GROUND CONTROL AND IMPROVEMENT

GROUND CONTROL AND IMPROVEMENT

PETROS P. XANTHAKOS
LEE W. ABRAMSON
DONALD A. BRUCE

WILEY

Ground Control and Improvement

Authorized reprint by Wiley India Pvt. Ltd., 4436/7, Ansari Road, Daryaganj, New Delhi – 110002.

Reprint: 2018

Printed at: Rajeev Book Binder

ISBN: 978-81-265-7433-9

CONTENTS

PREFACE xxi

1 GROUNDWATER LOWERING AND DRAINAGE TECHNIQUES 1

by Lee W. Abramson

1-1 Common Reasons to Lower Groundwater Levels / 1
Construction Dewatering / 3
Permanent Dewatering / 4
1-2 Design Input Parameters / 5
Existing Groundwater Levels and Fluctuations / 7
Depth of Required Groundwater Lowering / 7
Zone of Groundwater Lowering / 7
Permeability / 7
Transmissibility / 9
Storage Capacity / 9
Groundwater Quality / 10
1-3 Investigation Methods / 10
Borings / 10
Grain Size Distribution / 12
Permeability Tests / 17
Pump Tests / 19
1-4 Theoretical Background / 22
Soil and Rock Permeability / 22

Flow Nets / 23
Aquifer Characteristics / 24
Pump Theory / 27
1-5 Estimation of Flow Rates / 34
Darcy's Law / 34
Well Formulas / 35
Two-Dimensional Flow Nets / 35
1-6 Dewatering Methods / 38
Ditches and Trenches / 38
Sumps / 40
Weeps and Horizontal Drains / 40
Well Points / 41
Deep Wells / 42
Other Methods / 43
1-7 Design of Dewatering Systems / 43
Well Size / 43
Well Spacing / 45
Pump Size / 45
1-8 Plumbing / 48
Headers / 49
Valves / 49
Pipes and Connectors / 49
1-9 Discharge / 50
Storm Drains / 51
Sanitary Sewers / 51
Bodies of Water / 51
Recharge / 51
Treatment / 52
1-10 Effects of Adjacent Structures / 52
Consolidation / 52
Pumping of Fines / 53
Settlement / 53
Angular Distortion / 53
Rotting and Corrosion of Submerged Structures / 54
1-11 Impermeable Barriers / 54
Grouting / 54
Sheet Piles / 54
Slurry Walls / 56
Compressed Air / 56
Ground Freezing / 57

1-12 Case Histories / 58
Dewatering of Sand and Gravel for Construction of a 200-ft-Diameter Clarifier Tank / 58
Dewatering for a Powerhouse Excavation / 62
Dewatering for Construction of a 50-Story Building / 66
Excavation of a Slipway in Sand Below the Groundwater Table / 67
Deep Wells Used to Dewater Ground for New Sewer Tunnel / 70
Power and Signal Trench Dug with the Aid of Well Points / 71
Recovery of Chlorinated Hydrocarbon Solvents Using Deep Wells / 72

References / 73

2 UNDERPINNING **75**

by Petros P. Xanthakos

2-1 Basic Principles / 75
Definitions / 75
Underpinning Grouping / 76
General Requirements / 77

2-2 Conventional Underpinning Techniques / 78
Pit or Pier Underpinning / 79
Pile Underpinning / 82
Grouted Piles / 82

2-3 Shoring and Temporary Support / 91
Grillages / 94

2-4 Underpinning Applications and Examples / 95
Remedial Underpinning / 95
Underpinning for Buildings / 99
Underpinning for Tunneling / 102

2-5 Design Considerations for Underpinning / 104
Assessment of Existing Structures / 104
Pit and Pile Underpinning / 108
Root Pile Design / 110

2-6 Slurry Walls in Lieu of Underpinning / 115
Examples and Applications / 115
Factors Affecting Choice / 117
Observed Performance / 119

2-7 Prefounded Columns / 120
2-8 Intermittent Lateral Underpinning / 124
Horizontal Walls / 124
Vertical Rib Walls / 125
Construction Considerations / 127
2-9 Examples of Element Wall Underpinning / 128
2-10 Effect of Surcharge Loading / 131
2-11 Topics Relevant to Analysis / 140
Inclined Walls / 140
Effect of Soil Strength / 141
References / 141

3 EXCAVATION SUPPORT METHODS 145
by Petros P. Xanthakos

3-1 Convergence of Support–Control Process / 145
Example of Convergence / 145
The Rationale of Convergence / 151
3-2 Ground Response in Braced Excavations / 152
Theoretical Aspects / 152
Effects Inherent in Construction / 154
3-3 Supports for Braced Excavations / 159
Soldier Pile Walls / 159
Steel Sheet Pile Walls / 161
Bracing Systems / 162
3-4 Special Problems in Excavations / 170
Braced Cuts in Clay / 170
Pore Pressure Dissipation During Excavation in Clay / 176
3-5 Design Options in Collapsible Soils / 179
3-6 Excavation Supported by Temporary Walls / 181
Support Requirements / 183
3-7 Shotcrete as Structural Support / 186
Conventional Shotcrete / 186
Steel-Fiber-Reinforced Shotcrete / 188
3-8 Structural Performance of Shotcrete / 193
Plate Tests / 193
Rebound Tests / 194
Pullout Test / 197

3-9 Shotcrete Linings in Rock Openings / 198
Support Mechanism in Tunnels / 198
Laboratory Model Tests / 200
Field Observations / 200
Design Procedure / 204
The Convergence-Confinement Method / 207
3-10 Ground Rock Tunnel Interaction / 209
3-11 Trends in Control and Improvement Techniques with Particular Applications / 215
Stability Aspects / 216
Grouting Techniques for Structural Repairs and Seepage Control / 217
Seepage Control by Concrete Diaphragms / 218
3-12 Blasting and Excavation / 219
3-13 Soil-Cement Structural Walls / 224
References / 228

4 SOIL COMPACTION AND CONSOLIDATION 234
by Lee W. Abramson
4-1 Introduction / 234
Conventional Compaction / 238
Preloading / 238
Consolidation Drainage / 239
Dynamic Compaction / 240
Compaction Grouting / 241
Vibrocompaction / 242
Stone Columns / 243
Combined Methods / 244
Costs / 244
4-2 Uses and Applications / 245
Densification / 246
Increased Rate of Consolidation / 248
Settlement Reduction / 248
Increased Bearing Capacity / 252
Slope and Embankment Stability / 252
Reduction of Liquefaction Potential / 253
4-3 Principles of Behavior / 253
Decreased Porosity / 254

Increased Drainage / 254
Increased Shear Strength / 254
4-4 Theoretical Background / 255
Compaction / 255
Consolidation of Fine-Grained Soils / 257
Deformation of Cohesionless Soils / 263
Bearing Capacity / 266
Liquefaction Potential / 268
4-5 Design Considerations / 270
General Soil Characteristics / 270
Density / 271
Shear Strength / 271
Rate of Consolidation / 271
Settlement / 273
Bearing Capacity / 273
Slope Stability / 274
Liquefaction Potential / 275
4-6 Design Fundamentals / 275
Densification of a Cohesionless Soil Deposit / 275
Consolidation of a Cohesive Soil Deposit / 277
4-7 Construction Methods / 281
Compaction / 282
Dynamic Compaction / 282
Compaction Grouting / 283
Sand Drains / 284
Wick Drains / 284
Vibrocompaction / 285
Vibroreplacement (Stone Columns) / 287
4-8 Geotechnical Verification Testing / 288
Standard Penetration Tests (SPTs) / 288
Cone Penetrometer Tests (CPTs) / 288
Other Tests / 289
4-9 Performance Monitoring / 292
Optical Survey Techniques / 292
Settlement Plates and Deep Settlement Markers / 293
Piezometers / 294
4-10 Case Histories / 294
Preloading Used to Improve Site of Veterans Administration Complex in Tampa / 294

Preloading of Silty Sand Required for Sewage Treatment Plant / 297
Consolidation Drainage by Gravel Drains in San Francisco Bay Mud / 300
Sand Drains Used to Accelerate Settlement of I-95 Interchange Approach Embankments / 302
Dynamic Compaction of a Fill to Build an Executive Park / 307
Compaction Grouting of Loose Sand Beneath a Dam to Prevent Liquefaction / 311
Vibrocompaction Used to Densify Granular Backfill Behind Bulkhead / 316
Foundation Improvement with Stone Columns for Naval Housing Facility / 318
Several Methods Used at One Site with Variable Soil Conditions / 320
References / 325

5 IN SITU GROUND REINFORCEMENT 331

by Lee W. Abramson

5-1 Introduction / 331
Description / 331
Advantages / 332
Disadvantages / 332
5-2 Uses and Applications / 336
Excavation Support / 336
Slope Reinforcement / 338
Slope Stabilization / 339
Retaining Wall Repair / 341
5-3 History and Development / 342
France / 342
Germany / 345
United States / 345
5-4 Theoretical Background / 348
Nails / 349
Facing / 349
5-5 Design Considerations / 350
Wall Configuration / 352
Deflections / 353

Design Life / 355
Drainage / 355
5-6 Design Methods / 356
Davis Method / 359
Modified Davis Method / 361
German Method / 361
French Method / 363
Kinematical Method / 365
Computer Methods / 366
Design Method Inconsistencies / 366
5-7 Soil Nail System Design / 366
Empirical Methods / 367
Global Stability / 371
Internal Stability / 371
Corrosion Protection / 379
Facing / 381
Drainage / 382
Aesthetic Facades / 384
5-8 Construction Methods / 384
Nail Driving / 384
Nail Drilling / 385
Nail Materials / 385
Grouting / 386
Corrosion Protection / 387
Shotcreting / 388
Drainage / 389
5-9 Geotechnical Investigation and Testing / 389
Geologic Exploration / 389
Laboratory Testing / 390
Field Testing / 390
5-10 Performance Monitoring / 391
Optical Survey Techniques / 391
Inclinometers / 391
Strain Gages / 391
Load Cells / 392
Earth Pressure Cells / 392
5-11 Case Histories / 392
Nails Used in Stuttgart in Place of
Tieback Anchors / 392

Soil Nailing in Seattle Glacial Soils / 393
Soil Nailing Used in the Replacement of SR|504 Near Mt. St. Helens / 395
Soil Nailing Used in Loess for Washington Building Excavation / 399
Soil Nailing Used to Stabilize a Highway Embankment in Australia / 401
References / 402

6 SMALL-DIAMETER CAST-IN-PLACE ELEMENTS FOR LOAD-BEARING AND IN SITU EARTH REINFORCEMENT **406**

by Donald A. Bruce

6-1 Introduction / 406
6-2 Load-Bearing Pin Piles / 407
Historical Background and Characteristics / 407
Construction / 408
Design / 413
Case Histories and Performance / 431
Overview / 471
6-3 In Situ Earth Reinforcing—Type "A" Walls / 471
Definition / 471
Historical Background and Applications / 473
Construction / 473
Design / 473
Case Histories / 481
Overview / 489
References / 489

7 Permeation Grouting **493**

by Donald A. Bruce

7-1 Background to Chapters 7 and 8 / 493
Hydrofracture Grouting / 493
Compaction Grouting / 494
Permeation Grouting / 495
Jet Grouting / 495
7-2 Permeation Grouting / 498
Historical Development / 498
Applications / 498

Grouts / 501
Design / 528
Construction / 540
Evaluation of Results / 566
Cost Considerations / 570
Overview / 571
References / 573

8 JET GROUTING **580**

by Donald A. Bruce

8-1 Historical Development / 580
8-2 General Operational Features of the Three Generic Methods / 585
8-3 Applications / 592
Veritical/Subvertical Applications / 593
Horizontal/Subhorizontal Applications / 620
8-4 Design Aspects / 625
Preliminary Site Investigation and Testing / 626
Selection of Grout / 626
Selection of Jet Grout Parameters with Respect to Soil Types / 627
Characteristics of Jet Grouted Soils / 632
8-5 Quality Control and Assessment / 640
General Principles / 640
Estimating the Composition of Cuttings and Soilcrete / 641
Quality Assessment by the Energy Approach / 643
8-6 Summaries of Major Test Programs / 650
Varallo Pombia, Italy / 651
Sao Paulo, Brazil (F1) / 651
Norfolk, Virginia / 655
Volgodansk, C.I.S. / 659
Singapore / 661
Sao Paulo, Brazil (F2) / 665
Osaka, Japan / 669
Casalmaiocco, Italy / 673
Monte Olimpino, Italy / 676
8-7 Overview / 679
References / 679

9 REHABILITATION OF AGING ROCK SLOPES 684

by Lee W. Abramson

9-1 Introduction / 684
9-2 Exploration Methods / 684
Review of Existing Data / 685
Site Reconnaissance / 685
Geotechnical Investigation / 686
9-3 Stereographic Projection Plots / 692
9-4 Geotechnical Reports / 694
Data Reports / 695
Interpretive Reports / 696
9-5 Theoretical Background / 697
Intact Rock / 697
Discontinuous Rock Masses / 697
Shear Strength of Joints / 698
Effects of Water / 699
Factors of Safety / 699
9-6 Stability Analyses / 701
Circular Analysis / 702
Irregular Surface Analysis / 703
Planar Analysis / 704
Two-Block Analysis / 707
Wedge Analysis / 708
Multiple Plane Analysis / 710
Toppling / 711
Other Numerical Methods / 715
9-7 Potential Causes of Failure / 715
Weathering / 716
Hydrostatic Pressure and Freeze/Thaw Cycles / 717
Seismic Events / 718
Creep / 718
Progressive Failure / 719
9-8 Rating Systems / 720
9-9 Remediation Criteria / 722
9-10 Remediation Alternatives / 723
Do Nothing / 724
Facility Relocation / 724
Removal of Unstable Rock / 724

Catchment / 725
Flatten Slope / 727
Buttresses / 728
Surface Protection / 729
Reinforcement / 729
Drainage / 732
9-11 Estimation of Costs / 733
9-12 Program Planning / 734
Classification of Problem Areas / 734
Prioritization of Remediation Program / 735
9-13 Monitoring and Maintenance / 735
Geotechnical Instrumentation Monitoring / 735
Slope Maintenance Programs / 736
9-14 Case Histories / 736
Interstate 40 Sterling Mountain Portal Failure / 737
Wire Netting Used in Montana and Nevada for Rockfall Protection / 739
Shot-in-Place Rock Buttress Used to Repair Landslide / 740
Deteriorating Rock Slope Remediated Above Transit Tunnel Portal / 742
Rock Slope Failure in Singapore / 746
Woodstock Rock Slope Failure in New Hampshire / 746
Massive Rock Slide Derails Train in Pittsburgh / 749
Cedar Canyon Landslide Destroys State Highway 14 in Utah / 751
References / 753

10 VERTICAL SCREENS 755
by Petros P. Xanthakos

10-1 Introduction / 755
10-2 Earth Cutoffs / 756
Clay Mixes / 756
Composition and Permeability of Backfill / 758
Design Considerations / 758
Blowout Tests / 760
Composite Permeability of Earth Cutoff / 765
Control Limits of Slurry / 768

10-3 Earth Cutoffs for Pollution Control / 770
Applications / 770
Effect of Pollutant Attack on Cutoff Integrity / 772
Compressibility Characteristics / 775
10-4 Clay-Cement-Bentonite Mixes / 778
Applications / 778
Fundamentals of Clay-Cement Mixes / 779
Cement-Bentonite Mixes / 780
10-5 Plastic Concrete Cutoffs / 782
Applications / 782
Selection of Modulus / 783
Elasticity and Strength of Plastic Concrete / 784
Examples of Plastic Concrete Mixes / 786
Durability of Plastic Concrete / 790
10-6 Permeable Treatment Beds / 790
Applications / 791
Design and Construction Considerations / 792
Materials Analysis / 793
10-7 Cement-Bentonite Cutoffs (Solidified Walls) / 795
Characteristics / 795
Proportioning the Mix / 796
Properties of Mix / 797
Control of Setting Time / 800
Mixing Procedures / 800
Strength and Permeability of Set Mix / 801
Resistance to Deterioration / 801
The Slurry Replacement Method / 802
Design Principles / 803
Examples / 804
10-8 Injected Screens / 808
Construction Procedure / 808
Grout Mix Design / 810
Flow Properties of Grout / 811
Penetrability and Strength of Grout / 812
Applications / 813
10-9 Impermeable Membranes / 815
Chemical Resistance / 818
Installation / 819
Example of Environmental Application / 819

10-10 Interceptor Trenches / 820
Construction Requirements / 820
Design Considerations / 821
Advantges and Disadvantges / 822
10-11 High-Resistance Noncorrosive Cutoffs / 822
References / 824

11 ARTIFICIAL GROUND FREEZING **827**

by Petros P. Xanthakos

11-1 Introduction / 827
Background / 827
Basic Processes / 828
11-2 Sand-Ice Systems / 828
Mechanical Properties and Creep / 828
Further Studies of Creep Behavior / 832
Triaxial Tests of Frozen and Unfrozen Sands / 833
Applications / 836
11-3 Clay-Ice Systems / 837
General Principles / 837
Rheological Model of Laterally Stressed Frozen Soil / 838
11-4 Seismic Effects and Dynamic Response of Frozen Ground / 841
Seismic Response and Dynamic Behavior / 842
Parameter Effects on Dynamic Properties of Frozen Soils / 845
Frozen Clay Under Cyclic Axial Loading / 846
11-5 Design Requirements for Artificial Freezing in Temporary Ground Support / 848
11-6 General Description of Freezing Systems / 849
System Components / 849
Review of Freezing Procedures / 851
11-7 Design Considerations / 852
11-8 Basic Design Parameters / 854
Thermal Properties / 854
Hydrologic Properties / 856
Mechanical Properties / 856
Geometry and Capacity of Freezing System / 858

11-9 General Design Approach / 859
Thermal Considerations / 859
Effect of Mechanical Parameters / 863
Ground Movement and Groundwater / 864
Selection of Freezing System / 866
11-10 Construction and Monitoring Program / 868
General Approach / 868
Monitoring / 869
11-11 Shaft Design and Construction / 872
Design Principles / 873
Thermal Aspects / 875
Relevant Design Data and System Components / 877
Shaft Construction Inside Frozen Earth / 879
Thawing and Abandonment Stage / 880
Examples of Shaft Freezing / 882
11-12 Temporary Tunnel Support by Artificial Ground Freezing / 883
Tunnel Freezing Studies / 884
Design Aspects of Case Studies / 885
Structural Investigation / 889
Observed Performance / 891
References / 896

INDEX 905

PREFACE

A 1978 survey by the Committee on Placement and Improvement of Soils of the Geotechnical Division of the ASCE focused on possible future advances in this field. Participants were asked to identify long-range developments and to provide an assessment of their importance, feasibility, and probable time of occurrence. The consensus of opinion was that emphasis should continue on densification, admixture, reinforcement, moisture control, grouting, and regulation.

Motivated by similar considerations and an assessment of the current state of the art, the authors have selected 11 topics for discussion. Some of these topics are on traditional techniques, whereas others represent recent developments. Nonetheless, they all have demonstrated high capability, desirablity, and feasibility. Furthermore, the progress associated with them has been mainly technological rather than conceptual.

The book is a synthesis of the beliefs of its authors. The choice of material has been narrowed, however, to include techniques that have been tested in applications of ground support, control, and improvement schemes. Whereas this synthesis is the result of a logical assessment and merging of principles and concepts, the process does not stop here. Thus, impressive future advances in these and other areas may occur in a clear and consistent format, and progress will continue to evolve so that new developments will result from ideas and concepts that have not yet been foreseen.

The subject matter has been developed independently, but an effort has been made to produce a unified technology that can be used directly in new construction or in rehabilitation programs. The underlying principle is that demand in ground engineering has gone beyond the stage of a single application, and practice has moved into the realm of multiple uses and purposes that require a wide variety of construction controls. In this context, the authors caution about problems related to the placement of foundations and supports on poor soils and deteriorated rocks. Often, these cases will signify the absence of ground control and may reflect uncertainties in design criteria. Thus, ground control and improvement should be considered as a formidable supplement that balances the support requirements or that may eliminate them completely.

The convergence between support and control or improvement options must, however, be based on the explicit understanding of ground response to an externally applied action. This response may be specific or random, rapid or slow, or temporary or permanent. Its forecast in engineering terms is the determining factor for the use of ground controls or for the exclusive reliance on ground supports. More often, this convergence will provide the optimum solution to most ground engineering problems, and, since this trend is expected to continue, planning to mitigate the risks of underground construction by combining artificial support with control and improvement techniques should become a common engineering approach.

What is unique about this book is the up-to-date information, data compilation, and synthesis of material that it provides. The book has two main purposes: (1) to enable practicing engineers to make the best decision, in technical and economic terms, regarding ground engineering problems or when alternative schemes are formulated and evaluated, and (2) to provide the necessary information and credible data for a complete design.

During the planning and development of the text, each author wrote his own content and discussion in light of his own background and design experience, so the scope of the book may vary. However, the authors made every effort to coordinate the end result and to ensure that unnecessary gaps and overlapping would be avoided. Cross-references between chapters and sections are provided where indicated, but readers are encouraged to consult the index in order to find all pertinent information. Because this is a textbook, not a handbook, each chapter describes the process, articulates the design philosophy, and presents the design methodology, and is supplemented by examples and case studies.

Chapter 1 discusses groundwater lowering and drainage techniques for projects that require construction or permanent dewatering. The text presents the details of dewatering methods and discusses the theoretical principles and the design of dewatering systems, while also articulating the effects on adjacent structures. Dewatering techniques are examined in conjuction with other controls, such as impermeable barriers, and criteria are developed for selecting a scheme in which dewatering is combined with other methods to produce a unified system.

Underpinning is the subject of Chapter 2. In the traditional context, underpinning involves the addition of structural foundation units to give extra support to structures at or below grade. In the technical context, underpinning is the insertion of a new foundation or support below an existing one for the transfer of load to a lower level, but, in a broader sense, it may also refer to the lateral protection of a foundation by a retention system, the strengthening of ground beneath, or both. The decision to underpin, protect laterally, or strengthen the ground depends on such various interrelated factors as cost, technical expediency, and associated risks. These concepts are discussed in detail and are illustrated by examples and case histories.

Chapter 3 reviews excavation support methods. The associated support systems are primarily intended for temporary use and may or may not become part of the permanent structure. The rationale of convergence of the support control process is emphasized and articulated by examples. The text discusses ground response in supported excavations and gives a comparative review of artificial supports. Special problems are commonly encountered for excavations in clays, particularly those in which the effects of anisotropy must be considered, and in collapsible soils. In addition to the conventional ground support systems, the discussion includes shotcrete, grouting applications, blasting, and special systems, such as the soil-cement structural walls.

In Chapter 4, it is demonstrated that, with the availability and popularity of soil compaction and consolidation techniques, there are no longer unacceptable construction sites. Three tested and promising techniques supplement the theory of soil compaction and consolidation: vibro techniques, dynamic compaction, and compaction grouting. Dynamic compaction improves weak soils by controlled high-energy tamping. In this case, a comprehensive understanding of soil behavior is vital to a successful application. Compaction grouting involves the injection of material under high pressure to compact and densify loose soil beneath distressed structures. The application offers economic advantages when a thin, loose, deep stratum, overlain by a very dense stratum, requires densification. Vibrocompaction densifies granular soil by rearranging loose grains into a denser array. The stone column technique, or vibroreplacement, enhances displacement and drainage to improve weak ground.

In situ ground reinforcement is the subject of Chapter 5. The main topic is soil nailing, a process whereby steel rods or nails are installed in a cut face of original ground and are connected by steel mesh or shotcrete facing to support the soil near the cut. The steel nails reinforce and strengthen the ground, and, as the latter deforms, the nails share the loads, gradually becoming more stressed in tension. Uses and applications of soil nailing are mentioned for excavation support, slope reinforcement, slope stabilization, and retaining wall repair. The text traces the history and development of soil nailing, reviews the theoretical principles, and presents the design considerations. Corrosion protection and construction methods are discussed in detail.

A different form of in situ earth reinforcement is discussed in Chapter 6. This technique involves small-diameter cast-in-place elements. The broad usage of this system is reflected in the wide variety and range of names such as minipiles, micropiles, root piles, pali radice, needle piles, pin piles, and so on, all of which are used to describe basically a special type of small-diameter bored pile. Pin piles may be used as load-bearing elements. Reticulated micropiles may be installed in various configurations to produce a stable block of reinforced soil that can act as a coherent retaining structure. Soil doweling is another application whereby dowels are used to reduce or stop downslope movement on well-defined shear surfaces. The text reviews the theoretical background, origin and development, construction procedures, and uses and applications of soil doweling.

Chapters 7 and 8 summarize the most recent practice in permeation and jet grouting. In some respects, the techniques involved in permeation grouting are the oldest and best researched. In practical terms, the intent of permeation is to introduce grout into soil pores without any essential change in the original soil volume and structure. The method can be applied to both rocks and soils, but clearly the soil properties and pore geometry are the determining factors of the success of the application. On the other hand, jet or replacement grouting is the most recent category of ground treatment, and it is only since the early 1980s that the various derivatives of the method have reached its economic and operational potential. Jet grouting can be executed in soils with a wide range of characteristics, such as granulometrics and permeabilities, and, in fact, limitations to the applicability are usually imposed by strength parameters or by economic considerations. These chapters present the theoretical principles, design considerations, construction requirements, performance data, and applications.

Aging rock slopes show stability problems, and this is documented in mine quarries, water resource projects, and transportation facilities. A rehabilitation program requires a detailed assessment of site conditions, potential modes of failure, failure causes, slope condition rating methods, and monitoring and maintenance procedures. These principles are interfaced with rehabilitation methods and are discussed in Chapter 9, along with practical problems and appplications.

Vertical screens, which are discussed in Chapter 10, cover a broad area of protective or remedial systems that include continuous earth, semirigid and rigid cutoff walls, plastic barriers, permeable treatment beds, synthetic membranes with overlapping or interlocking sheet-pile sections, and clay-cement grouts injected under pressure into preformed narrow slots. The advantages of the technique are fully realized if the screens satisfy certain requirements, namely: (1) the insertion of the screen is not inhibited by site and ground conditions, even in very mobile formations; (2) the barrier is made continuous but is flexible enough to deform, if necessary; (3) the construction is rapid and has a low cost; and (4) the groundwater flow or level is not required to be altered. These principles are discussed in the chapter, along with the various types of screens.

Chapter 11 reviews the fundamentals of ground freezing. This technique is applied in the mining and construction industries in two basic modes: (1) as a supplementary or emergency procedure for stabilizing excavations and installations where the more traditional supports are used, and (2) as a primary independent construction method for stabilizing underground openings. Although freezing can be used practically in any type of soil that has pore water, it is more often used below the groundwater table. The principles of freezing are reviewed in this chapter, modeled, and quantified, and emphasis is placed on freezing design, applications, and examples. Among the various uses, shaft construction and temporary tunnel support by artificial ground freezing are given special consideration.

Inasmuch as the text focuses on the structural and geotechnical aspects of ground engineering, it should apply equally to these two fields. The text should also be of interest to construction engineers, contractors, planners, and administrators. In general, the book is oriented toward the needs of practicing engineers, but the material may be rearranged to fit one or two courses at the graduate level.

PETROS P. XANTHAKOS
LEE W. ABRAMSON
DONALD A. BRUCE

Great Falls, Virginia
January 1994

CHAPTER 1

GROUNDWATER LOWERING AND DRAINAGE TECHNIQUES

1-1 COMMON REASONS TO LOWER GROUNDWATER LEVELS

Common reasons to lower groundwater levels are for construction excavations and for permanent structures that are below the water table and are not waterproof or are waterproof but are not designed to resist the hydrostatic pressure. Permanent dewatering systems are far less commonly used than temporary or construction dewatering systems. When construction below the water table is planned, choices for dealing with this problem include construction "in the wet" (i.e., with water or some other type of fluid remaining in the excavation during construction, see Figure 1-1), use of cutoff walls, which limit inflow into the excavation (Figure 1-2), or lowering of the groundwater levels to reduce the hydraulic head and hence the inflow into the excavation (Figure 1-3). Even when cutoff walls are used, dewatering within the confines of the cutoffs may still be required to improve the stability of working areas, but probably to a lesser extent.

This chapter deals primarily with techniques used for lowering groundwater levels and related issues, such as effects on adjacent structures and the use of impermeable barriers in combination with groundwater lowering techniques. In this chapter, as well as in others in the book, the reader will find overlap between subjects. For instance, grouting can be used in conjunction with dewatering to reduce the quantity of water inflow into the excavation. Grouting is addressed in Chapter 7. Drainage trenches, cutoff walls, and ground freezing are often used in conjunction with or in place of dewatering systems and are addressed in Chapters 3, 10, and 11, respectively. The art of ground improvement is now beyond the stage of unique application to special individual problems and has moved into the realm of

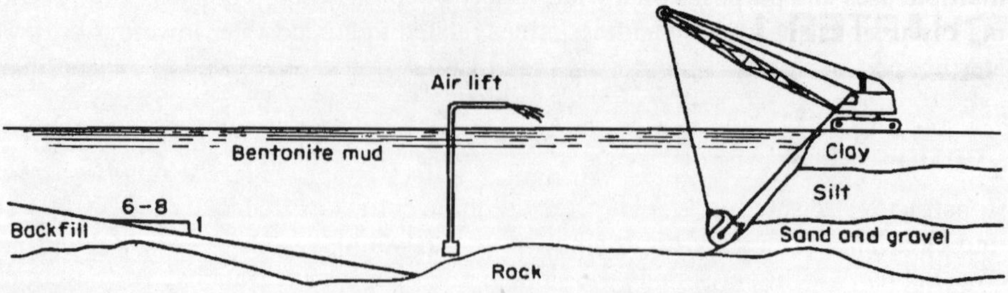

Figure 1-1 Construction in the wet. (From Xanthakos, 1979, by permission of McGraw-Hill.)

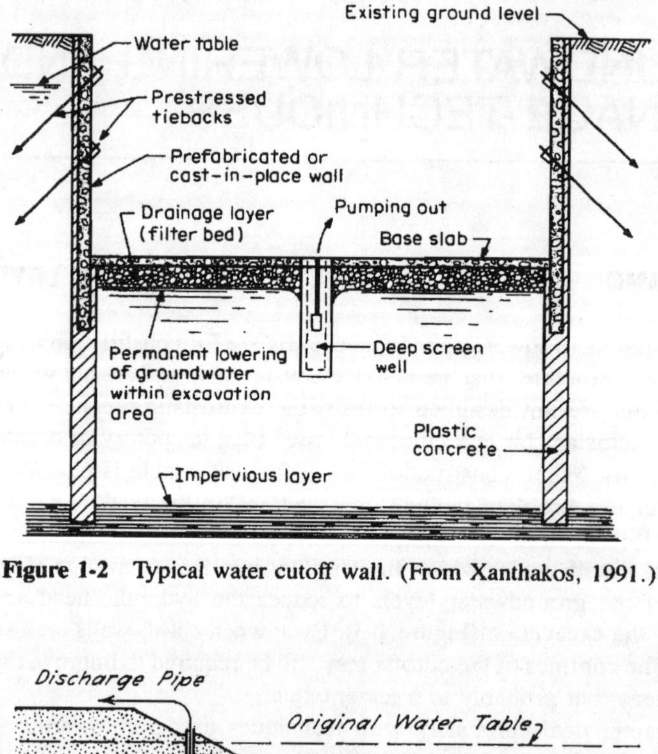

Figure 1-2 Typical water cutoff wall. (From Xanthakos, 1991.)

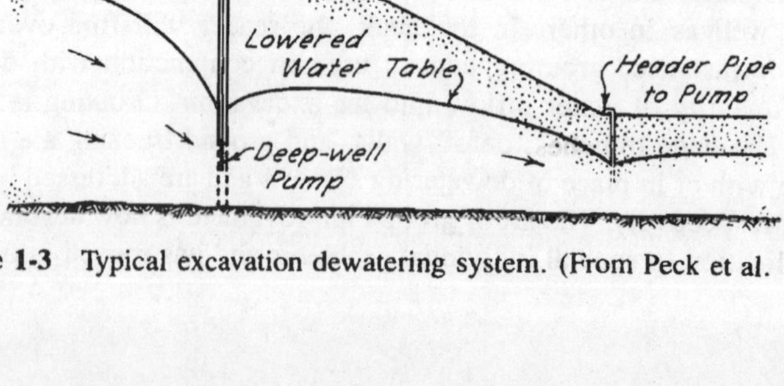

Figure 1-3 Typical excavation dewatering system. (From Peck et al. 1974.)

multiple uses and purposes on a wide variety of construction problems. The following discussions specifically address issues related to groundwater lowering or dewatering; pertinent information related to other subjects is referenced as appropriate.

Construction Dewatering

Construction dewatering is most often used by contractors to decrease water inflow into excavations, thereby improving working conditions in the excavation and increasing the stability of soils in the sides and base of the excavations. The height of groundwater above the base of the excavation (i.e., the anticipated head and thus the amount of inflow) will dictate the methods used. A contractor's experience and available equipment will also affect the methods chosen. Dewatering is a necessary evil. It is avoided to the extent possible because of cost, disruption to other construction tasks, schedule constraints, discharge and disposal requirements, sensitivity to discharge water quality, potential effects on water supply, and potential effects on adjacent structures. Often construction below the water table cannot be avoided and will cost less than other alternatives such as impermeable excavation supports (e.g., slurry walls).

The most common dewatering methods chosen by contractors are: sumps, trenches, and pumps; well points; and deep wells with submersible pumps.

Briefly, the first method involves handling minor amounts of water inflow into an excavation by channeling the water to trenches and sumps and then pumping the sumps out with a submersible pump as necessary to keep the excavation bottom dry and stable. This method is usually used where the height of groundwater above the

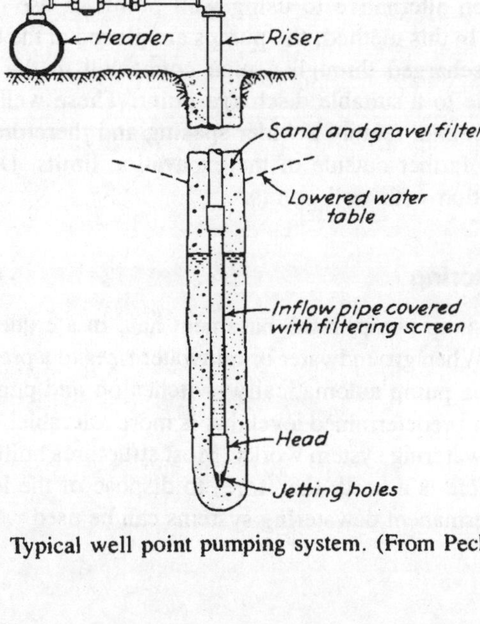

Figure 1-4 Typical well point pumping system. (From Peck et al., 1974.)

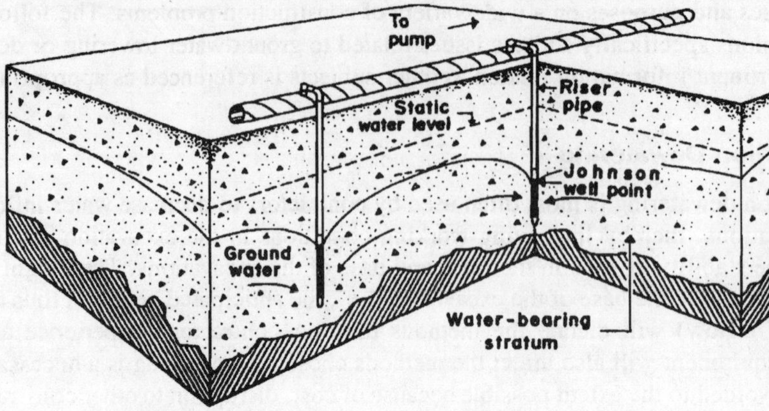

Figure 1-5 Typical well point system. (From Johnson, 1975.)

excavation bottom is relatively small (5 ft or less) and the surrounding soil mass is relatively impermeable (clayey soil for instance).

The well point method involves multiple closely spaced wells connected by pipes to a strong pump that can suck the water out of the ground through the well points via the "header" manifold pipe, through the pump, and out of the discharge end of the pump (Figure 1-4). Multiple lines or stages of well points are required for excavations that extend more than about 15 to 20 ft below the groundwater table (Figure 1-5). Ejectors or eductors can be used to enhance the capabilities of a well point system but require careful design for maximum efficiency at the anticipated head and discharge conditions.

The most common alternative to using well points is use of deep wells with submersible pumps. In this method, the pumps are placed at the bottom of the wells and the water is discharged through a pipe connected to the pump and run up through the well hole to a suitable discharge point. These wells are usually more powerful than well points, require a wider spacing and therefore fewer well holes, and can be installed farther outside of the excavation limits. Deep wells are used alone or in combination with well points.

Permanent Dewatering

Anyone who has a sump pump in the basement has, in a crude way, a permanent dewatering system. When groundwater or rainwater rises to a predetermined level in the basement sump, a pump automatically switches on and pumps the water level back down to another predetermined level that is more tolerable. This is more or less how a permanent dewatering system works. Most structures built below the groundwater level leak. There is a need, therefore, to dispose of the leakage with sumps and pumps. Also, permanent dewatering systems can be used where the structure is

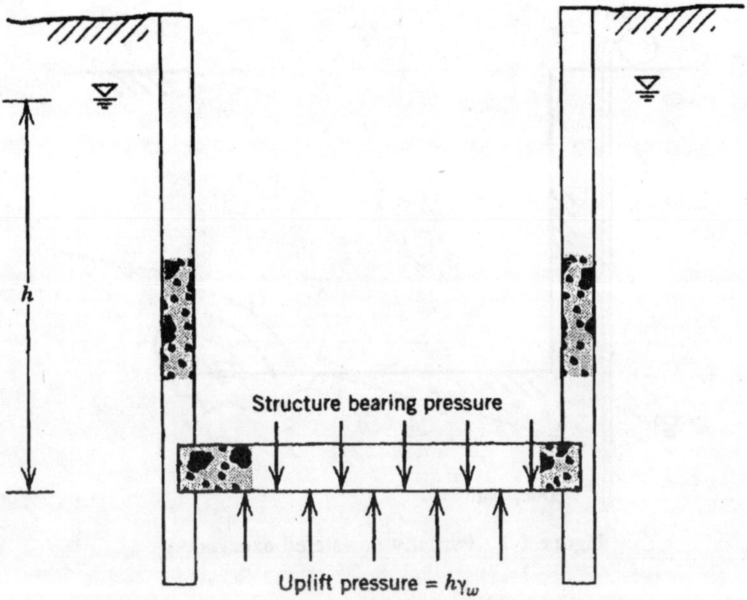

Figure 1-6 Buoyancy effects on underground structure.

far below the prevailing groundwater level outside the structure and where it is more economical to dewater than to design the structure walls and slabs to resist the water pressure outside and also to counteract the buoyancy effects (Figure 1-6). In this case, the higher initial capital cost of building a stronger and heavier structure must be traded off against the future operating and maintenance costs.

1-2 DESIGN INPUT PARAMETERS

The most important input parameters for selecting and designing a dewatering system are the height of the groundwater above the base of the excavation and the permeability of the ground surrounding the excavation. To know the depth of groundwater lowering, one must know what the prevailing groundwater levels are at the site and the depth of excavation. The groundwater level is usually lowered to at least 2 ft below the bottom of the excavation. The field permeability of the ground must be known to estimate the amount of pumping, or flow rate, that will be required to attain the required groundwater level. Also of importance is the shape of the dewatered zone, often referred to as the cone of depression. The cone of depression must encompass the excavation limits or the excavation bottom will be only partially dewatered (Figure 1-7). Water quality must be known too. The amount of discharge from a dewatering system is usually significant. If treatment will be required, this must be known in advance of construction.

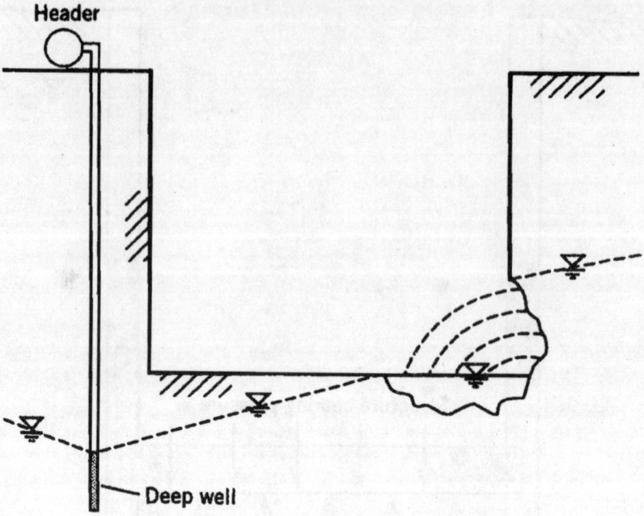

Figure 1-7 Partially dewatered excavation.

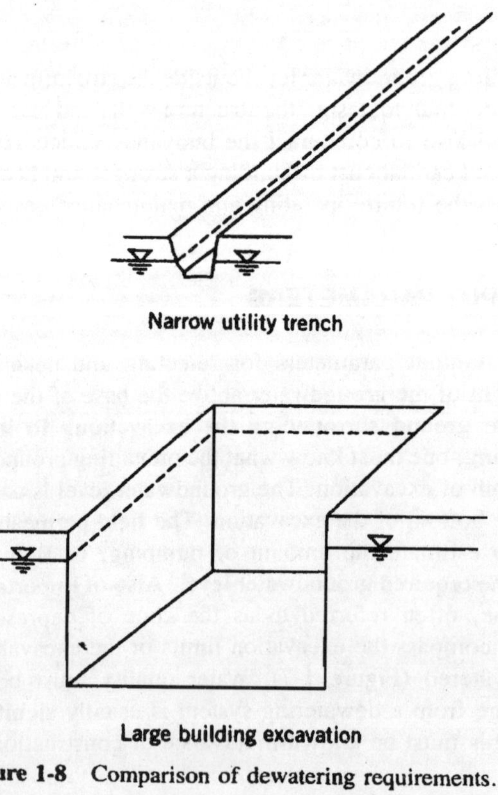

Figure 1-8 Comparison of dewatering requirements.

Existing Groundwater Levels and Fluctuations

A good starting point in assembling the information necessary to select and design a dewatering system is to determine where the prevailing groundwater level is at the site. This is usually accomplished with observation wells or with piezometers, instruments installed in boreholes to sense piezometric surfaces in the ground. If the site is near a large body of water such as a river, lake, or ocean, chances are that the groundwater level is at or near the river, lake, or sea level. Also, variations in water levels due to floods, storms, control structures, and tides must also be considered. Another important consideration is the presence of artesian (high pressure) and perched water, which can exhibit piezometric levels different from that expected. Any offsite pumping for water supply, hazardous waste remediation, or other construction projects should be known and evaluated for possible effects on the dewatering system being considered.

The most conservative approach would anticipate the highest water level possible during construction. One might also want to weigh the effects of assuming a lower water level with the potential economic impacts of that lower level being exceeded during the construction duration. That analysis usually results in the decision to be on the safe side by assuming the high water level. The potential costs of excavation flooding, construction delays, or the emergency mobilization of additional equipment usually far exceed the initial costs of mobilizing larger equipment or a few extra wells.

Depth of Required Groundwater Lowering

The required depth of groundwater lowering is usually related to the bottom of the excavation. The water level should be lowered to about 2 to 5 ft below the base of the excavation. If the absolute bottom of the excavation is not known, a conservative (i.e., lowest possible) estimate should be made. This should include any overexcavation required for footings, slabs, and shafts. The maximum depth of groundwater lowering is then the difference between the prevailing groundwater level and the required level during construction.

Zone of Groundwater Lowering

The zone of groundwater lowering involves not only the depth but the three-dimensional shape required. For instance, the requirements for a thin, linear, utility trench will be different from those for a large, square parking structure (Figure 1-8). The limits of the excavation must be known or estimated. For a long linear excavation, the entire site may not need to be dewatered at the same time.

Permeability

Permeability or hydraulic conductivity is the rate of water movement through the ground at a hydraulic gradient of one. Common values for a variety of soil and rock types are shown in Table 1-1. Of the parameters needed for dewatering system

TABLE 1-1 Permeability Values for Common Soils and Rocks

Formation	Value of k(cm/sec)
River Deposits	
Rhone at Genissiat	Up to 0.40
Small streams, eastern Alps	0.02–0.16
Missouri	0.02–0.20
Mississippi	0.02–0.12
Glacial Deposits	
Outwash plains	0.05–2.00
Esker, Westfield, Mass.	0.01–0.13
Delta, Chicopee, Mass.	0.0001–0.015
Till	Less than 0.0001
Wind Deposits	
Dune sand	0.1–0.3
Loess	0.001 ±
Loess loam	0.0001 ±
Lacustrine and Marine Offshore Deposits	
Very fine uniform sand, C_u = 5 to 2	0.0001–0.0064
Bull's liver, Sixth Ave., N.Y., C_u = 5 to 2	0.0001–0.0050
Bull's liver, Brooklyn, C_u = 5	0.00001–0.0001
Clay	Less than 0.0000001

	k(cm/sec)	Intact Rock	Porosity n(%)	Fractured Rock
Practically im-	10^{-10}	Massive low-porosity	0.1–0.5	
permeable	10^{-9}	rocks	0.5–5.0	
	10^{-8}			
	10^{-7}			
Low discharge,	10^{-6}	Sandstone	5.0–30.0	
poor drainage	10^{-5}	Weathered granite		Clay-filled joints
	10^{-4}	schist		
	10^{-3}			
High discharge,	10^{-2}			
free draining	10^{-1}			Jointed rock
	1.0			Open-jointed rock
	10^{1}			
	10^{2}			Heavily fractured rock

Source: Terzaghi and Peck, 1967 (Wiley).

selection and design, permeability is probably the most elusive and hardest to predict for the field case. Common methods for estimating permeability include empirical formulas, laboratory permeability tests, borehole packer tests, and field pump tests. The reliability and cost of these methods increases more or less in the order given. Field pump tests are the most reliable method but also the most costly. It is easy to say that permeability may vary by one or two orders of magnitude. Pump discharge rates are proportional to the coefficient of permeability. Pump discharge rates, however, must be predicted within more refined limits than plus or minus two orders of magnitude. So one of the most important parameters needed for dewatering analysis is one of the hardest to predict.

Transmissibility

The coefficient of transmissibility indicates how much water will move through the formation. It is defined as the rate at which water will flow through a vertical strip of the formation 1 ft wide and extending through the full saturated thickness under a hydraulic gradient of 1 or 100 percent. It can be calculated by multiplying the coefficient of permeability by the thickness of the formation. Common values range between 1000 and 1 million gallons per day per foot. The higher the value is, the more water will flow through the formation. Also, formations with higher transmissibility values will exhibit shallower cones of depression (less drawdown) that extend farther from the well (larger radius of influence) than formations with lower values at the same pumping rate (Johnson, 1975). The coefficient of transmissibility can be determined at a site by conducting a field pump test and recording the relationship between pumping rate and drawdown, as discussed later.

Storage Capacity

The coefficient of storage indicates how much can be removed from the formation by pumping. It is defined as the volume of water released from storage per unit of surface area of the formation per unit change in head. For a nonartesian groundwater table, the storage coefficient is the same as the specific yield of a formation. This is a dimensionless parameter often ranging between 0.1 and 0.35. The coefficient of storage of a site is a function of transmissibility and can be determined from a field pump test if the coefficient of transmissibility and the time-drawdown relationship is recorded.

Groundwater Quality

Requirements on the disposal of water from construction sites have become restrictive during recent years due to heightened awareness of water quality. It is more costly now to dispose of construction water and water quality requirements on the effluent are more stringent than ever. Therefore groundwater quality must be determined before a dewatering effort is undertaken. The quantity of discharge as well as the quality must be known ahead of time so that discharge facility requirements are known and any needed treatment can be planned far ahead of time.

The actual water quality requirements governing a construction project vary from location to location as well as from agency to agency. The city sewer company may have different requirements for discharge into a sanitary or storm sewer than the Army Corps of Engineers or Environmental Protection Agency may have for ocean or river discharge. The concept of putting the same kind of water back to where it came from no longer passes as a justification for minimal processing and handling. Once the water is removed from the ground, it is subject to requirements that are more stringent than the ones governing its condition in situ. It may have to be disposed of in a cleaner condition than it left the ground!

Another dilemma is determining who has jurisdiction over the groundwater and discharge and what quality requirements must be met. Governing agencies having similar charges have a variety of names from place to place. For each dewatering project, there is no alternative to contacting the city, state, and federal agencies each time to ascertain the requirements for a specific project. Also, the regulations are changing almost continuously. The requirements now may be different than they were last time. On a more positive note, the agencies tend to have knowledgeable people who are anxious and willing to help, not only with their own regulations but also in pointing us in the right direction for determining other agencies' regulations.

1-3 INVESTIGATION METHODS

There are certain geotechnical investigation methods that should be used when planning or designing a dewatering program. Firstly, conventional borings should be conducted to characterize the subsurface profile in terms of soil and rock types (i.e., gravel, sand, silt, clay, fractured basalt, massive limestone, etc.), lateral and vertical extent and variability of zones and layers, and the location and variability of the groundwater table including artesian and perched groundwater conditions. One common characteristic of soils that relates to dewatering is the grain size distribution. For some soils this is a reliable predictor of permeability. Sometimes laboratory and field tests are run to estimate the permeability of the soil and rock. Finally, the most reliable, and unfortunately the most costly, test method for dewatering programs is a full-scale field pump test. On large projects, field pump tests are usually cost effective.

Borings

Many geotechnical references (Hvorslev, 1949; Peck et al., 1974; Sowers, 1979; Hunt, 1984) discuss the wide variety of geotechnical investigative methods available. The staple of these methods is conventional borings, usually drilled using hollow-stem augers or casing and a roller bit. The most common methods of soil

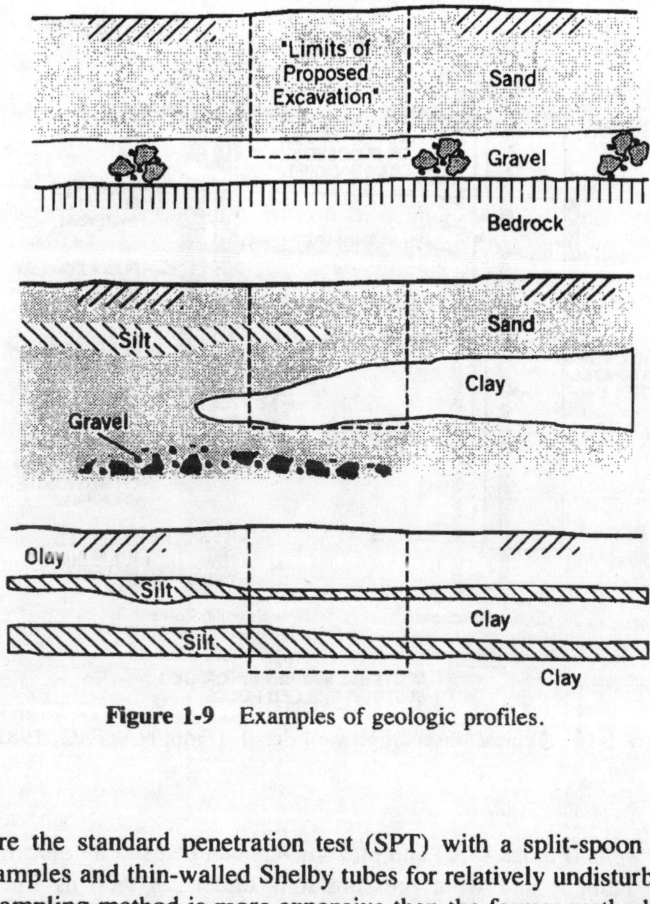

Figure 1-9 Examples of geologic profiles.

sampling are the standard penetration test (SPT) with a split-spoon sampler for disturbed samples and thin-walled Shelby tubes for relatively undisturbed samples. The latter sampling method is more expensive than the former method and is most commonly used for strength testing when required. Continuous rock coring is usually the method of sampling used in bedrock.

The three most important pieces of information that must be obtained from a boring program for dewatering are: (1) the lateral and vertical extent and variation of the soil and rock deposits at the site; (2) the hydraulic characteristics of each soil and rock deposit; and (3) the level and characteristics of the groundwater table. Examples of geologic site profiles are given in Figure 1-9. Dewatering requirements for these example sites would be quite different from site to site.

Important considerations in conducting a boring program include:

- Understanding the geologic origins of the site
- Careful field classification of the samples
- Retention of representative samples for testing
- Installation of observation wells or piezometers

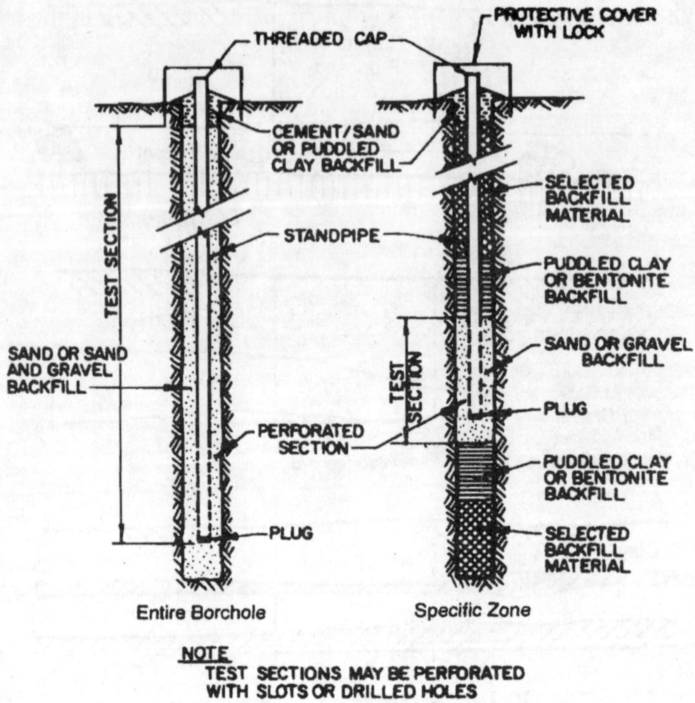

Figure 1-10 Typical observation well detail. (From NAVFAC, 1982.)

Observation wells (Figure 1-10) and piezometers can be used for sensing where the groundwater table is and what pressure it is under, as well as for performing borehole field permeability tests. Borehole field permeability tests (Figure 1-11) are conducted by adding water to the well (falling head test) or baling water out of the well (rising head test) and timing how long it takes for the well to reestablish an equilibrium condition. The foregoing test methods are used in soil zones. In rock, permeability tests can be run by using inflatable packers (Figure 1-12) to seal a zone of rock off and by pumping water into the sealed-off zone.

Grain Size Distribution

Grain size distribution of soil deposits affects their permeability and therefore is of primary concern to predicting water inflow into an excavation. Relative permeabilities of a variety of soil and rock types are given in Table 1-1. The amount of fines in the soil has a significant effect on permeability. A sand with 10 percent fines (10 percent passing a No. 100 sieve) could have 100 to 1000 times lower permeability than a cleaner sand with no fines (Bush, 1971).

The most common way of running a grain size distribution test in the laboratory is according to ASTM Test Method D-422 (ASTM, 1990). These are done with different sizes of sieves and a vibratory shaker for coarse-grained soils and a hydrometer for fine-grained soils. Typical grain size distribution curves are shown in Figure 1-13. Since there is usually variability from sample to sample, the grain size distribution limits or bounds will often be shown for different zones. Suitable dewatering and other treatment methods are shown according to grain size distribution in Figure 1-14.

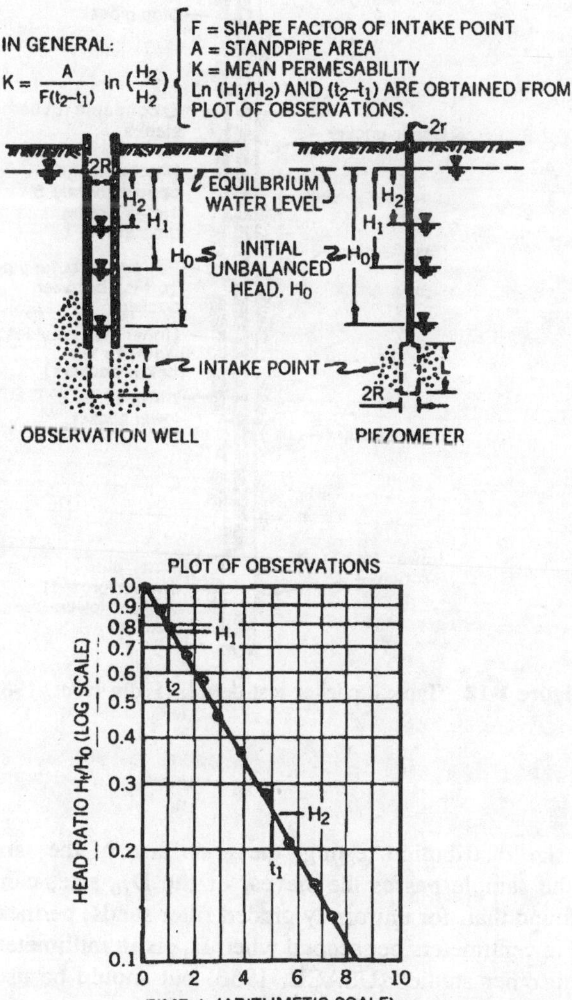

Figure 1-11 Typical borehole permeability test. (From NAVFAC, 1982.)

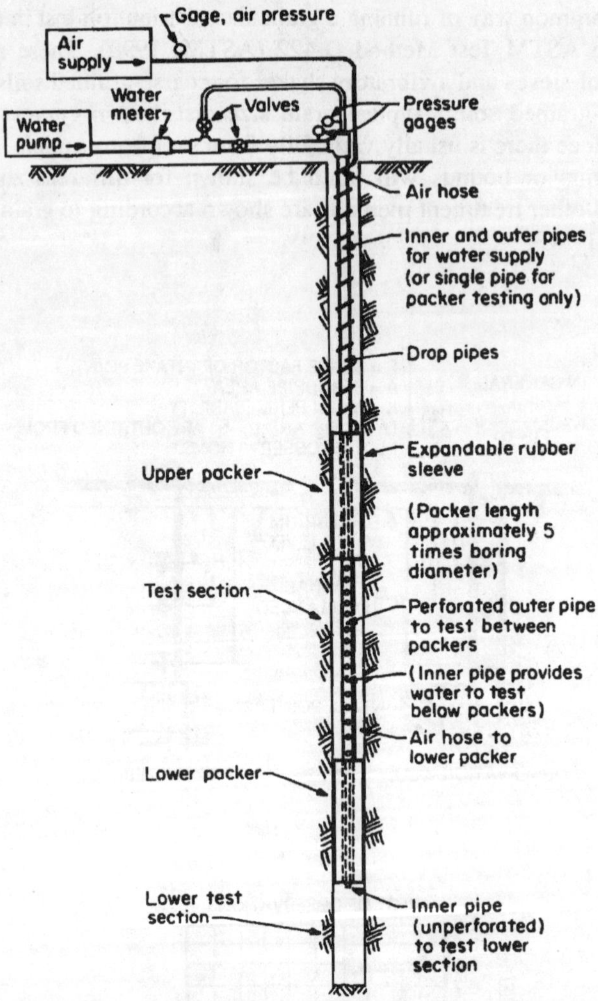

Figure 1-12 Typical packer test detail. (From Hunt, 1984.)

From grain size distribution testing, the grain size of the sample where only 10 percent of the sample passes the sieves, or the D_{10} size, can be determined. Hazen (1893) found that, for uniformly graded filter sands, permeability is roughly equal to $(D_{10})^2$ in centimeters per second when D_{10} is in millimeters. This relationship agrees with other studies (USACE, 1956) but should be used with caution, especially in fine-grained soils and in soil deposits with a high degree of variability.

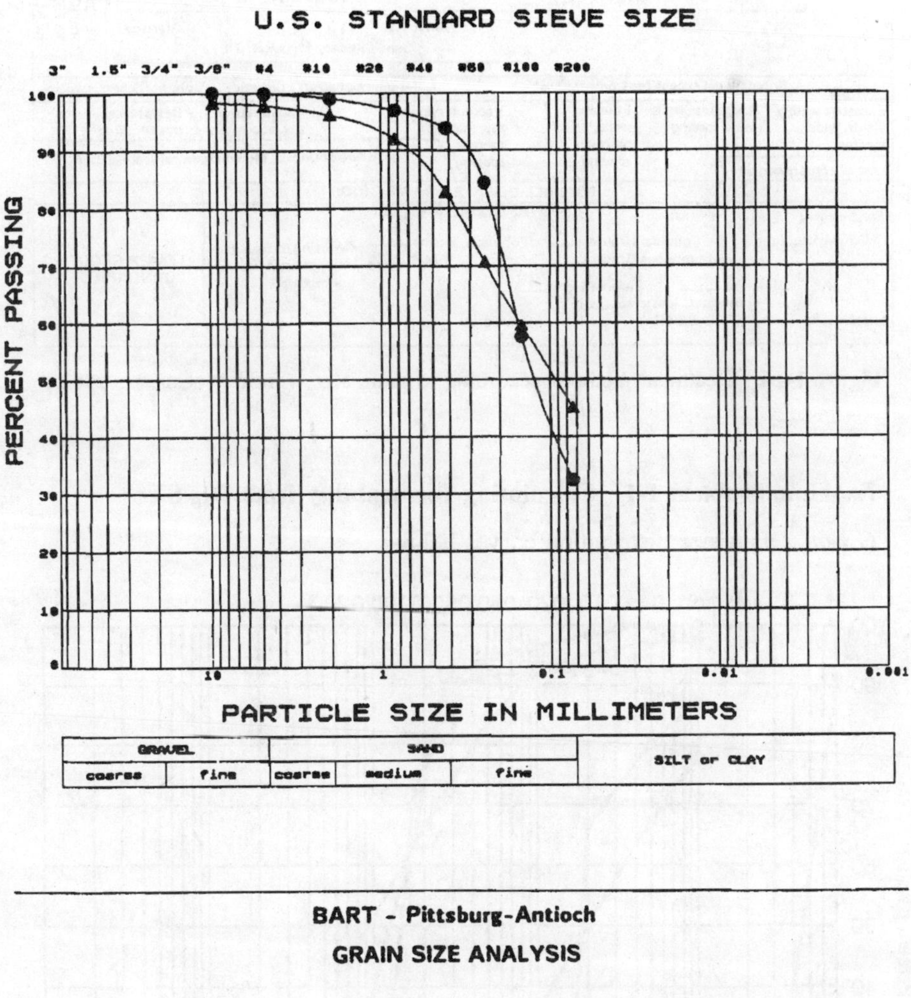

Figure 1-13 Example of grain size distribution curves.

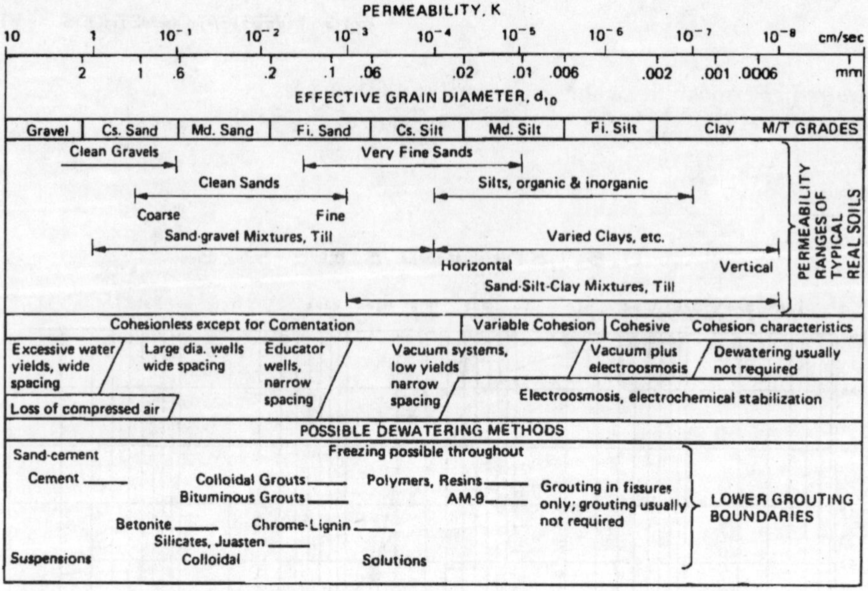

Figure 1-14 Treatment methods according to grain size. (From McCusker, 1982.)

Example Problem 1-1 Calculating Permeability from D_{10} Size

Given: Grain size distribution curves below:

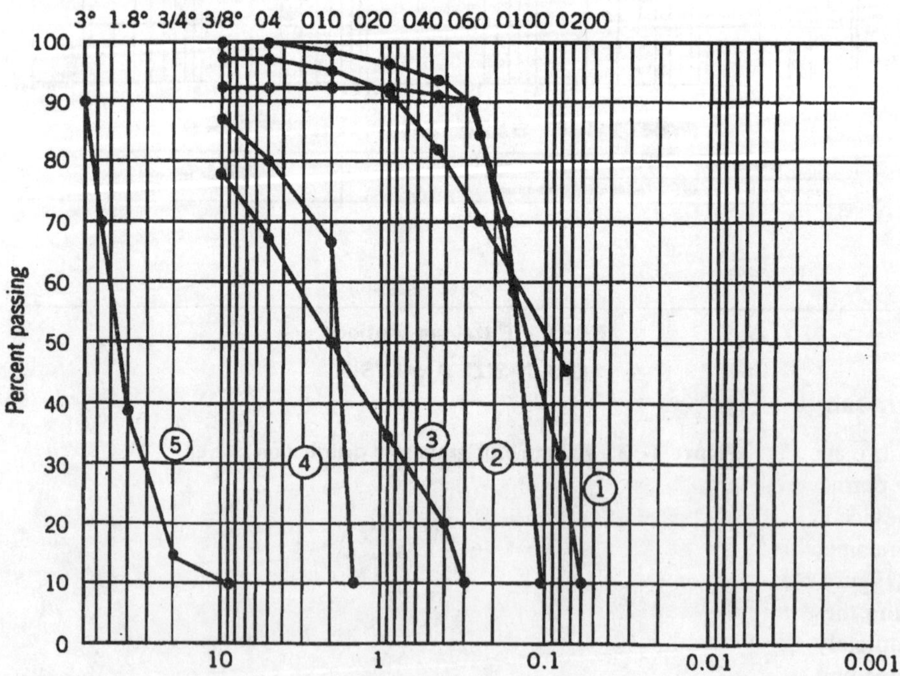

Required: Permeability values for the five soil samples.
Solution: Permeability, k(cm/sec) = D_{10}^2 (mm)

Curve Number	Soil Type	D_{10} (mm)	D_{10} Particle Size	k(cm/sec)
1	Sand	0.06	Silt	3.6×10^{-3}
2	Sand	0.1	Fine sand	1.0×10^{-2}
3	Gravel	0.3	Fine sand	9.0×10^{-2}
4	Gravel	1.5	Medium sand	2.3×10^{0}
5	Gravel	9.0	Fine sand	8.1×10^{1}

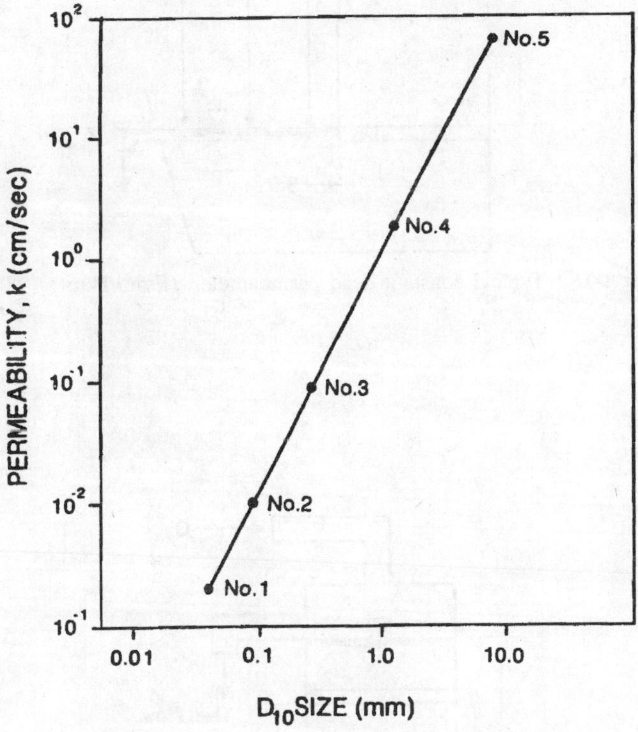

Permeability Tests

A more accurate method of determining soil permeability is by conducting laboratory permeability tests on representative samples obtained from the boring program. A device called a permeameter is used for testing (ASTM, 1990). A constant-head permeameter (Figure 1-15) is used for sands and gravels. A falling-head permeameter (Figure 1-16) is used for silts and clays. Because of changes imposed on the soils during the sampling, transportation, and preparation processes, samples are never completely undisturbed. Therefore, laboratory test results can be unreliable and misleading (Carson, 1961).

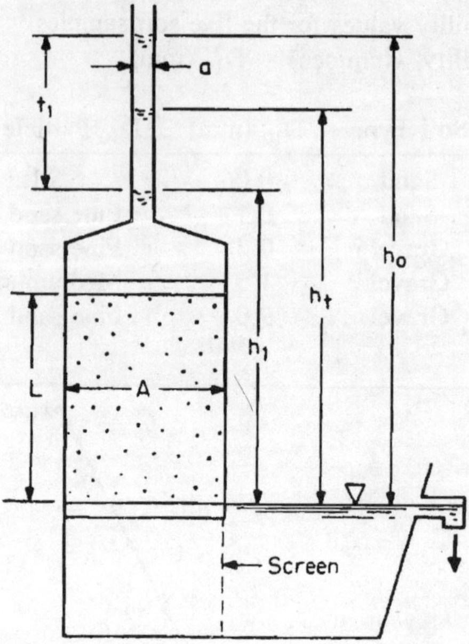

Figure 1-15 Typical constant-head permeameter. (From Hunt, 1984.)

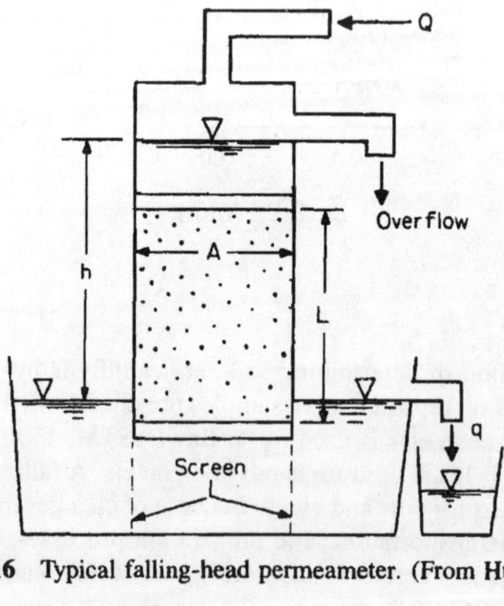

Figure 1-16 Typical falling-head permeameter. (From Hunt, 1984.)

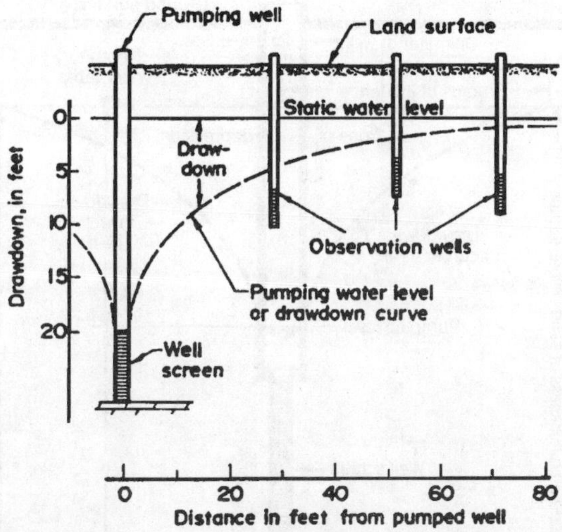

Figure 1-17 Typical pump test setup. (From Johnson, 1975.)

Pump Tests

The best method for predicting field permeability rates at a site is by conducting a full-scale field pump test where a test well, similar to the anticipated dewatering wells, is installed and pumped for a duration of time to predict flow rates and cone of depression geometry (Figure 1-17). The test well should penetrate the aquifer if practical and should be located near the center of the project site. A constant pump rate should be used and continued until equilibrium or static levels are reached in the observation wells. Two other pump rates can be tried to verify the results of the first pump test. The results should be analyzed using both equilibrium and non-equilibrium formulas.

The equilibrium well formula according to Johnson (1975), the one that applies to most groundwater conditions (Figure 1-18), is

$$Q = \frac{k\,(H^2 - h^2)}{1055 \log R/r} \tag{1-1}$$

where Q = pumping rate, gal/min
k = permeability, gal/day/ft^2
H = aquifer thickness, ft
h = depth of drawdown in the well, ft
R = cone of depression radius, ft
r = well radius, ft

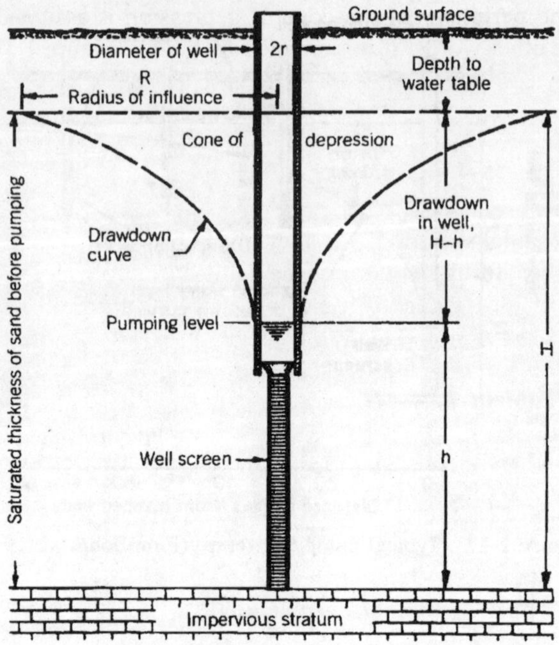

Figure 1-18 Equilibrium well formula. (From Johnson, 1975.)

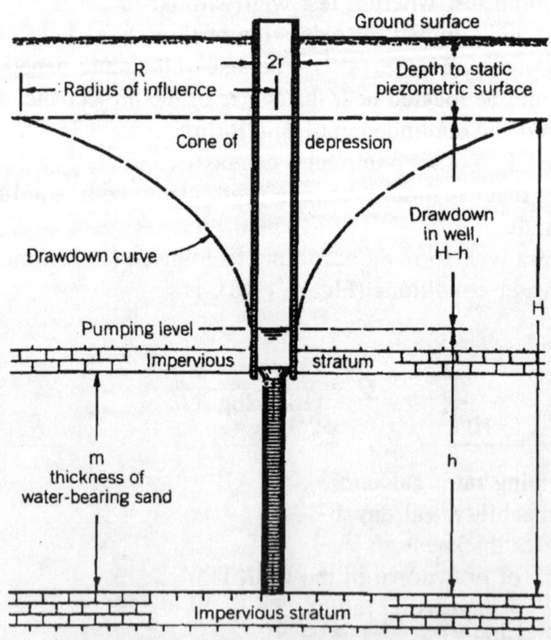

Figure 1-19 Pumping from an artesian aquifer. (From Johnson, 1975.)

Recharge at the periphery of the cone of depression is assumed. If the aquifer is confined, or in other words, if it is an artesian aquifer (Figure 1-19), the equilibrium well formula becomes

$$Q = \frac{km\,(H - h)}{528 \log R/r} \tag{1-2}$$

where the terms are as defined above except

m = aquifer thickness, ft

H = static head at the bottom of the aquifer, ft

These equations are frequently used to determine the field permeability from a pump test.

Example Problem 1-2 Calculating Permeability from Pump Test

Given

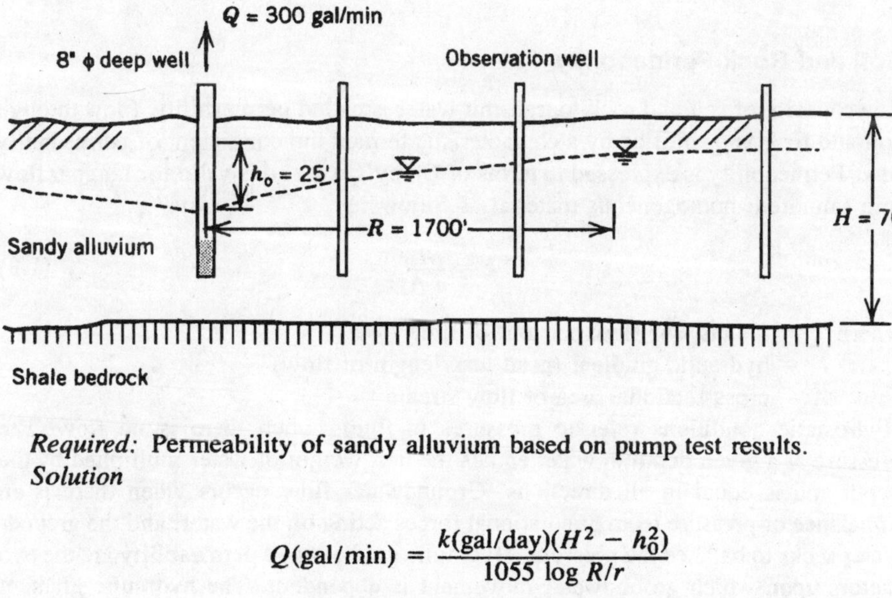

Required: Permeability of sandy alluvium based on pump test results.

Solution

$$Q(\text{gal/min}) = \frac{k(\text{gal/day})(H^2 - h_0^2)}{1055 \log R/r}$$

or

$$k = \frac{1055\,Q \log R/r}{H^2 - h_0^2} = \frac{1055 \times 300 \times \log 1700/0.33}{(70^2 - 25^2)}$$

$$= 275 \text{ gal/day} = 1.3 \times 10^{-2} \text{ cm/sec}$$

The nonequilibrium well formula, developed by Theis (1935), takes into account the effect of time on pumping. By use of this formula, the drawdown can be predicted at any time after pumping begins. Using this method can eliminate the need to reach a static condition in the observation wells during a pump test, thereby reducing the time and cost of the test. Also, only one observation well is required to develop the site hydraulic characteristics from a pump test, instead of the two needed for the equilibrium well formulas given above. While the Theis non-equilibrium formula is useful in running field pump tests, it is not often used in conjunction with dewatering calculations.

1-4 THEORETICAL BACKGROUND

Dewatering involves theory dealing with fluid flow through soil and rock media, aquifer properties, and hydraulic flow through pumps and pipes. To understand the requirements of a dewatering system and response characteristics to pumping, one must understand the concepts of permeability, transmissibility, storage, specific capacity, pump hydraulics, and flow thr ugh pipes. These concepts are discussed below.

Soil and Rock Permeability

The capacity of soil and rock to transmit water is called permeability. Flow through soil and rock is quantified by a characteristic termed the coefficient of permeability or k. Permeability is expressed in terms of Darcy's law and is valid for laminar flow in a saturated, homogeneous material as follows:

$$k = \frac{q}{i\,A} \tag{1-3}$$

where q = quantity of flow per unit of time
i = hydraulic gradient (head loss/length of flow)
A = cross-sectional area of flow stream

Hydrostatic conditions refer to pressures in fluids when there is no flow. The pressure at a given depth in water equals the unit weight of water multiplied by the depth and is equal in all directions. Groundwater flow occurs when there is an imbalance of pressure from gravitational forces acting on the water, and the groundwater seeks to balance the pressure. Hydraulic gradient and permeability are the two factors upon which groundwater movement is dependent. The hydraulic gradient between two points on the water table is the ratio between the difference in elevation of the two points and the distance between them. It reflects the friction loss as the water flows between the two points. Flow condition nomenclature is illustrated in Figure 1-20.

Flow in soil is affected by the grain size distribution and the dependent volume of voids through which water can pass. Since soil formations are often stratified and consist of alternating layers of coarse-grained and fine-grained soil, horizontal permeability may be greater than vertical permeability.

Flow in rock generally follows the path of joints, partings, shear zones, and faults of the formation. The intact rock is generally much less permeable than the jointing except in the case of highly porous rock such as coral. Joint conditions that

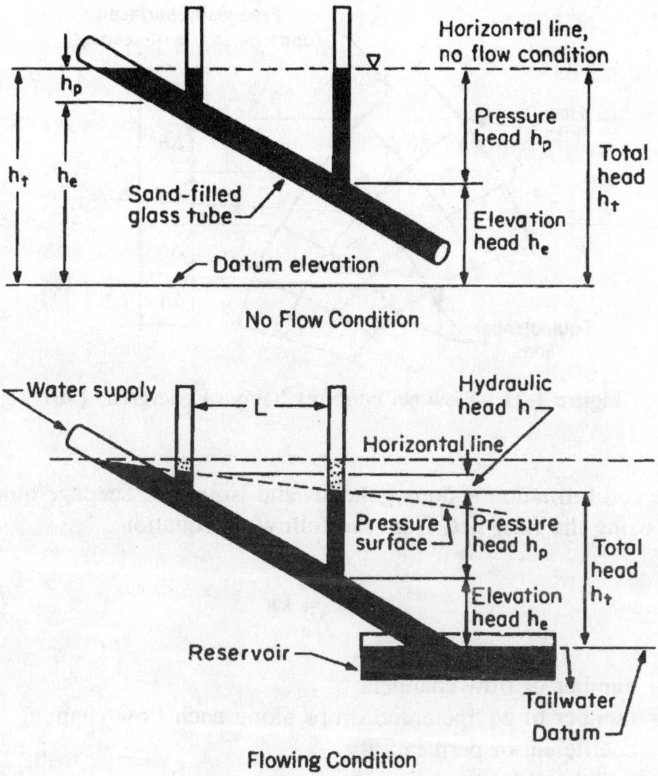

Figure 1-20 Hydraulic flow nomenclature. (From Hunt, 1984.)

affect rock mass permeability include spacing, orientation, continuity, interconnectivity, aperture width, and filling characteristics. Sedimentary rock formations often exhibit stratification of water flow, especially when coarse-grained rock such as sandstone is interlayered with fine-grained rock such as claystone and shale. Rock joints often stop at fine-grained rock boundaries, causing water to flow along bedding until another geologic feature permits flow across bedding.

Flow Nets

Flow through a soil medium may be represented by a flow net: a two-dimensional graphical presentation of flow consisting of a net of flow lines and equipotential lines, the latter connecting all points of equal piezometric level along the flow lines (Figure 1-21). Flow-net construction is accomplished by trial and error. The flow zone, bounded by the phreatic surface and an impermeable stratum, is subdivided on a scaled drawing of the problem area as nearly as possible into equidimensional quadrilaterals formed by the flow lines and equipotential lines crossing at right angles. The basic assumptions in flow-net construction are that Darcy's law is valid

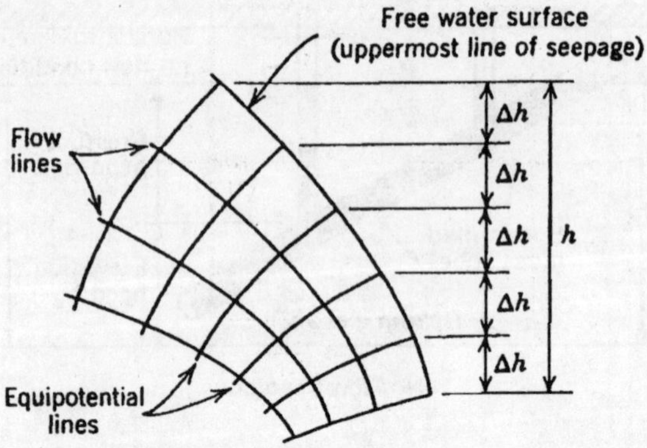

Figure 1-21 Flow net concepts. (From Cedergren, 1967.)

and that the soil formation is homogeneous and isotropic. Seepage quantity can be calculated, using the flow net, from the following equation:

$$q = \frac{N_f}{N_e} kh \qquad (1\text{-}4)$$

where N_f = number of flow channels
N_e = number of equipotential drops along each flow channel
k = coefficient of permeability
h = total head loss

Aquifer Characteristics

Other terms of interest when contemplating a dewatering program include transmissibility, storage, specific capacity, and radius of influence. The transmissibility, T, of an aquifer can be described as the ease with which water moves through a unit width of aquifer (Figure 1-22) and is defined as follows:

$$T = kB \qquad (1\text{-}5)$$

where k = coefficient of permeability
B = thickness of aquifer

If T is being determined from a pump test, then

$$T = \frac{Q}{d} \qquad (1\text{-}6)$$

where Q = pumping rate
d = change in drawdown per log cycle

The storage coefficient, C_s, is defined as the volume of water released from

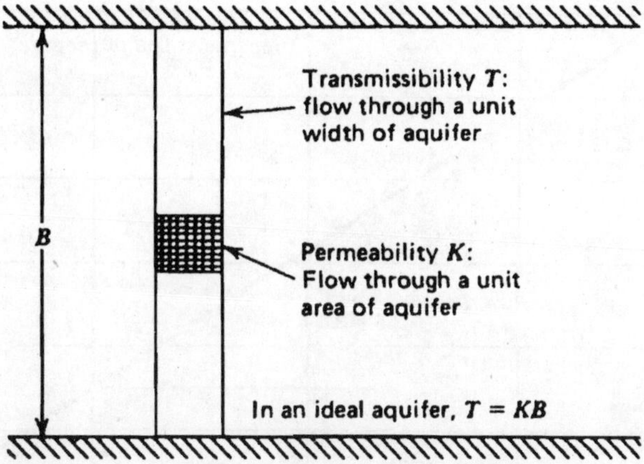

Figure 1-22 Aquifer characteristics. (From Powers, 1992.)

storage, per unit area, per unit reduction in head. In the average water table aquifer, C_s approaches 0.2 as water drains by gravity from the pores (Powers, 1992). In a confined aquifer, the pores remain saturated, but there is nevertheless a small release from storage when the head is reduced, due to the elasticity of the aquifer, and the compressibility of water. For confined aquifers, C_s is on the order of 0.0005 to 0.001. In rock aquifers, C_s can be lower than the above values by several orders of magnitude because of low effective porosity and rigid aquifer structure. If C_s is being determined from a pump test,

$$C_s = \frac{Tt_0}{r^2} \tag{1-7}$$

where T = transmissibility
t_0 = zero drawdown intercept (Figure 1-23)
r = distance of measurement from pumping well

The specific capacity of a well at time t, q_s, is defined as

$$qs = \frac{Q}{d} \tag{1-8}$$

where Q = pump rate at time t
d = drawdown at time t

The variables defined above are interrelated as shown in Figure 1-24.

An ideal aquifer has no recharge within the zone of influence of pumping. But, as illustrated in Figure 1-25, most natural aquifers are constantly discharging and being recharged. When dewatering begins, natural discharge from the aquifer diminishes. Recharge usually increases. For mathematical convenience, we say that the sum of the recharge from all the sources act as an equivalent single source, large in capacity, acting on a vertical cylindrical surface at distance R_0 from the

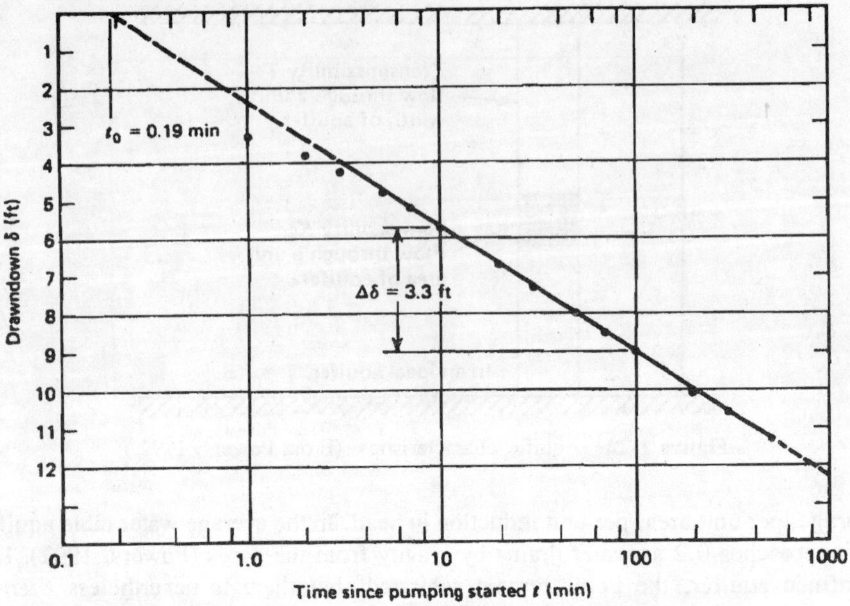

Figure 1-23 Zero drawdown intercept from pump test. (From Powers, 1992.)

center of pumping. R_0 is called the equivalent radius of influence. Dewatering volume varies inversely as the log of R_0. The only reliable indication of R_0 is from a properly conducted pump test (Powers, 1992). Lacking that, a rough guide to total recharge, and to the probable equivalent R_0, can be inferred from soil borings, permeability estimates, areal geology, and surface hydrology. Other ways to estimate R_0 include

$$R_0 = r_s + (Tt/C_s)^{1/2} \tag{1-9}$$

where r_s = equivalent radius of the pump array
T = transmissibility
t = pumping time
C_s = storage coefficient

and

$$R_0 = 3\ (H - h)\ k^{1/2} \tag{1-10}$$

where H = initial head
h = final head
$H - h$ = amount of drawdown, ft
k = coefficient of permeability, microns/sec

Water table aquifers can be analyzed using two-dimensional computer models such as FLOW-PATH and MODFLOW developed by the U.S. Geological Survey. A pump test is still recommended to define the characteristics of the aquifer and to

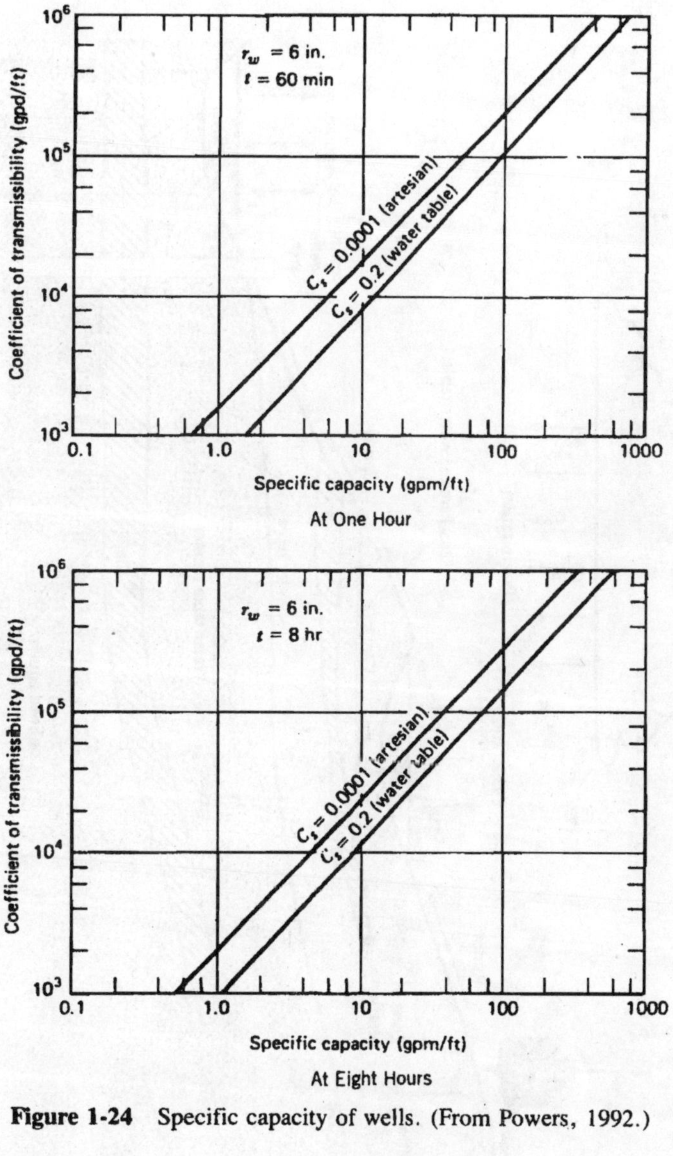

Figure 1-24 Specific capacity of wells. (From Powers, 1992.)

calibrate the model. Even with a computer model, several iterations are required to model the aquifer correctly. Figure 1-26 illustrates the type of output one can generate with such a program.

Pump Theory

Compared to the complexities of soils and groundwater, the pump is a rather straightforward mechanical device, whose performance should be predictable and

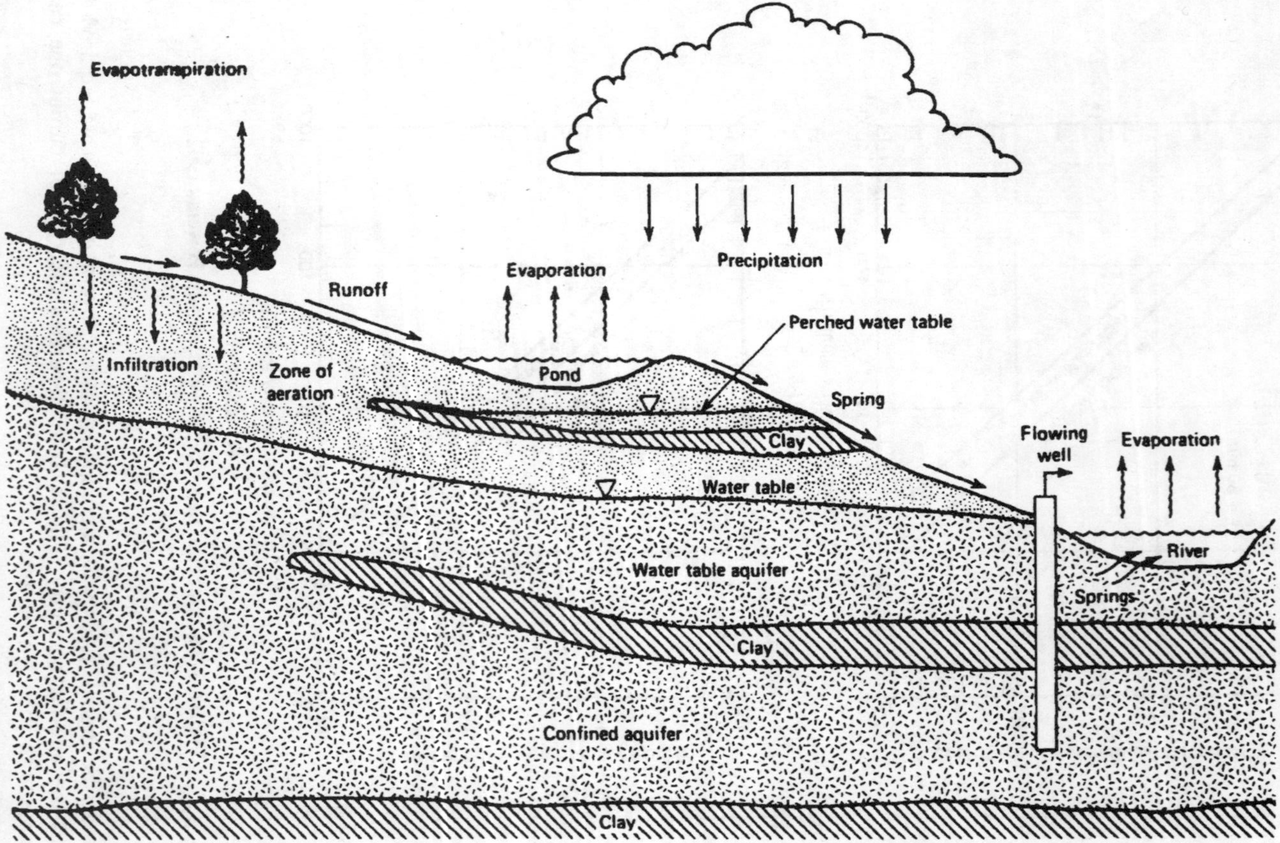

Figure 1-25 Natural aquifer characteristics. (From Powers, 1992.)

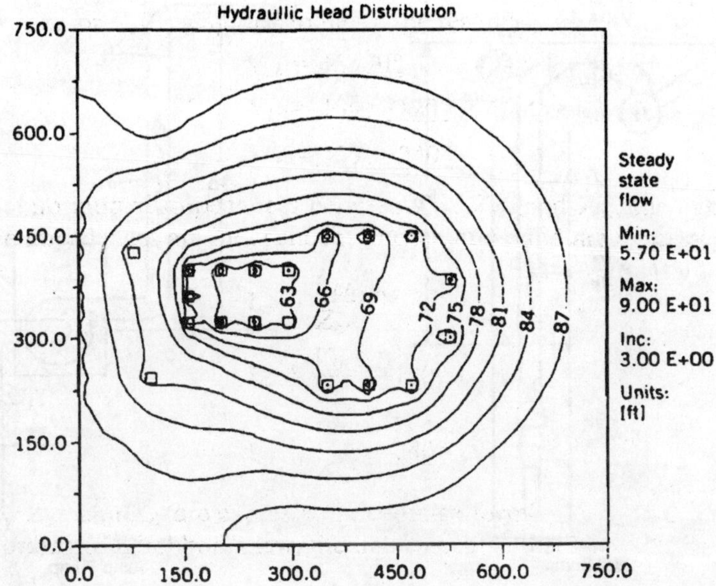

Figure 1-26 Computer modeling of well drawdown. (From Powers, 1992.)

reliable. The work a pump must accomplish, termed the water horsepower (WHP), is the product of the volume pumped times the total dynamic head (TDH) on the unit. TDH is the sum of all energy increase, dynamic and potential, that the water receives. Figure 1-27 illustrates the calculation of TDH in various pumping applications.

The well pump in Figure 1-27 faces a static discharge head h_D from the operating level in the well to the elevation of final disposal from the discharge manifold. In addition, the pump must provide the kinetic energy represented by the velocity head h_v. And it must overcome the friction f_1 in the discharge column and fittings and f_2 in the discharge manifold.

$$\text{TDH} = h_D + h_v + f_1 + f_2 \tag{1-11}$$

The velocity head, h_v, is calculated at the point of maximum velocity by

$$hv = \frac{v^2}{2g} \tag{1-12}$$

where v = water velocity
g = acceleration of gravity

The sump pump in Figure 1-27 faces a discharge head h_D, plus a suction head h_s, plus the velocity and friction heads. For the well point pump in Figure 1-27, it is not

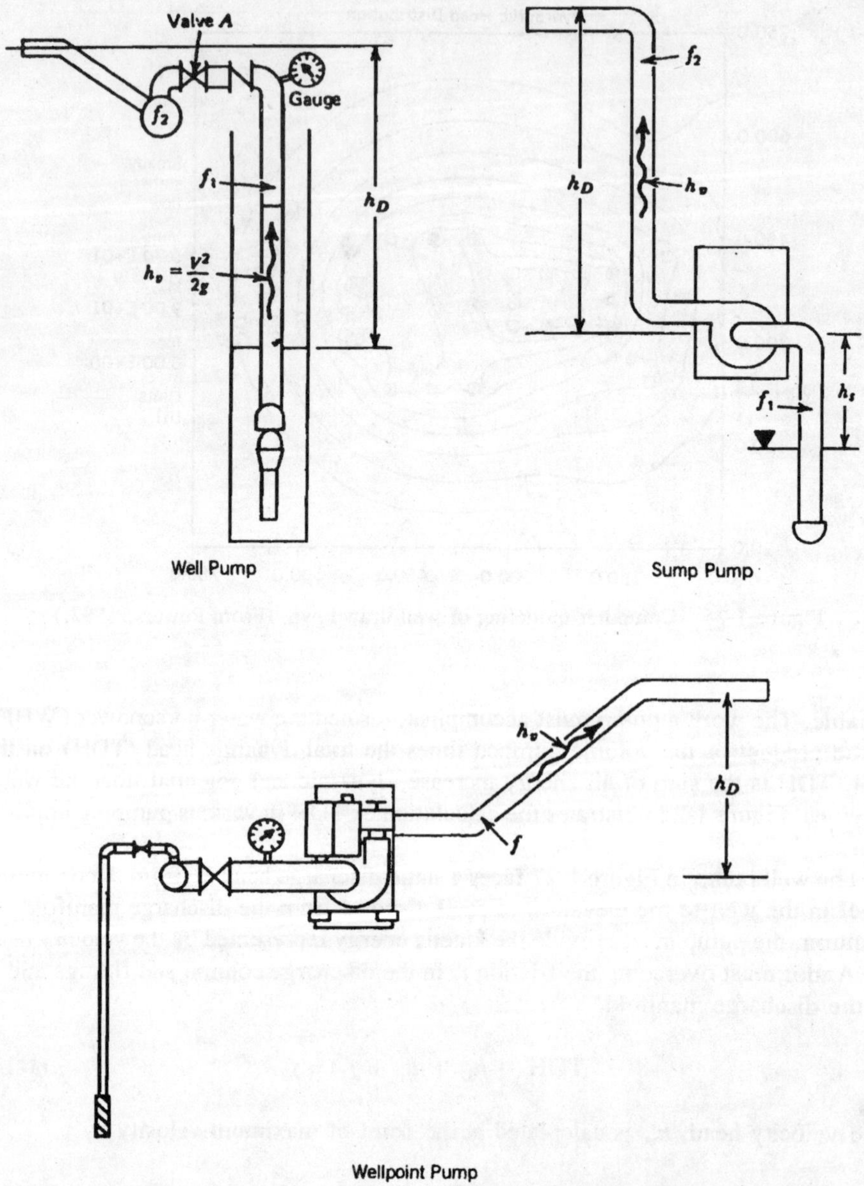

Figure 1-27 Variables in pump performance curves. (From Powers, 1992.)

possible to measure the suction head h_s. An approximate value can be estimated for h_s as equal to the maximum operating vacuum of the well point pump, usually 28 ft (Powers, 1992).

Figure 1-28 shows the basic performance curve of a centrifugal well point pump. The head-capacity curve shows the capacity of the pump at various values of TDH.

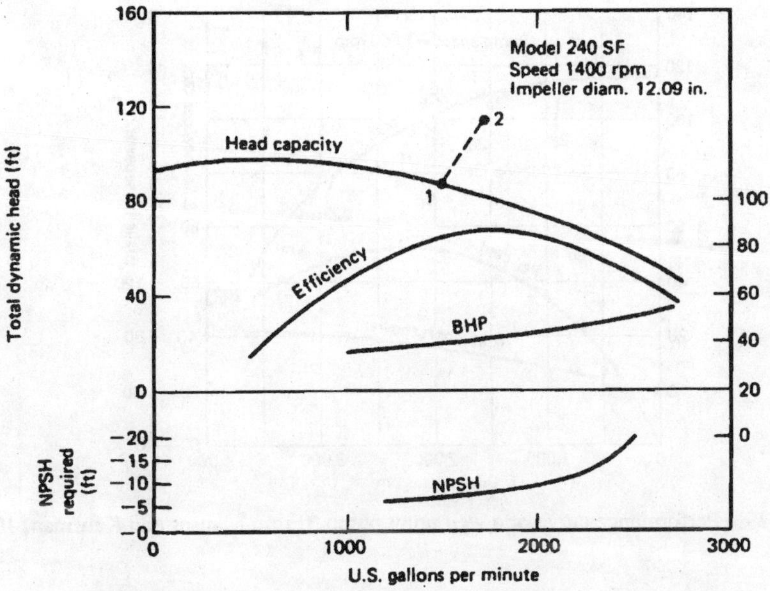

Figure 1-28 Total dynamic head of pump. (From Powers, 1992.)

The water horsepower (WHP) the pump is producing is the product of head and capacity with appropriate conversion factors.

$$\text{WHP} = \frac{\text{TDH(ft)} \times Q(\text{gal/min})}{3960} \tag{1-13}$$

The brake horsepower (BHP) is the amount of power that must be applied to the pump. It is greater than the WHP by the amount of hydraulic and mechanical losses in the pump. The efficiency, e, of the pump is

$$e = \frac{\text{WHP}}{\text{BHP}} \tag{1-14}$$

The BHP required by the pump is therefore

$$\text{BHP} = \frac{\text{TDH} \times Q}{3960 \times e} \tag{1-15}$$

Figure 1-29 shows a performance curve for a well point pump operating at a speed of 1600 revolutions per minute (rpm) for various suction lifts. Figure 1-30 shows a family of curves indicating the performance of one diameter impeller at various speeds. Figure 1-31 shows the performance at 1150 revolutions per minute (rpm) with various diameters.

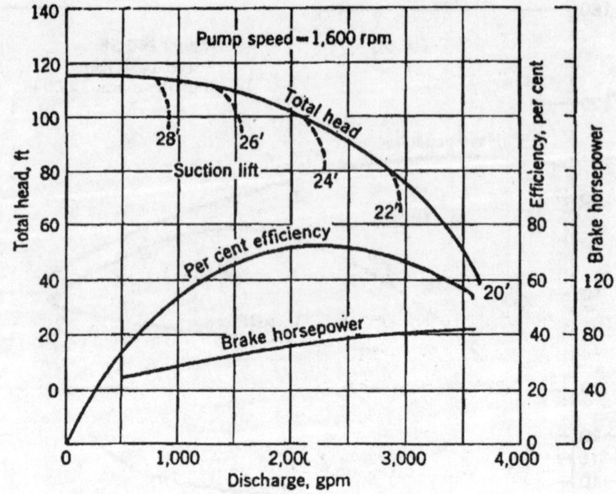

Figure 1-29 Performance curve of a well point pump. (From Mansur and Kaufman, 1962.)

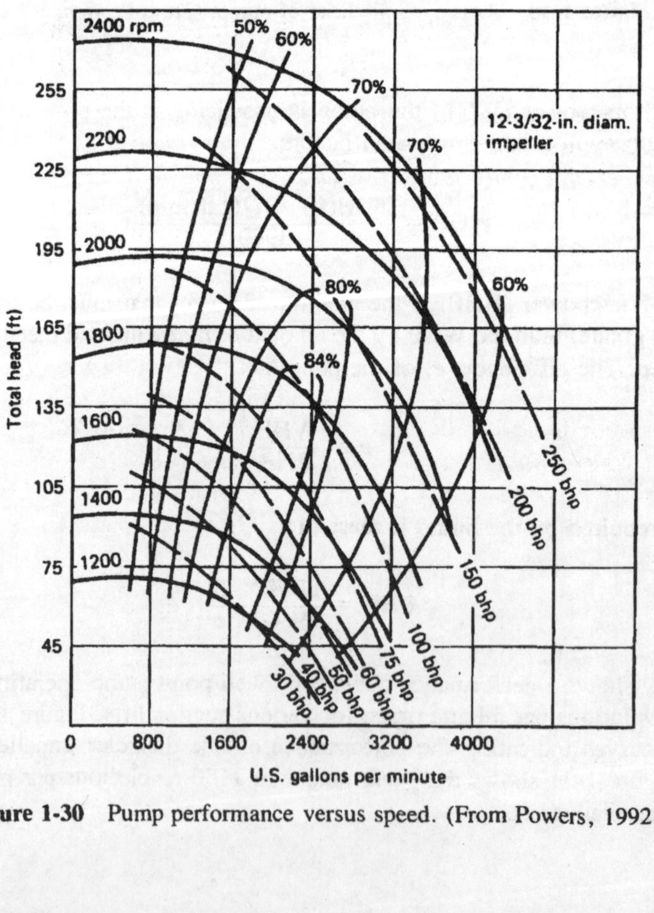

Figure 1-30 Pump performance versus speed. (From Powers, 1992.)

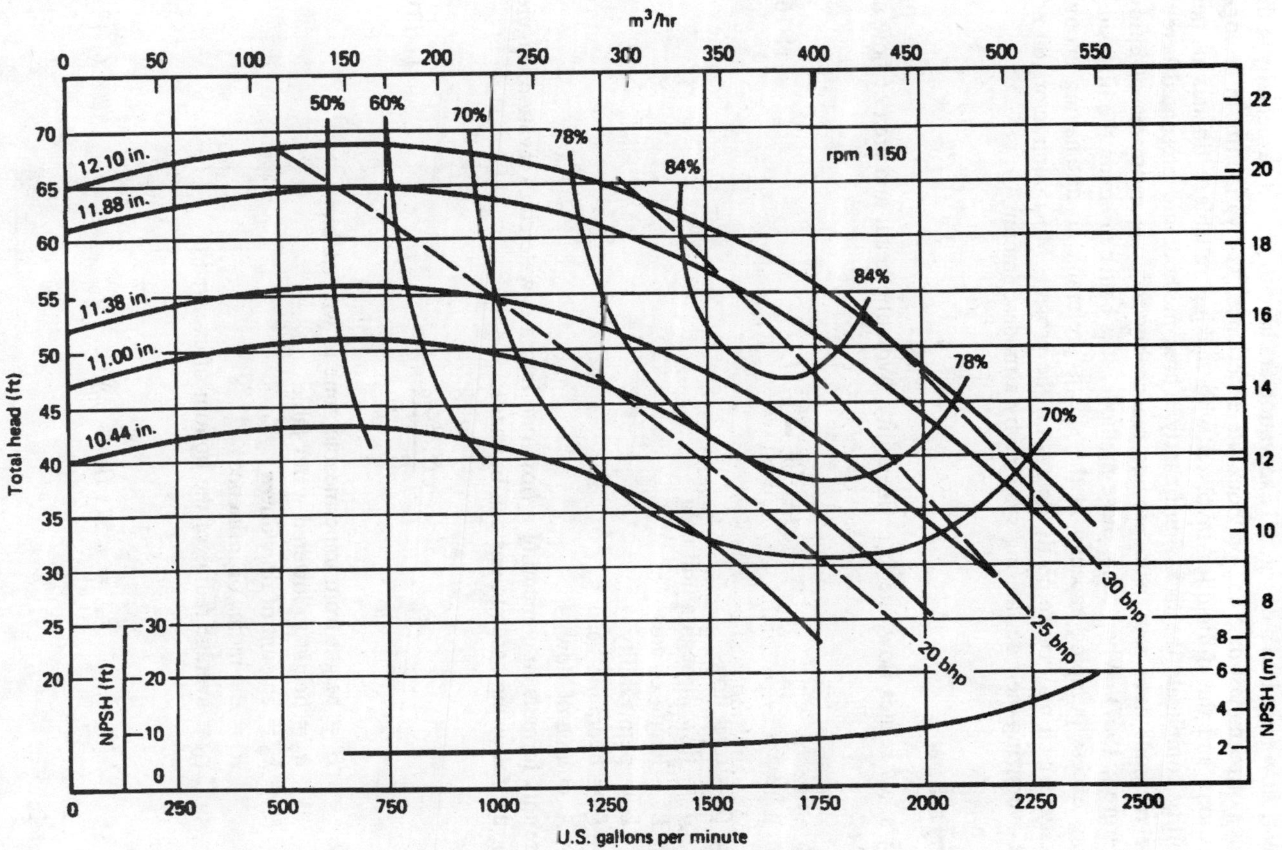

Figure 1-31 Pump performance versus impeller size. (From Powers, 1992.)

1-5 ESTIMATION OF FLOW RATES

At best, flow rates into a dewatered excavation are hard to predict, especially in the absence of pump test data and previous site experience. The variability in geologic formations and the difficulties involved with making reliable estimates of permeability complicate the task significantly. Three methods are presented herein. These methods have proven to provide reasonably accurate estimates of inflow (Cedergren, 1967). When using these methods, it is better to err on the high side. Having excess pumping capacity at the site is preferable to not having enough capacity, which may cause costly delays on the project. Most contractors do not mind oversizing their dewatering systems by a modest amount.

Darcy's Law

Darcy's law relates flow rate to permeability, hydraulic head, and area of flow as follows:

$$Q = qL = kiAL \tag{1-16}$$

where Q = total flow
q = flow through a unit area
L = length of area
k = permeability
i = hydraulic head
A = area of flow

To use this formula in computing a flow rate into a dewatered excavation (Figure 1-32), the following assumptions can be made:

$$i = \frac{H - h_0}{R - r_0} \tag{1-17}$$

where H = height from impermeable zone to water table
h_0 = height to lowered water table
$H - h_0$ = amount of drawdown
R = the radius of influence
r_0 = average radius of the bottom of excavation

and

$$AL = 1.571\ (H + h_0)\ (R + r_0) \tag{1-18}$$

Therefore

$$Q = 1.571\ k\ \frac{(H - h_0)(H + h_0)(R + r_0)}{(R - r_0)} \tag{1-19}$$

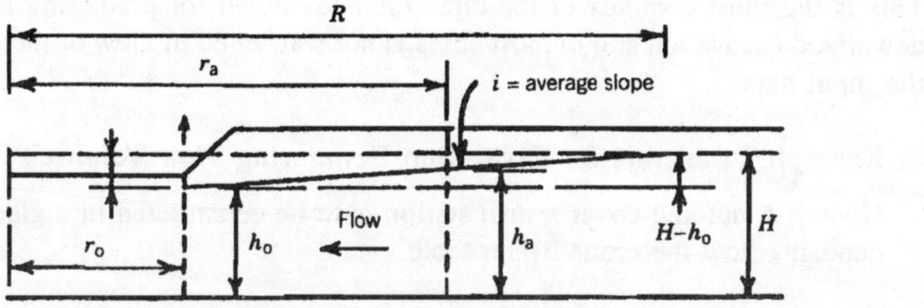

Figure 1-32 Darcy's law. (From Cedergren, 1967.)

Well Formulas

Another way to compute inflow into a dewatered excavation is by using the well formulas given above under the discussions of pump tests. Simplified, the nonartesian equation would appear as follows for this case:

$$Q = 1.366\,k\,\frac{H^2 - h_0^2}{\log(R/r_0)} \tag{1-20}$$

Two-Dimensional Flow Nets

Flow nets are a graphical representation of water flow (Figure 1-33) whereby, in the direction of flow, lines are drawn to designate individual flow channels (n_f) and perpendicular to the direction of flow, cross-flow lines, called equipotential lines, are drawn to designate head drops (n_d). A classic publication discussing the construction of flow nets is Casagrande (1940). After a flow net is constructed for the dewatered excavation, the amount of inflow can be calculated with the following equation:

$$Q = 3.14\,k\,(H - h_0)\,(R + r_0)\,n_f/n_d \tag{1-21}$$

where n_f = number of flow channels in flow net
n_d = number of head drops in flow net

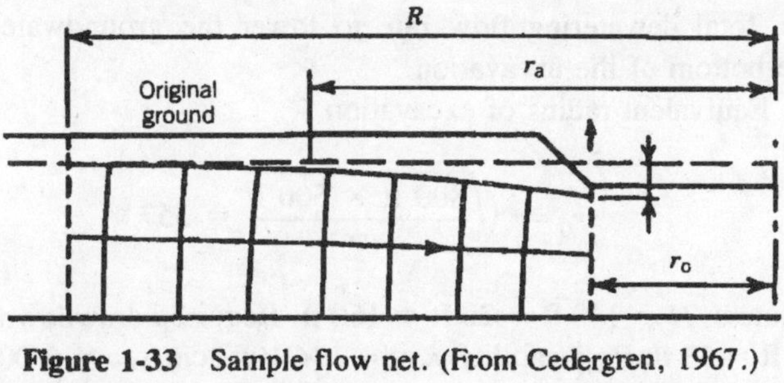

Figure 1-33 Sample flow net. (From Cedergren, 1967.)

This is the most complex of the three methods given for predicting inflow into a dewatered excavation and in most cases is not warranted in view of the reliability of the input data.

Example Problem 1-3 Excavation Dewatering Flow Volume

Given: A cut-and-cover transit station is to be constructed in a glacial outwash deposit below the groundwater table.

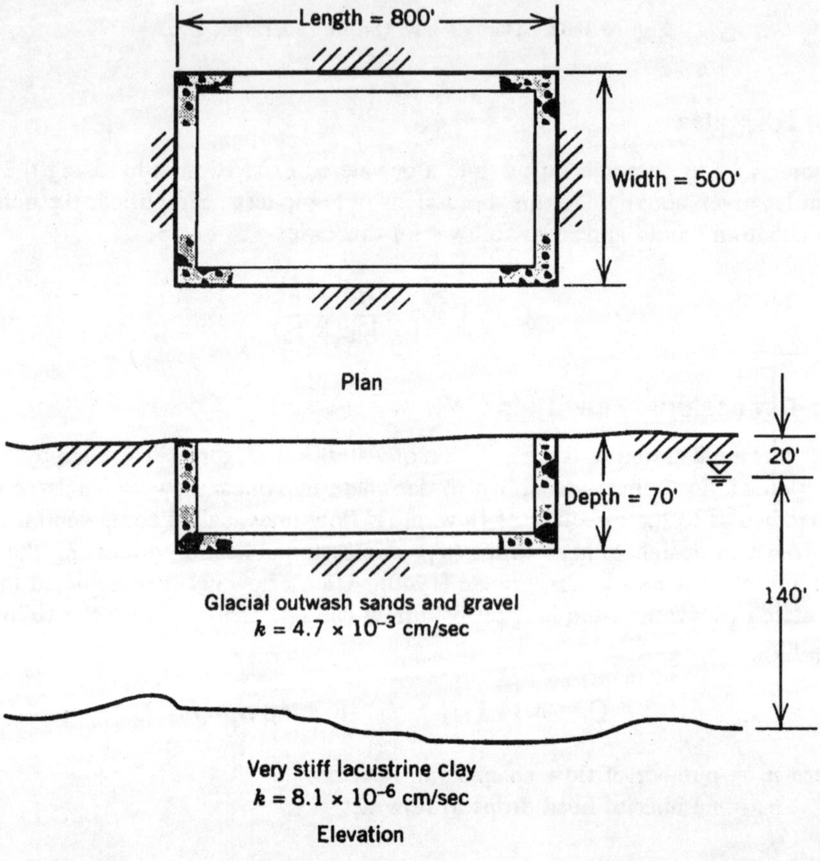

Required: Total dewatering flow rate to lower the groundwater table to 5 ft below the bottom of the excavation.
Solution: Equivalent radius of excavation

$$r_0 = \sqrt{\frac{800 \text{ ft} \times 500 \text{ ft}}{\pi}} = 357$$

Aquifer thickness, $H = 140$ ft + 20 ft = 160 ft. Required drawdown, $h_0 = 70$ ft + 5 ft − 20 ft = 55 ft. Permeability, $k = 4.7 \times 10^{-3}$ cm/sec = 0.00925 ft/min. Radius of dewatering, R, is unknown. Assume 1500 ft. Using Darcy's law,

$$Q = 1.57\,k\,\frac{(H - h_0)(H + h_0)(R + r_0)}{(R - r_0)}$$

$$= 1.57 \times 0.00925\,\frac{(160 - 55)(160 + 55)(1500 + 357)}{(1500 - 357)}$$

$$= 533 \text{ ft}^3/\text{min} \quad \text{or} \quad 3998 \text{ gal/min}$$

Using the simple well formula,

$$Q = \frac{1.37\,k\,(H^2 - h_0^2)}{\log\,(R - r_0)}$$

$$= \frac{1.37 \times 0.00925(160^2 - 55^2)}{\log\,(1500/357)}$$

$$= 459 \text{ ft}^3/\text{min} \quad \text{or} \quad 3442 \text{ gal/min}$$

Using a flow net,

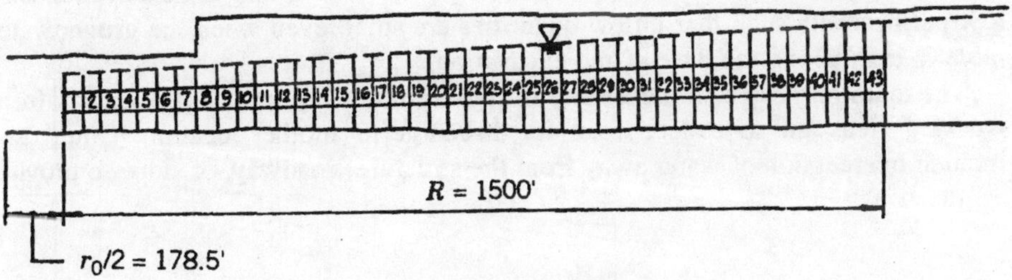

Number of flow drops, $n_d = 43$. Number of flow channels, $n_f = 3$.

$$Q = 3.14\,k\,(H - h_0)(R + r_0)\,\frac{n_f}{n_d}$$

$$= 3.14 \times 0.00925 \times (160 - 55)(1500 + 357)\,\frac{3}{43}$$

$$= 395 \text{ ft}^3/\text{min} \quad \text{or} \quad 2963 \text{ gal/min}$$

Based on the three alternative methods, the flow can be expected to be between 3000 and 4000 gal/min.

A pump test indicates that the field permeability rate $k = 9.2 \times 10^{-4}$ cm/sec and the radius of influence $R = 2200$ ft. The new solutions based on the pump test results are:

Method	Darcy's Law	Well Formula	Flow Net
Q (gal/min)	668	532	544

Say 500 to 700 gal/min.

1-6 DEWATERING METHODS

Dewatering methods include ways of passively accumulating excess water inflow into an excavation and ways of actively lowering water levels such that inflow into an excavation is within tolerable limits. The first category generally includes trenches, ditches, sumps, weeps, and drain pipes. Submersible pumps are usually used to transport water out of the collection points to a discharge facility. It is noted here that when a structure is built below the groundwater table and is not watertight, it can become a dewatering facility of sorts by attracting water and possibly lowering the surrounding groundwater table. More sophisticated methods are used to deliberately lower groundwater levels, including well points and deep submersible pumps. Other methods exist in the literature but are seldom used in practice.

Ditches and Trenches

Ditches and trenches are used frequently in excavations where the groundwater table is near the base of the excavation naturally or due to other groundwater lowering methods. Less often they are used when the ground surrounding the excavation is of low permeability such that inflow quantities are small even when the groundwater table is high above the base of the excavation.

The main purpose of trenches and ditches is to convey water inflow away from working areas and to collect water for discharge to another location. Other uses include interception of water away from the structure, roadway, or slope to provide

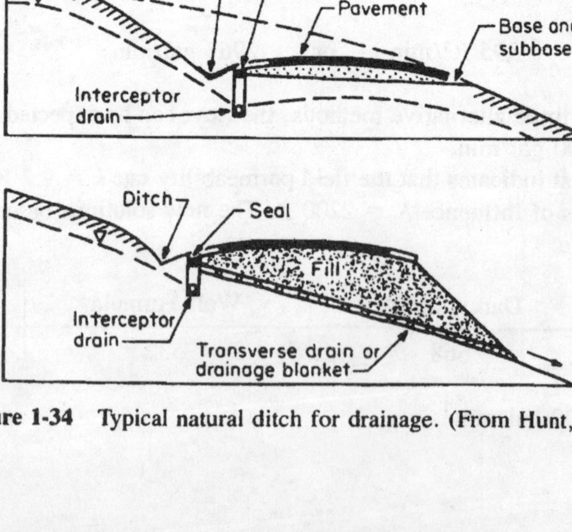

Figure 1-34 Typical natural ditch for drainage. (From Hunt, 1984.)

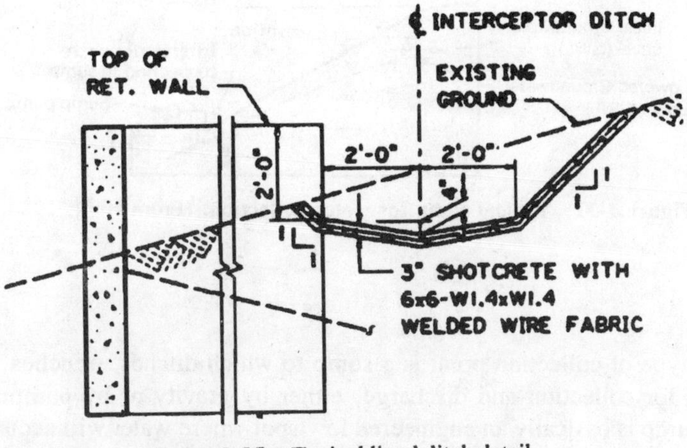

Figure 1-35 Typical lined ditch detail.

a dry surface or prevent the presence of unwanted water accumulations. The ditches and trenches can be in natural ground (Figure 1-34) or lined with engineered materials like shotcrete or concrete (Figure 1-35). There are usually collection points (drop inlets), pipes, culverts, and other discharge facilities to handle the accumulated water (Figure 1-36).

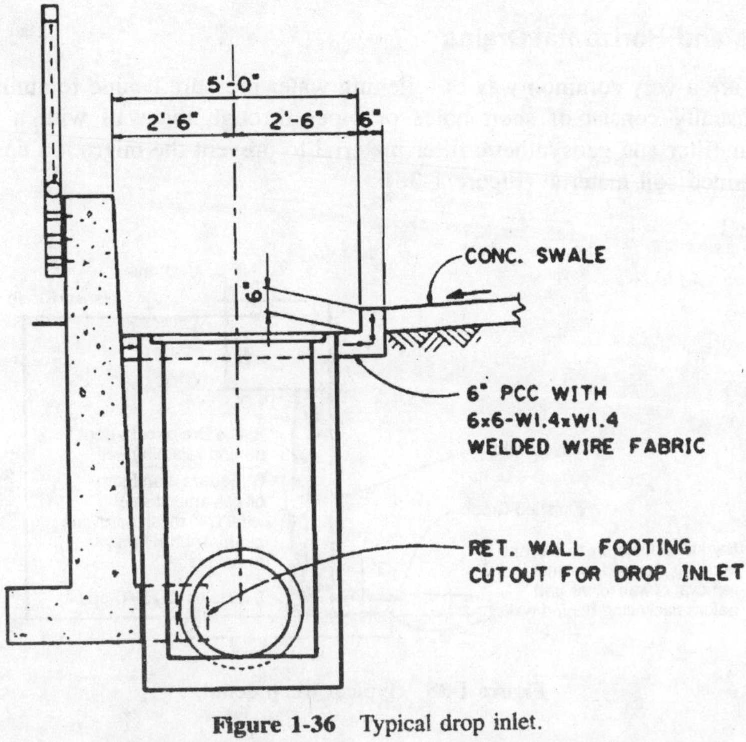

Figure 1-36 Typical drop inlet.

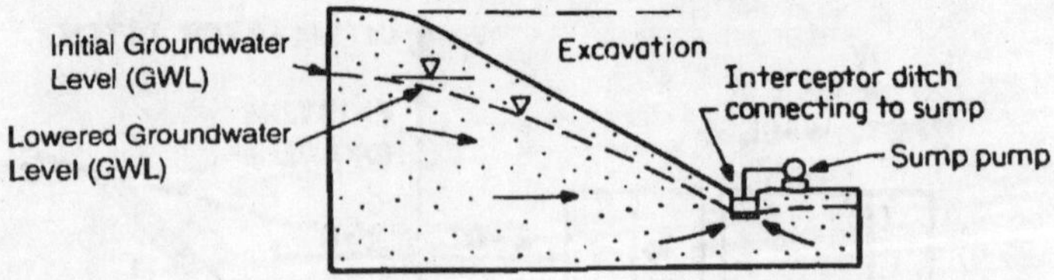

Figure 1-37 Typical sump for water collection. (From Hunt, 1984.)

Sumps

A special type of collection point is a sump to which ditches, trenches, weeps, or pipes lead for collection and discharge, either by gravity or by pumping (Figure 1-37). A sump is basically an engineered low spot where water will accumulate and flow to other more desirable locations. If discharged by gravity, a pipe will conduct water away from the sump and at a downward slope to a discharge point. If discharge is by submersible pump, the sump will require electrical power, an automated trigger mechanism, and a discharge pipe in addition to the pump.

One very important but elusive parameter for sizing pumps and discharge pipes is the magnitude of inflow expected. Since inflow often cannot be predicted with a high degree of certainty, it is usually estimated on the high side so that excess capacity is built into the system within practical limits.

Weeps and Horizontal Drains

Weeps are a very common way of relieving water pressure behind retaining walls. They usually consist of short holes or pipes through the wall with a pervious granular filter and geosynthetic filter material to prevent the migration and loss of fine-grained soil material (Figure 1-38).

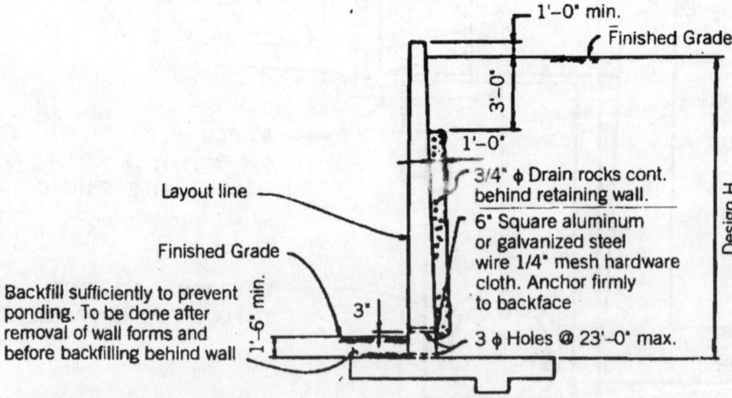

Figure 1-38 Typical weep detail.

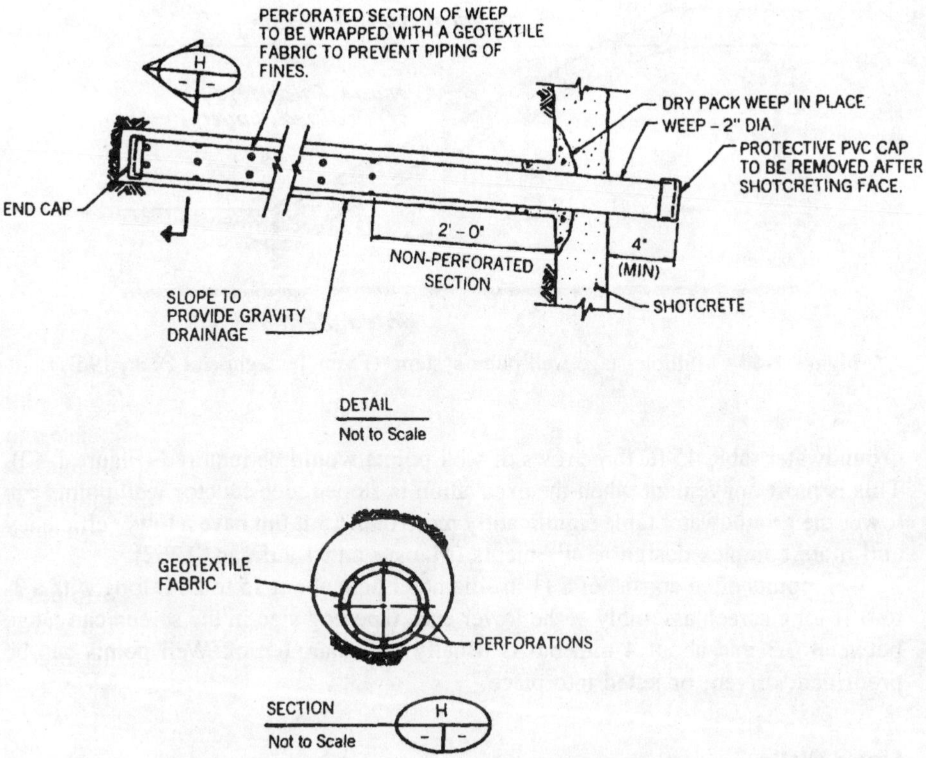

Figure 1-39 Typical horizontal drain detail.

Horizontal drains are basically smaller in diameter and longer than weeps and are installed into a soil or rock formation to prevent the buildup of water pressure and to direct water flow to a more desirable location (Figure 1-39). They are often used to reduce excess pore pressures in the vicinity of a landslide. Horizontal drains are usually between 100 and 200 ft long and consist of PVC pipe with holes and geosynthetic filter wrap.

Both weeps and drains are usually directed into a ditch, drop inlet, or other type of drainage way to carry the accumulated water away from the outlet area. Sumps and pumps may also be used.

Well Points

Well points are perforated pipes placed into the ground and connected to a pump by a header pipe to pull water out of the ground, thereby lowering the groundwater table. Normally, well points are put in a line or ring with 3 to 12 ft spacing between center-to-center. The header is usually between 6 and 12 in. in diameter. The header pipe is connected to a combined vacuum and centrifugal pump. One line of well points is adequate for about 15 ft of groundwater lowering. Therefore, to lower the

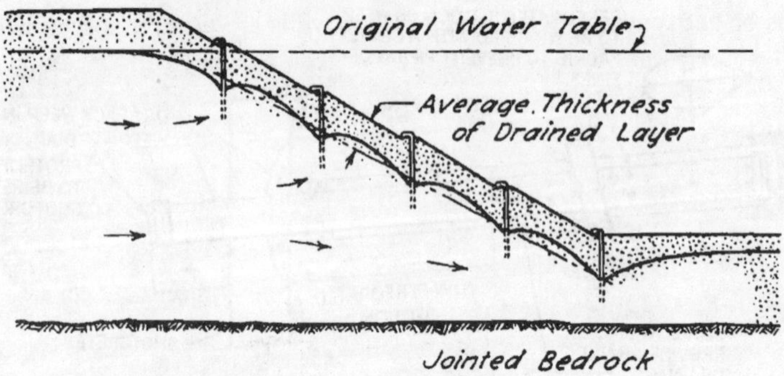

Figure 1-40 Multiple-stage well point system. (From Terzaghi and Peck, 1967.)

groundwater table 45 ft, three rows of well points would be required (Figure 1-40). This is most convenient when the excavation is sloped. Jet-eductor well points can lower the groundwater table significantly more than 15 ft but have a lower efficiency and more complex design requirements (Mansur and Kaufman, 1962).

Well points often consist of a $1\frac{1}{2}$-in.-diameter pipe about 15 to 20 ft long with a 3- to 4-ft-long screen assembly at the lower end. Opening size in the screen can range between 0.3 and about 4 mm but is usually less than 1 mm. Well points can be predrilled, driven, or jetted into place.

Deep Wells

For large excavations or where the depth of excavation below the groundwater table is more than 30 or 40 ft, or where artesian pressure in a deep aquifer beneath an excavation must be reduced, it may be more desirable to use deep wells and turbine pumps with or without well points as an accessory (Figure 1-41). Large-diameter deep wells are also suitable for lowering the groundwater table where the formation becomes more permeable with depth. If a more permeable zone underlies a less permeable zone, drainage in the upper zone can be enhanced with the installation of

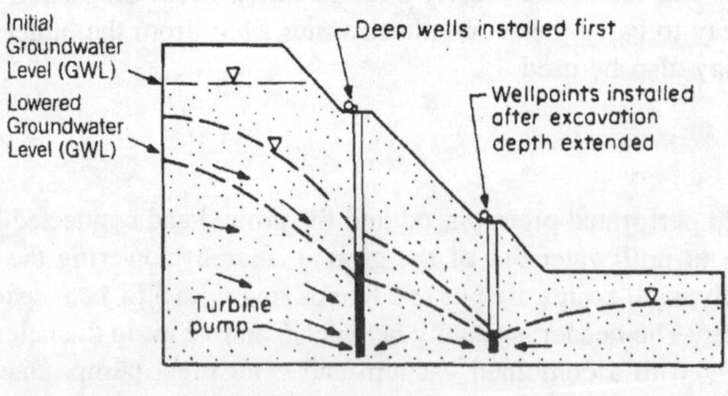

Figure 1-41 Dewatering with deep wells and well points. (From Hunt, 1984.)

sand drains or geosynthetic wick drains. Deep wells are usually spaced between 20 and 200 ft apart center-to-center and are between 6 and 18 in. in diameter. Well screens may be 20 to 75 ft long and surrounded by a sand-gravel mixture filter. Pumping is generally done with a submersible or vertical turbine pump installed near the bottom of the well.

Other Methods

Other methods for groundwater lowering exist but are seldom used. These include vacuum systems, which consist of well points that are airtight and sealed in boreholes. This technique is used in fine-grained soils, which tend to hold water in the void spaces between soil grains by capillary forces. Since it is difficult to make an airtight seal at the top of a borehole, this method has limited effectiveness and requires excess capacity to overcome the variability and ineffectiveness of each well point installation.

Electro-osmosis is a method of drainage that increases the amount of drainage toward well points or wells through the application of electrical current in the ground. This technique was largely developed by Casagrande (1952). By installing two electrodes, an anode (+) and a cathode (−), in saturated ground, water will flow from the anode toward the cathode. If well points or wells are used as the cathodes, water flow will be increased. While this approach is novel and technologically interesting, the additional complexity and cost of the dewatering system is usually not warranted.

1-7 DESIGN OF DEWATERING SYSTEMS

The design of a dewatering system starts with estimation of the inflow quantities using methods described above. In addition to knowing the permeability of the formation, the depth and plan dimensions of the excavation must by known. Duration of dewatering as it relates to power consumption, operating and maintenance costs, equipment rental, discharge facility usage costs, and so on, are important for determining which system is the most economical. Then selection of a system that will perform economically for the parameters and constraints set forth can be accomplished. Size and spacing of the wells or well points and pump sizes are key outputs in a design analysis.

Well Size

The size of well points varies by manufacturer but is usually between about 1 and 3 in. in diameter. The well point size characteristics are often a matter of preference. Other design characteristics can be obtained from manufacturer's information (Figure 1-42). A single well point seldom discharges more than about 10 to 20 gal/min.

The size of a deep well is implicitly required in the design, since the discharge capacity of the well is a function of the size, as well as speed, number, type and size

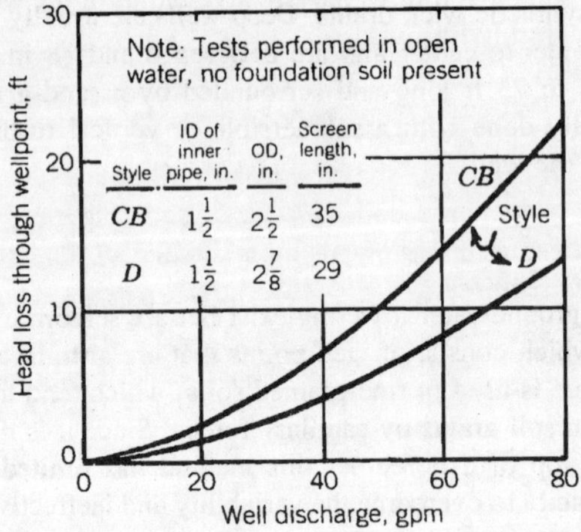

Figure 1-42 Well point design characteristics. (From Mansur and Kaufman, 1962.)

of impellers, and total dynamic head. Table 1-2 can be used to estimate the capacity of different sizes of wells. For a required inflow into an excavation, there is more than one potential solution since a greater number of smaller wells can do the same job as a smaller number of larger wells.

The rate of flow into a pumped well or well point depends upon the area and permeability of the ground immediately outside of the well and upon the hydraulic gradient causing the flow. Bush (1971) gives the following formula for approximating flow into a well:

$$Q = 44\, k^{1/2} r_w h_0 \tag{1-22}$$

where k = permeability, ft/min
r_w = effective radius of the well, ft
h_0 = depth of immersion of well, ft

TABLE 1-2 Capacities of Common Deep Well Pumps

Pump Bowl Size (in.) (Minimum i.d. of Well Pump Will Enter)	Preferred minimum i.d. of well (in.)	Approximate Maximum Capacity (gal/min)
4	5	90
5⅝	6	160
6	8	450
8	10	600
10	12	1,200
12	14	1,800
14	16	2,400
16	18	3,000

Source: From Mansur and Kaufman, 1962.

It follows then that for a given flow, the effective radius of the well or well point can be approximated with this formula.

Well Spacing

Charts can be used to determine the spacing of well points based on the amount of groundwater lowering required and the type of soil (Figures 1-43 and 1-44). For deep wells, the spacing required equals the perimeter of the excavation divided by the number of wells required.

Pump Size

Centrifugal pumps are usually used for well point and deep well collection systems. Detailed data can be found in textbooks on hydraulics and catalogues issued by pump manufacturers. The selection of a pump and motor depends on the required discharge, total dynamic head, suction lift, air-handling capacity, power available,

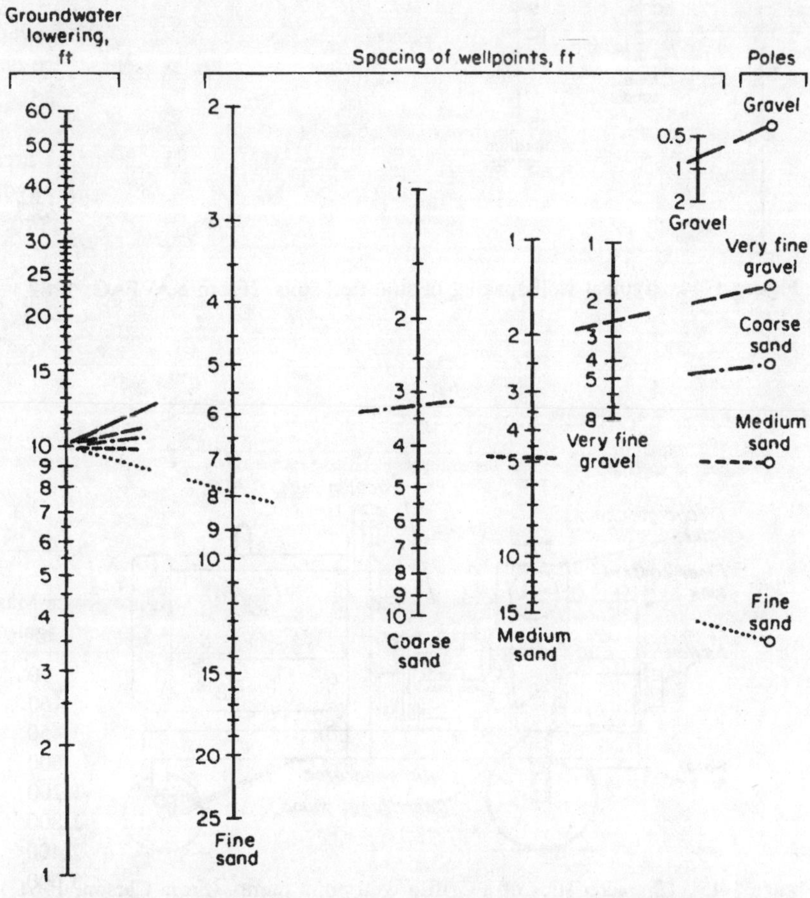

Figure 1-43 Typical well spacing in granular soils. (From NAVFAC, 1982.)

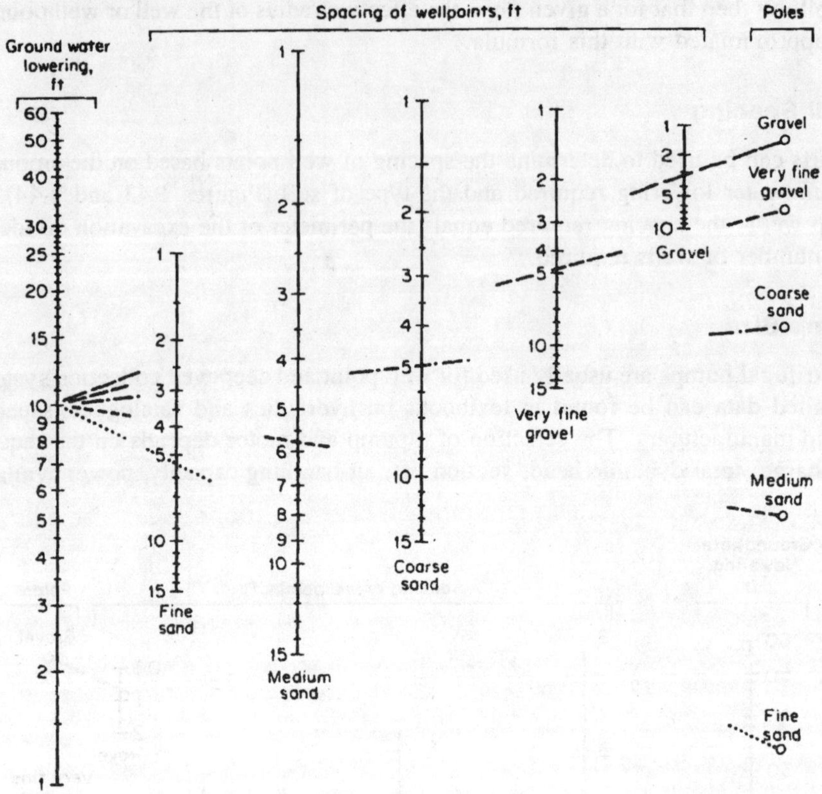

Figure 1-44 Typical well spacing in stratified soils. (From NAVFAC, 1982.)

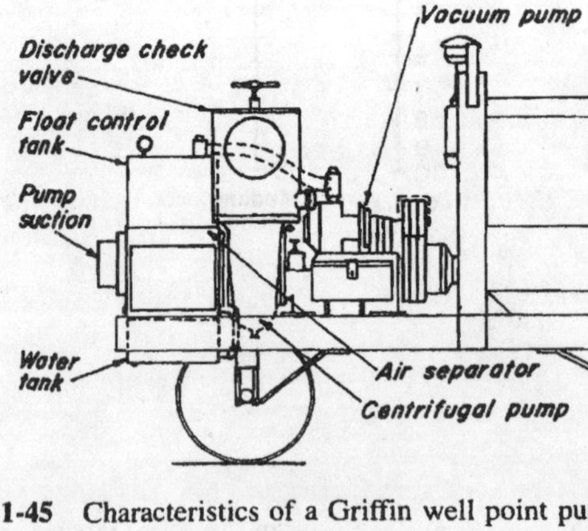

Figure 1-45 Characteristics of a Griffin well point pump. (From Carson, 1961.)

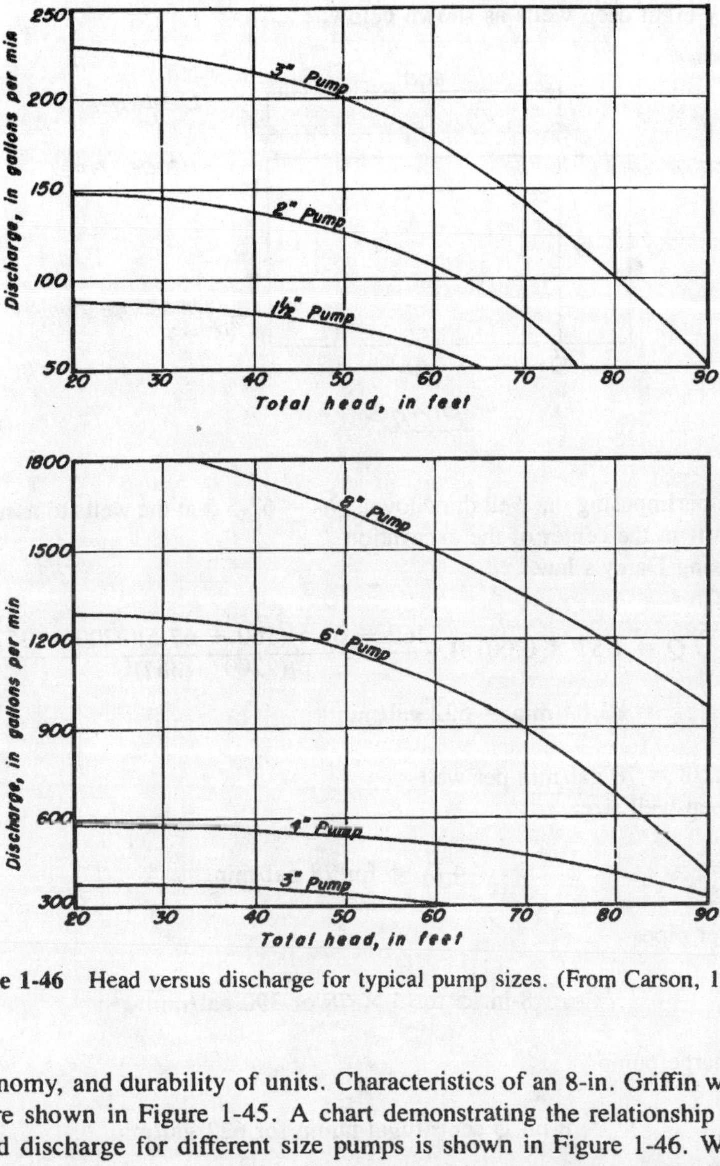

Figure 1-46 Head versus discharge for typical pump sizes. (From Carson, 1961.)

fuel economy, and durability of units. Characteristics of an 8-in. Griffin well point pump are shown in Figure 1-45. A chart demonstrating the relationship between head and discharge for different size pumps is shown in Figure 1-46. Well point pumps may be driven by gasoline, diesel, or electric power and are often 1500-gal/min, self-priming, centrifugal type. They can have up to a 4000-gal/min capacity.

Example Problem 1-4 Sizing of Dewatering System

Given: The excavation and conditions described in Example Problem 1-3.
Required: The dewatering system design.
Solution: Because the depth of drawdown is greater than 15 ft and the excavation walls will be vertical, a deep well dewatering system is required.

Try eight deep wells as shown below:

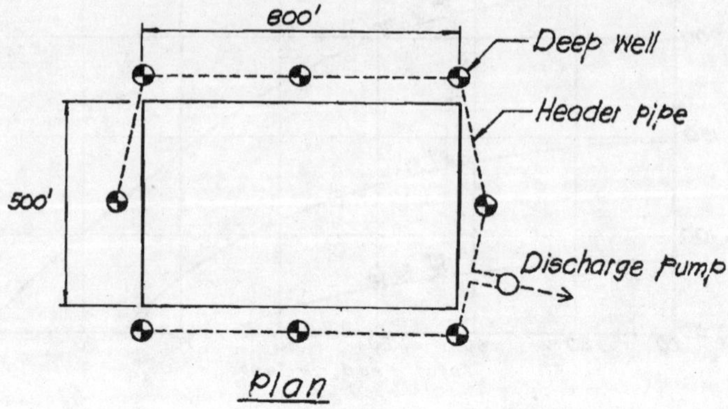

By superimposing the well drawdowns, $h_0 = 67.5$ ft at the wells to assure that $h_0 = 55$ ft in the center of the excavation.

Using Darcy's law,

$$Q = 1.57 \times 0.00181 \frac{(160 - 67.5)(160 + 67.5)(2200 + 357)}{(2200 - 357)}$$

$$= 83 \text{ ft}^3/\text{min} = 622 \text{ gal/min}$$

or 622/8 = 78 gal/min per well.

Deep well size:

4-in. ϕ for 78 gal/min

Header pipe:

8-in. ϕ for 5 × 78 or 390 gal/min

Discharge pump:

8-in. ϕ centrifugal pump for 622 gal/min

Additionally, valves at each deep well and header pipes are needed, as are swing joints and other associated hardware.

1-8 PLUMBING

Although well points and pumps are the essential components of dewatering systems, connecting pipes and valves are important parts of the system also (Figure

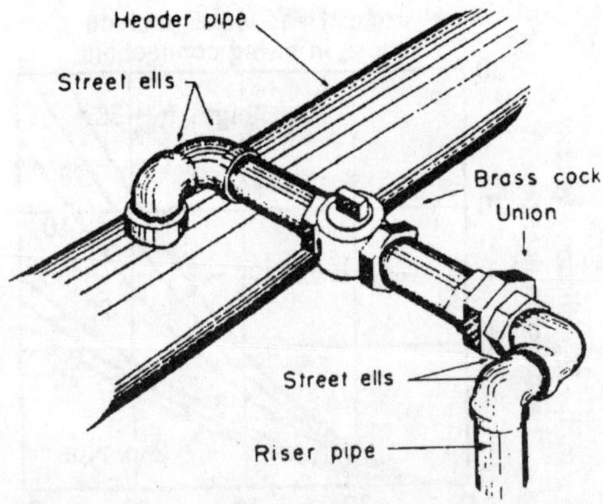

Figure 1-47 Swing connector for dewatering systems. (From Johnson, 1975.)

1-47). They must be sized correctly to be compatible with the other parts of the system, must be reliable, must cause minimal friction losses, and must be easy and fast to assemble.

Headers

Header pipes, or "headers," usually consist of plain-end, standard weight steel pipe to which multiple inlets are welded. PVC pipe is sometimes used instead of steel. The inlets are located on 2- to 3-ft centers, depending on the size and length of pipe section. Well point spacing is a multiple of 2 or 3 ft, depending on the header pipe configuration or vice versa. Therefore, a 6-ft well point spacing would use every other or every third inlet point for a 3- and 2-ft inlet spacing, respectively.

Valves

Valves are required throughout the system to isolate certain parts of the system, for priming parts of the system, and for finding leaks. As a minimum, there are usually main valves at the pump and header as well as individual valves at each well or well point. Gate valves are most often used because of their low cost, availability, and low head loss when fully open.

Pipes and Connectors

Various pipes, joints, elbows, tees, and other connectors are required to assemble the system completely. One important criterion is that no more than absolutely necessary are used because each one represents frictional head losses that diminish the capacity of the system, or in other words, will necessitate a larger system and higher costs. Connecting pipes should be as short as possible. This means the layout

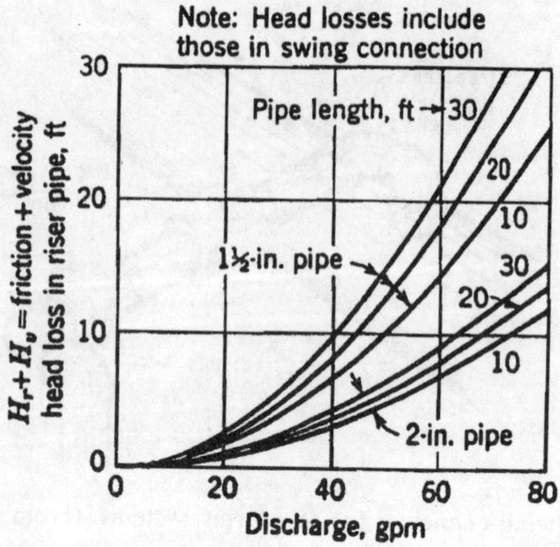

Figure 1-48 Frictional head losses in pipes. (From Mansur and Kaufman, 1962.)

of the system has to be well thought out to optimize the proximity of the pumping apparatus relative to the wells or well points and the discharge points. Typical values of frictional head loss for different size pipes are shown in Figure 1-48.

1-9 DISCHARGE

The discharge of drainage water is a major issue for permanent facilities and construction sites because of the quantities of water involved and strict requirements on the quality of the discharged water. Once water is removed from the ground and collected, it is subject to stricter requirements than if it were just left in place. The owner or contractor responsible for collecting the water is also responsible for the quality of the discharge regardless of the type of pollutants in the water or how they got there. Government regulations are in place that dictate effluent quality requirements as well as treatment, remediation, and monitoring needs. Permits that spell out dewatering and drainage plans in detail are generally required before discharge is approved by the governing agencies.

After the amount, duration, and means of discharge are ascertained, the receptacle into which discharge will occur must be determined. The choices are usually storm drains, sewers, or bodies of water. The last choice is seldom used because of the extremely strict quality and environmental protection requirements that must be met. The use of storm drains or sewers usually has a cost associated with it and also has strict quality requirements. Often storm drains and sanitary systems are combined. Treatment and recharge may be required if the water quality is poor, water supply is affected, or lowering of the groundwater table will have detrimental effects on adjacent properties.

Storm Drains

Storm drains are frequently available in most cities. With permission of the owner and governing agencies, water can be discharged using available inlets and manholes. Strict water quality requirements can be anticipated, with destination points of the effluent being a major concern. If the storm drain discharges into a body of water, governing water quality requirements will be those of the body of water, not the storm drain itself.

The size and capacity of the storm drain will govern the amount of discharge it can receive. It must have excess capacity to permit additional discharge beyond its operating requirements. Also, in most cases, the storm drain must not be pressurized or, in other words, must flow by gravity.

Sanitary Sewers

Sanitary sewers are another potential receptacle for dewatering or drainage discharge. To use a sanitary sewer for discharge, one must know whether the flow is going toward a treatment plant or away from a treatment plant (i.e., is the effluent flowing through the pipe or box raw sewage or treated water?). If the flow is going toward a plant, increases of the plant usage may be a concern of the owner. If the flow is going away from a plant, increases in discharge flow and effects on treated water quality may be of concern. Size and capacity will be of concern for a sanitary sewer. In addition the contents usually must flow by gravity for connection to a dewatering system.

Bodies of Water

The types of water bodies that can be used for discharge include canals, rivers, lakes, and oceans. One major advantage of using a body of water for discharge is that generally large quantities can be disposed of. Because of environmental concerns including preservation of wildlife, wetlands, recreational usage, historical artifacts, water supply, power generation facilities, transportation routes, and so on, regulations governing discharge into bodies of water are quite restrictive. In addition to very strict water quality requirements, extensive studies and time consuming environmental impact evaluations need to be made. It is normally more expedient to select another type of discharge location.

Recharge

Sometimes a requirement for permission to dewater is that the discharge be reinjected into the ground without pollutants to prevent such effects as subsidence, rotting and corrosion, and changes in water supply. Some form of treatment will be required before doing so, as described in the next section. Recharge must be done far enough away from the excavation and in such a way that the dewatering effect is still achieved. The design of a recharge system is similar to a dewatering system with consideration given to pumping capacity, water flow distribution, and avoiding the pumping of fine soil particles by use of filters. Recharge systems are usually used only in extreme circumstances, such as construction near a historical landmark that is prone to settlement, a water supply system serving a protected species, or some such specialized situation.

Treatment

As a minimum, water from a construction site usually requires removal of soil particles (sedimentation) and oil separation (skimming) before it can be discharged. There are mobile treatment plants that can be transported to construction sites for the treatment of more deleterious pollutants. Flow quantities must be estimated very carefully because of the additional processes involved and dewatering costs are obviously much higher when treatment is required. Mobile treatment plants are particularly useful on sites where no treatment was anticipated and fast reaction is required to prevent delays on the project. Alternatively, a more permanent facility may be required because of high flow quantities, long project durations, or long-term treatment requirements for drainage systems in the permanent facility.

1-10 EFFECTS ON ADJACENT STRUCTURES

Adjacent structures can be affected by changes in groundwater levels. The most common effect is settlement or subsidence due to increased effective soil stress in a compressible soil zone due to groundwater lowering. Differential settlement can also cause angular distortion of a structure. Loss of fine soil particles due to pumping can also induce settlement as ground is lost to the voids created. Another potential hazard to adjacent structures is exposure to the air of previously submerged subsurface features, such as wooden piles; such exposure accelerates the rotting process.

Consolidation

Soil consolidation is the compression of soil due to the delayed volume change associated with pore pressure dissipation in response to a change in soil stress. Consolidation settlement occurs predominantly in fine-grained compressible soils that cannot dissipate pore pressure changes quickly due to low permeability. The lowering of a groundwater table can increase the effective stress on a compressible layer by changing its stress state from submerged and buoyant to saturated and not buoyant. This change in stress is first taken on by the water in the pores of the soil. As the water is squeezed out of the pores and the soil grains start to bear the load, a volume change occurs that may cause settlement. For instance, a 15-ft-thick clay layer ($C_c = 0.36$, $e_0 = 1$) at ground surface would consolidate about 10 in. if the groundwater table was lowered from ground surface to the base of the clay layer. If the clay layer was 25 ft below ground surface and the water table was lowered from the top of the clay layer to the bottom of the clay layer, the settlement would be about 2 in. In addition to the settlement of overlying structures, this phenomenon can also cause additional loads on deep pile foundations by inducing additional downward vertical load, called negative skin friction, on the piles.

If consolidation of soils nearby is anticipated, and if such consolidation may have adverse effects on adjacent structures, settlement monitoring points should be placed on the suspect structures to document the rate and magnitude of any settlement that occurs. It is also a good idea to perform preconstruction surveys of nearby structures to document preexisting conditions. If observations indicate that intolerable levels of settlement are occurring, pumping should be halted until new estimates are made based on actual field observations and preventative or remedial measures are planned and implemented.

Pumping of Fines

Another potential hazard to structures adjacent to a dewatering site as well as to the excavation supports is the pumping or piping of fine-grained soil particles. If this occurs in significant quantities, voids may be created that are filled by overlying soils, creating depressions at the ground surface. If the dewatering system is designed, operated, and monitored properly, with suitable screens and filters in place, the piping of fines should not occur. If observations indicate a high amount of fine soils in the discharge water, pumping should be stopped until the source of these fines can be determined and corrected.

Settlement

Both of the phenomena above, as well as elastic response of soils, can cause settlement adjacent to dewatered excavations. It should be noted here too that there will likely be settlement due to the excavation itself, which must be considered in the context of the overall effects on adjacent structures that construction may cause. The total anticipated settlement around the dewatered excavations should be estimated and evaluated with respect to the allowable settlements that adjacent structures may tolerate. As a rough guide, masonry walls can tolerate about 1 to 2 in. of total settlement; framed buildings can tolerate about 2 to 4 in. of total settlement (Sowers, 1979).

Angular Distortion

Angular distortion represents the differential vertical movement between two points divided by the horizontal distance between the points. Most structures can tolerate modest amounts of settlement if it is uniform across the structure. It is often the differential settlement, which can be expressed as angular distortion, that causes damage to structures. A further consideration is whether this damage is architectural or structural. According to Cording et al. (1978), architectural damage can begin at an angular distortion of about 1/1000 for brick bearing walls and 1/650 for infilled frame structures. Again, preconstruction surveys and settlement instrumentation will give indications of angular distortion damage to adjacent structures caused by dewatering.

Rotting and Corrosion of Submerged Structures

It is well known that rotting and corrosion do not normally occur without the presence of oxygen. When dewatering is planned near other structures, consideration should be given to whether these structures have any submerged wooden or metallic components that could be exposed to air as a result of the dewatering program. If so, the possible deterioration and corrosion of these components must be considered to verify that the duration of dewatering and any possible future effects of construction, such as permanent lowering of the groundwater table, will not have detrimental effects on the components.

If lowering the groundwater table may cause significant amounts of rotting and corrosion, actions will have to be taken to prevent or mitigate these effects. For instance, a structure supported on wooden piles might need to be underpinned prior to construction, or exposed concrete reinforcement might need to be sealed.

1-11 IMPERMEABLE BARRIERS

Often excavation supports and ground treatment are combined with dewatering plans to work as a system. This should be done whenever possible. For instance, grouting can be used to reduce soil permeability and thereby water inflow into the excavation. Barrier or cutoff walls can be placed deep enough to reduce or eliminate the need for a dewatering system (Figure 1-49). Still other techniques, like compressed air in tunnels or ground freezing, can eliminate the need for dewatering completely. These techniques are discussed in greater detail in later chapters. The possible relationship to dewatering, however, is discussed below.

Grouting

Grouting can be used both to reduce the permeability of the ground and to strengthen it. A method such as jet grouting, for instance, does both. Types of grouting include permeation with cement or chemicals, contact and pressure grouting to fill voids with neat cement, compaction grouting with neat cement to densify the ground, and jet grouting, which mixes or completely replaces the soil with cement. By decreasing the permeability of the soil or rock, dewatering needs can be reduced or eliminated. A cost study can be made to determine the best combination from effectiveness and economic standpoints (Figure 1-50).

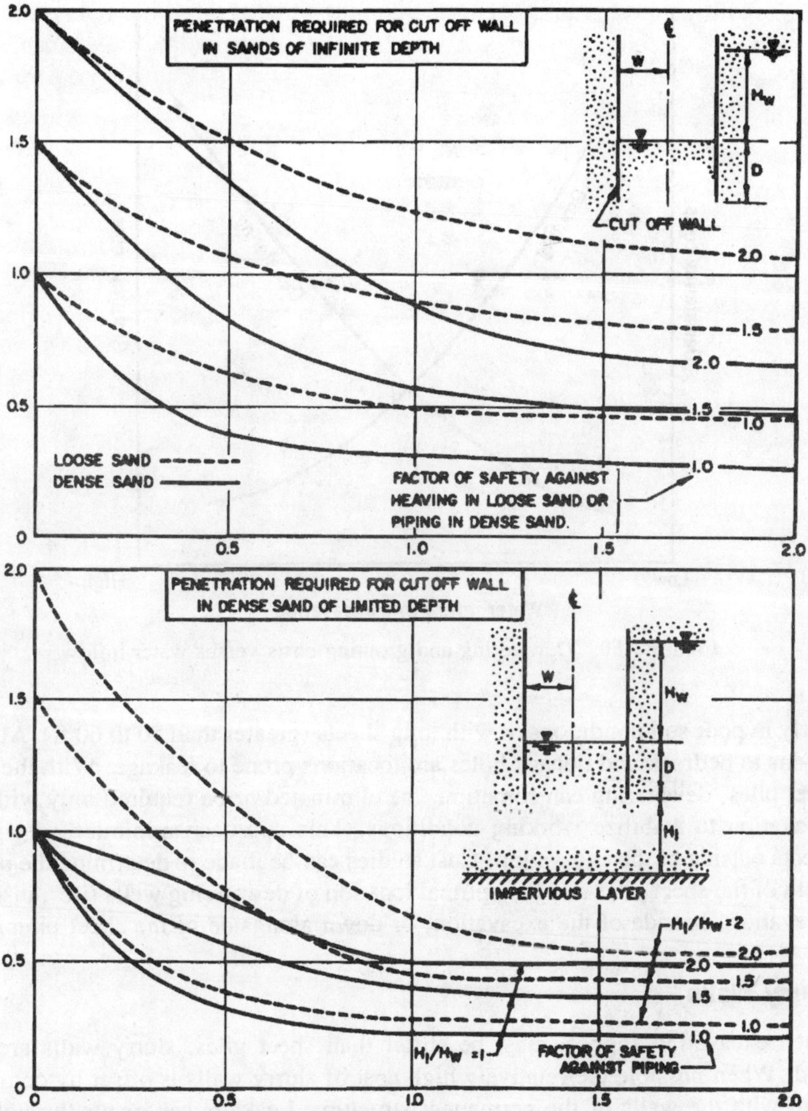

Figure 1-49 Sheet pile cutoff chart. (From NAVFAC, 1982.)

Sheet Piles

Interlocking steel sheet piles are frequently used for excavation support when a flexible system can be used but when the high water inflow amounts possible with soldier piles and lagging systems cannot be tolerated. Sheet pile support systems are relatively impermeable, although leakage between the sheets is common partic-

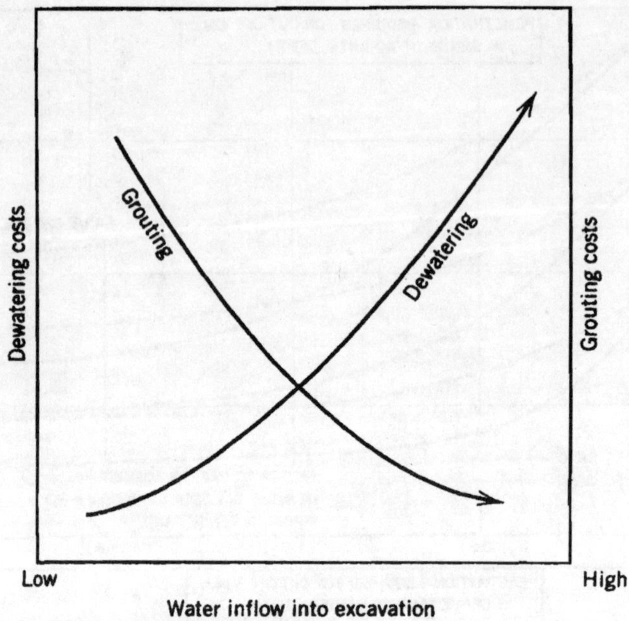

Figure 1-50 Dewatering and grouting costs versus water inflow.

ularly in poor soil conditions or with long sheets (greater than 50 to 60 ft). Also, the toe-ins to bedrock and tieback holes are locations prone to leakage. With the use of sheet piles, dewatering can sometimes be eliminated or be required only within the excavation to stabilize working conditions. This approach minimizes dewatering effects outside of the excavation. Cost studies can be made to determine the optimal depth of the sheet piles and the optimal location of dewatering wells (i.e., inside the excavation, outside of the excavation, or down alongside of the sheet piles).

Slurry Walls

When excavation support must be stiffer than sheet piles, slurry walls are often used. When possible the relatively high cost of slurry walls is offset by combining them with the walls of the permanent structure. Leakage can occur through construction joints and tieback holes not made watertight, and where the slurry wall panels toe into bedrock. So like sheet piles, slurry walls are not completely impermeable. Like sheet pile systems, dewatering wells can be inside, outside, or within the slurry wall panels. Trade-off studies should be made to develop the optimum design in combination with or without dewatering systems. An innovative system using grouting, slurry walls, and dewatering is shown is Figure 1-51.

Compressed Air

Compressed air methods have been used historically in caisson and tunnel construction below the groundwater table. The combination of high water pressures and

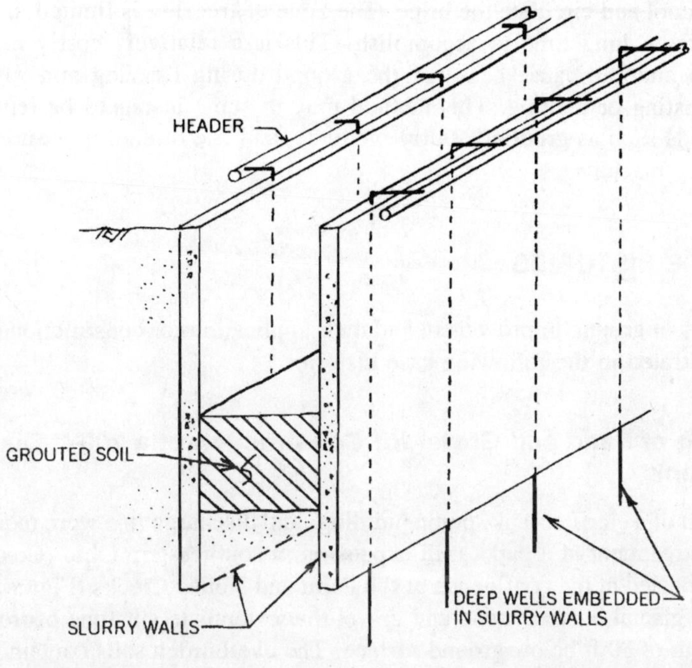

Figure 1-51 Excavation with slurry walls, grouting, and dewatering.

highly permeable or very weak soils often obviated the need for compressed air assistance in the past. The air inside of a caisson or tunnel is compressed to about 2 or 3 atmospheres, 1 or 2 atmospheres above normal atmospheric pressure, to apply additional support pressure to the sides of the excavation or to overcome the water pressure outside of the excavation. One atmosphere of pressure is equal to about 14.7 lb/in.^{2} (at sea level) or about 34 ft of water pressure. In recent times, however, the costs and health risks associated with compressed air work has all but eliminated it as a viable construction method. Advancements in marine construction, including the use of remotely operated equipment, slurries, tremie methods, divers, and so on, have replaced compressed air caisson construction techniques for the most part. In tunneling, earth pressure balance tunnel boring machines are in common use, which pressurize the tunnel face but not the remainder of the tunnel behind the tunnel heading. The workers can therefore work in normal atmospheric pressures. Also, ground improvement methods like grouting are now frequently used to stabilize weak ground and decrease water inflow.

Ground Freezing

Another technique used to stabilize the ground around an excavation and eliminate water inflow is ground freezing. The technique, although somewhat uncommon, is self-explanatory. Water in the ground is frozen by circulating brine at subfreezing temperatures through several pipes placed in the ground. A large refrigeration plant is set up to cool and circulate the brine. The zone of freezing is limited in size and freezing takes a long time to accomplish. This is a relatively costly method of construction and can cause heave of the ground during freezing and subsidence during defrosting or melting. This method may in some instances be replaced by other methods such as grouting, slurry methods, and use of earth pressure balance tunnel boring machines.

1-12 CASE HISTORIES

The methods of ground improvement and their application for construction dewatering are illustrated in the following case histories.

Dewatering of Sand and Gravel for Construction of a 200-ft-Diameter Clarifier Tank

Construction of a clarifier tank, pump building, and discharge line were required for wastewater treatment at a paper mill expansion in southeastern Ohio (Cox, 1984). The site is located at the confluence of the Paint and Honey Creeks (Figure 1-52) in a deposit of glacial outwash sand and gravel that extends to siltstone bedrock at an average depth of 80 ft below ground surface. The overburden soils contain between about 16 to 28 percent of gravel, 61 to 80 percent of sand, and 4 to 10 percent of fines.

The clarifier tank is circular in shape and 200 ft in diameter (Figure 1-53). The tank has vertical side walls and a sloping floor. The base of the tank is 20 ft below groundwater level. In addition to the tank, a rectangular pump building supported on a mat foundation was built next to the tank. Also, a trench was required for placement of the discharge line.

Parameters used to design the dewatering system included the following:

- $k = 6 \times 10^{-1}$ in./sec
- Saturated thickness of aquifer = 70 ft
- Transmissivity = 200,000 gal/(day)(ft)
- Storage coefficient = 0.2
- Well capacities = 1000 to 1200 gal/min
- Depth of dewatering = 20 ft

Other wells were in place for water supply and were incorporated into the new dewatering system.

Nine wells were installed to bedrock. Twelve hundred gal/min pumps were installed in the wells. The telescopic screens used were 18 in. in diameter and 30 ft long. This design was formulated based on an initial design and one test well. Obtaining sufficient electrical power for the pumps and routing of pump discharge lines so they did not interfere with construction were a couple of problems that occurred during construction.

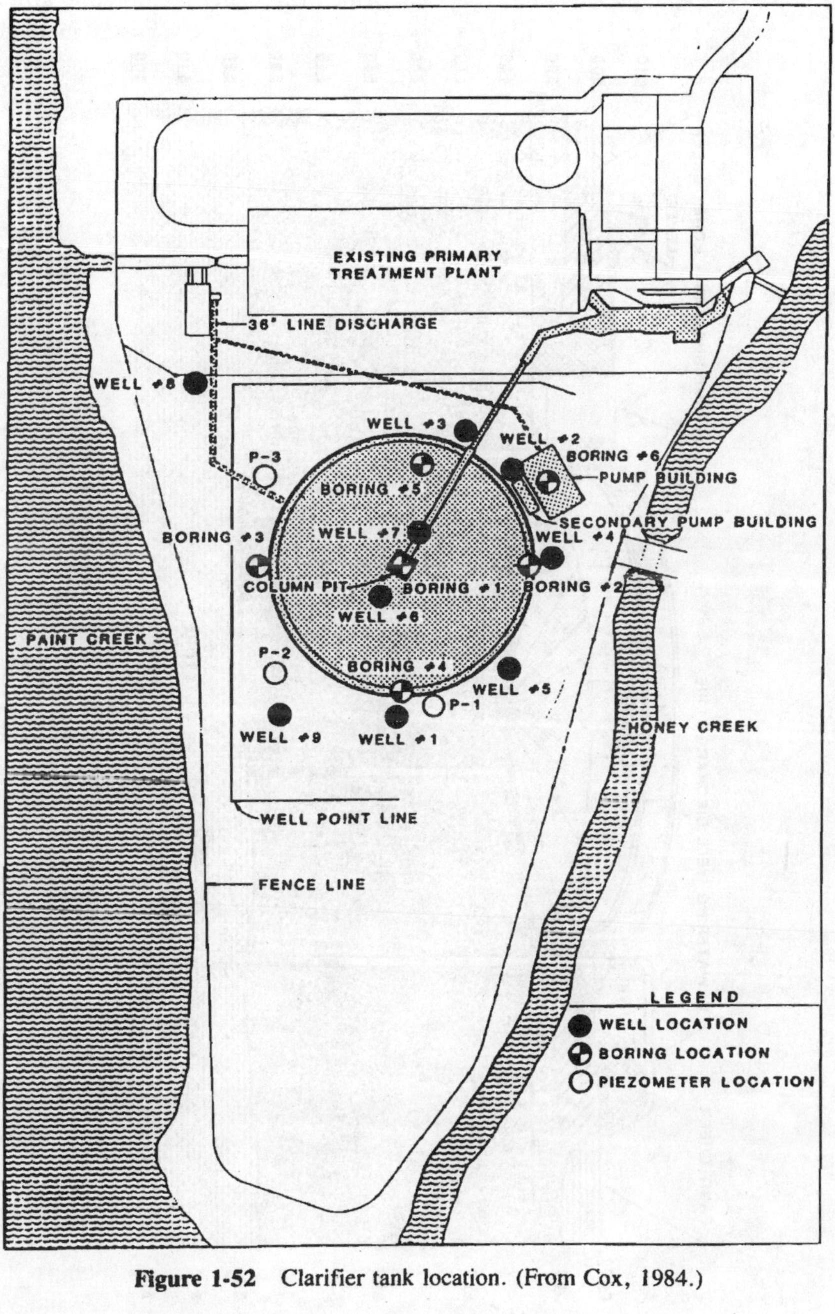

Figure 1-52 Clarifier tank location. (From Cox, 1984.)

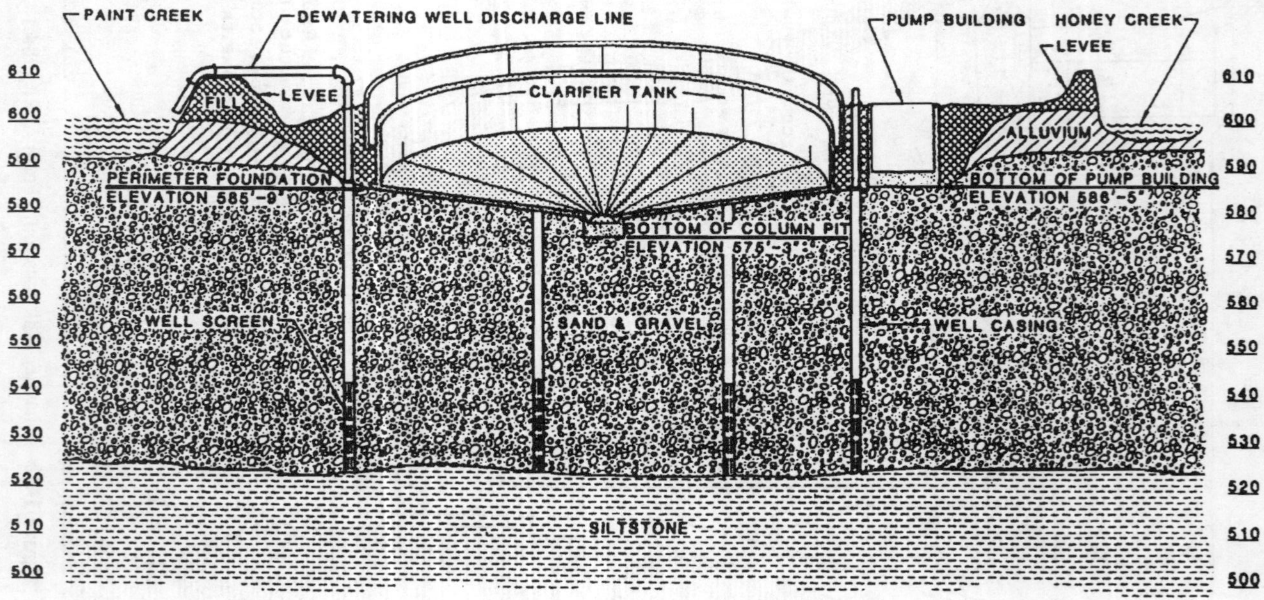

Figure 1-53 Clarifier tank details. (From Cox, 1984.)

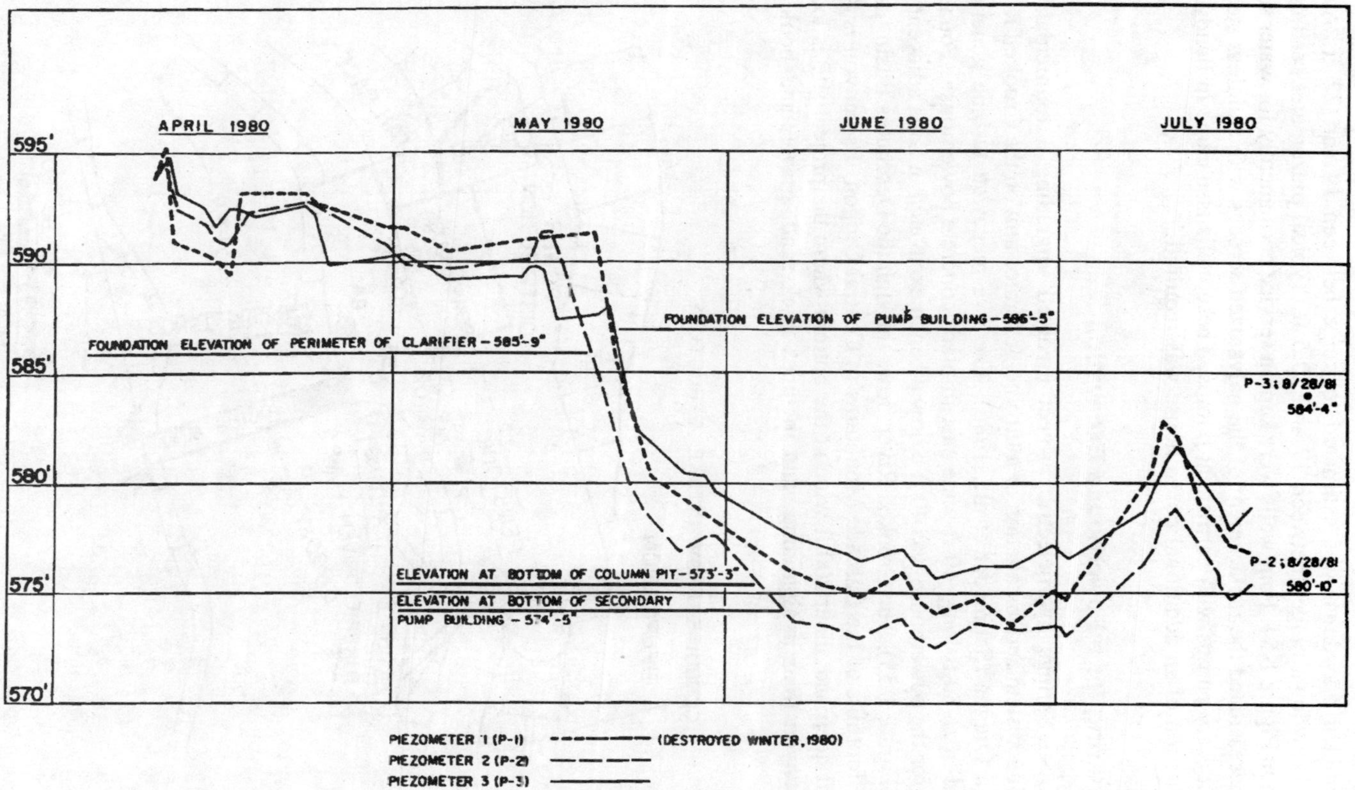

Figure 1-54 Clarifier tank construction piezometer readings. (From Cox, 1984.)

Predicted drawdown was estimated to range between 14 and 22 ft. Average actual drawdown ranged between 17 and 20.5 ft. Typical piezometer readings are shown in Figure 1-54. Four wells were kept in service permanently for water supply. This necessitated the replacement of the galvanized screens with stainless steel and of temporary pumps with permanent pumps. These wells also came in handy when construction of an additional pump house was required.

Dewatering for a Powerhouse Excavation

The Opal Springs hydroelectric project involved upgrading a municipal water pumping system and installation of a turbine-generator unit on the Crooked River in Central Oregon (Schroeder et al., 1986). The five megawatt turbine generator is housed in an 80-ft-long, 40-ft-wide cast-in-place concrete powerhouse. Foundation grade for the powerhouse was 40 ft below a bench excavated in talus adjacent to the river (Figure 1-55). The Crooked River flows through the Deschutes Basin just east of the foothills of the Cascade Mountains in Central Oregon. Its canyon is 800 to 1000 ft deep and about 500 ft wide at the bottom where the project site is located. The canyon floor is aggrading, and includes thick sand, gravel, and boulder al-

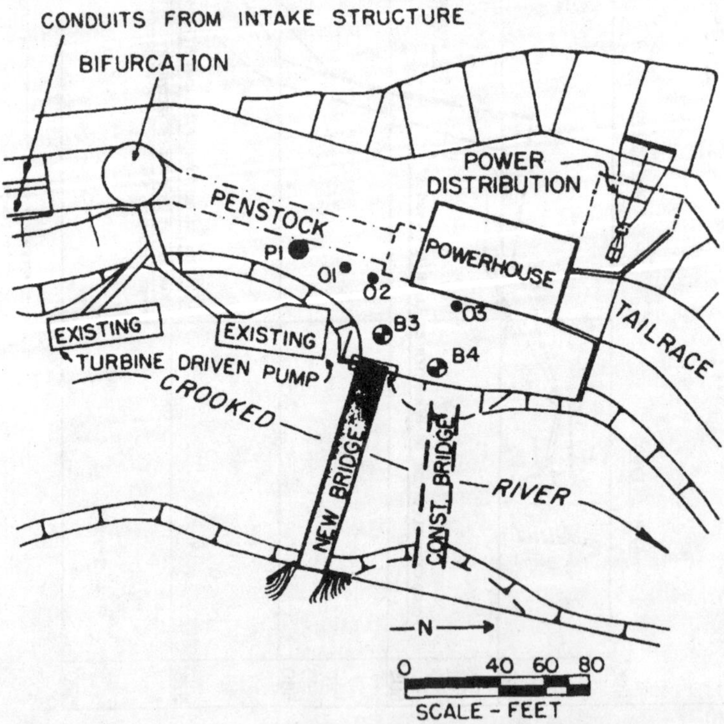

Figure 1-55 Powerhouse excavation location. (From Schroeder et al., 1986, by permission of ASCE.)

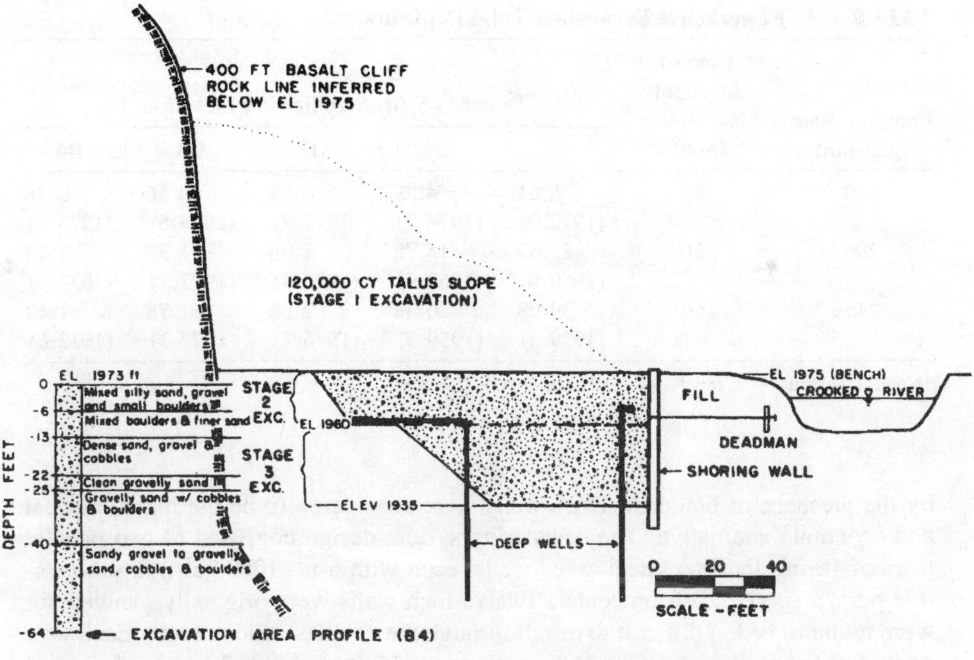

Figure 1-56 Powerhouse excavation geology. (From Schroeder et al., 1986, by permission of ASCE.)

luvium and mixed talus material from the nearby basalt formation. The boulders were up to 3 ft in diameter. The thickness of these deposits is greater than 64 ft (Figure 1-56). Constant head permeability tests in the borings indicated permeability values between 0.2 and 0.3 cm/sec. During construction of the original turbine-driven pump pad nearest the powerhouse, large boulders, flowing sand, and local inflows from springs on the order of 3000 gal/min had been encountered. Especially difficult dewatering requirements were therefore anticipated for the proposed 120,000-yard^3 excavation.

Preconstruction pump tests were run to provide a basis for planning the dewatering system. A 12-in. diameter pump well was used (P1 in Figure 1-55) with the lower 20 ft of the well casing being perforated. The well depth was 45 ft below the proposed excavation depth. Three short-term tests were run by pumping the test well with a 7.5-HP pump at rates of 100, 200, and 245 gal/min until the four observation well water levels stabilized (O1, O2, O3, and B4 in Figure 1-55). Total drawdown achieved is summarized in Table 1-3. A permeability rate of 0.15 cm/sec and transmissivity rates of between 3000 and 120,000 gal/(day)(ft) were calculated from the tests. A radius of influence of about 300 ft for 24 ft of drawdown was observed. This was extrapolated to 625 ft for the required 50 ft of drawdown.

Possible groundwater control options included slurry walls, caisson construction methods, and deep well dewatering. The first alternative would have required specialized methods and equipment. The second method would have been hampered

TABLE 1-3 Powerhouse Excavation Total Drawdown[a]

Pumping Rate (gal/min)	Time of Apparent Stabilization (min)	Drawdown (ft)(Stabilization Elevation, ft)				
		P1	O1	O2	O3	B4
100	40	6.61 (1972.0)	4.89 (1974.8)	0.99 (1977.9)	0.56 (1978.5)	0.38 (1973.6)
200	230	27.69 (1950.9)	15.78 (1963.9)	2.03 (1976.9)	1.38 (1977.7)	1.13 (1972.9)
245	160	39.08 (1939.5)	20.38 (1959.3)	2.68 (1976.2)	1.78 (1977.3)	1.43 (1972.6)

[a]Note: 1 gal/min = 3.785×10^{-3} m³/min; 1 ft = 0.305 m.

by the presence of boulders. Deep wells were considered to be the most practical and economic alternative. The dewatering system design consisted of two parallel lines of 10-in.-diameter, steel-cased wells, each with 6-in. 1000-gal/min submersible pumps spaced 20 ft on center. Twelve-inch wells were originally planned but were found to be too difficult to install through the cobbles and boulders. Each well had a 5-ft-long tail pipe casing at the bottom and 20 ft of slotted 8-in. stainless steel well screen (Figure 1-57). Discharge was measured using two V-notch weir boxes at the end of the discharge headers. Two backup pumps were available and backup power was provided by a portable 500-kilowatt generator. As a last resort if all backup systems failed, the excavation would be flooded to prevent the development of a quick condition in the excavation bottom subsoils. Dewatering was required for five months.

The well lines were activated sequentially in stages. Sumps were constructed along the river, within the excavation, and at a spring location to control seepage not mitigated by the deep wells. The distribution of measured discharge from the dewatering wells and sumps was as follows:

- Cliffside wells—2400 gal/min
- Riverside wells—640 gal/min
- Riverside ditch sump—2700 gal/min
- Excavation sump—500 gal/min
- Spring sump—2400 gal/min

The total measured discharge of 8640 gal/min compared favorably with the design estimate of 8,820 gal/min. Sixty-five percent of the flow came from the cliff side of the excavation.

Performance of the dewatering system was good in general and accomplished the difficult dewatering task well. Three pumps burned out because of low flow conditions and were replaced with 200-gal/min pumps. Pump wiring was damaged four

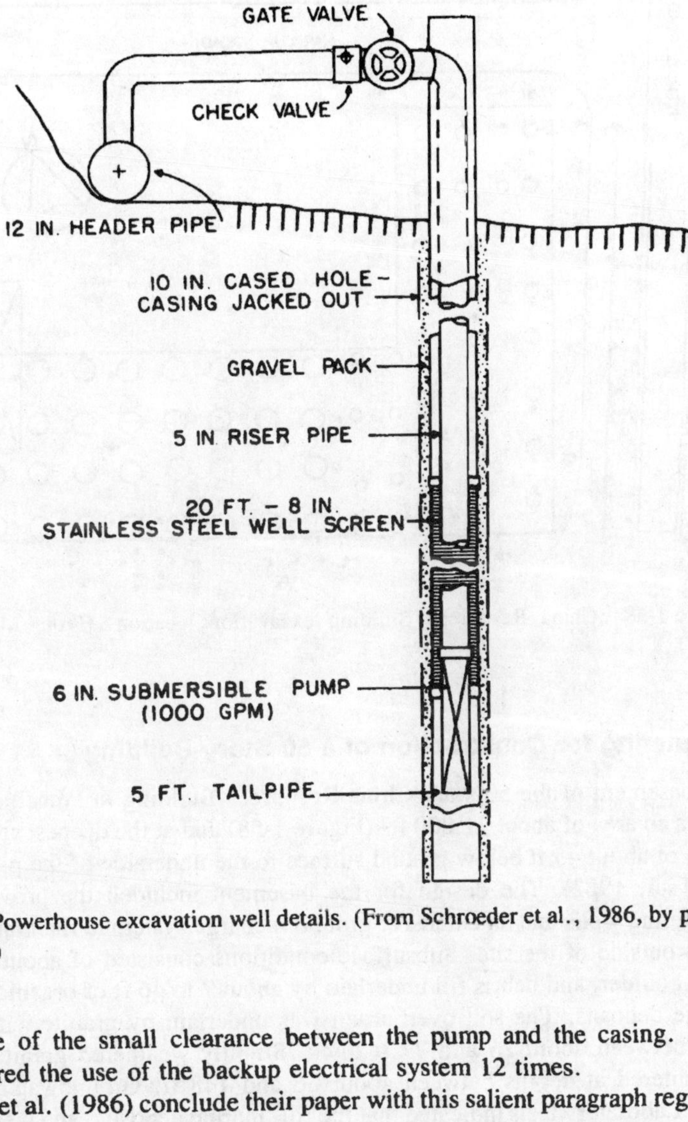

Figure 1-57 Powerhouse excavation well details. (From Schroeder et al., 1986, by permission of ASCE.)

times because of the small clearance between the pump and the casing. Power failures required the use of the backup electrical system 12 times.

Schroeder et al. (1986) conclude their paper with this salient paragraph regarding dewatering work:

> Dewatering a deep excavation in heterogeneous alluvium provides the opportunity to use the observational design method. A final design cannot be achieved until the system is in place and operating successfully. Since this method requires modifications, the total costs of accomplishing a required result are not precisely predictable. We must, therefore, approach such tasks with a flexible plan of action to accommodate the inevitable changes that will be required. Extensive analysis requiring radical simplification or idealization of ill-defined geologic conditions can be misleading and futile, if relied on exclusively for dewatering planning.

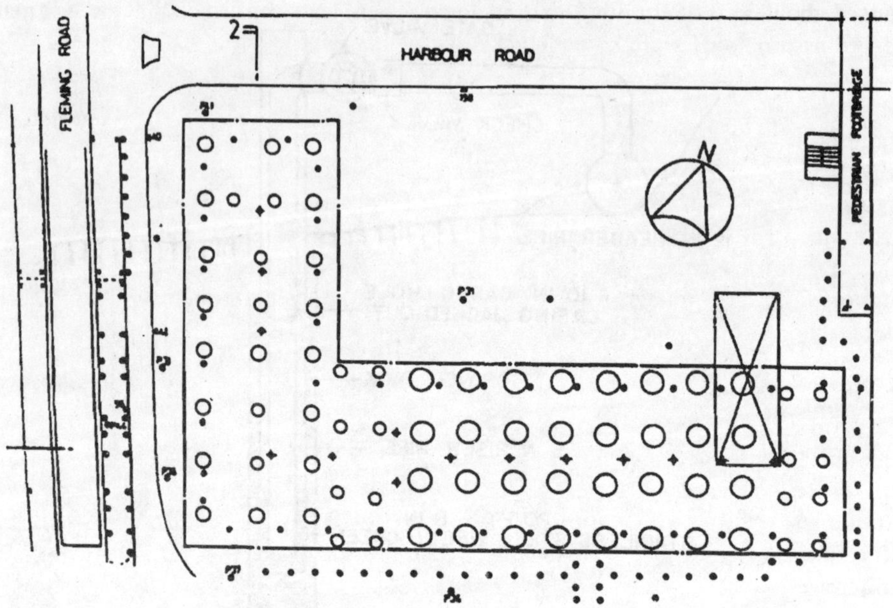

Figure 1-58 China Resources Building excavation location. (From Morton and Tsui, 1982.)

Dewatering for Construction of a 50-Story Building

The basement of the 50-story China Resources Building in Wanchai, Hong Kong, covers an area of about 71,000 ft^2 (Figure 1-58) and at the deepest section reaches a depth of about 45 ft below ground surface to the underside of the pile cap (Morton and Tsui, 1982). The design for the basement included the provision for deep dewatering wells and an extensive ground recharge system to maintain groundwater levels outside of the site. Subsurface conditions consisted of about 13 to 46 ft of sand, boulder, and debris fill underlain by about 7 to 46 ft of organic clay and sand marine deposits. The soil overburden was underlain by granite with a weathered zone between about 26 and 79 ft thick. Slightly weathered granite bedrock was encountered at depths between about 66 and 148 ft. Groundwater was near sea level. Laboratory tests indicated that the fill, marine deposits, and weathered granite contained 0, 20 to 60, and 20 to 35 percent fines, respectively.

The dewatering system design consisted of fourteen 8-in.-diameter wells to bedrock at the center of the excavation and forty 8-in.-diameter wells to a depth of about 66 ft along the perimeter of the sheet-pile supported excavation site. A pump test using four deep well pumps was conducted to check capacity and zone of influence and drawdown. One well operating at pump rates of 8, 24, and 32 gal/min caused 26 ft of equilibrium drawdown in the wells after 10 hr of pumping. Drawdowns of about 3 and 7 ft were obtained at distances of about 79 and 26 ft, respectively. Four wells operating at a stabilized pump rate of about 50 gal/min caused about 33 ft of equilibrium drawdown. These tests confirmed that the original design assumptions were of the right order of magnitude.

To avoid settlement of adjacent properties, a recharge system was installed. The recharge system consisted of seventy-five 4-in.-diameter boreholes to a depth of about 66 ft with 2-in.-diameter PVC well screens surrounded by a filter of sand and gravel. The wells were connected to a header and tank with water level at about 23 ft above sea level. A constant average recharge rate of about 0.3 gal/min per well was maintained. This recharge rate was effective in mitigating settlement of adjacent sites due to dewatering and the dewatering system allowed the deep, sheet pile supported excavation to proceed in the dry in spite of the poor quality fill soils near the surface and the high groundwater table.

Excavation of a Slipway in Sand Below the Groundwater Table

The construction of a slipway at Paradeep, a port situated in Orissa on the coast of Bay of Bengal, involved excavation through sandy soil up to a depth of 40 ft below ground level (Banerjee and Sen, 1970). The subsoil in the area consists primarily of silty fine sand with the groundwater level at about 16 ft below ground surface. Concrete sheet piles were driven into the ground to act as the side retaining walls of the slipway, but excavation within the sheet pile walls was necessary for the construction of the bottom reinforced concrete beams and cross frames (Figure 1-59). The proposed lowering of the groundwater table was 17 ft. Therefore, a single-stage, vacuum-pump-driven well point system was selected for dewatering.

A design permeability value of 3×10^{-2} cm/sec was used, based on falling head borehole permeability tests and grain size distribution analyses (Hazen's formula). The radius of influence, R, was estimated based on the formula

$$R = C' (H - h_w)k^{1/2} \tag{1-23}$$

where C' = 3 for gravelly soils
H = height above impervious stratum of original groundwater table, ft
h_w = head at well above impervious stratum, ft
k = soil permeability, 10^{-4} cm/sec

Accordingly, the radius of influence was calculated to be 800 ft.

Other estimated parameters used in the design are as follows:

- Drawdown at well point = 32 ft
- Well point spacing = 5 ft
- Maximum discharge = 650 gal/min (for 67 well points)

Three-foot-long, $1\frac{1}{2}$-in.-diameter, self-jetting well point screens were installed to a depth of 22.5 ft below sea level. A maximum of 40 lb/in.^{2} of water pressure was required to jet the well points into place. Eight-inch-diameter header pipes were used to connect the well points to two 40-HP pumps. The layout is shown in Figure

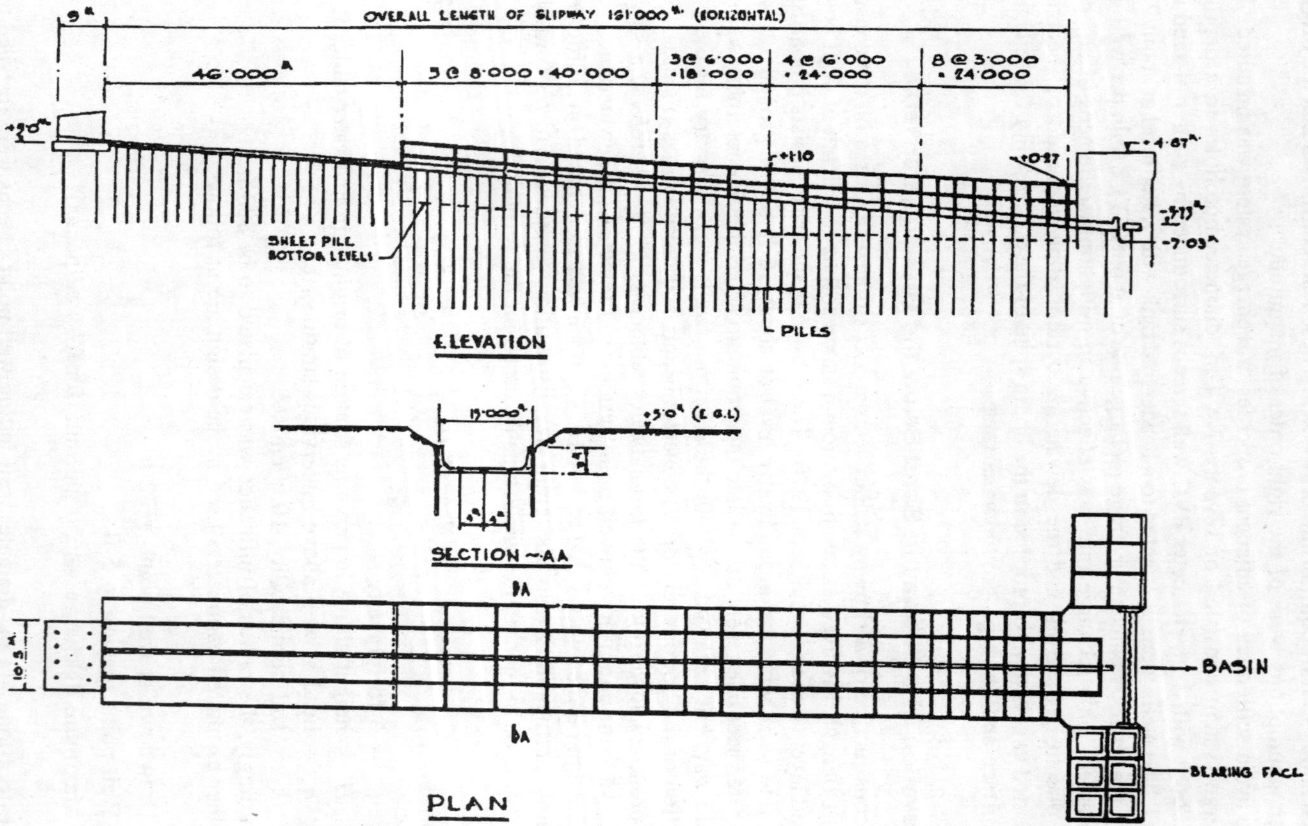

Figure 1-59 Paradeep slipway location plan. (From Banerjee and Sen, 1970.)

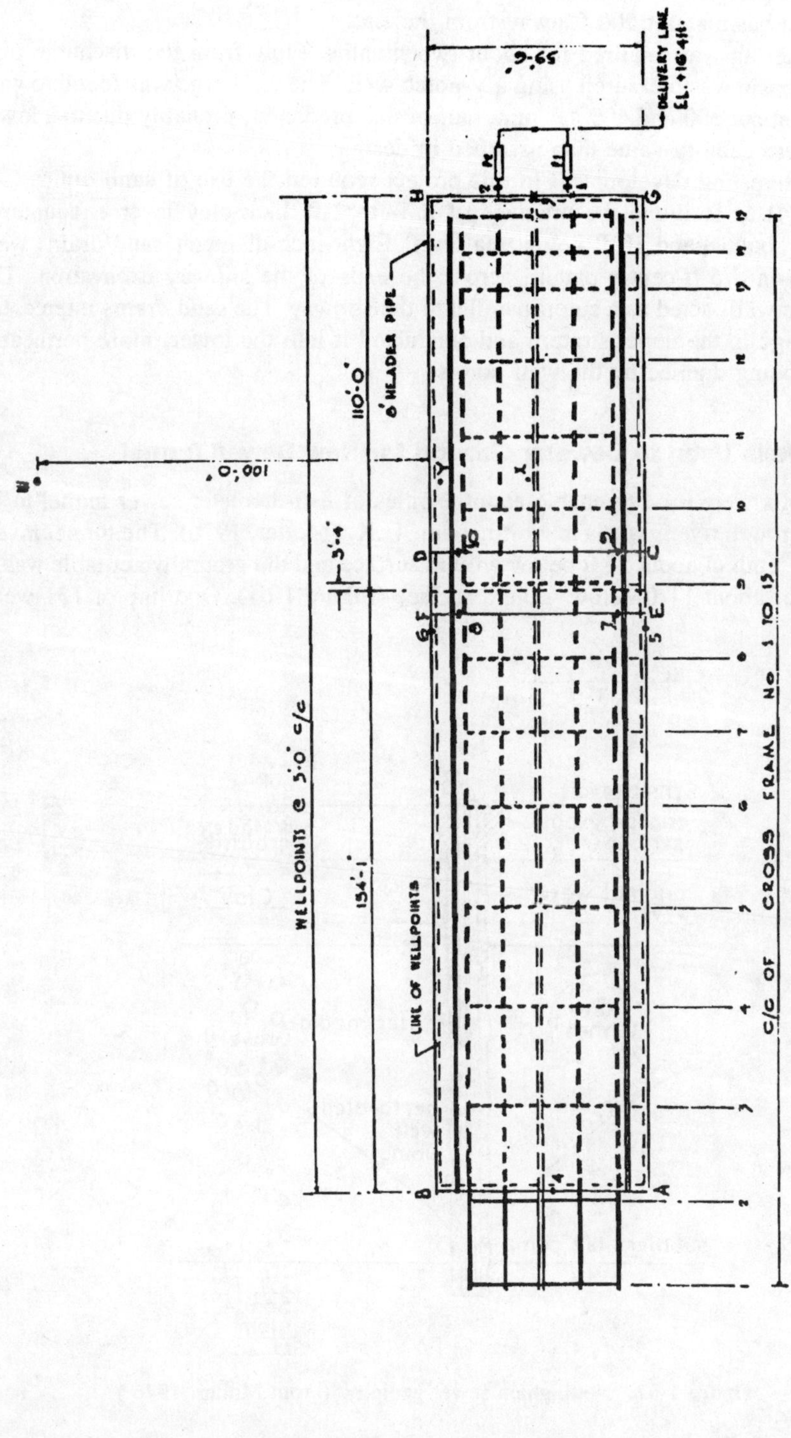

Figure 1-60 Paradeep slipway dewatering plan. (From Banerjee and Sen, 1970.)

1-60. The pumps were connected to a common 6-in.-diameter discharge line leading to the basin about 500 ft away from the site.

Dewatering was required for about two months. Flow from the discharge pipe near the basin was measured using a V-notch weir. The discharge was found to vary between about 300 and 400 gal/min, half of that predicted, probably due to a lower in situ permeability value than assumed in design.

An interesting development in this project required the use of sand drains (see Chapter 4) to facilitate the drainage of a 1- to 2-ft-thick clay layer encountered between about 8 and 10 ft below sea level. Eight-inch-diameter sand drains were installed on a 5-ft-center pattern across the ends of the slipway excavation. The sheet pile walls acted as a cutoff parallel to the slipway. The sand drains intercepted the seepage in the upper stratum and conducted it into the lower, more permeable stratum being drained by the well points.

Deep Wells Used to Dewater Ground for New Sewer Tunnel

Deep wells were used to enable about 2 miles of 8-ft-diameter sewer tunnel to be driven through river gravels in Nottingham, U.K. (Moller, 1976). The tunnel invert was at a depth of about 25 ft below ground surface and the groundwater table was at a depth of about 11 ft below ground surface (Figure 1-61). One line of 131 wells

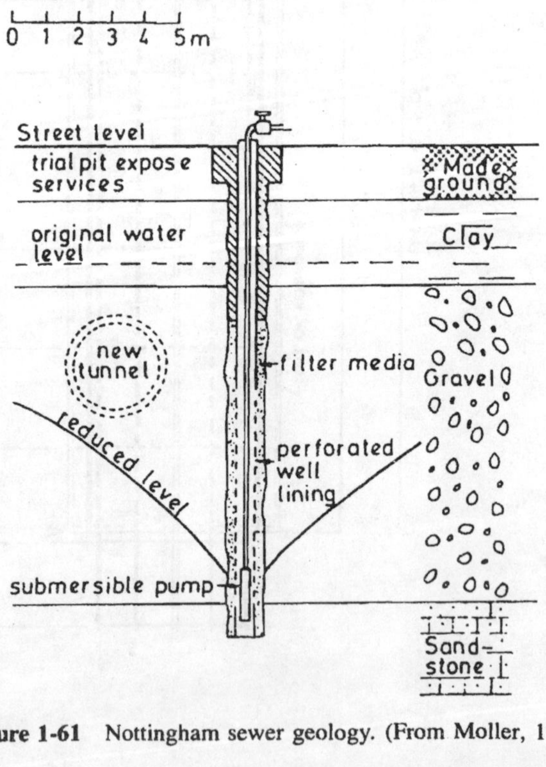

Figure 1-61 Nottingham sewer geology. (From Moller, 1976.)

was installed at about 100-ft intervals to a depth of about 43 ft below ground surface into the underlying sandstone bedrock. The submersible pumps were leap frogged as the tunnel construction advanced. At the peak, 65 pumps were in operation at one time. The groundwater level was maintained below tunnel invert throughout sewer construction. After an overall pumping period of 18 months, there were no reports of settlements to buildings or services.

Power and Signal Trench Dug with the Aid of Well Points

The installation of power and signal cables to provide additional services at Esso's Antwerp refinery required an extensive system of well point dewatering to control inflows during trench excavation and pipework construction (Anonymous, 1976). The project involved over 1400 miles of wire to be installed in about 34 miles of trench. The site is located on a bank of the River Scheldt where soil conditions consist of 6 to 13 ft of dense sand overlying alluvial silt and stiff clays. The groundwater table is located between about 1 and 3 ft below ground surface. Eight hundred well points were jetted into place ahead of the trench excavations with depths of between about 5 and 7 ft. The well points were connected to a 7200-ft-

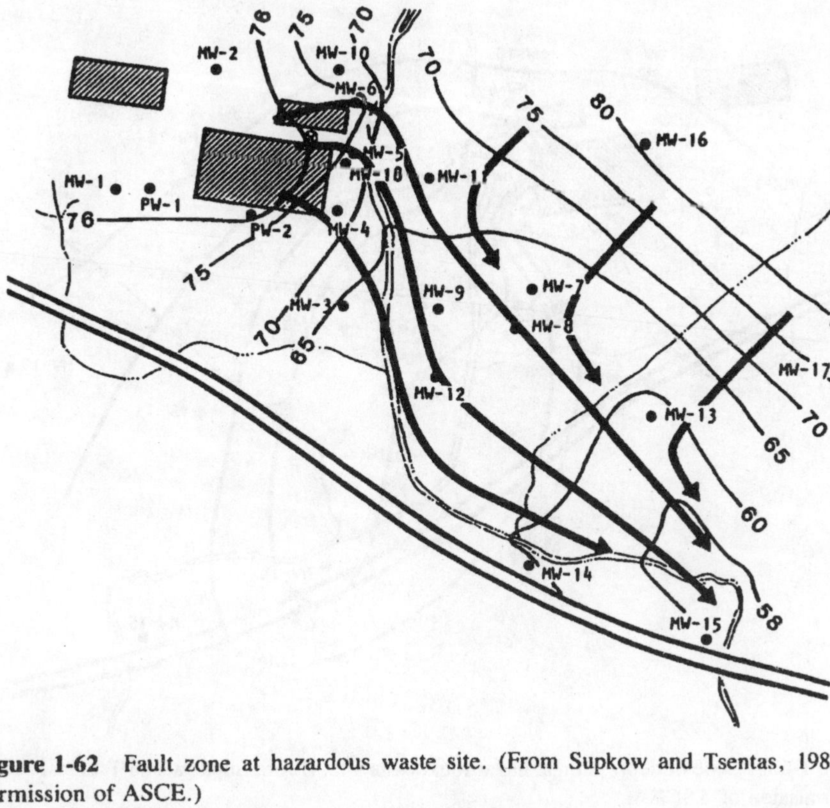

Figure 1-62 Fault zone at hazardous waste site. (From Supkow and Tsentas, 1985, by permission of ASCE.)

long header pipe, which was powered by 20 vacuum pumps with suction heads ranging between 3 and 6 in. Up to 2300 ft of trench was opened at one time and maximum pumping rates were on the order of 2000 gal/min for that length of trench.

Recovery of Chlorinated Hydrocarbon Solvents Using Deep Wells

An application of dewatering that has recently grown significantly is the use of wells to recover contaminants that have leaked into the ground. One such case involved the recovery of pure tetrachloroethylene degreasing solvent, also known as perchloroethylene or "perc," from a fractured bedrock aquifer. The site is in the northeastern United States and consists of gently rolling land with low relief (Supkow and Tsentas, 1985). The bedrock is composed of interbedded siltstones and shales with occasional sandstones having very little primary permeability. Secondary permeability through cracks and fractures near the top of bedrock provides for most of the groundwater movement in the rock. The groundwater table is generally less than about 20 ft below ground surface. Natural recharge occurs by rainfall percolation through the 10-ft-thick residual soil horizon.

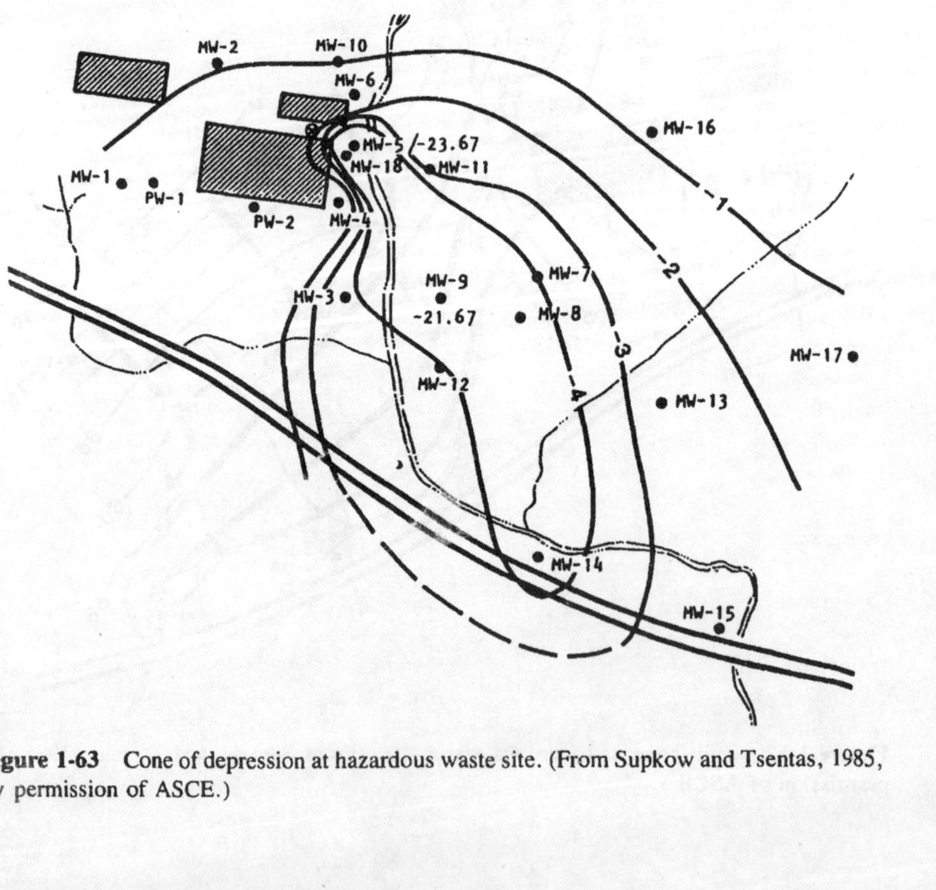

Figure 1-63 Cone of depression at hazardous waste site. (From Supkow and Tsentas, 1985, by permission of ASCE.)

Tetrachloroethylene is heavier than water and therefore tends to sink to the bottom of an aquifer. Because of its low solubility in water, removal of tetrachloroethylene is more difficult and time consuming than removal of contaminants having a high solubility in water. Also, tetrachloroethylene tends to displace water from mineral surfaces in the bedrock and becomes adsorbed preferentially.

In response to the spill, an array of monitoring wells was installed to characterize the groundwater table and flow, extent of contamination, and rate of contaminant migration. The wells were 4 in. in diameter and between 40 and 120 ft deep. Submersible pumps were installed in each of the wells to facilitate purging and sampling operations. The results of monitoring indicated that the groundwater tended to concentrate in a northwest-southeast trending zone, which was suspected of being a fault zone (Figure 1-62). Wells within the fault zone had specific capacities in the range of 1 to 1.5 gal/min/ft of drawdown, whereas wells outside of the fault zone had specific capacities in the range of 0.3 to 0.7 gal/min/ft of drawdown.

Two of the monitoring wells were converted to recovery wells by installing permanent pumps. Pumping of these two recovery wells at a combined rate of about 18 gal/(min)(ft) produced an elongated cone of depression, as shown in Figure 1-63. The water from the recovery wells was pumped to a municipal wastewater treatment plant. Following an initial rise in concentration shortly after the groundwater recovery operation started, the tetrachloroethylene concentration gradually declined over time until the end of the remediation program.

REFERENCES

American Society for Testing and Materials (ASTM), 1990. "Annual Book of ASTM Standards," Section 4, Vol. 04.08—Soil and Rock; Dimension Stone; Geosynthetics, Philadelphia.

Anonymous, 1976. "Dewatering at Antwerp Refinery," *Ground Engineering*, Vol. 9, No. 5, July, p. 4.

Banerjee, A. and S. Sen, 1970. "Study on Dewatering by Well Point," 2nd Southeast Asian Conference on Soil Engineering, June, Singapore, pp. 547–558.

Bush, R. Y., 1971. "Construction Dewatering," Section 29 in *Handbook of Heavy Construction*, J. A. Havers and F. W. Stubbs, Jr., Eds., 2nd ed., McGraw-Hill, New York, pp. 29-1–29-21.

Carson, A. B., 1961. *General Excavation Methods*, McGraw-Hill, New York, pp. 170–235.

Casagrande, A., 1940. "Seepage Through Dams," *Contributions to Soil Mechanics*, Boston Society of Civil Engineers.

Casagrande, L., 1952. "Electro-Osmotic Stabilization of Soils," *J. Boston Soc. Civil Eng.*, Jan.

Cedergren, H. R., 1967. *Seepage, Drainage, and Flow Nets*, Wiley, New York.

Cording, E. J., T. D. O'Rourke, and M. Boscardin, 1978. "Ground Movements and Damage to Structures," Proc. Int. Conf. on Evaluation and Prediction of Subsidence, S. K. Saxena, Ed., January, American Society of Civil Engineers, New York, pp. 516–537.

Cox, G. C., 1984. "Dewatering of the Construction Site for a 200-Foot Diameter Clarifier

Tank Founded on Sand and Gravel," Int. Conf. on Case Histories in Geotechnical Engineering, S. Prakash, Ed., January, University of Missouri—Rolla, pp. 1319–1327.

Hazen, A., 1893. "Some Physical Properties of Sands and Gravels," Massachusetts State Board of Health, 24th Annual Report.

Hunt, R. E., 1984. *Geotechnical Engineering Investigation Manual*, McGraw-Hill, New York.

Hvorslev, M. J., 1949. "Subsurface Exploration and Sampling of Soils for Civil Engineering Purposes," American Society of Civil Engineers, New York, Nov.

Johnson Division, UOP Inc., 1975. *Ground Water and Wells*, 4th printing, Saint Paul, Minn.

Mansur, C. I. and R. I. Kaufman, 1962. "Dewatering," Chapter 3 of *Foundation Engineering*, G. A. Leonards, Ed., McGraw-Hill, New York, pp. 241–351.

McCusker, T. G., 1982. "Soft Ground Tunneling," *Tunnel Engineering Handbook*, J. O. Bickel and T. R. Kuesel, Eds., Van Nostrand Reinhold, New York, pp. 35–69.

Moller, K., 1976. "Groundwater Control for Underground Construction," *Ground Engineering*, Vol. 9, No. 3, Apr., pp. 43–45.

Morton, K. and P. Tsui, 1982. "Geotechnical Aspects of the Design and Construction of the Basement of the China Resources Building, Wanchai," 7th Southeast Asian Geotechnical Conf., Hong Kong, Vol. I, pp. 529–543.

Naval Facilities (NAVFAC) Engineering Command, 1982. *Soil Mechanics Design Manual 7.1*, NAVFAC DM-7.1, May, Department of the Navy, p. 7.1–319.

Peck, R. B., W. E. Hanson, and T. H. Thornburn, 1974. *Foundation Engineering*, 2nd ed., Wiley, New York.

Powers, J. P., 1992. *Construction Dewatering — New Methods and Applications*, 2nd ed., Wiley, New York.

Schroeder, W. L., V. W. Rybel, and L. Cochran, 1986. "Dewatering for Opal Springs Powerhouse Excavation," *J. Construction Eng.*,ASCE, Vol. 112, No. 3, Sept., pp. 440–451.

Sowers, G. F., 1979. *Introductory Soil Mechanics and Foundations: Geotechnical Engineering*, 4th ed., MacMillan, New York.

Supkow, D. J. and C. Tsentas, 1985. "Migration of Chlorinated Hydrocarbon Solvents in a Fractured Bedrock Aquifer," Geotechnical Aspects of Waste Management, Foundations and Soil Mechanics Group, New York Metropolitan Section, ASCE, Dec. 5 pp.

Terzaghi, K. and R. B. Peck, 1967. *Soil Mechanics in Engineering Practice*, Wiley, New York.

Theis, C. V., 1935. "The Relation Between the Lowering of the Piezometric Surface and the Rate and Duration of Discharge of a Well Using Ground Water Storage," *Trans. Amer. Geophys. Union*, Washington, D.C., pp. 519–524.

United States Army Corps of Engineers (USACE), 1956. "Investigation of Underseepage and Its Control, Lower Mississippi River Levees," Waterways Experiment Station, TM 3-424, Vicksburg, Miss., Oct.

Xanthakos, P. P., 1979. *Slurry Walls*, McGraw-Hill, New York.

Xanthakos, P. P., 1991, *Ground Anchors and Anchored Structures*, Wiley, New York.

CHAPTER 2

UNDERPINNING

2-1 BASIC PRINCIPLES

Definitions

Underpinning relates to the practice of foundation analysis and construction. As a general method it involves the addition of structural foundation units to give extra support to structures at or below grade. Underpinning aspects represent therefore an old building art, but until the proliferation of procedures and standards for underground work the application was mostly remedial, its purpose being to increase the support of settling or deteriorating structures. In the technical context underpinning is the insertion of a new foundation or support below an existing one for the transfer of load to a lower level. In a broader sense underpinning may also refer to the lateral protection of a foundation, the strengthening of ground beneath, or both. The decision to underpin, protect laterally, or introduce ground strengthening depends on various interrelated factors such as cost, technical expediency, and associated risk of each alternative.

Underpinning analysis is commonly based on accepted procedures for structural and foundation work. A typical field operation involves the insertion of new load bearing units, usually in stages, and the eventual transfer of load. Where it is feasible, a foundation or building may be surrounded or confined by a rigid lateral support combined with grouting of the zone beneath. In other cases, the new work may accelerate the renewal of unstable circumstances unless the structural frame of the existing building is reinforced or some load is removed from the foundation before underpinning. Since the process commonly is applied to existing structures, some of which may be old and difficult to preserve, it suggests the value of engineering familiarity with older types and forms of construction.

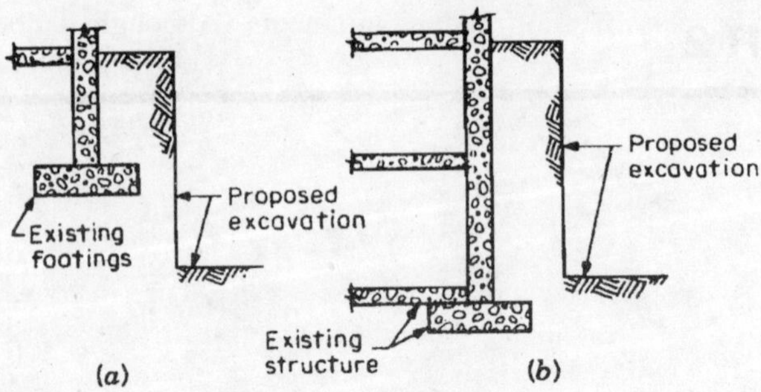

Figure 2-1 Excavation adjacent to existing structures. (*a*) Base of excavation below existing footing. (*b*) Base of excavation above existing footing.

Underpinning may be remedial or precautionary. The former adds foundation capacity to a structure that is deteriorating or is inadequately supported. The latter is necessary where foundation capacity must be adjusted so that greater loads can be sustained, or where changes in ground conditions are anticipated. For example, foundation adjustment may become necessary as a result of (1) a new activity such as vibrations near the structure, and (2) nearby excavation or tunneling likely to adversely affect ground strength and cause ground loss or movement.

We make use of these concepts for the structures shown in Figure 2-1. The need for underpinning or some form of protection is obvious in Figure 2-1 *a*. The case shown in Figure 2-1 *b* has an implicit sensitivity in spite of the fact that the excavation stops above the existing footing. In this instance removal of overburden can impart to the ground a loss of strength with a corresponding decrease of foundation capacity for the footing.

In general, underpinning is accompanied by some settlement or lateral movement. Even if a building rests on deep foundation, it still may register effects associated with horizontal extension. If design and field work are satisfactory, a settlement range of $\frac{1}{4}$ to $\frac{1}{2}$ in. is normal during the underpinning phase and must be added to the settlement associated with ground movement. As long as the settlement is uniform or varies within the foregoing range, damage is unlikely. Useful data on settlement and lateral movement of underpinned structures are given by Ware (1974), and are compiled from the Washington, D.C. area.

Underpinning Grouping

The procurement of design standards and construction specifications has revealed a tendency to become more codified. A great deal of this effort has followed the direction of engineering rationalization. Essentially, this applies to techniques and procedures that contractors have found to work well (White, 1976). Beginning first with the Bay Area Rapid Transit (BART) in San Francisco, the trend has continued with underground work in New York, Washington, D.C., Atlanta, and Baltimore, and has produced the concept of two distinct categories of structures with corresponding specific classification and requirements. Although it does not express underpinning in terms of a qualitative index, it provides a convenient approach. Thus, we introduce the following grouping:

- Category 1 structures, for which underpinning is mandatory and is designed by the engineer.
- Category 2 structures, which the contractor can choose to support temporarily, underpin, or both, or not to support or underpin if they are not likely to be affected by the work. Any of these options may be exercised at the contractor's discretion.

Obviously, in the latter case the decision rests solely with the contracting party, which assumes responsibility for the results (WMATA, 1973).

The technical and economic implications of this grouping are explicit. For buildings and structures where damage in absolute terms can be extensive, a safe and conservative design is prepared by the engineer. During the bidding stage contractors develop and maintain the same perceptions of project risk and requirements, and uniform fair competition is thus preserved. However, the approach is not always consistent with the objectives since it cannot adequately address the following concerns: (a) the owner is basically deprived of the opportunity of a lower underpinning cost for a different option that might result from the contractor's (or the underpinning specialist's) own experience and ingenuity; (b) the incentive of having the work done by a specialist diminishes, whereas the burden is placed upon the owner for adequate supervision and inspection and without risk sharing; and (c) in case of unforseen events such as varying conditions implying changes and revisions in the initial plans, the owner is isolated and in a difficult position to negotiate change orders.

These implications can be avoided by placing structures and buildings in Category 2. This trend has become obvious for small and relatively unimportant buildings. By rational extension, Category 2 includes also buildings located beyond a relevant influence zone with reference to the limits of the proposed excavation as well as structures with deep foundations relative to the same influence zone (New York City Transit Authority, 1974). If the owner chooses to place a building in Category 1, the disadvantages can be mitigated by including a value engineering clause that provides the opportunity for an independent assessment (White, 1976).

General Requirements

An underpinning scheme is planned in conjunction with the proposed construction project, and with the intent, if feasible, to incorporate the underpinning system in the lateral support of the permanent construction. The decision to underpin must have a technical merit, but often reflects a subjective judgment based on risk assessment that quantifies site and ground conditions, construction practices, public attitudes, and effects of jurisdictional authority.

Existing Structures These are studied to determine structural systems, possible existing defects, foundation type and loads, soil bearing pressure, soil characteristics and groundwater level, and excess foundation capacity. When necessary, weak portions of a building, such as masonry walls, might have to be repaired or strengthened before underpinning starts. For the purpose of expediency and economy, advantage may be taken of arching action and of the ability of a structure to withstand moderate overloading. Columns supported centrally on large footings do not have to be shored when digging is along an edge and involves only a small percentage of the total footing area, since a large portion of the column load is supported by the soil directly under the column.

From this analysis, we establish the engineering requirements for a safe underpinning operation, and define the constraints necessary to preserve overall structural integrity.

Effects from Operations Temporary conditions that arise during the underpinning process deserve consideration. Underpinning elements normally are embedded within an earth volume that is susceptible to vertical and horizontal deformation. Ground movement may be transmitted to the new structural elements, or impose additional load on them. The latter can thus be subjected to downdrag forces, lateral loads, and/or displacements. Certain incidental factors are likely to enhance ground loss. For example, lagged underpinning pits for pier construction have the same contingencies as soldier pile walls. The potential for ground loss also exists when blow conditions are manifested in open shafts or open-ended piles below the groundwater table.

Composite Movement We know from experience that vertical settlement can indicate the extent of potential damage to buildings and structures, especially if it occurs as differential settlement. Likewise, horizontal displacement can cause damage, and for underpinned structures it is of greater concern (Febesh, 1975). The amount and distribution of movement in a soil mass adjacent to an excavation is governed by the soil type, stiffness of support, and construction procedures, including the sequence of excavation. An explicit understanding of the mechanics of ground behavior permits the rational assessment of effects on adjacent buildings, and provides the basis for formulating techniques that can reduce movement. This places emphasis on a quantitative evaluation of displacement patterns as the basis for the decision to underpin. If the cost of underpinning is high compared to the value of the structure and if there is no danger of collapse, we may consider the do-nothing option, accept the risk of not underpinning, and remedy any damage after completion of the work.

2-2 CONVENTIONAL UNDERPINNING TECHNIQUES

Probably the oldest, and a fairly simple type of underpinning, is pit or pier underpinning. In Chicago, and until the advent of machine drilling, hand-dug foundation piers of circular configuration were common, and they are still in use. Their main advantage is that they require minimum headroom and can be sunk directly beneath footings and walls. Because of the danger of loss of ground and consequent settlement where soils are saturated, this method usually is restricted to dry soil. In water-bearing formations nonconducive to excavation, either open or closed-ended jacked pipe or H-section pile provide a good choice. These types are reviewed in this section.

Pit or Pier Underpinning

Pits or piers generally are square or rectangular excavations filled with concrete. Horizontal wood sheeting commonly is used to retain the sides of the earth cut. Interestingly, since the technique was introduced, the construction procedure has not changed significantly. The same method used for access below a foundation also provides the basis for other underpinning procedures. When piers must be placed close together, a continuous wall can be constructed instead. However, the underpinning wall should be build in sections to avoid undermining the existing foundation. The sequence requires alternate sections, with the gaps filled-in later. Pier underpinning is installed as shown in Figure 2-2.

Construction Sequence This involves the following steps:

1. Beginning at the existing foundation level, excavate an approach pit directly adjacent to the footing as shown in Figure 2-2 *a*. For access, the pit should be 4 ft long, 3 ft wide, and 4 ft deep, but in stable ground it can be as large as a square with 10-ft sides, and 6 ft deep.
2. Brace the four sides of the approach pit with wood sheeting fit tight against the ground. Pack behind the sheeting boards to establish full and firm contact with the earth face, as shown in Figure 2-2 *b*.
3. Excavate beneath the existing footing to the same depth as the approach pit, and likewise brace the face with tightly fit sheeting boards, as shown in Figure 2-2 *c*.
4. Continue excavation beneath the footing in layers just deep enough to install one ring of wood sheeting at a time, pack and brace as before, as shown in Figure 2-2 *d*.
5. When the intended underpinning depth is reached, remove the loose material from the bottom and pour concrete to fill the pit to within 2 to 3 in. of the underside of the footing, as shown in Figure 2-2 *e*.
6. When the concrete has set, usually 24 to 48 hr, drypack or ram into the annulus space a moistened mixture of cement and sand, or insert a set of plates and wedges for full contact and bearing.

Experience suggests that the completed piers should be laterally braced if the ground (soil) is excavated on one side to a depth exceeding 6 ft. An example of pit

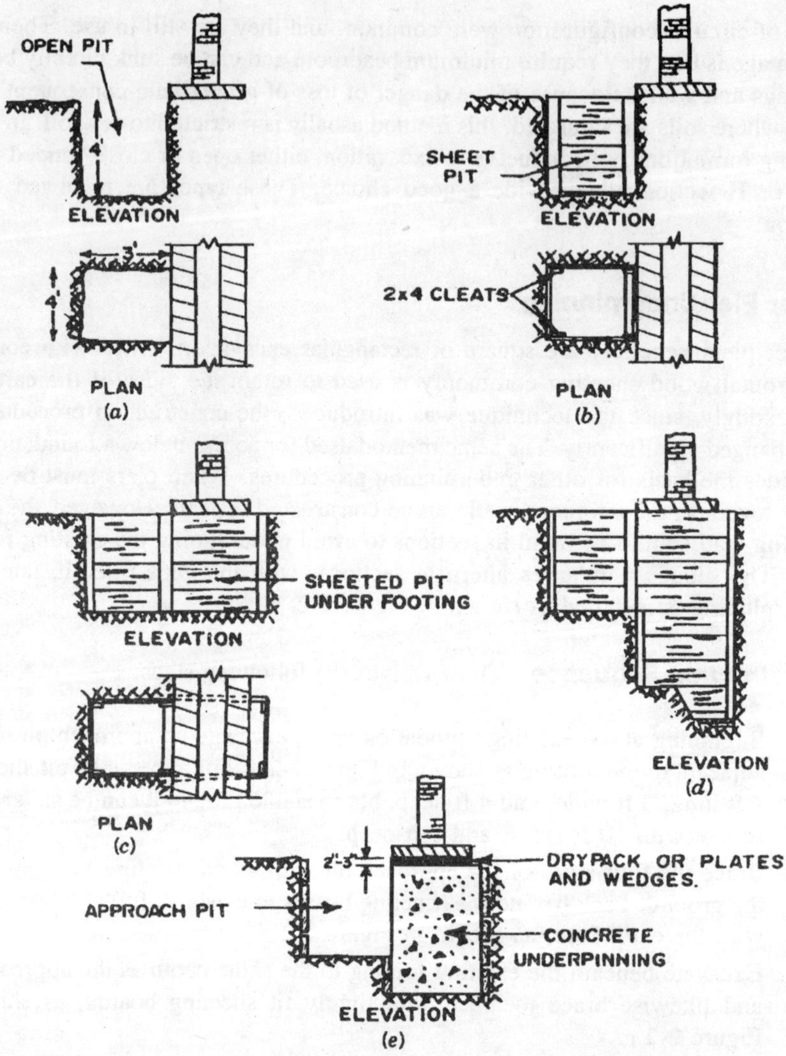

Figure 2-2 Pit or pier underpinning; typical installation sequence.

underpinning is the work done in the restoration of the White House, where a cellar and subcellar were added.

Load Transfer The materials filling the gap between the underside of the footing and the top of the concrete pier should develop a compressive strength equivalent to the soil strength in the initial configuration. For carefully packed and rammed materials this strength normally would be achieved. In certain conditions, attainable settlement as a result of the load transfer at this location may be excessive and thus unacceptable. In this case, jacks will have to be inserted and the loads maintained on them, an arrangement that will allow the concrete pier to settle while the underpinned structure is maintained at its original level.

At the base of the concrete pit, the soil-bearing capacity will be adequate as long as the same bearing area is available, the ground is not disturbed, and the soil strength is unchanged. When necessary, underpinning pits can be enlarged or belled at the bottom, especially in cohesive soils.

Pit Size and Details The thickness of the sheeting generally is independent of the pit depth, since the soil stresses are assumed to be distributed according to an arching effect. An exception to the spacing and sizing guidelines is noted for shallow pits (< 10 ft deep), or in firm cohesive soils where excavation and concreting can be completed in a single shift. In these instances sheeting requirements are less critical (Goldberg et al., 1976).

In soils other than soft clay, 2-in.-thick wood planks installed horizontally can be used as sheeting for pits up to 5-ft. square and regardless or depth. Board width is 8, 10, or 12 in., according to the type of soil. The material usually is untreated wood, but occasionally treated wood may be specified because of concern over future deterioration. The boards are installed one at a time, and soil is repacked to fill any voids. In running sands, contractors use hay stuffed behind the boards to block the flow. Corners of the sheeting typically are nailed.

The selection of pit size is based on the following considerations: (a) the pit must be large enough for easy operation and access to the footing; (b) it should be sufficiently small, however, to inhibit the in-place boards from deflecting excessively before the concrete is poured; and (c) the pit size should not undermine the footing but should allow it to span over the opening and without overstressing the soil. Common sizes of excavated pits are 3 × 4 ft or 3 × 5 ft. Square pits with sides 4, 5, or 6 ft are not unusual. Larger pits will require a thicker sheeting and supplementary bracing.

Pit spacing and sequence are coordinated with the existing foundation type, size, and load so that the latter is carried and distributed safely. The sequence occasionally may specify primary and secondary pits at selected spacing, particularly for continuous footings and until the entire underpinning scheme is in place.

If the pit depth does not accommodate manual soil removal, a scaffold can provide an intermediate level for two-stage removal. For great depths several intermediate tiers are necessary.

Maintenance Underpinning Shallow concrete piers can be used for maintenance underpinning. This is economical for keeping a building in a near level position and despite the subsidence that may occur, for example, as a result of tunneling in soft ground. Examples presented in subsequent sections demonstrate this procedure.

Ground Conditions A pit should be excavated and concreted as quickly as possible to reduce the risk of damage or settlement. During the operation, loss of ground can occur as an outflow of running soils that are difficult to drain. Ground loss can also result from moving squeezing soils, particularly cohesive materials that have a stability number >5. In both instances, the potential for ground loss will remain and persist while the soil face is exposed for the placement of lagging, and thereafter by soil moving through open lagging or by peal-off into an overcut behind the lagging.

Pit or pier underpinning is particularly sensitive to groundwater conditions, and thus best suited to dry ground. If the excavation level is below the groundwater table, a different underpinning scheme is indicated and should be considered. A difficult condition is manifested in sand or gravel formations that are stratified with impervious layers tending to support perched groundwater levels even after dewatering. In this case, insufficient dewatering can result in soil erosion if the water is allowed to flow into the pit through open lagging.

Alternatively, in water-bearing soil with depth not exceeding about 5 ft, vertical sheeting can be driven to cutoff the water. For this purpose, light steel or tongue-and-groove wood sheeting is suitable. The sheeting should be driven below the bottom of the pit a sufficient distance to prevent bottom heave because of hydrostatic pressure. With water cutoff the pit can be pumped dry and excavation continued.

Pile Underpinning

Pile underpinning is indicated if a good bearing stratum is at a relatively great depth, where groundwater prohibits excavation, or where the elements to be underpinned are mainly columns that carry considerable load. These factors, individually or in combination, can render pit underpinning uncertain or too costly. The piles can be installed either directly under or alongside a footing. In the latter case, the load is transferred to a beam connecting two piles and inserted horizontally under the footing, or to a bracket attached to a single pile. The piles have normal configuration and are selected from standard structural shapes, but because they are installed from inside structures they require special techniques. They may be jacked in place, or they may have to be driven, in which case the pile section and the hammer must work within the available headroom. The requirement for short sections implies structural types that can readily be spliced and practically precludes timber piles.

Steel H beams or pipe sections are suitable for use as underpinning, and in open- or close-ended configurations. Piles that will encounter relatively small resistance during driving are preferred, however, and the choice is thus confined to H beams and open-ended pipe sections. Close-ended pipes are suitable for penetration through soft materials, and for sites where displacement and vibrations from pile driving can be tolerated.

If the excavation is carried out close to an underpinned structure, the underpinning scheme can be designed to become part of the lateral support. In this case the piles are treated as soldier beams with lagging, and support lateral loads in addition to the axial loads from the foundation.

Jacked Piles These require normal headroom, and can be placed under a footing. The installation is better accommodated by hydraulic jacks that are light and easy to handle in a confined pit. Usual capacity of commercially available equipment is 40 to 60 tons, but this must be consistent with the intended use of the footing as jack reaction. Jack loads are monitored, therefore, to ensure that excessive upward force on the foundation is not induced. Coated or lubricated piles can reduce resistance to penetration.

Removal of soil from within becomes easier if a pipe is open-ended. In soft soils, a plug is formed with cinders, sand, or lean concrete, all allowing the pile to be advanced through soil layers without accepting earth materials inside. Where strong resistance is encountered, jacking is helped with ground excavation from within and below the tip of the pile (Bathel, 1968). This excavation or cleaning-out is carried out with the use of suitable tools, including dwarf orange peel buckets, post-hole diggers, and augers (hand-operated, suction pumps, and jets). Churn-drilling equipment can be used where large gravel, cobbles, and boulders are present. This method is fairly common in pile underpinning in the Washington, D.C. area (White, 1976). Efforts to produce powered augers have not been particularly successful because these machines are difficult to set, and require wide and deep pits for headroom and clearance.

Jacked pile installation usually involves the following steps:

1. Excavate and brace the approach and access pit under the footing. Attach a steel bearing plate to the underside of the existing footing, and establish full contact with drypack or mortar. This plate will accept and distribute the jacking force to the footing as an upward load.
2. Position a steel pipe section or H-beam 4 to 5 ft long in the bottom of the pit and place a steel bearing plate on top. Set the jack on top of the plate, and shim the space between the jacks and the plate at the underside of the footing to ensure full contact between the pipe, the jacks, and the footing.
3. Begin jacking the pile into the ground, and when its top reaches the bottom of the pit remove the jacking equipment. Clean the pipe if necessary, and add a jacking sleeve and the next pipe section. Repeat jacking, cleaning, and blocking until the intended foundation level is reached.
4. Clean out the pipe, and add more sleeve and pipe to bring the top to within 1 to 1.5 ft of the underside of the footing. Fill the pipe with concrete.

A typical jacked pile installation is shown in Figure 2-3.

Pretest Piles These were developed by Spencer, White, and Prentis (White, 1975). When the pile is in place and its tip is at the desired depth, a load test is carried out using two jacks placed on top of the pile. The jacks overload the pile by 50 percent of the design load, the latter varying from 40 to 100 tons, in a gradual application to detect any settlement. In addition, the intent is to prevent the rebound of the pile when jacking stops, and subsequent settlement when the load of the structure is transferred to the pile. Thus, when overloading is completed and settlement does not occur, a wedging beam is inserted between the two plates and tightly fit while the jacks continue to apply the test load. This insertion inhibits rebound

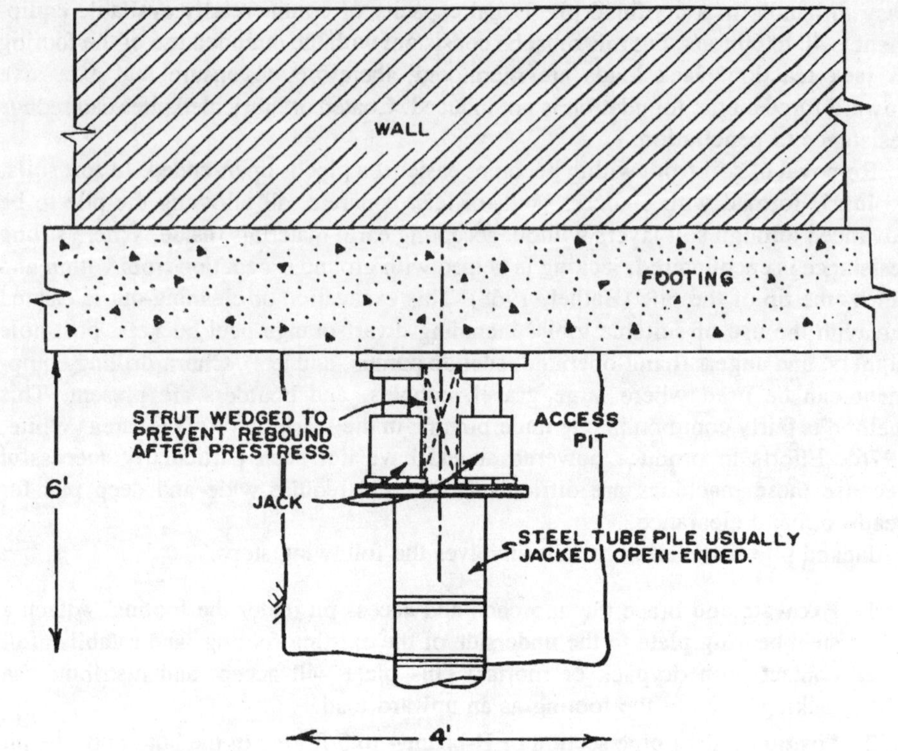

Figure 2-3 A typical jacked pile installation.

and subsequent settlement of the pile under hysteresis effects (White, 1975). When the transfer of load is accomplished, the jacking pit is backfilled to within 2 in. of the underside of the plate on top of the pile, the wedging strut and plates are encased in concrete, and the approach pit is backfilled as required. Figure 2-4 shows how pretest piles were used for underpinning existing structures during construction of a subway in New York City.

In certain cases pretest piles may be more economical than pit underpinning, even where groundwater problems do not exist. White (1976) cites an example that involves a small footing carrying a load of 100 tons (200 kips). For a pit well, it may be necessary to install first a column pick-up system, a preliminary support, a combination of needle beams, grillage beams, and screw jacks before the concrete element is sunk. This can be costly and space consuming. As an option, two pretest piles can be installed sequentially and without preliminary support, and may cost less.

Driven Piles These are usually 12 to 14 in.-diameter steel pipe $\frac{3}{4}$ in. thick. Commonly, they are driven open-ended to reduce vibrations, and in lengths determined by available headroom.

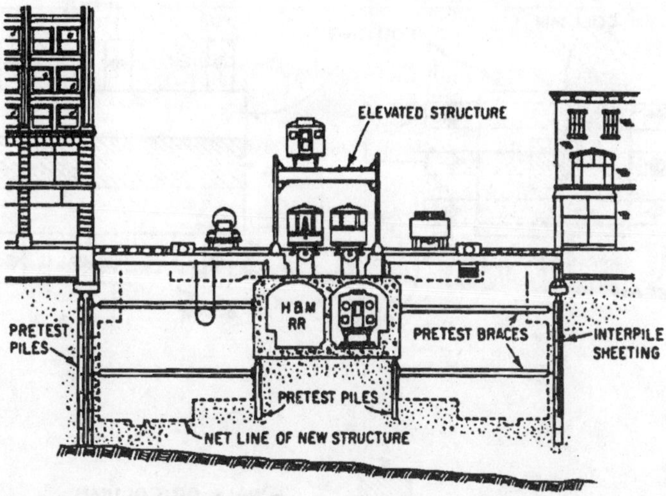

Figure 2-4 Pretest piles supporting existing buildings, old elevated railway, and Hudson and Manhattan Railroad Tunnel during construction of subway under Avenue of the Americas, New York City.

Piles can be driven using conventional hammers or drop weights. With hammers the actual energy utilized is limited by the available headroom, which restricts the drop height. Physically, the pit must be deep enough to accommodate the pile section, the hammer, and the associated appurtenances necessary to support the hammer. Piles installed below foundations usually are tested by jacking against the foundation, and the load transfer is accomplished as in jacked piles.

For open-ended piles reduction of side friction and base bearing during installation improves advancing rates, and requires only periodic soil cleaning from within. Pipe sections are spliced using tight-fitting sleeves fastened to the outside face to avoid inside interference with cleaning or removal of obstructions. The sleeves are preattached to the section, and help keep it in correct alignment. Cleaning methods usually involve mechanical tools or air and water jetting. With the latter, care is necessary to avoid cleaning below the bottom of the pipe. In most cases, a positive hydrostatic pressure can prevent bottom blowout during driving and cleaning.

The installations become more economical if the piles can be located outside the building. Frequent applications for driven pile underpinning are for columns of elevated transit sections. If these columns are within the net boundary of the subway structure, the load in the final stage can be transferred to the superstructure. If piles can be driven on both sides of a footing, the latter is intercepted by transverse beams, as shown in Figure 2-5*a*.

In this case, the construction involves the following steps:

1. Excavate to the base of the existing footing. If existing headroom is not available, excavate and brace a pit at each pile location.

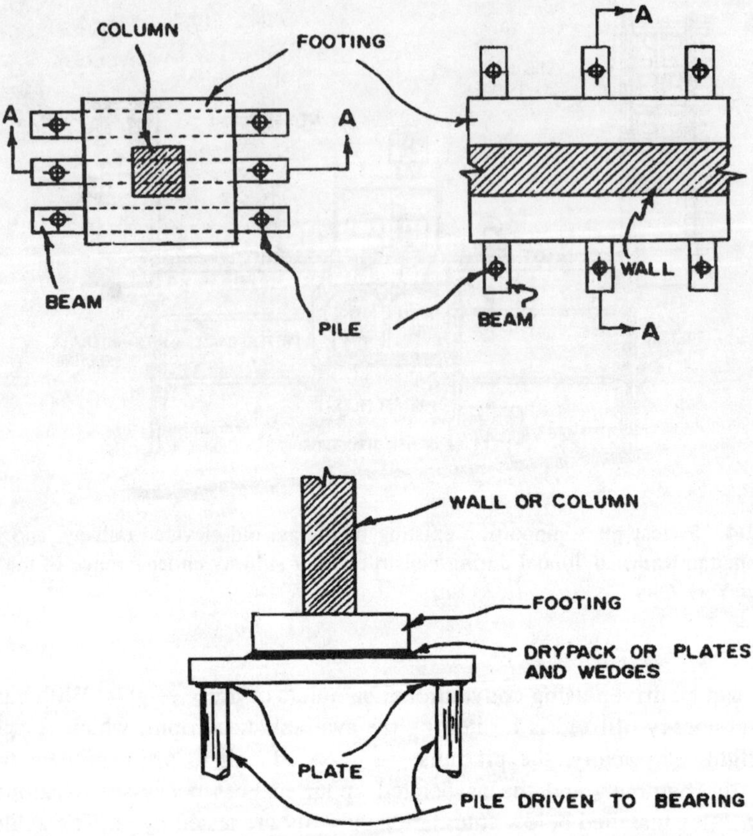

(a)

Figure 2-5 (*a*) Piles driven alongside footing, load transferred by beams. (*b*) Steel pile with steel bracket.

2. Install piles on both sides.
3. Excavate and brace a cross trench for one beam under the footing. Install the beam, and transfer the load by drypack, plates, and wedges. The load transfer can be made at the bottom of the footing, at the top of the piles, or at both locations.
4. Complete the work in phases, inserting one beam at a time and then transferring the load in a similar manner.

The beams are steel sections, reinforced or posttensioned concrete, and are encased in concrete one at a time or after the entire footing is underpinned.

If piles can be driven only on one side of the footing, the connection is provided with a bracket, as shown in Figure 2-5*b*. Since the load in this case is applied eccentrically, the pile sections must also resist bending moments in addition to the axial load. Piles on one side only are particularly suitable for exterior walls or

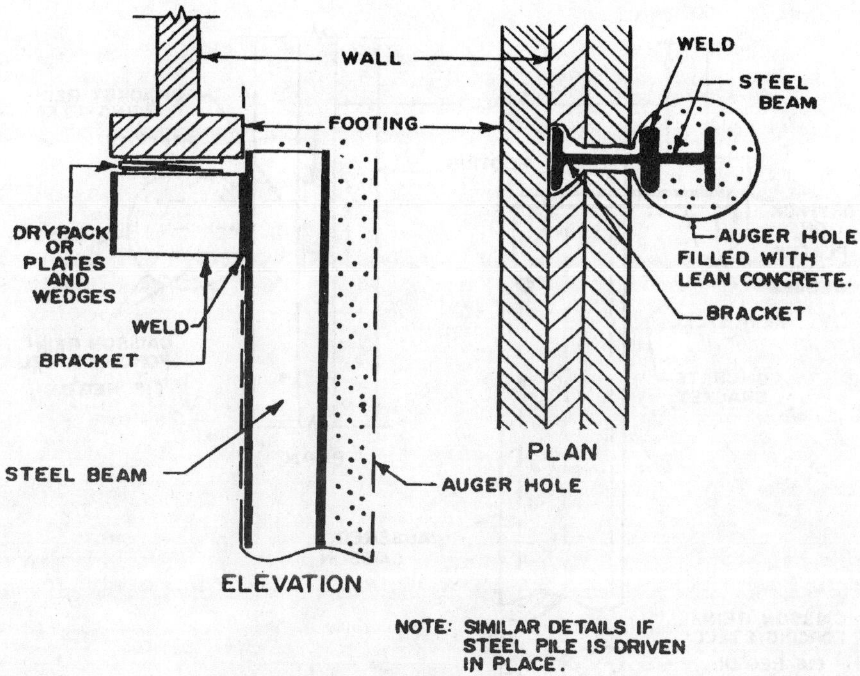

(*b*)

Figure 2-5 *continued.*

continuous footings, and where the bracket can be used without the risk of shearing effects on the footing. The installation involves the following steps:

1. Excavate to the base of the existing footing. If possible, undercut footing so that the pile can be inserted close to the wall to minimize eccentricity. Special offset driving brackets can be fabricated to accommodate close driving.
2. Install piles to the intended depth.
3. Excavate a sheeting pit beneath the footing in line with the pile. Weld bracket to the pile, and transfer the load by drypack, plates, and wedges. Encase bracket and top of pile in concrete.

Augered Caissons These can be used where the available space does not preclude the drilling equipment necessary for the work, but preferably outside a building. The caissons transfer the load to the ground by shaft resistance, base bearing, or both. An example of augered concrete caisson is shown in Figure 2-6, and requires the following installation sequence:

1. Auger a vertical hole directly adjacent to the footing.
2. Insert a steel beam into the hole and fill with concrete. Alternatively, the steel beam is omitted and conventional reinforcement is provided.

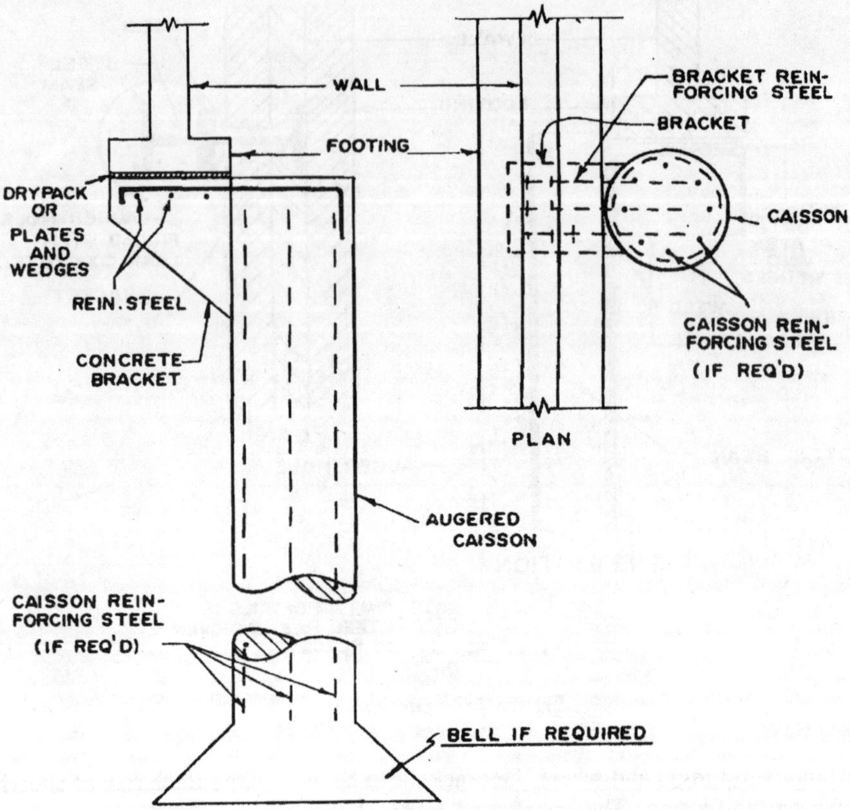

Figure 2-6 Augered concrete caisson with concrete bracket.

3. Excavate a pit under the footing, install reinforcement, and form bracket by poured concrete. Stop bracket 2 to 3 in. below the underside of the footing, and after the concrete has set place drypack to fill the gap.

A variance from the procedure of Figure 2-6 is shown in Figure 2-7. An auger hole is excavated in conjunction with a steel pile installed in the adjacent slot. The slot is cut under the footing for the entire depth of the hole, and is sheeted as the excavation progresses. After the pile is inserted the slot is concreted, and the load transfer is completed with the use of drypack, steel plates, and wedges.

Battered Piles When a proposed excavation encroaches onto private property or must be carried out inside the property line, preexcavated battered piles, also called slant drilled piles, can be used, as shown in Figure 2-8. The underpinning is thus confined to the area outside the excavation zone. The hole is drilled at a batter or slant beginning as close to the footing as practicable, and continues until the bearing stratum is reached. Underpinning is completed by cutting a slot below the existing

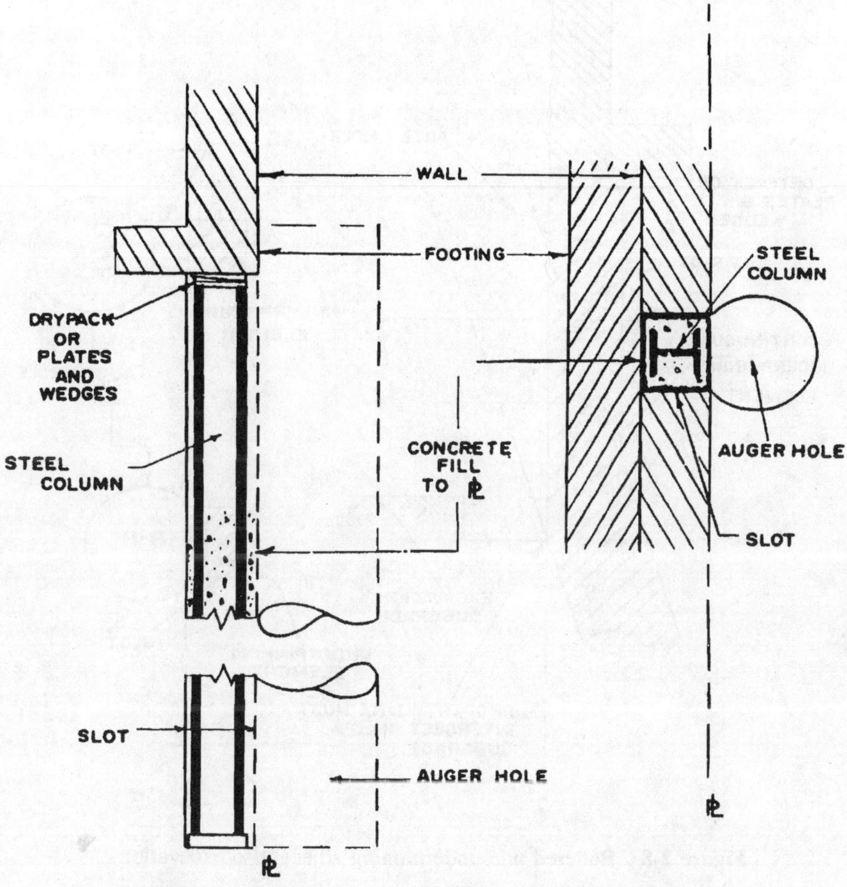

Figure 2-7 Augered hole with pile installed in adjacent slot.

foundation, as shown, and connecting to the slant pile reinforced concrete properly doweled into the pile element.

Grouted Piles

Unlike caissons, hollow-stem augered piles have a relatively small diameter, usually 12 to 16 in. A continuous-flight hollow-shaft auger is rotated into the ground until it reaches the required depth, and as it is withdrawn high-strength mortar is placed under pressure to form a compact concrete element. If the installation is inside buildings, special low-headroom equipment is used. The slender piles can be installed next to a footing or through it. The load transfer is accomplished by beams or brackets, and by making the piles integral wit! the footing. The piles can be installed at close spacing, and the process gener :es nominal vibration and soil heave. If the auger is withdrawn too quickly, some : oughing can occur, or soil may

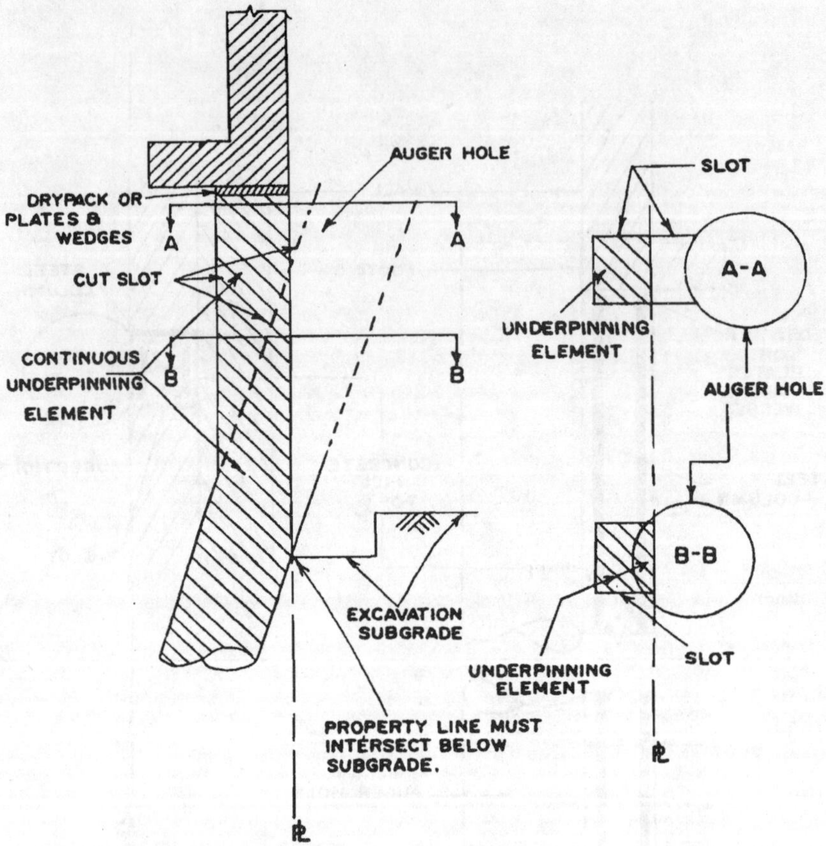

Figure 2-8 Battered pile underpinning adjacent to excavation.

fall into the hole before the mortar is placed. If this occurs, the resulting discontinuous element may not be detected until the loads are imposed on the underpinning.

Root Piles This method is popular in Europe where it was developed (Bares, 1974). Root piles are combined as a group to resist vertical loads or to provide lateral support. Their group action satisfies the underpinning concept, or may serve to strengthen an existing foundation. Typical underpinning schemes using root piles are shown in Figure 2-9. The piles provide direct foundation support for the structure shown in Figure 2-9*a*, and serve as underpinning for the wall shown in Figure 2-9*b*.

For usual applications involving nominal loads, the root pile diameter can be as small as 3 to 5 in. Where necessary the diameter can be as great as 16 in., and with nominal reinforcement. Initially, a hole is drilled with rotary or percussive tools and with the use of a casing. A steel bar is then inserted, and the hole is grouted as the

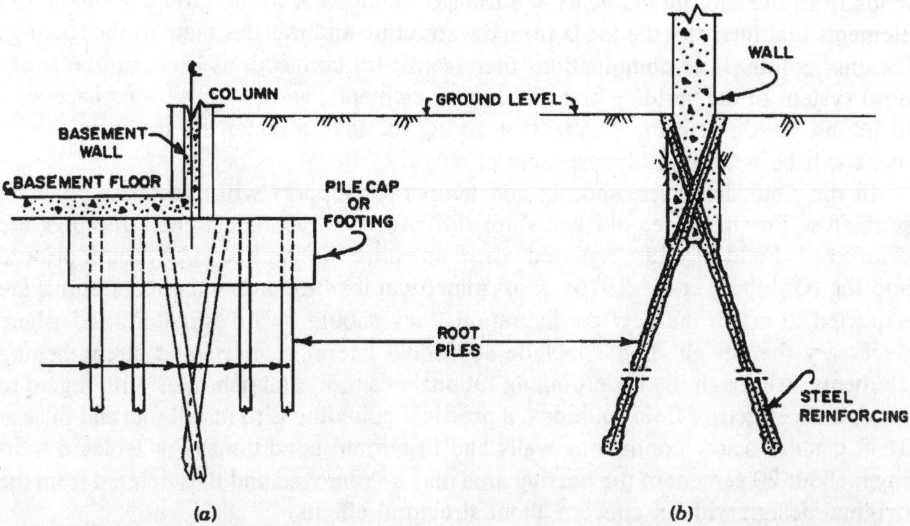

Figure 2-9 Typical underpinning schemes using root piles. (*a*) Direct support of a foundation (either friction piles or end-bearing piles). (*b*) Underpinning of a continuous wall.

casing is withdrawn. This operation is essentially similar to a ground anchor with a bar tendon except for the direction of the applied load. Grout pressure is quoted between 70 and 100 lb/in.^{2} With larger diameters (8 to 12 in.) the pile is usually reinforced with a spiral cage inserted before concreting. A wide range of uses and applications is reported in several European countries (Lizzi, 1970, 1974). Examples in the United States are reported by Bares (1974) and ENR (1972).

Root piles carry relatively small loads, 10 to 15 tons, and for a normal underpinning scheme several units are required. Pile inclination can vary within a practical range, and depends on design requirements. Piles battered as shown in Figure 2-9 will convert lateral loads into axial forces.

2-3 SHORING AND TEMPORARY SUPPORT

The need for shoring (preliminary) or temporary support during underpinning usually is based on the structural sensitivity or integrity of the building to be underpinned. Other factors to be considered are: (a) probable arching effect along the excavation zone; (b) acceptable increase (temporary) in the soil bearing pressure; and (c) the extent and potential effects of underpinning. It is quite difficult, and often impracticable, to predict the loads expected to act on temporary supports or shoring, and this uncertainty should be reflected in a conservative factor of safety combined with movement control and monitoring of the shored sections.

From the design standpoint temporary supports require analysis of structural and geotechnical data, and encompass the following: (a) new footings to receive the loads from the shoring elements and transfer them back to the ground; (b) shoring elements that intercept the loads from the structure and transfer them to the footings (beams, columns, or combinations therefrom); (c) connections between the structural system of the building and the shoring elements; and (d) provisions necessary to inhibit elastic deformations so that during the load transfer the structural settlement will be within an inconsequential range.

In the field, however, shoring and temporary support will encounter practical problems. For instance, old walls are difficult to shore unless their footings are reinforced. Field cases are reported where an entire footing had to be rebuilt prior to shoring (Goldberg et al., 1976). If asymmetrical loading or load concentrations are expected to act in the new configuration, they should be investigated, and where necessary the design should include additional lateral supports and strengthening elements. Although the underpinning rationale cannot establish rules with regard to acceptable effects of field solutions, a practical guideline is to install pits and piles at 16-ft centers below continuous walls and in normal conditions. For isolated footings, about 20 percent of the bearing area may be removed and thus deleted from the original design without concern about structural effects.

Needle Beams These are installed horizontally to transfer the load of a wall or column to either or both sides of its foundation, to permit digging of underpinning pits. These beams usually are more expensive than shores. Needles are steel wide-flange beams, and sometimes plate girders, used in pairs with bolts and pipe spreaders between the beams. This arrangement provides resistance to lateral buckling and torsion. The needles may be prestressed with jacks, since this can eliminate settlement that may occur when the load is applied. Typical needle beam shoring arrangements for continuous walls and for individual columns are shown in Figure 2-10.

The load from steel columns may be transmitted through brackets to the needles. For masonry walls the needles can be inserted through niches. The load should be transferred from the masonry to the needles through thin wood fillers that crush when the needles deflect and maintain nearly uniform bearing. Wedges commonly are inserted under the ends of the needles to shift the load from the underpinned member to the beams. Where only limited settlement can be tolerated, the shoring scheme may include several concrete pads and jacks at the support points to control and adjust structural movement. At best, and in favorable conditions, the shoring system can consist of timber members.

Inclined Shoring Shores can be installed vertically or can be inclined; in the latter case they provide both bearing and lateral support. Inclined shores placed only on one side of a wall require support at the base for vertical as well as horizontal forces. One method is to brace the shores against the opposite wall at the floor.

This type of support is particularly applicable where access is limited, where needle beams must be exceptionally long or deep, and where lateral support is needed. Geometric and structural schemes are developed and tailored to the actual field conditions, but essentially depend on the structure type, excavation sequence,

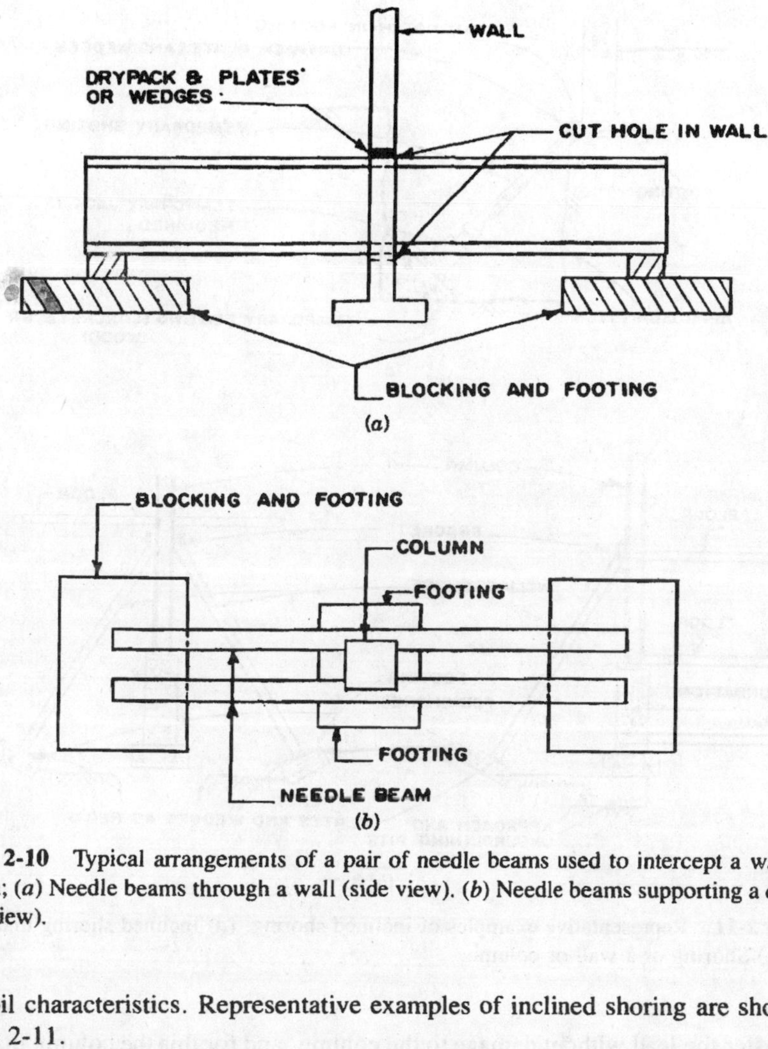

Figure 2-10 Typical arrangements of a pair of needle beams used to intercept a wall or a column; (*a*) Needle beams through a wall (side view). (*b*) Needle beams supporting a column (plan view).

and soil characteristics. Representative examples of inclined shoring are shown in Figure 2-11.

Good bearing is essential at the top and bottom of the shores. One method of providing bearing at the top is by cutting a niche to attach a steel bearing plate against the upper face. An alternative to the plate is a Z shape made by removing diagonally opposite half flanges from an H beam. When the top of the shore is cut to fit between flange and web of the Z, movement of the shore is restrained. However, for weak masonry walls, the load should be distributed over a larger area.

Both the vertical and the horizontal loads carried by an inclined shore are resisted by the shore footing, which preferably should be constructed perpendicular to the shore. The lateral load capacity is a function of the shoring inclination and the vertical load transmitted to the member. The footing must be sized and arranged to develop equal bearing capacity and passive resistance at the soil interface. Where steel or cast iron columns must be intercepted, the connections should be designed

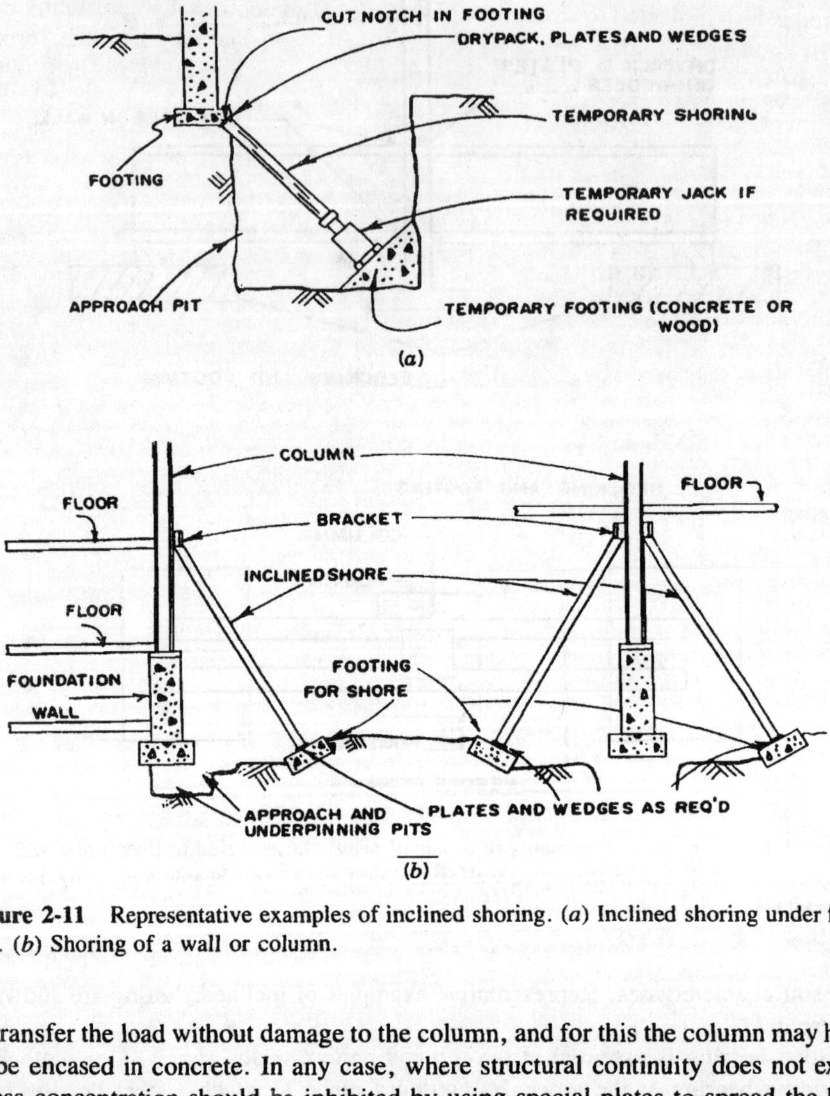

Figure 2-11 Representative examples of inclined shoring. (*a*) Inclined shoring under footing. (*b*) Shoring of a wall or column.

to transfer the load without damage to the column, and for this the column may have to be encased in concrete. In any case, where structural continuity does not exist, stress concentration should be inhibited by using special plates to spread the load transfer.

Grillages

These provide considerably more bearing area on the ground than needles; hence, they are often used as an alternative to needles and shores for closely spaced columns. A grillage may be installed directly on soil to support and tie together two or more adjoining column footings, or it may rest on a basement floor. As preliminary support, grillage may consist of two or more steel beams tied together with bolts and pipe spreaders, or of composite steel-and-concrete arrangements. The method is also indicated to strengthen or repair existing footings by reinforcing them and by increasing the bearing area. Alternatively, grillages may take the form of dowels or of encircling concrete or steel-concrete beams, in which case they should have adequate cross bracing to resist torsion and buckling.

2-4 UNDERPINNING APPLICATIONS AND EXAMPLES

Remedial Underpinning

This work becomes necessary for deteriorating foundations or where loss of foundation support has occurred and must be restored. An example is the circular tank shown in Figure 2-12. The tank is used for storing liquid ammonia, and has a capacity of 10,000 tons. Its diameter is 40 m (131 ft), and its height is 17.5 m (57 ft). The tank initially was supported by concrete piles, placed concentrically and capped with a flexible reinforced concrete raft, serving also as the bottom slab for the structure. The piles penetrated 26 m (85 ft) into soft marine clay, and 8 m (26 ft)

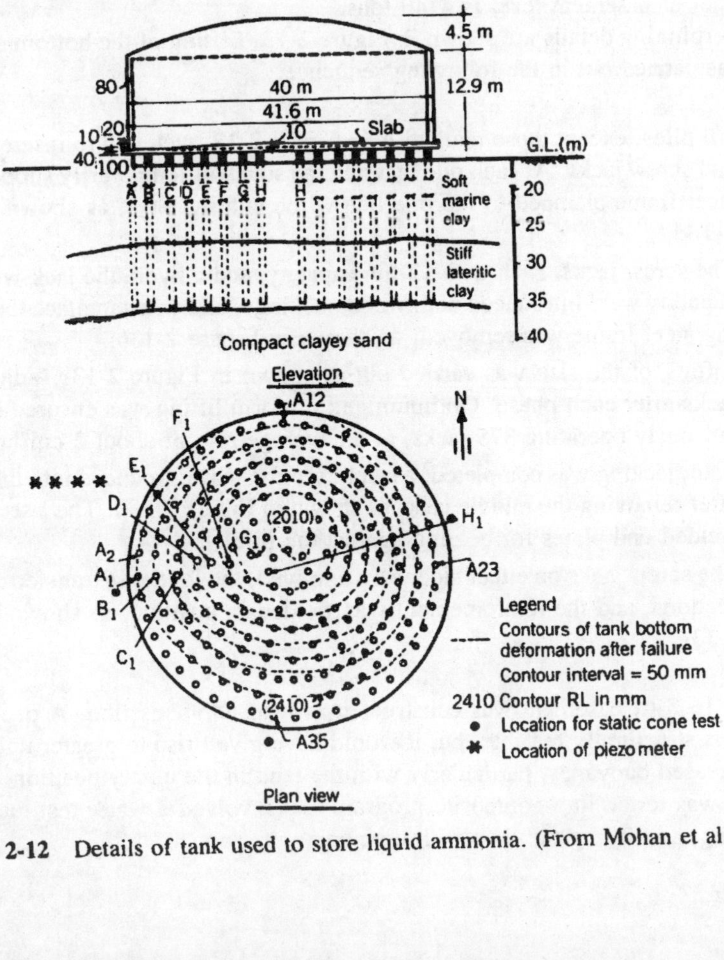

Figure 2-12 Details of tank used to store liquid ammonia. (From Mohan et al., 1978.)

into underlying stiff clay, and had an initial safe load capacity of 88 tons. Besides the storage load, an additional dead weight of 1600 tons is imposed on the foundation by the structure.

During service the tank experienced a series of problems. On initial testing approximately 8100 tons of water were pumped into the tank. As the load was completed, cracks appeared on the outer piles. Following this incident, the tank was emptied, and a geotechnical investigation was carried out. The tank was restored to normal use after constructing a watertight basement below the main structure, repairing the defective piles, and raising the deformed tank bottom to its original configuration (Mohan et al., 1978).

Underpinning Schemes Two remedial concepts are presented in Table 2-1. The first solution is moving the tank to a new foundation, while the second focuses on remedial underpinning, with four potential alternatives. The final choice was remedial underpinning combined with a partially compensated foundation and raising of the tank bottom to its original level. Partial compensation was achieved by constructing a 5-m (16.5-ft) deep chamber beneath the tank, fairly watertight and with its bottom slab rigidly connected to the piles. As a result of this work, total load relief at basement level is 4740 tons.

Underpinning details are shown in Figure 2-13. Lifting of the bottom slab of the tank was carried out in the following sequence:

1. All piles, except those marked A in Figure 2-12, were cut to insert hydraulic and screw jacks. At each pile location the slab was temporarily supported by a steel frame clamped to the pile below the cutting level, as shown in Figure 2-13*a*.
2. The screw jacks, each with 12-ton capacity, and a hydraulic jack with 30-ton capacity were introduced with wood packings, and when in place the supporting steel frame was removed, as shown in Figure 2-13*b*.
3. Lifting of the slab was carried out as shown in Figure 2-13*c*, adjusting the jacks after each phase. Optimum and uniform lifting was ensured by simultaneously operating 375 jacks, at an average rate of about 2 cm/hr.
4. After jacking was completed, a steel I section was introduced into the pile gap after removing the middle jack, as shown in Figure 2-13*d*. The I sections had welded end plates for bearing against the pile.
5. The screw jacks on either side were removed, the load was transferred to the I sections, and the reinforcement was welded in position, as shown in Figure 2-13*e*.

A 5-m (16.5-ft) basement was constructed as load compensation. A deeper basement was structurally feasible, but it would have given rise to greater uplift forces and increased buoyancy, particularly with the tank in the empty position. The new scheme was tested in a composite program that involved a water test loading and unloading, and filling the tank with ammonia.

TABLE 2-1 Remedial Schemes for Foundation Restoration

Alternative	Steps Involved	Cost (millions of U.S. dollars)	Time (months)	Remarks
1. Shifting tank to new foundation	Either fully compensated foundation or cellular raft after ground consolidation or piled foundation	1.28	13	Auxiliary services already built were also to be shifted
2. Remedial underpinning involving partially compensated foundation and:				
Providing additional camber in tank bottom	Cutting bottom steel plates of cup and shell, padding fresh concrete to achieve desired camber, and rewelding bottom	0.76	10	This measure would have reduced the tank capacity to 7,000 tons (62.23 MN)
Sinking exterior piles	Sinking exterior piles till desired camber is achieved	0.68	12	Higher time factor and cost
Lowering of tank bottom slab	Cutting piles and lowering tank bottom slab at outer periphery using a battery of jacks	0.68	11	Do
Lifting of tank bottom slab	Jacking tank bottom slab to its original configuration	0.58	9	Measure was finally adopted

Source: From Mohan et al. (1978).

Load settlement curves obtained at various points on the basement floor are shown in Figure 2-14, and lead to the following observations:

1. During the water loading test, measurable settlements are absent until the load reaches 5000 tons.

2. Tank bottom slab and basement settlements are almost the same near the tank center. However, along the periphery the settlement is larger at the basement than the corresponding points on the tank bottom slab, probably because a special construction detail allowed unrestricted movement of the basement slab at this location.
3. For a water load of 11,600 tons the maximum differential settlement of the basement is 23 mm (about 1 in.), and the maximum differential settlement for the tank bottom is 34 mm (1.3 in.).
4. The settlements at the tank bottom periphery and at tank center vary from 63 to 90 percent.
5. The load/settlement of the structure during the ammonia loading appears to be erratic and nonspecific, probably reflecting the effect of higher rates in tank filling.

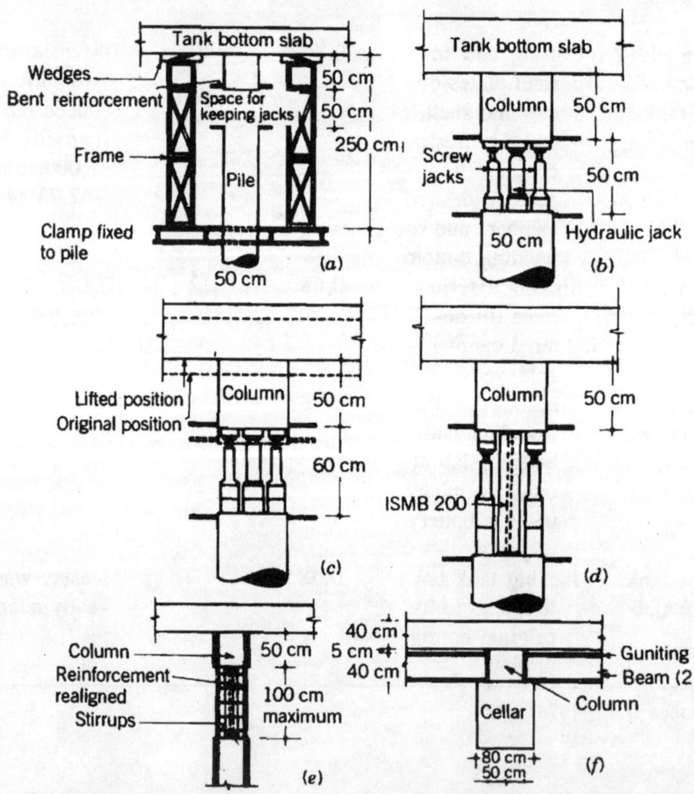

Figure 2-13 Details showing the lifting of tank bottom slab, tank of Figure 2-12. (*a*) Supports for slab and cutting pile. (*b*) Housing of hydraulic jacks and screw jacks. (*c*) Lifted bottom slab position. (*d*) Placement of ISMB and removal of jacks. (*e*) Realigning of reinforcement. (*f*) Construction of RCC collar and ISMC in position. (From Mohan et al., 1978.)

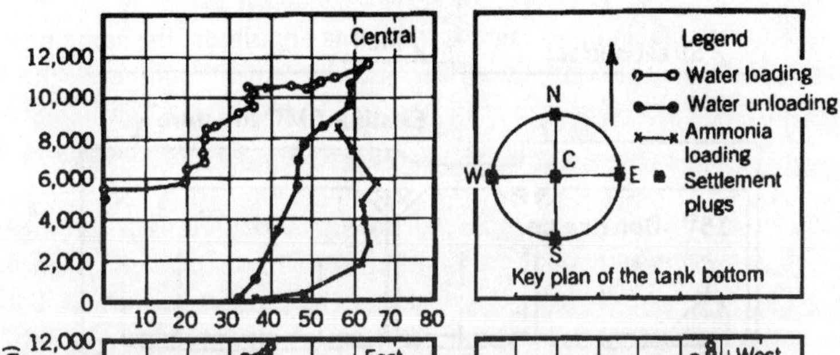

Figure 2-14 Load versus settlement diagrams for the tank structure of Figure 2-12, after construction of basement. (From Mohan et al., 1978.)

Underpinning for Buildings

As mentioned in Section 2-1, risk mitigation often provides a value engineering clause, commonly used as an adjunct to the normal contractual procedures to improve response to the underpinning invitations for bids. Two typical examples are given by White (1976).

Case 1 The new subway along Crystie Street, Lower East Side in Manhattan, New York City, passes underneath an existing-track subway. At the site, the ground consists of sand with a high water table. The owner presented a plan in the contract documents to support existing structures and buildings using riveted plate girders that would receive the loads from the steel walls and interior columns. Inside the old structure, the installation would probably face critical time constraints, limited working hours (during nights only), and corresponding delays and overtime costs, all intended to minimize disruptions to transit service.

The underpinning subcontractor proposed an alternative, combining pit and pre-

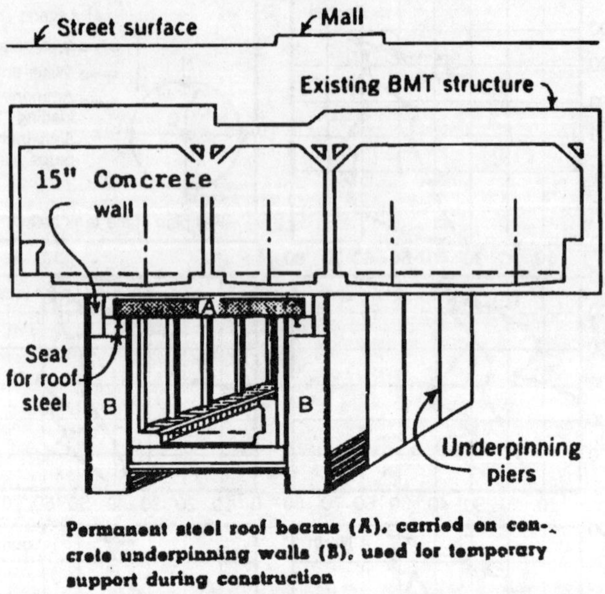

Figure 2-15 Underpinning scheme for Chrystie Street subway, New York City. (From White, 1976.)

test pile underpinning that eliminated the need to enter active subway areas. The owner accepted the proposed option under a note on the plans that allowed consideration of alternatives (Goldfinger, 1960). The final underpinning scheme is shown in Figure 2-15.

Case 2 Part of the excavation for the Washington, D.C. metro was adjacent to the Fine Arts Building on Seventh and "G" Streets. The underpinning was designed and detailed on the plans and the bidding documents. This building has cut-stone wall facing dating to 1836. The plans specified structural steel needles and grillages to be followed by large pits excavated to subgrade, intended to remain for as long as the building walls were to be suspended. During this period, large settlement cracks appeared in the gallery rooms, which were closed to the public.

In this instance, and despite a value engineering clause, submission of alternate underpinning designs was not encouraged. It appears that simple pit underpinning in short sections and without the use of needling might have prevented the problem of differential settlement (Mergentime, 1972). This was based on a comparison of this case with the successful pit underpinning of the White House in 1950 (Prentis and White, 1950).

Case 3 The Monadnock Building in Chicago is one of the city's oldest and tallest buildings with all-masonry construction. Underpinning of parts of the structure was carried out on two separate occasions, first in 1940 and again in 1967. The earlier underpinning was in conjunction with the construction of the Dearborn Street sub-

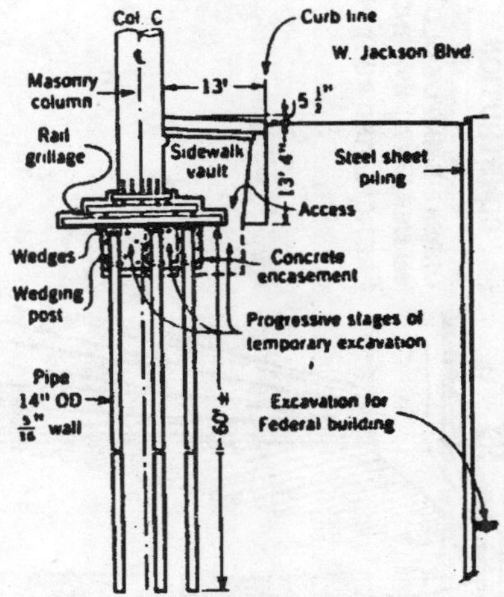

Section through underpinned column adjacent to excavation for the Federal building.

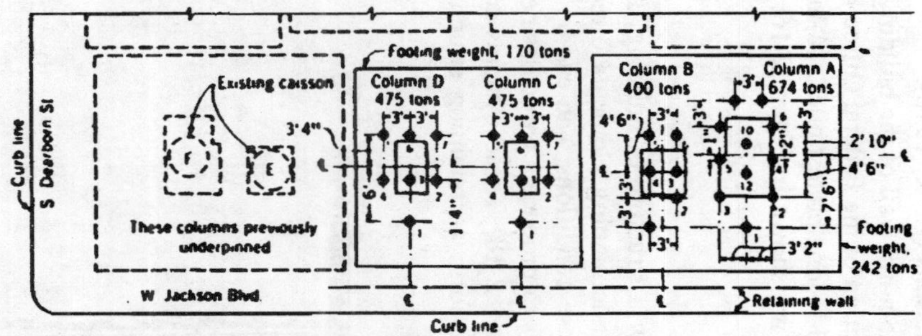

A small excavation permitted moving under old footings to install Pile No. 1 and pick up 80 tons of the load using the pretest system of transfer. Then the other piles were installed progressively from carefully sheeted excavations while part of the load was still carried by the spread footing.

Figure 2-16 Pretest pile underpinning of the Monadnock Building, Chicago, 1967. (From White, 1976.)

way along the east side of the building, and required extensive use of grillages, needle beams, mats, screw jacks, and caissons (ENR, 1940).

The underpinning in 1967 protected the north wall of the building along Jackson Boulevard during excavation and construction of the Federal Office Building across the boulevard. This operation involved 60-ft-long pretest piles, 14-in. o.d., consisting of $\frac{5}{16}$-in.-thick pipe sections. Details are shown in Figure 2-16.

Underpinning for Tunneling

Where tunnel alignment is directly below a structure or building, it is not always feasible or practical to insert vertical underpinning elements directly below the foundation. Root piles are suitable in these conditions, and can be used as reticulated walls formed by combining the piles to produce a particular network, such as the examples shown in Figures 2-17 and 2-18. These systems are designed to function as load-bearing elements or perform as retaining structures. (See also Chapter 3.)

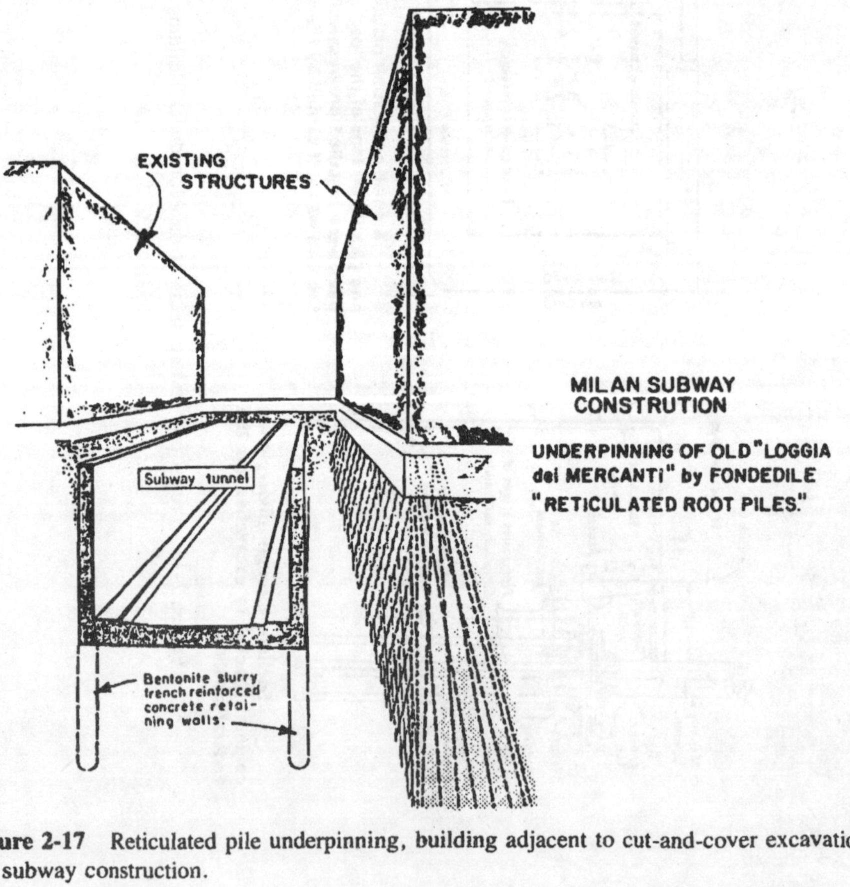

Figure 2-17 Reticulated pile underpinning, building adjacent to cut-and-cover excavation for subway construction.

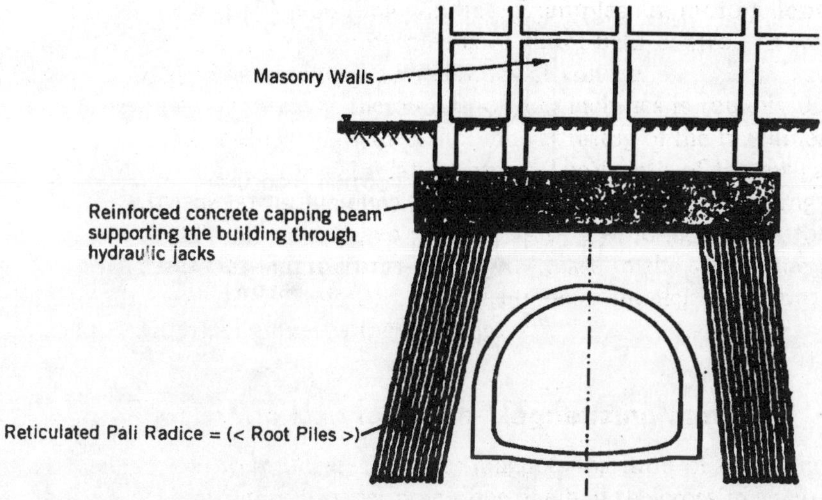

Figure 2-18 Reticulated pile underpinning, building above a subway tunnel.

When a tunnel is planned beneath a building, the structure can be maintained in the same elevation by disconnecting the columns from the footings, and jacking the columns. This operation is shown in Figure 2-19. A bracket is first installed and connected to the column, the anchor bolts are removed to disconnect the column from the footing, and jacks are placed as shown in Figure 2-19*a*. As tunnel excavation approaches, the jacks are activated while the load continues to act on the footing. As the footing begins to settle, the load is dissipated and is picked up by the jacks, thus maintaining the columns elevation. When the construction is completed and overall settlement of the footing has occurred, the base plate is reshimmed, as shown in Figure 2-19*b*, the anchor bolts are retightened, and the jacks are removed.

Figure 2-20 shows the application of the so-called pipe shield method. A series of contiguous horizontal pipe tunnels are drilled, 3 to 4 ft in diameter. The pipe tunnels are reinforced and concreted to provide a protective roof or shield cover above the proposed tunnel.

For the example shown in Figure 2-20, the jacking pits are excavated from the surface (Zimmerman, 1969). The construction is carried out in the following sequence.

① Underpin bridge superstructure with steel piles and jacks to adjust for settlement.
② Construct jacking pits on each side of the highway embankment, jack 1.2-m (4-ft) diameter pipes, and complete by concreting the pipes.
③ Construct tunnels below pipes (3-m wide, 2-m high). Concrete each tunnel before excavating the adjacent one.
④ Construct walls of highway tunnel.

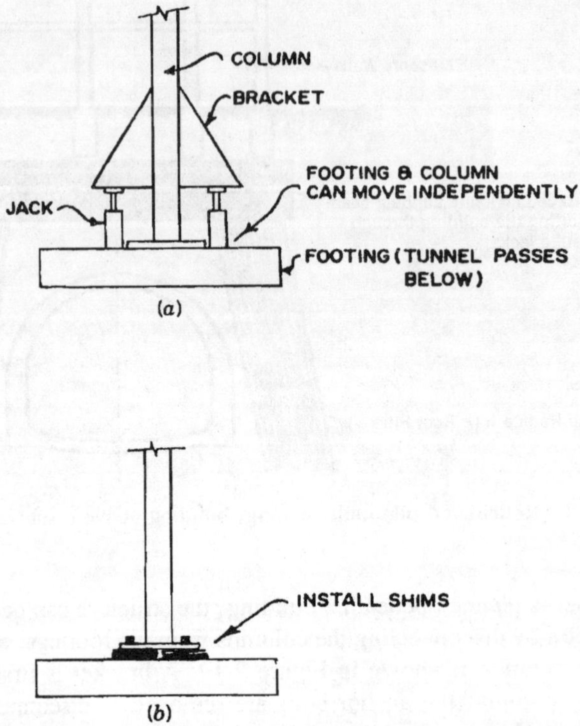

Figure 2-19 Diagramatic presentation of column jacking to prevent structure settlement during tunnel construction. (*a*) Brackets installed, anchor bolts loosened, jacks installed. (*b*) Shims installed, anchor bolts tightened.

A case reported by Maidl and Nellessen (1973) involved a subway passing beneath a heavy building, but site conditions did not permit excavation of a jacking pit from the surface. In this case, a primary drift tunnel was advanced, and the pipe shields were jacked transversely. Excavation was carried out below the pipe shield roof using secondary drift tunnels and general excavation.

2-5 DESIGN CONSIDERATIONS FOR UNDERPINNING

Assessment of Existing Structures

As we mentioned in reviewing the general requirements, the usual approach is to compile a record of buildings and structures to be underpinned. This inventory should be complete and should include the structure type, condition, loads, and materials data.

The intent of underpinning is to control settlement associated with new construction. Since the settlement that a building or structure can tolerate is not predictable

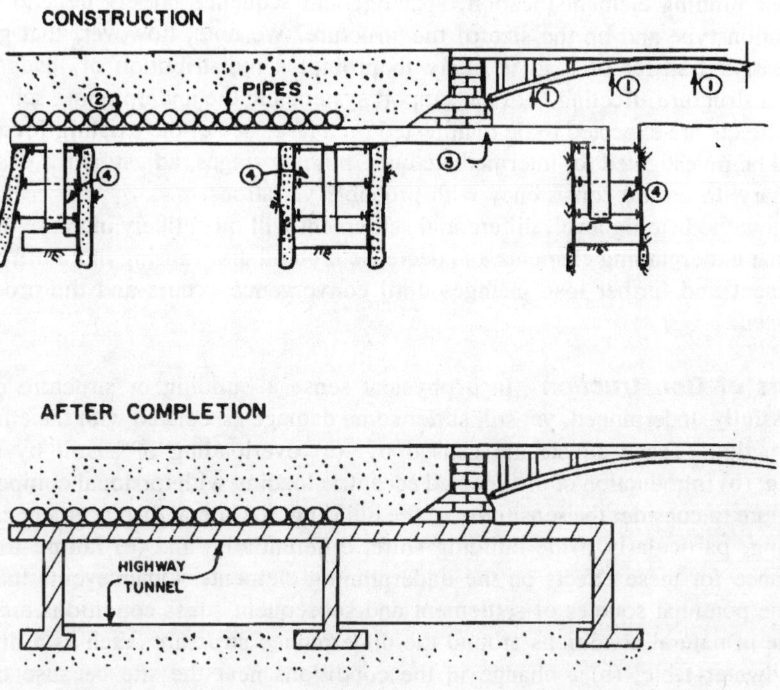

Figure 2-20 Pipe shield technique. (From Zimmerman, 1969.)

within the scope and the means of engineering analysis, correlation with documented history and performance records under similar conditions is essential. Useful conclusions can be drawn by studying buildings with known settlement records, and this reference can provide the basis of a safe and economical underpinning design.

For example, by observing a 5-story bearing-wall building, Terzaghi (1938) concluded that a stiff wall takes more differential settlement than an interrupted or broken wall, and heavy structures can undergo differential settlement without damage if they are separated into small rigid units. In general, a different tolerance is stipulated for low and tall buildings, and for wall-bearing or framed buildings. A different tolerance is accepted for structures resting on granular and plastic soils. Interestingly, a building can suffer more damage by a change in length from 100 ft to 100 ft, 1 in. than by a 1-in. settlement (White, 1976). Likewise, the time-rate settlement pattern is equally essential because a building may need more time to adjust to new support conditions, and will show distress if the settlement is sudden. Very rigid structures are prone to damage by differential settlement, and this conclusion is documented by field experience.

Estimation of Loads Estimating actual loads acting on foundation elements is difficult but essential. If the building record and plan examination does not produce these data, a structural analysis is mandatory.

Underpinning element location, spacing, and sequence largely depend on the foundation type and on the size of the structure. We note, however, that gradual progressive transfer of load is likely to prompt a redistribution of loads in the original structure affecting both the support system and the underpinning scheme. If these effects are expected to be manifested on a large scale, the existing foundation should be investigated for intermediate underpinning stages, adjusting the design, if necessary, to ensure consistency with probable variations.

Below the bearing level, differential settlement will most likely increase the load on some underpinning elements and decrease it on others, causing more differential settlement and further load changes until convergence occurs and the process is stabilized.

Effects of Construction In a physical sense a building or structure can be successfully underpinned, yet still suffer some damage associated with the effects of construction. These effects are caused by: (a) overloading of a soil by spread footing; (b) introduction of unexpected eccentric loading with torsional components; (c) failure to consider the sensitivity of the building; (d) lateral displacement causing cracking, particularly if the building shifts differentially; and (e) failure to make allowance for these effects on the underpinning elements. Other events that may become potential sources of settlement and subsequent stress conditions are: (a) a change in natural conditions around the underpinned structure, such as a different groundwater table; (b) a change in the conditions near the site because of new construction, excavation, surcharge load removal, or installation of a new foundation at a lower level; or (e) events such as blasting, vibrations, impact, and pile driving.

Settlement of underpinned buildings adjacent to excavations is documented in the engineering literature. Peck (1969) describes the settlement of structures underpinned on piles and adjacent to cut-and-cover excavation. Settlement probably caused by downdrag on new underpinning elements is reported by O'Rourke and Cording (1974). The Norwegian Geotechnical Institute (NGI, 1962) has documented a case history of an underpinned structure that moved considerably during construction of a cut-and-cover tunnel.

A considerable record exists regarding subsidence caused by tunneling in soft ground. This subsidence is manifested at ground surface as a trough, generally several inches along the tunnel centerline and extending for some distance beyond the tunnel zone, gradually diminishing to zero. A building or wall located in the trough zone will probably sustain some damage as a result of settlement and lateral movement toward the tunnel. Field investigations show that horizontal movement can be the same as the settlement along the side of the tunnel.

An example is shown in Figure 2-21. The small hexagon on the plot of soil movements represents a point on the ground surface 20 ft from the subway centerline and under 38 ft of cover. This point settled approximately 2 in. and also moved horizontally toward the tunnel by the same amount. The tunnel has a inside diameter of 21 ft, and was shield-driven through drained granular river terrace deposits (White, 1976). In this case, full underpinning to the depth of the invert was

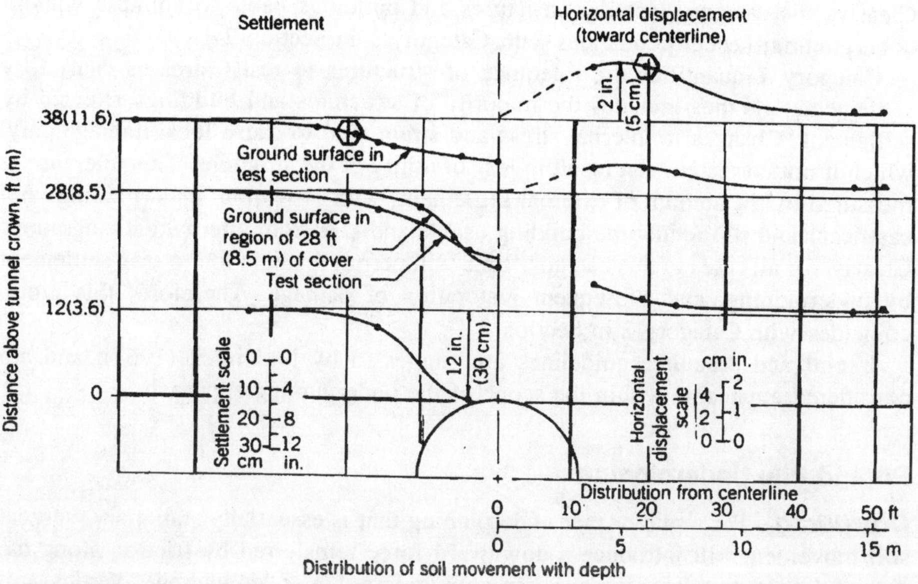

Figure 2-21 Distribution and soil movement associated with tunneling for the Washington, D.C., metro. (From White, 1976.)

quite expensive. Instead, hydraulic jacks were placed on top of shallow concrete piers or footings reacting against the wall or columns above, and kept the buildings fairly level as tunnel excavation approached. This procedure, described as maintenance underpinning in Section 2-2, is not expected to eliminate movement completely, and some form of cracking still may occur.

Maintenance underpinning was introduced to 35 buildings in San Francisco during construction of the BART project (Kuesel, 1972). These structures were equipped with column pickups for adjustments and leveling operations. However, with large settlement or lateral movement, tipping of the piles can occur, rendering the use of jacks impractical and often difficult. In some instances, maintenance underpinning is carried out with the use of pretest piles.

Criteria of Tolerable Conditions Underpinning must be planned to prevent damage, but indiscriminate elimination of damages is not considered practical or economically justifiable. An equally viable approach is to limit the extent of damage but accept some form of damage provided structural collapse and loss of utility is not involved. With this criterion we consider two categories of tolerance.

Category 1 is for tolerance of structures to settlement caused by external stresses only. It applies to structures and buildings that are not noticeably affected internally by external underpinning failure until the structure itself collapses. In this group a building must have a design and configuration that provides considerable structural resistance to internal stresses, the underlying condition being that settlement occurs so quickly that there is no time to register and adjust to changes in internal stress.

Clearly, this category includes structures and buildings liable to collapse without underpinning; hence it coincides with Category 1 in Section 2-1.

Category 2 quantifies the tolerance of structures to resist stresses caused by settlement, and thus includes the majority of structures and buildings affected by settlement. Changes in internal stress and strain tend to cause local damage only, which if not corrected, can result in loss of utility of the structure. This tolerance is measured by the amount of external settlement, with or without underpinning, that can occur and still render the building usable and safe from internal disintegration. Clearly, for all cases in this category time will permit remediation to stop settlement by underpinning, and subsequent restoration of damage. Therefore, this group coincides with Category 2 in Section 2-1.

Useful and articulate guidelines are suggested by McKinley (1964), and are considered applicable within the scope of the underpinning approach.

Pit and Pile Underpinning

Downdrag For a pile or pier underpinning that is essentially stationary, vertical soil movement will introduce a downward force transferred by friction along the side of the element. This force is commonly referred to as "downdrag." Vertical soil displacement may be the result of consolidation of compressible soils or it may be associated with nonvolumetric vertical strain within an earth mass influenced by an excavation. This action develops negative friction forces on the perimeter of the pile and tends to carry it further into the ground. The pile may not move, or it may move but not as much as the consolidation movement. In either case, the maximum frictional strength of the soil is developed, and this force may be large enough to cause the pile to settle or to overstress the pile material.

Negative skin friction may also be developed by the lowering of the groundwater, often specified to ensure a dry excavation. The increase in effective stress causes consolidation with associated settlement and friction along the piles.

In cohesive soils, the magnitude F_n of negative friction is taken as

$$F_n = p\, L_s\, s_u \tag{2-1}$$

where L_s = depth of soil within element zone over which settlement occurs
p = perimeter of pile
s_u = shear strength of soil in the zone of settlement

For granular soils

$$F_n = \tfrac{1}{2}\, L_s^2\, p\, \gamma\, K f \tag{2-2}$$

where K = pressure coefficient ($K_a < K < K_O$)
f = friction coefficient (upper limit of tan ϕ for smooth piles and tan ϕ for rough cast-in-place elements)
γ = unit weight of soil

Displacement Sources and causes of settlement and movement that are unique to underpinning were reviewed in the foregoing sections. This displacement is associated with the sequence and nature of the operation. Settlement may also occur as elastic deformation due to a load increase, or as inelastic deformation such as creep and shrinkage of the concrete used to form the underpinning elements. A further cause of movement is strain within the bearing stratum.

Lateral soil movement, particularly the type associated with adjacent excavation, will induce the same tendency to the underpinning group. Since this movement must be inhibited, it is necessary to introduce restraint such as lateral bracing and thus impose horizontal forces on the underpinning. In this case, selecting the underpinning scheme is the most important and critical part of the design.

Load Transfer This may be accomplished by friction, base bearing, or a combination, depending on whether the base rests on a suitable firm stratum, on the element size, type, and depth, and on the method of installation. All these factors, however, must be taken into account in a rational manner if we are to apply them to a given problem.

For friction piles in clay the adhesion between the soil and the element will provide the load resistance. When pile underpinning is driven or jacked in clay, the operation tends to remold the material, and this either reduces the shear strength or destroys is entirely. With time, however, a portion or all the cohesion is regained, particularly in soft clays that tend to remain in better contact with the element. The same operation in stiff clays tends to cause disturbance that may result in separation of the element and the soil, so that adhesion development is this case is less than the cohesion.

For augered caissons, particularly with a bell or pit underpinning, more reliance should be placed on base bearing, although some allowance may be made for adhesion of cohesive soils to the sides of the shaft according to bearing capacity theories.

Since underpinning is a group of elements rather than a single element, we must consider the associated interaction. A group of elements will stress the soil to a greater depth than an individual element. Thus, given the load per pile, the settlement of a group of piles will be greater than the settlement of a single pile. This distinction is essential when specifying the preloading of pile underpinning and the locked-in load. Since piles usually are preloaded individually rather than in groups, the resulting interaction from this sequence must be understood. For example, an initially installed and loaded pile will undergo elastic shortening and cause some compression of the bearing stratum. Installation and preloading of adjacent piles will cause more strain on the bearing layer and some relief of load in the initial piles. Thus, group piles that have been preloaded and locked-off separately may settle more than predicted under the full structural load.

When several piles in an underpinning scheme are clustered, we can reasonably expect that the soil pressure zones developed as soil resistance will overlap. With sufficient overlap, either the soil can fail or the group can settle excessively. A decision can be made on the significance of this action after considering the relevant factors, namely pile spacing and capacity, strength characteristics of the supporting ground, and sequence of preloading. As a practical rule, group action may be disregarded if the pile spacing is greater than three pile diameters, or if elements bear on competent granular soils and rock.

Laterally Loaded Underpinning Lateral loading conditions are induced by eccentric loading, such as the underpinning schemes shown in Figures 2-5 *b*, 2-6, and 2-8, or where the underpinning elements are braced at or near the top to inhibit movement. The resulting combination is an element acted upon by axial load, bending moments, and shear stresses. Suitable structural configurations to resist this action include steel H-beam sections with high moment of inertia in the direction of bending, and large-diameter reinforced concrete caissons. The literature contains a considerable record on procedures and methods of analysis, and these need not be repeated here.

Laterally loaded piles can be analyzed as infinite beams embedded in an elastic medium. This requires the consideration of a horizontal modulus of subgrade reaction k_n in conjunction with an appropriate differential equation. We may write the developed passive earth pressure p_p as

$$p_p = k_h\, y \tag{2-3}$$

where y is the associated displacement. Equation (2-3) generally is valid subject to the following observations: (a) the modulus of horizontal subgrade reaction k_h will decrease with time in soils for which consolidation theory applies as the consolidation takes place; (b) lateral deflection will increase as consolidation takes place; (c) in sands, both subgrade reaction modulus and deflection are independent of time since the deflection occurs instantaneously; (d) in sands, the modulus of subgrade reaction increases proportionally with depth; and (e) in clays, the modulus of subgrade reaction is essentially the same both horizontally and vertically, and independent of depth, except for normally consolidated clays and silts.

Root Pile Design

Analysis and design of root pile underpinning (see also Section 2.2) requires essentially the same procedures that apply for friction and bearing piles, but with the empirical coefficients modified for the construction effects of this operation. The load transfer can be by friction, end bearing, or both, if a compatible mechanism is established in terms of the associated displacement. As mentioned, usual root pile capacity is 10 to 15 tons, but loads up to 40 tons can be transferred with larger diameter piles.

Table 2-2 shows results of load tests on root piles, from both published and unpublished sources (Bares, 1974). Since the test load shown in the table is not necessarily the failure load, ultimate pile capacity, and by analogy the factor of safety, is not available. However, Table 2-2 provides settlement data that can be used to develop an approximate relation between pile geometry, load, and settlement.

Assuming that the load transfer is primarily by friction (which is fairly reasonable considering the small bearing area) and that the settlement is inversely proportional to the average skin friction, a procedure can be developed (Goldberg et al., 1976) for estimating pile settlement modulus. By reference to Figure 2-22, we obtain

$$\delta = k_s \frac{P}{DL'} \tag{2-4}$$

where δ = settlement, in.
P = load, tons
D = pile diameter, ft
L' = pile length (effective) in load transfer zone, ft
k_s = settlement modulus

Rewriting Eq. (2-4), we obtain

$$k_s = \frac{\delta_{max} DL'}{P_{max}} \tag{2-5}$$

where δ_{max} is the observed settlement under maximum load P_{max}.

The settlement modulus values shown in Table 2-2 generally are less than 0.1 in.-ft^2/ton, although slightly higher values are reported in some cases. In clayey soils, the pile settlement modulus can be higher (> 0.7 in.-ft^2/ton in one case), and this suggests that in cohesive soils the settlement should be carefully analyzed. In most cases, a load test should be mandatory to assess settlement characteristics.

Example Assuming a settlement modulus of 0.12 in.-ft^2/ton, we can estimate the settlement of a 25-ton pile with an effective transfer length of 20 ft. The pile diameter is 5 in.

From Eq. (2-4) we obtain directly

$$\delta = 0.12 \frac{25}{0.42 \times 20} = 0.36 \text{ in.}$$

The cases described in Table 2-2 involve a gradual load transfer to the root pile system as the original foundation support is removed in the process of excavation. This process is different from the conventional underpinning methods where the load transfer is attained after installation by jacks or wedges.

Inclined Root Piles The inclined pile configuration shown in Figure 2-9*b* imparts to the wall stability against overturning and lateral translation. This is important if the underpinning system is subjected, besides vertical loads, to the action of lateral loads and bending moment near its top. We may analyze the behavior of root

TABLE 2-2 Results of Load Tests on Pali Radice (Root Piles)

Case Number	Nominal Diameter D (in.)	Length L (ft)	Assumed Effective Length L' (ft)	Maximum Test Load P (tons)	Settlement at Maximum Load ρ (in.)	Settlement Modulus k^a(in. ft^2/ton)	Soil Type[b]	Location
1	4	21	21	22	0.04	0.013	G	School Building, Milan, Italy
2	4	40	40	22	0.16	0.097	C	Olympic Swimming Pool, Rome
3	12	90	90	50.6	0.32	0.570	G	Bausan Pier, Naples
4	4	49	20	19.8	0.08	0.0270	Si, G	Italian State Railroad, Rome
5	4	52	42	17.6	0.09	0.072	G	Bank of Naples
6	8.5	99	66	108	0.22	0.087	G	Corps of Engineers, Naples
7	5	65	24	50	0.32	0.062	G	Washington, D.C., Subway
8	9	19.5	10	45	0.45	0.075	G	Queen Anne's Gate, London
9	7	28	18	50	0.30	0.063	G	Queen Anne's Gate, London
10	4	52.8	52.8	23.1	0.236	0.1798	C-G	Salerno-Mercatello Hospital, Salerno-Mercatello
11	8	82.5	43	108	0.472	0.125	G	Marinella Wharf, Port of Naples, Naples
12	8	47.5	47.5	59.4	0.035	0.0187	G	Main Switching Plant, Genoa
13	8	73	73	62.5	0.065	0.0506	G	Mobil Oil Italiana, Naples
14	8	66	66	58.7	0.037	0.0247	G	Railway Terminal, Naples (Corso A. Lucci)

15	8	63	63	56.5	0.065	0.0483	G	Plant (Brindisi)
16	8	60.5	60.5	56.5	0.028	0.0200	G	Plant (Brindisi)
17	8	73.5	73.5	27.5	0.252	0.4490	C	Special Foundations for Transmission (Electrical Towers between Garigliano-Latina)
18	8	66	66	24.2	0.386	0.7018	C	Special Foundations for Transmission (Electrical Towers between Garigliano-Latina)
19	8	66	66	48.5	0.205	0.1860	C	Special Foundations for Transmission (Electrical Towers between Garigliano-Latina)
20	8	99	66	110.2	0.213	0.0850	G	Belt (Expressway) East-West, Naples
21	8	99	66	88.2	0.127	0.0634	G	Belt (Expressway) East-West, Naples
22	8	59.5	59.5	68.3	0.061	0.0354	G	Swimming Pool—Scandone Pool, Naples
23	4	33	33	21.5	0.087	0.0445	G	Casa Albergo in Viace Piave
24	8.5	82.5	82.5	69.7	0.148	0.1241	G	Port of Naples
25	8.5	82.5	82.5	69.7	0.150	0.1258	G	Port of Naples

Source: From Bares (1974).

[a] $k = \rho/(P/L'D) = L'D/P$.

[b] G = granular; C = clay; Si = SiH

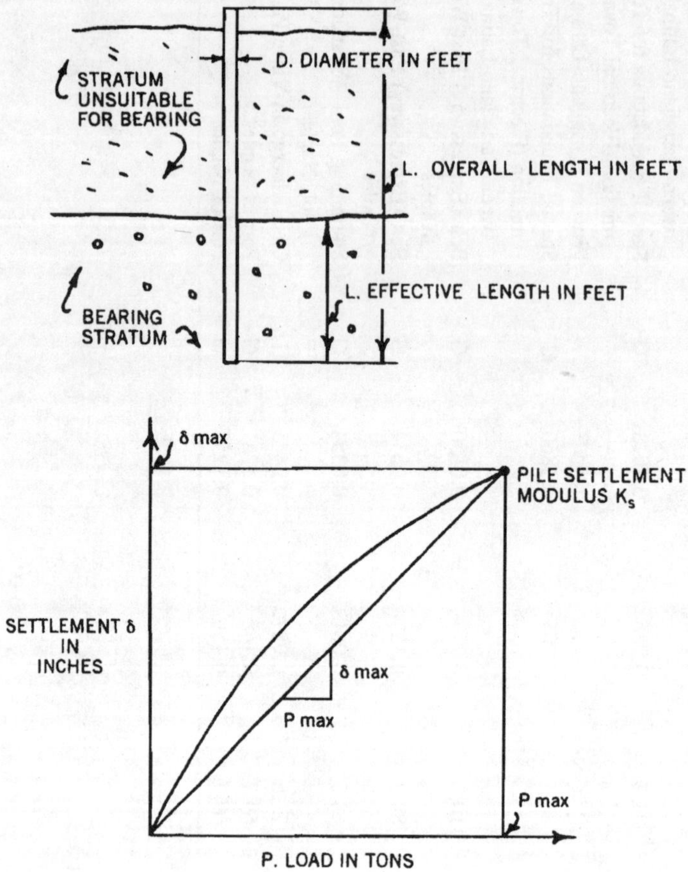

Figure 2-22 Procedure for estimating root-pile settlement modulus.

piles by reference to Figure 2-23*a*, subjected to the action of the lateral force *F*. As shown in Figure 2-23*b*, this force is resisted by the root pile system as an axial force *T*. From the geometry, we can easily obtain $T = F/2 \sin(\alpha/2)$. Along the front pile this load is compression whereas the back pile is in tension. This system is equivalent to a system of ground anchors in compression and in tension, respectively, and given the relatively small diameter of the piles, it can be analyzed as a group of anchors. Interestingly, the steel bar in the back pile will resist all the tension.

Reticulated Root Piles The reticulated pile walls shown in Figures 2-17 and 2-18 serve as underpinning and simultaneously resist lateral loads. In this case, the underlying principle is to engage an earth mass in a desired interaction by installing a specific root pile network at close spacing, and with a specific batter and orientation. The resulting scheme is a lattice structure that encompasses the soil and responds monolithically. System design is essentially similar to the analogy of

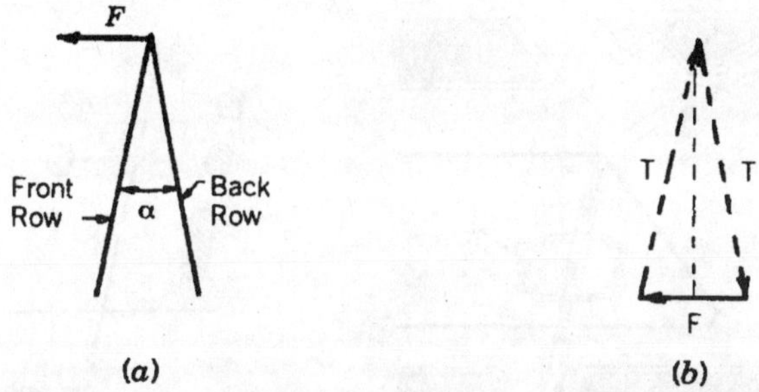

Figure 2-23 Inclined root piles. (*a*) Pile configuration and lateral load. (*b*) Force equilibrium triangle.

gravity walls, and involves analysis of overturning moments, determination of position and magnitude of the vertical reaction at the base, and estimation of horizontal shear through and below the monolith. The application shown in Figure 2-18 indicates that some arching is developed over the tunnel and contributes to stability.

2-6 SLURRY WALLS IN LIEU OF UNDERPINNING

Examples and Applications

An example of lateral protection of a structures is shown in Figure 2-24 (Xanthakos, 1979, 1991). This work is along Subway Line 6 of the Paris Metro, and became necessary prior to excavation on both sides of the subway for the foundations and basements of a 60-story tower. This excavation is 18 m (59 ft) deep. As shown in Figure 2-24, the ground beneath the subway was strengthened by cement-grout injection carried out from the original ground and from the first excavation level.

The retaining system consists of diaphragm walls built in intermittent panels, tied together by means of four levels of prestressed anchors (Xanthakos, 1991). These panels serve also as foundation elements for the new tower. The subway and the ground beneath are thus confined within the geometry of the diaphragm walls, but the ground outside this zone was excavated on both sides of the subway for the basement portion of the new construction. The lateral version of the underpinning scheme is, therefore, permanent. Strip piles are also used as the main foundation for the rest of the tower. These were excavated and cast from the final basement level. The piles are 50 m (164 ft) deep, and transfer the loads to a thick hard chalk layer with unconfined compressive strength of 25 to 40 kg/cm^2.

A somewhat different example is shown in Figure 2-25 (Fenoux, 1971). On one side of the construction, underpinning was carried out under the edge of the existing building in a sequential placement of four rows of horizontal strips, cast in place and anchored individually. On the left side, the excavation is laterally protected by a diaphragm wall that also protects the adjacent building, treated as a surcharge load.

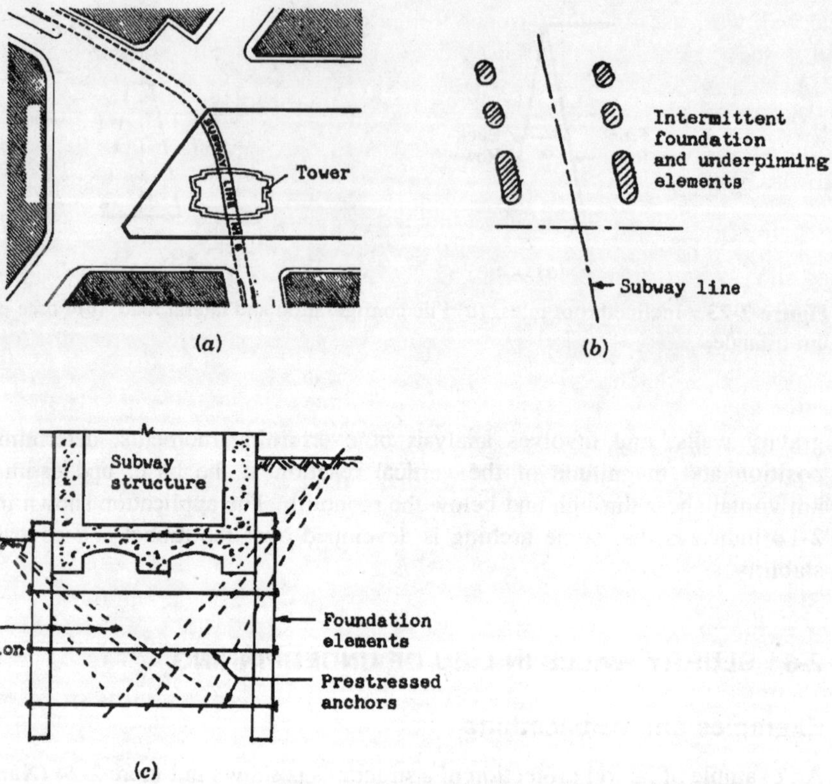

Figure 2-24 Use of anchorages and strip panel walls to underpin an existing subway structure. (*a*) Site plan. (*b*) Partial foundation and underpinning plan. (*c*) Typical section. (From Xanthakos, 1991.)

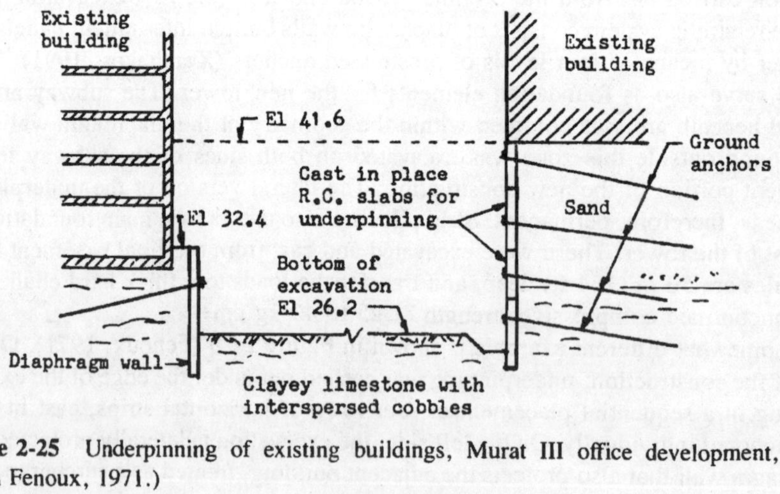

Figure 2-25 Underpinning of existing buildings, Murat III office development, Paris. (From Fenoux, 1971.)

The wall was cast in a slurry trench from a level just above the existing foundation, and is supported by prestressed anchors near the top and by embedment into firm soil at the bottom.

An inclined secant pile wall constructed to protect buildings and structures directly adjacent to, and often encroaching onto, the excavation line of the Munich Subway is shown in Figure 2-26*a* (Weinhold, 1969). Compared to the cost and time necessary for conventional underpinning, this solution was structurally adequate, and had a cost slightly higher than the cost of a vertical wall. About 600 m (1970 ft) of wall was built using a Benoto rig and without lowering the groundwater table (Xanthakos, 1979, 1991). The rather steep wall inclination was dictated by a tight geometric alignment of the new construction and the limited clearance. Construction was staged as follows: (a) installation of the pile wall inclined as shown; (b) installation of steel H-columns using the prefounded column concept; (c) placement of temporary decking at street level; (d) excavation to just above existing foundations to place struts as uppermost bracing; (e) excavation to first anchor level and installation of anchors, followed by excavation and installation of anchors at second level; (f) prestressing of anchors at both levels; and (g) excavation to final level.

This work exemplifies the principles of hybrid construction (lateral support by strut bracing, anchorages, and wall embedment in firm soil). The introduction of top and bottom bracing, before excavation begins, restrains wall movement and implies a design based on lateral earth stresses at rest combined with full hydrostatic pressure and surcharges imposed by the continuous footing load.

Loads and support conditions are simulated in the analytical model shown in Figure 2-26*b*. The wall is elastically embedded in the subgrade, a condition approached by the horizontal modulus of subgrade reaction. At the anchorage points elastically yielding supports are simulated, and as a statical system the wall is treated as a continuous beam.

Factors Affecting Choice

Slurry walls are good substitutes for conventional underpinning, particularly if they are incorporated in the final structure, or where the excavation requires control of groundwater and ground movement. The concept of a low-cost solution is enhanced if we consider the various wall systems and configurations available with this technique. The cost is also influenced by: (a) the required embedment below excavation level for statical support and control of movement or seepage at this location; (b) the nature of ground to be excavated, especially the presence of boulders and hard obstructions; (c) overall stability during trenching, concreting, and general excavation; and (d) relevant site factors such as mud disposal, time restrictions, and available working space (Xanthakos, 1979, 1994). Secant pile walls formed in prebored holes stabilized with bentonite slurry are good choices in unfavorable and adverse ground conditions, and thus provide a higher factor of safety.

Since the intent is to intercept loads and also limit displacement, slurry walls can satisfy these requirements in properly engineered excavations. The arrangement shown in Figure 2-24 can limit movement within the confined ground mass except

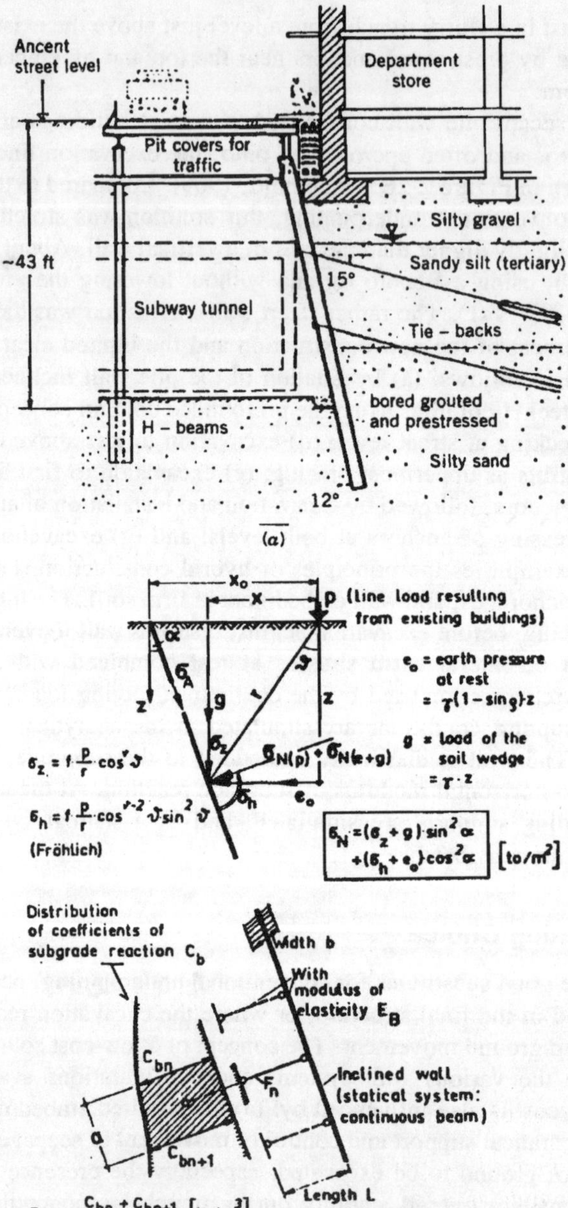

Figure 2-26 (*a*) Excavation for subway construction in Munich, and underpinning of buildings with inclined secant wall. (From Weinhold, 1969.) (*b*) Analytical model and simulation of load and support conditions.

for displacement related to the elastic stretching of anchors. Likewise, the scheme shown in Figure 2-25 is adequate in reducing movement provided the anchors are monitored and restressed if necessary. The inclined wall shown in Figure 2-26 *a* has its top and bottom braced prior to excavation; hence, movement is controlled accordingly. We can state, therefore, that in choosing a slurry wall two considerations are important: the effect of surcharge loads from foundations, and the mechanics of the general excavation.

Observed Performance

The literature contains a well-documented record of wall performance in a variety of conditions, and for various soil types. This record includes the effect of construction operations, soil strength, wall stiffness, anchor stiffness and prestressing, and interior bracing. The reader is referred to the bibliography at the end of this chapter.

Excavations supported by slurry walls designed to function as underpinning have been monitored by Ware et al., (1979); their paper includes data on four field-instrumented buildings. The work monitored was for the Federal Center of the Washington, D.C. Metro, and required an excavation 1065 ft long and 65 ft deep. Vertical and horizontal building movement was monitored by instrumentation at critical points. The plans and contract documents stipulated soldier piles with lagging as excavation support, and jacked pile or pit pier underpinning for the buildings. As an option, the contractor was allowed to bid on a slurry wall.

The four buildings monitored vary in height from four to six stories, and have either marble or brick facing. Exterior wall loads are estimated at 31 to 50 kips/ft of building face. Contract bids for the slurry wall option were \$2 to \$3 million less than the conventional underpinning scheme, and were accepted by the owner. The slurry wall consists of composite steel-and-concrete panels 7 ft long, 33 in. wide, and 80 ft deep. The joints between panels are formed by steel H-beams. The excavation was braced at four levels by struts.

At the site, the ground consists of fill in the upper layers, underlain by stiff to medium-stiff silty clay ($c = 1$ to 2 kips/ft^2) and compact brown fine to medium sand with traces of gravel ($\phi = 32°$ to $34°$) and $\gamma = 130$ lb/ft^3. The depth of these layers varies from 25 to 30 ft. Below these strata, the profile shows a layer of hard red-brown and gray plastic clay ($c = 2$ to 5 kips/ft^2) interbedded with layers of sand and some silt, cobbles, and boulders ($\phi = 32°$ to $38°$). This plastic clay extends 10 to 15 ft below the base of excavation.

Instrumentation The slurry wall and the building were monitored with inclinometers and conventional optical surveys. Inclinometer casings installed outside the wall provided data on ground movement during panel excavation and concreting. During panel excavation under slurry maximum lateral movement was $\frac{3}{4}$ in., reduced to $\frac{1}{4}$ in. after panel concreting.

Inside the wall panels inclinometer casings were attached to H beams installed during excavation, and representative readings were taken as excavation was progressing. Final displacements at two critical locations are presented in Figures 2-27 and 2-28. The solid line represents maximum displacement during excavation whereas the dashed line indicates total movement after pouring the invert slab and removing the lowest strut level to build the sidewalls (Ware et al., 1979).

Instrumentation Results The first deep excavation was completed in front of the Terminal Refrigeration Building, which has eight stories and is the tallest in the group. This building is also the heaviest, with exterior walls applying a load of 53 kips/ft. From Figure 2-27 we note that inward lateral movement reached 0.8 in., slightly higher than $\frac{3}{4}$ in., the maximum specified. For the same cross section, lateral wall movement along the H.E.W. building is much less, and in fact near the top the wall moved toward the ground. The H.E.W. building has four stories, and its external walls apply a load of 31 kips/ft. This pattern of movement clearly shows the effect of support stiffness and strut bracing. Translation caused by heavier lateral loads on the left side transmits these loads through the struts to the right side so that the movement there is stopped or reversed until the loads are balanced.

The same pattern is evident in the cross section of Figure 2-28, although it occurs to a lesser extent. The FBI Building on the left induces a load of 34 kips/ft of exterior wall, and although data for Federal Building No. 8 are not available, we can assume that the loads there are less.

The Terminal Refrigeration Building, being the heaviest, underwent vertical settlement that reached $\frac{1}{2}$ in. maximum. The remaining three buildings were subjected to heave which, although small, was measurable. Likewise, this pattern of settlement is consistent with the overall lateral movement, but is also associated with unloading of ground resulting from excavation.

2-7 PREFOUNDED COLUMNS

The use of underpinning implies that an underground structure cannot be built until some form of excavation is carried out. By analogy, underpinning may not be necessary if an underground structure or the below-ground portion of a building can be built either before or simultaneously with an excavation.

Using a deep basement or a deep excavation level to support superstructure loads has a technical appeal, since the intensity of load on the ground below that level is reduced by the weight of ground excavated, and this provides the basis for the design of compensated foundations.

Theoretically, therefore, the deeper the basement the cheaper the foundation, but in reality the deeper the excavation the more expensive the ground support and the higher the cost of excavation.

Thus, building a deep basement for the purpose of supporting superstructure loads is rare and unorthodox, although a deep basement provides a better load transfer. The latter is a real advantage for relatively small loads and where a foundation raft is used.

A further consideration relates to architectural and structural trends, which tend

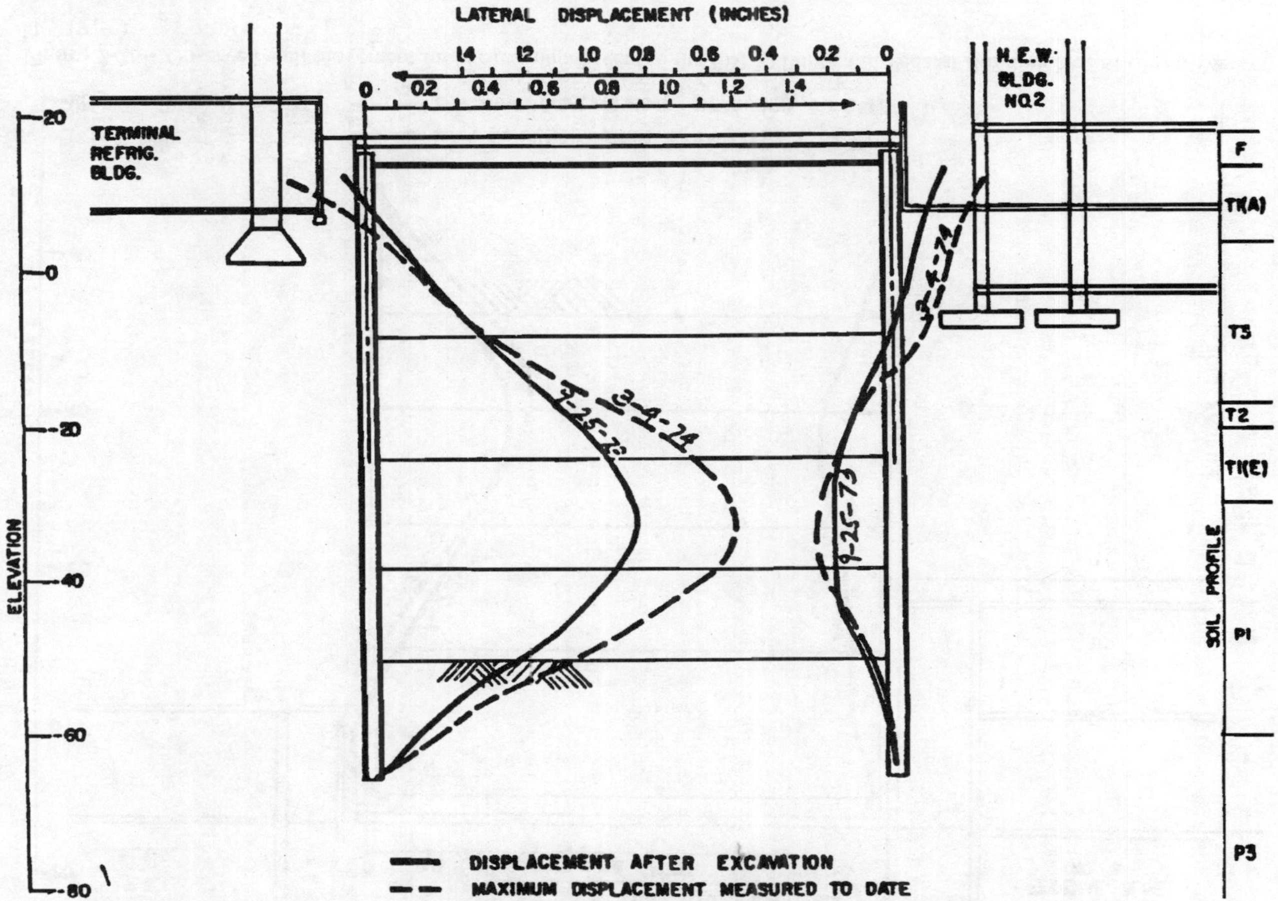

Figure 2-27 Observed wall movement for excavation adjacent to H.E.W. Building, and Terminal Refrigeration Building. (From Ware et al., 1979.)

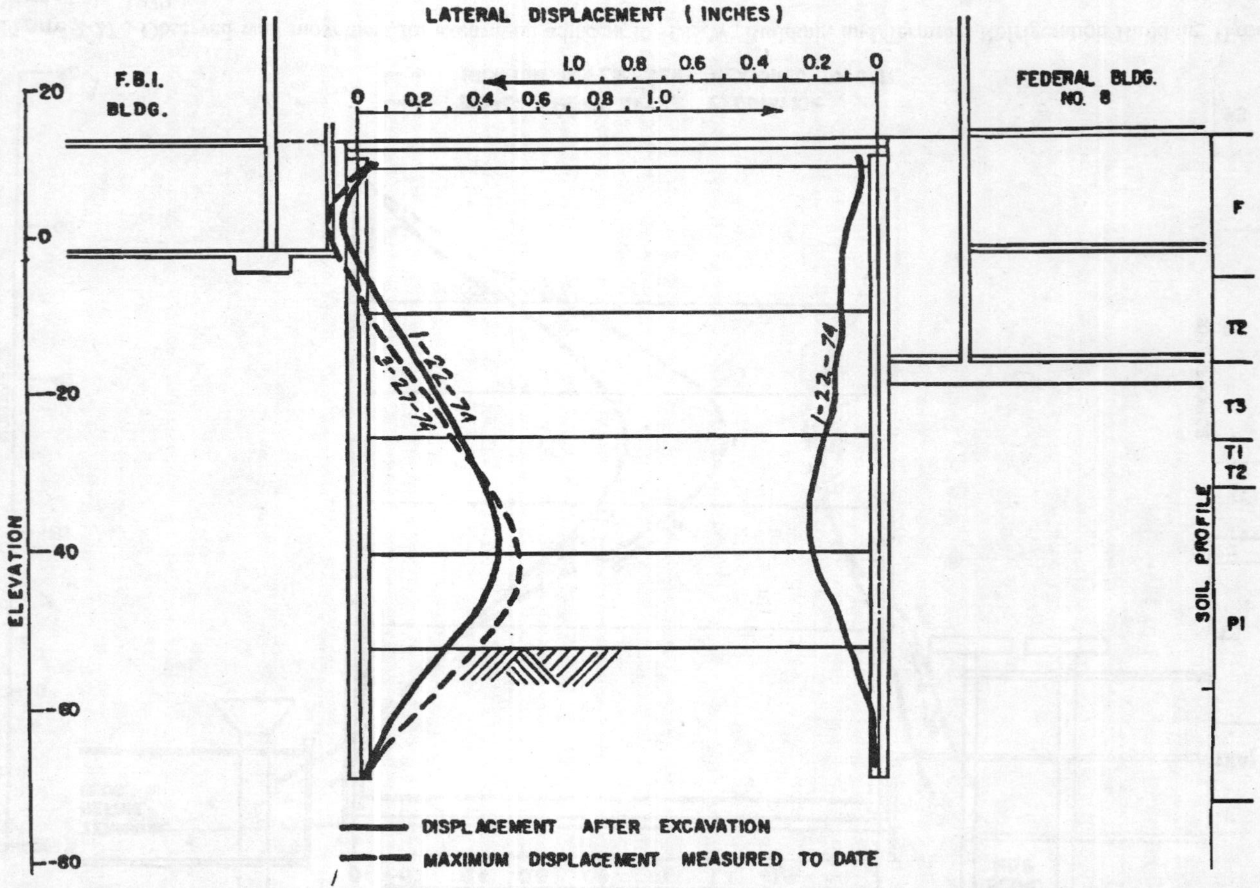

Figure 2-28 Observed wall movement for excavation adjacent to the FBI Building and Federal Building No. 8. (From Ware et al., 1979.)

to dictate heavier column loads because of the larger panels used to provide unobstructed floor space. In other instances, the main superstructure is formed as a central core and is used only above a part of the usable underground space. These arrangements predetermine the load distribution through the basement substructure, and affect the load-transfer scheme. In some cases conventional column-and-beam construction allows the loads from the periphery of the superstructure to be transferred to load-bearing walls in the basement, but a high-rise tower supported on a central core cannot derive these benefits unless the basement is locally deepened at the core.

Alternate Construction Methods We distinguish two conventional methods. In the first case, the foundation, usually extended to rock or hardpan, is inserted from the base of the general excavation, after allowance for the underground portion of the building. In the second instance, deep foundations are inserted from ground surface prior to general excavation, but are terminated at the level of the basement raft. If the formation includes clayey soils and natural water table, unloading during excavation will probably cause upward swelling, subjecting the foundation elements to considerable tension along the shaft, applied as negative friction. Concrete subpiers subjected to this action have failed unless they were heavily reinforced.

An alternate design and construction method is the use of prefounded columns. Beginning at ground level, and in a single operation, we construct the foundation element and the basement portion of the column as shown in Figures 2-29 *a* and *b*. In this example, a casing is sunk to just above the basement level as the hole is bored, and the excavation is continued below this level under bentonite slurry. When

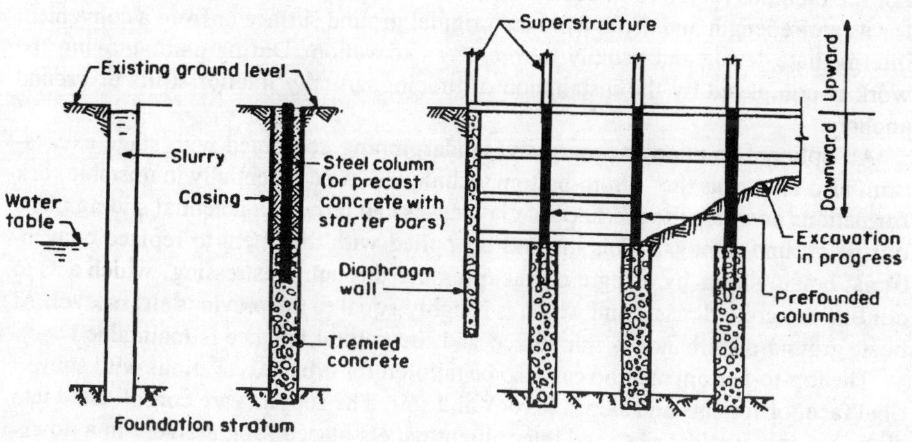

Figure 2-29 Prefounded columns and use in downward construction. (*a*) Drilling hole under slurry. (*b*) Lowering and positioning column and concreting the foundation element. (*c*) Simultaneous construction of superstructure and substructure by upward and downward process.

boring is completed, the steel column is lowered into the casing, positioned, and firmly held in place. Concrete is tremied into the hole and up to the intended excavation level, while the column maintains sufficient embedment for load transfer. On completion of the operation the basing is withdrawn, and the hole above the concrete is backfilled with granular material.

The downward construction is carried out as shown in Figure 2-29*c*, essentially a stage underpinning, and is completely simultaneous with the erection of the superstructure. The method eliminates the need for a temporary ground support, and produces overall time savings. One of the benefits is the early release of parts of the building for service and use, before the entire construction is completed. An example is the Haussman-Mogador nine-level underground parking garage in Paris, built beneath a new department store. The method reduced the total project time by one-third, and allowed the garage levels to become operational as they were completed (Fenoux, 1971).

Unlike the composite steel-and-concrete prefounded column shown in Figure 2-29, rectangular piles can be built in bentonite slurry, and concreted to ground level. The superstructure columns are erected on top of the piles, whereas the frame and floors for the underground portion are suspended or connected to the piles as excavation progresses. A notable example of prefounded rectangular piles is the Manhattan Building in Brussels. This building has 31 stories above ground, and 7 basement levels for parking and services.

2-8 INTERMITTENT LATERAL UNDERPINNING

For the methods reviewed in Sections 2-6 and 2-7 the ground support is constructed for its entire length and depth from the original ground surface or from a convenient intermediate level, and usually before any excavation. During earth moving the work is completed by the installation of bracing, usually interior struts or ground anchors.

An optional solution is progressive underpinning combined with stage excavation, also known as the "top-to-bottom technique," used essentially in unstable rock formations or where the presence of clay lenses can provide a potential sliding plane during ground removal. This method is applied with the intent to replace conventional bracing units by using a contact pressure through prestressing, which acts to confine the ground mass, and which is roughly equal to the previous stress level. At best, ground disturbance is minimized and the natural balance is maintained.

The top-to-bottom method can also be tailored for urban excavations with subvertical face, or for cuts and slopes in rock and soil. The supports are consolidated into slab sections, vertical ribs, and intermittent panels placed progressively in a downward process. These supports are also referred to as "element walls."

Horizontal Walls

Figure 2-30*a* shows a partially completed wall stabilized by the prestressing action of the elements. These units confine and support the soil locally, and act as cover to

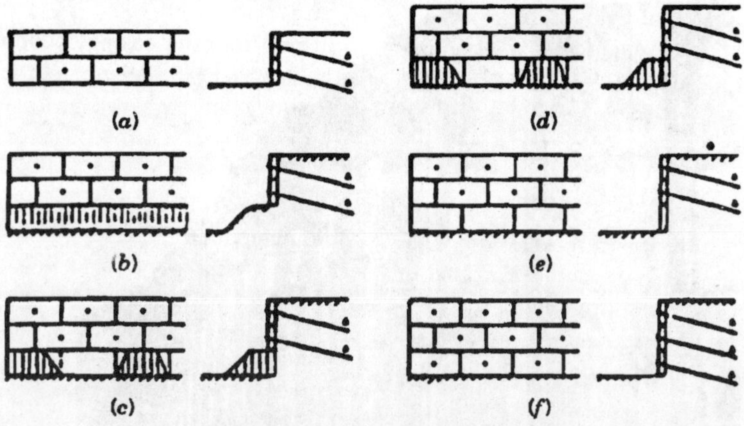

Figure 2-30 Typical construction procedure for a horizontal element wall.

distribute the pressure uniformly. The excavation continues with a cut slope and berm as shown in Figure 2-30*b*. The slope is then removed in front of two or three alternate panels, and new elements are installed and prestressed as shown in Figure 2-30 *c* through *e*. The strip is completed with the intermediate panels sometimes bonded with the primary units by steel dowels. The same procedure is repeated for the next lower strip.

The procedure is claimed to be unrestricted and completed quickly since the panels can be designed and sized to fit the ground conditions. Their structural integrity and fine appearance make them suitable for permanent construction. For large projects productivity is improved if several crews work simultaneously to maximize the number of units installed daily. Site nonideality, which characterizes most rock types, is not necessarily an impedement to progress, although the size and shape of wall units must be consistent with such rock characteristics as seams, fractures, and discontinuities to ensure a generally stable face. Properly sized and installed panels provide protection against ground loosening or crumbling of loose rock materials resulting from weathering because of rain, snow, or frost action. Likewise, the panels can be placed in cohesionless soils provided the groundwater does not induce unstable conditions.

A typical panel is 1.5 to 3 m (5 to 10 ft) high, and 3 to 4 m (10 to 20 ft) long, but the in situ optimum size and shape are determined by the local conditions and stability requirements. One or two anchors per panel are sufficient. Unlike other retaining walls, embedment below final excavation level is not necessary except for a stripe foundation constructed beneath the wall to provide a base support.

Vertical Rib Walls

If heavy concentrated or continuous loads act within a ground mass, a safe solution is to stabilize and simultaneously underpin. The horizontal strip wall is modified in this case with emphasis on vertical staging rather than strip progression. The construction begins within a narrow vertical zone where concrete ribs or columns are

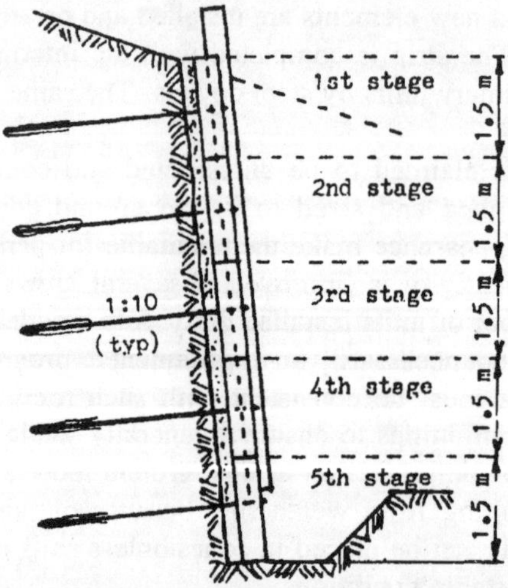

(b)

Figure 2-31 Construction of a retaining wall by progressive underpinning, Wingreis, Switzerland. (*a*) View of the wall showing the vertical tie beams and the filler concrete in between. (*b*) Section through the wall showing cover with concrete cladding. (VSL-Losinger.)

placed and properly anchored, and the open ground between them is protected by a gunned concrete layer. A cladding or outer facing is finally applied, serving both an aesthetic and structural functions. Usually the ribs are 3 m (10 ft) high, and are anchored at both ends before the next phase is carried out. A ground support constructed in this manner is shown in Figure 2-31, and includes vertical ribs and concrete filler. The ribs in this example are spaced at 4-m (13-ft) intervals.

For a cut that is unusually high and relatively close to existing heavy loads, a good choice is to stagger the ribs for each consecutive stage to ensure the same degree of structural uniformity and protection.

Construction Considerations

If a cut or slope is in unstable equilibrium, the continuous rigid wall should be approached with caution, particularly in view of considerable scatter in soil and rock properties. Conversely, the independent action of an element wall ensures the ground-support interaction in the area prestressed by the element. As the gradual removal of ground mass is followed by immediate protection, overall stability is afforded with a higher factor of safety. A system made up of disconnected components is free to move when each component is acted upon by the prestressing of anchors so that it does not experience secondary stresses related to nonuniform strain. Differential deformations and movement are confined essentially to the area acted upon by the individual units, and this behavior imparts to the system stable ground response. In this sense, local variations in ground conditions become less relevant, and an extensive geologic investigation can be avoided. However, during construction a mandatory observation and monitoring program is carried out to quantify ground response and complete the interpretational analysis of ground stability. If anchor restressing is necessary, the effect on neighboring units is minimal so that an in situ measured factor of safety is obtained for each panel and structural section (Xanthakos, 1991).

For a moderate cut inclination, horizontal strips are suitable. If the strip width can be made relatively large, cast-in-place concrete is practical and also allows the bearing face to be adapted to the excavated profile and surface irregularities for full contact. With steep slopes, the use of small-height tie ribs in prefabricated sections ensures better stability, particularly if the ribs are stressed following their placement.

The construction shown in Figure 2-32 requires relatively stable ground where

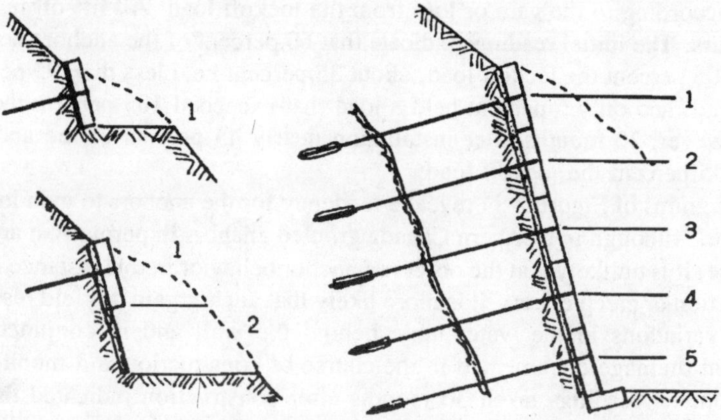

Figure 2-32 Anchored wall built in horizontal strips in a downward construction process; this process is feasible in relatively stable ground.

the risk of collapse does not exist. For the uppermost cut (step 1) the material is removed in a single run along the entire length of the slope, and a filler concrete layer is placed against the face to even irregularities and produce a smooth surface for the structural panels. The panels are cast and stressed before the next phase. Stages 2 through 5 are completed in a different manner, one similar to the wall shown in Figure 2-30. The excavation is now carried out for the entire length of the slope in one pass but at an inclination close to the angle of repose, up to the anchored slabs of the row above. Along the same strip the slope is cut to the final support inclination at alternate panels in a checkerboard pattern. The intervening material is removed after each alternate slab is in place and anchored.

2-9 EXAMPLES OF ELEMENT WALL UNDERPINNING

Element Wall for Rogers Pass Ventilation Building The nine-mile-long Rogers Pass tunnel is probably the longest rail tunnel in North America. Ventilation is provided by a mid-tunnel shaft. For the ventilation station at ground surface a 26-ft-deep excavation was required (Abramson and Hansmire, 1988). At the site the overburden consists of bouldery colluvium that inhibited construction of a soldier pile and lagging wall. The soil is dense, poorly graded gravelly sand with some cobbles and boulders, and with an angle of internal friction between 35° and 40° and negligible cohesion.

Details of the element wall for this excavation are shown in Figure 2-33. The top portion was constructed in 30-ft sections as shown in the elevation, with three point supports. The sequence followed excavation, forming and casting-in-place the concrete wall, installing the ground anchors, backfilling behind the wall, and tensioning the anchors. The two lower panels were installed in strip excavation.

Wall and anchor performance was documented over a period of 30 months by lift-off tests, inclinometer readings, water seepage measurements, and visual observations. Anchor load variation is shown in Figure 2-34. In this diagram, anchors are grouped according to the gain or loss from the lockoff load. All lift-off tests lasted for 24 hours. The initial readings indicate that 60 percent of the anchors held a load of 95 to 105 percent the lockoff load, about 35 percent held less than 95 percent the lockoff load, and only 5 percent held a load that exceeded 105 percent the lockoff load. However, 30 months after installation nearly 45 percent of the anchors exceeded 105 percent the lockoff load.

The diagrams of Figure 2-34 reveal a tendency for the anchors to gain load in the springtime. Although in Northern Canada grouted anchors in permafrost are routine experience, it is unlikely that the observed anchor behavior in this instance is related to frost action or creep effects. It is more likely that anchor gain or yield results from seasonal variations in the water table behind the wall and in conjunction with insufficient drainage documented in the course of construction and monitoring.

Inclinometer readings taken 30 months after construction indicated that maximum lateral wall deflection ranged from $\frac{1}{16}$ to $\frac{1}{4}$ in. This movement is clearly inconsequential, and probably associated with the stiff soils and a construction

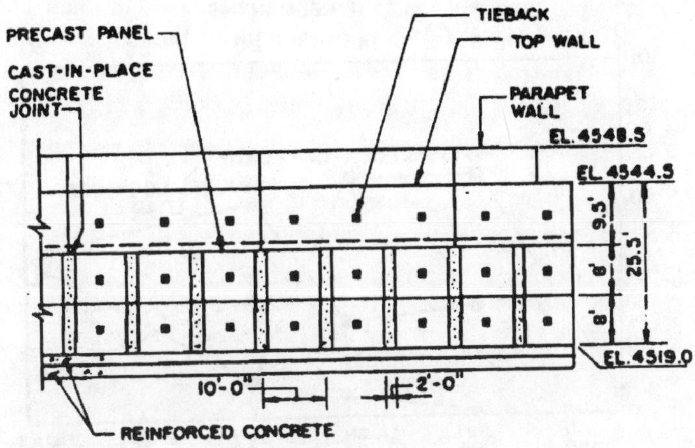

(a)

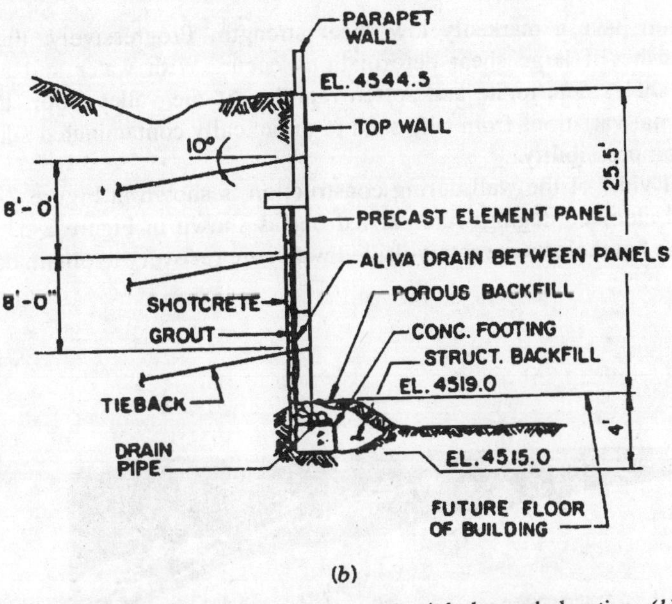

(b)

Figure 2-33 Element wall for Rogers Pass. (*a*) Partial plan and elevation. (*b*) Typical wall cross section.

procedure that allowed the ground to be readjusted before the wall was in place. The design was optimized by incorporating the element wall into the permanent building structure.

Element Wall at Flachau, Austria This wall supports a cut slope where the Takern motorway crosses a pronounced geological fault zone. The outcropping rock is in a deeply weathered graywacke zone in which the products of cohesive clay

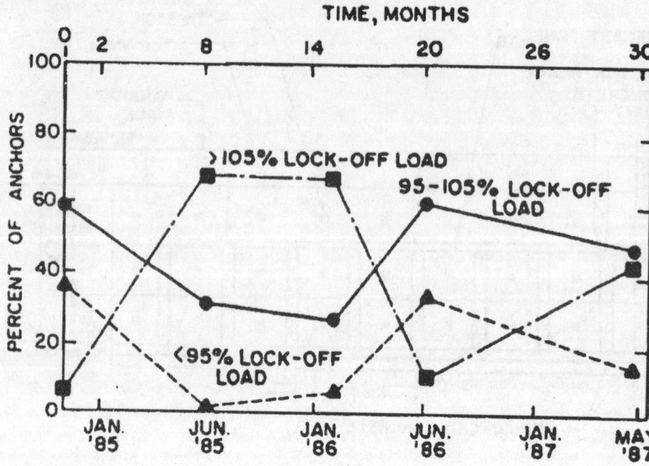

Figure 2-34 Variation in anchor load with time, Roger Pass element wall.

decomposition have a markedly low shear strength. Progressively, this strength decreases further if large shear deformations occur. With water and air gaining access, the shales deteriorate and soften rapidly. At the valley floor, the ground exhibits normal variations from silty sand to organically contaminated silty clays of very high compressibility.

A general view of the wall during construction is shown in Figure 2-35. Earth removal and element staging were carried out as shown in Figure 2-32. About 5 percent of the boreholes were rotary-drilled with core recovery to obtain data on soil

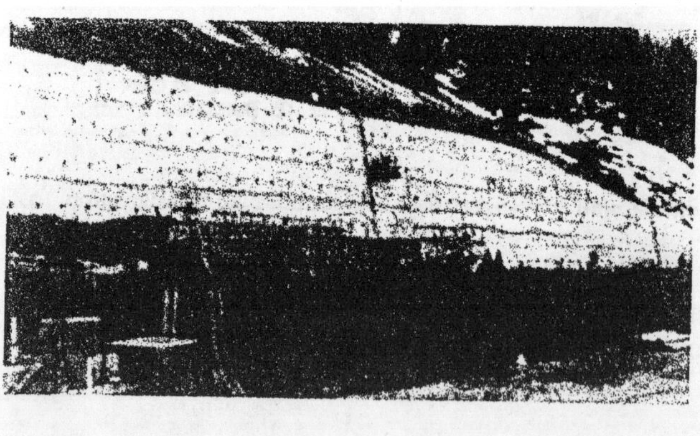

Figure 2-35 General view of element wall during construction.

types and to estimate fixed anchor length. The remaining boreholes were formed using down-the-hole hammers and casing (VSL-Losinger, 1978).

The first stage of the construction was completed with the minimum acceptable prestressing load, and consequently the minimum number of anchor units, although the design included contingencies for additional anchors. A protracted and intense precipitation during the winter and spring following construction resulted in anchor overloading, and the slope began to move. This movement was intercepted and stopped by installing extra anchors.

Retaining Walls at Lake Biel, Switzerland These walls support a cut necessary to widen the highway along the shores of Lake Biel. Along the upper part of the cut individual prestressed rock anchors support the face and are fixed in concrete blocks. The lower part of the excavation is protected by element walls placed along the new road alignment.

Where the rock is relatively stable at the surface, the element wall consists of vertical ribs 3 to 4 m (10 to 13 ft) high and anchored with two or three anchors. The ribs were constructed of cast-in-place concrete poured against the rock surface, and with a connecting cladding, essentially similar to the details shown in Figure 2-31.

Where the rock is less firm, the excavation was limited to horizontal strips 1.5 m (5 ft) deep. Between the vertical ribs a drainage layer was provided by placing filter concrete, and the entire assembly was covered with concrete cladding.

2-10 EFFECT OF SURCHARGE LOADING

For uniform surcharge load acting at the ground surface the resulting lateral stress is usually obtained by applying an appropriate coefficient K_A or K_0 that converts the surcharge into an equivalent lateral earth load. The effect of concentrated loads acting at the surface can be considered in terms of elastic theory, combined with empirical data, if available, and modified where necessary to account for increased stiffness of rigid walls (Newmark, 1942; Spangler, 1951; Terzaghi, 1954). Other investigators have considered the soil elastic but introduced a modulus of elasticity increasing linearly with depth (Turabi and Balla, 1968). Although comparison of measured stresses from the field cases available with stresses calculated from elastic theory shows surprisingly good agreement, engineers are cautioned to expect a possible deviation of ± 30 percent or even higher.

Surcharge loading from construction operations at the ground is conveniently considered as a distributed surface surcharge of the order of 300 lb/ft^2, which represents materials storage and general equipment. The surcharge is usually applied along a strip 20 to 30 ft from the excavation face. Concentrated loads from heavy equipment (trucks, cranes, etc.) are assumed to be covered in the 300 lb/ft^2 surcharge provided they are kept 20 ft from the wall (Goldberg et al., 1976). Within this strip, concentrated loads are far more critical than surcharge loading, and should be analyzed separately.

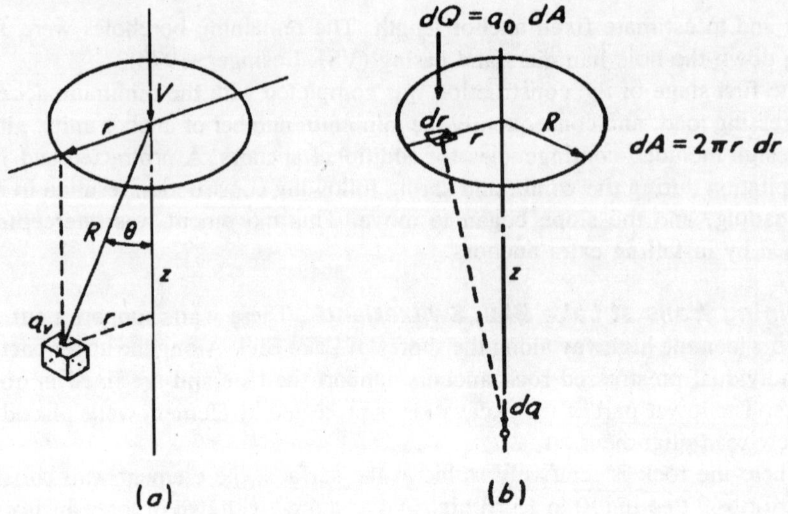

Figure 2-36 (*a*) Intensity of pressure *q*, based on Boussinesq approach. (*b*) Pressure at a point of depth *z* below the center of a circular area acted upon by intensity of pressure q_0.

Surcharge Distribution by Elastic Theory Tests by Gerber (1929), Terzaghi (1954), and Spangler and Mickle (1956) indicate that the lateral stress can be estimated for various surcharge loads by using modified versions of elastic theory analysis. We begin by considering the general equation for lateral pressure:

$$\sigma_h = \frac{V}{2\pi z^2}\left[3 \sin^2 \theta \cos^2 \theta - \frac{(1-2\mu)\cos^2 \theta}{1+\cos \theta}\right] \tag{2-6}$$

where the symbols *V*, *z*, and θ correspond to Figure 2-36. If we set $x = mH$, $z = nH$, $r = x$, and take Poisson's value of $\mu = 0.5$, Eq. (2-6) becomes

$$\sigma_h = \frac{3V}{2\pi} \cdot \frac{m^2 n^2}{(m^2 + n^2)^{5/2}} \tag{2-7}$$

When computing lateral earth stresses against rigid walls, Eq. (2-7) needs certain adjustments to make its theoretical derivation consistent with measured test values.

Point Load This may be a concentrated load acting on a finite area, but for simplicity we assume it to be a point load. The practical value of this assumption is that the estimated lateral stress for an area or a point load is essentially the same if the distance to the face is relatively large compared to the size of the loaded area. Likewise, the difference is small if a point load is considered when this distance is greater than twice the average dimension of the loaded area. The practicality of this assumption therefore suggests its validity, and it is customary to treat an isolated footing or a heavy load concentration as a point load.

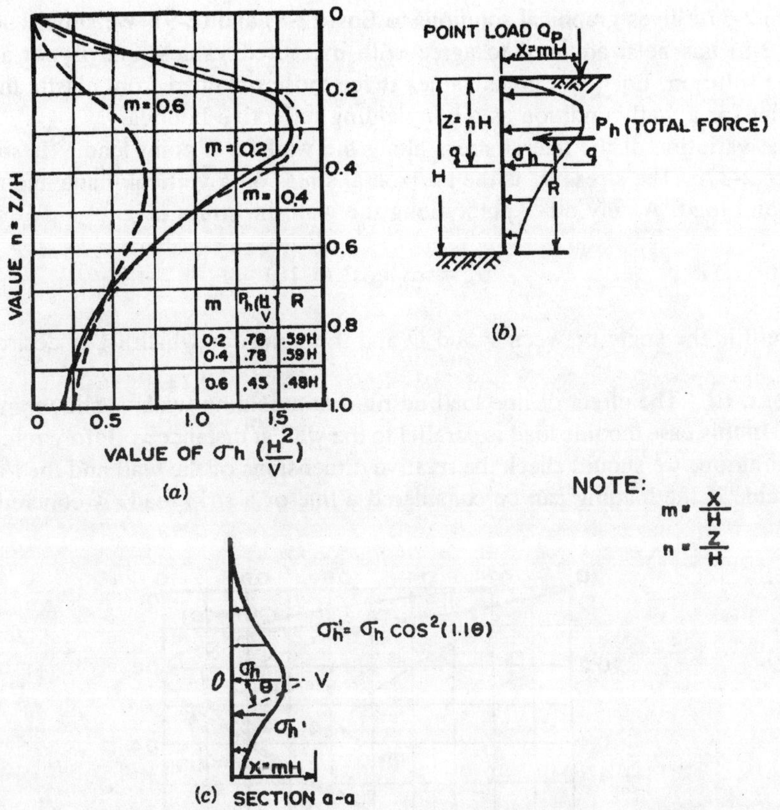

Figure 2-37 Lateral stresses on the face of a nonyielding wall from a point load. (*a*) Stress variation in terms of *m* and *n*. (*b*) Load and stress geometry. (*c*) Lateral stresses at points along the wall on each side of a perpendicular from the point load. (From Terzaghi, 1954; NAVFAC, 1971.)

In their work, summarized by Terzaghi (1954), Spangler and Gerber have shown that the magnitude and distribution of lateral stresses change very little from that determined by elastic analysis until the point load is at distance $x < 0.4H$, where x and H are as shown in Figure 2-37*b*. Thus, for $m \leq 0.4$,

$$\sigma_h = \frac{0.28V}{H^2} \cdot \frac{n^2}{(0.16 + n^2)^3} \tag{2-8}$$

where σ_h = horizontal stress at depth $z = nH$, as in Eq. (2-6)
V = magnitude of point load, as in Eq. (2-6)
n = z/H, $m = x/H$, as before

Likewise, for $m > 0.4$,

$$\sigma_h = \frac{1.77V}{H^2} \cdot \frac{m^2 n^2}{(m^2 + n^2)^3} \tag{2-9}$$

Figure 2-37*a* gives graphical solutions to Eqs. (2-8) and (2-9). We should note that Eq. (2-8) has been adjusted to agree with measured values, and is not a purely elastic solution. Eq. (2-9) gives values twice those obtained from elastic theory to account for a wall condition as a nonyielding reflective boundary.

The variation of the lateral stress along the wall for a point load V is shown in Figure 2-37*c*. The stress σ_h is the horizontal stress on a vertical plane and through the point load. At any other point along the wall the stress is

$$\sigma_h' = \sigma_h \cos^2 (1.1\theta) \tag{2-10}$$

where θ is the angle between V and O and the point for which σ_h' is desired.

Line Load The effect of line load on rigid nonyielding walls is shown in Figure 2-38. In this case the line load is parallel to the wall at distance x. Before referring to this solution, we should check the relative dimensions of the wall and the load area to decide if the loading can be considered a line or a strip load. A concrete block

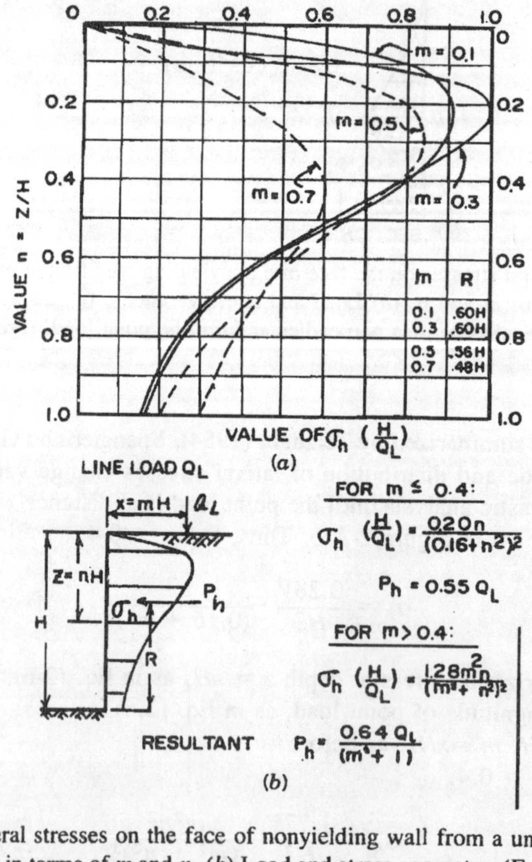

Figure 2-38 Lateral stresses on the face of nonyielding wall from a uniform line loading. (*a*) Stress variation in terms of *m* and *n*. (*b*) Load and stress geometry. (From Terzaghi, 1954; NAVFAC, 1971.)

wall or fence could be considered a line load, and a continuous strip footing parallel to the excavation fits the same condition.

Likewise, for $m \leq 0.4$,

$$\sigma_h = \frac{Q_L}{H} \cdot \frac{0.203\, n}{(0.16 + n^2)^2}, \qquad P_h = 0.55\, Q_L \tag{2-11}$$

where all notation corresponds to Figure 2-38. Likewise, for $m > 0.4$,

$$\sigma_h = \frac{4}{\pi} \cdot \frac{Q_L}{H} \cdot \frac{m^2 n}{(m^2 + n^2)^2}$$
$$P_h = \frac{0.64\, Q_L}{(m^2 + 1)} \tag{2-12}$$

Again, for $m \leq 0.4$, the lateral stresses predicted from elastic theory are too high. Therefore, Eq. (2-11) is the modified version that corresponds to measured values. For a line load at distance $x < 0.4H$ there is little change in the magnitude and distribution of lateral stresses from the values obtained for $x = 0.4H$, and the same is true for the resultant P_h until $x > 0.4H$. However, when we use the foregoing equations a stress discrepancy may arise at the base of the wall.

Strip Load A strip load has a uniform intensity within a finite width, such as a highway, railroad, earth embankment, and so on, which is parallel to the excavation support.

A solution is presented by Terzaghi (1954), somewhat modified for this case, as follows:

$$\sigma_h = \frac{2q}{\pi} (\beta + \sin\beta \sin^2\alpha - \sin\beta \cos^2\alpha) \tag{2-13}$$

where all symbols correspond to Figure 2-39, and the angle β is expressed in radians.

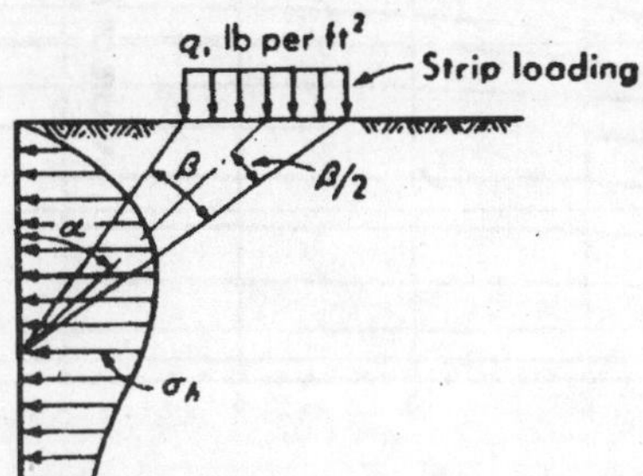

Figure 2-39 Lateral stresses on the face of nonyielding wall from a strip load.

Irregular Area Loading Approximate solutions by elasticity equations for evaluating lateral stresses against a wall caused by a vertical loading system have been developed by Newmark (1942). This investigator used elastic theory to construct an influence chart. A chart thus derived can be used similarly to the Bousinesq or Westergaard influence charts merely by plotting the shape of the load to the scale of the chart for the desired depth.

The method provides a rapid solution for irregularly shaped areas, such as a loading of limited or irregular dimensions acting in elastic half-space. Although the charts usually are developed for Poisson's ratio $\mu = 0.5$, they can be converted for other μ values. The lateral stresses computed by this method should be doubled for rigid walls.

Figure 2-40 shows an influence chart that can be used to evaluate lateral stresses against a rigid wall from a rectangular loading (Sandhu, 1974). The assumed Poison's ration μ is 0.5. Alternatively, an irregular-shaped surcharge loading can be resolved into several component loads, which are then treated as point loads, and the corresponding lateral stresses are more readily calculated.

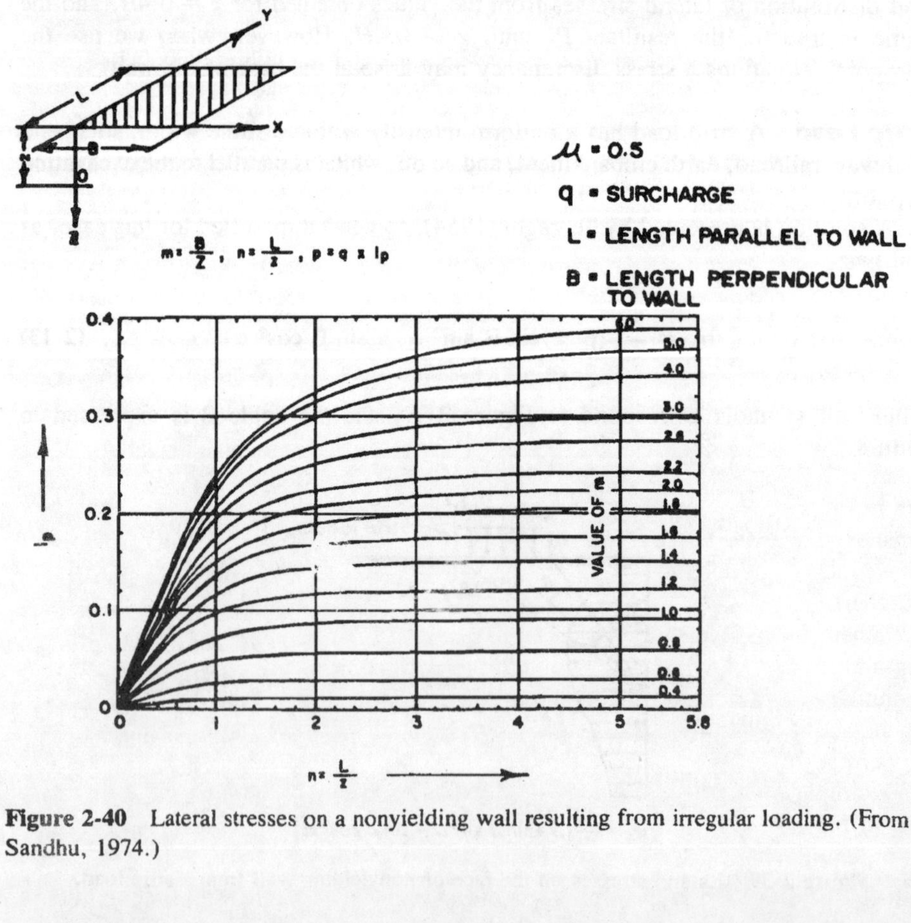

Figure 2-40 Lateral stresses on a nonyielding wall resulting from irregular loading. (From Sandhu, 1974.)

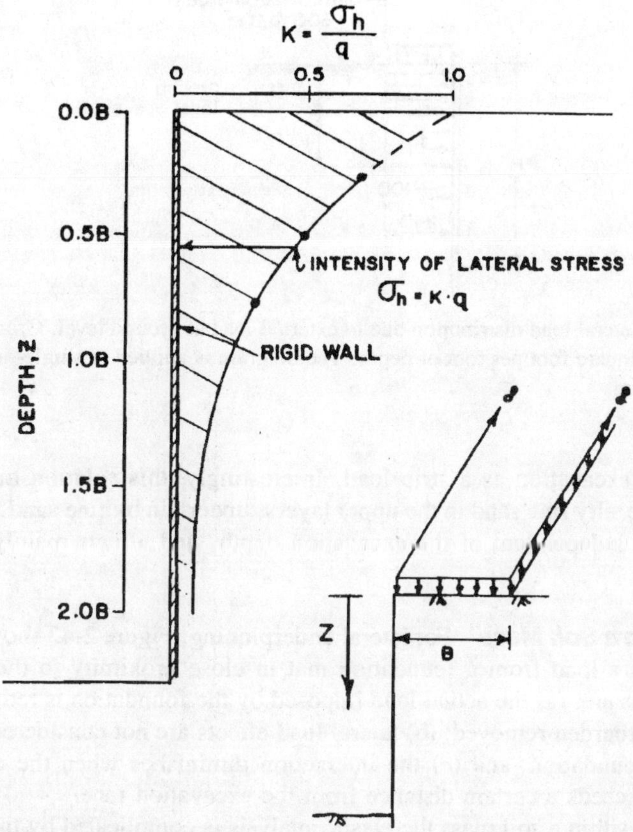

Figure 2-41 Elastic solution for lateral stresses on rigid wall from uniform surcharge of width *B* and infinitely long. (From Sandhu, 1974.)

Uniform Area Loading A solution based on elastic theory is shown in Figure 2-41 for a uniform area loading of width B infinitely long, and for a rigid wall. This is a partial application of the strip load distribution expressed by Eq. (2-13) and shown in Figure 2-39. We can note that the stress influence below a depth of $1.5B$ diminishes and can be ignored.

As we mentioned at the beginning of this section, a second approach is to convert the surcharge into an equivalent lateral stress by applying an earth pressure coefficient. Where the design implications can be significant, the two methods (elastic theory and limiting equilibrium) should be compared for results, considering also the boundary conditions. If earth pressure coefficients are used, it is prudent to apply the active condition K_A to temporary surcharges during construction, and the K_O to permanent surcharge loads.

An example of lateral stresses resulting from an external load at the surface is shown in Figure 2-42, and presumably it represents the effect of traffic and construction equipment. It is assumed to result from a surcharge of 600 lb/ft² applied

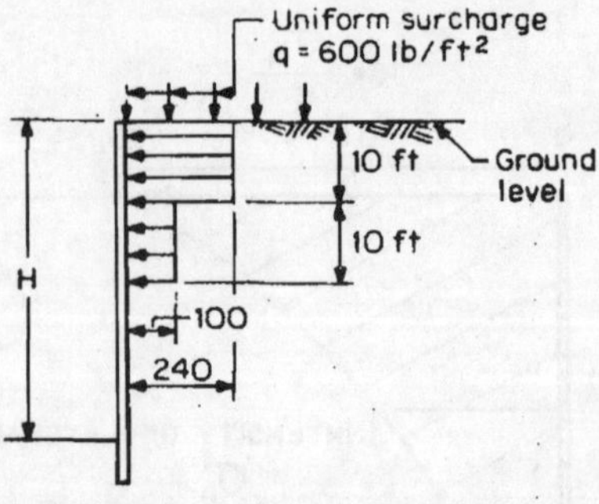

Figure 2-42 Lateral load distribution due to external load at ground level. Gradients shown are pounds per square foot per foot of depth. The diagram is applied to single- and multiple-braced excavation.

alongside the excavation as a strip load. Interestingly, this solution has been accepted in loose silty fine sand in the upper layers, underlain by fine sand. The lateral distribution is independent of the excavation depth, and affects mainly the upper wall section.

Loads Within a Soil Mass For lateral underpinning, Figure 2-43 shows a simple distribution of a load from a foundation mat in close proximity to the wall. The essential points are: (a) the actual load imposed by the foundation is reduced by the weight of overburden removed; (b) lateral load effects are not considered above the level of the foundation; and (c) the interaction diminishes when the edge of the loaded area exceeds a certain distance from the excavation face.

For a load within a soil mass the elastic analysis is complicated by the extension of the elastic soil medium above the plane of load application, and this makes the simple load distribution more convenient to use. Figures 2-44 *a* and *b* show two simple methods for distribution of loads within a soil mass. In Figure 2-44*a* the load applied by the foundation is distributed first vertically, according to an inclination

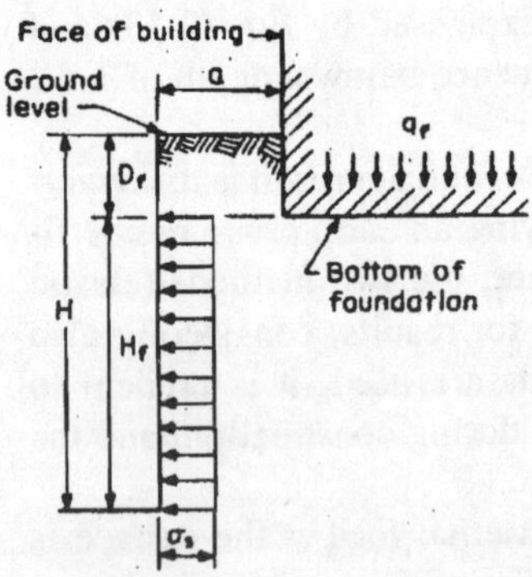

$$q_f = \text{total live- and dead-load foundation pressure, lb/ft}^2 = \frac{\text{building load in width } H_f}{\text{area of width } H_f}$$

$$q_n = q_f - \text{weight of overburden} = q_f - \gamma D_f$$

$$\sigma_s = \begin{cases} 0.5q_n\left(1 - \dfrac{a}{1.5H_f}\right) & \text{for } 0 < \dfrac{a}{H_f} < 1.5 \\ 0 & \text{for } \dfrac{a}{H_f} > 1.5 \end{cases}$$

Figure 2-43 Lateral load stresses distribution resulting from a nearby foundation mat.

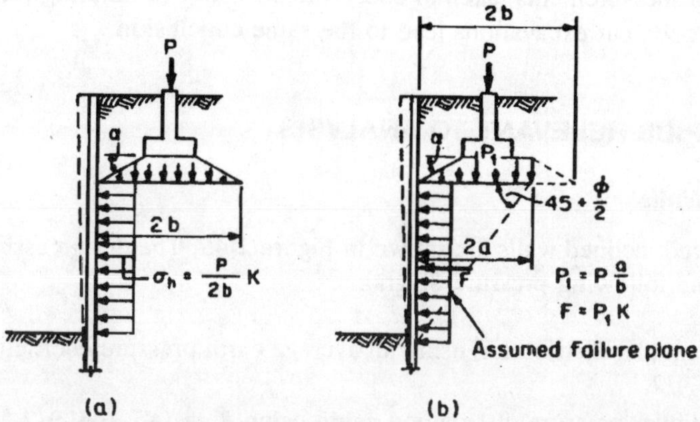

Figure 2-44 Lateral stress distribution due to external load within a soil. (*a*) Vertical and horizontal distribution according to an inclination angle. (*b*) Distribution taking into account the zone of influence within the failure wedge.

angle α, and then horizontally as shown. The coefficient *K* is the ratio of horizontal to vertical earth stress. In Figure 2-44*b* the inclination angle is considered within the failure wedge, and only the portion of load within this zone is assumed to cause lateral stresses.

Limitations of Analysis From the foregoing it appears that the lateral stress pattern caused by external loads acting at the surface or within a soil mass is based on elastic or limit theory, and very often there is an arbitrary crossover and shifting from one method to the other as the analysis considers more and different types of load. An approximate and perhaps more realistic approach is to establish the conformity between soil deformation and wall movement by iteration. An example is the solution suggested by Haliburton (1968), which assumes a discontinuous elastic-plastic deformation to represent the soil, and a multistrutted wall. This approach has been followed by other investigators using finite element techniques, and although the work is in the right direction it still is not completely satisfactory because: (a) total wall movement depends not only on the type and position of the bracing but also on the excavating sequence, time, and other related factors; (b) the deformation modulus at a given state of the soil varies with previous deformations; (c) the support sometimes moves toward the ground and sometimes away from it, particularly with element walls; and (d) in most instances there is a change and reversal from loading to unloading as external loads are first applied and then removed.

Tests on small-scale models simulating a strutted excavation and carried out to assess the lateral effect of foundations show that the pattern and distribution of external loads depend on the bracing position and excavation sequence in much the same manner as earth stress distribution does (Breth and Wanoscheck, 1972). The magnitude of these stresses is influenced by the position of the external load and its proximity to the wall, but its distribution hardly follows an elastic or limit theory. This distribution is decisively governed by the excavation and bracing process.

Recent field measurements taken to check the influence of building loads on walls supporting open cut excavations lead to the same conclusion.

2-11 TOPICS RELEVANT TO ANALYSIS

Inclined Walls

Two anchored inclined walls are shown in Figure 2-45. The design assumptions are based on the following pressure diagrams:

1. Rectangular distribution using an average earth pressure coefficient $K = (K_O + K_A)/2$.
2. Triangular pressure distribution again using $K = (K_O + K_A)/2$.
3. Triangular pressure distribution as in assumption 2, but multiplied by a reduction factor F.

Assumptions 1 and 2 apply mainly to walls with positive inclination, such as the supports shown in Figure 2-26. Assumption 3 is for walls negatively inclined, and applies readily to most element walls. The accompanying reduction factor is $F =$ (area ACD)/(Area ABD).

Results from model tests indicate that the trapezoidal earth pressure diagram is more likely to develop behind positively inclined walls if they are somewhat restrained against lateral movement, especially near the top. For walls with negative inclination the reduction factor is more applicable if we can assume that the failure plane does not change appreciably (Schnabel, 1971). For example, a negative slope of 12–2 in soil with $\phi = 30°$ gives a reduction factor 0.646. For an assumed vertical cut 25 ft deep, the resulting earth pressure is 0.625 kip/ft², but on the sloped wall the same pressure is reduced to $0.646 \times 0.625 = 0.404$ kip/ft². Sloped element

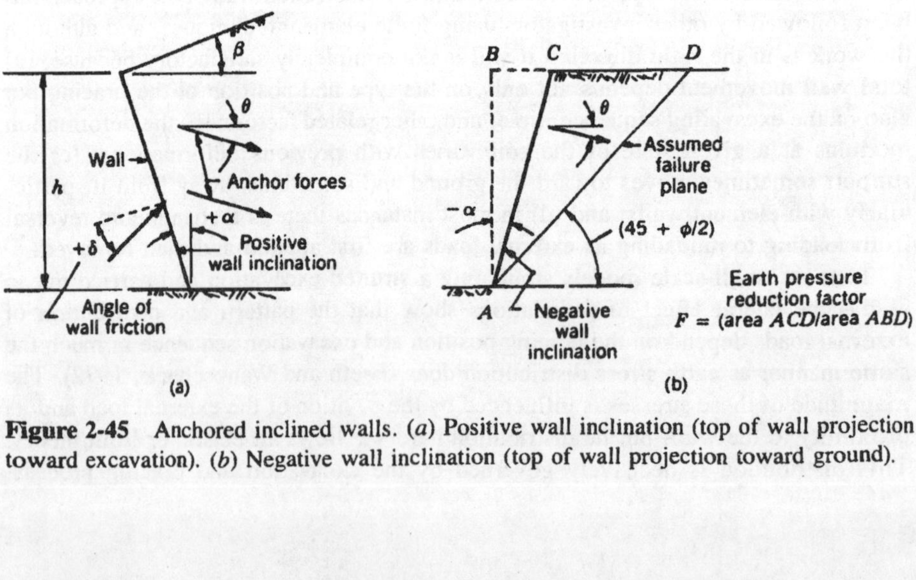

Figure 2-45 Anchored inclined walls. (*a*) Positive wall inclination (top of wall projection toward excavation). (*b*) Negative wall inclination (top of wall projection toward ground).

walls are thus particularly effective for supporting deep excavations or cuts, and the resulting economy relates to the smaller lateral forces that must be resisted.

As an excavation progresses positively inclined walls undergo more movement near the top; hence, it is important to provide a nonyielding lateral support at this location. Walls with negative inclination register the largest lateral movement near the base. The load on struts or anchors changes as excavation proceeds. More excavation causes load increase for positively inclined walls, but for walls with negative inclination the effect is an actual loss of load.

Effect of Soil Strength

For steep slopes of considerable height the effect of soil strength is significant, especially combined with pore pressure. Conventional stability analysis provides a general index only, and serves mainly to establish limiting values. Refined procedures can yield a deceptively high degree of accuracy that may not exist in practice and in spite of the theoretical basis. The margin of error increases if idealized assumptions are applied to a site characterized by a wide scatter of soil parameters.

The marked influence of soil strength (expressed by the parameters ϕ and c) on the results of stability analyses is demonstrated by a simple example. For the element wall shown in Figure 2-35, a change of only 1° in the angle of internal friction ϕ changes the anchor force necessary for stability with $F_s = 1$ by almost 1 MN/m of fixed anchor length. For the same wall, a variation in the cohesion c of 1 N/mm^2 (145 lb/in.^{2}) changes the required anchor force by 1 to 1.5 MN/m of fixed anchor length. In this example, a recheck of the soil parameters after completion of the wall showed that the angle ϕ was overestimated only by 1 to 1.5°, yet it caused an increase of the necessary anchor force by a factor of 2 in the final configuration.

REFERENCES

Abramson, L. W. and M. H. Hansmire, 1988. "Three Examples of Innovative Retaining Wall Construction," 2nd Int. Conf. Geot. Eng., June 1–5, Univ. of Missouri, St. Louis, paper 6.11.

Bares, F. A., 1974. "Use of Pali Radice (Root Piles) for the Solution of Different Foundation Problems," ASCE/EIC/RTAC Joint. Transp. Eng. Meeting, Preprint MTL-41, Montreal.

Bares, F. A., 1975. Personal communications.

Bathel, R., 1968. "Underpinning a Landmark", *Civ. Eng., ASCE*, Vol. 38, No. 8, Aug., pp. 67–68.

Bishop, A. W., 1955. "The Use of the Slip Circle in the Stability Analysis of Slopes," *Geotechnique*, London, Vol. 5, No. 1, pp. 7–17.

Breth, H. and H. R. Wanoscheck, 1972. "The Influence of Foundation Weights upon Earth Pressure Acting on Flexible Strutted Walls," Proc. 5th Eur. Conf. Soil Mech. Found. Eng., Madrid, Vol. 1.

Chugaev, R. R., 1964. "Stability Analysis of Earth Slopes", USSR All Union Scientific

Research Inst. of Hydraulic Engineering. (Translated from Russian, Israel Program for Scientific Translation, Jerusalem, Israel, 1966.)

ENR, 1940. "Underpinning Beats Subway Threat to Famous Chicago Building," *Eng. News Record,* Vol. 125, No. 13, Sept., pp. 64–68.

ENR, 1972. "Small-Diameter Piles Protect Tunnel, Buildings," *Eng. News Record,* April 20, pp. 22–23.

Febesh, M., 1975. "Underpinning and Lateral Movement," ASCE, New York Metropolitan Section, New York.

Fellenius, W., 1936. "Calculations of the Stability of Earth Dams," Trans. 2nd Congr. on Large Dams, Washington, D.C., Vol. 4, pp. 445–459.

Fenoux, Y., 1971. "Deep Excavations in Built-Up Areas," Soletanche Enterprises, Paris.

Gerber, E., 1929. "Unterschungen uber die Druckverteilung im ortlich belasteten Sand," Technische Hochschule, Zurich, Switzerland.

Goldberg, D. T., W. E. Jaworski, and M. D. Gordon, 1976. "Lateral Support Systems and Underpinning," Federal Highway Administration, Office of Research and Development, Washington, D.C.

Goldberg, D. T. et al., 1978. "Slurry Wall Test Panel Report," MBTA Red Line Extension, Davis Square to Alewife, Boston.

Goldfinger, H., 1960. "Permanent Steel Used as Subway Shoring," *Civ. Eng.*, Mar.

Haliburton, T. A., 1968. "Numerical Analysis of Flexible Retaining Structures," *ASCE J. Soil Mech. Found. Div.*, Vol. SM 6, Nov.

Janbu, N., 1957. "Earth Pressures and Bearing Capacity Calculations by Generalized Procedures of Slices," Proc. 4th Int. Conf. Soil Mech. Found. Eng., Vol. 2, pp. 207–212.

Kenney, T. C., 1956. "An Examination of the Methods of Calculating the Stability of Slopes," thesis presented to the Univ. of London at London, England in partial fulfillment of the requirements for the degree of Master of Science.

Lizzi, F., 1970. "Reticolo Di Pali Radice," Italian Geotech. Association Meeting, Bari.

Lizzi, F., 1974. "Special Patented System of Underpinning by Means of Pali Radice (Root Piles)," Spec. Conf. on Subway Constr. in Built-up Areas, Fondedile Corp.

Maidl, B. and H. G. Nellessen, 1973. "Reinforcement in Tunnel Construction as a Result of the New DIN 1045," *Die Bautechnik,* Vol. 50, No. 11, pp. 361–370.

McKinley, D., 1964. "Field Observations of Structures Damaged by Settlement," Proc. Design of Found. for Control of Settlement, ASCE, June, Northwestern Univ., Evanston, Ill., pp. 311–330.

Mergentime, C. E., 1972. "Underpinning Massive Structures," *Civ. Eng.*, Mar. pp. 40–41.

Mohan, D., G. R. S. Jain, and R. K. Bhandari, 1978. "Remedial Underpinning of Steel Tank Foundation," *ASCE J., Geotech. Div.*, May, pp. 639–655.

Morgenstern, N. R., and V. E. Price, 1965. "The Analysis of the Stability of General Slip Surfaces," *Geotechnique,* London, Vol. 15, No. 1, pp. 79–93.

NAVFAC DM-7, 1971. Design Manual—Soil Mechanics, Foundations and Earth Structure, Dept. of the Navy, Naval Facilities Engineering Command, Washington, D.C.

New York City Transit Authority, 1974. "Underpinning Section UP, Field Design Standards," Eng. Dept., New York.

Newmark, N. M., 1942. "Influence Charts for Computation of Stresses in Elastic Foundations," *Univ. of Illinois Bulletin,* Vol. 40, No. 12.

Norwegian Geotechnical Institute (NGI), 1962. "Measurement of Settlement Adjacent to Excavation," Report No. 7, Oslo, 111 pp.

O'Rourke, T. D. and E. J. Cording, 1974. "Observed Loads and Displacements for a Deep Subway Excavation", Proc., Rapid Exc. and Tunnel Conf., San Francisco, Vol. 2, p. 1305.

Peck, R. B., 1969. "Deep Excavations and Tunneling in Soft Ground," State-of-the-Art Report, 7th ICSMFE, Mexico City, pp. 225–290.

Prentis, E. A. and L. White, 1950. *Underpinning—Its Practice and Applications*, 2nd ed., Columbia Univ. Press, New York.

Saito, J., Y. Goto, and H. Sato, 1974. Stability of Trench Against Dynamic Loads during Slurry Excavation Kajima Corp., Spec. Bull., Tokyo.

Sandhu, B. S., 1974. Earth Pressure on Walls Due to Surcharge, *Civil Eng., ASCE*, Vol. 44, No. 12, Dec., pp. 68–70.

Sarma, S. K., and M. V. Bhave, 1974. "Critical Acceleration Versus Static Factor of Safety in Stability Analysis of Earth Dams and Embankments," *Geotechnique*, London, Vol. 24, No. 4, pp. 661–665.

Schnabel, H., 1971. "Sloped Sheeting," *Civ. Eng.*, Feb., pp. 48–50.

Schneebeli, G., 1964. "Le Stabilite des Tranchees Profondes Forees en Presence de Boue," *Houille Blanche*, Vol. 19, No. 7, pp. 815–820.

Skempton, A. W. and J. N. Hutchinson, 1969. "Stability of Natural Slopes and Embankment Foundations," Proc. 7th Int. Conf. Soil Mech. Found. Eng., State-of-the-Art Volume, pp. 291–340.

Spangler, M. G., 1951. "Soil Engineering", art. 21.18, International Textbook, Scranton, Pa.

Spangler, M. G. and J. Mickle, 1956. "Lateral Pressure on Retaining Walls Due to Backfill Surface Loads," Highway Res. Board Bull. 141, pp. 1–18.

Spencer, E., 1967. "A Method of Analysis of the Stability of Embankments Assuming Parallel Interslice Forces," *Geotechnique*, London, Vol. 17, No. 1, pp. 11–26.

Terzaghi, K., 1938. "Settlement of Structures and Methods of Observations," *Trans., ASCE*, Vol. 103, p. 1432.

Terzaghi, K., 1954. "Anchored Buluheads," *Transactions*, ASCE, Vol. 119, Paper 2720, pp. 1243–1324.

Turabi, D. A. and A. Balla, 1968. Distribution of Earth Pressure on Sheet Pile Walls, ASCE Soil Mech. Found. Div., vol. SM6, Nov.

VSL-Losinger, 1978. "Soil and Rock Anchors, Examples from Practice," VSL Report, Berne, Switzerland.

Ware, K. R., 1974. "Underpinning Methods and Related Movements," Rapid Excav. Tunnel Conf., San Francisco, Vol. II.

Ware, K. R., R. N. Evans, and J. G. Beck, 1979. "Use of Slurry Walls in Lieu of Underpinning," Proc. Slurry Walls for Underground Transp. Facilities, U.S. Dept. of Transp., Cambridge, pp. 315–343.

Weinhold, H., 1969. "Inclined Walls for the Munich Subway," 7th Int. Conf. Soil. Mech. Found. Eng., Specialty Sessions 14 and 15, Mexico City, pp. 102–103.

White, R. E., 1975. "Underpinning," *Foundation Engineering Handbook*, Van Nostrand Reinhold, New York, pp. 626–648.

White, R. E., 1976. "Underpinning for Transportation Tunnels," Proc. Underground Construction Problems, Techniques and Solutions, U.S. Dept. of Transportation, UMTA, Chicago.

WMATA, 1973. "Design Standards and Specifications," Section 210, Washington Metropolitan Area Transit Authority, Washington, D.C.

Xanthakos, P. P., 1974. "Underground Construction in Fluid Trenches," Colleges of Engineering, Univ. of Illinois, Chicago.

Xanthakos, P. P., 1979. *Slurry Walls*, McGraw-Hill, New York.

Xanthakos, P. P., 1991. *Ground Anchors and Anchored Structures*, Wiley, New York.

Xanthakos, P. P., 1994. *Slurry Walls as Structural Systems*, McGraw-Hill, New York, 855.

Xanthakos, P. P., 1994. Theory and Design of Bridges, Wiley, New York, 1443 pp.

Zimmerman, H., 1969. "Crossing Construction for North-South Highway in Cologne," *Strasse Brucke Tunnel*, Vol. 21, No. 5 and 6, pp. 113–120 and 151–156.

CHAPTER 3

EXCAVATION SUPPORT METHODS

3-1 CONVERGENCE OF SUPPORT-CONTROL PROCESS

Example of Convergence

Free Excavation Figure 3-1*a* shows an excavation carried out with free slopes. As long as these remain stable, the excavation is completed safely. In this example, however, the work is performed in water-bearing ground, and thus necessitates a suitable control system in the form of an impervious curtain. This curtain surrounds the site completely and ensures the completion of construction in dry conditions.

As the excavation continues, the adjacent ground mass begins to deform until it reaches self-equilibrium. If the impervious curtain is located within the zone of ground prone to movement, the screen may not possess sufficient strength to resist these deformations without cracking, and some form of damage is thus likely. The associated loss of efficiency must be considered in the design, and will affect the selection of the defense system. Suitable materials and methods must be selected, designed, and detailed in conjunction with relevant mechanical and hydrogeologic characteristics of the ground.

If the work is in soft ground (as opposed to rock), a curtain consisting of plastic materials will probably have an elastic modulus that will allow the system to deform and adapt to ground deformation without cracking. Walls of this type are reviewed in Chapter 10. They usually are sufficiently impermeable, and their increased plasticity enhances deformability, although it results in a lower mechanical strength.

Figure 3-1*b* shows a section for the excavation of a two-story basement for a building. This construction is carried out about 10 m (33 ft) from a river. The excavation is 7 m (23 ft) deep in sandy-gravelly alluvia, with the lower 5 m (16.5 ft) below the groundwater table. On the three sides bounded by streets the excavation is

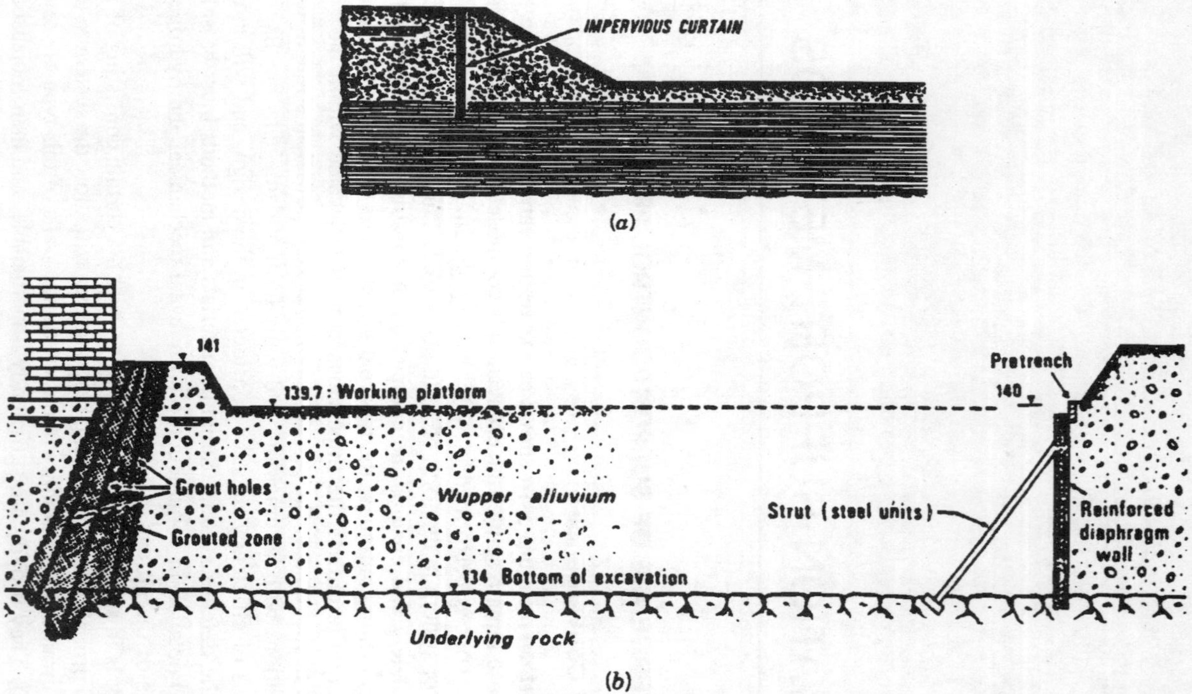

Figure 3-1 (*a*) Use of flexible impervious curtain to ensure a dry excavation. (*b*) Section through an excavation for a two-level building basement.

protected by a slurry wall combined with a free slope in the upper section extended to just above the water table. Along the adjacent building stability is ensured by underpinning preceded by consolidation grouting as shown. The slurry wall is braced internally by inclined steel struts.

Interestingly, this work was completed in 1960–1961, an therefore marks the transition from the more conventional techniques to the use of consolidation grouting and slurry walls.

Excavation Supported Entirely with Diaphragm Walls Figure 3-2 shows an excavation 17 m (56 ft) deep through noncohesive soils (mainly sandy-gravelly alluvia) and a formation of water-bearing coarse limestone. At the raft level the excavation is almost 5 m (16.5 ft) below the average groundwater table.

For the lateral support shown, the feasibility of the excavation is ensured if groundwater is prevented from entering the construction area, and if the bottom raft is stable against uplift pressures. The artificial supports are therefore supplemented

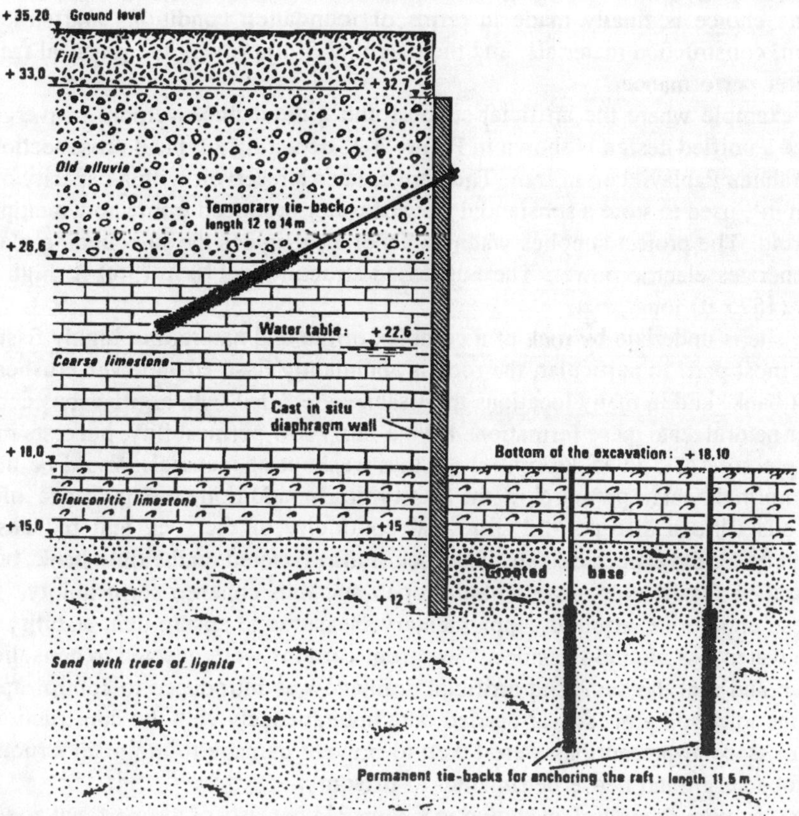

Figure 3-2 Cross section of an urban excavation showing the grouted base for groundwater control.

by a grouted base fixed at a level where active uplift is counteracted by the effective weight of the overburden. The grouted zone is 3 m thick and bears against the bottom section of the diaphragm wall. It is located in sand but its upper boundary coincides with the lower boundary of the limestone layer.

Grouting was carried out from level +32.7 through vertical drillings arranged on a grid 2 × 2 m. Grout was injected through the lower 3-m section of each drilling. The first grouting stage was completed with the intent to inhibit unnecessary loss of grout along the sand-limestone interface during subsequent injections and thus form a compact grouted base. The grout consisted of a clay-cement mix.

A second grout injection was applied to the sand layer, again using a clay-cement system. The treatment was completed by a third stage, in which a bentonite gel was injected to achieve higher penetrability in the critical zone.

Example of Dam Design and Construction Initial selection of a dam site and type of dam typically requires reference to water resource development and management criteria. The physical factors associated with this process are hydrologic, geologic, topographic, and biologic circumstances articulated by engineering analysis. The choice is finally made in terms of foundation conditions and structure location, construction materials, and the potential of improving the physical factors for better performance.

An example where the artificial support and ground improvement converge to produce a unified design is shown in Figure 3-3, depicting a typical cross section of the Farahnaz Pahlavi dam in Iran. The dam creates a reservoir with a capacity of 85 million m^3, used to store a substantial flow of water produced from snow melting in the spring. The project supplies water to Teheran, irrigates the surrounding plains, and generates electric power. The buttressed structure is 110 m (360 ft) high and 480 m (1575 ft) long.

The site is underlain by rock of a complex origin and nature, but highly fissured for the most part. In particular, the rock is abundantly fissured and even crushed on the left bank, and in many locations the fissures are filled with argillaceous debris. In their natural state these formations have a fairly low permeability, but tests made under pressure of 5 to 15 kg/cm^2 show that washout of materials would be likely under normal water pressure in the reservoir. In addition, the presence of argillaceous fillings disrupts the structural continuity of the rock and reduces its strength under the expected loads. The rock is also fissured on the right bank, but at this location the fissures are essentially open and thus enhance permeability.

These conditions required supplementary measures to ensure dam stability and reduce water loss around the sides. Two grout curtains were injected to consolidate the rock and seal the dam, and artificial drainage was introduced to control uplift. Since the injection process was carried out simultaneously with the construction of the dam, special precautions were taken to prevent damage to the main structures and probable ground heave detrimental to masonry.

The upstream grout curtain shown in Figure 3-3 consists of a superficial zone 10 to 15 m (33 to 50 ft) deep, formed by four rows of grout holes. Hole spacing is 2.8 m (9 ft), occasionally reduced to 1.4 m (4.6 ft). The injection pressure was 12 to 14

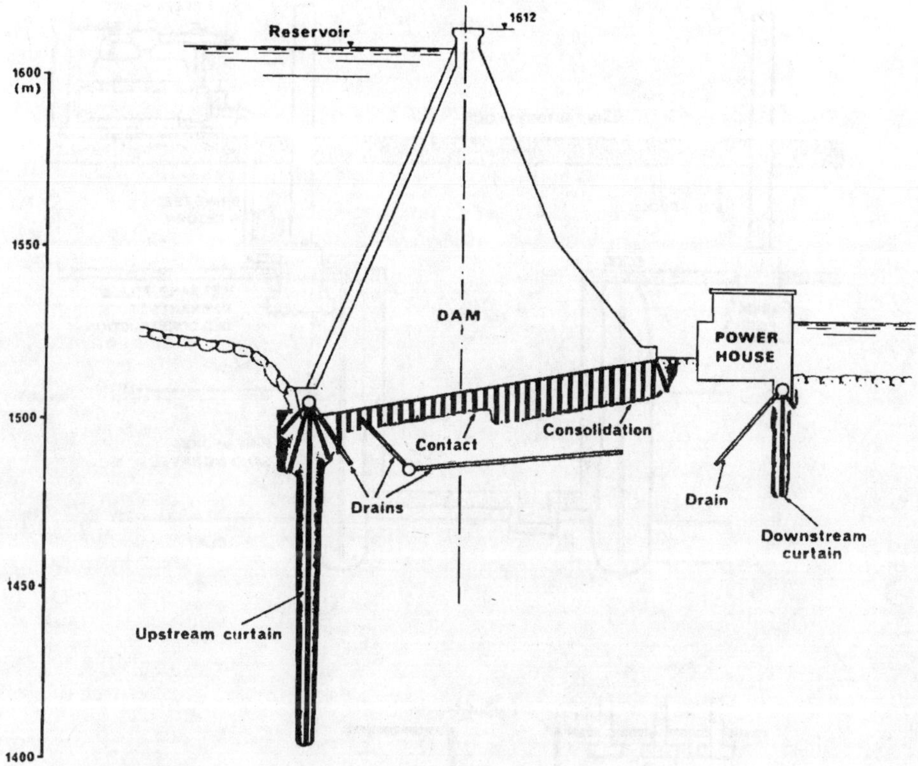

Figure 3-3 Typical cross section, Farahnaz Pahlavi dam, Iran.

kg/cm^2 in order to exceed the hydrostatic pressure of the reservoir water at full stage. A second zone was formed by injecting grout to produce a curtain 80 to 100 m (260 to 330 ft) deep in a single row of grout holes spaced 2.3 m (7.5 ft) apart. A primary hole drilled every 14 m confirmed the depth of the curtain.

The downstream grout curtain protects the dam from water circulation originating downstream. It is built with a single row of grout holes, and its average depth does not exceed 30 m (100 ft).

The direct result of this treatment is the consolidation of rock, and a 25 percent increase in its mechanical strength. The injection also ensures a full contact between rock and the concrete structure above. Outside the consolidated areas, a rock-concrete bond treatment was applied to the foundation by means of grout holes 5 m (16.5 ft) deep injected at low pressure (5 kg/cm^2).

A full-scale drainage system installed in the foundation protects the dam from unwarranted uplift pressures. The drain network is particularly dense underneath the right bank to stop water circulation before it reaches the downstream area.

Unsupported Excavation As mentioned in Chapter 11, artificial ground freezing can be used as structural support and for groundwater control. A direct applica-

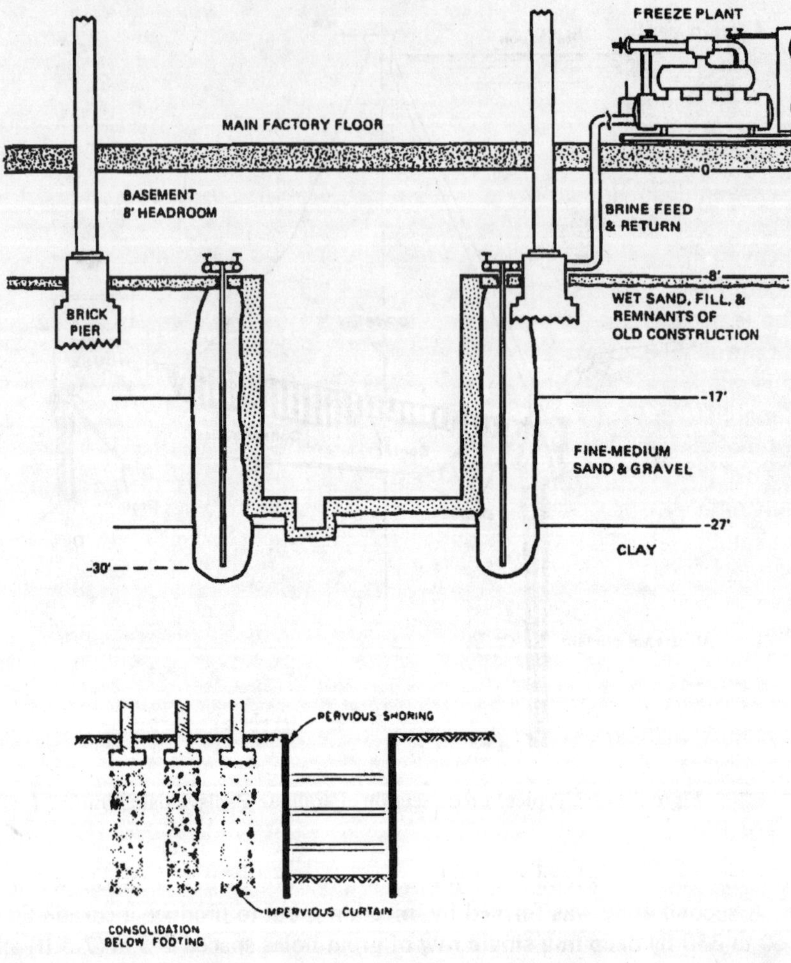

Figure 3-4 (*a*) Section through an existing basement showing excavation in frozen ground. (*b*) Underpinning by grouting.

tion is shown in Figure 3-4*a*, and involves excavation below the level of an 8-ft-high factory basement.

The work had to be completed without settlement or ground loss in the saturated sand and gravel layers underlying the building. In addition, plant operations had to continue without disruptions while construction continued as scheduled.

Working in the basement with rotary and percussive drilling tools, 28 vertical freeze pipes were installed in the clay bed below the sand-gravel layers. A packaged refrigeration plant was assembled and placed in a convenient location on the main floor of the building. Calcium chloride brine was circulated at temperatures close to $-30°C$ to form a frozen cofferdam that was structurally stable and that had a watertight anchor into the clay.

After a prefreezing period of three weeks, the excavation was carried out in the dry and without adverse effects. After pouring the concrete floor, the concrete walls were formed and placed in direct contact with the frozen ground.

Underpinning by Grouting If a slurry cutoff wall can be built to protect a deep excavation, as shown in Figure 3-1a, the associated low cost will make this option economically attractive. If a diaphragm wall can ensure the watertightness of the excavation while serving also as ground support, the result is a marked improvement of the technical aspects of the project. Where the feasibility of this construction is questionable and the planning must address dewatering as well as underpinning problems, an impermeable consolidated gravity earth wall may be considered. This situation is illustrated in Figure 3-4*b*. In view of the wide range of strengths that are attainable using suitable grouting techniques and products, considerable in-ground support capacity can be imparted to superficial foundations. If the mass of soil to be consolidated is easily accessible, grouting may be competitive with conventional underpinning methods (see also Chapter 2). The combined effects of consolidation and watertightness will be advantageous for reducing water flow as well as altering earth pressure loading conditions and surcharge on the lateral support system.

The Rationale of Convergence

Problems arising from the placement of foundations and supports on poor soils and deteriorated rocks often signify the absence of ground controls, and also indicate uncertainty in design criteria. Conversely, the application of ground controls and improvement techniques should be accepted as a formidable supplement that balances structural requirements of the artificial supports or eliminates these supports altogether. Whereas this contribution can improve structural safety and functional performance, it is not always considered in quantitative terms and is often diagnosed as an independent remedy.

The convergence between support and control requirements must recognize ground response to any externally applied action. This response may be specific or random, rapid or slow, and temporary or permanent. For example, a structure placed in clay soils will undergo settlement associated with distinct physical phenomena: (a) shear strains that develop simultaneously with the change in load (immediate or initial effects); (b) time-dependent shear strains (creep); (c) time-dependent volumetric changes resulting from the dissipation of excess pore pressure (consolidation); and (d) time-dependent volume changes after excess pore pressures are essentially dissipated (secondary compression).

A forecast of this response in engineering terms is the main determining factor for or against the introduction of controls, or for the exclusive use of ground supports. Ground response is also recognized in a quantitative analysis, particularly in a time-dependent fashion, where it can supplement the total capacity of the support system; examples are situations where some ground deformation is encouraged and allowed to occur under controlled conditions to mobilize ground strength and thus reduce the requirements of the artificial support. A well-known ground-structure interaction clearly demonstrates the effect of ground and support stiffness on the support load: the stiffer the support is relative to the ground, the greater will be the support load.

The interpretation and analysis of ground behavior around an excavation or an underground opening requires good judgment, particularly where it involves time-dependent response. In order to reach an optimum level of convergence by reducing the implicit overdesign of the support, it is essential to understand the ground-support interaction and how control measures will alter this process.

3-2 GROUND RESPONSE IN BRACED EXCAVATIONS

Theoretical Aspects

In general, excavation is the removal of a mass of soil and water in a process that causes unloading. Where it must be done in the dry and other controls are not feasible or available at the site, it is customary to lower the groundwater level outside and below the construction area (see also dewatering techniques in Chapter 1). Total stress release and absence of pore pressures normally result in movement of the surrounding soil, and control of this movement is a basic requirement in engineering an excavation. Ground response is influenced by three main factors, closely interrelated; these are the soil properties, control of groundwater, and the time element.

If an instantaneous earth removal could be made, the soil around it would strain under undrained conditions. If the same excavation could be made at an infinitely slow rate, the surrounding soil would strain under fully drained conditions. In most practical cases excavations in sands and gravels are typically assumed to be drained, and those in clay essentially undrained except near the clay boundaries; excavations in silts usually the assumed partially drained.

The Role of Groundwater If the groundwater level is unchanged, the water continues to act against the support and contributes to the total lateral stress. Conversely, a decrease in groundwater pressure can cause an increase in effective earth stress with subsequent settlement of the ground. Water flowing into an excavation area can cause practical construction problems. Commonly, we choose to estimate the total lateral thrust acting against a support from two components: effective earth stress and pore water pressure.

If a watertight wall surrounds an excavation and penetrates an impervious formation below the base, the water conditions can be assumed static, resulting in a simple water pressure distribution. With more conventional temporary supports ideal water cutoff is unlikely, and a more complex pattern of water inflow should be expected (Lambe, 1970). The latter case warrants the analysis of pore water flow. If a two-dimensional steady-state flow can be assumed, pore pressures may be estimated from flow nets or from finite element solutions. The complexity of this analysis becomes obvious, however, if relevant factors must be considered such as in situ permeability, pore fluid flow through the wall, pore pressure caused by changes in total stress, probable flow parallel to the wall, and actual time necessary to complete the construction including effects on consolidation. In the usual case, pore water pressures are likely to be less than static, but dynamic pressures can develop following certain events, for example, broken water mains outside the excavation.

The vertical effective stress σ'_v in the soil outside an excavation can be estimated assuming geostatic conditions as

$$\sigma'_v = \sigma_v - u \tag{3-1}$$

where σ_v is total overburden stress and u is the pore pressure near the support. Figure 3-5 shows the distribution of vertical stresses for static pore pressure conditions and for a groundwater level about 6 ft below surface. Two horizontal stress diagrams are shown in Figure 3-5*c* and correspond to $0.4\,\sigma'_v$ and $0.3\sigma'_v$, respectively. The distribution diagram based on $\sigma'_{ho} = 0.4\sigma'_v$ approximates horizontal stresses for the K_0 state and for normally consolidated soil. The distribution $\sigma'_{ho} = 0.3\sigma'_v$ represents the active state for a soft soil.

It is apparent that pore water pressure can be a significant component of the total thrust acting on a support. Outside the excavation, this relation is articulated by noting that decrease in pore pressure results in increase in the effective stress leading to settlement.

Stresses and Strains Near an Excavation Figure 3-6 shows stress-strain patterns experienced by two soil elements *A* and *B* near an excavation (Lambe, 1970). The effects of the excavation are: (a) the reduction of total vertical and horizontal stresses; (b) changes in equilibrium pore water pressure; and (c) the clear dependence of wall movement on the horizontal thrust acting on the wall.

The stress paths apply to soils normally consolidated to an anisotropic stress state. Since excavation causes unloading, it also alters the boundary pore pressure on the inside. If it remains long enough to allow steady-state seepage to develop, the equilibrium pore pressures essentially are the same as obtained from a flow net.

The changes in stress and strain caused by excavation and steady-state seepage are summarized in Table 3-1. Soil element *A* shows a tendency for settlement outside the excavation, but the soil below the base tends to expand. According to these strains we should expect an increase in shear strength at element *A* and a slight decrease in shear strength for element *B*.

Stress paths and associated horizontal strains are shown in Figure 3-7 for an element outside the support. The numbers 1 through 5 characterize the sequential changes in stress and strain. According to this pattern, the soil is initially consolidated to the stress system indicated by point 1, unloaded to the stress-strain system shown by point 2, and so on. The variation in horizontal stress at element *A* can cover a wide range, depending on the inward and outward wall movement.

The stress-strain plots of Figures 3-6 and 3-7 demonstrate soil behavior following

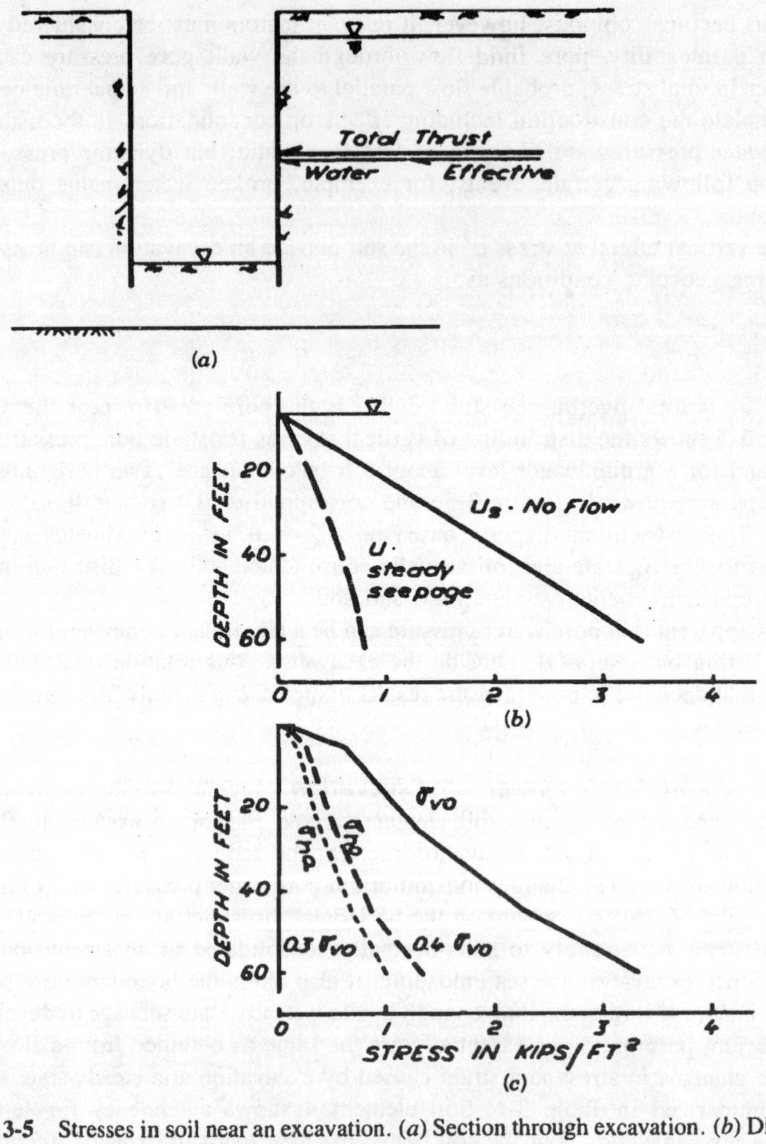

Figure 3-5 Stresses in soil near an excavation. (*a*) Section through excavation. (*b*) Distribution of water thrust. (*c*) Distribution of effective soil thrust.

an excavation. These plots, however, do not include the effect of friction at the wall-soil interface, and ignore changes caused by the excavation sequence and bracing.

Effects Inherent in Construction

Besides soil properties, groundwater, and time effects, factors affecting ground response are the dimensions of the excavation, the support system, excavation and

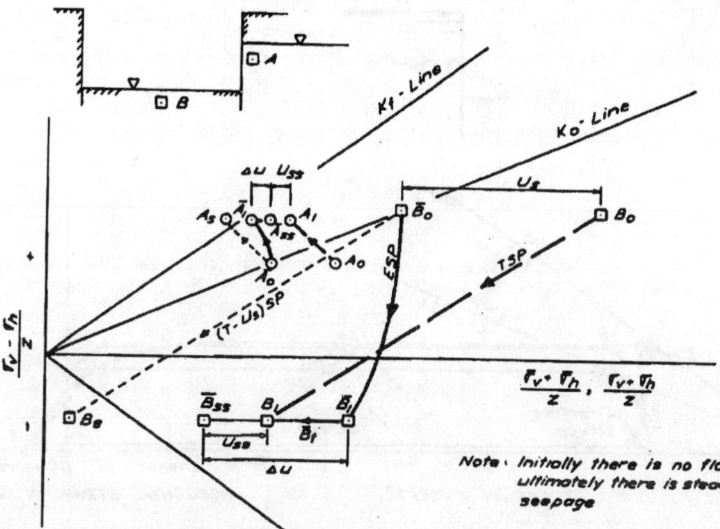

Figure 3-6 Stress paths for soil elements near an excavation. (From Lambe, 1970.)

bracing sequence, the presence of structures and foundations in the immediate vicinity, and transient surcharge loads. The first comprehensive investigation and state-of-the-art report on these effects is given by Peck (1969).

Incidental Factors In built-up areas engineering an excavation normally requires the selection of ground support and suitable control measures to ensure the stability of the cut and confine ground movement to an inconsequential range. For a safe design of ground support, apparent pressure envelopes may be used alternating with stress-strain relations. In contrast, settlement patterns associated with braced cuts are summarized and predicted in a general manner. The broad range of perfor-

TABLE 3-1 Stresses and Strains for Two Soil Elements near an Excavation

	Soil Element A	Soil Element B
Initial (static) pore pressure, u_s	$A_o\overline{A}_o$	$B_o\overline{B}_o$
Pore pressure at steady-state flow, u_{ss}	$A_1\overline{A}_{ss}$	$B_1\overline{B}_{ss}$
Pore pressure upon unloading	Decreases	Decreases
Pore pressure during consolidation	Decreases	Increases
Strain upon unloading	Vertical compression	Vertical extension
Strain during consolidation	Vertical compression	Vertical extension
Undrained shear strength during consolidation	Increases	Decreases

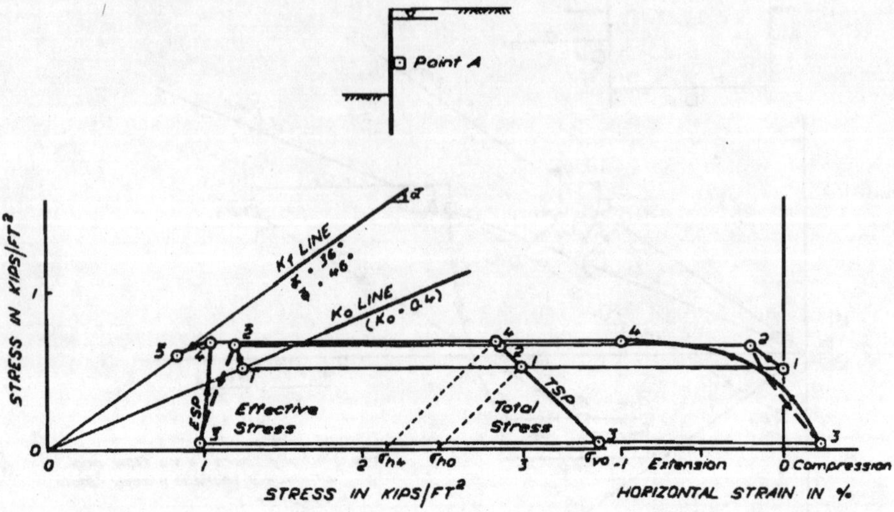

Figure 3-7 Stress-strain behavior for an element outside an excavation; horizontal stress paths. (From Lambe, 1970.)

mance patterns that result from different scenarios of construction methods and soil conditions often inhibit the decision on the degree and kind of protection and the associated controls.

O'Rourke (1981) has studied ground movement in braced excavations by correlating its pattern and magnitude to relevant aspects and phases of the construction. The stress-strain relationship and groundwater flow patterns discussed in the preceding section form the conceptual basis for much of the analysis, but are interfaced with physical activities and construction events to investigate the composite effects.

Figure 3-8 shows a typical settlement and lateral movement for a braced cut at the Embarcadero III site in San Francisco. This cut was excavated through a 22-ft (6.5-m) layer of granular fill and sand, and 23 ft (7 m) of soft to medium clay. The bottom of the cut is underlain by 35 ft (11 m) of medium clay followed by deposits of interbedded dense sand and stiff clay. The ground support is provided by a sheet pile wall, braced internally (Drossel, 1975; Tait and Taylor, 1975).

In this instance ground movement was strongly governed by construction activities related to site preparation and pile driving. About 800 timber piles had to be removed and replaced by prestressed concrete piles driven in situ. Sheet piling was installed along the perimeter of the cut using vibratory hammers. The plots of Figure 3-8 indicate that about 30 percent of the total movement occurred from day 90 to day 170, and in spite of the fact that the excavation was only 11 ft deep. In some instances, the dependence of movement on a specific construction activity is apparent; for example, when timber piles were removed near the property boundary, the associated movement caused cracking of the adjacent street pavement.

Site preparation effects are also highlighted in construction examples of drilled

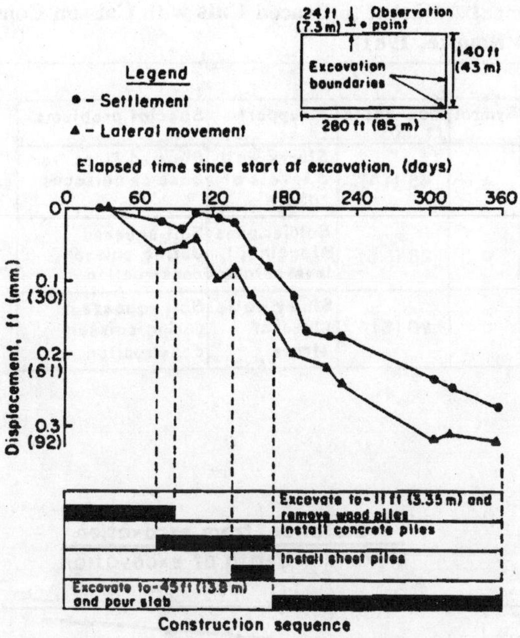

Figure 3-8 Ground movement record for a 45-ft (14-m) deep cut in soft to medium clay. (From O'Rourke, 1981.)

shafts summarized in Table 3-2 (Baker and Khan, 1971; Lukas and Baker, 1978). These shafts were first drilled to depths of 20 to 35 feet (6 to 10.5 m) from the construction level to install steel linings, and then further extended by augering and advancing a second steel casing left in place. The drilled subpiers are supported on limestone bedrock typical in Chicago, approximately 110 ft (34 m) from ground level.

The settlement data of Table 3-2 confirm the construction difficulties. For case 1 most of the settlement was caused by dewatering problems in deposits of sand, silt, and gravel directly overlying the bedrock. Pumping was necessary to prepare the shaft bottom for concreting, and in this process sand-silt particles were pumped out with the water. For cases 2 and 3 considerable settlement was caused by clay squeezing into the shafts during excavation and into annular space around the shaft after lining installation. In all three cases, movement associated with drilled shaft construction was between 50 and 70 percent the total settlement.

General Effects Ground response in braced excavations can be articulated for three distinct stages: (a) initial excavation before bracing; (b) excavation to subgrade after the top braces are installed; and (c) removal of the braces.

An excavation is usually carried down to a certain level (usually 15 to 20 ft) before the first bracing is installed. During this stage the deformation of the ground

TABLE 3-2 Settlement Adjacent to Braced Cuts with Caisson Construction Difficulties (From O'Rourke, 1981).

Case	Symbol	Max. depth ft (m)	Support	Special problems
1	▲	45 (14)	Slurry wall; 3 levels of rakers	Pumped fines from base of caissons
2	●	28 (8.5)	Soldier piles &lagging; 1 level of rakers	Soil squeeze during caisson construction
3	○	26 (8)	Slurry wall; 1 level of struts	Soil squeeze during caisson construction

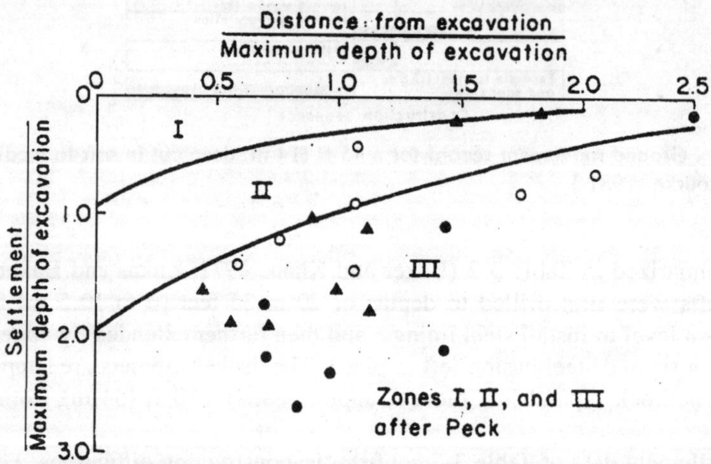

Note: The settlements and distances in the sketch are plotted in dimensionless form as fractions of the maximum excavation depth. Zones of settlement, delineated by Peck (1969) for various soil types and excavation conditions, are shown for comparison in the accompanying graphs.

support resembles a free cantilever. The horizontal strains reflect this mechanism by developing a triangular contour pattern that decreases with depth and distance from the support.

After the uppermost bracing is installed, this portion of the wall is restrained against lateral movement. If necessary, preloading can be used at this level to control further movement, although it will cause a corresponding recompression of the soil analogous to passive resistance. In the deeper portion of the cut the wall continues to bulge inwards and cause tensile strains. Installation of intermediate bracing stops this movement.

As the bottom braces are sequentially removed to construct the underground structure, further inward movement occurs at these locations. When the upper braces are removed and with the lower wall section braced by the permanent structure, the support reverts to the cantilever pattern. Movement stops when the upper portion of the excavation is braced.

3-3 SUPPORTS FOR BRACED EXCAVATIONS

Soldier Pile Walls

Soldier piles and lagging represent a time-tested technique for the support of excavations, and where necessary they can be successfully combined with control measures. The piles are usually spaced 6 to 10 ft apart, and lagging is inserted to span this distance. The piles receive and carry the full lateral load. Typical versions are shown in Figure 3-9. Lagging may be omitted in hard clays, soft shales, or where cohesion and natural cementation binds the soil and the piles are inserted at relatively close spacing. In this case control measures may be required to prevent erosion and spalling of the face (Shannon and Strazer, 1970; Clough et al., 1972). Erosion or raveling caused by drying of the exposed face can be inhibited by spraying the exposed soil.

Lagging Wood lagging is the most commonly used material. It can be installed to bear directly against the soil side or wedged as shown in Figure 3-9*a*. In the latter case the soil is squeezed to develop full-pressure contact that prevents its loosening. In the United States the most common wood used for lagging is construction grade lumber, usually rough-cut. Strength properties and physical characteristics are listed in standard manuals (Goldberg et al., 1976).

Earth pressure measurements show that lagging installed in the conventional manner and in reasonably competent soils is not likely to receive the full active pressure mobilized behind it. This pressure concentrates on the much stiffer soldier pile, and less pressure is transmitted to the lagging. This redistribution essentially constitutes arching and is related to the construction procedure. If the lagging is supported on the front flange, a slight overcut in the soil profile behind the lagging is always necessary for the placement of the boards; thereafter the soil moves to fill the annular space and arching is developed at the ends. This response means that

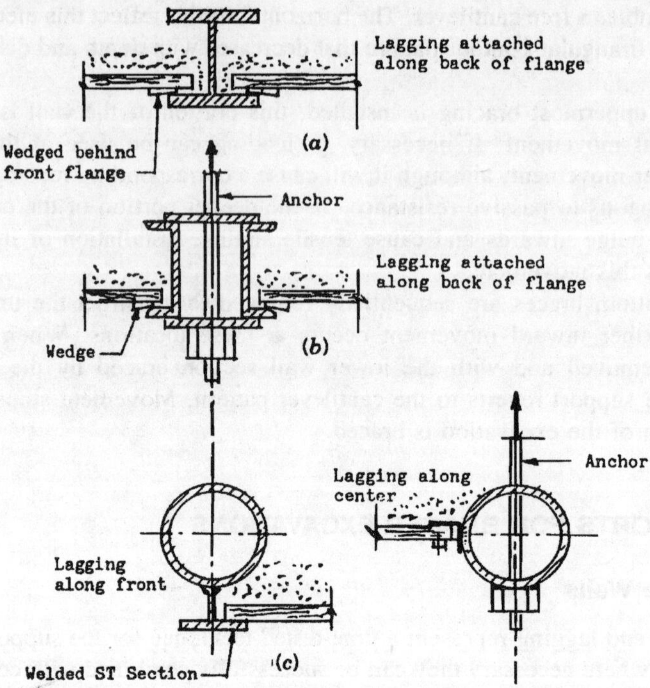

Figure 3-9 Various types of soldier piles. (*a*) Wide flange or H-pile. (*b*) Channel section. (*c*) Pipe section.

earth pressures on lagging are relatively unaffected with depth, and larger lateral loads in deeper cuts are transmitted through the soldier piles. Recommended design pressures and thickness of the lagging are reviewed by Goldberg et al. (1976).

Loss of Ground The installation procedures can contribute markedly to ground loss. A typical example is the soil response behind the lagging and the flexural deflection of the boards with increased pile spacing. In order to control the latter, board thickness should be chosen to limit deflection to less than 1 in. Movement caused by overcut is best controlled by packing of soil behind the lagging.

Loss of ground is also likely in soft clays and loose soils of low plasticity below the water table. The fast exposure of these soils by excavating below installed lagging provides the opportunity for deformation. Examples of complete failure in soft sensitive clays are reported by Broms and Bjerke (1973). In the same context, stress relief associated with arching can be uncertain and is unlikely in very soft soils or in soils prone to plastic creep. Conversely, where soils are difficult to drain, dewatering in advance of excavation is indicated. Dry cohesionless soil can cause problems, particularly in hot arid areas, and in this case the remedy is to moisten the face by spraying while placing the boards.

Preexcavation to install the soldier piles is a further cause of material loss. This loss can occur from suction effects manifested during withdrawal of the auger or in the form of collapse of soil into the hole.

Surface Water and Groundwater In water-bearing formations of cohesionless materials the groundwater should be drained before excavating to expose the face. A suitable groundwater control scheme is determined by the depth of excavation below the water table, the soil permeability, and the presence of any underlying or interbedded layers (see also Chapter 1).

In soils that drain slowly the excavation is slowed down by a corresponding slow advancement, usually 1 ft at a time. In these conditions contractors usually choose to slope the bottom in a V profile to collect surface drainage and also to produce a depressing effect on the phreatic surface at the side of the cut. Groundwater control is more difficult if impermeable layers are interbedded with pervious formations. In this case the groundwater tends to flow longer above the impervious layers.

Steel Sheet Pile Walls

The most common forms of steel sheeting are Z-shaped or arch-shaped interlocked sections. Where a higher bending resistance is necessary, Z sections are more suitable because of their greater moment of inertia. Domestic sheet pile configurations and sections are shown in Figure 3-10, and relevant data are summarized in Table 3-3.

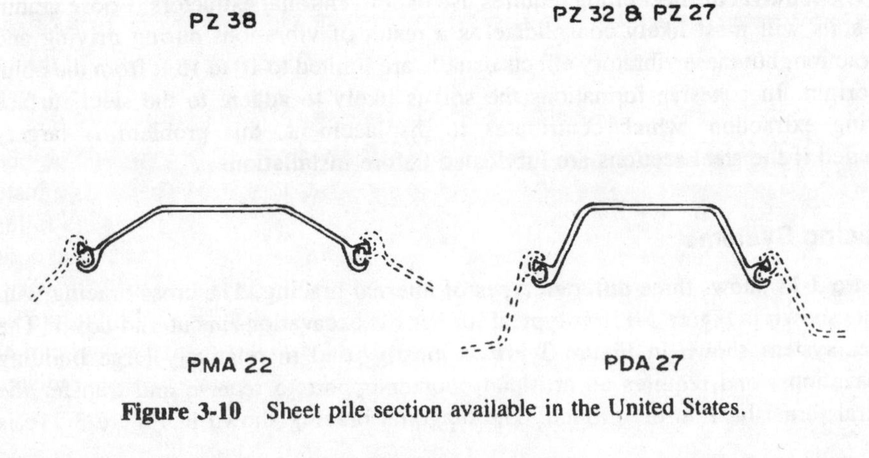

Figure 3-10 Sheet pile section available in the United States.

TABLE 3-3 Data for Sheet Pile Sections Available in the United States

Section	Dimension (in.) *D*, depth	*L*, length	Weight (lb/ft²)	Moment of Inertia (in.⁴/ft)	Section Modulus (in.³/ft)
PMA 22	$3\frac{1}{2} \times 2 = 7^{a}$	19.6	22.0	16	5.4
PDA 27	$5 \times 2 = 10$	16	27.0	40	10.7
PZ 27	12	18	27.0	183	30.2
PZ 32	11.5	21	32.0	220	38.3
PZ 38	12.0	18	38.0	281	46.8

[a]Single pile is $3\frac{1}{2}$ in. deep. As driven, wall is 7 in. deep.

Steel sheet pile walls are chosen where soldier piles are not suitable, for example, in soft clays, saturated silts, and loose silty or clayey sands. If the ground is hard or contains boulders, driving sheet piles is difficult and often impracticable. Depth limitations are imposed by site conditions and available headroom, and in a congested urban site the installation may not be possible because of the danger of cutting utilities.

Interlocked sheet piling is effective in cutting off concentrated water flow through pervious formations within or below the excavation and in protecting against a blow condition or other form of ground loss. The seal obtained at the base of the excavation ensures groundwater control for the construction period. This, however, does not imply that the installation will prevent the general lowering of the piezometric level and the accompanying consolidation in relatively pervious soils.

Ground Response Groundwater leakage and loss of ground can occur if the sheeting is forced out of the interlocks following hard driving or as a result of misalignment. The likelihood of this problem is particularly strong in dense soils or in the presence of boulders and obstructions.

The removal of sheet piling requires use of conventional extractors. Loose granular soils will most likely consolidate as a result of vibrations during driving and extraction, but these vibratory effects usually are limited to 10 to 15 ft from the point of origin. In cohesive formations the soil is likely to adhere to the steel surface during extraction, which contributes to displacement; this problem is largely avoided if the steel sections are lubricated before installation.

Bracing Systems

Figure 3-11 shows three different types of internal bracing. The cross bracing with struts shown in Figure 3-11*a* is typical for narrow excavations in cut-and-cover. The raker system shown in Figure 3-11*b* is mostly used in relatively large building excavations, and requires an artificial countersupport to receive and transfer the lateral thrust back to the ground. The diagonal bracing shown in Figure 3-11*c* is

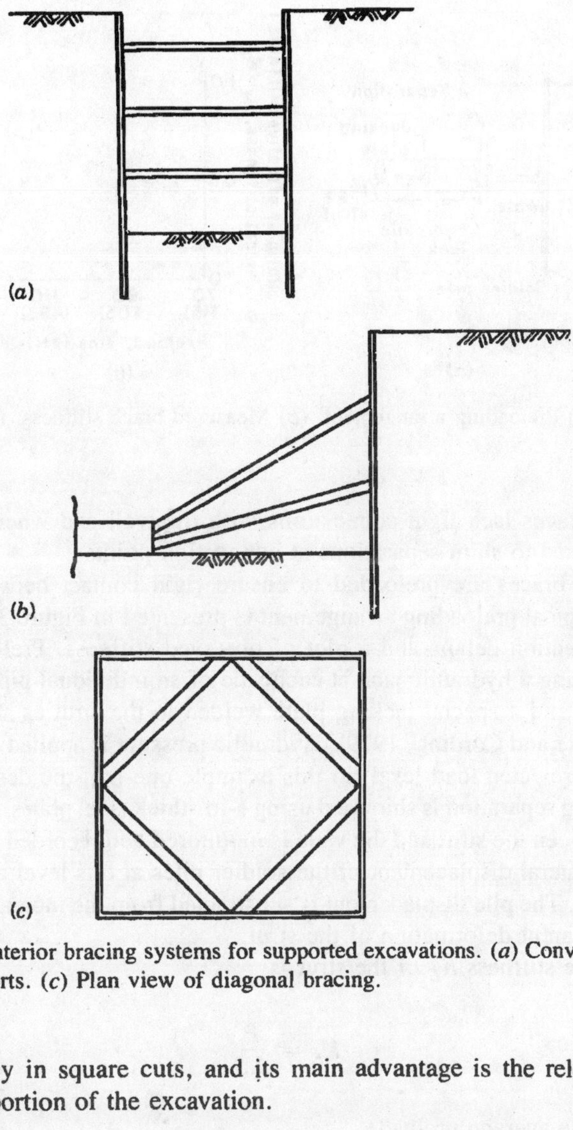

Figure 3-11 Interior bracing systems for supported excavations. (*a*) Conventional bracing. (*b*) Raker supports. (*c*) Plan view of diagonal bracing.

used essentially in square cuts, and its main advantage is the release of a large unobstructed portion of the excavation.

Effect of Brace Stiffness At each level of brace application secondary wall movement occurs as a result of compression at loose connection points, horizontal bending of wale beams, and elastic shortening of the braces. In practice, the effective stiffness of a brace (expressed as the ratio of preload to apparent deformation) may be markedly less than its ideal elastic stiffness (O'Rourke, 1981). Studies by Palmer and Kenney (1972) during open cut excavation for the Oslo subway show that the effective brace stiffness was only $\frac{1}{50}$ the ideal elastic stiffness. Other studies (Scott et al., 1972; Jaworski, 1973) confirm that significant horizontal displacement

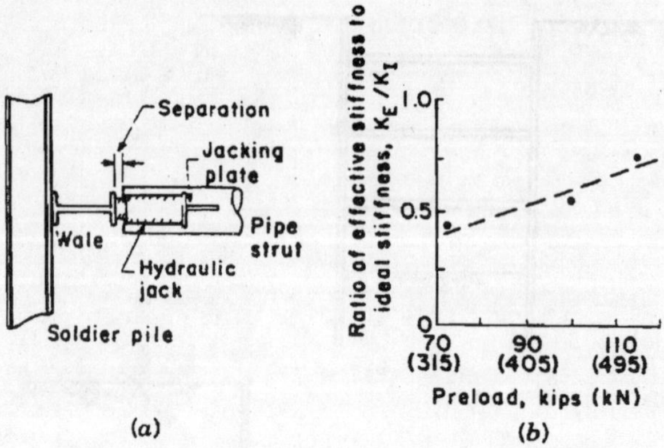

Figure 3-12 (*a*) Preloading arrangement. (*b*) Measured brace stiffness. (From O'Rourke, 1981.)

occurs when braces lack tight connections with the wall and where compressible materials are used to shim separations at connection points.

In practice, braces are preloaded to ensure rigid contact between interacting members. A typical preloading arrangement is presented in Figure 3-12, and shows the brace connection details and a plot of observed stiffness. Preloading is introduced by inserting a hydraulic jack at each side of an individual pipe strut between the wale beam and a special jacking plate welded to the strut as shown in Figure 3-12a (O'Rourke and Cording, 1979). Hydraulic pressure is applied until the strut is stressed to the selected load level (in this example one-half the design strut load), and the resulting separation is shimmed using $\frac{1}{8}$-in.-thick steel plates. The increase in separation between the strut and the wale is monitored and recorded during preloading, as is the lateral displacement of the soldier piles at this level and on opposite sides of the cut. The pile displacement is substracted from the increase in separation to give the apparent deformation of the strut.

The effective stiffness K_E of the strut is

$$K_E = \frac{P}{\Delta S} \tag{3-2}$$

where P = the average preload
ΔS = the average apparent deformation at each strut level

Conversely, the ideal elastic stiffness K_I of the strut is

$$K_I = \frac{EA}{L} \tag{3-3}$$

where E = the modulus of elasticity of the strut material
A = the cross-sectional area of the strut
L = the length of the strut.

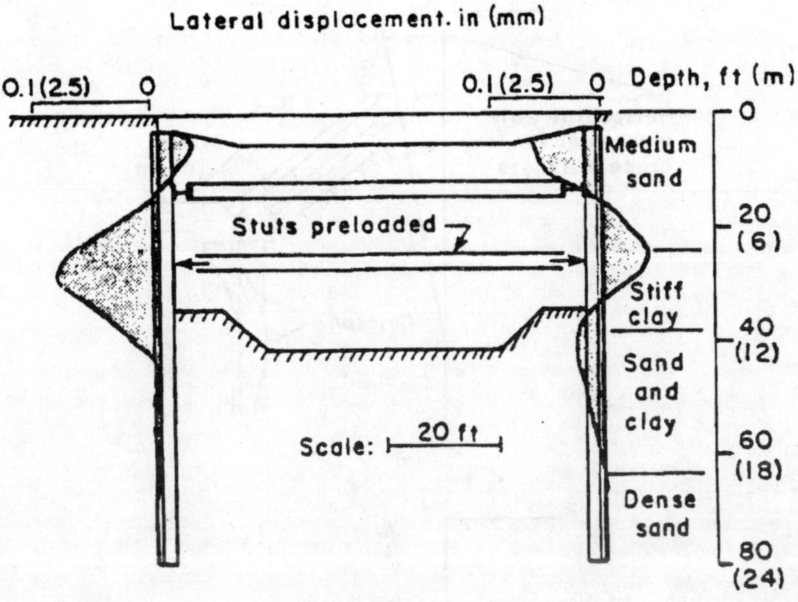

Figure 3-13 Wall displacement associated with the preloading of struts, cut-and-cover excavation. (From O'Rourke, 1981.)

In Figure 3-12*b* the ratio K_E/K_I is plotted versus the preload. The data are obtained from measurements at three separate strut levels from cut-and-cover construction for the Washington, D.C. metro (O'Rourke and Cording, 1979). The relatively high percentage of ideal stiffness attained at each strut level documents the effectiveness of preloading in developing rigid supports. The data show that strut stiffness increases with preload, but very high preload levels are not necessarily more advantageous. In practice, they may create locally stiffened sections of the supporting wall and induce relatively large pressure concentrations as excavation continues.

Figure 3-13 shows wall displacement at soldier pile profile (shaded area) caused by preloading a strut level. Wall movement toward the ground confirms that preloading is effective in closing separations within the bracing system. This movement, however, reaches a maximum of 0.1 in. at the level of the preloaded strut, and is distributed to within 10 ft (3 m) of the elevation of the jacking operation. It follows, therefore, that the preloading of struts in soldier pile walls will not compensate for or regain previous soil movement toward the excavation. The relatively small stiffness of the composite system confines this effect to the soldier piles; hence, preloading should be applied only to the extent necessary to produce a stiff bracing combination.

Effect of Rakers The use of preload in conjunction with rakers does not always provide control of movement. For example, a raker transmitting lateral loads to one

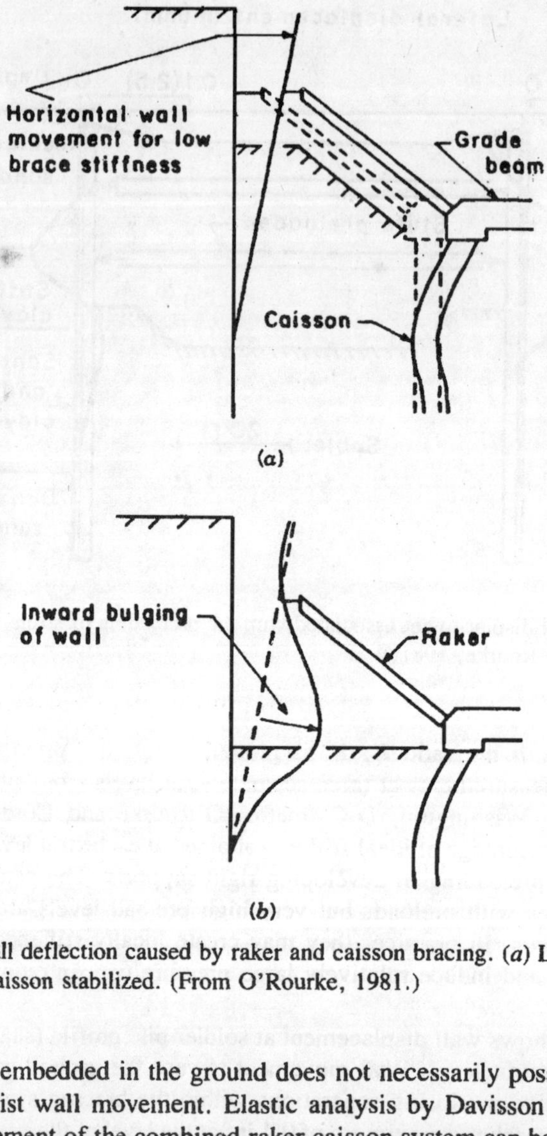

Figure 3-14 Wall deflection caused by raker and caisson bracing. (*a*) Lateral deformation of caisson. (*b*) Caisson stabilized. (From O'Rourke, 1981.)

or two caissons embedded in the ground does not necessarily possess the restraint necessary to resist wall movement. Elastic analysis by Davisson and Gill (1963) shows that movement of the combined raker-caisson system can be as high as 3 in. (7.6 cm) for braced cut loads and construction conditions typical in Chicago.

An idealized wall deformation associated with lateral loading and movement of caisson is shown in Figure 3-14*a*; it is apparent that at this stage the wall approaches cantilever behavior. As more construction is completed and the foundation slabs are in place, the caisson is further restrained and eventually stabilized, making the raker system more effective. Figure 3-14*b* shows wall movement with the excavation completed and the caisson under stable conditions. Although the volumes of incremental displacement in Figure 3-14*a* and *b* are essentially the same, the patterns of movement are markedly dissimilar.

Unsupported Slopes and Berms Examples of unsupported slope constructed as part of excavation programs are shown in Figure 3-15. In Figure 3-15*a* the excavation is made as a combination of cut slopes and braced cut. The unsupported slope begins at the top of the braced support and continues to ground line. Any failure occurring at this stage will prompt a large soil mass to fall into the excavation area. In Figure 3-15*b* the slope is made part of the bracing system, and any instability or failure will deprive the wall of lateral support.

In most large excavations berms are left in place either as part of a sequential construction or as designated lateral bracing. Usually, however, sloughing, seepage activity, creep, and construction operations lead to gradual distortion and eventual deterioration of berms.

Figure 3-16 shows wall behavior for an 82-ft-deep braced excavation in sand and interbedded stiff clay. The wall consists of I solder piles (W 24 × 130), spaced 7.5 ft apart, and timber lagging. The piles are 90 ft long. The struts were preloaded to one-half their design load. Wall displacements and excavation levels have their reference to the time of installation and preloading of the fourth strut level. The progressive wall movement shown in Figure 3-16a occurs as the excavation continues from the 58-ft level to the final 82-ft level (O'Rourke, 1981). The curves of the movement profile give the total incremental volume per unit length for the days indicated. Apparently, from day 0 to day 44 almost half the total movement occurred as ground response to excavation at the center of the cut. Between day 44 and day 147 a berm was left in place as shown, and no additional excavation was carried out during this period. Nonetheless, wall movement continued and reached almost 80 percent of the total movement. Excavation was resumed on day 147, changing the movement profile as shown.

Figure 3-16*b* shows the change in average load for the fourth strut level as a function of time since installation and preloading. The load increased, as the central portion was deepened with the berm in place, from the initial preload of 90 kips to 180 kips.

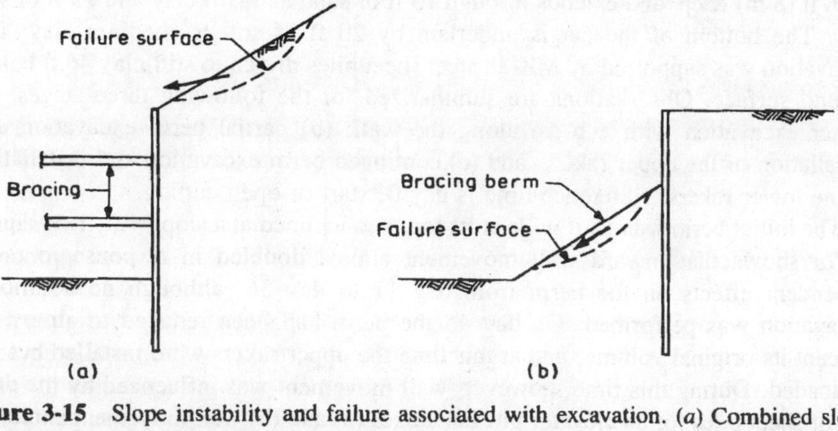

Figure 3-15 Slope instability and failure associated with excavation. (*a*) Combined slope and braced wall. (*b*) Braced wall supported by berm.

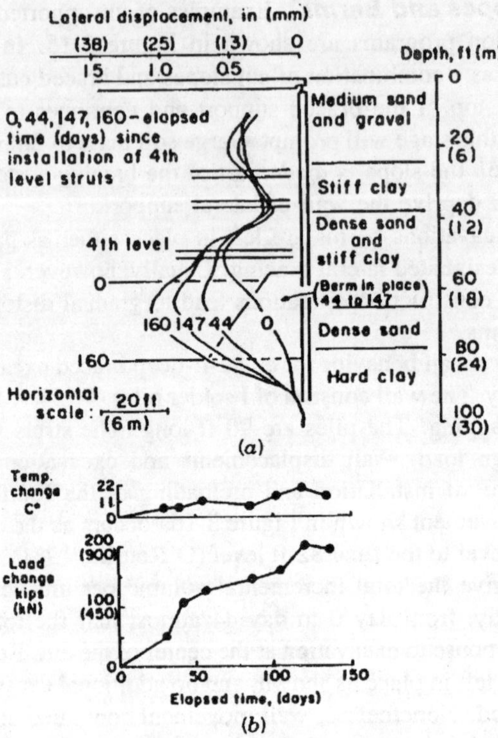

Figure 3-16 Observed lateral displacement and strut loads in conjunction with the use of berm. (*a*) Lateral displacement profiles. (*b*) Fourth level strut load. (From O'Rourke, 1981.)

Berms for excavation in soft to medium clay exhibit a higher dependence of time versus movement because of distinct creep effects. This is shown in the examples of Figure 3-17, giving time-displacement data from an excavation in Chicago. This cut is 26 ft (8 m) deep, and extends through 13 ft of sand and stiff clay and 13 ft of soft clay. The bottom of the cut is underlain by 20 ft of soft to medium clay. The excavation was supported by MZ-38 steel sheet piles driven to stiff clay 46 ft below ground surface. Observations are summarized for the following three stages: (a) center excavation with a b rm along the wall; (b) partial berm excavation and installation of the upper rake and (c) continued berm excavation and installation of the lower rakers. Reference time is day 0, start of open cutting.

The initial berm was 15 ft wide at its top, and inclined at a slope 4.0–1.5. Figure 3-17*a* shows that inward wall movement almost doubled in response to time-dependent effects on the berm from day 11 to day 36, although no additional excavation was performed. On day 45 the berm had been reduced to almost 70 percent its original volume, and at this time the upper rakers were installed but not preloaded. During this time, however, wall movement was influenced by the driving of sheet piles for an elevator pit. For stages (a) and (b) wall movement exhibited essentially a cantilever behavior with maximum displacement 5 in. As the berm was

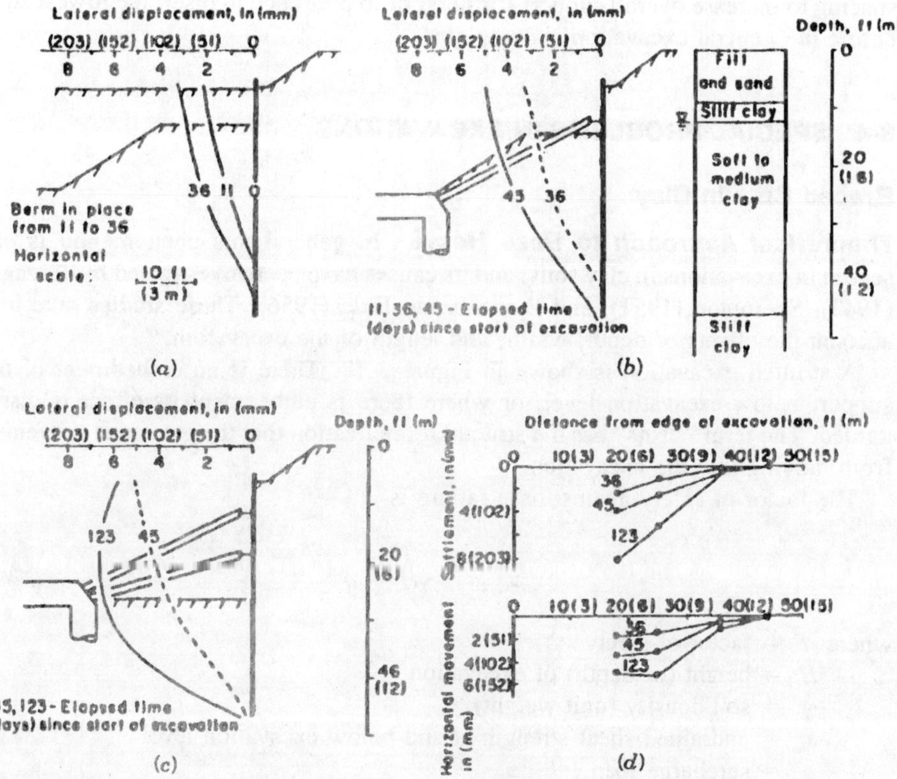

Figure 3-17 Observed displacement for raker and berm excavation in soft to medium clay. (*a*) Excavation for berm. (*b*) Installation of upper raker. (*c*) Excavation to subgrade. (*d*) Ground surface movements. (From O'Rourke, 1981.)

removed to install the bottom rakers, the wall deflected with an inward bulging, as shown in Figure 3-17*c*, and had a maximum displacement 7 in., occurring at the base of excavation. Settlement and horizontal displacement at street level adjacent to the cut are shown in Figure 3-17*d*.

The foregoing results articulate the sensitivity of excavation to raker and berm bracing in soft and medium clay. Guidelines for sizing berms in these conditions are proposed by Clough and Denby (1977), based on finite element analysis. The time dependence of berm deformations because of creep and gradual attrition from construction events contributes to displacement. A practical guideline is to choose construction methods that minimize the time elapsed from excavation at the center of the cut to the installation of stiff braces.

Excavation Below Lower Braces Limiting the excavation depth below the lowest bracing level is essential to controlling movement, particularly with flexible supports and deep cuts. The usual procedure is to reduce the vertical bracing spacing to increase overall support stiffness, or to pretrench to insert the lowest strut before the general excavation is completed.

3-4 SPECIAL PROBLEMS IN EXCAVATIONS

Braced Cuts in Clay

Theoretical Approach to Base Heave In general this phenomenon is observed in excavations in clay soils, and its causes have been investigated by Terzaghi (1943), Skempton (1951), and Bjerrum and Eide (1956). These studies take into account the effects of depth, width, and length of the excavation.

A strutted excavation is shown in Figure 3-18. There is no embedment of the support below excavation level, or where there is embedment its effect is disregarded. The term "struts" has the structural implication that the ground is prevented from moving into the excavation.

The factor of safety against base failure is

$$F = N_c \frac{s_u}{\gamma H + q} \tag{3-4}$$

where F = factor of safety
H = height (or depth) of excavation
γ = soil density (unit weight)
s_u = undrained shear strength at and below excavation level
q = surcharge load
N_c = stability number

Values of N_c are estimated from the diagram of Figure 3-18 using $D = H$, and then are entered in Eq. (3-4). The results are not completely accurate, although satisfactory for practical solutions and where the cut is temporary. Theoretically, base heave is small and inconsequential if $(\gamma H + q)/s_u < 6$. If the excavation support is anchored, the foregoing analysis is not entirely valid, and large movements have been documented in this case if the factor $(\gamma DH + q)/s_u$ is close to 4 (Stille, 1976).

Observed Field Behavior Predictive techniques on base heave patterns based on field performance are suggested by Mana and Clough (1981) for braced sheet pile or soldier pile walls. In these instances the first strut level should be inserted before the excavation depth exceeds $2s_u/\gamma$ to avoid unnecessarily large initial movement. Unlike the stability number approach, which ignores ground conditions below excavations, the potential for base heave is defined in terms of its own factor of safety considering also the presence and the stabilizing influence of underlying strong layers below the base of excavation.

Case histories for which data have been obtained are summarized in Table 3-4. This table also shows basic soil characteristics, minimum factors of safety against base heave, and lateral wall movement and surface settlement. The factors of safety are computed using the Terzaghi (1943) approach illustrated in Figure 3-19. In all

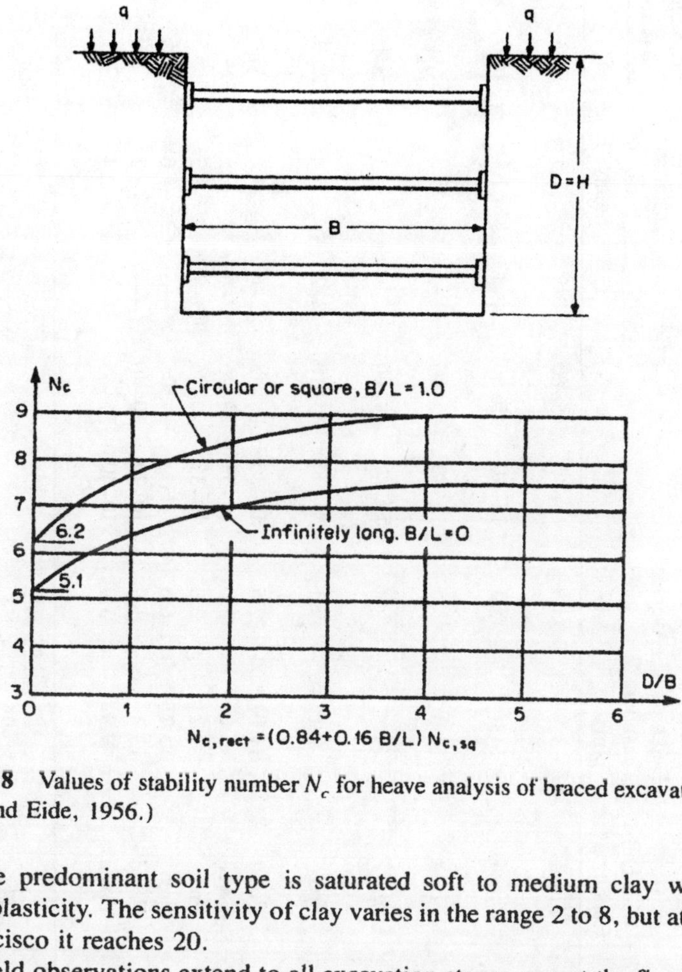

Figure 3-18 Values of stability number N_c for heave analysis of braced excavations. (From Bjerrum and Eide, 1956.)

cases, the predominant soil type is saturated soft to medium clay with low to medium plasticity. The sensitivity of clay varies in the range 2 to 8, but at one site in San Francisco it reaches 20.

The field observations extend to all excavation stages except the first (cantilever stage), and the movement at each stage is correlated to the appropriate base heave factor of safety. Where time-dependent movement occurs at any excavation depth, the final observed value is used. With the foregoing criteria, the ratio of maximum observed lateral movement to the corresponding excavation depth is plotted versus the factor of safety against base heave in Figure 3-20. Ignoring scattered results, the plots reveal a specific pattern that shows a tendency for movement to increase rapidly if the factor of safety is below 1.4 to 1.5. At higher values, the movement ratio is essentially constant and close to 0.5%.

Interestingly, the examples analyzed in Figure 3-20 include free-end and fixed-end bottom conditions, and apparently these conditions have little effect on wall behavior. Thus, lateral movement changes very little between walls with only just a tip below excavation level and sheet pile walls with significant embedment below this level. Maximum lateral movement is manifested at or just below the base and irrespective of wall penetration.

TABLE 3-4 Data from Case Histories; Base Heave and Movement for Excavations in Clay

Case Number	Location	Wall Type	Wall End Condition[b]	Final Depth of Excavation (m)	Clay Properties: Average s_u, tons/m²	w_p (%)	w (%)	s_t	Minimum Factor of Safety, FS	Movement (cm) δ_h	δ_v
1	San Francisco, Calif.	Sheet pile	Fixed	13.5	3.5	15–40	45–60	4–8	1.3	19.3	15.2
2	San Francisco, Calif.	Sheet pile	Fixed	14.0	3.5	15–40	45–60	4–8	1.3	15.0	14.4
3[a]	San Francisco, Calif.	Sheet pile	Fixed	14.0	3.5	15–40	45–60	4–8	1.3	10.2	—
4	San Francisco, Calif.	Sheet pile	Free	9.1	2.4	35–60	55–90	4–8	1.3	3.8	—
5	San Francisco, Calif.	Sheet pile	Free	9.1	2.1	35–60	75–100	8–20	1.0	25.4	—
6	Oslo, Norway	Sheet pile	Fixed	11.0	2.5	15–30	20–45	2–6	1.0	23.5	23.3
7	Oslo, Norway	Sheet pile	Fixed	11.0	3.0	10–35	20–45	2–6	1.1	14.2	14.0
8	Oslo, Norway	Sheet pile	Free	9.2	3.0	10–35	20–45	3–7	1.3	18.5	11.6
9	Boston, Mass.	Sheet pile	Free	15.2	6.6	11	30	—	1.6	11.5	—
10	Chicago, Ill.	Soldier pile	Fixed	13.4	1.9	10–20	20–40	—	1.2	8.9	8.0
11	Bowline Point, N.Y.	Sheet pile	Fixed	9.8	3.9	10–40	35–65	4–8	2.4	5.1	—

Source: From Mana and Clough, 1981.

[a]Local unusual construction effects. Affected data not used.

[b]Fixed end refers to wall tip embedded in underlying stiff layer; free end, tip in soft to medium clay.

Note: s_u = undrained shear strength; w_p = plasticity index; w = water content; s_t = sensitivity.

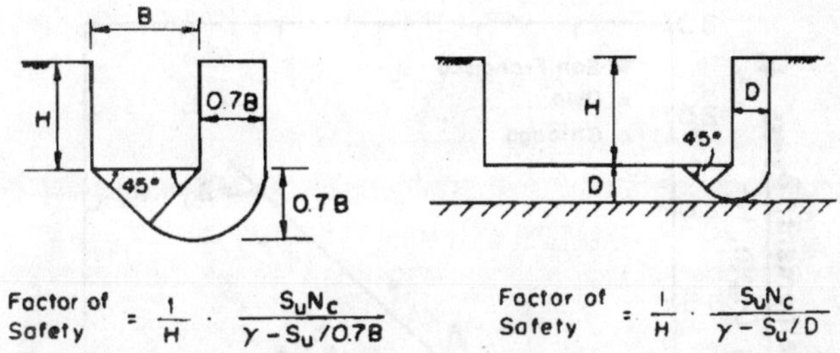

Figure 3-19 Method of base heave analysis. (From Terzaghi, 1943.)

The dependence of surface settlement on lateral movement can be established in general terms, and is likewise related to the factor of safety. The ratio of maximum settlement to excavation depth is plotted versus lateral movement in Figure 3-21. Although the results are not entirely accurate, they confirm that settlement is 50 to 100 percent the lateral movement. The analysis does not consider effects from unusual or special construction procedures, and does not articulate the sensitivity of the support system to such effects.

Effects of Anisotropy Anisotropy, as a difference in response to loading in the vertical and horizontal direction, is manifested in most natural clays, and its effect on the undrained behavior of soft to medium clays is now recognized in most analyses. In braced excavations the significance of anisotropy reflects physical differences in soil behavior; behind the wall (active side) the soil is acted upon by vertical gravity, but the soil on the passive side (inside the excavation) is loaded horizontally by the thrust of the wall.

Beginning with the Terzaghi base heave theory, Clough and Hansen (1981) have extended the analysis to include anisotropic behavior. For these effects to be consid-

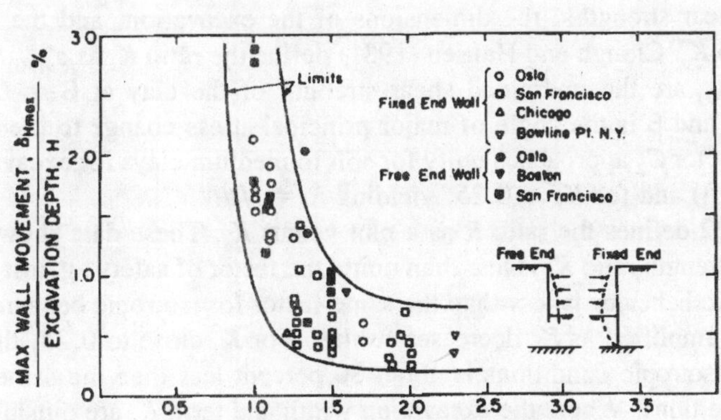

Figure 3-20 Relation between factor of safety against base heave and percentage of lateral movement. (From Mana and Clough, 1981.)

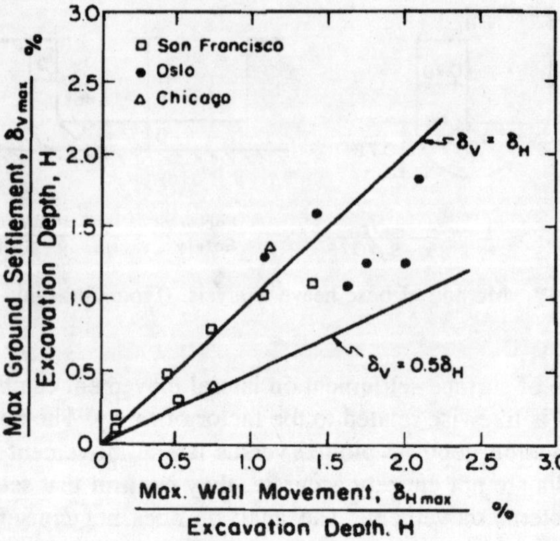

Figure 3-21 Relation between maximum observed settlement and maximum wall movement. (From Mana and Clough, 1981.)

ered, it is essential to introduce a strength variation along potential failure planes so that stress reorientation is applicable. Furthermore, it is useful to consider the ratio of the factors of safety for the anisotropic and the isotropic stage. This ratio, R, is given as

$$R = C_1 C_2 \tag{3-5}$$

where $C_1 = N_c^*/N_c$ (ratio of the anisotropic bearing capacity factor to the bearing capacity factor normally used for isotropic soils), and C_2 is given in terms of the undrained shear strengths, the dimensions of the excavation, and the anisotropic strength ratio K_s. Clough and Hansen (1981) define the ratio K_s as $s_{u(90)}/s_{u(0)}$ where $s_{u(0)}$ and $s_{u(90)}$ are the undrained shear strength of the clay at $\beta = 0°$ and $90°$, respectively, and β is the angle of major principal stress change to the vertical. In reality, the factor C_2 approaches unity for soft to medium clays for excavation width > 15 m (50 ft) and for $K_s \geq 0.25$, yielding $R = N_c^*/N_c$.

Figure 3-22 defines the ratio R as a plot versus K_s. These data show that if the anisotropic strength ratio K_s is less than unity, the factor of safety against base heave for anisotropic behavior is less than the same factor for isotropic behavior, and this difference is amplified as K_s decreases further. For K_s close to 0.25, the factor of safety for anisotropic conditions is about 30 percent less than the same factor for isotropic conditions. Where the excavation width and term K_s are outside the range $C_2 = 1$, the influence of anisotropy is further accentuated. These results suggest that base heave factors determined from conventional analysis overestimate the factor of safety for the anisotropic form, and the error becomes more critical as the degree of

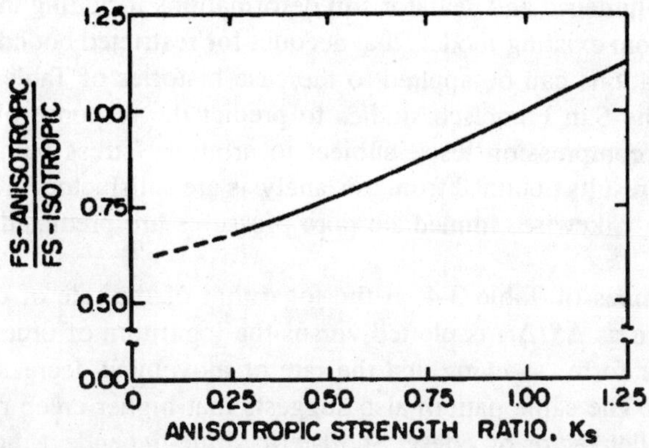

Figure 3-22 Ratio of factor of safety (FS) against base heave for anisotropic soil to that for isotropic soil for wide excavation. (From Mana and Clough, 1981.)

anisotropy increases. In more practical terms, the decrease in the factor of safety will be within 10 percent its isotropic value, unless the passive strength of the clay is less than 80 percent its active strength.

Likewise, movement can be doubled, earth pressure distribution altered, and strut loads markedly increased as soil behavior shifts from isotropic to anisotropic. The potentially detrimental effects that may result from these conditions appear most strongly if the factor of safety against base heave is reduced below 1.4. These phenomena are better understood by initially testing the compression (active) and extension (passive)modes to determine the degree of anisotropy. This information can be used in an isotropic bearing failure analysis as suggested by Clough and Hansen (1981). Alternatively, results reported in literature can be combined with data from a specific project to establish a conservative value of undrained shear strength that balances anisotropic extremes. This strength is then introduced in an isotropic heave analysis to predict the factor of safety against base heave.

Time Effects Several walls of Table 3-4 were monitored for certain periods of time while the excavation depth did not change. These projects were in the San Francisco Bay.

In this area, the assumption is often made that time-dependent effects in excavations are caused by undrained creep in the bay mud. However, a general constitutive model that can adequately predict the time-dependent deformation of cohesive soils under arbitrary three-dimensional stress is not as yet available. Such a model is sought, for example, in studies of the stand-up time of tunnels excavated in squeezing ground, but could also be useful in a variety of ground engineering problems (Kavazanjian and Mitchell, 1980). Considerable information on time-dependent behavior is available for certain cases, such as one-dimensional compression, plane strain, and undrained triaxial creep.

The general theory is formulated on the concept of pseudolinear elasticity. Representations for volumetric and deviator soil deformations including the effect of time are developed from existing models that account for restricted boundary conditions. These representations can be applied to the case histories of Table 3-4, and more particularly to the San Francisco studies to predict the response of remolded bay mud in triaxial compression tests, subject to arbitrary stress paths and drainage conditions. The results obtained from this analysis are satisfactory (Kavazanjian and Mitchell, 1980). Likewise, immediate pore pressures are predicted with sufficient accuracy.

For the examples of Table 3-4, if the logarithm of the rate of observed lateral movement (taken as $\Delta S/\Delta t$) is plotted versus the logarithm of time t, the resulting path has a linear form, showing that the rate of movement decreases rapidly with increasing time. The same pattern also suggests that higher creep rates are associated with a smaller factor of safety. Short-term, time-dependent movement should be expected for braced excavations in clay, particularly with reduced safety factor.

Pore Pressure Dissipation During Excavation in Clay

In Section 3-2, the undrained condition assumed for excavations in clay reflects an end-of-construction stage. The analysis, also known as total stress approach, does not articulate actual shear strength and deformation parameters. For example, the presence of a drainage layer in the clay matrix can enhance consolidation during excavation, and in this case the fallacy of idealized assumptions is obvious (Lambe and Turner, 1970; DiBiagio and Roti, 1972).

The occurrence of consolidation around an excavation is a complex phenomenon, and its quantification a difficult problem. As earth removal begins and continues, the drainage boundaries are likely to change, and this will allow consolidation to occur at increasing rates. Conversely, the presence of impervious barriers in the path of flow impedes the process and complicates consolidation further.

These factors can be analyzed within the matrix of finite element solutions. Osaimi and Clough (1979) have extended the scope of this problem to include sequential excavations carried out in nonlinear soil mediums supported by artificial wall systems. This program considers deformation and pore pressures, and gives predictions for total, effective, and pore water pressures for each construction stage as well as for periods following construction. In this context, consolidation patterns are studied in a broad range of excavation problems.

An example of supported excavation in linear elastic low-permeability soil ($k = 1 \times 10^{-8}$ cm/sec) is given by Osaimi and Clough (1979). The support is a 2-ft thick concrete diaphragm wall penetrating the full depth of the clay and assumed to be hinged at the level of a rigid base. As expected, the deformations in the soil mass were reduced with the wall in place. Along the face of the slope movement was about two-thirds of that without the wall. However, negative pore pressures were larger when the wall was present since this system is impervious and inhibits lateral drainage. Pore pressures in the soil mass directly below the excavation in front of the wall were basically similar to those that would be expected for simple one-dimensional vertical drainage. Since consolidation is retarded by the presence of an impervious wall, undrained loading conditions can reasonably be assumed in the analysis.

Two-Dimensional Excavation with Nonlinear Soil Behavior Figure 3-23 shows the excavation profile and finite element mesh for a cut in homogeneous saturated soil deposit underlain by an impervious hard clay layer. The soil is assumed to have a basic elastic modulus E = 100,000 lb/in.^{2} (4800 kN/m^2), and a density γ = 120 lb/ft^3. The excavation rate is taken as 0.5 ft/day (0.15 m/day), implying that about 60 days are required to complete the cut.

For nonlinear soil behavior, the clay properties were modified to represent a nonlinear elastic material according to the Duncan–Chang (1970) modeling criteria. Drained parameters are as follows: cohesion $c' = 0$; modulus number $K_t = 100$ or 300; friction angle $\phi = 30°$; modulus exponent $n = 0.5$; Poisson's ratio $\mu' = 0.3$; failure ratio $R_f = 0.85$; and coefficient of lateral earth pressure $K_0 = 0.8$. Soil permeability is considered in the range of 2.8×10^{-5} ft/day.

Since the soil has no drained cohesion and is internally unstable in a vertical excavation, the cut is supported by a fully penetrating 2-ft-thick wall likewise

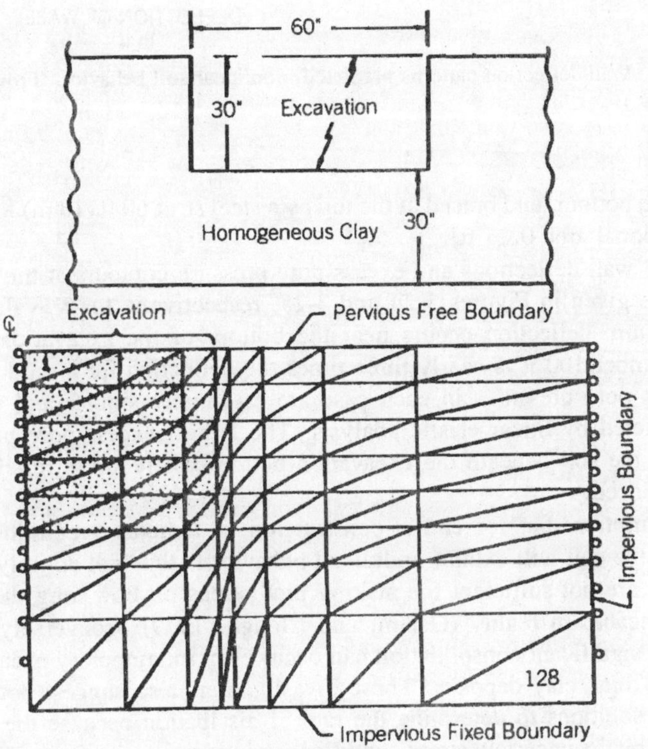

Figure 3-23 Finite element mesh and geometry of problem used in two-dimensional analyses. (From Osaimi and Clough, 1979.)

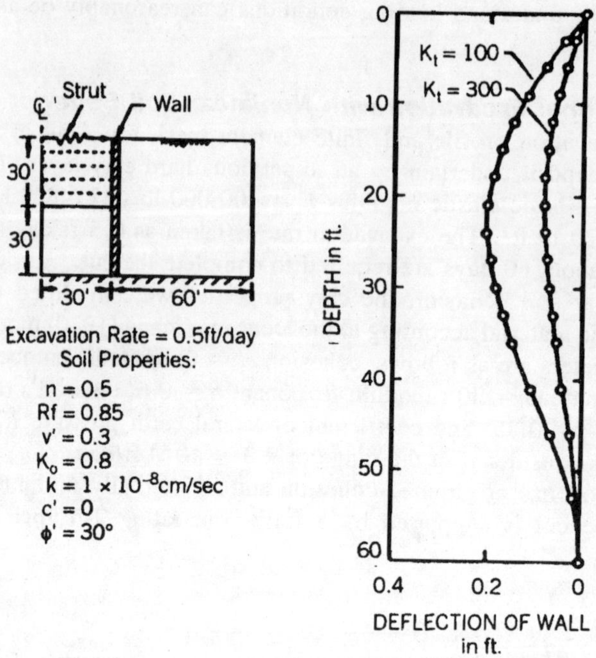

Figure 3-24 Wall deflection patterns predicted, nonlinear soil behavior. (From Osaimi and Clough, 1979.)

hinged at the bottom, and braced at the top by a steel strut 60 ft (18 m) long and with a cross-sectional area 0.25 ft².

Predicted wall deflections and excess pore pressure contours at the end of construction are given in Figures 3-24 and 3-25, respectively, for K_t values 100 and 300. Maximum deflection occurs near the bottom of the excavation, and for a modulus number 100 it is nearly three times the deflection for a modulus number 300. Excess pore pressures in each case are essentially similar and close to the values predicted by linear elastic analysis. The largest excess pore pressures are observed in the soil beneath the excavation bottom where stress relief effects are most pronounced.

The assumption that for end-of-construction or temporary conditions in clay excavations the soil will exhibit undrained behavior is thus not entirely valid. The present data are not sufficient for making predictions on how long the undrained case is applicable in reality (Osaimi and Clough, 1979). Conversely, field data indicate that significant consolidation can occur even in temporary excavations and in relatively thick clay deposits. These investigations also suggest not relying on closed-form solutions to determine the rate of dissipation because the problem is complicated by the uncertain stress conditions around an excavation and the moving drainage boundaries during earth removal.

In more specific terms, for the example of Figure 3-23 consolidation occurred at

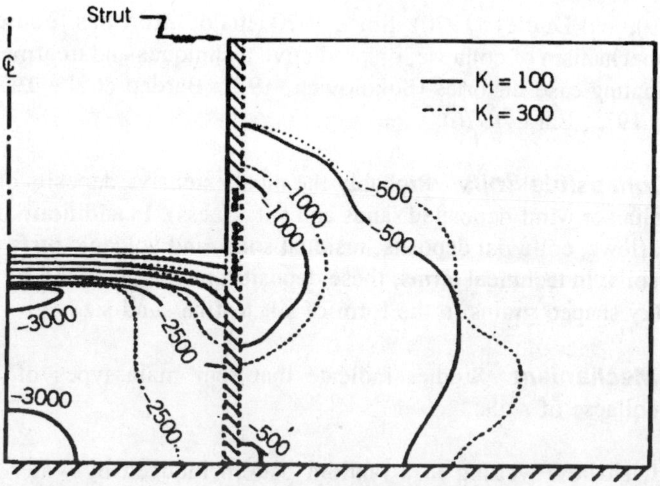

Figure 3-25 Contours of negative excess pore pressure, in pounds per square foot, at end of excavation, nonlinear soil behavior. (From Osaimi and Clough, 1979.)

the end of construction, although it was relatively small and hardly affected the undrained behavior assumption. Considerable consolidation should be expected in practical cases where field permeabilities are higher, deposit thickness smaller, construction periods longer, and the support is less impervious than a diaphragm wall. As a practical matter, nearly 2000 days are required for a 30-ft excavation in low-permeability clay to attain 50 percent consolidation, but more rapid dissipation rates may be expected in many practical situations. Consideration of nonlinear soil behavior for excavations supported by diaphragm walls is not indicated since it is likely to have only a small effect on pore pressure dissipation compared to the results of linear elastic soil response. Conversely, deformations are better predicted by nonlinear analysis since the choice of a single elastic modulus to characterize stress-strain relations is difficult and at best uncertain.

3-5 DESIGN OPTIONS IN COLLAPSIBLE SOILS

By definition metastable or collapsible soils are unsaturated soils undergoing a radical particle rearrangement and considerable volume loss upon wetting with or without additional loading (Clemence and Finbarr, 1981). Difficulties associated with the use of these soils as foundation support have long been recognized, but until recently concern was limited because such soil deposits were located mainly in arid regions with modest economic development potential. With recent advances in irrigation these regions have been made available for industrial development and associated construction, and collapsible soils are becoming ideal candidates for control measures and ground improvement.

Comprehensive reviews on the state of the art are given by Northen (1969),

Sultan (1969), and Dudley (1970). Since 1970, major effort has focused on determining the mechanism of collapse, on predictive techniques and treatment methods, and on evaluating case histories (Sokolovich, 1971; Barden et al., 1973; Jennings and Knight, 1975; Bara, 1976).

Types of Collapsible Soils Probably the most extensive deposits of collapsible soils are aeolian or wind-deposited sands and silts (loess). In addition, alluvial flood plains, mud flows, colluvial deposits, residual soils, and volcanic turfs can produce collapsible soils. In technical terms, these deposits are characterized by loose structures of bulky shaped grains in the form of silt to fine sand size.

Collapse Mechanism Studies indicate that four main types of wetting can trigger the collapse of soils:

1. Local, shallow wetting in a random pattern induced by water sources can cause settlement in the upper layer of soil below the wetted zone.
2. Intense, deep local wetting of soil can occur from continuous discharge of industrial effluents or irrigation, and may result in a rise in the groundwater table. In this case, the entire zone of collapsible soil may become saturated, leading to settlement that is extremely uneven and dangerous.
3. Slow and relatively uniform rise of the groundwater level may occur following gradual but steady flow from water sources outside the collapsible soil area. The associated settlement in this case is slow and uniform.
4. Gradual increase in moisture content can result from condensation of steam and accumulation of moisture because of changes in evaporation conditions. The associated effect is a partial weakening of the internal cohesion of the soil, and as a result the settlement is incomplete and slow. Collapse, however, may be triggered by water alone, or by saturation and loading acting together.

According to Barden et al. (1973), appreciable collapse of a soil requires three conditions: (a) an open, potentially unstable, partly saturated structure; (b) a stress component application of an intensity high enough to develop a metastable condition; and (c) a strong initial bonding or cementing action to stabilize intergranular contacts, which is diminished upon wetting and causes collapse.

Design and Treatment Methods In many instances, deep foundations (piles, caissons, slurry wall panels) are required and used to transmit the loads to a suitable bearing level below the collapsible soil deposit. Where loads can be supported on shallow foundations on or above the collapsible soil, continuous strip footings are safer and more economical than individual footings. Results from laboratory or field tests can be used to predict the settlement.

Pretreatment techniques reviewed in this book are often indicated and feasible to apply to collapsible soil deposits. They may be used to either stabilize the unstable deposit or cause its collapse and subsequent readjustment prior to construction. The

TABLE 3-5 Treatment Methods for Collapsible Foundation Soils

Depth of Subsoil Treatment Desired (m)	Foundation Treatment Method
	Current and past methods:
0–1.5	Moistening and compaction (conventional extra-heavy, impact, or vibratory rollers)
1.5–10	Overexcavation and recompaction (earth pads with or without stabilization by additives such as lime or cement); vibroflotation (free-draining soils); rock columns (vibroreplacement); displacement piles; injection of silt or lime; or ponding or flooding (if no impervious layers exist)
Over 10	Any of the aforementioned, or combinations of the aforementioned, where applicable; ponding and infiltration wells; or ponding and infiltration wells with the use of explosives.
	Possible future methods:
	Heat treatment to solidify the soils in place; ultrasonics to produce vibrations that will destroy the bonding mechanisms of the metastable soil; chemical additives to strengthen the bonding mechanism of the metstable soil structure (possibly electrochemical methods of application); or use of groutlike additives to fill the pore spaces before solidification

Source: From Bara, 1976.

method and extent of treatment will depend on the depth of the collapsible stratum and on the support requirements of the new construction.

Table 3-5 gives a summary of treatment methods (Bara, 1976). In the context of treatment and improvement technologies reviewed in this book, a separate assessment is necessary before selecting heat treatment and ultrasonics. Extensive studies have been carried out in Russia with chemical stabilization techniques (Sokolovich, 1971), and include: (a) gaseous silicatization of sandy and loessial soils; (b) strengthening of carbonate cements by polymers; and (c) chemical strengthening of alluvial soils by clay-silicate solutions.

Pretreatment is more feasible if collapsible soils are detected and identified during initial soil investigations.

3-6 EXCAVATION SUPPORTED BY TEMPORARY WALLS

Figure 3-26 shows section through building for the Beaubourg Cultural Center in Paris (Corbett and Stroud, 1974). This construction required an excavation 122 × 154 m (400 × 505 ft), and 16.5 m (54 ft) below street level, deepened in certain sections to 20.5 m (67 ft). The aboveground structure forming the main center consists of steel framing. The loads are resisted by twin columns 6 m apart grouped

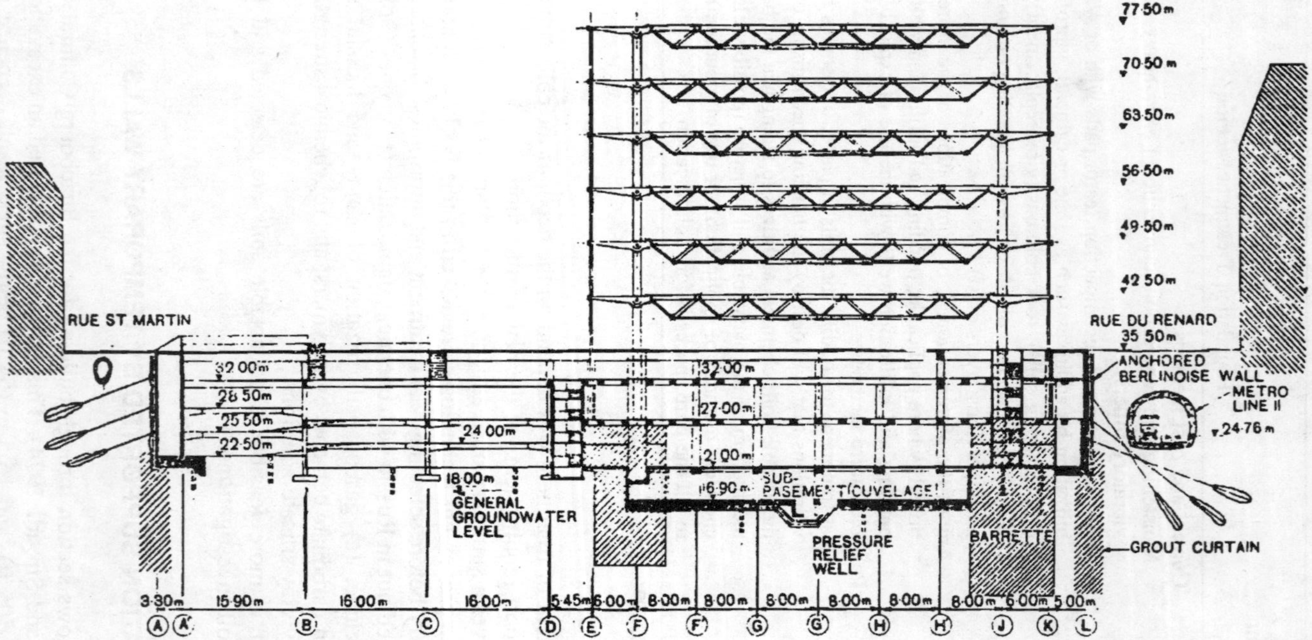

Figure 3-26 Section through building; Beaubourg Cultural Center, Paris. (From Corbett and Stroud, 1974.)

at 12.8-m centers along the long side of the building. The column loads are transferred to reinforced concrete walls at a +27-m level, as shown in Figure 3-26.

A temporary retaining wall was built along Rue du Renard, shown in the right side of the building cross section. This project was complicated by the presence of metro line 11 running parallel to the wall as shown.

Ground Conditions The geologic sequence at the site consists mainly of alluvial deposits of the Seine River overlying soils and rocks of Eocene age. It is difficult to differentiate between old and new alluvia since they both occur as dense gravelly sands. Soil properties were established from site investigations, and relevant design parameters are shown in Table 3-6.

Older buildings in the immediate vicinity had one or two basement levels, and many collapsed during demolition and were backfilled with masonry rubble. The normal groundwater level is approximately 10 m (33 ft) below ground level, and somewhat lower than the Seine River. Under flood conditions, the water table may rise by almost 14 m (45 ft).

Support Requirements

Although the construction was feasible as open excavation with free slopes along a large section of its perimeter, this option was abandoned for several reasons. First, the usable space at the bottom of the excavation would be reduced significantly, imposing particular restrictions along the right side of the construction by reference to Figure 3-26. Secondly, free slopes would affect the positioning and installation of the barrette piles (special linear foundation elements built by the slurry trench method). Because of requirements relating to accurate positioning and workmanship of installation, the barrettes were installed from the bottom level of the excavation, and space had to be available for cranes, grabs, and ancillary equipment. Finally,

TABLE 3-6 Soil Profile and Properties; Beaubourg Center, Paris

		Properties in Terms of Effective Stress	
Type of Soil or Rock	Thickness (m)	Cohesion c' (kPa)	Angle of Internal Friction ϕ (degrees)
Remblais (fill, old foundations)	6	0	35
Alluvions modernes et anciennes (new and old alluvium)	6	0	35
Marnes et Caillasses (marls and stony deposits)	5–14	50	28
Calcaire Grossier (coarse limestone)	15	—	—

Source: From Corbett and Stroud (1974).

the form of superstructure was critical to the overall structural scheme and impacted on all construction phases preceding the erection of the superstructure. An appropriate ground support in this case was a vertical temporary retaining wall extending over the entire excavation depth and free of interior obstructions. The presence of metro line 11 imposed additional stability requirements.

Support Systems

Diaphragm walls were considered and eventually rejected because the depth of the guide walls through old basements and debris would require extensive preparatory work and preliminary excavations within a strutted trench (Corbett and Stroud, 1974).

A second system under consideration was the so-called "puits blinde," very common in France, which is a modified version of the soldier pile wall. A series of shafts are drilled and concreted. The soil between the shafts is excavated in a second series of pits and concreted in sections to produce a continuous wall, usually braced with ground anchors. This method was rejected as being labor intensive and too slow for this project.

The temporary support system finally selected was a soldier pile and lagging wall. The steel piles were inserted in drilled holes concreted to the basement level. However, unlike the more conventional soldier beam walls where a horizontal wale beam is used at each bracing level, the piles were anchored individually and without lateral continuity. In this manner, the temporary wall provided a back shutter for placing the concrete of the permanent wall.

Three levels of multistrand prestressed anchors were installed and steeply inclined to avoid the metro tunnel. Design loads and anchor inclination are shown in Table 3-7. Design criteria included the assumptions that the wall moves sufficiently to mobilize total active pressure and 50 percent of the total passive pressure, and that there is no friction at the wall-soil interface. Under these conditions vertical loads induced by anchor inclination had to be resisted by the foundation element at each pile. The arch of the metro tunnel was considered rigid and the thrust at the springing was estimated as 390 kN/m. For design purposes, the lateral pressure distribution was assumed uniform, as shown in Figure 3-27.

TABLE 3-7 Design Loads and Anchor Inclination; Temporary Wall for Beaubourg Center, Paris

Level	NGF (m)	Force (kN)	Inclination to Horizontal (degrees)
Top	31.50	883	55
2nd	28.00	1180	35
3rd	25.00	1180	20

Source: From Corbett and Stroud (1974).

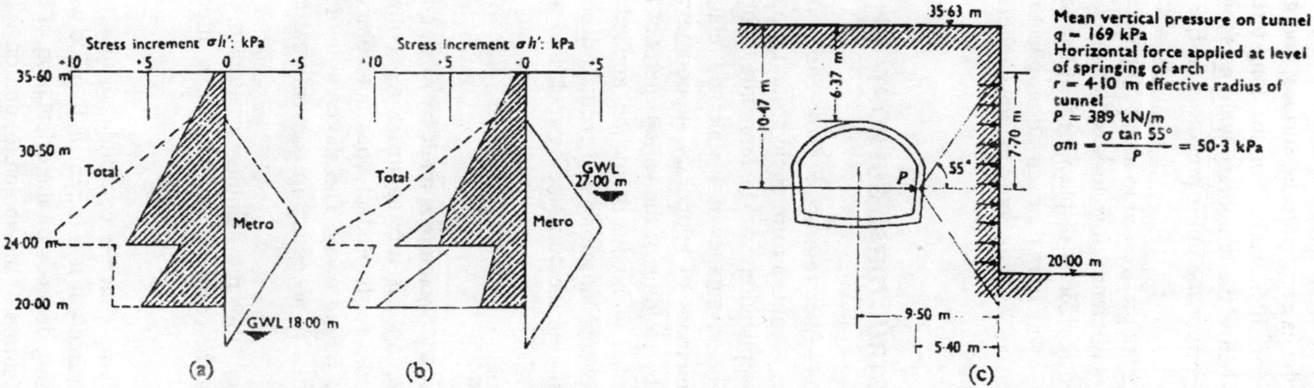

Figure 3-27 Lateral pressure distribution from metro tunnel, excavation for Beaubourg Center, Paris. (From Corbett and Stroud, 1974.)

During construction the groundwater level generally remained below excavation level. In the final stags, however, the excavation was taken down to the +15.5 m level of 2.5 m (8 ft) below the existing phreatic surface, and groundwater lowering was achieved by a system of filter wells. For the long-term condition, a grout curtain cutoff was installed below the temporary wall, as shown in Figure 3-26, to inhibit base instability and relieve the uplift pressures on the base slab.

Monitoring With all anchors prestressed to about 50 percent of the design load, the soil deformability was sufficiently low to keep lateral wall movement within acceptable limits. Measured horizontal displacement at the top of the wall ranged from −6 to +20 mm (minus away from the excavation, plus toward the excavation), with 6 mm average. Base heave was attributed to stress relief following excavation and ranged from 3 to 4 mm.

3-7 SHOTCRETE AS STRUCTURAL SUPPORT

Conventional shotcrete, steel-fiber reinforced shotcrete, and shotcrete with rock bolts or ground anchors, has become a formidable ground support system during the last few decades. Viable applications are in tunneling, where shotcrete is often combined with grouting. The advantages of this method relate to ground response; whereas more traditional methods of temporary support invariably tend to cause some loosening and voids by yielding of the various parts of the support system, a thin layer of shotcrete together with bolts or anchors applied to a rock face immediately after excavation prevents loosening and reduces decompression to a large degree. This process transforms the surrounding rock into a self-supported arch.

Conventional Shotcrete

Structural Characteristics Shotcrete is made by mixing sand, gravel, cement, and water, and then projecting this mix against a receiving surface using compressed air. A portion of the fresh product does not adhere to the surface, but rebounds and is wasted by falling away. The shotcrete hardens in place and thus eliminates the formwork that is necessary in conventional cast-in-place concrete (Mahar, 1975).

As a construction material, shotcrete exhibits variations in its characteristics, and its quality, as measured by the compressive strength, is affected by the following factors:

1. Shotcrete is a low-slump concrete consolidated upon impact. It must, however, set and harden rapidly if it is to remain in place without sloughing. The actual rate of hardening depends on the orientation of the receiving surface, the water content or slump of the shotcrete, and on water flowing along the placement area.

2. Special admixtures are often used to accelerate the setting and hardening process of shotcrete for overhead placement, in running water conditions, and where temperatures are low to moderate. These agents produce the expected benefits but also result in a loss of strength of the set shotcrete, usually 20 to 30 percent. Organic accelerators are claimed to reduce setting time but without adverse effects on strength.
3. Where shotcrete is sprayed directly overhead, a higher accelerator content is required compared to spraying vertical surfaces. The extra benefit reflects better adhesion and less sloughing.
4. The quality of shotcrete depends on the quality of constituent materials but is also influenced by each operation for preparation and placement. Major variations can result from the moisture content present in the aggregates, the attainable degree of mixing, water quantity, material velocity and impact compaction, and angle of application.

Quality Control Requirement The problem of sampling shotcrete articulates the method of placement. Sampling is most effectively done by removing cores from hardened panels or from in situ lining (Mason and Lorig, 1981). This sampling is convenient and representative, but results are difficult to correlate with other concrete sampling methods, mainly because of the difficulty of interpreting test data or simply because of poor sampling techniques. For example, shotcrete placed overhead typically attains lower strength than the same material placed on vertical surfaces (Mason, 1970).

Quality control specifications are based on the statistical variation of data interpretation, preconstruction testing, and testing during construction. They also require closely specified compatibility setting times, and correlation of core sample strength with the design strength.

Tests from the Atlanta Research Chamber Useful data have been compiled from the Atlanta Research Chamber (Mason and Lorig, 1981), and are directly applicable to tunnels and underground openings. This procedure recognizes that structural requirements specify in situ shotcrete that develops a strength equivalent to that of standard test cylinders, but that in reality test values of shotcrete samples are quite different.

The design strength f'_c is used as a basis, with the intent to utilize a predetermined fraction to resist an assumed loading condition. For cast-in-place concrete, the in situ concrete meets or exceeds the design strength if samples cast and cured separately (for example standard cylinders) comply with the strength requirements, also taking into account the variability of concrete production. However, shotcrete samples such as typical drilled cores have been shown to exhibit lower strengths than cast cylinders because of internal damage in the process of coring and inherent sample size differences.

Shotcrete in tunnels is used for initial support and final lining, or as initial support with a final lining placed inside the shotcrete layer. In other cases, shotcrete is used as a protective semistructural final lining, and examples are water, utility, and hydroelectric diversion tunnels where high compressive strength is not as essential. The variety in shotcrete uses and applications imposes a corresponding variation in quality assurance requirements and rational performance.

Mason and Lorig (1981) give the following summary from the Atlanta Research Chamber tests:

1. Cement-accelerator compatibility tests are not conclusive. Given the confirmed poor reproducibility of compatibility testing, the practice of specifying field dosages to comply with laboratory tests is unfounded. Results are not necessarily reproducible in the field, and this makes it difficult for the accelerator cement to fit between specified limits, even if the accelerator dosage rate is varied. For compatibility testing, the water-cement ratio appears to be a critical factor (Parker et al., 1975; Mahar et al., 1976).
2. The mix used for the tests complied with CN 120 gradation specifications except for the maximum particle size requirement. In addition, good quality shotcrete was obtained by using an aggregate mix that exhibited a gradation wall outside the specified limits. Shotcrete quality, as measured by compressive strength, was excellent although not entirely consistent and in spite of the poor gradation compared to CN 120 specifications. It is premature to conclude that the inclusion of aggregate gradation limits is not necessary to ensure quality shotcrete, but the introduction of performance specifications that do not mention gradation should not be excluded.
3. The CN 120 specifications require an overdesign factor of 140 percent of the specified core strength. However, differences in sampling cores and cast cylinders yield an overdesign factor less than that. For example, 4000-lb/in.^{2} cores with a length-to-diameter ratio of 2 may have an actual overdesign factor close to 120 percent, and this may be insufficient to compensate for anticipated strength loss caused by the use of inorganic accelerators.
4. A specific test program was developed and recommended for preconstruction test panels, and for testing during construction.

Steel-Fiber-Reinforced Shotcrete

Cording et al. (1981) have compiled results of tests on steel-fiber-reinforced shotcrete. Preconstruction testing was carried out in two stages, first to determine the compressive strength variation with time, and then to compare the strength variation of different mixes. Testing in the Atlanta Research Chamber was carried out to determine the structural behavior of fiber-reinforced shotcrete placed in situ under conditions close to actual construction.

Fibers The fibers used in these tests are circular in cross section, 0.025 cm (0.01 in.) in diameter, and approximately 2.5 cm (1 in.) long. They are made in U.S.

markets and are designated as U.S. fibers. A standard fiber-shotcrete batch contained 660 lb of cement, 1790 lb of fine aggregate, 1300 lb of coarse aggregate, and 115 lb of fibers. The mix was designed to have a water-cement ratio 0.45.

Compressive Strength Results Compressive strength results are plotted with reference to time for Sigunit and Dryshot admixtures in Figures 3-28*a* and *b*, respectively.

From Figure 3-28*a* it appears that fibrous shotcrete with normal 1.5 to 3 percent Sigunit dosages sets quickly and gains considerable strength in the first 8 to 10 hr. A test on fibrous shotcrete with 3 percent Sigunit resulted in an average f'_c close to 1500 lb/in.^{2} at 8 hr. Excepting samples tested when an adequate fiber factor was not available (hatched area in the diagrams), fibrous shotcrete with Sigunit accelerator shows a gain of strength with time that is consistent with specification requirements.

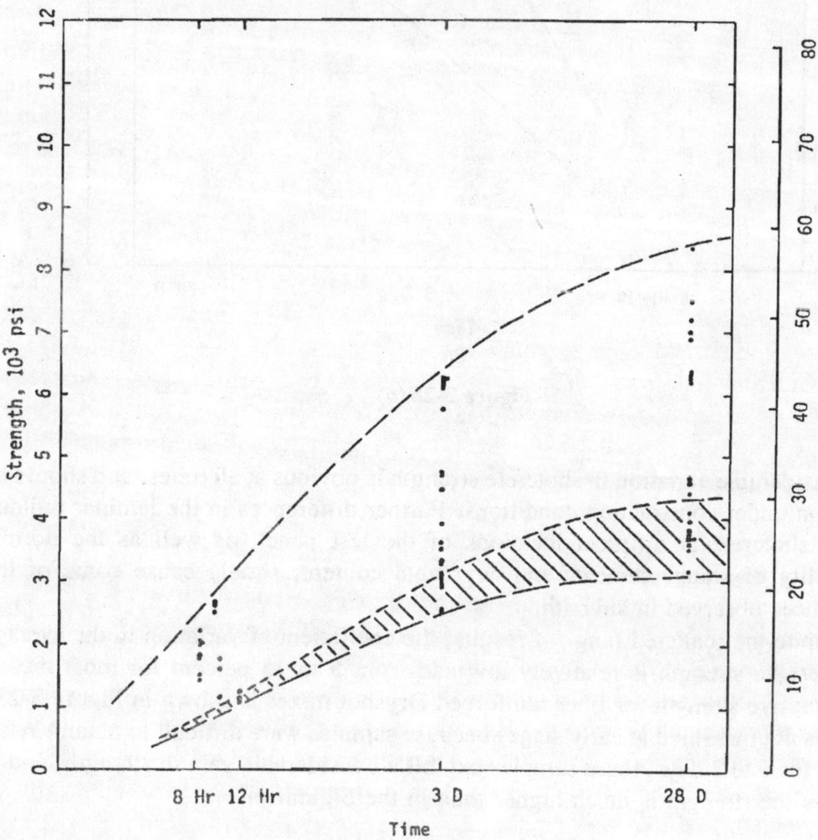

(*a*)

Figure 3-28 Compressive strength versus time for fiber-shotcrete. (*a*) Sigunit mixes. (*b*) Dryshot mixes. (From Cording et al., 1981.)

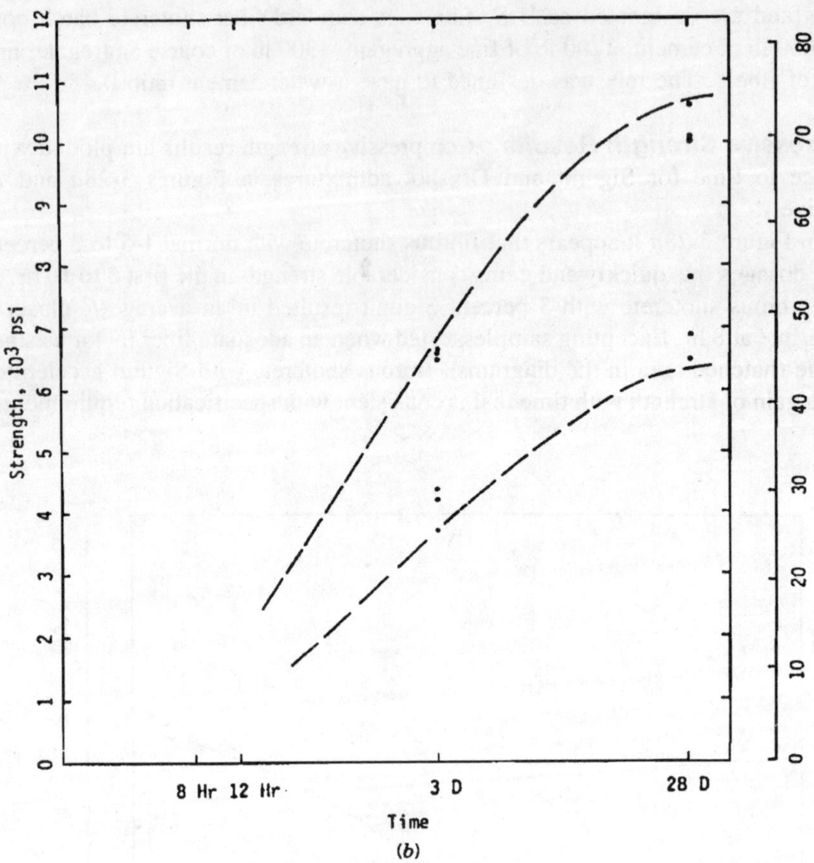

Figure 3-28 (*b*)

Considerable variation in shotcrete strength is obvious at all times, and should be common under construction conditions. Further differences in the laminar buildup of the shotcrete at different locations of the test panel, as well as the normal variability of water, cement, and aggregate content, should cause some of the differences observed in strer gth.

Despite the scattered rang f results, the coefficient of variation in the average compressive strength is relatively low and from 5 to 15 percent for most mixes. Compressive strength for fiber reinforced Dryshot mixes is shown in Figure 3-28*b* but was not measured at early stages because samples were difficult to obtain. After curing for 1 to 2 days, these samples exhibit a considerable gain in strength, and at 28 days the strength is much higher than in the Sigunit mixes.

Development of Flexural Strength Cording et al. (1981) also report tests on the flexural strength of fibrous shotcrete according to ASTM C78-64. The specimens were tested using third-point loading, as shown in Figure 3-29. In order to

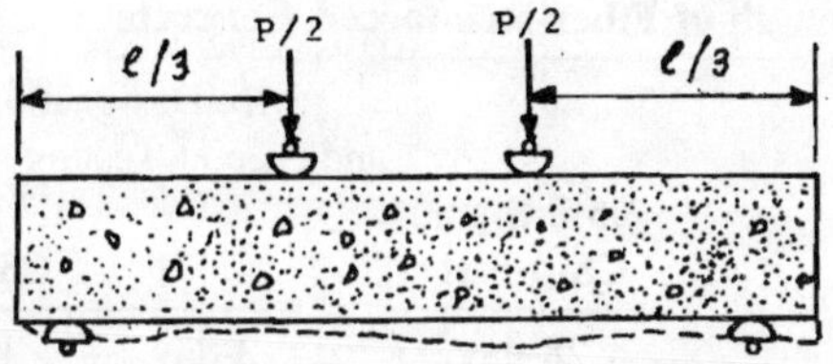

Figure 3-29 Shotcrete beam elevation and loading for flexural test. (From Cording et al., 1981.)

simulate field conditions, either the front or the back side of the beam was the tension side.

For the beam arrangement shown in Figure 3-29, the maximum flexural stress or modulus of rupture is

$$\sigma_f = \frac{Pl}{bd^2} \tag{3-6}$$

where σ_f = flexural stress (modulus of rupture)
P = total applied load (maximum)
l = span length
b, d = width and depth of section, respectively

Results of the tests are given in Table 3-8. The 28-day flexural strength for the 2.5 percent Sigunit mix is 23 percent (average) higher than that of the mix with 1.5 percent Sigunit. Interestingly, there is very little increase in flexural strength from 3 to 28 days in the mix with 1.5 percent Sigunit. This shows that flexural strength is developed very rapidly in the first days after shooting, a conclusion confirmed by other investigators (Parker et al., 1975).

Flexural strength is plotted versus compressive strength in Figure 3-30, which also shows the line corresponding to a flexural ratio σ_f/σ_c = 0.19. This ratio is reasonably useful in averaging data obtained from previous field tests (Parker et al., 1975).

The flexural ratio σ_f/σ_c is plotted versus the compressive strength σ_c in Figure 3-31. It appears that approximately one month after shooting the ratio σ_f/σ_c for the mixes tested is 0.13 to 0.17. Interestingly, the ratio for ordinary concrete with similar compressive strength is 0.11 to 0.14. The slightly higher ratio for shotcrete may be explained by the higher cement content for the in situ shotcrete.

Tests essentially similar to the foregoing (Parker et al., 1975) show a tendency for the flexural ratio to decrease with time, and this is evident from the plot of Figure 3-31. Results of tests from the Atlanta Chamber confirm this tendency. For example, young shotcrete (8 to 12 hr) should have a flexural ratio almost twice the ratio of older cured shotcrete.

TABLE 3-8 Flexural Strength of Fiber-Reinforced Concrete

	Mix Designation and Age at Testing (days)			
	2.5% Sigunit CS-4-F-H[a]	1.5% Sigunit CS-8-F-V[a]		
	28-day	3-day	10-day	28-day
Flexural strength, σ_f (lb/in.^{2})	1005			
	742	705	714	861
	1241	881	839	871
	1257			
Average compressive strength f'_c (lb/in.^{2})	6150	4600	6000[b]	6480
Flexural ratio = σ_f/f'_c	0.16			
	0.12	0.15	0.12	0.13
	0.20	0.19	0.14	0.13
	0.20			

Source: From Cording et al. (1981).

[a] S = admixture type = Sigunit
C = crushed stone mix
4, 8 = day of shooting
F = fiber mix
V, H = panel shooting position (vertical, horizontal).

[b] Value interpolated from compressive strength tests at 3 days and 28 days.

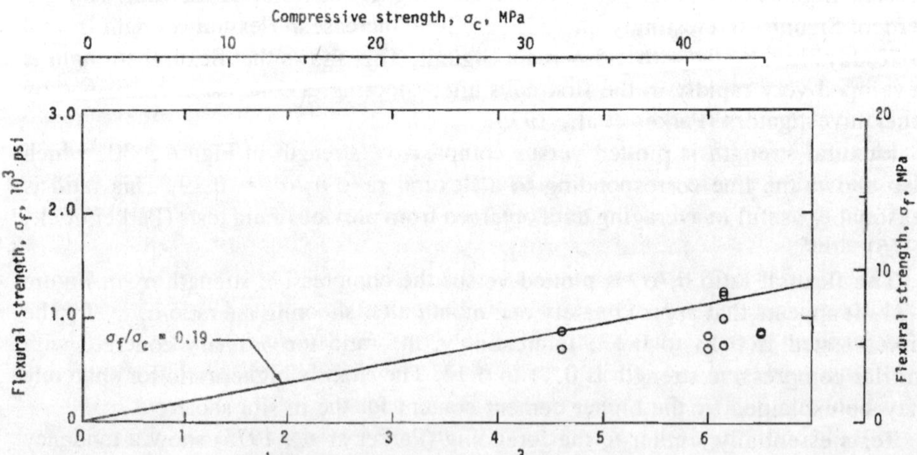

Figure 3-30 Flexural strength versus compressive strength; 10- to 28-day-old shotcrete. (From Cording et al., 1981.)

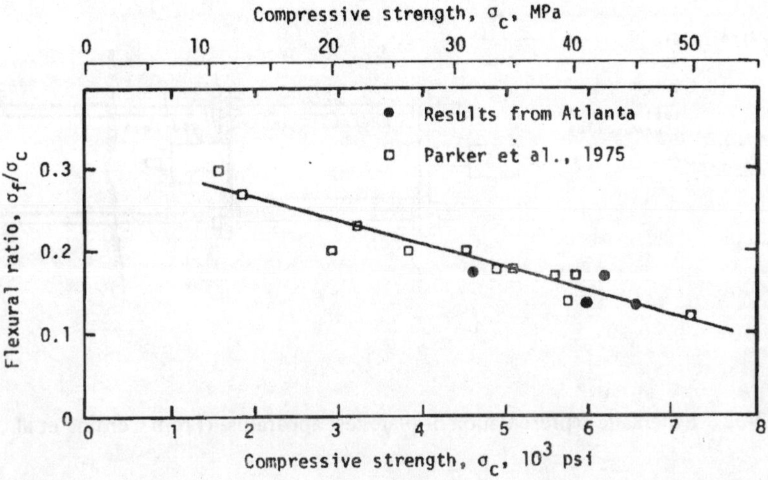

Figure 3-31 Average flexural ratio versus average compressive strength. (From Cording et al., 1981.)

Tests performed by the University of Illinois show that fibrous shotcrete has a peak flexural strength essentially comparable to conventional shotcrete. However, conventional shotcrete exhibits brittle behavior in bending, whereas fibrous shotcrete develops considerable ductility. This makes it particularly suitable as support of temporary loads, for example in the case of loosening ground.

3-8 STRUCTURAL PERFORMANCE OF SHOTCRETE

Plate Tests

Figure 3-32 shows the geometric configuration and equipment used for plate tests in shotcrete. A steel plate 2-ft square and 2-in. thick is placed in contact with rock, and is covered by a shotcrete layer wider than 2 ft but extending about 8 ft on either side of the plate. After the shotcrete has set, the plate is pulled at its center with a hydraulic jack and the failure load is measured.

Cording et al. (1981) have carried out plate tests on conventional and fibrous shotcrete. Figure 3-33 shows the configuration of the tested layers devised so that for each shotcrete type (conventional or fibrous) a flat and an arched layer were obtained. The results are tabulated in Table 3-9.

Shotcrete layers in flat configurations fail by loss of adhesion, and the adhesive strength between shotcrete and rock is higher than observed in laboratory tests (values of a_0 obtained in planar laboratory tests are close to $0.05 f'_c$). However, both the 4-in. and the 8-in. thick arched shotcrete layers in these tests failed in shear. The shear strength $0.10 f'_c$ is consistent with laboratory tests.

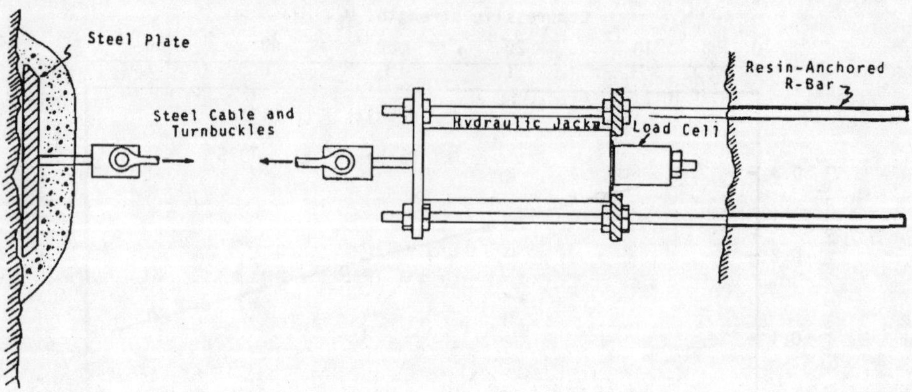

Figure 3-32 Schematic representation of plate test apparatus. (From Cording et al., 1981.)

These results suggest that natural surface irregularities in dry and clean rock increase the adhesive strength beyond the laboratory range. This effect is stronger in arched configurations where compressive stresses tend to develop across irregularities. In this case failure occurs not only as loss of adhesion but also as shear failure along irregularities in the rock and in the shotcrete.

The addition of fiber reinforcement increases the ductility of the shotcrete layers, although it does not necessarily increase their ultimate capacity. Cording et al. (1981) report that visual observations during the tests showed that the flat shotcrete layers with fiber reinforcement developed a series of visible cracks and moved 1 to 2 in. before failure. The flat shotcrete layers without fiber reinforcement remained brittle, however, and failed without warning.

Rebound Tests

In these tests shotcrete is shot against a rock surface for a certain period, and any material that fails to adhere to the receiving surface is collected in special nets and screens. Cording et al. (1981) report rebound tests on a 6-ft-wide and 10-ft-high strip of rock until the receiving surface was covered with a 3-in. thick fibrous shotcrete layer. Shotcrete lost by rebound was collected in a clean tarpaulin assembled and placed in front of the test panel.

During the test, 2530 lb of dry mix with fiber were shot in 10 minutes. The rebound collected on the tarpaulin weighed 553 lb, for an average rebound rate 553/2530 = 22 percent and material delivery rate 253 lb/min. Steel fibers in the rebound material were essentially straight after shooting. Fiber content by weight was 3.8 percent before shooting and 4.6 percent in the rebound material. A comparison of the gradation curves indicated that the rebound mix had 70 percent gravel, compared to 60 percent for the original mix. The cement content of the rebound was considerably lower as compared to the initial mix, but its water content was markedly higher.

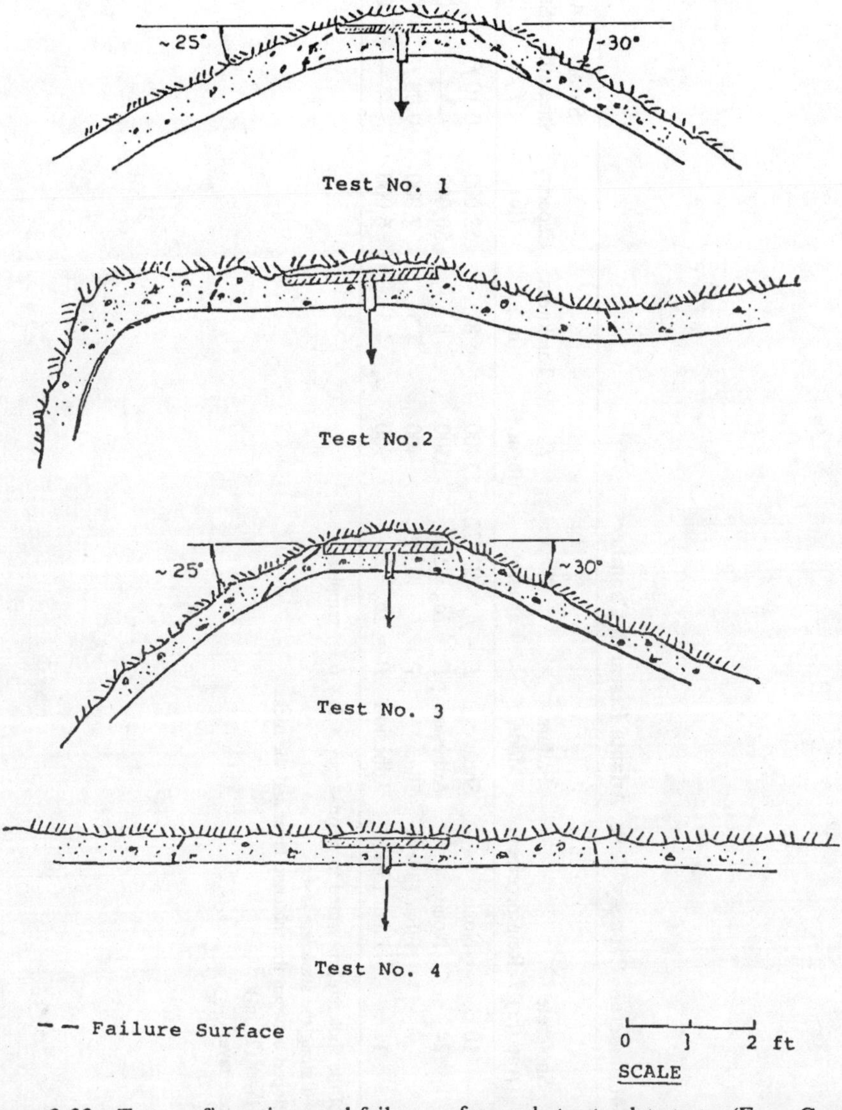

Figure 3-33 Test configurations and failure surfaces, shotcrete plate tests. (From Cording et al., 1981.)

Results from a series of field rebound tests on 27 different mixes applied under variable conditions show that the prime factor controlling the average rebound is the total thickness of the shotcrete layer placed in a single lift (Parker et al. 1975). The relationship between average rebound, rebound rate, and layer thickness is shown in Figure 3-34. The rate of rebound drops when an initial critical layer thickness is reached (phase 1), and then becomes nearly constant with thickness (phase 2).

TABLE 3-9 Summary of Shotcrete Capacity Tests, Atlanta Research Chamber[a]

Test Number	Configuration	Shotcrete Age (hr)	Reinforcement	Failure Mode	Capacity	f'_c (lb/in.^{2})	Thickness (in.)	Capacity (lb)	Shear Strength, f_d	Adhesive Strength, a_0
1	Arched	10	None	Shear	$P = f_d \cdot H \cdot 2L$	1400	8	52,000	$0.10 f'_c$	—
2	Flat	24	None	Adhesion	$P = 2a_0 \cdot 2L$	3500	8	50,000	—	$0.15 f'_c$
3	Arched	7	Fiber (3%)	Shear	$P = f_d \cdot h \cdot 2L$	400	4.5	7,000	$0.08 f'_c$	—
4	Flat	11	Fiber (3%)	Adhesion	$P = 2a_0 \cdot 2L$	900	4.5	6,000	—	$0.07 f'_c$

[a] f'_c = compressive strength of the shotcrete, measured in prismatic, 3 × 3 × 6 in. samples.
f_d = shear strength developed along the shotcrete layer.
a_0 = adhesive strength developed between the shotcrete layer and the rock.
L = width of the shotcrete layer (24 in.)
H = thickness of the shotcrete layer

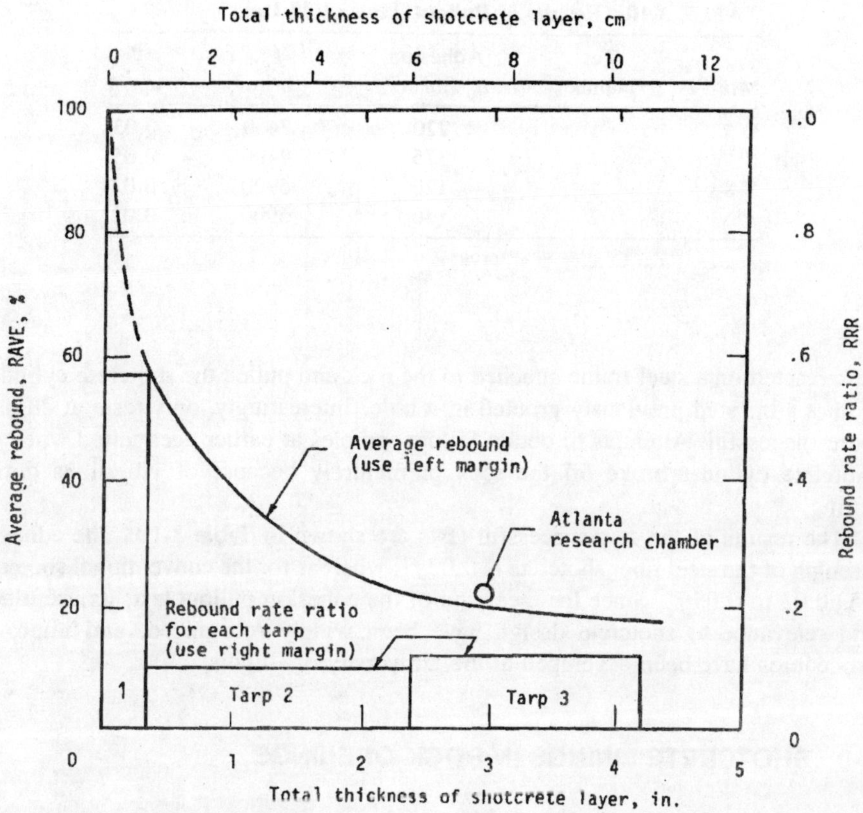

Figure 3-34 Change in measured rebound values with thickness. (From Parker et al., 1975.)

Average rebound (total weight rebounded divided by total weight shot) is reduced gradually and at a rate depending on the degree of the initial loss. For the test conditions of Figure 3-34, the dependence of the average rebound on the initial loss (phase 1) ceased to exist when the shotcrete layer became 4 in. thick. Interestingly, the results of tests from the Atlanta Chamber agree with the findings reported by Parker et al. (1975).

Pullout Tests

Like the plate tests reviewed at the beginning of this section, pullout tests are intended to measure the adhesion developed between shotcrete and rock. The test procedure is described by Cording et al. (1981).

These investigators report results of pullout tests carried out in the field. During the test a 3-in. diameter cylinder was isolated from the remaining shotcrete by drilling through it and into the receiving rock using a coring bit. A hollow hydraulic

TABLE 3-10 Results of Pullout Tests at 28 Days

Mix	Test Number	Adhesion, a_0 (lb/in.^{2})	f'_c, (lb/in.^{2})	a_0/f'_c
D-7	1	220	7400	0.03
	2	375	7400	0.05
S-8-F	1	129	5900	0.02
	2	130	5900	0.02

Source: From Cording et al. (1981).

jack reacted on a steel frame attached to the rock and pulled the shotcrete cylinders from a $\frac{3}{8}$-in. stud previously grouted in a hole. Interestingly, only tests at 28 days were successful. Attempts to obtain failure samples at earlier ages failed when the shotcrete cylinder broke off the rock prematurely because of vibrations during coring.

The results of the few successful tests are shown in Table 3-10. The adhesive strength of the steel fiber shotcrete is $0.02 f'_c$, whereas for the conventional shotcrete it is 0.03 to $0.05 f'_c$. Since the inception of the adhesion pullout test, its usefulness and relevance to shotcrete design have been widely recognized, and improved procedures have been developed at the University of Illinois.

3-9 SHOTCRETE LININGS IN ROCK OPENINGS

Support Mechanism in Tunnels

In tunnels and other underground openings, the shotcrete adheres directly to rock and prevents its loosening. When applied to a rock surface shotcrete is forced into fissures, open joints, and seams and provides the same binding action as mortar in a stone wall. In addition, it hinders water seepage and prevents piping of joint filling materials, especially if it is applied before any water forced into the surroundings returns to the rock surface.

The most important feature of shotcrete as support against loosening and stress-rearrangement pressures is its interaction with neighboring rock. A shotcrete layer applied immediately after opening a new rock face acts as a tough surface that improves stability even in weak rock. This close interaction enables the adjoining blocks to maintain their original undisturbed state, and prompts an arch action where the shotcrete absorbs the tangential stresses close to the surface. Disintegration normally begins by the opening of a minute surface fissure, but if the resulting movement can be prevented by applying a shotcrete layer, the rock behind the shotcrete remains stable.

A thin layer of shotcrete can prevent rock falls. Holmgren (1975) has demonstrated that the carrying capacity for thin unreinforced layers of shotcrete depends mainly on the adhesion developing on a strip of about 3 cm ($1\frac{1}{4}$ in.). As an example, we consider the rock pyramid shown in Figure 3-35, with a base 1 m^2 and a height 1 m, weighing almost 1 ton. Assuming an adhesion strength of 10 kg/cm^2 (140 lb/in.^{2}), which is consistent with the plate load tests discussed in Section 3-8, the carrying capacity is estimated as $10 \times 4 \times 100 \times 3 = 12{,}000$ kg, or twelve times the weight of the loose pyramid. The same investigator suggests that a theoretical model proposed by Alberts and Backstrom (1971) is basically correct, and in practice adhesion failure is predominant.

In loosening ground (where rock loads develop from the self weight of individual rock blocks that tend to loosen from the roof and the walls of the opening) the intent is to provide an artificial support before loosening occurs. Very small movement will most likely relieve high pressures, but it is seldom necessary to make allowance in the design if the construction is in blocky loosening rock. Conversely, for tunnels in squeezing ground (where high pressures are generated on the tunnel support with time as the ground creeps under the effect of natural stresses), the design allows a certain controlled movement in order to relieve these stresses and reduce the support requirements; an example is the New Austrian Tunneling Method (NATM) where this philosophy is demonstrated (Rabcewicz, 1965; Xanthakos, 1991).

In the United States, full-scale field studies have been carried out beginning with the Washington metro tunnels, in blocky and seamy foliated gneisses and schists where the boundaries of rock wedges are formed by continuous, planar joints and

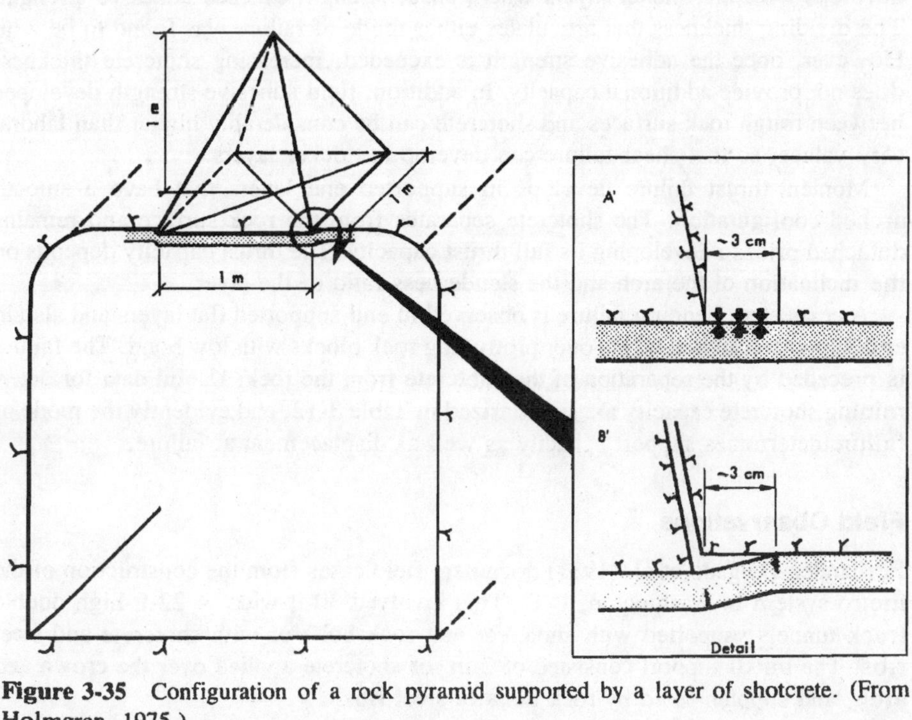

Figure 3-35 Configuration of a rock pyramid supported by a layer of shotcrete. (From Holmgren, 1975.)

shears (Fernandez-Delgado et al., 1981). In these tests, shotcrete-rock adhesion was often poor, particularly along sheared surfaces and foliation planes. Rock fallout and overbreak was monitored, and rock displacements were related to cracks and strains in the shotcrete lining. Tensile fractures were observed on portions of the shotcrete membrane covering protruding rock blocks (Mahar et al., 1972). Similar conditions have been noted by other investigators in loosening ground where bond development is poor. In a survey of Swedish and Norwegian tunnels (Cecil, 1970), adhesion failure was observed on clay-coated bedding plane surfaces. Brekke (1972) has documented adhesion failure where rotation and displacement of isolated blocks tended to puncture and break the shotcrete lining.

Laboratory Model Tests

Table 3-11 summarizes the principal failure modes observed in laboratory tests (Fernandez-Delgado et al., 1981). The main variables influencing failure are: (a) the geometric configuration of the rock surface; (b) the end conditions of the shotcrete layer; (c) the thickness of the shotcrete; and (d) the bond strength developed at the shotcrete-rock interface.

Adhesive failure typically develops in poorly bonded shotcrete layers with free ends, and irrespective of rock geometry and shotcrete thickness. With increasing bond strength, shear failure occurs with the thinner layers but adhesive failure still develops with the thicker layers where shear strength exceeds adhesive strength. The dividing thickness that articulates either mode of failure was found to be 2 in. However, once the adhesive strength is exceeded, increasing shotcrete thickness does not provide additional capacity. In addition, field adhesive strength developed between rough rock surfaces and shotcrete can be considerably higher than laboratory values, so that shear failure can develop in thicker layers.

Moment-thrust failure develops in supported end layers that have a smooth arched configuration. The shotcrete separates from the rock surface and remains detached prior to developing its full thrust capacity. The thrust capacity depends on the inclination of the arch and the slenderness ratio of the layer.

As expected, bending failure is observed in end-supported flat layers and also in end-supported arched layers over protruding rock blocks with low bond. The failure is preceded by the separation of the shotcrete from the rock. Useful data for determining shotcrete capacity are summarized in Table 3-12, and evidently the mode of failure determines support capacity as well as displacement at failure.

Field Observations

Fernandez-Delgado et al. (1981) document field cases from the construction of the metro system in Washington, D.C. This involved 30-ft-wide × 22-ft-high double track tunnels supported with shotcrete and rock bolts or with shotcrete and steel ribs. The initial support consisted of 2 in. of shotcrete applied over the crown and arch, and supplemented by rock bolts or steel ribs.

The rock is foliated schists and gneisses, with foliation dipping 50° to 70° to the

TABLE 3-11 Failure Modes of Shotcrete Layers; Shotcrete Linings in Loosening Rock

		Poor Bond		Good Bond	
		$h < 2$ in.[a]	$h > 2$ in.	$h < 2$ in.	$h > 2$ in.
	Planar free end	Adhesion	Adhesion	Shear	Adhesion
	Planar supports at 8 ft	Bending	Bending	Shear	Bending
	Arched smooth surface free end	Adhesion	Adhesion	Shear	Adhesion
	Arched smooth surface supported end	Moment thrust	Moment thrust	Shear	Moment thrust
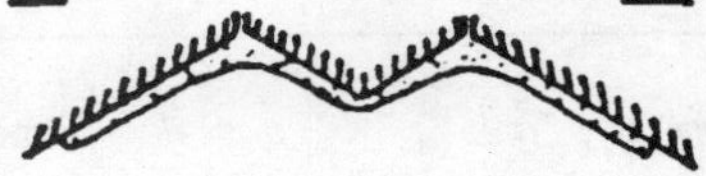	30° arched irregular surface free end	Adhesion	Adhesion	Shear	Adhesion
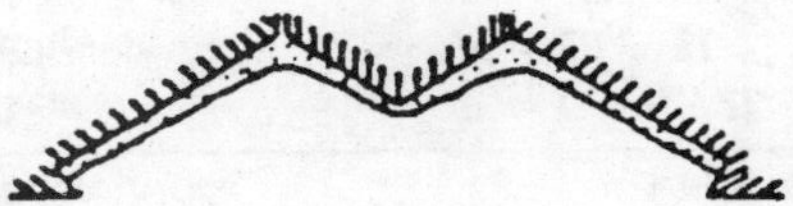	30° arched irregular surface supported end	Bending (apex)[b]	Bending (apex)[b]	Shear	Moment thrust
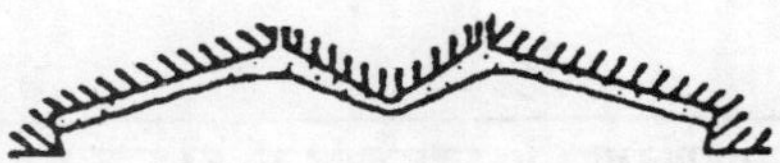	15° arched irregular surface supported end	Bending (apex)[b]	Bending (apex)[b]	Shear	Moment thrust

Source: From Fernandez-Delgado et al. (1981).

[a] h = Shotcrete thickness. [b] If enough shotcrete is placed to fill in the recess and develop a smooth arch configuration, the failure mode changes to moment thrust.

TABLE 3-12 Capacities of Unreinforced Shotcrete Layers Tested in the Laboratory

Failure Mode	Capacity		θ 0°	θ 15°	θ 30°	Range of Typical Displacements at Failure (in.)
Shear	$P = f_d \cdot h \cdot 2L$	f_a	0.05	$0.05 f'_c$	$0.10 f'_c$	0.004–0.07
Adhesion	$P = 2a_0 \cdot 2L$	$a_0{}^a$	0.05	$0.05 f'_c$	$0.10 f'_c$	0.02–0.05
Moment thrust	$P = 2T_c \cdot \sin\theta \cdot f'_c \cdot h \cdot L$	T_c	—	0.30	0.45–0.60	0.15–0.25
Bending	$P = B_c f'_c h^2 L/d^b$	B_c	0.12[c] 0.32[c]	0.12[c] 0.32[c]	0.25	0.02–0.20

[a] a_0 values are for good bond.

[b] d = distance between rock bolts or any other support element when planar conditions are present

d = half of the moving block width in the case of an arched configuration, θ 71 30°

[c] Using a 4 × 4 × ⅛-in. mesh as reinforcement:

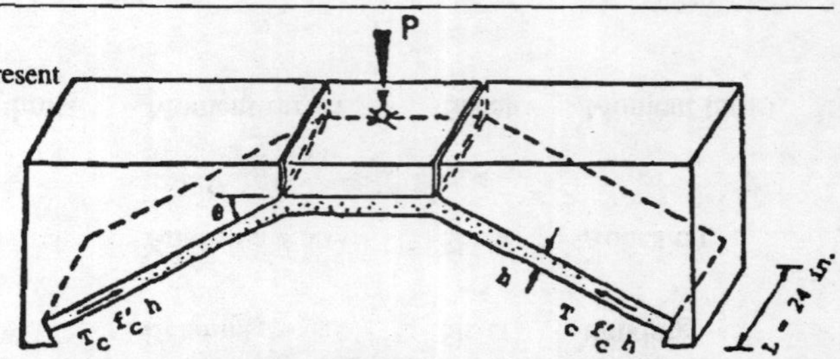

P = total load

f_d = shear strength developed along the shotcrete layer

$2L$ = total length of contact between the movable and fixed blocks (48 in. in the tests)

h = thickness of the layer

a_0 = adhesive strength developed between the shotcrete layer and the rock

T_c = dimensionless thrust coefficient, given by the ratio of the axial layer load at failure to the maximum compressive strength times the cross-sectional area of the layer

f'_c = compressive strength of the shotcrete, measured in prismatic, 3 × 3 × 6-in. samples

B_c = a dimensionless bending coefficient

west and striking north-south, nearly parallel to the long axis of most stations. Major geologic features dictating the type of support were the shear zones parallel to the steeply dipping foliation, and the sets of joints and shears that were planar, slick, and occasionally coated with a clay gouge. This configuration produced blocky rock, with blocks typically 2 to 6 ft in size. Typical shotcrete compressive strength at 28 days was 5000 lb/in^2.

Case A Figure 3-36*a* shows a rock wedge movement in a section where steel ribs were used as part of the permanent support. The block is about 8 × 8 ft, and was bounded by joint surfaces coated with clay and chlorite. This condition facilitated block movement and also inhibited the development of good adhesion. The movement was 0.2 in., and the rock eventually settled onto the steel supports.

The factor of safety is the maximum support capacity per unit length of liner divided by the weight of the rock wedge, or

$$F_s = \frac{P}{W} \tag{3-7}$$

where $P = B_c f_c' h^2 d / L$ (from Table 3-12)
$W = 8 \times 8 \times 160 = 10{,}240$ lb

A value of $B_c = 0.25$ is appropriate for this configuration. In addition, the following assumptions are made: (a) a portion of the weight of the rock wedge acted on the shotcrete liner a few hours after shooting; (b) measured rock wedge movement after two days was 0.1 in.; (c) the compressive strength of shotcrete at the time of loading was 2000 lb/in.^{2} and (d) about 25 percent of the wedge weight was active at this time. From these, we estimate

$$P = 0.25 \times 2000 \times 6.25 \times \tfrac{1}{4} = 780 \text{ lb}$$

and $F_s = P/W = 780/2560 = 0.33$. This low factor of safety is consistent with the observed behavior.

Case B The sliding rock wedge in this example occurred in a tunnel section supported with shotcrete and rock bolts, 15-ft long and on 4-ft centers, fully grouted with resin. Rock bolts were installed within 2 ft of the tunnel face. In one location, continuous joints combined to form a rock wedge, 8 × 10 ft in size, hanging in the tunnel arch as shown in Figure 3-36*b*. Both joints were coated with soft clay gouge containing water, and exhibited minor slickensides. Bolts installed as shown prevented further overbreak along the joint planes. The entire wedge consisted of several blocks, 1.5 to 2 ft in size, separated by several tight joint planes.

Rock movement close to 0.03 in. was recorded between the crown and the 40-ft anchor as the heading was advanced 5 ft beyond the extensometer, and essentially it involved separation along the continuous joint surfaces that bounded the wedge. Less than 0.002 in. of movement was recorded behind the 2L plane. The measured

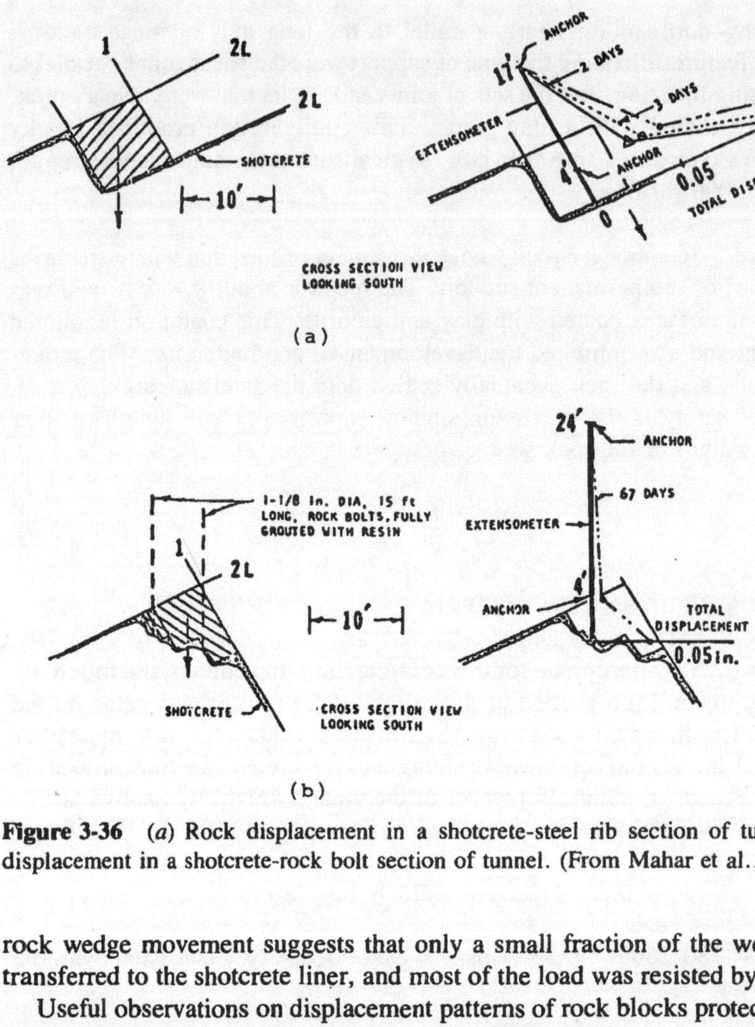

Figure 3-36 (*a*) Rock displacement in a shotcrete-steel rib section of tunnel. (*b*) Rock displacement in a shotcrete-rock bolt section of tunnel. (From Mahar et al., 1972.)

rock wedge movement suggests that only a small fraction of the wedge load was transferred to the shotcrete liner, and most of the load was resisted by the rock bolt.

Useful observations on displacement patterns of rock blocks protected with shotcrete are presented in Figure 3-37 (Cording, 1974). Rotation, outward displacement of a block, and slippage along the shotcrete-rock interface were observed in tunnel sections where the shotcrete was placed on irregular surfaces bounded by clean smooth planar and often slick joint surfaces. These results suggest that tensile stresses are developed in shotcrete placed over the smooth surfaces of protruding blocks. This shotcrete behavior has been modeled in finite element studies, and confirms the need of tensile reinforcement in the shotcrete for tunnel sections with irregular and smooth or slick surfaces.

Design Procedure

Fernandez-Delgado et al. (1981) offer guidelines for estimating shotcrete requirements. These guidelines consider the effect of natural geologic discontinuities on

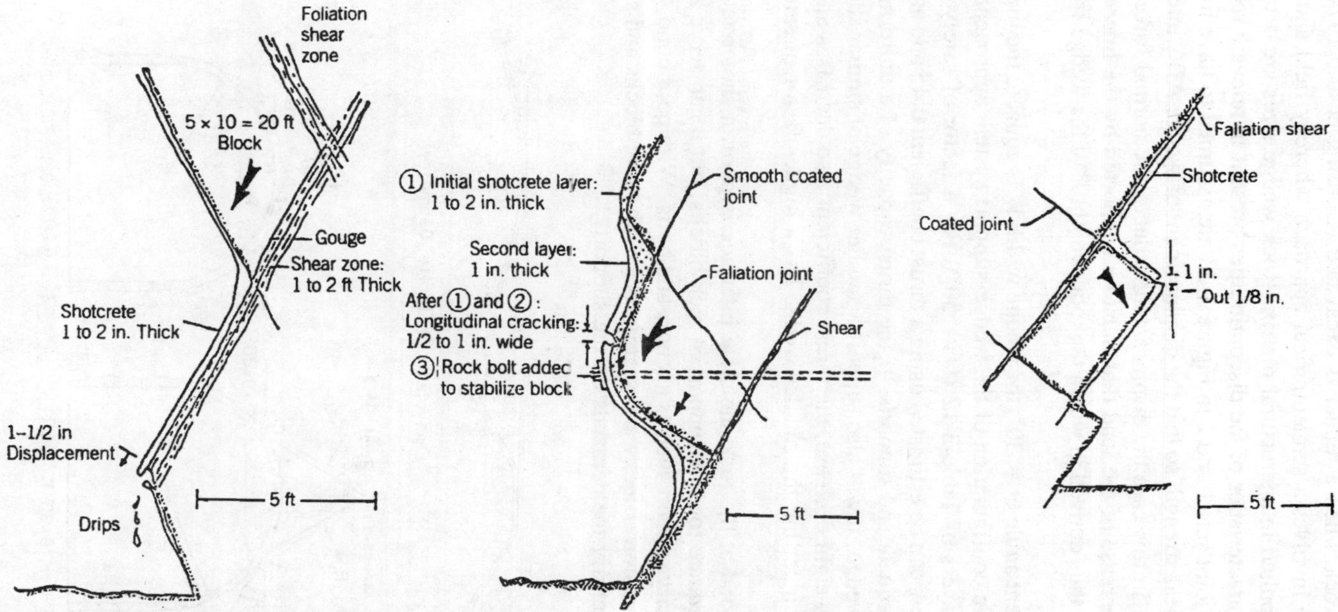

Figure 3-37 Displacement patterns, rock blocks supported with shotcrete, Washington, D.C. metro. (From Cording et al., 1974.)

the behavior of rock when a tunnel is excavated (Cording et al., 1974). A basic design procedure involves a semiempirical approach whereby field and laboratory test results are applied in a structural matrix. Rock wedge sizes are determined by the geometric characteristics of the discontinuities present in relevant rock masses.

For example, joint sets 1 and 2 in Figure 3-38 are assumed to have irregularities and negligible shear strength so that the displacement of wedge ABC into the tunnel is not precluded. In this case the displacement reduces the normal force P_n, with a corresponding increase of the load that must be supported by the liner. A limiting load condition is thus derived in which the total weight P of the wedge W will act on the support.

For an abutment angle $\Theta = 30°$, the wedge width W is equal to the tunnel radius, and the geometric configuration of the wedge-support system approaches the steep smooth-arched layers of Table 3-12 ($\Theta = 30°$). For a factor of safety of 1.0, the shotcrete thickness can be estimated using a thrust coefficient 0.45 and assuming the layer to have adequate end supports. If the thrust capacity of a uniform lining can support a large wedge, it will also support a smaller wedge of similar shape. As the wedge size and weight decrease, the thrust coefficient also decreases and the thrust acts at a flatter angle, but the smaller wedge has a higher factor of safety.

Bond If the bond at the rock-shotcrete interface is poor, a shotcrete lining is in most cases inadequate for sole support of rock blocks, but it can serve as a supplement to the primary support such as anchors and bolts. With poor bond, shotcrete is likely to fail in tension. Blocky and seamy rock with large blocks and slickensided surfaces will intensify the demands on the support.

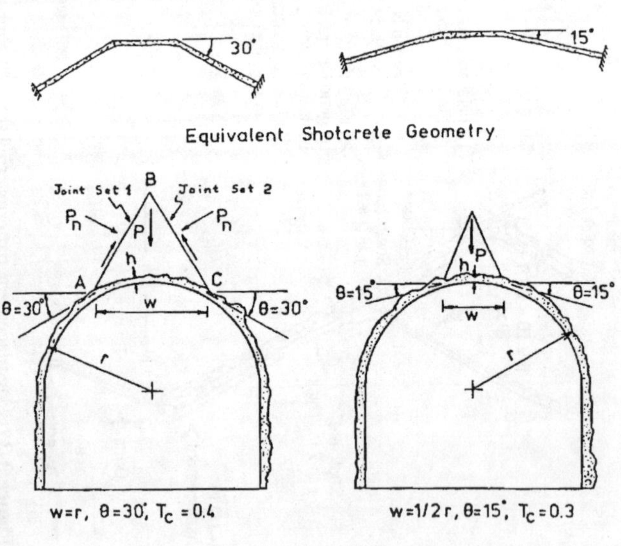

$$P = 2\,T_c\,[\sin\theta]\,[h]\,[f'_c \cdot L]$$

Figure 3-38 Cross section. Triangular moving wedge in smooth surface tunnel. (From Cording et al., 1974.)

The size of a two-dimensional (long) wedges that can, with good bond, be supported with a 2-in. thick shotcrete layer is shown in Table 3-13. In this case, the rock surface in the vicinity of the wedge is assumed planar or protruding rather than arched. The values given in Table 3-13 are not design parameters since they do not include load and safety factors.

For preliminary estimates, the block size that can be supported at the tunnel heading can be determined from early shotcrete strength. At 2 hours, f'_c = 400 lb/in.^{2}, and the maximum wedge weight for adhesive strength 0.05 f'_c would be 1200 lb/ft.

Reinforcement Wire mesh can increase the support capacity by limiting crack widths and by providing post-crack resistance that is essential for a ductile failure. Steel-fiber-reinforced shotcrete is particularly effective in the tunnel heading because it ensures ductility that cannot otherwise be provided. According to the results of laboratory tests and field tests in the Atlanta Research chamber, fibrous shotcrete allows substantial deformation and visible cracking to occur prior to structural failure, and this is a marked improvement from the brittle failure of plain shotcrete.

The Convergence-Confinement Method

Unlike the foregoing applications, where the primary function of shotcrete is to support gravity loads of the loosened rock above the opening, shotcrete is used in

TABLE 3-13 Two-Dimensional Wedge Sizes for Adhesion Failure of Well-Bonded Shotcrete Layer, 2 in. or More Thick

	Wedge Shape		Maximum Wedge Width, w (ft)	Maximum Cavern Radius if $w \leqq R/2$ (ft)
7-Day (3000-lb/in.^{2} Shotcrete: Maximum Wedge Weight = 9000 lb/ft for a_o = 0.05 f'_c				
Deep wedge		15°	7.6	15
Intermediate wedge	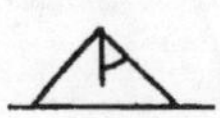	45°	15	29
Shallow wedge	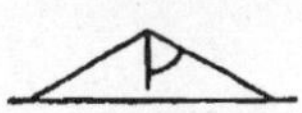	60°	19	39
7-hr (1000-lb/in.^{2} Shotcrete: Wedge Weight = 3000 lb/ft for a_o = 0.05 f'_c				
Deep wedge		15°	2.5	5
Intermediate wedge		45°	5	10
Shallow wedge		60°	6.5	13

plastic-squeezing ground to stabilize the surrounding rock after excavation and maintain its shear resistance. In addition to the requirement for a quick sealing of the rock surface as a practical matter, the convergence-confinement approach has emerged from the following developments: (a) the reciprocal relationship of required lining resistance and deformations (Fenner, 1938); (b) the finding that the time-dependent behavior of the rock mass is fundamental for predicting the behavior of the tunnel structure; (c) the development of the shear failure theory for tunnels under high overburden; (d) the introduction of the so-called semirigid linings combined with a semiempirical design approach using in situ measurements as part of the analysis; and (e) the incorporation of rock in the carrying support system.

Plasticity is commonly experienced in squeezing ground when the gradual depletion of strength drives the zone of broken rock deeper into the tunnel until the supports undergo a gradual buildup of pressure. In this case two distinct forms of behavior may be exhibited, as shown in Figure 3-39. If the rock tends to arch and the support can provide sufficient resistance to stop deterioration, the inward displacement of the walls will decrease with time, approaching an asymptote (stable condition). If the installation of the support is delayed, or if the rock induces very high loads, the inward movement will accelerate, thus creating unstable conditions.

In general, the plastic region is assumed to have been reached when a developing stress approaches a certain fraction of the rock strength (Deere et al., 1969; Rabcewicz, 1969; Ward, 1978). Irrespective of this definition, when the tangential stresses become greater than about one-half the unconfined compressive strength, cracks will begin to form.

Basic Features of Method In addition to the foregoing principles, certain basic features must be taken into consideration. These are: (a) the appropriate geomechanical ground behavior; (b) a suitable statical shape and tunnel configuration; (c) prevention of unfavorable stresses and deformations by means of suitable supports installed in a proper sequence; and (d) optimization of the support resistance as

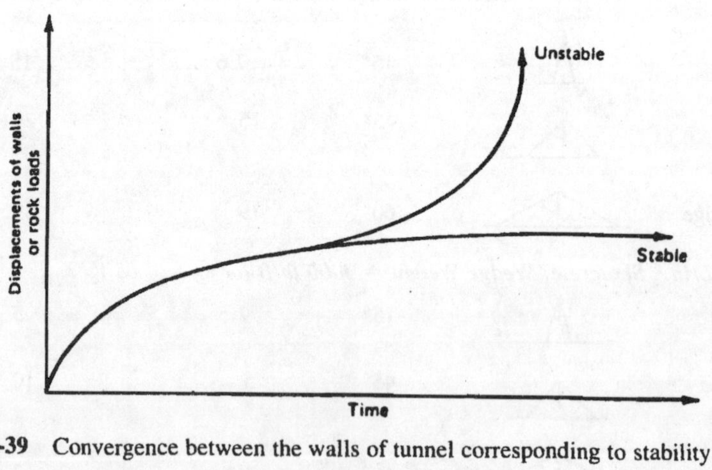

Figure 3-39 Convergence between the walls of tunnel corresponding to stability and instability.

a function of allowable deformations. A tunnel thus becomes a composite structure consisting of rock, supports, and strengthening elements such as shotcrete, anchors, steel ribs, and so on (Xanthakos, 1991). In excavating a tunnel the primary aim is to transform the natural state of equilibrium into a new secondary stability. In this process rock deformations must be controlled so that they remain small enough to prevent undue loss of rock strength, yet become large enough to activate the rock to a load-bearing ring in order to reduce artificial support requirements.

Functions of Shotcrete In this matrix the shotcrete essentially forms a skin around the opening and enables the rock to form a composite load-carrying ring. When the shotcrete is well bonded to rock, a close interaction results and the two materials tend to act as one unit. Where there is a tendency toward movement along joints or bedding planes, the shearing resistance provided by the shotcrete will inhibit this movement. Likewise, if the shear stresses along rock discontinuities become large enough, the shotcrete may follow the rock response and fail in shear.

Using appropriate shear failure theories, Rabcewicz (1969, 1970) assumes that failure takes place along the Mohr planes perpendicular to the direction of the arching thrust. In this case the shearing resistance provided by the shotcrete is

$$p_i^s = \frac{\tau_s d}{(b/2)\sin\alpha} \tag{3-8}$$

where p_i^s = shear resistance of shotcrete lining
d = thickness of shotcrete
b = height of shear zone (see Xanthakos, 1991)
α = angle of shear resistance between rock and shotcrete
τ_s = shear strength of shotcrete

Additional resistance can be provided by wire mesh in the shotcrete as follows:

$$p_i^m = \frac{A_m \tau_m}{(b/2)\sin\alpha} \tag{3-9}$$

where p_i^m = shear resistance of steel wire mesh
A_m = area of reinforcing per unit length of tunnel
τ_m = shear strength of steel wire mesh

3-10 GROUND-ROCK TUNNEL INTERACTION

Section 3-9 demonstrates the usefulness of (support pressure)–(tunnel convergence) relationships in improving our understanding of rock-support interaction and in estimating the support requirements. As the preexisting natural rock support is removed during excavation, the associated displacements in the rock surrounding the opening cause a redistribution of stresses. In the simple example shown in Figure 3-40, the ground-support reaction curve (in this case a straight line) correlates radial displacement with support pressure. The support is installed just as the

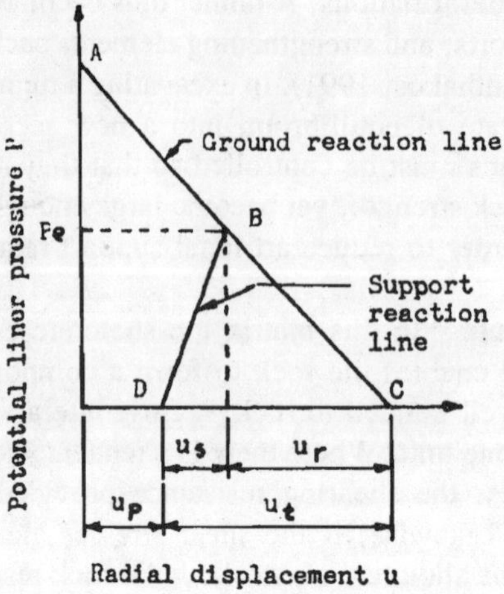

Figure 3-40 Ground-support reaction diagram; elastic conditions. (From Ranken et al., 1978.)

radial wall displacement reaches a value u_p, and at this point the pressure acting on it is zero. As the face of the excavation advances, pressure builds up on the support that deflects according to line *DB*. Equilibrium is reached at *B* when the support pressure reaches a maximum value p_e corresponding to displacement u_s.

Ground-support interaction diagrams have been proposed to help us understand the support mechanism of openings in rock (Rabcewicz, 1964, 1965; Daemen, 1975; Ward, 1978; Muiz Wood, 1979; Ladanyi, 1980; Lombardi, 1980; Xanthakos, 1991). Procedures therefrom can be used quantitatively in the selection and dimensioning of tunnel support (Ladanyi, 1974; Daemen, 1975; Ward, 1978; Hoek and Brown, 1980). Essentially, the design must predetermine the ground response curve for the rock mass, stress regime, and tunnel geometry. The usual approach follows closed-form solutions for problems of simple tunnel geometry and hydrostatic in situ stresses, but finite element techniques are necessary for problems involving more complex excavation scenarios and stress fields (Daemen, 1975, 1977; Lombardi, 1980). Invariably, it is necessary to introduce relevant assumptions about the stress-strain behavior and strength criteria of the rock mass.

Alternatively, it is possible to introduce more complex (and realistic) models of rock mass behavior into ground engineering solutions. Brown et al. (1983) present solutions to a simple axisymmetric problem obtained using nonlinear peak and residual rock strength criteria with an improved treatment of the influence of volumetric strains in the rock mass surrounding the excavation.

This approach is illustrated by solving a simple axisymmetric problem. Consider a tunnel of radius r_i driven in homogeneous, isotropic, initially elastic rock mass, and subjected to a hydrostatic stress field p_0, as shown in Figure 3-41. The support system is assumed to provide a uniform radial support pressure p_i. The stress induced in the rock as a result of excavation may exceed the yield strength of the

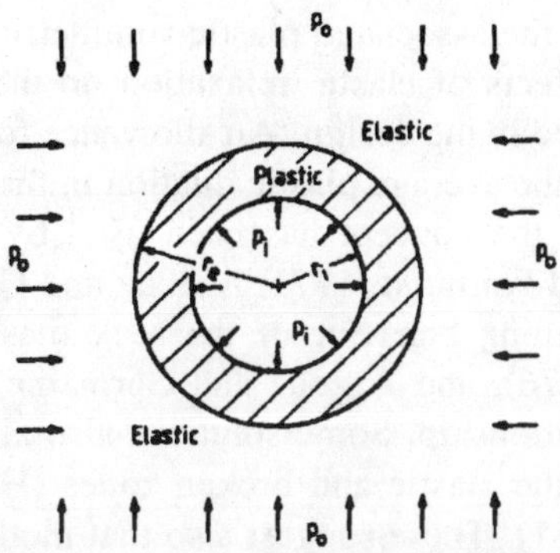

Figure 3-41 A typical axisymmetric tunnel problem

rock mass, and in this case a plastic zone of radius r_e will form around the tunnel. Outside the boundary defined by r_e the rock is assumed to remain elastic.

Solutions can be obtained that include the influence of the proximity of the tunnel face or construction procedures on ground-support interaction (see also Egger, 1974, 1980; Panet, 1976; Schwartz and Einstein, 1980; Kaiser, 1981; Tanimoto et al., 1981). For the present analysis, plane strain conditions are assumed with strain increments occurring in the plane of Figure 3-41. Because of axial symmetry, the tangential and radial stresses σ_θ and σ_r in the rock mass around the tunnel are the principal stresses. If the axial stress is the intermediate principal stress, then σ_θ and σ_r become the major and minor principal stresses σ_1 and σ_3, respectively. The simplicity of the problem is enhanced further if the influence of the rock weight in the plastic zone on tunnel displacements and support pressures is ignored.

Based on these assumptions, Brown et al. (1983) develop solutions that can be used as aids to tunnel design. The elastic-plastic analysis is approximate since it does not reflect the incremental plasticity (Dragon and Mroz, 1979) and does not allow for time-dependent effects (Ladanyi, 1974, 1980).

Comparison of Methods A complete review of solutions obtained for axisymmetric tunnel problems is given by Brown et al. (1983). These include elastic-plastic, elastic-brittle-plastic, and elastic-strain softening behavior. Elastic-plastic stress calculations for determining support pressures in the problem of Figure 3-41 have been presented by Fenner (1938), later improved by Morrison and Coates (1955). Notable contributions were made by Kastner (1949) and by Labasse (1949), who presented solutions for the case of a nonhydrostatic pretunneling stress field and addressed the issue of rock-support interaction. These solutions also recognized that in calculating displacements and tunnel stability it is necessary to consider the influence of volume increase associated with material failure and plastic deformation. An empirical nonlinear strength criterion for rock masses was developed and applied by Hoek and Brown (1980) and is incorporated in the present analysis.

Early solutions assume that the rock in the plastic zone will deform at constant volume. However, even with a strength reduction allowed for in the plastic zone, no consideration is given to the associated plastic volumetric strains in the failing rock mass, although some effects of elastic relaxation on the radial displacements are recognized and considered in the design. An allowance for the influence of nonelastic strains by estimating the average plastic dilation in the rock mass has been made by Labasse (1949), and the concept has been used by others (Lombardi, 1970, 1979, 1980; Daemen and Fairhurst, 1971; Azzouz and Germaine, 1981). Solutions allowing for strain-softening behavior of the rock mass are presented by Egger (1974, 1980), Panet (1976), and Azzouz and Germaine (1981), generally using a trilinear stress-strain relationship. Some solutions also allow for different values of the elastic constants in the elastic and broken zones (Hobbs, 1966; Hendron and Aiyer, 1972; Kaiser, 1981). These suggest also that modulus reductions associated with progressive failure of the rock alone can account for the displacements observed in many cases at the tunnel boundaries.

Behavioral Model Hoek and Brown (1980) present the following empirical peak strength criterion for rock masses:

$$\sigma_1 = \sigma_3 + (m\sigma_c\sigma_3 + s\sigma_c^2)^{1/2} \tag{3-10}$$

where σ_1 = major principal stress at failure

σ_3 = minor principal stress

σ_c = uniaxial compressive strength of intact rock

m, s = constants depending on the nature of rock mass and the extent to which it has been disturbed before subjected to the failure stresses σ_1 and σ_3

The parameters m and s vary with rock type and the rock mass quality Q (Barton et al., 1974).

Equation (3-10) is used for the initial rock mass strength. In the broken or plastic zone, the parameters m and s are reduced to m_r and s_r so that the residual strength of the broken rock mass is now

$$\sigma_1 = \sigma_3 + (m_r\sigma_c\sigma_3 + s_r\sigma_c^2)^{1/2} \tag{3-11}$$

The strength criterion given by Eqs. (3-10) and (3-11) has the advantage that it is based on one simple material property (σ_c) and rock mass quality data that are available during the investigation and construction phase of a tunneling project.

The idealized stress-strain relationship used in a closed-form analysis is shown in Figure 3-42. For the more complex model shown in Figure 3-43, a closed-form solution cannot be obtained, and a stepwise calculation procedure must be used in this case. In both instances the rock is assumed linearly elastic until the initial strength for the appropriate value of σ_3 is reached. Thereafter, the strength immediately drops to the residual in the simple model of Figure 3-42, while it gradually, with increasing strain, assumes the more complex model of Figure 3-43. Compressive stresses and direct strains are taken as positive.

When the stresses are reduced, some elastic volume increase will occur, but this

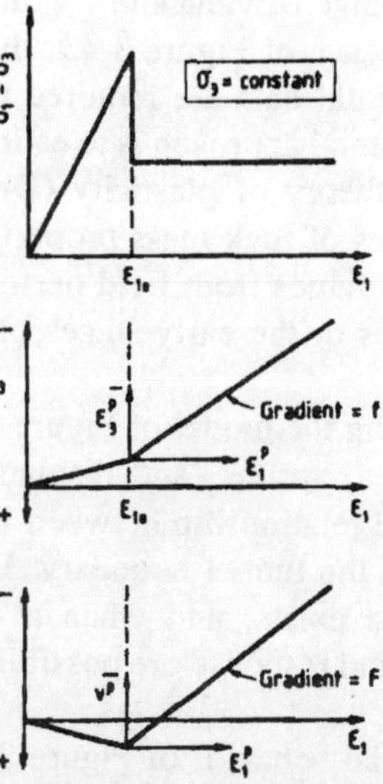

Figure 3-42 Material behavior model used in closed-form solution.

effect is not evaluated explicitly, although it is subsumed in the gradients of the post-peak ϵ_3 versus ϵ_1 and v versus ϵ_1 curves (Figures 3-42, 3-43). In this case, ϵ_1 and ϵ_3 = major and minor principal strains; and v = volumetric strain in the rock mass surrounding the tunnel. Thus, the strain components ϵ_1^p, ϵ_3^p, and v^p are total post-peak rather than plastic strain increments. The gradients f and F in Figure 3-42 are

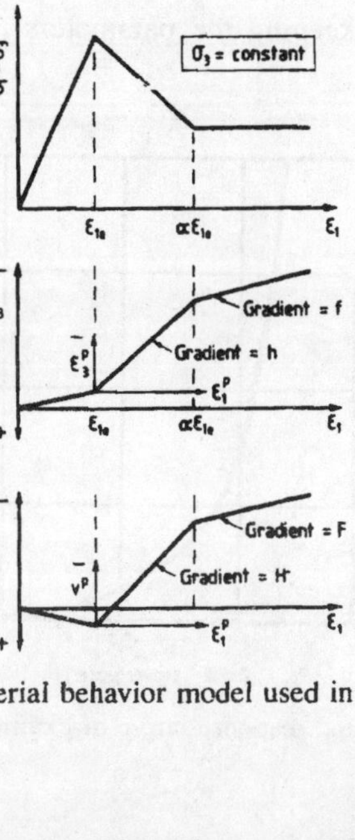

Figure 3-43 Material behavior model used in stepwise solution.

independent of σ_3 over the range of values of σ_3 that apply in the broken zone. In the stepwise calculation sequence of Figure 3-43, the parameters $\alpha, f, h, F,$ and H may vary with σ_3. Experimental data are required to determine these parameters. An alternative to the experimental approach is to estimate these parameters using the associated flow rule of the theory of plasticity (Drucker and Prager, 1952). The selection of appropriate values of rock mass properties, such as the parameter f, is aided by back-calculation of values from field performance data obtained by monitoring either trial excavations or the early stages of full-scale construction (John, 1977; Lombardi, 1977).

Closed-form solutions using the model of Figure 3-42 are given by Brown et al. (1983), and apply to stresses, strains, and displacements. The ground response curve is determined from the relationship between internal support pressure p_i and the radial displacement u_i at the tunnel boundary. Equations are available for two cases: when no broken zone exists, and when a broken zone exists. The steps required to calculate the ground response are possible with a programmable calculator.

Solutions using the material behavior of Figure 3-43 must take into account the possible existence of three different zones around the tunnel: (a) an elastic zone remote from the tunnel; (b) an intermediate plastic zone where stresses and strains fall on the strain-softening portion of Figure 3-43; and (c) an inner plastic zone in which stresses are limited by the residual strength of the rock mass. Likewise, Brown et al. (1983) give expressions for stresses, strains, and displacements.

Example A 35-ft-diameter highway tunnel is driven in a fair-to-good quality limestone at a depth of 400 ft below the surface. The hydrostatic in situ stress is $p_0 = 480$ lb/in.^{2} The following material property data are given for the rock mass: $\sigma_c = 4000$ lb/in.^{2}; $m = 0.5$; $s = 0.001$; $E = 2 \times 10^5$ lb/in.^{2}; $\nu = 0.25$; $m_r = 0.1$; $s_r = 0$; $h = 2.0$; $f = 1.2$; and $\alpha = 3.5$.

The example is solved keeping the parameters f and h constant. Figure 3-44

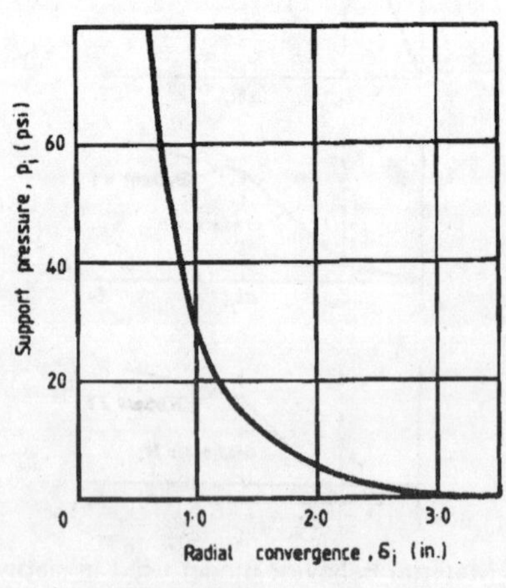

Figure 3-44 Computed ground response curve of example. (From Brown et al., 1983.)

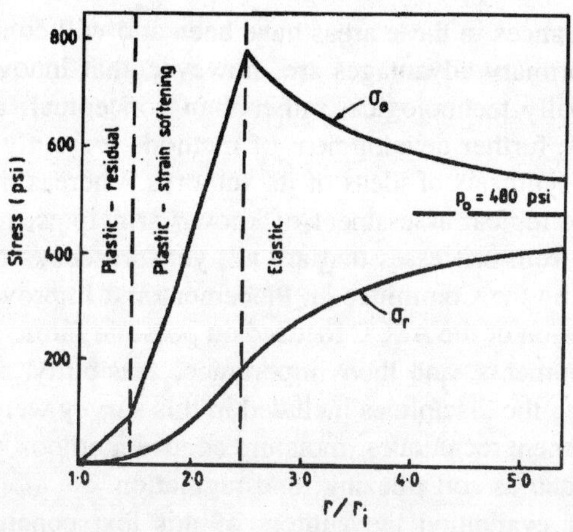

Figure 3-45 Development of tangential and radial stresses with distance from tunnel wall, design example. (From Brown et al., 1983.) σ_θ = tangential stress; σ_r = radial stress.

shows the ground response curve, and Figure 3-45 shows the development of tangential and radial stresses in the rock with distance from the tunnel wall. A maximum radial convergence 3.5 in. is calculated for $p_i = 0$, and at this stage an internal support pressure 10 lb/in.^{2} is required (for $p_i = 0$) and can be readily applied with pattern rock bolts or anchors limiting the convergence to 1.6 in. A supplementary support of wire mesh and shotcrete will stabilize small rock blocks developed at the excavation surface when the rock mass is broken and at residual strength. In this case, the effect of gravity on the broken rock along the roof will probably require support pressures greater than 10 lb/in.^{2} to control roof displacement.

Figure 3-45 shows the difference in stress distribution, particularly for σ_θ, in the three zones surrounding the tunnel. Since $s_r = 0$, the residual strength of the broken rock is zero when $p_i = \sigma_3 = 0$. Thus, at the tunnel boundary, $\sigma_\theta = 0$ when there is no internal support pressure. The tangential stress increases at a greater rate with distance from the tunnel wall in the strain-softening plastic zone than in the residual plastic zone, and reaches a maximum at the elastic-plastic boundary. The radial stress increases throughout the three zones as shown.

The foregoing review highlights the procedures that are normally required to articulate ground response in rock excavations and identify ground treatment and ground support options. For additional reading reference is made to Xanthakos (1991) and Goodman (1989).

3-11 TRENDS IN CONTROL AND IMPROVEMENT TECHNIQUES WITH PARTICULAR APPLICATIONS

The disciplines presented in this book as viable solutions to ground engineering problems are tested and promising methodologies for ground support, control, and improvement. Advances in these areas have been and will continue to be rapid and impressive. The primary advantages are, however, that innovations and advances have been principally technological rather than conceptual, and this trend is expected to dominate further development of methods presently available.

This book is a synthesis of ideas of its authors. Whereas the choice of material reflects merely the logical assessment of known principles, many future developments may result from processes that are not yet foreseen.

A 1978 survey by the Committee on Placement and Improvement of Soils of the Geotechnical Division of the ASCE focused on possible future advances, significant long-range developments, and their importance, feasibility, and probable time of occurrence. Among the disciplines included in this survey were densification methods, soil reinforcement techniques, moisture control methods, grouting procedures, related methods such as soil freezing, and regulation.

From a similar evaluation the authors of this text concluded that the eleven disciplines that make up the topics of this book are consistent in that they have demonstrated high capability, high desirability, and high feasibility. Hence, future advances should be expected to occur in a clear and consistent format. These methodologies may be used singly or in combinations either in new construction or in rehabilitation of existing structures.

One example of where the foregoing principles have found direct application is the vast program of dam reassessment, remedial work, and rehabilitation currently under way in the United States. Some of these techniques apply to a particular type of dam. For example, rock anchors can be used only in a concrete dam, whereas diaphragm walls and vertical screens are feasible only in earth embankments. Grouting for seepage control is common to both types but remedial foundation grouting for concrete dams is invariably applied to rock. Such grouting for embankment dams usually consists of soil treatment methods. The same methods are used to control foundation soil liquefaction, a problem arising typically in embankment dams. Alternatively, other geotechnical processes can be used for soil improvement, such as vibroflotation, blasting, compaction piles, and dynamic consolidation, but it is inconceivable that these could be applied to the foundations of an existing dam.

Bruce (1990) gives a comprehensive summary of dam rehabilitation and remedial work involving specialist techniques for strengthening and for improving function and performance. A brief review with illustrations is presented in this section.

Stability Aspects

A typical problem in concrete gravity dams is stability against overturning and sliding. The use of prestressed rock anchors provides additional resistance to overturning and improves resistance against sliding (Xanthakos, 1991). Notable examples are the Cheurfas dam in Algeria, and the Tarbela dam in Pakistan, where anchors were installed to provide a zone of compressed rock to enhance resistance to the dynamic forces induced by the water during operation of the spillway/flipbucket structure. A relatively new concept was demonstrated with the Stewart Mountain Dam, located on the Salt River in Arizona. Anchors were designed to provide overall stability and continuity to the arch structure and the left thrust block during a projected major earthquake.

For this project the maximum credible earthquake has an estimated magnitude 6.75 10 miles away, capable of producing a site acceleration of 0.34 g. The associated dynamic effects could shake the structure with a corresponding loss of stability and probable failure. Rock anchors were used to restore the structural integrity and increase the factor of safety against dynamic effects. Three-dimensional finite element studies show that anchors develop resisting forces that have a positive effect on the arch by providing additional constraint across the horizontal construction joints and by compressing the arch horizontally.

Grouting Techniques for Structural Repairs and Seepage Control

Grouting, discussed in Chapters 7 and 8, is recognized and routinely used in ground engineering. Major dams commonly incorporate this technique as an integral activity and part of the design. In remedial construction, however, there appears to be less evidence of constant use, although more recently this misconception tends to be recognized considering the effectiveness of remedial grouting. Thus, this technique may now be used in the context of three particular applications: (a) for seepage control in concrete dams (involving concrete and rock grouting); (b) for seepage control in embankment dams (involving rock and soil grouting); and (c) for miscellaneous applications including void filling, consolidation grouting, and slabjacking.

Water seeping through or under a concrete dam may be the cause of considerable uplift pressure. Foundation scour resulting from piping was the cause of failure of the St. Anthony Falls lower dam powerhouse on the Mississippi River in Minneapolis. For embankment dams, seepage can be equally dangerous, especially where sufficient internal filters have not been provided. Many of the most damaging dam failures have been caused by seepage-induced piping, and typical examples are the Teton dam and the Quail Creek dam.

It may appear that insufficient design considerations with respect to drilling and grouting parameters and procedures have rendered the application in some instances less effective with time. For example, Petrovsky (1982) discusses the leaching of cementitious compounds from grout curtains as a function of water-cement ratios, a problem also addressed by Houlsby (1982). The question of grout permanence is even more important for curtains in alluvials formed with the earlier silicate based chemicals alone, but this is less of a concern in the United States since such technologies have not been routinely applied to U.S. dams.

A classic example favoring grouting galleries in dams and abutments, especially for taller structures on relatively poor foundations, is the Hoover (Boulder) dam, a 222-m-high arch-gravity structure founded on imperfectly cemented volcanic breccia. The original grout curtain was extended from 40 m to 130 m deep in the valley bottom and to 90 m into the abutments after the detection of high seepage volumes and pressures.

A review by Davidson (1990) of five case histories of remedial grouting under embankment dams resulted in the following comments: (a) grouting can be successful in reducing seepage volume but may not significantly reduce piezometric levels, especially in soils and rocks with open but ungroutable pores or fissures; (b) grouting may be only a temporary solution if the real cause is solution of a soluble horizon; (c) attacking the seepage problem upstream with a clay blanket may be more cost effective; (d) although in the past state-of-practice methods and materials may have been used in original solutions, they may now be judged ineffective and inappropriate in view of current knowledge; and (e) the extent of knowledge of the foundation and embankment materials as well as the construction history will determine the accuracy of the analysis of the seepage problem.

A comprehensive review of grouting application and case histories is given by Bruce (1990) for concrete dams and earth embankments.

Seepage Control by Concrete Diaphragms

Various types of cutoffs are reviewed in Chapter 10. Concrete cutoffs built by the slurry trench method are discussed by Xanthakos (1979). These systems are now commonly incorporated in dam construction on alluvial foundations, since they have been tested and are known to provide effective seepage barriers. In the last decade the use of concrete diaphragms to repair existing embankment dams has been expanded in North America.

The interlocked bored pile wall, constructed by overlapping large-diameter piles (Xanthakos, 1979), may be used in special remedial applications as well as in new

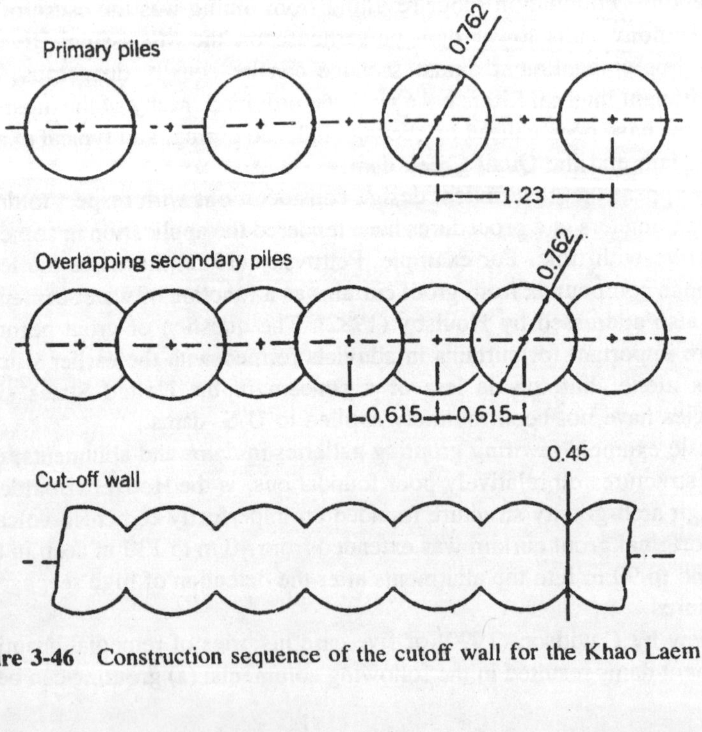

Figure 3-46 Construction sequence of the cutoff wall for the Khao Laem dam.

dam construction. An example is the Khao Laem dam in Thailand (Watakeekul and Coles, 1985). This is a concrete-faced rock fill dam 114 m high, founded on limestone containing major karstic features to a depth of 60 m, and other highly permeable zones to a depth of 180 m. Although conventional grouting was provided to this depth, the dam was reinforced with a cutoff concrete wall of interlocking bored piles. The construction sequence of this wall is shown in Figure 3-46.

3-12 BLASTING AND EXCAVATION

Blasting in underground construction is no longer a mere appendage to plans and specifications for a major project. Considering the scope, cost, and safety aspects of this work, blasting must be well designed and controlled. In addition, precision blasting indicates more severe restrictions than those associated with cautious blasting. A considerable amount of research has taken place in the last decades to warrant the technical requirements of blasting and place this technique on a higher technical level. New products have been introduced, making the use of explosives more effective and safer, and allowing better control of blast effects. Rapid advances in instrumentation accommodate almost automatic monitoring of routine blasting, and allow more aspects of blasting to be quantified and understood.

Although the trend is toward more restrictive limitations, recent research shows that some existing limitations are unduly restrictive under certain conditions. Such factors can be combined to increase the cost of construction unnecessarily, or entirely prohibit the work from being done. Conversely, blasting typically faces the problem of public sensitivity and the adverse reaction to the dynamic and sound effects that accompany this operation. In this context, there is no suitable substitute for advance preparation, dissemination of information, preblast inspection, and good public relations.

Importance of Specifications Contract formats typically should encourage good safety practices and tend to penalize carelessness that may lead to damage and injury. A good specification should warn bidders of the degree of caution to be exercizsed during blasting and excavations in an urban setting. For underground chambers, the specifications should indicate the special controlled blasting techniques required for all perimeter surfaces. The scope becomes more definitive if limits are placed specifically on hole dimensions, spacing, and explosive charge concentrations (Oriard, 1981). If unique methods are required, they should be included in the construction documents, and explicitly stated.

Drilling For controlled blasting in structural excavations, drilling is the first input to the operation. In recent years, there have been notable advances in drilling equipment, particularly for large-hole drilling aboveground. Many operations use blast holes in the range of 10 to 15 in. in diameter, and often larger, advanced as rapidly as small-diameter drills. A notable change in drilling has been the increasing use of hydraulically operated drills. These drills have demonstrated penetration rates much higher than those for ordinary pneumatic drills, but require extensive care and maintenance. Some drills are fully articulated and semiautomatic in the sense that they can maintain positions parallel to the original setting when properly operated.

Explosives Products The most widely accepted changes relate to the development of less sensitive, safer blasting agents as opposed to traditional explosives. More choices have also become available for detonators, mainly among the nonelectric types. Among these are nonelectric caps made with the delay element inside the cap, initiated with low-energy detonating chords.

Optimum Results and Fracture Control The final results of blasting in underground chambers may be predicted by the techniques used to develop the final perimeter surfaces. The final walls are better preserved by: (a) careful drilling of perimeter holes; (b) light loading of perimeter holes with explosive charges that are not concentrated or fully coupled to the walls of the drill holes; and (c) making certain that blasting in the row next to the perimeter is incapable of shattering to the perimeter and beyond.

Consideration should be given to newly developing techniques for use in fracture control. It is well-known from fracture mechanics that less energy is required to extend an existing crack than to form a new one. The first crack to form is where some flaw or stress concentration exists. Such flaws can be introduced in the desired plane by cutting small notches into the walls of the drill holes. If a high-pressure water jet is used, a narrow slot can be cut in the form of a true crack or rock joint. The full effectiveness of this simple technique is demonstrated, however, with isotropic rocks. A natural tendency for the rock to break along foliation planes will resist breaking perpendicular to these planes. The important influence of foliation on crack formation is demonstrated from actual cases.

For small-burden blasting such as smooth blasting, the notching or hole scribing technique may result in a reduction in charge concentration to about $\frac{1}{4}$ or $\frac{1}{5}$ of the normal condition with a simultaneous increase in hole spacing to almost twice that of the normal condition. For optimum results, the notch or slot in the drill hole should be sharp, well-formed, and in the plane of the desired breakage pattern. In addition, a suitable kind of stemming should be provided in order to contain the explosive gases long enough to allow the cracks to propagate.

Control of Vibrations and Airblast Overpressure Probably the best indicator of the potential for damage to structures from blasting vibrations is peak particle velocity. Some standards refer to poor plaster in a residential structure as the index of damage. A peak particle velocity of 2 in./sec is rather widely accepted as the safe limit, with the probability of damage increasing with particle velocities exceeding this limit. Limitations for concreted and engineered structures appear to be more liberal, with particle velocity in the range of 4 to 8 in./sec. It is beyond the scope of this text to comment on the ramifications of these standards, including observational procedures used to conclude when damage has occurred and what its extent may be.

This issue is often overdiscussed under arbitrary assumptions without proper verification.

A review of response spectra theories demonstrates the relevance of the time history of vibrations as it relates to the natural frequency of a given structure. This analysis is typically necessary for large critical structures in anticipation of structural survival following significant vibrations such as earthquakes. However, blasting vibrations cover a relatively narrow range of frequencies, and thus it is acceptable to use a single value of particle velocity to standardize the damage potential for a given structure. Others contend that this view is too simple and can give rise to problems at either end of the frequency spectrum. For the case of low-frequency vibrations on a structure with low natural frequency, the criterion may not be conservative, whereas for high-frequency vibrations it may be too conservative with associated unnecessary cost increases.

A simple way to limit ground vibrations generated by blasting is to limit the quantity of explosives detonated at any given instant of time (Oriard, 1981). This quantity is usually referred to as the charge weight per delay, under the assumption that all detonators of the same nominal delay interval will detonate simultaneously. Recent advances have increased the number of delay intervals available to explosives users, and the change is for both electric and nonelectric delays. Proper use allows a nearly unlimited choice of delays, but some previous experience is helpful in planning and executing the more complicated systems.

A development of parallel criteria is noted for airblast overpressures. Early research led to the conclusion that an overpressure of 1 $lb/in.^2$ could almost cause some windows to break, so that an acceptable limit would be in the range of 0.1 to 0.5 $lb/in.^2$ Some plaster damage has been reported even with relatively low airblast overpressures. This, combined with public complaints, has resulted in limits of the order of 0.01 $lb/in.^2$, or even lower. This conservative approach reflects also the fact that better instruments for detecting overpressures now indicate that earlier instruments did not register portions of the very low frequency energy.

Some of the measures that control ground vibrations have a similar function in the control of airblast overpressures. For example, delay detonators can serve the purpose of limiting both effects, although not to the same degree.

Precision Blasting The definite relationship between charging and permitted impact described as shock-wave-velocity is the normal criterion for defining blasting limitations to protect buildings and structures. In some instances there is a restriction based on minimum shock wave acceleration, and this is prevalent in connection with flexible structures and where delicate mechanical and electrical equipment is present. The fact that the limitation is defined as maximum acceleration implies that the relationship between charging and restriction is known and must be continuously determined. It also implies that the choice of blasting method must be adapted to the significance of the restrictions.

Precision blasting, on the other hand, involves, in addition to velocity and acceleration limitations, a relation between maximum acceleration and charge levels, as well as a relationship between bench height and burden, and also between cost and the acceleration limitation. It can be demonstrated by the following example.

Consider the plan of power stations O-1 and O-2, shown in Figure 3-47. The blasting volumes are as follows: 125,000 m^3 of benching, 20,000 m^3 of tunneling, and 5,000 m^3 of trenching (Widmark and Platzer, 1976). The allowable blasting acceleration is 0.7 g at 13 properly located points. There are also restriction on maximum velocity (35 mm/sec or 1.4 in./sec), for covering, for firing, and for high tension lines.

The restriction 0.7 g is significant for work scheduling and planning, whereas other criteria and limitations affect the planning only marginally. A maximum charge level 0.004 $kg/m^{1.5}$ is calculated to correspond to the acceleration limit. An adequate relationship between charge level and attained maximum acceleration is obtained after a period of trial blasting. For this case, the simple relationship is

$$a = (\text{constant})v \pm 0.1 \text{ g} \tag{3-12}$$

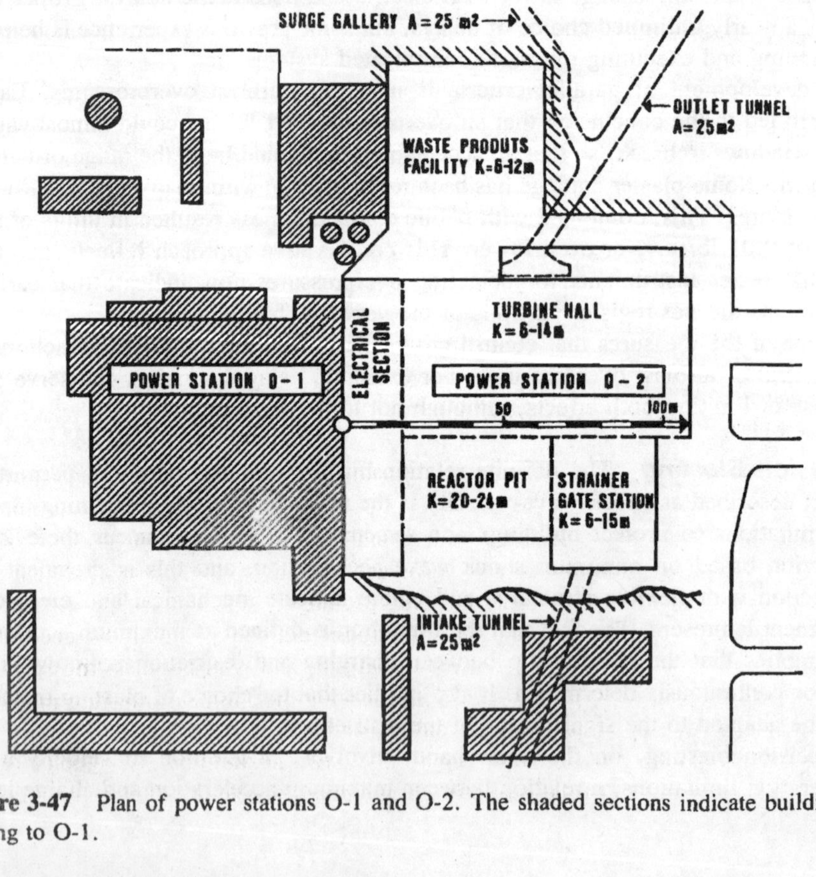

Figure 3-47 Plan of power stations O-1 and O-2. The shaded sections indicate buildings belong to O-1.

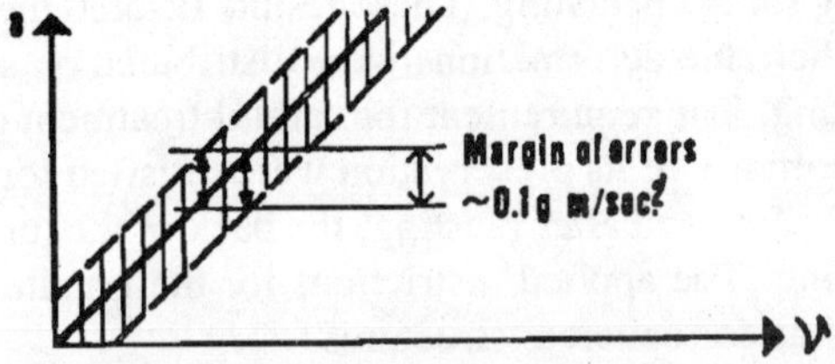

Figure 3-48 Graphical presentation of relationship between maximum acceleration and change level (a versus ν).

where a = acceleration (m/sec^2), and
ν = charge level factor.

After more than 1000 rounds, it became obvious that Eq. (3-12) was satisfied by almost 99 percent of the measurements. The relationship between a and ν is shown graphically in Figure 3-48.

The planning of the bench blasting must also consider optimum data regarding economy, restrictions, and time. The analysis shows that the choice of bench heights is relevant to optimum economic results and blasting cost. For this example, the bench heights were varied between 3 and 6 m, according to the distance to power plant number O-1. As the volume of benching was substantially increased during the working period, parts of the reactor pit were charged as doubled benches with separate column charges. The relationship between permissible charge and specific cost is shown graphically in Figure 3-49. The detonators have 25-msec delays.

Another objective of precision blasting is to avoid damage to surrounding rock. For the benching it was necessary to use a method of contour-blasting that would not damage the rock mass and would also allow satisfactory production rates. A series of test blastings showed that presplitting was not applicable in this case because of the acceleration limits. Smooth blasting worked well in the tunneling but required a

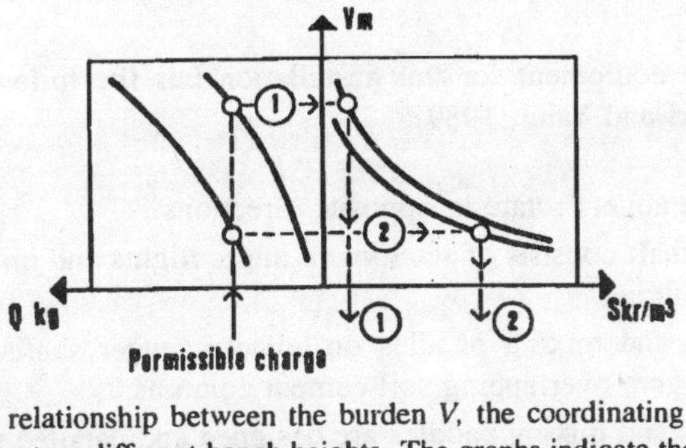

Figure 3-49 The relationship between the burden V, the coordinating charge Q, and the specific cost of k for two different bench heights. The graphs indicate that if the permissible charge amount is given, then the combination of bench height and burden in alternative 1 results in a significantly lower specific cost than alternative 2.

high amount of drilling for the benching. These results favored the use of a modified presplitting method where the deformations were distributed on several intervals in the initial part of blasting. The requirement for careful treatment of the surrounding rock mass and the importance of its preservation were satisfied for benching close to the excavated tunnel. Before and after blasting, the back of the tunnel was examined by painting and wedging. The applied restrictions for the granite rock were almost four times as rigorous as for concrete structures.

3-13 SOIL-CEMENT STRUCTURAL WALLS

This system represents a relatively new concept developed independently in Japan (Taki and Yang, 1989). It consists of mixing in situ soils with cement grout using multiaxis augers and mixing paddles to construct overlapping soil-cement columns. The resulting applications are as cutoff walls for groundwater control and structural diaphragm walls for excavation supports.

The installation is feasible in a broad range of soils including: soft to very stiff and low to highly plastic clays and silts; loose to dense sand, gravel, and cobble; and soft rock. During the process, soils are broken up by the cutting heads of the multiaxis auger and mixed with cement grout in situ, in a pugmill fashion by sections of auger flight and paddles mounted on the multiaxis auger shafts. Lightweight H piles are inserted into the columns before the hardening of the soil-cement for reinforcement. The resulting structural combination can resist lateral earth pressures, and when the wall is properly braced laterally it can support excavations of considerable depth.

Origin and Development The concept appears to have originated from the construction of mixing in-place single soil-cement piles, using single-axis earth augers, in 1954. Single piles often result in incomplete overlapping because of limitations of the single-axis equipment. The multiaxis machine has solved this problem, and thus provides continuity in the soil-cement columns. The insertion of steel H piles imparts to the system a structural configuration.

Equipment The equipment for this installation has the following operational characteristics (Taki and Yang, 1989):

1. The adjacent augers rotate in opposite directions.
2. Each auger shaft consists of sections of auger flights and mixing paddles for in situ soil mixing.
3. Auger flight and mixing paddles on adjacent auger shafts overlap during operation to give overlapping soil-cement columns.
4. Auger flights and mixing paddles are designed and detailed to accommodate various soil conditions.

5. The auger mixes soil with grout at its original depth uniformly and continuously but without the traditional auger that moves the soil upward.
6. Tie bands are used to maintain the rigidity of the auger shafts and the space between individual auger shafts for better operation and quality control.

Comparison with Conventional Slurry Wall Construction In principle, the application is completely independent of stability requirements. There is no need to use slurry, and because most of the in situ soil is used as construction material, the volume of soil spoils for disposal is smaller than the by-products of slurry wall excavations. Likewise, the construction time is shorter, and the in situ conversion of natural soil into a construction system results in higher production rates. The process is also compatible with urban requirements and ideal for use in noise- and vibration-sensitive areas.

Mechanical Characteristics These depend essentially on the physical properties of the in situ soil materials, the soil-cement mixing rates, and extent of mixing.

As in conventional cement technology, the strength of soil-cement is influenced by the method of sample preparation and testing conditions. The compressive strength of the soil-cement mix is used as the basic characteristic for design and quality control. According to tests, the curing period affects the development of strength, and increasing the curing period results in higher strength. The 28-day strength is almost twice the 7-day strength for either sand or clay, as shown in Figure 3-50. Likewise, within the working stress range, the soil-cement material is

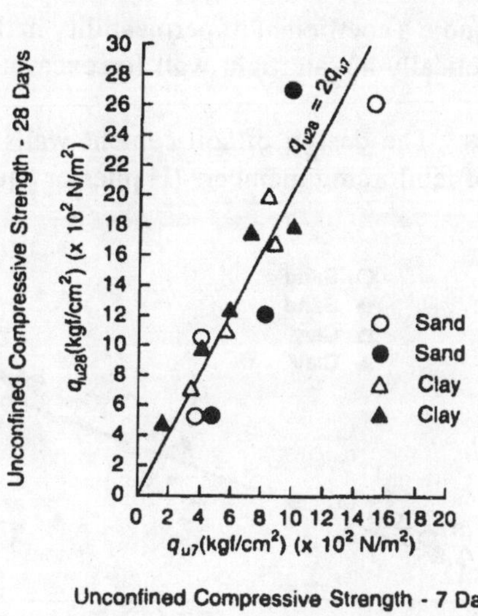

Figure 3-50 Relation between unconfined compressive strengths with different curing times.

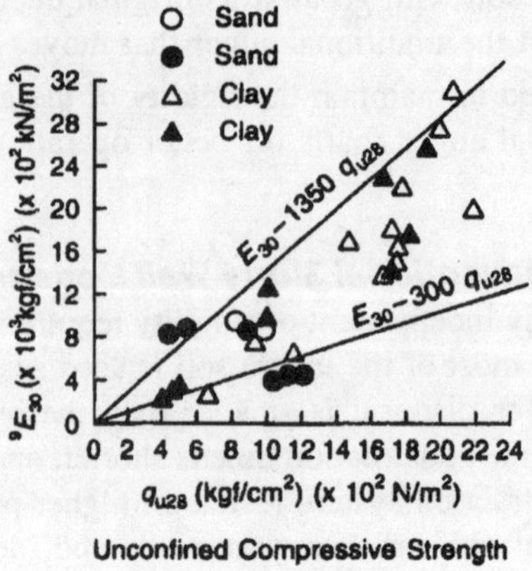

Figure 3-51 Relationship between modulus of elasticity and unconfined compressive strength.

linearly elastic, as evidenced by Figure 3-51, where the unconfined compressive strength is plotted versus the modulus of elasticity. The tensile strength of the mix is low and generally ignored in structural computations.

The relationship between cohesion and unconfined compressive strength is shown in Figure 3-52. The function of shear strength and unconfined compressive strength is shown in Figure 3-53. It follows that for design purposes the shear strength of the soil-cement material can be taken as one-third the unconfined compressive strength.

Taki and Yang (1989) quote a coefficient of permeability in the range 1×10^{-6} to 1×10^{-7} cm/sec, or practically a watertight wall for excavation purposes.

Design Considerations The design of soil-cement walls involves essentially two steps: (a) the design of reinforcing members (H piles or equivalent members) to

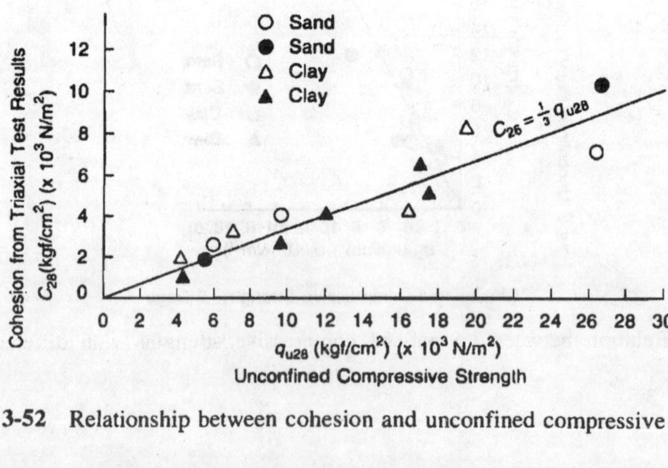

Figure 3-52 Relationship between cohesion and unconfined compressive strength.

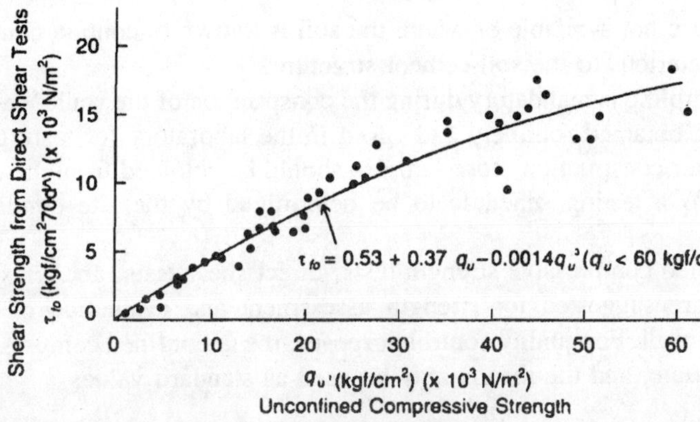

Figure 3-53 Relationship between shear strength and unconfined compressive strength.

resist moments, shears, and so on, as in conventional soldier pile walls, and (b) the design of the soil-cement elements to resist and transfer horizontal earth stresses to the H members.

If the reinforcement member (H beam) is installed in every column, as shown in Figure 3-54*a*, it is only necessary to consider the punch-through shear force Q in calculating the shear stresses. Where the reinforcement member is not installed in every soil-cement column, the soil-cement element may be analyzed using the hypothetical model shown in Figure 3-54*b*. In this case, in addition to the punch-through shear stresses, the analysis may consider compressive stresses along a hypothetical parabolic arch with a configuration formed as shown.

A strength test and a permeability test are commonly performed to obtain strength parameters and provide the basis for a semiempirical design. Prior to construction, tests may be performed on samples prepared in the laboratory using in situ soil mixed with cement. This test is indicated where previous data for the side

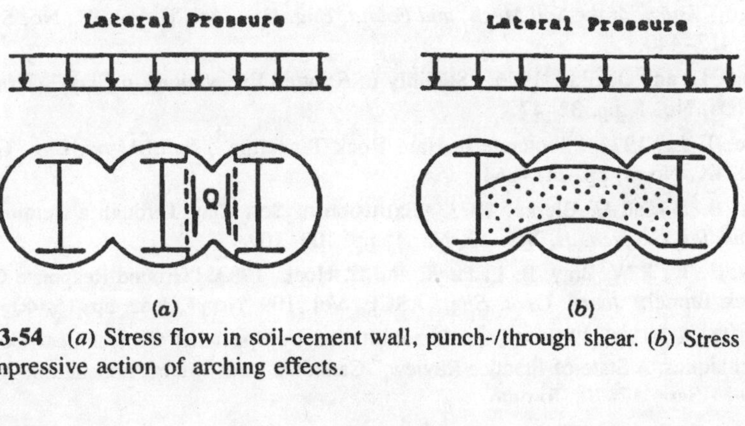

Figure 3-54 (*a*) Stress flow in soil-cement wall, punch-/through shear. (*b*) Stress analysis and compressive action of arching effects.

conditions are not available or where the soil is known to contain constituents that may be deleterious to the soil-cement structure.

Field sampling is mandatory during the construction of the wall. Wet soil-cement samples are obtained routinely and cured in the laboratory for testing and quality control. After construction, core samples should be obtained from the exposed wall according to a testing schedule to be determined by the site conditions during excavation.

Unconfined compressive strength tests, direct shear tests, and triaxial compression tests are suggested for strength assessment and evaluation of the finished cement-soil wall. For quality control purposes, the unconfined compressive strength test is adequate, and the results may be used as standard values.

REFERENCES

Alberts, C. and S. Backstrom, 1971. "Instant Shotcrete Support in Rock Tunnels," *Tunnels and Tunneling*, Jan.

Amberg, W. A. and G. Lombardi, 1974. "Une Methode de Calcul Elasto-Plastique de l'Etat de Tension et de Deformation Autour d'une Cavite Souterraine," Advances in Rock Mechanics, Proc. of the 3rd Congress of the International Society for Rock Mechanics, Vol. 2, Part B, National Academy of Sciences, Washington, D.C., pp. 1055–1060.

Azzouz, A. S., and J. T. Germaine, 1981. "Behavior of Cylindrical Openings in Strain-Softening Ground," Rock Mechanics from Research to Application, Proc. of the 22nd U.S. Symp. on Rock Mechanics, Massachusetts Institute of Technology, Cambridge, Mass., pp. 258–263.

Bara, J. P., 1976. "Collapsible Soils," ASCE Annual Conv. Exposition, Sept., Philadelphia, Pa.

Barden, L., A. McGown, and K. Collins, 1973. "The Collapse Mechanism in Partly Saturated Soil," Engineering Geology, Amsterdam, The Netherlands, pp. 49–60.

Barton, N., R. Lien, and J. Lunde, 1974. "Engineering Classification of Rock Masses for the Design of Tunnel Support," *Rock Mechanics*, Vol. 6, No. 4, Vienna, Austria, Dec., pp. 189–236.

Baker, C. N., and F. Khan, 1971. "Caisson Construction Problems and Correction in Chicago," *Journ. of the Soil Mech. and Found. Eng. Div.*, ASCE, Vol. 97, No. SM2, Feb., pp. 417–440.

Bjerrum, L. and O. Eide, 1956. "Stability of Strutted Excavations in Clay," *Geotechnique*, Vol. 6, No. 1, pp. 32–47.

Brekke, T. L., 1972. "Shotcrete in Hard Rock Tunneling," *Bull. Assoc. Eng. Geologists*, Vol. IX, No. 3, pp. 241–264.

Broms, B. B. and H. Bjerke, 1973. "Extrusion of Soft Clay Through a Retaining Wall," *Canadian Geotech. J.*, Vol. 10, No. 1, pp. 103–109.

Brown, E. T., J. W. Bray, B. Ladanyi, and E. Hoek, 1983. "Ground Response Curves for Rock Tunnel," *Journ. Geot. Eng.*, ASCE, Vol. 109, No. 1, Jan., pp. 15–40.

Bruce, D. A., 1990. "Major Dam Rehabilitation by Specialist Geotechnical Construction Techniques: A State-of-Practice Review," Canadian Dam Safety Association, 2nd Annual Conf., Sept. 17–20, Toronto.

Cecil, S. O., 1970. "Correlations of Rock Bolt-Shotcrete Support and Rock Quality Parameters in Scandinavian Tunnels," Ph.D. Thesis, Univ. of Illinois, Urbana.

Christian, J. T., 1989. "Design of Lateral Support Systems," Design, Construction and Performance of Deep Excavations in Urban Areas, Proc. BSCES/ASCE Seminar, Boston, pp. 1–26.

Clemence, S. P. and A. O. Finbarr, 1981. "Design Considerations for Collapsible Soils," *ASCE Geotech. Journ.*, March, pp. 305–317.

Clough, G. W. and G. M. Denby, 1977. "Stabilizing Berm Design for Temporary Walls in Clay," *Journ. of the Geot. Eng. Div.*, ASCE, Vol. 103, No. GT2, Feb., pp. 75–90.

Clough, G. W. and L. A. Hansen, 1981. "Clay Anisotropy and Braced Wall Behavior," *ASCE Geotech. J.*, July, pp. 893–913.

Clough, G. W. and T. D. O'Rourke, 1990. "Construction Induced Movements of in situ Walls," Design and Performance of Earth Retaining Structures, Proc. of ASCE Specialty Conf., Ithaca, N.Y., pp. 439–470.

Clough, G. W., P. R. Weber, and J. Lamont, Jr., 1972. "Design and Observation of a Tied-Back Wall," ASCE Specialty Conf. on Performance of Earth and Earth-Supported Structures, Purdue Univ., June, Vol. I, Part 2, pp. 1367–1389.

Corbett, B. O. and M. A. Stroud, 1974. "Temporary Retaining Wall Constructed by the Berlinoise System at Beaubourg Center, Paris," Proc. Diaphragm Walls Anchorages, Inst. Civ. Eng. London.

Cording, E. J., 1974. "Geologic Considerations in Shotcrete Designs," ASCE-ACI, SP-45, pp. 175–199.

Cording, E. J., J. Mahar, and G. Fernandez-Delgado, 1981. "Steel-Fiber-Reinforced Shotcrete," The Atlanta Research Chamber, U.S. Department of Transportation, UMTA Repot No. GA-06-0007-81-1. Also p. VIII-18.

Daemen, J. J. K., 1975. "Tunnel Support Loading Caused by Rock Failure," Technical Report MRD-3-75, Missouri River Division, U.S. Corps. of Engineers, Omaha, Neb.

Daemen, J. J. K., 1977. "Problems in Tunnel Support Mechanics," *Underground Space*, Vol. 1, No. 3, pp. 163–172.

Daemen, J. J. K. and C. Fairhurst, 1971. "Influence of Failed Rock Properties on Tunnel Stability," Dynamic Rock Mech., Proc. 12th Symp. Rock Mech., AIME, New York, pp. 855–875.

Davidson, R. R., 1990. "Rehabilitation of Dam Foundations," Internal Report available from Woodward Clyde Consultants, Denver.

Davisson, M. T. an H. J. Gill, 1963. "Laterally Loaded Piles in a Layered Soil System," *J. Soil Mech. Ground. Eng. Div.*, ASCE, Vol. 89, No. SM3, May, pp. 63–94.

Deere, D., R. Peck, J. Monsees, and B. Schmidt, 1969. "Design of Tunnel Liners," Univ. of Illinois, Urbana, Feb.

DiBiagio, E. and J. A. Roti, 1972. "Earth Pressure Measurements on a Braced Slurry-Trench Wall in Soft Clay," Proc. 5th Europ. Conf. on Soil Mech. Found. Eng., Vol. 1, pp. 473–484.

Dragon, A. and Z. Mroz, 1979. "A Model for Plastic Creep of Rock-Like Materials Accounting for the Kinetics of Fracture," *Int. J. Rock Mech. Mining Sciences*, Vol. 16, No. 4, Aug., pp. 253–259.

Drossel, M. G., 1975. "Braced Sheet Piling Shores High-Rise Building Excavation in Poor Soil," Construction Methods and Equipment, Vol. 57, No. 8, pp. 38–40.

Drucker, D. C. and W. Prager, 1952. "Soil Mechanics and Plastic Analysis or Limit Design," *Quart. Appl. Math.*, Vol. 10, No. 2, July, pp. 157–165.

Dudley, J. H., 1970. "Review of Collapsing Soils," *J. Soil Mech. Found. Div., ASCE*, Vol. 96, No. SM3, Proc. Paper 7278, May, pp. 925–947.

Duncan, J. M. and C. Y. Chang, 1970. "Nonlinear Analysis of Stress and Strain in Soils," *J. Soil. Mech. Found. Div., ASCE*, Vol. 96, No. SM5, Proc. Paper 7513, Sept., pp. 1629–1653.

Egger, P., 1974. "Gebirgsdruck im Tunnelbau und Stutzwirkung der Ortsburst bei Uberschreiten der Gebirgsfestigkeit," Advances in Rock Mech., Proc. 3rd Congr. Int. Soc. Rock Mech., National Academy of Sciences, Washington, D.C., Vol. 2, Part B, pp. 1007–1011.

Egger, P., 1980. "Deformations at the Face of the Heading and Determination of the Cohesion of the Rock Mass," *Underground Space*, Vol. 4, No. 5, Mar.–Apr., pp. 313–318.

Fenner, R., 1938. "Untersuchungen zur Erkenntnis des Gebirgsdruckes," *Gluckauf*, Vol. 74, Essen, Germany, pp. 681–695 and 705–715.

Fernandez, G., J. W. Mahar, and H. W. Parker, 1976. "Structural Behavior of Thin Shotcrete Liners Obtained from Large Scale Tests," Shotcrete for Ground Support, ASCE and ACI SP-54, pp. 399–442.

Fernandez-Delgado, G., E. J. Cording, J. W. Mahar, and M. L. Van Sint, 1981. "Thin Shotcrete Linings in Loosening Rock," The Atlanta Research Chamber, Applied Research for Tunnels, UMTA Report No. GA-06-0007-81-1, Mar.

Goldberg, D. T., W. E. Jaworski, and M. D. Gordon, 1976. "Lateral Support Systems and Underpinnings," Federal Highway Administration, U.S. Department of Transportation, Washington, D.C.

Goodman, R. E., 1989. *Rock Mechanics*, 2nd ed., Wiley, New York.

Hendron, A. J. and A. K. Aiyer, 1972. "Stresses and Strains Around a Cylindrical Tunnel in an Elasto-Plastic Material with Dilatancy," Technical Report No. 10, Missouri River Division, U.S. Corps of Engineers, Omaha, Neb.

Hobbs, D. W. 1966. "A Study of the Behavior of Broken Rock under Triaxial Compression and Its Application to Mine Roadways," Int. J. of Rock Mech. Mining Sciences, Vol. 3, March, pp. 11–43.

Hoek, E. and E. T. Brown, 1980a. "Empirical Strength Criterion for Rock Masses," *J. Geotech. Eng. Div., ASCE*, Vol. 106, No. GT9, Proc. Paper 15715, Sept., pp. 1013–1035.

Hoek, E. and E. T. Brown, 1980b. Underground Excavations in Rock, The Institution of Mining and Metallurgy, London.

Holmgren, J., 1975. "Plane Shotcrete Layer Subjected to Punch Loads," Swedish Rock Mech. Res. Found., Stockholm.

Houlsby, A. C., 1982. "Optimum Water–Cement Ratios for Rock Grouting," as Reference 23, pp. 317–331.

Jaworski, W. E., 1973. "An Evaluation of the Performance of a Braced Excavation," Ph.D. Thesis, MIT, Cambridge, Mass.

Jennings, J. E. and K. Knight, 1975. "A Guide to Construction on or with Materials Exhibiting Additional Settlement Due to 'Collapse' of Grain Structure," 6th Regional Conf. for Africa on Soil Mech. Found. Eng., Sept., pp. 99–105.

John, M., 1977. "Adjustment of Programs of Measurements Based on the Results of Current Evaluation," *Field Measurements in Rock Mechanics*, K. Kovari, Ed., Vol. 2, A. A. Balkema, Rotterdam, Holland, pp. 639–656.

Kaiser, P. K., 1981. "A New Concept to Evaluate Tunnel Performance—Influence of Excavation Procedures," Rock Mechanics from Research to Application, Proc., 22nd U.S. Symp. Rock. Mech., Massachusetts Institute of Technology, Cambridge, 1981, pp. 264–271.

Kastner, N., 1949. "Uber den echten Gebiergsdruck beim Bau tiefliengender Tunnel," *Osterreich Bauzeitscarift*, Vol. 10, No. 11, Vienna, Austria.

Kavazanjian, E. and J. K. Mitchell, 1980. "Time-Dependent Deformation Behavior of Clays," *ASCE Geotech. J.*, June, p. 611–630.

Labasse, H., 1949. "Les Pressions de Terrains dans les Mines de Huiles," *Revue Universelle des Mines*, Series 9, Vol. 5, No. 3, Liege, Belgium, Mar., pp. 78–88.

Ladanyi, B., 1974. "Use of the Long-Term Strength Concept in the Determination of Ground Pressure on Tunnel Linings," Advances in Rock Mechanics, Proc. 3rd Congr. Int. Soc. Rock Mech., National Academy of Sciences, Washington, D.C., Vol. 2, Part B, pp. 1150–1156.

Ladanyi, B., 1980. "Direct Determination of Ground Pressure on Tunnel Lining in a Nonlinear Viscoelastic Rock," Underground Rock Engineering, Proc. 13th Canadian Rock Mech. Symp., The Canadian Institute of Mining and Metallurgy, Montreal, Canada, pp. 126–132.

Lambe, T. W. 1970. "Braced Excavations," Proc. ASCE Specialty Conf., Lateral Stresses and Earth Ret. Struct., Cornell Univ., Ithaca, N.Y., pp. 149–218.

Lambe, T. W. and C. K. Turner, 1970. "Braced Excavations," Design of Earth-Retaining Structures, ASCE, pp. 149–218.

Lombardi, G., 1970. "Influence of Rock Characteristics on the Stability of Rock Cavities," *Tunnels and Tunnelling*, Vol. 2, No. 1, London, Jan.–Feb., pp. 19–22; Vol. 2, No. 2, Mar.–Apr., pp. 104–109.

Lombardi, G., 1977. "Long-Term Measurements in Underground Openings and Their Interpretation with Special Consideration to the Rheological Behavior of Rock," *Field Measurements in Rock Mechanics*, K. Kovari, Ed., Vol. 2, A. A. Balkema, Rotterdam, Holland, pp. 839–858.

Lombardi, G., 1980. "Some Comments on the Convergence-Confinement Method," *Underground Space*, Vol. 4, No. 4, Jan.–Feb., pp. 249–258.

Lowrance, W. W., 1976. *Of Acceptable Risk-Science and the Determination of Safety*, W. Kaufmann, Los Altos, Calif., 180 pp.

Lukas, R. G. and C. N. Baker, 1978. "Ground Movement Associated with Drilled Pier Installations," Proc. ASCE Conv. Preprint 3266, Pittsburgh, Pa., April.

Mahar, J., 1975. "Shotcrete Practice in Underground Construction," Federal Highway Administration, U.S. Department of Transportation, Office of Res. and Development, NTIS PB 248 675.

Mahar, J. W., F. L. Gau, and E. J. Cording, 1972. "Observations during Construction of Rock Tunnels for the Washington, D.C. Subway," Proc. North Amer. Rapid Exc. and Tunneling Conf., Vol. 1, pp. 659–681.

Mahar et al., 1976. "Shotcrete Practice in Underground Construction," U.S. Department of Transportation, Report FRA-OR&D 75-90.

Mana, A. I. and G. W. Clough, 1981. "Predictions of Movement for Braced Cuts in Clay," *ASCE Geotech. J.*, June, pp. 759–777.

Mason, E. E., 1970. "The Function of Shotcrete in Support and Lining of the Vancouver Railway Tunnel," in *Rapid Excavation, Problems and Progress*, D. M. Yardley, Ed., The American Inst. of Mining Metalurgical and Petroleum Engineers, Inc.

Mason, R. and L. Lorig, 1981. "Conventional Shotcrete," The Atlanta Research Chamber, U.S. Department of Transportation, UMTA, Report No. GA-06-0007-81-1.

Morrison, R. G. K. an D. F. Coates, 1955. "Soil Mechanics Applied to Rock Failure in Mines," *Canadian Mining and Metallurgical Bulletin*, Vol. 48, No. 523, Montreal, Canada, Nov., pp. 701–711.

Muiz Wood, A. M., 1979. "Fourteenth Sir Julius Wernher Memorial Lecture—Ground Behaviour and Support for Mining and Tunnelling," *Tunnelling '79*, M. J. Jones, Ed., The Institution of Mining and Metallurgy, London, pp. xi–xxii.

Northen, R. D., 1969. "Engineering Properties of Loess and Other Collapsible Soils," 7th Int. Conf. Soil. Mech. Found. Eng., pp. 445–452.

Nussbaum, H., 1972. "Recent Development of the New Austrian Tunnelling Method," ASCE Nat. Meeting of Structural Engineering, Cleveland, Ohio, Apr.

Oriard, L. L., 1981. "Modern Blasting in an Urban Setting, The Atlanta Research Chamber, U.S. Department of Transportation, UMTA, Report No. GA-06-0007-81-1.

O'Rourke, T. D., 1981. "Ground Movement Caused by Braced Excavations," *ASCE Geotech. J.*, Sept., pp. 1159–1177.

O'Rourke, T. D., 1989. "Predicting Displacements of Lateral Support Systems," Design Construction and Performance of Deep Excavations in Urban Areas, Proc. BSCES/ASCE Seminar, Boston, pp. 1–35.

O'Rourke, T. D. and E. J. Cording, 1979. "Observed Loads and Displacements for a Deep Subway Excavation," Proc. RETC, Vol. 2, San Francisco, CA, pp. 1305–1326.

Osaimi, A. E. and G. W. Clough, 1979. "Pore Pressure Dissipation During Excavation," *ASCE Geotech. J.*, Apr., pp. 481–498.

Palmer, J. H. L. and T. C. Kenney, 1972. "Analytical Study of a Braced Excavation in Weak Clay," Canadian Geot. J., Vol. 9, pp. 145–164.

Panet, M., 1976. "Analyse de la Stabilite d'un Tunnel Creuse dans un Massif Rocheux en Tenant Compte du Comportement apres la Rupture," *Rock Mechanics*, Vol. 8, No. 4, Vienna, Austria, Nov., pp. 209–223.

Parker, H. W., G. Fernandez, and L. J. Lorig, 1975. "Field Oriented Investigation of Conventional and Experimental Shotcrete for Tunnels," U.S. Department of Transportation, Report FRA OR&D [illegible]6, Aug.

Peck, R. B., 1969. "Deep Excavations and Tunneling in Soft Ground," State-of-the-Art Report, 7th ICSMFE, Mexico City, pp. 225–290.

Petrovsky, M. B., 1982. "Monitoring of Grout Leaching at Three Dams Curtains in Crystalline Rock Foundations," as Reference 23, pp. 105–120.

Rabcewicz, L. V. 1964. "The New Austrian Tunneling Method," *Water Power* (Nov., Dec.).

Rabcewicz, L. V., 1965. "The New Austrian Tunneling Method," *Water Power*, Jan.

Rabcewicz, L. V., 1969. "Stability of Tunnels under Rock Load," *Water Power* (June, July, August).

Rabcewicz, L. V., 1970. "Die halbsteife Schale als Mittel zur empirisch-wissenschaftlichen Bemessung von Hohlraumbauten," *Rock Mechanics*, Suppl. IV.

Ranken, R., J. Ghaboussi, and A. Hendron, 1978. "Analysis of Ground Liner Interaction for Tunnels," UMTA-IL-06-0043-078-3, Springfield, VA, NTIS PB 294818, Oct.

Schwartz, C. W. and H. H. Einstein, 1980. "Simplified Analysis for Ground-Structure Interaction in Tunnelling," The State of the Art in Rock Mechanics, Proc. 21st U.S. Symp. Rock Mech., D. A. Summers, Ed., Univ. of Missouri—Rolla, pp. 787–796.

Scott, J. D., N. E. Wilson, and G. E. Bauer, 1972. "Analysis and Performance of a Braced Cut in Sand with Large Deformations," *Canadian Geot. J.*, Vol. 9, No. 4, pp. 384–406.

Shannon, W. L. and R. J. Strazer, 1970. "Tied-Back Excavation Wall for Seattle First National Bank," *Civ. Eng., ASCE*, Vol. 40, Mar., pp. 62–64.

Shotcrete Strength Testing: Comparing Results of Various Specimens, 1976. T. Rutenbeck, ACE Publication SP-54, Shotcrete for Grounded Support, Eng. Found. Conf.

Skempton, A. W., 1951. "The Bearing Capacity of Clays," Building Research Congr., Div. 1, pp. 180–189.

Sokolovich, V. E., 1971. "New Development in the Chemical Strengthening of Ground," *Osnovaniya, Fundamenty i Mekhanika Gruntov,* No. 5, Sept.–Oct., pp. 26–28.

Stille, H., 1976. "Behavior of Anchored Sheet Pile Walls," Thesis, Royal Institute of Technology, Stockholm.

Sultan, H. A., 1969. "Collapsing Soils, State-of-the-Art," 7th Int. Conf. Soil Mech. Found. Eng., No. 5.

Tait, R. G. and H. T. Taylor, 1975. "Rigid and Flexible Bracing Systems on Adjacent Sites," *Constr. Div.,* ASCE, Vol. 101, No. C02, pp. 365–375.

Taki, O. and D. S. Yang, 1989. "Excavation Support and Groundwater Control Using Soil-Cement Mixing Wall for Subway Projects," Proc. RETC, Los Angeles, June 11–14.

Tanimoto, C., S. Hata, and K. Kariya, 1981. "Interaction Between Fully Bonded Bolts and Strain Softening Rock in Tunnelling," Rock Mechanics from Research to Application, Proc. 22nd U.S. Symp. Rock Mech., Massachusetts Institute of Technology, Cambridge, pp. 347–352.

Terzaghi, K., 1943. *Theoretical Soil Mechanics,* Wiley, New York, 510 pp.

United States Steel Corporation, 1975. Steel Sheet Piling Design Manual, United States Steel, Pittsburgh, Pa.

Use of Shotcrete for Tunnel Lining, 1976. U.S. Bureau of Reclamation, Engineering and Research Center, Contract Report S-76-4, State-of-the-Art Review on Shotcrete. Published by U.S. Army Engineer Waterways Experiment Station.

Ward, W. H., 1978. "Eighteenth Rankine Lecture—Ground Supports for Tunnels in Weak Rocks," *Geotechnique,* Vol. 28, No. 2, London, June, pp. 133–170.

Watakeekul, S. and A. J. Coles, 1985. "Cutoff Treatment Methods in Karstic Limestone," Proc. 15th ICOLD Congr., Lausanne, Vol. 3, pp. 17–38.

Widmark and Platzer Intern., 1976. "Precision Blasting: Swedish Underground Construction Mission, Stockholm, Sept.

Xanthakos, P. P., 1979. *Slurry Walls,* McGraw-Hill, New York.

Xanthakos, P. P., 1991. *Ground Anchors and Anchored Structures,* Wiley, New York.

Xanthakos, P. P., 1994. *Slurry Walls as Structural Systems,* McGraw-Hill, New York.

CHAPTER 4

SOIL COMPACTION AND CONSOLIDATION

4-1 INTRODUCTION

Considering the fact that one of the earliest soil compaction methods originated with herding sheep back and forth across newly placed ground (the original version of a sheep's foot roller), soil compaction and consolidation methods have evolved to more sophisticated techniques involving mechanical vibrators for densification and replacement of loose deposits (vibrocompaction), the dropping of large weights for deposits requiring high amounts of compactive energy (deep dynamic compaction), installation of vertical drains to speed consolidation of cohesive soils (wick drains), and grouting to densify loosened ground far below ground surface (compaction grouting). Most of these newer techniques, in contrast to conventional compaction methods, are performed by specialty contractors with specialized equipment and experience. However, the conditions requiring these techniques are quite common and can be detected readily with conventional equipment and testing.

With the availability and popularity of these techniques, there are no longer unacceptable building sites. Soft fine-grained marine deposits, liquefiable sands, sanitary landfills, and sinkholes are just some of the deposits that can be improved using dynamic compaction, wick drains, vibrocompaction, and compaction grouting (Table 4.1). In a most unusual application, dynamic compaction was used to consolidate low-level nuclear waste on a 58-acre site operated by the Department of Energy in 1989 and 1990 (Schexnayder and Lukas, 1992). This process was used to reduce future settlement without exposing workers to radiation.

The need to compact a soil deposit is nothing new in geotechnical engineering. Embankments and fills are commonly placed in 8- to 12-in.-thick lifts and then compacted to a specified density. This is one of the more common construction

TABLE 4-1 Recent Ground Improvement Densification Examples

Name of Project	Soil Type	Ground Improvement Method[a]					Reference
		WD	DC	CG	VC	SC	
Kwang-Yang Steel Mill Complex, Korea	Silt and clay	x					Shin et al. (1992)
Salem/Hope Creek Nuclear Power Station, N.J.	Fill					x	DeStephen et al. (1992)
Two grain silos, Turkey	Fill				x	x	Ergun (1992)
Seventh Street Marine Terminal, Oakland, Calif.	Fill					x	Egan et al. (1992)
Hospital, San Francisco, Calif.	Silty sand			x			Graf (1992)
Offshore platform yard, Trondheim, Norway	Sand and silt		x		x		Senneset and Nestvold (1992)
Industrial mill, Tenn.	Clay			x			Brill and Darnell (1992)
Storage tanks, U.S.	Miscellaneous			x			Berry and Buhrow (1992)
Landfill lining, Fla.	Sinkholes			x			Schmertmann and Henry (1992)
Wanaque Filtration Plant, N.J.	Shot rock fill			x			Chastanet and Blakita (1992)
St. John's River Power Park, Jacksonville, Fla.	Sand		x	x			Welsh et al. (1987); Schmertmann et al. (1986)
Bolton Hill Tunnel, Baltimore, Md.	Sand and gravel			x			Baker et al. (1983)
Steel Creek Dam, Steel Creek, S. Car.	Clayey sand		x			x	Dobson (1987)
LNG Tank, Vancouver, B.C.	Silty sand					x	Dobson (1987)
Bridge approaches, Bay area, Calif.	Clay	x					Hoover (1987)

(continued)

TABLE 4-1 (*Continued*)

Name of Project	Soil Type	Ground Improvement Method[a]					Reference
		WD	DC	CG	VC	SC	
Port of Portland, Portland, Ore.	Sand				x		Leycure and Schroeder (1987)
Kings Bay Naval, Kings Bay, Ga.	Sand		x	x	x		Hussin and Ali (1987)
Pinopolis West Dam, Charleston, S. Car.	Sand			x			Salley et al. (1987)
Interstate 70, Glenwood Canyon, Colo.	Talus		x				AASHTO (1990)
Several cases	Miscellaneous					x	AASHTO (1990)
Several cases	Miscellaneous	x					AASHTO (1990)
K-D Tool Co., Walterboro, S. Car.	Sand		x				Welsh (1986)
14-story building, Minneapolis, Minn.	Peat					x	Venema et al. (1989)
Large facility, North England	Mixed glacial					x	Slocombe and Moseley (1991)
Warehouse, Southwest England	Clay					x	Slocombe and Moseley (1991)
Large retail units, South England	Fill, clay, sand					x	Slocombe and Moseley (1991)
Gilberton Power Project, Pa.	Coal waste					x	Davie et al. (1991)
Fallon Naval Air Station, Fallon, Nev.	Sand					x	Hayden and Welch (1991)
Several cases	Miscellaneous					x	Hussin and Baez (1991)
Port of Kismayo, Somalia	Sand fill				x		Castelli (1991)
Interstate 90, Seattle, Wash.	Glacial sand					x	Allen et al. (1991)
Freedom Business Center, King of Prussia, Pa.	Sandy clay		x				Welsh (1988)

(*continued*)

TABLE 4-1 (*Continued*)

Name of Project	Soil Type	Ground Improvement Method[a] WD	DC	CG	VC	SC	Reference
Townhouses, United States	Sandy silt		x				Byle et al. (1991)
Downtown tunnel project, Seattle, Wash.	Glacial sand			x			Hayward Baker (1990)
Riverview Executive Park, Trenton, N.J.	Mixed			x			Partos et al. (1989)
Regency Hotel, Atlantic City, N.J.	Sand				x		Partos et al. (1989)
Franklin Mills Mall, Philadelphia, Pa.	Fill			x			Partos et al. (1989)
Lee Boulevard, Chester Co., Pa.	Clay		x				Partos et al. (1989)
Jebba Hydroelectric, Nigeria	Sand				x		Mitchell and Welsh (1989)
Several cases	Sinkholes		x	x		x	Henry (1989)
Several cases	Miscellaneous		x	x	x	x	Hayward Baker (1991)
U.S. 71, Fayetteville, Ark.	Sanitary landfill			x			Welsh (1983)
U.S. 41/I-164 interchange, Evansville, Ind.	Sanitary landfill			x			Wardlaw (1986)
Several cases	Miscellaneous	x			x	x	GKN (1989)
Richmond Freeway, Richmond, B.C.	Sand					x	GKN (1987)
Wastewater treatment facility, Modesto, Calif.	Sand				x		GKN (1988)
Tapia water reclamation facility, Calabasas, Calif.	Silty sand					x	Hayward Baker (1990)
Several cases	Sinkholes					x	Henry (1986)

[a] WD = wick drains
DC = dynamic compaction
CG = compaction grouting
VC = Vibrocompaction
SC = Stone columns

activities, especially on land development, water retention, flood control, and transportation projects. The amount and rate of compaction of the new soils are often essential design parameters on large projects that are schedule driven. For a large embankment requiring great quantities of borrow material, 10 ft of settlement could have a significant impact. For an area requiring fill, the time for settlement of underlying foundation soils to cease could be critical to the schedule of construction activities. Here, preloading of the fill and accelerated consolidation drainage with wick drains might be advantageous.

The other common areas requiring ground improvement include cases where existing soils at the site are not strong enough to support the imposed loads, will settle or subside excessively, or might liquefy during an earthquake. Soft fine-grained soil and loose coarse-grained deposits are frequent candidates for these kinds of ground improvement. Dynamic compaction and vibrocompaction methods can be used to densify the soils or replace them with columns of stronger soil materials. These and other methods such as compaction grouting have been used on other nonsoil deposits as well, like sanitary landfills and sinkholes.

Standard penetration tests (SPTs) and cone penetrometer tests (CPTs) are frequently used field test methods to ascertain the extent of ground improvement needed. In the laboratory, identification, shear, and consolidation tests are used to characterize the in-place soils and to engineer the ground improvement method selected. Settlement markers and piezometers are often used to monitor the effectiveness of ground improvement. Sometimes, before and after field testing is conducted.

Conventional Compaction

Conventional compaction, particularly of a new fill or embankment, is usually carried out with heavy, vibratory, steel-drum rollers. Densification of clean sands can be accomplished to depths of about 6 ft, with the most effective vibration frequencies being between 25 and 30 hertz. For most projects, fill lift thicknesses of 8 to 12 in. are specified. Thicker lifts can be used with heavier equipment. No organic matter, refuse, or expansive materials are used in the fill. The degree of compaction is normally specified in terms of a minimum percentage of the maximum dry density obtained from laboratory testing of the fill soils (Figure 4-1). For most projects, 90 to 95 percent of the maximum density is adequate except under pavements, slabs, footings, and other structural elements. In these cases, 100 percent maximum dry density may be required. Particle sizes should be less than 3 in. within 18 in. of any structural element. Additionally, field moisture contents within 2 to 4 percent of optimum are specified. The American Society for Testing and Materials has standard guidelines and test methods for investigation and quality control.

Preloading

Advance consolidation of compressible clay deposits below a fill embankment or structure may be required in some cases to reduce future settlement and increase

Soil texture and plasticity data

No.	Description	Sand	Silt	Clay	LL	PI
1	Well-graded loamy sand	88	10	2	16	N.P.
2	Well-graded sandy loam	72	15	13	16	N.P.
3	Med-graded sandy loam	73	9	18	22	4
4	Lean sandy silty clay	32	33	35	28	9
5	Lean silty clay	5	64	31	36	15
6	Loessial silt	5	85	10	26	2
7	Heavy clay	6	22	72	67	40
8	Poorly graded sand	94	– 6 –		N.P.	–

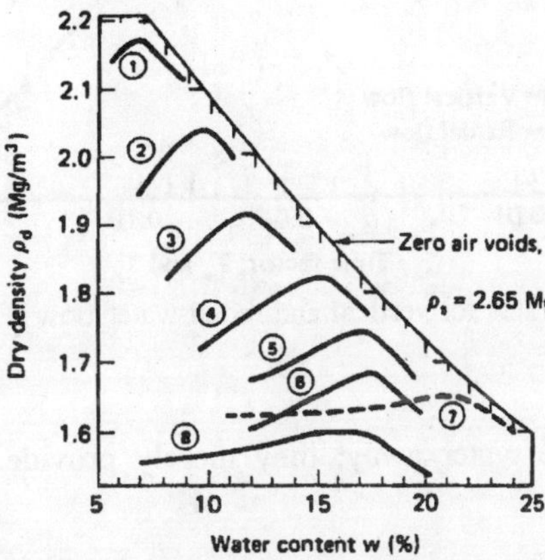

Figure 4-1 Moisture-density compaction curve. (From R. D. Holtz and W. D. Kovacs, © 1981. Reprinted by permission of Prentice Hall, Englewood Cliffs, N.J.)

strength. Usually, additional fill is placed above the final grade in an amount to compensate for the future loads of the structure. After primary consolidation is complete as verified by careful settlement and pore pressure monitoring, the extra fill is removed and the structure is constructed. Soil strength generally increases with decreases in pore pressure.

Consolidation Drainage

For a well-drained deposit, consolidation of the underlying clay layer may not take long. To accelerate drainage in poorly drained deposits, sand drains or wick drains can be installed. As the consolidation process is governed by the rate of excess pore pressure dissipation, shortening the length of pore water flow paths greatly reduces the consolidation time (Figure 4-2).

The concept of installing some sort of vertical drainage system in soft soil deposits to accelerate consolidation began in the 1920s with the sand drain (Welsh et al., 1987). This type of artificial consolidation drainage endured until the beginning of the 1970s, when developments in plastics and fabric filters led to the development of prefabricated wick drains.

Sand drains are placed in predrilled holes. Wick drains can be drilled or driven into place using specialized equipment. As Hoover (1987) points out, wick drains

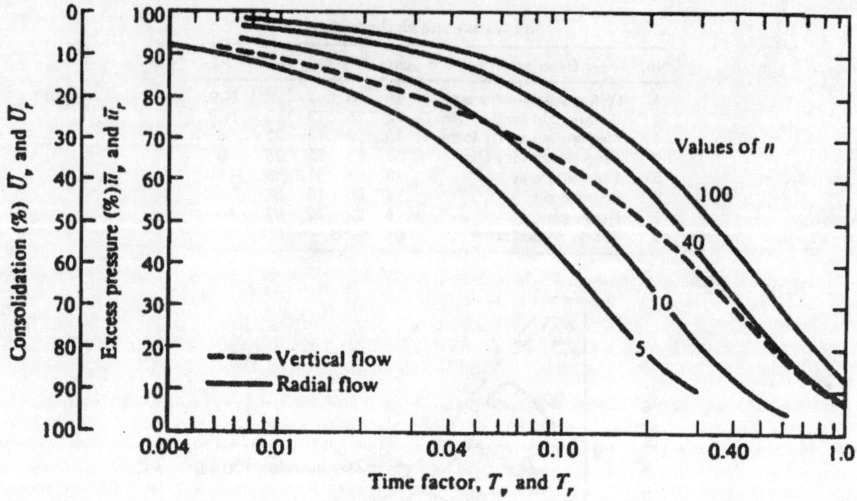

Figure 4-2 Consolidation rates for vertical and radial water flow (From Cedergren, 1975).

do not really "wick" the water away; they merely provide paths for increased vertical drainage.

Dynamic Compaction

Dynamic compaction is in some ways similar to what is accomplished with a vibratory roller. A vibratory roller uses an eccentric weight rotating at high frequency within the steel drum to impart energy to the soil, causing it to densify. Dynamic compaction also uses a weight to impart energy to the soil, again causing it to densify. The differences are that dynamic compaction uses much bigger weights, up to 20 tons, which are repeatedly dropped onto the soil from heights of up to 100 ft with a crane (Figure 4-3). The amount of energy imparted to the soil and the consequential depth of improvement is much greater than that which results from a vibratory roller. Deposits up to 35 ft deep have been improved using this technique.

Dynamic compaction is one of the oldest forms of ground improvement in existence. The Romans reportedly utilized a variation of this technique and it was used in the United States as early as 1871. There are reports of its use in Germany and China in the 1930s and 1940s (Welsh, 1986).

Dynamic compaction can be used to densify loose sand deposits, often of coastal, glacial, and alluvial origins. It has also been used to densify fills, mine refuse, collapsible soils, sanitary landfills, and soils loosened by underlying sinkholes and mining activities. The method works best on granular deposits where the degree of saturation is low, the permeability of the soil mass is high, and drainage is good (AASHTO, 1990).

Dynamic compaction is not appropriate for saturated clayey soils. Intermediate soils may benefit from dynamic compaction but will require more time for drainage

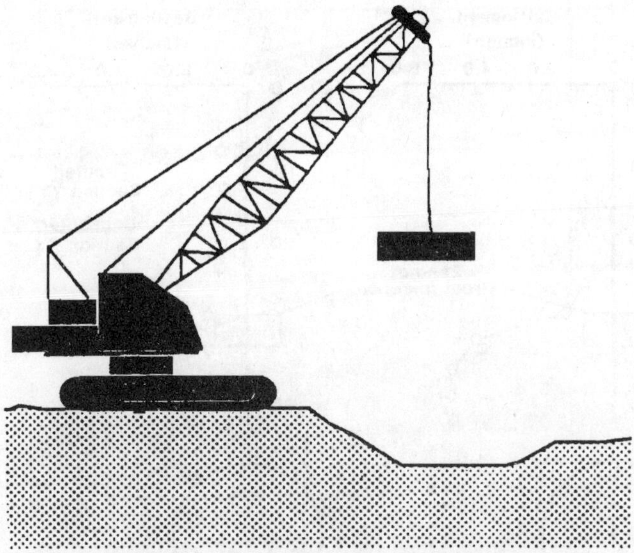

Figure 4-3 Dynamic compaction equipment.

and possibly multiple passes with the equipment. Sufficient time between passes should be allowed so that excess pore water pressures dissipate.

Compaction Grouting

Compaction grouting consists of the injection of low-slump (less than 2 in.) soil cement mortar grout under high pressure (500 to 1000 lb/in.^{2}) to compact and displace the adjacent soils. The grout does not penetrate soil pores but displaces the subsurface soils by forming a homogeneous grout bulb near the grout pipe tip. The purposes of compaction grouting are to densify subsurface soil zones prior to construction and to compensate for soil loss or settlement after construction. The most common applications are for arresting foundation settlements, compensating for loss of ground due to tunneling, slab and foundation jacking, densification of liquefiable soils, and treatment of sinkhole problems (AASHTO, 1990). Compaction grouting offers definite economic advantages when a thin, loose, deep strata, overlain by a very dense strata, requires densification.

Compaction grouting was invented in the United States and has been used over the last 40 years (Warner, 1982). Use on the Bolton Hill Tunnel (Figure 4-4) in Baltimore (Baker et al., 1983) was one of the first major projects to bring this technique to the public eye.

The most costly and disruptive aspect to compaction grouting is the great number of injection points required, especially if used in an urban environment. Grout pipes are normally designed to be installed and injected on a primary and secondary spacing with final grid spacing between 6 and 12 ft. The vertical spacing between injection points varies between 1 and 3 ft (Welsh et al., 1987).

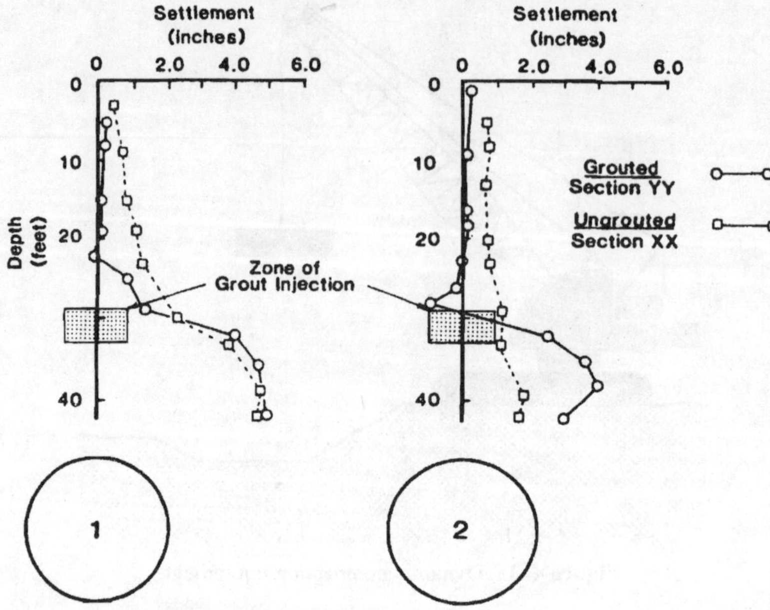

Figure 4-4 Compaction grouting of the Bolton Hill Tunnel. (Reprinted from *Underground Space*, Vol. 7, W.H. Baker, E.J. Cording, and H.H. McPherson, "Compaction Grouting to Control Ground Movements During Tunneling," pp. 205–212, Copyright 1983, with permission from Pergamon Press Ltd, Headington Hill Hall, Oxford OX3 OBW, UK.

Vibrocompaction

Vibrocompaction is "a method of deep densification of in situ granular soils by means of rearranging loose cohesionless grains into a denser array [Figure 4-5] by insertion of a vibratory probe" (AASHTO, 1990). The vibrating probe, assisted by water or air jets, is used to make a hole in the ground and then compact the ground in place (Figure 4-6) by multiple insertions of the vibrating probe. As discussed by Mitchell (1986), vibrocompaction was patented in Germany in 1930 and introduced in the United States in 1941.

The vibrocompaction method works best in granular soils with typically less than 15 percent fines (Figure 4-7). Soils of this type are commonly found in coastal, alluvial, glacial, and fill deposits. Also, there have been successful uses on cinders, slag, bottom ash, mineral tailings, coral, and aluminum ore (AASHTO, 1990). Sand or gravel backfill is usually added at the ground surface as the vibrator is intermittently withdrawn and repenetrated to ensure densification effort throughout the penetration depth.

Vibrocompaction can be used to depths up to about 100 ft (Dobson, 1987) and can be effective up to 13 ft from the vibrator, depending upon the type of soil and vibrating power (Mitchell, 1981). Typical depths range from 10 to 50 ft.

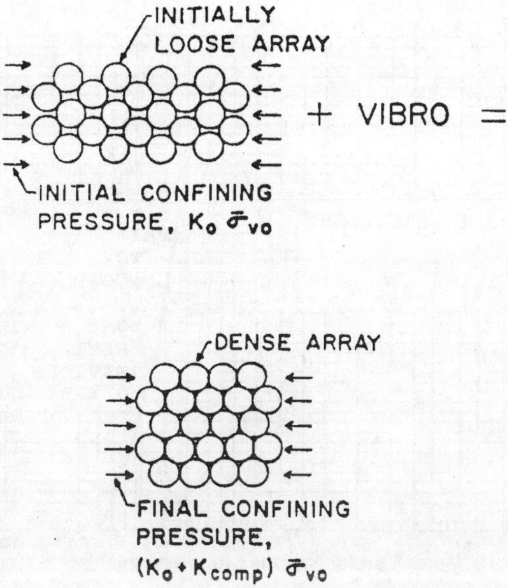

Figure 4-5 Rearrangement of soil grains by vibrocompaction (From AASHTO, 1990).

Stone Columns

In granular soils that exhibit some cohesive characteristics and have a fines content ranging between about 15 and 25 percent, a stone backfill is used during vibrocompaction to enhance displacement and drainage and subsequently assist in the densification process (Figure 4-8). This process is referred to as vibroreplacement. The

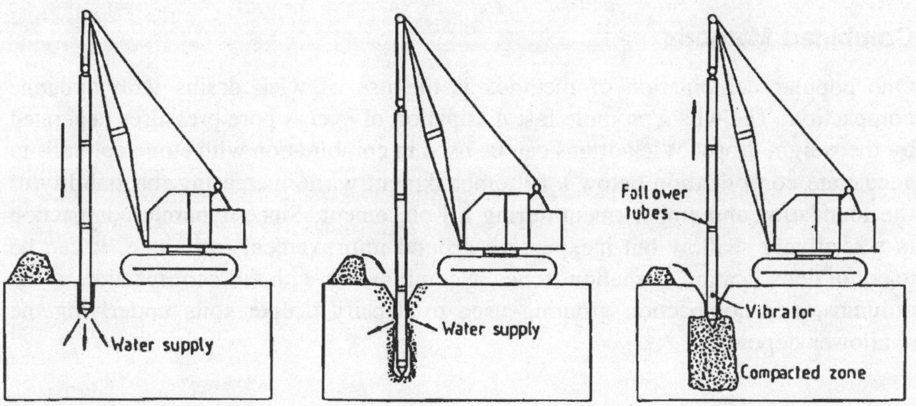

Figure 4-6 Vibrocompaction process. (From Welsh et al., 1987. Reproduced by permission of ASCE.)

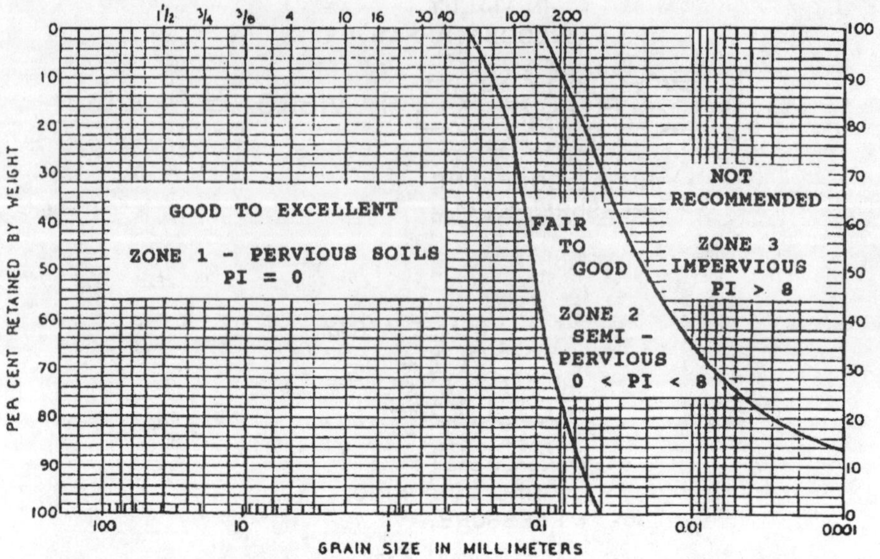

Figure 4-7 Grain size distribution for use of vibrocompaction. (From Welsh et al., 1987. Reproduced by permission of ASCE.)

increased drainage also contributes to the protection of the subsurface mass from liquefaction in a seismically active region. In more cohesive soils, the stone backfill in its densified state performs as a structural reinforcement element to increase the bearing capacity of the mass, and it greatly reduces total and differential settlements. The most recent improvement in this technique is the bottom-feed method, in which the stone backfill is fed via a hopper through a supply tube directly to the bottom of the vibrator (Welsh, 1986).

Combined Methods

One popular combination of methods is the use of wick drains with dynamic compaction. The wicks promote fast dissipation of excess pore pressures generated by the weight drops. Wick drains can be used in combination with stone columns to accelerate consolidation below a fill embankment while increasing the stability of the foundation and embankment during fill placement. Since dynamic compaction is a relatively shallow but inexpensive ground improvement technique, it can be used for densification of shallow zones in combination with vibrocompaction, stone columns, and compaction grouting used to densify deeper soils underlying the shallower deposits.

Costs

According to Welsh et al. (1987), "Compaction grouting is normally more costly than dynamic compaction and vibrocompaction, and on par with vibroreplacement"

Figure 4-8 Typical stone columns (from GKN Hayward Baker, Inc., 1989).

for new construction. Figures given in Welsh (1986), AASHTO (1990), and the authors' experience indicate the following rough order of magnitude estimates for ground improvement in 1993 U.S. dollars per cubic yard of soil treated:

Conventional compaction	1.0 – 5.0
Dynamic compaction	0.5 – 3.0
Vibrocompaction	1.0 – 7.0
Stone columns	5.0 – 11.0
Compaction grouting	15.0 – 100.0

Costs will be higher on small projects and lower on major projects.

4-2 USES AND APPLICATIONS

Most of the uses and applications of the ground improvement methods described above are related to densification of sandy soil or consolidation of clayey soil.

Densification and consolidation may be desired simply to close large voids in the soil matrix, for instance, over a landfill or sinkhole, or to increase the strength of the soil in situ. Dense granular soils have higher strengths (Figure 4-9), higher bearing capacities (Figure 4-10), lower permeabilities (Figure 4-11), and lower liquefaction potential (Figure 4-12) as compared to loose deposits. Consolidation of fine-grained soils brings about increases in strength (Figure 4-13), bearing capacity (Figure 4-14), and slope stability (Figure 4-15). Also, a single method often brings about multiple ground improvement results. For instance, ground improvement with stone columns promotes drainage and consolidation of fine-grained deposits as well as strengthening the deposit with high-strength soil inclusions.

Densification

Densification usually refers to making a predominantly granular deposit more compact. The density of granular deposits is most often detected with the standard penetration test (SPT) blow counts (*N*). These blow counts are then related to terms describing the density of the deposit as shown in Table 4-2. Other parameters, such

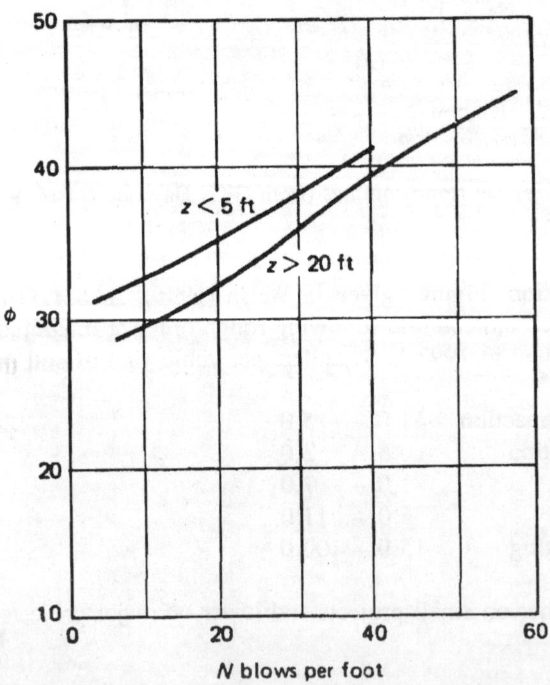

Figure 4-9 Strength versus density relationship for sands. (Reprinted with the permission of Macmillan Publishing Company from *Introductory Soil Mechanics and Foundations: Geotechnical Engineering*, 4th ed. by G. F. Sowers. Copyright © 1979 by Macmillan Publishing Company.)

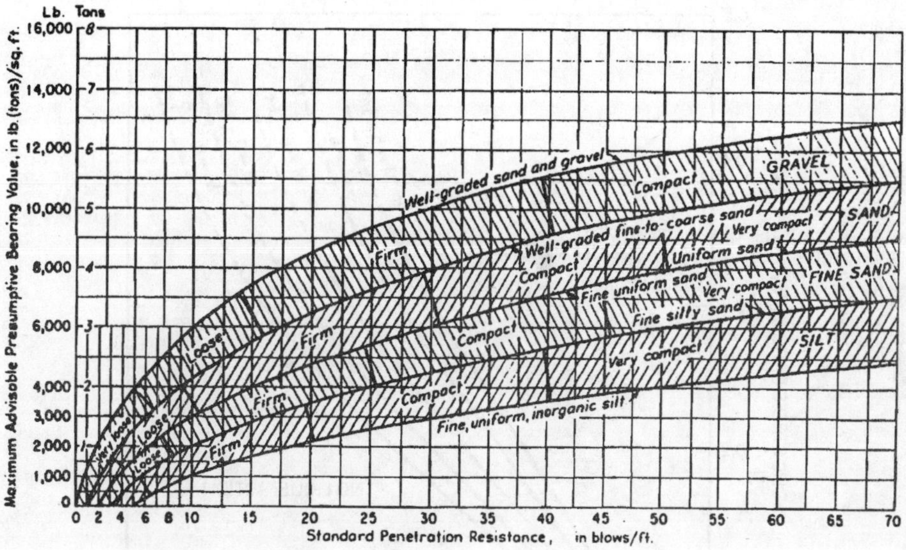

Figure 4-10 Bearing capacity versus density relationship for sands (from Hough, 1969).

as the relative density and angle of internal friction, can be related to the blow count. As discussed later, other field and laboratory test methods can be used to confirm or more accurately estimate soil density and strength. Ways to increase the density of granular soil deposits include conventional vibratory compaction, vibroflotation, dynamic compaction, and compaction grouting (Figure 4-16).

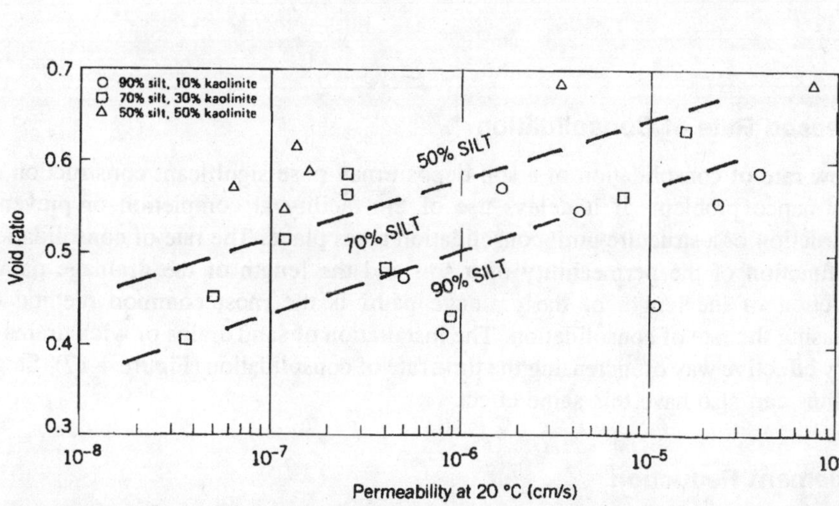

Figure 4-11 Permeability versus density relationship. (from R. D. Holtz and W. D. Kovacs, © 1981. Reprinted by permission of Prentice Hall, Englewood Cliffs, N.J.)

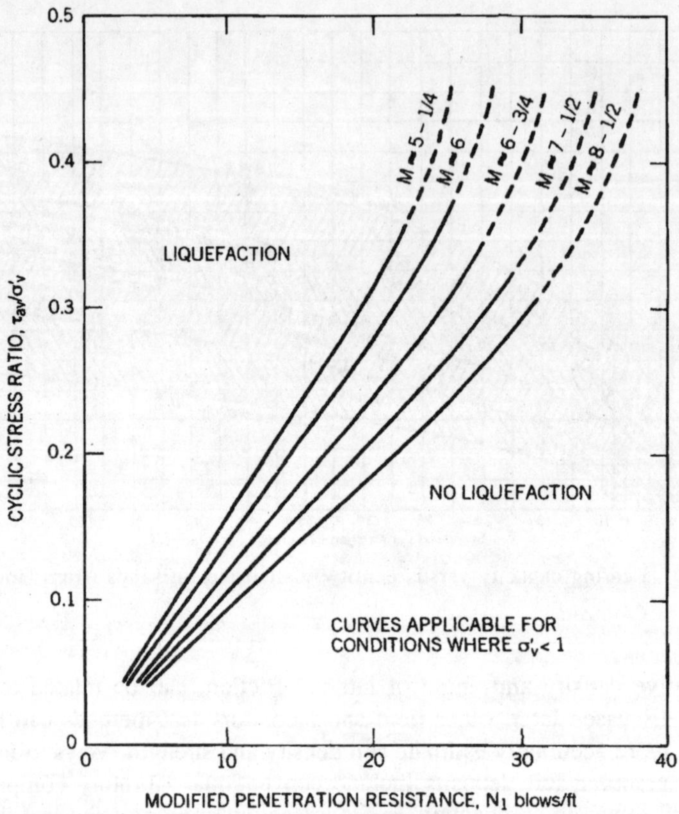

Figure 4-12 Liquefaction potential versus corrected blow counts (from Seed and Idriss, 1982).

Increased Rate of Consolidation

A slow rate of consolidation of a soil deposit may pose significant construction or performance problems if it delays use of the facility at completion or prevents construction of a structure until consolidation takes place. The rate of consolidation is a function of the permeability of a soil and the length of the drainage paths. Reduction of the length of the drainage paths is the most common method of increasing the rate of consolidation. The installation of sand drains or wick drains is a very effective way of increasing the time rate of consolidation (Figure 4-17). Stone columns can also have this same effect.

Settlement Reduction

By densifying the granular soil below an embankment or structural foundation, the elastic settlements that take place upon loading can be reduced. The relative in-

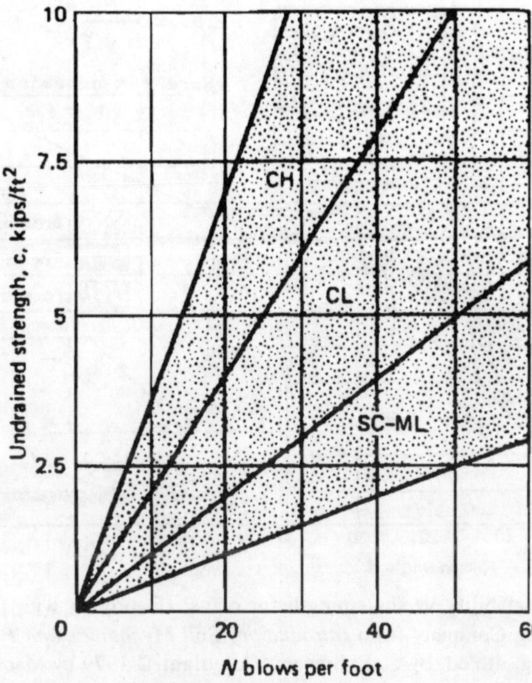

Figure 4-13 Strength versus density relationship for clays. (Reprinted with the permission of Macmillan Publishing Company from *Introductory Soil Mechanics and Foundations:* Geotechnical Engineering, 4th ed. by G. F. Sowers. Copyright © 1979 by Macmillan Publishing Company.)

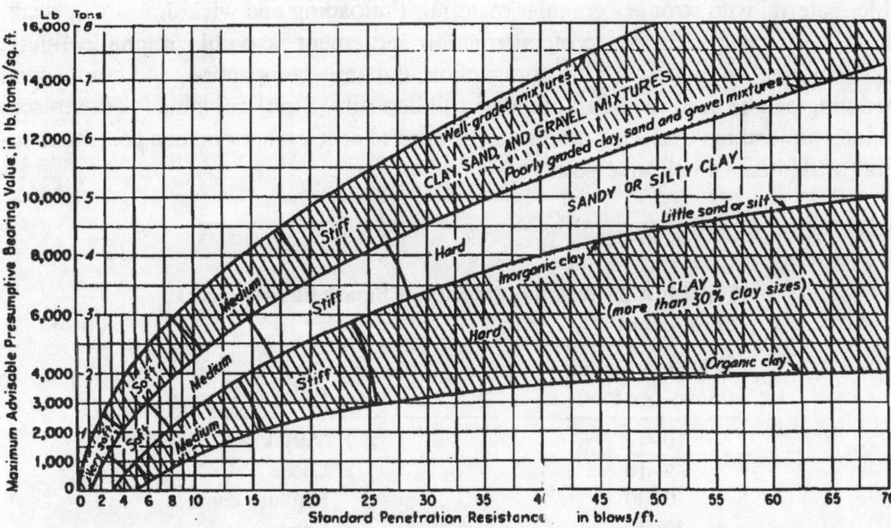

Figure 4-14 Bearing capacity versus density relation.hip for clays (from Hough, 1969).

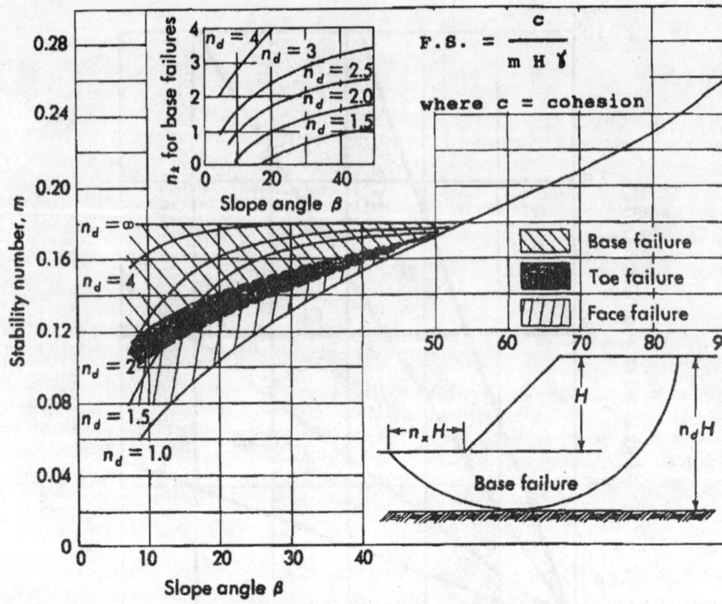

Figure 4-15 Slope stability versus strength for clays. (Reprinted with the permission of Macmillan Publishing Company from *Introductory Soil Mechanics and Foundations:* Geotechnical Engineering, 4th ed. by C. F. Sowers. Copyright © 1979 by Macmillan Publishing Company.)

crease in modulus for increasing degrees of density is shown in Figure 4-18. For cohesive soils, total settlement can be reduced by replacing areas of the compressible material with stronger granular material. Preloading and wick drains can accelerate settlement such that postconstruction settlement is within tolerable limits. Conventional compaction, vibrocompaction, dynamic compaction, and compaction grouting can be used to stiffen granular soil deposits. Stone columns (vibroreplacement), preloading, and wick drains can be used to accelerate or reduce postconstruction settlement of cohesive soils.

TABLE 4-2 Relative Density of Sands Based on SPT Blow Count

SPT Blow Count (blows per foot)	Relative Density
0–4	Very loose
4–10	Loose
10–30	Medium dense
30–50	Dense
Over 50	Very dense

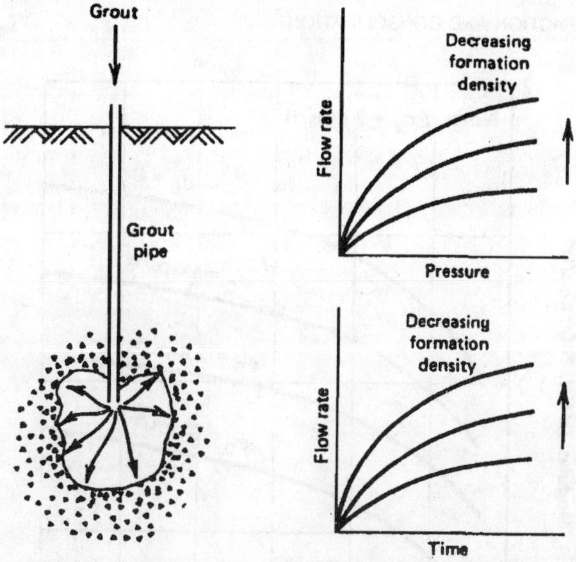

Figure 4-16 Compaction grout bulb. (From R. M. Koerner, *Construction and Geotechnical Methods in Foundation Engineering*, © 1984. Reproduced with permission of McGraw-Hill.)

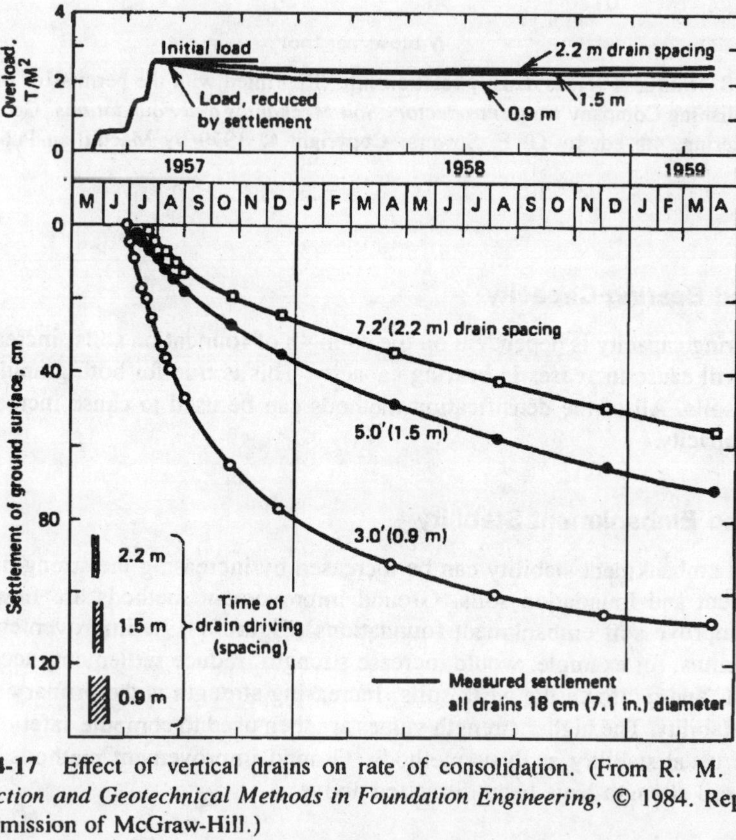

Figure 4-17 Effect of vertical drains on rate of consolidation. (From R. M. Koerner, *Construction and Geotechnical Methods in Foundation Engineering*, © 1984. Reproduced with permission of McGraw-Hill.)

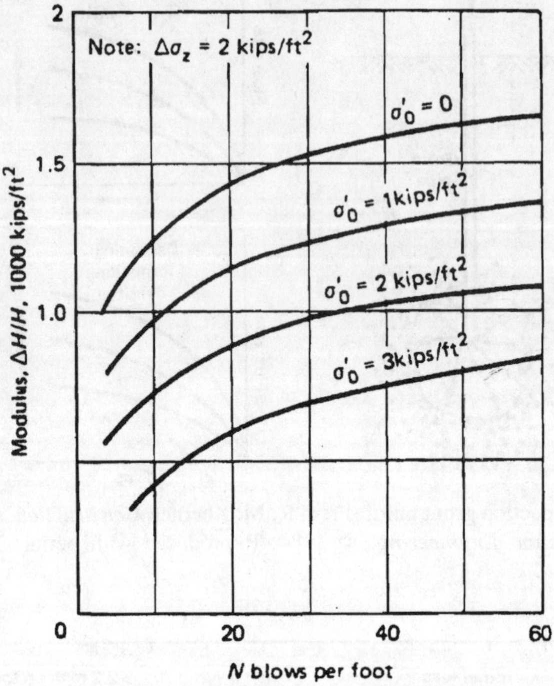

Figure 4-18 Stiffness versus density relationship. (Reprinted with the permission of Macmillan Publishing Company from *Introductory Soil Mechanics and Foundations:* Geotechnical Engineering, 4th ed. by G. F. Sowers. Copyright © 1979 by Macmillan Publishing Company.)

Increased Bearing Capacity

Since bearing capacity is dependent on the strength of foundation soils, increases in strength will cause increases in bearing capacity. This is true for both granular and cohesive soils. All of the densification methods can be used to cause increases in bearing capacity.

Slope and Embankment Stability

Slope and embankment stability can be increased by increasing the strength of the embankment and foundation soils. Ground improvement methods are frequently used to improve soft embankment foundations (Figure 4-19). Improvement with stone columns, for example, would increase strength, reduce settlement, accelerate settlement, and reinforce the weak soils. Increasing strength is the primary way to increase stability. The higher strength values are then used to compute safety factors using the usual stability analysis methods. Ground improvement methods for increasing soil strength have been discussed above.

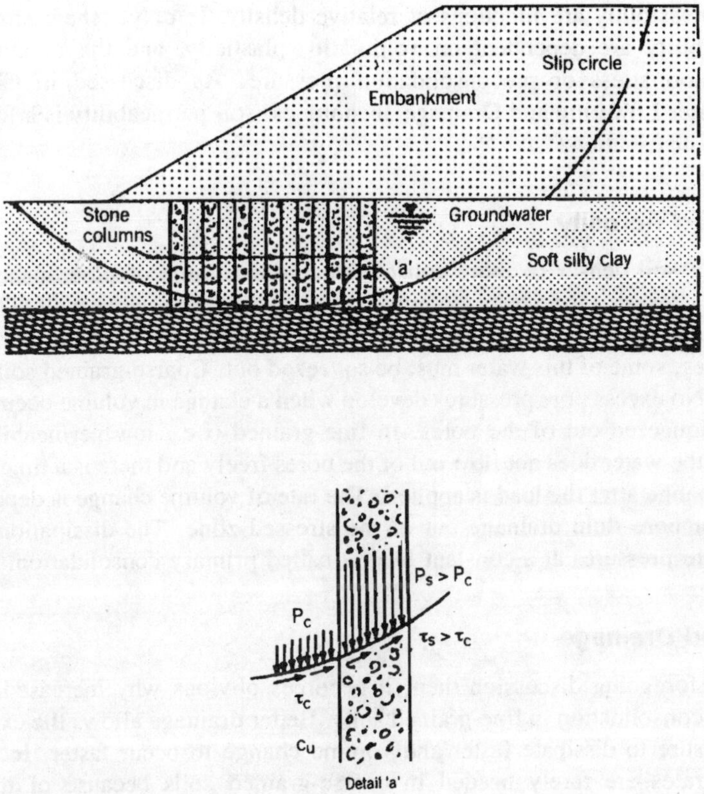

Figure 4-19 Embankment foundation improvement with stone columns. (From Munfakh et al., 1987. Reproduced by permission of ASCE.)

Reduction of Liquefaction Potential

Liquefaction potential of a granular deposit is largely dependent on fines content of the soil, the location of the groundwater table, and density as defined by modified blow count (N) values. Liquefaction potential is most often reduced by increasing the density of the soil. Hence the ground improvement methods used for increasing density, as described above, can also be used for liquefaction reduction.

4-3 PRINCIPLES OF BEHAVIOR

As with most soil mechanics problems, the principles of behavior are related to the interaction of soil grains and the spaces or pores between them. The porosity of a soil, n, is the ratio of pore volume to total volume. The ratio of pore volume to solids volume is known as the void ratio, e. The relative density of a cohesionless soil, D_r, may be expressed in terms of void ratio or dry unit weight, which is basically a function of the void ratio and the water content. In sand, shear strength and compressibility are functions of relative density. In clays, shear strength and compressibility are dependent on void ratio, plasticity, and the maximum past consolidation stress or preconsolidation pressure. As discussed in Chapter 1, Groundwater Lowering and Drainage Techniques, soil permeability is a function of grain size distribution.

Decreased Porosity

To make a soil denser or more consolidated, the volume of pore space must be decreased. Porosity and void ratio must therefore be decreased as well. This means that the soil particles must be in a more compact configuration, and if there is water in the pores, some of this water must be squeezed out. Coarse-grained soils are free draining. No excess pore pressures develop when a change in volume occurs and the water is squeezed out of the pores. In fine-grained (i.e., low permeability) soils however, the water does not flow out of the pores freely and there is a time delay for volume change after the load is applied. The rate of volume change is dependent on the rate of pore fluid drainage out of the stressed zone. The dissipation of these excess pore pressures at a constant load is called primary consolidation.

Increased Drainage

From the foregoing discussion then, it becomes obvious why increased drainage promotes consolidation in fine-grained soils. Better drainage allows the excess pore fluid pressure to dissipate faster and volume change to occur faster. Increases in drainage rates are rarely needed in coarse-grained soils because of their free-draining characteristics (i.e., higher permeabilities).

Increased Shear Strength

For cohesionless soils, strength is expressed in terms of the angle of internal friction, or ϕ angle, which is dependent on density or void ratio, gradation, grain shape, and grain mineralogy. For cohesive soils, strength is expressed in terms of a cohesion or c value. This value is dependent on the effective overburden stress and preconsolidation pressure. The plasticity of the soil also plays a role in the effective shear strength of a clay.

To increase the strength of a sandy soil, one must increase the density. To increase the strength of a clayey soil, one must increase the maximum past consolidation stress or decrease the void ratio. In practice, many soils possess both frictional strength (ϕ) and cohesion (c). These soils are frequently referred to as c-ϕ soils and do not neatly conform to either purely sandy or purely clayey soil behavior characteristics. Most ground improvement methods can be analyzed and verified using field and laboratory tests that are not soil-type dependent (e.g., standard penetration, cone penetrometer, pressuremeter, geophysical, consolidation, direct shear, triaxial, and load tests). So these common types of intermediate soils do not present an insolvable analytical problem.

4-4 THEORETICAL BACKGROUND

The ground improvement methods being discussed in this chapter primarily deal with reducing settlement, increasing bearing capacity, and reducing liquefaction potential. To understand how these methods should be used, one must first understand how the soil tends to behave. In the following paragraphs are brief discussions of the theories behind compaction, settlement, bearing capacity, and liquefaction of fine-grained and coarse-grained soils. For more in-depth treatments, the reader is directed to basic soil mechanics and foundation engineering texts, of which there are many.

Compaction

The behavior of any soil is influenced to a considerable extent by its relative looseness or denseness. In this respect, however, a distinction is necessary between coarse-grained cohesionless soils and cohesive materials. In a mass of coarse-grained soil, most of the grains touch several others in point-to-point contact and efforts to densify the mass can reduce the void ratio only through rearrangement or crushing of particles. On the other hand, the densification of fine-grained soil, especially clay, depends on other factors, such as cohesion and the presence of water films on the particle surfaces.

The void ratio or porosity of any soil usually does not in itself furnish a direct indication of its behavior under load or during excavation. Of two coarse-grained soils at the same void ratio, one soil may be in a dense state whereas the other may be loose. Thus, the relative density of a coarse-grained material is much more significant than the void ratio alone. The relative density can be expressed numerically by the density index, I_d, defined as

$$\text{Density index, } I_d = \frac{e_{max} - e}{e_{max} - e_{min}} \tag{4-1}$$

where e = actual void ratio
e_{min} = void ratio in the densest state
e_{max} = void ratio in the loosest state

Hence, $I_d = 1$ for a very dense soil and 0 for a very loose soil.

In practice, the relative density of granular soils is usually judged indirectly by penetration or load tests, because direct measurement of the void ratio of a soil in the field is not convenient. However, if e is known, the values of e_{min} and e_{max} can be determined in the laboratory.

For a soil containing appreciable amounts of silt or clay, the density index loses its significance because the values of e_{max} and e_{min} have no definite meaning. Yet many construction operations deal with such materials. Moreover, the beneficial effects of compacting soils have been demonstrated by long experience. The need for a method of defining the degree of compaction led in the early 1930s to the development in California of a laboratory compaction test. This test has been refined and standardized by the American Society for Testing and Materials (ASTM) and the American Association of State Highway Officials (AASHO) as the moisture-density relations test, more widely known as the Standard Proctor test.

The Standard Proctor test is conducted by tamping soil into a cylinder having a fixed volume. Tamping is carried out with a tamper of constant weight and height of fall. Soil samples of progressively higher water contents are tested until the weight of the moist soil that can be packed into the mold has reached a maximum and starts to decrease. For each test the dry unit weight and moisture content of the soil is determined. The results are plotted as shown in Figure 4-20. The modified Proctor test is carried out similarly except with a different weight tamper and height of fall. The ordinate of the peak of the curve is designated the maximum dry density, γ_{max}, or 100 percent compaction, and the abscissa the optimum water content, w_{opt}.

The results obtained from the laboratory tests are used as guidelines in the field. Different curves would be obtained in the field depending on such variables as type, weight, and number of passes of compaction equipment, or thickness of layers being compacted. Commonly, specifications require that dry densities be obtained in the field that are at least equal to 95 percent of maximum dry density determined on the basis of the laboratory tests. The type of test used must be specified since two different moisture-density relations for the same soil can be obtained. Percent compaction can be determined in the field using the sand cone method, water balloon

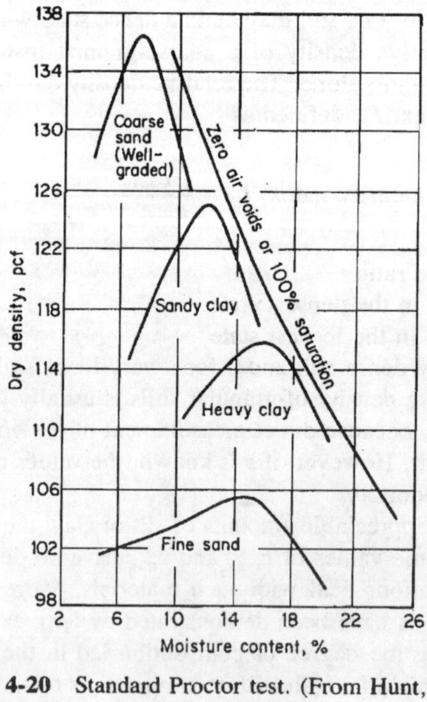

Figure 4-20 Standard Proctor test. (From Hunt, 1986.)

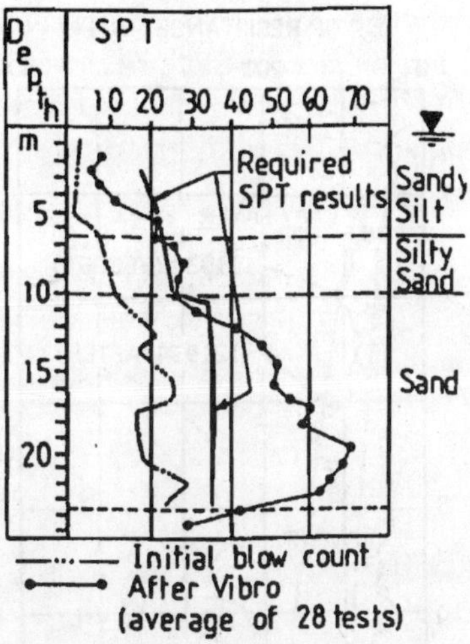

Figure 4-21 Standard penetration resistance before and after compaction. (From Dobson, 1987. Reproduced by permission of ASCE.)

method, or nuclear moisture and density meters. Also, standard penetration tests (SPTs) and cone penetrometer tests (CPTs) can be used to measure the density of the soil before and after compaction (Figure 4-21 and 4-22).

Consolidation of Fine-Grained Soils

The theory of consolidation for fine-grained soils first evolved from observations of new structure settlements that occurred long after construction. Fine-grained soils did not appear to follow typical stress-strain behavior of engineering materials because of the time lapse of settlement after application of the load. Terzaghi was the first soil mechanic to explain this phenomenon on a scientific basis in 1919 (Terzaghi, 1943). His studies indicated that when a load is applied to a fine-grained soil, the increase in stress is first shared by the soil grains and the water contained within the pore space between the soil grains. Over time, some of the water gets squeezed out of the pores (the amount of pore space decreases) until the soil particle skeleton is able to support its share of the increased stress. The resulting decrease in pore space results in compression of the soil layer and settlement.

The most common way of studying and discussing consolidation theory is with a laboratory device called a confined compression oedometer, or consolidation test device (Figure 4-23). During this test, a sample is placed in a test apparatus that prevents the sample from experiencing lateral displacements and allows drainage of water upward and downward as the soil particles squeeze together. This process is known as one-dimensional consolidation, which often represents conditions approached in the field. The relations among vertical pressure, settlement, and time can be investigated using this type of test.

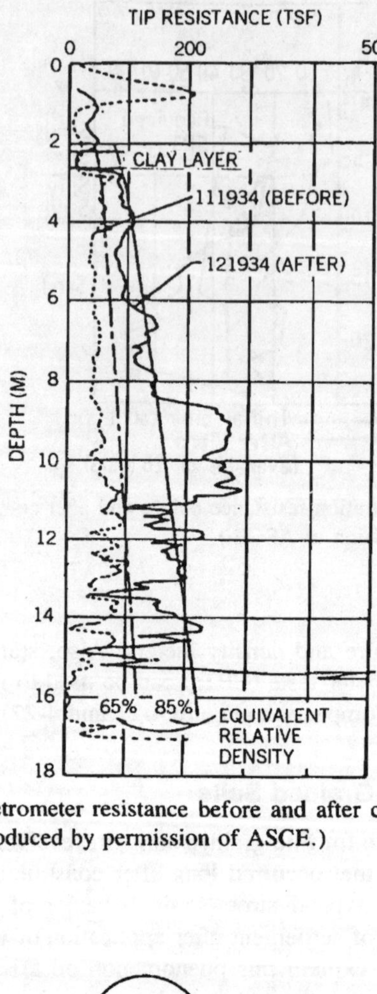

Figure 4-22 Cone penetrometer resistance before and after compaction. (From Schmertmann et al., 1986. Reproduced by permission of ASCE.)

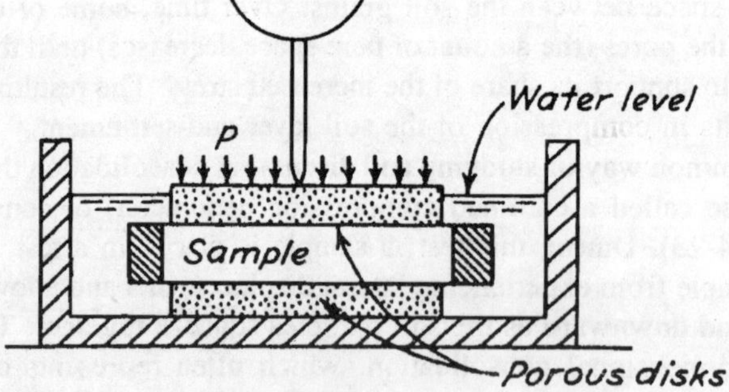

Figure 4-23 Consolidation test apparatus. (From Peck et al., 1974.)

A consolidation test is run by increasing pressure on the sample in steps. Each step load is held constant until deformation practically ceases (Figure 4-24). The time it takes for this to occur depends on how fast water can be squeezed out of the pores between the soil grains. The results of the test are plotted by using the final void ratio corresponding to each increment of pressure as a function of the accumulated pressure. It is convenient to plot the pressure to a logarithmic scale. The plot is then known as an *e*-log *p* curve (Figure 4-25).

Most soils have been preloaded or overconsolidated, meaning that they have had pressures acting on them greater than the pressure under which the soils are now in equilibrium. The ratio of preconsolidation pressure to current pressure is known as the overconsolidation ratio. Cassagrande (1936) developed a method to estimate the preconsolidation pressure of a sample from consolidation test results (Figure 4-26). Once this preconsolidation pressure is known, settlements can be predicted from an *e*-log *p* curve by assuming that any load change which results in a pressure less than the preconsolidation pressure falls on the flat portion of the curve, the reload/unload portion, and that any load change which results in a pressure greater than the

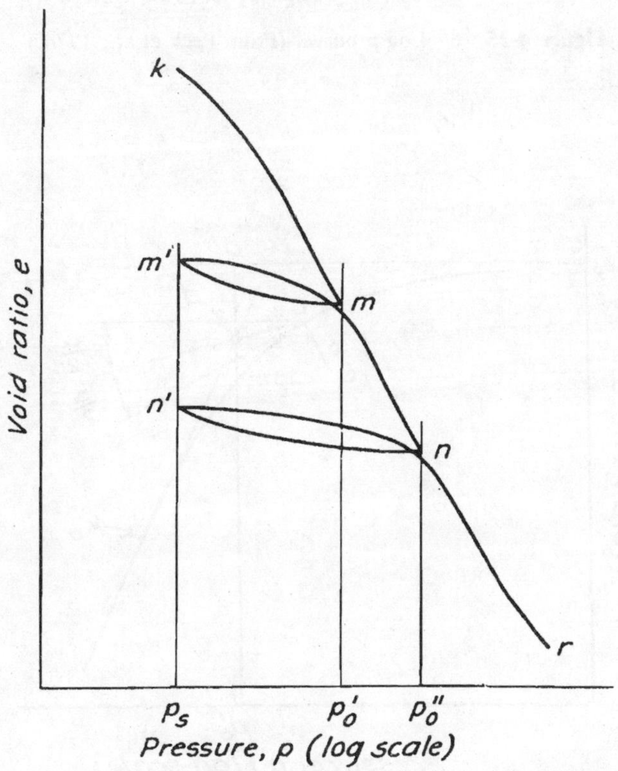

Figure 4-24 Deformation during consolidation test. (From Peck et al., 1974.)

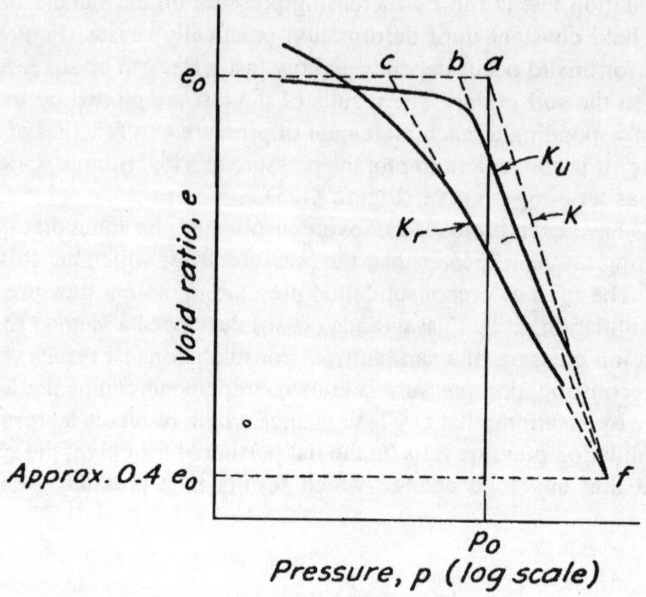

Figure 4-25 e - Log *p* curve. (From Peck et al., 1974.)

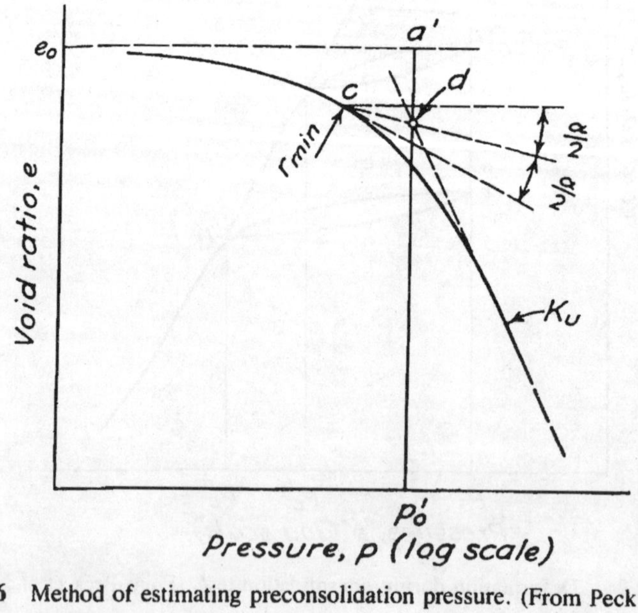

Figure 4-26 Method of estimating preconsolidation pressure. (From Peck et al., 1974.)

preconsolidation pressure falls on the steep or virgin portion of the curve. The slopes of these lines are used to calculate settlements.

As mentioned earlier, settlement of a fine-grained soil layer is attributable to the change in pore space or void ratio in response to a change in stress. The general equation for calculating settlements due to consolidation of a fine-grained soil layer is

$$S = \frac{C_c}{1 + e_0} H \log_{10} \frac{p_0 + \Delta p}{p_0} \tag{4-2}$$

where S = amount of predicted settlement of soil layer
C_c = compression index (slope of the e-log p curve)
e_0 = initial void ratio
H = thickness of compressible soil layer
p_0 = initial vertical pressure
Δp = change in vertical pressure

The value of C_c will depend on whether the stress change is on the reload or virgin portion of the e - log p curve. If p_0 is less than the preconsolidation pressure but $p_f = p_0 + \Delta p$ is greater than the preconsolidation pressure, two calculations must be made. The first calculation would be for the stress increase between the initial stress and the preconsolidation pressure using the value of C_c from the flat portion of the curve; the second calculation would be for the stress increase between the preconsolidation pressure and the final stress level using the value of C_c from the steep portion of the curve.

The time it takes for settlement to occur can also be predicted from the results of consolidation tests. This is an important consideration when decisions are being made about whether to preload a site, whether to employ ground improvement methods, and when to time construction of settlement sensitive structures. Fine-grained soils are relatively impervious. Water in the pores is more or less trapped there. When a change in stress occurs, the water cannot squeeze out of the pores immediately. Pressure develops in the pore water. This increased pressure eventually causes the water to flow out of the pores, rapidly at first and then more slowly. The soil particles move closer as a result and settlement occurs. The rate of settlement is rapid at first and slower as the pore water pressure equilibrates to a new stable pressure.

The distribution of pore water pressure throughout the sample layer is not constant. The pores closest to the drainage exit consolidate the fastest; the pores farthest from the drainage exit consolidate the slowest (Figure 4-27). However, after sufficient time the pore pressures stabilize. This point is known as 100 percent consolidation or the end of primary consolidation. To calculate the time of consolidation, one must know the coefficient of consolidation c_v:

$$c_v = \frac{k}{m_v \gamma_w} \tag{4-3}$$

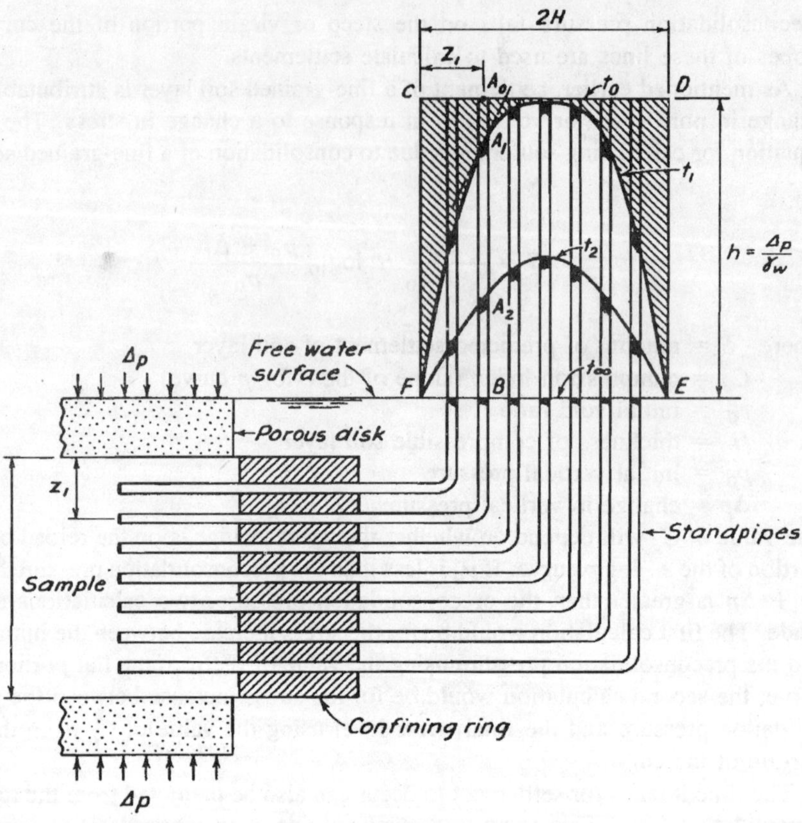

Figure 4-27 Pore pressure distribution during consolidation. (From Peck et al., 1974.)

where k = coefficient of permeability
m_v = coefficient of volume compressibility
δ_w = unit density of water

With the value of c_v, one can calculate the time of consolidation with the following equation:

$$T_v = \frac{c_v}{H^2}\, t \tag{4-4}$$

where T_v = dimensionless number called the time factor
H = half-thickness of layer for double drainage
t = time corresponding to the degree of consolidation U_z

The value for T_v may be found in Figure 4-28 for different degrees of consolidation (U_z).

In advanced stages of consolidation, the time-settlement curve does not actually approach a horizontal asymptote at 100 percent consolidation as is shown in Figure

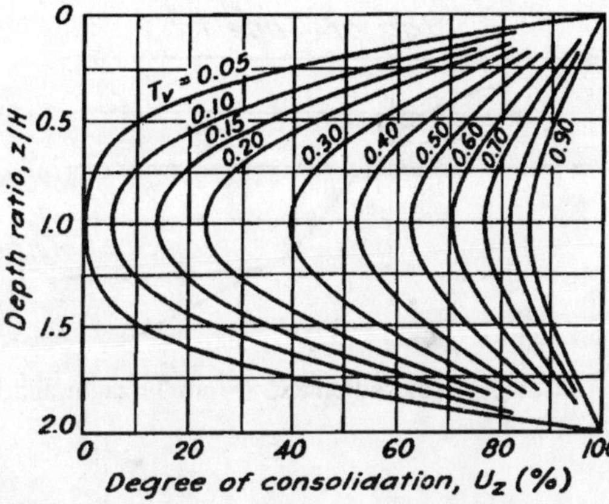

Figure 4-28 Degree of consolidation. (From Peck et al., 1974.)

4-29. Instead, the curve approaches an inclined tangent with a nearly constant slope when plotted to a logarithmic scale (Figure 4-30). The additional settlement that takes place after primary consolidation occurs is designated as secondary settlement. The amount of secondary settlement that may occur is defined by the slope of the e - log t curve (Figure 4-31). The amount of secondary settlement is often small and thus can often be neglected.

Deformation of Cohesionless Soils

Cohesionless soils are difficult to sample in an undisturbed state. Sometimes consolidation tests are run on reconstituted samples at the in situ relative density. More often, the compressibility is measured in other ways, such as with a pressuremeter, and settlement calculations are made using the deformation modulus obtained from in situ testing. One such method (SOLS SOILS, 1975) that uses the pressuremeter modulus for calculating footing settlements is

$$S = \frac{p}{4.5E_c} [2B_0 (\lambda_2 \frac{B}{B_0})^\alpha + (\alpha\lambda_3 B)] \tag{4-5}$$

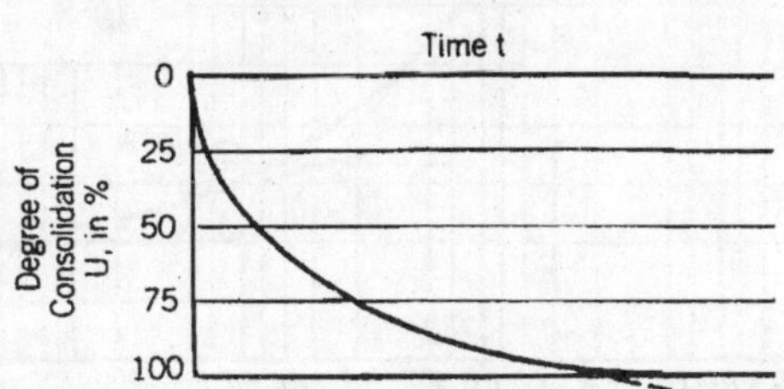

Figure 4-29 Time versus settlement curve. (From Terzaghi and Peck, 1967.)

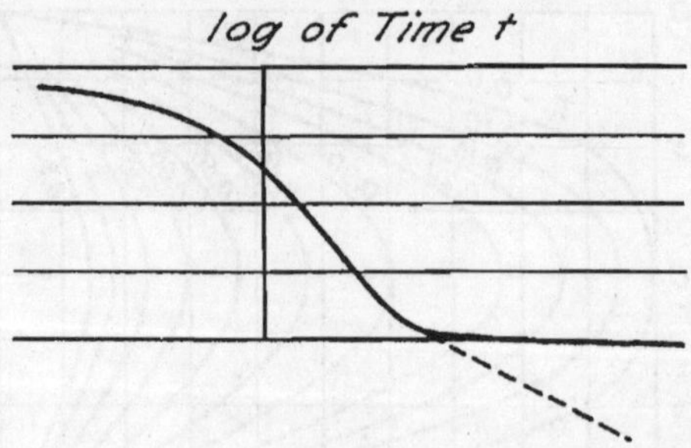

Figure 4-30 Log of time versus settlement. (From Terzaghi and Peck, 1967.)

where p = contact pressure at base of footing
E_c = pressuremeter modulus
B_0 = reference width equal to 30 cm
B = footing width
λ_2, λ_3 = shape factors (Figure 4-32)
α = soil structure factor (Figure 4-33)

Schmertmann (1970) gives the following relationship for settlement occurring in a number of cohesionless soil layers beneath a footing. For each layer, the estimated settlement is

$$S = C_1 C_2 \, \Delta p \, \left(\frac{I_z}{E_s}\right) \Delta z \tag{4-6}$$

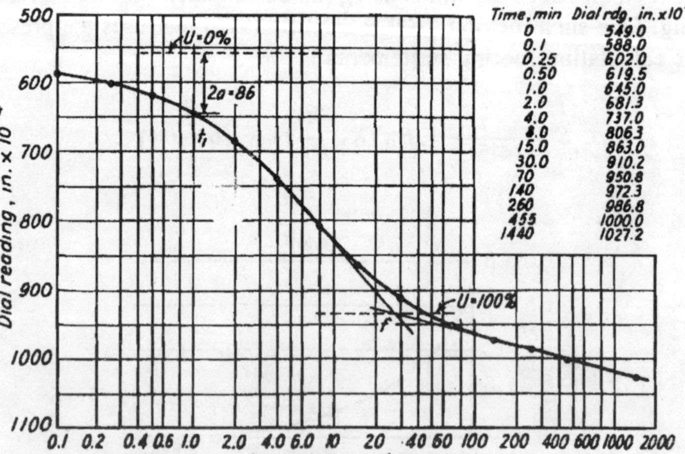

Figure 4-31 e - log t curve. (From Peck et al., 1974.)

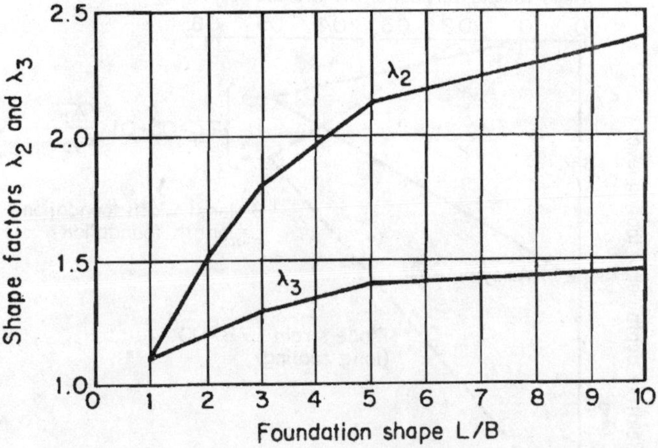

Figure 4-32 Shape factors for calculating settlement. (From Hunt, 1986.)

where

$$C_1 = 1 - 0.5\left(\frac{\bar{p}_0}{\Delta p}\right) \tag{4-7}$$

$$C_2 = 1 + 0.2\log\left(\frac{t}{0.1}\right) \tag{4-8}$$

and where p_0 = effective overburden pressure
Δp – footing pressure
t = settlement period (years)
I_z = strain influence factor (Figure 4-34)

	Peat		Clay		Silt		Sand		Sand and Gravel	
	E_c/P_L	α	E_c/P_L	α	E_c/P_L	α	E_c/P_L	α	E_c/P_L	α
SOIL MATERIAL										
Overconsolidated or very dense			16	1	14	0.67	12	0.5	10	0.33
Normally consolidated or dense		1	9–16	0.67	8–14	0.5	7–12	0.33	6–10	0.25
Underconsolidated or loose			7–9	0.5	5–8	0.5	5–7	0.33		

ROCK CONDITIONS			
Wide joint spacing	Moderately close joint spacing	Close spacing	Very close spacing, low strength
α = 0.67	α = 0.5	α = 0.33	α = 0.67

Figure 4-33 Soil structure factors for calculating settlement. (From Hunt, 1986.)

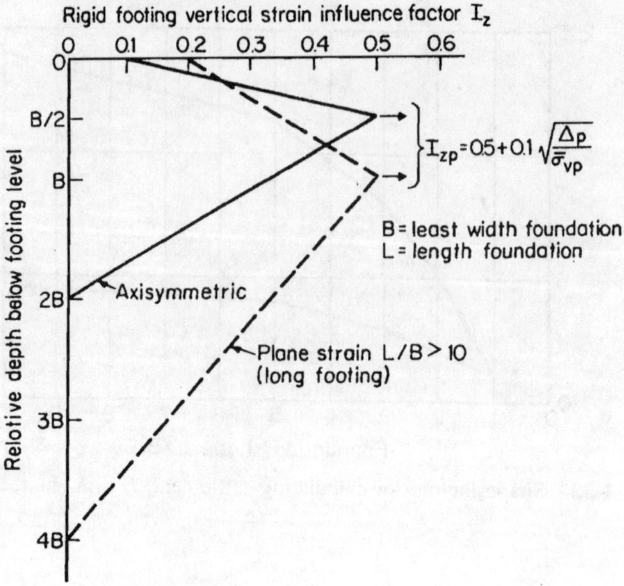

Figure 4-34 Strain influence factors for calculating settlement. (From Hunt, 1986.)

E_s = soil modulus
Δz = layer thickness

To arrive at the total predicted settlement, the settlements of each layer are summed together.

Hunt (1986) gives a simplified version of this method as follows:

$$S = 0.7\,\Delta p \quad \frac{B}{N} \tag{4-9}$$

where S is in inches
Δp is in tons per square foot
N = standard penetration test (SPT) blow count

Bearing Capacity

Bearing capacity theory is based on solutions developed for the rigid-plastic solid of the classical theory of plasticity, in which the solid is assumed to exhibit no deformation prior to shear failure and plastic flow under constant stress after failure. The problem is illustrated in Figure 4-35. A closed form equation for solving this problem in soil with drained loading conditions and a long rectangular footing has been found by Terzaghi (1943):

$$q_b = cN_c + qN_q + 0.5\,\delta\,BN \tag{4-10}$$

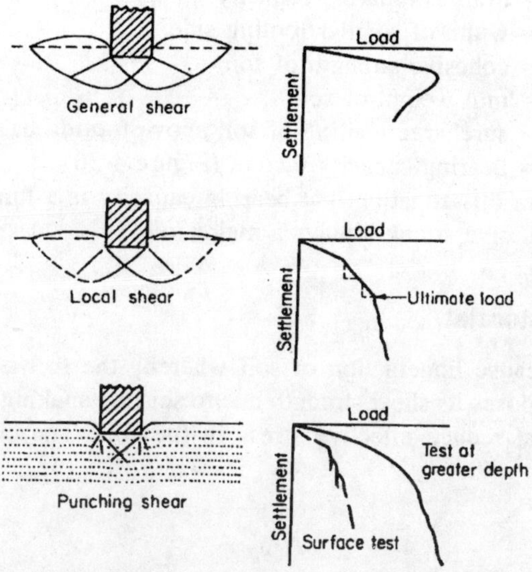

Figure 4-35 Bearing capacity failure. (From Hunt, 1986.)

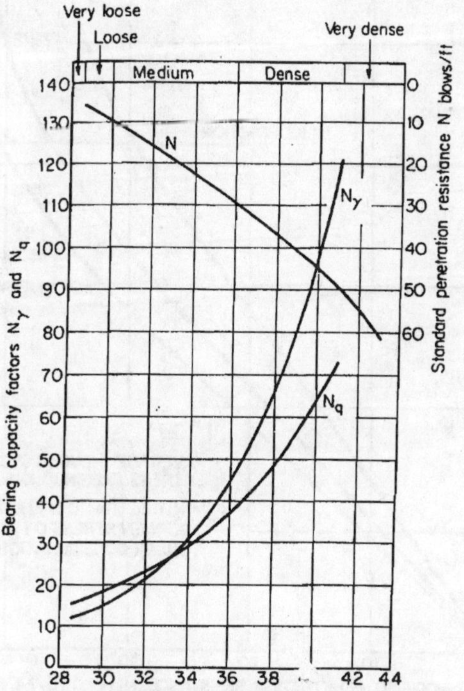

Figure 4-36 Bearing capacity factors. From Hunt, 1986.)

where q_b = ultimate bearing capacity
B = width of smaller footing side
c = cohesive strength of soil
δ = unit weight of soil
q = surcharge loading of soil above footing base
N_c, N_q, N = bearing capacity factors (Figure 4-36)

It can be seen from this equation that bearing capacity is a function of soil shear strength where cohesive strength plays a major role.

Liquefaction Potential

Earthquakes can cause liquefaction of soil whereby the soil is subjected to high shear strains and looses its shear strength due to seismic shaking and the buildup of pore pressures that reduce effective stress in the soil. The character of ground

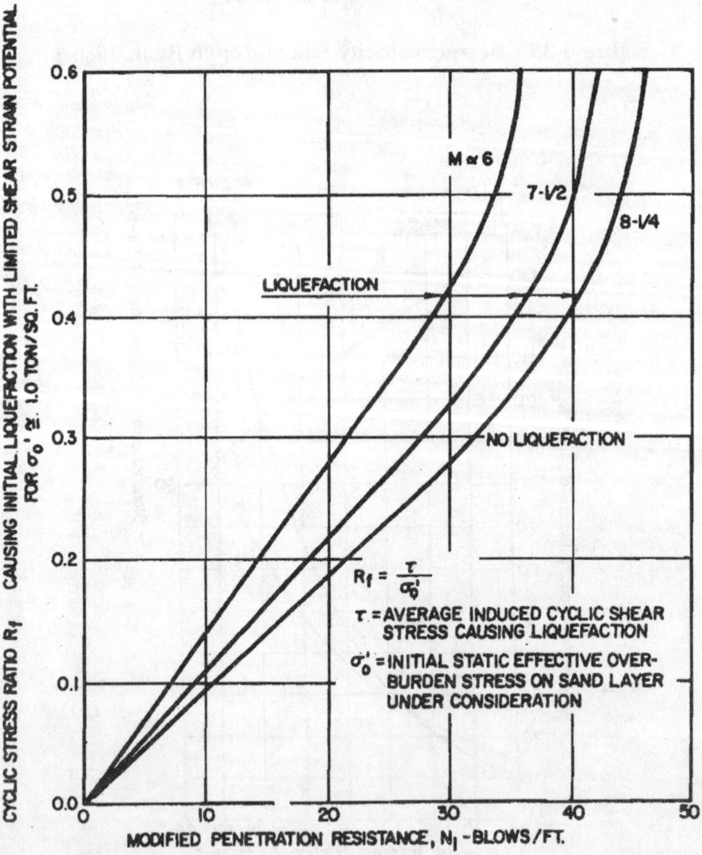

Figure 4-37 Cyclic stress ratio versus blow count. (From NAVFAC, 1983.)

motion (acceleration and frequency content), soil type, and in situ stress conditions are the three primary factors controlling the development of liquefaction. Dense gravelly soils are less likely to liquefy than loose sandy soils. Fines content improves the susceptibility of soils to liquefaction. Also, low water table and high effective stress makes liquefaction less likely. Case histories indicate that liquefaction normally occurs within a depth of 50 ft or less (NAVFAC, 1983).

Liquefaction potential can be evaluated using empirical methods by comparing the probable earthquake and in situ density of a site to previous case histories. The first step is to determine the cyclic stress ratio at various depths of interest, using the following equation:

$$R_i = 0.65\, a_{\max} \frac{\sigma_0}{\sigma_0'} r_d \tag{4-11}$$

where R_i = cyclic stress ratio in the field for design earthquake
$a_{\max}$ = peak surface acceleration in g's
σ_0' = effective overburden stress on layer

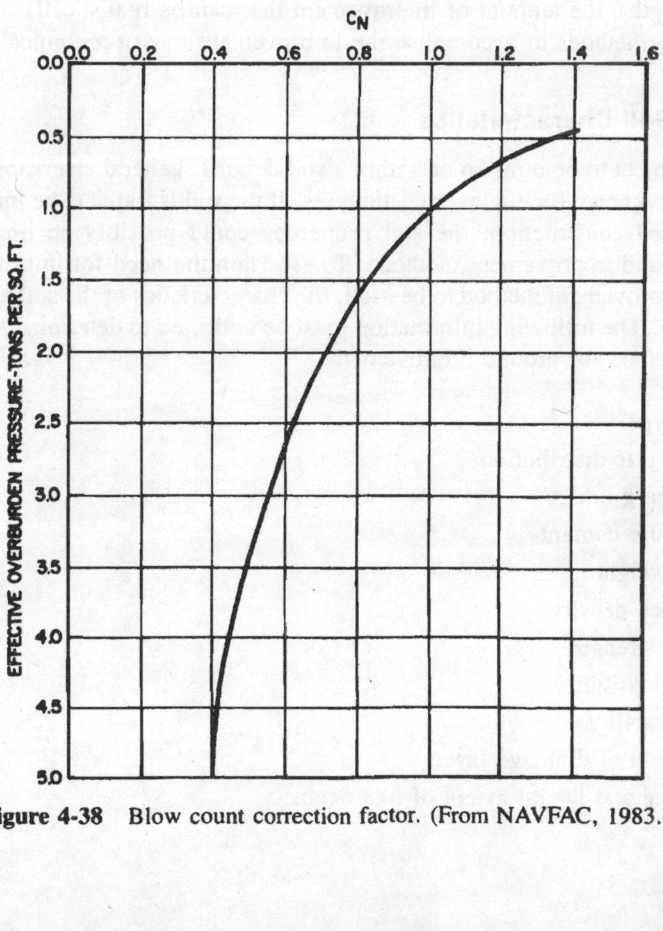

Figure 4-38 Blow count correction factor. (From NAVFAC, 1983.)

σ_0 = total overburden stress on layer
r_d = stress reduction factor equal to 1 at ground surface and 0.9 at a depth of 30 ft

Next, corrected blow counts (N_1) are calculated and used to find the cyclic stress ratio required for liquefaction, R_f, using Figure 4-37. N_1 equals $C_n N$ using the correction factors in Figure 4-38. If R_i is greater than R_f, there is a high potential for liquefaction. These can be plotted with depth to identify the zone of possible liquefaction where ground treatment may be required.

4-5 DESIGN CONSIDERATIONS

Ground improvement by compaction or consolidation is generally due to increased density in granular soils and increased drainage to facilitate consolidation in fine-grained soils. Increased density and consolidation generally cause increases in shear strength and hence, increased bearing capacity and decreased liquefaction potential. To decide on a ground improvement program, it is necessary to understand the soil characteristics that can be modified (i.e., consolidation properties, density, and shear strength), the amount of improvement that can be realistically attained, and the optimal methods to accomplish the improvement most economically.

General Soil Characteristics

For any project to be built on or within a soil deposit, general characteristics of the soil must be known for engineering analyses. If the soil is found to be inadequate for the proposed construction, the soil properties could possibly be improved with certain ground improvement methods. To ascertain the need for improvement and the best improvement method to be used, the characteristics of the soil are of central importance. The following information must be collected to determine the need and optimal method for ground improvement:

- Soil type
- Grain size distribution
- Atterberg limits
- Moisture content
- Unit weight
- In-place density
- Shear strength
- Compressibility
- Permeability
- Location of drainage layers
- Vertical and lateral extent of like deposits

- Location of groundwater table
- Presence of cobbles, boulders, and so on.

All of these properties are determined generally during a standard geotechnical exploration program. Most of them must be known to ascertain the acceptability of the site for the proposed construction. If the site is found not to be suitable, a form of ground improvement may be applicable. If that is the case, the soil properties are used again to determine the applicability of different improvement methods and the amount of improvement that can reasonably be expected.

Density

A common problem with new building sites is that the soil is not dense enough to support the new structure loads without excessive settlement, bearing capacity failure, or high liquefaction potential. The density of some soil deposits and even landfills and karst terrains can be increased by using certain ground improvement methods. In addition to soil stiffness, soil density, especially for granular (cohesionless) soils, is related to the angle of internal friction, generally the measure of granular soil strength. Density can also be correlated with permeability, compressibility, small-strain shear modulus, and cyclic shear strength. The relationships of granular soil density with compressibility are shown in Figure 4-39.

A most common and convenient way to evaluate the density of a soil deposit in the field is with the standard penetration test (SPT) blow count. The relative density of a soil deposit is a function of the in situ void ratio and the unit dry weight. As discussed below, other test methods include cone penetrometer tests (CPTs) pressuremeter tests, geophysical tests, and the like.

Shear Strength

When a site is found to be incapable of supporting the stresses imposed by changes to the site due to excavation, embankment filling, changes in the groundwater regime, new structure loads, or design seismic loadings, it may be possible to increase the shear strength of the soil. The shear strength of a granular soil was discussed above. The shear strength of a cohesive soil is important too with regard to bearing capacity and compressibility. The shear strength of a clay, the cohesion value, is related to the standard penetration resistance and the void ratio. Change in void ratio is normally how settlement calculations are related to changes in stress (Figure 4-40).

Rate of Consolidation

When a site underlain by clayey soil is to be loaded with an embankment or structure, consolidation will occur. The time it takes for primary consolidation to occur will depend upon the rate of consolidation. The rate of consolidation of a clay

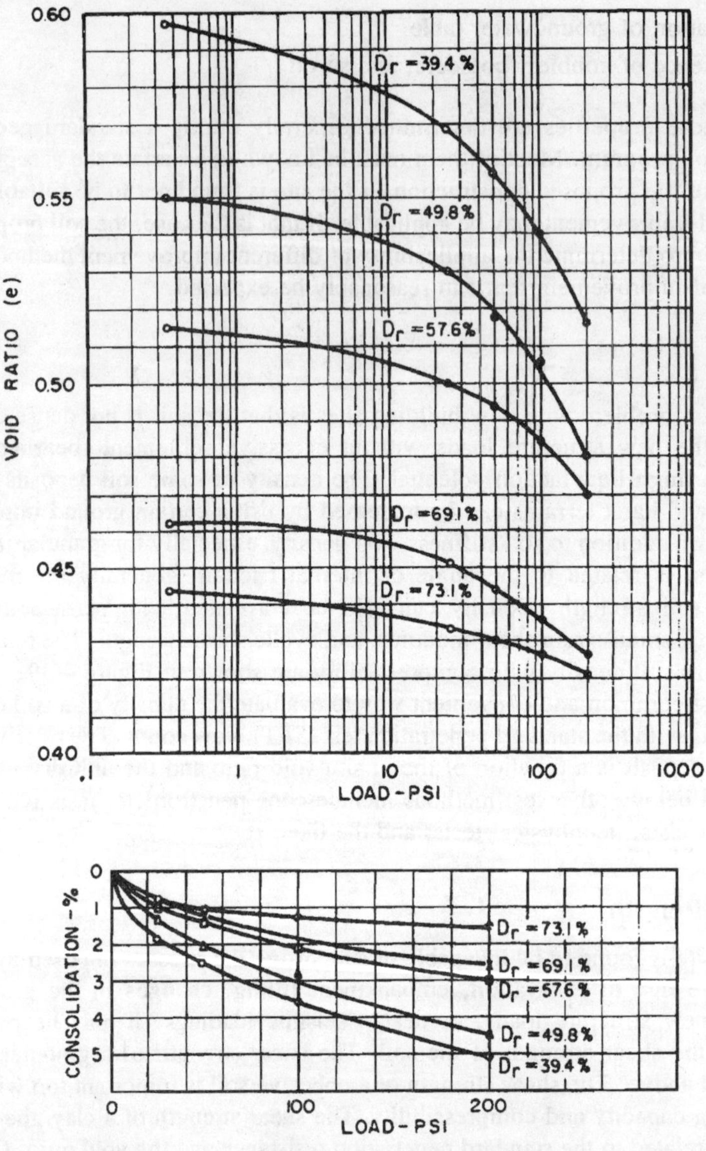

Figure 4-39 Compressibility versus density. (From Hilf, 1975.)

layer is a function of the permeability of the soil and the distance to drainage layers. To increase the rate of consolidation, it is easiest to decrease the distance to drainage layers. With vertical drains, drainage tends to occur in the horizontal direction to the drain and then vertically up the drain. Not only are the drainage paths shorter, but horizontal permeability of a clay deposit is often greater than the vertical permeability. The faster excess pore pressures can be dissipated through drainage, the faster the process of consolidation can occur.

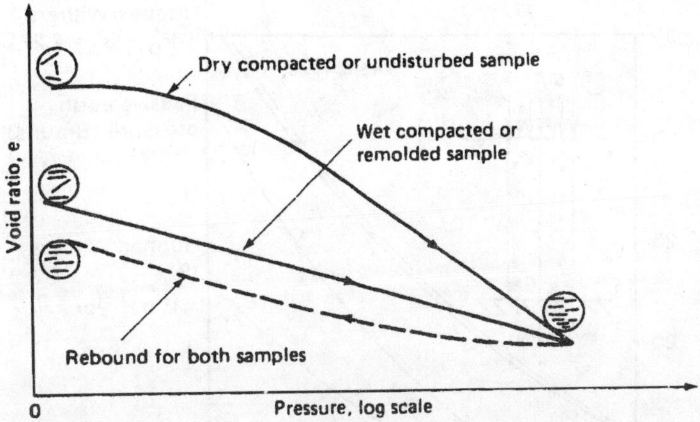

Figure 4-40 Typical consolidation curve (from R. D. Holtz and W. D. Kovacs, © 1981. Reprinted by permission of Prentice Hall, Englewood Cliffs, N.J.)

Settlement

Some sites may simply be judged to produce settlements that are too high for the structures being built to tolerate. In that case, there are three ways to reduce settlement. The first way involves reducing the compressibility of a deposit so that the total magnitude of settlement is less when the structure loading is applied. The second way is to introduce into the deposit stiff soil elements that will behave in a composite manner with the in situ soils by reinforcing the weak soils with stiffer load sharing elements. The third way is to preload the deposit so that much of the total settlement has occurred prior to application of the structure loading.

In a granular deposit, the way to reduce settlement is by increasing density, as discussed above. In a cohesive deposit, the best way to reduce settlement is by preloading. Preloading the site with a surcharge generally takes a significant amount of time. One way to accelerate the process is with vertical drains. Otherwise, stone columns can be constructed within the cohesive deposit to share the structure loads and reduce the pressure transmitted to the surrounding compressible soils (Figure 4-41). In extreme cases, all three of these techniques (preloading, vertical drains, and stone columns) can be used in combination.

Bearing Capacity

Rather than excessive settlements, new embankment or structure loads might be in excess of what the foundation soils can realistically be relied upon to support without the chance of a bearing capacity failure. By increasing the shear strength of a soil, the bearing capacity is also increased. Ways to accomplish this in granular soils include densification by dynamic compaction, vibrocompaction, or compaction grouting. In cohesive soils, modest increases in shear strength will occur with

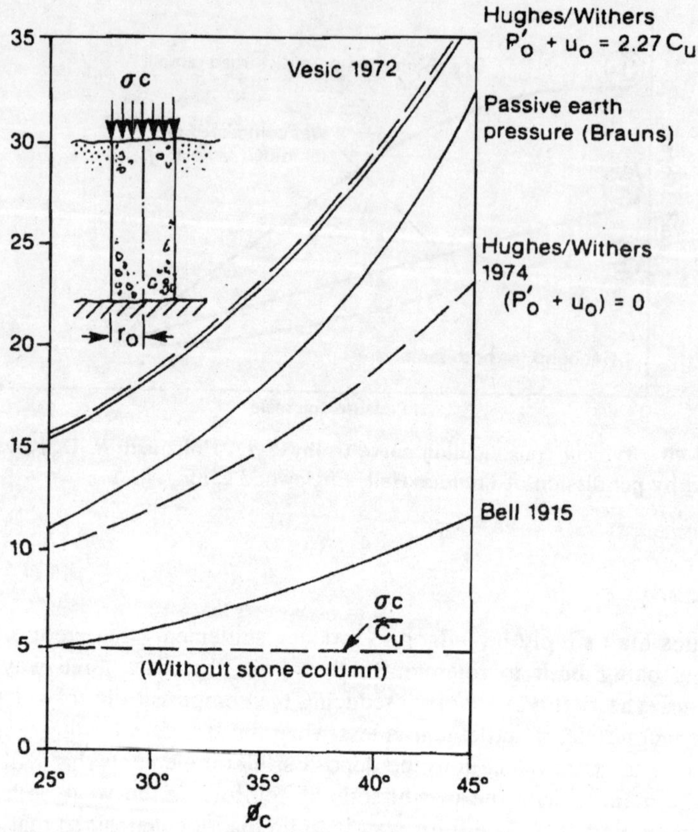

Figure 4-41 Ultimate capacity of stone columns. (From Munfakh et al., 1987. Reproduced by permission of ASCE.)

consolidation. However, the most effective method for increasing bearing capacity of clayey soils is probably through reinforcement with stone columns.

Slope Stability

Occasionally, ground improvement methods are used to increase the stability of a slope that is actively moving or is expected to move upon the application of new loads. By increasing the shear strength of a soil slope, one also increases the stability of that slope. If the in situ soil shear strength cannot be improved, stronger soil columns can be placed within the soil mass to reinforce it across potential failure planes. Another factor that can affect the stability of a soil slope is drainage. The installation of vertical drains may be all that is required to stabilize the slope by reducing excess pore pressures.

Liquefaction Potential

Liquefaction potential is dependent on the magnitude of a probable earthquake, the fines content of the soil, the density of the soil, and the location of the groundwater table. Granular deposits are by far most prone to liquefaction. The only characteristic of a granular soil that is commonly modified to reduce liquefaction potential is density. Little control over the other characteristics is practical. Methods for increasing soil density have been discussed above and include dynamic compaction, vibrocompaction, and compaction grouting.

4-6 DESIGN FUNDAMENTALS

Because of the specialty nature of ground improvement and the use of proprietary systems, detailed design work is generally not performed. Specifications are written on a performance basis, and include any necessary restrictions and guidelines. Performance is evaluated in terms of a measure of improvement. How that improvement is to be measured will be specified. For instance, Welsh (1986) gives potential measures of ground modification in granular soils in terms of standard penetration test (SPT), cone penetration test (CPT), pressuremeter test (PMT), and dilatometer test (DMT) results, as shown in Table 4-3. Design work chiefly centers around what type of and how much improvement is required, what potential methods can be used to accomplish the needed improvement, and what requirements will be specified to accomplish the improvement.

Densification of a Cohesionless Soil Deposit

Densification is generally required on a site to reduce settlement, increase bearing capacity, or reduce liquefaction potential. The need for and amount of ground improvement is identified through conventional geotechnical analyses. As a stan-

TABLE 4-3 Potential Improvement of Granular Soils

	Dynamic Compaction	Vibrocompaction	Stone Columns	Compaction Grouting
SPT (blows/ft)	3–5 × 25 Max	3–5 × 25 Max	2–4 × 15 Max	3–5 × 25 Max
CPT (tons/ft^2)	80–150	80–150	80–150	80–150
PMT (K_0)	0.6–1.3	0.6–1.3	0.6–1.3	0.6–1.3
DMT (tons/ft^2)	500–1000	500–1000	500–1000	500–1000

Source: From Welsh (1986). Reproduced by permission of ASCE.

dard matter on a project, settlement, bearing capacity, and liquefaction calculations are ordinarily carried out after determination of reliable soil property values. The resulting quantities are then compared to the allowable values and demands of the facility. If the estimated settlement is higher than allowable limits, if the allowable bearing capacity is lower than the bearing pressure of the structure, or if the seismically induced cyclic stress ratio is higher than cyclic shear ratio capacity, then ground modification is a candidate remedy if one of the methods can bring the site to within allowable limits.

To evaluate the use of ground improvement for densification, the following information must be known:

- In-place density of the soil
- Grain-size distribution
- Required density for satisfactory performance
- Net differences if in-place density is too low
- Depth and extent of zone requiring improvement

Some of this information can be plotted versus depth (Figure 4-42). This method of presentation is useful for evaluating the amount and location of improvement needed as well as the amount of improvement achieved after construction.

The amount of improvement needed can be expressed in terms of the following measurements (AASHTO, 1990):

- Minimum or average percent relative density
- Minimum or average percent of maximum density
- Minimum or average SPT blow count
- Minimum or average cone penetration resistance
- Minimum size of gravel or sand added (vibro methods)
- Minimum load-bearing requirement

These quantities can be readily measured in the field and laboratory using standard methods.

The steps to designing a ground improvement program for the purposes of increasing soil density are to:

1. Carry out a conventional geotechnical exploration.
2. Synthesize and prepare the data obtained.
3. Perform standard foundation analyses including settlement, bearing capacity, and liquefaction potential estimates.
4. Check the values obtained against project requirements.
5. If density is found to be too low, consider improvement.
6. Determine the amount and location of improvement needed.

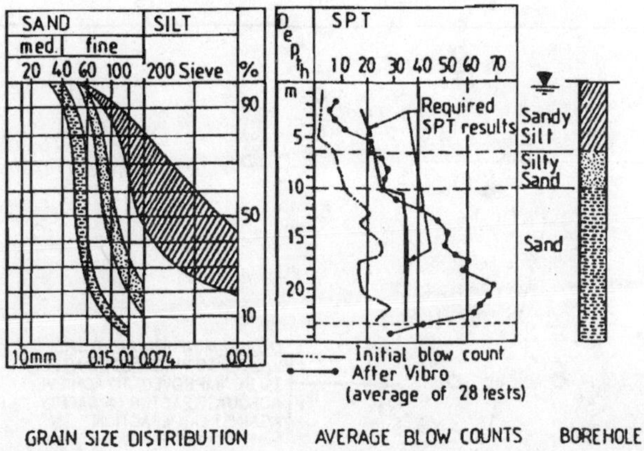

Figure 4-42 Soil density versus depth. (From Dobson, 1987. Reproduced by permission of ASCE.)

7. For shallow problems, consider conventional compaction and dynamic compaction. For deep problems consider vibrocompaction, vibroreplacement (stone columns), or compaction grouting.
8. Evaluate the amount of improvement that can be expected from candidate improvement methods and evaluate whether the amount of improvement is acceptable (Figure 4-43).
9. If ground improvement is a viable solution, prepare the necessary contract documents. If not, consider a deep foundation or another site.
10. Contract documents should clearly detail the expectations of the ground improvement method(s).

It is important to provide the contractor with good quality geotechnical information. The specifications should provide for processes by which the contractor's experience is evaluated, proposed methods are preapproved, and practical verification of the work is carried out.

Consolidation of a Cohesive Soil Deposit

The variables involved in the design of a ground improvement program related to consolidation of a cohesive soil deposit typically are how much settlement will occur and how fast. Methods used to increase the magnitude and rate of settlement include:

- Surcharging
- Vertical drains (sand or wicks)
- Dewatering through vertical drains

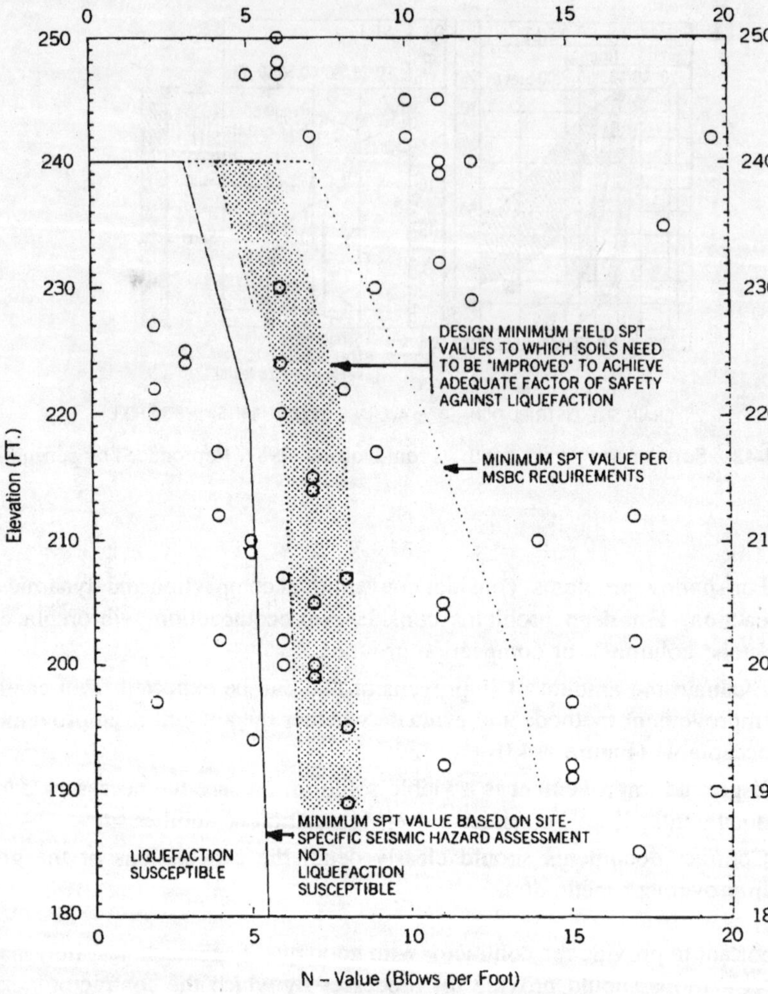

Figure 4-43 Typical ground improvement profile. (From La Fosse and von Rosenvinge, 1992. Reproduced by permission of ASCE.)

These methods can be used singly or in combination. If surcharging is used, care must be taken to make sure that shear failure of soft foundation soils does not occur under the increased loading. Dewatering methods can be used in combination with vertical drains to increase water flow into the drains. Dewatering methods are covered in Chapter 1, Groundwater Lowering and Drainage Techniques.

Estimating the amount of settlement that may occur in a compressible deposit is carried out using traditional formulas dealing with one-dimensional consolidation theory. The time-rate of consolidation can be calculated for a variety of surcharge heights. If the time required for completion of primary and secondary consolidation is too long, vertical drains can be used to increase the rate of consolidation. A

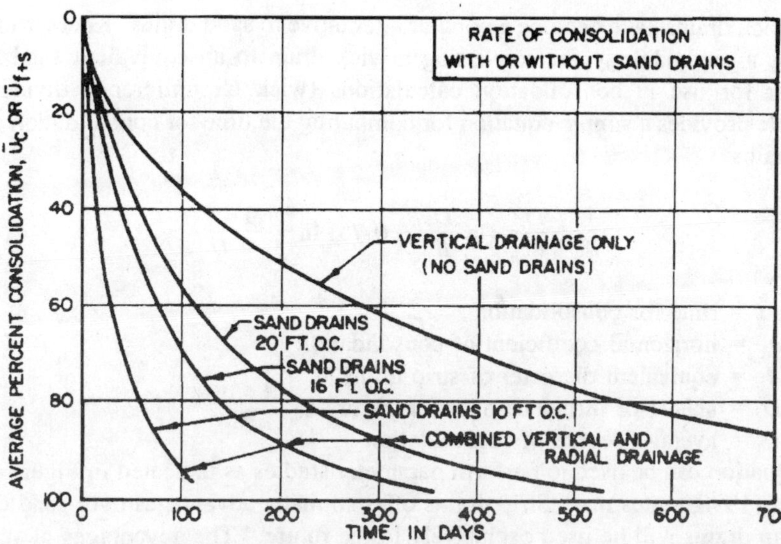

Figure 4-44 Rate of consolidation with and without sand drains. (From NAVFAC, 1982.)

variety of drain types and spacings can be evaluated to find the most cost-effective layout (Figure 4-44). Surcharge heights can be varied along with drain spacings to reach the optimum design (Figure 4-45).

Most of the literature dealing with vertical drain design (NAVFAC, 1982) for use in accelerating consolidation is based on the characteristics of sand drains. Geocom-

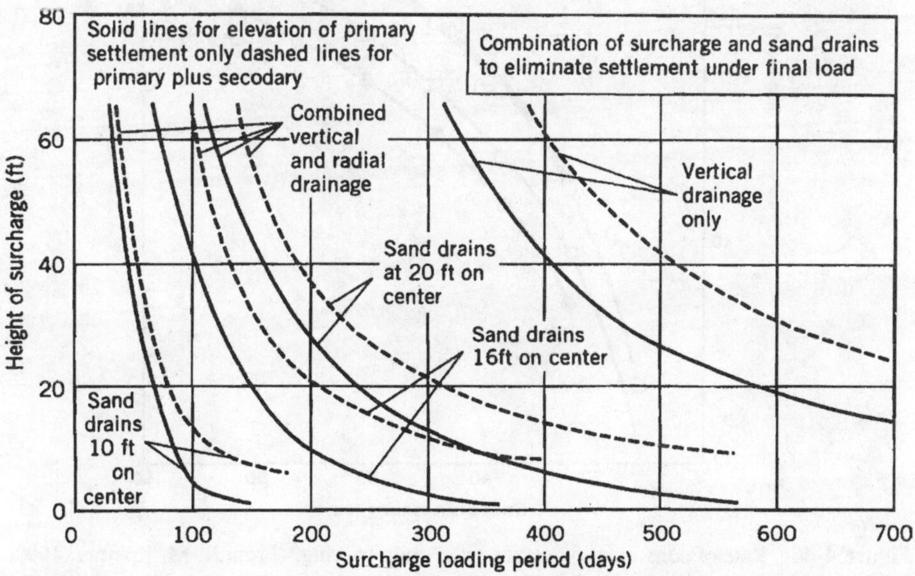

Figure 4-45 Time of surcharging with and without sand drains. (From NAVFAC, 1982.)

posite wick drains have become a popular alternative to sand drains. Koerner (1990) provides a methodology for converting a wick drain to an equivalent sand drain diameter for use in consolidation calculations (wick circumference divided by 3.14). He provides a simple equation for computing the time for consolidation using wick drains:

$$t = \frac{D^2}{8c_h}\left(\ln \frac{D}{d} - 0.75\right) \ln \frac{1}{1-U} \tag{4-12}$$

where t = time for consolidation
c_h = horizontal coefficient of consolidation
d = equivalent diameter of strip drain
D = sphere of influence of the wick drain
U = average degree of consolidation

This equation can be used to perform parameter studies as indicated in Figure 4-46. Koerner (1990) states that "Strip drains offer so many advantages over sand drains that strip drains will be used exclusively in the future." The advantages of strip or wick drains are:

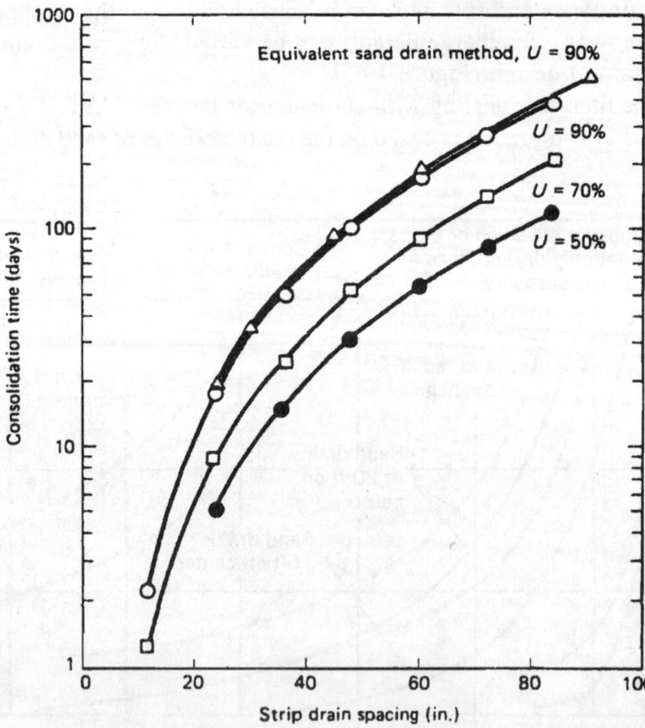

Figure 4-46 Rate of consolidation versus wick drain spacing. (From R. M. Koerner, 1990, Designing with *Geosynthetics*, © 1990. Reprinted by permission of Prentice Hall, Englewood Cliffs, N.J.)

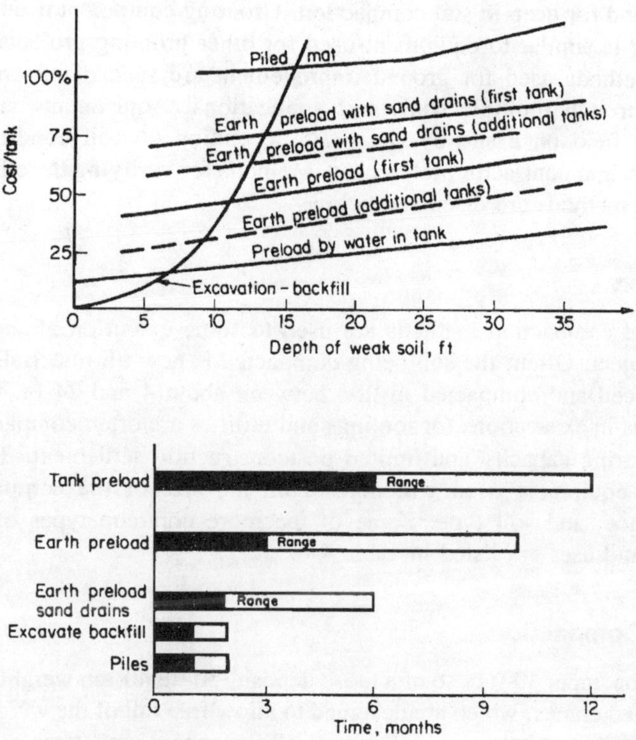

Figure 4-47 Economic comparison of ground improvement alternatives. (From Hunt, 1986.)

- Unrestricted water flow through the drain
- Relatively small installation equipment required
- Simple, straightforward, fast, and clean installation
- Reinforcement of the weak soil

The use of surcharging and vertical drains to accelerate consolidation is quite versatile and can be tailored to a variety of site conditions and constraints (Figure 4-47).

4-7 CONSTRUCTION METHODS

The construction methods used for ground improvement by compaction and consolidation range from using conventional compaction equipment in conjunction with vertical drains to more specialized methods using large weights, vibroflots, and grouting equipment. Most of the specialized methods require the use of a large crane modified for the specific use such as dropping heavy weights for dynamic compaction, driving wick drains into the ground for drainage, or lowering vibrators into the ground for deep in situ compaction. Grouting equipment used for compaction grouting is similar to equipment used for other grouting projects. The equipment and methods used for ground improvement are specialized and have been developed through multiple trials and applications. Adjustments are frequently made in the field on a site by site basis depending on soil conditions, project requirements, and contractor preferences. Methods for verifying the effectiveness of construction methods are discussed below.

Compaction

Conventional compaction methods are used to some extent on almost every construction project. Often, the soil being compacted is new fill materials imported to the site, placed and compacted in lifts between about 4 and 24 in. thick. Newly exposed soils in excavations for footings and utilities are often compacted to insure adequate bearing capacity and limited postconstruction settlement. The choice of compaction equipment used will depend on the size of the compaction effort, working space, and soil type. Some of the more common types of compaction equipment and uses are listed in Table 4-4.

Dynamic Compaction

To densify the upper 40 ft or so of a loose deposit, 30- to 40-ton weights are dropped from modified cranes, which are designed to allow free fall of the weight from up to 100 ft above the ground surface (Figure 4-48). Standard cranes are not designed to repeatedly drop heavy weights. The crane must be modified so that most (80 to 95 percent) of the potential energy is realized at the point of impact. Modifications to the crane boom, cables, drums, clutch, and outriggers are usually necessary to attain these potential energy levels in a safe manner.

TABLE 4-4 Types of Conventional Compaction Equipment

Equipment Type	Use	Comments
Sheepsfoot roller	High-speed production in fine-grained soils	Compaction depends on unit pressure and roller speed
Rubber-tired rollers	Granular soils	Very light weight to 200 tons; self-propelled or towed
Vibratory compactors	Granular soils	Towed, self-propelled, or hand-held; compaction depends on frequency and energy of vibrations
Air tamps	Low production; restricted access; shallow lifts	Compaction by reciprocating blows powered by compressed air
Steel-wheel rollers	Provides a smooth sealed surface; shallow lifts	Self-propelled

Source: From Sain (1983).

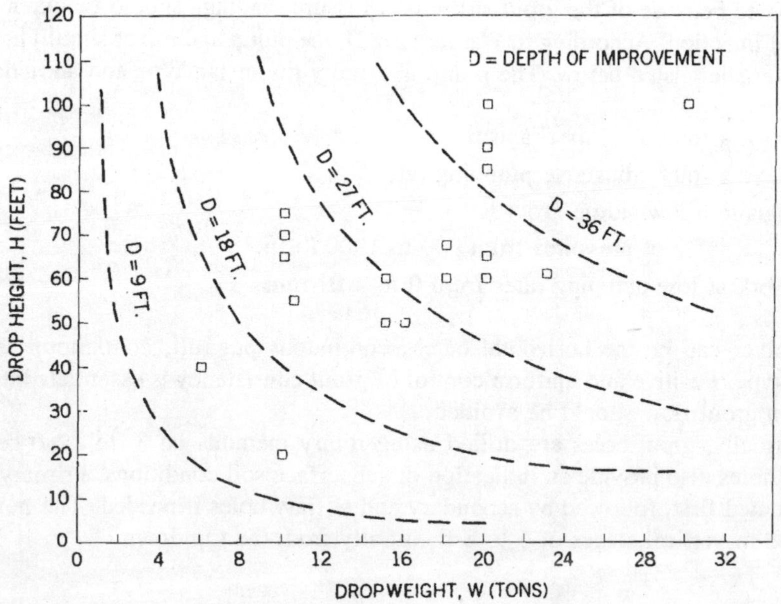

Figure 4-48 Depth of improvement using dynamic compaction (data from AASHTO, 1990).

Ground improvement programs using dynamic compaction are usually planned in phases with separate plans for the "primary pass," the "secondary pass," the "tertiary pass," and the "ironing pass." Each pass typically has a prespecified weight, drop height, number of drops, and drop pattern spacing. At a single site, multiple passes may be required with different drop weights and drop heights. This may require the use of different cranes at the same site.

After site improvement, earth moving equipment may be required to level the site and fill in the craters. At soft sites or landfills, it might be necessary to lay down a granular working pad prior to construction for the purposes of facilitating movement of the crane and distributing the dynamic load (weight drops). Other ground improvement methods may be used in conjunction with this method to facilitate relief of excess pore pressures with wick or sand drains and ground improvement at greater depths with vibrocompaction or compaction grouting.

During dynamic compaction, special procedures are required to prevent damage of the crane equipment and injury to the crane operator. In addition to these procedures, off-site vibrations and flying debris should be of concern during construction.

Compaction Grouting

Compaction grouting requires grout pipes, hoses, valves and fittings, a pump, and mixer. Because of the very stiff nature of the type of grout and the relatively high grouting pressures used, special equipment is required. The pipes, hoses, valves, and fittings are generally $1\frac{1}{2}$ to 2 in. in diameter. High pressure losses should be anticipated because of the grout stiffness. Pressure readings should be taken at the point of injection. According to Warner (1982), the pump and mixer should have the characteristics listed below. The pump is usually the piston type and should:

- Have a force feed mechanism
- Have a fully adjustable pumping rate
- Handle a low slump grout
- Be capable of pressures from 600 to 1000 lb/in.^{2}
- Work at low pumping rates from 0 to 2 ft^3/min

The mixer can be the horizontal batch, continuous pugmill, continuous belt, or auger type. Positive and uniform control of grout consistency is essential. Interruptions in grout flow should be avoided.

Normally, grout holes are drilled using rotary methods on 8- to 12-ft-centers. These holes also provide an indication of subsurface soil conditions. Primary holes are grouted first, followed by secondary and tertiary holes if needed. The holes are grouted in vertical stages of 5 to 8 ft, usually from the top down.

Sand Drains

Sand drains consist of columns of sand placed in vertical holes in compressible soil at sufficiently close spacing so that the horizontal drainage path for consolidation is less than the vertical path. They can be installed within mandrel-driven or conventionally driven pipes or rotary- or auger-drilled boreholes (Figure 4-49). The sand can be placed in the hole or pipe by gravity or by jetting it into place. Closed-mandrel driven pipe sand drains are the most common type (NAVFAC, 1982). Sand drains can be from 6 to 30 (usually 16 to 20) in. in diameter and are typically spaced between 5 and 20 (usually 6 and 10) ft apart.

Wick Drains

Wick drains are inserted into the ground using a crane- or backhoe-mounted mast containing a hollow mandrel or lance through which the drain is threaded (Figure 4-50). The mandrel and wick drain are inserted into the ground to the required depth and then the mandrel is withdrawn, leaving the drain in place. The drain is anchored into the ground by an anchor plate, rod, or other special device. The mandrel can be pushed into place using hydraulically driven cables or chains or can be driven into the ground with a vibratory hammer. The drains are from about 2 to 5 in. wide (Figure 4-51) and are installed on a spacing from about 1 to 12 ft.

The equipment and qualified personnel for this work can be supplied by a specialty contractor, although it would not be impossible for a general contractor to perform the work with the proper equipment and supervision. The size and type of mandrel equipment is an important construction consideration. A practical depth limitation on wick drains is about 100 ft.

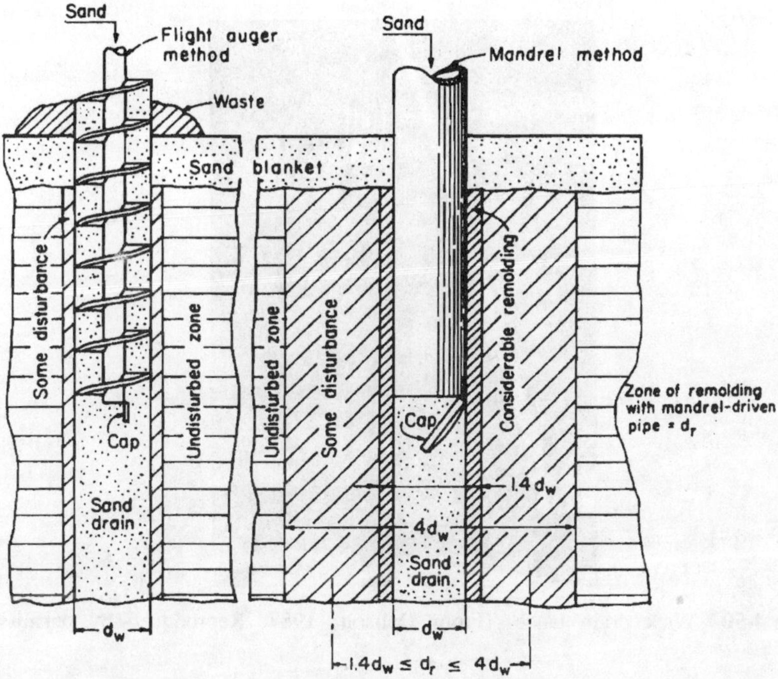

Figure 4-49 Sand drain installation. (From Landau, 1966.)

Adequate subsurface exploration should be carried out to determine installation costs, predrilling or jetting needs, and the possibility of encountering obstructions. These same borings can be used for design of the wick drain system, which requires knowledge of the extent and characteristics of the subsurface conditions and the amount and rate of settlement.

Vibrocompaction

Vibrocompaction consists of first boring a hole by air or water jetting a vibrating probe into the predominantly granular soil to the required improvement depth. Once the probe reaches the bottom, the jets are turned off and wash water is circulated up the hole so that it remains open and sand or stone backfill materials can be fed into the hole and compacted by the vibrating probe at the bottom of the hole. The probe is raised in small increments and momentarily held in place as the materials around and in the hole (the backfill) are compacted. Backfill is fed into the top of the hole until the probe reaches the ground surface again. A variation to this is to use a "bottom-feed" probe, which conveys the backfill material down a delivery pipe directly to the vibrating tip as opposed to down the sides of the borehole from the collar.

The main pieces of equipment are the electrical or hydraulic vibrator, follower

Figure 4-50 Wick drain lance. (From Dobson, 1987. Reproduced by permission of ASCE.)

pipes, and a 30- to 100-ton capacity crane to support the string. Supporting equipment usually includes a generator or drive unit, high-pressure high-volume water pump, and a front-end loader for the backfill.

The vibrator probe is usually about 30 in. in diameter and 7 to 16 ft long (Figure 4-52). The vibrator shown in the figure transmits radial vibrations at the nose of the probe. Centrifugal forces in excess of 20 tons can be generated at frequencies ranging from 1500 to 3000 revolutions per minute (Dobson, 1987). The probe weighs between 10 and 20 tons. Other operating parameters according to Mitchell (1986) are:

- 3- to 6-ft/min probe sinking rate
- 1-ft/min withdrawal/compaction rate

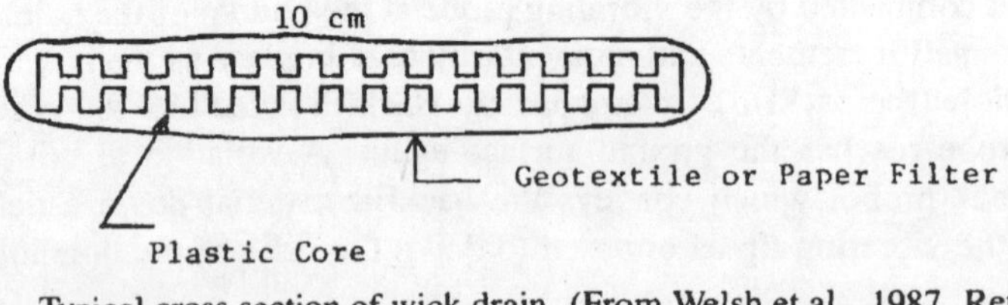

Figure 4-51 Typical cross section of wick drain. (From Welsh et al., 1987. Reproduced by permission of ASCE.)

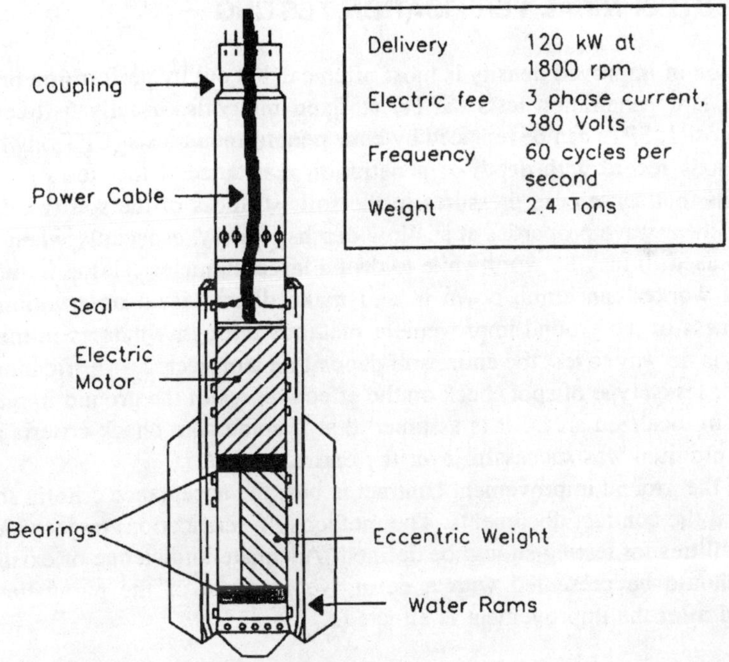

Figure 4-52 Typical vibroflotation vibrator. (From Venema et al., 1989. Reproduced by permission of ASCE.)

- Up to 115 lb/in.^{2} water pressure
- Up to 600-gal/min water flow
- Up to 16-ft^3/ft backfill consumption
- 3 to 20 yards^3 of improved soil/ft

Some probes transmit vibrations vertically down the follower pipes and probe. This method has a quicker cycle time, but requires a closer pattern spacing (AASHTO, 1990).

Vibroreplacement (Stone Columns)

The process of vibroreplacement is essentially the application of the vibrocompaction process to fine-grained in situ soils. The same equipment is used and is described above. The backfill is composed predominantly of crushed rock or stone instead of sand. The major difference in construction methods is that the probe is worked up and down in the hole so that the stone penetrates the soft cohesive soils around the hole. The backfill is not simply used to fill the void left by the probe and densification process as it is with vibrocompaction. During vibroreplacement, the stone backfill actually mixes with the surrounding soils as well as filling the probe hole. This process takes longer than vibrocompaction and uses more stone backfill.

4-8 GEOTECHNICAL VERIFICATION TESTING

Verification of improved density is most often carried out by performing before and after standard penetration tests (SPTs) at fixed intervals, usually 5-ft-centers. In some deposits, SPTs can be replaced by cone penetrometer tests (CPTs), which give a continuous record with depth of penetration resistance. Other tests can also be used, tests that sense pore pressures in the soil, stiffness of the soil (moduli), and dynamic shear wave properties at shallow depths. Rarely, especially when grouting has been used, it may be worthwhile to drill a large-diameter caisson in the ground so that a worker can climb down in and make direct visual observations of the effectiveness of the ground improvement methods through windows in the casing.

There is no way to test the entire soil deposit on a project. So verification testing is more or less a type of spot check on the effectiveness of the ground improvement methods in localized areas. It is assumed that if some spot check criteria are met, then the program was successful over the entire project.

When the ground improvement contract is bid, the acceptance criteria should be detailed in the contract documents. The methods of verification, target values, and responsibilities for testing should be defined. Adequate knowledge of existing conditions should be presented with a definitive measure of the parameters to be compared after the improvement is effected.

Standard Penetration Tests (SPTs)

A split-spoon sampler is used to take SPT samples according to methods documented by the American Society for Testing and Materials (ASTM). Most split spoons are 18 in. long, although some are 24 in. long. The SPT is conducted by driving the split spoon into soil that has not yet been disturbed by the drilling process. The sampler is driven 18 in. with a 140-lb hammer blow falling 30 in. The number of blows it takes to drive the sampler each 6-in.-increment is recorded. The first 6-in.-drive for seating the sampler is usually affected by cuttings at the bottom of the borehole and the number of blows is neglected. Blow counts from the second and third 6-in.-increments are added together and the sum equals the SPT resistance, N, or blow count. This number is expressed in units of blows per foot. A disturbed sample is obtained for visual classification of the soil.

Standard penetration tests are most effective in granular deposits, although correlations with blow counts in cohesive deposits do exist but are less reliable. There is quite a lot of variability in SPTs and one should not rely on their accuracy beyond plus or minus 5 to 10 blows/ft. In other words, an after-treatment blow count of less than 5 more than the before-treatment blow count might not be a reliable indicator of improvement. However, changes in penetration resistance of 10 or 15 blows/ft would reliably indicate a change in density.

Cone Penetrometer Tests (CPTs)

A cone penetrometer test (CPT) is a method by which a cone-shaped steel tip is forced into the ground by connecting rods and hydraulic equipment at the ground surface. A continuous record of point resistance and side friction is obtained as the cone is jacked through the soil (Figure 4-53). No samples are recovered. This test method is particularly good for detailing soil stratification in thick deposits of weak to moderately strong soils. Shear strength and relative density are correlated to the CPT bearing capacity. Comparisons of CPT resistance before and after treatment are very illustrative (Figure 4-54).

Other Tests

The two test methods given above are the most common methods used to verify ground treatment effectiveness. Other test methods, however, are used regularly and include:

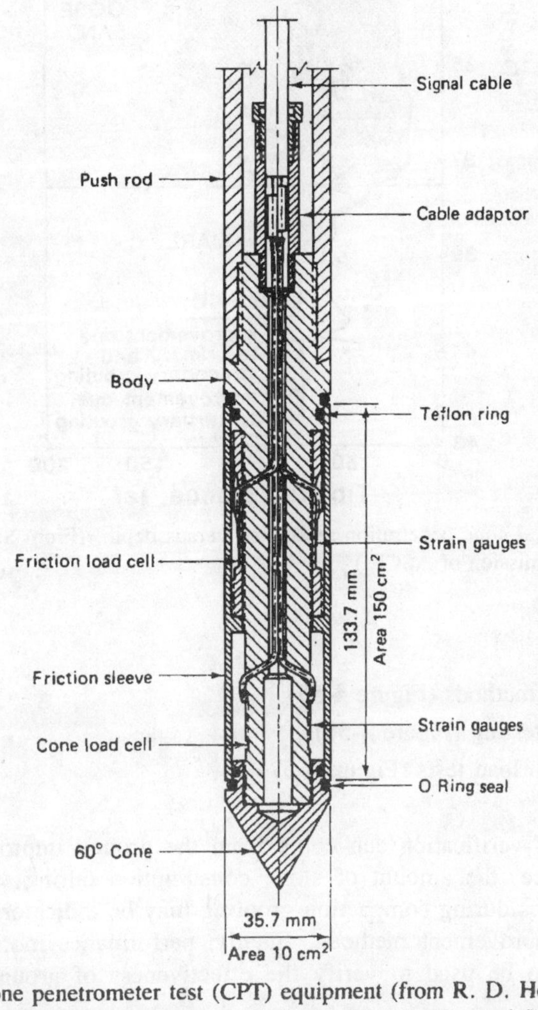

Figure 4-53 Cone penetrometer test (CPT) equipment (from R. D. Holtz and W. D. Kovacs, © 1981. Reprinted by permission of Prentice Hall, Englewood Cliffs, N.J.)

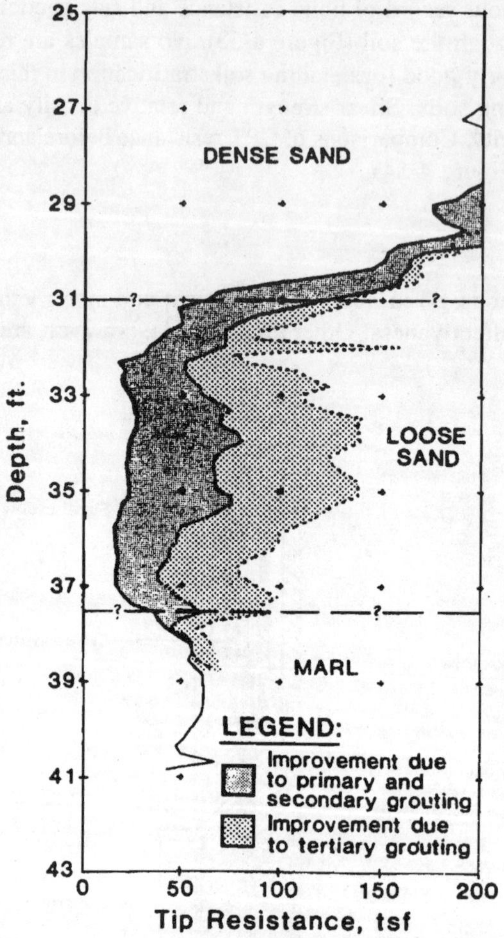

Figure 4-54 Typical cone penetration resistance versus depth. (From Salley et al., 1987. Reproduced by permission of ASCE.)

- Geophysical methods (Figure 4-55)
- Dilatometer testing (Figure 4-56)
- Stone column load tests (Figure 4-57)

Other methods of verification can come from the ground improvement method itself. For instance, the amount of stone consumption during vibroflotation or reduced grout takes during compaction grouting may be indicators of the success of the ground improvement methods. Finally, performance monitoring, as discussed below, can be used to verify the effectiveness of ground improvement methods.

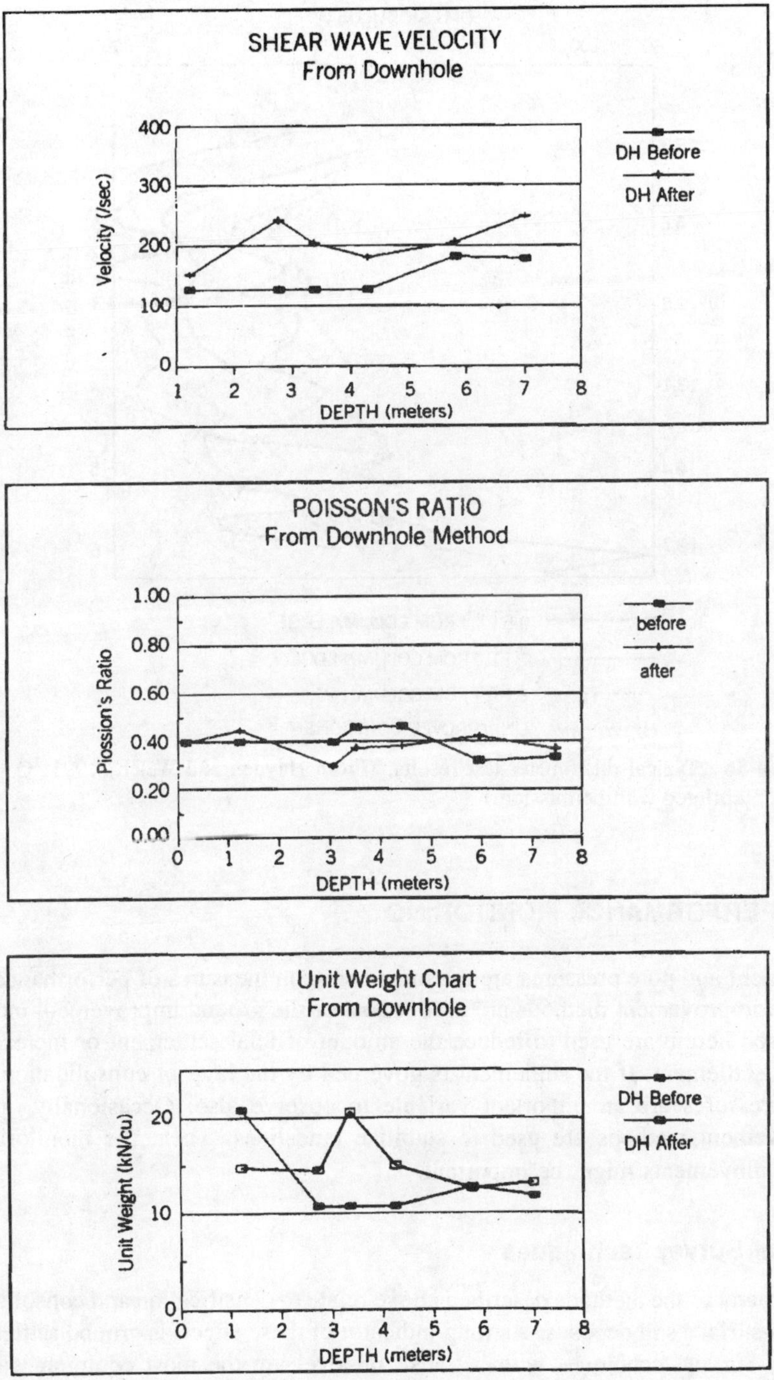

Figure 4-55 Typical geophysical testing results. (From Byle et al., 1991. Copyright ASTM. Reprinted with permission.)

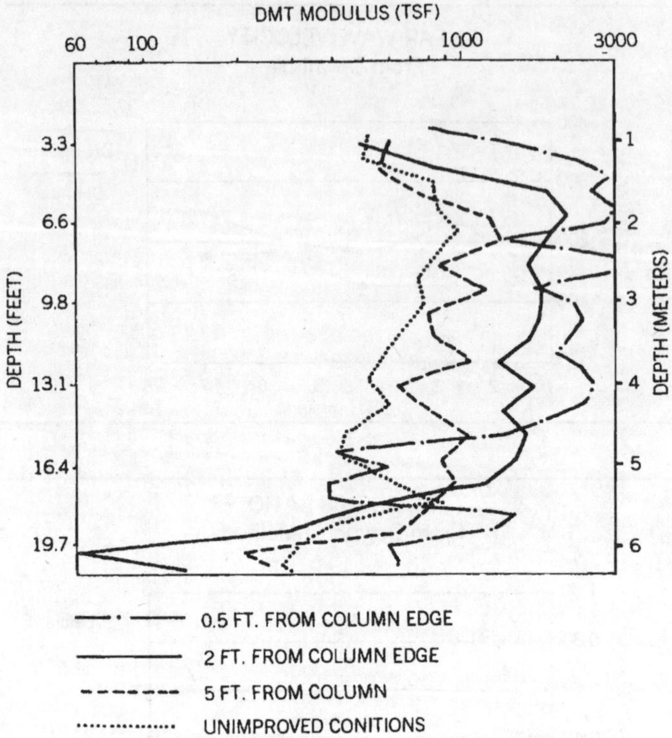

Figure 4-56 Typical dilatometer test results. (From Hayden and Welch, 1991. Copyright ASTM. Reprinted with permission.)

4-9 PERFORMANCE MONITORING

Settlement and pore pressures are the most common measures of performance after ground improvement methods are used. Most of the ground improvement methods discussed herein are used to reduce the amount of total settlement or increase the rate of settlement. If the settlement is governed by the laws of consolidation, then pore pressures are an important variable to observe also. Occasionally, ground improvement methods are used to stabilize landslides. Then, the monitoring of lateral movements might be important.

Optical Survey Techniques

Since many of the methods described above relate to densification and consolidation of near-surface soil deposits, a strong indicator of these effects is ground settlement. Optical survey techniques with a transit or level are the most common ways of observing the amount and rate of ground settlement that is taking place before and after ground improvement. The same types of survey equipment are commonly used

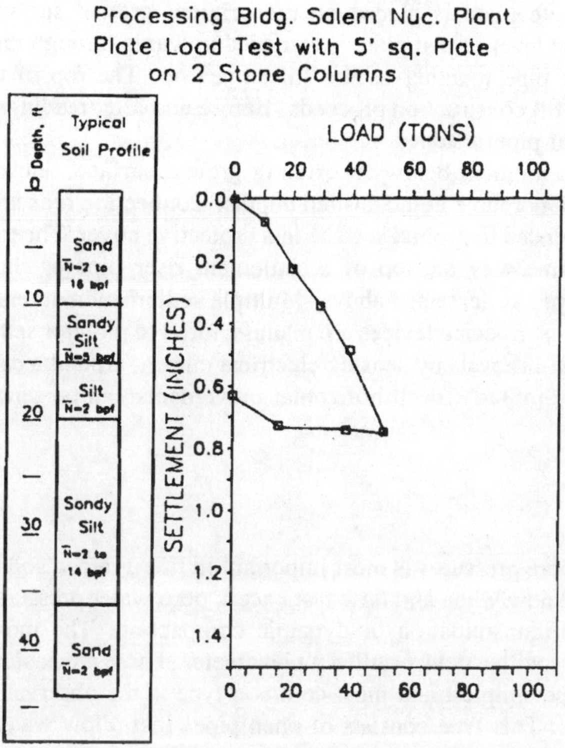

Figure 4-57 Typical stone column load test results. (From Hussin and Baez, 1991. Copyright ASTM. Reprinted with permission.)

on a construction site for layout. One need only have a reliable bench mark to start from and monitoring points chosen that can be repeatedly checked on. Survey of the monitoring points can often be done during the course of other surveying activities. If horizontal measurements are required, the same procedures can be followed.

The routine gathering of the survey data should be accompanied with clear lines of responsibilities for processing and transmitting the data in a timely manner. Someone should be reviewing the data regularly and action plans should be identified for when unusual readings or trends develop. The frequency of readings should be well understood so that missed readings do not occur at critical times during project development. The readings are usually plotted against time to observe any changes that are taking place and at what rate.

Settlement Plates and Deep Settlement Markers

The survey methods discussed above are used for measurements at the ground surface. To monitor subsurface settlements, settlement plates or deep settlement markers must be utilized. Settlement plates are used in embankments. A steel, wood, or concrete plate is placed on the original ground surface before filling commences. The level of that plate is extended upwards through the fill by connecting sections of pipe together as the fill comes up. The top of the riser pipe is surveyed as the fill construction proceeds. Before and after readings are taken every time a section of pipe is added.

To monitor settlement below an existing ground surface, anchor points are installed through a borehole at the desired depths. Connecting rods are attached to the anchors and extended to ground surface in a protective cover. The top of the rods are surveyed the same way the top of a settlement riser pipe or shallow settlement marker is surveyed as described above. Multiple settlement markers can be installed at different depths. Special devices are manufactured to monitor settlement electronically at frequent intervals by sensing electrical current around a casing with induction coil rings. Similarly, deep horizontal movements can be sensed using inclinometer casings.

Piezometers

Monitoring of pore pressures is most important in fine-grained soil deposits when it is important to know when and how fast excess pore water pressure is dissipated in connection with consolidation or dynamic compaction. The monitoring of pore pressures is done with a device called a piezometer. There are basically two types of piezometers. The simplest and most common type is the observation well or open pipe piezometer. This type consists of open pipes that allow water to filter in and stabilize at the prevailing groundwater level outside of the pipe. The open pipe allows someone to drop a tape measure or other sensing device down to the water level to measure and record the depth of the water level below ground surface. This type of piezometer was discussed earlier in Chapter 1.

The second type of piezometer is closed to the atmosphere and senses the pore pressure in the ground through a filter and diaphragm arrangement (transducer). Movement of the diaphragm can be sensed pneumatically by pressure gages or electrically by strain gages that are connected to the ground surface read-out device by tubes or wires (Figure 4-58). Single- or multiple-point devices can be used. The piezometers can be installed in boreholes or driven into the ground.

4-10 CASE HISTORIES

The methods and applications of ground improvement methods used for compaction and consolidation of soil are illustrated in the following case histories.

Preloading Used to Improve Site of Veterans Administration Complex in Tampa

The site for a Veterans Administration hospital complex in Tampa, Florida was underlain by compressible clayey sand and limestone solution channels (Wheeless

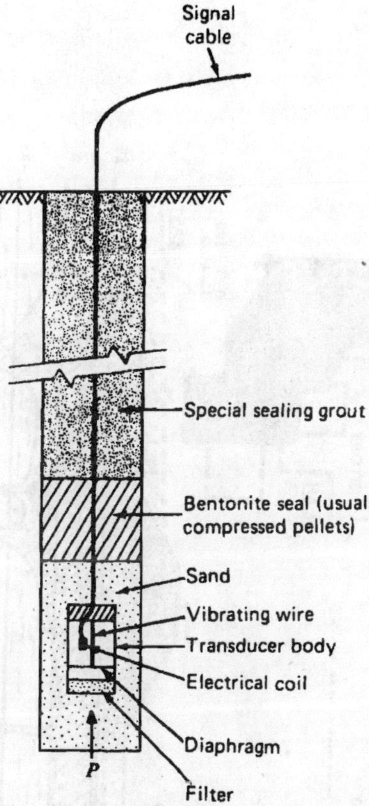

Figure 4-58 Typical piezometer equipment. (From Dunnicliff and Green, 1988.)

and Sowers, 1972). These subsurface conditions are depicted in Figure 4-59. A combination of grouting, preloading, and a compensated mat foundation was used to overcome the potential for excessive building settlement after construction. The eight-story-high hospital was a 240 × 260-ft rectangle in plan with three-story-high wings on all sides. Maximum column loads for the hospital and wings were 600 and 300 kips, respectively. For mat foundations, the maximum bearing pressures were about 1900 and 850 lb/ft^2, respectively.

Subsurface conditions consisted of clayey sand near the ground surface, becoming sandy clay at depths of about 15 to 20 ft below ground surface. Standard penetration resistances ranged between 5 and 20 blows/ft. Typical void ratios ranged between 0.7 and 1.4. Local lenses of very soft clay associated with erosion into the limestone typically had poorer engineering properties. These clay zones had penetration resistances between 1 and 3 blows/ft and void ratios exceeding 1.5. Tampa Limestone underlay the overburden and consisted of chalky, poorly consolidated calcareous deposits containing lenses of hard, calcareous clays and inclusions of hard reprecipitated calcite. The groundwater level was about 15 ft below ground surface.

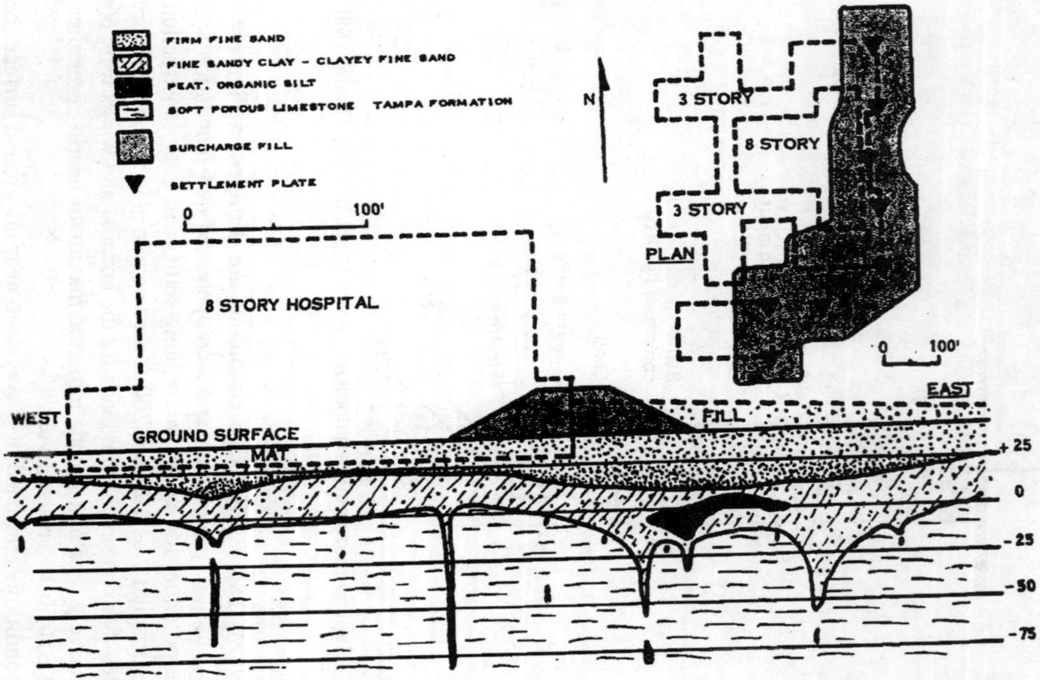

Figure 4-59 VA Hospital subsurface conditions. (From Wheeless and Sowers, 1972. Reproduced by permission of ASCE.)

A mat foundation was selected because it would produce the least stress concentration in the soft limestone (as compared to driven piles for instance) and would bridge over localized weak areas that might develop in the future. The chances of tilting of the mat foundation due to fill settlement and large-scale subsidence due to sinkhole development were lessened by a preload fill and cap grouting, respectively.

The preload fill was 26 ft high. Between 3 and 4 in. of settlement were predicted from laboratory test results. The preload fill was expected to accelerate the settlement to within four months, the time at which mat construction was to start. This is about twice as fast as settlement would have occurred without the preload fill. Settlement plates were used to monitor progress of the settlement. Observed settlement was between $2\frac{1}{2}$ and $3\frac{1}{2}$ in., very close to predicted values. However, settlement occurred faster than predicted (Figure 4-60) suggesting that sandy seams provided faster drainage than the 20- ft thickness of clay suggested.

Preloading of Silty Sand Required for Sewage Treatment Plant

A major sewage treatment plant was built along the Willamette River in Milwaukie, Oregon (Figure 4-61). The work involved 80,000 yards³ of excavation on the river slope and 200,000 yards³ of embankment on the 4-acre site (Schroeder and Worth, 1972). The subsurface conditions consisted of loose, poorly graded, fine alluvial sand and silty sand with occasional lenses of silt and some gravels. The sand was

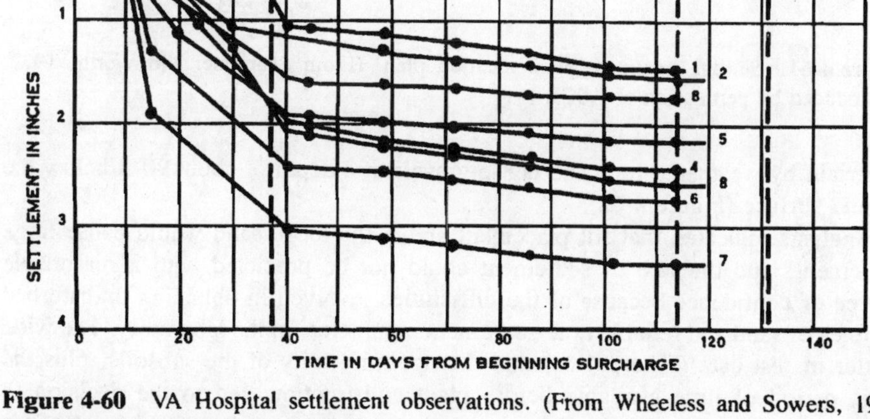

Figure 4-60 VA Hospital settlement observations. (From Wheeless and Sowers, 1972. Reproduced by permission of ASCE.)

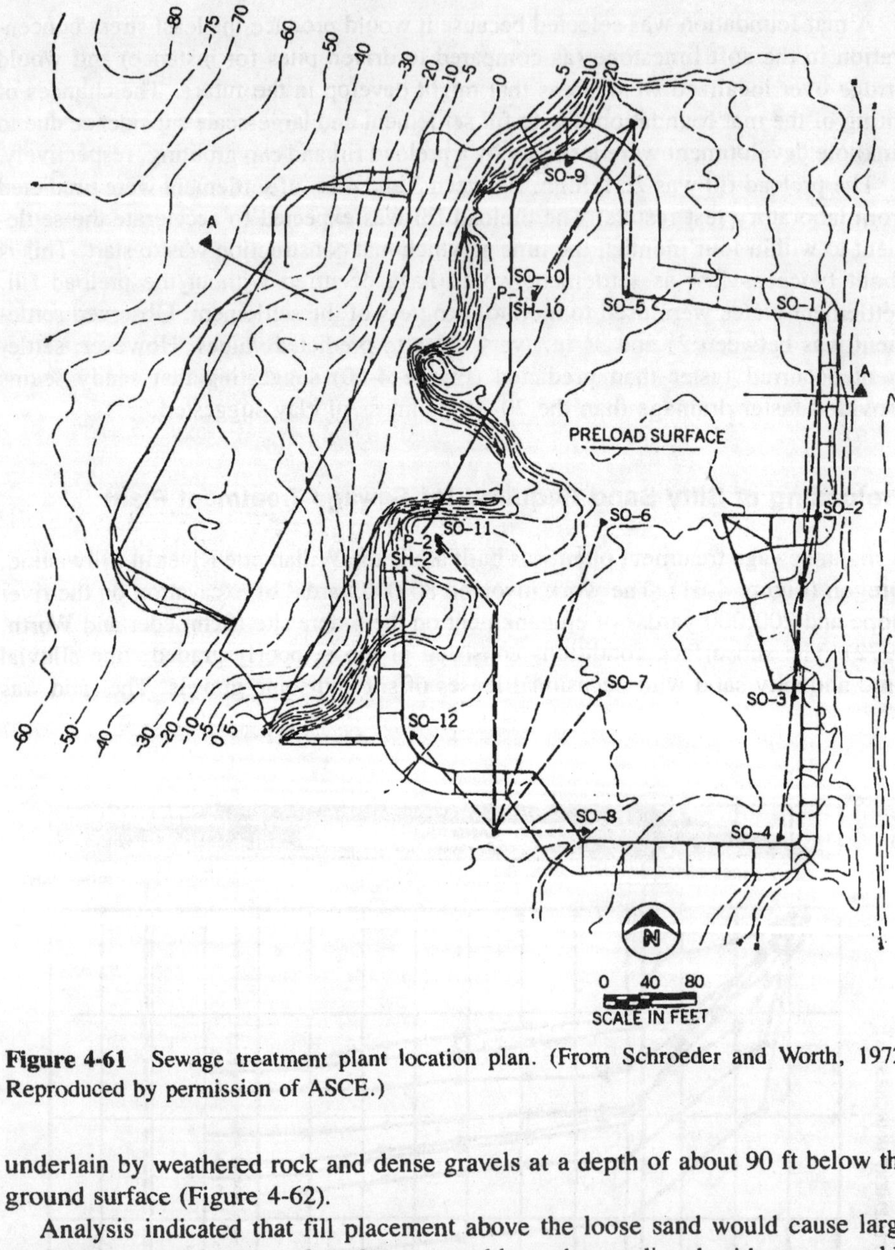

Figure 4-61 Sewage treatment plant location plan. (From Schroeder and Worth, 1972. Reproduced by permission of ASCE.)

underlain by weathered rock and dense gravels at a depth of about 90 ft below the ground surface (Figure 4-62).

Analysis indicated that fill placement above the loose sand would cause large settlements and the rate of settlement could not be predicted with a reasonable degree of confidence because of the difficulties involved in obtaining undisturbed samples of sand and reliable time-settlement parameters in the laboratory. The wide scatter of test data (Figure 4-63) and the nonuniformity of the subsoils, plus the short period of time available before plant construction, led to the decision to preload and surcharge the site. Instrumentation was used to monitor response of the loose sand during fill placement.

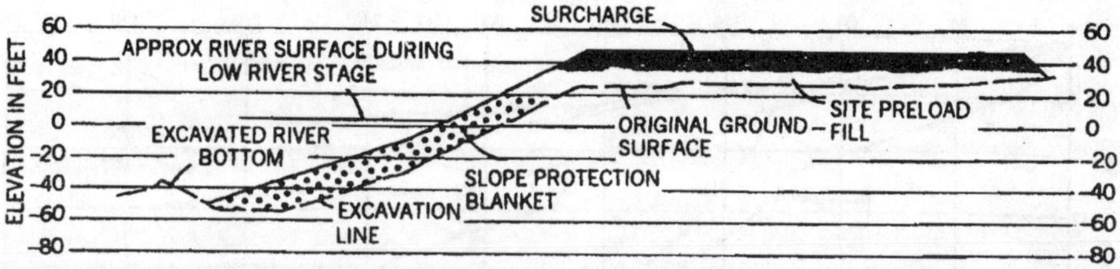

Figure 4-62 Sewage treatment plant subsurface conditions. (From Schroeder and Worth, 1972. Reproduced by permission of ASCE.)

Results of the instrumentation program are shown in Figures 4-64, 4-65, and 4-66. This case history demonstrates that consolidation of loose silty sand can be analyzed using conventional theoretical analysis. However, the input parameters needed for good settlement predictions are almost impossible to obtain using conventional field and laboratory testing. The use of the observational method during

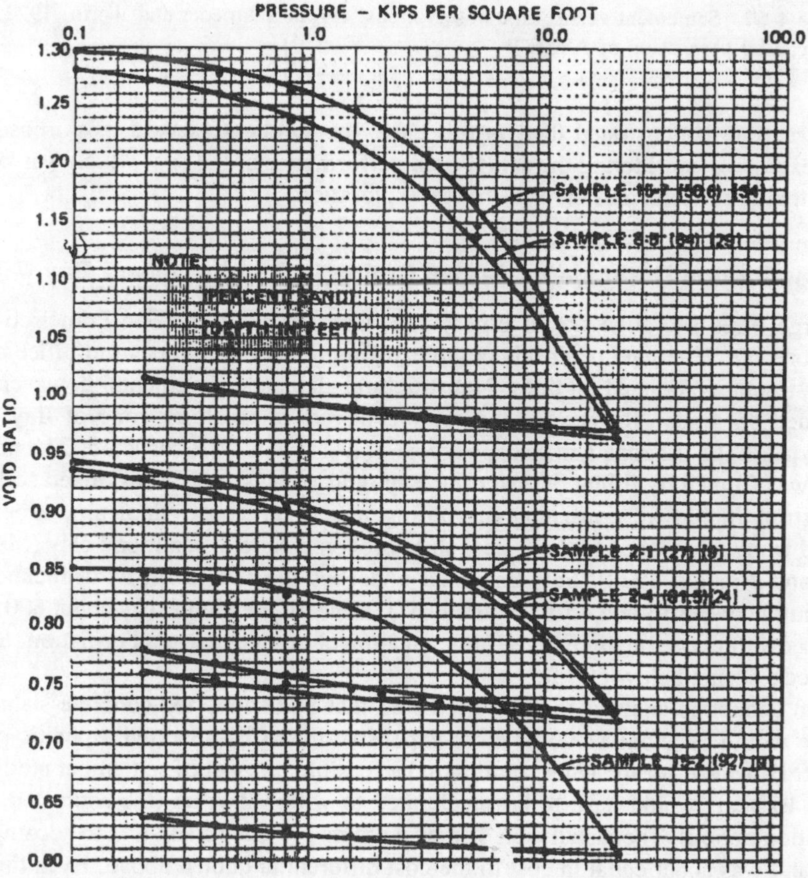

Figure 4-63 Scatter in laboratory data. (From Schroed :r and Worth, 1972. Reproduced by permission of ASCE.)

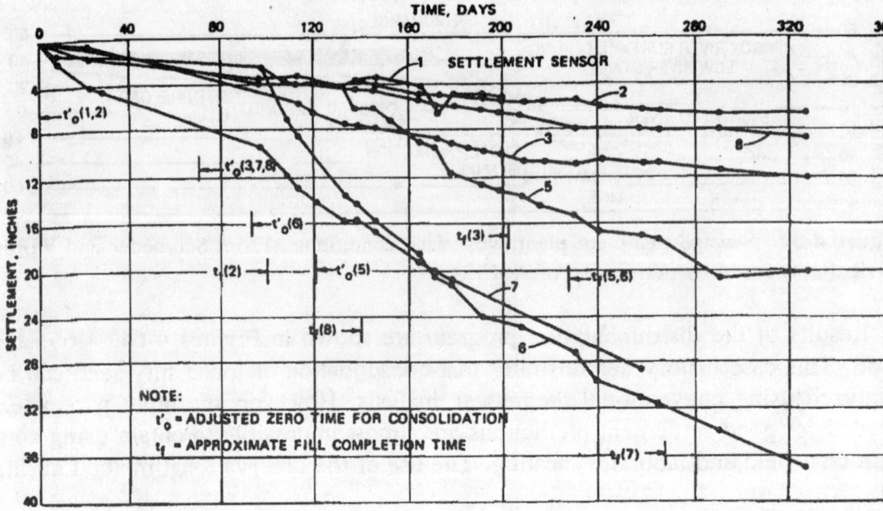

Figure 4-64 Settlement versus time observations. (From Schroeder and Worth, 1972. Reproduced by permission of ASCE.)

construction demonstrated that, in this case, the surcharge served its purpose but was excessive considering the actual field performance of the site, which was better than that predicted from the subsurface exploration.

Consolidation Drainage by Gravel Drains in San Francisco Bay Mud

A large filled shoreline site in Emeryville, California, on the San Francisco Bay supports the Watergate peninsula complex of multistory apartment and office complex structures founded on piles. Nonstructural slabs and pavements, however, are founded on grade. To increase the settlement rate of the thick zone of Bay mud underneath the fill zone and slabs, vertical gravel drains were installed. The peninsula was formed by filling out over the Bay mud with 20 to 25 ft of mixed soil and industrial rubble fill consisting of roofing paper, tar paper, linoleum, asbestos, wood from demolished houses, steel slag, broken concrete, sand, and clay (Margason and Arango, 1972). Soft Bay mud underlies the rubble fill with thicknesses varying between 15 and 45 ft (Figure 4-67). Beneath the mud exists about 600 ft of layered stiff clays, medium dense silts, and gravel of the Alameda Formation. Shale bedrock occurs below 600 ft.

An economic analysis in 1969 predicted that pile-supported concrete slabs-on-grade would be about $1.00/ft^2 more expensive than fill-supported asphaltic pavement. However, for the latter solution to be feasible, the 4 ft of settlement predicted over 10 years would have to be accelerated to within 2 years (the estimated construction period). The installation of gravel drains provided a solution to accomplish this and was about equal in cost to the cost differential quoted above. Even though the alternatives had equivalent costs, the gravel drain solution was selected for scheduling reasons.

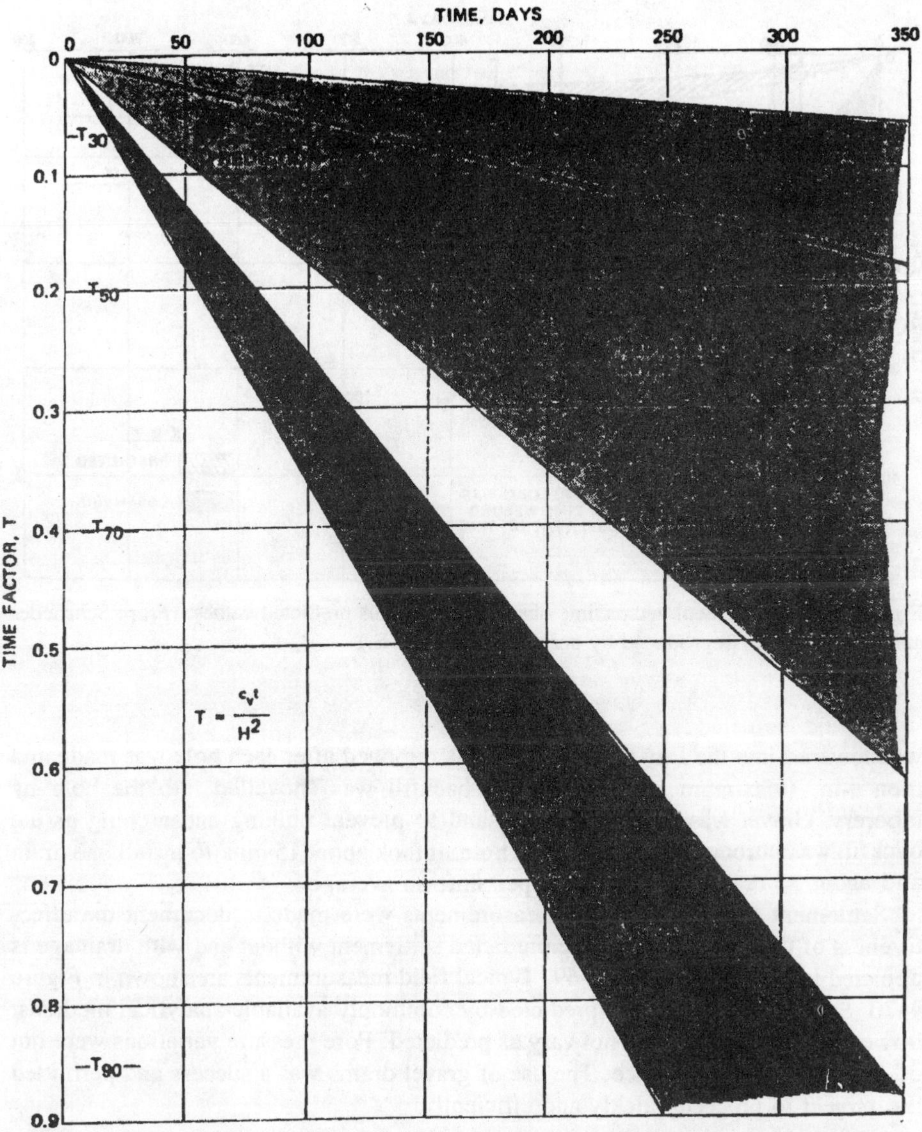

Figure 4-65 Settlement rate observations. (From Schroeder and Worth, 1972. Reproduced by permission of ASCE.)

Over 3000 pea-gravel-filled, 12-in.-diameter drains were installed at a horizontal spacing of between 10 and 14 ft and a depth up to 65 ft. A modified Dutch jetting probe was used, which consisted of a 50-ft-long, 6-in.-diameter, heavy walled pipe with a 12-in.-diameter serrated jetting ring at the base suspended by a crane. The jetting ring was supplied by salt water from the Bay delivered by an 800-gallons/minute pump. The rubble fill was predrilled and cased before the probe

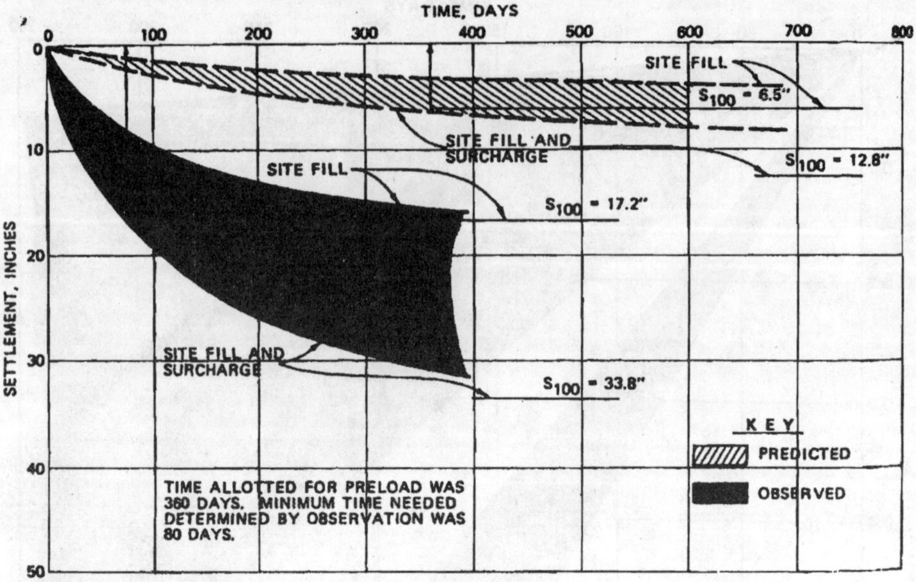

Figure 4-66 Settlement versus time observations versus predicted values. (From Schroeder and Worth, 1972. Reproduced by permission of ASCE.)

was lowered into the Bay mud. The jet was removed after each hole was made and then $\frac{3}{8}$-in. (maximum size) pea gravel backfill was shovelled into the hole by laborers. Gravel was used instead of sand to prevent bulking and arching as the backfill was introduced into the probe hole. It took about 15 min. to install one drain and about 30 drains were installed per shift on average.

Settlement and pore pressure measurements were made to document the effectiveness of the gravel drains. The predicted settlement without and with drainage is depicted in Figures 4-68 and 4-69. Typical field measurements are shown in Figure 4-70. Settlement occurred as predicted by commonly available analytical methods; however pore pressures did not vary as predicted. Pore pressure variations were not of consequential importance. The use of gravel drains was a success and permitted the project to proceed quickly and efficiently.

Sand Drains Used to Accelerate Settlement of I-95 Interchange Approach Embankments

Improvements to the Interstate Route 95 Interchange in Portsmouth, New Hampshire required stabilization of the soft clay subsoils beneath the approach embankments requiring several million cubic yards of fill material. A combination of surcharge fills, stabilizing berms, staged construction, and the installation of vertical sand drains was required (Ladd et al., 1972). The interchange project included the construction of five bridges and approach embankments up to 35 ft in height

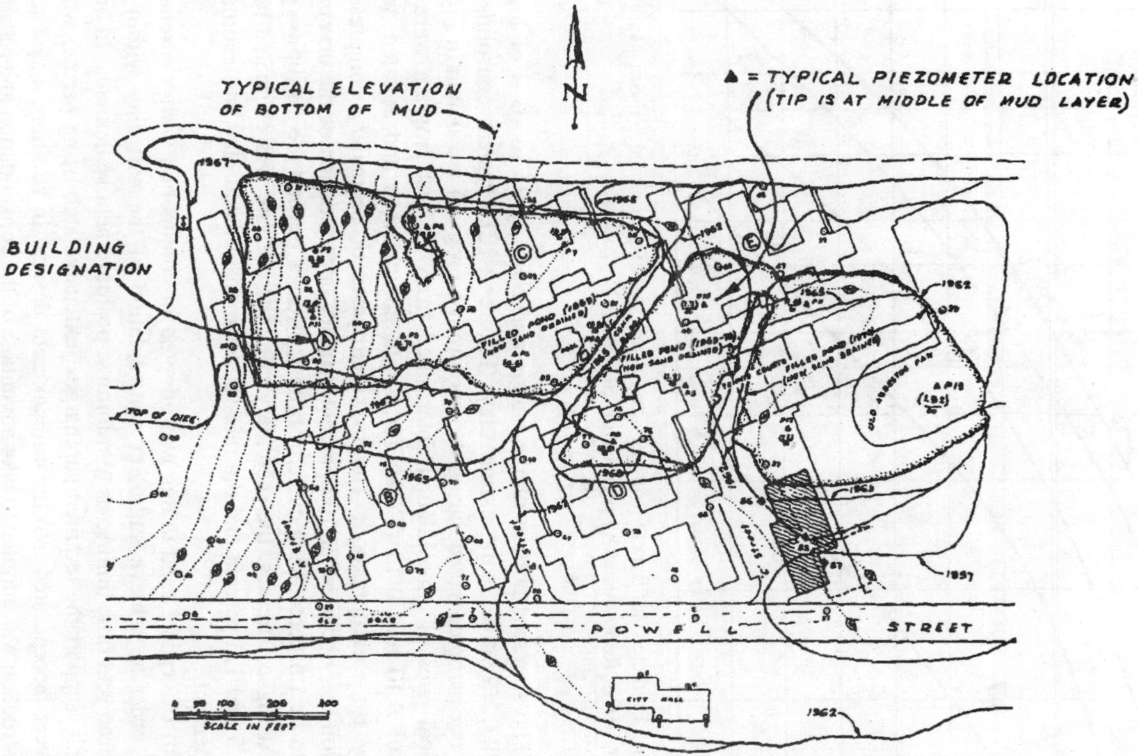

Figure 4-67 Watergate complex subsurface conditions. (From Margason and Arango, 1972. Reproduced by permission of ASCE.)

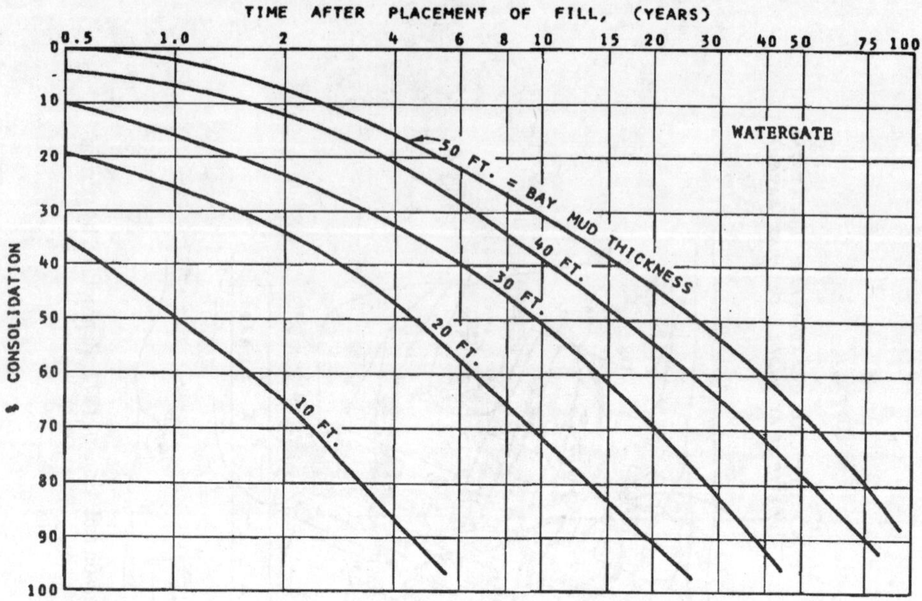

Figure 4-68 Predicted settlement without sand-drain-assisted drainage. (From Margason and Arango, 1972. Reproduced by permission of ASCE.)

(Figure 4-71). A major portion of the site is underlain by 35 to 40 ft of highly compressible, sensitive, soft marine clay (Figure 4-72). Typical consolidation curves for two samples tested during the exploration program are shown in Figure 4-73. Typical rates of consolidation determined from the test program were between about 0.10 and 0.15 ft^2/day for virgin compression and equaled about 1.0 ft^2/day for recompression. The maximum rate of secondary compression (compression that occurs after 100 percent primary consolidation of the clay) was estimated to be 1.5 percent/log cycle of time at stresses beyond the maximum past pressure. At higher stresses, the secondary compression rate was estimated to be between 0.5 and 1.0 percent and the rate for overconsolidated clay was estimated to be 0.1 percent.

A system of vertical sand drains was adopted in conjunction with a surcharge scheme in order to achieve sufficient consolidation of the soft clay within the available three year time limit so as to minimize postpavement settlements. In order to adequately support the required surcharges, extensive stabilizing berms, where possible, were incorporated into the embankment design. In addition, a system of staged construction was adopted to take advantage of the consolidation and resulting anticipated strength increase that would occur in the sand drain areas beneath the central portions of the embankments.

For design purposes, target limits for postpavement settlement were set consisting of less than 2 in. for embankments within several hundred feet of the bridge abutments and 6 in. for embankments not in the vicinity of bridge abutments. Based

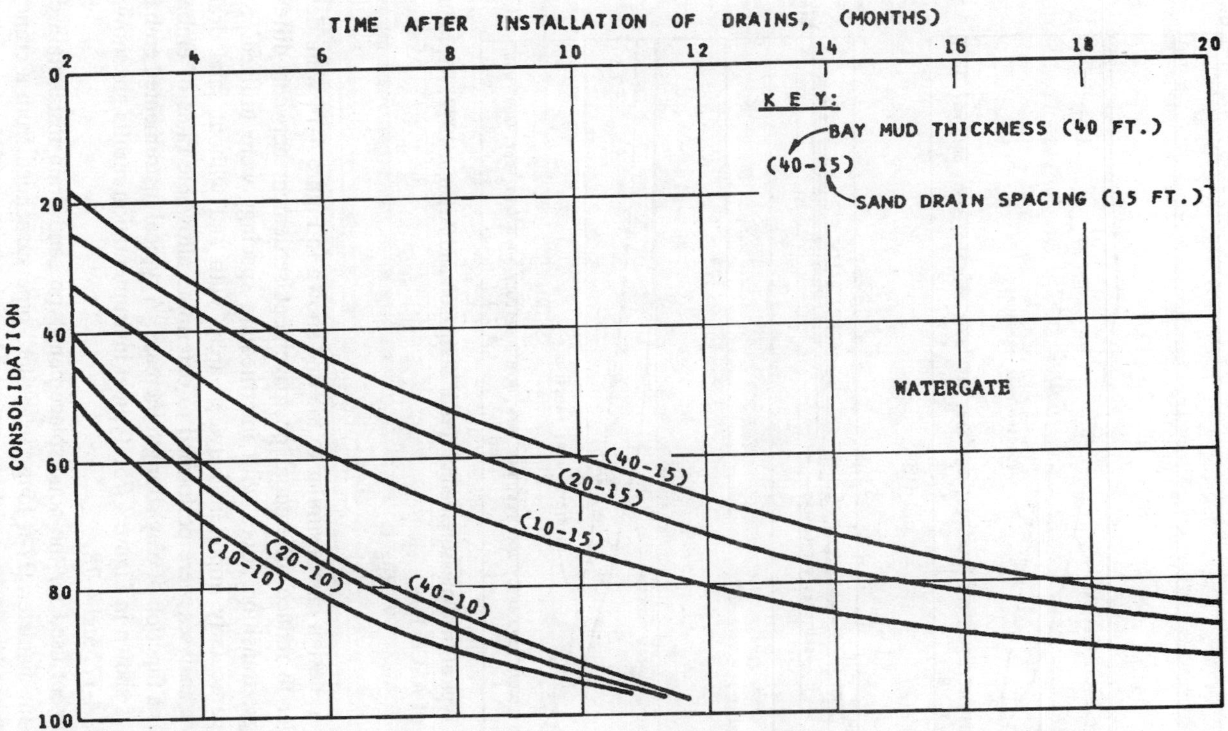

Figure 4-69 Predicted settlement with sand-drain-assisted drainage. (From Margason and Arango, 1972. Reproduced by permission of ASCE.)

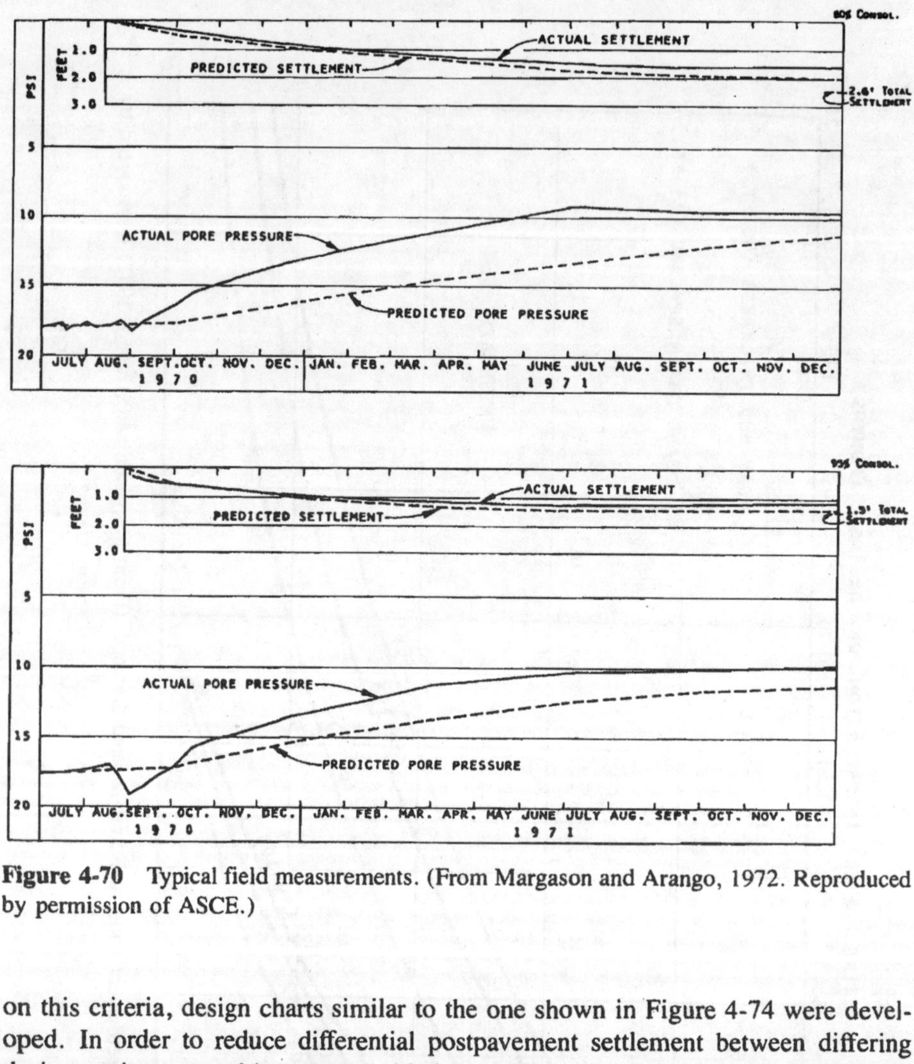

Figure 4-70 Typical field measurements. (From Margason and Arango, 1972. Reproduced by permission of ASCE.)

on this criteria, design charts similar to the one shown in Figure 4-74 were developed. In order to reduce differential postpavement settlement between differing drain spacings, transition zones with intermediate spacings were utilized. Such zones were particularly important between areas with and without sand drains. Slope stability analyses were performed to determine allowable rates of embankment filling as a function of clay strength increase. A typical instrumented embankment section is shown in Figure 4-75. Typical instrumentation results are shown in Figures 4-76, 4-77, and 4-78.

The equipment used for the nondisplacement-type sand drain installation consisted of a 5-in.-diameter, 27-ft-long "jet-bailer" pipe suspended from a crane. A 12-in.-diameter tooth cutting head was attached to the bottom of the jet-bailer. A 6-in.-diameter high-pressure pump supplied water to the jet nozzle at the center of the cutting head. After the sand drain hole was made, concrete sand was placed using shovels to backfill the hole. Sand drain spacings ranged between 9.0 and 16.2 ft on center.

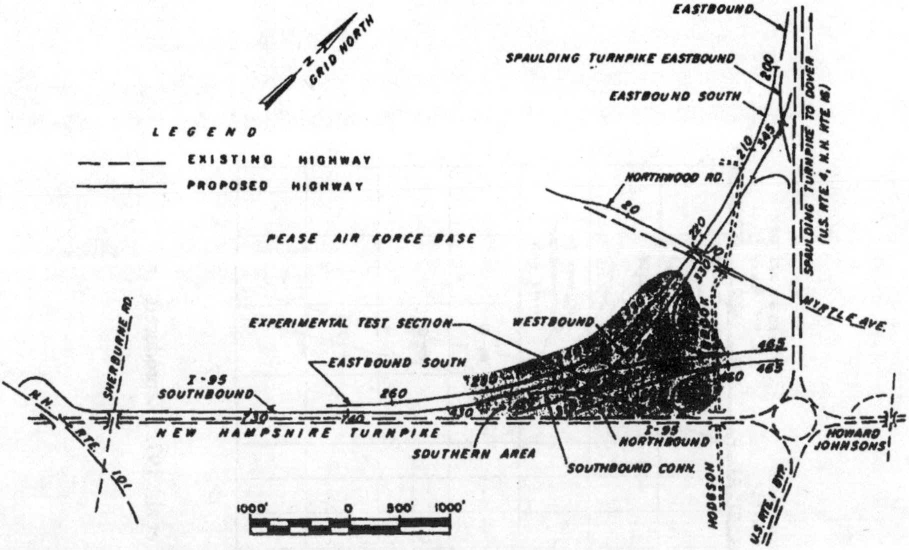

Figure 4-71 I-95 Interchange site plan (From Ladd et al., 1972. Reproduced by permission of ASCE.)

Dynamic Compaction of a Fill to Build an Executive Park

Riverview Executive Park consists of three steel frame and glass curtain wall office buildings located along the Delaware River in Trenton, New Jersey (Partos et al., 1989). The buildings are each 20,000 ft^2 in size, four stories high, and have maximum column loadings of about 550 kips. Subsurface conditions consist of about 13 to 20 ft of slag, cinder, rubble, and soil fill having blow counts between 9 blows/ft and 50 blows/0 in. (Figure 4-79). The fill is underlain by a soft, sandy silt layer up to 4 ft in thickness. The fill and/or silt is underlain by an 8- to 12-ft-thick layer of medium to very dense sand and gravel. This layer is underlain by weathered mica schist at depths greater than about 35 ft below ground surface. Groundwater is encountered near the bottom of the fill. The soil and groundwater were found to contain heavy metals, petroleum hydrocarbons, and PCBs.

Based on the subsurface conditions encountered, it was decided that the fill was not suitable to support spread footing foundations without the probability of large differential settlements. Replacement of the fill was not economical, and it was feared that deep foundations such as piles might encounter buried obstructions. Deep dynamic compaction was proposed, therefore, to improve the bearing capacity and settlement characteristics of the fill. A design bearing capacity value of 4000 lb/ft^2 was selected.

The deep dynamic compaction program consisted of dropping an 18.5-ton weight from a height of 85 ft. The number of drops per location was decided to be seven, based on a test section. The drops were made on a 10.6-ft grid pattern. The column locations were subjected to six additional drops. Three in. of penetration was targeted for the last drop.

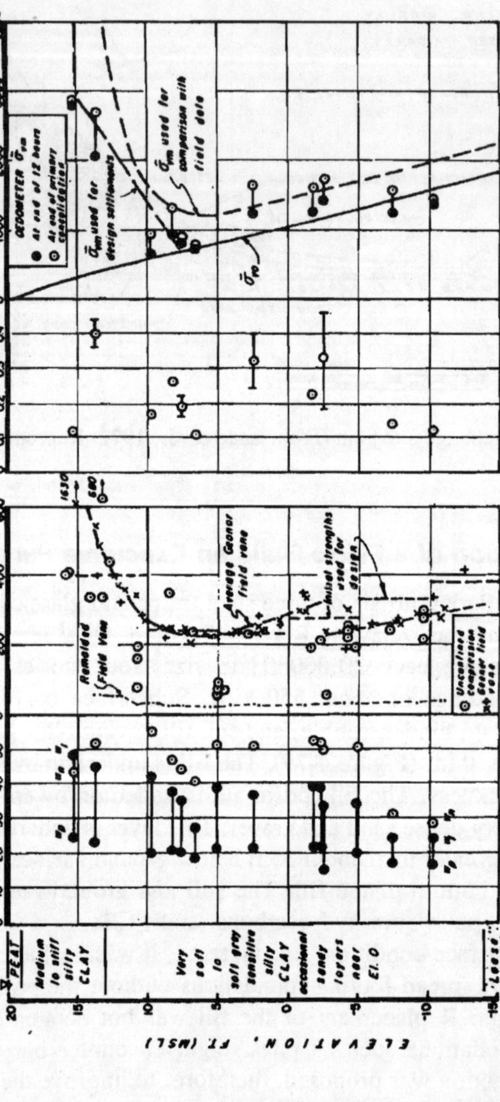

Figure 4-72 I-95 interchange subsurface conditions. (From Ladd et al., 1972. Reproduced by permission of ASCE.)

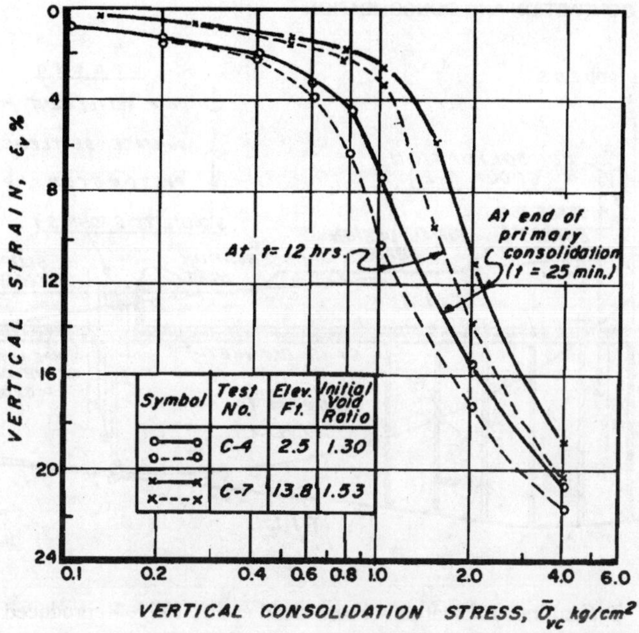

Figure 4-73 Typical consolidation curves. (From Ladd et al., 1972. Reproduced by permission of ASCE.)

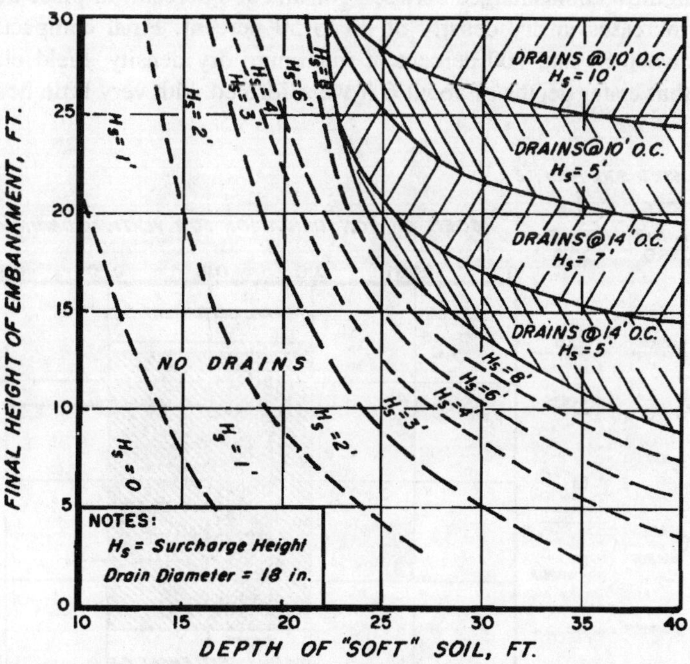

Figure 4-74 Settlement design charts. (From Ladd et al., 1972. Reproduced by permission of ASCE.)

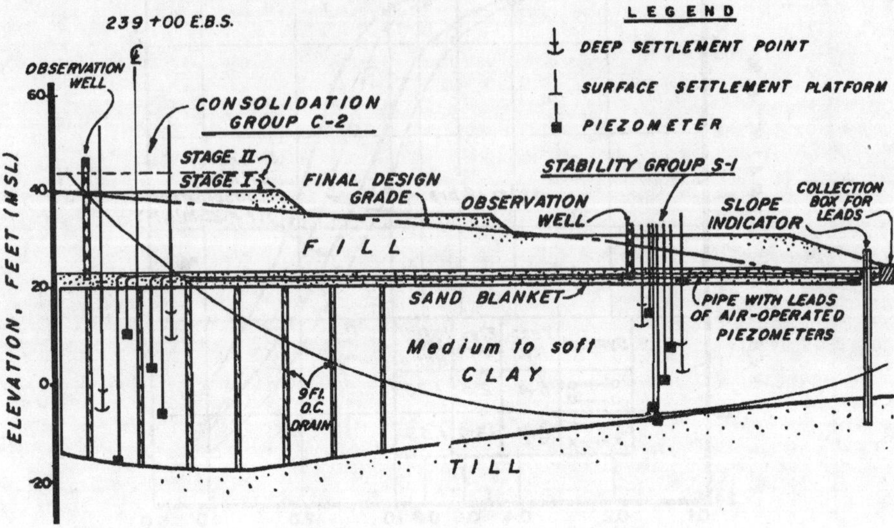

Figure 4-75 Typical instrumentation. (From Ladd et al., 1972. Reproduced by permission of ASCE.)

Results of the program were verified using SPTs and are shown in Figure 4-80. Increases in blow counts ranged between 100 and 500 percent. In-place density tests indicated increases in dry density of up to 50 percent. Final compaction values ranged between 89 and 100 percent of maximum dry density. Field observations indicated that crater depths of about 6 ft were formed with very little heave notice-

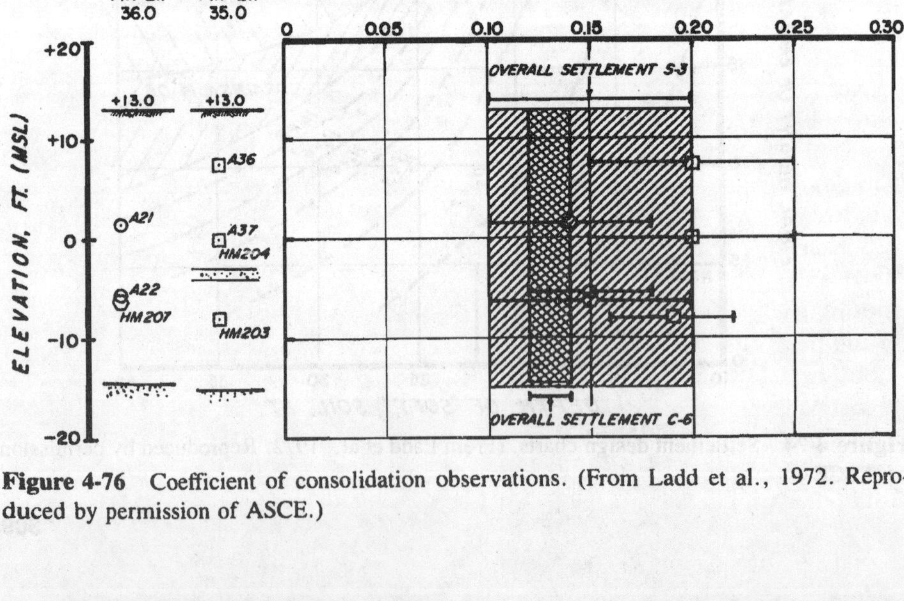

Figure 4-76 Coefficient of consolidation observations. (From Ladd et al., 1972. Reproduced by permission of ASCE.)

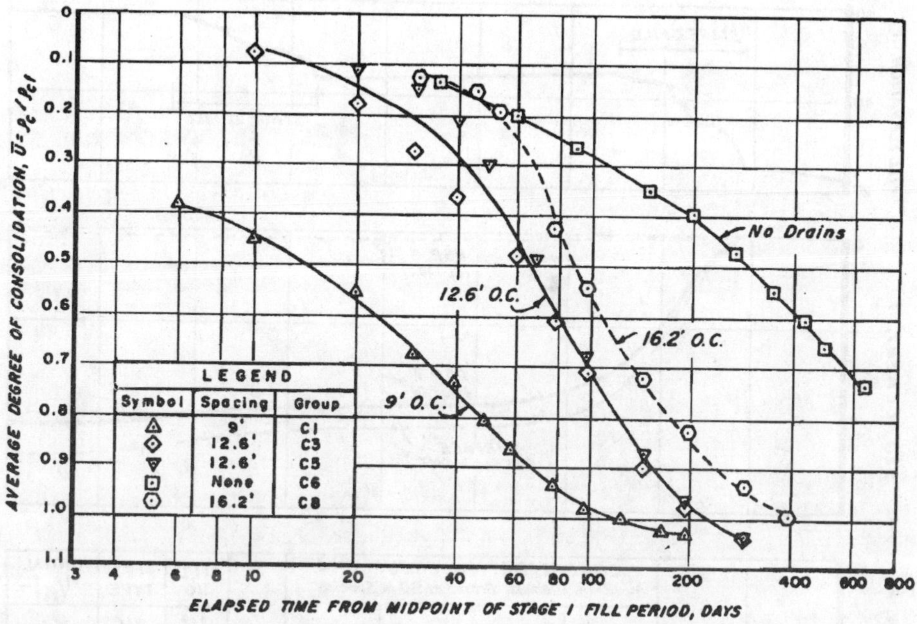

Figure 4-77 Degree of consolidation observations. (From Ladd et al., 1972. Reproduced by permission of ASCE.)

able. The site was lowered on the average by approximately 1.5 ft after completion of deep dynamic compaction. The structures have been completed since 1987, and no unacceptable settlements have been noted.

Compaction Grouting of Loose Sand Beneath a Dam to Prevent Liquefaction

The Pinopolis West dam consists of a 6600-ft-long homogeneous, rolled earthfill structure built in 1940 as part of the Santee Cooper Hydroelectric and Navigation Project (Salley et al., 1987). The dam is 70 ft high and impounds Lake Moultrie approximately 30 miles northwest of Charleston, South Carolina (Figure 4-81). The dam is underlain by a 4- to 8-ft-thick layer of very loose sand (about 4 blows/foot average), which could liquefy during an earthquake in this seismically active area. It was determined that blow counts of between 11 and 20 were required to provide a 1.25 factor of safety against liquefaction during an earthquake. Among the dam improvements considered to remedy the potential for liquefaction, compaction grouting was tested to see if the loose sand layer could be densified using in situ ground improvement techniques.

Subsurface conditions at the site consisted of about 20 ft of sand fill (placed for the test pad), between about 2 and 12 ft of dense to very dense sand, between about 4 and 8 ft of very loose to loose sand, and between about 6 and 8 ft of Cooper Marl, a very stiff to hard clayey silt (Figure 4-82). The piezometric water level is located about 6 ft above the loose sand layer. Corrected blow counts for the loose sand were

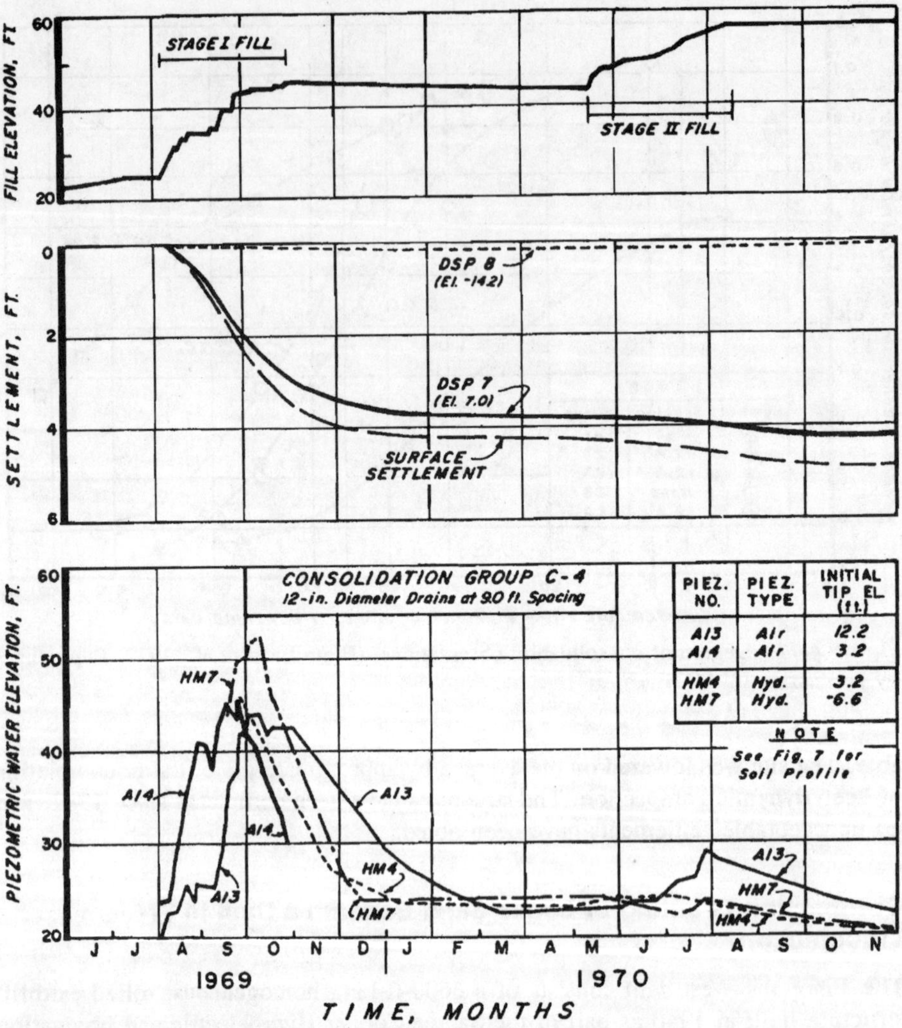

Figure 4-78 Field instrumentation data observations. (From Ladd et al., 1972. Reproduced by permission of ASCE.)

about 4 blows/ft. Cone penetrometer tip resistance ranged between about 18 and 27 tons/ft^2. In-place dry densities ranged between about 85 and 100 lb/ft^2. Void ratios ranged between 0.65 and 0.94.

Four instrumented pads were constructed for the testing, with the configuration shown in Figure 4-83. At each pad, compaction grouting was initiated on 12-ft grid patterns with secondary and tertiary grout stages closing the grid to 6 ft. Anticipated grout take of about 10 percent of the formation volume was calculated on the basis of increasing the in-place dry density from 95 to 105 lb/ft^3. For the 12-ft primary spacing, this equates to about 7 ft^3 of grout per foot of loose sand formation. The

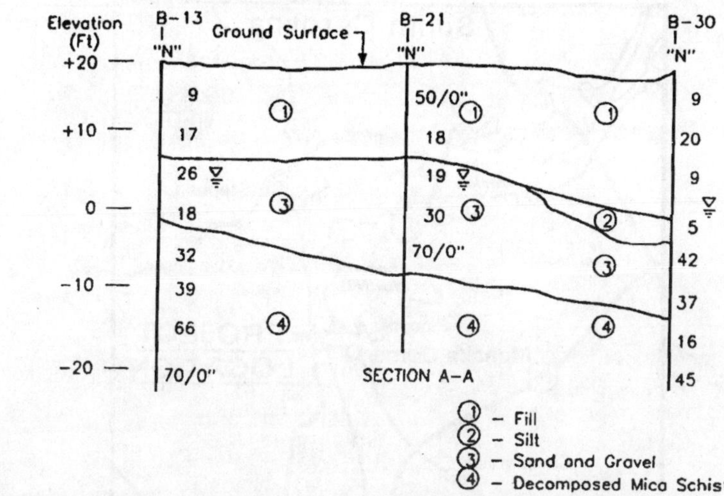

Figure 4-79 Riverview Executive Park subsurface conditions. (From Partos et al., 1989. Reproduced by permission of ASCE.)

primary grout holes were expected to take 50 to 75 percent of the grout take with heave of the ground being a limiting factor.

The work was accomplished by installing the grout hole casings with a track-mounted, air rotary, percussion hammer drill rig down to the marl layer. Grout was delivered to the injection point under pressure by a hydraulically powered dual piston pump. The grout mix consisted of 200 lb of Portland cement, 1650 lb of sand, 1,340 lb of flyash, 60 gal of water, and 46 ounces of pozzalin per cubic yard of grout. Grouting of a pipe continued until a predetermined amount of grout was

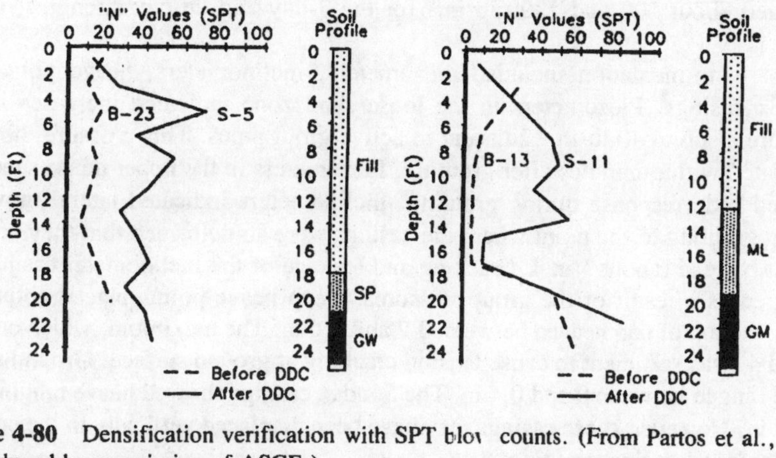

Figure 4-80 Densification verification with SPT blow counts. (From Partos et al., 1989. Reproduced by permission of ASCE.)

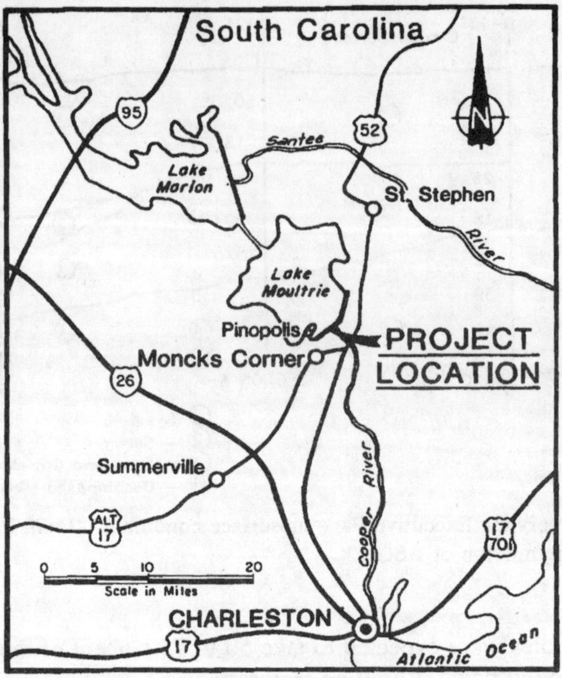

Figure 4-81 Pinopolis West dam site plan. (From Salley et al., 1987. Reproduced by permission of ASCE.)

injected (based on thickness of loose sand), the stroke pressure could not be kept below 300 lb/in.^{2} (the pressure at which maximum grout takes could be achieved), or the tip of the grout hole casing was above the loose sand layer (Figure 4-84). The grout was injected at a rate of approximately 2 ft^3/min. Grout cylinder strengths averaged about 600 and 1300 lb/in.^{2} for the 7-day and 28-day strengths, respectively.

Test instrumentation included piezometers, inclinometers, heave points, and Sondex casings. Piezometers in the loose sand zone indicated increases in pore pressure of up to 40 lb/in.^{2} adjacent to active grout pipes. This pressure, however, dissipated within minutes after grouting. Piezometers in the upper dense sand layer showed little response during grouting. Inclinometers indicated lateral movement during grouting to the point where the casings were so deformed that they could no longer be read (about 1 in.). Cracking and leakage of the inclinometer casings also occurred as a result of the grout pressures. Deep heave points placed about 25 ft below the top of pad heaved between 0.2 and 1.6 in. The maximum, which occurred at Pad 4, was sufficient to cause tension cracking at ground surface. Ground surface heave ranged between 0 and 0.4 in. The Sondex casings showed heave ranging from 1 to 2 in. However, these casings may have been displaced vertically in the borehole and damaged during grouting.

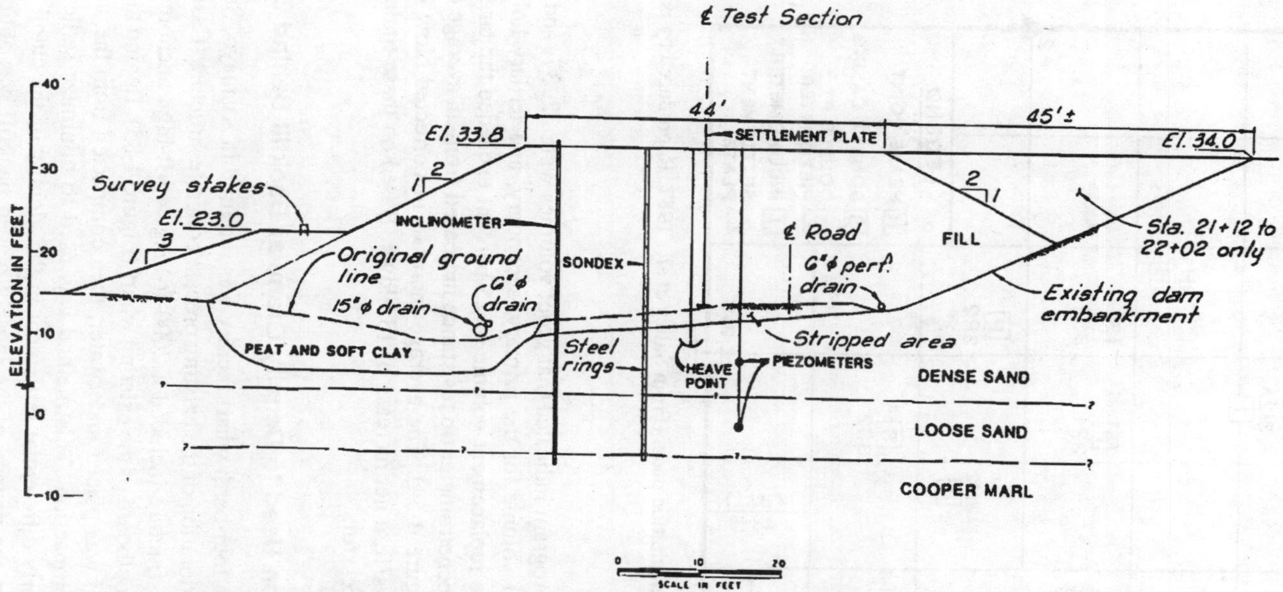

Figure 4-82 Pinopolis West dam subsurface conditions. (From Salley et al., 1987. Reproduced by permission of ASCE.)

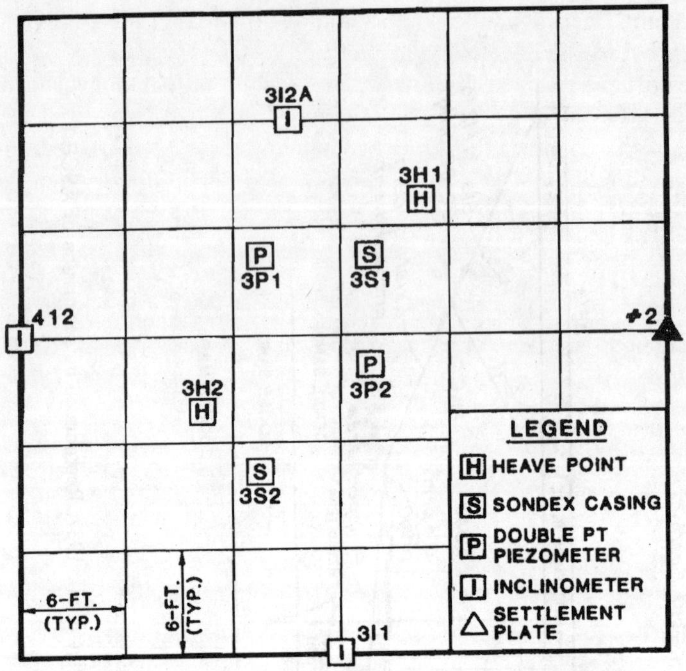

Figure 4-83 Instrumentation pads. (From Salley et al., 1987. Reproduced by permission of ASCE.)

Results of the program indicated that grout volumes were 22, 37, and 41 percent of the replacement volume for the primary, secondary, and tertiary holes, respectively. The average replacement volume of 25 percent exceeded the target value of 10 percent. Cone penetrometer tip resistance increased from an average of about 24 to 83 tons/ft^2 (Figure 4-54). The average increase in corrected blow counts was from 4 to 17 blows/ft, a net increase of 13 blows. Based on these findings the test program was successful.

Vibrocompaction Used to Densify Granular Backfill Behind Bulkhead

Rehabilitation of a four-berth wharf facility was required in Kismayo, Somalia, due to premature deterioration of the existing precast concrete structure (Castelli, 1991). The replacement structure included a 2000-ft-long, 45-ft-high anchored sheet pile bulkhead located outboard of the existing wharf (Figure 4-85). Behind the bulkhead, a granular backfill was placed underwater, then compacted from the surface using deep vibratory compaction. This method was used to minimize settlement of the backfill, particularly differential settlement in the vicinity of the cutoff piles where wide variability in the density of the loosely placed backfill was anticipated. The intent of ground improvement was to minimize the need for future maintenance of the wharf's rigid concrete pavement and avoid possible disruption to the surface drainage system. Deep dynamic compaction also increased the passive soil resistance for support of the deadman anchorage.

Two types of underwater fill material were specified, including sand fill behind the bulkhead and a select fill at the deadman for increased passive soil resistance. The sand fill used in construction consisted of a uniformly graded, medium- to fine-sized beach sand (Figure 4-86). The minimum and maximum dry densities of the sand fill were 101 and 108 lb/ft³, respectively. The select fill was a sand-gravel mix with about 30 percent gravel size. The minimum and maximum dry densities of the select fill were 104 and 121 lb/ft³.

Twin, horizontally oscillating vibroflots with water jetting proved to be the most effective way of achieving the specified 80 percent density limit. A probe grid spacing of 6 to 7 ft was used with average probe penetration of about 14 ft. A

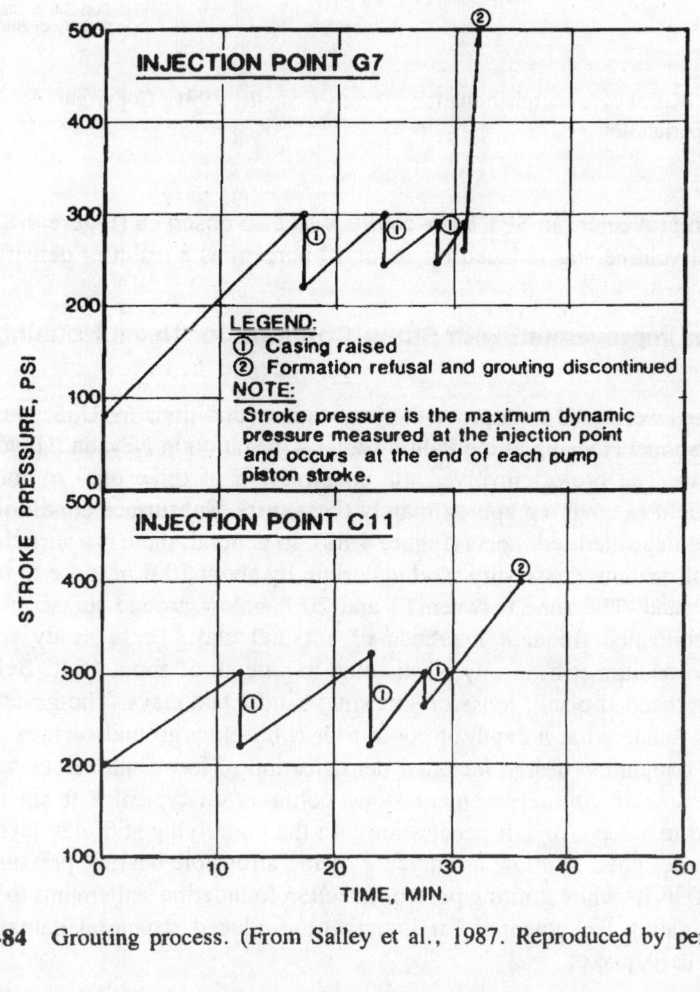

Figure 4-84 Grouting process. (From Salley et al., 1987. Reproduced by permission of ASCE.)

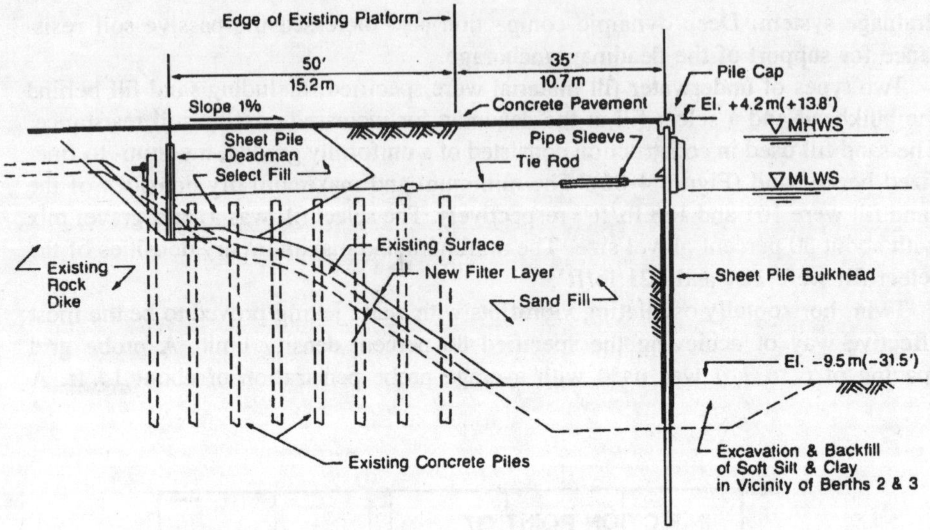

Figure 4-85 Kismayo port rehabilitation. (From Castelli, 1991. Copyright ASTM. Reprinted with permission.)

significant improvement in SPT blow counts was also observed (Figure 4-87). The sand backfill volume was reduced by about 10 percent as a result of densification.

Foundation Improvement with Stone Columns for Naval Housing Facility

Stone columns were used to improve the foundation conditions for Unaccompanied Enlisted Personnel Housing at the Fallon Naval Air Station in Nevada (Hayden and Welch, 1991). The project involved the construction of three one- to four-story masonry structures covering approximately 65,000 ft^2. Subsurface conditions consisted of lake-deposited sediments (Figure 4-88). In general, there is a surficial layer of 3 to 4 ft of medium dense silty sand underlain by about 10 ft of loose to medium dense clean sand. The zone between 13 and 20 ft below ground surface was also sandy but contained frequent interbeds of silt and clay. These sandy soils are underlain by medium stiff, highly plastic clay to a depth of about 45 ft. Below the clay was interbedded black, loose or soft silts, sands, and clays. The groundwater table was encountered at a depth of about 5 or 6 ft below ground surface.

Building foundation design included densification of loose subsurface sands by the construction of vibroreplacement stone columns on typical 8-ft square grid spacings and to a depth of 1-ft penetration into the underlying stiff clay layer. This spacing was designed to allow an increase of the allowable bearing pressure from 3000 to 4500 lb/ft^2 while limiting postconstruction foundation settlements to about $\frac{3}{4}$ in. and mitigating the potential for liquefaction-induced structural damage as a result of an earthquake.

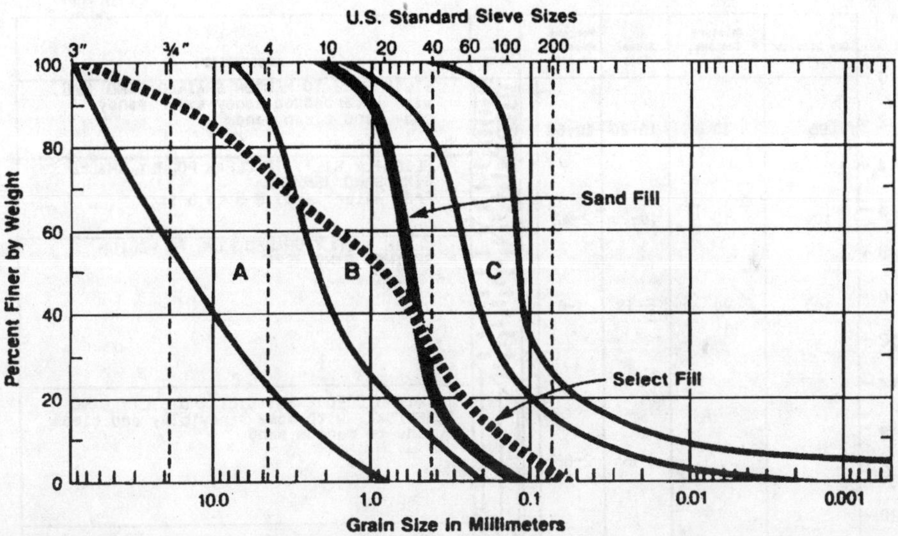

Figure 4-86 Gradation of sand fill. (From Castelli, 1991. Copyright ASTM. Reprinted with permission.)

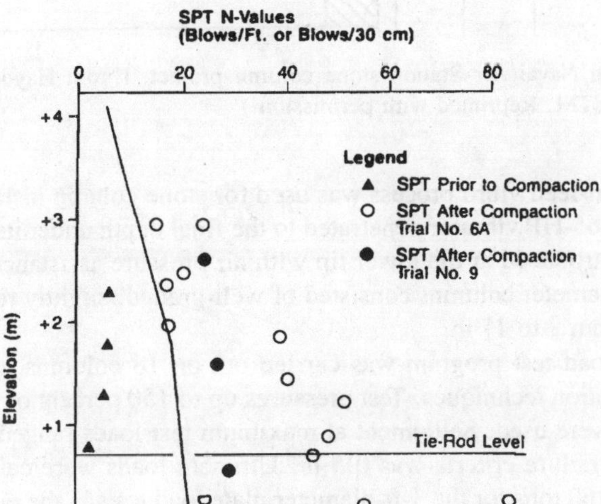

Figure 4-87 Increase in SPT blow counts. (From Castelli, 1991. Copyright ASTM. Reprinted with permission.)

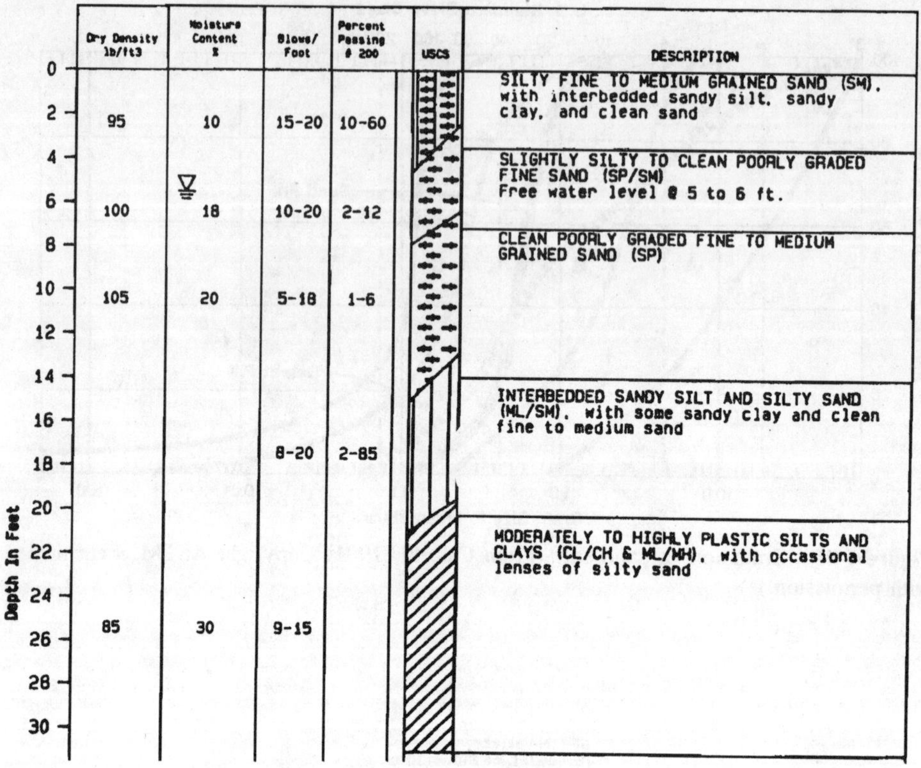

Figure 4-88 Fallon Naval Air Station stone column project. (From Hayden and Welch, 1991. Copyright ASTM. Reprinted with permission.)

The dry, bottom-feed vibro process was used for stone column installation. With this method, the 165-HP vibrator penetrated to the final depth under its own weight. Stone was then introduced at the lower tip with air pressure assistance. Stone used to form the 3-ft-diameter columns consisted of well-graded, slightly rounded gravel ranging in size from $\frac{3}{8}$ to $1\frac{1}{2}$ in.

A short-term load test program was carried out on 18 columns to confirm the design and installation techniques. Test pressures up to 150 percent of the allowable bearing pressure were used. Settlement at maximum test loads ranged between 0.1 and 0.49 in. The failure criteria was 0.5 in. Ultimate loads were calculated to be between 125 and 160 tons for the 3-ft-diameter plate load tests in the range of clayey to sandy near-surface soils. Typical load test results are shown in Figure 4-89. Cone penetrometer and dilatometer tests were run before and after ground improvement to predict bearing capacity and settlement response of the foundations (Figure 4-90).

Several Methods Used at One Site with Variable Soil Conditions

The Trident Atlantic Coast Strategic Submarine Base was constructed in Kings Bay, Georgia. The new facility supports a squadron of Trident submarines as well as crew

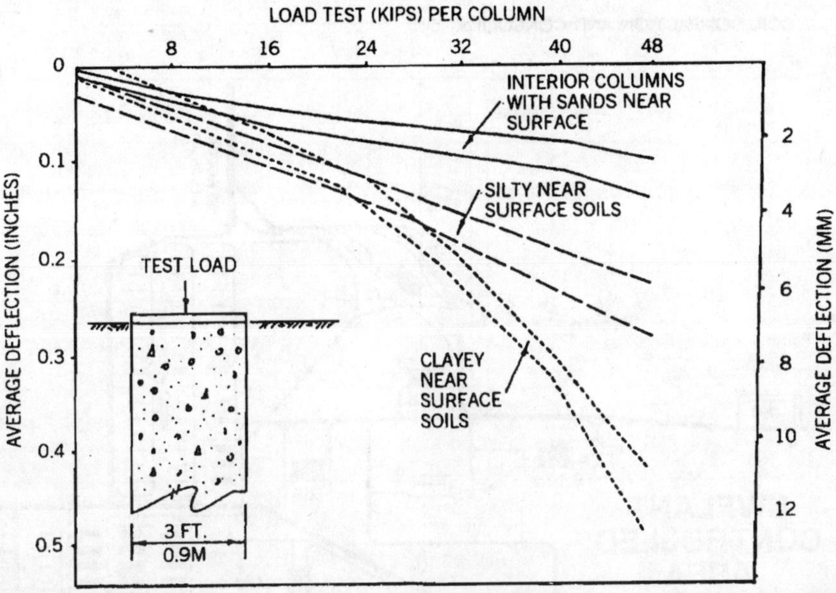

Figure 4-89 Load test results. (From Hayden and Welch, 1991. Copyright ASTM. Reprinted with permission.)

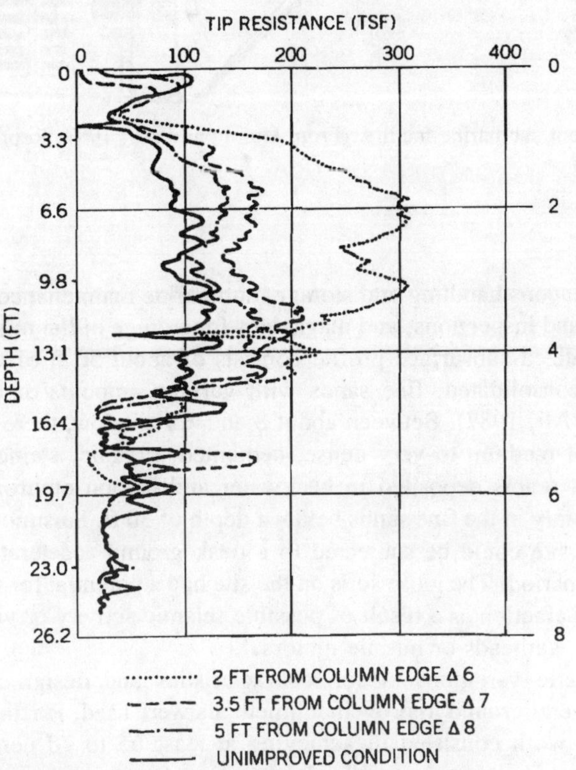

Figure 4-90 Increase in cone penetrometer resistance. (From Hayden and Welch, 1991. Copyright ASTM. Reprinted with permission.)

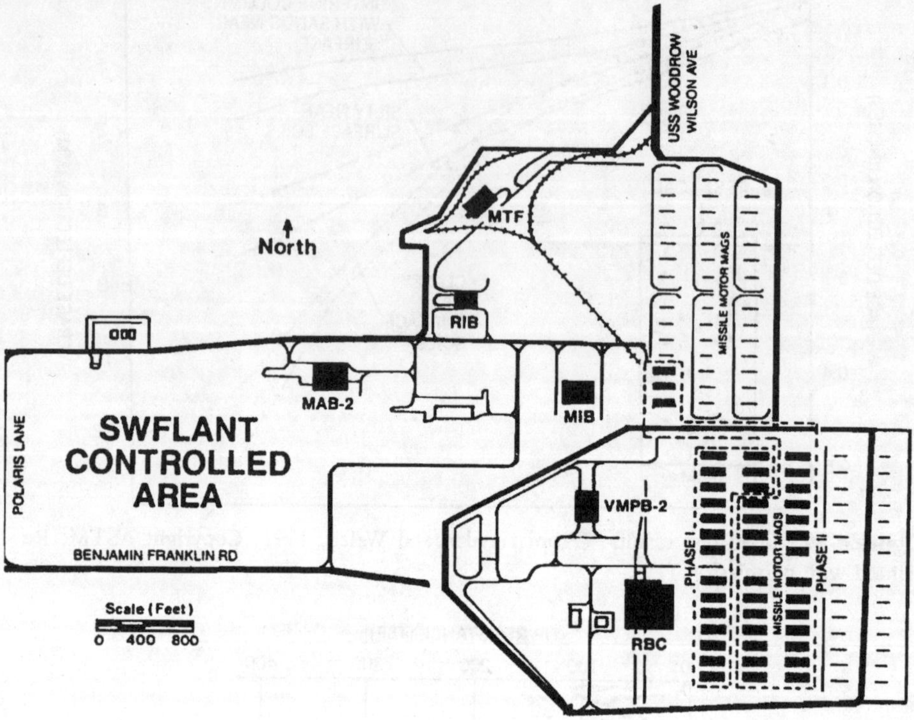

Figure 4-91 Trident submarine facility. (From Hussin and Ali, 1987. Reproduced by permission of ASCE.)

training, small weapons handling and storage, submarine maintenance and repairs, missile assembly and inspections, and magazines for storage of the missiles (Figure 4-91). The generalized subsurface profile consists of about 50 ft of very loose to dense, normally consolidated, fine sands with varying amounts of organics and fines (Hussin and Ali, 1987). Between about 8 and 15 ft below, there is a hardpan layer consisting of medium to very dense, cemented, organic, stained fine sand. Thin clay and silt seams deposited in backwater and lagoon environments were encountered randomly in the fine sands below a depth of 30 ft. Seismic risk analysis indicated that the site could be subjected to a peak ground acceleration of 0.1 g during a 250-year period. The loose soils on the site had a potential for unacceptable settlement and liquefaction as a result of possible seismic activity or vibrations due to sudden blast of warheads or missile motors.

Because of subtle variations in soil characteristics and design requirements across the site, several ground improvement methods were used. Performance criteria applied to the work consisted of achieving at least 65 to 70 percent relative density in the case of cohesionless soils. In the case of cohesive soils, the criteria was improvement of the soil profile to allow a maximum of $\frac{1}{2}$-in. total settlement.

Field verification testing consisted of electronic cone penetrometer testing, dilatometer testing, and standard penetration testing. The structures involved and the methods of improvement used are listed below:

- Reentry Body Complex (RBC)—vibroreplacement (VR).
- Vertical Missile Packaging Building (VMPB-2)—vibroreplacement (VR).
- Missile Inspection Building (MIB)—vibroreplacement (VR).
- Motor Transfer Facility (MTF)—compaction grouting (CG).
- Motor Assembly Building (MAB-2)—deep dynamic compaction and compaction grouting (DDC/CG).
- Radiographic Inspection Building (RIB)—deep dynamic compaction (DDC).
- Missile Motor Magazines (MMM)—vibroreplacement and compaction grouting (VR/CG).

The incredible aspect of this project is that it required one of the highest concentrations of equipment and materials for deep soil improvement ever used in the United States.

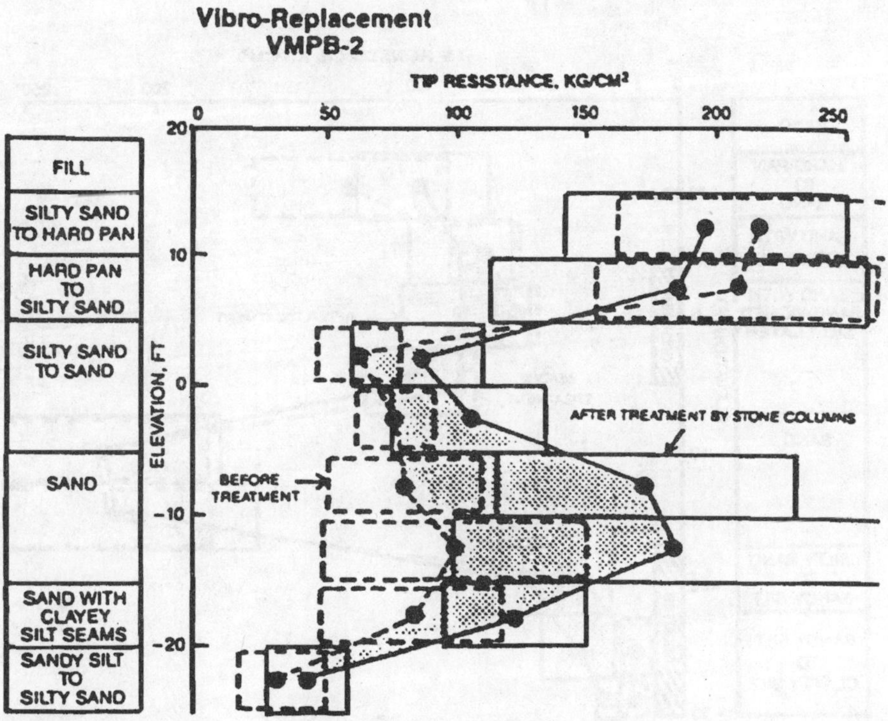

Figure 4-92 Increases in Cone penetrometer resistance. (From Hussin and Ali, 1987. Reproduced by permission of ASCE.)

The variety of equipment used included four vibro units, each with a 60 ton crane, a generator, high pressure water pump, and a 2½ yard³ front-end loader. The dynamic compaction unit consisted of a 200-ton Olympus Crane to drop a 32-ton weight. A dozer was used to fill in the resulting craters. The compaction grouting unit consisted of a 60-ton crane, a vibratory hammer with power pack, a specially designed mobile batch plant, a specially designed high-volume/high-pressure grout pump, and a 2½ yard³ front-end loader.

The materials used for the project were sand, stone, and grout. The sand was a native fine sand from local borrow pits. The stone was a coarse granite ballast with a maximum grain size of 2 in. The grout was batched on the site and consisted of silty fine sand, cement, additives, and sufficient water to achieve a 2-in. slump.

The methods used for the variable site conditions allowed for comparisons to be made between the improvement methods. Representative plots of cone penetrometer tip resistance before and after improvement are shown in Figures 4-92, 4-93, and 4-94. It was concluded that vibroreplacement was not effective for improving silty sand beneath the hardpan and cohesive soils with more than 12 percent fines.

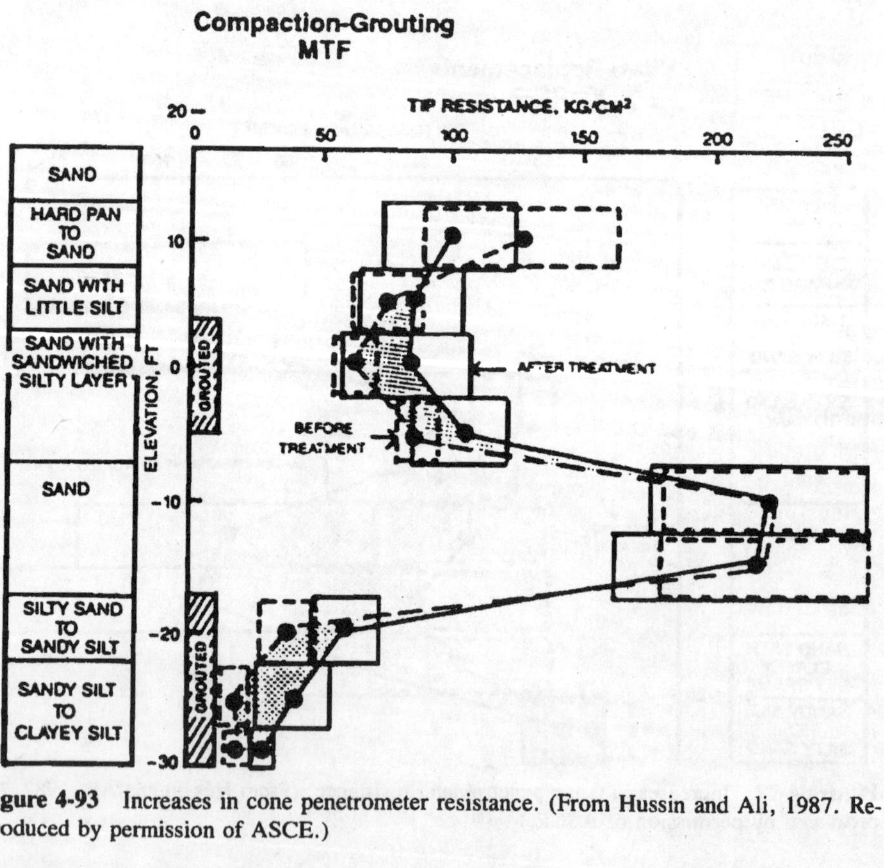

Figure 4-93 Increases in cone penetrometer resistance. (From Hussin and Ali, 1987. Reproduced by permission of ASCE.)

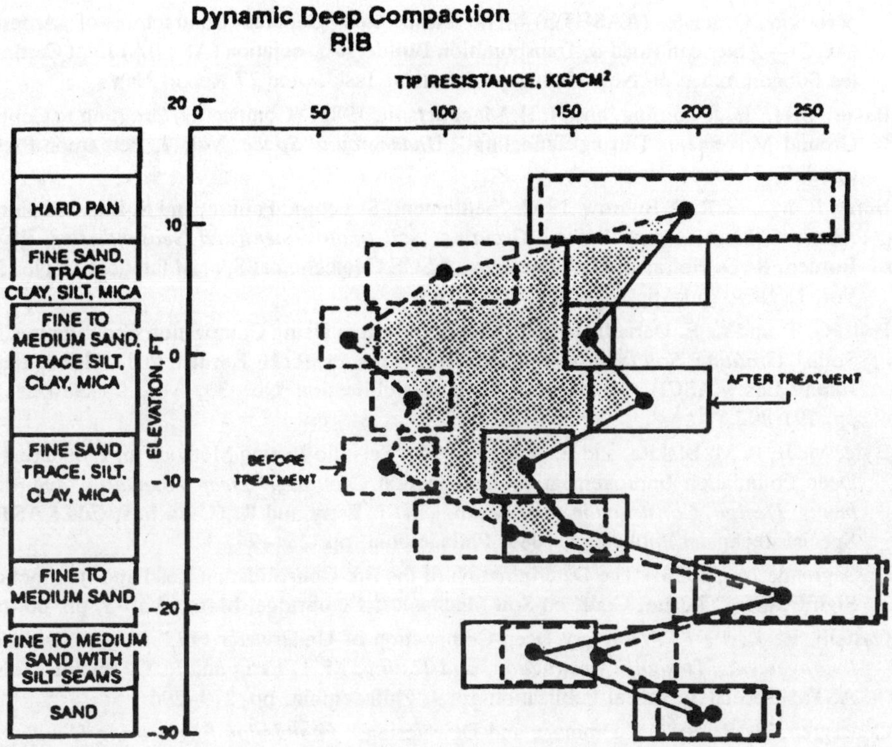

Figure 4-94 Increases in cone penetrometer resistance. (From Hussin and Ali, 1987. Reproduced by permission of ASCE.)

Vibroreplacement was more effective than deep dynamic compaction in sands greater than 25 ft deep.

Deep dynamic compaction in granular soils was observed to depths of 35 ft in fine sands using the 32-ton weight and a free-fall drop height of 100 ft. No improvement was observed in cohesive soils.

Compaction grouting showed the most significant improvement of deep loose or soft sandy silt and silty sand layers. The mean cone penetrometer tip resistance values increased by as much as 100 percent.

REFERENCES

Allen, T. M., T. L. Harrison, J. R. Strada, and A. P. Kilian, 1991. "Use of Stone Columns to Support I-90 Cut and Cover Tunnel," *Deep Foundation Improvements: Design, Construction, and Testing,* M. I. Esrig and R. C. Bachus, Eds., American Society for Testing and Materials Special Technical Publication 1089, Philadelphia, pp. 101–115.

American Association of State Highway and Transportation Officials (AASHTO), 1990. "In Situ Soil Improvement Techniques," American Association of State Highway and Transportation Officials (AASHTO)—The Associated General Contractors of America (AGC)—American Road & Transportation Builders Association (ARTBA) Joint Committee Subcommittee on New Highway Materials, Task Force 27 Report, Jan.

Baker, W.H., E. J. Cording, and H. H. MacPherson, 1983. "Compaction Grouting to Control Ground Movements During Tunneling," *Underground Space*, Vol. 7, Pergamon Press, pp. 205–212.

Berry, R. M. and R. P. Buhrow, 1992. "Settlement, Structural Failure, and In-Place Repair of Above Ground Storage Tanks," *Grouting, Soil Improvement and Geosynthetics*, R. H. Borden, R. D. Holtz, and I. Juran, Eds., ASCE Geotechnical Special Publication No. 30, Vol. 1, New York, pp. 240–251.

Brill, G. T. and K. E. Darnell, 1992. "Retention System Using Compaction Grouting in Clay Soils," *Grouting, Soil Improvement and Geosynthetics*, R. H. Borden, R. D. Holtz, and I. Juran, Eds., ASCE Geotechnical Special Publication No. 30, Vol. 2, New York, pp. 791-802.

Byle, M. J., P. M. Blakita, and E. Winter, 1991, "Seismic Testing Methods for Evaluation of Deep Foundation Improvement by Compaction Grouting," *Deep Foundation Improvements: Design, Construction, and Testing*, M. I. Esrig and R. C. Bachus, Eds., ASTM Special Technical Publication 1089, Philadelphia, pp. 234–247.

Cassagrande, A., 1936. "The Determination of the Pre-Consolidation Load and Its Practical Significance," 1st Int. Conf. on Soil Mechanics, Cambridge, Mass., Vol. 3, pp. 60–64.

Castelli, R. J., 1991. "Vibratory Deep Compaction of Underwater Fill," *Deep Foundation Improvements: Design, Construction, and Testing*, M. I. Esrig and R. C. Bachus, Eds., ASTM Special Technical Publication 1089, Philadelphia, pp. 279–296.

Cedergren, H. R., 1975. "Drainage and Dewatering," *Foundation Engineering Handbook*, H. F. Winterkorn and H. Y. Fang, Eds., Van Nostrand Reinhold, New York, p. 229.

Chastanet, J. D. and P. M. Blakita, 1992. "Wanaque Filtration Plant Subgrade Stabilization—A Case History," *Grouting, Soil Improvement and Geosynthetics*, R. H. Borden, R. D. Holtz, and I. Juran, Eds., ASCE Geotechnical Special Publication No. 30, Vol. 1, New York, pp. 265–274.

Davie, J. R., L. W. Young, M. R. Lewis, and F. J. Swekosky, 1991. "Use of Stone Columns to Improve the Structural Performance of Coal Waste Deposits," *Deep Foundation Improvements: Design, Construction, and Testing*, M. I. Esrig and R. C. Bachus, Eds. ASTM Special Technical Publication 1089, Philadelphia, pp. 116–130.

DeStephen, R. A., D. W. Kozera, and F. J. Swekosky, 1992. "Design of Floating Stone Columns in Hydraulic Fill," *Grouting, Soil Improvement and Geosynthetics*, R. H. Borden, R. D. Holtz, and I. Juran, Eds., ASCE Geotechnical Special Publication No. 30, Vol. 2, New York, pp. 829–841.

Dobson, T., 1987. "Case Histories of the Vibro Systems to Minimize the Risk of Liquefaction," *Soil Improvement—A Ten Year Update*, J. P. Welsh, Ed., ASCE Geotechnical Special Publication No. 12, Apr., New York, pp. 167–183.

Dunnicliff, J. and G. E. Green, 1988. *Geotechnical Instrumentation for Monitoring Field Performance*, Wiley, New York, 577 pp.

Egan, J. A., R. F. Hayden, L. L. Scheibel, M. Otus, and G. M. Serventi, 1992. "Seismic Repair at Seventh Street Marine Terminal," *Grouting, Soil Improvement and Geosynthetics*, R. H. Borden, R. D. Holtz, and I. Juran, Eds., ASCE Geotechnical Special Publication No. 30, Vol. 2, New York, pp. 867–878.

Ergun, M. U., 1992. "Design and Performance of Two Port Silos on Improved Ground,"

Grouting, Soil Improvement and Geosynthetics, R. H. Borden, R. D. Holtz, and I. Juran, Eds., ASCE Geotechnical Special Publication No. 30, Vol. 2, New York, pp. 843–854.

GKN Keller Canada, Ltd., 1987. "Case History—Richmond East-West Freeway, Richmond, British Columbia," GKN Keller Canada, Ltd., Vancouver, B.C.

GKN Hayward Baker, 1988. "Case History—Vibro Compaction, Wastewater Treatment Facility, Modesto, California," GKN Hayward Baker, Odenton, Md.

GKN Hayward Baker, 1989. "Case Histories," GKN Hayward Baker, Odenton, Md.

Graf, E. D., 1992. "Earthquake Support Grouting in Sands," *Grouting, Soil Improvement and Geosynthetics*, R. H. Borden, R. D. Holtz, and I. Juran, Eds., ASCE Geotechnical Special Publication No. 30, Vol. 2, New York, pp. 879–888.

Hayden, R. F. and C. M. Welch, 1991. "Design and Installation of Stone Columns at Naval Air Station," *Deep Foundation Improvements: Design, Construction, and Testing*, M. I. Esrig and R. C. Bachus, Eds., ASTM Special Technical Publication 1089, Philadelphia, pp. 172–184.

Hayward Baker Inc., 1990. "Ground Modification Specialists," Hayward Baker Inc., Odenton, Md.

Hayward Baker Inc., 1991. "Solving Problems Underground," Hayward Baker Inc., Odenton, Md.

Henry, J. F., 1986. "Low Slump Compaction Grouting for Correction of Central Florida Sinkholes," Proc. Conf. on Environmental Problems in Karst Terrains and Their Solutions, Bowling Green, Ky., Oct.

Henry, J. F., 1989. "Ground Modification Techniques Applied to Sinkhole Remediation," Engineering and Environmental Impacts of Sinkholes and Karst, Proc. 3rd Multi-Disciplinary Conf. on Sinkholes, B. F. Beck, Ed., Balkema, Rotterdam, The Netherlands.

Hilf, J. W., 1975. "Compacted Fill," *Foundation Engineering Handbook*, H. F. Winterkorn and H. Y. Fang, Eds., Van Nostrand Reinhold, New York, pp. 259 and 264.

Holtz, R. D. and W. D. Kovacs, 1981. *An Introduction to Geotechnical Engineering*, Prentice-Hall, Englewood Cliffs, N.J., pp. 115, 120, 212, 580.

Hoover, T., 1987. "Caltrans' Wick Drain Experiences," *Soil Improvement—A Ten Year Update*, J. P. Welsh, Ed., ASCE Geotechnical Special Publication No. 12, Apr., New York, pp. 184–196.

Hough, B. K., 1969. *Basic Soils Engineering*, 2nd Ed., Ronald, New York.

Hunt, R. E., 1986. *Geotechnical Engineering Analysis and Evaluation*, McGraw-Hill, New York.

Hussin, J. D. and S. Ali, 1987. "Soil Improvement at the Trident Submarine Facility," *Soil Improvement—A Ten Year Update*, J. P. Welsh, Ed., ASCE Geotechnical Special Publication No. 12, New York, pp. 215–231.

Hussin, J. D. and J. I. Baez, 1991. "Analysis of Quick Load Tests on Stone Columns: Case Histories," *Deep Foundation Improvements: Design, Construction, and Testing*, M. I. Esrig and R. C. Bachus, Eds., ASTM Special Technical Publication 1089, Philadelphia, pp. 185–198.

Koerner, R. M., 1984. *Construction and Geotechnical Methods in Foundation Engineering*, McGraw-Hill, New York.

Koerner, R. M., 1990. *Designing with Geosynthetics*, Prentice Hall, Englewood Cliffs, N.J.

Ladd, C. C., J. J. Rixner, and D. C. Gifford, 1972. "Performance of Embankments with

Sand Drains on Sensitive Clay," Performance of Earth and Earth-Supported Structures, ASCE, June, New York, pp. 211–242.

La Fosse, U. and T. von Rosenvinge, 1992. "Densification of Loose Sands by Deep Blasting," *Grouting, Soil Improvement and Geosynthetics*, R. H. Borden, R. D. Holtz, and I. Juran, Eds., ASCE Geotechnical Special Publication No. 30, Vol. 2, New York, pp. 954–968.

Landau, R. E., 1966. "Method of Installation as a Factor in Sand Drain Stabilization Design," Highway Research Record No. 133, Highway Research Board, Washington, D.C., pp. 75–96.

Leycure, P. and W. L. Schroeder, 1987. "Slope Effects on Probe Densification of Sands," *Soil Improvement—A Ten Year Update*, J. P. Welsh, Ed., ASCE Geotechnical Special Publication No. 12, Apr. New York, pp. 197–214.

Margason, E. and I. Arango, 1972. "Sand Drain Performance on a San Francisco Bay Mud Site," Performance of Earth and Earth-Supported Structures, ASCE, June, New York, pp. 181–210.

Mitchell, J. K., 1981. "Soil Improvement—State-of-the-Art Report," Proc. 10th Int. Conf. on Soil Mechanics and Foundation Engineering, June, Stockholm, pp. 509–565.

Mitchell, J. K., 1986. "Practical Problems from Surprising Soil Behavior, the Twentieth Terzaghi Lecture," *J. Geot. Eng. ASCE* vol. 112, no. 3, pp. 255–289.

Mitchell, J. K. and J. P. Welsh, 1989. "Soil Improvement by Combining Methods," Proc. 12th Int. Conf. on Soil Mechanics and Foundation Engineering, Aug., Rio de Janeiro, Balkema, Rotterdam, The Netherlands, pp. 1393–1396.

Moorhouse, D. C. and G. L. Baker, 1968. "Sand Densification by Heavy Vibratory Compactor," *Placement and Improvement of Soil to Support Structures*, ASCE, Aug., New York, pp. 379–388.

Munfakh, G. A., L. W. Abramson, R. D. Barksdale, and I. Juran, 1987. "In-Situ Ground Reinforcement," *Soil Improvement—A Ten Year Update*, J. P. Welsh, Ed., ASCE Geotechnical Special Publication No. 12, Apr., New York, pp. 167–183.

Naval Facilities (NAVFAC) Engineering Command, 1982. *Soil Mechanics*, Design Manual 7.1, NAVFAC DM-7.1, May, Department of the U.S. Navy.

Naval Facilities (NAVFAC) Engineering Command, 1983. *Soil Dynamics, Deep Stabilization, and Special Geotechnical Construction*, Design Manual 7.3, NAVFAC DM-7.3, Apr., Department of the U.S. Navy.

Partos, A., J. P. Welsh, P. W. Kazaniwsky, and E. Sander, 1989. "Case Histories of Shallow Foundations on Improved Soil," Proc. Foundation Engineering Congr., Evanston, Ill., ASCE, June, pp. 313–327

Peck, R. B., W. E. Hanson, a T. H. Thornburn, 1974. *Foundation Engineering*, 2nd ed., Wiley, New York.

Sain, C. H., 1983. "Earthwork," *Standard Handbook for Civil Engineers*, 3rd ed., F. S. Merritt, Ed., McGraw-Hill, New York, Chapter 13, pp. 13-1–13-41.

Salley, J. R., B. Foreman, J. Henry, and W. H. Baker, 1987. "Compaction Grout Test Program—West Pinopolis Dam," *Soil Improvement—A Ten Year Update*, J. P. Welsh, Ed., ASCE Geotechnical Special Publication No. 12, Apr., New York, pp. 245–269.

Schexnayder, C. and R. G. Lukas, 1992. "Dynamic Compaction of Nuclear Waste," *Civ. Eng. ASCE*, New York, Mar., pp. 64–65.

Schmertmann, J. H., 1968. Discussion of "Sand Densification by Heavy Vibratory Compactor" by M. C. Moorhouse and G. L. Baker, *Placement and Improvement of Soil to Support Structures*, ASCE, Aug., New York, pp. 389–391.

Schmertmann, J. H., 1970. "Static Cone to Compute Static Settlement Over Sand," *Proc. ASCE J. Soil Mech. Found. Eng. Div.*, Vol. 96, No. SM3, May, p. 1045.

Schmertmann, J., W. Baker, R. Gupta, and K. Kessler, 1986. "CPT/DMT QC of Ground Modification at a Power Plant," Use of In Situ Tests in Geotechnical Engineering, Proc. In Situ '86, ASCE, New York, June, pp. 985–1001.

Schmertmann, J. H. and J. F. Henry, 1992. "A Design Theory for Compaction Grouting," *Grouting, Soil Improvement and Geosynthetics*, R. H. Borden, R. D. Holtz, and I. Juran, Eds., ASCE Geotechnical Special Publication No. 30, Vol. 1, New York, pp. 215–228.

Schroeder, W. L. and E. G. Worth, 1972. "A Preload on Silty Fine Sand," Performance of Earth and Earth-Supported Structures, ASCE June, New York, pp. 379–394.

Seed, H. B. and I. M. Idriss, 1982. *Ground Motions and Soil Liquefaction During Earthquakes* (in revision), Earthquake Engineering Research Institute, Berkeley Calif. 134 pp.

Senneset, K. and J. Nestvold, 1992. "Deep Compaction by Vibro Wing Technique and Dynamic Compaction," *Grouting, Soil Improvement and Geosynthetics*, R. H. Borden, R. D. Holtz, and I. Juran, Eds., ASCE Geotechnical Special Publication No. 30, Vol. 2, New York, pp. 889–901.

Shin, E. C., B. W. Shin, and B. M. Das, 1992. "Site Improvement for a Steel Mill Complex," *Grouting, Soil Improvement and Geosynthetics*, R. H. Borden, R. D. Holtz, and I. Juran, Eds., ASCE Geotechnical Special Publication No. 30, Vol. 2, New York, pp. 816–828.

Slocombe, B. C. and M. P. Moseley, 1991. "The Testing and Instrumentation of Stone Columns," *Deep Foundation Improvements: Design, Construction, and Testing*, M. I. Esrig and R. C. Bachus, Eds., ASTM Special Technical Publication 1089, Philadelphia, pp. 85–100.

SOLS SOILS, 1975. "The Menard Pressuremeter: Interpretation and Application of Pressuremeter Test Results to Foundation Design," SOLS SOILS No. 26, Paris.

Sowers, G. F., 1979, *Introductory Soil Mechanics and Foundations: Geotechnical Engineering*, 4th ed., MacMillan, New York, 621 pp.

Terzaghi, K., 1943. *Theoretical Soil Mechanics*, Wiley, New York.

Terzaghi, K. and R. B. Peck, 1967. *Soil Mechanics in Engineering Practice*, Wiley, New York.

Venema, T. P., C. R. Waletzko, and F. J. Swekosky, 1989. "Ground Modification for a 14-Story Structure Over a Deep Peat Deposit," Proc. Foundation Engineering Congr., Evanston, Ill., ASCE, June, pp. 392–405.

Wardlaw, K. R., 1986. "Ground Modification with Crushed Stone—Dynamic Deep Compaction Presents a High Market Potential," *Stone Review*, Oct./Nov., pp. 11–14.

Warner, J., 1982. "Compaction Grouting—The First Thirty Years," Proc. Conf. Grouting in Geotechnical Engineering, W. H. Baker, Ed., ASCE, New York, pp. 694–707.

Welsh, J. P., 1983. "Dynamic Deep Compaction of Sanitary Landfill to Support Superhighway," Proc. 8th Europ. Conf. on Soil Mechanics and Foundation Engineering, Helsinki, May, Balkema, Rotterdam, The Netherlands, pp. 319–321.

Welsh, J. P., 1986. "In Situ Testing for Ground Modification Techniques," Use of In Situ Tests in Geotechnical Engineering, Proc. In Situ '86, ASCE, New York, June, pp. 322–335.

Welsh, J. P., 1988. "Sinkhole Rectification by Compaction Grouting," Proc. Conf. on Geotechnical Aspects of Karst Terrains, May, Nashville, pp. 115–132.

Welsh, J. P., R. D. Anderson, R. P. Barksdale, C. K. Satyapriya, M. T. Tumay, and H. E. Wahls, 1987. "Densification," *Soil Improvement—A Ten Year Update*, J. P. Welsh, Ed., ASCE Geotechnical Special Publication No. 12, Apr., New York, pp. 67–97.

Wheeless, L. D. and G. F. Sowers, 1972. "Mat Foundation and Preload Fill, VA Hospital Tampa," Performance of Earth and Earth-Supported Structures, ASCE, June, New York, pp. 939–952.

CHAPTER 5

IN SITU GROUND REINFORCEMENT

5-1 INTRODUCTION

In situ ground reinforcement is used to strengthen the ground in place to allow slopes to stand at steeper angles, to prevent further movement along preexisting failure planes, and to reinforce the ground after a cut slope has been made. In recent years, it has proven to be cost effective to reinforce the ground and help it to support itself rather than try to support the entire earth load with a gravity wall (Figure 5-1) or buttress. Many ground reinforcement methods utilize this concept, including mechanically stabilized embankments (MSE) (Figure 5-2), fabric walls (Figure 5-3), tunnel reinforcement (Figure 5-4), landslide stabilization (Figure 5-5), and, one of the newest methods to gain acceptance in practice, soil nailing (Figure 5-6). Various types of soil reinforcement techniques are discussed in Christopher et al. (1990), Mitchell and Villet (1987), and Walkinshaw (1990).

Description

Many earth reinforcement techniques involve the inclusion of metal strips, welded wire fabric, geosynthetics, tree branches, twine, and so on, in engineered fill embankments. The contrast in construction sequences between a mechanically stabilized embankment and a soil nailed cut is shown in Figure 5-7. Soil nailing, the main topic of this chapter, is a process whereby steel rods or "nails" are installed in a cut face in original ground on a pattern varying with wall height; the nails are connected by steel mesh or shotcrete facing to hold soil near the cut face in place between them. Sometimes precast or cast-in-place concrete facades are used to improve the appearance of the shotcrete facing for permanent facilities.

Figure 5-1 Comparison of soil nailed wall to gravity wall.

The steel nails reinforce and strengthen the ground by helping the soil to resist deformations that tend to take place as a result of the excavation process. As the ground deforms, the nails share the load with the soil, gradually becoming more stressed in tension from an initially nominal stress level at installation. As the soil deforms due to excavation, the nails become tensioned to the extent necessary to arrest the deformations and stabilize the soil (i.e., to prevent slope failure). The nails therefore provide passive resistance to the soil, unlike tiebacks, which are pretensioned at installation (Weatherby, 1982; Xanthakos, 1991).

Advantages

Several authors, including Bruce and Jewell (1986), Mitchell and Villet (1987), Elias and Juran (1991), Fannin and Bowden (1991), have discussed the advantages of soil nailing. These authors consistently agree that the advantages of soil nailing are:

- Cost savings of 10 to 30 percent as compared to tieback walls may be realized.
- Mobile, small scale, and quiet construction equipment can be used.
- Rapid and flexible construction methods, accommodating variations in soil conditions and work progress, can be used.
- Only small movements will be required to mobilize the reinforcement action.
- Surface settlement can be mitigated by tensioning the upper levels of reinforcement.
- The entire system is flexible and can tolerate large horizontal and vertical movements.
- The system of reinforcement is redundant; a weak nail will not cause failure of the entire wall system.

Disadvantages

Of course there are disadvantages that must be mentioned as well. However, as the reader will see, the disadvantages are also common to other reinforcement systems and are minor as compared to the advantages. The disadvantages of soil nailing are:

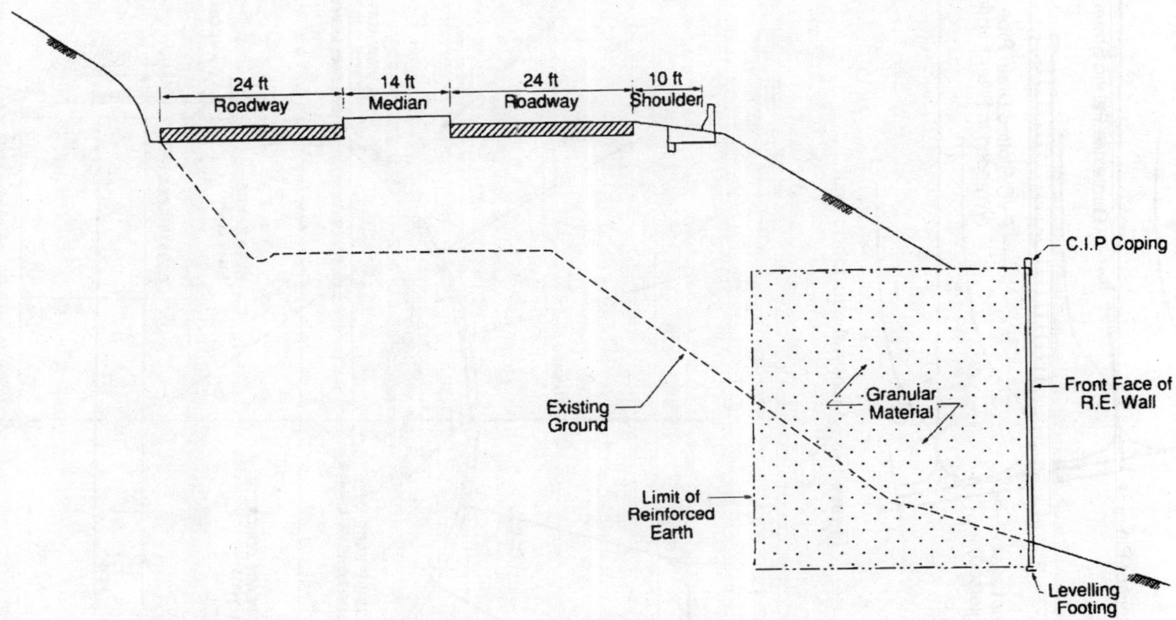

Figure 5-2 Mechanically stabilized embankment (MSE).

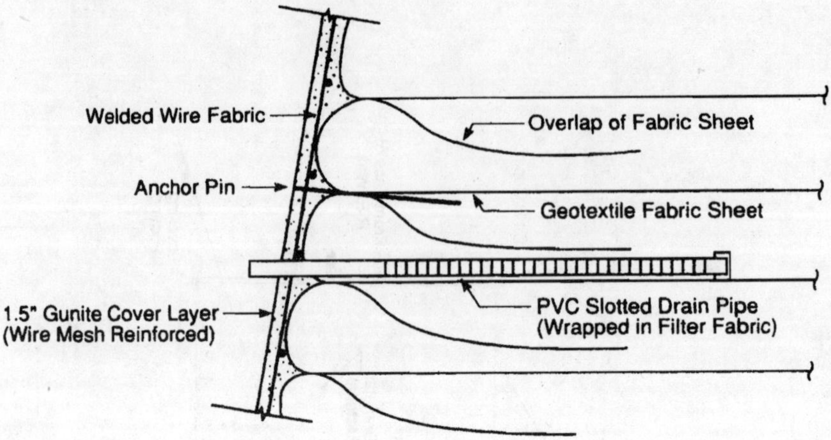

Figure 5-3 Geosynthetic fabric wall.

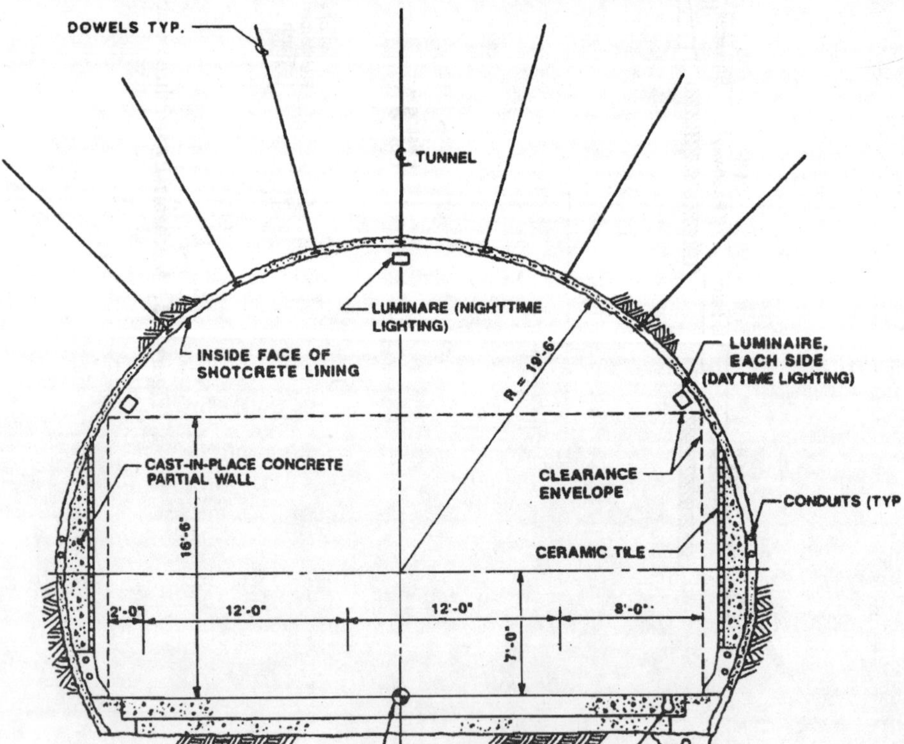

Figure 5-4 Tunnel reinforcement.

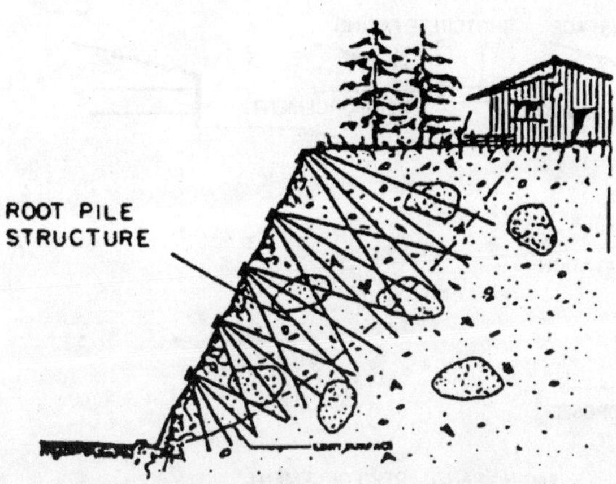

Figure 5-5 Landslide stabilization. (From Mitchell and Villet, 1987.)

- The ground must be strong enough for 3- to 8-ft-high cuts to remain stable for at least a few hours to allow time for reinforcement installation.
- Water cannot be flowing heavily out of the face of the excavation since that would prevent the application of shotcrete facing.
- Reliable drainage systems are difficult to construct.
- Very soft clays are not suitable for this type of reinforcement.
- Permanent or temporary underground easements may be required; interference with nearby utilities may occur.

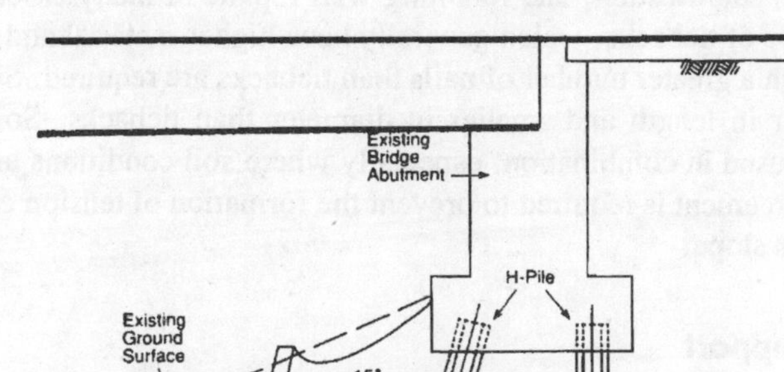

Figure 5-6 Soil nailed wall.

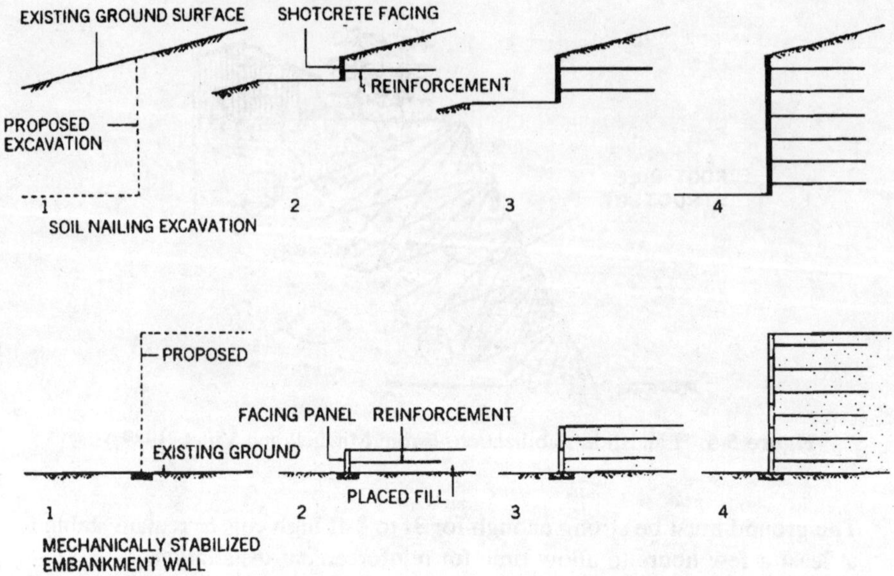

Figure 5-7 Comparison of a mechanically stabilized embankment to a soil nailed wall. (From Bruce and Jewell, 1986.)

5-2 USES AND APPLICATIONS

The uses and applications of soil nailing include excavation support, slope reinforcement, slope stabilization, and retaining wall repair. In many cases, soil nails can take the place of tiebacks, which generally have higher material and installation costs even though a greater number of nails than tiebacks are required. Soil nails are generally shorter in length and smaller in diameter than tiebacks. Soil nails and tiebacks can be used in combination, especially where soil conditions are such that tensioned reinforcement is required to prevent the formation of tension cracks at the top of the cut or slope.

Excavation Support

Deep open cut excavations in soil generally require lateral bracing. A deep cut is one greater than about 5 or 10 ft in depth. This lateral bracing can be accomplished with struts, rakers, tiebacks, or nails (Figure 5-8). Struts and rakers are often undesirable methods of bracing because they interfere with work within the excavation and construction area. Soil nails can be used for the support of an excavation if some amount of lateral deflection can be tolerated. The excavation can be any size and shape with nails. This is not so with struts. Very wide spans require a high number of very large struts. Two examples of soil nail supported building excavations are shown in Figures 5-9 and 5-10. Bruce and Jewell (1986, 1987a, b) provide other examples of nail supported excavations.

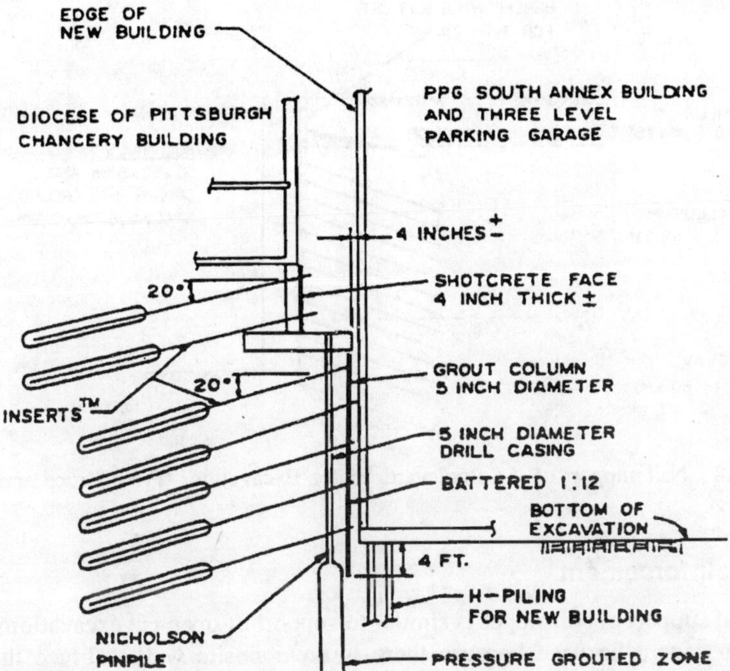

Figure 5-8 Lateral bracing for an open cut excavation. (From Nicholson, 1986.)

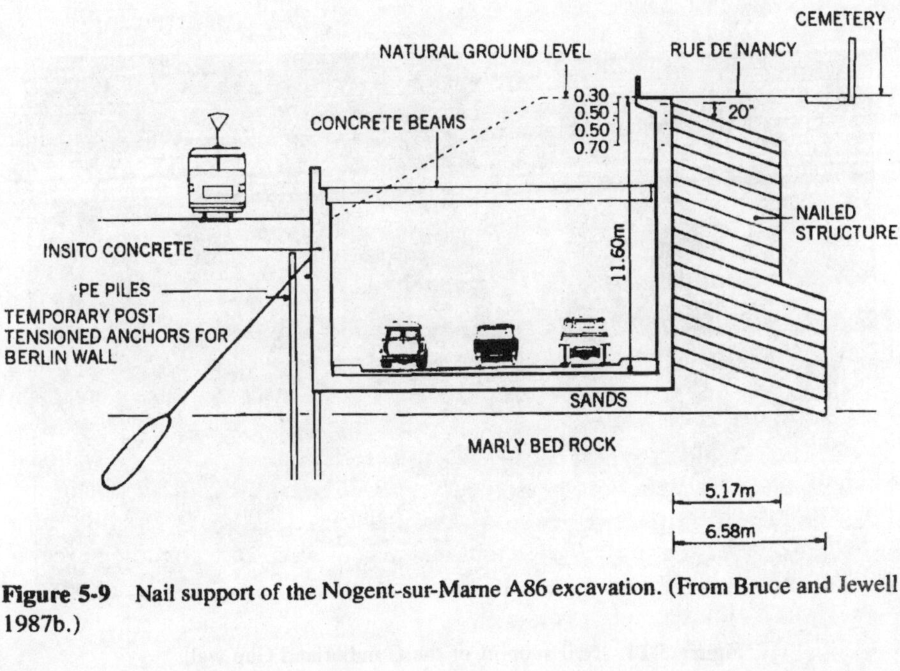

Figure 5-9 Nail support of the Nogent-sur-Marne A86 excavation. (From Bruce and Jewell, 1987b.)

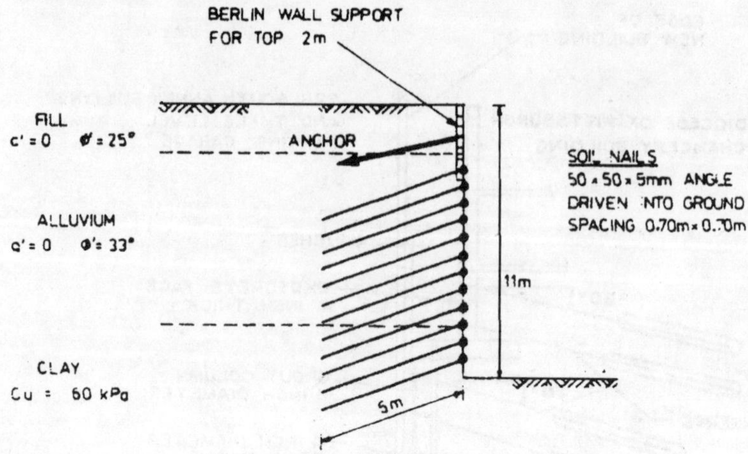

Figure 5-10 Nail support of the Boulevard Victor Excavation. (From Bruce and Jewell, 1987b.)

Slope Reinforcement

The lateral support of soil slopes is similar to support of open cut excavations except that struts cannot be used because there is no opposite wall to brace the struts against. Usually, tiebacks are used to reinforce slopes. However, soil nailing is becoming a viable option to tieback support. Nicholson and Boley (1985) describe the use of soil nails for permanent support of a colluvial highway slope in Kentucky (Figure 5-11). Boyce and Abramson (1992) describe the combined use of dowels

Figure 5-11 Nail support of the Cumberland Gap wall.

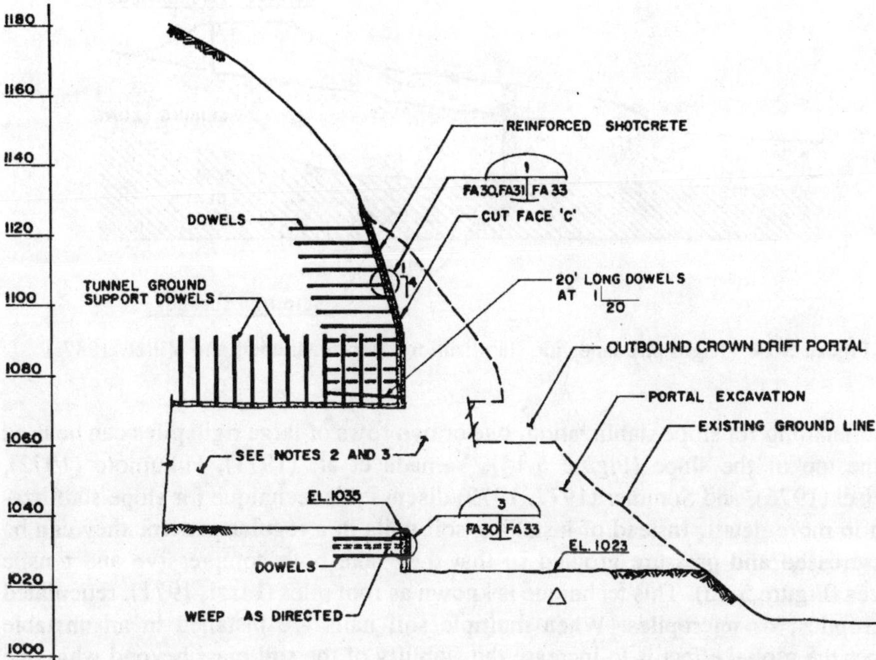

Figure 5-12 Nail support of the Interstate Route H-3 tunnel portals.

and tiebacks in extremely weathered basalt "saprolite" at a highway tunnel portal in Hawaii (Figure 5-12).

Slope Stabilization

Slopes can be stabilized with soil nails when the safety factor for slope failure is too low or after a failure surface has formed and is creeping downhill at a very slow rate. An imagined example of this situation is depicted in Figure 5-13. Alternative

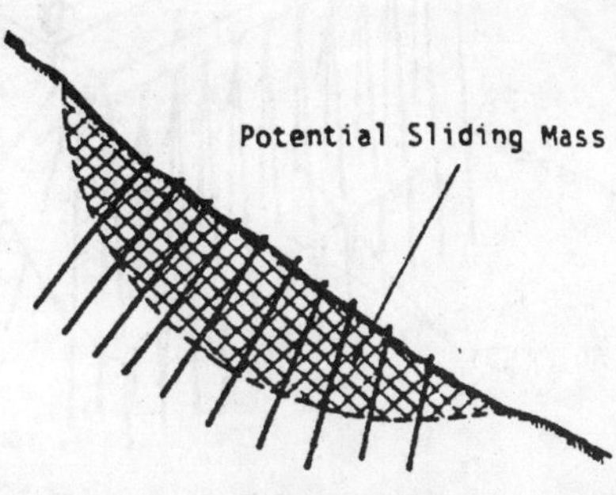

Figure 5-13 Nail support of a creeping landslide. (From Munfakh et al., 1987. Reproduced by permission of ASCE.)

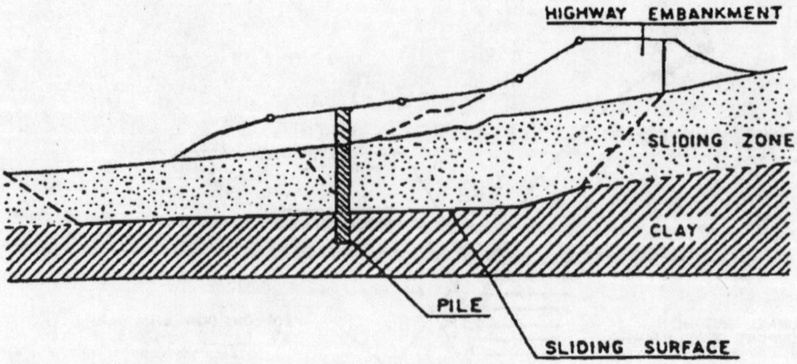

Figure 5-14 Rigid pile landslide stabilization. (From Mitchell and Villet, 1987.)

to soil nailing for slope stabilization, one or two rows of large rigid piles can be used at the toe of the slope (Figure 5-14). Yamada et al. (1971), Fukumoto (1972), Kerisel (1976), and Sommer (1977, 1979) discuss this technique for slope stabilization in more detail. Instead of installing soil nails in a regular pattern, they can be crisscrossed and pressure grouted so that they take both compressive and tensile forces (Figure 5-15). This technique is known as root piles (Lizzi, 1971), reticulated micropiles, or micropiles. When multiple soil nails are installed in an unstable slope, the global effect is to increase the stability of the soil mass beyond what one would expect by multiplying the resistance of one nail by the number of nails installed. The soil/nail interaction must account for the increased efficiency of the nails.

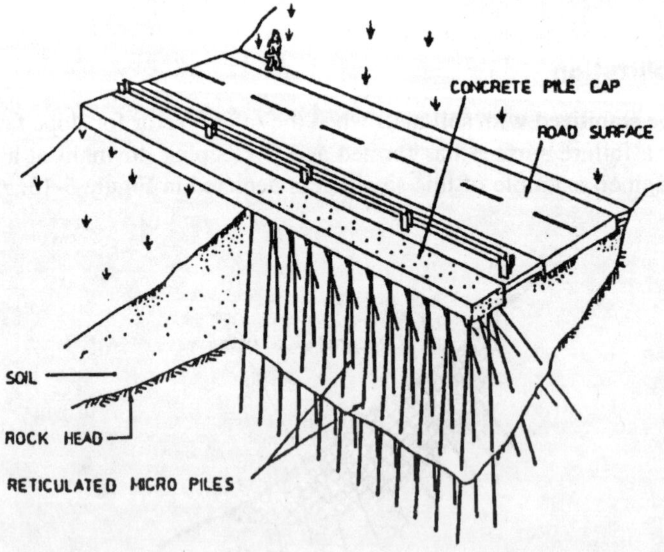

Figure 5-15 Reticulated micropile landslide stabilization. (From Bruce and Jewell, 1986.)

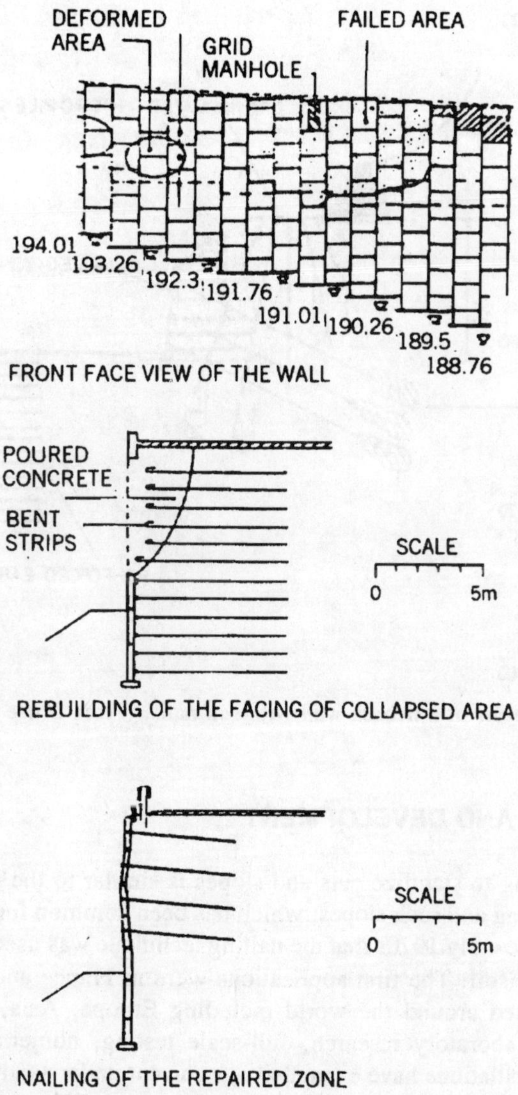

Figure 5-16 Repair of a reinforced earth wall with soil nails. (From Bruce and Jewell, 1986.)

Retaining Wall Repair

Repairs of a reinforced earth wall (Figure 5-16), a tieback wall (Figure 5-17), and a masonry wall (Figure 5-18), all using soil nails, are reported by Bruce and Jewell (1986, 1987a, b). These applications demonstrate how flexible and innovative soil nailing can be.

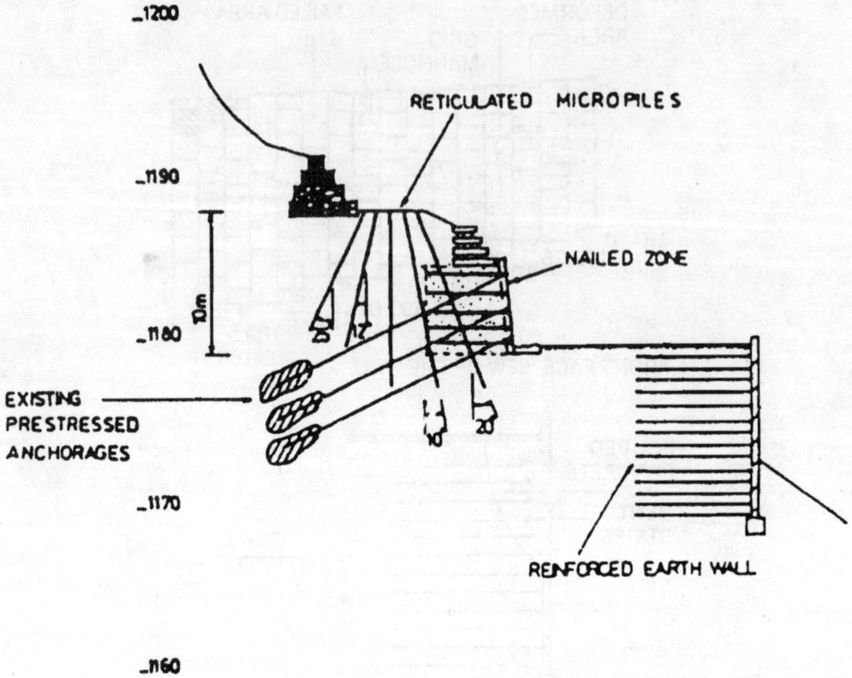

Figure 5-17 Repair of a tieback wall with soil nails. (From Bruce and Jewell, 1986.)

5-3 HISTORY AND DEVELOPMENT

Use of soil nailing to stabilize cuts and slopes is similar to the use of untensioned dowels in tunneling and rock slopes, which has been common for several years. But it was not until the early 1970s that the nailing technique was used for reinforcement of excavations in soil. The first applications were in France and Germany. Today, soil nailing is used around the world including Europe, Asia, England, and the United States. Laboratory research, full-scale testing, numerical modeling, and instrumented installations have expanded our current understanding of the behavior and performance of soil nailed walls. Literature from France, Germany, and the United States has been the most extensive.

France

The French have probably done more to advance soil nailing state-of-the-art than any other country in the world. The first applications of this technique were for inclined slopes along the French railroad system in the early 1970s (Figure 5-19). Rabejac and Toudic (1974) and Hovart and Rami (1975) describe these applications. Other more recent projects are described by Louis (1979), Schlosser and Juran (1979), Guilloux and Notte (1983), and Schlosser (1983).

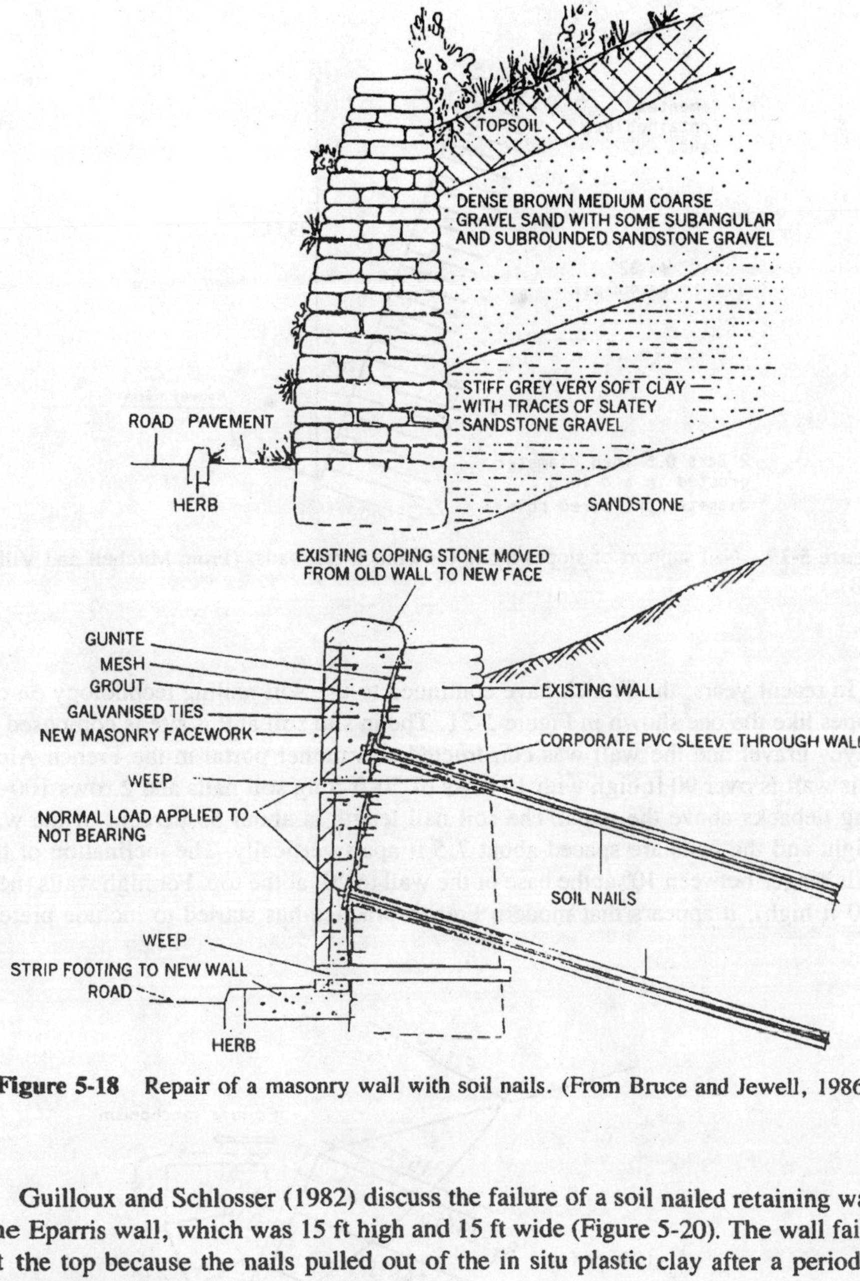

Figure 5-18 Repair of a masonry wall with soil nails. (From Bruce and Jewell, 1986.)

Guilloux and Schlosser (1982) discuss the failure of a soil nailed retaining wall, the Eparris wall, which was 15 ft high and 15 ft wide (Figure 5-20). The wall failed at the top because the nails pulled out of the in situ plastic clay after a period of heavy rainfall.

The French have put great effort into measuring and analyzing the tensile, shear, and bending stresses in soil nail inclusions and the effects on the in situ soil mass (Munfakh et al., 1987). As described below in Section 5-6, Design Methods, the French design method is based to a large extent on these phenomena.

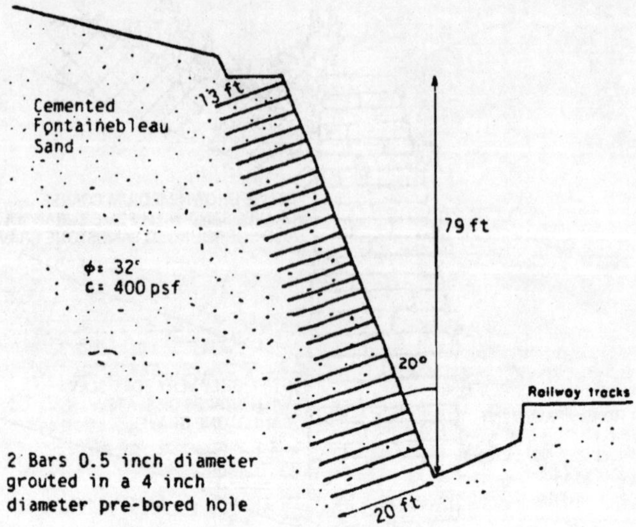

Figure 5-19 Nail support of slopes along the French railroads. (From Mitchell and Villet, 1987.)

In recent years, the French have continued to use soil nailing technology on cut slopes like the one shown in Figure 5-21. The in situ soil at this site is composed of clayey gravel and the wall was constructed at a tunnel portal in the French Alps. This wall is over 90 ft high with 10 rows of 50-ft-long soil nails and 2 rows 100-ft-long tiebacks above the nails. The soil nail length is about 55 percent of the wall height and the nails are spaced about 7.5 ft apart vertically. The inclination of the nails ranges between 10° at the base of the wall to 20° at the top. For high walls (near 100 ft high), it appears that modern French practice has started to include preten-

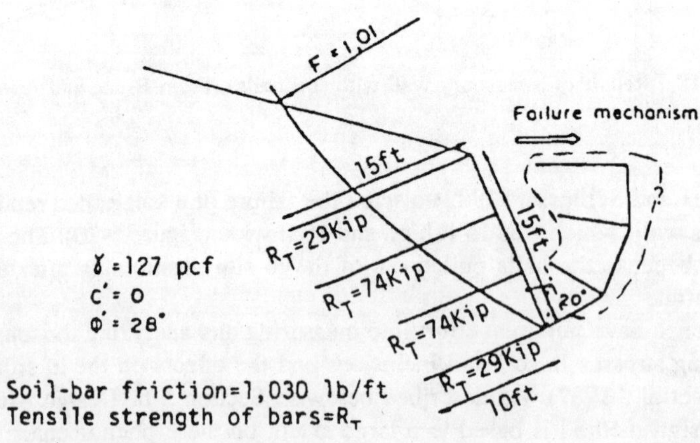

Figure 5-20 Eparris wall failure. (From Mitchell and Villet, 1987.)

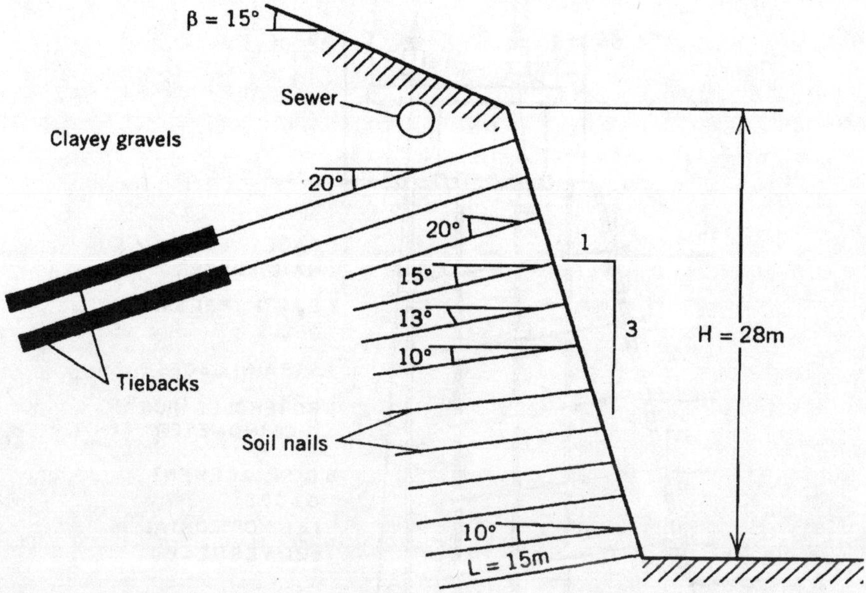

Figure 5-21 Nail support for the Mur de la tete Nord du tunnel de Cotiere in France. (From Walkinshaw, 1991.)

sioned soil anchors at the top of the cuts to limit excessive horizontal slope movement and to prevent the formation of tension cracks above the cut slope.

Bruce and Jewell (1986) estimate that there are about 50 soil nailing projects per year in France. About 10 percent of these are permanent applications. Average project sizes range between about 10,000 and 20,000 ft². France has a nationally organized research program on soil nailing, called "Programme Clouterre," in which the École National des Ponts et Chausees and the Centre d'Études et de Recherches du Batiment et des Travaux Public are taking part.

Germany

The Germans were not far behind the French in transferring what they knew about tunnel reinforcement to soil cut slopes. Some of the published literature from Germany by Gassler (1977), Stocker et al. (1979), and Gassler and Gudehus (1981) discusses their experiences with instrumented walls and full-scale load tests (Figure 5-22). Extensive measurements of nail forces, earth pressures behind the wall facing, and ground movements were made. Bruce and Jewell (1986) estimate that the current estimated level of soil nailing activity is about 25 percent that of France.

United States

Soil nailing came to the United States from Europ · in 1976 with the first application of this technology to the foundation excavation :or the Good Samaritan Hospital

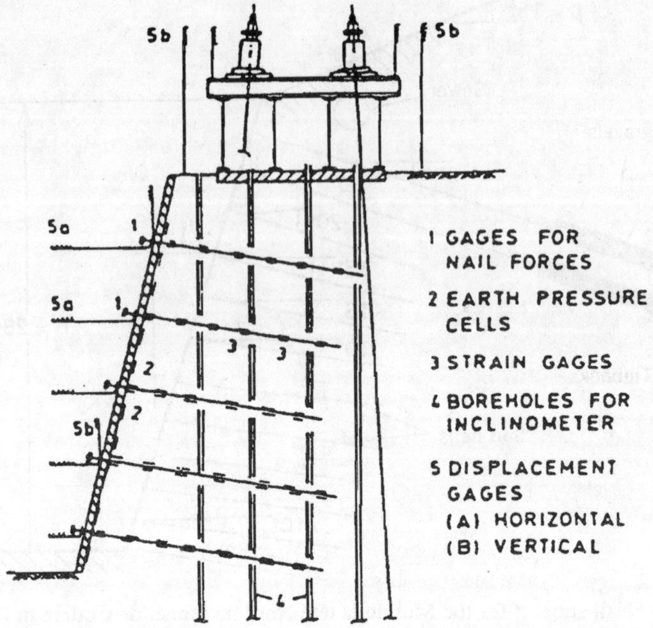

Figure 5-22 Typical German load test facilities. (From Mitchell and Villet, 1987.)

Extension in Portland, Oregon (ENR, 1976). Shen et al. (1981) discuss monitoring of the Portland project as well as centrifuge testing, finite element analyses, and a full-scale instrumented test section. The next significant published case history about soil nailing in the United States was for the foundation excavation of the PPG Industries headquarters in Pittsburgh, Pennsylvania (Nicholson and Boley, 1985).

The first time that soil nailing was used on a publicly funded project in the United States was in 1985 on the Cumberland Gap Tunnel Project (U.S. Highway 25E). The in situ soils were composed of residual soils derived from the shale, sandstone, and siltstone bedrock that had weathered in-place (Nicholson, 1986). The height of the wall was typically 30 to 40 ft (Figure 5-23). The nails were typically drilled and grouted in a 4 ½-in.-diameter hole on 5-ft centers, 20 to 30 ft long, and inclined 15° below horizontal (Figure 5-24). The wall was instrumented with slope inclinometers, strain gages, and load cells. Pull-out tests were conducted on the nails.

Berry (1991) cites new Occupational Safety and Health Administration (OSHA) regulations for prevention of trench deaths as a reason to use soil nailing for an excavation at the University of Tennessee. Excavations between 22 and 32 ft deep were made for renovation and expansion of the university's engineering campus Dabney Hall. Design details include:

- 5 × 5-ft nail spacing
- 4- to 6-in.-diameter No. 8 thread bar nails
- 18- to 32-ft-long nails (0.8 to 1.0 × excavation height)

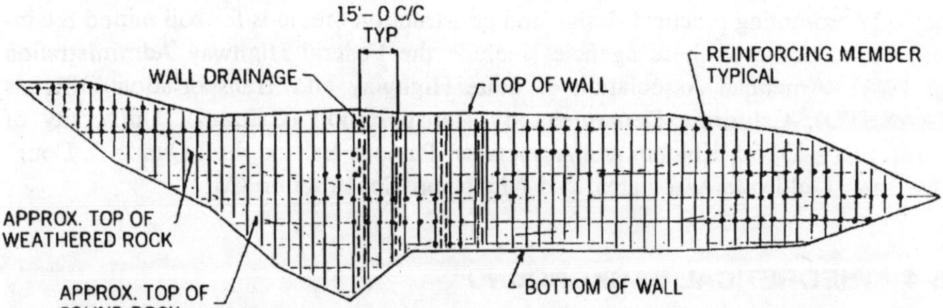

Figure 5-23 Elevation of Cumberland Gap wall. (From Nicholson, 1986.)

- 15° angle below horizontal
- 3-in.-thick shotcrete facing

This is just one example of the use of soil nailing rapidly gaining in popularity in the United States.

Representatives from the public and private sectors in the United States are

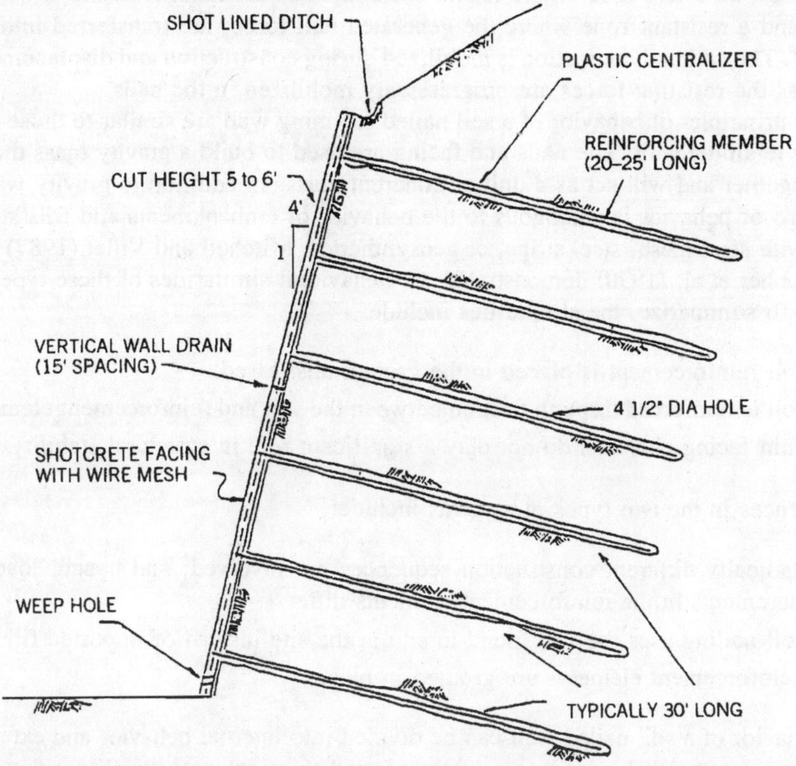

Figure 5-24 Cross section of Cumberland Gap wall. (From Nicholson, 1986.)

actively promoting practical design and construction methods for soil nailed retaining walls. Some of these agencies include the Federal Highway Administration (FHWA), American Association of State Highway and Transportation Officials (AASHTO), California Department of Transportation (Caltrans), University of California—Davis, Kulchin and Associates, Parsons Brinckerhoff Quade & Douglas, Inc., Golder & Associates, and Nicholson Construction Co.

5-4 THEORETICAL BACKGROUND

Soil nailed structures rely on the transfer of tensile forces generated in the nails in an active zone to a resistant zone through friction or adhesion mobilized at the soil/nail interface and passive resistance developed on the surface perpendicular to the direction of soil/nail relative movement (Elias and Juran, 1991). The frictional interaction between the ground and the nails restrains ground movement during and after construction. The resisting tensile forces mobilized in the nails induce an apparent increase of normal stresses along potential sliding surfaces, increasing the overall shear resistance of the native ground. Nails placed across a potential slip surface can resist the shear and bending moment through the development of passive resistance. The locus of maximum tensile forces in the nails separates the nailed soil mass into two zones: an active zone where lateral shear stresses are mobilized and stress the nails, and a resistant zone where the generated nail forces are transferred into the ground. The soil nail interaction is mobilized during construction and displacements occur as the resisting forces are progressively mobilized in the nails.

The principles of behavior of a soil nailed retaining wall are similar to those of a gravity retaining wall. The nails and facing are used to build a gravity mass that is held together and will act as a unit or coherent mass. In addition to gravity walls, this type of behavior is analogous to the behavior of embankments and fills stabilized with steel mesh, steel strips, or geosynthetics. Mitchell and Villet (1987) and Christopher et al. (1990) demonstrated the behavioral similarities of these types of walls. To summarize, the similarities include:

- Soil reinforcement is placed in the ground unstressed.
- Soil is reinforced through friction between the soil and reinforcement element.
- Thin facing elements do not play a significant role in structural stability.

Differences in the two types of systems include:

- Radically different construction sequences are involved, and hence, loading increments in the reinforcement elements differ.
- Soil nailing uses the soil found in situ at the site instead of imported fill.
- Reinforcement elements are grouted in place.

Behavior of a soil nailed wall can be divided into internal behavior and external behavior. Internal behavior, more often referred to as internal stability, relates to internal mechanisms of behavior including the in situ soil properties, stresses within the structure, and characteristics of the nails and facing. External behavior, or external stability considerations, for a soil nailed retaining wall are similar to those for all types of retaining walls, namely, sliding, overturning, bearing capacity, and failure slip surfaces beyond the reinforced soil mass.

Nails

For internal stability, the nails must be strong enough not to fail in tension and be long enough not to pull out of the ground when they become loaded (Figure 5-25). Also, the nails must be spaced closely enough to "knit" the soil together so that it acts as a coherent mass. For external stability, this mass must be sized large enough (i.e., the nails must be long enough) such that the reinforced soil mass will not slide, tilt, or fail within a larger slip surface (Figure 5-26).

Facing

The wall facing, usually shotcrete, primarily holds the soil in place at the cut face surface between the nails. It also protects the near-surface soil from excessive distortion, erosion, and weathering upon excavation. Typically, the facing or, at least part of the facing, is constructed (Figure 5-27) immediately after excavation of

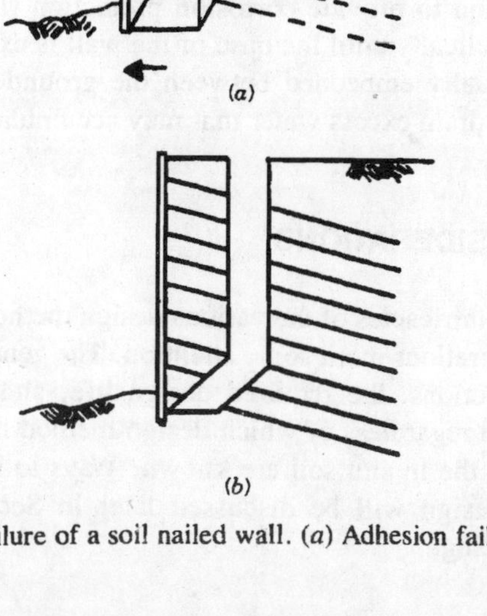

Figure 5-25 Internal failure of a soil nailed wall. (*a*) Adhesion failure. (*b*) Tension failure.

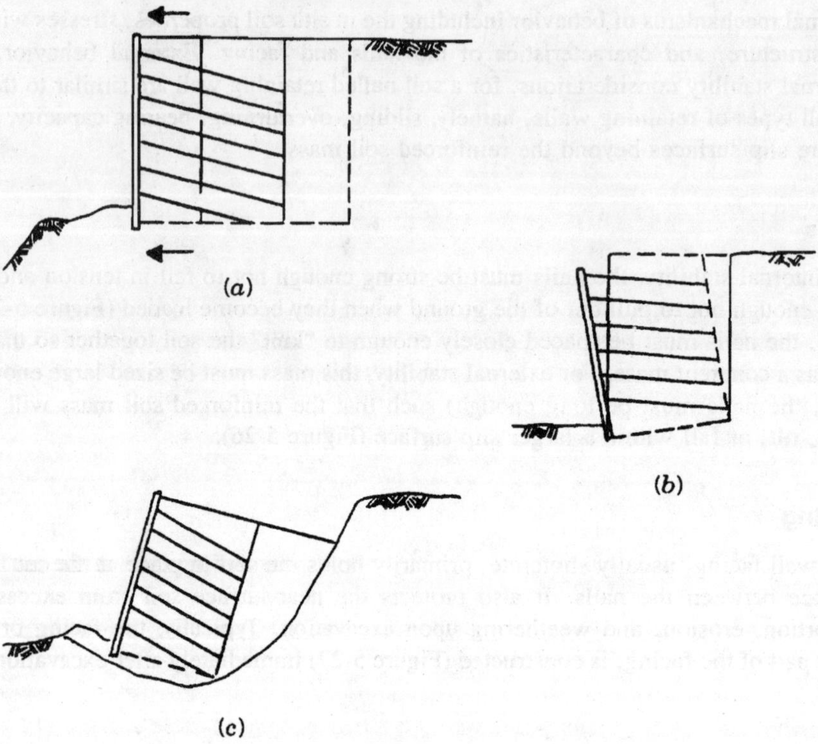

Figure 5-26 External failure of a soil nailed wall. (*a*) Sliding. (*b*) Tilt/bearing failure. (*c*) Slip failure.

a lift (about 5 to 8 ft in height). This initial facing layer then holds the soil in place while the row of nails is installed to reinforce the soil mass. Once the row of nails is installed, a second shotcrete facing layer is usually applied to provide continuity of the facing and nails and to provide corrosion protection (Figure 5-28). Then, the process is repeated cyclically until the base of the wall is excavated and reinforced. Drainage fabric is usually embedded between the ground and the shotcrete, and weeps are installed to drain excess water that may accumulate behind the shotcrete.

5-5 DESIGN CONSIDERATIONS

Before discussing the intricacies of the various design methods in detail, other more general design considerations merit some attention. The general configuration of the wall, allowable deflections, the required design life, and drainage requirements need to be determined regardless of which design method is utilized. It is assumed that characteristics of the in situ soil are known. Ways to accomplish that for soil nail retaining wall design will be discussed later in Section 5-9, Geotechnical Investigation and Testing.

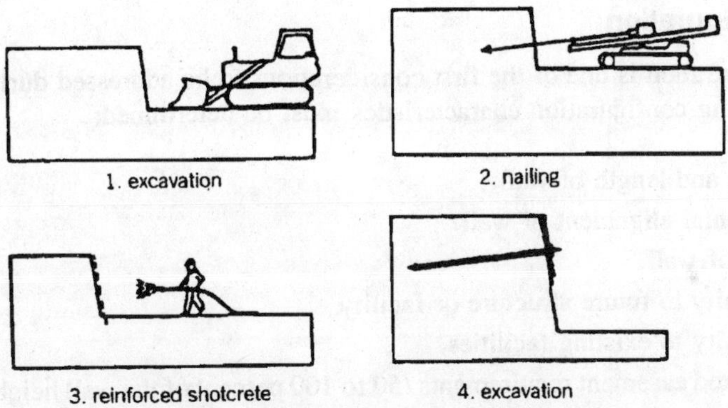

Figure 5-27 Construction sequence of soil nailing.

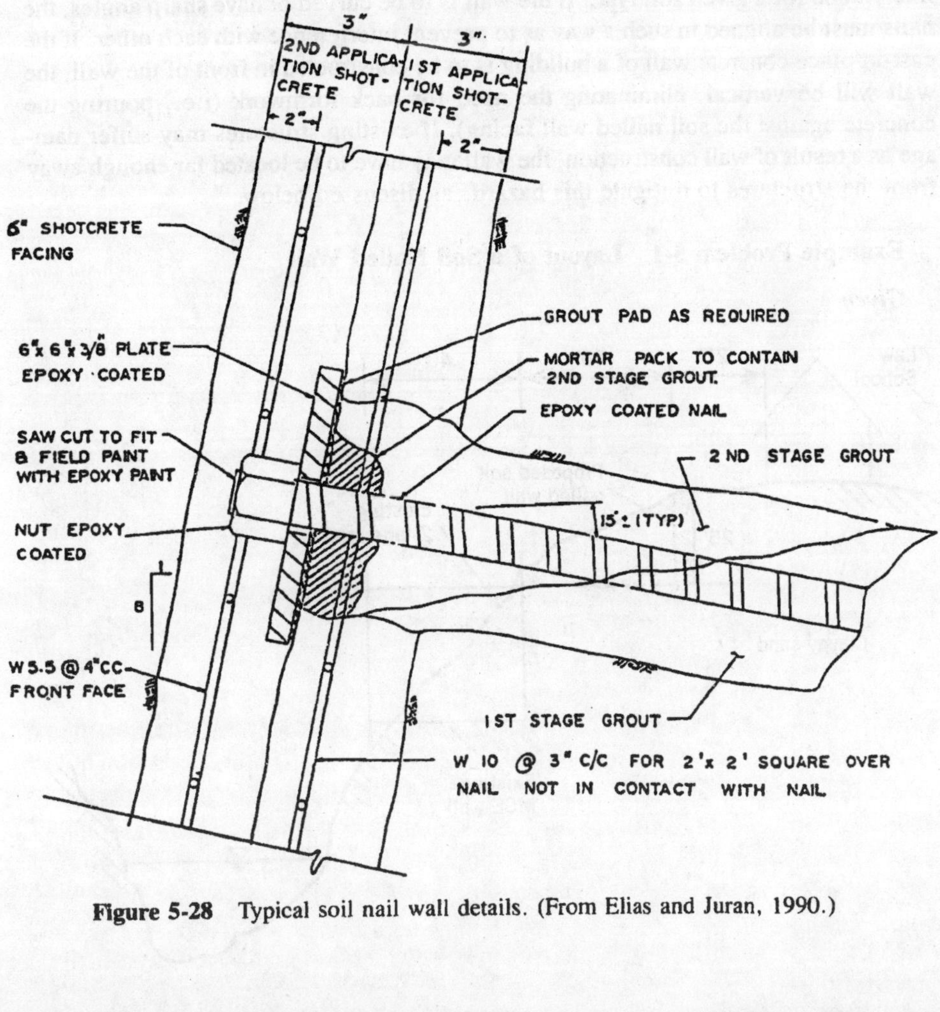

Figure 5-28 Typical soil nail wall details. (From Elias and Juran, 1990.)

Wall Configuration

Wall configuration is one of the first considerations to be addressed during design. The following configuration characteristics must be determined:

- Height and length of wall.
- Horizontal alignment of wall.
- Slope of wall.
- Proximity to future structure or facility.
- Proximity to existing facilities.
- Estimated easement requirements (50 to 100 percent of the wall height as a first guess).

These elements of design will dictate the space available for the wall and the space required for the wall. The higher and steeper the wall is, the longer the nails are likely to be for a given soil type. If the wall is to be curved or have sharp angles, the nails must be aligned in such a way as to prevent interference with each other. If the cast-in-place concrete wall of a building is to be constructed in front of the wall, the wall will be vertical, eliminating the need for back formwork (i.e., pouring the concrete against the soil nailed wall facing). If existing structures may suffer damage as a result of wall construction, the wall may have to be located far enough away from the structures to mitigate this hazard, as discussed below.

Example Problem 5-1 Layout of a Soil Nailed Wall

Given

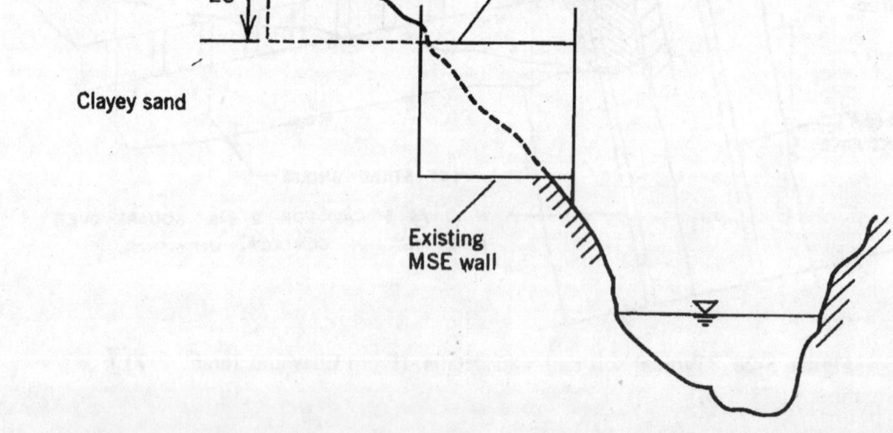

Required: Wall configuration requirements to expand to a four-lane road.
Solution: Height of wall = 25 feet. Alignment will be parallel to the road. Slope of wall: Vertical to avoid easement requirements; check property line. Proximity to Law School: 25 ft. Check for potential of building settlement and distortion.

Deflections

An excavation support system using soil nails and shotcrete should be considered to be a flexible-type system. Abramson and Hansmire (1988) made a comparison of tieback wall and soil nail wall deflections due to similar height excavations in similar soils (Figure 5-29). A soil nail and shotcrete wall is not a rigid system, like for instance a slurry wall. Soil nailed walls can be expected to deform laterally from about one-tenth to one-third percent of the excavation height (0.001 to 0.003 × excavation height) according to Bruce and Jewell (1987b). By comparison, Goldberg et al. (1976) report lateral deflections of the same order of magnitude for soldier pile and wooden lagging, steel sheet pile, and slurry walls in sand, gravel, and very stiff to hard clay. Distortions of nearby structures due to construction of a soil nailed wall may be intolerable, especially in urban environments with sensitive facilities nearby.

Horizontal wall deflections are typically related to settlement or subsidence at the ground surface, as shown in Figure 5-30. Ground surface settlement attenuates with distance from the excavation, as shown in Figure 5-31. If there are existing structures near the proposed excavation, potential settlement and angular distortion of these structures must be evaluated. Differential settlement due to varying wall

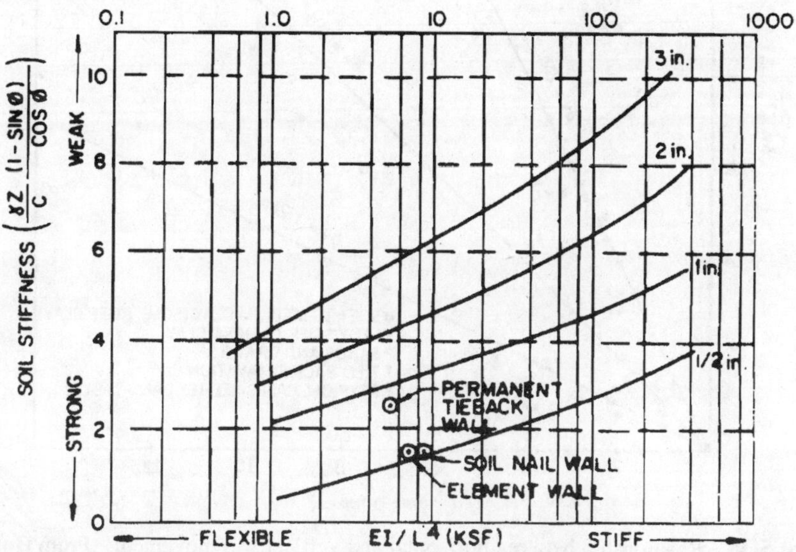

Figure 5-29 Wall deflections as a function of soil and wall stiffness. (From Abramson and Hansmire, 1988.)

heights should also be looked at. It may be necessary to locate the wall far enough away from the affected structure to eliminate the potential for distortion or the structure may have to be underpinned, as discussed in Chapter 2, Underpinning.

Example Problem 5-2 Estimation of Wall Deflections

Given: Problem given in Example Problem 5-1.
Required: Estimated settlement and angular distortion of Law School building.
Solution: A first estimate would be that maximum vertical settlement will equal maximum horizontal movement, which can range between 0.1 to 0.3 percent of the wall height (see Figure 5-30).

$$\begin{aligned}\delta_v = \delta_h &= 0.001 \text{ to } 0.003\, H \\ &= 0.001 \text{ to } 0.003 \times 25 \text{ ft} \times 12 \text{ in.} \\ &= 0.3 \text{ to } 0.9 \text{ in.}\end{aligned}$$

Say $\frac{1}{4}$ to 1 in. (maximum). The maximum would be expected to occur at the wall facing.

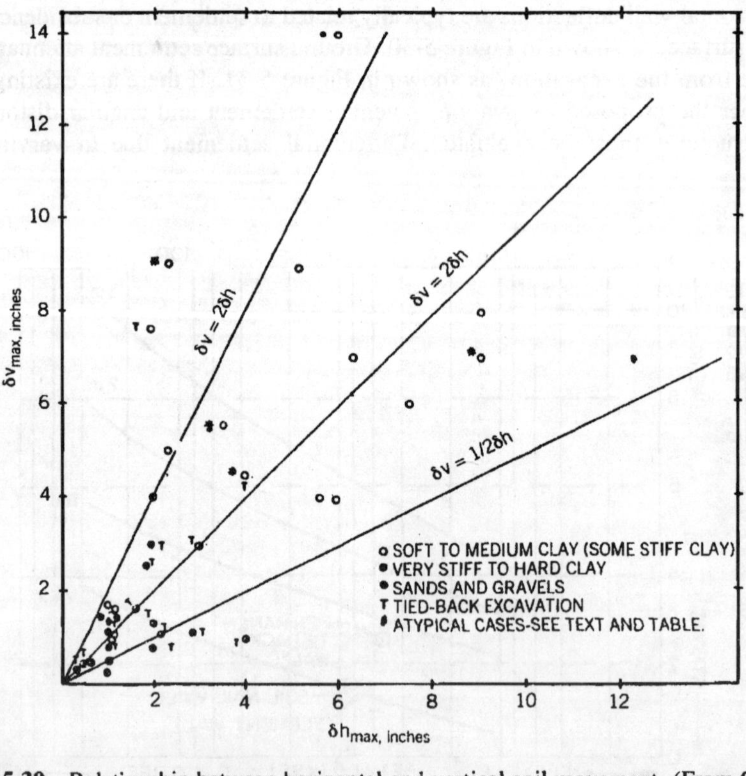

Figure 5-30 Relationship between horizontal and vertical soil movement. (From Goldberg et al., 1976.)

At the Law School building line the settlement should attenuate to about one-third of the maximum:

$$\frac{\delta_v}{3} = 0.1 \text{ to } 0.3 \text{ in. (Figure 5-31)}$$

This should be tolerable for the building. Angular distortion should be less than 0.3 in./25 ft, or 0.001. The building should tolerate this too. (See Figure 5-31.)

For conservatism, it may be prudent to use a soil nail design that is on the stiff side of usual practice to minimize building settlement (see Figure 5-29).

Design Life

The design life of a soil nailed wall is usually divided into two broad categories: temporary or permanent. If the wall is to be used for temporary excavation support, little concern is normally given to steel corrosion, cracking of the shotcrete, and sophisticated drainage features. However, if the wall is to be a permanent structure, the following factors must be considered during design:

- Corrosion protection of the nails and facing reinforcement.
- Structural and aesthetic performance of the wall facing.
- Drainage behind the wall facing with highly permeable, well connected drains or drainage fabric, weeps, and discharge outlets as discussed below.

Drainage

Poor drainage behind a retaining wall may cause the following problems:

- Excess horizontal loading
- Piping and erosion of soil fines
- Freeze/thaw pressures
- Accelerated steel corrosion
- Wall cracking or tilting
- Reduced sliding resistance
- Reduced slope failure resistance

Three words suggest the way to avoid these problems: drainage, drainage, drainage.

With the availability of several suitable types of drainage fabrics (geosynthetics) and industry experience with weeps and wall drainage piping, it is now easier than ever to specify high-quality wall drainage systems with a high degree of confidence. The costs of these systems are relatively low as compared with the total wall cost and can be installed with little additional effort during construction. Most of the potential problems with soil nail wall performance relate to drainage and can be mitigated easily with appropriate forethought during design and construction.

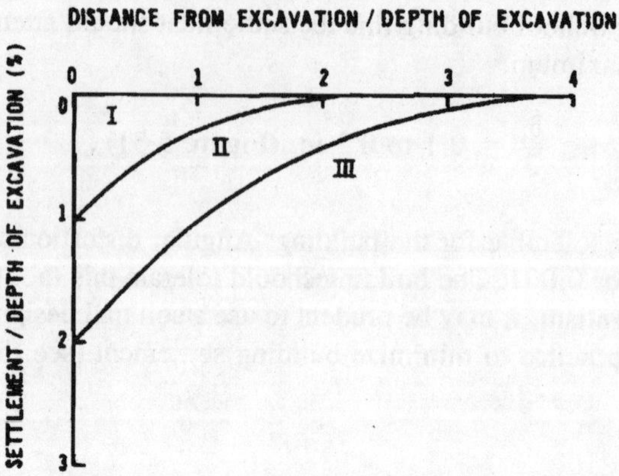

Figure 5-31 Ground settlement versus distance from the excavation. (From Peck, 1969.)

5-6 DESIGN METHODS

The first three soil nailing design methods to emerge in the literature in the 1970s and 1980s were: the Davis (and modified Davis) method, the German method, and the French method. These methods are referred to as limit analysis design methods (Elias and Juran, 1991). Critical potential failure surfaces must be assumed and the analyses are predicated on global or partial factors of safety. Since in reality, failure of a soil nailed wall would be progressive and initiated at the top of the slope with pull-out of the top rows of nails, a global factor of safety (equal for all nails) does not accurately predict the behavior of nails in different rows during failure. The basic assumptions of the different design approaches are discussed in Elias and Juran (1991) and summarized in Table 5-1.

A more complex and cumbersome method of analysis is based on the behavior of mechanically stabilized embankments. This kinematical method, described by Juran (1977), considers kinematically admissible displacement failure modes in a limit analysis framework. The kinematical method places undue emphasis on nail stiffness and is difficult to use.

It is the authors' opinion that with appropriate knowledge, understanding, and experience with any of the methods, reasonable results can be obtained by making the appropriate assumptions. These necessary assumptions may not be identical for each method. None of the methods is superior to the others and a good design with one method should stand up to scrutiny with another method. Using two independent design methods is a good way to provide an independent check on the design.

As with most engineering calculations today, computer software has been developed to perform soil nail design studies. Older traditional slope stability computer programs can be used to check the external stability of a wall design. A few computer programs have been developed especially for soil nail design. More sophisticated studies can be conducted with finite element programs (Shen et al., 1981) if desired.

TABLE 5-1 Assumptions of Different Design Methods

Features	French Method (Schlosser, 1983)	German Method (Stocker et al., 1979)	Davis Method (Shen et al., 1981)	"Modified" Davis (Elias and Juran, 1991)	Kinematical Method (Juran et al., 1990)
Analysis	Limit moment Equilibrium Global stability	Limit force Equilibrium Global stability	Limit force Equilibrium Global stability	Limit force Equilibrium Global stability	Working stress Analysis Local stability
Input material properties	Soil parameters (c, ϕ') limit nail forces Bending stiffness	Soil parameters (c, ϕ) lateral friction	Soil parameters (c, ϕ') limit nail forces Lateral friction	Soil parameters (c, ϕ') limit nail forces Lateral friction	Soil parameters ($c/(\gamma H)$, ϕ') Nondimensional bending stiffness parameter (N)
Nail forces	Tension, shear, moments	Tension	Tension	Tension	Tension, shear, moments
Failure surface	Circular, any input shape	Bilinear	Parabolic	Parabolic	Log-spiral
Failure mechanisms	Mixed[a]	Pull-out	Mixed	Mixed	Nonapplicable
Safety factors[b]					
Soil strength $F_{c'} F_{\phi-}$	1.5	1 (residual shear strength)	1.5	1	1
Pull-out resistance $F_{p'}$	1.5	1.5 to 2	1.5	2	2

(*continued*)

TABLE 5-1 (*Continued*)

Features	French Method (Schlosser, 1983)	German Method (Stocker et al., 1979)	Davis Method (Shen et al., 1981)	"Modified" Davis (Elias and Juran, 1991)	Kinematical Method (Juran et al., 1990)
Tension bending[c]	Yield stress Plastic moment	Yield stress	Yield stress	Yield stress	Yield stress Plastic moment
Design output	GSF[d] CFS[e]	GSF CFS	GSF CFS	GSF CFS	Mobilized nail forces CFS
Groundwater	Yes	No	No	No	Yes
Soil stratification[f]	Yes	No	No	No	Yes
Leading[f]	Slope, any surcharge	Slope surcharge	Uniform surcharge	Slope, uniform surcharge	Slope
Structure geometry[f]	Any input geometry	Inclined facing Vertical facing	Vertical facing	Inclined facing Vertical facint	Inclined facing Vertical

From Elias and Juran (1991).

[a] Mixed failure mechanisms: limit-tension force in each nail is governed by either its pull-out resistance factored by the safety factor or the nail yield stress, whichever is smaller.

[b] Definitions of safety factors used in this analysis:

For soil strength, $F = c/c_m$, $F_\phi = (\tan \eta)/(\tan \phi_m)$; where c and ϕ are the soil cohesion and friction angle, respectively, while c_m and ϕ_m are the soil cohesion and friction angle mobilized along the potential sliding surface.

For nail pull-out resistance, $F_P = f_1/f_m$, f_1 and f_m are the limit interface shear stress and the mobilized interface shear stress, respectively.

[c] Recommended limit nail force.

[d] GSF: Global safety factor.

[e] CFS: Critical failure surface.

[f] Present design capabilities.

Davis Method

The Davis method (Shen et al., 1981) assumes a parabolic failure surface that passes through the toe of a vertical wall (Figure 5-32). A slope stability analysis using the method of slices is used to evaluate the contribution of the nails to overall stability. Components of the tensile forces in the nails are considered parallel and perpendicular to the failure surface. Two conditions are considered in the analysis:

1. The failure surface extends beyond the reinforced zone.
2. The failure surface is entirely within the reinforced soil mass.

The solutions for analysis of these two conditions contain the factors of safety and therefore must be solved by iteration. Shen et al. (1981) developed a computer program to solve the problem readily.

For the first condition, the force equilibrium equations for Element 1 are:

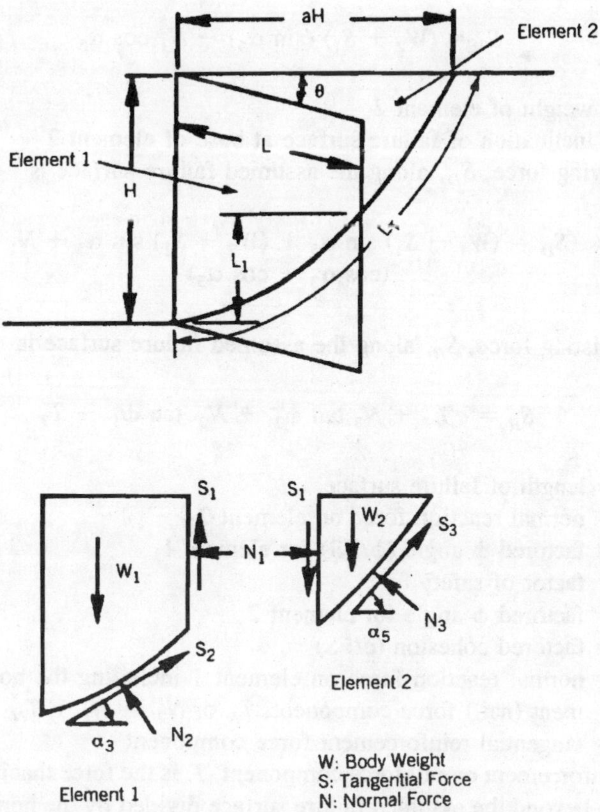

Figure 5-32 University of California—Davis design method. (From Mitchell and Villet, 1987.)

$$N_2 = (W_1 - S_1)(\cos \alpha_3) - N_1 \sin \alpha_3 \tag{5-1}$$

$$S_2 = (W_1 - S_1)(\sin \alpha_3) + N_1 \cos \alpha_3 \tag{5-2}$$

where W_1 = weight of element 1
S_1 = vertical tangential force between elements 1 and 2
α_3 = inclination of failure surface at base of element 1
N_1 = horizontal side force between elements 1 and 2 or

$$\tfrac{1}{2} K_0 (H - L_1)^2 \tag{5-3}$$

The force equilibrium equations for element 2 are:

$$N_3 = (W_2 + S_1)(\cos \alpha_5) + N_1 \sin \alpha_5 \tag{5-4}$$

$$S_3 = (W_2 + S_1)(\sin \alpha_5) - N_1 \cos \alpha_5 \tag{5-5}$$

where W_2 = weight of element 2
α_5 = inclination of failure surface at base of element 2

The total driving force, S_D, along the assumed failure surface is

$$S_D = (W_1 - S_1) \sin \alpha_3 + (W_2 + S_1) \sin \alpha_5 + N_1 (\cos \alpha_3 - \cos \alpha_5) \tag{5-6}$$

The total resisting force, S_R, along the assumed failure surface is

$$S_R = c' L_T + N_3 \tan \phi_{2'} + N_{2'} \tan \phi_{1'} + T_T \tag{5-7}$$

where L_T = length of failure surface
N_3 = normal reaction force on element 2
$\phi_{1'}$ = factored ϕ angle (ϕ/FS) for element 1
FS = factor of safety
$\phi_{2'}$ = factored ϕ angle for element 2
c' = factored cohesion (c/FS)
$N_{2'}$ = normal reaction force on element 1 including the normal reinforcement (nail) force component, T_N or $N_{2'} = N_2 + T_N$
T_T = tangential reinforcement force component

The total reinforcement or nail force component, T, is the force that is representative of the length beyond the assumed failure surface divided by the horizontal distance between nails.

Modified Davis Method

Elias and Juran (1991) proposed modifications to the Davis method that allow input relative to the pull-out resistance of the nail, multiple nail lengths, inclination of the wall face, a sloping bench above the wall, and factored soil strength input parameters. Others have proposed modifications to the original Davis method, including practitioners at the University of California—Davis, the California Department of Transportation, and Golder & Associates (Chassie, 1993).

German Method

The German method proposed by Stocker et al. (1979) and Gassler and Gudehus (1981) uses a force equilibrium analysis assuming a bilinear failure surface (Figures 5-33 and 5-34). The shearing resistance of the soil, as defined by Mohr–Coulomb's failure criterion, is assumed to be entirely mobilized along the potential failure surface. Only reinforcement tension forces are considered, as with the Davis method. The global factor of safety for this method is the ratio of the sum of the available resisting nail forces to the total required nail forces to maintain equilibrium. The available resisting nail forces are the ones available beyond the assumed failure surface. The total force required to maintain limit equilibrium is readily obtained by

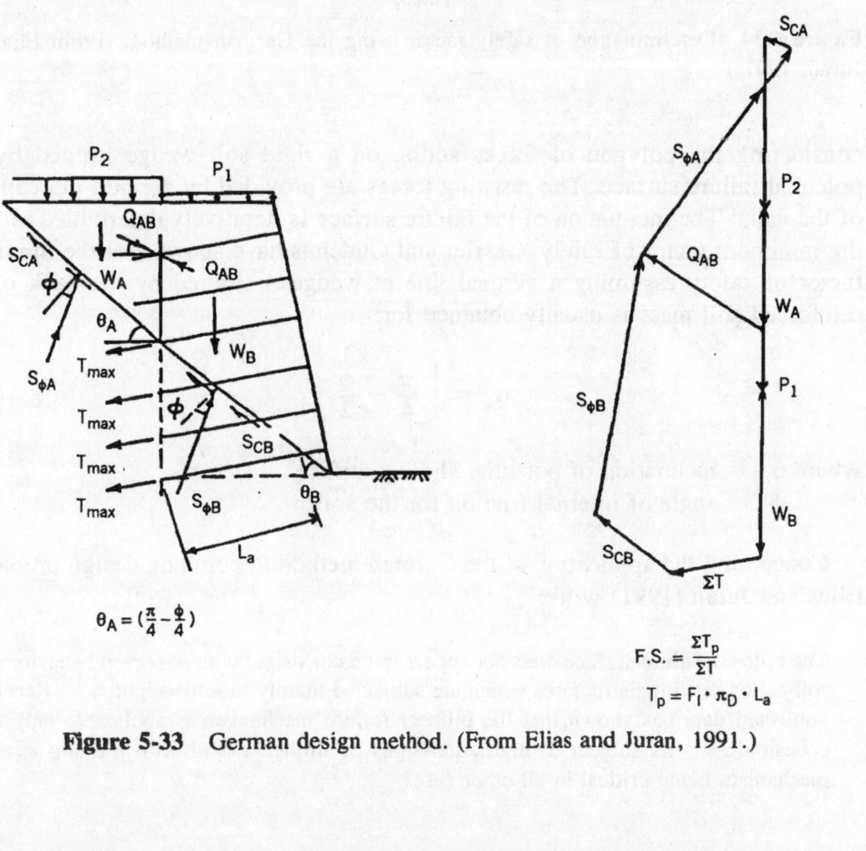

Figure 5-33 German design method. (From Elias and Juran, 1991.)

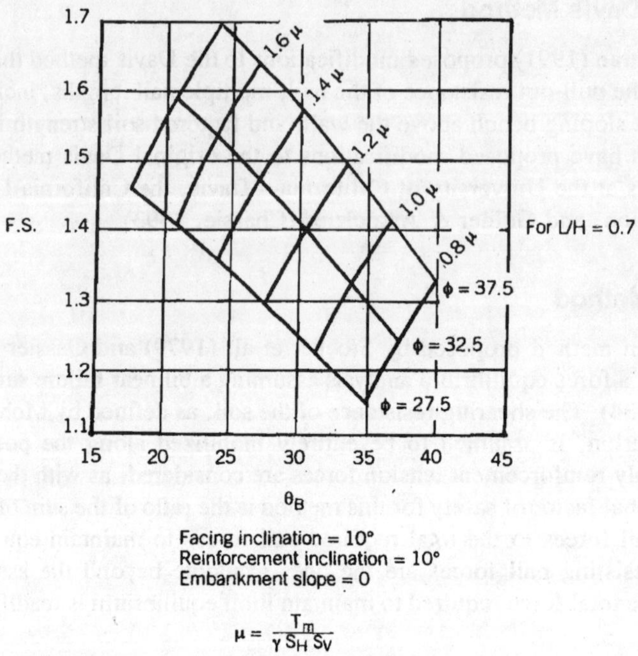

Figure 5-34 Determination of safety factor using the German method. (From Elias and Juran, 1991.)

considering the polygon of forces acting on a rigid soil wedge limited by the potential failure surface. The resisting forces are provided by the pull-out capacity of the nails. The inclination of the failure surface is iteratively determined to yield the minimum factor of safety. Gassler and Gudehus have shown that the minimum factor of safety assuming a vertical line at wedge *A* limited by the back of the reinforced soil mass is usually obtained for:

$$\sigma_a = \left[\frac{\pi}{4} - \frac{\phi}{2} \right] \tag{5-8}$$

where σ_a = inclination of potential sliding surface
ϕ = angle of internal friction for the soil

Concerning the application of the German method to common design problems, Elias and Juran (1991) write:

> The bilinear failure surface does not appear to be consistent with observed behavior of soil nailed retaining structures which are subjected mainly to self weight. . . . Recent published data has shown that the bilinear failure mechanism is applicable only in cohesionless soils subject to high surcharges of limited extent with the slip circle mechanism being critical in all other cases.

French Method

Either circular or noncircular failure surfaces can be used with the French method (Schlosser, 1983), which can be solved with the method of slices like the Davis method. The reinforced soil mass is treated as a composite material. The key difference of this method as compared to the other methods is that four failure criteria are considered, as shown in Figure 5-35. Each soil nail is evaluated with respect to the four failure criteria addressing different failure modes of the nail itself, the soil around the nail, and the nail/soil interface. Guidelines for these evaluations are summarized below and elaborated on in Mitchell and Villet (1987) and Elias and Juran (1991).

The shear resistance of the soil is evaluated on the basis of the traditional Mohr–Coulomb failure criteria with the angle of internal friction, ϕ, and the cohesion

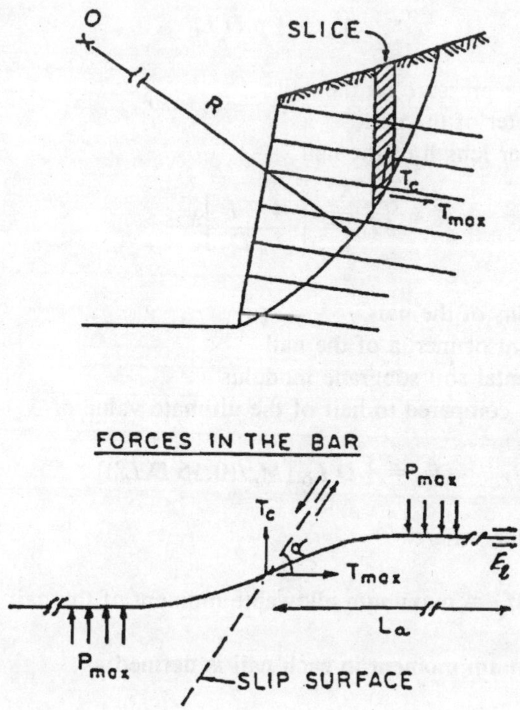

FAILURE CRITERIA

Shear resistance of the bar $T_{max} \leq A_S \cdot Fy$, $T_c \leq R_c = A_S \cdot Fy$

Soil bar friction $T_{max} \leq \pi D \tau_{ult} L_a$

Normal lateral earth thrust on the bar $\rho \leq \rho_{max}$

Shear resistance of the soil $\tau < c + \sigma \tan \phi$

Figure 5-35 French design method. (From Elias and Juran, 1991.)

value, c, at the base of each slice being the input material properties. The soil is failing if the mobilized shear stress is greater than the normal stress multiplied by tan ϕ plus the cohesion. A factor of safety of 1.5 is acceptable.

The tensile nail force is calculated from the pull-out resistance of the nail beyond the assumed failure surface. The nail is pulling out of the soil beyond the assumed failure surface if the mobilized tensile force is greater than the surface area of the bar beyond the failure surface multiplied by the maximum allowable skin friction (ultimate skin friction divided by a factor of safety). A factor of safety of 1.5 is acceptable.

The soil/nail interaction failure criterion is similar to estimating the capacity of a laterally loaded pile using a load versus deformation soil/structure interaction analysis. The allowable bending moments and shear forces of the nails are compared with the mobilized shear and bending forces. The nail fails if the mobilized forces are greater than the allowable forces. The induced shear force in each nail is defined as

$$V_0 = \tfrac{1}{2}\, p\, D\, L_0 \tag{5-9}$$

where p = passive pressure on the nail
D = diameter of the nail
L_0 = transfer length of the nail

$$L_o = \left[\frac{4\,E\,I}{k_h\,D}\right]^{0.25} \tag{5-10}$$

where E = modulus of the nail
I = moment of inertia of the nail
k_h = horizontal soil subgrade modulus

The value of V_0 is compared to half of the ultimate value or

$$V_0 = \tfrac{1}{2}\, D\, L_0\, [M_p/(0.16\, D\, L_0^2)] \tag{5-11}$$

whichever is smaller, where

M_p = maximum allowable moment of the nail

The induced maximum moment in each nail is defined as

$$M_{max} = 0.16\, p\, D\, L_0^2 \tag{5-12}$$

This is compared to the maximum allowable moment of the nail, M_p. A factor of safety of 2.0 is acceptable for computation of allowable shear and bending forces.

Finally, the combined tensile and shear strength of the nail is considered using Tresca's failure criterion:

$$\frac{T^2}{R_n^2} + \frac{V^2}{R_c^2} < 1 \tag{5-13}$$

where V = mobilized shear in the nail

$$V = \frac{R_c}{[1 + 4\tan^2(1.57 - \alpha)]^{0.5}} \tag{5-14}$$

T = mobilized tension in the nail

$$T = 4\,V\tan(1.57 - \alpha) \tag{5-15}$$

R_n = tensile strength of the nail, f_y

R_c = shear strength of the nail, $f_y/2$

α = relative angle between the nail and the slope failure surface

Factors of safety of 1.3 and 1.5 are recommended for temporary and permanent walls, respectively.

Kinematical Method

Elias and Juran (1991) proposed the kinematical method, which is based on "a limit analysis associating a kinematically admissible displacement failure mode as observed on model walls with a statically admissible limit equilibrium solution." This method differs from the others mentioned above in that it does not utilize the method of slices for solution of the problem and treats the soil nailed wall more or less like a mechanically stabilized embankment. It assumes a "quasi-rigid body rotation," defined by a circular or log spiral assumed failure surface (Figure 5-36). Otherwise, this method is similar to the French method in that it uses the Mohr–Coulomb failure criterion for the soil and the Tresca failure criterion for the nail. This method is explained fully in Elias and Juran (1991).

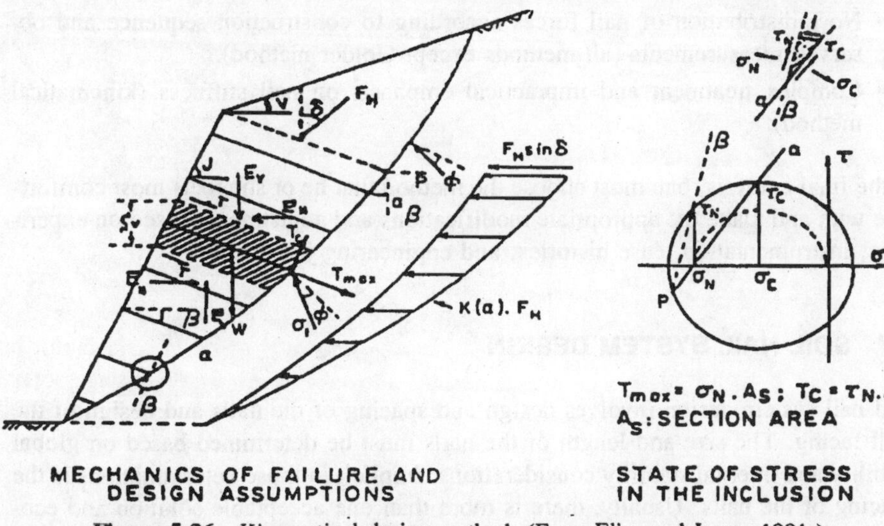

Figure 5-36 Kinematical design method. (From Elias and Juran, 1991.)

Computer Methods

If it were not for the common availability of computer programs to solve soil nailing design problems, design calculations would be prohibitively arduous and time consuming. There are computer methods that can be and should be used to design soil nailed retaining walls. Valuable time can be spent on optimizing a design by trial and error instead of on time consuming hand calculation of single alternatives.

Available computer programs include: "homemade" programs that solve soil slope stability problems, modified to account for the presence of reinforcement; public-domain and commercially available programs that solve soil slope stability problems using the method of slices and that are or can be modified to include reinforcement; and specialized soil nailing computer programs like the program by Shen et al. (1981) with appropriate modifications, a program by Caltrans (1991) called SNAIL, and a program by Golder & Associates called GOLDNAIL. A particular advantage of the Caltrans program is that tiebacks and nails can be combined if appropriate. The French (TALREN) and Germans also have similar programs, but they are more difficult and costly to obtain. For detailed studies, finite element methods can also be applied to the problem.

Design Method Inconsistencies

None of the methods described above solve the soil nailed wall problem without inconsistencies in the input parameters, analytical methods, and comparisons to observed behavior. Some of these inconsistencies according to Walkinshaw (1992) include:

- Improper cancellation of interslice forces (Davis method).
- Lateral earth pressures inconsistent with nail force and facing pressure distribution (all methods).
- No redistribution of nail forces according to construction sequence and observed measurements (all methods except Golder method).
- Complex treatment and impractical emphasis on nail stiffness (kinematical method).

In the final analysis, one must choose the method that he or she feels most comfortable with and make the appropriate modifications and adjustments based on experience, instrumentation, case histories, and engineering judgment.

5-7 SOIL NAIL SYSTEM DESIGN

Soil nail system design involves design and spacing of the nails and design of the wall facing. The size and length of the nails must be determined based on global stability and internal stability considerations. Implicit in those determinations is the spacing of the nails. Usually, there is more than one acceptable solution and economics may govern the final choice of configuration. Corrosion protection should be considered during design if appropriate.

Design of the wall facing (materials, thickness, and reinforcement) is dependent on the nail forces assumed. The stiffness of the ground is a factor, particularly if the facing is to be modeled as a beam on an elastic foundation. During design, it must also be decided whether the facing is to serve a temporary or permanent function. The materials and factors of safety used in the analysis would be different for these cases.

Empirical Methods

Bruce and Jewell (1987b) derived parameters that can be used as a first cut at design based on published case histories. These parameters include:

$$\text{length ratio} = L/H = 0.5 \text{ to } 0.8 \text{ for drilled and grouted nails in granular soils} \quad (5\text{-}16)$$

$$= 0.5 \text{ to } 0.6 \text{ for driven nails in granular soils}$$

$$= 0.5 \text{ to } 1.0 \text{ for moraine and marl}$$

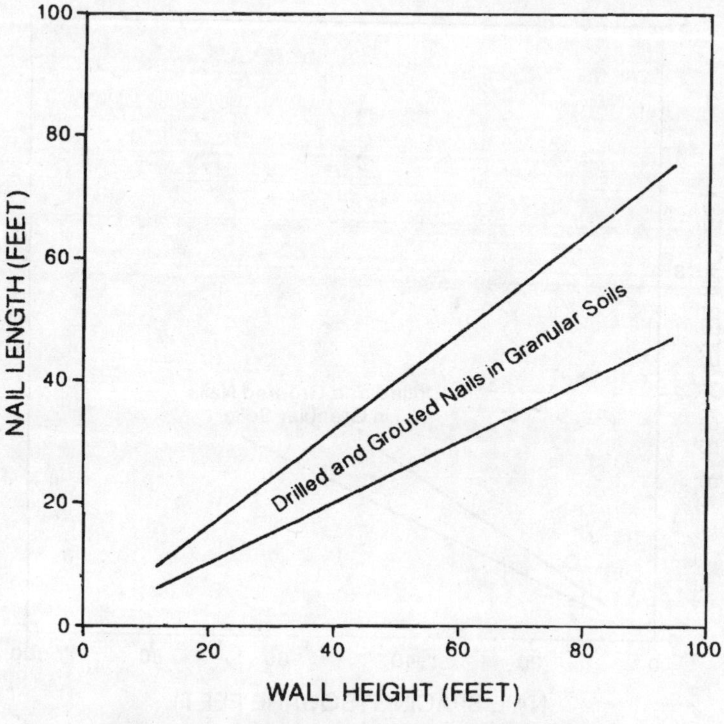

Figure 5-37 Empirical soil nailing length ratio.

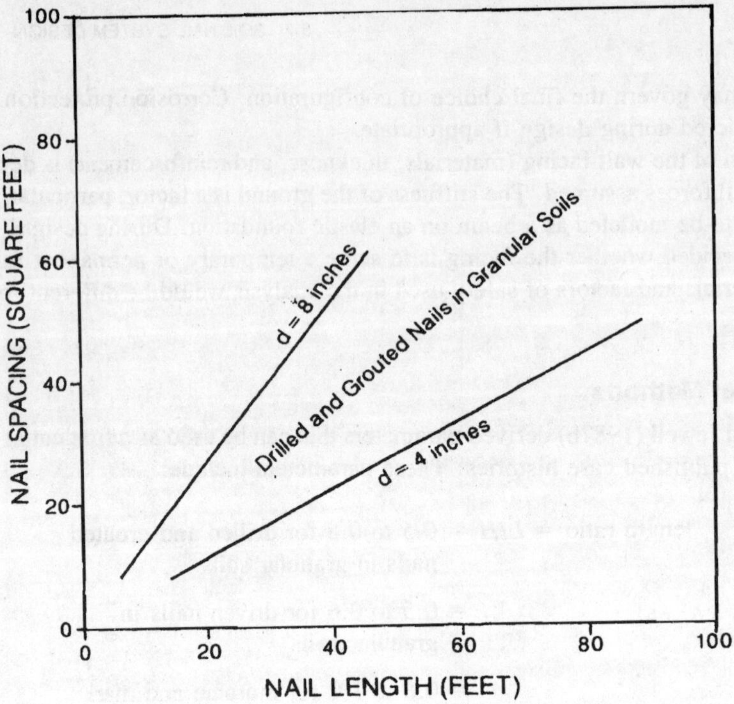

Figure 5-38 Empirical soil nailing bond ratio.

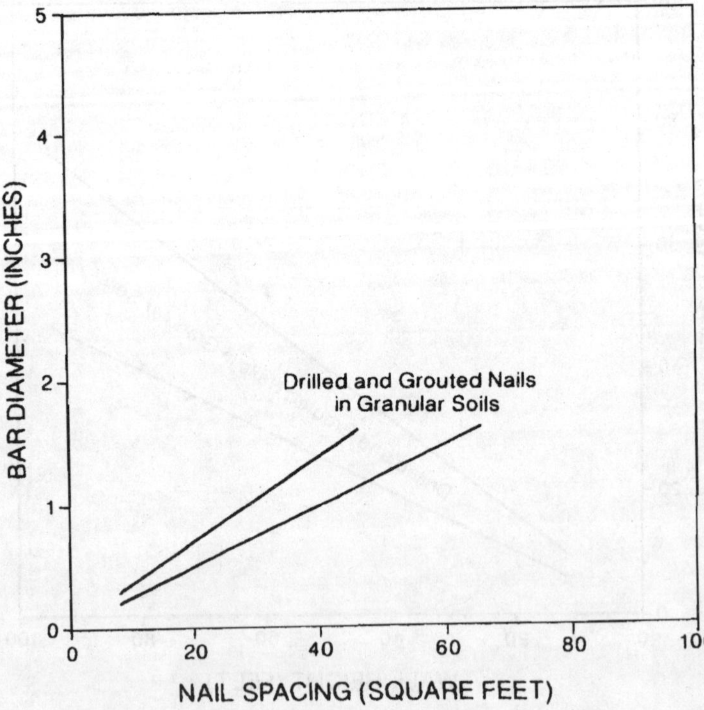

Figure 5-39 Empirical soil nailing strength ratio.

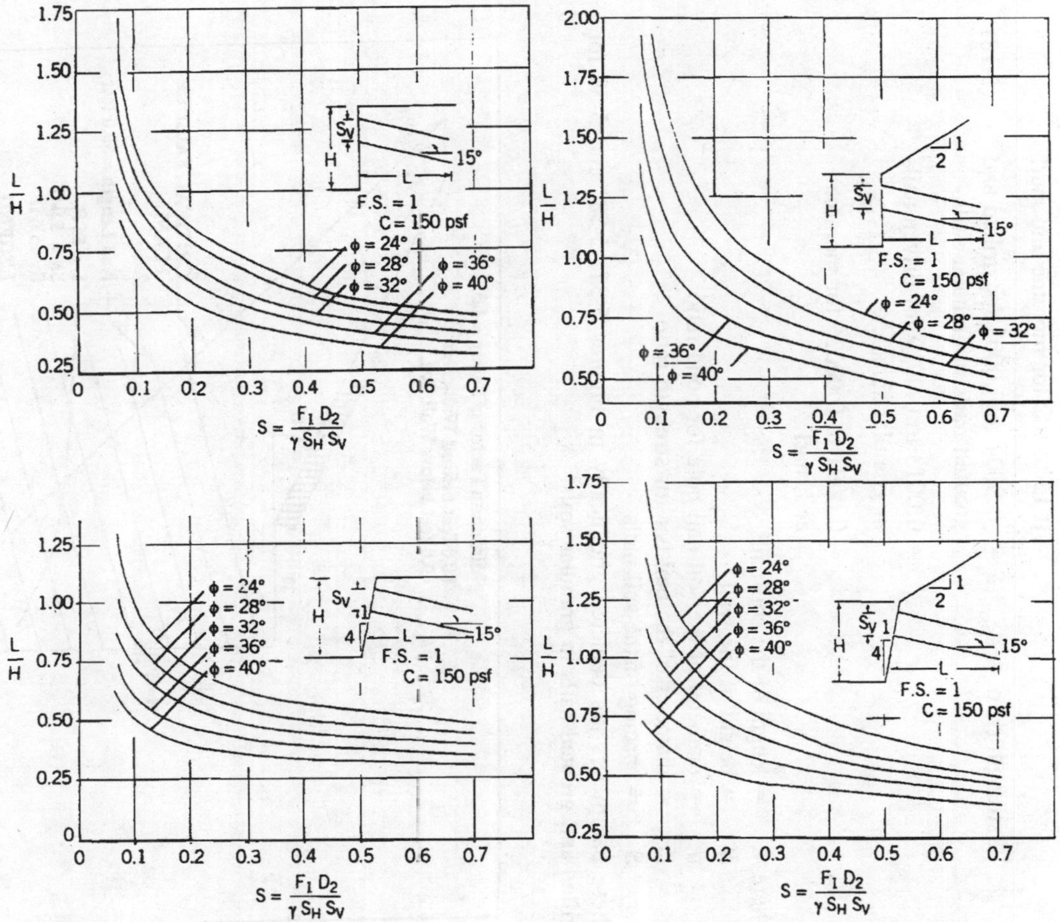

Figure 5-40 Modified Davis method design charts. (From Elias and Juran, 1991.)

$$\text{bond ratio} = dL/S = 0.5 \text{ to } 0.6 \text{ for drilled and grouted nails in granular soils} \tag{5-17}$$

$$= 0.6 \text{ to } 1.1 \text{ for driven nails in granular soils}$$

$$= 0.15 \text{ to } 0.2 \text{ for moraine and marl}$$

$$\text{strength ratio} = d_{bar}{}^2/S = 0.0004 \text{ to } 0.0008 \text{ for drilled and grouted nails in granular soils} \tag{5-18}$$

$$= 0.0013 \text{ to } 0.0019 \text{ for driven nails in granular soils}$$

$$= 0.0001 \text{ to } 0.00025 \text{ for moraine and marl}$$

where L = length of the soil nails
H = height of the wall
d = diameter of the soil nail hole for bond ratio
d_{bar} = diameter of the nail bar for strength ratio
S = spacing of the soil nails

These parameters are depicted graphically in Figures 5-37, 5-38, and 5-39 for drilled and grouted nails in granular soils.

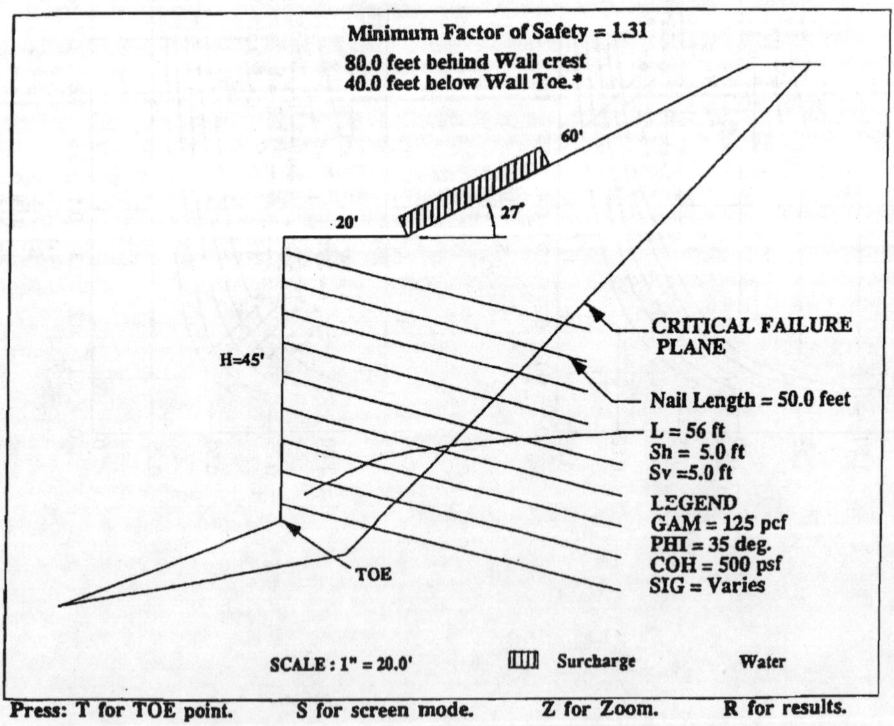

Figure 5-41 SNAIL computer program output.

Global Stability

In general, detailed design of the soil nail system should begin with an analysis of global stability. This can be done using whatever soil slope stability analysis method is available. The modified Davis method is one of the simplest to use and easiest to obtain. Design charts for using this public-domain method are given in Figure 5-40. The Caltrans computer program is another public-domain resource that is a handy way to compute global stability. Output from this program is depicted in Figure 5-41.

Internal Stability

Some of the methods that can be used to calculate global stability also output force distribution and failure modes of the nails (Figure 5-42). This is the most desirable

RESULTS OF THE MINIMUM SAFETY FACTORS
FOR THE FAILURE PLANES AT THE TOE AND BELOW

DISTANCE ALONG TOE SLOPE (ft)	MINIMUM SAFETY FACTOR	DISTANCE BEHIND WALL TOE (ft)	LOWER FAILURE PLANE ANGLE (deg)	LOWER FAILURE PLANE LENGTH (ft)	UPPER FAILURE PLANE ANGLE (deg)	UPPER FAILURE PLANE LENGTH (ft)
0.0	1.260	80.0	37.7	70.8	50.3	37.6

Reinf. Stress at Level

1 = 18.949 Ksi (Pullout Controls..)
2 = 31.061 Ksi (Pullout Controls..)
3 = 43.173 Ksi (Pullout Controls..)
4 = 47.619 Ksi (Yield Stress Controls)
5 = 47.619 Ksi (Yield Stress Controls)
6 = 47.619 Ksi (Yield Stress Controls)
7 = 47.619 Ksi (Yield Stress Controls)
8 = 47.619 Ksi (Yield Stress Controls)
9 = 47.619 Ksi (Yield Stress Controls)

Press <ENTER> to Continue ...

MOVING DOWN THE SLOPE ...

DISTANCE ALONG TOE SLOPE (ft)	MINIMUM SAFETY FACTOR	DISTANCE BEHIND WALL TOE (ft)	LOWER FAILURE PLANE ANGLE (deg)	LOWER FAILURE PLANE LENGTH (ft)	UPPER FAILURE PLANE ANGLE (deg)	UPPER FAILURE PLANE LENGTH (ft)
8.00	1.367	80.0	15.9	27.3	47.8	91.1

Reinf. Stress at Level

1 = 13.182 Ksi (Pullout controls..)
2 = 21.675 Ksi (Pullout Controls..)
3 = 30.168 Ksi (Pullout Controls..)
4 = 38.662 Ksi (Pullout Controls..)
5 = 43.885 Ksi (Yield Stress Controls)
6 = 43.885 Ksi (Yield Stress Controls)
7 = 43.885 Ksi (Yield Stress Controls)
8 = 43.885 Ksi (Yield Stress Controls)
9 = 43.885 Ksi (Yield Stress Controls)

Figure 5-42 Nail forces computed using SNAIL.

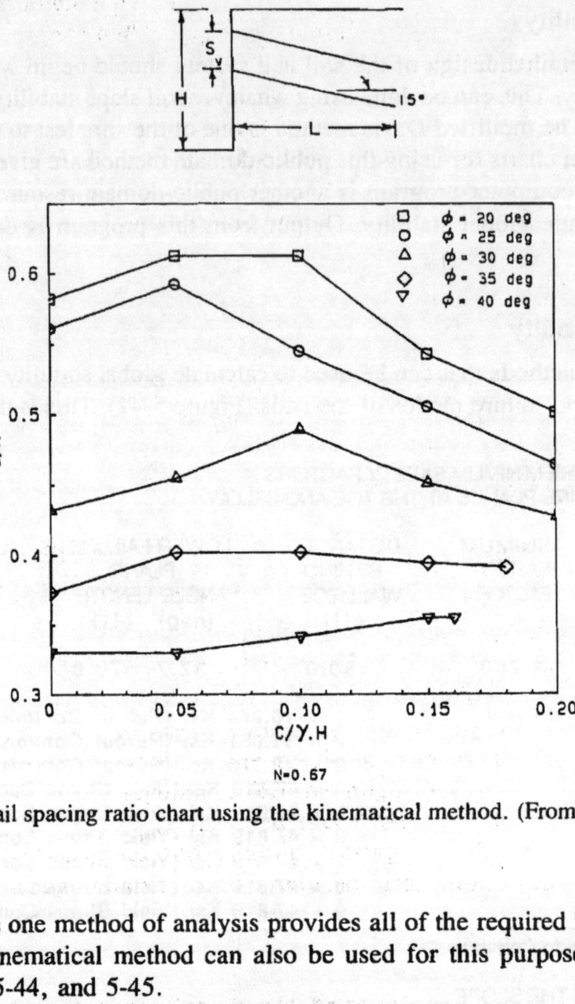

Figure 5-43 Nail spacing ratio chart using the kinematical method. (From Elias and Juran, 1991.)

situation where one method of analysis provides all of the required information for design. The kinematical method can also be used for this purpose, as shown in Figures 5-43, 5-44, and 5-45.

Example Problem 5-3 Soil Nail Design

Given: Problem given in Example Problem 5-1.
Required: Length (L), spacing (s), and size of nails (d).
Solution

Soil shear strength	$\phi' = 30°$
	$c_d = 150$ lb/ft²
Unit weight	$\gamma = 120$ lb/ft³
Ultimate bond stress	$F_1 = 4000$ lb/ft²

Assume nails will be drilled and grouted with a rig capable of drilling 6-in. diameter nail holes. Then, using empirical methods:

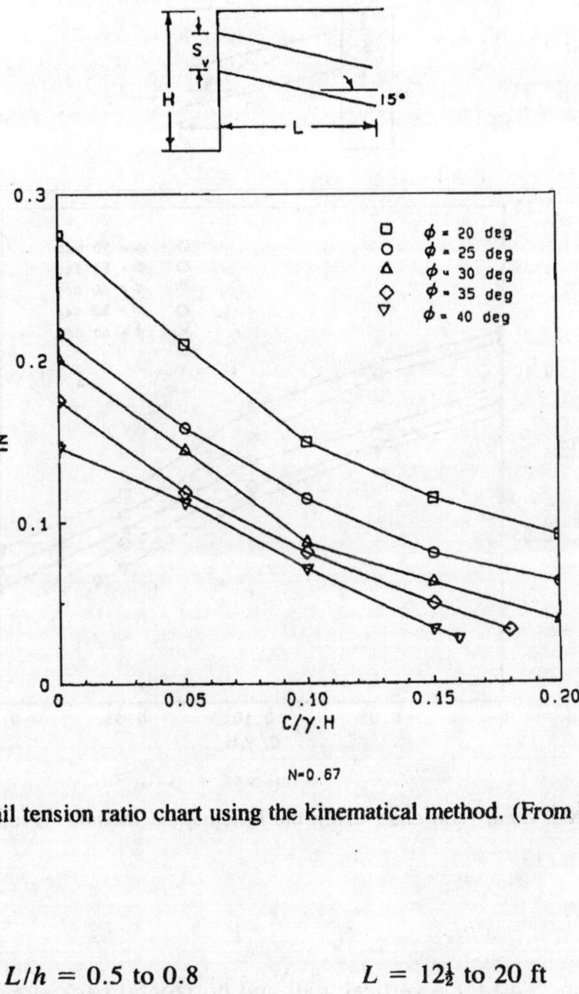

Figure 5-44 Nail tension ratio chart using the kinematical method. (From Elias and Juran, 1991.)

$$L/h = 0.5 \text{ to } 0.8 \qquad L = 12\tfrac{1}{2} \text{ to } 20 \text{ ft}$$

$$dL/s = 0.5 \text{ to } 0.6 \qquad s = 10 \text{ to } 20 \text{ ft}^2$$

$$d^2/s = 0.0004 \text{ to } 0.0008 \qquad d_{\text{bar}} = \tfrac{3}{4} \text{ to } 1\tfrac{1}{2} \text{ in. } \phi$$

Using the modified Davis method:

$$\phi_d = \frac{30°}{1.2} = 25°$$

$$F_{ld} = \frac{4000 \text{ lb/ft}^2}{2.0} = 2000 \text{ lb/ft}^2$$

$$\frac{C_d}{\gamma H} = \frac{150 \text{ lb/ft}^2}{120 \text{ lb/ft}^3 \times 25 \text{ ft.}} = 0.05$$

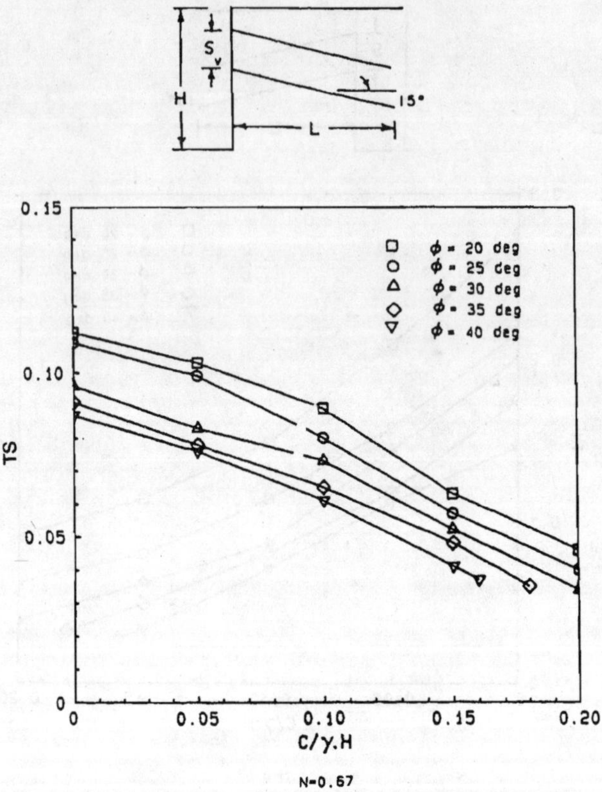

Figure 5-45 Nail shear ratio chart using the kinematical method. (From Elias and Juran, 1991.)

From Figure 5-40 for a vertical wall and horizontal back slope:

$$\frac{F_{ld}\, d}{\gamma\, S_h\, S_v} = \frac{2000\ \text{lb/ft}^2 \times \frac{1}{2}\ \text{ft}}{120\ \text{lb/ft}^2 \times 4\ \text{ft} \times 4\ \text{ft}} = 0.52$$

$$L/H = 0.53 \qquad L = 0.53 \times 25\ \text{ft} = 13\ \text{ft}$$

The following design seems reasonable:

nail length = 14 ft
spacing = 4 ft × 4 ft (S = 16 ft^2)
d_{bar} = 1.0 in. (No. 8 bars)

Check design using SNAIL.

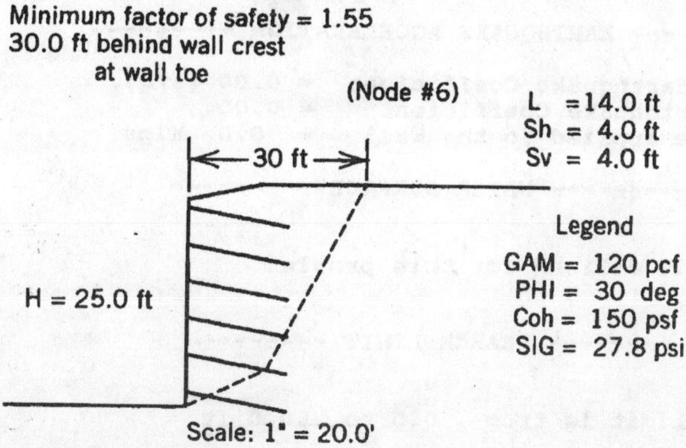

Press: N for new node. S for screen mode. Z for zoom.
R for results

```
*******************************************************
*      CALIFORNIA DEPARTMENT OF TRANSPORTATION        *
* DIVISION OF NEW TECHNOLOLY, MATERIAL & RESEARCH *
*        OFFICE OF GEOTECHNICAL ENGINEERING           *
* Reinforced Soil Systems & Earthquake Engineering*
*    Date: 02-27-1992              Time: 17:36:54     *
*******************************************************
```

```
--------- WALL GEOMETRY ---------

Vertical Wall Height                      =  25.00 ft
Wall Batter                               =   0.0 degrees
First Slope Angle-above the wall          =  10.0 degrees
First Slope Length from Wallcrest         =  10.00 ft
Second Slope Angle                        =   0.0 degrees
Second Slope Length from 1st Slope        = 100.00 ft
Third Slope Angle                         =   0.0 degrees
Third Slope Length from 2nd Slope         =   0.00 ft
Fourth Slope Angle                        =   0.0 degrees

--------- SLOPE BELOW THE WALL ---------

There is NO SLOPE BELOW THE TOE of the wall

--------- SURCHARGE ---------

There is NO SURCHARGE imposed on the system.

--------- SOIL PARAMETERS ---------

Unit Weight, GAM                    = 120.00 pcf
Friction Angle, PHI                 =  30.0  degrees
Cohesion, COH                       =  150.0 psf
Bond Stress, SIG                    =  27.80 psi
```

--------- EARTHQUAKE ACCELERATION ---------

Horizontal Earthquake Coefficient = 0.00 (a/g)
Vertical Earthquake Coefficient = 0.00
Horiz. Force applied to the Wall = 0.0 Kips

--------- WATER SURFACE ---------

NO Water Table defined for this problem.

--------- SEARCH LIMIT ---------

The Search Limit is from 0.0 to 50.0 ft

You have chosen TO LIMIT the search to the following nodes:

BEGIN search at node - 1
END search at node - 10

--------- REINFORCEMENT PARAMETERS ---------

Number of Reinforcement Levels = 6
Horizontal Spacing = 4.00 ft
Diameter of Reinforcement Element = 1.00 in
Yield Stress of Reinforcement = 60.00 ksi
Diameter of Grouted Hole = 6.00 in
Punching Shear Capacity = 45.00 kips

--------- (For ALL Levels) ---------

Reinforcement Lengths = 14.00 ft
Reinforcement Inclination = 6.0 degrees
Vertical Spacing to First Level = 2.00 ft
Vertical Spacing to Remaining Levels = 4.00 ft

DISTANCE ALONG TOE SLOPE (ft)	MINIMUM SAFETY FACTOR	DISTANCE BEHIND WALL TOE (ft)	LOWER FAILURE PLANE ANGLE (deg)	LOWER FAILURE PLANE LENGTH (ft)	UPPER FAILURE PLANE ANGLE (deg)	UPPER FAILURE PLANE LENGTH (ft)
Toe	2.898	5.0	74.6	18.8	89.9	7.8

Reinf. Stress at Level 1 = 20.702 Ksi (Yield Stress controls.)
2 = 20.702 Ksi (Yield Stress controls.)
3 = 20.702 Ksi (Yield Stress controls.)
4 = 20.702 Ksi (Yield Stress controls.)
5 = 20.702 Ksi (Yield Stress controls.)
6 = 20.702 Ksi (Yield Stress controls.)

DISTANCE ALONG TOE SLOPE (ft)	MINIMUM SAFETY FACTOR	DISTANCE BEHIND WALL TOE (ft)	LOWER FAILURE PLANE ANGLE (deg)	LOWER FAILURE PLANE LENGTH (ft)	UPPER FAILURE PLANE ANGLE (deg)	UPPER FAILURE PLANE LENGTH (ft)
NODE 2	2.458	10.0	56.1	16.1	85.7	13.4

```
Reinf. Stress at Level  1 = 14.011 Ksi (Pullout controls...)
                        2 = 14.983 Ksi (Pullout controls...)
                        3 = 15.955 Ksi (Pullout controls...)
                        4 = 22.946 Ksi (Pullout controls...)
                        5 = 24.406 Ksi (Yield Stress controls.)
                        6 = 24.406 Ksi (Yield Stress controls.)
```

DISTANCE ALONG TOE SLOPE (ft)	MINIMUM SAFETY FACTOR	DISTANCE BEHIND WALL TOE (ft)	LOWER FAILURE PLANE ANGLE (deg)	LOWER FAILURE PLANE LENGTH (ft)	UPPER FAILURE PLANE ANGLE (deg)	UPPER FAILURE PLANE LENGTH (ft)
NODE 3	1.943	15.0	48.1	18.0	77.4	13.7

```
Reinf. Stress at Level  1 =  0.365 Ksi (Pullout controls...)
                        2 =  3.999 Ksi (Pullout controls...)
                        3 =  7.633 Ksi (Pullout controls...)
                        4 = 20.306 Ksi (Pullout controls...)
                        5 = 30.882 Ksi (Yield Stress controls.)
                        6 = 30.882 Ksi (Yield Stress controls.)
```

DISTANCE ALONG TOE SLOPE (ft)	MINIMUM SAFETY FACTOR	DISTANCE BEHIND WALL TOE (ft)	LOWER FAILURE PLANE ANGLE (deg)	LOWER FAILURE PLANE LENGTH (ft)	UPPER FAILURE PLANE ANGLE (deg)	UPPER FAILURE PLANE LENGTH (ft)
NODE 4	1.683	20.0	33.8	14.4	66.9	20.4

```
Reinf. Stress at Level  1 =  0.000 Ksi
                        2 =  0.000 Ksi
                        3 =  0.000 Ksi
                        4 =  5.838 Ksi (Pullout controls...)
                        5 = 23.310 Ksi (Pullout controls...)
                        6 = 35.646 Ksi (Yield Stress controls.)
```

DISTANCE ALONG TOE SLOPE (ft)	MINIMUM SAFETY FACTOR	DISTANCE BEHIND WALL TOE (ft)	LOWER FAILURE PLANE ANGLE (deg)	LOWER FAILURE PLANE LENGTH (ft)	UPPER FAILURE PLANE ANGLE (deg)	UPPER FAILURE PLANE LENGTH (ft)
NODE 5	1.572	25.0	28.1	11.3	55.0	26.1

```
Reinf. Stress at Level  1 =  0.000 Ksi
                        2 =  0.000 Ksi
                        3 =  0.000 Ksi
                        4 =  4.701 Ksi (Pullout controls...)
                        5 = 18.083 Ksi (Pullout controls...)
                        6 = 38.175 Ksi (Yield Stress controls.)
```

DISTANCE ALONG TOE SLOPE (ft)	MINIMUM SAFETY FACTOR	DISTANCE BEHIND WALL TOE (ft)	LOWER FAILURE PLANE ANGLE (deg)	LOWER FAILURE PLANE LENGTH (ft)	UPPER FAILURE PLANE ANGLE (deg)	UPPER FAILURE PLANE LENGTH (ft)
NODE 6	1.550	30.0	30.7	10.5	45.5	30.0

```
Reinf. Stress at Level  1 =  0.000 Ksi
                        2 =  0.000 Ksi
                        3 =  0.000 Ksi
                        4 =  3.814 Ksi (Pullout controls...)
                        5 = 22.303 Ksi (Pullout controls...)
                        6 = 38.708 Ksi (Yield Stress controls.)
```

DISTANCE ALONG TOE SLOPE (ft)	MINIMUM SAFETY FACTOR	DISTANCE BEHIND WALL TOE (ft)	LOWER FAILURE PLANE ANGLE (deg)	LOWER FAILURE PLANE LENGTH (ft)	UPPER FAILURE PLANE ANGLE (deg)	UPPER FAILURE PLANE LENGTH (ft)
NODE 7	1.559	35.0	27.0	11.8	41.1	32.5

```
Reinf. Stress at Level  1 =  0.000 Ksi
                        2 =  0.000 Ksi
                        3 =  0.000 Ksi
                        4 =  0.000 Ksi
                        5 = 14.776 Ksi (Pullout controls...)
                        6 = 38.479 Ksi (Yield Stress controls.)
```

DISTANCE ALONG TOE SLOPE (ft)	MINIMUM SAFETY FACTOR	DISTANCE BEHIND WALL TOE (ft)	LOWER FAILURE PLANE ANGLE (deg)	LOWER FAILURE PLANE LENGTH (ft)	UPPER FAILURE PLANE ANGLE (deg)	UPPER FAILURE PLANE LENGTH (ft)
NODE 8	1.609	40.0	24.0	13.1	37.4	35.2

```
Reinf. Stress at Level  1 =  0.000 Ksi
                        2 =  0.000 Ksi
                        3 =  0.000 Ksi
                        4 =  0.000 Ksi
                        5 =  7.367 Ksi (Pullout controls...)
                        6 = 37.281 Ksi (Yield Stress controls.)
```

DISTANCE ALONG TOE SLOPE (ft)	MINIMUM SAFETY FACTOR	DISTANCE BEHIND WALL TOE (ft)	LOWER FAILURE PLANE ANGLE (deg)	LOWER FAILURE PLANE LENGTH (ft)	UPPER FAILURE PLANE ANGLE (deg)	UPPER FAILURE PLANE LENGTH (ft)
NODE 9	1.668	45.0	21.6	14.5	34.2	38.1

```
Reinf. Stress at Level  1 =  0.000 Ksi
                        2 =  0.000 Ksi
                        3 =  0.000 Ksi
                        4 =  0.000 Ksi
                        5 =  0.604 Ksi (Pullout controls...)
                        6 = 35.973 Ksi (Yield Stress controls.)
```

DISTANCE ALONG TOE SLOPE (ft)	MINIMUM SAFETY FACTOR	DISTANCE BEHIND WALL TOE (ft)	LOWER FAILURE PLANE		UPPER FAILURE PLANE	
			ANGLE (deg)	LENGTH (ft)	ANGLE (deg)	LENGTH (ft)
NODE 10	1.744	50.0	15.0	10.4	31.0	46.7

```
Reinf. Stress at Level 1 =  0.000 Ksi
                       2 =  0.000 Ksi
                       3 =  0.000 Ksi
                       4 =  0.000 Ksi
                       5 =  0.000 Ksi
                       6 = 27.089 Ksi (Pullout controls...)
```

* FOLLOWING IS LOCATION WHERE MAX. REINFORCEMENT FORCE OCCURS *

LOWER FAILURE PLANE		UPPER FAILURE PLANE		DISTANCE BEHIND THE WALL TOE (ft)	FACTOR OF SAFETY F.S.
ANGLE (deg)	LENGTH (ft)	ANGLE (deg)	LENGTH (ft)		
0.0	20.0	89.9	26.7	20.0	$999.000

Maximum Average Reinforcement Working Force = 8.066 kips/level

Corrosion Protection

Corrosion of loaded steel members is a serious concern in any structural system. Much has been learned over the years from related civil engineering applications including culverts, buried utilities, tunnels, pile foundations, bridges, marine structures, and water resource projects. Corrosion of steel is an electrochemical process. For it to occur, there must be a potential difference between two points, an anode and a cathode, that are electrically connected and immersed in an electrolyte. The anode is the point that has the most negative potential and is the area that becomes corroded through loss of metal ions to the electrolyte. Correlation of this theory with actual or potential of metals underground is complicated and difficult because of the many factors that singly or in combination affect the course of the electrochemical reaction. These factors not only determine the rate and amount of corrosion that occurs, but also the kind of corrosion.

Galvanic and stray current corrosion are two broad categories of corrosion. Galvanic corrosion results when a galvanic current originates between dissimilar metals or differential electrolytic concentrations. Stray currents are present in the soil as a result of electrical leaks, or failure to provide positive and permanent electrical grounding. In these systems, a portion of the current leaks into a structure and then discharges back into the soil at another point some distance away.

Steels are subject to five types of corrosion: uniform surface corrosion, pitting, bacterial corrosion, stress induced corrosion, and embrittlement corrosion. Uniform surface corrosion is caused by electrochemical reactions that proceed uniformly over the entire metal surface. Pitting corrosion of a metal is the result of intense localized attack in an electrolyte. Once initiated, the corrosion process within the pit produces a condition stimulating and sustaining further corrosion. Reduction of steel area at a pit, if allowed to continue, will lead to ductile failure of a stressed member. Bacterial corrosion is caused by the presence of sulfate-reducing anaerobic bacteria in deaerated soils such as wet clays, marshes, and organic soils below the groundwater table. Stress corrosion is a brittle failure that occurs when a normally ductile metal or alloy is subjected to tensile stresses above a threshold level in the presence of specific corrosive environments. Hydrogen embrittlement occurs when atomic hydrogen resulting from a corrosion reaction or cathodic polarization enters the metal lattice at cathodic sites, and upon reaching a void of favorable site, combines to form molecular hydrogen. The interstitial molecular hydrogen significantly reduces the ductility of the metal.

The type of corrosion protection chosen for a particular project will depend upon the:

1. Aggressiveness of the environment.
2. Relative costs of the various protection systems.
3. Bar type.
4. Installation methods.
5. Risk associated with failure.
6. Contractors' patents.

The corrosivity of the ground should be evaluated carefully during design. Soil corrosivity is influenced by:

1. The resistivity of the soil.
2. The pH of the soil.
3. The chemical composition of the groundwater and soil.
4. The permeability of the soil.
5. The groundwater table elevation and variations.
6. External electrochemical and physical factors such as stray currents.

Some requirements for soil nail corrosion protection systems include the following:

1. Effective life of the nails and corrosion protection should be at least equal to that required for the facility.
2. Corrosion protection should not have adverse effects on the environment.
3. The materials used should be mutually compatible with respect to deformability, permanence, and corrosion.
4. The materials used should be tough enough to withstand handling damage during manufacture, transport, storage, and installation.

Guidelines for tieback construction, especially the bonded zone of a tieback, are applicable to soil nails. For tiebacks as well as soil nails, corrosion protection should include the following considerations (Cheney, 1988):

- Full cement grout encapsulation around the nails ½ to 2 in. thick, minimum.
- Bar centralizers to ensure grout coverage.
- Topping off of the anchor head with cement mortar dry-pack.
- Epoxy-resin coated bars and maximum grout coverage for permanent installations and very aggressive environments.
- Avoidance of quenched and tempered prestressing steels.
- Avoidance of blast furnace slag fill, moist low pH soil, light ash industrial environments, and organic soils.

It is difficult or impossible to provide adequate corrosion protection with jet grouted and driven nails for permanent use.

Cement encapsulation of a soil nail for corrosion protection should be adequate if:

1. The tendon is electrically insulated from the structure.
2. The nail is in oxygen deficient ground.
3. The pH of the ground is greater than 4.5.
4. The resistivity of the soil is greater than 2000 ohm-centimeters.
5. Sulfides are not present in the soil surrounding the nail.

If any of these conditions are known not to exist, additional encapsulation using plastic pipes or sheaths may be required. Plastic acts as an additional barrier preventing:

1. The nail from being depassivated.
2. Sulfides from reaching the nail.
3. Oxygen from reaching the nail and creating or maintaining a local corrosion system in low resistivity soils.
4. Acid attack of the nail in low pH environments.
5. Long-line and stray-current corrosion systems.

Facing

Reinforced shotcrete facing can be designed using conventional reinforced concrete design methods for a two-way slab on an elastic foundation. Using American Concrete Institute (ACI) working stress design procedures, 4000-lb/in.^{2} shotcrete, and 60,000-lb/in.^{2} steel, the approximate thickness of the shotcrete, d, can be calculated according to the following formula:

$$d = [0.2\ T\ L]^{\frac{1}{2}} \tag{5-19}$$

where d = shotcrete thickness, in.
T = maximum nail force, kips
L = nail spacing, ft

Most often shotcrete is placed in two 2- to 4-in.-thick layers. The initial layer is applied immediately after the soil is exposed by excavation. The second layer is applied after the nails are installed. Two inches of shotcrete is the usual minimum cover over the nail heads.

The approximate amount of reinforcing steel in the shotcrete in the inside and outside faces, A_s, can be computed for each face as:

$$A_s = 0.0052\, d \text{ (in.}^2) \qquad (5\text{-}20)$$

About 2 in. of shotcrete cover should be provided for the steel on both faces. For shotcrete thicknesses between 3 and 8 in., plain wire (ASTM A82) sizes of W1 and W2.5, respectively, would be acceptable on 6-in. centers. Welded wire fabric (WWF) is often used. A common size would be specified as 6X6, W2/W2.

Example Problem 5-4 Wall Facing Design

Given: Problem given in Example Problem 5-1.
Required: Shotcrete thickness and reinforcing.
Solution: Maximum nail force:

$$T = \Pi\, d\, LF_1 = 88 \text{ kips}$$
$$\text{nail spacing} = 4 \text{ ft}$$

Shotcrete thickness:

$$d = \sqrt{0.2 \times 88 \times 4} = 8.4 \text{ in.}$$

Say two 4-in. layers.
Reinforcing:

$$A_S = 0.0052 \times 8.4 = 0.04 \text{ in.}^2$$

Use 6 × 6 in. WWF.
Required wire area: 0.02 in.^{2} or W 1.4 × W 1.4 (0.028 in.^{2})

Drainage

The primary methods of draining soil nailed walls are strip drains embedded in the shotcrete and weeps (Figures 5-46 and 5-47). The strip drains should be fabricated from suitably permeable geosynthetic or geocomposite materials. The strips should be no wider than one foot and interconnected to the extent possible. Manufacturers of such products are excellent sources of additional information. The weeps are

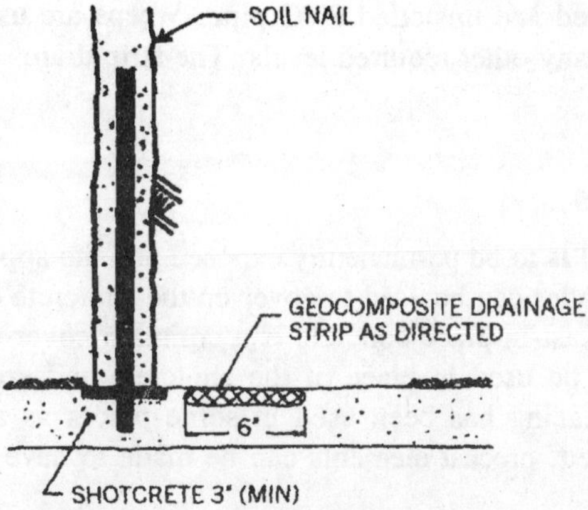

Figure 5-46 Typical strip drain detail.

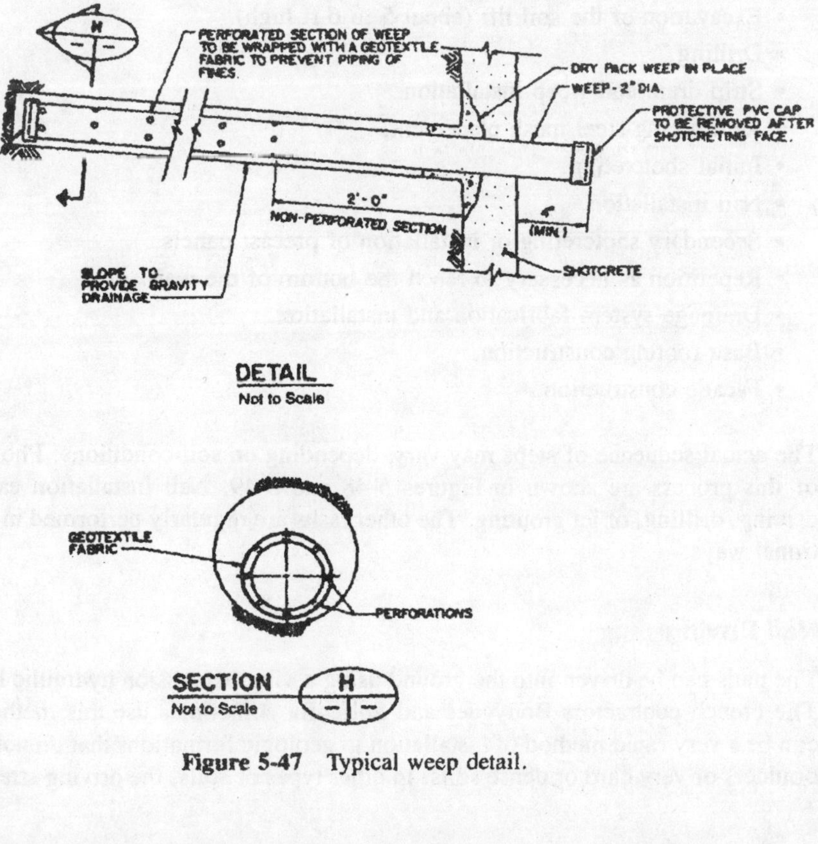

Figure 5-47 Typical weep detail.

fabricated from slotted and unslotted PVC pipe. Weeps are usually located at the base of the wall and any other required levels. The strip drains can be connected to the weeps.

Aesthetic Facades

If the soil nailed wall is to be permanently exposed and the appearance of shotcrete is objectionable, facades can be used to cover up the shotcrete or can even take the place of shotcrete. Cast-in-place concrete is a common cover-up material. Precast concrete panels can be used in place of the shotcrete and erected as the wall is constructed. Metal facing has been used in some places as a facade. If a more natural look is desired, precast elements can be made to have plantable boxes for vegetation.

5-8 CONSTRUCTION METHODS

The primary tasks of soil nail wall construction in the usual "top-down" sequence are:

- Excavation of the soil lift (about 5 to 6 ft high).
- Drilling.
- Strip drain and weep installation.
- Reinforcing steel mesh placement.
- Initial shotcreting.
- Nail installation.
- Secondary shotcreting or installation of precast panels.
- Repetition as necessary to reach the bottom of the wall.
- Drainage system fabrication and installation.
- Base footing construction.
- Facade construction.

The actual sequence of steps may vary, depending on soil conditions. Photographs of this process are shown in Figures 5-48 and 5-49. Nail installation can be by driving, drilling, or jet grouting. The other tasks are regularly performed in conventional ways.

Nail Driving

The nails can be driven into the ground using a vibropercussion hydraulic hammer. The French contractors Bouygues and Solrenfor sometimes use this method. This can be a very rapid method of installation in geologic formations that do not contain boulders or very hard or dense soils. In other types of soils, the driving stresses and

Figure 5-48 Construction of the Cumberland Gap wall.

vibrations can weaken the soils. The newest driving method, used by Louis (1984), employs a system that can grout the nail at high pressures while simultaneously driving it into the ground. The grout lubricates the nail as it is driven, improving drivability as well as the character of the soil near the nail.

Nail Drilling

The most common method of nail installation is by rotary drilling a 4- to 6-in.-diameter hole. The two most common drilling methods are auger drilling and duplex drilling, although percussive drilling with an air-track drill and other methods are possible. Auger drilling must be with a hollow stem auger if the soil is not strong enough to stand in the hole while the drill string is withdrawn and the nail is inserted and grouted. Even with hollow stem augers, care must be taken to avoid cohesionless soil from running in the hole and causing overexcavation, ground loss, and possibly settlement.

Duplex drilling consists of advancing an outer casing simultaneously with an inner drill string. This is a high-quality method, because disturbance to the in situ soil is minimized and the outer casing provides support to the drill hole until grouting is started and the casing is simultaneously removed. Drilling fluid is usually used and can consist of water or air. The use of bentonite is not recommended because of the possibility of leaving a smear in the hole, which can diminish the load transfer capacity between the grout and the soil.

Nail Materials

The nails are usually steel reinforcing bars or Dywidag bars. Common nail sizes range between $\frac{3}{4}$ and $1\frac{1}{4}$ in. in diameter. Couplers are discouraged but are often needed for nail lengths greater than about 35 ft. Plastic centralizers should be

Figure 5-49 Construction of the Interstate Route H-3 tunnel portals.

mounted to the bars at a maximum spacing of 10 ft. Epoxy coatings should be between 14 and 18 mils thick. Samples of the bars should be provided and tested to document quality and strength. The bars should be free of rust.

Grouting

Grouting should be carried out by starting at the bottom of the hole. The grout is introduced into the hole by gravity or very low pressures. The neat cement or sand cement grout should have a minimum cube strength of 3000 lb/in.^{2} A water-cement ratio of 0.4 to 0.5 is common. The grout pour should be continuous without interruption. Grout cubes are made to document the grout strength. Some contractors like to use the same mix for grouting as for shotcreting. This should be discouraged because the shotcrete mix does not flow around the nail centralizers well enough, possibly leaving voids in the grout column.

Corrosion Protection

For corrosion protection of soil nails, some methods utilized for permanent tiebacks should be considered. A soil nail is similar to the bonded zone of a tieback. A nail does not have a stressing or unbonded zone. Cheney (1988) provides relevant information regarding corrosion protection of permanent tiebacks. The important factors that can cause galvanic corrosion of steel, as cited by Cheney, are:

- Groundwater quality and flow rate
- Ground temperature
- Stress condition of the steel

Extremely acidic, alkaline, or salty groundwater promotes corrosion. High rates of groundwater flow promote corrosion. High ground temperatures promote corrosion. High stresses promote corrosion. Testing of the soil and groundwater should be conducted at the site to determine whether extremely corrosive conditions exist.

For corrosion protection, it is important that epoxy-coated steel bars are used and that there is a minimum of $1\frac{1}{2}$ in. of cement grout around the steel bar. The centralizers should insure this and counteract the weight of the bar, which makes it want to sit on the bottom of the hole in direct contact with the soil. Care in handling of epoxy-coated bars should avoid unusually harsh treatment of the bar coating and possible scratching of the bar. If the epoxy coating is noticeably damaged, it should be patched in the field or the entire bar replaced. Elias and Juran (1991) recommend the following requirements for most applications:

> A minimum grout cover of 1.5 inches shall be achieved throughout the grout zone. Centralizers must be placed at distances not exceeding 10 feet center to center, and the lowest spacer located a maximum of 1 foot from the bottom of the grouted drill hole. Centralizers should be made from a plastic material and not impede the free flow of grout. Centralizer diameter shall not be less than 1 inch smaller than the hole diameter.
>
> The nail structural section shall be fusion bonded epoxied, using an electrostatic process, to minimum thicknesses of 14 mils. This protection scheme for permanent routine structures could be increased to full encapsulation of the nail for those applications in aggressive regimes previously identified where field observations have indicated corrosion of similar structures or where field tests . . . have identified aggressive regimes. Full encapsulation is generally accomplished as with tiebacks, by a uniform corrugated plastic or metal tube. This tube must be capable of withstanding deformations associated with transportation, installation and passive stressing of the nail and transferring the load applied to the bar.

For fully encapsulated nails, the outer grout cover in most cases can be reduced to $\frac{1}{2}$ in. and epoxy coating is probably not necessary.

It is difficult to completely grout the drill hole. Because of the angle below the horizontal, there will be an ungrouted portion left at the top of the hole where the grout reaches the bottom of the hole collar. This portion of the hole should be dry-packed with a thick neat cement mortar so that the entire hole is completely filled with cementitious material. This part of the installation has a high probability of corrosion if not properly protected.

Shotcreting

Shotcreting is a specialized method of applying concrete to a soil face using a high-pressure spray. It can be premixed with water (wet mix) or combined with water at the nozzle (dry mix). Accelerator is usually added to the mixture at the nozzle. It can be batched at the site, delivered from a concrete plant in ready-mix trucks, or can be supplied in 94-lb bags. Competence of the nozzle operator should be demonstrated and is essential for a good installation. The shotcrete mix, equipment, and nozzle operators should be chosen to maximize early shotcrete strength, good homogeneity (i.e., no laminations), and minimum rebound.

Fiber reinforced shotcrete can be used in place of plain shotcrete reinforced with steel mesh. This should be an economic decision made by the designer or an option decided by the contractor. Fibers increase wear and tear on the equipment but save labor time and money by eliminating the need to hang steel mesh. The surface of fiber reinforced shotcrete can be sharper but this can be mitigated by a plain surface coat of shotcrete on top of the fiber shotcrete. The long-term corrosion potential of steel fibers is debatable and should be considered on a case by case basis.

Some common shotcrete mix parameters include:

- Minimum compressive strength:
  - 2600 lb/in.^{2} in 7 days
  - 4000 lb/in.^{2} in 28 days
- Maximum water-cement ratio = 0.4
- Air content = 4 to 8 percent
- Slump = 2 to 6 in.

Since it is difficult to make conventional concrete cylinder test samples with shotcreting equipment, quality control and field testing is usually done by laboratory trial mixes, field test panels, and test cores. The mixes tested should contain all admixtures planned to be used.

Construction joints in the shotcrete are necessary between vertical excavation lifts and other stoppages of work. The joints should be tapered to a thin edge ("feathered"). Before applying additional shotcrete, the joint should be wetted and cleared of any rebound, overspray, dirt, and debris.

Drainage

Strip drains must be installed from the top down. Continuity between strip segments must be provided. The strip drains must be securely held in place during shotcreting to avoid the creation of voids and laminations. A suitable connection to weeps and discharge pipes must be provided. Wet spots that show up in the shotcrete after construction should be relieved with weep drains.

5-9 GEOTECHNICAL INVESTIGATION AND TESTING

For soil nailed walls, a combination of conventional and specialized geotechnical investigation and testing is necessary. Conventional exploration (i.e., boring, sampling, and laboratory soil testing) is needed to characterize the soil deposit with respect to:

- Geologic origin and stratification
- Soil density and shear strength
- Groundwater level and flow regime
- Presence of running sands
- Presence of boulders and other geologic anomalies

Specialized testing is needed primarily to determine the ultimate pull-out resistance of prototype nails.

Geologic Exploration

A conventional boring program should be planned such that there are enough borings to construct a geologic profile along the entire wall alignment. Variations in geologic conditions perpendicular to the wall should also be ascertained. The borings should extend a minimum of 10 ft below the proposed wall foundation bottom and deeper if geologic conditions suggest the need for evaluation of weak soils at deeper levels.

In situ testing should include standard penetration tests (SPTs) at 5-ft intervals as a minimum. For clayey soils, SPT blow counts are not very reliable and other in situ testing methods may be warranted such as pressuremeter, vane shear, or cone penetrometer testing.

Representative samples should be obtained for field classification of the soils, pocket penetrometer or pocket shear vane testing, and running laboratory tests. Relatively undisturbed samples should be obtained for laboratory testing, if soil conditions permit, and should be sealed, stored, and transported to the laboratory in such a way as to minimize further disturbance.

If soil conditions permit, test pits can be excavated at the site to provide more representation of soil stratification and content, water inflow rates, stand-up time for excavated materials, and bulk samples. Personnel should not enter test pits greater than 5 ft deep unless shoring is in place. Test pits are usually constructed with a moderately sized backhoe machine, which can be rented with its operator for about $500/day. At an average rate of progress, between about 10 and 15 test pits can be excavated during an 8-hr day. Test pits are therefore a quick and economical way to get additional geologic information.

Laboratory Testing

Laboratory testing should be conducted to determine the following soil properties:

- Water content
- Unit weight
- Grain size distribution
- Atterberg limits
- Shear strength
- Corrosivity (resistivity, pH, sulfates, chlorides)

Soil shear strength can be evaluated with results from direct shear and triaxial shear tests.

Additional laboratory testing is normally not conducted for most projects but can include direct shear tests of the soil/grout interface and scale-model nail pull-out tests with a large shear box. Other less common tests include centrifuge modeling and direct shear tests perpendicular to prototype nails.

Field Testing

The best field test that can be conducted for a soil nailing project is a field pull-out test of prototype nails. It is desirable to conduct this test so that the results can be used during design. But most often it is not practical to mobilize the necessary equipment during design, and testing is conducted at the beginning of construction as a confirmation of design values. The tests should be taken to ultimate failure and results should be presented as load versus deflection plots. If the soils are possibly susceptible to creep, tests should be conducted to determine the critical creep tension value (Weatherby, 1982).

Elias and Juran (1991) have collected published nail pull-out values for comparison purposes. The ultimate pull-out resistance of a soil nail equals the ultimate friction value multiplied by the surface area of the grouted nail drill hole. The ultimate friction value is related to the angle of internal friction and overburden values in cohesionless soils and the cohesion value and alpha factor in cohesive soils. Ultimate friction value ranges given by Elias and Juran (1991) for rotary drilled and grouted nails in kips per square foot are:

Fill	0.4 to 0.6 kip/ft^2
Clay/silty clay	0.4 to 1.2 kips/ft^2
Sandy clay	4.0 to 6.0 kips/ft^2
Silt/loess	0.5 to 2.0 kips/ft^2
Sand with and without fines	1.7 to 5.0 kips/ft^2
Sand and gravel	6.0 to 9.0 kips/ft^2

However, values from site to site and driller to driller can vary significantly. That is why it is best to conduct representative field tests for prototypic conditions at each project site using the proposed in situ soils, nail materials, and drilling equipment.

5-10 PERFORMANCE MONITORING

It is desirable to monitor the performance of soil nailed walls in view of the uncertainties in design and construction. Of greatest interest are the global stability, as evidenced by wall and soil movements, and internal stability, as evidenced by the mobilized forces in the nails. Optical survey techniques and inclinometers are often used to monitor wall and soil movements. Strain gages and load cells are often used to monitor nail forces. Earth pressure cells can be used to monitor pressures on the wall facing.

Optical Survey Techniques

Optical survey techniques can be used to monitor vertical and horizontal movements of the wall and soil. Survey points should be installed at the ground surface above the proposed wall construction and on the wall facing as it is constructed. Deep settlement markers can be installed in boreholes within the soil mass for additional data. The coordinates and elevations of the points should be established as soon as possible and monitored at frequent and regular intervals during construction. Monitoring should continue after construction for as long as possible.

Inclinometers

Reading inclinometer casings are an effective way to monitor lateral deflections of the wall and soil mass. The inclinometers are usually installed by grouting the casing in a borehole drilled to a depth of at least 10 or 20 ft below the planned bottom of the wall. The time consuming process of reading inclinometers is accomplished by lowering a specially designed probe to the bottom of the hole and taking readings in 2-ft increments to the top of the hole. The inclinometer measurements yield a continuous profile of lateral movements with depth.

Strain Gages

Strain gages can be mounted to the nail bars at regular intervals to sense the strain distribution in the bar. Nail forces can be calculated from the strains if the modulus and the area of the bar are known. Many practitioners prefer vibrating wire type strain gages for this purpose (Abramson and Green, 1985) because of their reliability and resistance to damage and zero shift due to water ingress.

A nail installation with multiple strain gage mountings will have several wire attachments, and special care is required during installation to prevent damage to the gages and wiring. Also, facilities for reading the gages must be provided including waterproof junction boxes, read-out devices, and telephonic connections if remote monitoring is desired. Automatic data loggers and scanners can dramatically simplify the strain gage monitoring process.

Load Cells

Load cells can be used to monitor the loads in the nails at the wall facing. This will provide information related to wall facing loads, and hence pressure at the nail location, but will not yield information about maximum load in the bar and load distribution. Electrical resistance strain gaged load cells are common but vibrating wire and pneumatic load cells are also available. A high-quality data logger can be used to record load cell readings as well as the strain gage readings. Load cells on the Cumberland Gap wall alerted the Federal Highway Administration to the very high freeze/thaw pressures that the wall facing was experiencing (Juran and Elias, 1987).

Earth Pressure Cells

Earth pressure cells have been used in the past to monitor earth pressures within embankments and at the facing of walls (Abramson, 1983). This technique is highly unreliable, because it is extremely difficult, if not impossible, to install the earth pressure cells so they sense the actual earth pressure in the ground without causing arching and altering the state of stress. The use of earth pressure cells behind the face of a soil nailed wall should only be considered for very unique studies and should be well backed up with load cell and deflection measurements.

5-11 CASE HISTORIES

Case histories are not yet abundant in the literature. Most applications have been for temporary support of commercial excavations, which are not typically written up for publication. Only recently have applications extended to transportation structures and public facilities; detailed accounts of such applications are more widely published. Some of the case histories found in the literature are summarized below.

Nails Used in Stuttgart in Place of Tieback Anchors

Soil nailing was used for a 50-ft-deep excavation at the corner of a sloping site that was bordered by a city street on one side and by residential structures on the other, as depicted in Figure 5-50 (Stocker and Riedinger, 1990). The installation of long tieback anchors underneath these properties was not permitted. The soil consisted of about 2 to 6 ft of silt, sand, and cinder fill, underlain by about 26 ft of medium to stiff clayey, sandy silt and gravel. The overburden soils were underlain by Keuper Marl consisting of alternating layers of siltstone and claystone. The unit weights for the fill, silt, and marl were 122, 128, and 134 lb/ft^3, respectively. The angle of

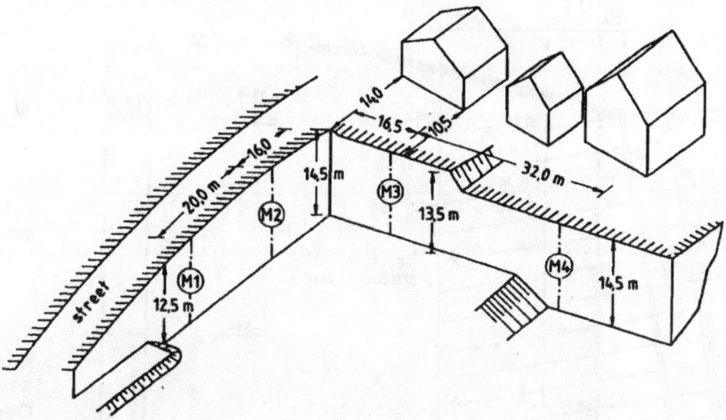

Figure 5-50 Stuttgart building site plan. (From Stocker and Riedinger, 1990. Reproduced by permission of ASCE.)

internal friction for the fill was 30°. The silt had an angle of internal friction of 27.5° and a cohesion value of between about 100 and 200 lb/ft². The marl had an angle of internal friction of 23° and a cohesion value greater than 1000 lb/ft².

The top two rows of nails consisted of 20-ft-long, $\frac{7}{8}$-in.-diameter deformed steel bars (Figure 5-51). The other rows consisted of 26-ft-long, 1-in.-diameter bars. The yield and ultimate strengths of the bars was about 60,000 and 72,000 lb/in.², respectively. Corrosion protection consisted of $\frac{1}{25}$-in.-thick PVC sleeves cement grouted to the entire length of the bars. The wall facing consisted of 10-in.-thick, steel-mesh-reinforced shotcrete. Excavation was carried out in 3- to $3\frac{1}{2}$-ft-high lifts.

The excavation support was instrumented and monitored with slope indicators, load cells, extensometers, and strain gages. Five percent of the nails were load tested to 45,000 lb. A plot of maximum nail forces after excavation indicated a potential failure zone such as that shown in Figure 5-52. Long-term monitoring over a period of 10 years indicated a significant load increase when the daily minimum temperature at ground surface fell below freezing. Maximum horizontal deformation as a function of excavation depth varied between 0.1 and 0.36 percent. Additional amounts of deformation over time ranged between 0.06 and 0.15 percent, probably due to continuing construction and loads from the new building. Deformations generally ceased by the end of the third year of monitoring. Deformations within the ground mass decreased with increasing distance from the wall face (Figure 5-53). The earth pressure distribution observed is more or less rectangular with a maximum of about 1000 lb/ft² (Figure 5-54).

Soil Nailing in Seattle Glacial Soils

Thompson and Miller (1990) reported on the use of soil nailing to support a Seattle area excavation in glacial soils. Two sides of a building excavation were temporarily shored with soil nails. The walls were 35 and 55 ft high and adjacent to city streets.

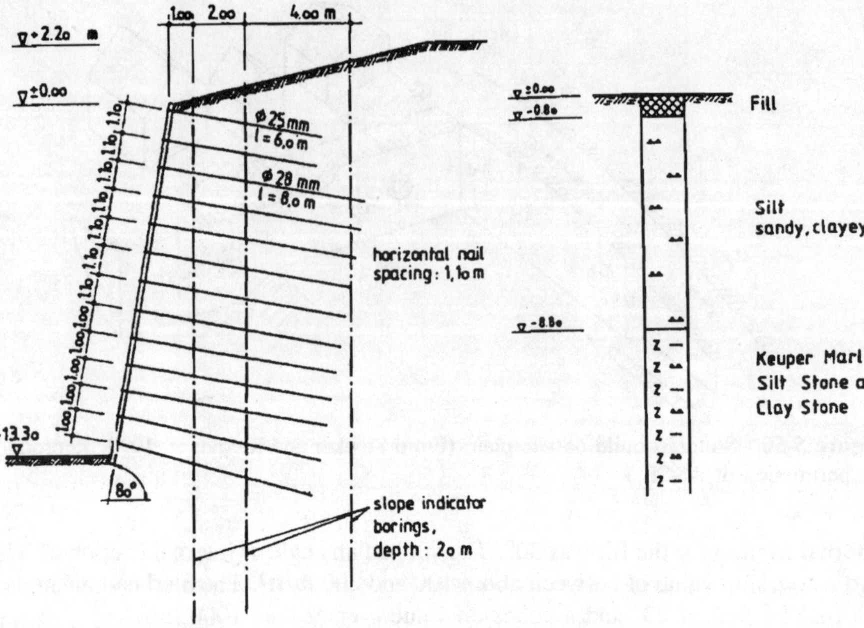

Figure 5-51 Excavation cross section. (From Stocker and Riedinger, 1990. Reproduced by permission of ASCE.)

Subsurface soil conditions consisted of 8 ft of fill underlain by very dense glacial outwash and lacustrine sand and silt. The contact between the outwash and the lake deposits occurred about mid-height on the low wall and at the base of the excavation on the high wall. Groundwater was below the base of the excavation.

The nails consisted of 1- to 1½-in.-diameter, 35-ft-long (maximum) Dywidag bars installed on a 6-ft-wide center to center pattern at 15° below horizontal typically

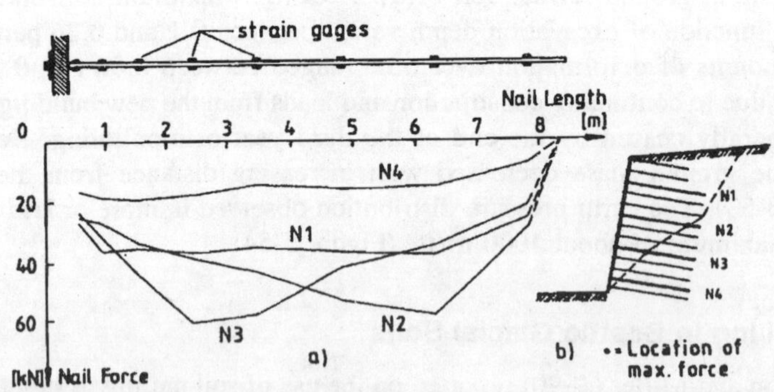

Figure 5-52 Observed maximum nail forces. (From Stocker and Riedinger, 1990. Reproduced by permission of ASCE.)

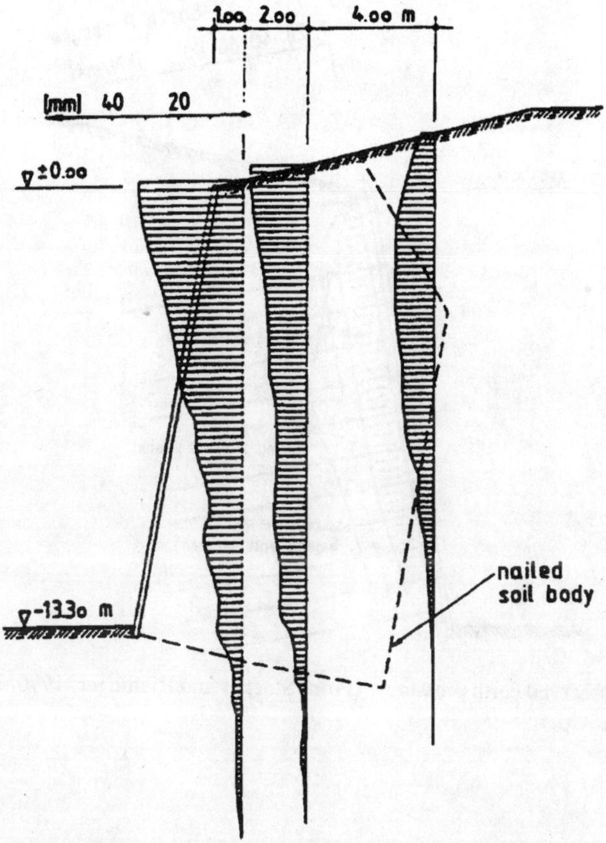

Figure 5-53 Observed ground deformations. (From Stocker and Riedinger, 1990. Reproduced by permission of ASCE.)

(Figure 5-55). Holes were drilled with an 8-in.-diameter continuous flight, hollow stem auger.

The wall was instrumented using strain gages on the bars and slope inclinometers. Wall deflections observed near the top of the wall were about 0.1 percent of the wall height (Figure 5-56). The measured load distribution in the nails compared to two design methods is shown in Figure 5-57. Comparisons of observations made to the results of studies using the finite element method are shown in Figure 5-58. There was also good agreement with the pressure distribution one might predict using conventional apparent earth pressure diagrams for braced excavations (Figure 5-59).

Soil Nailing Used in the Replacement of SR 504 Near Mt. St. Helens

Over 20 miles of State Highway Route 504 (SR 504) was destroyed as a result of the eruption of Mt. St. Helens on May 18, 1980 (Leonard et al., 1988). This road led to

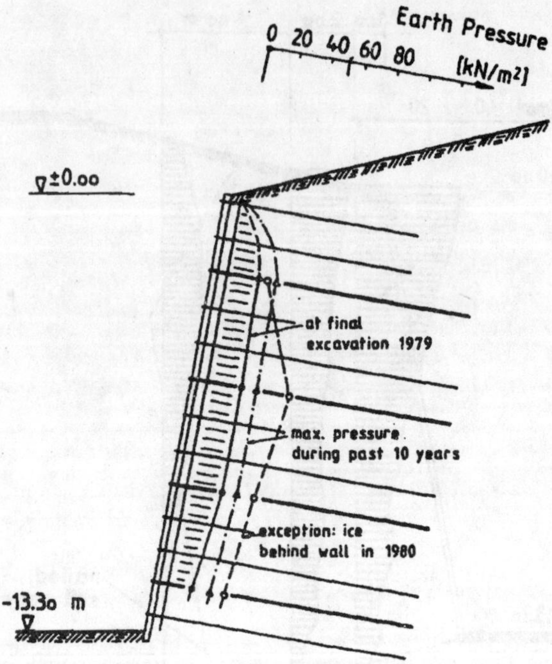

Figure 5-54 Observed earth pressures. (From Stocker and Riedinger, 1990. Reproduced by permission of ASCE.)

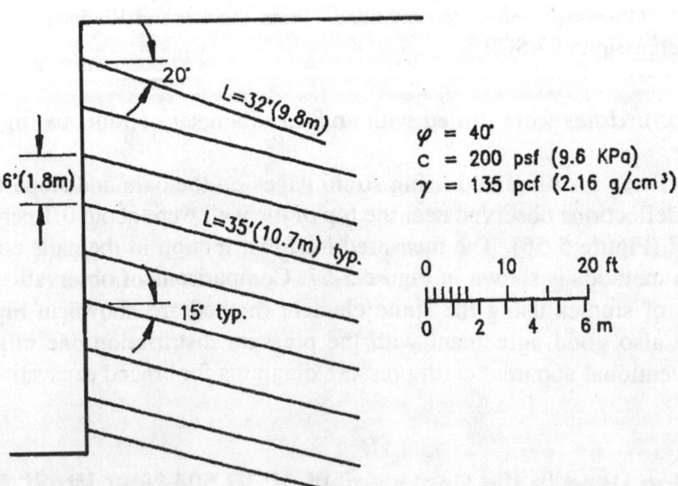

Figure 5-55 Seattle excavation cross section. (From Thompson and Miller, 1990. Reproduced by permission of ASCE.)

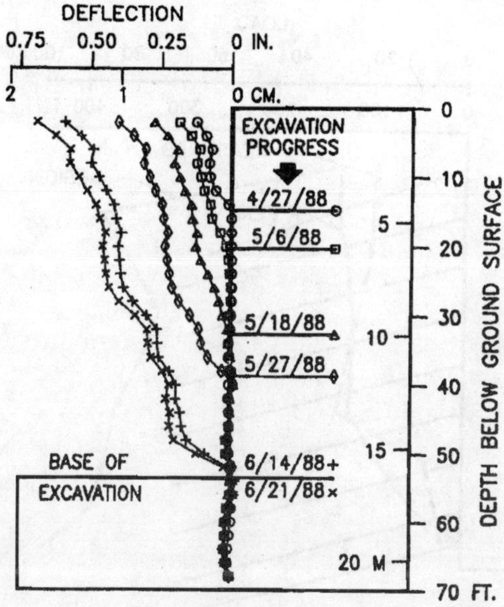

Figure 5-56 Observed wall deflections. (From Thompson and Miller, 1990. Reproduced by permission of ASCE.)

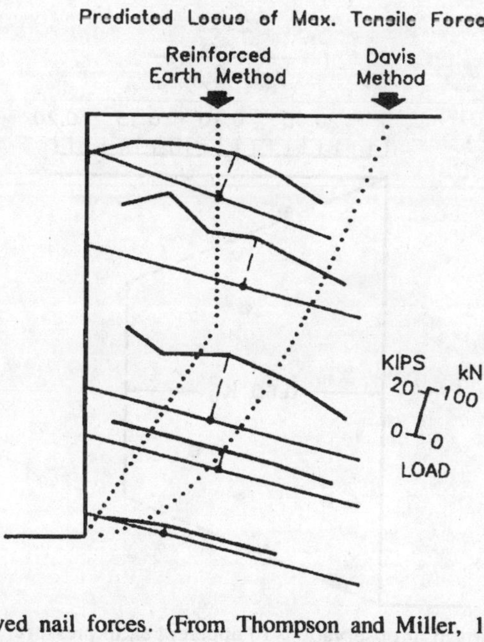

Figure 5-57 Observed nail forces. (From Thompson and Miller, 1990. Reproduced by permission of ASCE.)

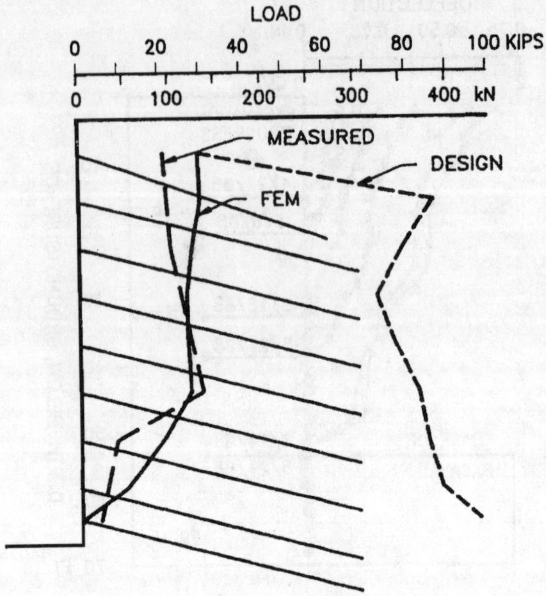

Figure 5-58 Comparison of observations to finite element studies. (From Thompson and Miller, 1990. Reproduced by permission of ASCE.)

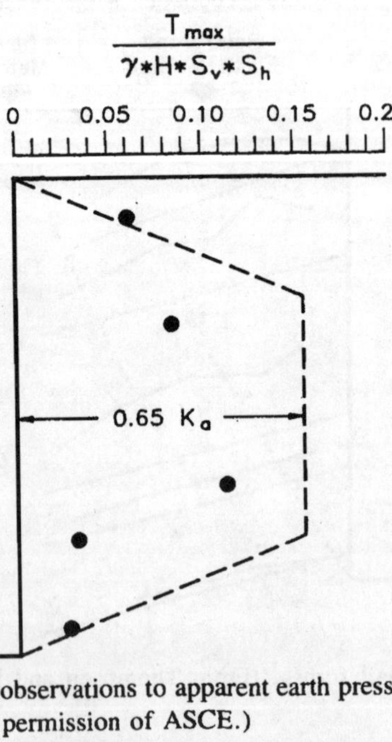

Figure 5-59 Comparison of observations to apparent earth pressure. (From Thompson and Miller, 1990. Reproduced by permission of ASCE.)

the mountain along the valley of the North Fork of the Toutle River before being covered with tons of ash and debris. The new road was located on the valley sides above any future debris flows emanating from the mountain. The cuts were on the order of 65 ft high in an area where the natural slopes range up to about 35°. The cut involved compact to dense silty sand and gravel colluvium, talus gravel, cobbles, and boulders, and poor to good quality basalt bedrock. The overburden thicknesses ranged between about 20 and 35 ft. The cuts therefore involved both soil and rock. The wall design included 1H–10V slopes with a bench at the rock surface (Figure 5-60), 25-ft-long nails on a 6-ft pattern, shotcrete initial support, concrete facade final support, and drainage fabric and weeps.

Soil Nailing Used in Loess for Washington Building Excavation

Another site in Washington that used soil nailing is the extension of an academic research building on the Washington State University campus (Ho et al., 1989). The excavation for the spread footing foundation was about 100 ft wide, 120 ft long, and between 25 and 40 ft deep (Figure 5-61). The excavation was within Palouse loess, a low plasticity soil with approximately 5 percent sand and 25 percent clay. Typical design parameters for this soil are an angle of internal friction of 20°, a cohesion value of 200 lb/ft^2, and a unit weight of 92 lb/ft^3.

The nails consisted of 27-ft-long, No. 10 (1.27-in.-diameter), threaded, Grade

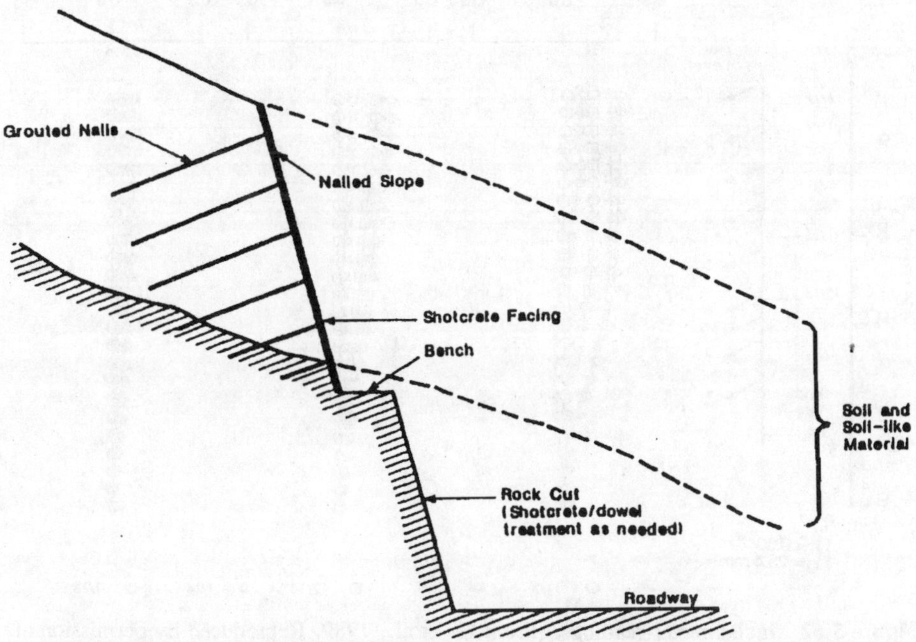

Figure 5-60 SR 504 highway cut cross section. (From Leonard et al., 1988.)

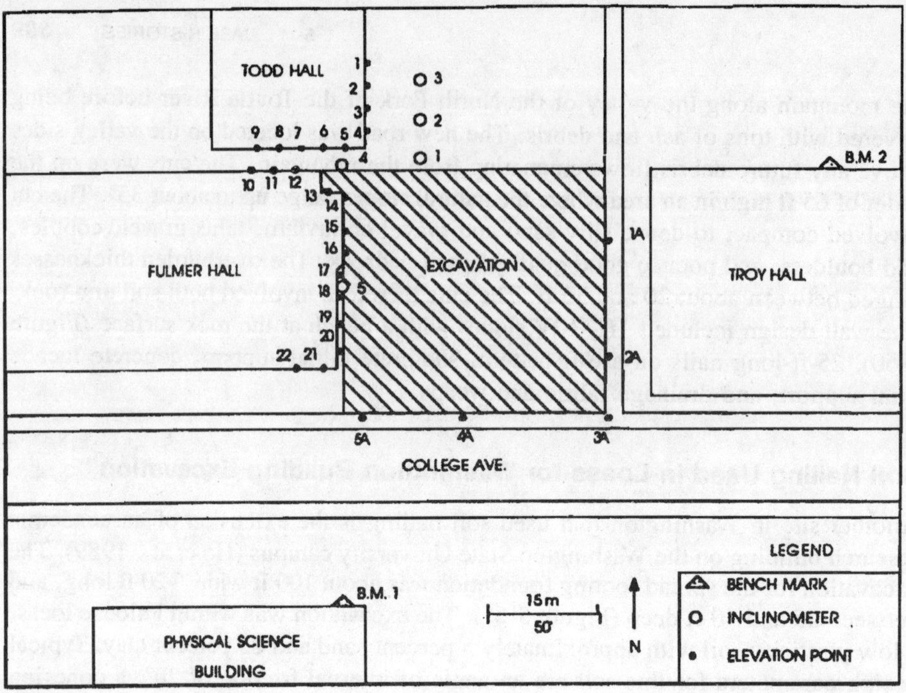

Figure 5-61 Washington State University excavation. (From Ho et al., 1989. Reproduced by permission of ASCE.)

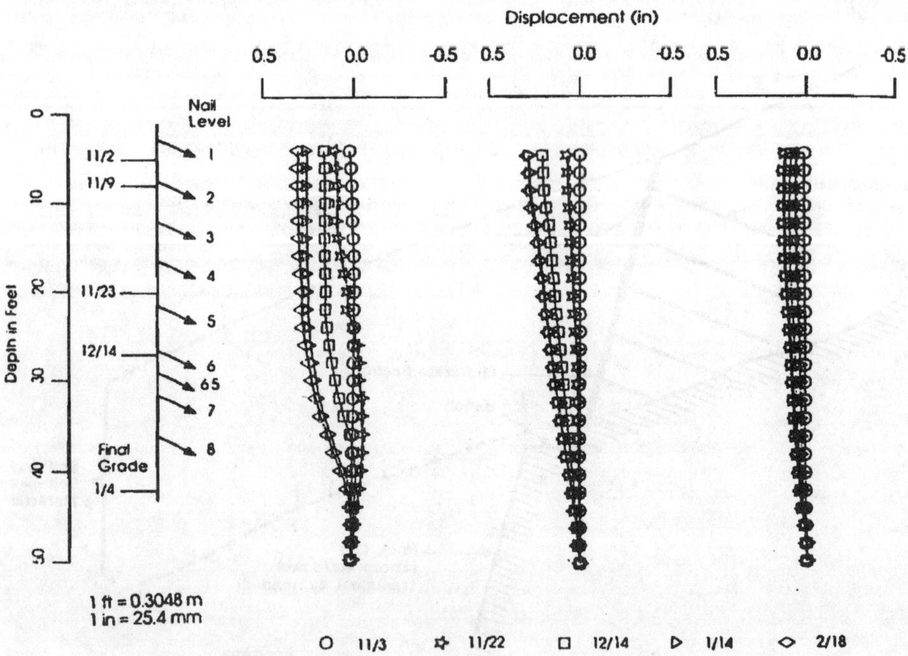

Figure 5-62 Inclinometer readings. (From Ho et al., 1989. Reproduced by permission of ASCE.)

60 Dywidag bars and were installed using augers on a 5-ft (horizontal) by 4.9-ft (vertical) pattern. Excavation was carried out in 5-ft lifts. After excavation and nail placement, steel mesh and Miradrain was fastened to the face and then shotcreted.

Instrumentation consisted of inclinometers, settlement points, and strain gages. Additionally, eight pull-out tests were conducted. The pull-out loads ranged between 29 to greater than 73 tons, with an average of 50 tons. These values convert to skin friction values between 0.51 and greater than 1.33 kips/ft^2 and an average of 1.1 kips/ft^2. A value of 1 kip/ft^2 was used in design. Inclinometer readings indicated lateral movements of 0.06 percent of the wall height (Figure 5-62) and the strain gages indicated a curved potential failure surface (Figure 5-63). There were no significant settlements adjacent to the site.

Soil Nailing Used to Stabilize a Highway Embankment in Australia

A highway embankment was stabilized using soil nailing at Mt. White, about 37 miles north of Sydney, Australia (Hausmann, 1992). The embankment, consisting of clayey silty sands and sandstone fill, threatened to fail after heavy rains in June 1989. Four benches were constructed on which two rows of nails were installed (Figure 5-64). A total of 400 nails were installed on a 6.5-ft-pattern horizontally and a 3.3-ft-staggered pattern vertically. The nails were $\frac{3}{4}$ in. in diameter and 39 ft long, installed at an angle of 10° below horizontal. The wall was faced with 6 in. of steel-mesh-reinforced shotcrete.

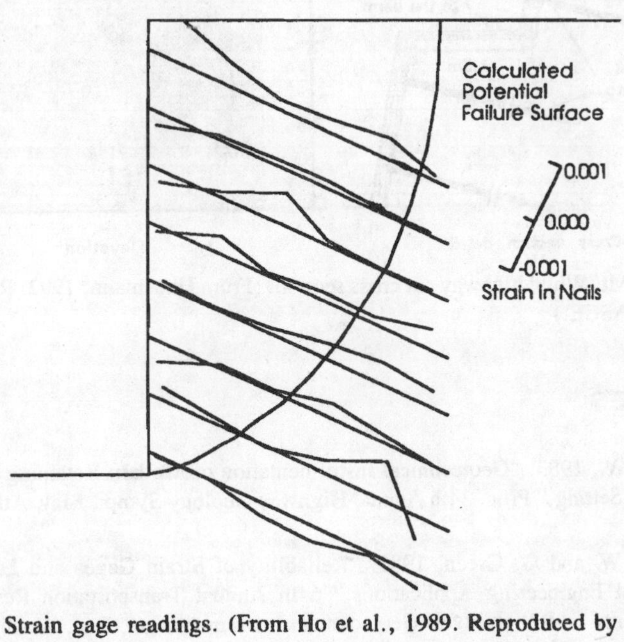

Figure 5-63 Strain gage readings. (From Ho et al., 1989. Reproduced by permission of ASCE.)

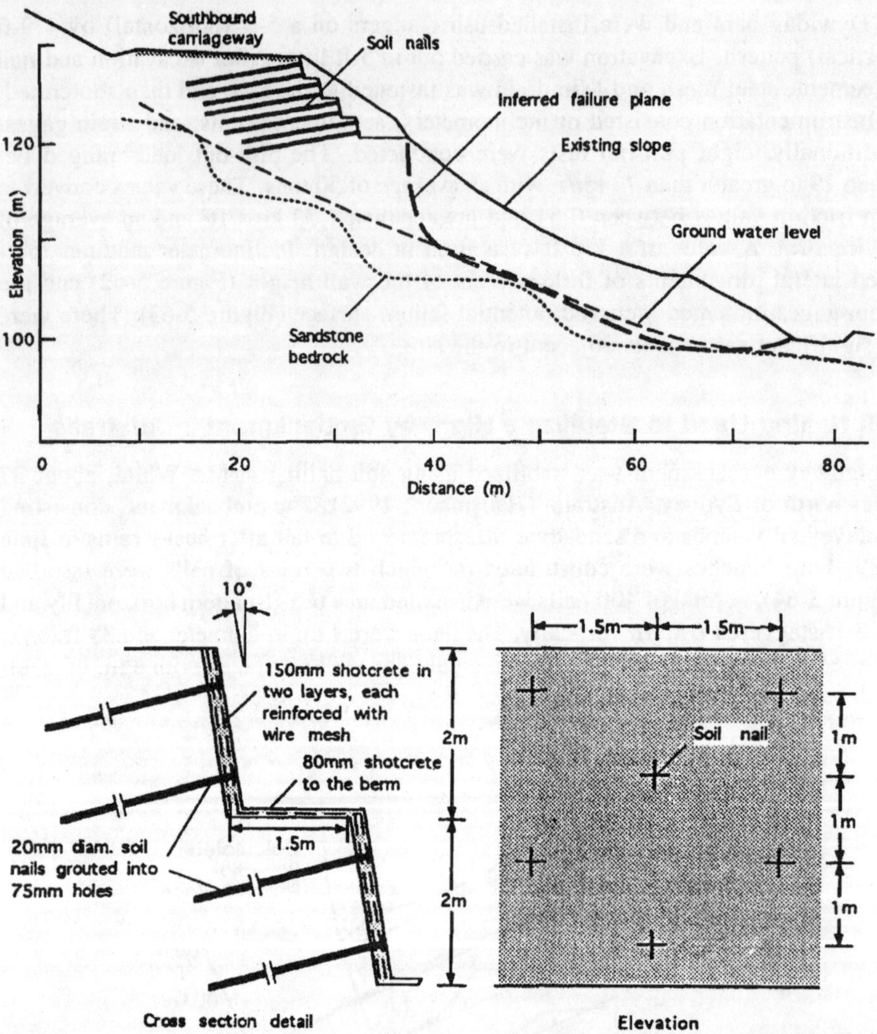

Figure 5-64 Mt. White highway cut cross section. (From Hausmann, 1992. Reproduced by permission of ASCE.)

REFERENCES

Abramson, L. W., 1983. "Geotechnical Instrumentation of Modern Retaining Wall Designs in an Urban Setting," Proc. 34th Annual Highway Geology Symp., May, Atlanta, pp. 87–127.

Abramson, L. W. and G. Green, 1985. "Reliability of Strain Gages and Load Cells for Geotechnical Engineering Applications," 64th Annual Transportation Research Board Meeting Symp. on Reliability of Geotechnical Instrumentation, Jan., Washington, D.C., TRB Publication No. 1004, pp. 13–19.

Abramson, L. W. and W. H. Hansmire, 1988. "Three Examples of Innovative Retaining Wall Construction," Proc. 2nd Int. Conf. on Case Histories in Geotechnical Engineering, June, St. Louis, Paper No. 6.11, pp. 1191–1200.

American Association of State Highway and Transportation Officials (AASHTO), 1990. "In Situ Soil Improvement Techniques," American Association of State Highway and Transportation Officials (AASHTO)—The Associated General Contractors of America (AGC)—American Road & Transportation Builders Association (ARTBA) Joint Committee Subcommittee on New Highway Materials, Task Force 27 Report, Jan.

Berry, R. M., 1991. "Excavating, OSHA-Style," *Civil Eng. ASCE,* Dec., pp. 61–63.

Boyce, G. M. and L. W. Abramson, 1992. "Slope Stabilization Methods Used on the Interstate Route H-3 Project," Proc. 28th Engineering Geology and Geotechnical Engineering Symp., Apr., Boise.

Bruce, D. A. and R. A. Jewell, 1986. "Soil Nailing: Application and Practice—Part 1," *Ground Engineering,* The Journal of the British Geotechnical Society, Vol. 19, No. 8, Nov., pp. 10–15.

Bruce, D. A. and R. A. Jewell, 1987a. "Soil Nailing: Application and Practice—Part 2," *Ground Engineering,* The Journal of the British Geotechnical Society, Vol. 20, No. 1, Jan., pp. 21–28.

Bruce, D. A. and R. A. Jewell, 1987b. "Soil Nailing: The Second Decade," Int. Conf. on Foundations and Tunnels, Mar., London.

Caltrans, 1991. "A User's Manual for the SNAIL Program, Version 2.02—Updated PC Version," Obtained from Mr. K. A. Jackura, California Department of Transportation, Division of New Technology, Material & Research, Office of Geotechnical Engineering, Sacramento, Oct.

Chassie, R. G., 1994. "FHWA Ground Nailing Demonstration Project, Guideline Manual and Workshop," United States Federal Highway Administration Publication, in print.

Cheney, R. S., 1988. "Permanent Ground Anchors," United States Federal Highway Administration, Publication No. FHWA-DP-68-1R, Mar., 136 pp.

Christopher, B. R., S. A. Gill, J.-P. Giroud, I. Juran, J. K. Mitchell, F. Schlosser, and J. Dunnicliff, 1990. "Reinforced Soil Structures, Vol. 1—Design and Construction Guidelines," United States Federal Highway Administration, Publication No. FHWA-RD-89-043, Nov.

Corte, J. F. and P. Garnier, 1984. "Transformation d'un mur ancre par clouage du sol," Proc. Int. Conf. In Situ Soil and Rock Reinforcement, Oct., Paris, pp. 327–332.

Elias, V. and I. Juran, 1991. "Soil Nailing for Stabilization of Highway Slopes and Excavations," United States Federal Highway Administration, Publication No. FHWA-RD-89-193, June.

Engineering News Record (ENR), 1976. "Sprayed Concrete Wall Cuts Overall Cost By 30% In Underpinning, Shoring," Aug. 19, p. 26.

Fannin, R. J. and R. K. Bowden, 1991. "Soil Nailing: An In-Situ Reinforcement Technique," *Geotechnical News,* June, pp. 32–34.

Fukumoto, Y., 1972. "Researches on the Behavior of Piles for Preventing Landslides," *J. Japanese Soc. Soil Mech. Found. Eng.,* Vol. 12, No. 2, pp. 61–73.

Gassler, G., 1977. "Large Scale Dynamic Test of In-Situ Reinforced Earth," Proc. DMSR 77 Conf., Karlsruhe, Vol. 2.

Gassler, G. and G. Gudehus, 1981. "Soil Nailing—Some Soil Mechanic Aspects of In-Situ

Reinforced Earth," Proc. 10th Int. Conf. on Soil Mechanics and Foundation Engineering, Stockholm, Vol. 3, Session 12, pp. 665–670.

Goldberg, D., W. Jaworski, and M. Gordon, 1976. "Lateral Systems and Underpinning, Volume 1—Design and Construction," United States Federal Highway Administration, Publication No. FHWA-RD-75-128, Apr.

Guilloux, A. and G. Notte, 1983. "Experiences on a Retaining Structure by Nailing," Proc. 8th European Conf. on Soil Mechanics and Foundation Engineering, Helsinki.

Guilloux, A. and Schlosser, F., 1982. "Soil Nailing Practical Applications," Symp. Soil and Rock Improvement Techniques, A.I.T.

Hausmann, M. R., 1992. "Slope Remediation," Stability and Performance of Slopes and Embankments—II, R. B. Seed and R. W. Boulanger, Eds., ASCE Geotechnical Special Publication No. 31, New York, pp. 1274–1317.

Ho, C. L., H. P. Ludwig, R. J. Fragaszy, and K. R. Chapman, 1989. "Field Performance of a Soil Nail System in Loess," Foundation Engineering: Current Principles and Practices, F. H. Kulhawy, Ed., ASCE, New York, Vol. 2, pp. 1281–1292.

Hovart, C. and R. Rami, 1975. "Elargissement de l'Emprise SNCF pour la Desserte de Saint-Quentin-Yvelines," *Revue Travaux,* Jan., pp. 44–49.

Juran, I., 1977. "Dimensionement Interne, Ovrages en Terre Armee," These de Docteur-Ingenieur, ENPC, Paris.

Juran, I. and V. Elias, 1987. "Soil Nailed Structures: Analysis of Case Histories," ASCE Geotechnical Special Publication No. 12, New York, pp. 232–244.

Juran, I., G. Baudrand, K. Farrag, and V. Elias, 1990. "Kinematical Limit Analysis for Design of Soil-Nailed Structures" *J. Geot. Eng.* vol. 116, no. 1.

Kerisel, J., 1976. "Theorie du Clouage et Application au Pont de Puteaux," Rapport Interne SIMECSOL, Paris.

Leonard, M., R. Plum, and A. Kilian, 1988. "Considerations Affecting the Choice of Nailed Slopes as a Means of Soil Stabilization," 39th Highway Geology Symp., Aug., Park City, Utah, pp. 288–302.

Lizzi, F., 1971. "Special Patented Systems of Underpinning and Subsoil Strengthening by Means of Root Piles (Pali Radice)," Conf. Donnee a l'MIT, Boston; Universite de L'Illinois; United States Bureau of Reclamation, Denver; and Club of Civil Engineers, Vancouver, Canada.

Long, et al., 1984. "Repair of a Reinforced Earth Wall," Proc. 1st Int. Conf. on Case Histories in Geotechnical Engineering, pp. 335–339.

Louis, C., 1979. "Control and Monitoring in Tunnelling," Report of French Mission to China, Simecsol.

Louis, C., 1984. "New Method of Sinking/Grouting Ground Anchors," French Patent 84-02742.

Mitchell, J. K. and W. C. B. Villet, 1987. "Reinforcement of Earth Slopes and Embankments," National Cooperative Highway Research Program Report No. 290, United States Transportation Research Board, National Research Council, Washington, D. C., June.

Munfakh, G. A., L. W. Abramson, R. D. Barksdale, and I. Juran, 1987. "In-Situ Ground Reinforcement," ASCE Geotechnical Special Publication No. 12, New York, pp. 1–66.

Nicholson, P. J., 1986. "In situ Earth Reinforcement at Cumberland Gap, U.S. 25E," ASCE and Pennsylvania Department of Transportation Joint Conference, Harrisburg, Pa., April, 21 pp.

Nicholson, P. J. and D. L. Boley, 1985. "Soil Nailing Supports Excavation," *Civ. Eng. ASCE*, Apr., pp. 45–47.

Peck, R. B., 1969. "Deep Excavations and Tunneling in Soft Ground: State of the Art Report," 7th Int. Conf. on Soil Mechanics and Foundation Engineering, Mexico City, Mexico, State of the Art Volume, p. 266.

Rabejac, S. and P. Toudic, 1974. "Construction d'un Mur de Soutenement entre Versailles-Chantiers et Versailles-Matelots," *Revue Generale des Chemins de Fer*, 93 e Annee, Apr., pp. 232–237.

Schlosser, F., 1983. "Analogies et differences dans le Comportement et le Calcul des Ouvrages de Soutenement en Terre Armee et par Clouage du Sol," *Annales de L'Institut Technique du Batiment et des Travaux Publics*, No. 418.

Schlosser, F. and I. Juran, 1979. "Design Parameters for Artificially Improved Soils," Proc. 7th Europ. Conf. on Soil Mechanics and Foundation Engineering, General Report, Session No. 8, Sept., Brighton, U. K.

Shen, C. K., L. R. Herrman, K. M. Romstad, S. Bang, Y. S. Kim, J. S. De Natale, 1981. "An In Situ Earth Reinforcement Lateral Support System," Department of Civil Engineering, Univ. of California, Davis, Report No. 81-03, U.S. National Science Foundation Report No. NSF/CEE-81059, Mar., 130 pp.

Sommer, H., 1977. "Creeping Slope in a Stiff Clay," Proc. 9th Int. Conf. on Soil Mechanics and Foundation Engineering, Specialty Session 10, Tokyo, pp. 113–118.

Sommer, H., 1979. "Stabilization of Creeping Slope in Clay with Stiff Element," Proc. 7th Europ. Conf. on Soil Mechanics and Foundation Engineering, Sept., Brighton, U. K.

Stocker, M. F. and G. Riedinger, 1990. "The Bearing Behavior of Nailed Retaining Structures," Design and Performance of Earth Retaining Structures, P. C. Lambe and L. A. Hansen, Eds., ASCE Geotechnical Special Publication No. 25, New York, pp. 612–628.

Stocker, M. F., G. W. Korber, G. Gassler, and G. Gudehus, 1979. "Soil Nailing," Proc. Int. Conf. on Soil Reinforcement, Paris, pp. 469–474.

Thompson, S. R. and I. R. Miller, 1990. "Design, Construction and Performance of a Soil Nailed Wall in Seattle, Washington," Design and Performance of Earth Retaining Structures, P. C. Lambe and L. A. Hansen, Eds., ASCE Geotechnical Special Publication No. 25, New York, pp. 629–643.

Walkinshaw, J. L., 1990. "Handout on Retaining Wall Alternatives," U.S. Federal Highway Administration—Region 9, San Francisco, May.

Walkinshaw, J. L., 1991. "A Summary Update of French Practice on Soil Nailing," U.S. Federal Highway Administration—Region 9, San Francisco, Aug.

Walkinshaw, J. L., 1992. Personal Communication, U.S. Federal Highway Administration—Region 9, San Francisco, Jan.

Weatherby, D. E., 1982."Tiebacks," United States Federal Highway Administration, Publication No. FHWA/RD-82/047, July, 249 pp.

Xanthakos, P. P., 1991. *Ground Anchors and Anchored Structures*, Wiley, New York.

Yamada, G., M. Watari, and S. Kobashi, 1971. "Phenomena and Countermeasures of Landslides," *Sankaido*, pp. 338–339.

CHAPTER 6

SMALL-DIAMETER CAST-IN-PLACE ELEMENTS FOR LOAD-BEARING AND IN SITU EARTH REINFORCEMENT

6-1 INTRODUCTION

Small diameter cast-in-place elements have been installed for piling and in situ earth reinforcement throughout the world following their inception in Italy almost half a century ago. This broad usage is reflected in the wide range of names applied internationally: minipiles, micropiles, root piles, pali radice, needle piles, pieu racine, Gewi pile, Wurzelpfahle, and Estaca Raiz. In North America, the term "Pin Pile^SM" has become increasingly popular during their 20 years or so of application. However, all these names basically refer to a "special type of small diameter bored pile" (Koreck, 1978).

Some practitioners have tried to classify these elements on the basis of diameter: for example, "minipiles" are sized between 150 and 250 mm, with "micropiles" being smaller. Others consider that a diameter of 100 mm defines the lower limit, recalling that "conventional" bored piles commence at about 300 mm. Even allowing for the fact that actual in situ diameters are usually markedly in excess of the nominal drill diameter, subdivisions based on this dimension focus on the wrong aspect of construction. Rather, we may classify as a Pin Pile any bored, cast-in-place pile that can be installed using conventionally sized drilling and grouting equipment such as that used for ground anchorages or ground treatment. This approach is supported by the strong similarities that exist between certain aspects of Pin Pile and anchorage design, construction, and performance. We are therefore considering a fully grouted element of maximum diameter 300 mm, and maximum length 50 m. In the vast majority of cases, these typical dimensions are barely approached.

This method of categorization therefore excludes other types of small-diameter elements often loosely considered as Pin Piles, such as displacement piles of the types driven by hammers, jacks, or vibrators, although such elements can provide attractive solutions under certain conditions, where design capacities are low, and the geology and the structure can accommodate such vigorous installation techniques.

By far the greater number of Pin Piles have been installed to act as *conventional piling*, to accept direct structural loadings and transfer them to deeper, more competent horizons. The balance, although normally installed in exactly the same fashion, have been used to act as *in situ reinforcement* in slope stability applications. The differentiation between these two groupings is fundamental, as their use, design, and performance are completely different. And yet, until recently (Bruce, 1989a) this fundamental subdivision has rarely, if ever, been drawn, to the confusion of interested clients and the loss of potential contractors. This situation may even have been fostered by certain European specialty companies in the earlier days of importing these techniques, in their desire to maintain some form of proprietary mystique. This misguided strategy is reflected by the fact that today this North American market is served almost exclusively by domestic contractors who have made significant advances in the technology, as described below.

6-2 LOAD-BEARING PIN PILES

Historical Background and Characteristics

In 1950, at a time when contemporary construction codes (such as in Germany) were stipulating minimum diameters of 400 mm for cast-in-place piles, the Italian specialty contractor Fondedile began the commercial exploitation of root piles (pali radice) in the restoration and strengthening of historic buildings. Excellent illustrations were provided by Lizzi (1982).

The aim was to provide a support system that would, firstly, not adversely affect the structure; secondly, be capable of installation in restricted working areas; and, thirdly, accept load with minimal deflections. The piling community was reportedly initially skeptical but within a short period had been favorably impressed by the technical and economic potential of the method. This confidence was further strengthened by an ever growing pool of successful (and low cost) field tests, high profile applications, and aggressive promotion.

From the middle 1960s, similar systems became popular in Germany mainly associated with the underground construction of roads, subways, and metros. However, it was with the expiration of some of the original Italian patents in the early 1970s that the technique really developed internationally and new proprietary methods were developed. These included the Gewi pile (Dywidag), the Tubfix Micropile (Rodio), and the Menard Minipile, while in North America, the first applications of the Nicholson Pin Pile were undertaken. A significant later technical development was the use of preloading facilities in certain types of pile (Nicholson Compressed Anchor Pile, Rodio Ropress Pile) to eliminate any structural settlement during the action of load transfer from structure to pile. Examples of these preloaded pile systems are provided in case histories later in this section.

Throughout the world, Pin Piles are now used and regarded as a reputable construction tool of exceptional value and potential. While the underpinning of Europe's historic structures continues apace (e.g., Gouvenot, 1975; Herbst, 1982, Attwood, 1987; Doornbos, 1987), especially in cities now being impacted by new underground constructions, the expanding cities of the Far East (Bruce and Yeung, 1983; Mitchell, 1985), such as Hong Kong and Singapore, are also major markets as sophisticated construction continues in soft soils with high water tables in highly congested and populous locations. In North America, the refurbishment of industrial, transportation, and commercial structures provides constant challenges, while the redevelopment of the older cities, especially on the East Coast, ensures that the intensity of Pin Pile construction there is at least as great as that anywhere in the World. To illustrate the growing use of Pin Piles in the United States alone, a survey in 1988 (Bruce, 1988–1989) listed 25 projects completed between 1978 and 1988 by one company. A later, similar review (Bruce, 1992a) listed 20 new projects undertaken in the subsequent two years by the same firm.

In summary, applications are related to the prevention or arrest of structural movements generated by

- Adjacent underground excavation
- Changes in the groundwater level
- Changes to existing foundation loads
- Deterioration or inadequacy of existing foundations
- Imposition of machine vibrations
- Especially difficult and/or unpredictable foundation materials

Additional general overviews of Pin Piles have been provided by several authors, including Koreck (1978), Weltman (1981), Herbst (1982), Lizzi (1982, 1985), Bruce and Yeung (1983), and Bruce et al. (1985).

Construction

As with the later sections on Design, and Testing/Performance, this section concentrates on the general details and approaches. Particular details are provided later in the descriptions of individual case histories.

The most common basic method of installing Pin Piles is shown in Figure 6-1. Older variants using compressed air to pressure the grout or a vibrated mandrel (displacement pile) are described by ASCE (1987) but are rarely if ever used in developed countries. Likewise, the "expanded base" pile (Lizzi, 1982) and the Menard inflatable cylinder pile (Mascardi, 1982) are never seen nowadays.

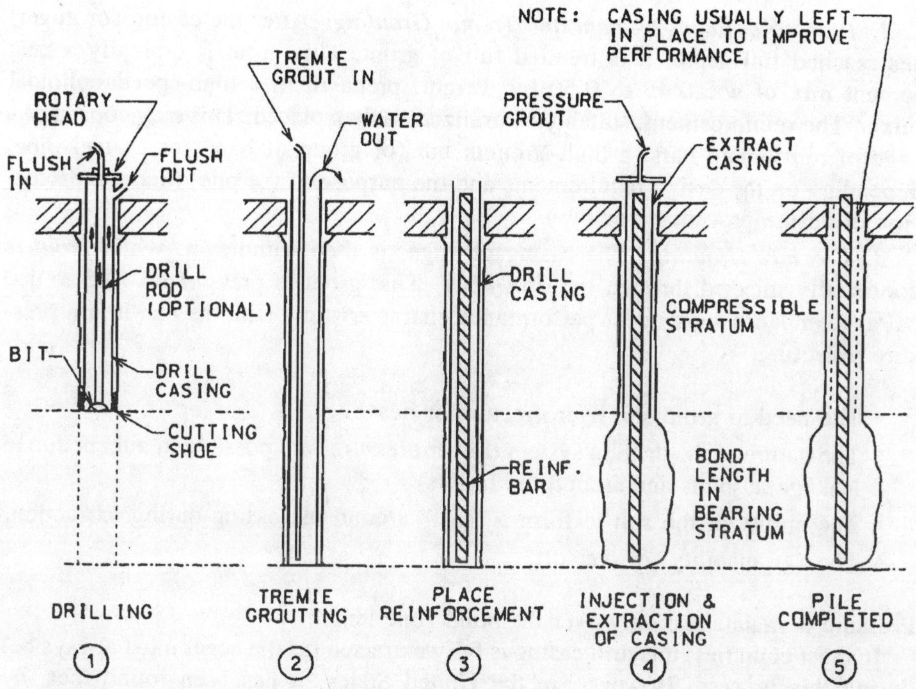

Figure 6-1 Stages in the construction of a typical Pin Pile in soil. (From Bruce, 1989a; after Koreck, 1978.)

The successive construction steps are:

1. Drilling. A drilling method is chosen to ensure the minimum practical disturbance or upheaval to the structure or the soil. Frequently a different system may be necessary to penetrate through any existing structure from that to be used in the soils below. For soil drilling, some type of duplex method (Bruce, 1989b) is common, although in certain conditions the use of a single casing is permissible. Water or foam flush is typical: air flushing is typically not allowed. In certain soil conditions (e.g., clays) or where fluid spoils cannot be tolerated within the structure being underpinned for environmental reasons, a hollow stem auger can be used, although subsequent grout/soil bond capacity may be impacted adversely as a result of lateral decompression of the surrounding soil. This is a clear reminder that the critical design aspect of pile-soil bond capacity is highly sensitive to constructional method, and especially the drilling and grouting techniques.

Contemporary drilling rigs for such work are diesel-hydraulically or electro-hydraulically powered, track mounted, and extremely powerful for their compact size. Many have dimensions allowing them to pass through very narrow openings and operate in less than 3 m of headroom. Such rigs are highly maneuverable and capable of drilling at any angle through rock, soil, and obstructions. They can commence drilling within 0.3 m of existing structures.

2. *Placing of Reinforcement and Tremie Grouting.* After the casing (or auger) has reached full depth, it is tremied full of grout. This grout is typically a neat cement mix of w:c 0.45 to 0.50 by weight, prepared in a high-speed colloidal mixer. The reinforcement, suitably centralized, is then placed. This may consist of a cage of reinforcing bars, a high-strength bar (or group of bars), or a steel pipe, depending on the design requirements and the purpose of the pile. (See the discussion of design procedures below.)

3. *Pressure Grouting.* The casing or auger is then withdrawn, while grout is continually injected through the drill head. This grout is pressurized (0.5 to 1.0 MPa) to enhance subsequent performance characteristics, with the maximum pressure reflecting:

- The need to avoid soil hydrofracture or heave.
- The nature of the drilling system (lower pressures are possible in augers due to leakage at joints and around the flights).
- The ability of the soil to form a "seal" around the casing during extraction.
- The "groutability" of the soil.

Pressure is maintained only over the bond zone length.

In most countries, this drill casing is fully extracted (as the auger must always be) during this process. However, in the United States, it has been found that, by leaving the casing in place through the zones above the pressured zone, the Pin Pile performance is greatly enhanced, both vertically and laterally. This option also prevents wasteful travel of grout into these often permeable upper horizons, while it also provides excellent corrosion protection to the interior of the pile in what is usually the most vulnerable zone.

4. *Connection to Structure.* Adequate bond or connection must always be provided between the pile and the structure in order to properly transfer load. Where piles have been installed through existing structures (Figure 6-2), the (normally smooth) structural interface can be roughened to give additional mechanical interlock in order to ensure adequate bond. Such a proprietary system, named Ankerbonder, has been used with success on a major underpinning project for cooling towers in England (Anonymous, 1987). Employing vibrating air-driven pistons with tungsten carbide tips, the head is lowered into the hole in the structure and rotated slowly. The resultant grooved configuration is claimed to increase ultimate structure/pile bond capacity by up to 10 times. Other, nonproprietary options also exist, based on the same principle of "roughening up" the interface.

In other applications, horizontal posttensioning of the foundation beam has been undertaken to provide a "clamping" effect to guarantee structure/pile continuity.

The use of high-strength, nonshrink grouts for the grouting of the upper pile section within the structure is also common.

5. *Corrosion Protection.* When piles are required to act in tension, or when they must be installed in particularly aggressive conditions, attention must be paid to the corrosion protection of the load-bearing steel element and to the chemistry of the

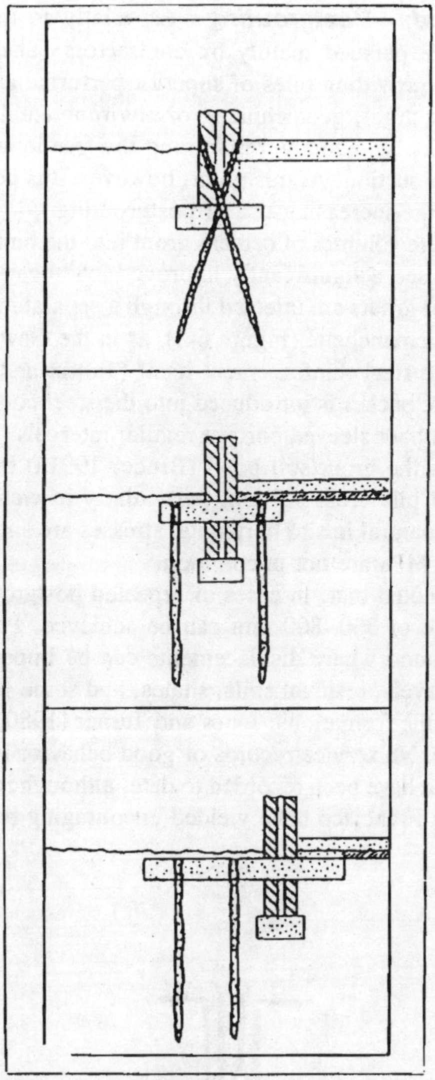

Figure 6-2 Basic methods of load transfer from Pin Piles to structure. (From Bruce, 1988. Reproduced by permission of Thomas Telford Publications.)

cement. Similar to ground anchorages (FIP, 1986), protection in the form of a corrugated sheath can be used, with centralizers to ensure a minimum grout cover of 20 mm. The protection of left-in-place drill casings is more difficult, but these can be coated with an anticorrosion compound or can themselves be protected by an outer casing. In general, however, the normal action of Pin Piles in compression aids corrosion protection, whereas the opposite is true of anchorages that act in tension.

Development Trends—Postgrouting Especially in the construction of Pin Piles, innovations are pursued mainly by contractors. These developments have been directed toward providing piles of superior performance more cheaply and in more challenging structural, geotechnical, or environmental settings. The benefits of two trends in particular, namely, reinforcing the free length and preloading, are illustrated later in this section. At this point, however, it is convenient to focus on a third development—the increasing use of postgrouting.

By injecting discrete volumes of cement grout into the bond zone after the initial, or primary, grout has set, a significantly improved load-bearing performance can be provided. The cement grouts are injected through a separate grouting tube, that is, a sleeved pipe or tube à manchette (Figure 6-3), as in the Gewi pile system of Herbst (1982) or through the steel reinforcement itself (Tubfix and Ropress piles). In the latter case, the double packer is introduced into the steel core pipe, and the grout is ejected through the rubber sleeved ports at regular intervals (Figure 6-4). Postgrouting greatly improves the grout/soil bond (Bruce, 1991a) but, in addition, it may increase the nominal pile cross section, particularly in weaker soil layers or near ground level, where natural in situ horizontal stresses are small. Postgrout pumping pressures for over 4 MPa are not uncommon.

Mascardi (1982) noted that, in cases of repeated postgrouting, an effective pile diameter in the range of 300–800 mm can be achieved. Postgrouting tends to be most effective in ground where displacements can be imparted relatively quickly, such as sands and gravels, residual soils, shales, and some weaker sedimentary and low-grade metamorphic formations. Jones and Turner (1980) also noted a favorable response in stiff clay. No service records of good behavior in very soft nonconsolidated clay or soft peat have been recorded to date, although recent tests conducted in the Bay Mud of San Francisco have yielded encouraging results.

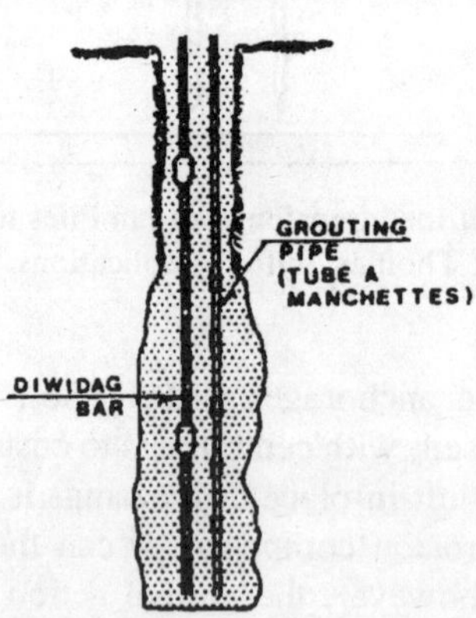

Figure 6-3 Gewi minipile. (After Mascardi, 1982.)

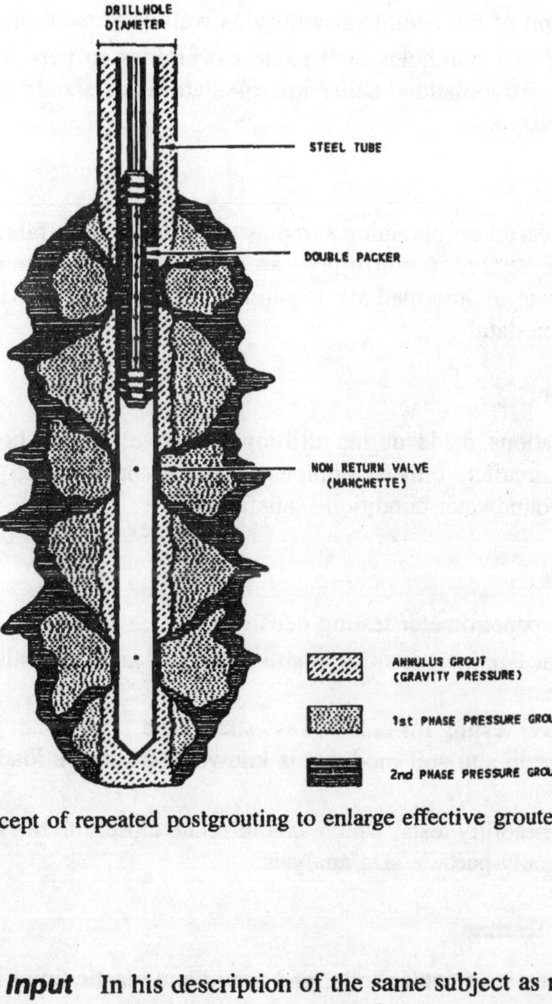

Figure 6-4 Concept of repeated postgrouting to enlarge effective grouted diameter. (From Mascardi, 1982.)

Design

Geotechnical Input In his description of the same subject as related to ground anchorages, Littlejohn (1990) states "The ground is one structural component of the . . . system, and the importance of a good quality site investigation cannot be overstated. Lack of adequate information on the ground remains the most common cause of . . . failures. . . ." This dictum applies equally to the requirements for Pin Pile design, and the following synopsis of Littlejohn's work provides a valuable checklist of desiderata.

General

- Detailed knowledge of soil conditions, especially in the bond zone (e.g., the presence of thin partings of silt or sand in a clay).
- Classification of the soil with respect to shear strength, compressibility, density, groundwater conditions, and chemical analysis.

- Consideration of the lateral variability as well as vertical changes.
- Location of key boreholes at the site extremities to permit interpolation as opposed to extrapolation. Other intermediate holes should have a maximum spacing of 20 m.

Sampling

- Emphasis placed on obtaining samples that can identify fabric or structure of the founding stratum (undisturbed samples at 1.5 m maximum intervals, with other samples at intermediate locations). Samples to provide standard soil classification data.

Groundwater

- All observations made during drilling site investigation holes carefully recorded. Thereafter, only standpipes or more sophisticated piezometers can measure groundwater conditions satisfactorily.

Field Testing

- Static cone penetrometer testing density.
- SPT tests (at 1-m intervals) for in situ density of granular soils, in situ strength of cohesives.
- Pressuremeter testing for radial stress-strain data. This is not yet common, but the apparent in situ soil modulus is known to influence load-transfer mechanisms.
- In situ permeability tests, which should relate closely to the soil classification data, especially particle size analysis.

Laboratory Testing

- Granulometry of granular soils, and liquid and plastic limits of cohesive soils for every stratum.
- Shear tests on granular soils, provided they are conducted at the density and stress level approximating that in situ. For cohesive soils, use the triaxial test whose drainage condit ons best reflect the project construction parameters such as rate of loading.

Chemical Testing

- To examine the ground or groundwater to determine potential aggressivity to the constituents of the pile. Sulfate and chloride contents dictate the choice of cement and are quantified in Table 6-1. For aggressivity toward metals, redox potential and resistivity are useful indicators (Table 6-2). If the site is adjacent to an electrical installation, the ground should be checked for stray currents.

TABLE 6-1 Aggressivity of Groundwater with Respect to Cement

Groundwater Environment	Remarks on Aggressivity
Very pure water ($CaO < 300$ mg/liter)	Such waters dissolve the free lime and hydrolyse the silicates and aluminates in the cement.
$pH < 6.5$	Acid waters attack the lime in the cement, but pH values of 9 to 12 are passivating
Selenious water (SO_3) > 0.5 g/liter (stagnant); > 2 g/liter (flowing) Magnesium water (SO_3) > 0.25 g/liter (stagnant); > 0.1 g/liter (flowing)	These sulfates react with the tricalcium aluminate to form salts that disarrange the cement by swelling

Source: From Littlejohn (1990).

Investigation During Construction

- Especially when dealing with potentially variable glacial soils, as much information as is reasonably and economically possible should be recorded during production drilling and grouting. Production drill rigs, of course, are not geared for precise site investigation, but, although they may miss subtle changes, they can provide useful overall data. Likewise, any other construction activities on site that may impact Pin Pile performance should be recorded (e.g., dewatering, blasting, superstructure changes).

Design Procedures and Considerations

General Observations The basic philosophy of Pin Pile design differs little from that required for any other type of pile: the system must be capable of sustaining the anticipated loading requirements within acceptable settlement limits, and in such a

TABLE 6-2 Corrosiveness of Soils Related to Values of Resistivity and Redox Potential

Corrosiveness	Resistivity ($\Omega \cdot$ cm)	Redox Potential (Corrected to pH = 7) (mV), Normal Hydrogen Electrode
Very corrosive	< 700	< 100
Corrosive	700 –2000	100 –200
Moderately corrosive	2000–5000	200–400
Mildly corrosive or noncorrosive	> 5000	> 430 if clay soil

Source: From Littlejohn (1990), after King (1977).

fashion that the elements of that system are operating at safe stress levels. In detail, attention must be paid analytically to settlement, bursting, buckling, cracking, and interface considerations, whereas, from a practical viewpoint, corrosion protection, and compatibility with the existing ground and structure (during construction) must be regarded. The system must, of course, also be economically viable.

Reference must always be made to local construction regulations or building codes (e.g., Commonwealth of Massachusetts, 1984) although the special and often novel demands of Pin Piling may not always be specifically or adequately addressed. In that event, sensible interpretation or extrapolation is essential by all parties, backed up by appropriately rigorous preproduction field testing if possible.

Whereas the design of a conventional pile is normally governed by the external (i.e., ground related) carrying capacity, Pin Pile design is frequently controlled by the internal design (i.e., the selection of the pile components). This reflects both the relatively small cross section available, and the unusually high grout/ground bond capacities that can usually be mobilized, as a consequence of the Pin Pile installation methods. This anticipates the point that Pin Piles typically transfer load to the ground through skin friction as opposed to end bearing: a pile 200 mm in diameter with a 5-m-long bond zone has a peripheral area 100 times greater than the cross-sectional area. This mode of load transfer directly impacts performance, in that the pile movements needed to mobilize lateral frictional resistance are of the order of 20 to 40 times *less* than those needed to mobilize end bearing.

Occasionally in the United States, Pin Piles are designed as simple struts between the structure and a particularly resistant bedrock surface. In such cases, and assuming the rock mass has sufficient "punching" resistance, the internal pile design obviously governs maximum capacity.

Internal Pin Pile Design

REINFORCING STEEL The purpose of the Pin Pile, its working load, and its permitted elastic deflection dictate the nature of the reinforcement, the primary load resisting element. For example, relatively small-capacity Pin Piles designed to act only in compression (e.g., Lizzi, 1982) usually comprise either a "cage" of high-yield rebars supported by helical reinforcement, or a very limited number of high-strength bars. When such piles have to act in tension, then the latter solution is adopted.

For the higher capacity Pin Piles increasingly becoming popular, where deflections must be minimized, or where significant lateral stresses have to be resisted, steel pipes or casings are common. Such elements have a high radius of gyration and a constant section of modulus in all directions and so have intrinsically good column properties. Frequently, however, simple economic logic will dictate the selection of the reinforcement if various options potentially satisfy the loading criteria.

Regarding working stresses in the steel, examples are given in the discussion of case histories later in this section. These confirm that there is no standard approach nationwide, as a consequence of the absence of a national code, and the presence of various and differing local regulations and contractor preferences.

As a guideline, Tables 6-3 to 6-5 summarize details of various reinforcing steels currently available.

GROUT Most commonly, the grout consists of simply cement and water. The type of cement will be dictated by the nature of the groundwater or by the strength/time requirements. Water should always be fresh and potable but with a chloride ion content of less than 500 mg/liter. As for all materials, these basic components must be in accordance with national or local standards.

Cement grouts should be sufficiently fluid to allow efficient pumping and injection, and sufficiently stable to resist displacement and erosion after injection.

The principal variable affecting the properties of cement grouts is the water-cement ratio (*w:c*). The amount of water determines the rate of bleeding, subsequent plasticity, and ultimate strength of the grout. The extent to which these, and also fluidity, are related to *w* is shown in Figure 6-5 (neat Type 1 cement). Excess water causes bleeding, low strength, increased shrinkage, and poor durability. Typically, *w* values of 0.45 to 0.55 are used for Pin Piling.

Fine sands can be added to neat cement-water suspensions to form an economical

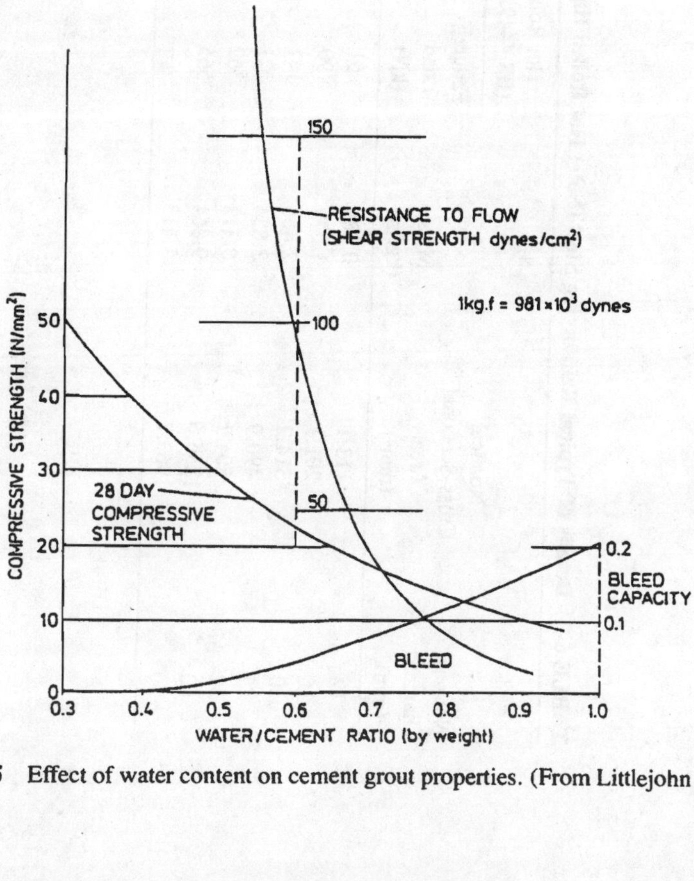

Figure 6-5 Effect of water content on cement grout properties. (From Littlejohn and Bruce, 1977.)

TABLE 6-3 Details of Typical Reinforcing Steel (Note: Hot Rolled High Yield [BS 4440]: 410 N/mm²)

Nominal Diameter (mm)	Nomical Cross-Sectional Area (mm²)	Mass (kg/m)	Hot Rolled Mild Steel Bar (BS 4449: Fy = 250 N/mm²)		Cold Worked High Yield Bar (BS 44461: Fy typically 425 N/mm²)	
			Failure Load (kN)	Characteristic Strength (kN)	Failure Load (kN)	Characteristic Strength (kN)
12	113.1	0.888	51	28	—	—
16	201.1	1.579	90	50	111	93
20	314.2	2.466	141	78	173	134
25	490.9	3.854	221	123	270	209
32	804.2	6.313	362	201	442	342
40	1256.6	9.864	565	314	691	534
50	1963.7	15.413	—	—	1080	834

TABLE 6-4 Details of High-Strength Bars

Details of Dywidag Bars[a]

Steel Grade Yield/Ultimate (N/mm²)	Number of Bars	Nominal Diameter (mm)	Ultimate Load $F_z = B_z \cdot A$ (kN)	Working Load: Ult. Load 1.6 (kN)	Ult. Load 1.7 (kN)	Ult. Load 1.8 (kN)	Ult. Load 2.0 (kN)	Yield Load $F_s = B_s \cdot A$ (kN)	Working Load: 0.75 Yield Load (Yield Load/ 1.33) (kN)	0.6 Yield Load (Yield Load/ 1.67) (kN)	Yield Load/ 1.75 (kN)
835/1030	1	26.5	568	355	334	316	284	460	345	276	263
835/1030	1	32.0	828	518	487	460	414	671	503	403	384
835/1030	1	36.0	1048	655	617	583	524	850	637	510	486
1080/1230	1	26.5	678	424	399	377	339	595	446	357	340
1080/1230	1	32.0	989	618	582	549	495	868	651	521	496
1080/1230	1	36.0	1252	783	736	696	626	1099	824	659	628

Details of Gewi Bars

LOAD CHARACTERISTICS OF THE GEWI-STEEL

GEWI Threadbar	Diameter 50 mm (kN)	Diameter 51 mm (kN)	Diameter 57 mm (kN)
Ultimate load	980	1,113	1,443
Yield load	824	795	1,013
Workingload at *yield load* 1.75	470	454	579
Area (cm²)	19.63	20.27	25.81

[a]Designation according to S1-standard: F = force, A = cross-sectional area of steel.

TABLE 6-5 Types of Pile Reinforcement as Used in Pieux IM by Soletanche Ltd.

Type	Type of Reinforcement		Characteristics		Nominal Capacity	
	Dimensions (mm)	Elastic Limit (N/mm²) σ_{ea}.	Minimum Hole Diameter (mm)	Steel Section (mm²) S_a	$\frac{2}{3}\sigma_{ea} S_a$ (k/N)	$\frac{1}{2}\sigma_{ea} S_a$ (kN)
Profile	IPE 100 × 55 × 4	240	150	1,000	160	120
Tubes	46/60	390	100	1,200	310	230
		530			420	310
	70/89	390	120	2,300	600	450
		530			820	620
	97/114	390	150	2,800	730	550
		530			1,000	750
	109/127	390	170	3,400	880	660
		640			1,200	900
	157/178	390	200	500	1,300	980
		530			1,760	1,320
Bars and cages	20T	400	60 to 250 depending on the reinforcement	300	80	60
	32T	400		800	210	160
	40T	400		1,300	340	260
	26DY	800		500	280	210
	33DY	800		800	450	330
	36DY	800		1,000	530	400
	6 Nr. 32T	400	150	4,800	1,290	970
	4 Nr. 36T	800	150	4,000	2,120	1600

grout, particularly where a high-solids, low-water grout with relatively high frictional shear strength is required. Sand is chosen as for concrete in relation to durability, shrinkage, and alkali reaction, and in general, hard bulky crushed rock is preferred to flat, angular, or flaky material, which gives poor fluid handling properties. Evenly graded sands are preferred (5 mm down to 75 μm) and for long pumping distances the maximum size should ideally be reduced to 0.5 mm and the maximum sand/cement ratio limited to 3 to maintain the particles in suspension and avoid segregation. Rarely, however, does the ratio exceed 1.5, and 1.0 is most common, giving characteristic strengths of the order of 30 to 35 MPa. In confirmation, Koreck (1978) advocates a minimum cement content of 600 kg/m^3 (equivalent to a sand/cement ratio of 2.3 at $w = 0.50$), and Lizzi (1982) advocates 600 to 800 kg of cement per cubic meter of sieved sand: "Therefore a high strength concrete." In general, however, the use of sanded grouts is growing less common, either dictated by local tradition or in cases where excessive neat grout takes are anticipated (e.g, when injecting Karstic limestone formations, or when using larger diameter hollow stem augers during drilling.)

Admixtures can be added in relatively small quantities to modify grout properties (Table 6-6), especially to prevent shrinkage, to allow reduction in w (while maintaining fluidity/pumpability), to accelerate or retard setting, and to prevent bleeding (thereby discouraging corrosion). No admixture containing more than a total of 0.1% (by mass) of chlorides, sulfides, or nitrates should be used (Littlejohn, 1990). Most commercial admixtures are compatible with Type I and III Portland cements,

TABLE 6-6 Common Cement Grout Admixtures

Admixture	Chemical	Optimum Dosage (% cement wt.)	Remarks
Accelerator	Calcium chloride	1–2	Accelerates set and hardening
	Sodium silicate	0.5–3	Accelerates set
	Sodium aluminate		
Retarder	Calcium lignosulfonate	0.2–0.5	Also increases fluidity
	Tartaric Acid	0.1–0.5	
	Sugar	0.1–0.5	
Fluidifier	Calcium lignosulfonate	0.2–0.3	Entrains air
	Detergent	0.05	
Air Entrainer	Vinsol resin	0.1–0.2	Up to 10% of air entrained
Expander	Aluminum powder	0.005–0.02	Up to 15% preset expansion
	Saturated brine	30–60	Up to 1% postset expansion
Anti-Bleed	Cellulose ether	0.2–0.3 (for $w < 0.7$)	Equivalent to 0.5% of mixing water
	Aluminum sulfate	Up to 20% (for $w < 5$)	Entrains air

Source: From Littlejohn (1982).

but many are incompatible with high alumina and super sulfated cements, which are in any case extremely rare in Pin Piling practice. Admixtures should not be regarded as a replacement for good grouting practice and must not be used indiscriminately.

Bearing in mind the various kinds of batching plant and mixing techniques, trial mixes must be carried out and the following data should be recorded for review:

- Type of cement.
- Water-cement ratio (by Baroid Mud Balance).
- Admixture or filler concentration.
- Flow reading (through flowmeter, flow cone, or viscometer).
- Crushing strengths (minimum two cubes at 3, 7, 14, and 28 days).
- Notes on amount of free expansion, shrinkage, bleed, and final setting time.

As for ground anchorages, a characteristic strength of 40 N/mm^2 at 28 days is an accepted target, while bleed should be less than 4 percent at final set.

GROUT-STEEL BOND This interface is fundamental in that it is a mechanism of load transfer from steel to ground, and it also acts to promote the composite action of the internal pile components. A great volume of research has been conducted into the nature, distribution, and controls over grout-steel bond characteristics (e.g., Littlejohn and Bruce, 1977), and so there is no shortage of regulated values or field data available. Bond stresses are assumed to be uniformly distributed along the element. Bond values have typically been generated in tension testing and can be regarded as at least equally valid for the compressional sense, as in most Pin Piles.

In the majority of cases, the grout-steel bond consideration does not govern the design: internal load capacity, or (less frequently) grout-ground capacity are the principal controls. As for the case with ground anchorage tendons, bond strength can be significantly affected by the surface condition of the reinforcement. A film of rust is not necessarily harmful, but pitting or the presence of loose debris or lubricant materials is not acceptable.

COMPOSITE ACTION For Pin Piles in compression, loaded across their full section, the load will be resisted jointly by the steel and the grout, in a certain proportion. This concept of composite action is clearly beneficial in optimizing internal pile design, in that it reduces the steel requirement. However, under some regulations (e.g., Hong Kong, 1976) it is disallowed despite the clear theoretical and practical evidence to the contrary: "When subjected to compressive stress, the cement grout cylinder also participates in the carrying of the load. This results in lower settlement values for the compression pile" (Dywidag, 1983).

It would seem that most American designs do now acknowledge this, and it is typical to find designs where the steel is permitted to accept load to a certain percentage of its yield point, and the grout can contribute a certain percentage of its *unconfined* compressive strength. More sophisticated analyses consider the strain compatibility between the steel and the relatively stiff grout. When the Pin Pile reinforcement is a pipe or casing, then the internal grout is very effectively *con-*

fined, and so it may be argued that it is capable of safely sustaining an even higher percentage of its unconfined strength.

STABILITY OF PIN PILES Where the soil modulus is less than 0.5 MPa, mathematical models can be called upon to investigate the stability of Pin Piles with respect to buckling and bursting resistances. Regarding the former, early work by Bjerrum (1957) is supported by the detailed analyses of Mascardi (1970, 1982) and Gouvenot (1975). These authors conclude that only in soils of the very poorest mechanical properties, such as loose silts, peat, and nonconsolidated clay, is even a possibility of failure through insufficient lateral restraint feasible.

Similarly, bursting can be equally discounted in conducting analysis to relevant codes for reinforced columns and piers. Where the possibility does exist, additional lateral restraint can be provided by increasing the thickness of the grout annulus, by modifying the grouting design and methods, or by maintaining a sacrificial casing through the suspect horizons.

External Pin Pile Design

GROUT-GROUND BOND As described in the discussion on Construction in Section 6.2, the drilling and grouting methods used in Pin Pile construction act to produce excellent bond characteristics in the load-transfer zone. Analogies can be drawn with ground anchorage practice, albeit for interfaces in the opposite sense of shear, and so we have access to a wealth of published data (e.g., Littlejohn and Bruce, 1977; Littlejohn, 1990; Xanthakos, 1991), and codes or regulations (e.g., FIP, 1982; PTI, 1986; BS, 1989).

For typical Pin Piles, injected at pressures of less than 1 MPa in cohesionless soils, a typical approach (BS, 1989) is

$$Tf = Ln \tan \phi'$$

where Tf = ultimate load holding capacity
L = bond length
ϕ' = angle of shearing resistance
n = factor that apparently takes into account the drilling technique, depth of overburden, pile diameter, grouting pressure, in situ stresses, and dilation characteristics

Field experience (Littlejohn, 1970) indicates*:

coarse sands and gravels
($k > 10^{-4}$ m/sec): n = 40 to 60 tonnes/meter
fine-medium sands
($k = 10^{-4}$ to 10^{-6} m/sec): n = 13 to 16 tonnes/meter

*These figures were initially recorded in normally consolidated materials for borehole diameters of about 100 mm, and where this diameter varies significantly, n is modified in the same proportion.

For piles grouted under gravity head, the following two relationships may be used, particularly valid as they relate to standard penetration test/N values:

1. $\tau_{ult} = 0.007N + 0.12$ MPa (Suzuki et al., 1972)
2. $\tau_{ult} = 0.01N$ MPa (Littlejohn, 1970)

Thus, for N values consistently 200 or above, and for a 220-mm hole, loads of around 40 tonnes/meter can be mobilized, with a theoretical safety factor of 3.

Especially in less competent materials, the magnitude of skin friction can be strongly influenced by the grouting pressure (Figure 6-6): PTI (1986) indicates an enhancement potential of 20 to 50 percent in both cohesive and cohesionless soils.

In addition, pressure grouting may increase the nominal cross section, particularly in the weaker soil layers or near ground level, where natural in situ horizontal stresses are small.

Where the bond zone is founded in rock, the bond is assumed to be uniformly distributed over the entire interface, even though this is unlikely to be the case where E grout/E rock is less than 10 (Littlejohn and Bruce, 1977). The working load, LW, is calculated from

$$LW = \frac{\pi DL}{sf} \tau_{ult}$$

where D = diameter
L = length

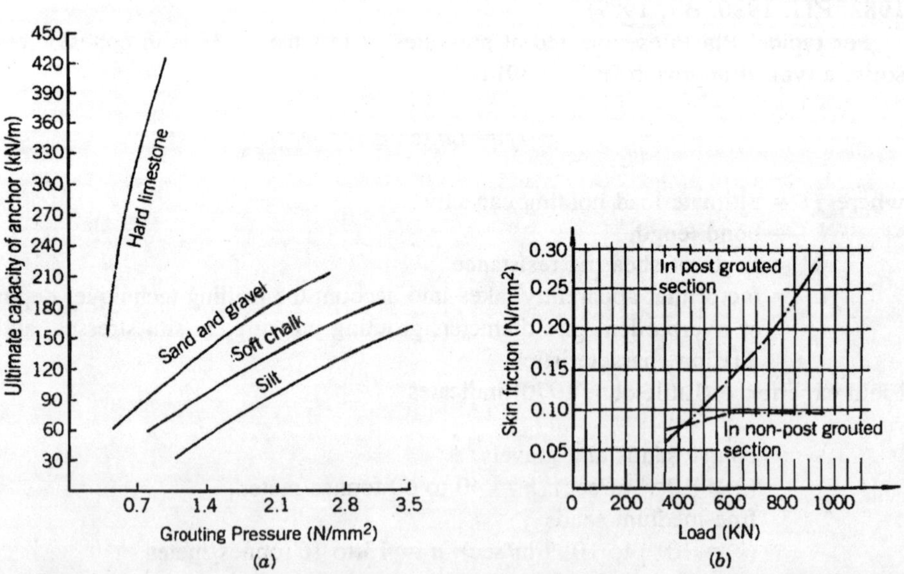

Figure 6-6 (*a*) Influence of grouting pressure on ultimate load holding capacity. (From Littlejohn and Bruce, 1977.) (*b*) Effect of postgrouting on skin friction. (From Herbst, 1982.)

τ_{ult} = ultimate skin friction
sf = safety factor

Similarly, for *cohesive soils* (grouted at gravity pressure), undrained shear strengths are used to estimate capacities:

$$LW = \pi DLCu\alpha$$

where LW, D and L are as above and
Cu = average undrained shear strength over the bond length
α = the adhesion factor

It is common to select $\alpha = 0.45$ in conventional large-diameter bored piling. However, due to their size, and construction procedures, Pin Piles are often designed satisfactorily with α values of 0.6 to 0.8.

A slightly different approach was described by Lizzi (1985), who advocated a "simple empirical formula" for the ultimate load P_{ult} (in kilograms):

$$P_{ult} = DLKI$$

where D = the nominal diameter, in cm, of the pile, that is, the drilling diameter
L = the length of the pile, in cm
K = a coefficient that represents, in kg/cm^2, the average interaction between the pile and the soil for the whole length (From the physical point of view, it can represent the pile/soil adherence, or the shear stress induced in the soil by the pile, or the cohesion of the soil, etc.)
I = a nondimensional coefficient of form that depends on the nominal diameter of the pile.

Values of K and I are as follows:

VALUES OF K (KG/CM^2)

Soil	K
Soft soil	0.5
Loose soil	1.0
Soil of average compactness	1.5
Very stiff soil, gravels, sands	2.0

VALUES OF I

Diameter of the Pile	I
D = 10 cm	1.00
D = 15 cm	0.90
D = 20 cm	0.85
D = 25 cm	0.80

The choice of a suitable value for τ_{ult} is often related to local knowledge and the back analysis of successful applications. Equally, trusted relationships such as the following can be employed, for competent rock with 100 percent core recovery:

$$\tau_{ult} = \frac{UCS}{10}_{rock} \text{ to a maximum value of 4.0 Mpa}$$

Thus, even if a strong granite of unconfined compressive strength far greater than 40 MPa is proved, the unconfined value of 4.0 MPa is still selected as τ_{ult}. Applying a safety factor of 3, a working value of τ_w of 1.3 MPa would be chosen. Even if a typical permissible local value of 1 MPa were to be enforced, for a 220-mm-diameter hole drilled 2000 mm* into such rock, the allowable working load could be

$$LW = \pi DL \tau_w$$

$$= \frac{\pi \times 220 \times 2000 \times 1}{1000} = 138 \text{ tonnes}$$

This underlines again that it is internal load-carrying capacity that limits individual pile rating. Conversely, a Pin Pile of 55 tonnes working load, with similar geometry, would mobilize an average τ_w value of less than 0.4 MPa, a figure to be regarded as conservative when compared with values of rock-grout bond (Table 6-7) used in practice in ground anchors (Littlejohn and Bruce, 1977). It will be noted that these calculations implicitly assume that the grout transferring load from steel to ground can readily sustain the required shear stresses.

For weak rocks (unconfined compressive strength less than 7 MPa) Littlejohn (1990) recommends that the ultimate bond should not exceed the minimum shear strength, based on shear tests on representative samples. The degree of weathering is a major factor that can affect both the ultimate bond and the load/displacement characteristics. This is seldom quantified, but for design in weak weathered rock, Littlejohn (1990) notes that SPT data can predict ultimate bond, for example, ultimate bond (kN/m^2) = 10N (where N = blows/0.3 m) for stiff to hard chalk.

As a final point, grout-ground bond potential can be very sensitive to construction techniques and quality. Special preproduction pile testing is therefore extremely valuable—and often essential—to demonstrate the adequacy of the design assumptions as married to the installation parameters.

Spacing of Pin Piles: Group Effect The design of underpinning systems incorporating Pin Piles usually dictates the need for groups of closely spaced piles. With conventional piles, there is usually a compromise to be resolved between the desire to select a close pile spacing, thus minimizing the size and cost of the pile cap and,

*It is noteworthy that such a minimum embedment length is often specified in order to avoid terminating the pile in a boulder: few boulders are found over 2000 mm in diameter and consisting of very fresh bedrock material.

TABLE 6-7 Rock/Grout Bond Values for Rock Anchors

Rock Type	Working Bond (N/mm^2)	Ultimate Bond (N/mm^2)	Factor of Safety	Source
Igneous				
Medium hard basalt		5.73	3.4	India—Rao (1964)
Weathered granite		1.50–2.50		Japan—Suzuki et al. (1972)
Basalt	1.21–1.38	3.86	2.8–3.2	Britain—Wycliffe-Jones (1974)
Granite	1.38–1.55	4.83	3.1–3.5	Britain—Wycliffe-Jones (1974)
Serpentine	0.45–0.59	1.55	2.6–3.5	Britain—Wycliffe-Jones (1974)
Granite and basalt		1.72–3.10	1.5–2.5	USA—PCI (1974)
General			3	Britain—Littlejohn (1972)
Competent rock (where $UCS > 20$ N/mm^2)	Uniaxial compressive strength ÷ 30 (up to a maximum value of 1.4 N/mm^2)	Uniaxial compressive strength ÷ 10 (up to a maximum value of 4.2 N/mm^2)		
Weak rock	0.35–0.70			Australia—Koch (1972)
Medium rock	0.70–1.05			
Strong rock	1.05–1.40			
Wide variety of igneous and metamorphic rocks	1.05		2	Australia—Standard CA35 (1973)
Wide variety of rocks	0.98			France—Fargeot (1972)
	0.50			Switzerland—Walther (1959)
	0.70			Switzerland—Comte (1965)
		1.20–2.50		Switzerland—Comte (1971)
	0.70		2.25 (Temporary) 3 (Permanent)	Italy—Mascardi (1973)
	0.69	2.76	4	Canada—Golder Brawner (1973)

(*continued*)

TABLE 6-7 *(Continued)*

Rock Type	Working Bond (N/mm²)	Test Bond (N/mm²)	Ultimate Bond (N/mm²)	s_m (test)	s_f (ultimate)	Source
	1.4	4.2			3	USA—White (1973)
			15–20 percent of grout crushing strength		3	Australia—Longworth (1971)
Concrete			1.38–2.76		1.5–2.5	USA—PCI (1974)

Recommendations for Design

Rock Type	Working Bond (N/mm²)	Test Bond (N/mm²)	Ultimate Bond (N/mm²)	s_m (test)	s_f (ultimate)	Source
Igneous						
Basalt	1.93		6.37		3.3	Britain—Parker (1958)
Basalt	1.10	3.60				USA—Eberhardt and Veltrop (1965)
Tuff	0.80					France—Cambefort (1966)
Basalt	0.63	0.72				Britain—Cementation (1962)
Granite	1.56	1.72				Britain—Cementation (1962)
Dolerite	1.56	1.72				Britain—Cementation (1962)
Very fissured felsite	1.56	1.72				Britain—Cementation (1962)
Very hard dolerite	1.56	1.72				Britain—Cementation (1962)
Hard granite	1.56	1.72				Britain—Cementation (1962)
Basalt and tuff	1.56	1.72				Britain—Cementation (1962)
Granodiorite	1.09					Britain—Cementation (1962)
Shattered basalt		1.01				USA—Saliman and Schaefer (1968)
Decomposed granite		1.24				USA—Saliman and Schaefer (1968)
Flow breccia		0.93				USA—Saliman and Schaefer (1968)
Mylontised prophyrite	0.32–0.57					Switzerland—Descoeudres (1969)
Fractured diorite	0.95					Switzerland—Descoeudres (1969)
Granite	0.63	0.81				Canada—Barron et al. (1971)

Source: Littlejohn and Bruce (1977).

on the other hand, the need to maintain a certain minimum interpile spacing so as to avoid the "group effect" necessitating a reduction in the nominal capacity of each pile. For example, "CP 2004 Foundations" (BS, 1972) states that, for "friction piles, the spacing center-to-center should be not less than the perimeter of the pile; with piles deriving their resistance mainly from end bearing the spacing center-to-center should be not less than twice the least width of the pile."

Conversely, the group effect paradigm for Pin Piles is the opposite. Lizzi (1982) and others, such as ASCE (1987), refer to the "knot effect" whereby a "positive" group effect is achieved in the loading of the soil-pile system. For example, Plumelle's (1984) full-scale testing yielded the results that are shown in Figure 6-7 and that confirmed Lizzi's earlier model tests (see Figure 6-8). The latter noted that the increase was proportionally greater in sand than in the cohesive pozzolanic material that allowed interaction in even the Group a arrangement.

In current practice, there is no evidence that this positive "group effect" is being

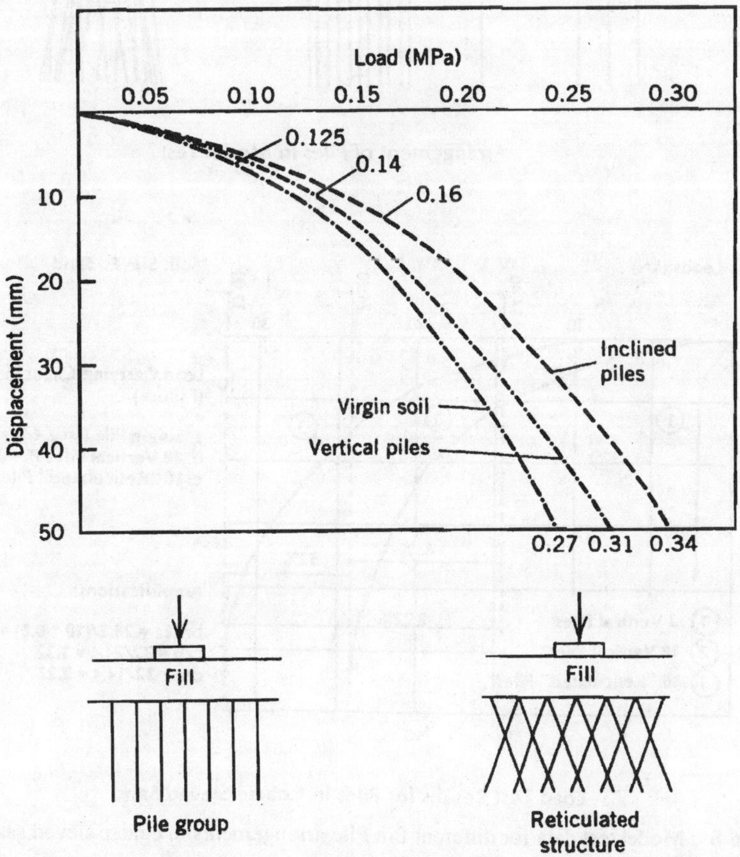

Figure 6-7 Field test data for different Pin Pile arrangements. (From Plumelle, 1984.)

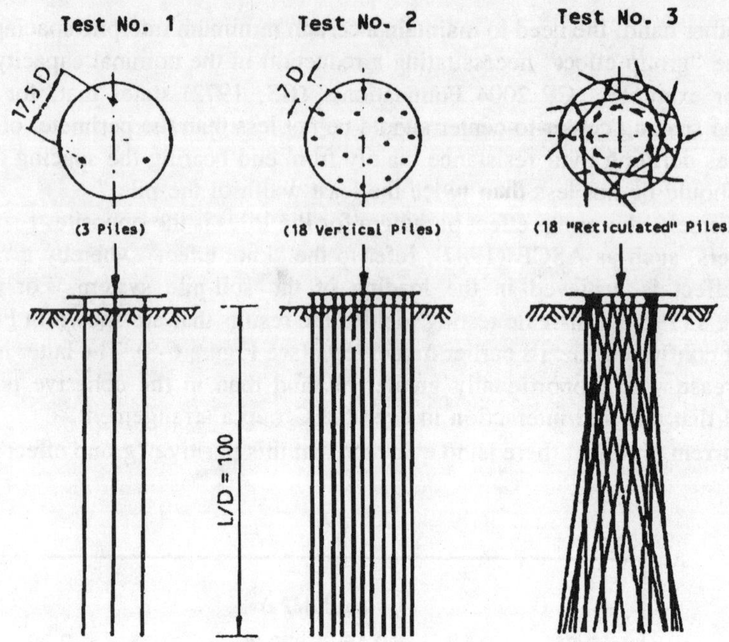

Arrangement of Piles in Model Test

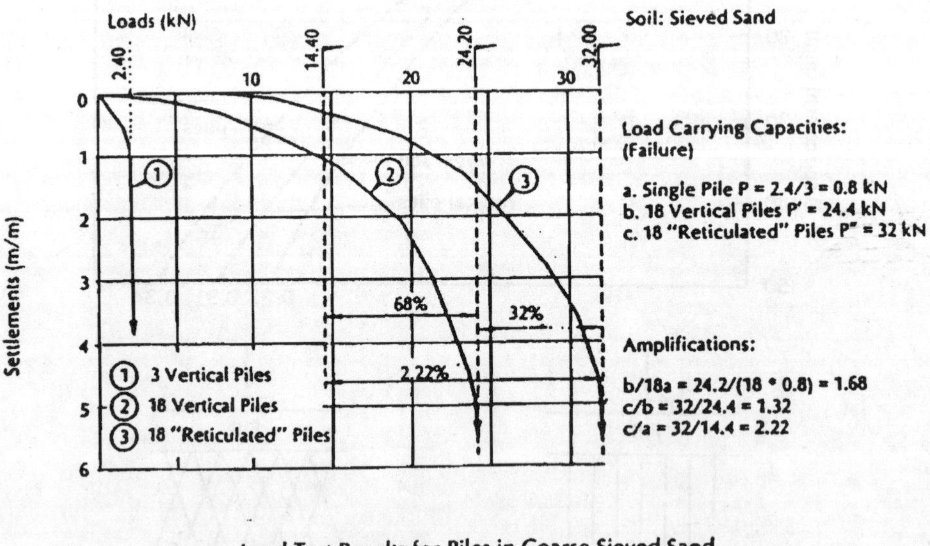

Load Test Results for Piles in Coarse Sieved Sand

Figure 6-8 Model test data for different Pin Pile arrangements in coarse sieved sand. (From Lizzi, 1978.)

routinely exploited in Pin Piles for axial load-bearing applications. On the other hand, there are no examples of accommodations for any negative group effect.

Case Histories and Performance

As an introduction to U.S. practice and developments over the last 15 years or so, Table 6-8 summarizes details of projects executed by one contractor—Nicholson Construction. These examples exclude the frequent cases where Pin Piles have been used as reaction elements in the course of large rock anchor tests, or where they have been installed as in situ reinforcement (Section 6.3), or where they have been used as simple pins to stabilize the toes of sheet pile walls.

Table 6-8 illustrates several points common to Pin Pile projects:

- The wide range in the scope of individual projects.
- The range in working loads.
- Their installation in virtually every ground condition.
- The relatively narrow range in dimensions.
- The typical applications of restricted headroom and access conditions, within existing structures and operating industrial facilities.
- The common use of a permanent steel casing from the surface to the load transfer zone.
- The excellent load holding performance, with minimal settlements.

The following seven major case histories have been selected from those listed in Table 6-8 to illustrate the details of design, construction, and performance outlined in the preceding sections of this chapter.

1. *Boylston St., Boston, Mass.:* Classic example of medium-capacity piles installed in very restrictive access and headroom conditions.
2. *Coney Island, N.Y.:* Installation and testing of several thousand Pin Piles in an operating transit repair facility in very restrictive conditions.
3. *Warren County, N.J.:* High-capacity Pin Piles installed in karstic limestone in place of caissons to support a new bridge pier.
4. *Postal Square, Washington, D.C.:* Medium-capacity Pin Piles in typical rehabilitation application with special testing of structure/pile contact.
5. *Presbyterian University Hospital, Pittsburgh, Pa.:* Innovative use of Pin Piles—subsequently exposed for the upper 22 ft of their lengths—to support a functioning hospital building.
6. *Pocomoke River Bridge, Md.:* Underpinning of an old delicate bridge using preloaded Pin Piles.
7. *Augusta, Ga.:* Postgrouted Pin Piles in operational "clean" factory environment.

TABLE 6-8a Some Pin Pile Projects Executed in the United States, 1978–1988

Location	Location/ Application for Foundations Being Underpinned	Ground Conditions	Installation Conditions	Load (tons) Working/Test	Number of Production Piles
Appollo, Pa.	New tank in existing wastewater treatment plant	Loose fill with concrete obstructions over clay over medium to very dense sands with silt and gravel	Plant measured 38′ × 48′ in plan. Maximum headroom 18′	10/20	45
Brookgreen Gardens, S.C.	Supported masts of suspended net forming "natural" aviary in swamp, with minimal damage to environment	Loose sands and organics over medium-dense sand	Natural cypress swamp	55 generally (15 for center pile)	25
Neville Island, Pa.	Existing dust collector structure on rapidly compacting soil	Loose fill over compact sand and gravel	10′ to 16′ headroom	30/60	32
Providence, R.I.	Test to assess viability of underpinning existing granite block seawall	Quay, bearing on silt, sand and till overlying sandstone bedrock	Open air	55/110	1 (Test)
Trafford, Pa.	New printing press in existing building	Loose cinder fill over silty clay and weathered shale bedrock	14′ headroom	10/20	20
Warwick, N.Y.	Existing gymnasium building (use of preloaded piles)	Loose sandy silt and glacial till becoming denser with depth	Minimum headroom 20′	27.5/55	62
Monessen, Pa.	Existing operating coke battery, emission control facility	Fill over clayey sand and gravel	19′ to 25′ headroom	50/100 (comp) 35 or (tension) 45	102
Mobile, Al.	Two existing sodium hydroxide storage tanks under which wood piles had failed.	Soft organic silt and clay over dense sand with gravel	Very restricted access. 8′ to 15′ headroom. Caustic chemical spills	34 54	171 7
Burgettstown, Pa.	Existing gantry runway	Slag, silty sandy clay and shales over sandstone and limestone	Maximum headroom 24′. Soil saturated with sulfuric acid	10	20

(continued)

TABLE 6-8a *(Continued)*

Total Length Installed (ft)	Individual Length (ft) Typical/ Range	Nominal Drilled Diameter in Bond Zone (in.)	Construction Data: Reinforcement and Casing	Construction Data: Grouting	Test Performance/ Special Notes
1,350	30	5	#11 rebar in lower 20′ + 5″ casing in upper 15′	Type I, $w = 0.5$, maximum pressure 100 psi	Test data on 2 piles: Total displacement at 20 tons—0.049″ and 0.077″ Permanent displacement after—0.008″ and 0.022″ respectively
1,174	30 to 35 for verticals, 55 for rakers	5	#9 rebar full length, 5″ casing in upper 20′	Type I, $w = 0.5$ maximum pressure 120 psi	Award winning solution to unique set of problems
928	29	5	#9 rebar in lower 16′, 5″ casing in upper 20′	Type I, $w = 0.5$, maximum pressure 100 psi	Test data on 1 pile: Total displacement at 60 tons—0.078″ Permanent displacement after—0.01″ (allowable 0.60″)
65	65	6	5″ casing for 57′	Type I, $w = 0.45$, gravity fill	Test data on 1 pile. Total displacement at 110 tons—0.70″ Permanent displacement after—0.03″
720	36	5	5″ casing full length	Type II, $w = 0.5$ maximum pressure 100 psi	Test data on 1 pile. Total displacement at 20 tons—0.055″ Permanent displacement after—0.005″
4,030	65	5	2 No. 0.6″ diameter strands (for preloading 5″ casing in upper 40′)	Tupe I, $w = 0.45$, maximum pressure 120 psi	Test data on 2 piles: Total displacement at 55 tons—0.188″ and 0.249″ Permanent displacement after—0.002″ and 0.005″ respectively
6,330	55 and 65	5	#7 rebar full length, 5″ casing for all except lower 10′	Type II, $w =$ 0.45, maximum pressure 100 psi	Test data on 1 pile. Total displacement at 100 tons—0.312″ Permanent displacement after—0.008″
9,600	56 (range 46 to 60)	5	5″ or 6⅝ for full length except lower 8′	Type I, $w = 0.5$, maximum pressure 80 psi	Piling part of major overall structural repair.
400		6⅝			
640	32	4 (for 3′ rock socket)	3½″ casing full length	Type II, $w =$ 0.45, maximum pressure 40 psi	—

(continued)

TABLE 6-8a *(Continued)*

Location	Location/ Application for Foundations Being Underpinned	Ground Conditions	Installation Conditions	Load (tons) Working/Test	Number of Production Piles
Dunbar, Pa.	Addition to water treatment plant	Fill over fine sand and sandstone	Open air	45	7
Pittsburgh, Pa.	Existing structure adjacent to deep excavation	Fill and fine alluvials over dense sands and gravels with trace silt	Open air	50	21
Pittsburgh, Pa.	Existing parking garage	Fill and alluvials over sandstone/ siltstone bedrock	8′ to 10′ headroom	55	46
Aliquippa, Pa.	New emission control building at existing coke battery	Slag fill over dense sand and gravel	25′ headroom	50/100 (comp) 75/150 (tension)	31 8
Jeanette, Pa.	New machine in existing building	Fill, silt, and clay over bedrock	20′ headroom	Total of 150 tons of structural weight supported	27
Appollo, Pa.	New nuclear power structure in existing building	Loose fills with clay over medium sands with gravel	20′ headroom	10	24
Marion, Ind.	Existing body stamping plant	Silty sand over rock	18′ headroom	60	24
Alcoa, Tenn.	New building in existing mill	Limestone	Open air	70/140	
Washington, D.C.	Existing structure at Castle Building, Smithsonian Institute	Fill over dense sands with gravel	Very restrictive access and hole entry conditions	50/100	21
Pittsburgh, Pa.	Restoration of existing Timber Court Building	Sands and gravels, over sandstone bedrock	10′ headroom	50	15
Warren Co. N.J.	New bridge pier	Karstic limestone with voids and gouge	Open air, small area	100/224	24

TABLE 6-8a (*Continued*)

Total Length Installed (ft)	Individual Length (ft) Typical/ Range	Nominal Drilled Diameter in Bond Zone (in.)	Construction Data: Reinforcement and Casing	Construction Data: Grouting	Test Performance/ Special Notes
179	26 (ranges 25 to 26)	5	#6 rebar for lower 10′, 5″ casing for upper 20′	Type III, w = 0.45, gravity pressure	——
630	30	5	5″ casing for upper 20′	Type I, w = 0.5, maximum pressure 60 psi	Piles installed in conjunction with subhorizontal soil nails for excavation stability.
1,980	43 (range 38 to 44)	5	5″ casing to rock head	Type I, w = 0.45, gravity pressure	——
2,170 600	70 75	5 5	#6 rebar for lower 25′ 5″ casing for upper 50′	Type I, w = 0.45, maximum pressure 120 psi	Test data on 1 pile: Total displacement at 100 tons—0.2″ Permanent displacement after—0.02″
945	35	5½	5½″ casing full depth	Type I, w = 0.45, gravity pressure	——
552	23	5½	#7 rebar full depth, 5½″ casing for upper 18′	Type II, w = 0.45, maximum pressure 150 psi	——
1,680	70	7	7″ casing for upper 50′, #11 rebar for lower 25′	Type I, w = 0.45, maximum pressure 50 psi	——
	40	5½			Total displacement at 140 tons—0.459″ Permanent displacement after—0.078″
1,580	75 (range 69 to 77)	5½	#11 rebar full depth, 5½″ casing between footing and bond zone	Type I, w = 0.5, maximum pressure 140 psi	1 Piles combined with subhorizontal soil nails to stabilize excavation adjacent to structure. 2 Data on Test Pile 2: Total displacement at 100 tons—0.653″ Permanent displacement after—0.078″
1,050	70	5½	5½″ casing full length	Type I, w = 0.45, gravity pressure	——
1,889	78 (ranges 44 to 200)	8½	7″ casing full length	Type III, w = 0.5, maximum pressure 50 psi	Test data on 1 pile: Total displacement at 205 tons—0.40″ Permanent displacement after—0.07″

(*continued*)

TABLE 6-8a (*Continued*)

Location	Location/ Application for Foundations Being Underpinned	Ground Conditions	Installation Conditions	Load (tons) Working/Test	Number of Production Piles
Kingsport, Tenn.	New storage tank in existing building	Silts and sands over limestone	11′ headroom	40/80	115
Boylston St. Boston, Mass.	Existing building being redeveloped	Soft fills and organics over medium dense sand	Minimum headroom 8′ in very restrictive basement conditions	40/92 (comp) 12/27 (tension)	262
Ann St., Pittsburgh, Pa.	To support new soldier beams for new retaining wall	Weathered shale and sandstone over competent sandstone	Open air	48/68 (comp) 8/12 (lateral)	86
Coney Island, N.Y.	Rehabilitation of existing repair shop	Fill and organic silt over dense sands	Minimum headroom 8′. Very difficult access in fully operational facility	15/30 and 30/60	2300 1900
Cleveland, Ohio	New addition to existing control building	Slag fill and soft silty clay over shale bedrock	Open air but difficult access due to ongoing steel plant operations	60	45

TABLE 6-8b Details from Pin Pile Projects Completed Since late 1988 by Nicholson Construction (to End 1990)

Location	Application	Ground Conditions	Installation Conditions	Load (tons) Working/ Test
Cleveland, Ohio	Foundations for new electric furnace in existing building	Slag fill, soft silty clay over shale bedrock	Low headroom	35/——
Boston, Mass.	Underpinning of existing building being redeveloped	Fill and soft clay over bouldery till	Restricted access, with 8′ minimum headroom	60/120
Cleveland, Ohio	Support to spread footings of existing pipe bridge, already settled 18″	Stiff clay	Very difficult access to and under bridge	$12\frac{1}{2}$/25

TABLE 6-8a (*Continued*)

Total Length Installed (ft)	Individual Length (ft) Typical/ Range	Nominal Drilled Diameter in Bond Zone (in.)	Construction Data: Reinforcement and Casing	Construction Data: Grouting	Test Performance/ Special Notes
4,025	35	5½	#8 rebar in lower 15 ', 5½" casing to bedrock	Type I, $w = 0.45$, gravity pressure	No measureable permanent displacement after testing to 80 tons
7,070	27	5½	#8 rebar full length, 5½" casing in upper 19′	Type II, $w = 0.5$, maximum pressure 60 psi	Total data on 2 piles: total displacement at 92 tons—0.44" and 0.34" Permanent displacement after—0.25" and 0.16"
1,000	11.5	6	#11 high-strength rebar full length	Type I, $w = 0.45$, gravity pressure	1 Piles subjected to vertical, lateral, and moment testing 2 Compression test data on 6 piles: Total displacement at 68 tons—0.059 to 0.099" Permanent displacement after—0.006 to 0.020"
80,500 85,500	35 45	6⅝ 7⅝	#6 rebar full length, # 9 rebar full length	Type I, $w = 0.45$ maximum pressure 60 psi	Extensive test program
6,390	142	6½ (for 5′ rock socket)	7" casing to rock head, #8 rebar for 5′ rock socket and 10′ into casing	Type I, $w = 0.45$, gravity pressure	——

Source: From Bruce (1988).

TABLE 6-8b (*Continued*)

Number of Production Piles	Total Length Installed (ft)	Typical Pile Length (ft)	Nominal Drill Diameter of Bond Zone (in.)	Interior Pile Composition	Notes
12	1500	125	5½	5½ casing plus 1–1⅜ rebar in bond zone	——
97	4,850	50	7	2" high-yield rebar	Two tests to 120 tons: Total def. = 0.223" Permanent def. = 0.050".
4	280	70	5½"	5½" casing	——

(*continued*)

TABLE 6-8b (*Continued*)

Location	Application	Ground Conditions	Installation Conditions	Load (tons) Working/ Test
Montgomery, Co., Pa.	Foundations for new bridge abutment	Silty soil over Karstic limestone	Overhead power lines	77/235
Rome, Ga.	Support for foundations in operational paper mill	Fills over shales with quartzitic seams	Access through doorways; minimum 12′ headroom	95/190
Apollo, Pa.	Support for column foundations to permit excavation of hazardous waste	Low level radioactive fill and silty clay with rock fragments over siltstone and shale	Interior of operating steel mill, minimum 10′ headroom	60/——
Orangeburg, N.Y.	Foundations for exterior stairway for existing psychiatric center	Loose fill overlying very compact glacial till	14 × 14′ access to interior courtyard	5–38/20–75
Huddleston, Va	Foundations for new river bridge	15′ of alluvials and weathered rock over granite/gneiss	Good access, unlimited headroom	70–140
Pocomoke City, Md.	Replacement foundations for 60-year-old delicate bascule bridge	River bed silts and clays over 30′ dense fine-medium sands	Most from bridge deck, 4 from very limited access/headroom	50/100
Augusta, Ga.	Underpinning of footings subjected to additional loads in operational detergent factory	23′ clay over various medium-fine sands with interbedded clays	Very restricted access, minimum 8–10′ headroom	50/100
Baltimore, Md.	Intensive underpinning of historic 5-story building threatened by deterioration of original wood piles	Peats and clayey silt over silty fine sands	Very restricted access, 8–10′ headroom	70/250
Seattle, Wa.	Test program for underpinning of historic building	Sands and silts over fine and silty dense sands	Through concrete footings in old structure, headroom as low as 8′	70/135–150
Pittsburgh, Pa.	Supporting existing columns of operating hospital to	Siltstone, shale, claystone	Interior of very sensitive building with 10′	125/325

TABLE 6-8b *(Continued)*

Number of Production Piles	Total Length Installed (ft)	Typical Pile Length (ft)	Nominal Drill Diameter of Bond Zone (in.)	Interior Pile Composition	Notes
48	1,026	29–80	8½	9⅝″ casing to rock, 7″ casing full length	At test load: Total def. = 0.290″ Permanent def. = 0.020″
33	1,320	40	5½	5½″ casing	——
20	800	40	5½	5½″ casing	——
103	2,760	26–32	8	½″–1″ rebar	At test load: Total def. = 0.110″ Permanent def. = 0.020″
72	1,901	25	7	7″ casing plus 1⅜″ high-yield rebar in bond zone	——
52	5,200	100	7	7″ casing plus 1⅜″ high-yield rebar in bondzone	See text
143	5,291	37	7⅝	1⅜″ diameter high-yield rebar, plus 7″ casing in upper 10′ for lateral resistance	Routine use of post-grouting to enhance soil-grout bond
121	4,500	35–40	7	15–20 ′ of upper 7″ casing with 20–25′ of 1⅜ rebar in bond zone	Described in *Civil Engineering* in December 1990
4 (test)	140	30–40	5½	10–30′ casing, bond lengths with full length 1¼″ rebar	Excellent test data, including use of post-grouting
18	775	43	7	7″ casing	See text

(continued)

TABLE 6-8b *(Continued)*

Location	Application	Ground Conditions	Installation Conditions	Load (tons) Working/ Test
	permit adjacent and ulterior excavation		headroom	
Brooklyn, N.Y.	Temporary and permanent piles to support overhead roadway	Fine-medium glacial sands with silts and clays	Reasonable access, 16′ + headroom	60–100/120–250
Covington, Va.	Foundations for pipe bridge foundations for mill expansion	20′ soils over 15′ shale and limestone	Through and around existing foundations	100/——
State College, Pa.	New column foundations for fire damaged church	Clay over karstic limestone	Difficult access, low headroom	20–35/——
Memphis, Tenn.	Test pile for underpinning of major transport facility	Clayey fill over sanitary landfill over loose sand and stiff clay	Unrestricted	——/80
Washington, D.C.	Underpinning for new and existing foundations for historic, massive building being refurbished	Fill over various alluvial fine-medium sands with cobbly/clayey horizons	Existing basement with 8–17′ headroom in 3 areas	75/150
Pittsburgh, Pa.	Foundation for pedestrian bridge	Backfill over claystone	20′ headroom within 18″ of existing structure	75/——
Baltimore, Md.	Foundation for temporary highway bridge	25′ of alluvials and weakened material over schist	Unrestricted	75/——

Equally significant is that each of these projects featured a preproduction test program, thoughtfully organized and carefully recorded. Bearing in mind that every Pin Pile is *not* routinely tested prior to being put into service—unlike the case with prestressed ground anchorages—such test programs are a vital component of successful applications. To aid comparison, each of the following case histories is presented in the same format, and uses imperial units reflecting the national construction practice.

1. Boylston Street, Boston, Mass.

Background The properties at 739–749 Boylston Street in the Back Bay area of Boston, Mass., were completed in the "Chicago style" in 1906. These derelict commercial buildings, six and three stories high, were acquired for redeveloping

TABLE 6-8b *(Continued)*

Number of Production Piles	Total Length Installed (ft)	Typical Pile Length (ft)	Nominal Drill Diameter of Bond Zone (in.)	Interior Pile Composition	Notes
77	4,250	50–60	7	7″ casing	Excellent vertical and lateral testing with postgrouting
172	6,020	35	6	7″ casing to rock, 3 ea 1¾″ rebar in bond zone	—
50	1,750	35	5½	5½″ casing to rock, 1″ rebar in bond zone	—
1	130	130	5	5″ casing to top of bond zone, 3 ea ⅝ diameter rebars below	At failure load: Total def. = 1.060″ Permanent def. = 0.094″
609	37,500	51–58	7	25–30′ casing plus 25′ of 1⅜ rebar in bond zone	See text
12	540	45	6½	5½″ casing	—
4	100	25	7	7″ casing	—

Note: Grouting conducted with or without excess pressure with neat cement grouts, Type I or II, at w/c = 0.45.

and refurbishing: the former, for example, will have retail space on the basement and first floors, with office space and a mechanical penthouse level above.

The structure was founded originally on pile caps bearing on timber piles. To accommodate the increased loadings from the new construction, additional support was required under enlarged pile caps (Figure 6-9). The engineer foresaw piles of working loads 20 tons (compression) and 6 tons (tension), but accepted the contractor's alternative design offering a cased pile with working loads of 40 tons and 12 tons, respectively.

Piling had to be executed from within the partially demolished basement of the structure (approximately floor elevation +8 ft) about 10 ft below existing sidewalk elevation, giving a minimum working headroom of 8 ft.

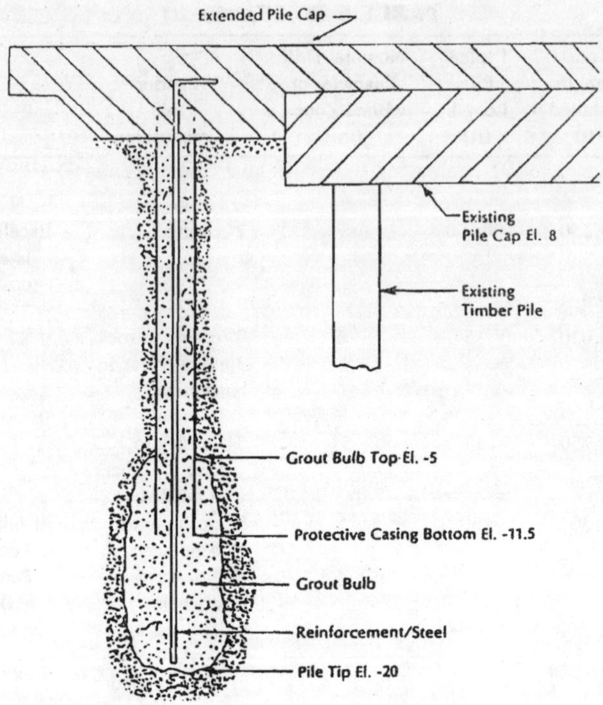

Figure 6-9 General arrangement of minipiles, Boylston St., Boston, Mass. (From Bruce, 1988. Reproduced by permission of Thomas Telford Publications.)

Site and Ground Conditions Access was awkward and restricted, and the position of several piles had to be adjusted slightly to accommodate particular site conditions.

The fill consisted of saturated loose grey-brown fine sand and silt, and overlaid soft grey organic silt with traces of shells, sand, and gravel. The founding layer occurred at about −4 ft and was 18 to 24 ft thick throughout the site. It comprised medium dense/dense fine medium sand with a trace of silt. Pile lengths were maintained within this horizon so as not to perforate the underlying Boston Blue Clay.

Design Piles were designed on the basis of an ultimate load 2.3 times design working load (i.e., 92 tons in compression, 27 tons in tension).

The length of the load transfer zone was designed on the basis of analogous soil anchor experience and assumed $\phi = 35°$ for the sand, a bulb diameter of $7\frac{1}{2}$ in., and a grout pressure of 60 lb/in.^{2} in these soils.

Ultimate soil/grout bond (τ_{ult}) was estimated empirically from the relationship

$$\tau_{ult} = \text{grout pressure} \times \tan\phi$$

$$= 60 \times 0.7 = 42 \text{ lb/in.}^2$$

The required load transfer length (L) was calculated from

$$L = \frac{\text{load}}{\pi \cdot d \cdot T_{\text{ult}}}$$

where d is the bulb (bond zone) diameter.

Thus, for an ultimate load of 92 tons,

$$L = \frac{92 \times 2000}{\pi \times 7.5 \times 42} = 186 \text{ in.}$$

Further routine calculations using the provisions of the Commonwealth of Massachusetts Building Code (1984) demonstrated that:

- The use of 5½ in. casing of 0.362 in. wall thickness, and *fy* (minimum specified yield stress) = 55 kips/in.^{2} as the major load-bearing element was safe. (Allowable stress $\not>$ 35 percent *fy* casing.)
- The anticipated pile settlement at working load was acceptable.
- The compressive strengths generated in the grout of the bond zone were acceptable. (Allowable stress $\not>$33 percent *fc*.)
- The use of an internal 1-in.-diameter 60-kips/in.^{2} rebar would adequately transfer loads in the founding horizon. (Allowable stress $\not>$50 percent *fc*.)

The individual piles were as shown in Figure 6-9 and were arranged as in Figure 6-10.

Construction A diesel hydraulic track rig was used to install all 260 piles. The 5½-in. casing was first water flushed to about 8 ft below the surface, before being pushed for a short distance to locate accurately the top of the dense bearing stratum. Rotary drilling then resumed in the sand to full depth. Neat Type I grout of water-cement ratio about 0.5 was placed by tremie, followed by the rebar. Pressure grouting of the sand was carried out to a maximum of 60 lb/in.^{2} during extraction of the casing, for the 15 ft to 16 ft of bond zone. The casing was then pushed back down about 5 ft into this pressure grouted zone and left in place.

Grout takes generally ranged from 2.5 to 3.5 times nominal hole volume, confirming that the enhanced effective diameter of the bond zone had been achieved. Grout cubes at 14 days gave unconfined crushing strengths of over 6000 lb/in.^{2}

During drilling, wood piles or granite blocks in the fill were occasionally encountered but were accommodated by relocation or perseverance. Overall, four piles had to be replaced due to constructional problems, while the construction of an additional two piles lifted the contract total to 262.

Testing and Performance Prior to the production piling program, compressive and tensile load tests on two typical piles were conducted. Each pile was constructed as described above, except for the addition of a "tell tale" anchored near the tip and the placing of an outer steel liner around the 5½-in. casing above the bond

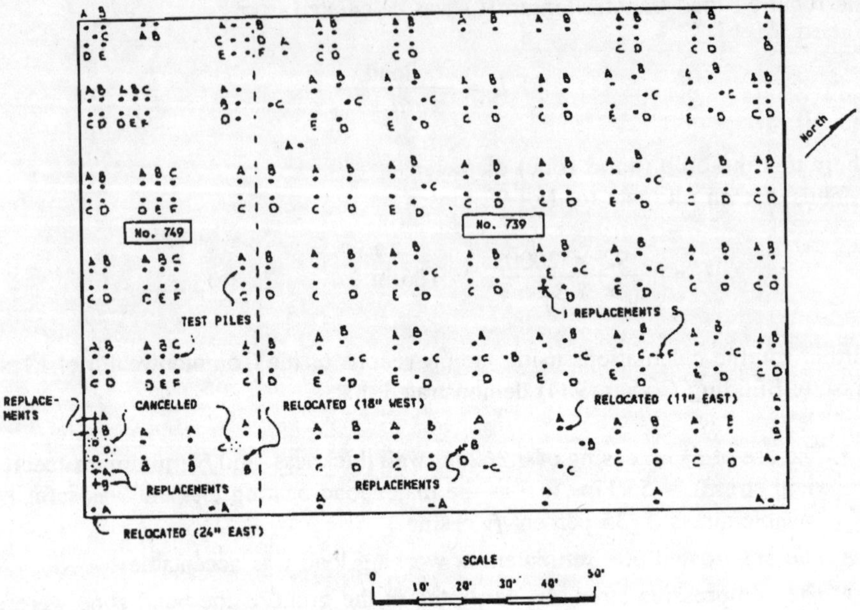

Figure 6-10 Plan of Pin Pile arrangement; Boylston St., Boston, Mass. (From Bruce, 1988. Reproduced by permission of Thomas Telford Publications.)

zone to prevent any load transfer in the upper soils. Reaction for each test pile was provided by adjacent ground anchors, and the tests were executed in accordance with the Massachusetts State Building Code and ASTM D1143. The data are summarized in Table 6-9, while the performance of TP2 (in compression) is shown in Figure 6-11, together with that of a timber pile, for comparison.

It was noteworthy that the elastic (recoverable) settlement at 80 tons was about half the total deflection, while no indication of pile or soil failure was evident from the butt or tip displacement curves. Furthermore, the net butt settlements were well below recommended Building Code criteria for maximum net settlements. The performance in tension was equally satisfactory.

Most of the major structural rebuilding work was completed in the eight-month period following completion of the Pin Piles. Readings were taken regularly of the pile cap deflections at 16 locations. The range of cap settlements during construction was 0.06 to 0.24 in. (average 0.16 in.) and entirely consistent with the test data of Table 6-9 (Total settlements of 0.34 to 0.44 in. at twice working load, without the benefit of existing timber piles).

2. Coney Island, N.Y.

Background The Coney Island Main Repair Facility of the New York Transit Authority has been in operation for over 70 years and is the largest of its kind in the world. It encompasses, including the rail yards, about 100 acres, of which 12 are

TABLE 6-9 Summary of Test Data on Test Piles (TP) 1 and 2, Boylston St., Boston, Mass.

	Butt (in.)		Tip (in.)	
	TP-1	TP-2	TP-1	TP-2
Compression test (to 80 t)				
Gross settlement	0.44	0.34	0.31	0.19
Net settlement (permanent)	0.25	0.16	0.25	0.16
Tension test (to 24 t)				
Gross heave	0.24	0.14	0.17	0.06
Net heave (permanent)	0.16	0.09	0.15	0.06

Source: From Bruce (1988).

covered building space. Constructed on the former Coney swamp, the repair shop was built on a loose fill surface with no pile support for the floors. The steel frame, columns, and outside walls were supported on piled foundations. Settlement had produced major underfloor voids, which had led to many floor collapses such as an 18-in. drop in the main shop in 1980.

During the original construction, the swamp filling had apparently created mud waves, resulting in uneven thicknesses of the soft organics underlying the structure. The subsequent settlement of the ground surface due to the loading by the fill and the structure had thus been irregular in magnitude across the site.

After "Years of Band-Aids" (Munfakh and Soliman, 1987) a $100 million repair program was initiated in 1984 coincident with the installation of new equipment that would alone have accelerated the settlement problem. Foundation repair had to be carried out in a fashion guaranteeing minimum disruption to continuing shop operation, as well as constituting a proven, compatible, and cost-effective solution.

Remedial options under consideration included compaction grouting, chemical grouting, and concrete filled steel shell piles. However, conventional Pin Piles proved to be the most attractive solution from all viewpoints, and a contract was let in early 1987 to install over 4200 piles.

Site and Ground Conditions Four distinct soil layers were identified under the slabs: fill, peat with organic silt, grey sand, and brown sand. Short- and long-term consolidation testing confirmed the organic layers to be the cause of the settlement. These strata experienced long-term secondary consolidation and peat/organic degradation, either from oxidation or micro organisms. Typically the medium dense, fine sands recognized as being adequate load-bearing materials commenced 10 to 25 ft below the surface. The piezometric level was at about −4 ft.

Access and headroom conditions were always restrictive and frequently obstructive, being as little as 8 ft. In addition, as the work was to be carried out in a busy,

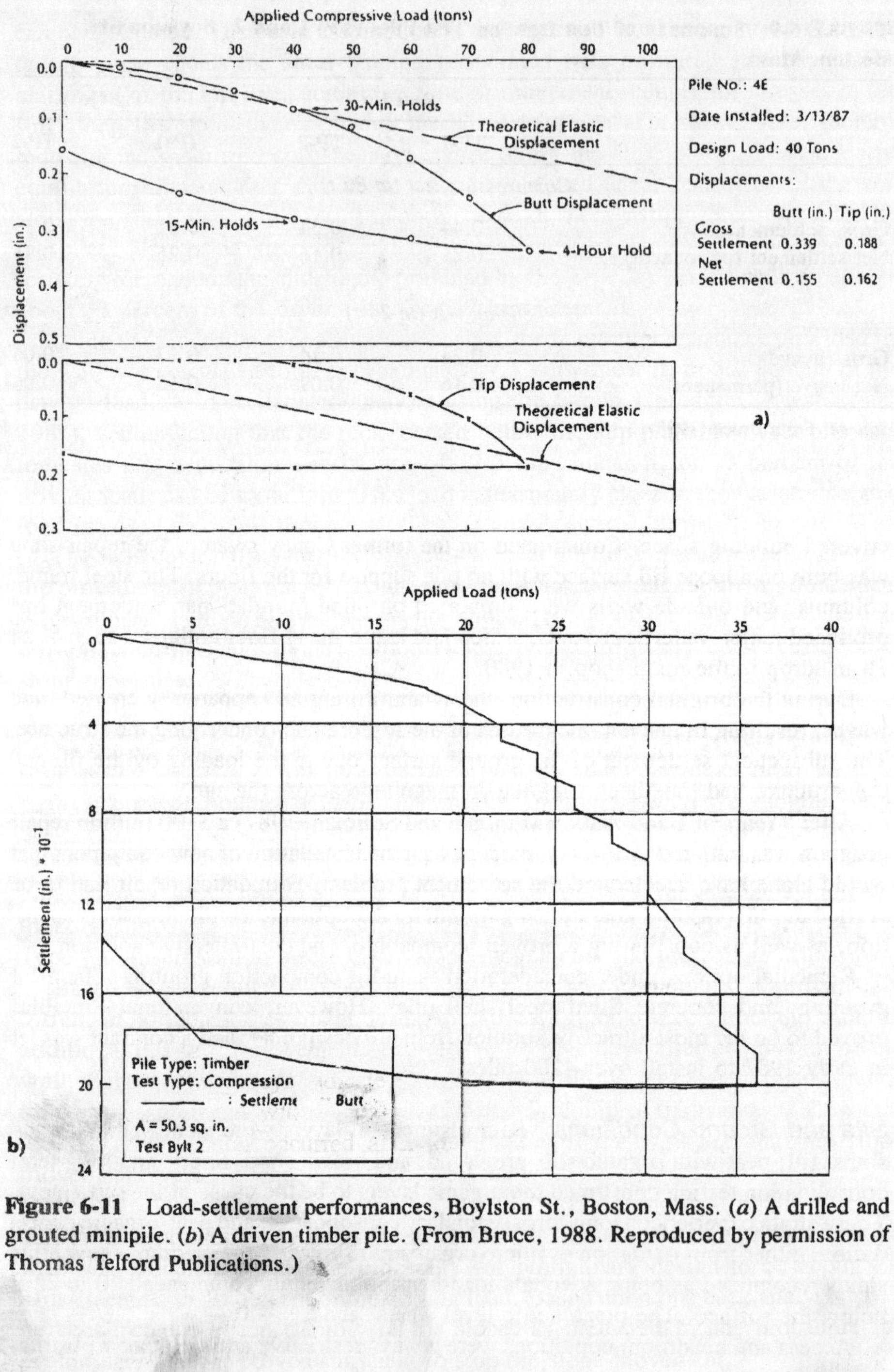

Figure 6-11 Load-settlement performances, Boylston St., Boston, Mass. (*a*) A drilled and grouted minipile. (*b*) A driven timber pile. (From Bruce, 1988. Reproduced by permission of Thomas Telford Publications.)

fully operational facility, in collaboration with other major structural repairs, it had to be executed in restricted "packages" in a piecemeal fashion.

Design Approximately 2300 15-ton piles and 1900 30-ton piles were required. The engineer's design allowed for the load to be taken on #6 bars, without the addition of sacrificial steel casing in the soft upper zones wherein resistance to buckling was analyzed and judged adequate. Standard design procedures, based on $\phi = 30°$, were used to arrive at total lengths of 35 and 45 ft for 15-ton and 30-ton piles, respectively, that is, 10 to 20 ft into the load-bearing sand.

Construction Before installing the piles, the existing underslab voids were filled with a lightweight foamed concrete. It was intended that its light weight would inhibit additional settlement and corresponding downdrag forces to the piles. The fill would also protect against erosion by blocking water flow through such voids.

The access and headroom restraints over much of the site demanded the use of specially constructed drilling equipment featuring short masts and remote power units. Whenever possible, conventional crawler mounted units were employed with special care having to be taken in all cases with exhaust fumes and drilling spoil disposal.

The 15-ton piles were drilled and cased to 6⅝-in. nominal diameter and the 30-ton piles to 7⅝-in. nominal diameter. Water flush was used. This casing was completely withdrawn during the pressure grouting of the sand using neat Type 1 with $w = 0.50$ to a maximum of 60 lb/in.^{2}, following the placing of the reinforcing bar (#6 or #9 rebar full length).

Load transfer to the existing slab structure was provided by an underreamed supporting zone formed under the slab (Figure 6-12).

Performance and Testing A program of 10 full-scale test piles was executed to verify assumptions regarding design and performance for the two pile types. PVC

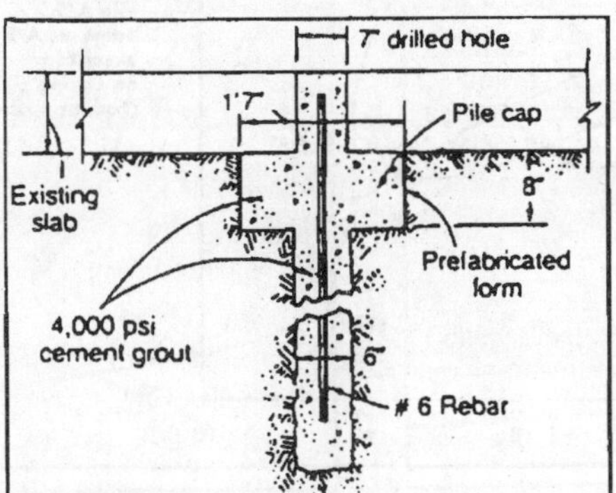

Figure 6-12 Schematic arrangement of minipile and existing base slab; Coney Island, N.Y. (From Munfakh and Soliman, 1987.)

liners were provided from the slab to the top of the sands to ensure transfer of load only in the lower horizons.

In the first three compression tests the load was applied directly to each pile via a beam/reaction anchor system. Munfakh and Soliman (1987) reported that the high concentration of stress crushed the top portion of each pile. The remaining test piles were given an enlarged cap, providing better load transfer to the grout and reinforcement. Load tests were run to twice working load in compression, and to 50 tons in tension.

The steel casing was left in place in one pile (number A/8) so that a performance comparison with the standard pile number A/9 could be obtained (Figure 6-13).

The first four piles (Table 6-10) experienced significant creep at maximum load

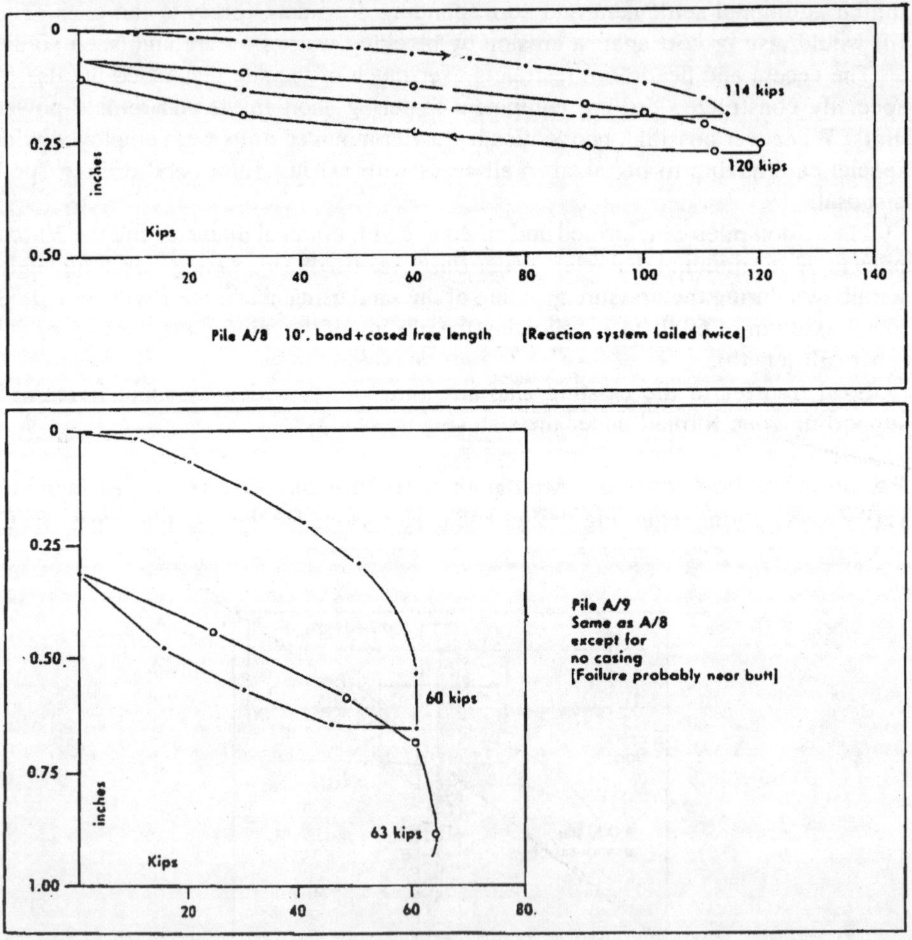

Figure 6-13 Comparative test performances of two 35-ft-long minipiles, with and without permanent casing; Coney Island, N.Y. (From Bruce, 1988. Reproduced by permission of Thomas Telford Publications.)

TABLE 6-10 Comparative Performance of 15-ton Working Load Piles, Coney Island, N.Y.[a]

Pile Number	Description	Ratio of Grout Volume To Hole Volume	Stiffness in Lineal Part[b] (tons/in.)	Maximum Load (tons)	Total Accumulated Deflection (in.)	Notes
A/3	Loaded annulus only	1.2	80	20 (F)	1.25	Failure premature and most probably due to crushing of pile head
A/4		3.7	85	31 (F)	0.65	
A/5	Loaded full section	2.5	95	29 (F)	0.75	Failure possibly due to soil/grout failure, although distress at head also noted
A/9		2.9	72	31 (F)	0.85	
A/7	Includes original concrete slap in cap	2.9	303	70 (F)	0.90	Soil-grout failure likely
A/10	Excludes concrete slab in cap	3.4	178	56	0.42	Test suspended upon failure of pile cap
A/8	With sacrificial casing for 25′	7.7	385	60	0.30	Test suspended when reaction pile pulled

Source: From Bruce (1988).

[a]All piles were $6\frac{7}{8}$ in. in diameter, 35 ft long, including 10-ft bond, and had a full length #6 rebar.

[b]A measure of pile stiffness obtained by dividing the maximum load over which displacement is relatively linear by the displacement of that load.

(up to 0.35 in. in 4 hr), whereas those tested through the cap had less (0.032 to 0.064 in. in 4 hr at 30 tons). The cased pile had less than half this amount of creep in 5 hr at 30 tons.

Such performances were acceptable to the structural designers and the benefits of the cased pile were not required in the production piles subsequently installed.

3. *Warren County, N.J.*

Background The I-78 dual highway was designed to cross the Delaware River between Pennsylvania and New Jersey (Warren County) on seven span, multigirder bridges. Generally, foundations on the Pennsylvania side incorporated driven H piles, whereas the river piers and the New Jersey piers were intended to be founded on solid rock. This proved to be practical except for pier E-6 on the eastbound structure, since the foreseen excavation for the footing to the planned elevation could not find rock head. Further excavation to an elevation 15 to 20 ft below revealed only random rock thicknesses of several feet and a highly irregular bedrock surface. The excavation was filled with lean mix concrete and the foundation design reassessed.

Various options were reviewed, including:

- Enlarged spread footings
- H piles in predrilled holes
- Elimination of the pier
- Relocation of the pier
- Deep bored piling.

Only the last option proved feasible and two alternates were considered:

1. Six large-diameter (36-in.) caissons, each of working load 360 tons.
2. 24 Pin Piles, each of nominal working load 100 tons (allowing an 11 percent redundancy to reflect somewhat the highly variable rock conditions).

Bids were solicited for each option, but due to the extremely onerous geological and programming restraints, only one contractor for each responded. The bid for the 36-in.-diameter caissons was essentially cost plus with a guesstimated price of about $1 million. Nicholson's fixed price offer for the Pin Piles was less than half that figure. The owner therefore decided on the latter option on grounds of cost, program time, and the ability to demonstrate the effectiveness of the system by a test pile installed in advance.

A further technical advantage was the action of Pin Piles in transferring load by skin friction as opposed to end bearing: the possibility of pile failure by punching through into any soft underbed immediately under founding level was therefore eliminated.

Site and Ground Conditions The bedrock was a Cambro-Ordovician dolomitic limestone referred to locally as the Allentown Limestone. It proved to be moderately to highly fissured, cherty, and very susceptible to karstic weathering. Major clay filled beds were intersected even over 100 ft below the surface, for example, 50 ft of soft brown silty clay below 106 ft at the location of pile 24. Dipping 55° to the southeast, the rock mass proved highly variable laterally and vertically. The shape of the solid bedrock surface, as revealed in site investigation holes, and by the subsequent pile drilling, is shown in Figure 6-14.

Design The owner's design regulations permitted:

- Maximum average rock-grout bond at working load (100 tons) of 50 lb/in.^{2}.
- Maximum allowable reinforcement steel stress at working load equivalent to 45 percent *fy*.

These factors led to the selection of:

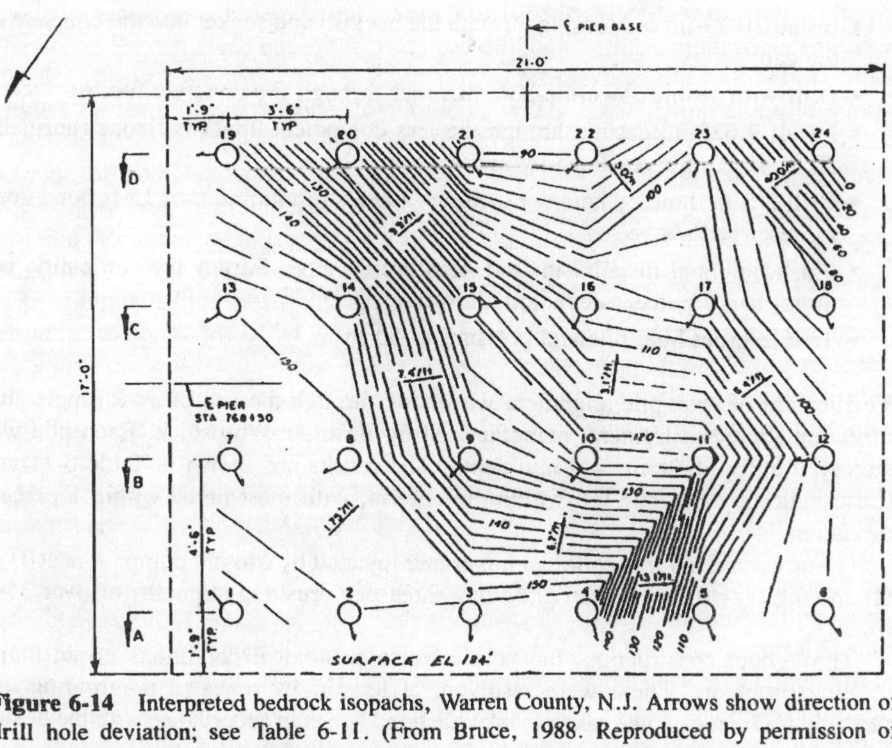

Figure 6-14 Interpreted bedrock isopachs, Warren County, N.J. Arrows show direction of drill hole deviation; see Table 6-11. (From Bruce, 1988. Reproduced by permission of Thomas Telford Publications.)

- A load-transfer zone, $8\frac{1}{2}$ in. diameter and 15 ft long in competent rock.
- Use of a 55-kips/in.^{2} low-alloy steel pipe of 7 in. o.d. and wall thickness of 0.408 in. as pile reinforcement.

Recognizing that the rock was likely to be very variable, provision was made to allow the 15-ft bond zone to not necessarily be continuous. In most piles this was subject to the following restrictions:

- The lower part of the zone was to contain at least 10 ft of continuous sound rock.
- Soft interbeds were to be less than 3 ft thick.
- A zone of acceptable load-bearing rock was to be at least 5 ft thick.
- Regrouting and redrilling of interbeds within the overall bond zone was to be undertaken.

Piles 1, 6, 17, 18, 19, 23, and 24 were required to have a continuous 15-ft-long bond zone.

Construction The sequence of drilling and installation was as follows:

- Install 10.75-in. o.d. casing through the backfill and socket into the concrete of the cap.
- Drill with 10-in. down-the-hole hammer through the concrete footing.
- Install 9.625-in. casing through the less competent upper horizons (normally 30 to 45 ft). Survey linearity and grout in place.
- Drill 8.5-in. hole by hammer or rotary to ensure minimum of 15-ft bond zone as described above.
- Flush hole and install 7-in. o.d. reinforcing pipe. Survey for verticality, not more than 2 percent deviation allowable.
- Tremie grout hole pile and pressure to 50 lb/in.^{2}

Verification of each pile alignment was made through the use of an R single shot direction survey instrument, manufactured by Eastman-Whipstock. Each pile was surveyed at top, bottom, and mid depth. The results are shown in Table 6-11 and these indicate every pile fell within the criteria, with most being within 1 percent deviation.

Grout was mixed in a colloidal mixer and injected by Moyno pump. A neat Type III mix of $w = 0.5$ was used providing three-day crushing strengths of over 3500 lb/in.^{2}

Throughout construction, the very adverse geological conditions posed major drilling problems. These were resolved, at length, by repeated pregrouting and redrilling. Great care was taken to provide bond zones in accordance with the design provisions. Figure 6-15 summarizes the actual total drilled lengths.

Regarding the anticipated caisson tip elevations, also shown in Figure 6-15,

TABLE 6-11 Borehole Deviation Data on Minipile Holes, Warren County, N.Y.

Pile	Length (ft)	Actual Drift (in.)	Ratio Actual to Allowable Deviation (Based on 2% Deviation)	Direction of Drift (see Figure 6-14)
1	44.0	4.6	44%	S 50°E
2	47.0	3.45	31%	N 45°W
3	46.0	6.13	57%	N 30°W
4	45.0	2.36	22%	N 85°W
5	93.0	1.95	9%	N 77°W
6	97.0	8.10	35%	S 85°W
7	49.0	6.11	53%	N 57°W
8	49.0	4.04	35%	N 05°E
9	67.0	8.21	70%	N 18°W
10	77.0	9.68	52%	N 13°W
11	77.0	5.64	30%	N 14°E
12	97.0	10.16	44%	N 32°E
13	49.0	5.13	43%	N 85°W
14	52.0	9.80	78%	N 75°E
15	80.0	5.03	26%	S 4°W
16	93.0	13.60	61%	N 20°W
17	96.0	2.00	9%	S 4°W
18	107.0	4.45	17%	N 70°W
19	60.0	5.91	41%	N 45°E
20	80.0	11.73	61%	N 10°W
21	108.0	11.20	44%	S 61°W
22	109.0	9.13	34%	S 12°E
23	98.0	10.26	44%	N 12°W
24	200.0	14.24	35%	– (22′ above base)
			Average = 40% i.e., average deviation of less than 1%	

Source: From Bruce (1988).

these would have been in all cases shorter than subsequently proved necessary to found safely the minipiles. Poor or voided rock was consistently found below these anticipated elevations, further supporting the decision to use minipiles.

Overall, the total drilled length of 1920 linear feet corresponded with the total foreseen quantity of 1710 linear feet. Variations from 43 ft less to 30 ft more with respect to foreseen were recorded on individual piles, highlighting the variability of the rock. Overall, a volume of grout equivalent to four times the nominal hole volume drilled was injected. Much of this was consumed in the zone above rock-head during pregrouting operations. The level of maximum takes corresponded with groundwater level.

Testing and Performance A separate test pile, 30 ft long with only 5.33 ft of bond was load tested in accordance with ASTM D1143 quick load test method to

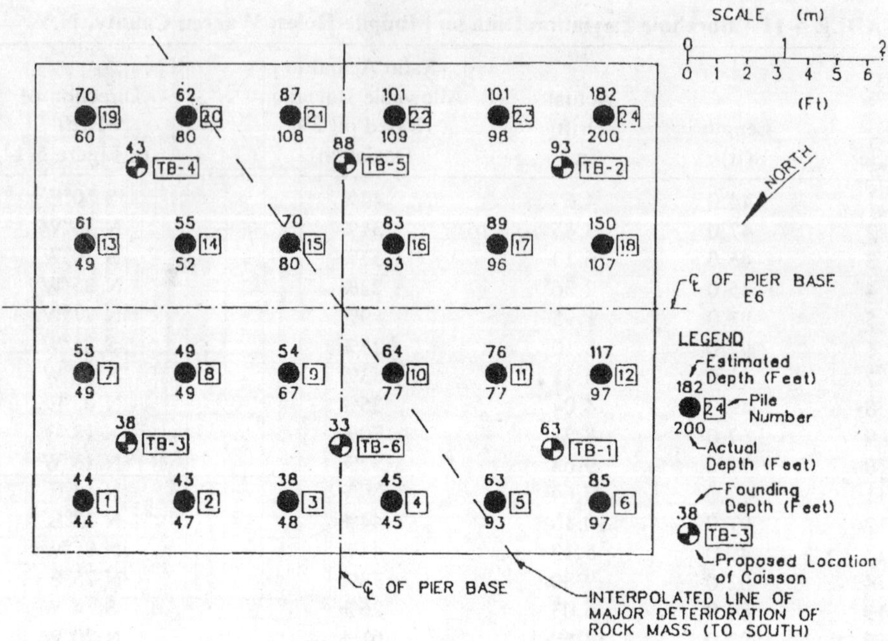

Figure 6-15 Actual Pin Pile lengths and foreseen caisson depths; Pier E6, I-78 Bridge, Warren County, N.J. (From Bruce, 1988. Reproduced by permission of Thomas Telford Publications.)

205 tons, using rock anchors as reaction. This particular short socket length was selected as at test load the average grout to rock and grout to steel bonds would be 304 and 250 lb/in.^{2}, respectively—both considered to be at or near ultimate values. An outer sleeve of PVC pipe extending to the top of the rock socket ensured load transfer only in the socket. A 6-in.-thick wooden plug was attached to the bottom of the steel pipe to ensure no load could be transferred in end bearing.

The data are presented in Figure 6-16. In summary, the total settlements recorded at each successive cycle to 205 tons were 0.367 and 0.373 in., respectively. Creep of 0.011 in. was recorded over 60 minutes hold at these loads. The permanent set after this operation was 0.07 in.

The next day testing was continued to higher levels, but at 224 tons the material of the upper casing began to fail. Until that point, the pile was performing exactly as it had during the previous testing sequence. Total displacement was 0.371 in. at 215 tons but 0.452 in. at 224 tons.

During installation of the reinforcing pipe in the last and deepest pile (No. 24), a thread parted and a 130-ft length of pipe fell into the 200-ft-deep hole. Borehole TV revealed the casing to be further ruptured 30 ft above the bottom of the hole, due to its impact with the bottom. After various attempts at recovery, it was decided to grout the pile, having previously suspended a 20-ft-long, $4\frac{1}{2}$ in. diameter, 150-kips/in.^{2} steel pin, with centralizers, from 62 to 82 ft below the top. The intention of

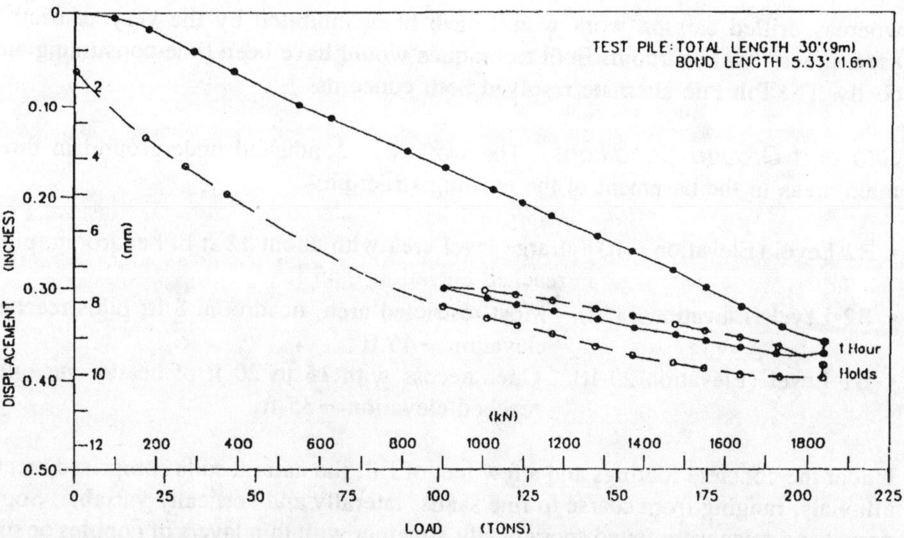

Figure 6-16 Load-displacement data, test pile at Pier E6, I-78 Bridge, Warren County, N.J. (From Bruce, 1988. Reproduced by permission of Thomas Telford Publications.)

this pin was to ensure effective load transfer across the upper discontinuity. A very rigorous extended load test was then executed to 170 tons. The performance of the pile proved excellent: the total displacement was 0.187 in. at 170 tons, 0.010 in. creep in 24 hr, permanent set 0.009 in. It was judged capable of safely performing its function in service.

The bridge is now complete and the performance of pier 6E has proved exceptional.

4. Postal Square, Washington, D.C.

Background The original portion of the massive Old Postal Building, Postal Square, was completed in 1911. A major extension followed in 1931. For many years it served as the main Post Office for Washington, D.C., being located adjacent to the Union Station on Massachusetts Avenue, a few blocks north of the Capitol Building. The Federal Government planned to remodel the existing structure by adding new office floors in the center court area and constructing mechanical space below the existing lowest basement elevation of +23 ft. This meant that existing foundations had to be upgraded and new columns added to support new interior framing. The existing supports are steel and concrete columns on large concrete footings, and 14-in.-square caissons end bearing on dense sands.

Originally a cumbersome underpinning scheme was envisaged, involving hand-dug support, massive spread footings, and large-diameter caissons, both hand-excavated and mechanically drilled. However, the hand work would probably have caused significant undermining of the existing footings, leading to settlement, whereas drilled caisson work would have been inhibited by the very restrictive access, and low headroom. Both techniques would have been time-consuming and costly. The Pin Pile alternate resolved both concerns.

Site and Ground Conditions The work was conducted underground in three main areas in the basement of the existing structure:

- B2 Level (Elevation 6 ft) Large level area with about 12 ft of headroom; piles reached elevation −45 ft
- B2 Level (Elevation 11 ft) Most restricted area, headroom 8 ft; piles reached elevation −45 ft
- B1 Level (Elevation 23 ft) Open access with 15 to 20 ft of headroom; piles reached elevation −35 ft

Under the concrete footings and a few feet of fill, the natural soils comprised recent alluvials, ranging from coarse to fine sands, laterally and vertically variable. Some gravel and mica were found sporadically, together with thin layers of cobbles or stiff clayey silt and silty sand in lower reaches. Typically the sands were dense to very dense. Natural groundwater level was at about Elevation −5 ft.

Design The overall design required 390 vertical piles in the B2 levels and 310 piles in the B1 level, each with a nominal working load of 75 tons. About 25 percent of the piles were installed in groups of 4 or 6 through 15 existing B2 (El 6) footings comprising 7 to 14 ft of concrete. Pile centers were within 20 in. of existing column faces.

Totals of 21 new reinforced concrete caps were cast in B2 (Elevation 6), 17 in B2 (Elevation 11), and 53 in B1. These featured standard (and several nonstandard, specially designed) plan geometries from 5 ft 4 in. × 4 ft 8 in. (3 piles) to 7 ft 6 in. square (9 piles). The minimum pile separation was 26 in. center to center, but was typically 30 in.

Construction Custom built, short mast diesel hydraulic track rigs were used to rotate 7-in.-diameter 0.5 in. wall N80 casing with water flush, to target depth. Type I grout of w/c = 0.45 was injected under excess pressures of 80 to 110 lb/in.^{2} during progressive extraction of the casing over the lower 25 ft. The casing was then reinserted 5 ft into this pressure grouted zone as permanent support. The lowermost 25 ft of pile was reinforced by Grade 60, $1\frac{3}{8}$-in.-diameter rebar in coupled lengths. For those holes through existing footings, an $8\frac{7}{8}$-in.-diameter down-the-hole hammer was used to penetrate until significant steel was encountered. Thereupon, the hole would be completed with an 8-in.-diameter core bit. Efficient load transfer between the casing of the pile and the footing was ensured by the use of a special nonshrink, high-strength grout. For the new pile caps, the Pin Pile casing was extended 4 in. up into the subsequent concrete, the horizontal reinforcing of which was fixed 2 in. above the top of the casing.

Testing and Performance Four special test piles (TP) were installed prior to construction of the production piles (Table 6-12). TP1 and TP2 were tested cyclically, yielding the analysis provided in Figure 6-17. TP3 and TP4 were also tested incrementally but progressively to maximum load in accordance with ASTM-1143. TP1 failed at the grout-soil interface, the founding horizons being on average finer and less dense than those for the other piles. Figure 6-17 also shows that the elastic compressions of TP1 and TP2 were similar at the failure load of the former. This shows that load must have been transferred to similar depths in both piles, despite the nominal difference in "free length" upon construction. The elastic performance of TP3 and TP4 was likewise similar, supporting the observation.

Table 6-12 also highlights higher total creep amounts in TP1 and TP2: largely a reflection that there were far more creep monitoring periods in the cyclic loading than in the progressive loading. This clearly impacts overall permanent displacement and is an important point to bear in mind when judging pile performance on this criterion.

A separate pull-out test was conducted in an existing column footing in the B2 (Elevation 6) level to quantify concrete-grout bond. A special element was grouted 4 ft 6 in. into an $8\frac{7}{8}$-in. hole drilled through the concrete. A special high-strength, nonshrink grout was used. After repeated cyclic loading to 525 kips (79 percent GUTS, or guaranteed ultimate tensile strength, and equivalent to 350 percent design load), the maximum uplift recorded was 0.005 in. reducing to 0.001 in. upon destressing. Assuming uniform bond distribution, an average grout-concrete bond of almost 350 lb/in.^{2} had therefore been safety resisted.

Following installation of the Pin Piles, the structural renovation has progressed, and the foundations have performed perfectly.

TABLE 6-12 Test Pile Data, Postal Square, Washington, D.C.

	TP1	TP2	TP3	TP4
Area	B2 (6 ft)	B2 (6 ft)	B2 (11 ft)	B1 (23 ft)
Total length (ft)	36	51	36	58
Bond length (ft)	25	25	25	25
Maximum test load (kips)	187.5 (failed)	300	300	300
Elastic extension at maximum load (in.)	0.173	0.461	0.383	0.374
Permanent deflection at end of test (in.)	0.313	0.291	0.187	0.113
Total cumulative creep (in.) during test	0.096	0.174	0.059	0.061

Source: From Bruce (1992a).

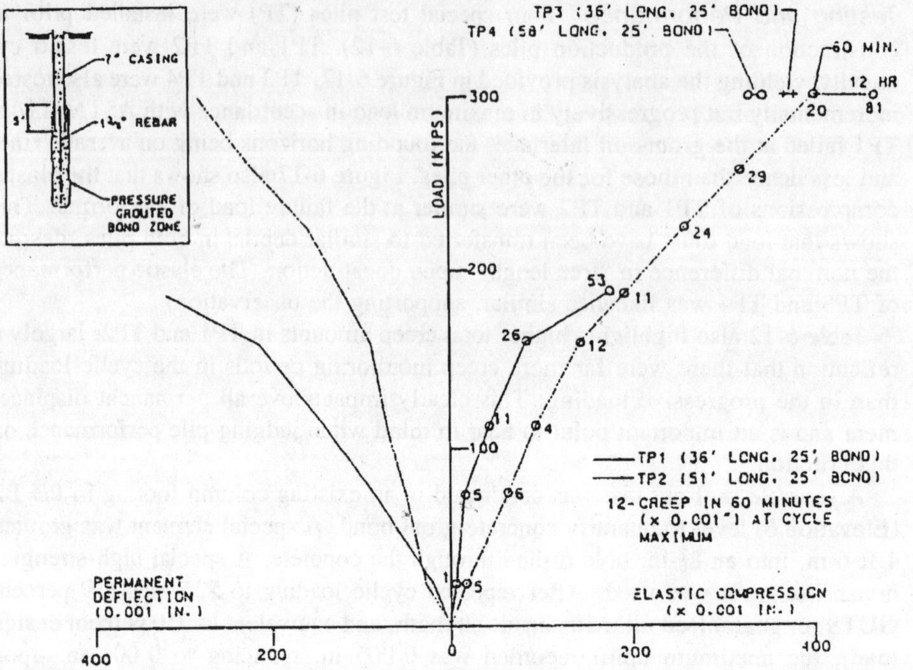

Figure 6-17 Elastic/permanent set performance of TP1, TP2, Postal Square, Washington, D.C. (From Bruce, 1992a.)

5. *Presbyterian University Hospital, Pittsburgh, Pa.*

Background The Presbyterian University Hospital complex already occupies two extremely congested city blocks, and so when the need for more facilities became apparent, the decision was made to vertically extend and laterally link several existing operational structures. Overall, 1.6 million ft^2 of new facilities were to be built in four major additions.

This highly delicate operation, conducted within a fully functional facility, necessitated some equally complex and innovative foundation engineering solutions involving excavation support and structural underpinning. One of the most dramatic operations was associated with the completion of a new Magnetic Resonance Imaging Center. The construction of a new elevator pit called for a 30-ft-deep excavation directly underneath three exterior column footings of the adjacent 13-story hospital structure. The pit, 60 × 32 ft in plan, was further bounded on two sides by five additional footings, and so these sides required anchored lateral support.

Historically, the support of columns in such circumstances has been achieved by conventional underpinning pits and needle beams. However, in this instance, the difficult access conditions, and the specified requirement to limit downwards movement of the columns to less than $\frac{1}{8}$ in. demanded a special solution, featuring high-capacity Pin Piles in rock.

Site and Ground Conditions The access was very restricted laterally and vertically (as low as 12 ft), and the work had to be conducted within the confines of a fully operational medical facility. The piles were installed through 3 ft 6 in. of existing nearby reinforced concrete footings case directly on fractured, fissile medium hard-hard siltstone, occasionally calcareous or limey.

Design and Construction At each existing footing, six Pin Piles (four working, plus two redundant) were installed in $8\frac{1}{2}$-in. holes drilled vertically by rotary percussive methods with air flush to the target depth (43 ft below the footing). Each pile had a design working load of 250 kips. The reinforcing element consisted of 7-in.-diameter, $\frac{1}{2}$-in. wall N80 casing placed full depth and then tremied full of neat cement grout of w/c = 0.45. The upper 23 ft of each pipe was greased on the outside to debond it from the surrounding grout in that region and so permit load transfer into the 10-ft-long bond zone. The suitability and security of this design had been proved in the earlier test program, described below. A structural steel jacking frame was then erected over the top of the piles and fastened to the existing steel column. Each of the steel columns—supporting an occupied hospital building—was then sequentially lifted off its existing spread footing by a distance of $\frac{1}{16}$ to $\frac{1}{8}$ in. This effectively preloaded the piles to prevent any later settlement of the building, and transferred the column loading into the bedrock, but 23 ft below. Excavation then proceeded, supported laterally by beams, shotcrete lagging, and prestressed rock anchors. As the excavation deepened, cross frames were welded to the Pin Piles to limit the unbraced lengths of these piles now exposed and acting as grout-filled steel columns.

Testing and Performance By the end of excavation, the foundations of the existing structure could be seen resting on the Pin Pile groups, 22 ft off the bottom of the excavation. During and after excavation, no movement of the structure was measured.

One of the most common problems foreseen for Pin Piles is their potential for buckling or bending, as an inferred consequence of their high slenderness ratio. This unique project—featuring Pin Piles with no surrounding ground to offer any lateral restraint—is proof that correctly designed and constructed Pin Piles can operate with surprising efficiency not only in the axial sense.

Clearly, testing of production piles was not possible in this instance, and so a full-scale test pile was installed beforehand at an adjacent location. Using identical construction methods, a Pin Pile with 20-ft bond was formed in the same geological stratum. The total length was 50 ft, including, therefore, 30 ft of debonded "free length." The casing was preassembled in the workshop and consisted of five separate lengths, hand tightened together. Two "telltales" were incorporated—one each at the top and bottom of the bond zone. A thick, soft wood plug was attached to the bottom of the reinforcing pipe to eliminate any possible end bearing contribution and to so allow only side shear to be mobilized. As part of the contract requirement, the pile was then tested to twice design working load (500 kips), according to ASTM-D1143-81 (modified to allow cycles at 25, 50, and 75 percent). Results are summarized in Table 6-13. At 250 kips, the elastic compression of 0.227 in. was exactly that predicted, while the permanent displacement of 0.052 in. was proved by the telltales to be due to some inelastic compression of the steel casing itself. While loading from 400 to 425 kips, a "bump" was recorded and the load dropped to 300 kips. Load was then increased to 500 kips when a further "bump" occurred. However, when the data from the cyclic loading and the telltales were analyzed, it became clear that:

1. The pile elastic deflection at 500 kips was exactly as predicted.
2. The apparently large permanent movement (Table 6-13) was due to an irreversible "one off" shortening of the steel pipe. The assembly records of the pile were then reviewed. It transpired that there had been several "unshouldered" hand-tightened joints between adjacent casing sections. It was suspected that each joint was unshouldered about $\frac{1}{4}$ to $\frac{1}{8}$ in. Thus, the sudden 0.471 in. permanent compression of the pile material was readily explained, and when subtracted from the permanent set of 0.503 in., gave a true movement of the pile tip into the rock of 0.032 in. at 500 kips—an outstanding performance. There was negligible creep at all load increments.

Thereafter, the pile was tested to a maximum load of 675 kips before it became clear that material failure of the steel casing was occurring. At this load, the steel had compressed 3.084 in. (from telltales), compared with the measured butt permanent displacement of 3.224 in. Thus, at 675 kips, a true permanent movement of the pile of 0.140 in. had been recorded, while analysis proved the perfect elastic performance of the pile with a calculated debonded length a few feet into the bond zone.

This project was therefore highly significant from several viewpoints:

- The excellent lateral and vertical performance in Pin Piles was demonstrated.
- The value of telltales in aiding understanding of internal pile performance was shown.

TABLE 6-13 Summary of Test Pile Performance, Presbyterian University Hospital, Pittsburgh, Pa.

Load Cycle Maximum (kips)	Total Butt Movement at Maximum (*A*) (in.)	Permanent Butt Movement at Subsequent Zero (*B*) (in.)	Therefore, Elastic Deflection at Maximum (*A*) − (*B*) (in.)	Apparent Bottom Telltale Movement (Relative to Butt) (in.)
125	0.127	0.042	0.085	0.037
250	0.279	0.052	0.227	0.044
375	0.448	0.077	0.371	0.063
300	0.663	0.329	0.334	0.295
500	1.020	0.517	0.503	0.471

- The warming fact that the boundaries of Pin Pile design are now those of the constituent materials—that is, independent of the surrounding ground properties—was underlined.

6. Pocomoke River Bridge, Md.

Background This 275-ft-long movable bascule pier drawbridge (Figure 6-18) was built over the Pocomoke River in 1921. Bascule Piers 3 and 4 were originally supported on wooden piles driven through the soft riverbed muds into the underlying compact sand. The support offered by these piles had been compromised by river scour that had exposed them in several places. The Pin Piles designed to stabilize the structure were remarkable on three counts:

1. They had to be installed through the structure and through the scour zone.
2. They had to provide support without allowing any additional structural settlement. This necessitated the use of preloading techniques.
3. A very intensive test program was required on special test piles to verify the concept and the performance.

Site and Ground Conditions In each of Piers 3 and 4 a total of 24 piles were drilled from the bridge deck. In addition, a further four piles were installed from the restricted access of the Control House of Pier 4 (8 × 8 ft plan, 14-ft headroom) (Figure 6-19). The riverbed materials into which the underpinning was installed comprised recent alluvial settlements. The founding horizon was dense medium to coarse sand, beginning about 60 ft below river surface level.

Design and Construction Each pile was installed as shown in Figure 6-20. The allowable stresses used in the design were 30 percent U.C.S. grout and 40 percent of the yield strength in both the casing and the epoxy coated rebar. To permit preloading of the pile, a tendon comprising 3 ea. 0.6-in.-diameter seven-wire strands was also installed through each hole, its 20-ft bond zone extending to 25 ft

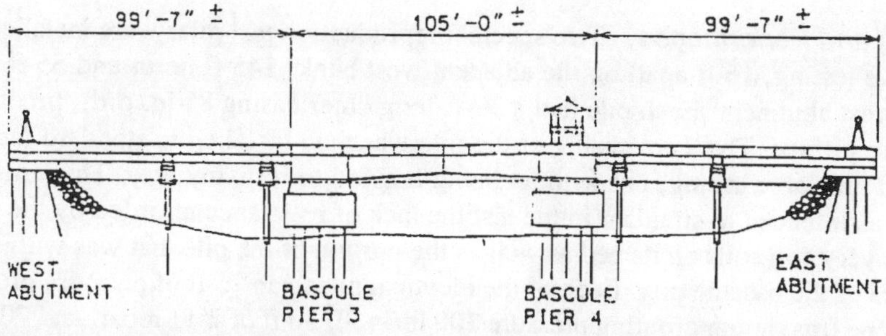

Figure 6-18 General configuration of Pocomoke River Bridge, Md. (From Bruce et al., 1990.)

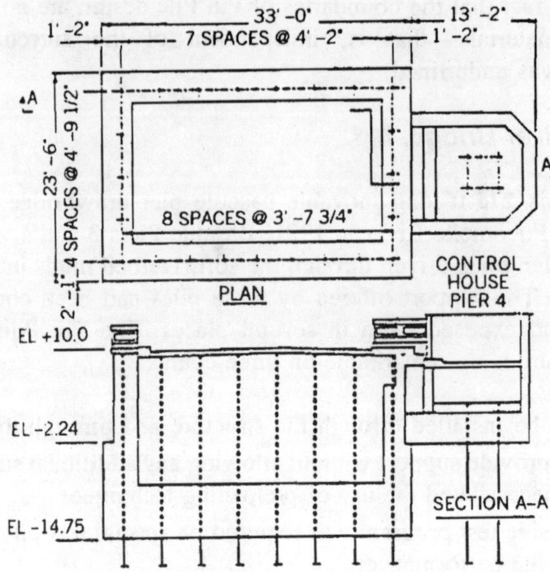

Figure 6-19 Plan and section of Bascule Pier #4, Pocomoke River Bridge, Md., showing Pin Pile locations (From Bruce et al., 1990.)

below the toe of the Pin Pile casing. Prior to drilling, grout filled bags had to be placed around the bases of the piers as framework for void filling grouting.

After the neat cement grout had reached 3500 lb/in.^{2}, the tendon was stressed against the top of the steel casing, to the design load of 82 kips. The annulus between casing and structure was then grouted with special high-strength grout. About 5 to 7 days later, the prestress was released at the tendon head, thereby allowing full structural load transfer to the pile but without obviously causing further pile compression. Slightly amended procedures had to be adopted in the restricted access of the Control House, but the same basic principles were followed, resulting in the perfect installation and functioning of all 52 Pin Piles on the project.

Testing and Performance Two special preproduction test piles were installed for intensive testing, 16 ft apart on the adjacent west bank, 145 ft north and 55 ft west of the west abutment. Each pile had a 30-ft-long outer casing $8\frac{5}{8}$-in. o.d., predrilled from the surface. The 7-in. casing of the pile was then installed in standard fashion through this large casing, but without being bonded to it in any way. This arrangement was intended to simulate in the test the lack of resistance afforded by the river and the very soft soils on its bed as well as the portion of the pile that was within the confines of the bascule pier. Each of the identical piles had 25 ft of pressure grouted bond zone (maximum grouting pressure 100 lb/in.^{2}), 30 ft of #11 rebar, and 70 ft of 7-in. casing (from surface to 5 ft into the bond zone). Soil anchors were installed to provide reaction to the test loads.

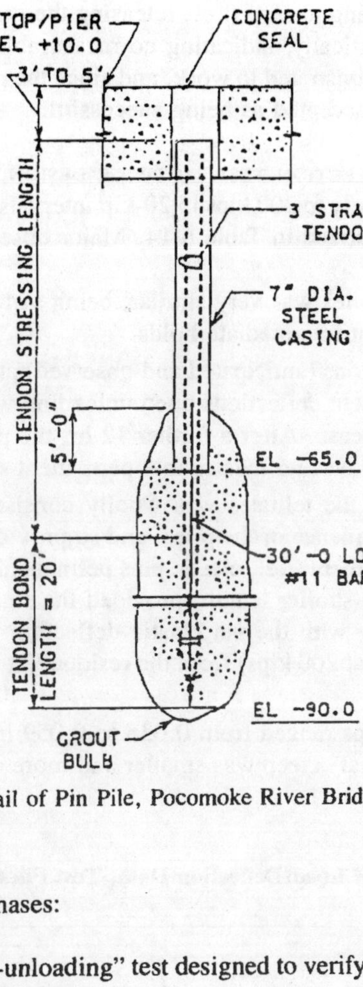

Figure 6-20 Typical detail of Pin Pile, Pocomoke River Bridge, Md. (From Bruce et al., 1990.)

The test had three phases:

Phase I: A "preload-unloading" test designed to verify the performance efficiency of the preloading system.

Phase II: A conventional pile load test to establish load-deflection performance within the scope of the specification (i.e., progressively to twice working load).

Phase III: On one pile, loading to failure.

The test was heavily instrumented, with load being measured independently by load cell and by hydraulic jack gage, and deflection monitored by dial gages supported from an independent reference beam and by piano wire and mirror scale. Dial gages were also used to indicate movements of telltales located at elevation −70 ft (i.e., at top of bond length) and at elevation −90 ft (bottom of bond length).

PHASE I TESTS (PRELOAD–UNLOADING TESTS) The anchor tendon in each pile was loaded to 82 kips, creating an elastic shortening of each pile by 0.123 and 0.137 in., respectively. Upon unloading to zero (i.e., releasing the prestressing load), the pile cap rebounded totally elastically, indicating no measurable permanent shortening. As the procedure was demonstrated to work, and since the performance was elastic, this phase of testing was accepted as being successful.

PHASE II TEST (LOAD/DEFLECTION TEST TO TWICE DESIGN WORKING LOAD) Each pile was loaded progressively to 200 kips in 20-kip intervals, each with a 5-min hold period. Details are summarized in Table 6-14. Major observations were:

- Performance of the piles was very similar, being virtually elastic, linear, and with minimal creep at intermediate holds.
- The total pile deflections (anticipated and observed) at 200 kips were less than ½ in., and the permanent deflections upon unloading were around 0.04 in. at 2 hr after final load release. After a further 12 hr, the piles had returned to full extension (i.e., there was no measurable permanent shortening).
- The performance of the telltales was wholly consistent. They reflected the internal elastic performance of the piles, and so provided movements less than the *total* pile displacement (i.e., elastic plus permanent). Predictably, the upper telltale, monitoring a shorter length, provided the smaller movements. These data compare closely with the net elastic deflection obtained by subtracting total cap movement (at 200 kips) from the residual (at zero), as shown in Table 6-15.
- Total creep at 200 kips ranged from 0.038 to 0.059 in. over 24 hr. However, the amount of "internal" creep was smaller and more uniform (0.021 to 0.033 in., average = 0.028 in.).

TABLE 6-14 Highlights of Load/Deflection Data, Test Piles 1 and 2, Pocomoke River Bridge, Md.

	Deflection at 200 Kips (in.)	Creep in 24 hr at 200 Kips (in.)	Permanent Displacement upon Unloading from 200 Kips Instantaneous/After 2 hr (in.)
		Pile 1	
Pile cap	0.442	0.038	0.044/0.020
Upper telltale	0.344	0.028	0.024/0.021
Lower telltale	0.374	0.031	0.095/0.093
		Pile 2	
Pile cap	0.437	0.059	0.047/0.027
Upper telltale	0.385	0.033	0.041/0.037
Lower telltale	0.420	0.021	0.067/0.063

Source: From Bruce et al., 1990.

TABLE 6-15 Comparison of Net and Measured Elastic Pile Performance, Pocomoke River Bridge, Md.

Pile Number	Net Elastic Deflection[a] at 200 Kips (in.)	Measured Elastic Deflection[b] at 200 Kips (in.)
1	0.398	0.374
2	0.390	0.420
Average	0.394	0.397

Source: From Bruce et al. (1990). Originally presented at the 7th Annual International Bridge Conference, June 18–20, 1990, Pittsburgh, Pennsylvania.

[a]Total deflection at 200 kips *less* permanent deflection at subsequent zero.
[b]From lower telltale.

- There was a time-related "rebound" evident in all points of measurement after unloading. Overall, this was 0.020 to 0.027 in. at the pile cap, including 0.002 to 0.004 in. of "internal" pile rebound.

PHASE III TEST (LOAD/DEFLECTION TEST TO FAILURE, TEST PILE 2) Once the required test to 200 kips was satisfied, an attempt was made to determine the ultimate skin friction. The hold down reaction system was sized for about 360 kips, which was initially felt to be sufficient to plunge the pile. Surprisingly, after four successive cycles to about 360 kips, the pile had not yet failed, despite a cumulative permanent displacement of 0.567 in. Thereafter, the test setup was overhauled, and the test rerun: a maximum load of 390 kips was reached before plunging of the pile was recorded.

Again the evidence of the telltales indicated virtually perfect elastic performance within the pile structure. The difference, at maximum load, between overall elastic performance (lower telltale) and total deflection was 2.874 in. − 1.128 in. = 1.746 in.: very close to the *measured* permanent set at zero load, of 1.712 in. The difference is probably due to the fact that the telltale was not exactly at the pile tip. Creep values were only significant from about 340 kips onwards.

As a final point, this project represented the second time that this particular preloaded Pin Pile concept had been used. Some years before a structure at the Mid Orange Correctional Facility, N.Y., had been saved by preloaded 55-kip piles (Bruce, 1988–1989). In both instances, the subsequent structural movements in service have been of the order of a few thousands of inches. These successes have recently encouraged the possible use of permanently preloaded piles in California to underpin transmission lower footings threatened by "rocking" actions produced by seismic events. The basic requirements of the problem can be readily satisfied with this technique:

- Resistance to uplift forces in the range of 100 to 400 kips.
- No additional compressive loads are imposed upon a new or existing footing.

- No structural movement within the design capacity of the system in service.
- Each pile installed is tested (during application of the prestress).

7. Augusta, Ga.

Background An existing soap manufacturing factory had been founded on both spread footings and driven pile foundations. Due to the planned heightening of the facility, certain footing capacities had to be upgraded in both the vertical and lateral senses. This necessitated the installation of 143 Pin Piles of nominal working load 50 tons through and adjacent to existing footings. Roughly half were vertical, with the remainder having a batter of 2 vertical to 1 horizontal.

Site and Ground Conditions The work had to be conducted within the fully functional facility where site cleanliness was paramount, and access was restrictive. Apart from a few inches of silty fill under the footings, the founding stratum consisted of a fine to coarse sand (N about 30) with lenses of clay, and underlain by a clay layer that dips across the site. The three test piles did, however, penetrate two feet into this underlying clay layer. The elevation of the water table was below the pile tips.

The typical soil conditions (variable by only 1 ft across the site) were:

0 to 3 ft, red clay
3 to 20 ft, sandy clay
20 to 23 ft, white sandy clay
23 to 34 ft, competent dense coarse sand
34 to 35 ft, pink sandy clay
35 to 37.5 ft, slick wet clay

Design The piles were designed to be 38 ft long within a $7\frac{1}{4}$-in.-diameter hole. A $1\frac{3}{8}$-in.-diameter, 150-kips/in.^{2} reinforcing bar was later specified as standard. Due to design changes prior to construction, however, the first two test piles were reinforced by a single #18 Grade 60 bar. Each pile also had a $5\frac{1}{2}$-in. o.d. steel casing installed in the top 10 ft of the pile to resist a 12-kip lateral design load (Figure 6-21). The grout was designed to provide a 28-day strength of 4000 lb/in.^{2} A maximum vertical deflection of 0.5 in. at 50 tons (after ultimate test load was reached and held) was specified and a maximum lateral deflection of 0.5 in. at 12 tons was anticipated.

Construction To minimize drill spoils in the manufacturing facility, drilling was conducted with hollow stem augers. Each reinforcing bar was placed in 8-ft-long sections to which were attached the 1-in. regrout tube. Primary grouting through the auger was conducted, but only to a maximum pressure of 40 lb/in.^{2} due to leakage between the auger sections and around the flights. This was a considerably lower grouting pressure than could have been applied with the typical rotary casing with water flush method of drilling.

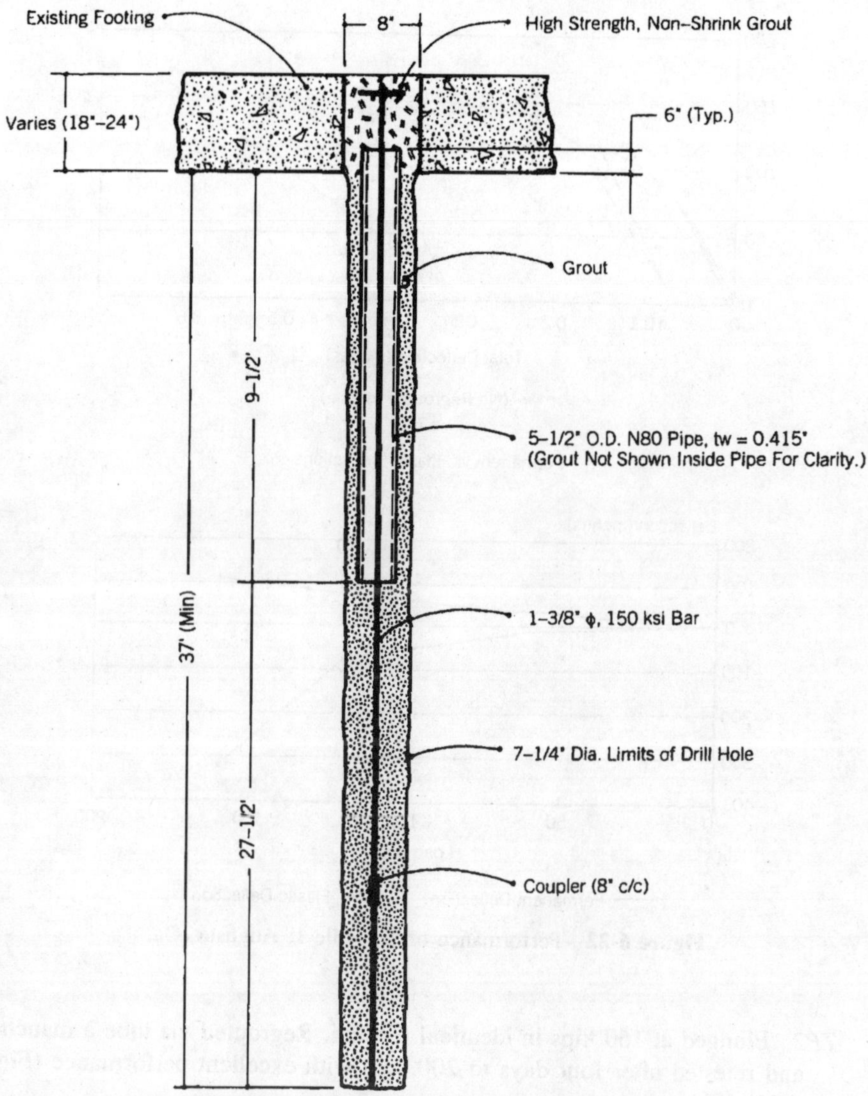

Figure 6-21 Pile arrangement; Augusta, Ga.

Testing and Performance The three test piles were generally tested cyclically using the ASTM D1143-81 quick test procedures to a target of 200 kips:

TP1: Failed at 160 kips (Figure 6-22). This was felt to be an atypically low value given the prevailing ground conditions, and was thought to be due to the low grout pressures during installation. Could not be regrouted due to blockage in tube. Test discontinued.

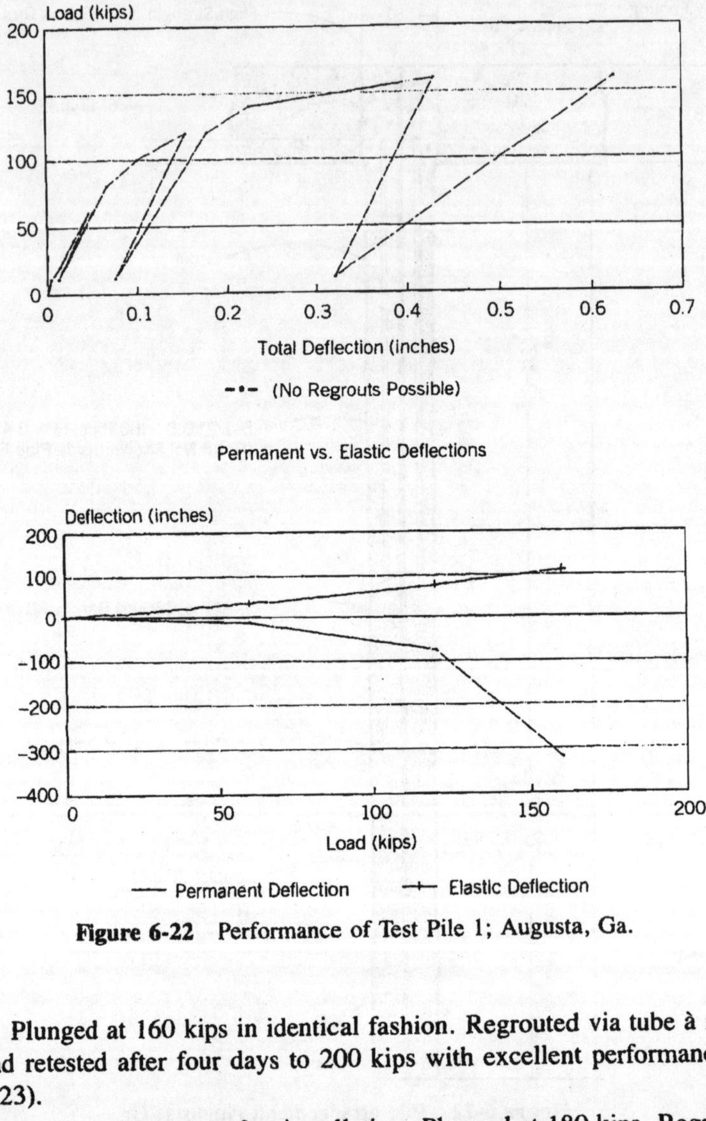

Figure 6-22 Performance of Test Pile 1; Augusta, Ga.

TP2: Plunged at 160 kips in identical fashion. Regrouted via tube à manchette and retested after four days to 200 kips with excellent performance (Figure 6-23).

TP3: Regrouted one day after installation. Plunged at 180 kips. Regrouted and retested successfully to 200 kips with excellent performance (Figure 6-24).

TP3 deflected elastically almost four times as much as the other two piles at equivalent loads. The only substantial difference between them was that TP3 had smaller diameter reinforcing.

Analysis of the loading data confirmed the mode of failure to be at the grout-ground interface. Postgrouting that interface appeared to impact performance in two ways:

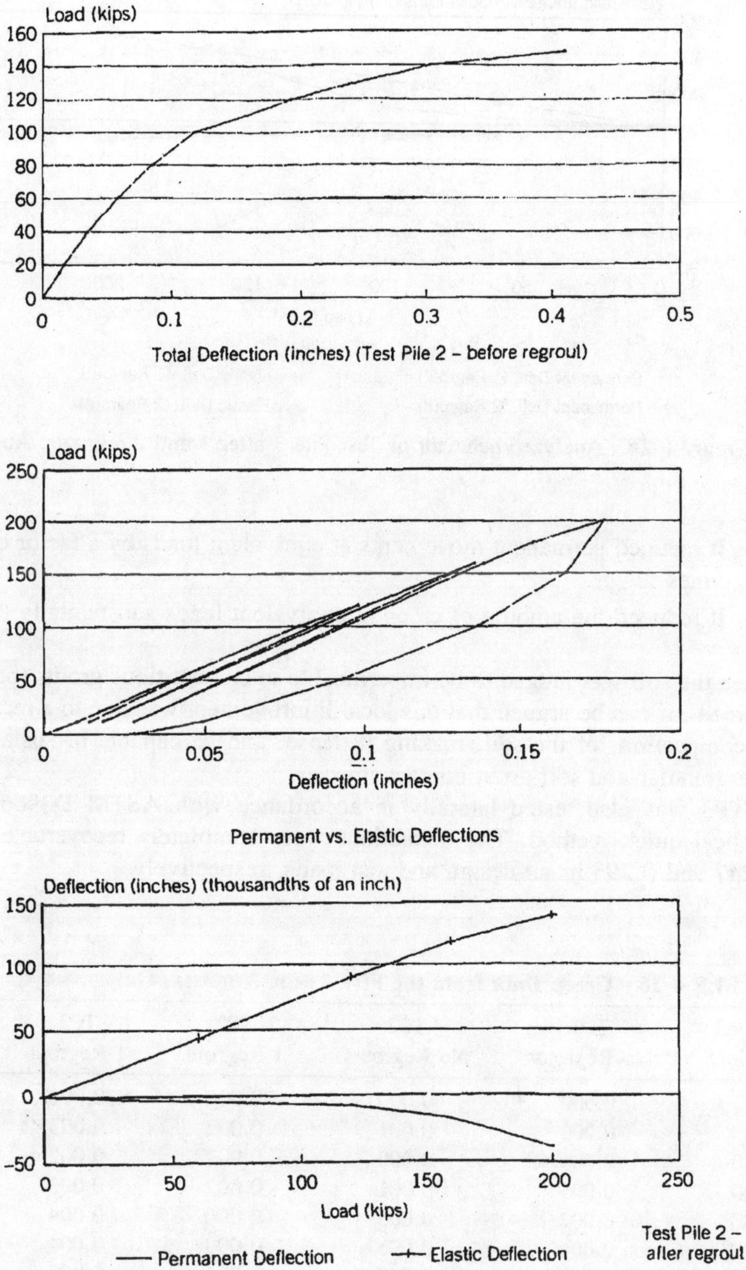

Figure 6-23 Performance of Test Pile 2, before and after regrouting; Augusta, Ga.

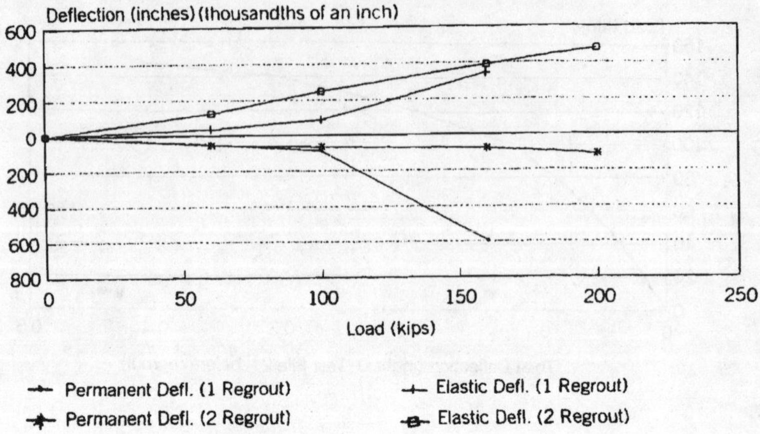

Figure 6-24 Analyzed behavior of Test Pile 3 after 1 and 2 regrouts; Augusta, Ga.

- It reduced permanent movements at equivalent loads by a factor of about five times
- It reduced the amount of creep at equivalent loads substantially (Table 6-16).

Since the soil was judged to be impermeable to cementitious grouts (being too fine grained), it can be argued that this local improvement was due to a recompression, or compaction, of the soil, making it denser and so capable of sustaining higher intergranular and soil-grout contact stresses.

TP3 was also tested laterally in accordance with ASTM D3966-81 using a cyclical quick method. The deflections were completely recoverable and totaled 0.247 and 0.293 in. at design and test loads, respectively.

TABLE 6-16 Creep Data from the Five Tests, Augusta, Ga.

Load (Kips)	TP1 No Regrouts	TP2 No Regrouts	TP2 1 Regrout	TP3 1 Regrout	TP3 2 Regrouts
13.6	0.000	0.000	0.000	0.001	0.000
20	0.000	0.001	0.002	0.003	0.004
40	0.000	0.000	0.002	0.002	0.001
60	0.001	0.001	0.002	0.003	0.001
80	0.002	0.005	0.000	0.004	0.000
100	0.004	0.005	0.000	0.004	0.000
120	0.010	0.015	0.001	0.028	0.002
140	0.017	0.032	0.002	0.032	0.002
160	0.045	Failed	0.000	Failed	0.002
180	Failed		0.004		0.004
200			0.012		0.007

Overview

These recent case histories illustrate the major characteristics of Pin Piles in general, and the specific nature of the U.S. market in particular. They confirm the trend toward designing for—and achieving—progressively higher unit loads, the advantage of postgrouting, the application of preloading principles, and the growing understanding of lateral loading behavior. Further research is obviously needed in certain directions, for example, composite action, and the "positive" group effect, while more construction-related developments, such as the exploitation of compaction grouting principles continue apace.

There is no doubt that the kind of demands that countries such as the United States will increasingly place on Pin Pile technology will continue to create an environment where both the application and development of Pin Piles will thrive.

6-3 IN SITU EARTH REINFORCING—TYPE "A" WALLS

Definition

The benefits of soil reinforcement in enhancing soil mass stability were first appreciated by the Babylonians 5000 years ago (TRB, 1987). Its principles have been exploited since then by a variety of organizations ranging from the Roman army to the U.S. government. The schemes have featured "bottom up" construction, utilizing carefully selected materials for reinforcement, facings, and fill. The current expression of the technique is in Reinforced Earth and its like.

In situ reinforcement, however, has a much shorter and less exotic history, although its exploitation of natural ground, in place, invites different dimensions of engineering originality and judgment. In situ reinforcement is proving increasingly popular in a wide range of applications for slope and excavation stability associated with deep foundation, tunneling, and highway construction. Three main categories can be identified (Figure 6-25):

1. *Soil Nailing*. This refers to reinforcing elements installed horizontally or subhorizontally into the cut face, as top-down staged excavation proceeds. The inserts improve the shearing resistance of the soil by being forced to act in tension.
2. *Reticulated Pin Piles*. These are similar inserts, but steeply inclined in the soil at various angles in planes, both perpendicular and parallel to the wall face. The overall aim is to provide a stable block of reinforced soil to act as a coherent retaining structure, holding back the soil behind, while providing resistance to shear across the failure plane.
3. *Soil Doweling*. This technique is applied to reduce or halt downslope movements on well-defined shear surfaces. The principle exploits the large lateral surface bearing area and high bending stiffness of the dowels, which are of far larger diameter than nails or Pin Piles (seldom greater than 6 inches). The use

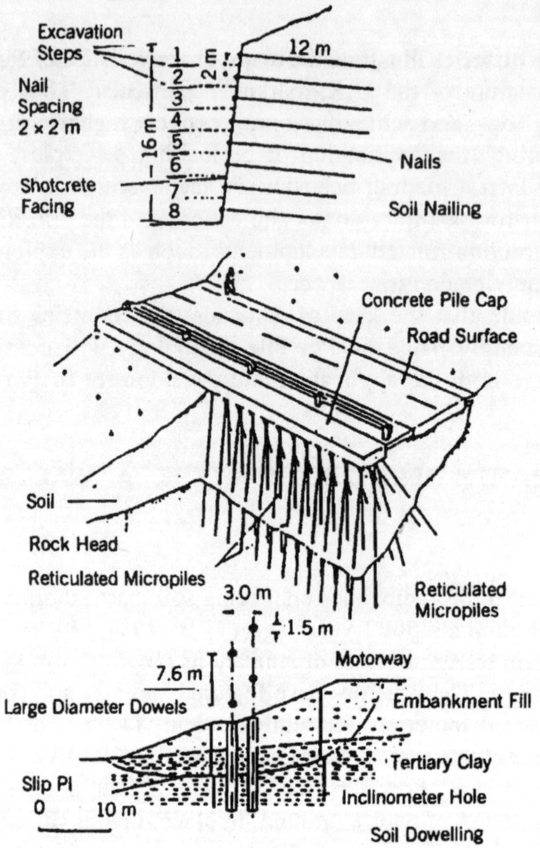

Figure 6-25 The family of in situ soil reinforcement techniques (From Bruce and Jewell, 1986–1987. Reproduced by permission of Thomas Telford Publications.)

of soil doweling is rare in urban environments, although it can prove attractive when combined with linked deep drainage in arresting massive land movements (e.g., in eastern Italy and southern California) (Bruce and Boley, 1987; Bruce and Bianco, 1991).

In terms of construction, the individual elements comprising reticulated Pin Pile arrays are no different from those described in Section 6-2 for direct axial load holding, and this explains their description in this chapter. However, their function, design, and performance when installed as in situ reinforcement are so different that their detailed description in a separate section is warranted. The balance of this chapter is therefore devoted to their application as in situ reinforcement. In the United States, such pile groups are called Type "A" insert walls because of the distinctive cross-sectional shape of the pile arrangement.

Historical Background and Applications

The earliest applications of Pin Piles were as conventional piles for direct underpinning. As development and testing of multiunit, three-dimensional arrays progressed and the concept of the "knot effect" was unveiled, very soon the advantages of this were applied to resolve slope stability problems, typically in rural areas (Figure 6-26). Then the value of such arrays in urban engineering applications, relating to tunneling works (both bored and cut-and-cover), deep excavations, and so on, became apparent and many excellent case histories have been reported, in particular throughout Western Europe (Figures 6-27 and 6-28). In each of these cases, the concept was to create protective structures in the ground to separate the foundation soil of a building from zones potentially subject to construction-related disturbance.

It should be emphasized that these in situ structures do not rely on intergranular improvement of the soil by permeation of grout. They rely, instead, on interaction between the soil and the inserts to create a coherent mass.

Construction

The individual inserts of a Type "A" wall are installed as described for load-bearing Pin Piles in Section 6-2's discussion of Construction. The overall construction sequence is illustrated in Figure 6-29. The concrete capping beam merits particular attention during design and construction. For relatively shallow slide planes (i.e., within about 20 ft of the ground surface), the beam provides added stiffness. For deeper slides, wall designs are being evaluated that do not require a cap beam or full extension of the pile reinforcement to the ground surface.

Design

Even Lizzi (1985), in introducing some basics of the approach, stated that the design is "not an easy task. In the very complex soil-pile interaction, there are many factors whose influence on the final behavior of the structure cannot be conveniently assessed." He cited the potential variations in the soil, in the piles, and the "practically unknown" relationship between the parameters. He concluded that designs should be based on "some simple assumptions" using the concept of reinforced soil, "and so similar to those currently used for reinforced concrete." In effect, the soil supplies the mass, while the inserts supply "lines of force" that allow the structure to resist compression, tension, and shear. The wall is thus conceived to physically prevent loss of soil from behind it, and to prevent sliding along potential failure planes.

Considering the case of a structure for "stiff or semi-rocky formations," Lizzi defined the problem as the calculation of the contribution made by the piles to the resistance of the natural soil. Having defined the "critical" surface, the design of the reinforcement follows: once a factor of safety *Fx* has been assessed, then

$$Fx = (R + R^1)/A \; : \, . \; 1$$

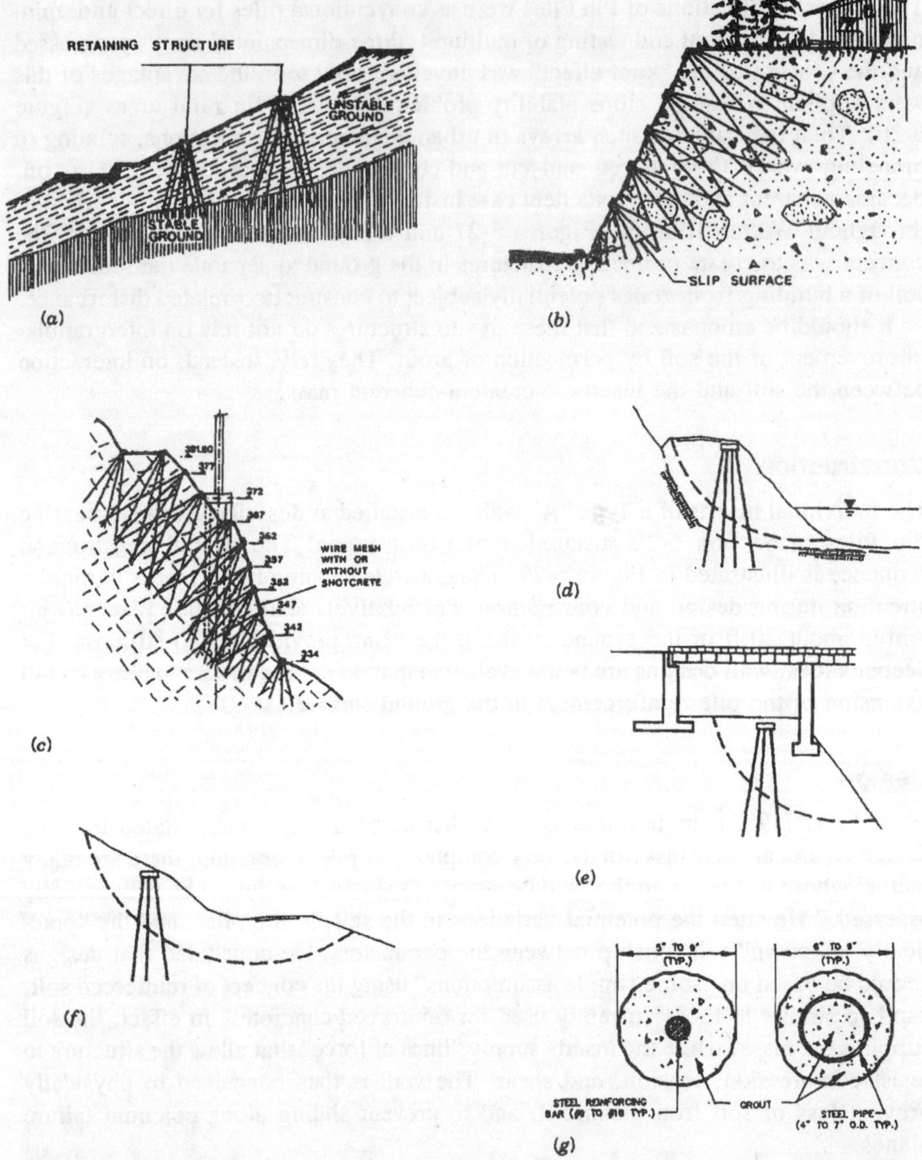

Figure 6-26 Applications of Type "A" walls in rural areas. (*a*) E. V. R., Rome (loose soil). (*b*) Southern highway, Italy (weathered formation). (*c*) Rodovia dos Immigrants, San Paulo (fissured rock). (*d*) Roadway slope. (*e*) Bridge foundation. (*f*) Slope above highway. (*g*) Typical pile sections for (*d*)–(*f*). [(*a*)–(*c*) from Lizzi, 1982. (*d*)–(*g*) from Pearlman et al., 1992. Reproduced by permission of ASCE.]

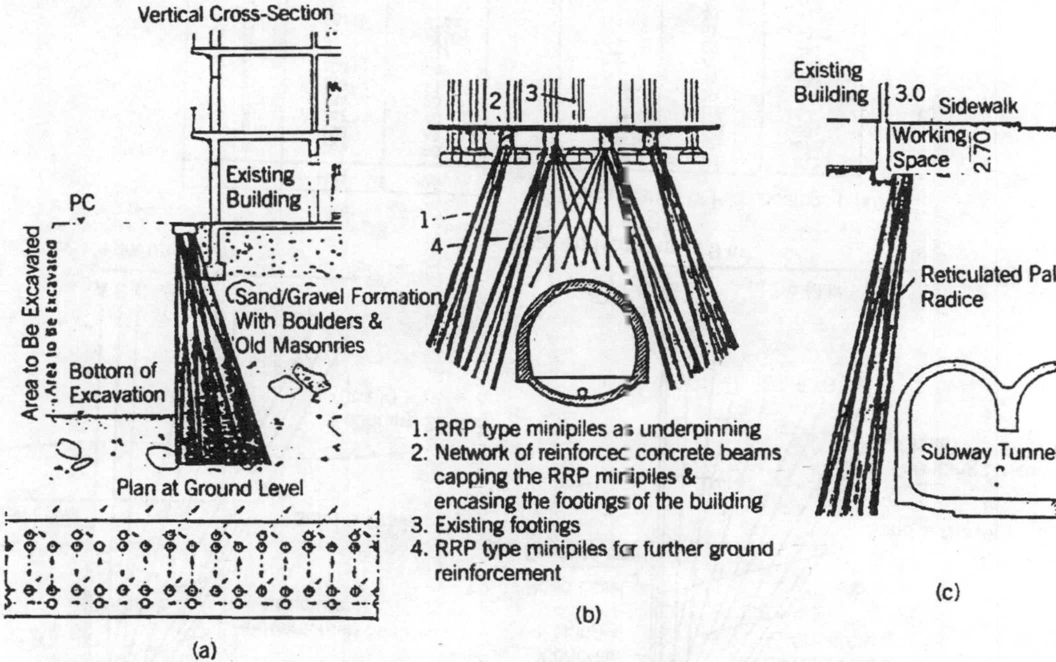

Figure 6-27 Applications of Type "A" walls in urban areas. (*a*) Cut-and-cover excavation. (*b*) and (*c*) Around bored tunnels. (After Lizzi, 1982.)

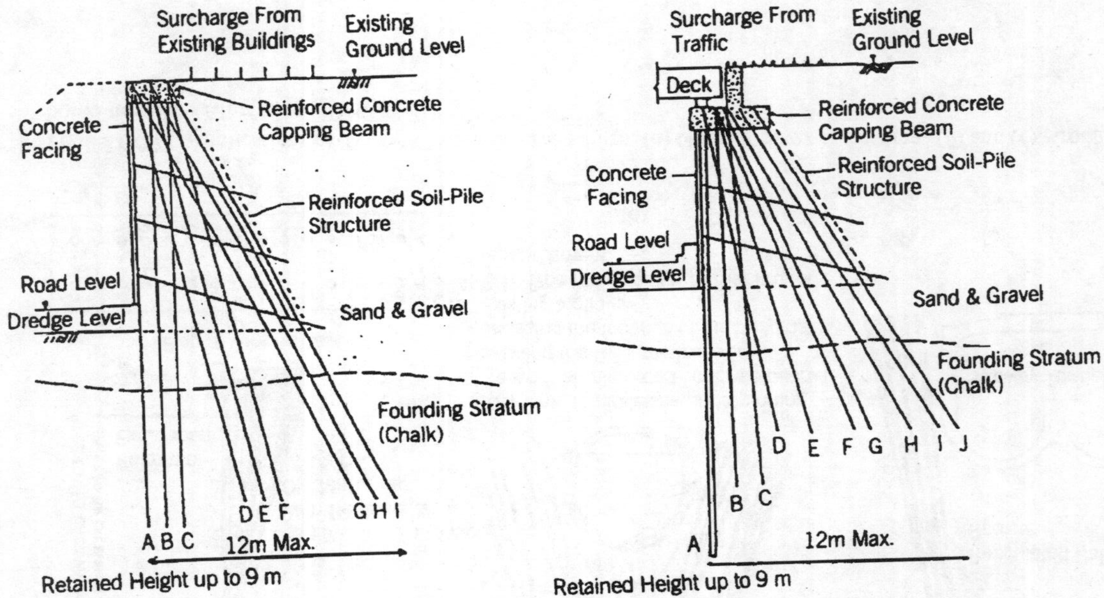

Typical Range of Pile Loadings (kN)

(a)

	Max	Min
Row A	165	90
B	153	65
C	142	40
D	134	23
E	118	−10
F	103	−43
G	87	−77
H	72	−110
I	56	−144

(b)

Row A	366	311
B	125	57
C	114	33
D	105	17
E	99	0
F	84	−33
G	77	−49
H	62	−81
I	51	−106
J	40	−130

Figure 6-28 In situ reinforcement for road widening project, Dartford, London. (*a*) Normal retaining wall. (*b*) Retaining wall serving as bridge abutment. (After Attwood, 1987.)

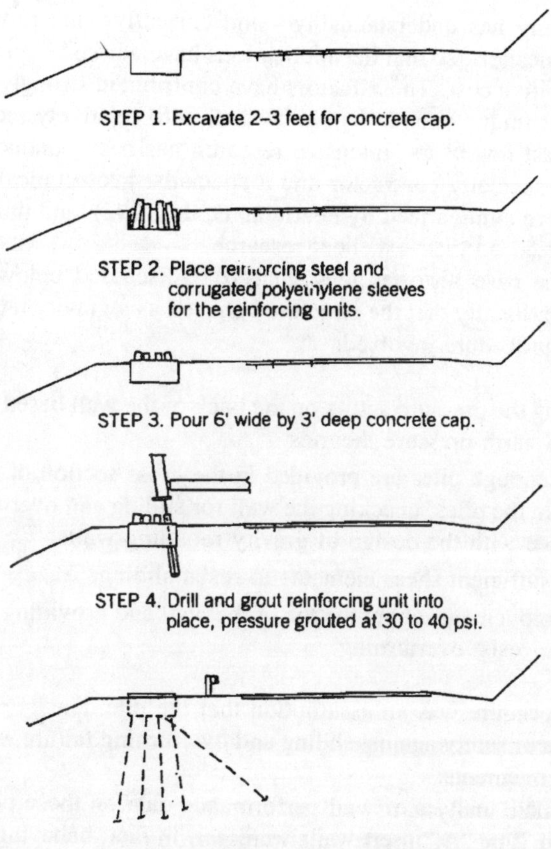

Figure 6-29 Typical steps in insert wall construction. (From Bruce, 1992b. Reproduced by permission of ASCE.)

where R = total resistive forces on the "critical surface"
A = driving forces on the same surface
R^1 = the additional shear resistance provided by the piles

R^1 depends either on the shear resistance of the piles or the resistance of the soil intersected by the piles, whichever is smaller. This approach is considered conservative, since it does not take into consideration the interaction developed between soil and pile.

Lizzi was still clearly wrestling with the details of his concepts, however, when he concluded that a better guide (to design) is the "extensive examination of works carried out," feeling that "it is not yet possible to have at our disposal an exhaustive means of calculation ready to be applied with safety and completeness" (1982). In addition, the ASCE Committee (1987) also alluded to the great reliance placed in designs on the soil/pile interaction "which is still subject to experience and intuition."

This uncertainty has understandably—and correctly—led to a high degree of conservatism in designs, so that the applications have worked extremely well, but at an almost prohibitive cost. These factors have contributed strongly to the very slow growth of the technique outside Italian borders until relatively recently.

Within the past few years, intensive research has been conducted by a consortium between a specialty contractor and a specialist geotechnical consultant. The findings have been summarized by Pearlman et al. (1992), and much of the following review is based on their excellent research.

Several of the case histories they analyzed (discussed below) were designed assuming conventionally that the structure behaved as a gravity retaining wall. This original design procedure involved:

- Determining the pressure acting on the back of the wall based on slope stability analyses or earth pressure theories.
- Assuming enough piles are provided in the cross section of the wall to retain soil between the piles, checking the wall for sliding and overturning stability in conformance with the design of gravity retaining walls.
- Providing sufficient shear elements to resist sliding.
- Summing movements about the toe of the wall and providing sufficient tension elements to resist overturning.

Basic to the procedure was an assumption that the pile density needed to achieve adequate factors of safety against sliding and overturning failure would be sufficient to stop slope movements.

From a detailed analysis of wall performance data on these projects, they concluded that such Type "A" insert walls were *not,* in fact, behaving as gravity walls. For example, wall movements seemed to be confined to a relatively thin and localized zone along the slide plane and additional slope movements were occurring after wall construction. Therefore, a new procedure was developed for preliminary design of these walls to better model the behavior of this relatively flexible slope stabilization system. The theoretical basis for the procedure has been verified by comparison with back analyses of the instrumented walls. In general, the new design procedure involves the following:

- Conducting stability analyses to determine the increase in resistance, along a potential or existing slip surface, that would be required to provide an adequate factor of safety.
- Checking the potential for structural failure of the piles due to loading from the moving soil mass.
- Checking the potential for plastic failure (i.e., flow of soil around the pile).

Typically, movement of marginally stable, noncreeping soil slopes occurs within a relatively thin zone that is subjected to large shear strains, not experienced within the materials above and below the zone of failure. The purpose of the Pin Piles is to connect the moving zone (above the slip surface) to the static zone (below the slip surface), and thus to increase the sliding resistance along the slip-plane

Because Pin Piles are relatively flexible, the maximum bending moments in the piles tend to develop relatively close to the slip-plane. Fukuoka (1977) devised a theory to evaluate the bending moments that develop in a pile oriented perpendicularly to a slip-plane, assuming a uniform velocity distribution of the soil above the slip-plane. Figure 6-30 is a chart for preliminary design of Type "A" insert walls. The chart was developed using the method described in Fukuoka (1977) and considers four sizes of pile elements. It should be noted that the ultimate horizontal resistance is either the load that causes yield stresses to develop on the outer edges of the steel pipes (i.e., Pile Types 1, 2, and 3) or the load that causes crushing of the concrete surrounding the centralized reinforcing bar (i.e., Pile Type 4). The ultimate horizontal load resistance of the piles is a function of the coefficient of subgrade reaction K_s of the soil or rock above and below the slip-plane. The K_s of the soil also has a significant effect on the amount of horizontal movement required before the pile reaches its ultimate horizontal resistance. Typical deflections and bending stresses along a pile are shown in Figure 6-31.

Plastic failure of soil around the piles can be analyzed using a procedure developed by Ito and Matsui (1975). The method is based on the fundamental consideration that soil deformation is restricted to a plane strain condition. Typically, this type of failure occurs if the soil above the slide plane is relatively soft and the piles are stiff and spaced far apart. This is usually not the case with relatively flexible Pin Piles, but may govern when stiffer pipe elements are employed. Based on the theory proposed by Ito and Matsui (1975), the predicted results for various pile spacings and soil conditions are plotted in Figure 6-32.

This procedure is useful in providing a preliminary estimate of the pile density and type of piles that are feasible for a particular application. The procedure is conservative for wall cross sections that use battered piles. Battering the Pin Piles with respect to the slide plane, and/or direction of slope movement tends to mobilize the axial resistance of the Pin Piles. Since the piles are typically small in diameter, their surface area to cross-sectional area ratio is relatively large. Hence they are very efficient at mobilizing skin friction, and typically have much higher axial capacity than lateral capacity.

The examples of Type "A" insert walls reviewed had used a centralized reinforcing bar in a grout-filled hole. As shown on Figure 6-30, piles constructed using a centralized #9 reinforcing bar (e.g., Type 4 piles) have significantly less resistance to horizontal loading than the piles installed with pipe reinforcement. However, once the concrete crushes, the reinforcing bar provides additional resistance to horizontal loading by the development of tensile forces in the bar across the slide plane. A Type "A" insert wall designed assuming the Pin Piles behave in this manner would therefore be more economical than a wall designed using the procedure proposed for the preliminary design. However, the additional movement of the slope needed to reach this condition may be intolerable in some cases.

It should be noted that these charts are for preliminary design only to establish general requirements for pile size and spacing. Final design of Type "A" insert walls

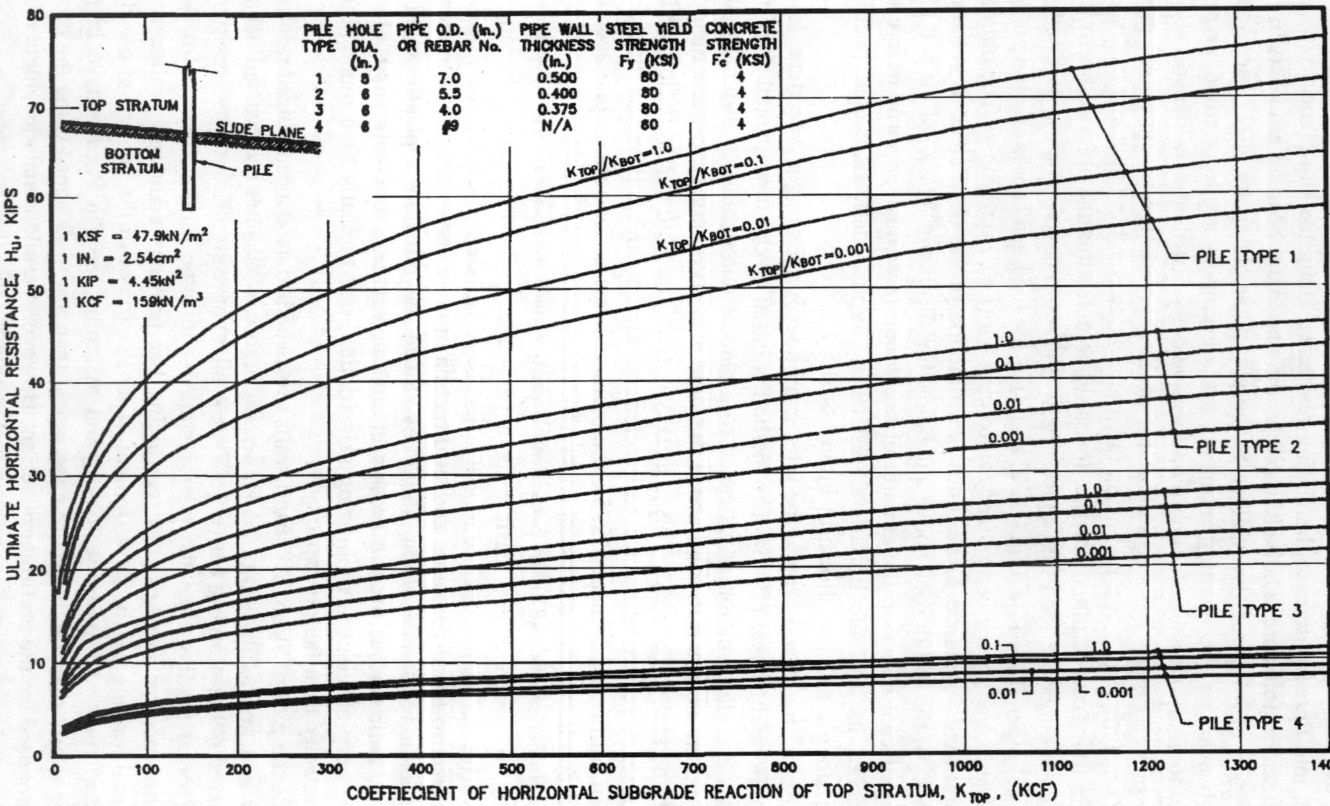

Figure 6-30 Preliminary design chart for ultimate horizontal resistance of piles (From Pearlman et al., 1992. Reproduced by permission of ASCE.)

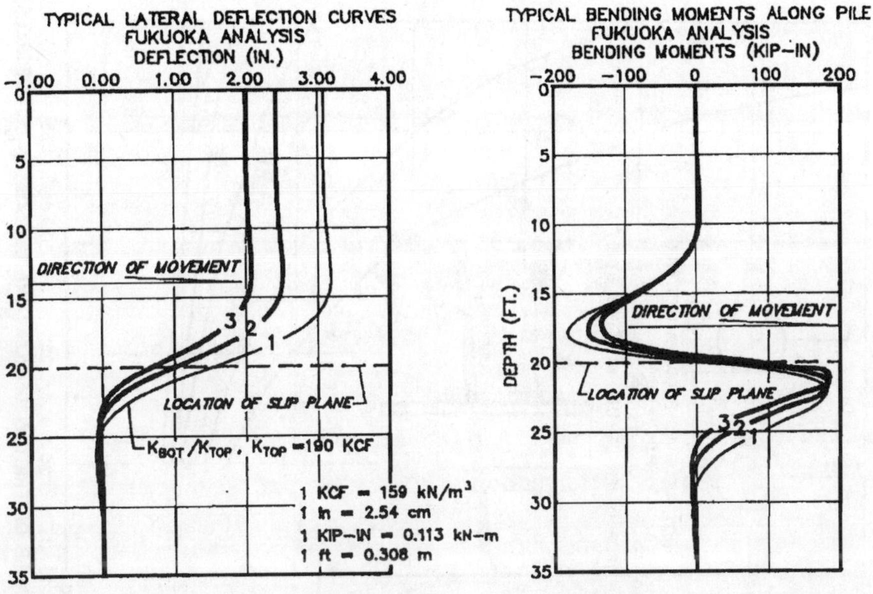

Figure 6-31 Typical results of analyses of Type 3 Pile (see Figure 6-30 for dimensions) (From Pearlman et al., 1992. Reproduce by permission of ASCE.)

requires consideration of various factors, including pile batter relative to the orientation of slope movement, the depth of the slide plane relative to the stiffness of the piles, and the additional capacity of the reinforcing bars after concrete crushing occurs.

Case Histories

Tables 6-17 and 6-18 summarize seven well documented and instrumented case histories of Type "A" insert walls and similar systems. (Other case histories have been alluded to, but the information presented in them was insufficient for current purposes.) It is noteworthy that the density of piles has reduced since the earliest installations from greater than two piles per foot to less than one pile per foot for more recently constructed walls. However, this reduction has not noticeably affected the performance of the walls with regard to limiting slope movements after wall construction. This observation suggests, and subsequent analyses later demonstrate, that the walls may have considerable reserve resistance if all possible soil-structure interaction modes are considered in design.

The information presented in the tables was used to evaluate the applicability of the proposed procedure for preliminary design of Type "A" insert walls. The following two projects are highlighted as they are generally similar in construction to the other walls, are well instrumented, and use a lower density of piles to control slope movements.

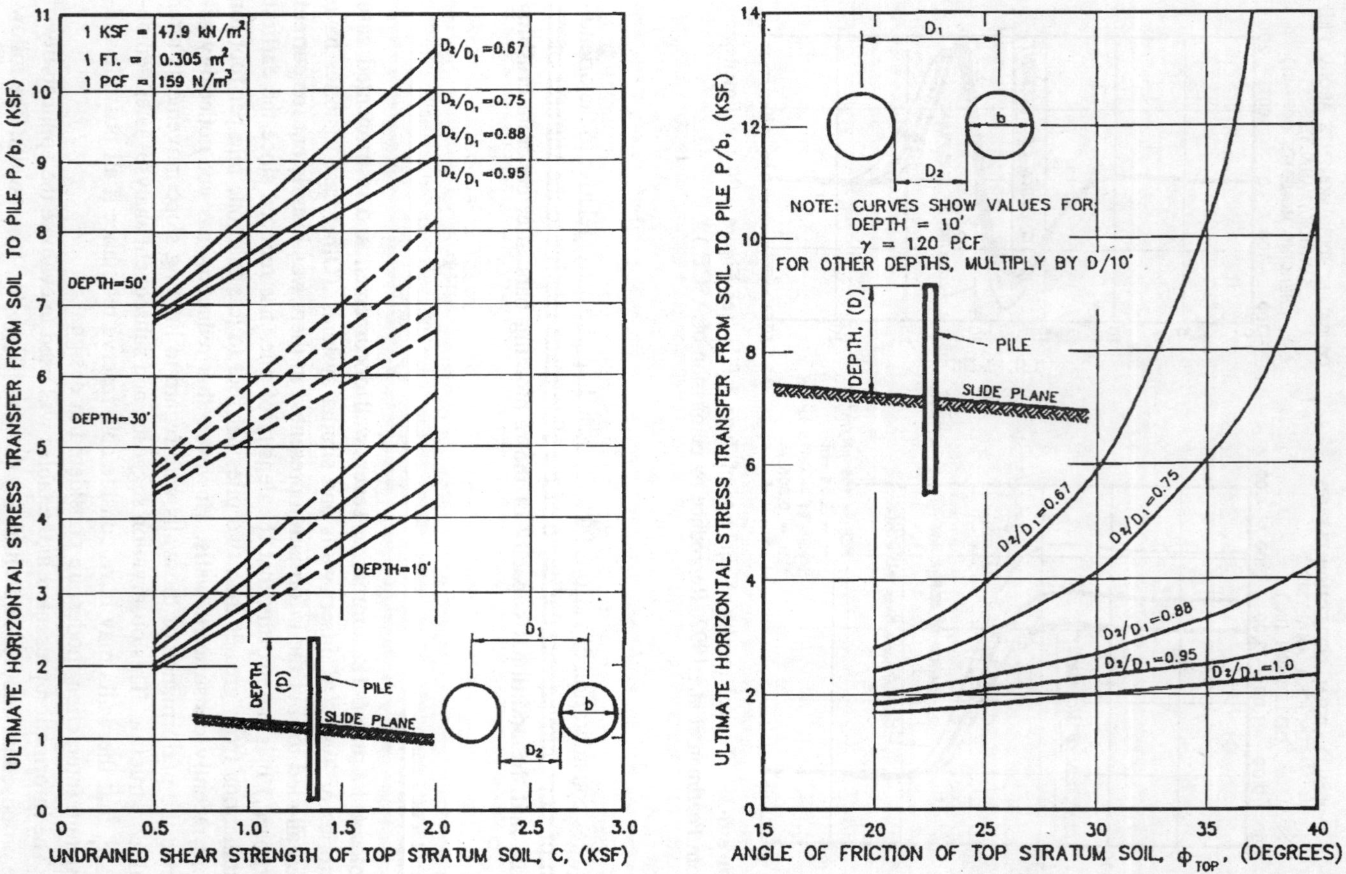

Figure 6-32 Ultimate stress transfer from soil to pile versus shear strength of soil (From Pearlman et al., 1992. Reproduced by permission of ASCE.)

TABLE 6-17 Summary of In Situ Wall Case Histories

Number	Project Name, Location, and Reference	Construction Date	Ground Conditions	Slope Geometry Upslope/Downslope	Depth to Slide at Wall (ft)
1	Forest Highway No. 7 Mendicino National Forest, Calif. Walkinshaw (1977); Palmerton (1984)	1977	Micaceous phyllite ($\phi = 13°$, $c = 500$ lb/ft^2) and schistose bedrock	2H–1V/2H–1V[b]	55
2	Route 23A Catskill State Park, N.Y. Murray (1984); Palmerton (1984)	1977	Moist, very dense glacial till with boulders ($\phi = 30°$, $c = 0$) and shale bedrock	2H–1V/2H–1.3V	10–26
3	PA-306 Monessen, Pa. Dash and Jovino (1980)	1979	Random fill and colluvium ($\phi = 17°$, $c = 100$ lb/ft^2, $\gamma = 134$ lb/ft^3) and weak shale	2.5H–1V/4H–1V	20
4	L.R. 69 Armstrong County, Pa. Earth, Inc. (1986); Dash (1987)	1985	Random fill, wet sandy clay, and hard clay ($\phi = 30°$, $c = 0$, $\gamma = 120$ lb/ft^3; $\phi = 17°$ along slide plane)	Level/1.2H–1V	34
5	Glady-Durbin Road No. 44 Randolph County, W.V. NCC[c] project files	1987	Sandy clay with rock fragments ($\phi = 30°$, $c = 0$, $\gamma = 120$ lb/ft^3) and sandstone bedrock	Level/1H–1V	16–40
6	S.R. 4023 Armstrong County, Pa. NCC project files	1988	Random fill, stiff colluvial clay with rock fragments ($\phi = 177°$, $c = 0$, $\gamma = 125$ lb/ft^3), and weathered claystone/ competent-sandstone	Level/2H–1V	23–35
7	Blue Heron Road Big South Fork River, Ky. COE[d] project files; NCC project files	1989	Medium stiff/stiff silty clay and shale bedrock ($\phi = 19°$, $c = 0$, $\gamma = 125$ lb/ft^3)	Level/1.5H–1V	25

Source: From Pearlman et al. (1992).

[a] 1 ft = 0.31 m; 1 lb/ft^2 = 47.9 N/m^2; 1 lb/ft^3 = 159 N/m^3.

[b] 2H–1V—2 horizontal to 1 vertical.

[c] NCC—Nicholson Construction Company.

[d] COE—U.S. Army Corps of Engineers.

TABLE 6-18 Summary of In Situ Wall Case Histories: Wall Geometry and performance[a]

Number	Cap Beam Geometry (L × W × T) (ft)	Pile Diameter (3) (in.)	Rebar Size (#)	Pile Density (pile/linear ft)	Pile Inclination (° from vertical)	Maximum Length (ft)	Maximum Embedment Below Slide (ft)	Lateral Displacement (in.)	Remarks
1	310 × 6 × 3 50 × 5 × 3	5 4.5/rock	9	2.33	19 to −19[b]	69 53	8	N.M.[e]	Cap constructed before pile construction
2	250 × 11 × 1.75	4	10	2.80	15 to −15	50	24–30	4 D.C.[f] 0.3 A.C.[g]	Cap constructed before pile construction
3	200 × 6 × 2.5	5	9	2.25	19 to −19	45	9	0.1–0.8 A.C.	Failure of downslope after construction; little wall movement

4	310 × 6 × 2.5	6	11/14	1.33	8 to −8	50	15	0.1 A.C.	Anchors inclined at 40° to vertical installed for rapid drawdown case
5	115 × 5 × 3 and 50 × 5 × 3	5.5/soil 4.5/rock	11/14/18	1.00	16 to −2	37[c] 53[d]	8	N.M.	Cap constructed before pile construction
6	122 × 4.6 × 3[c] 122 × 4.6 × 3[d]	5.5/soil 4.5/soil	11/14/18	1.25[c] 0.80[d]	26 to 0	60	17[c] 24d	1.0 D.C. 0.7 A.C.	Cap constructed after pile construction
7	34 × 5 × 3	5.5	11/14	0.75	19 to −5	35	10–15	0.6 D.C. 0.3 A.D.	Cap constructed after pile construction

Source: From Pearlman et al. (1992).

[a]1 ft = 0.31m; 1 in. = 2.54 cm.

[b]Upslope is + angle.

[c]Wall A.

[d]Wall B.

[e]N.M.—Not measured.

[f]D.C.—During construction.

[g]A.C.—After construction.

1. State Route 4023, Armstrong County, Pa. Portions of the two-lane roadway were constructed on a slope adjacent the Allegheny River. A 250-ft-long section of the road, and the railroad tracks located upslope, were experiencing damage caused by slope movements toward the river. A monitoring program initiated by the Pennsylvania Department of Transportation (PADOT) indicated a slip-plane was located about 25 to 35 ft below the road and that the slope was moving at a rate of up to 0.7 in./month.

PADOT designed a repair using prestressed rock anchors and tangent caissons extending into competent rock. The earth pressures used for the design were based on the results of stability analyses, for which the soil along the slip-plane was assigned a residual angle of friction. This design provided a minimum factor of safety with regard to the overall slope stability equal to 1.5 and 1.2 for the normal and rapid drawdown conditions, respectively. A postbid alternative based on a Type "A" insert wall design was accepted by PADOT with a resultant savings of about $1 million, compared to the lowest bid for the anchored caisson wall design.

In general, the wall consisted of four rows of Pin Piles extending across the slip-plane and into competent rock. The wall comprised two equal length sections designated as Wall A and Wall B. Wall A contained a higher density of piles than Wall B, because the top of a weathered rock dips to a lower elevation in the area of Wall A.

Performance monitoring consisted of: (1) strain gages on the reinforcement of the outermost pile rows; (2) inclinometers through and upslope from the wall; (3) fixed-end extensometers to measure average strain along the reinforcements; and (4) surface survey monuments. Due to space limitations, only the results of the slope inclinometer data for Wall A are described herein.

The maximum movement at select inclinometer locations along the length of the wall is plotted as a function of time in Figure 6-33. The data for inclinometers located relatively close to and within the wall indicate that up to 1.5 in. of horizontal movement occurred during construction, and up to 0.4 in. of movement occurred in the year following completion of construction. Inclinometer SI1W, located approximately 60 ft upslope from the wall, exhibited approximately 1.2 in. of additional movement in the first year after the completion of construction and 0.34 in. of movement in the second year following the end of construction.

Overall, the inclinometer data show that the wall significantly slowed the slope movements at the site from a rate of approximately 8 in. per year to a rate of 0.3 in. per year. Furthermore, the rate of movement decreased with time. The stabilizing effects of the relatively flexible pile elements appeared to become greater with increasing movement of the wall. If movements had continued and eventually reached an unacceptable magnitude, additional piles could have been installed.

The performance of the wall was back analyzed with a view to checking the revised design approach suggested in Section 6-3's discussion of Design. The location of the slip-plane was estimated based on inclinometer data, and the driving forces of the slide were calculated to be approximately 132 kips/linear ft.

Using equations developed by Leinenkugel (1976) and considered by Winter et al. (1983) for the stabilization of creeping clay slopes, the loads acting on the wall

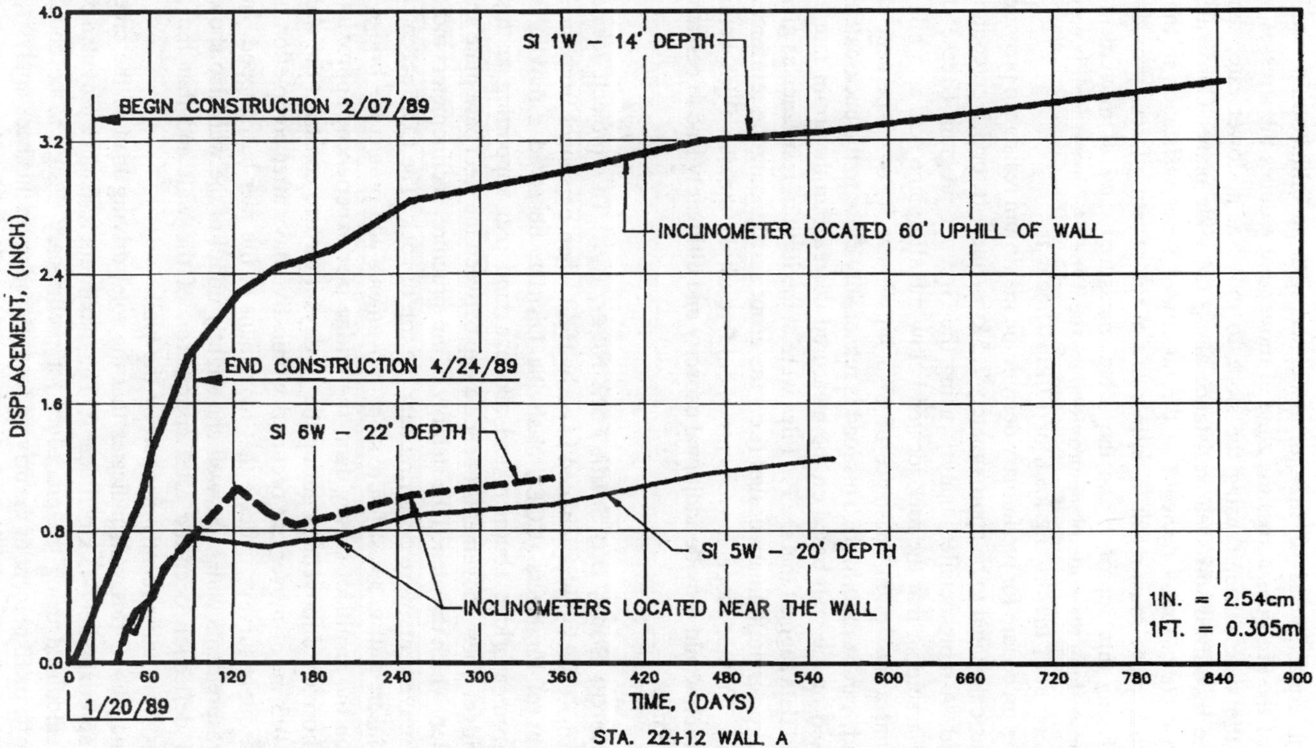

Figure 6-33 Maximum movement at select inclinometer locations with time; State Route 4023, Armstrong County, Pa. (From Pearlman et al., 1992. Reproduced by permission of ASCE.)

can be estimated. The basic assumption is that the mobilized shear stress along the failure plane equals the shear strength associated with an initial strain rate. The placement of the Pin Piles across the zone of movement reduces the stresses in the soil along the slide plane, causing the slope to move at a slower rate, thereby reducing the mobilized soil shear resistance along the slide plane. Consequently, equilibrium dictates that a decrease in the mobilized shear resistance in the soil along the slide plane must result in an increase of the resisting forces provided by the piles. Using this approach and the actual measured rates of movement, the resisting force along the slide plane provided by the Type "A" insert wall is equal to about 13 percent of the driving load, or 17 kips/linear ft.

Using the procedure for preliminary design, the maximum resistance provided by the wall was calculated to be approximately 7.5 kips/linear ft, or 5.7 percent of the driving load. A more detailed analysis using the Group 1 program (Reese et al., 1987), and assuming that the rock located below the slip-plane acts as a pile cap, indicates that a maximum resistance of 17.5 kips/linear ft, or 13 percent of the driving load, can be mobilized. This load is reasonably close to the back-calculated load imposed on the wall based on the measured decrease in the strain rate.

The field data suggest that the Pin Piles were at the limit of resistance as given in the procedure for preliminary design (i.e., the grout surrounding the reinforcement is crushing). Further slope movements should result in the Pin Piles acting in tension, which would provide additional capacity and ultimately lead to cessation of slope movement.

2. Blue Heron Road, Big South Fork River, Ky. This project, located in southeastern Kentucky, was initiated in early 1989 when personnel from the U.S. Army Corps of Engineers (COE), Nashville District, observed a moving slope downhill from a bridge abutment and above a land pier supporting an historic railroad bridge. These movements opened a gap between the steel base plate and the concrete pier, which threatened the stability of the structure. Inclinometers indicated that slope movements were occurring near the top of rock. The COE accepted the Type "A" insert wall to arrest future slope movements and protect the bridge pier.

Based on the results of survey data, the bridge pier experienced approximately 0.6 in. of horizontal movement during the 10 days it took to construct the wall; an additional 0.3 in. of movement occurred within five days after completion of the wall. Subsequent surveys indicate that movement of the pier has stopped. Inclinometer measurements within the wall and slightly uphill of the wall also show that 0.3 in. of deflection occurred after completion of the wall and that the slope movements were subsequently stopped by the wall.

Back-analysis of this wall indicates that the total driving load of the creeping slope was approximately 45 kips/linear ft. The maximum resistance provided by the wall was calculated using the procedure for preliminary design to be about 4.6 kips/linear ft, or 10 percent of the driving load. Based upon this back analysis, a 10 percent increase in the stability of the slope, as predicted by the procedure for preliminary design, proved sufficient to stop the slope movement.

Overview

Type "A" insert walls are used in applications where more traditional approaches (such as soil buttresses, anchored soldier pile walls, diaphragm walls, or conventional gravity retaining walls) may be possible but precluded by site or geological constraints or cost. For example, Type "A" walls are the least disruptive to the site, and may be constructed on extremely steep slopes using relatively small, lightweight drilling equipment. However, to date their use has been slowed in the United States by the absence of a rigorous design approach, which has in turn led to overconservative, expensive designs and a lack of confidence in the option overall. Given the significance and relevance of the most recent work by Pearlman and his co-workers, there is every reason to believe that the technique will now enjoy a major growth in popularity.

REFERENCES

Anonymous, 1987. "Core Drilling Through Concrete Pier Bases Preparatory to Mini Piling," *Geodrilling*, Aug., pp. 161–168.

ASCE Committee on Placement and Improvement of Soils, 1987. "Soil Improvement, A Ten-Year Update," Proc. Symp. at ASCE Convention, Apr. 28, Atlantic City, N.J.

ASTM—D1143-81, 1981a. "Method of Testing Piles Under Static Axial Compressive Load," Section 04, Vol. 04.08.

ASTM—D3966-81, 1981b. "Method of Testing Piles Under Lateral Loads," Section 04, Vol. 04.08.

Attwood, S., 1987. "Pali Radice: Their Uses in Stabilizing Existing Retaining Walls, and Creating Cast In Situ Retaining Structures," *Ground Engineering*, Vol. 20, No. 7, pp. 23–27.

Bjerrum, L., 1957. "Norwegian Experiences with Steel Piles to Rock," *Geotechnique*, Vol. 7, No. 2, pp. 73–96.

Bruce, D. A., 1988. "Developments in Geotechnical Construction Processes for Urban Engineering," *Civ. Eng. Practice*, Vol. 3, No. 1, Spring, pp. 49–97.

Bruce, D. A., 1988–1989. "Aspects of Minipiling Practice in the United States," *Ground Engineering*, Vol. 21, No. 8, pp. 20–33 and Vol. 22, No. 1, pp. 35–39.

Bruce, D. A., 1989a. "American Developments in the Use of Small Diameter INSERTS as Piles and In Situ Reinforcement," DFI Int. Conf. on Piling and Deep Foundations, May 15–18, London, 12 pp.

Bruce, D. A., 1989b. "Methods of Overburden Drilling in Geotechnical Construction: A Generic Classification," *Ground Engineering*, Vol. 22, No. 7, pp. 25–32.

Bruce, D. A., 1991. "The Construction and Performance of Prestressed Ground Anchors in Soils and Weak Rocks: A Personal Overview," DFI Conf., Chicago, Ill., Oct. 7–9.

Bruce, D. A., 1992a. "Recent Progress in American Pin Pile Technology," Proc. ASCE Conf., "Grouting, Soil Improvement and Geosynthetics," New Orleans, La., Feb. 25–28, pp. 765–777.

Bruce, D. A., 1992b. "Two New Specialty Geotechnical Processes for Slope Stabilization," ASCE Specialty Conf. on Stability and Performance of Slopes and Embankments—II, June 29–July 1, Berkeley, Calif., 15 pp.

Bruce, D. A. and B. Bianco, 1991. "Large Landslide Stabilization by Deep Drainage Wells," Proc. Int. Conf. on Slope Stabilization Engineering, Shanklin, Isle of Wight, Apr. 15–19, 8 pp.

Bruce, D. A. and D. L. Boley 1987. "New Methods of Highway Stabilization," Proc. 38th Annual Highway Geology Symp., May, Pittsburgh, Pa., pp. 57–72.

Bruce, D. A. and R. A. Jewell, 1986–1987. "Soil Nailing: Application and Practice," *Ground Engineering*, Vol. 19, No. 8, pp. 10–15, and Vol. 20 No. 1, pp. 21–32.

Bruce, D. A. and C. K. Yeung, 1983. "A Review of Minipiling, With Particular Regard to Hong Kong Applications," *Hong Kong Engineer*, June 1984, pp. 31–54, having been presented at the Institution in Nov. 1983.

Bruce, D. A., J. D. Ingle, and M. R. Jones, 1985. "Recent Examples of Underpinning Using Mini Piles," 2nd Int. Conf. on Structural Faults and Repairs, Apr.–May, London, 11 pp.

Bruce, D. A., S. L. Pearlman, and J. H. Clark, 1990. "Foundation Rehabilitation of the Pocomoke River Bridge, MD, Using High Capacity Preloaded Pin Piles," 7th Int. Bridge Conf., June 18–20, Pittsburgh, Pa.

British Standards Institution (BS), 1972. "CP 2004 Foundations," London.

British Standards Institution (BS), 1989. "Ground Anchorages," BS 8081, London.

Commonwealth of Massachusetts, 1984. State Building Code, 4th ed.

Doornbos, S., 1987. "The Renovation of the Amsterdam Conference Hall," Proc. Int. Conf. on Foundations and Tunnels, Mar. 24–26, London, pp. 61–67.

Dywidag, 1983. Munich, Germany. Promotional brochure.

Federation Internationale de la Precontrainte (FIP), 1982. "Recommendations for the Design and Construction of Prestressed Ground Anchorages, FIP 2/7," Cement and Concrete Association, Slough, England (under review).

Federation International de la Precontrainte (FIP) 1986. *Corrosion and Corrosion Protection of Prestressed Ground Anchorages: State of the Art Report*, published by Thomas Telford, London, 28 pp.

Fukuoka, M., 1977. "The Effects of Horizontal Loads on Piles Due to Landslides," Proc., Specialty Session 10, Ninth Int. Conf. on Soil Mechanics and Foundation Engineering, Tokyo, Japan, pp. 27–42.

Gouvenot, D., 1975. "Essaid de Chargement et de Plambement de Pieux Aiguilles," *Annales de ITdB et de Travaux Publics*, No. 334, Dec.

Herbst, T. F., 1982. "The GEWI Pile—A Solution for Difficult Foundation Problems," Symp. on Soil and Rock Reinforcement Techniques, AIT, Nov. 29–Dec. 3, Bangkok, Paper 1-10.

Hong Kong Building (Construction) Regulations, 1976. Hong Kong Government.

Ito, T. and T. Matsui, 1975. "Methods to Estimate Lateral Force Acting on Stabilizing Piles," *Soils and Foundations*, Vol. 15, No. 4, pp. 43–59.

Jones, D. A. and M. J. Turner, 1980. "Post-Grouted Micro Piles," *Ground Engineering*, Vol. 13, No. 6, pp. 47–53.

Koreck, H. W., 1978. "Small Diameter Bored Injection Piles," *Ground Engineering*, Vol. 11, No. 4, pp. 14–29.

Leinenkugel, H. J., 1976. "Deformations und Festigkeitsver Halten Bindiger Erdstoffe," Experimentelle Erges-Bnisse und Ihre Physikalische Deutung Veroffentil. Inst. f. Bodenmech. u. Felsmech. 66, Karlsruhe.

Littlejohn, G. S., 1970. "Soil Anchors," Ground Engineering Conf., Institution of Civil Engineers, London, pp. 33–44.

Littlejohn, G. S., 1982. "Design of Cement Based Grouts," ASCE Conf., Grouting in Geotechnical Engineering, New Orleans, La., Feb. 10–12, pp. 35–48.

Littlejohn, G. S., 1990. "Ground Anchorage Practice," ASCE Conf., Design and Performance of Earth Retaining Structures, Cornell University, Ithaca, NY, June 18–21, pp. 692–733.

Littlejohn, G. S. and D. A. Bruce, 1977. "Rock Anchors—State of the Art," Foundation Publications Ltd., Brentwood, Essex, England, 50 pp.

Lizzi, F., 1978. "Reticulated Root Piles to Correct Landslides," ASCE Conf., Chicago, Ill., October.

Lizzi, F., 1982. "The Pali Radice (Root Piles)," Symp. on Soil and Rock Improvement Techniques Including Geotextiles, Reinforced Earth and Modern Piling Methods, Dec., Bangkok, Paper D1.

Lizzi, F., 1985. "Pali Radice (Root Piles) and Reticulated Pali Radice," *Underpinning*, Surrey Univ. Press, pp. 84–151.

Mascardi, C. A., 1970. "Il Comportamento dei Micropali Sottoposti a Sforzo Assiale, Momento Flettente e Taglio," Verlag Leemann, Zurich.

Mascardi, C. A., 1982. "Design Criteria and Performance of Micropiles," Symp. on Soil and Rock Improvement Techniques Including Geotextiles, Reinforced Earth and Modern Piling Methods, Dec., Bangkok, Paper D-3.

Mitchell, J. M., 1985. "Foundations for the Pan Pacific Hotel on Pinnacled and Cavernous Limestone," Proc. 8th SE Asian Geotechnical Conf., Mar. 11–15, Kaula Lumpur, pp. 4-29–4-44.

Munfakh, G. A. and N. N. Soliman, 1987. "Back on Track at Coney," *Civ. Eng.*, ASCE Vol. 59, No. 12, pp. 58–60.

Pearlman, S. L., B. D. Campbell, and J. L. Withiam, 1992. "Slope Stabilization Using In-Situ Earth Reinforcements," ASCE Specialty Conf. on Stability and Performance of Slopes and Embankments—II. June 29–July 1, Berkeley, Calif., 16 pp.

Plumelle, C., 1984. "Amelioraton de la Portante D'un Sol Par Inclusions de Froupe et Reseaux de Micropieux," Int. Symp. on In Situ Soil and Rock Reinforcement, Paris, Oct. 9–11, pp. 83–90.

Post Tensioning Institute (PTI), 1986. "Recommendations for Prestressed Rock and Soil Anchors," Phoenix, 41 pp.

Reese, L., A. Awoshika, P. Lam, and S. Wang, 1987. "Documentation of Computer Program, GROUP1, Analysis of a Group of Piles Subjected to Axial and Lateral Loadings," Ensoft Inc., Austin, Tex.

Suzuki, I., T. Hirakawa, K. Morii, and K. Kanenko, 1972. "Developments Nouveaux dans les Foundations de Plyons pour Lignes de Transport THT du Japon," Conf. Int. des Grande Reseaux Electriques a Haute Tension, Paper 21-01, 13 pp.

TRB, 1987. "Reinforcement of Earth Slopes and Embankments," NCHRP Report 290, June.

Weltman, A., 1981. "A Review of Micro Pile Types," *Ground Engineering*, Vol. 14, No. 3, pp. 43–49.

Winter, H., W. Schwartz, and G. Gudehus, 1983. "Stabilization of Clay Slopes by Piles," Proc. 8th Europ. Conf. on Soil Mechanics and Foundation Engineerings, Helsinki, Vol. 2, pp. 545–550.

Xanthakos, P. 1991. *Ground Anchors and Anchored Structures*, Wiley, New York, 688 pp.

CHAPTER 7

PERMEATION GROUTING

7-1 BACKGROUND TO CHAPTERS 7 AND 8

Overviews of ground treatment typically identify four basic categories of soil grouting (Figure 7-1):

1. Hydrofracture (or claquage).
2. Compaction.
3. Permeation.
4. Jet (or replacement).

Hydrofracture Grouting

The ground is deliberately split by injecting stable but fluid cement-based grouts at high pressures (e.g., up to 4 MPa). The lenses and sheets of grout so formed are thought to increase total stress, fill unconnected voids, possibly consolidate the soil locally, and, conceptually, create mainly horizontal, impermeable barriers. However, hydrofracture grouting's effects are difficult to control, and the potential danger of damaging adjacent structures by the use of high pressure often proves prohibitive. It has not been common to find this technique alone deliberately exploited outside the French grouting industry, although some hydrofracture phenomena accompany most permeation grouting contracts either accidentally or in conjunction. Tornaghi et al. (1988) note that hydrofracture naturally occurs with conventional cement-based grouts in soils with a permeability of less than 10^{-1} cm/sec. In California, this technique is being promoted under the term "controlled fracture" grouting. Polypropylene fibers in the grout are claimed to impart significant tensile

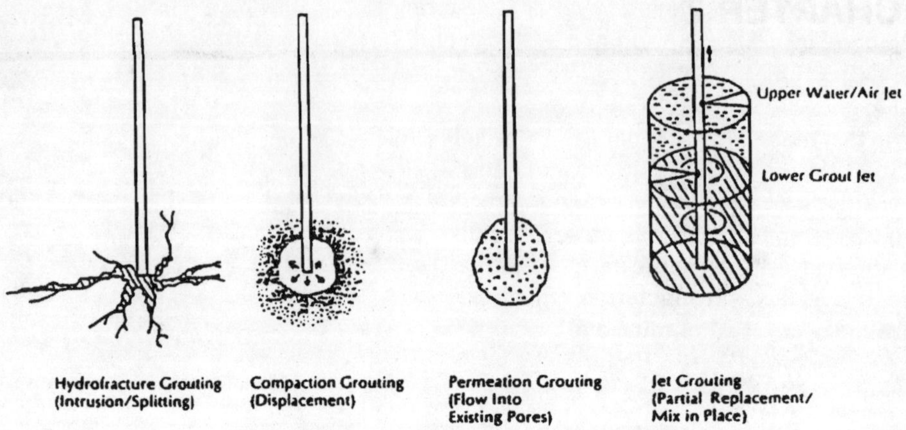

Figure 7-1 Basic categories of soil grouting.

and flexural strength to the grout lenses formed, so enhancing the potential of the technique in fill and slope stabilization, expansive soil treatment, and soft ground tunneling. Zuomei and Pinshou (1982) also reported an application of hydrofracture grouting in clay infilled karstic limestones. The aim was to increase the resistance of the clay against washout under high hydraulic gradients. More recently, Pototschnik (1992) described a successful application to reduce the settlement of a tall structure on fine, saturated sand, influenced by an adjacent new subway tunnel.

Compaction Grouting

This is a specialized "uniquely American" process (Baker et al., 1983) that has been used since the early 1950s and remains very popular in the United States. Very stiff, low-mobility soil-cement mixes are injected at high pressures (up to 3.5 MPa) at discrete locations to densify soft, loose, or disturbed soil. Unlike the case of hydrofracture grouting, the grout is intended to form a very dense and coherent bulb that does not extend far from the point of injection. Near-surface injections result in the lifting of the ground surface (the technique of slab jacking as described, for example, by Bruce and Joyce, 1983), and, indeed, the earlier applications were used exclusively on shallow foundations (Warner, 1982; Graf, 1992).

Although compaction grouting does have practical and technical limitations, its popularity continues to grow, in no small way due to its very active and professional promotion in the technical press and at geotechnical gatherings by speciality contractors. However, its possible application should be most carefully reviewed when dealing with tall structures or buildings that can tolerate only the smallest differential movements. Under such conditions, it is imperative to attack the cause of the settlements at the source, and prevent them from migrating away from the excavation. Permeation or replacement grouting may then be necessary. Good case histories and guidelines abound. Recent papers dealing with more novel applications include those by Salley et al. (1987), referring to liquefaction control measures at Pinopolis West Dam, S.C., and by Welsh (1988) for combating sinkhole settlements in karstic limestone topographies. Warner (1992) and Warner et al. (1992) provide fundamental reviews of mix design and rheology considerations. There is evidence also that the technique is being increasingly exploited in Western Europe, Japan, and Taiwan, largely based on U.S. expertise.

Akin to compaction grouting in principle is "squeeze grouting" (Greenwood and Hutchinson, 1982). In this variant, the fluid grout initially induces localized consolidation stresses centered on the injection point, but then acts on a larger ground volume as the hydrofracture mechanism progresses *in a controlled fashion.* Several examples are cited in mine and tunnel work, largely in Southern Africa. "Compensation grouting" (e.g., Essler and Linney, 1992) is another similar and recent offshoot of this concept and is beginning to find popularity in Western Europe and Canada in particular.

Permeation Grouting

In certain ways, the techniques involved in permeation grouting are the oldest and best researched. The intent of permeation is to introduce grout into soil pores without any essential change in the original soil volume and structure. The properties of the soil, and principally the geometry of the pores, are clearly the major determinants of the method of grouting and the materials that may be used (Figure 7-2). This method, as applied to both rocks and soils, is the prime subject of this chapter.

Jet Grouting

Jet, or replacement, grouting is the youngest major category of ground treatment. According to Miki and Nakanishi (1984), the basic concept was propounded in Japan in 1965, but it is generally agreed that it is only since the early 1980s that the various derivatives of jet grouting have approached their full economic and operational potential to the extent that today it is arguably the fastest growing method of ground treatment worldwide. Its development was fostered by the need to thoroughly treat soils ranging from gravels to clays to random fills in areas where major environmental controls were strongly exercised over the use of chemical (permeation) grouts and allowable ground movements.

Jet grouting can be executed in soils with a wide range of granulometries and permeabilities. Indeed, any limitations with regard to its applicability are imposed by other soil parameters (e.g., the shear strength of cohesive soils or the density of granular deposits) or by economic factors.

The ASCE Geotechnical Engineering Division Committee on Grouting (1980) defined jet grouting as a "technique utilizing a special drill bit with horizontal and vertical high speed water jets to excavate alluvial soils and produce hard impervious columns by pumping grout through the horizontal nozzles that jets and mixes with foundation material as the drill bit is withdrawn." Figure 7-3 depicts one particular

Type of soils	Coarse sands and gravels	Medium to fine sands	Silty or clayey sands, silts
Soil characteristics:			
Grain diameter	$D_{10} > 0.5$ mm	$0.02 < D_{10} < 0.5$ mm	$D_{10} < 0.02$ mm
Specific surface	$S < 100\ \text{cm}^{-1}$	$100\ \text{cm}^{-1} < S < 1000\ \text{cm}^{-1}$	$S > 1000\ \text{cm}^{-1}$
Permeability	$k > 10^{-3}\ \text{m s}^{-1}$	$10^{-3} > k > 10^{-5}\ \text{m s}^{-1}$	$k < 10^{-5}\ \text{m s}^{-1}$
Type of mix	Bingham suspensions	Colloid solutions (gels)	Pure solutions (resins)
Consolidation grouting	Cement ($k > 10^{-2}\ \text{m s}^{-1}$) Aerated mix	Hard silica gels: double-shot: Joosten (for $k > 10^{-4}\ \text{m s}^{-1}$) single-shot: Carongel Glyoxol Siroc	Aminoplastic Phenoplastic
Impermeability grouting	Aerated mix Bentonite gel Clay gel Clay/cement	Bentonite gel Lignochromate Light carongel Soft silica gel Vulcanizable oils Others (Terranier)	Acrylamide Aminoplastic Phenoplastic

(*a*)

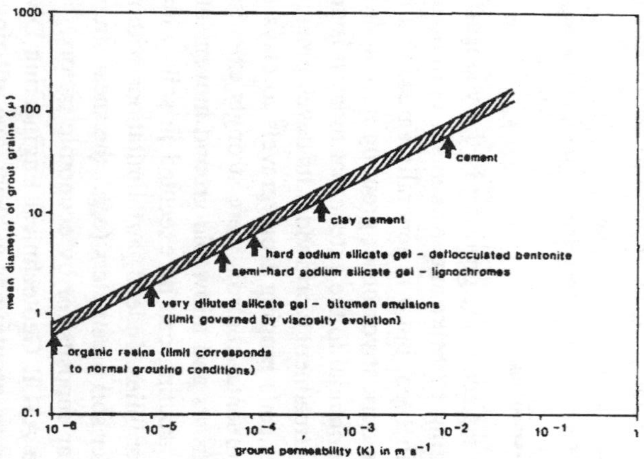

(*b*)

Rheological class	Particulate suspensions (Bingham fluids)			Solutions (Newtonian fluids)			Gaseous emulsions	
	Unstable	Stable		Colloidal solutions (evolutive)		Pure solutions (non-evolut.)		
Main types of grouts				Chemical grouts			Swelling grouts	
				Based on sodium silicate				
	Cement only	Cement with bentonite or clay	Deflocculated bentonite	High strength	Medium-low strength	Based on organic resins	Based on cement	Based on organic products
Fields of applicability	Fissured rock and masonry	╳	╳	Micro-fissured and porous rock			Large voids or cavities	Cavities with fast flowing water
		Granular soils						
		Prevailing gravel	Coarse sands	Medium-fine sands		Fine silty sands (sandy silts)		
Coefficient K of permeability (m/s)		$> 5.10^{-4}$	$> 5.10^{-5}$	$> 5.10^{-5}$	$> 1.10^{-6}$ ①	$> 1.10^{-6}$ ②	╳	
Specific surface S_0 (m²/l)		< 0.5	< 1.5	< 1.5	< 4	< 10		
Basic injection principles	High pressure	Controlled quantity and pressure					Low pressure (filling)	

① Limit governed by viscosity/time evolution ② Normal limit for uniform impregnation

(*c*)

Figure 7-2 Various classifications and guides to grout penetrability. (*a*) Grouting limits of common mixes. (After Caron, 1965.) (*b*) Limits of injectability of grouts based on the permeability of sands and gravels. (After Cambefort, 1977.) (*c*) Classification of grouts related to groutable media (From Gallavresi, 1992.)

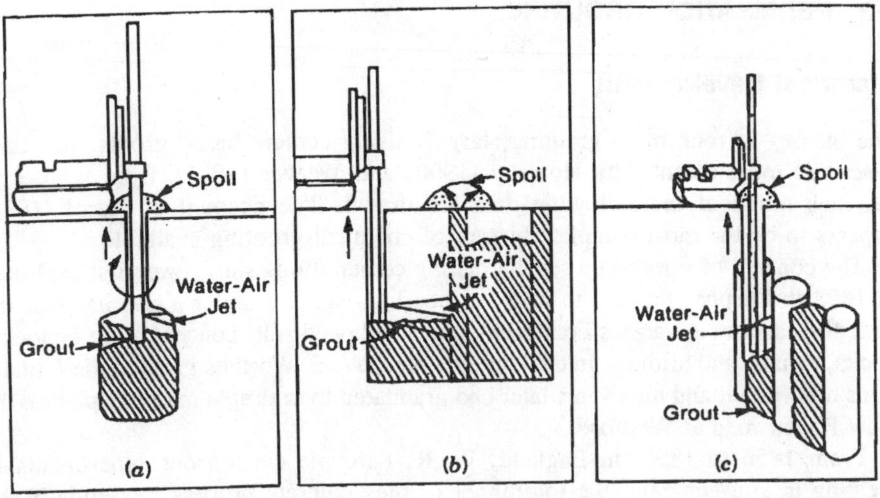

Figure 7-3 Jet grouting construction elements. (*a*) Column. (*b*) Panel. (*c*) Wing. (From ASCE, 1987. Reproduced by permission of ASCE.)

type in which the soil is jetted by an upper nozzle ejecting water at up to 50 MPa inside an envelope of compressed air at up to 1.2 MPa (i.e., a three-fluid system). The debris are displaced out of the oversized drill hole by the simultaneous injection of cement-based grout through a lower nozzle (at pressures up to 8 MPa). Other simpler variants (e.g., the one-fluid system) utilize grout jetting alone to simultaneously erode and inject, giving much more of a mix-in-place action. At the other extreme of complexity, the new Japanese Super Soil Stabilization Management (SSSMAN) system provides total (and verifiable) excavation of the soil prior to grouting or concreting. Clearly, each system has its own cost implications. It would seem that 45 m is the practical maximum depth of treatment.

In contrast to the sensitivity and sophistication of some aspects of permeation grouting, the principle of jet grouting stands as a straightforward positive solution, using only cement-based grouts across the whole range of soil types. It therefore has the potential of being "designer-driven" as a technology, unlike the case with other grouting methods. However, it must be emphasized that any system that may involve the simultaneous injection of up to three fluids at operating pressures of up to 50 MPa must be handled with extreme care and only in appropriate applications, circumstances, and ground conditions.

Applications of jet grouting have been reported throughout Western Europe, the Far East, Russia, South Africa, and South America. In addition, there is a small but growing market in North America, largely under the promotion of certain government agencies and specialist contractors, following a "slow and uncertain" start (Andromolos and Pettit, 1986). In Canada, numerous works have been conducted in the Montreal region, associated with deep excavations, while at John Hart Dam, B.C., jet grouting has been used through an existing dam to help create a seismic cutoff (Imrie et al., 1988).

Jet grouting is the subject of Chapter 8.

7-2 PERMEATION GROUTING

Historical Development

The history of rock mass grouting, largely using cement-based grouts, has been described most recently by Houlsby (1990) and Weaver (1991), from whose researches much of the following data are drawn. The account of Karol (1990) appears to be the most complete history of chemical grouting available.

The concept of injecting a self-hardening cementitious slurry was first exploited in 1802 in Dieppe, France, to improve bearing capacity under a sluice. Over the next 40 years or so, various French engineers followed suit, concentrating on locks, docks, canals, and bridges. In the United States, W. E. Worthen grouted the foundations of a flume, and nine years later had graduated to sealing a masonry pier on the New Haven road at Westford.

From 1856 to 1858, in England, W. R. Kinipple carried out experiments in creating in situ concrete (the forerunner of the "colcrete process": essentially the grout permeation of large, preplaced aggregate under water). True to imperial form, Kinipple regarded himself as the inventor of cement grouting, although by 1883 even he was still having difficulty convincing other engineers of the potential. However, applications did continue internationally and in 1876 the first dam grouting project was completed by T. Hawksley in Rochdale, England, and successful applications in French and German mines, London tunnels, and Maltese and Scottish docks soon followed.

At New Croton Dam, N.Y., the rock grouting process was applied systematically for the first time on a large scale in 1893, and remedial grouting projects continued on bridges, breakwaters, and dams in England and Egypt. By 1915 the first technical paper devoted to the grouting of a rock foundation under a dam (Estacada, Ore.) was published (Rands, 1915), and much interest resulted. The grouting at Hoover Dam between 1932 and 1935 is said to mark the beginning of systematic design of rock treatment in the United States (Glossop, 1961).

Since then, developments in rock fissure grouting have continued apace, with researches into drilling and grouting technologies, water testing, and materials developments being well documented (e.g., Simonds, 1947, 1958a,b; USCE, 1956; ASCE, 1982, 1985, 1992; Leonard and Grant, 1958; ICE, 1963, 1992; and ACI, 1984).

The use of chemicals in grouting evolved when attempted uses of cement grouts proved only partially successful. For example, as described by Littlejohn (1985), the progress of two deep shafts in finely fissured porous sandstone at Thorne in Yorkshire, England, was stopped in 1909 due to heavy water flows. The Belgian engineer, A. Francois, developed in 1911 a process called "silicatization" at a nearby colliery, and by 1913, he had solved the Thorne problem by injecting a solution of aluminum sulfate and sodium silicate *prior* to cement grouting. He originally thought that the chemicals were mere lubricants: he did not realize that they formed a chemical grout for filling fine pores and fissures, so allowing the cement grout to set normally without dissolution or displacement. This successful technique was then widely used in shaft sinking, in South Africa especially.

In 1925, Joosten patented the chemical grouting process now known by his name, and this permitted developments in grout chemistry that made it practical to permeate soils as fine as medium sands. The two components were injected successively during the placing (concentrated sodium silicate) and extraction (strong brine) of a perforated pipe, to permit a chemical reaction in situ. The search continued on commercial grounds for a "one shot" solution, capable of reacting in the ground after having been prepared from its components on the surface. These chemicals were based on diluted sodium silicate solutions, which had good waterproofing properties but low strength.

By 1933, Ischy had invented the tube à manchette system (see the discussion of Construction later in Section 7-2), a grout injection method ideally suited to the precise treatment of soils with great operational flexibility. This was used by the Rodio company at Bou-Hanifia Dam in Algeria under Terzaghi's advice, and facilitated over 5000 tons of sodium silicate injections. Thereafter, the approach to soil permeation was progressively enhanced and rationalized due to theoretical researches (e.g., Maag, 1938), and materials developments. These latter focused on lowering viscosity, increasing gel time control, increasing strength, and improving durability. In 1963, the ICE Conference in London reviewed the contemporary state-of-the-art and Table 7-1 illustrates the major commercial grout systems considered practicable at that time.

Since then, developments have continued principally into new materials, including those that are water reactive, elastic after gelling, highly durable, and environmentally compatible. Indeed, by 1983, Karol was able to list eight major research and review documents prepared directly by, or commissioned for, government agencies. These documents were in addition to fundamental, classic works by Cambefort (1977) and Caron (1982), as prime examples. Most recently, renewed attention has been devoted to the microfine cement-based grouts (e.g., DePaoli et al., 1992a, b) and the whole concept of grout rheology as related to efficiency of injection (Deere and Lombardi, 1985).

Regarding processes, the Japanese in particular have been active, bringing to commercial use a series of drill and grout systems, such as LAG, DDS, and SGR, (Bruce, 1989a), which have enjoyed considerable success on soft ground tunneling projects in the Far East, although they have received little attention elsewhere (see the discussion of Grouting in Soil under the Construction heading later in Section 7-2).

The interest shown in the recent literature (e.g., Karol, 1983, 1990) and at conferences (e.g., ASCE, 1992) confirms that permeation grouting remains a very dynamic, challenging, and evolving topic. Typically, developments originate with specialty contractors or materials suppliers, and are then explored further by universities and governmental agencies before entering general usage.

Applications

As noted in the AFTES (1991) report, permeation grouting has an extremely wide range of applications, which may be categorized according to general criteria such as:

TABLE 7-1 Classification of Grouts

Proprietary Name (if used)	Basic Composition	Type of Action	Strength of Gel or Cement (lb/in²)	Strength of Treated Soil (lb/in²)
Joosten I	Sodium silicate, calcium chloride (2F)[b]	V, A, 2P	—	Up to 1000
Joosten II	Sodium silicate, alkali dilution, calcium chloride (2F)	V, A, 2P	—	600
Joosten III	Sodium silicate, heavy metal salt, ammoniacal colloid (1F)	V, A, 2P	—	120
Guttman	Similar to Joosten II, sodium carbonate as alkali	V, A, 2P	—	200–700
Rodio	Sodium silicate, lime water (1F)	V, A, 2P	—	100
Langer	Sodium silicate, heavy metal salt, coagulant	V, A, 2P	—	
Polivka	Sodium silicate, sodium bicarbonate (1F)	V, A, 2P	—	70–100
	Silicate–ethyl acetate (1F)	V, A, 2P	20–100	Up to 300
	Resourcinol–formaldehyde (1F)	V, A, 2P	—	Up to 300
	Urea–formaldehyde (1F)	V, A, 2P	20–200	Up to 500
AM-9	Acrylamide (1F)	V, A, 2P	—	Up to 300
	Calcium acrylate (1F)	V, A, 2P	—	—
	Chrome–lignin (1F)	V, A, 2P	—	—
	Polyester (1F)	V, A, 1P or A, 1P	—	Up to 300
Polythixon	Polyurethanes (1F)	V, A, 1P	20–200	Up to 500

Source: After Skipp and Renner (1963).

[a]Literature gives little data on type and density of sands.

[b]1F = single-fluid; 2F = two-fluid.

[c]V = void filler; A = adhesive; 1P = single phase cement; 2P = two-phase cement.

- Purpose of the structure.
- Urban or nonurban location.
- Depth and dimensions of the structure.
- Type of ground.
- Purpose of the grouting, for example, strengthening or waterproofing.
- Temporary or permanent works.
- Preventative or remedial works.

The technical literature is full of examples of applications illustrating one or a combination of these criteria and one need only review the numerous conference proceedings and textbooks for appropriate case histories to cite in a wide range of applications.

Grouts

There are a number of comprehensive but basically similar classifications of grout types, e.g., FHWA (1976), Cambefort (1977), Karol (1982, 1990), Mongilardi and Tornaghi (1986), Shirlaw (1987), Naudts (1989), and AFTES (1991). This review draws freely from each but mostly from the AFTES document, both in structure and content. The author especially acknowledges the kind permission of the French Tunnelling Association (AFTES, 1991) to draw freely from their particular report.

There are fundamentally three categories (Figure 7-4), which can be listed in order of rheological performance and cost:

1. Particulate (suspension or cementitious) grouts, having a Binghamian performance.
2. Colloidal solutions, grouts that are evolutive Newtonian fluids in which the viscosity increases with time.
3. Pure solutions, nonevolutive Newtonian solutions in which the viscosity is constant until setting, within an adjustable period.

Category 1 comprises mixtures of water and one or several particulate solids such as cement, flyash, clays, or sand. Such mixes, depending on their composition, may prove to be stable (i.e., having little or no sedimentation or bleeding) or unstable,

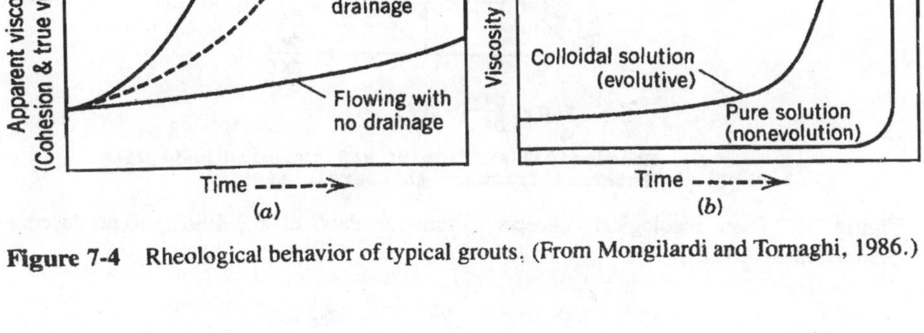

Figure 7-4 Rheological behavior of typical grouts. (From Mongilardi and Tornaghi, 1986.)

when left at rest. Stable thixotropic grouts have both cohesion and plastic viscosity increasing with time at a rate that may be considerably accelerated under pressure.

Category 2 and 3 grouts are commonly referred to as "chemical" grouts and are typically subdivided on the basis of their component chemistries, for example, silicate based (Category 2) or resins (Category 3). The outstanding rheological properties of Category 3 grouts, together with their low viscosities, permits permeation of soils as fine as silty sands ($k < 10^{-4}$ cm/s).

The various rheological concepts are illustrated in Figure 7-5.

Cementitious Grouts Due to their characteristics (including cost considerations), these grouts remain the most commonly used in both waterproofing and ground strengthening. The water to solids ratio is the prime determinant of their properties and characteristics including stability, fluidity, strength, and durability (Littlejohn, 1982). One can identify five broad classes:

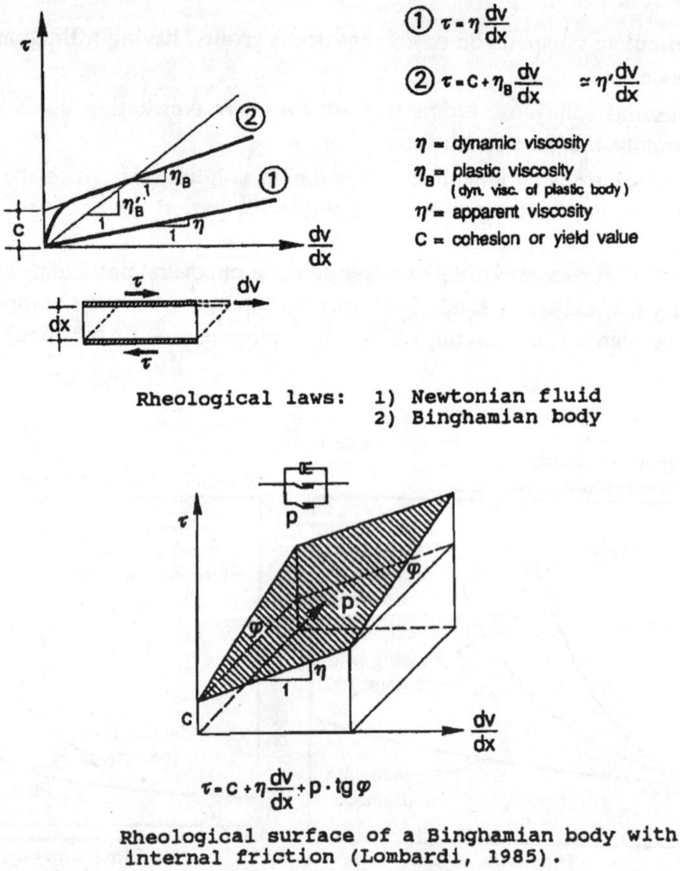

Figure 7-5 Basic rheological concepts. (From De Paoli et al., 1992a. Reproduced by permission of ASCE.)

1. Neat cement grouts.
2. Clay/bentonite cement grouts.
3. Grouts with fillers.
4. Grouts for special applications.
5. Grouts with enhanced penetrability.

Neat Cement Grouts As illustrated in Figure 7-6, such grouts are typically unstable, except at w/c ratios less then about 0.4 as prepared in high sheer mixers. Their mode of deposition in intergranular voids or fissures is akin to a hydraulic flushing action and so is sensitive to the dimensions of these spaces as well as the grout mix and injection parameters (see the discussions of Other Grouts and Design later in Section 7-2). Typically these grouts are associated with high strength and durability.

Clay/Bentonite Cement Grouts These cement suspensions are stabilized with a clay mineral to:

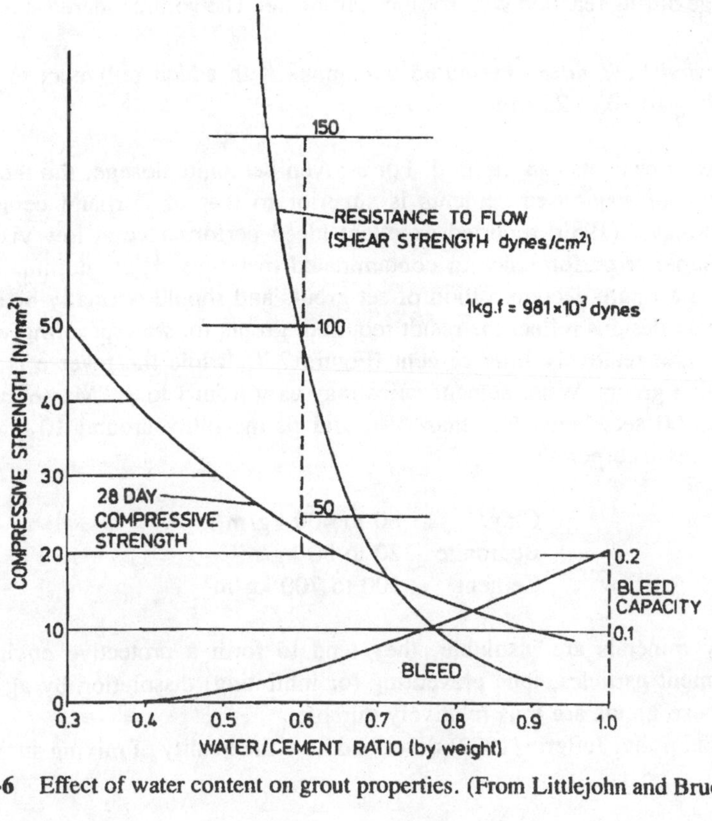

Figure 7-6 Effect of water content on grout properties. (From Littlejohn and Bruce, 1977.)

- Obtain homogeneous colloidal mixes with a wide range of viscosities.
- Reduce sedimentation (bleed).
- Decrease the setting time index and filtration tendencies.
- Increase the cement binding time.
- Improve penetrability and resistance to washout.
- Permit a wide range of mechanical strengths.
- Reduce permeability.

Common products used are

- *Natural Clay:* Economic, and swells when hydrated to as much as six times dry volume.
- *Natural Bentonites:* Montmorillonitic components predominate. Natural bentonites have remarkable colloidal properties in water and can swell to 18 times their original volume. The sodium bentonites unique to Wyoming give the best performance. Deere (1982) was among the first in the United States to describe the potential of such mixes in routine rock fissure grouting.
- *Permuted Bentonites:* Natural calcium bentonites that undergo an ion exchange during reaction with sodium carbonate. The volume increase is 10 to 15 times.
- *Activated Bentonites:* Permuted bentonites with added polymers to increase swelling to 10 to 25 times.

Most types of cements can be used. For a given bentonite dosage, the mechanical performance of slag-based cements is superior to that of Portland cement. For example, Weaver (1989) reported excellent bleed performance at low viscosities, and so a superior performance in contaminated materials. High alumina cements may lead to a future decomposition of set grouts and should normally be avoided.

Grout mix designs reflect the result required: grouts for waterproofing will have much clay and relatively little cement (Figure 7-7), while the reverse is true for strengthening grouts. Water-cement ratios may vary from 1 to 8, "Marsh" viscosity from 35 to 60 sec, bleed less than 5%, and permeability around 10^{-6} cm/sec. Typical mixes incorporate

Clay	80 to 400 kg/m^3
Bentonite	20 to 80 kg/m^3
Cement	100 to 700 kg/m^3

Since clay minerals are insoluble, they tend to form a protective environment around cement particles, thus preventing (or inhibiting) dissolution by aggressive waters. These grouts are thus relatively durable.

As a final point, Jefferis (1982) described how the quality of mixing such grouts

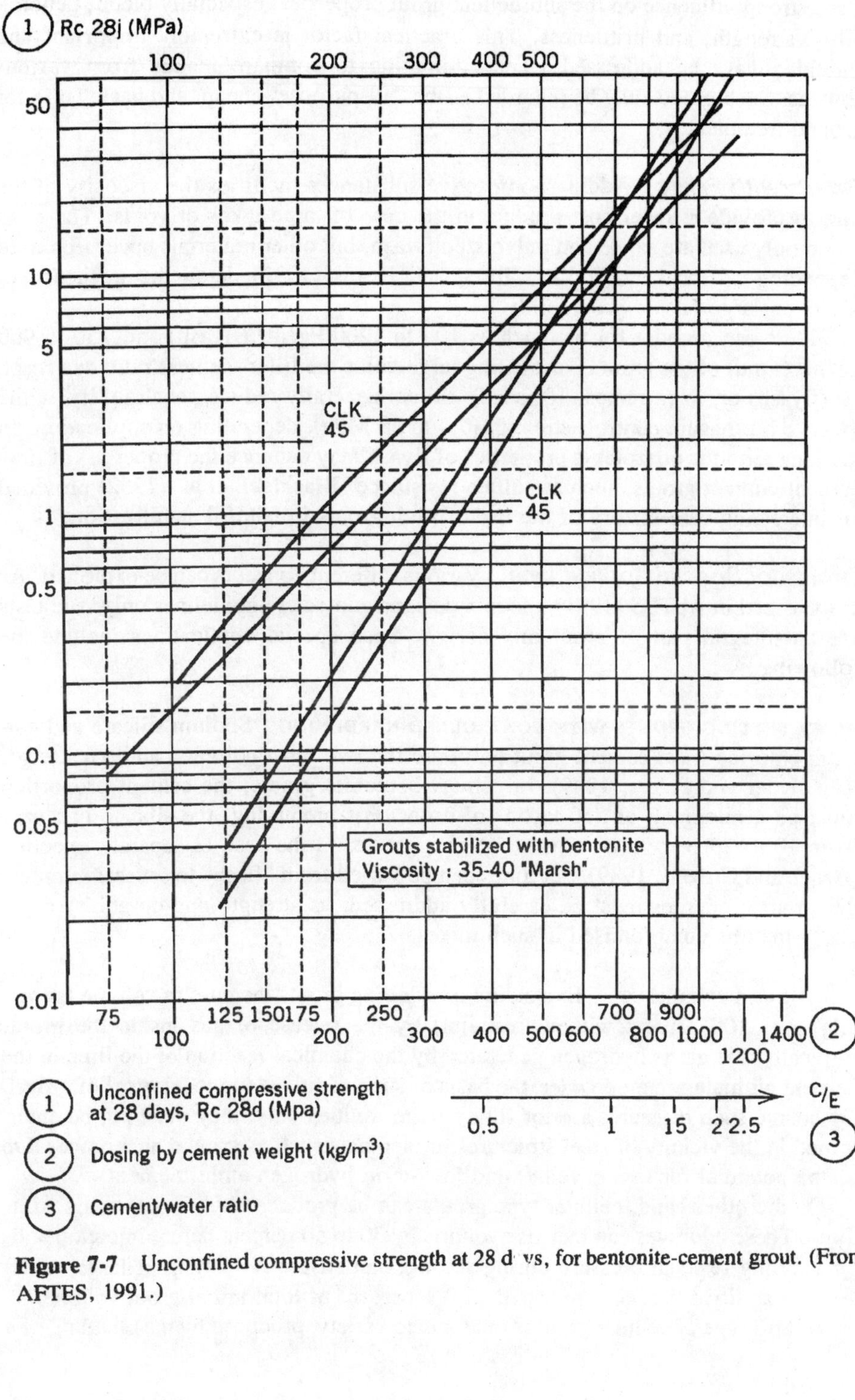

Figure 7-7 Unconfined compressive strength at 28 d vs, for bentonite-cement grout. (From AFTES, 1991.)

has a strong influence on the subsequent grout properties, especially bleed, penetrability, strength, and brittleness. This practical factor is extremely important and should always be addressed when attempting to compare results from various sources. Such grouts may be regarded as the "all-purpose, cheap, and basic" mix for ground treatment.

Grouts with Fillers Adding nonreactive substances modifies the viscosity of the mix to provide a lower cost product in the case of large takes or voids. The most commonly used are sands and pulverized flyash, but other materials have been used depending on local availability. Tosca and Evans (1992) detail the influence of fillers on large fissure groutability.

Fillers can account for as much as 100 to 1200 kg/m^3 (flyash) and 750 to 900 kg/m^3 (sand) of the grout. For treating large voids the filler–cement ratio can reach 20 (flyash) or, in extremes, 10 (sand). Fillers generally reduce penetrability, while the w/c ratio again controls strength (0.4 to 30 MPa), depending on application. In the long term the pozzolanic properties of flyash may enhance the properties of low-cement-content grouts, such as sulfate resistance. Blacklock et al. (1982) provided an interesting case history of the use of such grouts in landfill stabilization.

Grouts for Special Applications Various different concepts can be exploited. As emphasized in AFTES (1991), the mix designs quoted are guidelines only: site tests are an integral part of grout mix design. Such special applications include the following.

QUICK SETTING GROUTS WITH CONTROLLED HARDENING Sodium silicate and calcium chloride (neat cement grout only) are the two most common additives (e.g., Reifsnyder and Peters, 1989). In cement-bentonite grouts, the cement proportion must be a minimum of 250 kg/m^3 of grout. For premixing, the silicate can vary from 10 to 20 percent of cement weight, greater in the case of separate injection (Bruce and Croxall, 1989). Set times can be varied from "flash" to several minutes. Set grout properties must be carefully addressed, as strength and durability especially may be compromised in such mixes.

CELLULAR TYPE GROUTS Expanding or swelling grouts increase in volume (generally over 100 percent without restraint) by the release of gas inside the grout. Typically this gas is hydrogen generated by the chemical reaction of the lime in the cement with aluminum powder, the basis of such additives (up to 2 kg/m^3 of grout). Of course such measures are for filling large cavities only: they must not be entertained in the vicinity of steel structures or elements such as ground anchorages due to the potential for loss of bond, and long-term hydrogen embrittlement.

On the other hand, cellular type grouts can be produced by air entraining additives. These additives can increase volume by 30 to 50 percent before injection and, by exerting residual pressure during setting, can ensure full filling of large voids. Typical additive dosages are less then 0.1 percent of total initial grout volume.

A third type of cellular grout is the foamed variety, produced by the blending of a cement grout and a separately prepared foam. The advantages of these grouts over expanded grouts include their capacity for a higher cement content and for combining low density with high strength.

GROUTS WITH ENHANCED STRENGTH These grouts can be produced by: (1) adding a water reducing agent to permit the mixing and pumping of extremely low w/c ratio grouts; or (2) modifying the lime/silica ratio of the cement, by adding reactive siliceous products (e.g., silica fume) that give a pozzolanic-like compound with the lime of the cement. For example, ACI (1992) reports that the addition of 10 percent silica fume will double the compressive strength, but records also the practical difficulties in handling the bulk material. In some cases these additions will be supplemented with activators such as caustic soda or sodium carbonate.

GROUTS WITH IMPROVED RESISTANCE TO WASHOUT These grouts can be achieved by hardening (as described above for quick setting grouts), or by adding flocculating, coagulating, or thickening types of organic materials. These increase both viscosity and cohesion, which in turn modify grout rheology as well as the behavior at the grout/water interface. They are added in amounts equivalent to a few per thousand ratio of dry weight.

Grouts with Enhanced Penetrability Such grouts are becoming increasingly required to more thoroughly and cheaply fill atypically small pores or fissures while avoiding the typical problems of solution grouts (e.g., permanence, toxicity, strength, and cost). As described by DePaoli et al. (1992a, b), these significant investigations have proceeded along three major tracks:

1. *Improving the Rheological Properties.* Plastic viscosity, cohesion, and internal friction may be decreased by using deflocculating additives such as can be derived from natural organic products (polyacrylates, naphthalein sulfonates) or mineral products. Including 0.5 to a few percent of such fluidifiers will alone reduce Marsh viscosity from 55–60 sec to 32–35 sec.
2. *By Increasing Stability.* While rheological properties can be improved by simply increasing the water content, both bleed, and pressure filtration will increase, thus negating any real advantage during injection. Therefore, activators such as grain dispersants (peptizers) or water retaining polymers are being used. The former typically comprise 0.4 to 2.5 kg/m^3 of grout in cement bentonite mixes, while polymers vary from 0.1 to 5 kg/m^3 in neat cement, or cement-bentonite mixes.
3. *By Reducing Grain Size.* Until recently, the concept revolved around the supply of microfine *cements* (e.g., Clarke, 1982, 1984, 1987; Clarke et al., 1992; Shimada and Ohmori, 1982) who described materials with a mean grain size of 4 μm, and a maximum of 10 μm, capable of permeating fine sands ($k = 10^{-3}$ to 10^{-4} cm/sec). Regrinding can reduce by two or three times normal cement particles to a size of 5 to 15 microns. This corresponds to an increase in Blaine specific surface of 3500 to 8000 cm^2/g. In reground

dry cements care must be taken to prevent the selective elimination of certain components and so changes in chemistry. Such cements are typically expensive and, being hygroscopic, awkward to store and handle. These problems can be resolved by the new development of wet grinding the *mixed grout*, as in the CEMILL® process described below.

Figure 7-8 shows results recently obtained on unstable, stable (i.e., with bentonite added), and MISTRA® mixes (those containing bentonite and additives).

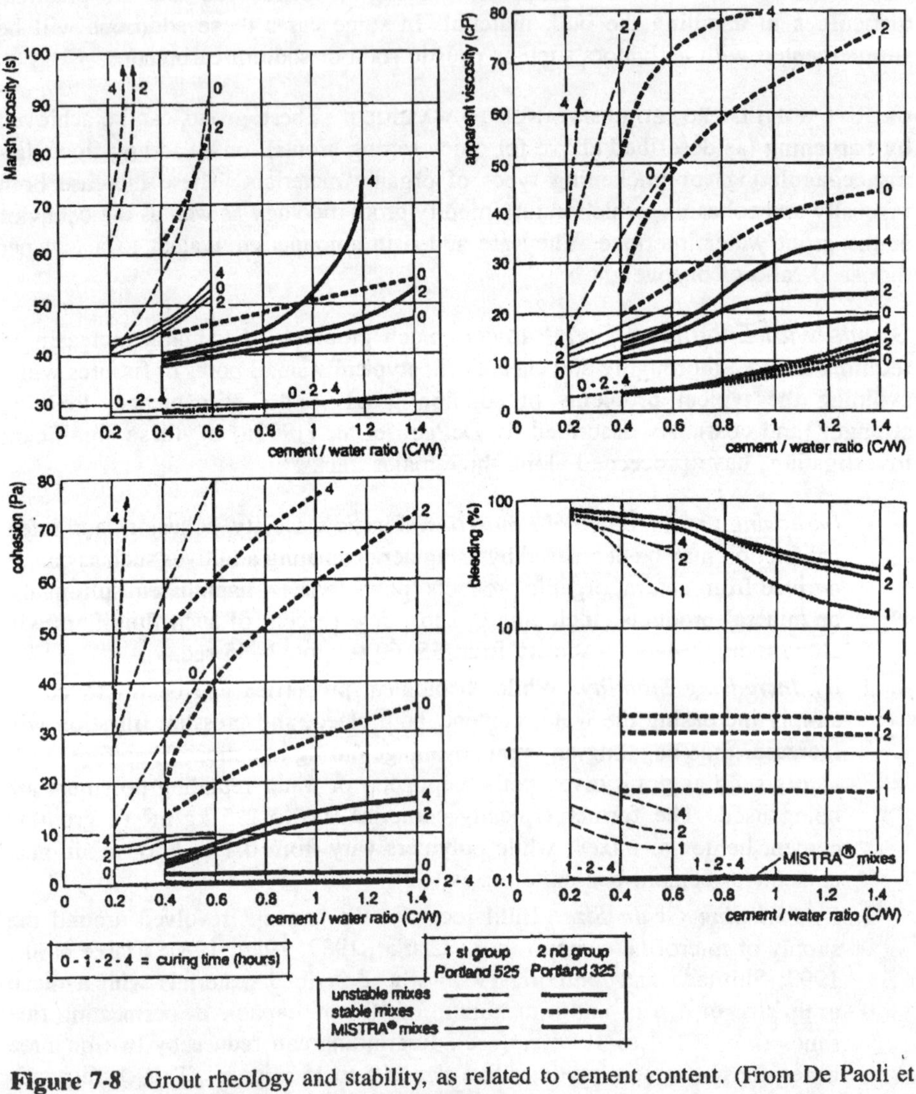

Figure 7-8 Grout rheology and stability, as related to cement content. (From De Paoli et al., 1992a. Reproduced by permission of ASCE.)

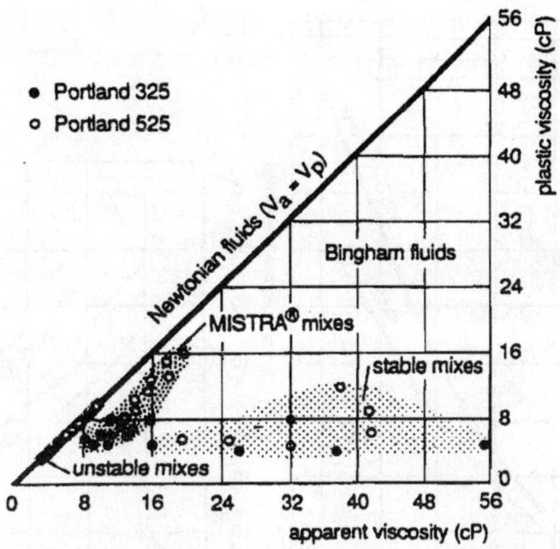

Figure 7-9 Relationship between plastic and apparent viscosities for the different types of mixes. (From De Paoli et al., 1992a. Reproduced by permission of ASCE.)

Marked improvements in viscosity, cohesion, and bleed can be noted for the MISTRA® grouts—all to the advantage of penetrability. Figures 7-9 and 7-10 illustrate these data in a different way, while Figure 7-11 shows strength and permeability data. Properties of MISTRA® grouts are summarized in Table 7-2, which confirms the achievement of:

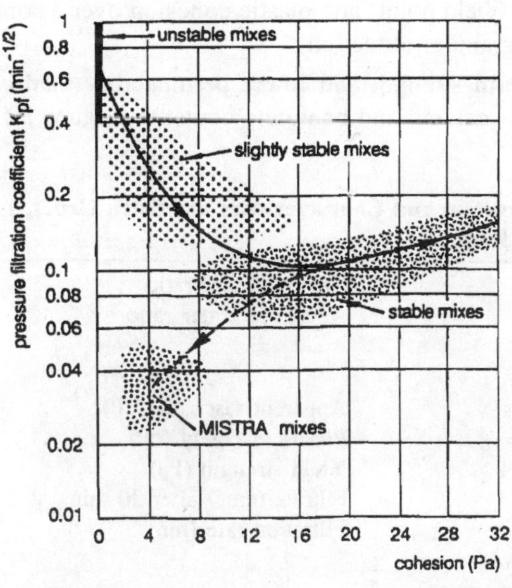

Figure 7-10 Relationship between stability under pressure and cohesion for the different types of mixes. (From De Paoli et al., 1992a. Reproduced by permission of ASCE.)

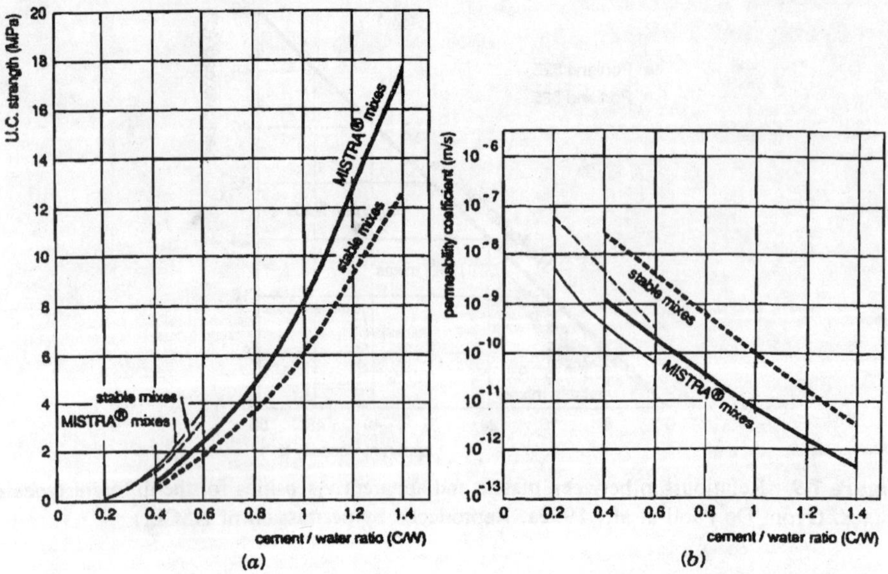

Figure 7-11 (*a*) Unconfined compressive strength, and (*b*) permeability of stable and MISTRA[R] mixes (at 28 days). (From De Paoli et al., 1992a. Reproduced by permission of ASCE.)

- No bleed, and low coefficient of pressure filtration (by lowering filter-cake permeability) (Figure 7-12).
- Low cohesion (yield point) and plastic cohesion over a controllable period (by reducing intergranular attraction).
- Higher long-term strength and lower permeability than conventional grouts with equivalent cement and bentonite contents (Figure 7-11).

TABLE 7-2 Composition and Characteristics of Mistra Grout, Lot 1PB, Passante Ferroviario, Milan, Italy

Composition	Cement/water ratio	0.35
	Additives/water ratio	0.04–0.05
Bleed capacity (%)		0–2
Marsh viscosity (sec)		33–37
Rheometer parameters	Apparent viscosity (cP)	8–12
	Plastic viscosity (cP)	5–8
	Yield strength (Pa)	1.5–5
Filter press test	Filtrate (cm^3) after 30 min.	36–72
at 0.7 N/MPa	Filtration rate ($mm^{-1/2}$)	0.016–0.032
UCS[a] (N/MPa) of grouted sand after 28 days		1.2—1.8

Source: From Mongilardi and Tornaghi (1986).

[a]UCS = unconfined compressive strength.

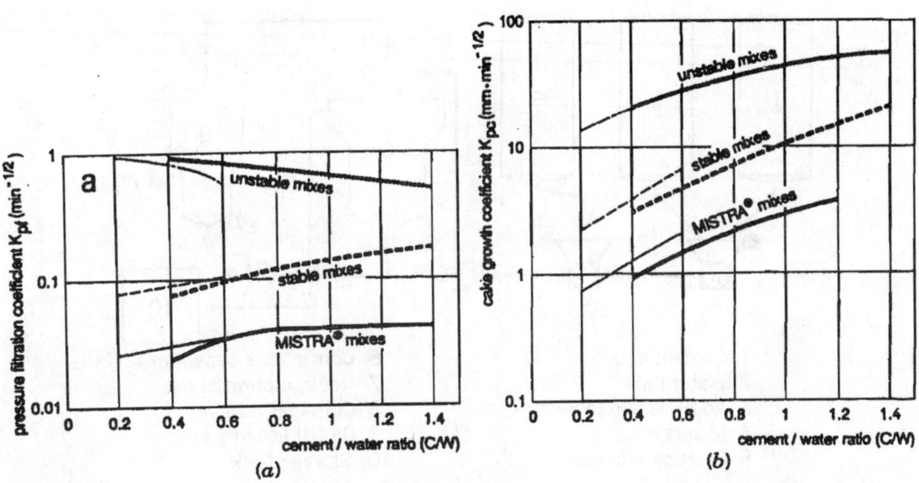

Figure 7-12 (*a*) Pressure filtration and (*b*) cake growth coefficients of different mixes, related to cement content. (De Paoli et al., 1992a. Reproduced by permission of ASCE.)

Figure 7-13 provides data on the various microfine materials, including the grouts produced by the CEMILL® process. This method can operate with both unstable and stable (i.e., with added bentonite) grouts, as shown in Figure 7-14. Experimental data are provided in Figures 7-15 to 7-18. As a consequence of these developments, the penetrability limits of the various families of grouts can be revised, as shown in Figure 7-19. It may be noted that soils with permeabilities as low as 5×10^{-3} cm/sec can be permeated with such grouts. (See also discussion of the Principles of Injection later in Section 7-2). As for the conventional cement-

	grain size (μmm)					
	D 95	D 85	D 60	D 50	D 15	D 10
CEMILL® 6	15.0	9.0	6.0	5.0	1.3	0.9
CEMILL® 9	9.0	5.5	3.5	2.5	0.6	0.4
CEMILL® 12	6.0	4.0	3.0	2.2	0.4	0.3
ONODA MC-500	8.0	60.0	4.5	4.0	2.5	2.0
Portland 525	40.0	22.0	11.0	8.0	2.5	2.0
bentonite	60.0	40.0	15.0	10.0	1.7	1.2

ⓐⓑⓒ sands for injection tests
ⓐ $\gamma - \gamma_{max} = 1.713$ g/cm³
ⓑ $\gamma - \gamma_{max} = 1.701$ g/cm³
ⓒ $\gamma - \gamma_{max} = 1.690$ g/cm³
ⓓ bentonite
ⓔ Portland 525 cement
ⓕ ONODA MC-500 cement
ⓖⓗⓘ CEMILL® mixes

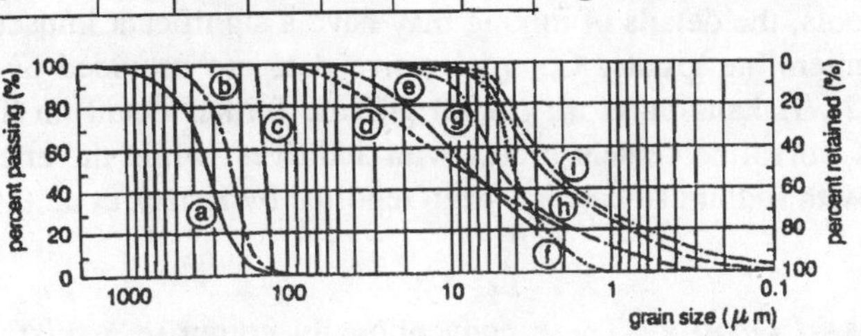

Figure 7-13 Grain size distribution curves for sands, dry materials, and grouts. (From De Paoli et al., 1992b. Reproduced by permission of ASCE.)

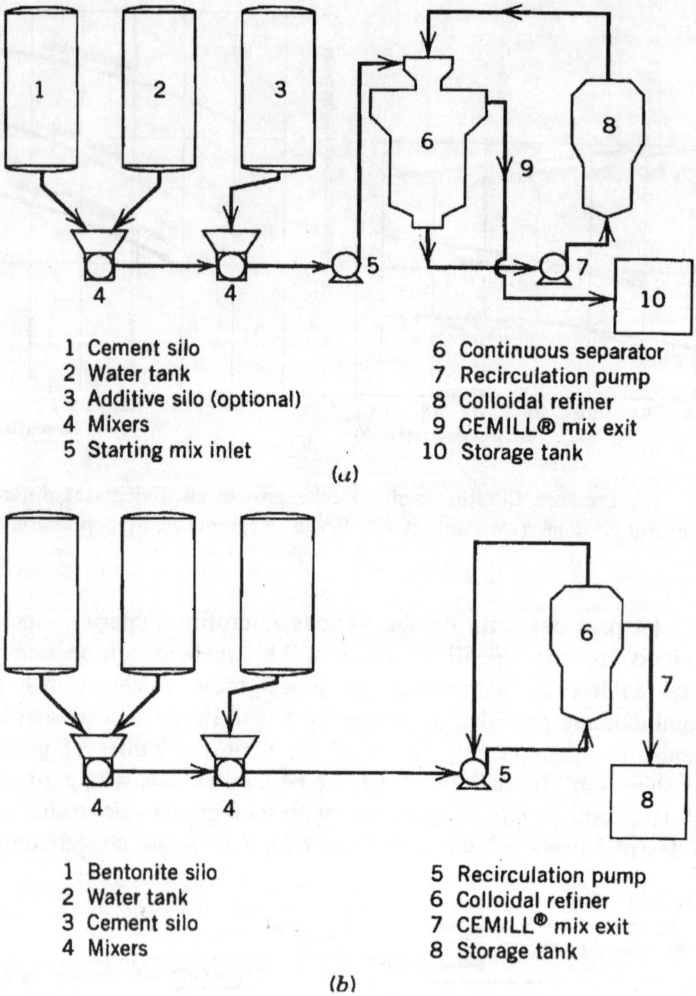

Figure 7-14 Layout of production plants. (*a*) For CEMILL®I (unstable grouts). (*b*) For CEMILL®S (stable grouts). (From De Paoli et al., 1992b. Reproduced by permission of ASCE.)

bentonite grouts, the details of mixing may have a significant impact on the rheology of the microfine grouts. Comprehensive data are provided by Schwarz and Krizek (1992). Hakansson et al. (1992) provide further detail on the rheological properties of microfine cement grouts with additives, while the effect of reacting such grouts with sodium silicate has been reported by Krizek et al. (1992) and Liao et al. (1992).

Silicate-Based Grouts These conventionally comprise a mixture of sodium silicate and reagent solutions, which change in viscosity over time to produce a gel. The sodium silicate is an aqueous solution of alkaline colloidal type silicium

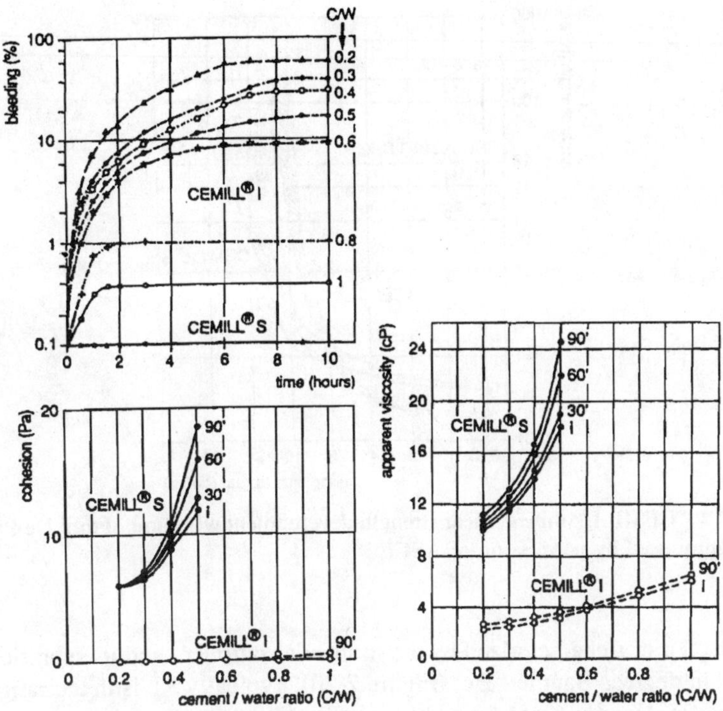

Figure 7-15 CEMILL® mixes: rheology and stability as a function of cement content. (From De Paoli et al., 1992b. Reproduced by permission of ASCE.)

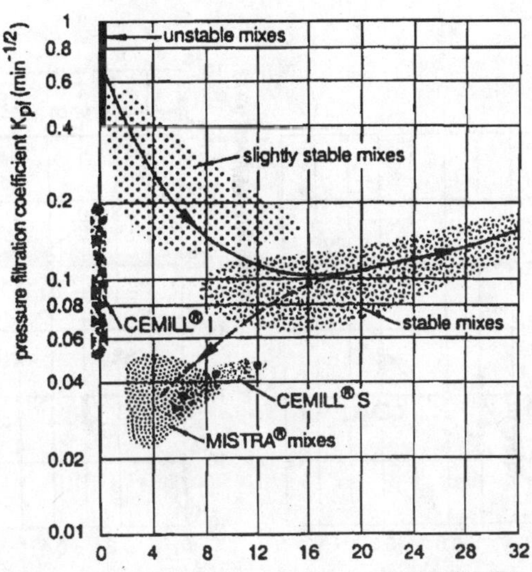

Figure 7-16 Relationship between stability under pressure and initial cohesion for CEMILL® and other mixes prepared with cements with traditional fineness. (From De Paoli et al., 1992b. Reproduced by permission of ASCE.)

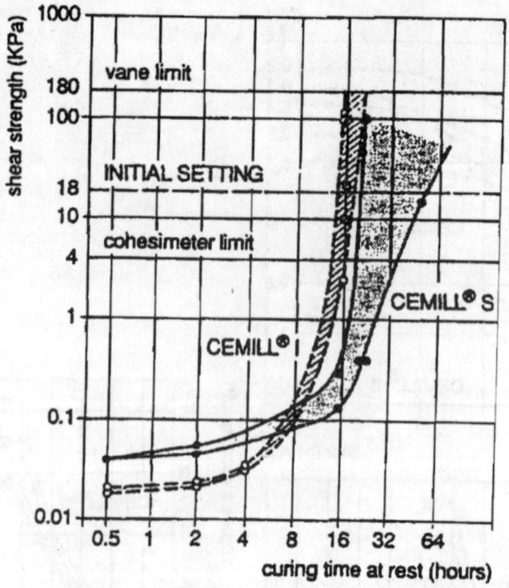

Figure 7-17 CEMILL® mixes: shear strength development with time. From De Paoli et al., 1992b. Reproduced by permission of ASCE.)

($nSiO_2 \cdot Na_2O$). It is characterized by the molecular ratio Rp, and its specific density, expressed in degrees Baumé (°Be) (Figure 7-20). Typically Rp is in the range 3 to 4, while specific density varies from 30 to 42°Be. Reagents may be organic or inorganic (mineral). The former exert a saponification hydraulic reaction that frees acids, and can produce either soft or hard gels depending on silicate and reagent

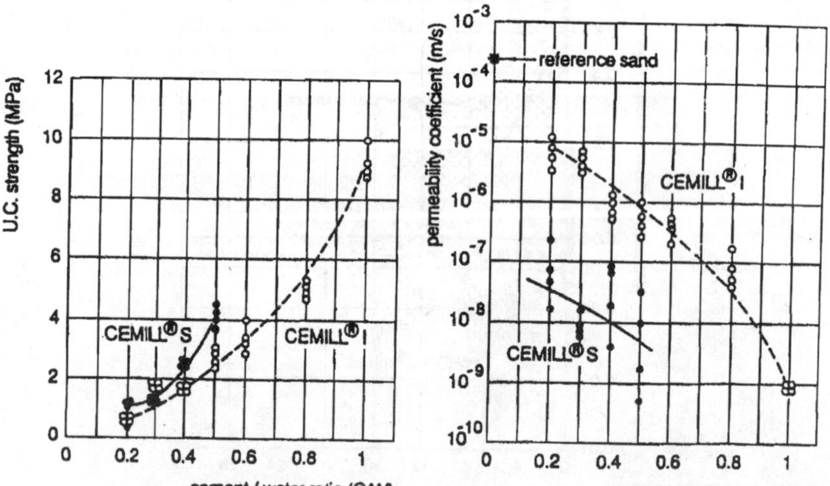

Figure 7-18 CEMILL® mixes. (*a*) Unconfined compressive strength. (*b*) Permeability. (From De Paoli et al., 1992b. Reproduced by permission of ASCE.)

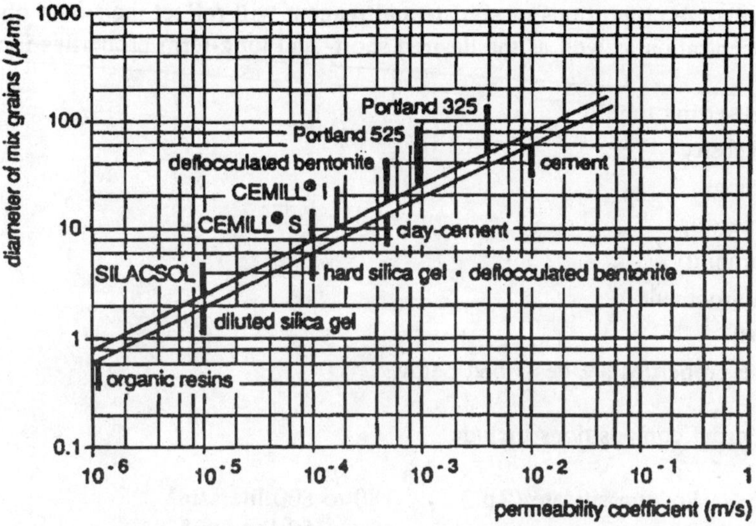

Figure 7-19 Penetrability limits of grouts according to Cambefort (1977) (lower) and the current authors (upper). (From De Paoli et al., 1992b. Reproduced by permission of ASCE.)

concentrations. Common types include monoesters, diesters, triesters, and aldehydes, while organic acids (e.g., citric) and esters are now much rarer. Inorganic reagents contain cations capable of neutralizing silicate alkalinity. In order to obtain a satisfactory hardening time, the silicate must be strongly diluted, and so these gels are typically soft and therefore of use only for (temporary) waterproofing. Typical reagents are sodium bicarbonate and sodium aluminate.

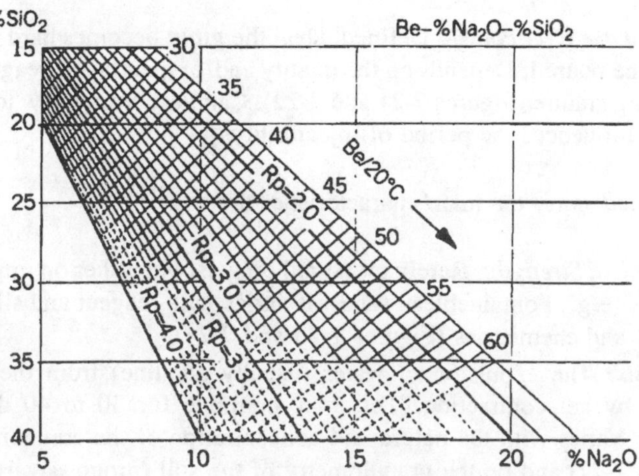

Figure 7-20 Sodium silicate: relation between degree Baumé and Na_2O and SiO_2 percentage for various molecular ratios (Rp). (From AFTES, 1991.)

The relative proportions of silicate and reagent will reflect their own chemistry and concentration as well as the desired short- and long-term properties including:

- Gel setting time
- Viscosity
- Strength
- Syneresis
- Durability
- Environment
- Cost

All these properties are described below.

Typical grout compositions include

Organic	Sodium silicate (Rp 3.3)	180 to 800 liters/m³
	Reagent	40 to 150 liters/m³
	Water	To make up 1 m³ of grout
Inorganic	Sodium silicate (Rp 3.3)	100 to 300 liters/m³
	Sodium aluminate (dry)	10 to 30 kg/m³
	Water	To make up 1 m³ of grout

In its *liquid state,* the main characteristics of a silicate grout are:

- *Density:* Linked to the silicate ratio.
- *Initial Viscosity:* Depends mainly on the silicate Rp and concentration.
- *Evolutive Viscosity:* Changes until gel point, and strongly influences grouting time.
- *Setting Time (Gel Point):* Defined when the grout becomes hard enough that it cannot be poured. Depends on the quality and/or quantity of reagent and varies with temperature (Figures 7-21 and 7-22). Can vary from a few to 120 min and clearly influences the period of injectibility.

In its *hardened state,* the main characteristics are:

- *Mechanical Strength:* Rarely measured on gels, but rather on impregnated soil samples (e.g., Fontainebleau sand). It varies with reagent and silicate concentrations and chemistries (Figures 7-23 and 7-24).
- *Syneresis:* The expulsion of water (usually alkaline) from the gel, accompanied by gel contraction. This may continue for 30 to 40 days after gel setting. Varies with the nature and concentration of the components (Figures 7-25 to 7-27) and on the granulometry of the soil (progressively less in finer soils).
- *Resistance to Washing Out:* Along with gel dissolution, depends on silicate

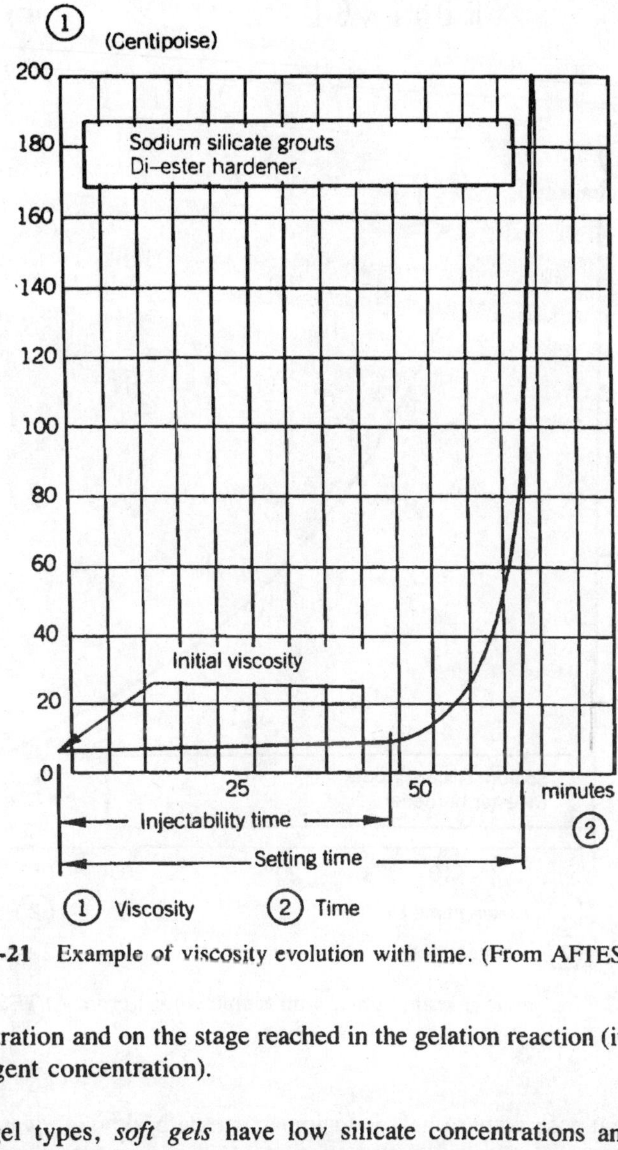

Figure 7-21 Example of viscosity evolution with time. (From AFTES, 1991.)

concentration and on the stage reached in the gelation reaction (itself linked to the reagent concentration).

Regarding gel types, *soft gels* have low silicate concentrations and usually an inorganic reagent. They have very low viscosity (close to water) and so are used for sealing fine sands or very fine rock fissures. *Hard gels* have higher silicate concentrations and organic reagents, the proportion of which is selected to achieve the best possible neutralization rate. Initial viscosities can reach 30 cP, and strengths can vary from 0.2 to 6 MPa.

Soils treated by soft gels can have permeabilities as low as 10^{-5} cm/sec, strengths of 0.2 MPa, and durabilities that vary greatly with soil grain size granulometry. In this context, grouts with sodium bicarbonate—which produces higher syneresis—are acceptably durable only in fine to very fine granular materials.

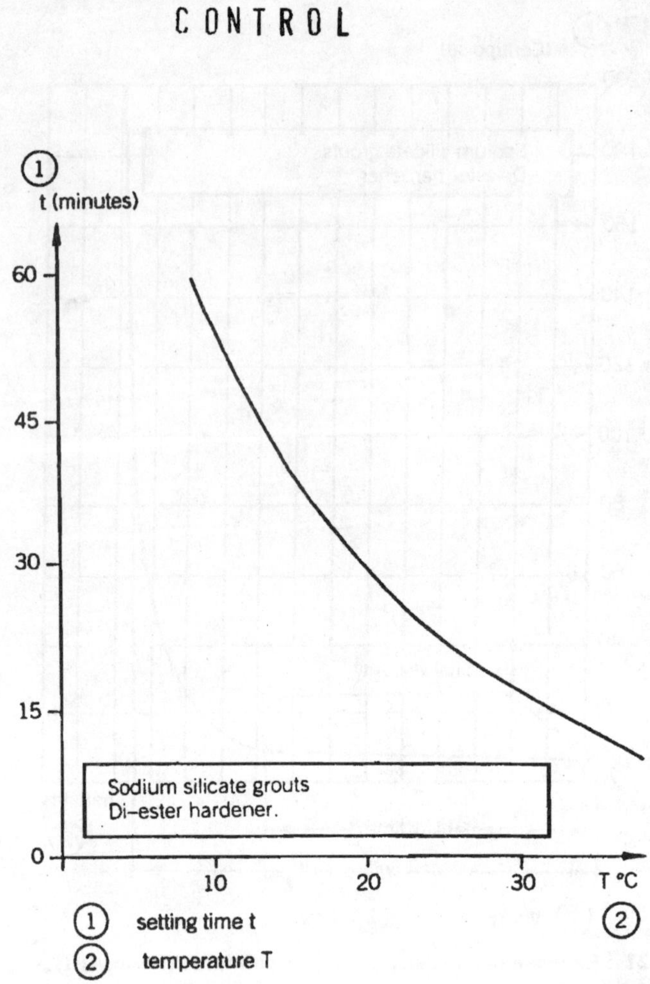

Figure 7-22 Variation in setting time, with temperature. (From AFTES, 1991.)

The main purpose of hard gels is to impart strength although waterproofing is also provided. Strength is controlled by both the solid and the grout, and other factors. Higher strengths are found in finer soils (Figure 7-28), while increasing density has a similar control. For the grout, the silicate Rp and concentration, and the nature and concentration of the reagent control the strength. In addition, the efficiency of pore filling, and the grouting pressure also influence the strength of the grouted soil.

Strengthening appears to be due primarily to an increased cohesion value as opposed to a change in the internal friction angle. The stress rate also is significant (Figure 7-29) in determining strength, although this has less influence in triaxial testing.

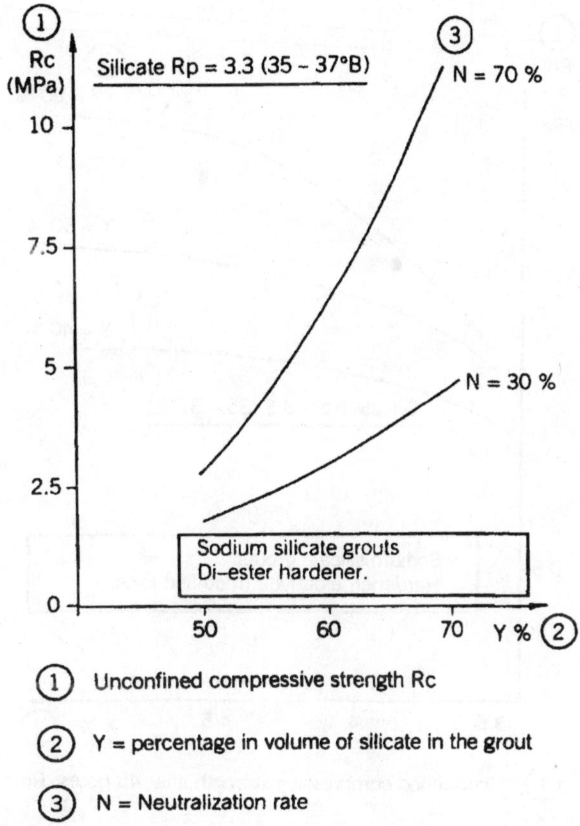

Figure 7-23 Unconfined compressive strength in Fontainebleau sand. Influence of volumetric silicate content and neutralization rate. (From AFTES, 1991.)

Immediate strength and resistance to creep increase with reagent content, and sensitivity to creep varies with the silicate concentration rate. Most recently, considerable progress has been made in Japan (Shimada et al., 1992) using carbon dioxide as a reagent, in the "Carbo rock" method. The gas neutralizes the alkalis, thus preventing environmental contamination during the precipitation of the silica gel (Figure 7-30). Depending on concentrations, gel times can be substantially varied. Treated soil unconfined strengths of 0.6 to 1.2 MPa are obtained, with residual permeabilities of 1.5 to 2.1×10^{-5} cm/sec.

Resins Resins are solutions of organic products in water or in a nonaqueous solvent capable of causing the formation of a gel with specific mechanical properties under normal temperature conditions and in a closed environment. They exist in different forms characterized by their mode of reaction or hardening:

- *Polymerization:* Activated by the addition of a catalyzing element (e.g., polyacrylamide resins).

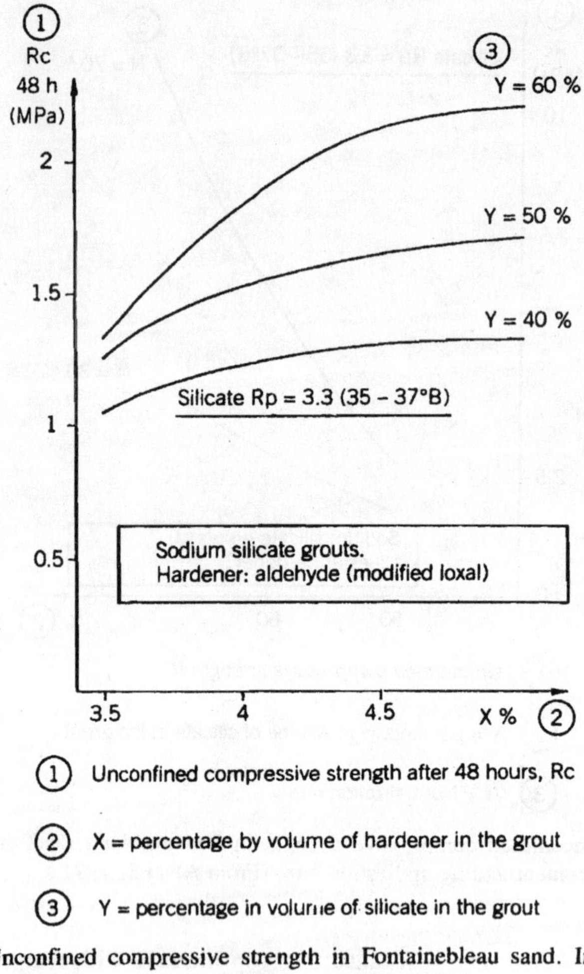

Figure 7-24 Unconfined compressive strength in Fontainebleau sand. Influence of volumetric content of silicate and hardener. (From AFTES, 1991.)

- *Polymerization and Polycondensation:* Arising from the combination of two components (e.g., epoxies, aminoplasts).

In general, setting time is controllable by varying the proportions of reagents or components.

Resins are used when cement or silicate grouts prove inadequate. Examples of such situations would include:

- Particularly low viscosity
- High rapid gain of strength (a few hours)
- Variable setting time (few seconds to several hours)

- Good chemical resistance
- Special rheological properties (pseudoplastic)
- Resistance to groundwater flow.

Resins are used for both strengthening and waterproofing in cases where durability is essential, and the above characteristics must be provided. Four categories can be recognized: acrylic, phenolic, aminoplastic, and polyurethane. Applications are summarized in Table 7-3.

Acrylic Resins Acrylic resins are monomers in aqueous solution. A polymerization and reticulization interaction is obtained by adding catalyzers (0.1 to 5%) (redox system). Accelerators can also be used in the same range of dosages to adjust setting time. Viscosities as low as water can be achieved. The set gel, depending on the degree of reticulation, will be more elastic or more plastic in place and will swell accordingly in the presence of water. Strengths of pure gels are low, but testing of grouted sand samples may yield up to 1.5 MPa. Modified acrylic resins can be produced with:

- Sodium silicate, to have low viscosity (2 cP), good mechanical properties, and expansion in water;
- Latex polymers, to have lowish viscosity (15 cP), good adherence, elasticity, and great resistance to extrusion under water pressure.

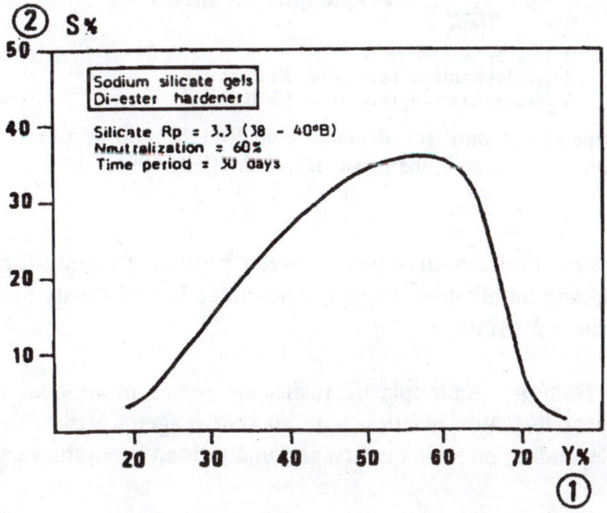

1 Y = volumetric percentage of silicate in the grout
2 S = syneresis rate

Figure 7-25 Syneresis in pure gel, depending on volumetric content of silicate. (From AFTES, 1991.)

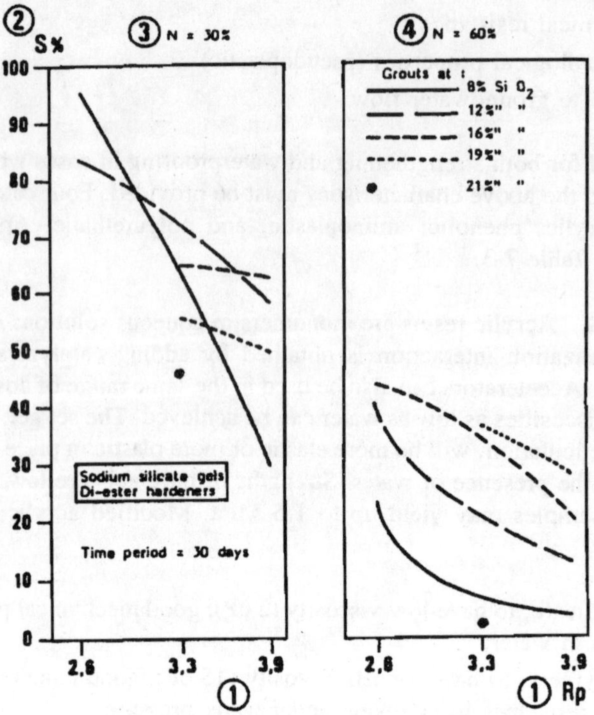

1 Rp : $\frac{SiO2}{NaO2}$, weight ratio of silicate

2 S = syneresis ratio
3 Neutralization rate N = 30
4 Neutralization rate N = 60 %

Figure 7-26 Syneresis in pure gel, depending on: weight ratio of silicate; neutralization rate; weight content of silicate in the grout. (From AFTES, 1991.)

Phenolic Resins Powders dissolved in water provide a phenol-formol polycondensation by adding an alkaline reagent. They have low viscosity and high treated soil strengths (over 2 MPa).

Aminoplastic Resins Aminoplastic resins are resins in aqueous solutions that undergo a polycondensation reaction with an acid reagent. Viscosities range from 10 to 100 cP, depending on resin quality, and unconfined strengths vary from 3 to 10 MPa.

Polyurethane Resins Polyurethane resins have two basic classes:

- *Water-Reactive:* Liquid resin, often in solution with a solvent or in a plasticizing agent, possibly with added accelerator, reacts with groundwater to provide either a flexible (elastomeric) or rigid foam. Viscosities range from 50 to 100 cP.

- *Two Components:* Two compounds in liquid form react to provide a rigid foam due to multiple supplementing with a polyisocyanate and a polyol. Such resins have viscosities from 100 to 1,000 cP and mechanical strengths as high as 2 MPa. An extremely thorough description of these grouts is provided by Naudts (1989), while a particularly illustrative case history of application was provided by Jiacai et al. (1982).

Other Grouts The following grouts are essentially composed of organic compounds or resins. In addition to waterproofing and strengthening, they also provide very specific qualities such as great strength, resistance to corrosion, or flexibility. Their use is limited by specific concerns such as toxicity, injection and handling difficulties, and cost. Major categories are bitumen, latex, polyesters, epoxies, furanic resins, silicones, and silacsols.

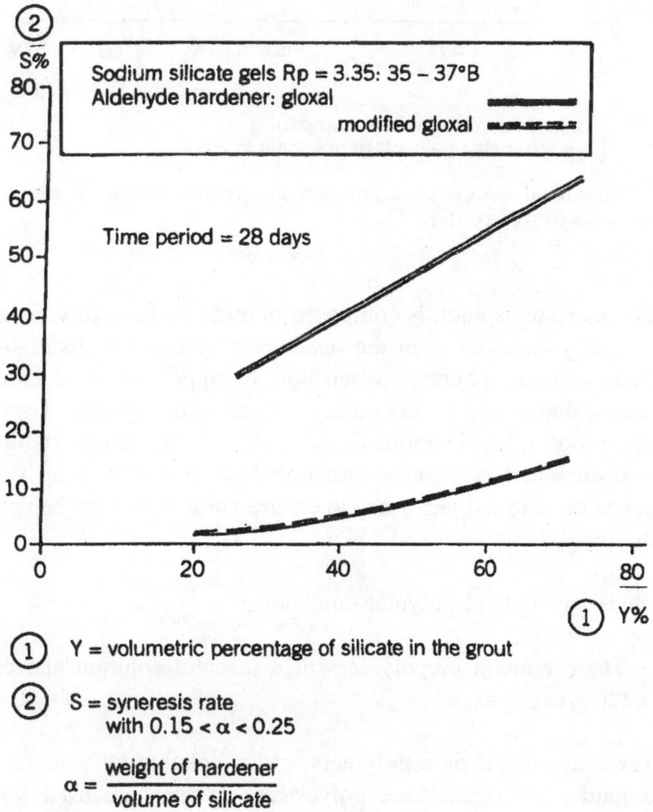

Figure 7-27 Syneresis in pure gel, depending on volumetric content of silicate. (From AFTES, 1991.)

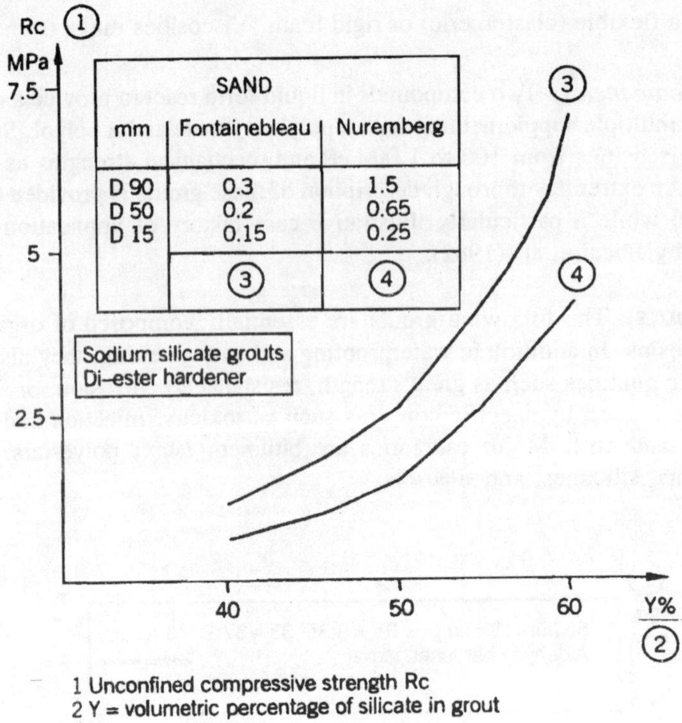

Figure 7-28 Unconfined compressive strength of grouted sands. Incidence of granular distribution. (From AFTES, 1991.)

Bitumen Bitumen, or asphalt, is composed of hydrocarbons of very high molecular weights, usually obtained from the residues of petroleum distillation. Asphalt may be viscous to hard, pourable when hot, or may consist of emulsions with restricted stability that coagulate in contact with mineral surfaces. They are used in particular waterproofing applications (Bruce, 1990a, b), remain stable with time, and have good chemical resistance, although later penetration by stable cementitious grouts has been found necessary to ensure good long-term behavior (Lukajic et al., 1985).

Latex Latex is a coagulant polymer emulsion.

Polyesters These contain prepolymers in a reactant solution and can be polymerized by adding catalysts.

Epoxies These are liquid pure polymers, cross linkable by reaction (poly addition) with a hardening agent. Like polyesters, epoxies are used for their high mechanical strength and good adhesive qualities (e.g., Bruce and DePorcellinis, 1991). They also have excellent chemical resistance.

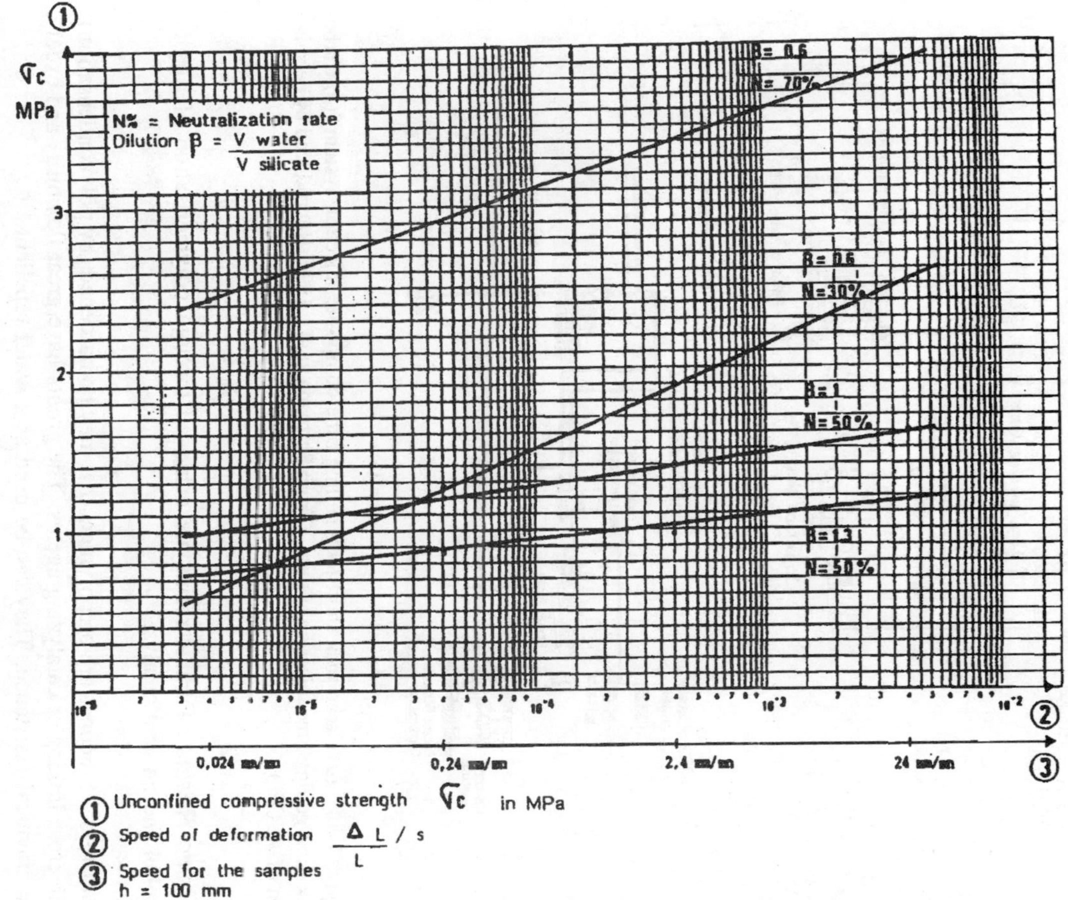

Figure 7-29 Rupture strength of a Fontainebleau sand injected with silicate gels. (From AFTES, 1991.)

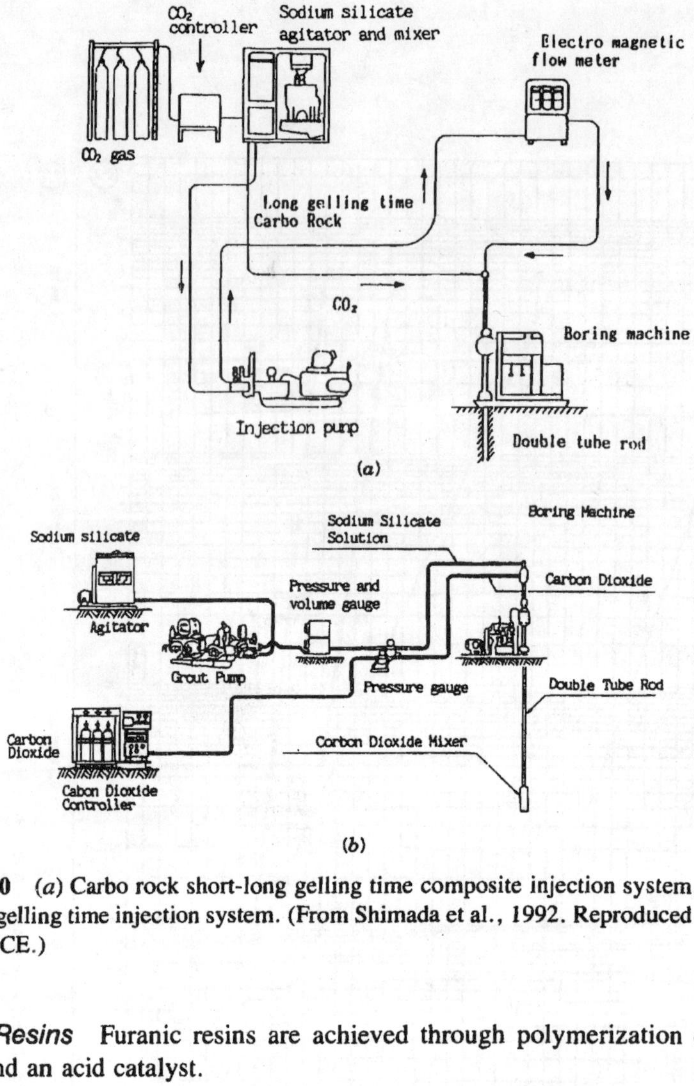

Figure 7-30 (*a*) Carbo rock short-long gelling time composite injection system. (*b*) Carbo rock short gelling time injection system. (From Shimada et al., 1992. Reproduced by permission of ASCE.)

Furanic Resins Furanic resins are achieved through polymerization of furylic alcohol and an acid catalyst.

Silicones Silicones are prepolymers that may be hardened (by polycondensation) with cross linking or catalyzing agents. The grouts have great flexibility and excellent chemical resistance. They can be used as a water repellent.

Silacsols Silacsols are formed by reaction between an activated silica liquor and a calcium-based inorganic reagent. Unlike the sodium silicates previously discussed—aqueous solutions of colloidal silica particles dispersed in soda—the silica liquor is a true solution of activated silica. The reaction products are calcium hydrosilicates with a crystalline structure similar to that obtained by the hydration and setting of cement: a complex of permanently stable crystals. This reaction is not therefore an evolutive gelation involving the formation of macromolecular aggre-

TABLE 7-3 Uses and Applications of Resins

Type of Resin	Nature of Ground	Use/Application
Acrylic	Granular, very fine Finely fissured rock	Waterproofing by mass treatment Gas tightening (mines, storage) Strengthening up to 1.5 MPa Strengthening of a granular medium subjected to vibrations
Phenol	Granular, very fine	Strengthening
Aminoplast	Schists and coals	Strengthening (by adherence to materials of organic origin)
Polyurethane	Large voids	Formation of a foam that forms a barrier against running water (using water-reactive resins) Stabilization or localized filling (using two-component resins)

Source: From AFTES (1991).

gates (Figure 7-31), but a direct reaction on the molecular scale, free of syneresis potential (Figure 7-32). This concept has been employed in Europe since the mid-1980s (Bruce, 1988) with consistent success in fine-medium sands. The grout is stable, permanent, and environmentally compatible. Other outstanding features, relative to silica gels of similar rheological properties, are:

- The far lower permeability (Figure 7-32).
- The far superior creep behavior of treated sands for grouts of similar strength (2 MPa) (Figure 7-33).
- Even if a large void is encountered, or a large hydrofracture fissure is created, a permanent durable filling is assured.

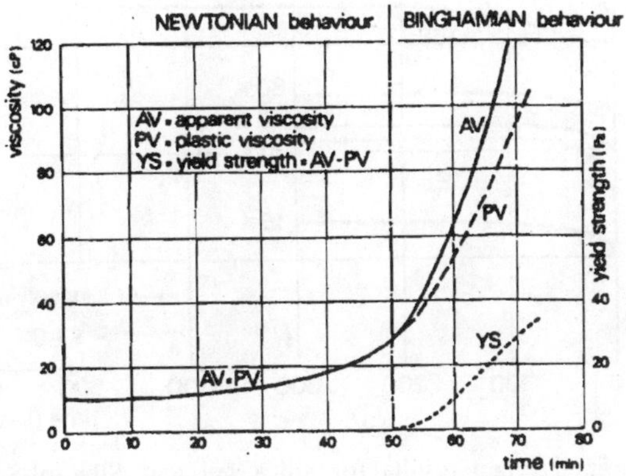

Figure 7-31 Typical viscosity-time behavior of Silacsol-S grout. (From Tornaghi et al., 1988.)

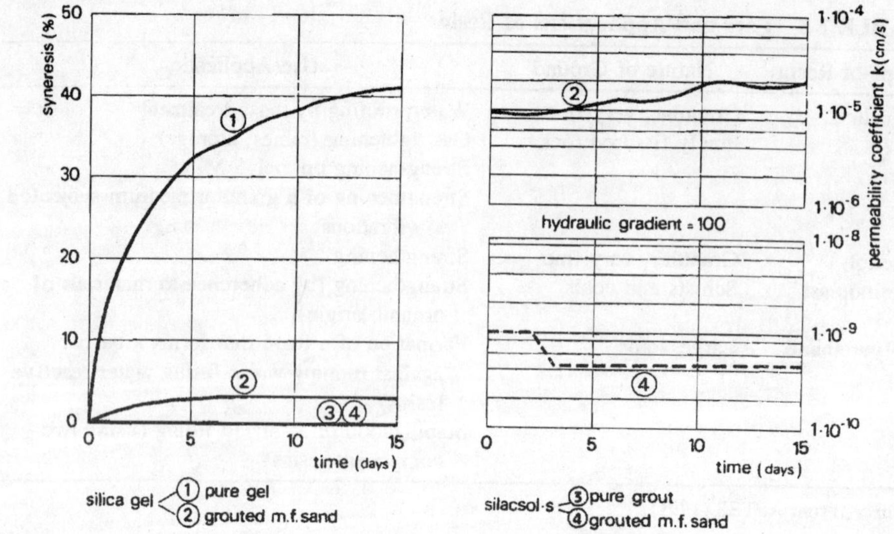

Figure 7-32 Effect of time on syneresis and permeability of typical chemical grouts. (From Tornaghi et al. 1988.)

Design

Geotechnical Parameters The overall objective of the site investigation is to obtain a detailed geotechnical classification of the different ground types, together with their locations, extents, and thicknesses.

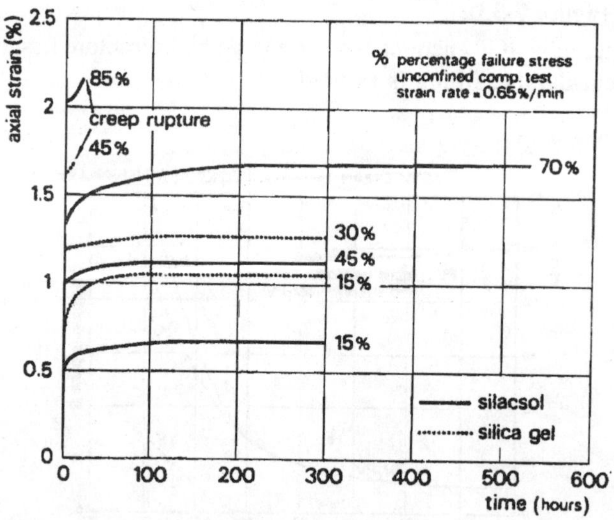

Figure 7-33 Unconfined creep results for silica gel and Silacsol-S grout (samples of grouted medium-fine sand). (From Tornaghi et al., 1988.)

The principal parameters to be obtained (AFTES, 1991) are

- The existence and size of voids
- The geometry of these voids
- The permeability
- The modulus of deformability
- Hydraulic gradients and chemistry of the groundwater

For rock masses, permeability (as measured by Lugeon tests) has two origins: fissure permeability and matrix (or material) permeability. The size of the fissures is typically the predominant factor (Table 7-4), and it is conventionally assumed that cement grouts will penetrate fissures of width at least three times the largest particle size. Littlejohn (1975) estimates this limiting width to be 160 μm for normally ground cements.

For granular soils, the principal parameters are granulometry and density, which define the size of the intergranular voids. The resultant coefficient of permeability largely controls grout choice (Table 7-5). For particulate grouts it sets practical lower limits for permeation, and for solution grouts it influences the rate of injec-

TABLE 7-4 Types of Grouts Used, Depending on the Type and Size of Fissures

Type and Size of Voids	Types of Grouts
Open voids; karsts; wide fissures.	Cement-based grouts with coarse fillers (gravels) Cement-based cellular grouts Quick setting grouts
Large fissures (average opening > 1 cm)	Cement grouts with fine fillers (ash, fine sand filler) Quick setting grouts Cement-based cellular grouts Bentonite, clay, cement grouts Polyurethane foam Carbamide
Average fissures (1 mm to 1 cm)	Pure cement or bentonite Clay-added grouts Synthetic foams Resins
Fine fissures (0.5 to 1 mm)	Special improved penetrability grout Silicate gel Acrylic resins
Very small fissures (less than 0.5 mm); porous material	Deflocculated bentonite Low-viscosity silicate gel Acrylic resins Phenol resins

Source: From AFTES (1991).

TABLE 7-5 Fields of Application of Grouts for Granular Soils

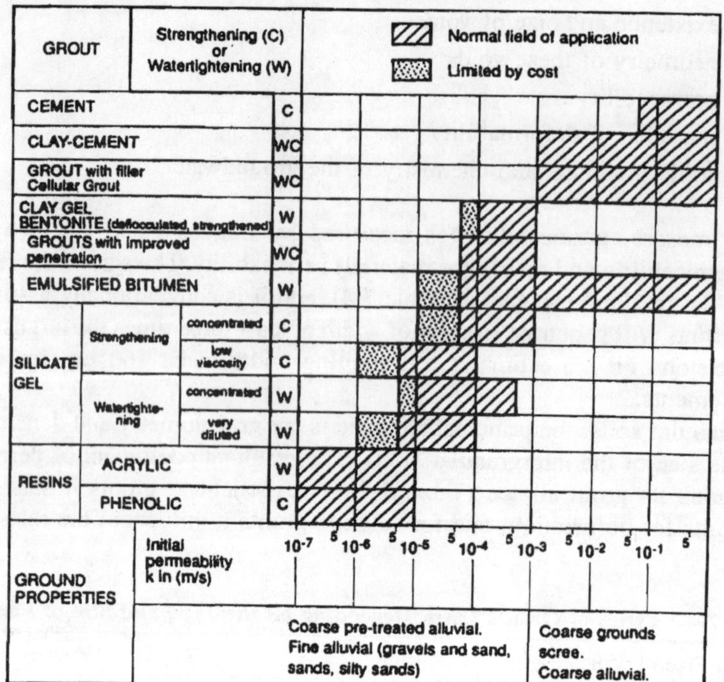

Source: From AFTES (1991).

tion. Porosity dictates grout consumption. For clean sands, a good approximation of k, in the absence of (Mandel–Lefranc) tests, can be calculated by Sherard's formula

$$K \text{ (m/sec)} = A \cdot 10^{-2} \cdot D_{15}^2$$

where A = 0.2 to 0.6 (usually 0.35)

D = size of the grid mesh (mm) permitting 15 percent of the sand to pass.

Groundwater details may influence choice of chemical formulation and the extent of treatment.

In summary, and as noted by Littlejohn (1985), a ground investigation is satisfactory only when it provides sufficient information to answer the following questions:

- Can the ground be grouted?
- What types and amounts of grout are required?
- What are the likely parameters of the treated soil?

Excellent examples of how to plan and execute grouting contracts can be found in the publications of Baker (1982) and Black et al. (1982), while Karol (1990) provides a comprehensive overview.

Grout Material Parameters As noted in the discussion of Grouts in Section 7.2, the penetrability of particulate grouts depends on

- Stability (i.e., bleed capacity)
- Pressure filtration
- Rheology (principally cohesion and viscosity, Figure 7-34)
- Grain size concentration

The solution grouts are evolutive Newtonian liquids during their period of practical injectability, when permeation occurs in accordance with Darcy's laws. Principal controls over penetration distance and grout characteristics are therefore

- Ground permeability and porosity.
- Initial grout viscosity (Figure 7-35) and its evolution (Figure 7-36). Deere and Lombardi (1985) note that cohesion determines distance of travel and viscosity determines the flow rate.
- Pressure (related to flow rate).
- Practical duration of injection.

Grouting Method Parameters Again the structure of AFTES (1991) provides a logical approach, identifying four main parameters:

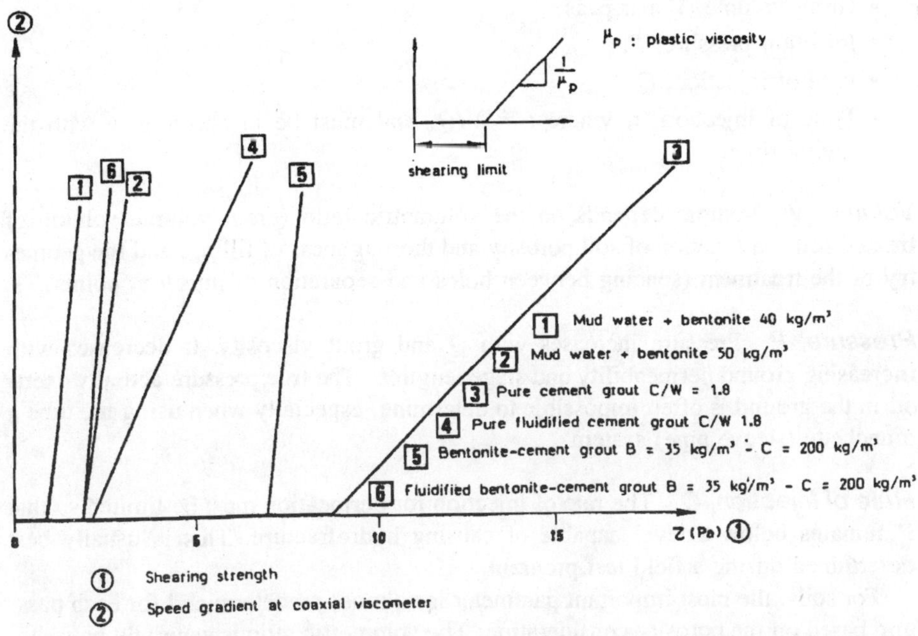

Figure 7-34 Rheological characteristics of grouts. (From AFTES, 1991.)

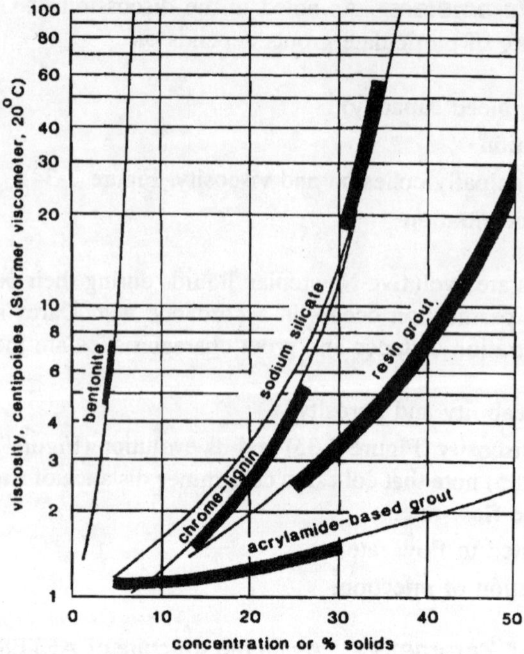

Figure 7-35 Viscosities of various grouts. Heavy lines indicate the solution concentrations normally used in the field. (From Karol, 1983. Reprinted by courtesy of Marcel Dekker, Inc.)

- Grout volume, V, per pass.
- Injection pressure, P.
- Rate of injection, Q.
- Time of injection, t, where $t = V/Q$, and must be in accordance with the setting time.

Volume, V Volume depends on the volumetric ratio (grout volume/volume of treated soil—a function of soil porosity and thoroughness of filling) and the geometry of the treatment (spacing between holes and separation of injection points).

Pressure, P Pressure increases with Q and grout viscosity. It decreases with increasing ground permeability and stage length L. The true pressure actually exerted in the ground is often impossible to determine, especially when using the tube à manchette (sleeve pipe) system.

Rate of Injection, Q The rate of injection for permeation must be limited so that P remains below a level capable of causing hydrofracture. This is usually best determined during a field test program.

For soils, the most important parameter is volume, predetermined for each pass, and based on the porosity consideration. The volumetric ratio is generally between:

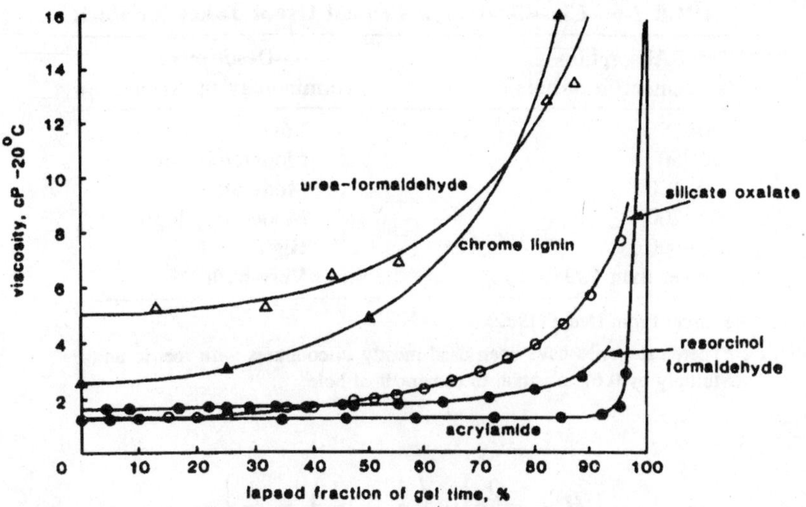

Figure 7-36 Growth of viscosity in period before gelation. [Modified after James, 1963. Reprinted from Thorburn and Hutchison, 1983, p. 250, by permission of Blackie Academic and Professional (an imprint of Chapman and Hall).]

15 and 45 percent for coarse sands and gravel; 5 and 25 percent for finer sands or fissured cohesives.

The rate of injection usually reflects permeability and may vary from 0.2 to 1.8 m^3/hr. Pressures may reach 1.5 to 3 times the total minor stress before hydrofracture is induced.

For rocks, injection pressure is paramount, and may excede 10 MPa, as determined by the nature of the rock mass and grout, and the geometrical and hydrogeological conditions. Cementitious grouts may experience a pressure filtration effect in contact with porous or microfissured rock, and this phenomenon is intensified as pressure increases. Flow rates may vary from 0.5 (fine fissures) to 2 m^3/hr (medium to large fissures). For stable grouts, volumetric ratios of 1 to 10 percent are typical of fine to medium size fissures, although many (e.g., Deere, 1982) use a quantity recording system based on dry weight of cement (Table 7-6).

Principles of Injection The theory of injection has been reviewed in detail by numerous sources including FHWA (1976), Bell (1982), and Karol (1990). Arguably the most lucid summary was provided by Littlejohn (1985), from whom much of the following is drawn.

Assuming permeation, grout advances through the ground under pressure, displacing air and water outwards and in a direction determined by ground permeability. In uniform isotropic soils, spherical flow is observed, and assuming Darcy's law and a Newtonian fluid, Raffle and Greenwood (1961) showed that the rate Q at a radius of penetration R is related to the hydraulic drilling head H as follows:

TABLE 7-6 Classification of Cement Grout Takes for Rock

Grout Absorption (kg cement/m of hole)[a]	Descriptive Terminology of Absorption
0–25	Low
26–50	Moderately low
50–100	Moderate
101–200	Moderately high
201–400	High
Greater than 400	Very high

Source: From Deere (1982).

[a]There categories have been used mostly in countries with metric units. Multiply by 0.67 to obtain lb cement/ft of hole.

$$H = \frac{Q}{4\pi k}\left[\mu\left(\frac{1}{r} + \frac{1}{R}\right) + \frac{1}{R}\right] \tag{7-1}$$

where k = ground permeability
μ = grout viscosity in centipoise
r = radius of spherical injection source (for a cylindrical injection source of length L
diameter D, $r = \frac{1}{2}\sqrt{LD}$, approximately.

The time for the grout to penetrate to radius R is given by

$$t = \frac{nr^2}{kH}\left[\frac{\mu}{3}\left(\frac{R^3}{r^3} - 1\right) - \frac{\mu - 1}{2}\left(\frac{R^2}{r^2} - 1\right)\right] \tag{7-2}$$

where n = porosity.

If the second component inside the main bracket is ignored, the relationship simplifies to the equation proposed by Maag (1938); namely,

$$t = \frac{\mu n}{3kHr}(R^3 - r^3) \tag{7-2a}$$

The results of equation (7-2) are shown in Figure 7-37, which permits rate of progress of injection to be estimated for different grout viscosities.

For Binghamian grouts under constant injection pressure, the opposing drag forces on the wetted surfaces of the ground structure gradually increase until the injection pressure is balanced, with no extra available to maintain viscous flow. According to Raffle and Greenwood, the pressure gradient (i) required to overcome the Bingham yield strength may be expressed as

$$i = 4\tau_s/d \tag{7-3}$$

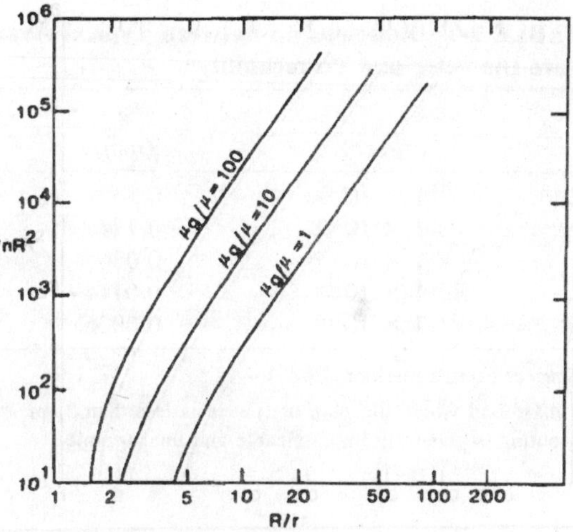

k = soil permeability
H = hydraulic head
t = time
n = porosity of soil
r = radius of source
R = radius of grout at time t
μ = viscosity of water
μ_g = viscosity of grout.

Figure 7-37 Dependence of penetration time on viscosity ratio. [After Raffle and Greenwood, 1961. Reprinted from Thorburn and Hutchison, 1983, p. 251, by permission of Blackie Academic and Professional (an imprint of Chapman and Hall).]

where τ_s = Bingham yield stress
d = effective diameter of the average pore

To maintain an advancing flow an additional pressure gradient is required. With reference to equation (7-3), the average pore diameter can be estimated from the Kozeny equation:

$$d = 2\sqrt{\frac{8\mu k}{\delta_w g n}} \tag{7-4}$$

where δ_w = density of water
g = acceleration due to gravity

Combining equations (7-3) and (7-4), Table 7-7 indicates typical average pore diameters for different permeabilities assuming a porosity of 25 percent. For the specific case of silts, experiments by Garcia-Bengochea et al. (1978) indicate that the predominant pore size is approximately equal to the effective size (D_{10}) of the soil.

TABLE 7-7 Relationship Between Typical Average Pore Diameter and Permeability[a]

k (m/sec)	d (mm)
1×10^{-2}	0.36
1×10^{-3}	0.174
1×10^{-4}	0.036
1×10^{-5}	0.0114
1×10^{-6}	0.0036

Source: From Littlejohn (1985).

[a]In Ground where the pore or fissure is less than 3μm, chemical grouting is generally impracticable and uneconomic.

For a ground permeability of 1×10^{-3} cm/sec and given a 5 percent bentonite solution (Figure 7-38), with a yield value of 2 MPa, then an injection head of 0.7 MPa is required for each meter of grout penetration just to overcome the inherent grout yield strength. For guidance, Table 7-8 shows approximate hydraulic gradients to maintain flow in Bingham-type fluids.

Where injection pressures must be limited to avoid ground disturbance and heave, then there is a limiting radius of penetration R_L given by

$$R_L = \frac{\delta_w g H d}{4\tau_s} + r \qquad (7\text{-}5)$$

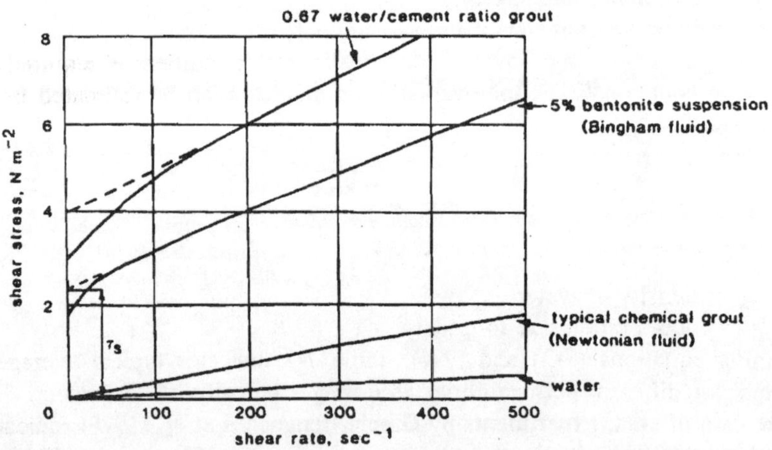

Figure 7-38 Flow properties of typical grouts. [After Bell, 1978. Reprinted from Thorburn and Hutchison, 1983, p. 257, by permission of Blackie Academic and Professional (an imprint of Chapman & Hall).]

TABLE 7-8 Hydraulic Gradient to Maintain Flow in Non-Newtonian Grouts

Soil Permeability (m/sec)	Yield Value (N/m^2)	Hydraulic Gradient
10^{-2}	1	1.2
	10	12
	100	120
	1000	1200
10^{-3}	1	4
	10	40
	100	400
	1000	—
10^{-4}	1	12
	10	120
	100	1200
	1000	—
10^{-5}	1	4.
	10	40.
	100	400.
	1000	—

Source: After Littlejohn (1985).

AFTES (1991) quotes the radius of injection R as given by

$$R = \sqrt{V/(n\pi L)} \tag{7-5A}$$

where V = injected grout volume per pass
n = soil porosity
L = thickness of injected section or pass length

From theoretical and experimental considerations, it can be seen that the penetration radius of a grout mainly depends upon:

- The nature of the ground (permeability k for soils or degree of fissuring for rocks).
- Injection pressure.
- Injected volume.
- Characteristics of the grout (viscosity, setting time).
- Desired efficiency (performance risks are often greater for waterproofing than for consolidation).

Based on these considerations, grout hole patterns can be designed (Figure 7-39), and typical spacings summarized (Table 7-9).

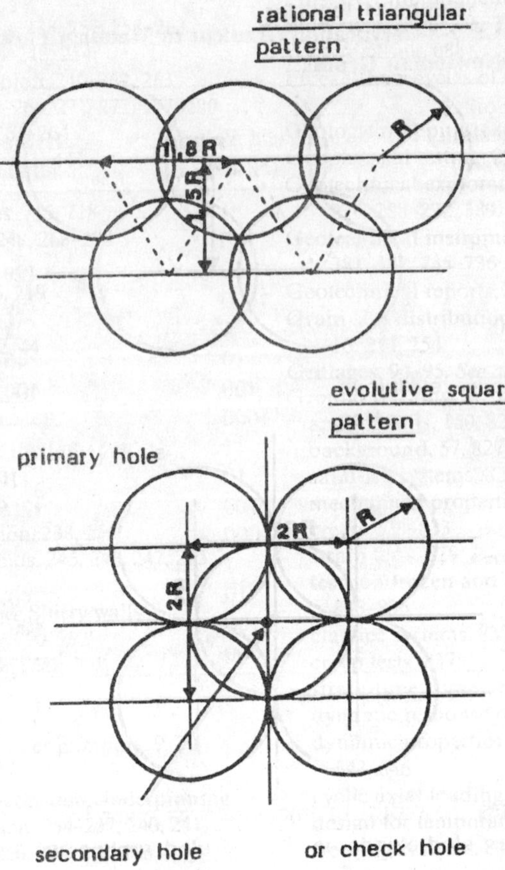

Figure 7-39 Standard pattern for ground treatment. (From AFTES, 1991.)

Although chemicals are marketed as pure solutions, they invariably contain particles up to 20 μm, say, which may block off fine pores in the ground. Additional chemical or centrifuge clarification may be needed prior to use. Also, silt impurities in commercial bentonites may have particles up to 50 μm in maximum dimension.

From Terzaghi's criterion for filter sizing, Mitchell (1981) defined the degree of groutability as follows:

$$N = (D_{15})\ \text{soil}/(D_{85})\ \text{grout} \tag{7-6}$$

$$N_c = (D_{10})\ \text{soil}/(D_{95})\ \text{grout} \tag{7-7}$$

With $N < 11$ or $N_c < 6$, injection would be impossible, while good results should be obtained with $N > 24$ or $N_c > 11$. However, according to Cambefort (1977), a more logical criterion of injectability should consider the dimensions of the voids as compared to those of the grains of the grout. Expressing the hydraulic radius of the

TABLE 7-9 Typical Drill Hole Spacings

Medium to be Injected	Description	Distance Between Holes (m)
Soil, depth < 25 m	Fine sand	0.8–1.3
	Sand, sand and gravel	1.0–2.0
	Gravel	2–4
	Gravel Sand and gravel ($kH > kV$)	Watertight ground 3–5
Rock depth < 25 m	Fine cracks	1–3
	Open cracks	2–4
Structures	Backing behind the vault	2–3
Cavities	Filling of large voids	3–15

Source: From AFTES (1991).

mean interstitial section as a function of the porosity and specific surface, and using the Kozeny formulae to determine permeability, he derived the relationship:

$$D \leq C \cdot K^{1/2} \tag{7-8}$$

where D is the average diameter of the grains of the suspension, C is a constant, and K is the permeability coefficient. More recently, Arenzana et al. (1989) have indicated, with specific reference to grouting with very thin microfine grouts (w/c = 4), the following permeation conditions apply at a pressure of 0.7 MPa:

$$D_{10\text{sand}} \geq 0.15 \text{ mm} \tag{7-9}$$

$$R_H = n / (1 - n)\, S_0 \cdot Y_d \geq 2 \text{ m} \tag{7-10}$$

where R_H = hydraulic radius, that is, the ratio between the surface available for the flow and the wet parameter

n, Y_d, and S_0 = porosity, dry density, and specific surface of the sand, respectively.

The experiments of DePaoli et al. (1992a, b) confirm the general validity of Mitchell's theories, although stable mixes and "lighter" mixes do perform markedly better near penetrability limits. Cambefort's limits were likewise confirmed, providing that the diameter values refer more to the D_{85} or D_{95} than to the D_{50} of the particles to be grouted. The experiments were able to confirm equation (7-9) only. These authors concluded that the most significant correlation would be between grout grain size (D_{85}) and soil pore diameter [D_{30} by Kozeny formula (7-4), or D_{10} by mercury porosimetry]. Permeation would occur for ratios over 4 and 5.5, respectively, but would not occur when these ratios were lower than 3 or 4. In the intermediate field, the grout concentration and stability are crucial controls.

Overview of Grout Selection The selection of the most suitable grout for a specific project still requires the judgment of an experienced engineer, who has to weigh the many factors—technical, economic, and practical—that have been touched upon, and that are summarized (Littlejohn 1985) as:

- Extent and quality of the site investigation (and especially permeability and porosity results).
- Optimum injection method and hole pattern.
- Availability of grout materials.
- Viscosity—time evolution of grout including sensitivity to temperature, dilution and mix proportioning errors.
- Stability of grout in situ.
- Degree of saturation of ground during service, including risk and effect of grout desiccation.
- Chemical composition of groundwater.
- Permanence of grout in situ [both Petrovsky (1982) and Deere (1982) describe leaching of *cementitious* grouts in dam curtains with time, while several authors, including Graf et al. (1982), describe similar observations in *chemically* treated soil].
- Toxicity of grout and chemical components and working environment [Ballivy et al. (1992) describe the impact of harsh service conditions on the grout in situ].
- Aggressivity of grout and chemical components toward plant and equipment.
- Residual permeability or strength of grouted ground.
- Degree of site supervision required, including sophistication of construction and monitoring systems.
- Overall cost including materials, mixing, and injection (see the discussion of Cost Considerations later in Section 7-2).

Works by various authors including FHWA (1976), Davidson and Perez (1982), Littlejohn (1985), Karol (1990), AFTES (1991), and several at the ASCE conferences of 1982 and 1992 provide detailed and systematic insight into each of the above factors.

Construction

Drilling

Rock The drilling of competent rock masses for grout hole purposes has long been a topic of strong international debate (e.g., Albritton, 1982). For historical reasons with commercial undertones, rotary percussion has been the most common choice of European-based specialists, whereas high-speed rotary drilling has been traditionally specified in North America. Deere (1982) provided early confirmation that rotary percussion has become the more common choice. Generically, one may classify rock drilling methods into three basic categories:

1. Low-speed rotary, for example, tricone or drag bit. (Augers are infrequently used, and only for drilling in claystones to moderate depths.)
2. High-speed rotary, either for core recovery or with "plug bits."
3. Rotary percussive, by top hammer.

The choice of drilling *method* is ideally linked to the geology (e.g., Figure 7-40), cost notwithstanding, although historical bias and local comfort levels are often a more common if unscientific determinant. Within each broad category there is, of course, a wealth of options with respect to equipment, flushing characteristics, and operational techniques.

Virtually all practitioners do agree, however, on the choice of flushing *medium:* water is preferred, although foam is gaining acceptance and popularity. Whereas water-based fluid flushes appear most adept at cleaning debris from fissures, and so promoting later, efficient grouting, air appears to have the opposite effect. Evidence seems to confirm that air promotes the wedging of rock fragments in fissures and so limits the potential for efficient grouting. In this regard, it is questionable whether recent North American trends to permit the use of (air powered) down-the-hole, rotary percussive methods are really wise, even allowing for the insistence on the use of an accompanying "water mist." Where the drilling is intended to penetrate large voids or other extremely permeable horizons, this is not such a crucial issue, however.

Boreholes in competent rock masses typically range from 38 to 76 mm in diameter, while depths in excess of 100 m are rare. Deviations as much as 1 in 30 may be expected, and will be worst when:

- The drilling equipment is light (small diameter, short rod length).
- The hole is long and/or inclined.
- The rock is heterogeneous (variable hardnesses, inclined fissures).
- The drilling method involves a top drive, rotary percussive hammer.

Deviations can be measured, as for rock anchorage holes (Bruce and Bianchi, 1992), but typically interhole spacings are selected to reflect and compensate this degree of uncertainty.

Detailed descriptions of rock drilling technology may be found in McGregor (1967), Acker (1974), Littlejohn and Bruce (1977), Houlsby (1990), and Weaver (1991).

Overburden and Soil This is a topic often overlooked in the technical literature but is, at the same time, potentially the greatest source of problems, and a contractor's most promising commercial advantage. Perhaps even more so than in rock drilling, there are paradigms and perceptions in soft ground drilling that often defy

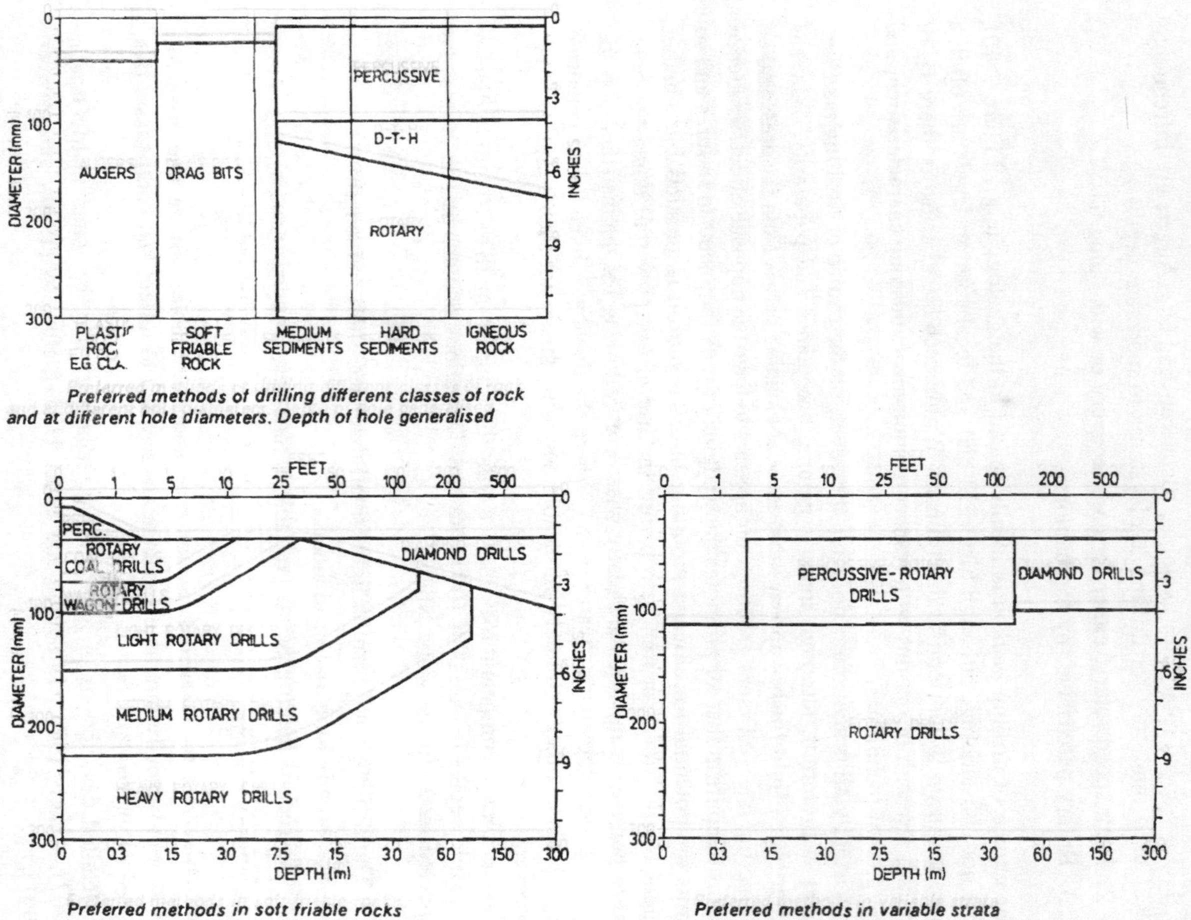

Figure 7-40 Typical drilling methods, related to geology. (After McGregor, 1967.)

engineering logic. In many instances, contractors will promote techniques and principles simply because they are well known to them, as opposed to advocating dispassionately the best and safest technical solution for the problem. This perceived mistrust is perhaps one of the main reasons that dam rehabilitation consultants, for example, are often disinclined to accept a drilling and grouting option for seepage control. Such engineers believe, often from bitter, personal experience, that the contractor may do more harm than good, causing hydrofracture, pneumatic fracture, or other unacceptable soil disturbance phenomena in the course of penetrating the soil or the structure.

More recently, this unacceptable trend has been challenged by the tendency of certain specialist contractors to be more scientific and pragmatic in their selection of overburden drilling methods. This was reflected in the publication of a generic classification of drilling methods, which attempts to cut across commercial lures and traditional preferences (Bruce, 1989b).

In some cases, the soil characteristics or hole geometry may permit "open holing" without temporary casing. In other cases it may be possible to temporarily stabilize holes by using a mud, polymer, or foam flush. In addition, there are other possible approaches to overburden drilling but they are synonymous with unacceptably high cost (e.g., diamond drilling) or restricted geological compatibility (e.g., vibrodriving).

Regarding "production" drilling methods, therefore, we can identify generically a six-fold classification, summarized in Table 7-10.

TYPE 1: SINGLE TUBE ADVANCEMENT In appropriate ground conditions with the proper equipment this is the cheapest and fastest method. There are two variants:

- Drive drilling
- External flush ("wash boring")

Drive Drilling This is a percussive system in which the steel casing is drilled with the leading end terminating in either a "knock off" drive shoe or bit. No flushing medium is used. A little rotation is necessary to prevent the string uncoupling during driving and to reduce deviation potential (recorded for the 76.1-mm size potentially being as much as 1 in 7.5). A standard range of sizes is shown in Table 7-11.

Rarely, however, are sizes over 101.6 mm practical, except in particularly loose, gravelly, or sandy conditions, and the 76.1-mm system appears to be optimum in terms of general cost effectiveness. Production figures of up to 250 m/shift are claimed for this size in "favorable" conditions, to maximum depths of 40 m.

External Flush Again, a single casing tube is advanced to the final depth. But in this case the drilling mechanism is rotary (not percussive), and the casing terminates in an open "shoe," or bit, and not a closed point. Water flush of suitable volume and pressure is passed continuously down through the casing during drilling and it washes debris out and away from the leading end. These water-borne debris

TABLE 7-10 Summary of Overburden Drilling Methods

Drilling Method	Principle	Common Diameters and Depths	Notes
1. Single tube advancement			
a. Drive drilling	Casing, with "lost point" percussed without flush	50–100 mm to 30 m	Hates obstructions or very dense soils
b. External flush	Casing, with shoe, rotated with strong water flush	100–200 mm to 50 m	Very common for anchor installation; needs high torque head and powerful flush pump
2. Rotary duplex	Simultaneous rotatin n and advancement of casing plus internal rod, carrying flush	100–200 mm to 60 m	Used only in very sensitive soil/site conditions; needs positive flush return; needs high torque
3. Rotary percussive concentric duplex	As 2, above, except casing and rods percussed as well as rotated	89–176 mm to 36 m	Useful in obstructed/bouldery conditions; needs powerful top rotary percussive hammer
4. Rotary percussive eccentric duplex	As 2, except eccentric bit on rod cuts oversized hole to ease casing advance	89–200 mm to 60 m	Obsolescent, expensive, and difficult system for difficult overburden; largely restricted to water wells
5. "Double head" duplex	As 2 or 3, except casing and rods rotate in opposite directions	100—150 mm to 60 m	Powerful, newer system for fast, straight drilling in worst soils; needs large hydraulic power
6. Hollow stem auger	Auger rotated to depth to permit subsequent introduction of tendon through stem	150–450 mm to 30 m	Hates obstructions, needs care in cohesionless soils; prevents application of higher grout pressures

TABLE 7-11 Standard Drive Drilling Casing Sizes

Casing Size		Recommended Casing Lengths
o.d. (mm)	i.d. (mm)	(Must be Portable by Not More Than 2 Men)
42.4	15	3.0 m
51.0	18	3.0 m
63.5	35	3.0 m
76.1	50	3.0 m
88.9	64	2.5 m
101.6	72	2.0 m
108.0	82	2.0 m
114.3	88	2.0 m
133.0	108	2.0 m
177.8	146	1.5 m

Source: From Hutte & Co., 1984.

then typically escape to the surface around the outside of the casing, but may equally be "lost" into permeable upper horizons. Depending on the purpose of the drilling, this system may be especially useful and economic or potentially very dangerous. For example, if ground anchorages are to be formed by subsequent pressure grouting through the same casing during its withdrawal, the washing will have helped promote a "bulb" of larger diameter, and so a fixed anchorage of higher load carrying potential. Conversely, if the drilling is to be conducted through highly sensitive soils under particularly delicate structures, the uncontrolled washing of the soil may prove inadvisable. However, even in this case, the system should not be rejected if good practice and common sense can together verify its applicability.

All sizes of casing can be used with this approach, but the most common applications of grout holes, anchoring, minipiling, and in situ reinforcement involve diameters of 100 to 200 mm.

TYPE 2: ROTARY DUPLEX In the most common situations, when ground conditions and job requirements combine to eliminate the "easy option" of single tube advancement, some method featuring the simultaneous advancement of internal rod (with bit), and external casing (with shoe) must be adopted. Such methods may collectively be referred to as "duplex."

The basic method, which most frequently carries the term "duplex," is purely rotary, and relies for its penetration performance on variations of rig thrust, head torque and rotational speed, and flushing characteristics, other factors being equal. The major components are illustrated in Figure 7-41 for a typical size and comprise:

- An outer casing (rotated), terminating in:
- A casing shoe (or crown).
- An inner-drill rod (rotated), terminating in:

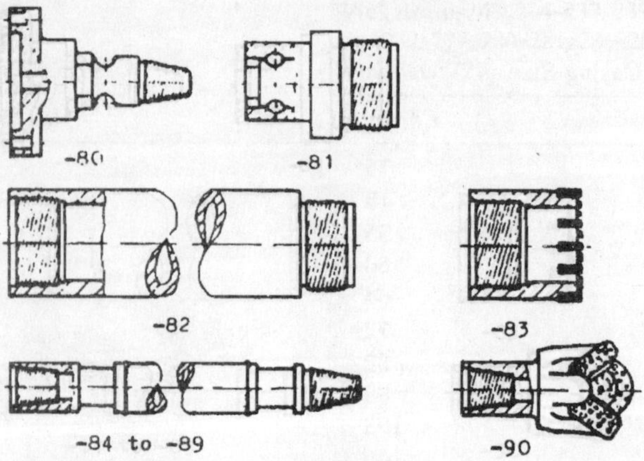

Order-no.	Name
178-80	transition flange with Wirth-thread 63,5
178-81	ejection flushing head 63,5 x 177,8
178-82	tube 177,8 x 900mm
178-83	casing crown button type 178 x ø 180mm
178-84	tube 63,5 x 900mm
178-85	tube 63,5 x 1500mm
178-86	tube 63,5 x 2000mm
178-87	tube 63,5 x 2500mm
178-88	tube 63,5 x 3000mm
178-89	tube 63,5 x 3500mm
178-90	rotary bit ø 149,2mm with transition to tube 63,5

Figure 7-41 Duplex drilling equipment, size 177.8. (From Hutte & Co., 1984.)

- A drill bit (usually tricone). Rods and casings connect to:
- A duplex head/transition flange, which is attached to the rotary head of the rig.

If hard obstructions are foreseen, it is common to exchange a down-the-hole hammer for the tricone bit, to hopefully fragment the obstruction and so permit the casing to be rotated down with less resistance (e.g., Bruce and Yeung, 1983). Equally, in especially difficult ground conditions, reverse circulation may be used. Duplex is most commonly used as a high production tool in soft and unstable ground conditions, typically powered by high torque hydraulic rotary heads. As a consequence, many contractors favor rather more robust systems than the one illustrated in Figure 7-41. However, where conditions are less onerous, or where environmental restraints are significant, standard flush coupled or jointed casing, or water well casing, with appropriate rod types, may be used, in accordance with local or national standards.

TYPE 3: ROTARY PERCUSSIVE CONCENTRIC DUPLEX This method, historically typified by the Atlas Copco OD72 System, is a duplex method wherein both rods and casings are simultaneously percussed and rotated. In its early years of use, it was driven by airpowered hammers with relatively restricted torque capacity. Therefore, the applicability was regarded as limited, and other methods, notably ODEX, with far less emphasis on rotational power and more on percussion, were developed. More recently, however, there is clear evidence of a resurgence of the method as a result of the increasing availability of higher torque hydraulic top hammers. By way of illustration, it may be noted that rotary percussive duplex was the preferred production drilling tool of major geotechnical contractors on Metro Tunnel related works in Hong Kong, where ground conditions were extremely onerous, featuring gritty decomposed granites with large fresh rock relicts. This market for grout hole installation alone was conservatively estimated at about 200,000 m of drilling per year, in the late 1970s and early 1980s.

Although the Atlas Copco System is available in only one size, other manufacturers can supply sizes as shown in Table 7-12.

The casings are, of necessity, special quality steel, and have modified rope threads and wall thicknesses of around 12 mm (as opposed to 6 mm for ODEX). One consequence is that the unit weight is high, and normally 2-m casing lengths are used in the larger sizes unless automatic rod handling is available. Drilling on with the rods into rock or other stable material is accomplished without the necessity of changing the bit. Both insert and button types are available for bits and casing shoes. As with other forms of concentric duplex, in especially sensitive ground, the bit can be retracted up into the casing behind the casing shoe, to minimize cavitation of the ground and promote good flush return. The opposite is done in particularly dense and competent ground. Flushing water is best introduced via an external flushing device and should have a minimum rate of about 100 to 150 liters/min at 1.5 to 2.0 MPa. To further improve flush return, sleeving can be inserted between adjacent couplers on the rod string to present a constant annular volume and reduce local "pressure drops" and resultant blockages.

TABLE 7-12 Standard Percussive Duplex Sizes

Casing			
o.d. (mm)	Minimum i.d. (mm)	Crown o.d. (mm)	Bit Diameter (mm)
88.9	64	95	60
101.6	72	107	70
108.0[a]	82	112	76
114.3	88	120	85
133.0[a]	108	140	105
177.8[a]	146	185	146

Source: From Hutte & Co., 1984.

[a]Denotes common sizes for double head drilling.

Assuming that sufficient torque (say to 6 kNm) is available at the hammer, and adequate pull-up force can be applied (say around 4 tonnes) then rotary percussive duplex may be regarded realistically as a viable and robust production method for holes to 60-m depth. Clearly, however, for the deeper drilling associated with water well drilling or mineral prospecting it may not be the most cost-effective option.

TYPE 4: ROTARY PERCUSSIVE ECCENTRIC DUPLEX Restricted in terms of torque availability and faced with an increasing demand for a system to reliably penetrate the difficult Scandinavian glacial deposits, Atlas Copco and Sandvik jointly developed the very successful ODEX system in 1972. This percussive duplex variant features a pilot bit with eccentric reamer, which cuts a hole of diameter slightly larger than the following casing. The manufacturers state that its performance is not impaired by gross changes in the ground from loose soil to fresh igneous rocks; the method cuts through obstructions or shoulders them aside. Early experience in Britain (Patey, 1977) also confirmed its ability to deal with artificial obstructions such as slag and other foundry spoil, typical of fill deposits in old industrialized areas. Good results in loose scree type deposits, rip-rap, and through old piled foundations have also been confirmed.

The principle of the operation is illustrated in Figure 7-42. In Figure 7-42a, the single piece pilot bit (concentric) is shown drilling beneath the casing; rotation has been applied to the rod string, swinging out the reaming device (eccentric), which enlarges the hole so facilitating the advancement of the casing (percussed only). The reamer is held in the correct position by stop lugs during drilling. Cuttings are transported upward past the guide device, inside the casing to exit via ports at the driving cap. Flush is usually water, although air can be used, and foam is common for depths over 30 m. When drilling is complete (Figure 7-42b) the rods are counter-rotated, so closing the reamer and permitting the withdrawal of the rod and bit assembly. Drilling on into rock must then be done with a suitable rock bit (Figure 7-42c). ODEX is available with both top hammer and down-the-hole options and selection reflects ground conditions, hole diameter, hole purpose, and the type of rig and head available. In the former case, part of the percussive energy is transferred from the top hammer, via a shank adaptor, to a driving cap above the casing. For down-the-hole drilling, the percussive energy is transferred to the casing from the hammer by a special "bit tube" with a driving (or impact) shoe. The casing is therefore pulled down, again without rotation, from its lower end. In both cases, however, the steel must be strong enough to resist the percussive energy of the hammer either in compression (top hammer), or in tension (down-the-hole). Also, where it is to be later extracted, the threaded casing must also have sufficient tensile strength, particularly in the threaded zones, and this parameter often dictates the practical depth to be drilled under any given conditions. Indeed, where ODEX 76 has been employed as a production drilling tool under adverse geological conditions, the typically thin-walled rotary casing of the standard systems has had to be modified by specialist contractors, within, of course, the limits imposed by the geometry of the other elements of the system.

Regarding the anticipated longevities of the key components of the ODEX and

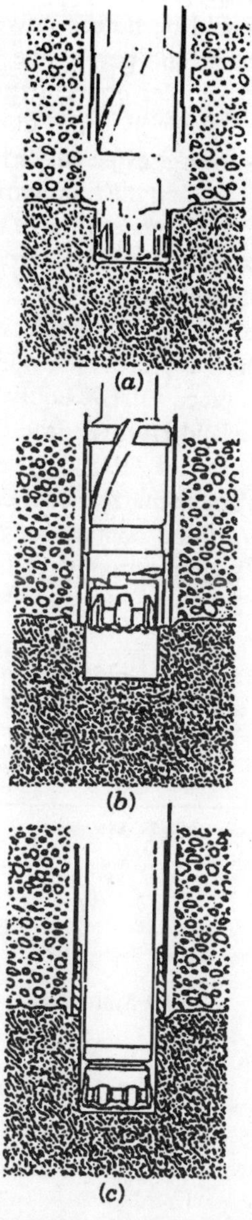

Figure 7-42 Operating principle of ODEX. (*a*) When drilling, the reamer of the ODEX bit swings out and drills a hole larger than the external diameter of the casing tube. (*b*) When the required depth has been reached, the drill is reversed and the reamer swings in to its minimum diameter, allowing the bit to be lifted up through the casing, which is left in the hole. (*c*) Drilling can continue with an ordinary DTH drill bit. (From Atlas Copco, 1976.)

OD systems (for comparison), Atlas Copco has published the indicative guidelines reproduced in Table 7-13. It should be noted, however, that the relatively recent developments of high torque rotary and percussive drill heads have breathed new life back into conventional and simpler concentric duplex systems, as described above. Therefore, the use of top drive eccentric duplex is becoming rarer. On the other hand, the demand remains strong, especially in the water well industry, for the drilling of large-diameter holes in which the casing may be left permanently. In such cases, the down-the-hold ODEX variants still have much to offer, especially when the driller has available only a standard medium-sized drill rig with rotary head, and has experience in down-the-hole drilling.

More recently, Halco has developed its own eccentric duplex system, Sim Cas. As the reaming device is only in two pieces, the operation is claimed to be simpler and more robust than the three-piece ODEX equivalent. Similar systems are also offered by Hutte and by Weaver and Hurt (Bulroc Overburden Drilling System).

In summary, a major attraction of ODEX type systems is that the effective efficient depth of penetration is not primarily dependent on drilling torque, since the presenter of the greatest steel/ground contact area, that is, the casing, does not require rotation. However, the system remains relatively sophisticated, and its suc-

TABLE 7-13 Indicative Guideline Longevities for Atlas Copco OD and ODEX Systems Components

Component	Anticipated Longevity (drilled meters)
ODEX System[a]	
Pilot bit	200–600
Reamer	100–300
Guide	400–1200
OD System	
Extension tube	1000–1500
Tube coupling	800–1000
Ring bit	150–400
Adaptor sleeve	1000–1200
Cross-bit	300–500
ODEX and OD Systems	
Shank adaptor, flushing head, driving cap	800–1000
Extension rods	1000–1500
Coupling sleeves	800–1000

Source: From Atlas Capco, 1976.

[a]The various items in ODEX systems are normally consumed in the following ratios: 1 guide to 2 pilot bits to 4 reamers.

cess is very sensitive not only to operator skill and expertise, but to the quality of the casing and its joints, and the efficiency of the flushing.

TYPE 5: DOUBLE HEAD DUPLEX DRILLING This rotary duplex method is claimed to be especially quiet, and to ensure minimal ground disturbance, and consistent cost-effective penetration to over 80 m in even the most difficult ground conditions. It is distinguished from conventional rotary duplex by the fact that the rods, and casings, are simultaneously rotated *but in opposite directions*. The inner drill rods, with right-hand rotation, carry either a down-the-hole hammer in hard conditions, or some form of rotary bit in soft ground. Typical rotary energy requirements are 2.5 kNm torque at 40 to 60 rpm.

The casing, with left-hand rotation, terminates in a substantial crown that cuts a slightly oversized hole, thus reducing casing/ground resistance. Rotational speeds are lower than in conventional duplex drilling (15 to 30 rpm) to the advantage of the torque availability (to 8 kNm). However, the benefits of the counter rotation are that the cutting action is enhanced, and the prospect of flush debris blockages in the casing/rod annulus is minimized due to the shearing action between its boundaries. (Water flush is typically 40 to 60 liters/min at 1.5 MPa.) In addition, the counter-rotation helps to offset natural tendencies for holes to deviate and, in conjunction with the stiff, thicker walled casing used, holes of exceptional straightness (say within 1 in 100) can be provided (Bruce and Kord, 1991).

This system is driven by special "double heads," with both Klemm and Krupp (Table 7-14) being prime examples. These heads can be mounted even on relatively small and mobile track rigs of sufficient hydraulic power. A particular feature is the facility to move the upper rotator coaxially (turning the rods) about 300 mm relative to the lower casing rotator. This affords the driller extra scope in selecting the relative advancement of rod bit and casing shoe in response to ground conditions. The lower rotator can also work in high gear (say 30 rpm, low torque range) or low gear (say 15 rpm, twice torque previously available). In addition, the upper rotator can be replaced with a rotary percussive head and the down-the-hole hammer omitted, as noted in Table 7-14. As with other percussive duplex variants, a retrievable underreamer can be used to precut the soil to a diameter just larger than the casing shoe.

Double-head duplex drilling is common on European sites with particularly difficult ground but restricted access. It was also used under similar conditions recently at the Hines Auditorium in Boston (Bruce, 1988–1989) while its use is growing—with the increasing popularity of diesel hydraulic track rigs—for anchor drilling on both coasts of the United States. In Canada, a project has recently been completed underground in North Ontario where the 133-mm casing was drilled, straight, to 60-m depth through loose mine backfill containing large boulders of over 300 MPa compressive strength, in headroom of 4 m, at outputs equivalent to over 50 m/shift (Bruce and Kord, 1991).

TYPE 6: HOLLOW STEM AUGER Auger drilling is a long established method of drilling cohesive soils containing the minimum of hard inclusions, and features the

TABLE 7-14 Specification for Double Head Drill, with Either Rotary or Rotary Percussive Option for Inner Drill String

Operating Method	Outer Casing Inner Rod	Option A Rotated[a] Rotated[a]	Option B Rotated[a] Percussed/Rotated[a]
Service weight including base plate	kg	630	700
Oil flow rate (front/rear rotary mechanism)	maximum liters/min	160/170	160
Oil flow rate (percussive mechanism for inner string)	maximum liters/min	——	——/85
Operating pressure (front/rear rotary/ percussion mechanism)	maximum bar	210/260/——	210/170/170
Torque (front/rear rotary mechanism)	maximum Nm	8,000/4,000	8,000/4,000
Number of revolutions[a] (front/rear rotary mechanism)	maximum rpm	110/145	110/110
Number of blows	maximum/min	——	——/1,800
Connection thread outer/inner drill strings		to be specified	to be specified
Hole diameter	nm	100/300	100–300
Flushing medium		air/water	air/water

Source: From Kruppo, 1983.

[a]Clockwise or counterclockwise, but inner and outer drill strings always counterrotating.

rotation of what is basically a continuous screw into the ground. The continuous flight auger may be in one part (as used in bored piling works) or in connecting sections. The basic method uses a solid stem (or core) to excavate the hole, which, when the auger is withdrawn, will remain open only due to the natural competence of the ground, and the absence of groundwater pressures. As noted earlier, such "open hole" methods are not the subject of this discussion.

Much recent development has focused on hollow stem augers, which permit water, and/or grout to be pumped to the bottom of the hole, allow placing of anchor bars or grout tubes, or enable drilling on into underlying strata for soil sampling or rock socketing. Generally, however, as emphasized by the range of standard sizes and the capacities of typical rotary head models, the whole concept of augering is still related to the larger diameter fields of cast in situ piles, prebored pile holes, and sand drains. In addition, lower capacity ground anchors on the West Coast and in certain parts of the Midwest have traditionally been installed in augered holes up to 400 mm in diameter. Common base machines are excavators, piling rigs, and crawler mounted cranes.

To reduce power requirements and allow adequate clearance for the flights, auger bits (or cutting heads) typically cut a hole 10 to 12 percent larger than the auger diameter. The pitch of the flights is 60 to 80 percent of the outside diameter of the auger to reduce the tendency of the cuttings to roll back down the hole. The leading auger section (0.2 to 0.5 m), fitted with the appropriate bit or drive shoe, is often armored to reduce wear on following flights. Expanding auger bits are available for use with continuous flight augers for boring inside casing. The auger bit has an outside diameter equal to the continuous flight auger, but expanding wings increase the cutting diameter to the outside diameter of the casing. During drilling, the auger is positioned so that the wings are just below the lower edge of the casing, which may then be advanced as cutting proceeds. Reversing the rotation causes the wings to fold back, enabling the auger and bit to be withdrawn without disturbing the casing.

Great care must always be exercised when augering holes in soils with poor cementation or limited cohesion. Continued rotation of an auger in such conditions will cause cavitation of the surrounding ground: the auger then effectively acts as a screw conveyor. Given only the lower pressures that can be applied when grouting through augers, there is then no guarantee that the voids will be completely filled and/or the soil redensified.

Hole Spacing and Location As a rough guide, AFTES (1991) produced Table 7-9 for typical grout hole spacings in different ground conditions. These are felt to be reasonable to depths of 25 m; thereafter the option to increase spacing (by making use of potentially higher grout pressures) should be balanced against the increased likelihood of borehole deviation. In all cases, borehole spacings must reflect the design assumptions of the grouting methods and the anticipated final result.

Patterns are determined by geometrical considerations related to the theoretical radius of effective treatment, and the type of grouting. Their three-dimensional layout will also reflect the restraints of the grouting, for example:

- Vertical (or parallel inclined) from the surface (Figure 7-39).
- Parallel or divergent holes along one or more parallel lines from the face or from an auxiliary underground structure.
- Radial holes from a structure.
- Holes drilled in overlapping cones from an advancing face.

Monitoring Production Drilling It is still relatively uncommon for detailed drill logs to be taken as production grout holes are formed. And yet, each hole drilled represents another opportunity to explore the ground, or assess the possible impact of any previous phase of grouting. Such data can even assist in refining grouting procedures and parameters.

In recent years, contractors in France and Italy have developed real time electronic monitoring of production drilling parameters (De Paoli et al., 1987). Such data have been used in deep oil well drilling for many years, and can be processed into a "drillability" factor, such as the specific energy that can then be interpreted as indicating various lithologies. Experimental and theoretical work has been conducted, based on the formula

$$e = \frac{F}{A} + \frac{2NT}{AR}$$

where e = the specific energy, kJ/m^3
F = the thrust, kN
N = the rotational speed, rev/sec
T = the torque, kNm
R = the penetration rate, m/sec
A = the cross section of the hole, m^2

Rowlands (1971) verified the equation's validity under ideal drilling conditions, including efficient hole flushing, no loss of drilling energy in the rods, and constant bit wear conditions. Regarding the parameters themselves:

1. *Penetration Rate*
   - In rock, the penetration rate is directly related to the mechanical characteristics such as hardness, abrasiveness, Young's modulus, sonic velocity.
   - In loose ground, the penetration rate indicates the relative ease of drilling.
   - Generally, it allows the identification, sizing, and analysis of fissures and cavities.
2. *Thrust or Hold Back on the Bit*. The measurement of the thrust pressure on the rod assembly completes the "speed" information in the search for cavities, since in the absence of ground reaction while passing through cavities, the thrust drops practically to zero. For certain types of hydraulic drill rigs, the

thrust measurement must be complemented by that of the holdback pressure applied to the drill string.

3. *Rotation Torque.* The torque used by the rig relates to the ground quality. This parameter clearly identifies grounds such as compact marls, gravels, and conditions that produce jerky rotation.
4. *Speed of Rotation.* The measurement of the speed of rotation of the rods completes the torque measurement data.
5. *Reflected Energy Characteristics.* In percussion drilling, the quantity of percussion energy that is reflected in the rod assembly is greater when the rock is harder. This parameter clearly identifies the hard and compact rocks, and coarse gravels and boulders.
6. *Drill Flush Pressure.* This parameter quantifies the pressure of the flushing medium. When the bit passes through a plastic formation (clay or marl), the pressure increases. On the contrary, formations that are very permeable, such as gravels, produce a drop in pressure. The measurement of pressure is also very useful for determining the existence and nature of any cavity infill material.
7. *Fluid Gain-Loss.* This parameter quantifies the losses and the gains of circulation fluid (water or mud). It identifies the zones of high permeability, such as gravels and very fissured zones.

Figure 7-43 shows the recorded parameters in a typical example, and the generation of the specific energy profile. These values are then related to the various lithologies by a statistical analysis, wherein all the specific energy values are represented as a histogram of frequencies (Figure 7-44). Each peak is then related to a lithological unit, based on general geological investigations at this, or other similar sites. Having performed this analysis on individual holes, and having defined the lithological groups statistically, it is possible to use the results obtained as the input data of more general computer programs. The variables are stored under a common format permitting interfacing with CAD facilities.

An excellent example was in the tunneling works for the Milan Metro (De Paoli et al., 1987), where close characterization of the highly variable alluvial soils was necessary to optimize subsequent grouting designs. As shown in Figure 7-45, the geological interpretation of the specific energy records can be made at 100 mm intervals. Such individual records can then be assessed longitudinally to illustrate the location of sand lenses needing particular attention. Continuing correlation of predicted, and observed, conditions permits the operation to be progressively refined during the work.

As AFTES (1991) notes, engineers are increasingly turning to real time, graphic recording of data with digital acquisition and storage of information, which can then be used in a microcomputer. However, even when the scale of the project or the level of sophistication of the contractor prevents automatic data recording and analysis, drill personnel should be encouraged to record manually as much information as possible, especially on the "easier" parameters such as penetration rate, hole

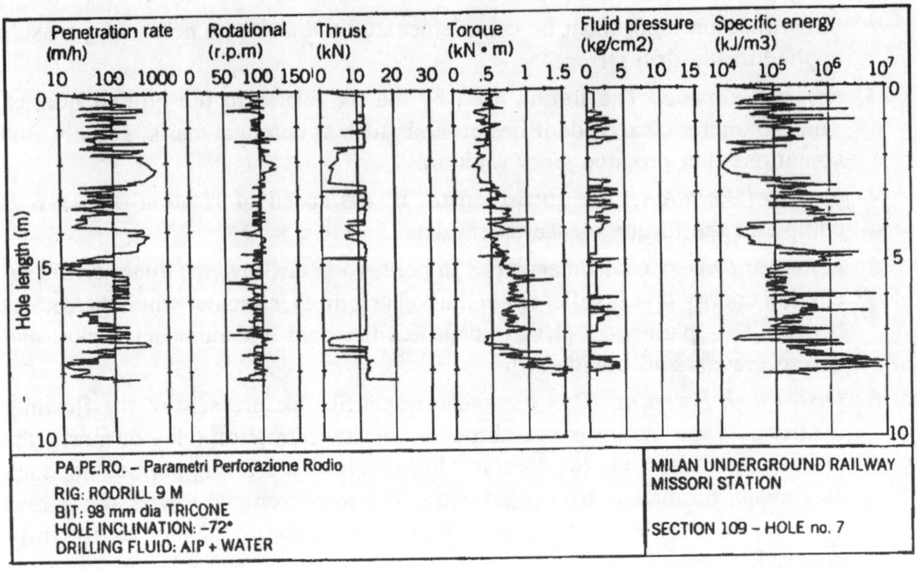

Figure 7-43 Plots reporting the quantities measured: rate of penetration, rotational speed, thrust, fluid pressure, specific energy. (From De Paoli et al., 1987.)

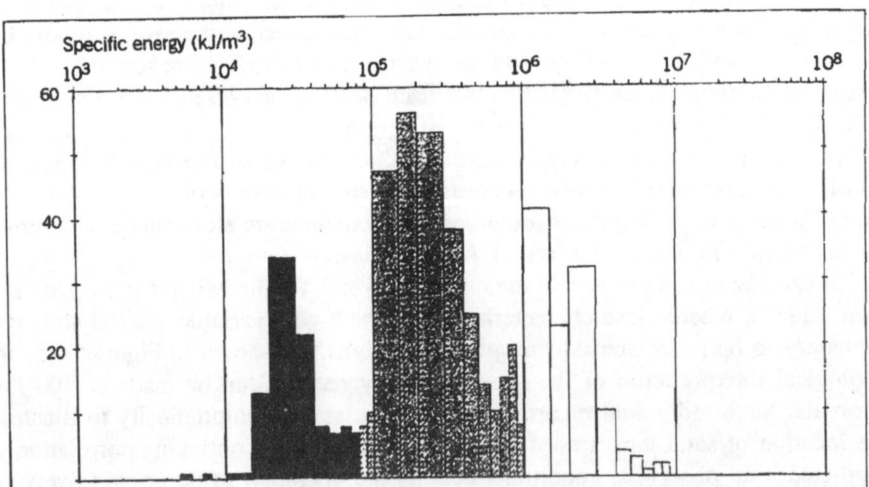

Figure 7-44 Frequency of specific energy values measured on a significant number of holes. Peaks represent typical soil layers. (From De Paoli et al., 1987.)

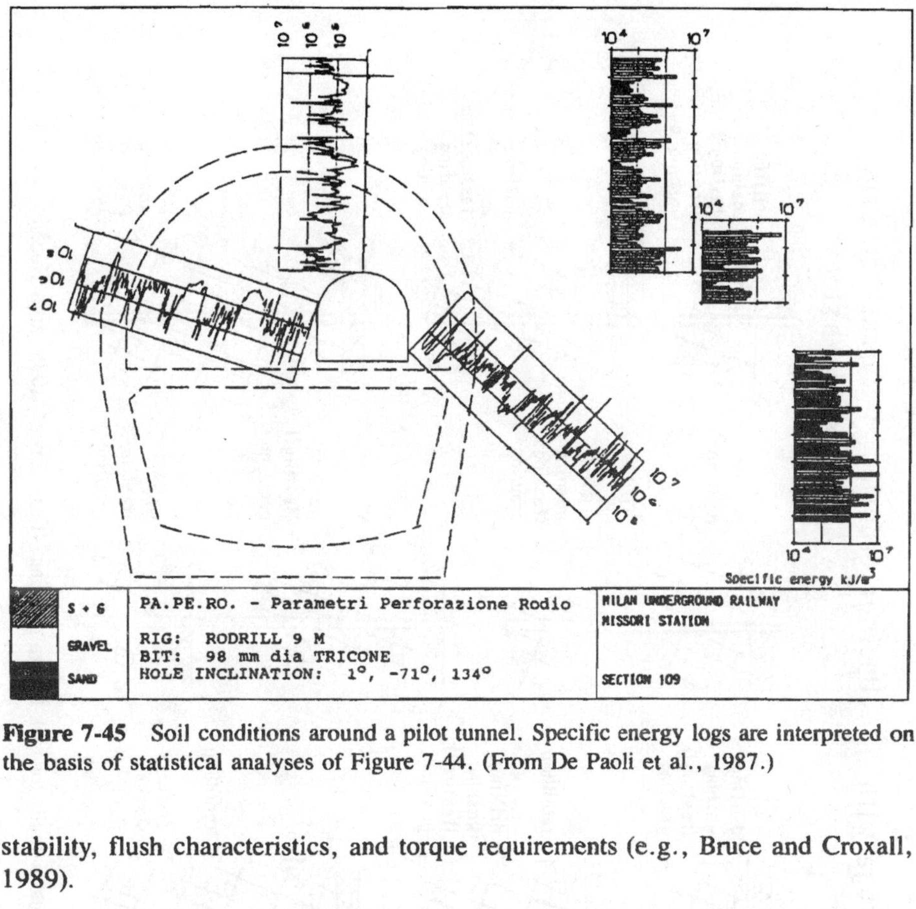

Figure 7-45 Soil conditions around a pilot tunnel. Specific energy logs are interpreted on the basis of statistical analyses of Figure 7-44. (From De Paoli et al., 1987.)

stability, flush characteristics, and torque requirements (e.g., Bruce and Croxall, 1989).

Grouting

Rock The standard, traditional methods of rock mass fissure grouting have long been used and are generally well known (e.g., Bruce and George, 1982). While it is clear from many recent conferences and textbooks that the details of fissure filling are rather more deceptively complex than often assumed, there is no doubt that a high level of rock grouting expertise and success is evident throughout the world. Very comprehensive treatments of the subject may be found in Houlsby (1982, 1990), Ewart (1985), and Weaver (1991), as well as in numerous technical papers in conference proceedings such as ASCE (1982, 1985, 1992).

In summary, depending largely on geological and economic factors, treatment is conducted in stages from the top of a hole down (descending stages) or from the bottom of a hole up (ascending stages) (Figure 7-46). In the former method, packers may be left at the top of each hole or placed at the top of each successive down stage. Simplistically, one may assume that if the rock mass is mechanically stable, permitting "open hole" drilling and disallowing leakage of grout around an inflated

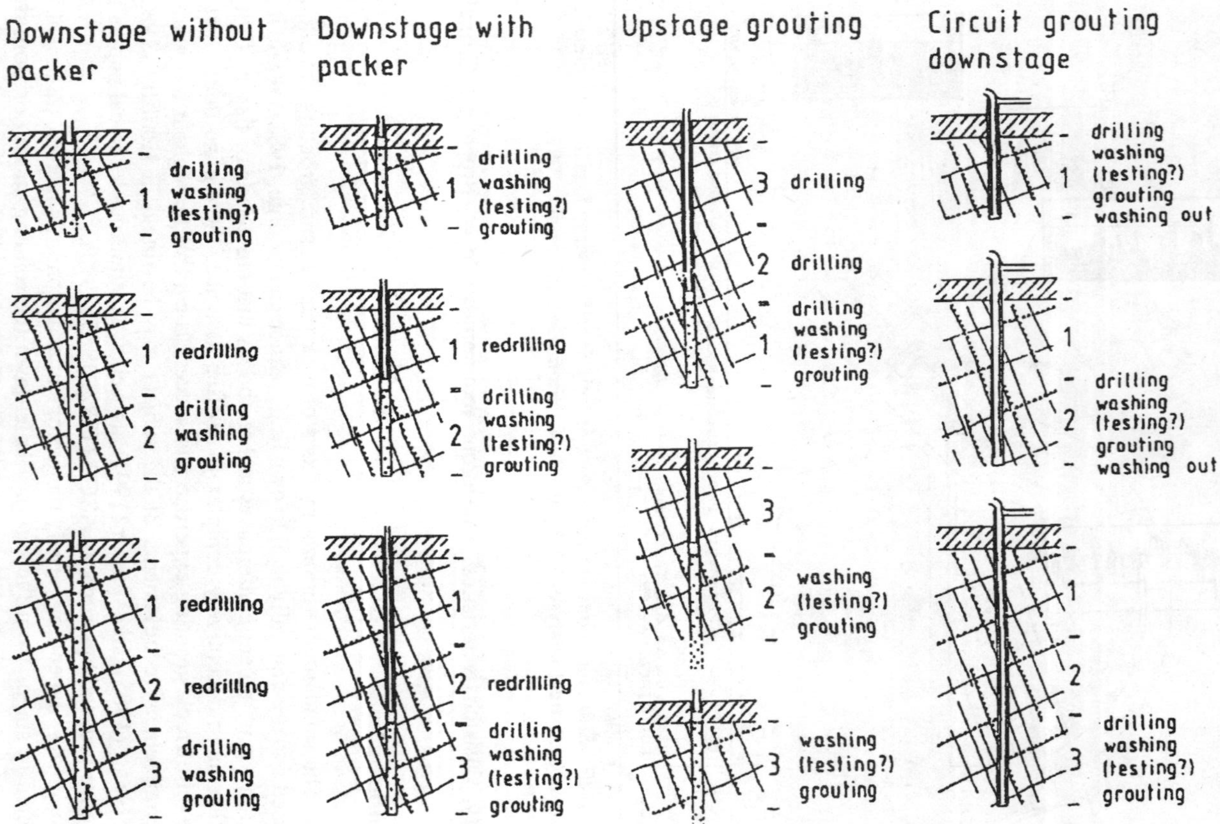

Figure 7-46 Conventional stage grouting methods for rock fissure grouting. (From Ewart, 1985, after Houlsby, 1982.)

packer, then ascending stage grouting is the most popular choice. This is reflected deep in the roots of U.S. practice where the earlier dams in particular had the advantage of being built on geologically sound sites due to the luxury of wide choice. The major advantages and disadvantages of each method are summarized in Table 7-15.

While changes in grouting materials and improvements in equipment continue, the main innovation in rock grouting methodology is the introduction of the multiple packer sleeved pipe (MPSP) system (Bruce and Gallavresi, 1988). Unstable rock masses will prevent "open holing" or efficient packer seating. As a consequence, stages may be incompletely grouted, if treated at all. In the late 1970s, Rodio developed the MPSP system to combat voided karstic limestone in Turkey (Oymapinar dam) and collapsing "sugary" limestone in Pakistan (Tarbela dam). These successes have led to widespread use of the system in especially difficult geological conditions.

MPSP owes much to the principle of the tube à manchette system, in that grouting of the surrounding rock is effected through the ports of a plastic or steel grout tube placed in a predrilled hole. However, unlike tube à manchette, no sleeve or annulus grout is used. Instead, the grouting tube is retained and centralized in each borehole by concentric collars, essentially fabric bags inflated in situ with cement grout. These collars are positioned along the length of each grout pipe, either at regular intervals (say 3 to 6 m) to isolate standard stage lengths, or at intermediate or closer centers to ensure intensive treatment of special or particular zones discovered during drilling. The system permits the use of all grout types, as dictated by the characteristics of the rock mass and the purpose of the ground treatment. The typical construction sequence is as follows (Figure 7-47):

1. The borehole is drilled by the most cost-effective method (usually rotary percussive) with water flush to full depth. Temporary casing may be necessary to full depth also, as dictated by the degree of instability of the rock mass. Typically borehole diameters are 100 to 150 mm.
2. The MPSP is installed. Pipe details can be varied with requirements, but a typical choice consists of steel pipe, 50 mm in diameter. Each 5-m long pipe may have three 80-mm-long, 4-mm-thick rubber sleeves equally spaced along the length, protecting groups of 4-mm-diameter holes drilled in the pipe. A concentric polypropylene fabric bag is sealed by clips above and below the uppermost sleeve in each length and is typically 400 to 600 mm long. For short drill holes, plastic pipes of smaller diameter may be used. The temporary drill casing is then extracted; any collapsing material simply falls against the outside wall of the MPSP tube.
3. Starting from the lowermost pipe length, each fabric bag is inflated via a double packer positioned at the sleeved port covered by the bag. A neat cement grout is used at excess pressures of up to 0.2 MPa to ensure intimate contact between bag and borehole wall. The material of the bag permits seepage of water out of the grout, thus promoting high early strength and no possibility of later shrinkage. Clearly the choice of the bag material is crucial

TABLE 7-15 Major Advantages and Disadvantages of Downstage and Upstage Grouting of Rock Masses

Downstage	Upstage
Advantages	
1. Ground is consolidated from top down, aiding hole stability and packer seating and allowing successively higher pressures to be used with depth without fear of surface leakage.	1. Drilling in one pass
	2. Grouting in one repetitive operation without significant delays
	3. Less wasteful of materials
2. Depth of the hole need not be predetermined; grout take analyses may dictate changes from foreseen, and shortening or lengthening of the hole can be easily accommodated	4. Permits materials to be varied readily
	5. Easier to control and program
	6. Stage length can be varied to treat "special" zones
3. Stage length can be adapted to conditions as encountered to allow "special" treatment	7. Often cheaper since net drilling output rate is higher
Disadvantages	
1. Requires repeated moving of drilling rig and redrilling of set grout; therefore, process is discontinuous and may be more time-consuming	1. Grouted depth predetermined
	2. Hole may collapse before packer introduced or after grouting starts, leading to stuck packers and incomplete treatment
2. Relatively wasteful of materials and so generally restricted to cement-based grout	3. Grout may escape upwards into (nongrouted) upper layers or the overlying dam, either by hydrofracture or bypassing packer; smaller fissures may not then be treated efficiently at depth
3. May lead to significant hole deviation	
4. Collapsing strata will prevent effective grouting of entire stage, unless circuit grouting method can be deployed	4. Artesian conditions may pose problems
5. Weathered and/or highly variable strata problematical	5. Weathered and/or highly variable strata problematical
6. Packer may be difficult to seat in such conditions	

Source: From Bruce and Gallavresi (1988).

to the effective operation of the system: the fabric must have adequate strength, a certain elasticity, and a carefully prescribed permeability.

4. Water testing may then be conducted if required through either of the two free sleeves between collars, again via a double packer. Tests show that a properly seated fabric collar will permit effective "stage" water testing at up to 0.4 MPa excess pressure.

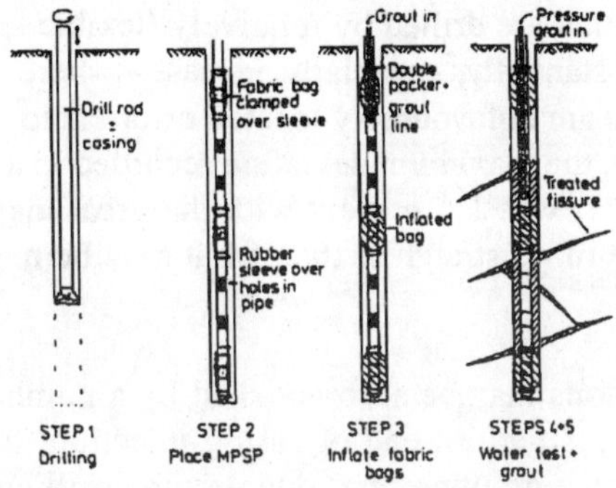

Figure 7-47 Steps in MPSP grouting. (From Bruce and Gallavresi, 1988. Reproduced by permission of ASCE.)

5. Grouting is executed in standard tube à manchette fashion from bottom up via the double packer (usually of the inflatable type). The grouting parameters are chosen to respect target volumes (to prevent potentially wasteful long-distance travel of the grout) and/or target pressures (to prevent potentially dangerous structural upheavals).

The following additional points are especially noteworthy regarding the MPSP system. First, if a hole has been grouted once, it generally cannot be regrouted: some of the pressure grout will remain in the annulus outside the tube and so form a strong sleeve grout preventing the opening of sleeves in contact. (The system does, however, allow different stages *in the same hole* to be treated at different times.) Thus the MPSP system accommodates the principles of conventional stage grouting, where split spacing methods are used: the intermediate secondary holes both demonstrate the effectiveness of the primaries and intensify the treatment by intersecting incompletely grouted zones. Analyses of water test records, grout injection parameters, reduction ratios, and so on will dictate the need for further intermediate grouting phases.

Second, in addition to the technical advantages of the system, there are significant logistical and work scheduling attractions. For example, the drilling and installation work can proceed regularly at well known rates of production, without requiring an integrated effort from the grouting crews (as in downstage grouting). In addition, the secure nature of the grout tube prevents the possibility of stuck packers, which is an unpleasant but unavoidable fact of life in upstage grouting in boreholes in most rock types. Grouting progress is therefore also more predictable and smoother, to the operational, technical, and financial advantage of all parties concerned.

A third point relates to the straightness of the borehole and thus the integrity and continuity of the ground treatment. The temporary drill casings often used in the hole drilling operations are thick-walled and robust (as described in the discussion of Drilling in Overburden and Soil earlier in Section 7-2). They therefore promote hole straightness, whereas the uncased boreholes that are common in normal stage grouting in rock, and that are drilled by relatively flexible small-diameter rods, are known to deviate substantially, especially in cases where fissures and/or softish zones in the rock mass are unfavourably located or oriented. By way of illustration, at Metramo dam, Italy, the maximum deviation recorded in a test block of 150 cased holes, each 120 m long, was 1.5 percent with the great majority being less than 1 percent. This is 2 or 3 times straighter than what may be reasonably expected with uncased boreholes.

Soil Permeation of soils may be accomplished by a number of systems, with the major groups being classified as: end of casing injection, tube à manchette, valve tube system, limited area grouting, and double tube drilling and seepage.

END OF CASING INJECTION When the ground is suspected of being very permeable and there is no recognized need for sophisticated multiphase or multimaterial injections in any one hole, then the simplest method is via "end of casing." In essence, the drill casing is installed to the final depth, and pressure grouting is conducted through it, as the casing is slowly withdrawn. All the forms of overburden drilling outlined under Drilling earlier in Section 7-2 can be used for this purpose. Typical examples would range from drive drilling (for shallow grouting of railway embankments) through percussive duplex (for deeper consolidation, as in mine shafts) to rotary duplex (for grouting anchors or pin piles). In addition, grouting through the drill rods, again during withdrawal, is often conducted for hole stabilization for watertightness, prior to redrilling. Compaction grouting is generally conducted by this method also.

TUBE À MANCHETTE It is generally recognized in Europe and North America that the most controlled method of overburden permeation is the tube à manchette (or sleeved pipe system, Figure 7-48). Essentially, it permits multiphase injections of various materials with a great degree of control over the grouting variables (Bruce, 1982). The method does, however, depend for its successful performance on the efficient and economic installation of the plastic or steel grouting tubes. In general, some form of duplex method is used to penetrate to the required depth. The inner rods are withdrawn, the casing topped up with bentonite-cement "sleeve" grout, the sleeved grouting pipe inserted, and the drill casing withdrawn. Recently, increasing use has also been made of hollow stem augers for this purpose, and in less cohesive soils, rotary methods with bentonite flush are common. Clearly, the casing must have sufficient bore to permit its extraction without damaging the tube or its rubber sleeves. However, too large an outside diameter will give an unacceptably thick annulus of sleeve grout, making a subsequent opening of the sleeves a question of very high initial rupture pressures. Usually an annulus of 20 to 30 mm is sought.

Despite the advances in other forms of soil grouting, permeation using the sleeved pipe system remains one of the most popular systems worldwide. Major recent applications include tunnels, for example, the Hong Kong Metro (Bruce and Shirlaw, 1985), the Cairo sewers (Greenwood et al., 1987), Milan Metro (Mongilardi and Tornaghi, 1986), deep excavations (Littlejohn et al., 1989), and dams (Bell,

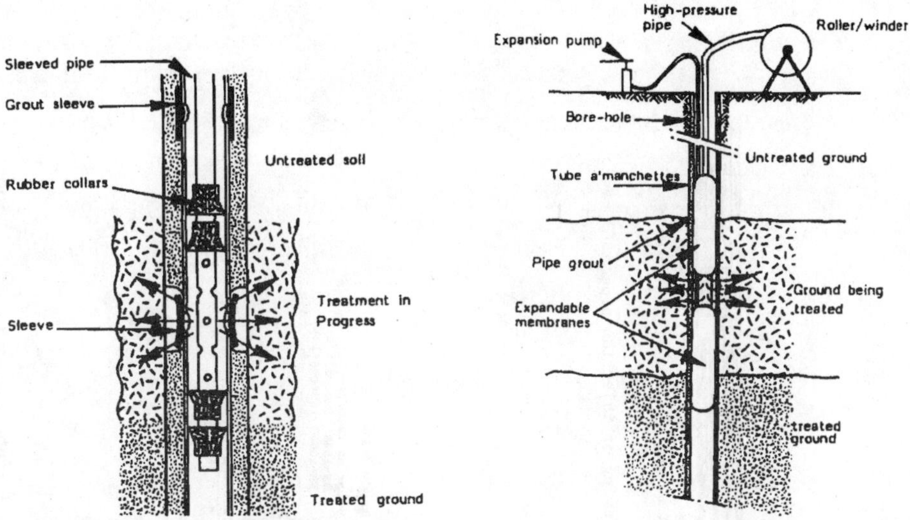

Injection with double collar packer in sleeved pipe: stage-up injection.

Grout injection in a tube à manchette (sleeve pipe), expandable with double packer: stage-up grouting.

Figure 7-48 Tube à manchette (sleeved pipe) grouting. (From AFTES, 1991.)

1982). In the United States, many examples can be cited of recent work in New York, Pittsburgh, and Baltimore, and ongoing work in the Los Angeles Metro (ASCE, 1992). Hydrofracture and compensation grouting are also conducted through sleeved pipes of this type.

VALVE TUBE SYSTEM In many ways similar to the tube à manchette system in terms of its grouting capabilities, this system, developed by Stabilator of Sweden in the middle 1960s, has one major difference. The steel grouting pipe, equipped with spring loaded grouting ports, doubles as the drill casing, and has a nonretrievable crown (or ring bit). The casing is not rotated during driving. Clearly the initial lineal cost of tube installed is relatively high, but this expense is claimed to be offset by the high rate of installation, in which no time need be spent annulus grouting or extracting temporary casings, as in the case of tube à manchette grouting. Several successful major applications have been recorded throughout the world with a particularly good description provided by Lamberton (1982).

LIMITED AREA GROUTING (LAG). In the last 40 years there has been a tremendous growth in tunneling and deep foundation projects in Japan. This is reflected in the high reputation currently held by the Japanese as soft ground tunnelers, and as developers of novel ground treatment systems, of which LAG (Tokoro et al., 1982) is one of the more common (Figure 7-49) in Southeast Asia.

It features the introduction by a small hollowhead rotary drill rig of a combined rod-casing assembly, followed by the injection of a flash-setting (less than 5 sec)

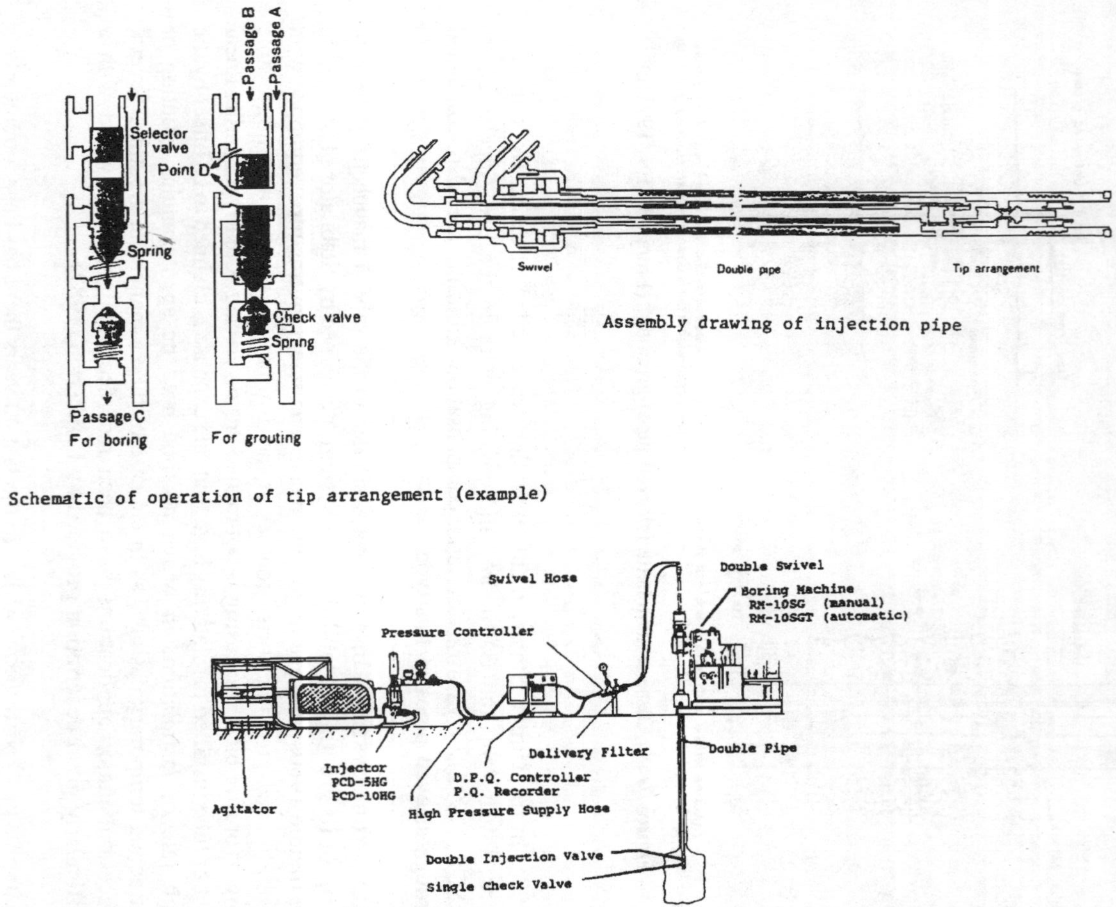

Figure 7-49 Operating principle of LAG grouting method. (From Bruce, 1989a, after Tokoro et al., 1982.)

grout via one exit port during rotated withdrawal of the string (20 rpm at about 2 m/min). With respect to Figure 7-49 passage A carries the base component (silicate solution), and passage B the reagent. These compounds are mixed and ejected only at the port, which during drilling is kept closed by a spring arrangement. A diameter of soil treatment of 0.6 to 1 m/hole is claimed.

Typical ground conditions suited to LAG are clays, silts, sands, and fine gravels. It is notable that the tube à manchette system is now relatively little used in Japan due to: (a) its relative cost and complexity; (b) its construction cycle time; (c) its potential for permitting dilution and dispersion of grout under dynamic groundwater conditions; (d) the presence of plastic or steel grouting types in the ground after completion of treatment; and (e) the possibility of water supply contamination due to comparatively large lateral grout travel resulting from high pressures and longish gel times.

Grouted ground strengths of 0.2 to 0.5 MPa are common with LAG, and this system accounts for 20 percent of the Japanese domestic market but a larger proportion of the work executed by their specialist companies elsewhere. The system is protected by at least six patents and one Association.

DOUBLE TUBE DRILLING AND SEEPAGE (DDS) The system is in some ways similar to LAG. It features the rotary insertion of a combined rod casing system (42 mm o.d.) with water flush. At the final depth a small plug is activated by grouting pressure against a retaining spring above the drill bit: this exposes six lateral nozzles through which the fast-setting (10 to 30 sec) grout components are ejected. As in LAG, the grout consists of a mix of silicate plus reagent.

These chemicals are prepared and delivered in separate passages in the drill string, with final mixing occurring only at the nozzles. No rotation is required during extraction. Water flush characteristics of 15 to 25 liters/min at 10 MPa during drilling give a diameter of influence of up to 1 m. Withdrawal rates of around 15 min/m are common, with grouting pressures of up to 1.5 MPa. In the mid 1980s, about 50 percent of the Japanese domestic chemical grouting market featured this system. Again, small hollow head drilling rigs (say up to 30 HP capacity) are adequate, especially as they are rarely expected to drill more than 20 m, and their quiet and vibrationfree operation makes them very popular in urban or underground grouting works. Recent developments in jet grouting have proved severe competition.

It should be noted that there are several other variants of this type, for example the "space grouting rocket system" (SGR) in Japan, where environmental and geotechnical considerations clearly favor this approach. However, their market share is relatively small, and the other systems described above would appear to be of far wider relevance outside that country.

As a final note, the practical aspects of mixing and injection, including equipment reviews, have been described by Deere (1982), Gourlay and Carson (1982), Littlejohn (1985), Houlsby (1990), Karol (1990), AFTES (1991), and Weaver (1991).

Evaluation of Results

During any test program, and during the production works, the drilling and grouting parameters should be carefully monitored and recorded. The significant parameters are described in earlier sections, while the means of actually varying and recording them are somewhat outside the scope of this chapter. The interested reader is referred to the works of Jefferies et al. (1982), Mueller (1982), USBR (1984, 1987), Fairweather (1987), Aberle et al. (1990), Houlsby (1990), AFTES (1991), and Weaver (1991), for detailed guidance in this respect.

However, the evaluation of the efficiency of the grouting works is within our scope, for it is an essential element of the logical "assess—design—build—verify" sequence of good grouting practice. Models of evaluation programs were provided by Baker (1982) and Davidson and Perez (1982).

AFTES (1991) notes there are two complementary methods of evaluation:

- During the grouting, for example, by noting grout interconnections, piezometric variations, ground uplift, and grout injection pressure–volume–time–composition characteristics.
- At the end of an intermediate phase or at the conclusion of the work, for example by testing the treated ground in relation to the desired objective such as permeability, or strength changes. In this regard it must be recalled that many such tests are usually highly local, and may not have the range or resolution to identify small—but potentially significant—defects in the treatment.

Figure 7-50 represents the main "global" test groups.

Geophysical and Related Methods The properties of the ground change during treatment, and several different geophysical principles have been exploited to investigate these variations. Excellent reviews were provided by Baker (1982) and Huck and Waller (1982), who described the following methods:

Acoustic Emission Monitoring of Injection Pressure (AEM) AEM may be used to detect structural distress in geotechnical materials. During grouting it can detect hydraulic fracturing and therefore aid control of this phenomenon. Indications of fracturing are bursts of microseismic noises heard by the system, denoted by increased acoustic emission count rates. Hydraulic fracturing can reduce grouting cost, but the critical initial pressure can vary by a factor of several times even in closely spaced holes. The sensor is placed in an inactive grout pipe at the approximate depth of injection. It can filter out frequencies below 1000 Hz including, therefore, most construction noise. After testing and calibration, the system is placed so that the grouter can see the recorded output. He can then increase the injection pressure to each injection point until fracture begins, and then decrease the pressure to a comfortable safety margin. He can also in theory track the flow of the grout through the foundation soils.

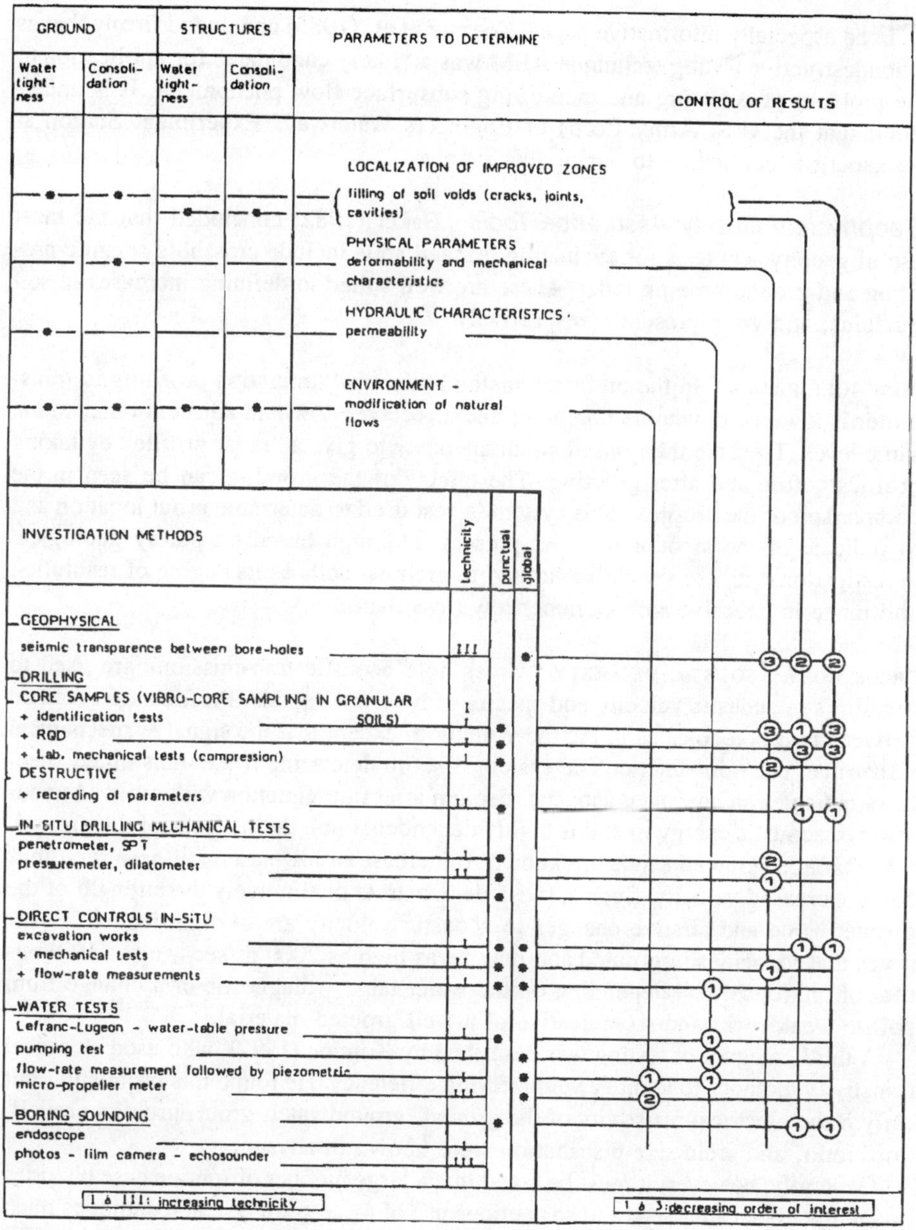

Figure 7-50 Methods of controlling the results of grouting treatment. (From AFTES, 1991.)

In an especially informative paper, Koerner et al. (1985) concluded firmly that as a nondestructive testing technique AEM was a "likely candidate" for application to the problem of detecting and monitoring subsurface flow phenomena. It is understood that the U.S. Army Corps of Engineers Waterways Experiment Station at Vicksburg is continuing to pursue this avenue.

Geophysical Quality Assurance Tools Baker (1982) concluded that the most useful geophysical tests for evaluating grouted soils include crosshole seismic profiling and ground probing radar. These are well suited to defining increases in soil modulus, and grout presence, respectively.

BOREHOLE RADAR In the preferred method of transillumination profiling, a transmitter is lowered down one borehole, and a receiver down an adjacent hole, to the same level. They are then raised simultaneously to give a "radar profile" by taking profiles before and after grouting. The effects of the grouting can be seen in the comparison of the profiles. This system is best used to determine grout location and to indicate of the amount of grout present, although based on purely geological investigations, its use should be limited to granular soils as its degree of resolution and range in cohesive soils is rather low (less than 3 m).

CROSS HOLE ACOUSTIC VELOCITY Cross hole acoustic transmissions are used to measure the acoustic velocity and spectra of received signals. Profiles are obtained between two boreholes as in the radar method, except that the signal is mechanical rather than electromagnetic. The system is set to determine if the transmitted spectrum indicates an improved acoustic medium after impregnation with grout. Attenuation of acoustic energy in soil is highly dependent upon the stiffness of the ground. For example, grouted sands are known to increase in low-strain stiffness, and thus show increased velocity. Such surveys demonstrate qualitatively the strength of the grouted zone and relative changes in acoustic velocity are of significance. Baker notes that velocities in grouted soil may be as high as 2000 m/sec—up to 10 times that of ungrouted soil depending on the water table—diagnostic of a change from soil to weak rock, and so indicative of a well grouted material.

A third category of testing was described by Komine (1992), who used electrical resistivity methods to identify void infilling efficiency. He found this promising, but only if the electrical resistivity of the ground, groundwater, grout, grouted ground, void ratio, and grain size distribution were known in advance.

Generally, however, it must be noted that a large number of routine case histories has not yet been amassed, and so cautious use of these geophysical techniques must currently be exercised. Such methods also typically involve exceptional skill and expertise in execution and analysis and this clearly will have a cost impact.

Direct Methods Involving Drilling

Core Sampling Core sampling of treated soil is often proposed, frequently attempted, but rarely conclusive. In addition to the fact that it is a very localized test, it is most commonly found that the action of drilling fragments the sample, even if the treatment had been efficient in situ.

The best chance of obtaining meaningful samples is in grouted rock and homogenous fine sand, provided that the core barrel is at least 100 mm in diameter and that a triple tube system is used. Endoscopes can be used to compare the nature of the grouted mass with that logged prior to treatment.

Production Grout Hole Drilling This type of Drilling was described in the Construction discussion earlier in Section 7-2. It can also give valuable information, especially if supplemented by an electronic drilling parameter acquisition system, and linked with systematic permeability testing.

SPT Penetrometer Testing This type of testing can be used to demonstrate effectiveness in conditions where similar tests have been conducted before grouting. Care should be taken, of course, when hydrofracture conditions are believed to exist, as misleadingly high values will be obtained when lenses or sheets of hardened grout are encountered.

Pressure Meter Testing This is another good "before and after" local test, giving localized deformation characteristics. In rock masses, special dilatometers, such as the Goodman Jack, are necessary, given the higher operational pressures and the lower radial deformations of the instrument.

Tests on Laboratory Samples Samples can be taken during the work or during excavation of the treated soil. In either case the quality and consistency of the sampling is paramount. The most common test is for unconfined compressive strength, although microscopic and petrographic testing can be done to demonstrate other aspects, and deformability testing is also feasible.

Direct In Situ Observations Inspection and measurements of excavated ground offer the best means of confirming the efficiency of grouting. When pretreating ground for tunneling or shaft sinking this opportunity is always present. In other applications, special provisions such as inspection shafts can be made. Observations should be made of the distribution and travel of the grout, and the homogeneity of the treatment in relation to geological variations.

It should be noted that these observations can be aided by:

- Coloring the grout (e.g., with fluorescein, methylene blue, eosine, rhodamine), if the nature and natural color of the soil allow such observations.
- Applying chemical indicators to the ground (e.g., phenol phthaleine) to highlight the location of the grout. This is especially useful in finer grained soils.

Excavation operations also represent a means of access to the treated zone from which surveys, in situ tests, and sampling for laboratory tests can be carried out.

Other groups of large-scale tests, such as plate loading tests, jacking tests, and shearing tests, can be used to determine deformability and rupture characteristics of the treated ground, but are often prohibitively expensive.

For waterproofing applications, pumpout tests can be performed directly on the excavations and the analyses will be enhanced by readings from astutely located piezometers.

Hydraulic Tests Water testing to investigate in situ permeabilities is an excellent method of demonstrating grouting effectiveness in waterproofing applications (Houlsby, 1990; Weaver, 1991). Such tests are usually Lefranc tests in soils, and Lugeon testing in rock and must be carried out in sufficient numbers to take into account the heterogeneities of the ground, and to detect any defects in the treatment.

As an example, Bruce and Millmore (1983) described a typical curtain grouting program in rock at Kielder dam, Scotland. Pregrouting multipressure Houlsby-type permeability testing was conducted in exploratory boreholes at approximately 40-m centers along the line of the curtain. Each subsequent individual grouting stage was then subjected to a single pressure water test prior to injection. Interposed posttreatment verification holes were then drilled and tested. In such cases, it is common to incline the verification holes within the plane of the curtain to investigate the possibility of untreated vertical fissures.

Piezometric observations are also very useful in providing a global verification of the data yielded by individual water tests in boreholes. Depending on the arrangement of the piezometers, the piezometric level across a curtain can be traced so as to indicate the effective width of the treatment and the prevailing hydraulic gradient (and any subsequent changes thereof).

Chemical or isotopic tracers can be used to investigate flow patterns and rates, while the newer, and sophisticated method of micropropellor meter measurements has recently been introduced by AFTES (1991), who confirm that "particular skill" is required on the part of the operators.

Cost Considerations

In practice, the choice of the grouting method and materials is initially influenced by technical considerations, but is finally dominated by economic concerns. It is extremely difficult and usually misleading to try to summarize the cost of ground treatment works: each project has its own set of determinant factors. However, Littlejohn (1985) produced a "relative cost of materials" table (Table 7-16) as a guide. Going one step further, AFTES (1991) produced a guide for the cost of supplying and using various grouts (Table 7-17). These are very useful works, but must be used with caution, and proper engineering and commercial judgement.

TABLE 7-16 Relative Material Costs of Grout Formulations

Formulation	Relative Cost of Materials
Cement–bentonite	
w/c = 3, 5% bentonite by weight of water	1.0
w/c = 2, 3% bentonite by weight of water	1.3
w/c = 1, 1% bentonite by weight of water	2.3
Cement	
(w/c = 0.5)	3.4
Silicate–bentonite	
20% bentonite, 7% silicate (by weight of water)	1.3
Silicate–chloride (Joosten)	4.0
Silicate–ester	
37% silicate, 4.4% ester (by volume)	5.0
47% silicate, 5.6% ester (by volume)	6.5
Silicate–aluminate	
46% silicate, 1.4% aluminate (by weight)	5.0
Phenol–formaldehyde	
13% (by volume)	10.5
19% (by volume)	15.3
Acrylate	
10% (by weight)	18.5
Resorcinol–formaldehyde	
21% (by volume)	23.0
28% (by volume)	31.0
Polyacrylamide	
5% (by volume)	20.0
10% (by volume)	40.0

Source: From Littlejohn (1985).

Overview

Of all the contemporary methods of ground treatment or support, permeation grouting is probably the most difficult to circumscribe and summarize. It touches upon many complex and evolving branches of science—from organic chemistry to fluid mechanics and from overburden drilling to geophysical surveying. In this chapter, the approach has been one of generic classifications, in order to give the reader both a survey of current practice, and a framework upon which to fix future knowledge and developments.

As noted throughout, the subject has a rich and expanding literature, and it is practically impossible to do full justice to the numerous works of theory, practice, and experiment that abound. The extensive references cited are by no means the only works deserving attention, and may not even be the best in their field. The reader is encouraged to pursue particular avenues: the authors of unwittingly unreferenced works are encouraged to be forgiving.

TABLE 7-17 Price of Supply and Use of Grouts (per m³ of Grout); Conditions are for the Year 1986[a]

Soil to be Grouted	Flow Rate (m^3/hr)	Percentage of Voids (%)	Distance Between Bore Holes (m)	Price of Supply and Use of Grouts (per cu. m of grout): Pure Cement	Clay Cement	Grout with Filler	Water Tightening Gel	Consolidation Gel
Gravels	0.8	30–45	2–4		1.5			
Sands, sands and gravel	0.4	30–45	1–2.5				2.7	3.2
Fine sands	0.2	30–45	0.8–1.5				3.5	4.2
Large voids and karsts	3	Volume of cavities	3–15			0.4		
Large cracks	1	5–20	2–4	1.1	1.0	0.8		
Fine cracks	0.5	1–10	1–3	1.5	1.5		2.3	3
Filling behind lining	1	Variable	2–3		1.4	1.4		
Filling behind lining	0.3	Variable	2–3		2.3			
Renewal of works	0.05	Variable	Variable	12	12			

Source: From AFTES (1991).

[a]As an example, for large sites in the Paris region, the coefficient 1 is valued at between 450 FF and 550 FF as of 1 January 1987 (prices exclusive of VAT). Under the same conditions, the price of drilling and equipment is between 150 FF and 250 FF/ml, or even 300 FF/ml in galleries.

REFERENCES

Aberle, P. P., R. L. Reinhardt, and R. D. Mindenhall, 1990. "Electronic Monitoring of Foundation Grouting on New Waddell Dam," ASCE Annual Conv., Nov. 5–8, San Francisco, Session T13, 18 pp.

Acker, W. L., 1974. "Basic Procedures for Soil Sampling and Core Drilling," Acker Drill Co., Inc. Scranton, Pa., 246 pp.

AFTES, 1991. "Recommendations on Grouting for Underground Works," *Tunnelling and Underground Space Technology,* Vol. 6, No. 4, pp. 383–461.

Albritton, J. A., 1982. "Cement Grouting Practices, U.S. Army Crops of Engineers," Proc. ASCE Conf., Grouting in Geotechnical Engineering, Feb. 10–12, New Orleans, pp. 264–278.

American Concrete Institute (ACI), 1984. "Innovative Cement Grouting," Publication SP-83, Collection of papers presented at 1983 Fall Conv., Kansas City, Mo., 174 pp.

American Concrete Institute (ACI), 1992. "Geotechnical Cement Grouting: State of Practice Report," Technical Committee 552, in finalization.

Andromalos, K. B. and P. J. Pettit, 1986. "Jet Grouting: Snail's Pace of Adoption," *Civ. Eng. ASCE,* Vol. 56, No. 12, pp. 40–43.

Arenzana, L., R. J. Krizek, and S. F. Pepper, 1989. "Injection of Dilute Microfine Cement Suspensions into Fine Sands," Proc. XII ICSMFE, Rio de Janeiro, Vol. 2, pp. 1331–1334.

ASCE Conference, 1982. "Grouting in Geotechnical Engineering," Feb. 10–12, New Orleans, 1017 pp.

ASCE Conference, 1985. "Issues in Dam Grouting," April 30, Denver, 165 pp.

ASCE Geotechnical Special Publication, No. 12, 1987. "Soil Improvement—A Ten Year Update," April 28, Atlantic City, N.J.

ASCE Conference, 1992. "Grouting, Soil Improvement and Geosynthetics," Feb. 25–28, New Orleans, 1451 pp.

ASCE Geotechnical Engineering Division Committee on Grouting, 1980. "Preliminary Glossary of Terms Related to Grouting," *J. Geotech. Eng. Div.*, ASCE, Vol. 196, No. GT7, July, pp. 803–805.

Atlas Copco, 1976, London, England. Trade literature.

Baker, W. H., 1982. "Planning and Performing Structural Chemical Grouting," Proc. ASCE Conf., Grouting in Geotechnical Engineering, Feb. 10–12, New Orleans, pp. 515–539.

Baker, W. J., E. J. Cording, and H. H. MacPherson, 1983. "Compaction Grouting to Control Ground Movements during Tunnelling," *Underground Space*, Vol. 7, pp. 205–212.

Ballivy, G., K. Saleh, T. Mnif, J. Maniez, L. M. Landry, and M. Nadeau, 1992. "Rehabilitation of Concrete Dams: Laboratory Simulation of Cracking and Injectability," Proc. ASCE Conf., Grouting, Soil Improvement and Geosynthetics, Feb. 25–28, New Orleans, Vol. 1, pp. 614–625.

Bell, L. A., 1978. "Alluvial Grouting," M.Sc. Dissertation, University of Durham, England.

Bell, L. A., 1982. "A Cutoff in Rock and Alluvium at Asprokremmos Dam," Proc. ASCE Conf., Grouting in Geotechnical Engineering, Feb. 10–12, New Orleans, pp. 172–186.

Black, J. C., A. Pollard, and G. P. Daw, 1982. "Hydrogeological Assessment and Grouting at Selby," Proc. ASCE Conf., Grouting in Geotechnical Engineering, Feb. 10–12, New Orleans, pp. 665–679.

Blacklock, J. R., R. C. Joshi, and P. J. Wright, 1982. "Pressure Injection Grouting of Landfills Using Lime and Fly Ash," Proc. ASCE Conf., Grouting in Geotechnical Engineering, Feb. 10–12, New Orleans, pp. 708–721.

Bruce, D. A., 1982. "Aspects of Rock Grouting Practice on British Dams," Proc. ASCE Conf., Grouting in Geotechnical Engineering, Feb. 10–12, New Orleans, pp. 301–316.

Bruce, D. A., 1988. "Developments in Geotechnical Construction Processes for Urban Engineering," *Civ. Eng. Practice,* Vol. 3, No. 1, Spring, pp. 49–97.

Bruce, D. A., 1988–1989. "Aspects of Minipiling Practice in the United States," *Ground Engineering,* Vol. 21, No. 8, pp. 20–33, and Vol. 22, No. 1, pp. 35–39.

Bruce, D. A., 1989a. "Contemporary Practice in Geotechnical Drilling and Grouting," Keynote Lecture, lst Canadian Int. Grouting Seminar, April 18, Toronto, 28 pp.

Bruce, D. A., 1989b. "Methods of Overburden Drilling in Geotechnical Construction—A Generic Classification," *Ground Engineering,* Vol. 22, No. 7, pp. 25–32. Also published in "Drill Bits—The Official Publication of the International Drilling Federation," Fall 1989, pp. 7, 8, 10, 11, 13, 14.

Bruce, D. A., 1990a. "Major Dam Rehabilitation by Specialist Geotechnical Construction Techniques: A State of Practice Review," Proc. Canadian Dam Safety Association 2nd Annual Conf., Sept. 18–20, Toronto, 63 pp. Also reprinted as Vol. 57 of the Institute for Engineering Research, Foundation Kollbrunner-Rodio, Zurich, Sept.

Bruce, D. A., 1990b. "The Practice and Potential of Grouting in Major Dam Rehabilitation," ASCE Annual Civil Engineering Conv. Nov. 5–8, San Francisco, Session T13, 41 pp.

Bruce, D. A., 1991. "The Construction and Performance of Prestressed Ground Anchors in Soils and Weak Rocks: A Personal Overview," Proc. 16th Annual Meeting, DFI, Oct. 7–9, Chicago, 20 pp.

Bruce, D. A. and R. H. Bianchi, 1992. "The Use of Posttensioned Tendons on Stewart Mountain Dam, Arizona: A Case Study Involving Precision Drilling," 2nd Interagency Symp. on Stabilization of Soils and Other Materials, Nov. 2–5, Metalrie, La., 15 pp.

Bruce, D. A. and J. E. Croxall, 1989. "The MPSP Grouting System: A New Application for Raise Boring," Proc. 2nd Int. Conf. on Foundations and Tunnels, London, Sept. 19–21, pp. 331–340.

Bruce, D. A. and P. DePorcellinis, 1991. "Sealing Cracks, in Concrete Dams to Provide Structural Stability," *Hydro Review,* Vol. 10, No. 4, pp. 116–124.

Bruce, D. A. and F. Gallavresi, 1988. "The MPSP System: A New Method of Grouting Difficult Rock Formations," ASCE Geotechnical Special Publication No. 14, "Geotechnical Aspects of Karst Terrains," pp. 97–114. Presented at ASCE Nat. Conv. May 10–11, Nashville, Tenn.

Bruce, D. A. and C. R. F. George, 1982. "Rock Grouting at Wimbleball Dam," *Geotechnique,* Vol. 23, No. 4, 14 pp.

Bruce, D. A. and G. M. Joyce, 1983. "Slabjacking at Tarbela Dam, Pakistan." *Ground Engineering,* Vol. 16, No. 3, pp. 35–39.

Bruce, D. A. and F. Kord, 1991. "A First for Kidd Creek," *Canadian Mining Journal,* Vol. 112, No. 7, Sept.–Oct., pp. 57, 59, 62, and 65.

Bruce, D. A. and J. P. Millmore, 1983. "Rock Grouting and Water Testing at Kielder Dam," *Quart. J. Engineering Geology,* Vol. 16, No. 1, pp. 13–29.

Bruce, D. A. and J. N. Shirlaw, 1985. "Grouting of Completely Weathered Granite with Special Reference to the Construction of the Hong Kong Mass Transit Railway," 4th Int. Symp., Tunneling 85, Mar. 10–15, Brighton, pp. 253–264.

Bruce, D. A. and C. K. Yeung, 1983. "Minipiling at Hong Kong Country Club." *Hong Kong Contractor,* Nov., pp. 13–18.

Cambefort, H., 1967. *Injection des Sols,* Editions Eyrolles, Paris, 2 vols.

Cambefort, H., 1977. "The Principles and Applications of Grouting," *Quart. J. Engineering Geology,* Vol. 10, No. 2, pp. 57–95.

Caron, C., 1965. "Physico-chemical Study of Silicagels, Ann-ITBTP-Essais Mesures, 81:447–484," Mar./Apr.

Caron, C., 1982. Background Talk: "The State of Grouting in the 1980's," Proc. ASCE Conf., Grouting in Geotechnical Engineering, Feb. 10–12, New Orleans, pp. 346–358.

Clarke, W. J., 1982. "Performance Characteristics of Acrylate Polymer Grout," Proc. ASCE Conf., Grouting in Geotechnical Engineering, Feb. 10–12, New Orleans, pp. 418–432.

Clarke, W. J., 1984. "Performance Characteristics of Microfine Cement," Preprint 84-023, ASCE Geotechnical Conf., May 14–18, Atlanta, Ga., 14 pp.

Clarke, W. J., 1987. "Microfine Cement Technology," 23rd Int. Cement Seminar, Dec. 6–9, Atlanta, Ga., 13 pp.

Clarke, W. J., M. D. Boyd, and M. Helal, 1992. "Ultrafine Cement Tests and Dam Test Grouting," Proc. ASCE Conf., Grouting, Soil Improvement and Geosynthetics, New Orleans, Vol. 1, Feb. 25–28, pp. 626–638.

Davidson, R. and J. Perez, 1982. "Properties of Chemically Grouted Sand at Locks and Dam No. 26," Proc. ASCE Conf., Grouting in Geotechnical Engineering, Feb. 10–12, New Orleans, pp. 433–449.

Deere, D. U., 1982. "Cement Bentonite Grouting for Dams," Proc. ASCE Conf., Grouting in Geotechnical Engineering, Feb. 10–12, New Orleans, pp. 279–300.

Deere, D. U. and G. Lombardi, 1985. "Grout Slurries—Thick or Thin?" Proc. ASCE Conf., Issues in Dam Grouting. April 30, Denver, pp. 156–164.

DePaoli, B., G. Viola, and A. Tomiolo, 1987. "The Use of Drilling Energy for Soil Classification," 2nd Int. Symp., FMGM87, Apr. 6–9, Kobe, Japan.

DePaoli, B., B. Bosco, R. Granata, and D. A. Bruce, 1992a. "Fundamental Observations on Cement Based Grouts (2): Microfine Cements and the Cemill[R] Process," Proc. ASCE Conf., Grouting, Soil Improvement and Geosynthetics, Feb. 25–28, New Orleans, Vol. 1, pp. 486–499.

DePaoli, B., B. Bosco, R. Granata, and D. A. Bruce, 1992b. "Fundamental Observations on Cement Based Grouts (1): Traditional Materials," Proc. ASCE Conf., Grouting, Soil Improvement and Geosynthetics, Feb. 25–28, New Orleans, Vol. 1, pp. 474–485.

Essler, R. D. and L. F. Linney, 1992. "Compensation Grouting Trial Works at Redcross Way, London," Proc. ICE Grouting in the Ground Conf. Nov. 25–26, London, 14 pp.

Ewart, F., 1985. *Rock Grouting,* Springer-Verlag, New York, 428 pp.

Fairweather, V., 1987. "Milan's Model Metro," *Civ. Eng.* Dec., pp. 40–43.

FHWA, 1976. "Grouting in Soils, Report No. FHWA-RD-76-26," prepared by Halliburton Services, June, 2 vols.

Gallavresi, F., 1992. "Grouting Improvement of Foundation Soils," Proc. ASCE Conf., Grouting, Soil Improvement and Geosynthetics, Feb. 25–28, New Orleans, Vol. 1, pp. 1–38.

Garcia-Bengochea, I., C. W. Lovell, and A. G. Altschaeffli, 1978. "Pore Distribution and Permeability of Silty Clays," *Proc. ASCE Geotech. Eng. Div.* Vol. 105, No. GT7).

Glossop, R., 1961. "The Invention and Development of Injection Processes, Part II: 1850–1960," *Geotechnique,* Vol. XI, No. 4.

Gourlay, A. W. and C. S. Carson, 1982. "Grouting Plant and Equipment," Proc. ASCE Conf., Grouting in Geotechnical Engineering, Feb. 10–12, New Orleans, pp. 33–48.

Graf, E. D., 1992. "Compaction Grout," Proc. ASCE Conf., Grouting, Soil Improvement and Geosynthetics, Feb. 25–28, New Orleans, Vol. 1, pp. 275–287.

Graf, T. E., G. W. Clough, and J. Warner, 1982. "Long-Term Aging Effects on Chemically Stabilized Soils," Proc. ASCE Conf., Grouting in Geotechnical Engineering, Feb. 10–12, New Orleans, pp. 470–481.

Graf, E. D., D. J. Rhoades, and K. L. Fa ıght, 1985. "Chemical Grout Curtains at Ox Mountain Dam," Proc. ASCE Conf., Issues in Dam Grouting, April 30, New York, pp. 99–103.

Greenwood, D. A., M. T. Hutchinson, and J. Rooke, 1987. "Chemical Injection to Stabilize Worker Logged Sand During Tunnel Construction for Cairo Wastewater Project." Proc. European Conf., Soil Mech. & Found Engg., Dublin, Eire, September, 8 pp.

Hakansson, U., L. Hassler, and H. Stille, 1992. "Rheological Properties of Microfine Cement Grouts with Additives," Proc. ASCE Conf., Grouting, Soil Improvement and Geosynthetics, Feb. 25–28, New Orleans, pp. 551–563.

Houlsby, A. C., 1982. "Cement Grouting for Dams," Proc. ASCE Conf., Grouting in Geotechnical Engineering, Feb. 10–12, New Orleans, pp. 1–34.

Houlsby, A. C., 1990. *Construction and Design of Cement Grouting,* Wiley, New York, 442 pp.

Huck, P. J. and M. J. Waller, 1982, "Grout Monitoring and Control," Proc. ASCE Conf., Grouting in Geotechnical Engineering, Feb. 10–12, New Orleans, pp. 781–791.

Hutte & Co., 1984, Munich, Germany. Trade literature.

Imrie, A. S., W. F. Marcusson, and P. M. Byrne, 1988. "Seismic Cutoff," *Civ. Eng.* ASCE, Vol. 58, No. 12, pp. 50–53.

Institute of Civil Engineers (ICE), 1963. "Grouts and Drilling Muds in Engineering Practice," Butterworths, England.

Institute of Civil Engineers (ICE), 1992. "Grouting in the Ground," Nov. 25–26. London.

James, A. N., 1963. "Discussion to Session 3—Grouting Symposium on Grouts and Drilling Muds in Engineering Practice," Butterworths, London, pp. 168–169.

Jefferies, M. G., B. T. Rogers, and D. W. Reades, 1982. "Electronic Monitoring of Grouting," Proc. ASCE Conf., Grouting in Geotechnical Engineering, Feb. 10–12, New Orleans, pp. 769–780.

Jefferis, S. A., 1982. "Effects of Mixing on Bentonite Slurries and Grouts," Proc. ASCE Conf., Grouting in Geotechnical Engineering, Feb. 10–12, New Orleans, pp. 62–76.

Jiacai, L., W. Baochang, C. Wengguang, G. Yuhua, and C. Hesheng, 1982. "Polyurethane Grouting in Hydraulic Engineering," Proc. ASCE Conf., Grouting in Geotechnical Engineering, Feb. 10–12, New Orleans, pp. 403–417.

Karol, R. H., 1982. "Chemical Grouts and Their Properties," Proc. ASCE Conf., Grouting in Geotechnical Engineering, Feb. 10–12, New Orleans, pp. 359–377.

Karol, R. H., 1983. *Chemical Grouting*. Marcel Dekker, New York, 327 pp.

Karol, R. H., 1990. *Chemical Grouting*, revised ed., Marcel Dekker, New York, 455 pp.

Koerner, R. M., R. M. Sands, and J. D. Leaird, 1985. "Acoustic Emission Monitoring of Grout Movement," Proc. ASCE Conf., Issues in Dam Grouting, April 30, Denver, pp. 149–155.

Komine, H., 1992. "Estimation of Chemical Grout Void Filling by Electrical Resistivity," Proc. ASCE Conf. Grouting, Soil Improvement and Geosynthetics, Feb. 25–28, New Orleans, pp. 372–383.

Krizek, R. J., H. Liao, and R. H. Borden, 1992. "Mechanical Properties of Microfine Cement/Sodium Silicate Grouted Sand ," Proc. ASCE Conf., Grouting, Soil Improvement and Geosynthetics, Feb. 25–28, New Orleans, pp. 688–699.

Krupp, 1983, Munich, Germany. Trade literature.

Lamberton, B. A., 1982. "Swedish Valve Tube Grouting," Proc. ASCE Conf., Grouting in Geotechnical Engineering, Feb. 10–12, New Orleans, pp. 907–922.

Leonard, G. K. and L. F. Grant, 1958. "Experience of TVA with Clay-Cement and Related Grouts," *ASCE J. Soil. Mech. Found. Div.* Vol. 84, No. SM1, Paper 1552.

Liao, H., R. J. Krizek, and R. H. Borden, 1992. "Microfine Cement/Sodium Silicate Grout," Proc. ASCE Conf., Grouting, Soil Improvement and Geosynthetics, Feb. 25–28, New Orleans, pp. 676–687.

Littlejohn, G. S., 1975. "Acceptable Water Flows for Rock Anchor Grouting," *Ground Engineering*, Vol. 8, No. 2, pp. 46–48.

Littlejohn, G. S., 1982. "Design of Cement Based Grouts," Proc. ASCE Conf., Grouting in Geotechnical Engineering, Feb. 10–12, New Orleans, pp. 35–48.

Littlejohn, G. S., 1985. "Chemical Grouting," *Ground Engineering*, Vol. 18, No. 2, pp. 13–16; No. 3, pp. 23–28; No. 4, pp. 29–34.

Littlejohn, G. S., and D. A. Bruce, 1977. "Rock Anchors—State of the Art," Foundation Publications, Essex, England, 50 pp.

Littlejohn, G. S., R. L. Newman, and C. T. Kettle, 1989. "Grouting to Control Groundwater During Basement Construction at St. Helier," *Ground Engineering*, Vol. 22, No. 1, pp. 22–31.

Lombardi, G., 1985. "The Role of Cohesion in Cement Grouting of Rock," 15th ICOLD, Lausanne, Q58, R13.

Lukajic, B., G. Smith, and J. Deans, 1985. "Use of Asphalt in Treatment of Dam Foundation Leakage, Stewartville Dam." Proc. ASCE Conf., Issues in Dam Grouting, April 30, Denver, pp. 76–91.

Maag, E., 1938. "Ueber die Vergestigung und Dichtung das Baugrundes (Injektionen)," Erdbaukers der ETH.

McGregor, K., 1967. *The Drilling of Rock*, Inst. ed. C.R. Books, Ltd., London, 306 pp.

Miki, G. and W. Nakanishi, 1984. "Technical Progress of the Jet Grouting Method and Its Newest Type," Proc. Int. Conf., In Situ Soil and Rock Reinforcement, Oct. 9–11, Paris, pp. 195–200.

Mitchell, J. K., 1981. "Soil Improvement—State of the Art Report," Proc. X ICSMFE, Stockholm, Vol. 4, pp. 509–565. 1.

Mongilardi, E. and R. Tornaghi, 1986. "Construction of Large Underground Openings and Use of Grouts," Proc. Int. Conf. on Deep Foundations, Sept., Beijing, 19 pp.

Mueller, R. E., 1982. "Multiple Hole Grouting Method," Proc. ASCE Conf. Grouting in Geotechnical Engineering, Feb. 10–12, New Orleans, pp. 792–808.

Naudts, A. M. C., 1989. "The Various Grouts and Their Applications," 1st Canadian Int. Grouting Seminar, April 18, Toronto, 25 pp.

Patey, D. R., 1977. "Grouting Old Mine Workings at Merthyr," *Ground Engineering*, Vol. 10, No. 8, pp. 24–27.

Petrovsky, M.B., 1982. "Monitoring of Grout Leaching at Three Dam Curtains in Crystalling Rock Foundations," Proc. ASCE Conf., Grouting in Geotechnical Engineering, Feb. 10–12, New Orleans, pp. 105–120.

Pototschnik, M. J., 1992. "Settlement Reduction by Soil Fracture Grouting," Proc. ASCE Conf., Grouting in Geotechnical Engineering, Feb. 10–12, New Orleans, pp. 398–409.

Raffle, J. F. and D. A. Greenwood, 1961. "The Relationship Between the Rheological Characteristics of Grouts and Their Capacity to Permeate Soils," Proc. 5th Int. Conf. on Soil Mechanics and Foundation Engineering, London, Vol. 2, pp. 789–793.

Rands, K. 1915. "Grouted Cut-Off for Estacada Dam," *ASCE Trans.*, Vol. 78, pp. 447–482. A discussion of the paper follows on pp. 483–546.

Reifsnyder, R. H. and J. F. Peters, 1989. "Sodium Silicate Grouts: The Solution to Difficult Subsidence Problems," Proc. Symp. on Evolution of Abandoned Mine Land Technologies, Riveston, Wyo., June 14–16, 6 pp.

Rowlands, D., 1971. "Some Basic Aspects of Diamond Drilling," Proc. 1st Australia New Zealand Conf. on Geomechanics, Melbourne.

Salley, J. R., B. Foreman, W. H. Baker, and J. F. Henry, 1987. "Compaction Grouting Test Program Pinopolis West Dam," Proc. ASCE Conv., Atlantic City, Special Publication 12, April 28, pp. 245–269.

Schwarz, L. G. and R. J. Krizek, 1992. "Effects of Mixing on Rheological Properties of Microfine Cement Grout," Proc. ASCE Conf., Grouting, Soil Improvement and Geosynthetics, Feb. 25–28, New Orleans, pp. 512–525.

Scott, R. A., 1975. "Fundamental Conditions Governing the Penetration of Grouts," *Methods of Treatment of Unstable Ground,* F. G. Bell, Ed., Newnes-Butterworths, London.

Shimada, S., M. Ide, and H. Iwasa, 1992. "Development of a Gas-Liquid Reaction Injection System," Proc. ASCE Conf., Grouting, Soil Improvement and Geosynthetics, Feb. 25–28, New Orleans, pp. 325–336.

Shimoda, M. and H. Ohmori, 1982. "Ultrafine Grouting Material," Proc. ASCE Conf., Grouting in Geotechnical Engineering, Feb. 10–12, New Orleans, pp. 77–91.

Shirlaw, J. N., 1987. "The Choice of Grouts for Hand-Dug Caisson Construction," *Hong Kong Engineer,* February, pp. 11–22.

Simonds, A. W., 1947. "Contraction Joint Grouting of Large Dams," *Proc. of the ACI,* Vol. 43, pp. 637–652.

Simonds, A. W., 1958a. "Present Status of Pressure Grouting Foundations," *ASCE J. Soil Mech. Found. Div.* Vol. 84, No. SM1, Paper 1544.

Simonds, A. W., 1958b. "Cement and Clay Grouting of Foundations, Present Status of Pressure Grouting Foundations," *J. of the Soil Mechanics & Foundations Division, ASCE,* No. SM1, Feb., p. 1544-1–11.

Skipp, B. O. and L. Renner, 1963. "The Improvement of the Mechanical Properties of Sands," Symp. on Grouts and Drilling Muds in Engineering Practice, Butterworths, London, pp. 29–35.

Thorburn, S. and J. F. Hutchison, Eds. 1983. *Underpinning*, Surrey Univ. Press, London, 296 pp.

Tokoro, T., S. Kashima, and M. Murata, 1982. "Grouting Method by Using Flash-Setting Grout," Proc. ASCE Conf., Grouting in Geotechnical Engineering, Feb. 10–12, New Orleans, pp. 738–752.

Tornaghi, R., B. Bosco, and B. DePaoli, 1988. "Application of Recently Developed Grouting Procedures for Tunnelling in Milan Urban Area," Proc. 5th Int. Symp. Tunnelling '88, London, April 18–21, 11 pp.

Tosca, S. Z. and J. C. Evans, 1992. "The Effects of Fillers and Admixtures on Grout Performance," Proc. ASCE Conf., Grouting, Soil Improvement and Geosynthetics, Feb. 25–28, New Orleans, pp. 337–349.

U.S. Bureau of Reclamation (USBR), 1984. "Policy Statements for Grouting," ACER Technical Memorandum No. 5, Sept., 65 pp.

U.S. Bureau of Reclamation (USBR), 1987. "Cement Grout Flow Behavior in Fractured Rocks," Report REC-ERC-87-7, June, 51 pp.

U.S. Corps of Engineers (USCE), 1956. "Pressure Grouting Fine Fissures," Tech. Report No. 6-437, October, W.E.S. Vicksburg, Miss.

Warner, J. F., 1982. "Compaction Grouting—The First Thirty Years," Proc. ASCE Conf., Grouting in Geotechnical Engineering, Feb. 10–12, New Orleans, pp. 694–707.

Warner, J. F., 1992. "Compaction Grout: Rheology vs. Effectiveness," Proc. ASCE Conf., Grouting, Soil Improvement and Geosynthetics, Feb. 25–28, New Orleans, pp. 229–239.

Warner, J. F., N. Schmidt, J. Reed, D. Shepardson, I. Lamb, and S. Wong, 1992. "Recent Advances in Compaction Grouting Technology," Proc. ASCE Conf., Grouting, Soil Improvement and Geosynthetics, Feb. 25–28, New Orleans, pp. 252–264.

Weaver, K. D., 1989. "Consolidation Grouting Operations for Kirkwood Penstock," Proc. ASCE Conf., Evanston, Ill., 2 Vols., June 25–29, pp. 342–353.

Weaver, K. D., 1991. *Dam Foundation Grouting*, ASCE Publications, 178 pp.

Welsh, J. P., (1988). "Sinkhole Rectification by Compaction Grouting," Proc. ASCE Conv., Nashville, May 9–13, and published in ASCE Geotechnical Special Pub. No. 14, pp. 115–132.

Zuomei, Z. and H. Pinshou, 1982. "Grouting of the Karstic Caves with Clay Fillings," Proc. ASCE Conf., Grouting, Soil Improvement and Geosynthetics, Feb. 25–28, New Orleans, pp. 92–104.

CHAPTER 8

JET GROUTING

Since jet grouting is still a relatively new technique, technical papers tend largely to be of the case history type, but with introductory sections summarizing the generalities of evolution, methodologies, advantages, applications, and so on. Much of the data are common to most papers and it would be repetitive to cite specific references for all such information. However, there are three particular sources that have proved of basic value, and from which much of the following is derived: Bruce (1988), Kauschinger and Welsh (1989), and Gallavresi (1992). The author acknowledges the permissions by Dr. Gallavresi, Prof. Kauschinger, Mr. Welsh, and their co-workers to recount so much of their writings in the following sections.

8-1 HISTORICAL DEVELOPMENT

Previous research and experience with high-pressure water cutting, for example in American coal mines, and conceptual developments in Britain (Nicholson, 1963) were seized upon by Japanese specialists in the mid-1960s. The original developments and studies using these principles to not only cut and erode, but to cement, soils were conducted about 1965 by the Yamakado brothers (Miki and Nikanishi, 1984). By early 1970, two competing forms of jet grouting had been developed (Figures 8-1 and 8-2 and Table 8-1).

The Chemical Churning Pile (CCP) method originally developed by Nakanishi and co-workers (Miki, 1973; Nakanishi, 1974) used chemical grouts as the jetting medium. These were ejected through nozzles 1.2 to 2.0 mm in diameter, located at the bottom of a simple drill rod that was rotated during injection. However, due to environmental concerns, chemical grouts were soon replaced by cement-based

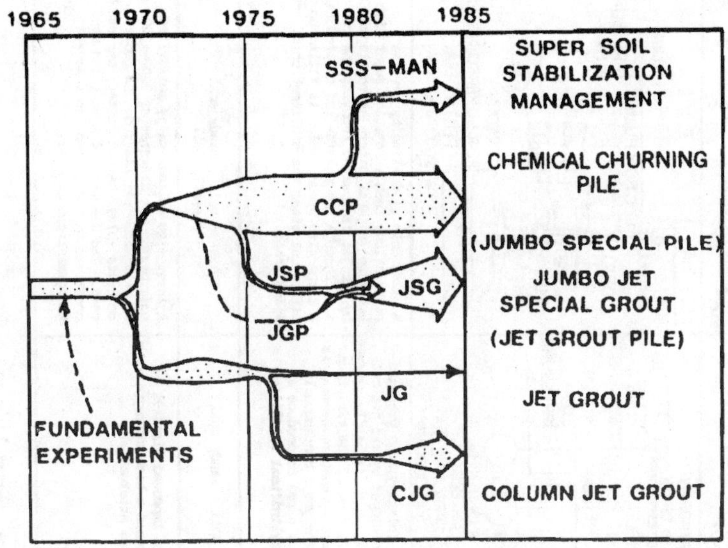

Figure 8-1 Development of jet grouting methods in Japan. (From Miki and Nakanishi, 1984.)

grouts, although for proprietary reasons the name CCP was preserved. In 1973, an Italian contractor became the first to form CCP columns using "ultra high pressures" (35 MPa) and high flow rates (100 to 250 liters/min), through two slightly larger nozzles (up to 2.4-mm diameter). This system is still widely used, for example, in South America (Guatteri et al., 1988).

By 1972, the CCP group in Japan had also developed the "Jumbo Special Pile" (JSP) method using compressed air as an envelope around the grout jet to give column diameters of 80 to 200 cm. Meanwhile, a "Jet Grout Pile" (JGP) method was being simultaneously developed by another independent group, and JSP and JGP merged around 1980 into the "Jumbo Jet Special Grout" (JSG) method.

The major rival group, headed by Yahiro (Yahiro and Yoshida, 1973, 1974; Yahiro et al., 1975), had also developed in 1970 the "Jet Grout" (JG) method, in which a horizontal high-speed water jet was used monodirectionally to excavate a panel, filled progressively from below with grout. By 1975, however, they had also adopted the idea of rotating the rods during withdrawal, giving birth to the "Column Jet Grout" (CJG) method, which also featured the use of compressed air to focus the upper, water jet.

What Kauschinger and Welsh (1989) refer to as the "last important advance" in jet grouting methodology was pioneered in 1980 by the CCP group in Japan, and first described in 1984 by Miki and Nakanishi. Responding to a request to make a deep, completely watertight wall, they developed the "Super Soil Stabilization Management Method" (SSS-MAN) (Figure 8-3). First, a pilot hole is drilled by reverse circulation, and then a rotating "super high pressure" air-enhanced water jet is lowered to excavate the soil to a greater diameter. The void is then surveyed with

Method	CCP : Mixing method using a high speed grout jet. ≡ F1	JSG : Mixing method using a high speed grout jet enveloped by air jet. ≡ F2	CJG : Half-replacement meyhod using a high speed water jet enveloped by air jet. ≡ F3	SSS-MAN : Reverse type replacement method using a high speed water jet enveloped by air jet.
Schematic Diagram	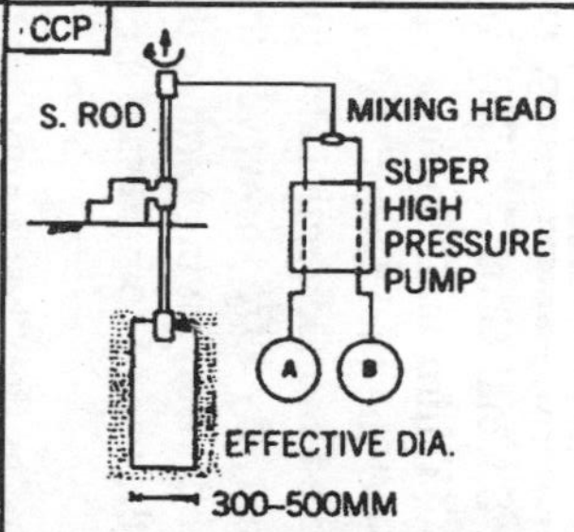	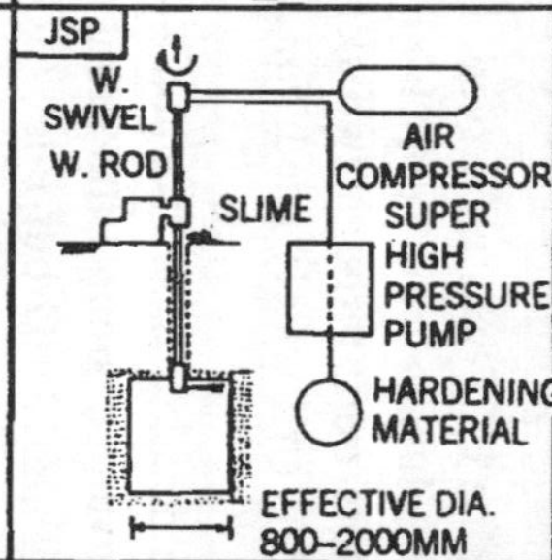	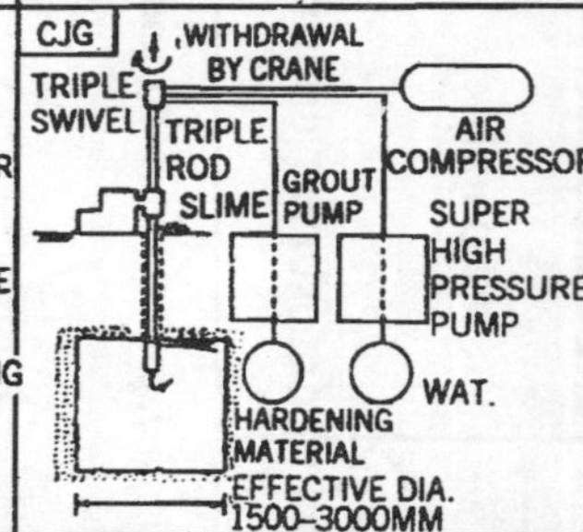	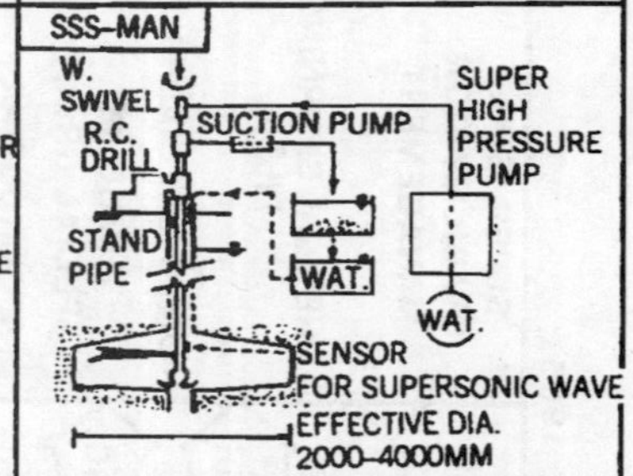
Outline of Method	Uplift of a rotating horizontal griut jet with a super high-pressure (20 MPa) mixes in situ soil with the grout and produces a cylindrically solidified body.	Uplift of a rotating high-speed horizontal grout jet enveloped by air jet mixes in situ soil with the grout and produces a large-size cylindrical solidified body.	Uplift of a rotating high pressure water jet enveloped by air jet cut in situ soil, which is halfly removed by the uplift flow of the air, then a space formed by that removement is mixed with the grout continuously supplied from a rod to produce a large-size cylindrical solidified body.	A pilot hole is drilled by reverse circulation, then a rotating super high pressure (60 MPa) water jet enveloped by air jet is lowered removing the cut soil through the reverse rod to produce a large-size space, which is filled up with grout after the confirmation of the space shape.
Fit Soil	cohesive soil (N<5) , sandy soil (N<15)	cohesive soil , sandy soil gravelly soil	cohesive soil , sandy soil	cohesive soil , sandy soil gravelly soil
Improvement efficiency	An improved diameter of 300-500 mm and a high strength with uniformity are attainable.	A large size improved diameter of 800-2000 mm and a high strength with uniformity are attainable.	A large -size improved diameter of 1500-3000 mm and a high strength with uniformity are attainable.	A large-size cylindrically solidified body having the diameter of 2000-4000 mm is attainable per one rod. As grout, concrete, clay, cement mortar, etc can be used. The diameter of a body can be confirmed on the ground, so a severe management can be done.

Figure 8-2 Illustrations of jet grouting methods used in Japan. (From Miki and Nakanishi, 1984.)

TABLE 8-1 Major Categories of Jet Grouting Methods

Original Japanese Name	Principle of Operation	Jetting Pressure (MPa)	Jetting Nozzle Diameter (mm)	Rotation Rate (rpm)	Anticipated Column Diameter (cm)	Notes
Jet Grout (JG)	Upper water and lower grout jet	20	?	None	—	Panels only, soon obsolete
Chemical Churning Pile (CCP)	Single grout jet	20–40	1.2–3.0	20	30–60	1. Chemicals now replaced by cement 2. Similar to Rodinjet 1 (Fl)
Jumbo Special Grout (JSG)	Single jet of grout enveloped in air	20	3–3.2	6	80–200	1. Originally called "jumbo special pile" (JSP) but name changed for patent reasons 2. Similar to Rodinjet 2 (F2)
Column Jet Grout (CJG)	Upper water and air jet and lower grout jet	40–50	1.8–3.0 (upper) 3.0–5.0 (lower) (8–9 mm in Kajima system)	5	150–300	1. Referred to as "half replacement" 2. Similar to Rodinjet 3 or Kajima/GKN Keller system (F3)
Mini Max (MM)	Like CCP but uses special "chemicolime" cement	20	1.2	20	80--160	Specially for very weak soil and organics (e.g., soft peaty clays under water)
Jumbo Mini Max (JMM)	As for MM except for addition of 20–40 cm wing jet	20	1.2	20	200	Specially for very weak soil and organics (e.g., soft peaty clays under water)
Super Soil Stabilization Management (SSS-Man)	Air water jet used to excavate volume completely underwater; this is then surveyed ultrasonically; if OK, then tremied full of desired material	20–60	2–2.8	3–7	200–400	1. To provide absolute control over shape and composition of column 2. Effective to over 70-m depth 3. "Complete replacement" 4. Most expensive technique, but ensures desired performance

Source: From Bruce (1988).

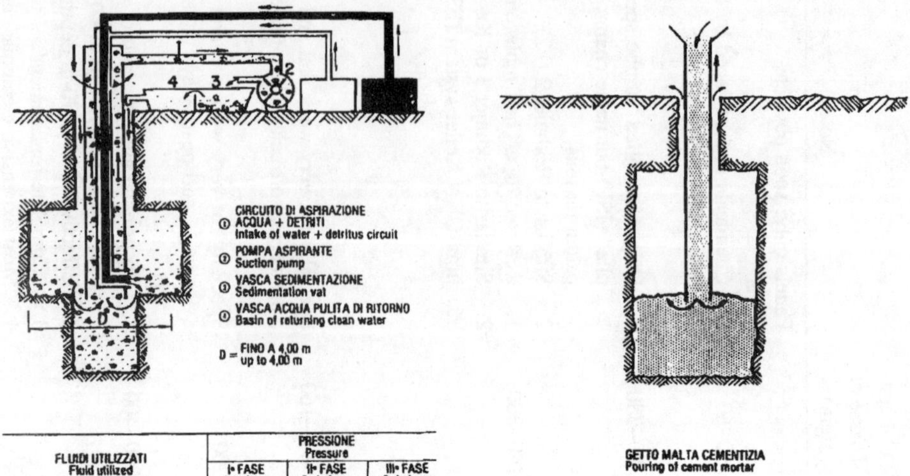

FLUIDI UTILIZZATI Fluid utilized	PRESSIONE Pressure		
	I• FASE 1st phase	II• FASE 2nd phase	III• FASE 3rd phase
ACQUA Water	20 MPa 200 BAR	40 MPa 400 BAR	60 MPa 600 BAR
ARIA COMPRESSA Compressed air		(7) ?	

Figure 8-3 Details of SSS-MAN method. (From Trevi, 1990.)

a sonic transducer (MASU—Mitsui Automatic Scanner of Underground) and further jetting conducted if required. The water-filled space is then backfilled by tremie placement of the selected grout or concrete.

By the late 1970s, most of the basic techniques were sought, under license, by groups of geotechnical contractors throughout the world, but initially and primarily in Germany, Italy, France, Singapore, and Brazil. As discussed in Section 8.3., this scope has widened considerably over the last decade to the extent that applications have been recorded throughout the world, on every continent.

In North America, the idea was first promoted in 1979, although by 1984 only a handful of small projects had been completed using either one-fluid or three-fluid systems. The slow acceptance of the method (Andromalos and Pettit, 1986) reflected a number of drawbacks including risk/legal concerns such as are associated with any novel method; inappropriate applications; initial technical problems leading to poor performance; and simply, a lack of commercial demand for the technique. In the last few years, however, a limited number of specialty contractors have conducted numerous, effective works (Burke et al., 1989) but typically and conservatively only for one purpose (structural underpinning) in one soil category (sands and gravels).

Despite all the ongoing developments and refinements, there are basically only three methods that are being used, and these may be generically classified as follows:

- *F1—One-Fluid System:* The fluid is grout, and in this system the jet simultaneously erodes and injects. It involves only partial replacement of the soil.

- *F2—Two-Fluid System:* This method uses a cement jet inside a compressed air cone. F2 gives a larger column diameter than F1 and gives a higher degree of soil replacement.
- *F3—Three-Fluid System:* Here, an upper ejection of water inside an air envelope is used for excavation, with a lower jet emitting grout for replacement of jetted soil.

These generic classes are now described in detail. Overall, on an international basis, the simplicity and versatility of the F1 method ensure that it is still the most common, although there are considerable national variations depending on the application of the jet grouting.

8-2 GENERAL OPERATIONAL FEATURES OF THE THREE GENERIC METHODS

Table 8-2 provides a general summary of operational parameters and grouted soil strengths. However, it must be noted that these data represent a typical range, and the details are dictated on any one project by a number of variables. For the F1 system, six jetting parameters must be determined: grout pressure, number and size of nozzles, w/c ratio, drill rod lifting rate, and rotational speed. For F2, there are two additional variables (air pressure and delivery rate), while for F3 there are ten parameters (the seven for F2 plus water pressure and water jet nozzle size and number).

F1 is the simplest and most straightforward method. The grout jet cuts, removes, and mixes the soil and so provides essentially a mix-in-place effect. In F2 and F3, the air also aids evacuation of the debris, but as the jet is rotated from vertical to horizontal, the air lifting efficiency decreases. Therefore, F1 is used almost exclusively for horizontal jet grouting. As noted by Kauschinger et al. (1992b), field data indicate that there is "significant" compaction of soil outside the perimeter of the jet grout column for a distance of about half a column diameter. As shown in Table 8-2, these column diameters are typically 40 to 60 cm (cohesive soils) and 50 to 120 cm (granulars). For the same amount of cement injected per volume treated, F1 produces the strongest soilcrete in granular soils. The technical limitations of the simple fluid system are strongly influenced by the horsepower and flow rate of the grout pump. The drill rods are usually 90 to 110 mm in diameter, and have a 10-mm wall thickness.

With the F2 method, the compressed air enhances the cutting effect of the jet to the extent that columns twice the diameter of F1 columns can be created. This enhancement is due to these factors:

- The air acts as a buffer between the jet stream and any groundwater present, permitting deeper penetration by the jet.
- The soil cut by the jet is prevented from falling back onto the jet, thus reducing the energy lost through the turbulent action of the cut soil.

TABLE 8-2 Typical Range of Jet Grouting Parameters and Soilcrete Formed Using the Single, Double, and Triple Fluid Systems

Jetting Parameter		F1	F2	F3
Injection pressure				
Water jet	(MPa)	PW[a]	PW	30–55
Grout jet	(MPa)	30–55	30–55	1–4
Compressed air	(MPa)	Not used	0.7–1.7	0.7–1.7
Flow rates				
Water jet	(liters/min)	PW	PW	70–100
Grout jet	(liters/min)	60–150	100–150	150–250
Compressed air	(m^3/min)	Not used	1–3	1–3
Nozzle sizes				
Water jet	(mm)	PW	PW	1.8–2.6
Grout jet	(mm)	1.8–3.0	2.4–3.4	3.5–6
Number of water jets		PW	PW	1–2
Number of grout jets		2–6	1–2	1
Cement grout W–C ratio			0.80–1 to 2–1	
Cement consumption	(kg/m)	200–500	300–1000	500–2000
	(kg/m^3)	400–1000	150–550	150–650
Rod rotation speed	(rpm)	10–30	10–30	3–8
Lifting speed	(min/m)	3–8	3–10	10–25
Column diameter:				
Coarse-grained soil	(m)	0.5–1	1–2	1.5–3
Fine-grained soil	(m)	0.4–0.8	1–1.5	1–2
Soilcrete strength				
Sandy soil	(MPa)	10–30	7.5–15	10–20
Clayey soil	(MPa)	1.5–10	1.5–5	1.5–7.5

Source: From Kauschinger and Welsh (1989).

[a]PW = Water jets only used during prewashing.

- The cut soil is more efficiently removed from the region of jetting by the bubbling action of the compressed air.

The major drawback with F2 is that the soilcrete has a higher air content, and so has the lowest strength for any of the systems. F2 also requires rather more complex hardware and techniques. For example, the pathway for conducting the air is a 5-mm-wide annulus between the inner rods (carrying the grout) and the outer rod. This must be kept open, or the process will revert to F1. On the other hand, Guatteri et al. (1988) claim that an operational advantage of F2 is that it permits slightly lower grouting pressures (25 to 33 MPa) to be used, and so causes less wear on the grout pumps.

F3 is the most complicated and slowest system, but permits virtually full replacement of the jetted soil, and provides the largest column diameters. There is no standard practice for selecting which passageway in the triple-rod system carries which medium. The only essential is that the air/water jet be located above the grout injection point. Parry-Davies et al. (1992) found by field observation in sandy, bouldery material that the most efficient geometry featured a 180° gap (in plan) between the direction of the upper (air/water) and lower (grout) nozzles, which had a vertical separation of 150 mm. One procedure is to use the central core to convey the grout (Figure 8-4), with the middle annulus carrying the water and the outer the air. Water can be used during drilling to permit self-jetting of the rods, in appropriate ground conditions. Again, a simple check ball is used to change the direction of flow for the grouting phase. An alternative method features the water in the core; this is mechanically the best arrangement since the highest pressure fluid (i.e., the water) is in the smallest diameter container and so generates the smallest hoop stresses. The air travels in the middle annulus, and the grout in the (larger) outer annulus. If the system is to be self-jetting, then special valving is necessary to seal the bottom of the monitor prior to grouting. (A simple check ball is not possible because the central passage does not typically pass uninterrupted to the bit).

Clearly, the diameter of the column formed and the strength of the cemented soil (soilcrete) are not related only to the grouting method. They are also strongly influenced by many other factors including: soil type, density, plasticity, water content, water table location, amount of cement injected, age of soilcrete, and the energy used to form the column, as discussed in later sections.

Most simply, there are two basic steps in the process: the drilling, and the subsequent grouting.

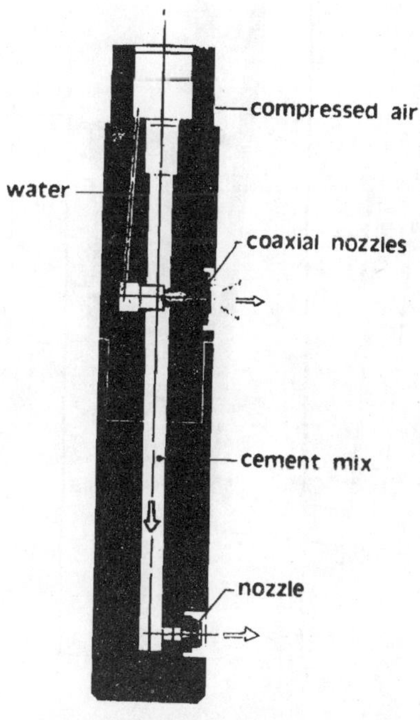

Figure 8-4 F3 monitor. (From Tornaghi and Perelli Cippo, 1985.)

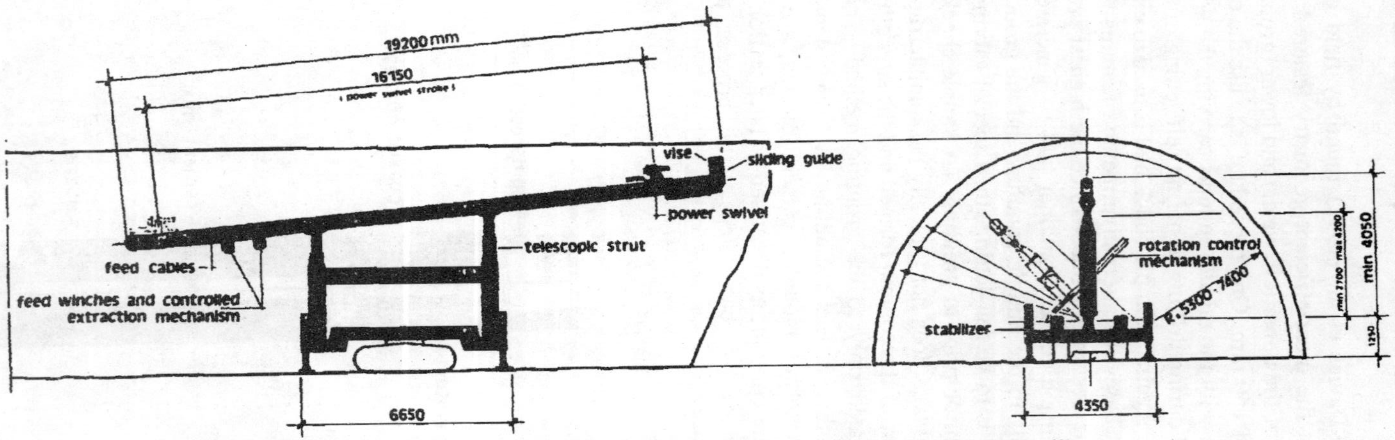

Figure 8-5 Early model drill rig for subhorizontal jet grouting. (From Tornaghi and Perelli Cippo, 1985.)

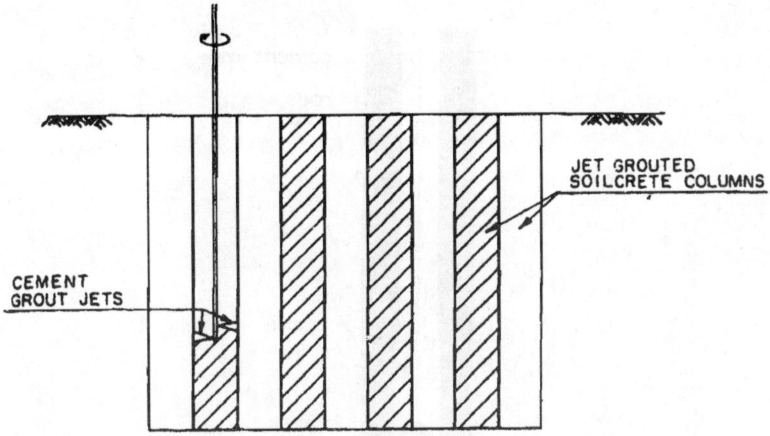

Soilcrete Wall Constructed by Single Stem Method

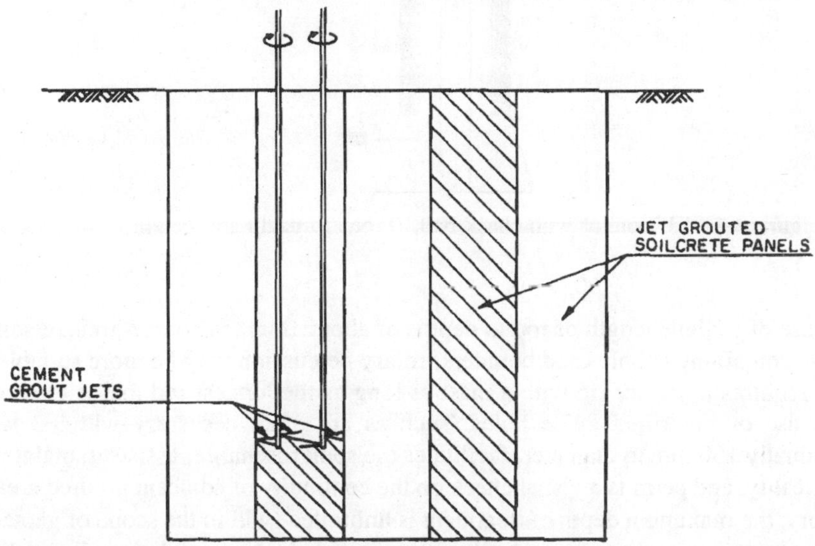

Figure 8-6 Soilcrete wall constructed by double-stem method, Norfolk, Virginia. (From Andromalos and Gazaway, 1989. Reproduced by permission of ASCE.)

The mode of drilling is selected according to soil conditions, general site features, and the design specifications in regard to hole length and inclination. Rotary drilling is preferred in medium- to fine-grained soils, and relatively small rigs can be used. Temporary drill casing may be needed, but more frequently uncased holes are drilled with direct circulation of water or bentonite slurry. If casing is used, it must obviously be withdrawn after the jetting rod is placed and grouting begins. The use of a rotary head with a hollow spindle running on a mast 4 to 5 m long permits

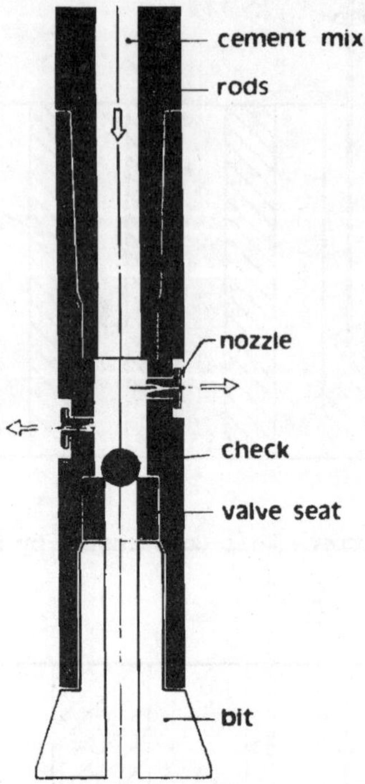

Figure 8-7 Fl monitor with check ball. (From Tornaghi and Perelli Cippo, 1985.)

the use of a single length of rod to depths of about 16 m. In coarse-grained soils, or those containing cobbles and boulders, rotary percussion may be more suitable, but this requires a heavier rig with a mast as long as the longest rod that is to be used. The use of predrilled guide holes (such as are often necessary with F3 work), nominally 150 mm in diameter, facilitates the spoil discharge, assists in maintaining verticality, and permits a visual check on the continuity of adjacent grouted areas. In theory, the maximum depth of treatment is unlimited (within the scope of geotechnical engineering applications), but the 70 m recorded when stabilizing slide debris in a railroad tunnel near Sempione in Italy is probably a practical maximum (Kauschinger and Welsh, 1989). Figure 8-5 shows the first generation of drill rig that was specially designed for subhorizontal percussive drilling for tunnel work. A system of hydraulic jacks allows the mast to be rotated axially over 180°, and inclined at up to 14° to the horizontal. Using such machines, all the holes necessary for the treatment of a tunnel section ahead of the excavation face can be drilled to a length of 16 m, with a single rod and with no interim setups of the rig. Later rigs are equipped with a movable subplatform permitting a more direct access of the operator to the heading and a greater versatility with respect to the treatment geometry.

Andromalos and Gazaway (1989) describe an interesting development in which the Fl process was conducted by twin 24-m rods (Figure 8-6) suspended on a crane equipped with a set of tubular leads.

The selection of a single rod or very long units (say 25 m) is advantageous not only to maximize productivity but, more importantly, to minimize interruptions (and hence the potential for blockages of nozzles and annuli) during injection. When using the F1 method, the same string of rods (typically 90 mm in diameter), is used for both drilling and grouting, whereas in F3, two strings are usually used consecutively given the delicacy and complexity of the injection string and monitor.

In F1 and F2 a check ball is introduced into the rod after drilling (Figure 8-7) to change the grout jet direction from axial to radial. Water jetting through the bottom of the rods during drilling improves the drilling rate and enhances verticality. Some systems also allow for horizontal water jetting during drilling.

Jet grouting uses heavy duty triplex piston pumps, of around 400 HP, often adapted from high-pressure oil field duties, to pump water or cement grouts. They typically have maximum pressure capacities of well over 70 MPa, although they operate considerably below this ceiling in jet grouting applications.

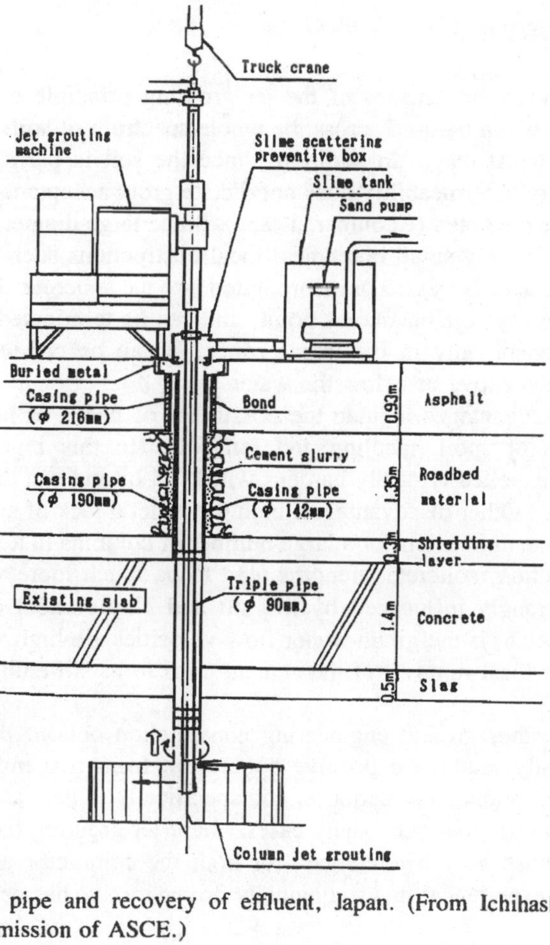

Figure 8-8 Guide pipe and recovery of effluent, Japan. (From Ichihashi et al., 1992. Reproduced by permission of ASCE.)

Grouts are prepared in automatic batch plants, specially designed to provide accurate proportioning and efficient colloidal mixing at rates adequate to meet the site demands: each drill rig may require up to 8 m^3 of grout per hour when jetting, and supply to the rig must not be interrupted during this operation.

During jetting, it is essential that the displaced slurry (and excess grout) can reach the surface with minimum impedence. If an annular pathway is not maintained, uncontrolled and usually unwanted ground heave or lateral movements will result. In the case of soft clays, heaves of up to 1 m have been reported (Kauschinger and Welsh, 1989). It may be necessary to design special procedures for preventing this, as well as for the handling and disposal of the spoil—especially important in urban areas and/or where the spoil may be contaminated or hazardous. Figure 8-8 is a good example of a system used during treatment of soils under an airport in Japan (Ichihashi et al., 1992). Burke et al. (1989) claim that F3 is the least likely to cause heave since "the water/soil/air mixture is of lower density and easily flows up the drill annulus as waste."

8-3 APPLICATIONS

One of the primary advantages of the jet grouting principle over other types of grouting is that it can be used across the whole spectrum of soils from the coarsest gravels to the finest clays. In addition, since the soil is partially or completely removed, its virgin permeability does not dictate grout acceptance and so costs can be more closely predicted (Coomber, 1985a). Since large diameter columns can be created from relatively small boreholes, local obstructions such as timber piles or large boulders can be bypassed or incorporated into the soilcrete. Jet grouting can be conducted from any suitable access point, and can be terminated at any elevation, providing treatment only in the target zones. It can be conducted vertically or (sub)horizontally, above or below the water table.

Potential disadvantages include the possibility of heave or lateral movements, and difficulties of spoil handling and removal. (In this regard, however, the grout/soil mix expelled typically hardens within 24 hr and can then be disposed of as solid residue.) Other disadvantages lie in the general lack of an economic means for measuring the dimensions of a large number of columns in a routine production project. In addition, soilcrete strengths tend to be much more variable than concrete, being strongly influenced by the silt and clay content of the native soil (Section 8-4). Also, if the groundwater flow velocities are high, the fluid soilcrete may experience local removal of the cement prior to its stiffening, and so unevenness in quality.

As for any other ground engineering construction option, the viability of jet grouting typically requires a positive conclusion from two independent studies: technical and economic. Furthermore, and specifically in the case of a newer technology such as this, there are many cases where jet grouting has been conducted simply to establish a "job reference" for both the contractor and the technique. Based on U.S. experience, this strategically driven motive has occasionally had the opposite, adverse effect, and the end result has been technically and economically unsatisfactory and strategically discouraging.

Nevertheless, jet grouting has fast become a common feature of geotechnical construction in countries with very active tunneling programs, and urban/industrial development and rehabilitation schemes. Other common applications to date have been the underpinning of existing structures, and impermeable cutoff walls for tunnels, open cuts, canals, and dams (ASCE, 1987).

Applications can be simplistically divided into vertical, and horizontal, jet grouting, although in each category the actual direction may be many degrees off these directions. Jet grouting is also used occasionally in the construction of soil anchorages (Anonymous, 1988), pinpiles (Trevi, 1990), or soil nails (Kauschinger et al., 1992a), but this study concentrates on the use for ground treatment per se. Many excellent case histories abound, but in the following, only recent and/or especially interesting and illustrative examples are referenced.

As a reminder of earlier cases, Table 8-3 provides some summary details. At that time, one of the two American examples, the test shaft at New Waddell dam, Arizona, was not entirely satisfactory in outcome due to "three deficiencies":

- Jet grout column diameter was not determined.
- No performance criteria were established prior to commencement.
- Optimization of drilling parameters was not carried out.

As a consequence, jet grouting was not adopted as the method of constructing the subsequent permanent cutoff under the dam, and indeed little jet grouting was conducted anywhere in the United States until the later 1980s.

Vertical/Subvertical Applications

Seepage Barriers, Lateral Retention, and Bottom Seals

General Comments Braced excavations in urban areas often have many utilities along the alignment. Typically these utilities can be rerouted prior to constructing the retaining wall. However, when this is not practical, jet grouting can provide both a seepage barrier, and an earth retention system by either of the following:

1. Jetting beneath the utility
2. Incorporating the utility into the wall by grouting all around it.

Sheet piles or secant walls used as seepage barriers as well as earth retention systems may have "windows." Jet grouting the soil behind such walls can seal these gaps (Figure 8-9). In such cases it may be more economic to conduct wing jet grouting as opposed to forming complete columns: twin jets are set 45° to 90° apart in plan and create a wedge-shaped body behind the wall as the string is lifted without rotation.

TABLE 8-3 Jet Grouting Projects (Up to 1986)

Project	Location	Nature of Problem	Soil Type	Columns			Grout		Source
				Rows	Diameter	Spacing	Mix	Pressure	
British Rail, Glasgow tunnel	Scotland	Track settlement	Saturated silt	2	6 ft (2 m)	6 ft (2 m)	Neat rapid-set cement	—	Welsh et al., 1986
Trent and Mersey Canal tunnel-renovation	England	Collapse of invert	Weathered marl	3	6 ft (2 m)	4–6 ft (1–2 m)	w/c = 1	—	Welsh et al., 1986
Oldenburg underpinning	Germany	Underpinning for basement construction	Silty and clayey sand and fill	—	—	—	—	—	Welsh et al., 1986
Stockton on Tees, new tunnel	England	Shaft excavation and correction to tunnel	Submerged silty sand	3	6 ft (2 m)	4–6 ft (1–2 m)	Cement PFA	—	Coomber, 1985b
Gateshead, new tunnel	England	Access shafts for bentonite TBM	Silty and clayey sands, gravel	—	6 ft (2 m)	4–6 ft (1–2 m)	Cement PFA	—	Coomber, 1985
Porto Tolle power plant	Italy	Cutoff wall for seepage and erosion prevention	Silty sand and organics	3	2 ft (0.6 m)	1–2 ft (0.3–0.6 m)	w/c = 2	4400 psi (30 MPa)	Aschieri et al., 1983
Research	Japan	Various methods	Cohesive, sandy and gravelly soils	—	1–13 ft (0.3–4 m)	—	—	2900–8700 psi (20–60 MPa)	Miki and Nakanishi, 1984
Halic sewer tunnel	Turkey	Eliminate water flow into open excavation	Alluvial sands	—	—	—	Neat cement	—	Abramson, 1986

Fatih sewer tunnel	Turkey	Eliminate water flow and increase stand-up time	Miscellaneous fill	11	6 ft (2 m)	2.5 ft (0.8 m)	Neat cement	5100 psi (35 Mpa)	Abramson, 1986
New Wadall dam	U.S.	Eliminate water flow into shaft	Sand, gravel, cobbles and boulders	1	2 ft (0.6 m)	1.7 ft (0.5 m)	w/c = 1	5100 psi (35 MPa)	ENR, 1986
Pump lift station	U.S.	Minimize subsidence	Saturated clays	—	—	—	—	—	Andromalos and Pettit, 1986
Milan Metro, Line 3	Italy	Increase stand-up and minimize settlement	Alluvium	1	—	1–2 ft (0.3–0.6 m)	—	—	Andromalos and Pettit, 1986
Milan–Rome railroad	Italy	Prevent scouring of bridge parts	Alluvium	1	—	—	—	—	Andromalos and Pettit, 1986
Europe Southern Observatory	Germany	Building expansion	Silt, sand, and gravel	1	—	—	—	—	Andromalos and Pettit, 1986
Singapore MRT tunnels	Singapore	Increase stand-up time	Marine clay	18	1.5 ft (0.4 m)	1–2 ft (0.3–0.6 m)	—	—	Martin, 1986
Building restoration	Japan	Underpinning	Sand and silt	1	8 ft (2.5 m)	—	w/c = 1	5800 psi (40 MPa)	Shibazaki and Ohta, 1982
Yokohama highway	Japan	Trench cutoff	Sand and silt	2	0.5 ft (0.2 m) Panels	4 ft (1.2 m)	Water cement bentonite	5600 psi (39 MPa)	ENR, 1984
Nuclear fuel reprocessing plant	Japan	Slope protection cutoff	Sand and gravel	1	0.5 ft (0.2 m) Panel	—	—	—	Yahiro et al., 1975

Source: From ASCE (1987).

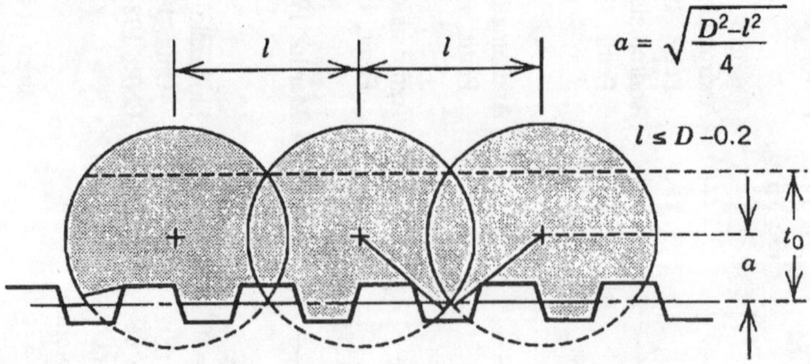

Figure 8-9 Location of jet grout columns behind sheet pile wall to create seepage barrier. (From Kauschinger and Welsh, 1989.)

Jet grouting has also been used extensively to create the side walls and bottom seals for shafts. Guatteri et al. (1988) reported on the construction of access shafts in Sao Paolo, Brazil, 8 m deep and 10 m in diameter. The water table was near the surface and the ground comprised soft marine clay in the upper 5 m with fine silty sand below. A double row of F2 columns was placed to create a wall about 2 m thick. The application was successful and the shaft remained open for about 12 months.

In cases where the depth to an impervious layer prohibits the driving of steel sheeting full depth, jet grouting can be used to form an impervious bottom seal at the bottom of the sheeting prior to excavation. A similar application was conducted in 1985 for part of a below grade expressway in Montreal, Canada, where it contacted an existing large sewer collector. A jet grouted wall was extended below a conventional slurry trench diaphragm wall, which could not penetrate through an existing concrete mat. Jetted columns also provided a roof over the subsequent excavation, which was conducted safely in dry conditions.

Details of Projects Aschieri et al. (1983) and Tornaghi and Perelli Cippo (1985) reported on the "typical and well documented" hydraulic cutoff, constructed around the existing Outlet Station of a thermal power plant at Porto Tolle, Italy, to prevent further seepage and internal erosion through loose medium-fine silty sands. Three rows of staggered, overlapping F1 columns were installed 17.5 m deep and penetrating about 3 m into a soft silty clay formation (Figure 8-10). The same authors described similar works for seepage control and earth retention for several shaft foundations for long viaducts along the Udine-Tarvisio motorway and adjacent railway. Some shafts 8 to 12 m in diameter were excavated to a depth of 22 m against a water head of 15 m in detrital soil of very variable granulometry. This involved two rows of F1 columns 26 m deep. Conversely, a single row was sufficient to allow the excavation of a circular service shaft 22 m deep, 4 m below the water table in Milan (Figure 8-11).

Mongilardi and Tornaghi (1986) describe how a jet grout curtain was installed as a "continuous retaining structure for the protection of old buildings overlying the new

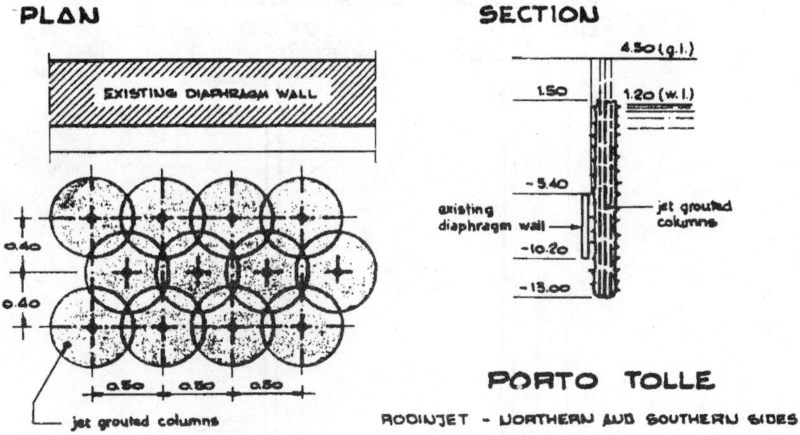

Figure 8-10 Scheme of Rodinjet treatment to form an impervious cutoff. (From Tornaghi and Perelli Cippo, 1985.)

Metro Tunnels in Milan" (Figure 8-12). Similar examples may be found throughout metro construction in the older cities of Western Europe, such as Vienna.

Regarding the bottom seals on shafts, Coomber (1985a) produced a useful table (Table 8-4) summarizing dimensions adopted to that time. It confirms that the weight of the plug, plus overlying soil to formation level, are inadequate to resist uplift forces. Coomber concluded that the majority of the resistance is developed by "vertically downward forces around the periphery of the grouted plug."

Figure 8-13 shows the successful results of a test shaft more recently constructed at Waggaman, La., in soft saturated clay. Excavation was carried out to a depth of 5.1 m with no measurable movement in the surrounding soils. The same company, Halliburton Services, also used F1 jet grouting as remedial shoring after the failure of a sheet piling wall during heavy rain. This grouting both reinforced the piling against lateral loads, and halted migration of soil and water. The soilcrete structure was 90 m long and keyed into a shale layer 7 m below the surface (Figure 8-14).

Miyasaka et al. (1992) described the design, construction, and performance of a grouted, gravity retaining wall as an earth support system under difficult site and access conditions. This was combined with a conventional diaphragm wall to provide a practical construction option (Figure 8-15).

Jet grouting has also been used in the sealing of existing dams. Panel jetting was conducted under the Brombach dam in Germany where the cutoff in weathered Keuper sandstones has successfully withstood hydraulic gradients over 100. Columns have been installed in the core of Villaneuva dam, Spain, to resist seepage, and under a concrete structure within the embankment of John Hart dam, Canada, as part of a cutoff aimed at improving liquefaction resistance (Imrie et al., 1988). Figure 8-16 also shows the application for a cutoff within the left bank of the Euphrates River in Iraq as a part of the channel management program.

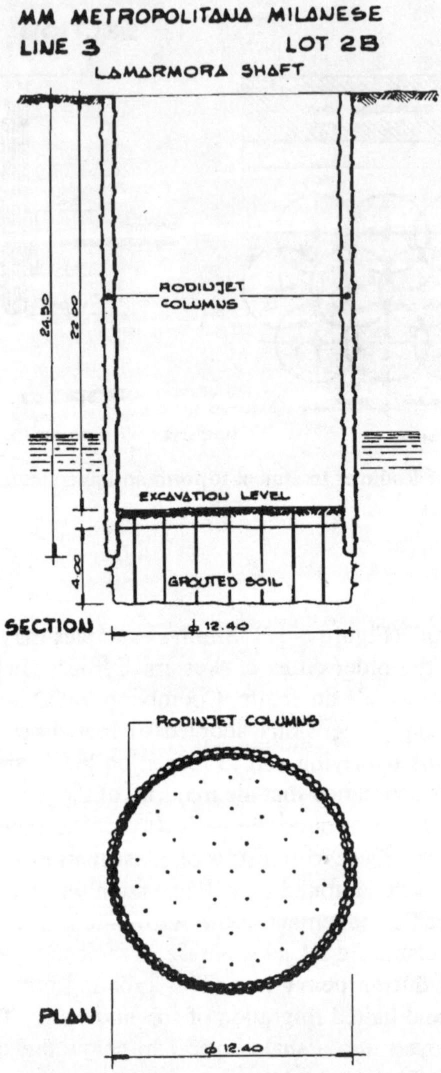

Figure 8-11 Scheme of Rodinjet treatment to protect deep shaft excavation. (From Tornaghi and Perelli Cippo, 1985.)

Tunnel Construction

General Comments Tunneling remains a prime source for jet grouting applications. In one concept, a "vault" (Figure 8-17) can be created around the perimeter, and if the tunnel is below the water table, a bottom plug can be placed. If the soils are particularly soft, and saturated, then full face grouting can be conducted. In addition to the precaution of running preproduction tests in such difficult conditions,

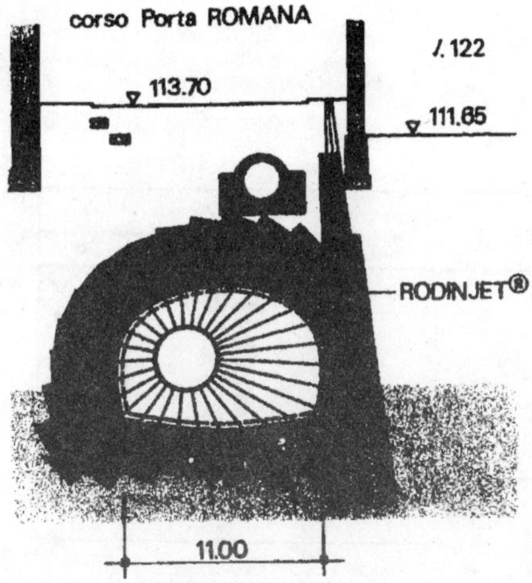

Figure 8-12 Milan Metro, Line 3/Lot 2B: radial conventional grouting from drift and subvertical jet grouting to protect an old building, Milan, Italy. (From Mongilardi and Tornaghi, 1986.)

one must also ensure that during excavation, steps must be taken to vent the large amount of heat generated during such works.

Other interesting tunnel applications include the formation of portals and starting blocks, and reaction blocks behind sheet pile walls for pipe jacking.

Details of Projects In contrast to the more numerous examples for new tunnel construction, Coomber (1985b) described two applications for remedial works. The first, in Glasgow, Scotland, was for a 25-m-long section of 100-year-old rail tunnel experiencing settlement due to the presence of saturated silt in a buried glacial hollow (Figure 8-18). Two rows of 2-m-diameter columns at 2-m centers beneath the center line of each track were installed to either bedrock or competent boulder clay. Mixing and ancillary equipment were located at the surface and grout, water, and compressed air were fed down via a purpose-drilled access shaft to the drilling locations.

Monitoring of tunnel movements and track deflections in service continued during and after jet grouting operations. These surveys showed that, although on completion of jet grouting, slab movement, unacceptable traffic load deflection, and sleeper debonding had all effectively ceased, some settlement of the side walls continued. Evidence also emerged to suggest that the original brick caisson support to the walls had been only at a few isolated points rather than at the expected close centers. It became apparent that problems had not been confined to the invert slab, as had originally been thought.

Subsequent contracts to strengthen the arch and drill pinpiles from the surface through the tunnel walls were therefore rapidly implemented. Tunnel performance has since been satisfactory throughout. Owner estimates showed the various reme-

TABLE 8-4 Jet Grout Treatment of Shaft Bases

Soil Treated	Z (m)	S (m)	H (m)	h (m)	d (m)
Alluvial silts/sands	10.5	14.5	7.0	2.0	2.0
	13.0	10.5	9.5	2.2	2.0
	7.5	10.5	4.0	Nil	1.5
Alluvial silty sand	10.6	3.5	6.9	Nil	2.0
Thanet sands	13.0	6.0	9.2	Nil	2.0
Gravelly sands	7.0	2.0	6.1	1.6	1.5
Alluvial clayey sands	10.0	8.0	9.0	Nil	2.6
Very soft clays	12.9	10.0	11.9	Nil	2.9
Loose sandy silt	14.7	5.0	12.4	Nil	3.0
Alluvial sands	13.4	7.5	9.6	Nil	2.5

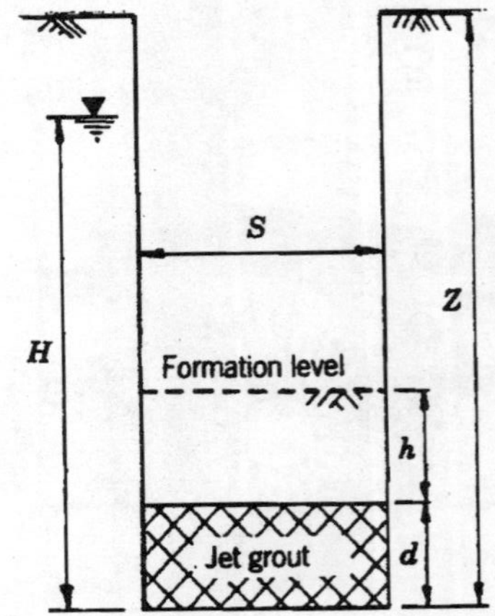

Source: From Coomber (1985a).

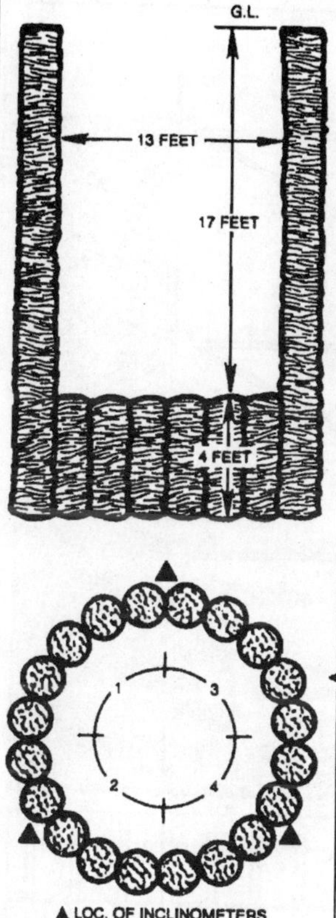

SHAFT DETAIL

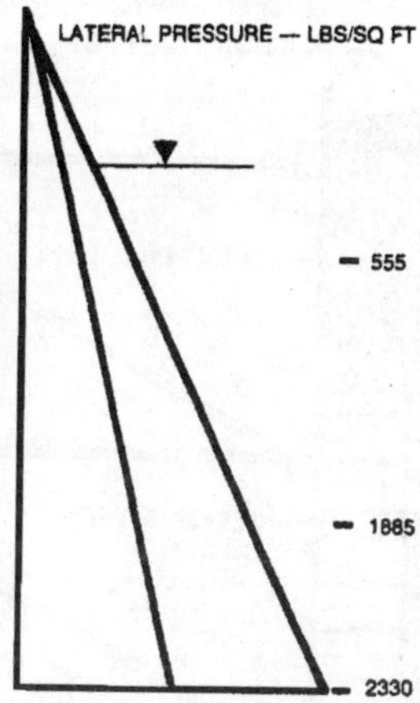

Sector Number	Cement Water Ratio	28 Day Engineering Properties: Compressive Strength (psi)	Elastic Modulus (psi)	Poisson's Ratio (unitless)
1	0.8::1	600	0.35×10^6	0.23
2	1.0::1	720	0.39×10^6	0.23
3	1.2::1	1530	0.77×10^6	0.28
4	1.4::1	1590	0.81×10^6	0.29

BORING LOG

STRATUM DEPTH IN FEET	VISUAL CLASSIFICATION	BLOWS PER FOOT
	MEDIUM STIFF TAN AND BROWN SILTY CLAY	9
2.0	MEDIUM STIFF TAN AND GRAY CLAY W/SILT AND SPECKS WOOD	6
5.0		6 3
11.5	SOFT TAN & GRAY CLAY W/MUCH SILT	4
15.0	VERY SOFT GRAY CLAY W/SILT & WOOD	2
18.5	VERY SOFT GRAY SILTY CLAY W/SAND	–
20.0	SOFT GRAY CLAY W/SILT	3

Figure 8-13 Shaft at Waggaman, Louisiana. (From Halliburton, 1987.)

dial works to have been completed at one-fifth of the cost of the only alternative course of action—tunnel closure and total reconstruction.

The second remedial example was a canal tunnel in Cheshire, England. A 12-m-long section of invert had subsided 450 mm and brick linings were damaged. Up to 6 m of soft clay-filled voids in the marl bedrock were found immediately beneath the invert. The proposed grout treatment was carried out in three phases: injection grouting to infill voids beneath and on either side of the tunnel in the marl; jet grouting to follow the first phase with 2-m-diameter columns to remove and replace the soft clay and silt infilling immediately below the tunnel invert; and secondary jet grouting along the center line to greater depths at selected locations to remove and replace silt and clay filling of critical voids. The extent of work is illustrated in Figure 8-19.

Access to the works—along a narrow private track—was extremely limited. The

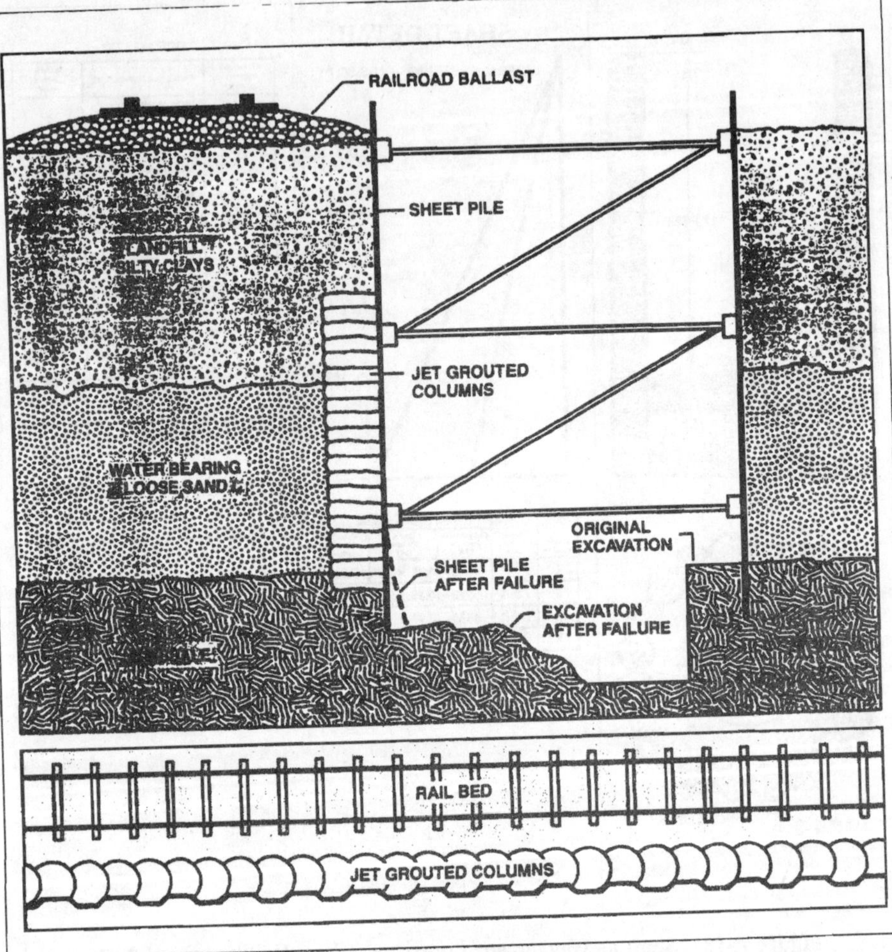

Figure 8-14 Jet grouting for construction shoring, Dallas, Texas. (From Halliburton, 1987.)

grout station was set up on the towpath adjacent to the tunnel portal, and drilling and jet grouting were carried out from a benched platform formed by excavation in the hillside above the tunnel center line. The jet grout mix consisted of equal weights of PFA and cement, with a water/solids ratio of 1. The work was satisfactorily undertaken within a tight schedule.

An early, major application of full face treatment (Tornaghi and Perelli Cippo, 1985) was on Contract 106 of the Singapore Mass Transit System between Dhoby Ghaut and City Hall Stations. Groundwater levels ranged from about 2 m below surface to less than 1 m in the Dhoby Ghaut Station area. Here a thin covering of fill overlay up to 7 m of peat and peaty clay, which in turn overlay up to 7 m of medium dense silty sands and stiff clays and then stiffer soils, grading into weathered sandstone or siltstone.

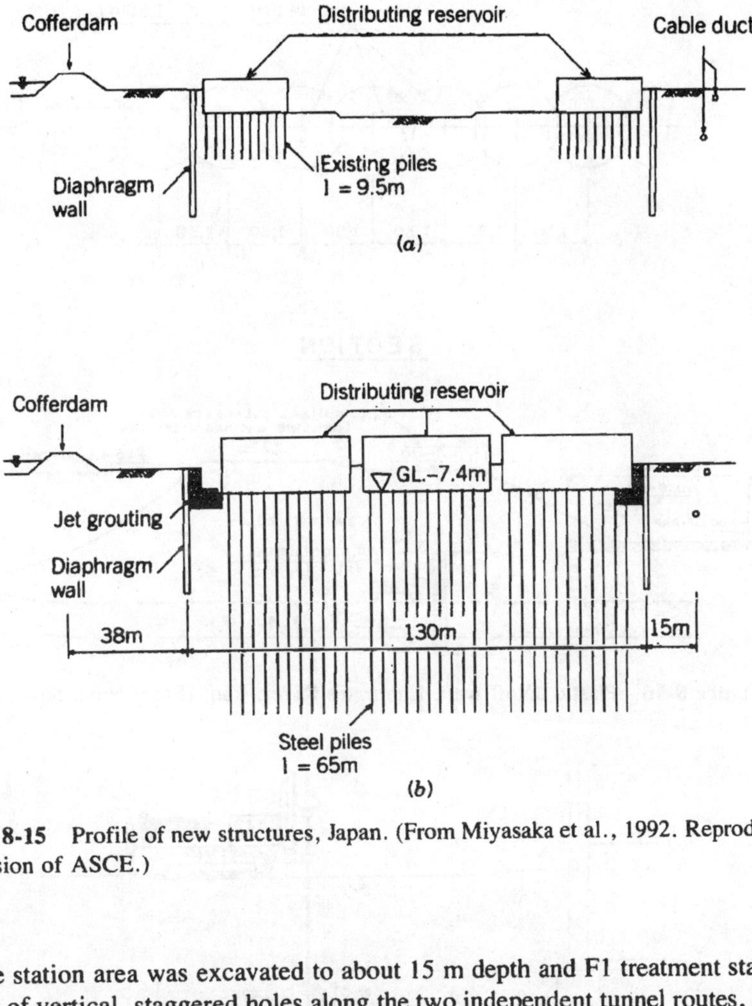

Figure 8-15 Profile of new structures, Japan. (From Miyasaka et al., 1992. Reproduced by permission of ASCE.)

The station area was excavated to about 15 m depth and F1 treatment started by means of vertical, staggered holes along the two independent tunnel routes, passing beneath roads and under (or very close to) buildings and public utility services. Without any soil improvement even shield tunneling was judged difficult and potentially unsafe, the permissible magnitude of vertical ground movements being restricted to a few centimeters.

Figure 8-20 shows the layout of the first section (about 70 m long), starting from Dhoby Ghaut Station, where soil treatment was required at variable depths according to the soil profile. Owing to the presence of soft highly plastic formations and to the shallowness of the tunnels, jet grouting from the surface was selected.

In line with the design specifications, the treatment had to be extended to the full excavation area above soft rock or very stiff clay, and had to create an arch of strengthened soil 1.5 to 3.0 m thick around the soft soil to be excavated. The thicker treatment was executed close to the station where the shield could not operate. To check the proposed solution and to set up the working program, a large-scale trial was carried out on site (as described in the discussion of Singapore in Section 8.6).

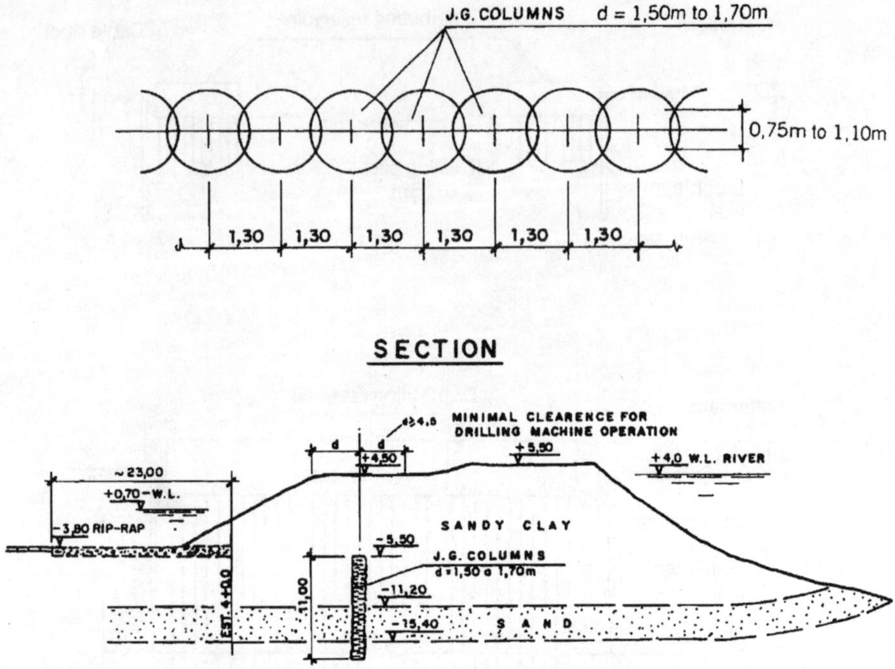

Figure 8-16 Plastic cutoff wall, Euphrates River, Iraq. (From Novatecna, 1988.)

Figure 8-17 Vertical jet grouting to create "vault" around perimeter of tunnel. (From Pacchiosi Drill, 1978.)

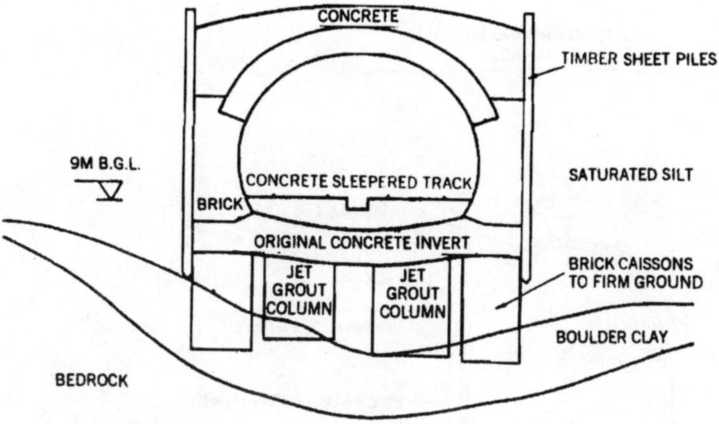

Figure 8-18 Cross section showing tunnel construction and jet grout treatment, Glasgow, Scotland. (From Coomber, 1985b.)

Figure 8-21 illustrates the application to repair a collapse that occurred in 1980 during driving of the Sanegran-Tiete-Andrade Gutierrez Tunnel, Brazil. A total of 6000 linear m of F1 columns was installed in 60 days with one rig to stop the inflow and settlements, and permit resumption of construction. This was the first F1 job conducted in South America (Guatteri et al. 1988). Thereafter, as the soil along the 12-km-long tunnel was very variable, jet grouting was used for the succeeding two years at ten different problem sections.

Structural Underpinning

General Comments This has become the major application of jet grouting in the United States, although slightly less common elsewhere. It can be conducted in low headroom conditions, and all three injection methods have been used, depending largely on the contractor's preference. Individual footings may be supported on groups of piles, or columns can be placed at wider spacings under continuous footings. Multiple rows of columns can be jetted at various angles to create an earth retaining structure (Figure 8-22) and so accommodate eccentric loads. Intimate contact can be established between column and footing by maintaining a positive head of grout above the footing to prevent shrinkage during hardening.

Where the rate of structural settlement is high, and ongoing during underpinning, jet grouting may not be the best choice as load will be applied to the columns before they reach their design strength (Bruce, et al., 1992).

Details of Projects The review paper by Welsh and Burke (1991) illustrates many of the various applications of jet grouting in citing 17 U.S. examples (Table 8-5). Of these, over half have been conducted in the last five years for structural underpinning—probably a fair indication of the major interest in the United States.

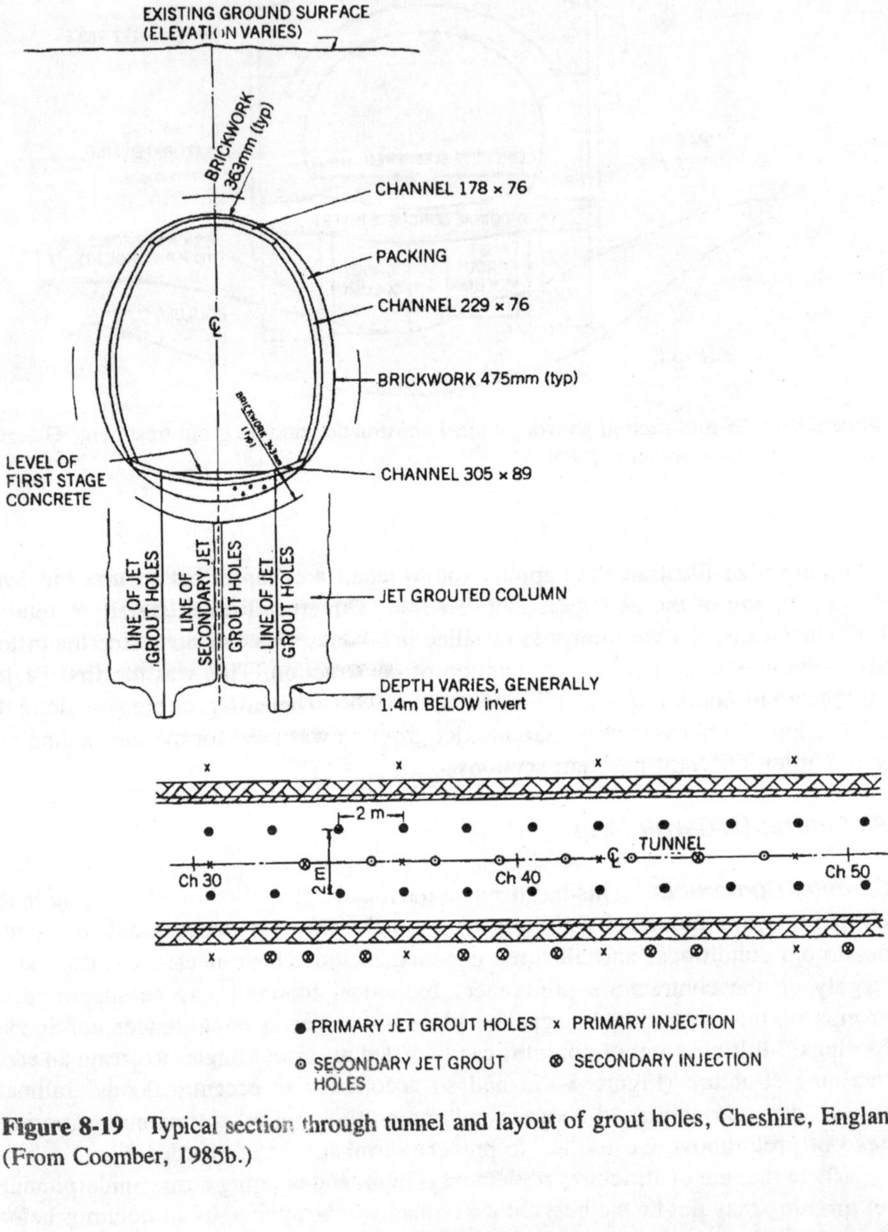

Figure 8-19 Typical section through tunnel and layout of grout holes, Cheshire, England. (From Coomber, 1985b.)

Burke et al. (1989) documented the underpinning of the Yale University Sterling Medical Library Building in New Haven, Connecticut. This building's expansion required that the Y-shaped, five-story, steel frame construction be underpinned to allow for filling in of the adjacent courtyards. Conventional pit underpinning was specified and bid. However, the general contractor was concerned about the dry sands ravelling during this method of excavation. After considering chemical grout

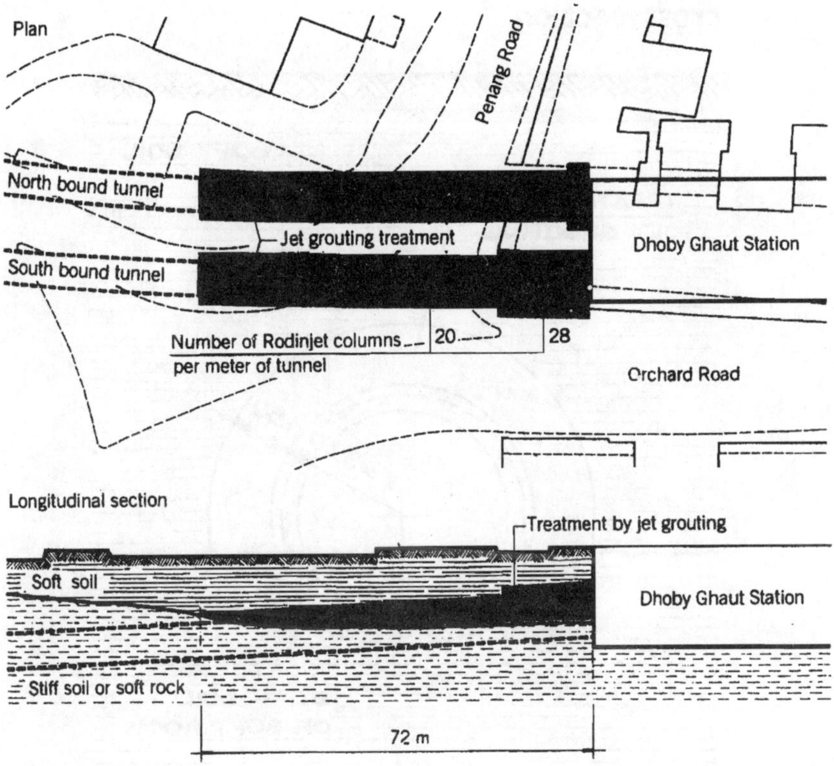

Figure 8-20 Singapore Mass Transit System: layout of Rodinjet treatment along two tunnel sections from Dhoby Ghaut Station toward Victoria Street. (From Tornaghi and Perelli Cippo, 1985.)

stabilization of the sand, the contractor determined that F3 jet grouting could not only underpin the structure, but also do it faster and more economically than the conventional method specified. It had been estimated that settlements of approximately 25 mm would occur during pit underpinning. However, the 45 tell-tales installed on the project indicated that the maximum recorded settlement after jet grouting was 4 mm, with the majority of readings being between 2 and 3 mm.

As Case Histories 6, 9, 10, 13, and 16 of Table 8-5 recall, jet grouting has also been used for excavation support. Often, renovation work and the expansion of facilities require completion of an addition while the existing facility remains operational. When the addition includes work adjacent to and beneath existing foundations, some means of supporting the structural loads as well as retaining the earth is required. In case history 16, soilcrete was used to support a heavy seven-story library in St. Paul, Minnesota, contain the earth pressure, and provide a back form for the new building wall construction: the silty sands and gravels were transformed into a soilcrete gravity support wall. This wall utilized an outside, half-column

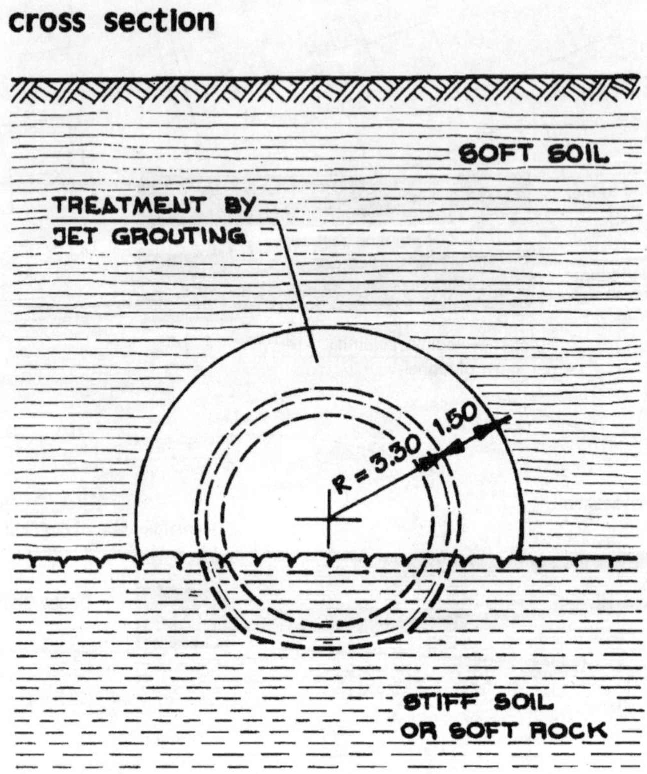

Figure 8-20 (Continued)

geometry to create a nearly flat face upon excavation, which required very little trimming prior to further construction activities.

Soilcrete Struts Soilcrete struts have also occasionally been used to internally support earth retaining structures such as in a cut and cover top down application when tiebacks cannot be used. The soilcrete is installed before any excavation begins, and so before any deformations or settlements can occur. Such temporary struts can also act as interim stable working platforms, especially useful if the excavation is in soft clays or silts. Struts can be created at different levels from the same borehole simply by selecting the elevations at which jet grouting is carried out.

Ground Improvement/In Situ Reinforcement Jet grouting has often been used to improve the overall bearing capacity of a large volume of soil. In this case the columns do not overlap, as this would be prohibitively expensive. Rather they work together with the ground similar to the action of stone or lime columns, and some inter-columnar soil improvement is achieved during the jetting (especially with F1).

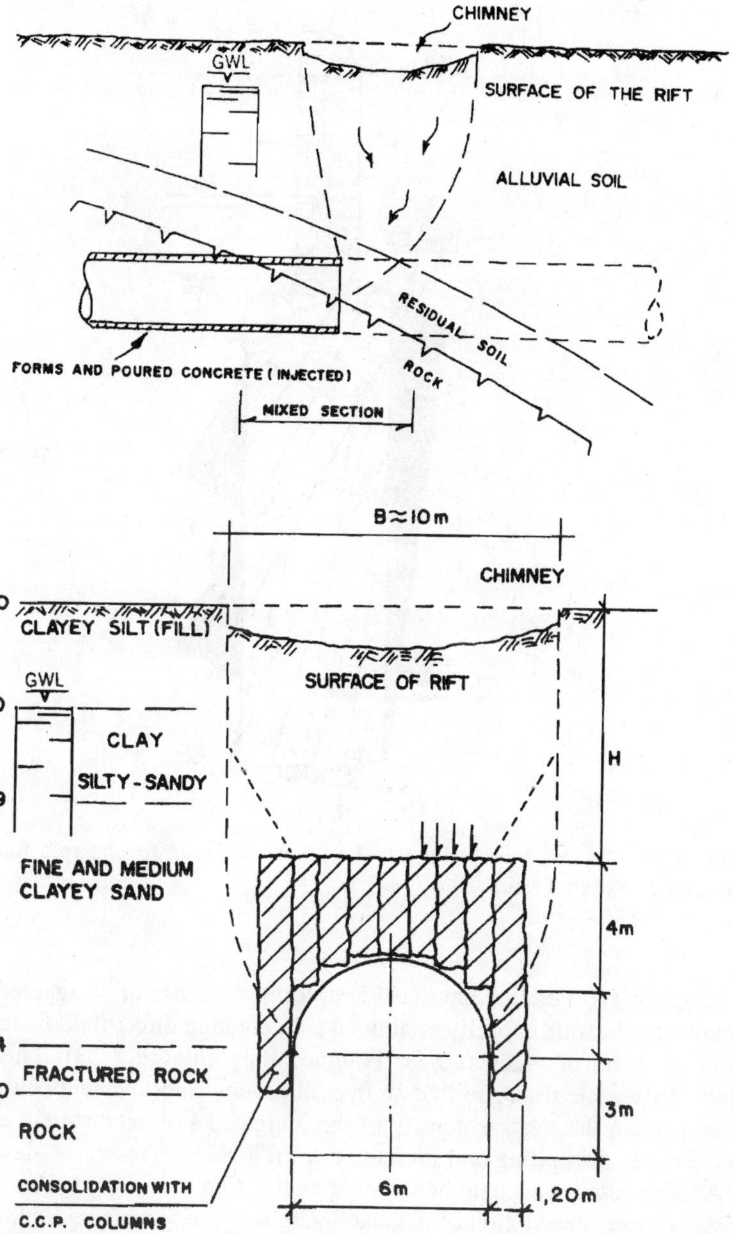

Figure 8-21 Repair of partial tunnel collapse with F1 columns, Sanegran Tunnel, Brazil. (From Novatecna, 1988).

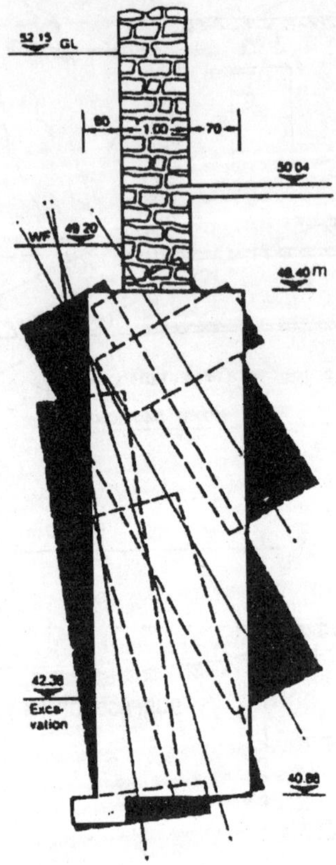

Figure 8-22 Typical cross section of jet grout wall used to underpin foundation walls. (From GKN Keller, 1985.)

Tornaghi and Perelli Cippo (1985) described the use of "staggered columns" to improve the bearing capacity of alluvial soil, creating direct shallow foundations for about 50 piers of viaducts near Foligno, Italy (Figure 8-23). The spacing was selected after site trials, so that an overall ground improvement could be achieved by increasing the relative density of the untreated soil between the columns. This was cheaper than piling and preferable from a seismic response viewpoint.

A more detailed account of a similar application was provided by De Paoli et al. (1989) in their description of the stabilization of peaty soils under 800 m of a new railway embankment near Como, Italy. These soils comprised highly compressible materials, principally fibrous peat with a silty-clayey matrix, and averaged 5 m thick. A previous track built close by had required releveling maintenance works on average every six months, and these seriously restricted the rail traffic along this very busy line. Following an extensive test, the F3 method was used to provide 2.1-m-diameter columns, at 2.75-m centers, extending into the underlying sands and

TABLE 8-5 Projects Executed in the United States by Hayward Baker

Case History Number and Location	Soil Type	Year Work Performed	Purpose
1. Charleston, S.C.	Fine sand	1980	Test program for liquefaction protection
2. Aberdeen, Ohio	Silty, fine sand seams between zones of sandy, silty clay	1980	Up to 30-ft-deep, 120-ft long panel cutoff wall formed
3. Cross, S.C.	Soft clay, silts, and sands	1980	Removal of infill soils and filling limestone cavities beneath caissons of power plant
4. Trussville, Ala.	Soft clay	1987	Stabilization of crown of hard rock tunnel by removing and replacement of soft clay
5. Jacksonville, Fla.	Fine sand	1988	Sealing 570 joints in precast concrete sheet piling up to 26 ft deep
6. New Haven, Conn.	Sand	1988	Excavation support and underpinning of existing 4-story library
7. Mt. Pocono, Pa.	Silty sand	1988	Underpinning for increased bearing capacity for new loads from building renovation
8. Kalamazoo, Mich.	Sand	1988	Underpinning for hospital renovation
9. Trumbull, Conn.	Silty sands with cobbles	1988	Underpinning 10 ksf structure load to rock to allow adjacent tunnel excavation
10. Kingsport, Tenn.	Silty sand with gravel	1989	Underpinning and permanent excavation support of pit walls; headroom restricted to 9 ft.
11. Danville, Pa.	Silty sand to clayey silt	1989	Underpinning for increased bearing capacity in headroom as low as 8.5 ft
12. Baltimore, Md.	Silts, clay, and sand	1989	Underpinning for settlement control in as low as 10-ft headroom
13. New Haven, Conn.	Clean sand	1989	Underpinning and 11-ft excavation support adjacent to 6-story hospital

(*continued*)

TABLE 8-5 (*Continued*)

Case History Number and Location	Soil Type	Year Work Performed	Purpose
14. North Elba, N.Y.	Silty sand with cobbles	1989	Underpinning to prevent frost heave of existing structure
15. Kingsport, Tenn.	Silty sand with gravel	1990	17-ft excavation support and groundwater control for pipeline installation along a river
16. St. Paul, Minn.	Silty sand with gravel and cobbles	1990	Underpinning and excavation support for 7-story library addition
17. Charleston, S.C.	Silts and sands	1990	Groundwater control and stabilization for WWTP addition

Source: From Welsh and Burke (1991).

gravels. This directly treated approximately 60 percent of the volume of the problem horizon (Figure 8-24).

Some cross sections of the embankment were monitored with: (a) strain meters (sliding micrometers) located both within and between columns of jet grouted peaty soil; (b) settlement plates on top of columns; and (c) inclinometers. With reference to the typical section of Figure 8-24, Figure 8-25 shows the vertical strains recorded by sliding micrometers SL1, SM1, and SM2 for columns 54/A, 54/C, and peaty soil between 53/C and 54/C, respectively, and the total settlements of columns 52A/C/E recorded by plates SP5/11/15, respectively.

About 130 days after completion of the first grouting, the maximum settlements recorded *in* the columns by sliding micrometers ranged from 3 mm (54/C) to more than 6 mm (54/A), while almost 9 mm was reached *between* columns 53/C and 54/C. The plates indicated settlements ranging between 2.4 mm (SP11) and 7.5 mm (SP15)—in good agreement with the sliding micrometer readings.

This proved that short-term compressibility was negligible within the columns and far reduced in the peaty soil between, fully vindicating the choice of the solution. Long-term monitoring continues. Data from the inclinometers indicate no significant horizontal displacements to date. Further information on the test program is given in the discussion of Monte Olimpino in Section 8.6.

Slope Stability Langbehn (1986) provided a detailed review of jet grouting's potential in slope stabilization. The concept is to place columns acting as dowels or shear pins across the shear zone, slip surface, or potential failure zone in a concentration large enough to prevent failure or continued failure and improve stability to an acceptable degree.

Columns can be installed over entire slopes or concentrated in a dense grid at the toe. The shear strength of jet grouted soil is typically about 10 to 15 percent of unconfined compressive strength. As this parameter depends on soil conditions and operational parameters, corresponding shear strengths in the range 0.4 to 0.8 MPa can be obtained in the most commonly found slope materials (silty clays and silty sands). Where tensile or bending stresses are likely, steel reinforcement can be placed. Configurations of column groups distributed over the slope were observed by Langbehn to provide approximately 10 percent higher factors of safety for comparable levels of treatment using dense column grids at the toe of the slope.

A 1985 project at Gela, Italy, consisted of the stabilization of 1500 m of a landslide slope threatening nearby buildings. The soils within the slide mass consisted of intermittent layers of sands, clayey sands, and clays. The slide was induced by the continued deterioration of the mechanical properties of the clayey sands at the base of the slide mass, caused by a fluctuating water table.

Several constraints were imposed on the work, necessitating the use of jet grouting for the stabilization. These constraints included:

1. A short time period in which to perform the work.
2. A fluctuating groundwater level.
3. A heterogeneous soil profile consisting of alternating layers with varying mechanical and granulometric properties.
4. The need to stabilize a precise volume of material without inducing surface heave or settlements in the area near the existing buildings.

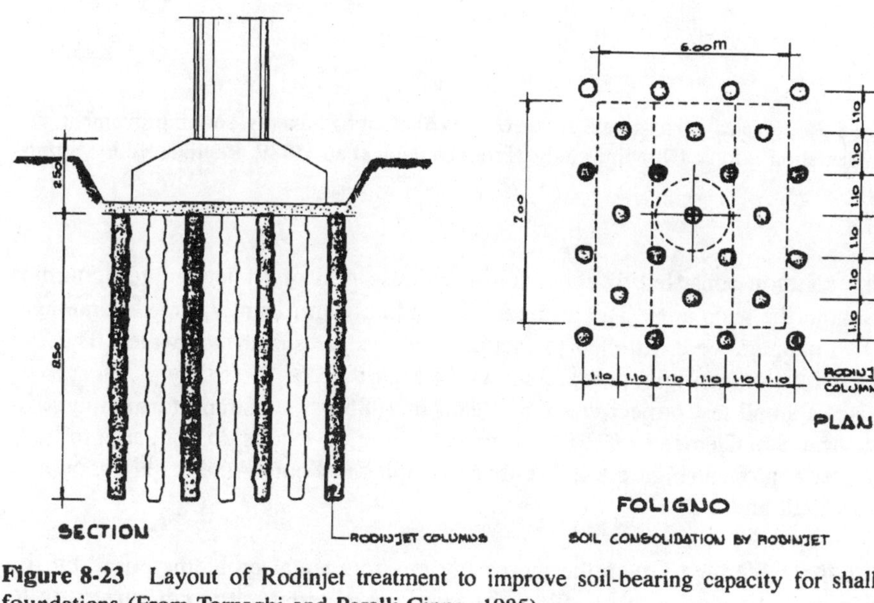

Figure 8-23 Layout of Rodinjet treatment to improve soil-bearing capacity for shallow foundations (From Tornaghi and Perelli Cippo, 1985).

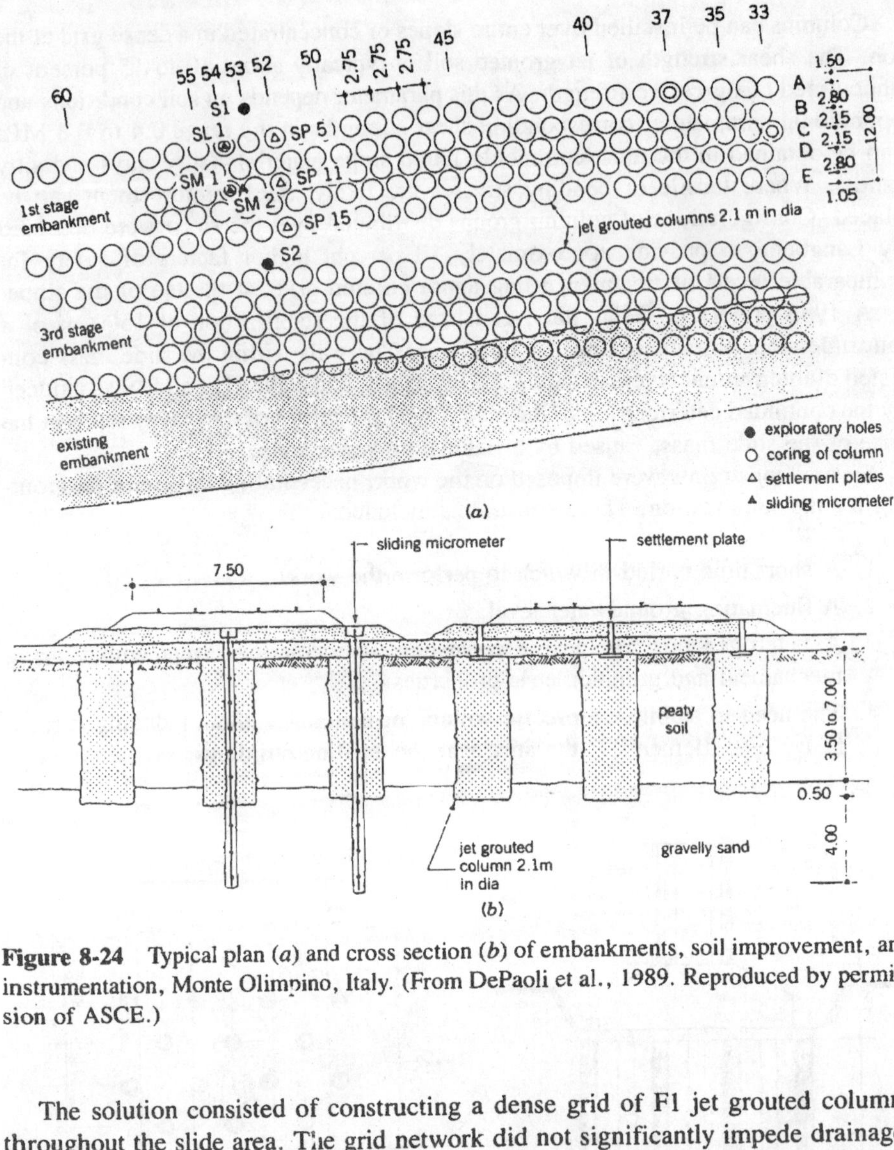

Figure 8-24 Typical plan (*a*) and cross section (*b*) of embankments, soil improvement, and instrumentation, Monte Olimpino, Italy. (From DePaoli et al., 1989. Reproduced by permission of ASCE.)

The solution consisted of constructing a dense grid of F1 jet grouted columns throughout the slide area. The grid network did not significantly impede drainage, and hydroaugers were installed to faciliate removal of subsurface water. The site and stabilization configuration are shown in Figure 8-26.

Also, a small test project was completed in 1985 in Capistrano formation sediments near San Clemente, Calif., but the encouraging results do not seem to have been later exploited (Figure 8-27). Other examples in South America are shown in Figures 8-28 and 8-29.

Hazardous Waste Given the increasing environmental problems posed by the containment, fixation, and handling of hazardous materials, it is not surprising to find jet grouting also entering this field. A typical example was provided by Hal-

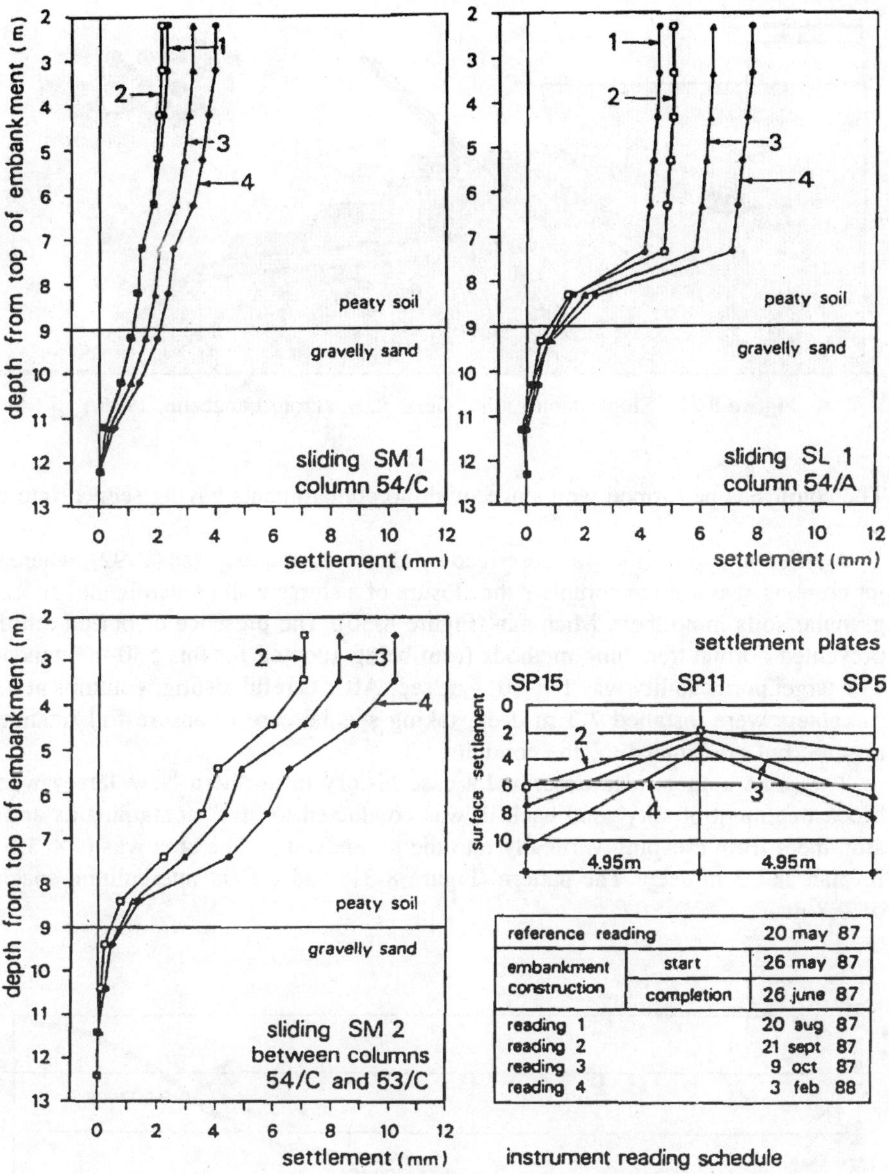

Figure 8-25 Preliminary results from settlement monitoring, Monte Olimpino, Italy. (From DePaoli et al., 1989. Reproduced by permission of ASCE.)

liburton (1987). An industrial gas producer in Oklahoma maintained a storage pit containing high pH, high sulfide water on an old dump site adjacent to a small stream. Regulatory agency instructions necessitated sealing the highly permeable soils on the side near the stream. A chemically resistant cement/ash grout was used in the F1 treatment of a 66-m length to a depth of 6 m into hard, dry clay aquiclude.

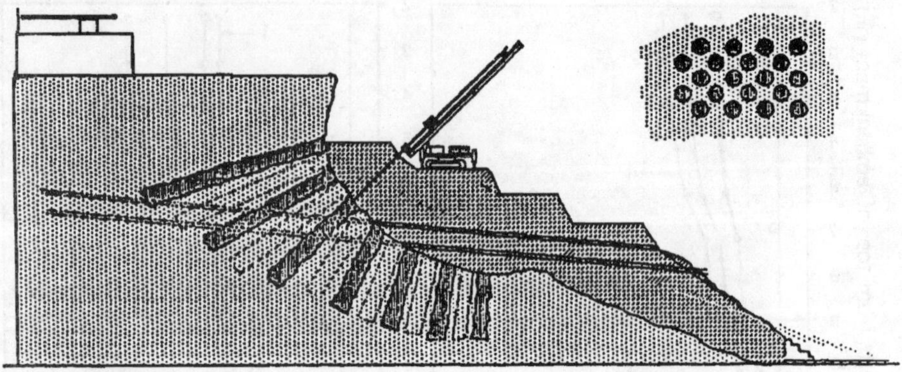

Figure 8-26 Slope stabilization, Gela, Italy. (From Langbehn, 1986.)

The cutoff has performed well since, with no contaminants having seeped into the river.

A different application was described by Gazaway and Jasperse (1992), where F1 jet grouting was used to complete the closure of a slurry wall contaminant barrier in granular soils in northern Michigan (Figure 8-30). The presence of buried utilities prevented normal trenching methods from being adopted for this 530-m^2 window. The target permeability was 1×10^{-8} m/sec. After careful testing, columns at 0.6-m centers were installed 7.3 m deep, taking special care to ensure full treatment around, but no damage to, the conduits.

The same authors also described a case history in northern New Jersey where block treatment of silty sand backfill was conducted to "fix" contaminants and to stop them from escaping vertically into the groundwater. The area was 6 × 3.6 m in plan and 3 m deep. The pattern (Figure 8-31) had a final intercolumn spacing of 0.8 m.

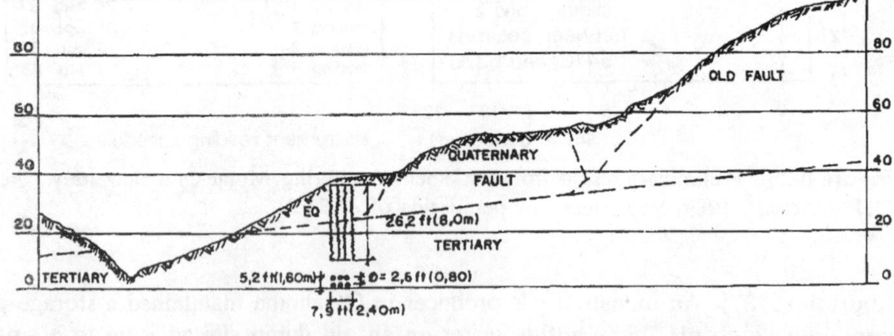

Figure 8-27 F1 columns for slope stability trial, San Clemente, California. (From Novatecna, 1988.)

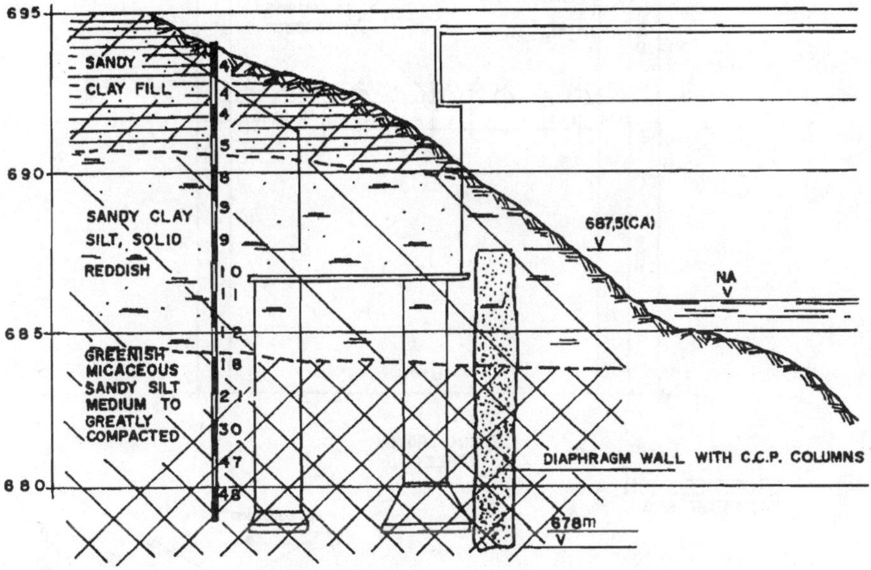

Figure 8-28 Protection by F1 columns of the abutments of railway and highway bridges over the Arrudas Stream, Belo Horizonte, Brazil. (From Novatecna, 1988.)

On both these projects, bentonite was incorporated into the grout to aid its waterproofing efficiency.

Miscellaneous Applications An excellent application was described by Andromalos and Gazaway (1989). An existing coal loading pier in Norfolk, Virginia was completed in 1961 for the purpose of loading oceangoing ships. The size of the ships that could be fully loaded was limited to the depth of the mud line adjacent to the pier: dredging would provide an increase in loading capacity and pier productivity. The pier was supported on concrete piles (Figure 8-32) and so there was a concern that their lateral stability would be significantly reduced by the sloughing of the silts and sands into the excavation. A soil retention system was necessary, and jet grouting was used to create a wall between the existing rows of concrete piles. The size of the project, together with its constraints on equipment access, led to the custom design and fabrication of the double stem F1 system previously shown in Figure 8-6. Over 1400 holes were involved at 560-mm centers. The quality of the soilcrete produced by this method was claimed by the authors to be superior to that obtained by the conventional single jet system. Details of the preliminary test program are provided in the discussion of Norfolk in Section 8.6.

A not dissimilar application was described by Parry-Davies et al. (1992). The F3 method was used to consolidate a marine, littoral boulder bed against scour beneath one of the columns of a newly constructed pier in Port Elizabeth, South Africa. Following intensive testing of parameters and effectiveness, the work was conducted under difficult ambient conditions, and involved a curtain of holes at 650-

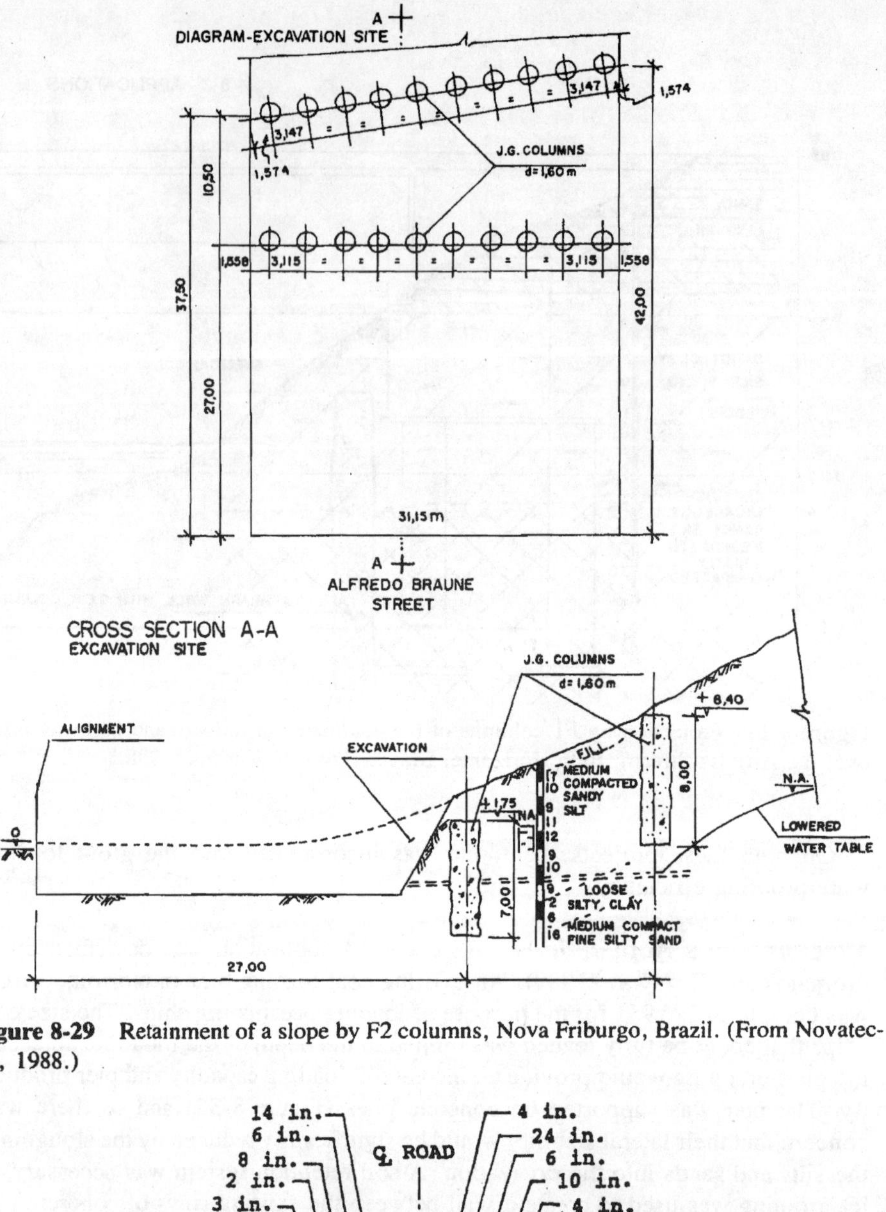

Figure 8-29 Retainment of a slope by F2 columns, Nova Friburgo, Brazil. (From Novatecna, 1988.)

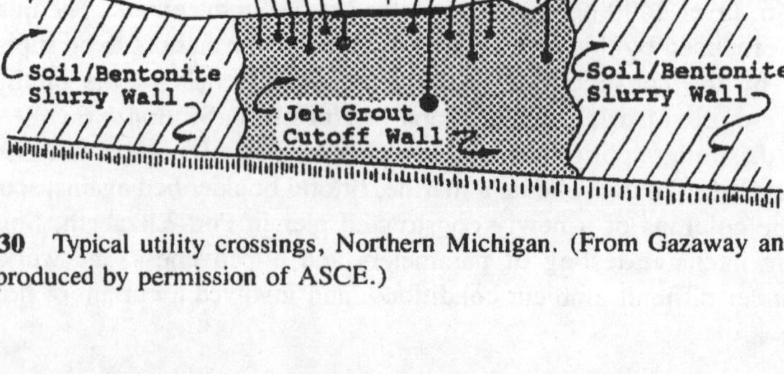

Figure 8-30 Typical utility crossings, Northern Michigan. (From Gazaway and Jasperse, 1992. Reproduced by permission of ASCE.)

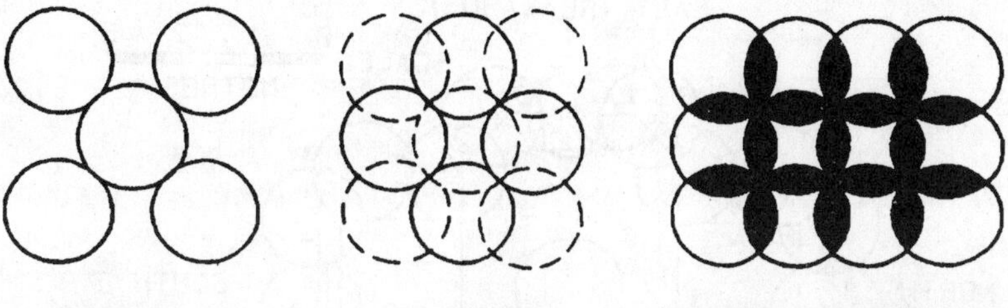

Figure 8-31 Typical block treatment pattern, New Jersey. (From Gazaway and Jasperse, 1992. Reproduced by permission of ASCE.)

mm centers around the footing (Figure 8-33). Holes were predrilled 125 mm in diameter with an eccentric duplex system prior to jetting. Internal "compartment" or consolidation treatment was then conducted, and posttreatment tests, coupled with the natural ocean activity, confirmed the excellence of the work.

Ichihashi et al. (1992) described how the bearing capacity of existing piles was increased by jet grout underpinning during the construction of a vehicular underpass in an international airport (Figure 8-34) in Japan. F3 jet grouting was also used to form two overlapping rows of columns as water cutoffs to prevent internal dewatering efforts causing drawdown, and structural settlement, outside the excavation. The layouts of the grout holes for cutoff and for underpinning are shown in Figures 8-35 and 8-36, respectively. Gauges showed 3 mm of "negative settlement" during

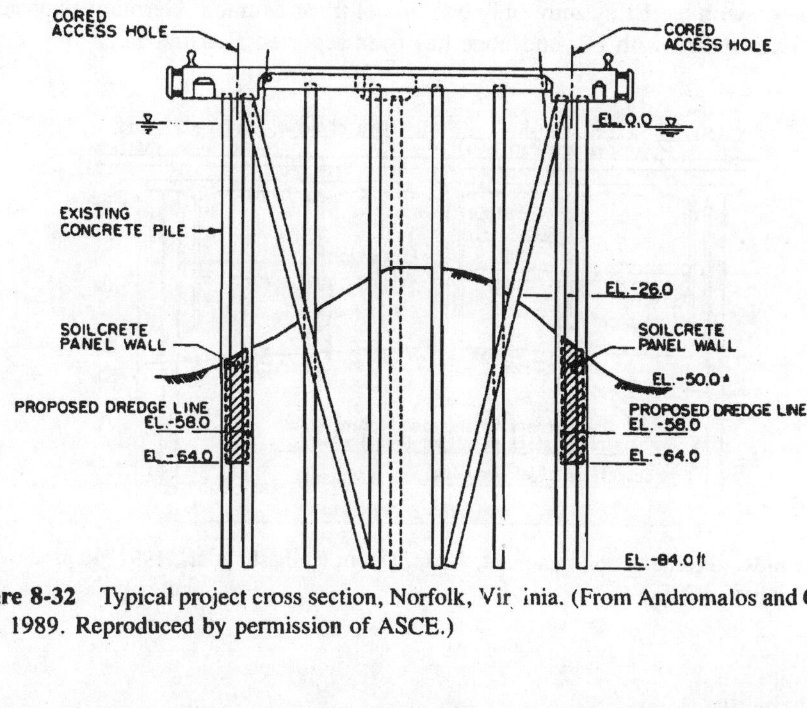

Figure 8-32 Typical project cross section, Norfolk, Vir, inia. (From Andromalos and Gazaway, 1989. Reproduced by permission of ASCE.)

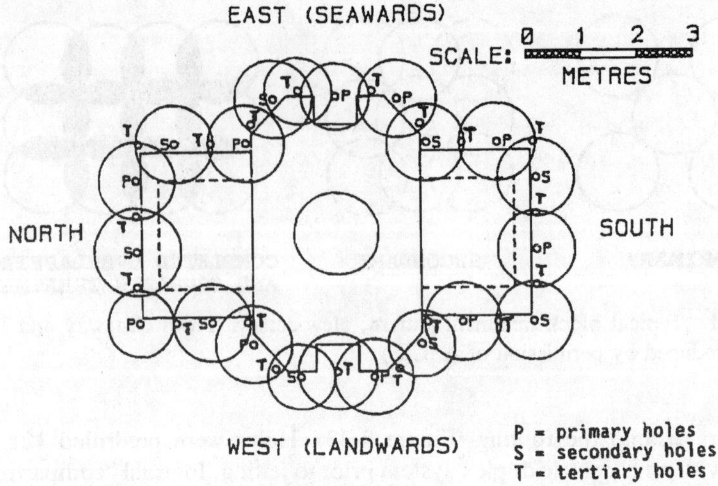

Figure 8-33 Theoretical grout travel based on grout acceptance, Port Elizabeth, South Africa. (From Parry-Davies et al., 1992. Reproduced by permission of ASCE.)

the subsequent tunnel excavation, while load tests on the supported piles proved "successful to clear the design goal." The permeability of the cutoff was negligible.

Horizontal/Subhorizontal Applications

General Comments Horizontal jet grouting for tunnel support represents one of the most popular applications worldwide. As noted above, this is almost exclusively conducted with the F1 system: only one tunnel (near Munich, Germany) appears to have been treated with F2, and none has been reported as using F3.

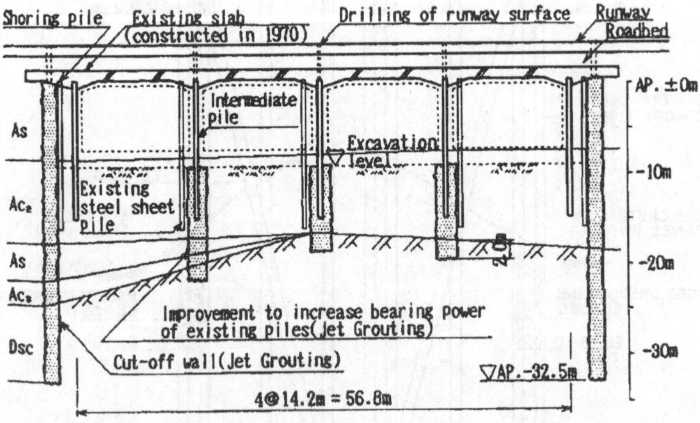

Figure 8-34 Profile of soil treatment, Japan. (From Ichihashi et al., 1992. Reproduced by permission of ASCE.)

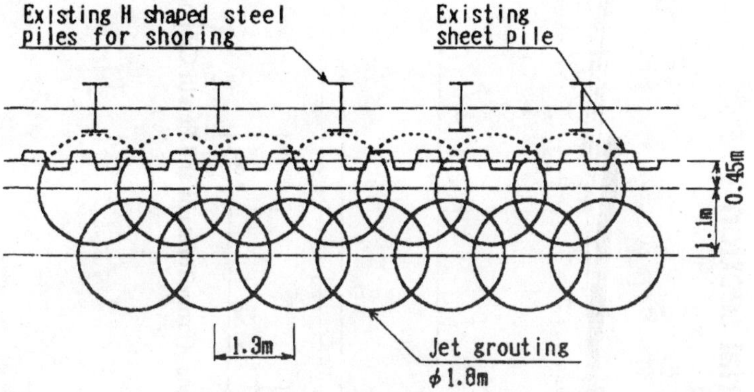

Figure 8-35 Alignment for water cutoff, Japan. (From Ichihashi et al., 1992. Reproduced by permission of ASCE.)

Columns are jetted about 8 to 10 m ahead of the face at angles of up to 15° above horizontal. A plug is placed after grouting to prevent egress of soilcrete during hardening. The face is temporarily supported by shotcrete, which also prevents the jetting spoil from eroding the face. This process is repeated until the perimeter has been completed from crown to springline. Inclined columns can then be formed at the invert to increase the bearing capacity of the soil at this point.

Details of Projects A typical example was described by Tornaghi and Perelli Cippo (1985). A railway tunnel, 12 m in diameter, was to be excavated through detrital soil at Moggio Udinese, Italy. Successive series of about 40 overlapping columns, 13 m long, were formed by holes drilled with an initial spacing of 0.45 m, permitting interim excavation in 10-m stages (Figure 8-37).

Mongilardi and Tornaghi (1986) describe similar applications on the Milan Metro, except that the jet grouting was used to permit the excavation of the 3-m-diameter pilot grouting tunnels (Figure 8-38) in 6-m stages. This method, used for

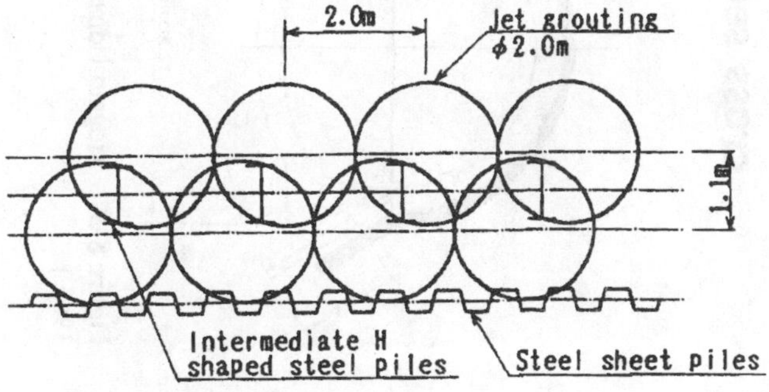

Figure 8-36 Alignment for bearing resistance, Japan. (From Ichihashi et al., 1992. Reproduced by permission of ASCE.)

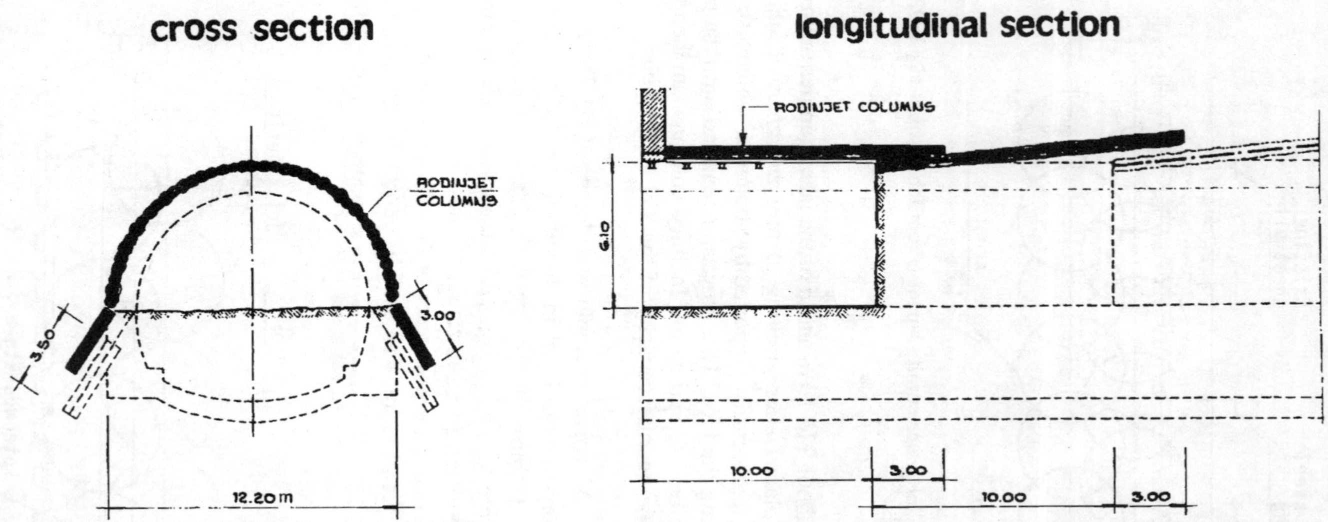

Figure 8-37 Moggio Udinese railway tunnel: scheme of Rodinjet treatment ahead of tunnel face. (From Tornaghi and Perelli Cippo, 1985.)

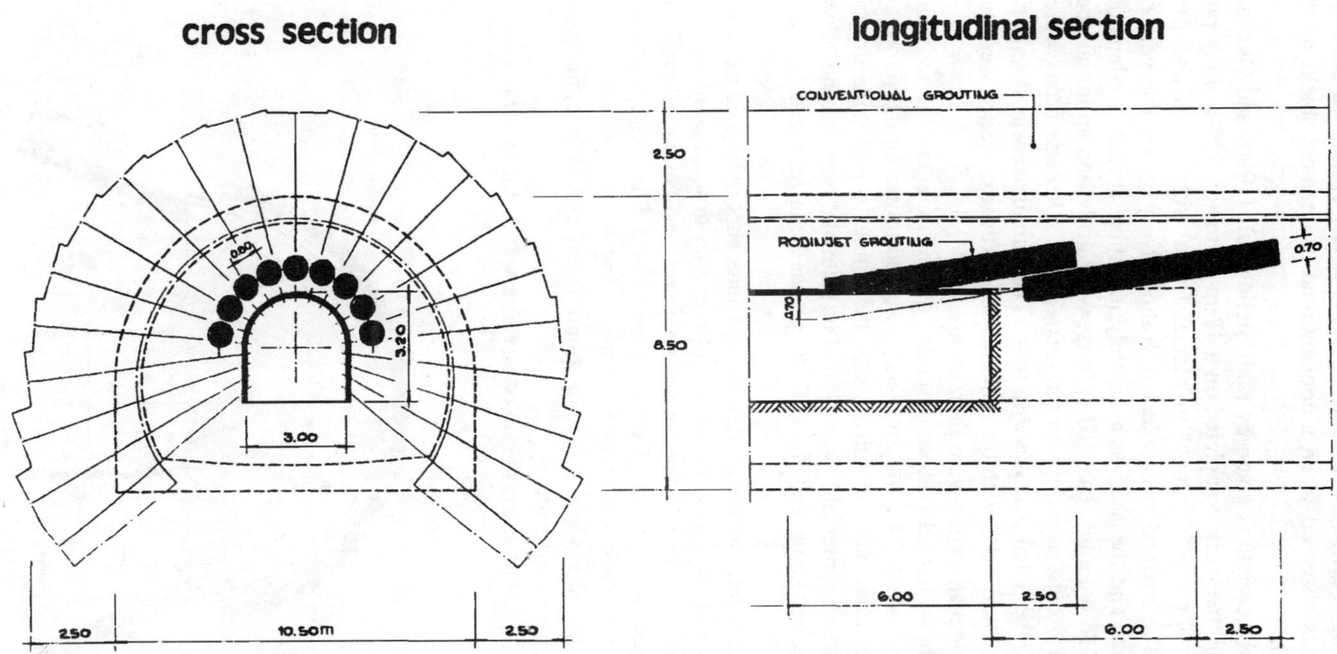

Figure 8-38 Milan Metro, Line 3: scheme of soil treatment for drift (subhorizontal Rodinjet columns) and for main bore (conventional grouting from drift. (From Mongilardi and Tornaghi, 1986.)

over 1300 linear m of drift on one contract alone, totally eliminated the problem of surface settlements observed when a conventional mechanical shield method had been earlier used.

A more recent example is given by both Stella et al. (1990) and Otto and Thut (1991), who detail the work completed on a 12-m-diameter, 750-m-long portion of the new Monte Olimpino 2 tunnel in glacial fluvial silty sands and gravels in northern Italy. The method was selected competitively given the very variable ground conditions, and the need to absolutely minimize surface settlements. Successive series of 35 columns, each 10 m long, were drilled on an initial spacing of 0.45 m and an 11° outwards slope to the tunnel axis. Alternate holes were drilled and grouted first; the jetted part was 7 m long. The remainder was simply plugged, thus providing a 1-m overlap with the previous series (the excavation steps were 6 m long). The intermediate holes were then completed.

Figure 8-39 shows in detail the cross section of the tunnel and the arch formed by the jet-grouted cylinders. The soil arch was designed to permit the removal of soil within the tunnel such that the soil outside the tunnel would undergo minimum deformations without appreciable variation of its state of stress. In other words, "the soil outside does not know that the soil inside has been removed" (Stella et al., 1990).

After the excavation, internal support, and face shotcreting of each 6-m-long heading, two parallel rows of subvertical reinforced jet-grouted columns spaced 0.6 m apart were created to form a continuous retaining wall below the prelined vault, as shown in Figure 8-39. Each of the columns was 5.5 m long. Their structural function was to withstand the earth pressure during the subsequent phase of the bottom excavation. The vault was then poured and carefully connected to the subvertical jet-grouted structure.

The overall average production rate of advance was 2.4 m per working day, and

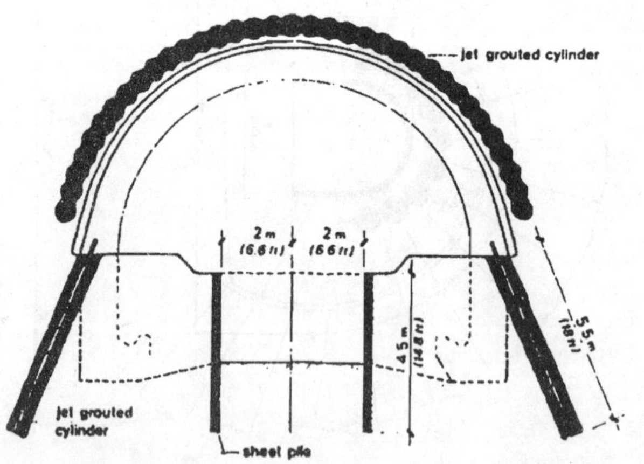

Figure 8-39 Cross section showing arch of jet grouted cylinders, Monte Olimpino, Italy. (From Stella et al., 1990. Reproduced by permission of ASCE.)

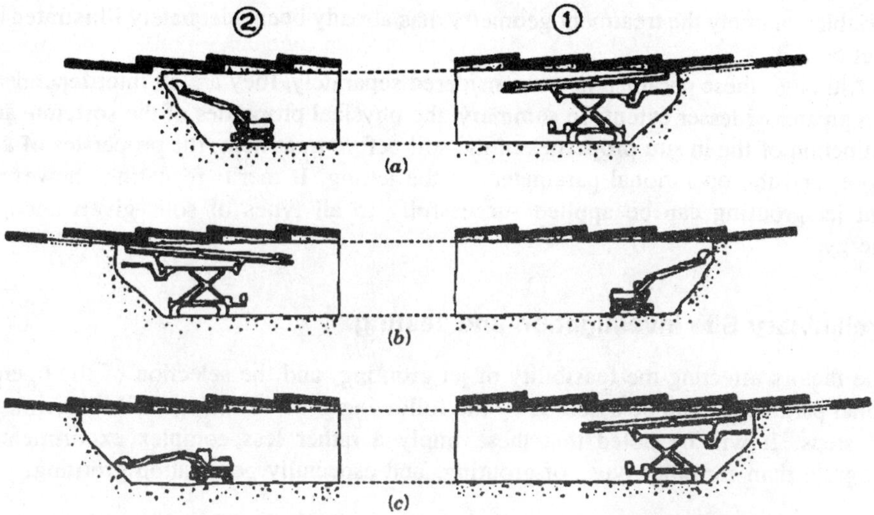

Figure 8-40 Monte Olimpino 2 Tunnel: sequence of jet grouting and excavation stages on two headings. (From Stella et al., 1990. Reproduced by permission of ASCE.)

"very few" problems were encountered with potentially problematical fine grained soils. The work was performed by alternatively shifting the rig between two opposite headings, and coordinating the jet grouting and excavation phases, as in Figure 8-40. Each column was left to cure for 24 hr. Close instrumentation of the soil above the tunnel proved that "the small tensile stress due to soil movement during excavation, and the fact that the net stress remained compressive indicates that the jet-grouting technique worked successfully in controlling ground movement by precompressing the overlying soil and providing support" (Stella et al., 1990).

8-4 DESIGN ASPECTS

Welsh and Burke (1991) referred to a "seven year hiatus" in records of their company's U.S. jet grout applications from 1980 to 1987. They ascribed this to the "concern of both specialty construction firms and consultants regarding satisfactory performance" of jet grouting. The situation was altered by the tremendous growth in European activity, especially with regard to equipment developments and fundamental field experiences. The current consequence is a relatively active domestic market.

At the root of the hiatus, however, was the lack of suitably researched design methods as directly related to local soil conditions. This symbionic relationship between theory and field experimentation has been deeply explored by specialists from each of the major contractors and the results of the most significant programs are detailed in Section 8.6. In this section, however, the impact of each of the major groups of variables is examined in broader, more generic fashion. One of these variables, namely the treatment geometry, has already been adequately illustrated in Section 8.3.

Although these parameters are considered separately, they are all interdependent to a greater or lesser extent. In summary, the physical properties of the soilcrete are a function of the in situ properties of the soil before treatment, the properties of the grout, and the operational parameters of the jetting. It merits repeating, however, that jet grouting can be applied successfully to all types of soil, given enough energy.

Preliminary Site Investigation and Testing

The factors affecting the feasibility of jet grouting, and the selection of the operational parameters can be assessed by the following general and specific experimental steps. It will be noted that these imply a rather less complex experimental program than for other types of grouting, and especially permeation grouting.

1. Detailed in situ tests, such as CPT or SPT, to estimate soil consistency or relative density.
2. Classification tests on representative soil samples for evaluation of the grain size distribution of cohesionless materials or the water content, bulk density, and Atterberg limits of cohesives.
3. Laboratory tests on trial grouts and soilcrete.
4. In situ jet grouting tests for verifying the operational parameters, and for providing further information for the final design.

Selection of Grout

The materials and their relative proportions are selected to meet the specified requirements of strength and permeability. They are bound by different and less restrictive criteria than those applying to permeation grouting and are selected primarily to provide high yields at minimum unit cost. They are almost exclusively cement based.

Regarding rheology, viscosity and rigidity should be low to allow an effective treatment to the greatest distance. When strength is the main criterion, a neat water-cement grout ($w/c = 0.6$ to 1.2; usually around 1) is selected depending on the soil granulometry, permeability, water content, and the average quantity of grout per unit volume of treated soil. In low-permeability soils, poor drainage of water from the grout will occur, so that the soilcrete is of lower strength than when a granular soil is treated, other factors being equal. The addition of bentonite (up to 5% of solids weight) may be appropriate to reduce the drainage effects in granular soils when permeability control is the main concern and high strength is not required. In this case the water : solids ratio can reach 4. Coomber (1985b) reports that fly ash has been used in ratios of cement : flyash = 1 : 1 to 10. Fillers coarser or more abrasive than ash are avoided because of wear on equipment and nozzles and the risks of blockages. Special additives or admixtures to control set time or to provide certain properties to the grout or soilcrete can be used but are relatively rare.

Selection of Jet Grout Parameters with Respect to Soil Types

The influences of nozzle size, pressure, type and quantity of grout, and monitor rotation and withdrawal rates have been closely researched, and are the subject of a large and complex literature. The general ranges are summarized in Tables 8-1 and 8-2.

Water flow rate and pressure, which dictate the energy imparted by the jet, together with withdrawal speed, are the most influential factors governing radial penetration. The decay of jet energy with distance from the nozzle is approximately exponential. Miki and Nakanishi (1984) published the data in Figure 8-41 for a stationary jet operating at a typical jetting pressure. This highlights both the restrictive and dominating effect of water (such as would occur in a saturated soil), but also the enhancement afforded by the air envelope (as in the F2 method). In very gravelly soils, however, it is the large size of particles relative to the jet diameter that controls the shape of the column: these particles will reflect energy and so locally reduce penetration. Diagrams similar to Figure 8-42 can be developed for all soil types, based on experiment. In reality, the true radius of a jet is reduced from the theoretical because of the dampening effect of the relatively dense water/soil slurry. This is particularly true in cohesive soils where considerable energy has to be consumed simply disrupting the soil structure and a minimum jetting pressure of 40 MPa may be needed (Coomber, 1985a) just to erode a soil of 0.04 MPa shear strength.

When the time of exposure to the jet is considered, a design curve of the type shown in Figure 8-43 is obtained. Increased pressure, or decreased lifting speed increases column diameter. Variation in column diameter rarely exceeds 10 percent except in very coarse soils. Broad recommendations for the withdrawal rate for different soil types and various strengths are shown in Figure 8-44.

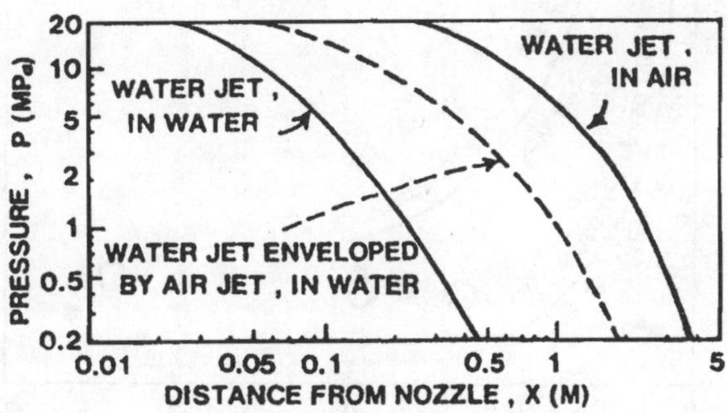

Figure 8-41 Pressure decrease of water jets along the axis of jet. (From Miki and Nakanishi, 1984.)

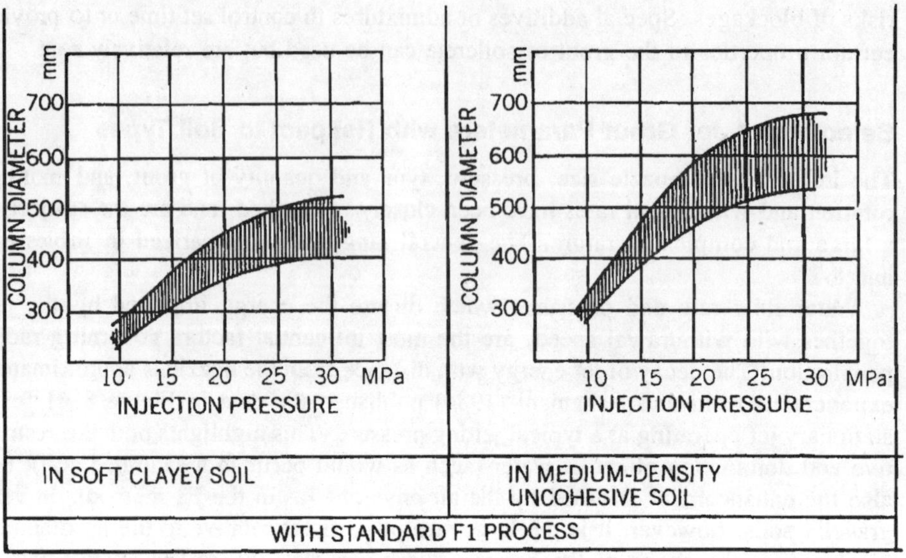

Figure 8-42 Diameter injection pressure relationship. (From Langbehn, 1986.)

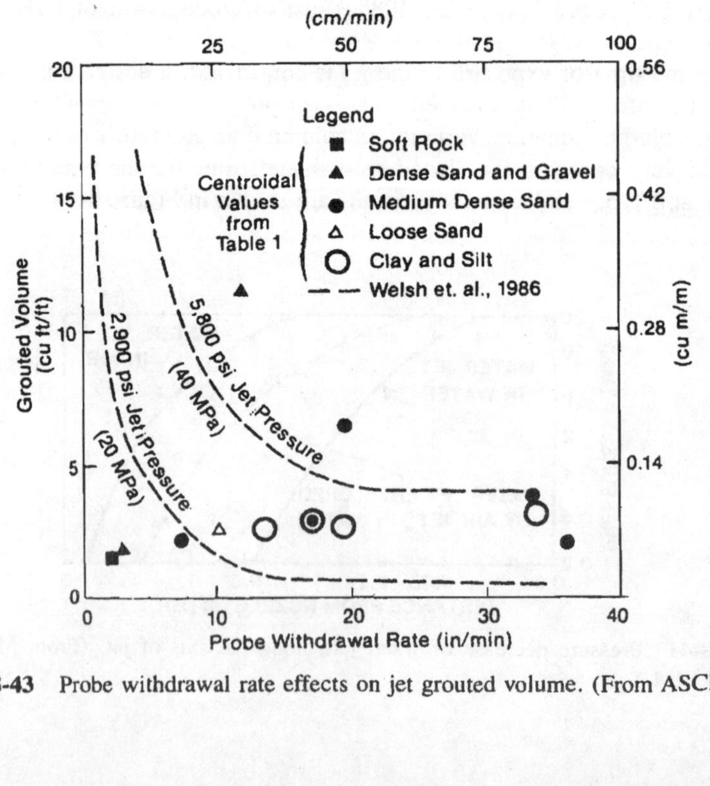

Figure 8-43 Probe withdrawal rate effects on jet grouted volume. (From ASCE, 1987.)

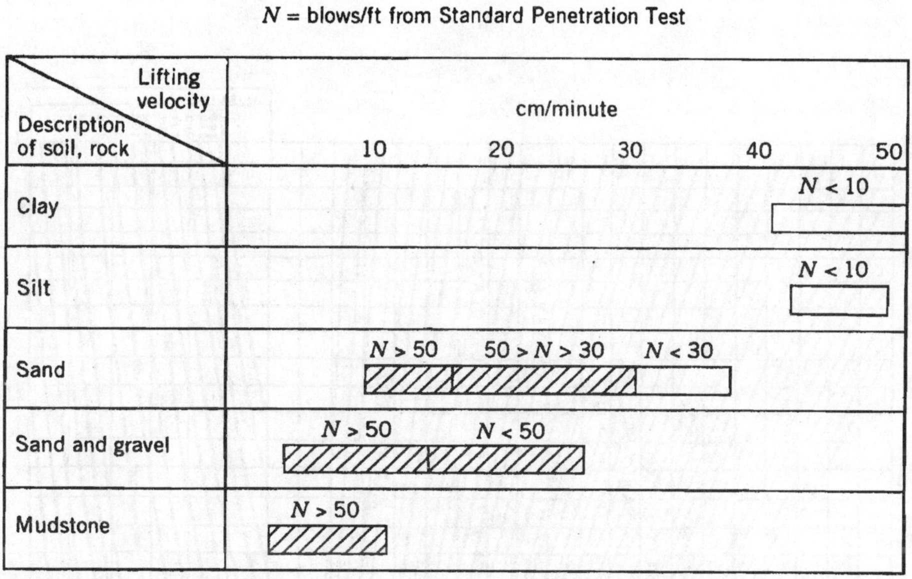

Figure 8-44 Relation between rate of withdrawal of monitor and soil type. (After Yahiro and Yoshida, 1973.)

With increasing experience and confidence, design charts such as Figure 8-45 can be developed to help identify the appropriate combination of process variables, as a first step in iterative refinement. Many equally acceptable solutions may be determined technically; the final choice is then based on economics, assuming field test data support the technical solution.

Miki and Nakanishi (1984) published empirically derived data on typical jet grout column diameters as related to SPT values (Figure 8-46). They showed that F3 creates columns three times larger than F1 regardless of soil type and density. Furthermore, their data show that there is generally only a 25 percent reduction in column diameter when passing from loose to very dense soil, all other parameters being equal.

As noted above, the jetting pressure should be related to soil type and density or consistency for optimal results, but typically ranges from 20 to 60 MPa. Friction losses between pump and nozzle may account for 5 to 10 percent of the pump pressure. In experiments conducted in the former Soviet Union, Broid et al. (1981) demonstrated that large-diameter (5 to 7 mm) and high-discharge (120 to 250 liters/min) nozzles doubled grouted volume in sandy soils at low pressures (7 MPa) as compared to Japanese experience. Considerable research remains to be conducted.

ASCE (1987) published a summary of grouting parameters recorded to that time (Table 8-6) and in summary quoted Welsh et al.'s (1986) rules of thumb regarding the relationship between soil type and soilcrete characteristics:

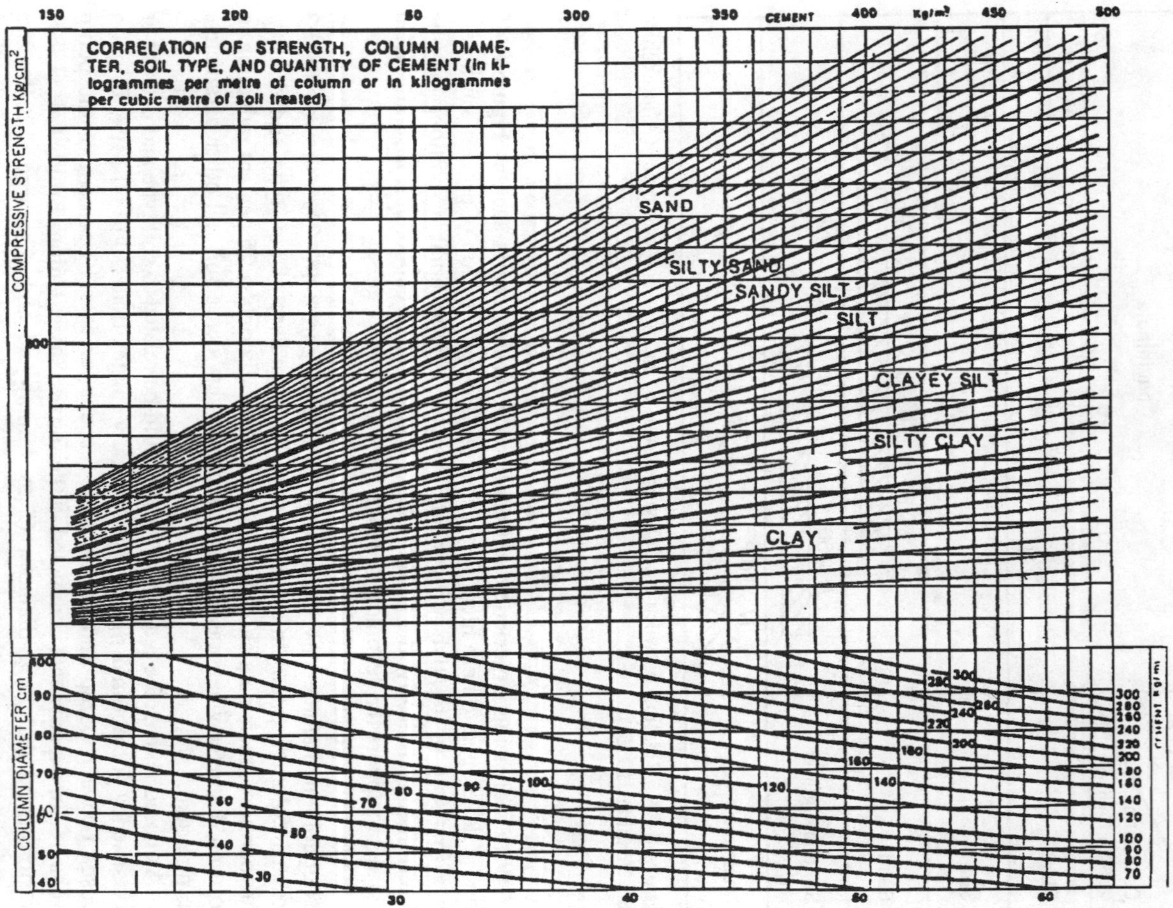

Figure 8-45 Fl design chart. (From Langbehn, 1986.)

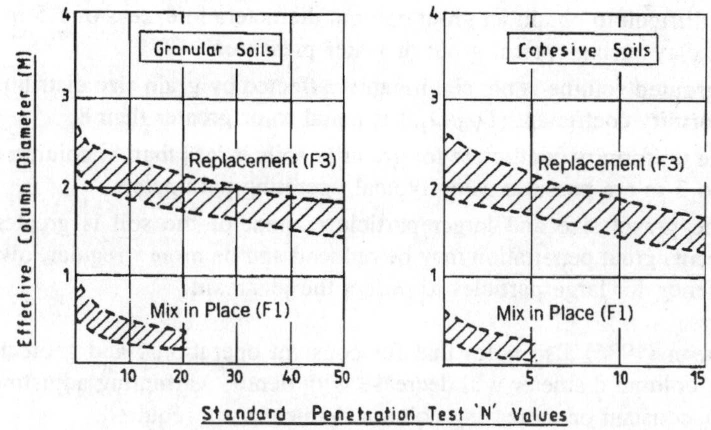

Figure 8-46 Column diameter related to in situ density. (After Miki and Nakanishi, 1984).

- For any soil, grout or water pressure and probe withdrawal rate are the most significant design factors impacting the volume of jet grouted soil.
- For any given grout pressure and withdrawal rate, jet grouted volume decreases with increasing clay content.
- As clay content increases, grout pressure must be increased and/or withdrawal rate decreased for a given jet grouted volume.

TABLE 8-6 Effect of Monitor Withdrawal Rate on Grouted Volume

Soil	Water/Grout Pressure (lb/in.^{2})[a]	Withdrawal Rate in./minute[a]	Grouted Volume ft^3/ft	Source
Soft rock	5700–7100	0.9– 3.7	1.5– 1.8	Yahiro et al., 1975
Dense sand	5700–7100	1.2– 4.6	1.6– 1.9	Yahiro et al., 1975
and gravel	5000–6000	12.0	3.1–19.6	ENR, 1986
Medium dense sand	5700–7100	5.2– 9.1	1.8– 2.6	Yahiro et al., 1975
	2900–5800	3.9–19.7	1.4–19.7	Welsh et al., 1986
	5600	36.0	2.0	ENR, 1974
	4,400	15.7	3.0	Aschieri et al., 1973
	5700–7100	19.7–47.2	0.9– 6.5	Yahiro and Yoshida, 1973
	800–1000	15.7–23.6	2.2–10.8	Broid et al., 1981
Loose sand	5700–7100	9.1–11.3	2.4– 2.7	Yahiro et al., 1975
Clay and silt	5700–7100	12.2–15.2	2.6– 2.9	Yahiro et al., 1975
	2900–5800	3.9–16.6	1.4–16.6	Welsh et al., 1986
	4400	15.7	3.0	Aschieri et al., 1983
	5700–7100	19.7–47.2	0.7– 5.8	
	800–1000	15.7–23.6	0.8– 4.3	Broid et al., 1981

Source: From ASCE (1987).

[a] 1,000 psi = 6.894 MPa; 1 in./minute = 2.54 cm/minute; 1 ft^3 = 0.028 m^3

- It is difficult to obtain jet grout column diameters in excess of 1.5 m in stiff to hard clays using typical grout or water pressures.
- Jet grouted volume is not significantly affected by grain size distribution if the uniformity coefficient (D_{60}/D_{10}) is equal to or greater than 8.
- If the uniformity coefficient for granular soils is less than 8, column diameters up to 3 m are possible using typical operating parameters.
- If the gravel size and larger particle content of the soil is greater than 50 percent, grout penetration may be reduced and be more irregular, owing to the tendency for large particles to reflect the jetstream.

Langbehn (1986) also notes that for constant operational and geotechnical parameters, column diameter will decrease with depth, warranting adjustment of the process if constant or increasing column diameters are required.

As a final point, it must be noted that a great deal of fundamental research has been conducted into the details of nozzle design, and the controls over the efficiency of the cutting action. These are described in detail by Kauschinger and Perry (1986), and Kauschinger and Welsh (1989). They calculated that a grout of w/c = 1 pumped through two nozzles, each 2 mm in diameter, at the ideal jet velocity of 250 m/sec would produce a flow rate of about 76 liters/min, necessitating a nominal 100 HP of driving power.

Characteristics of Jet Grouted Soils

Broadly, for a given cement in a given soil, the *strength* may be correlated statistically to the soilcrete w/c ratios, assuming a relatively homogeneous soilcrete is attained. For saturated cohesive soils, treated by F1, Figure 8-47 shows the ranges of soilcrete c/w as a function of grout content, for four different c/w contents and three different virgin soil water contents. This confirms the importance of cement content when appreciable drainage is not possible. On the other hand, in granular soils, the typically lower moisture content and the drainage effect during jetting are factors providing soilcrete w/c ratios even lower than that of the mix.

Figure 8-48, also for fine cohesive soils, shows unconfined compressive strengths (R) after one to two months, with two scales representing the statistical range of maximum frequency of the strength index:

$$R_o = R/(c/w)^n$$

n ranges from 1.5 to 3 but is typically assumed to be 2 in most inorganic soils.

The strength index *Ro* depends mainly on the type of cement, the principal properties of the soil, and on the age at test. For gravelly-sandy soils, and with high-strength cements, *Ro* in the long term can be 10 to 30 MPa for the F1 method. For highly plastic soils, it is difficult to provide strengths above 3 MPa unless very high cement contents are used, largely because the w/c ratio of the soilcrete will be greater than that of the grout. In most soils, the use of F2 and or F3 methods will

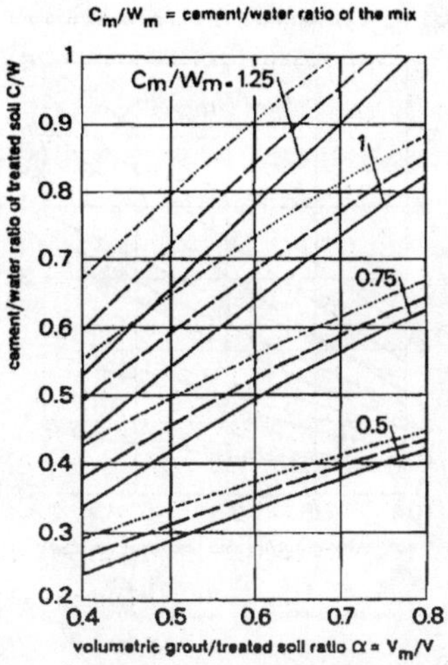

type	volumetric characteristics of native soil		
	saturated bulk density γ_t (t/m^3)	dry density γ_d (t/m^3)	water content W_t (%)
T1 ------	2.037	1.630	25
T2 — —	1.833	1.309	40
T3 ——	1.618	0.971	67

Figure 8-47 Saturated cohesive soils: influence of treatment on final cement-water ratio. (From Gallavresi, 1992. Reproduced by permission of ASCE.)

give larger column diameters, but strengths that are generally lower, and more difficult to estimate at the preliminary design stage. Figure 8-49 provides some average experimental data.

Kauschinger et al. (1992a) reported on Rodio's test program conducted at Casalmaiocco, Italy, in a site with three distinct soil types. The upper 4 m comprised silt (ML) near the surface, grading down to silty fine sand (SM). From 4 to 6 m there were significant silty layers and 5 percent gravel. Below 6 m, the sandy soils became less silty (<20 percent) and contained up to 50 percent gravel. All three injection methods were used, giving average column diameters of 0.8, 2, and 2 m, respectively. The corresponding cement injected was 200, 1000, and 1000 kg/m and so the amount of cement injected per volume of soilcrete was very similar (320 to 395 kg/m^3). Strengths at 220 to 240 days are shown in Figure 8-50. In the upper 4 m, F1 soilcrete was strongest, but between 4 and 6 m the difference was less as the soil became more gravelly. Below 6 m, the F1 soilcrete was very strong and the F2 method provided the weakest. Further details are provided in the discussion of Casalmaiocco in Section 8.6.

The F3 method was used in very dense alluvials by Ichihashi et al. (1992) in Japan, using the parameters of Table 8-7. The strength data shown in Figure 8-51 varied with the position of the core samples, but averaged 1.2 MPa for column A-1 and up to 6.3 MPa for the "modified specification" of columns A-2 and B.

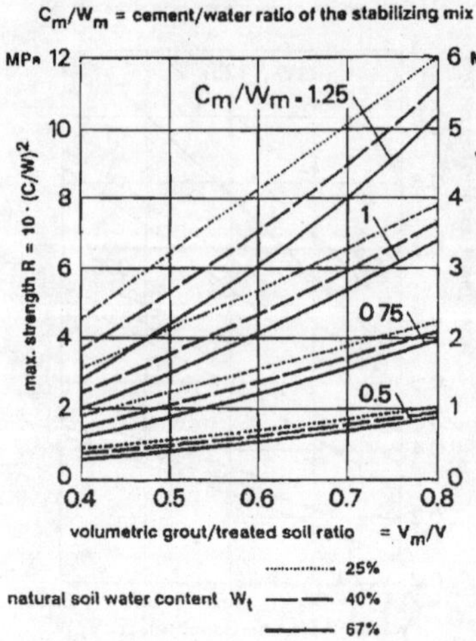

Figure 8-48 Cohesive saturated soils: influence of treatment on strength as a function of c/w (see Figure 8-47). (From Gallavresi, 1992. Reproduced by permission of ASCE.)

As an overview on the question of soilcrete strength, the major uncertainties in predicting soilcrete composition are related to the following (Kauschinger and Welsh, 1989):

1. The soil component of soilcrete is dictated by the in situ soil conditions, which may be very variable. Also, the grain size distribution of the soilcrete

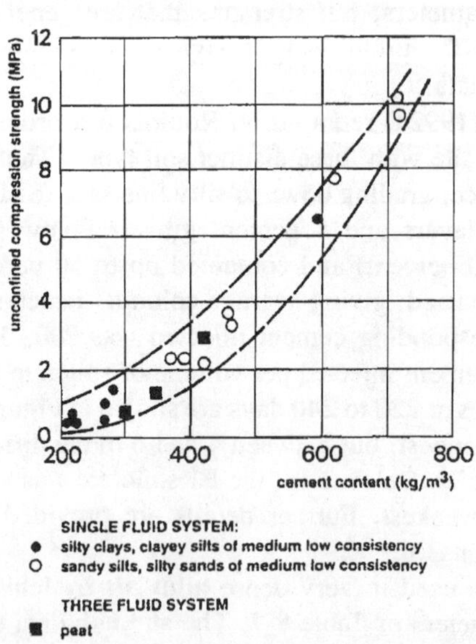

Figure 8-49 Strength as a function of cement content: mean experimental data from typical treatments. (From Gallavresi, 1992. Reproduced by permission of ASCE.)

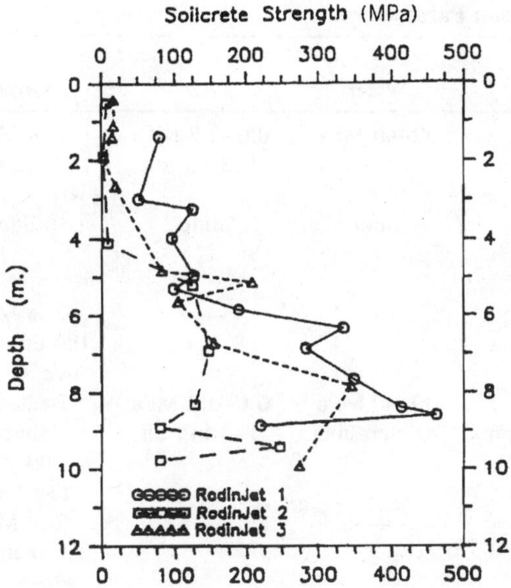

Figure 8-50 Soilcrete strengths for three jet grouting systems, Casalmaiocco test site. (From Rodio, 1987. Reprinted by permission of Ing. Giovanni Rodio & C., Milan, Italy.)

usually is *not* the same as the initial soil distribution because fine sand and silt particles are easily and preferentially eroded during jetting. However, removal of fines from the soilcrete is beneficial for strength increase.

2. The final water content of the soilcrete in the column is influenced by four major factors, which include:
   (a) The in situ water content of the soil: the higher the soil water content, the higher the w/c ratio of the soilcrete. High soil water content can be partially offset by using a thicker grout.
   (b) The w/c ratio of the injected grout specified.
   (c) Permeation of water out of the column, due to a gradient set up by residual jetting pressure in the column. This factor is especially important in sandy soils.
   (d) Self-weight consolidation of the wet soilcrete, which causes a reduction in the w/c of the soilcrete.
3. The soilcrete's w/c ratio and cement factor in situ are a function of the amount of cement and water injected, and the proportion of soil, water, and cement ejected in the cuttings. The volume of soilcrete created has a strong influence on both parameters. Only the amount of cement injected is under direct control. Mass balance approaches have been advocated (Rodio, 1983; Kauschinger and Welsh, 1989; Kauschinger et al., 1992a) for estimating the amount of soil, water, cement, left in situ, given measurements for the injected grout and cement content of the cuttings.

TABLE 8-7 Injection Parameters

Column	Type	Water	Air	Grout	Lifting Speed
A-1	Primary injection	30–40 MPa 70 liters/min	0.6–0.7 MPa 1–3 m/min	2–4 MPa 180 liters/min w/c = 100%	5 cm/min
A-2	Primary injection	ditto	ditto	ditto	ditto
	Secondary injection	——	——	2–4 MPa 180 liters/min w/c = 50%	10 cm/min
B	Primary injection	30–40 MPa 70 liters/min	0.6–0.7 MPa 1–3 m/min	Bentonite slurry 300–400 liters/min	5 cm/min
	Secondary injection	——	——	2–4 MPa 220 liters/min w/c = 50%	7.5 cm/min

Source: From Ichihashi et al. (1992).

4. Soilcrete is a nonhomogeneous mixture. The mixing of the in situ soil with grout is primarily a function of the jet velocity that breaks up the in situ soil, and the number of rod rotations (number of passes of the jet stream past a soil element). The nonhomogeneity can be caused by several factors:
   (a) Poor control over jetting parameters such as variable jetting pressure, worn nozzles, and lift rate, and poor control over step size. This points to the need for real-time data acquisition.
   (b) Insufficient mixing of in situ soil with grout. Field experiments indicate that about four to six revolutions of the drill string are needed for proper mixing of the grout and soil. Insufficient mixing will cause the soilcrete strength to be highest along the central axis and lower along the outer perimeter of the column. Improper mixing is most prevalent in clayey soils, where clay balls may become mixed into the soilcrete matrix.
   (c) Nonhomogeneous soil with depth.

Other minor influences affecting soilcrete strength are related to groundwater chemistry and seepage velocities through the region to be jet grouted. However, the influence of these parameters can be taken into account by testing the groundwater and calculating the seepage velocities. Appropriate action can easily be taken. For instance, if the seepage velocities are above 6 cm/sec (a recommended limit for normal grout), waterproofing additives can be mixed with the grout prior to injection.

Regarding the relationships between strength and other parameters, Gallavresi

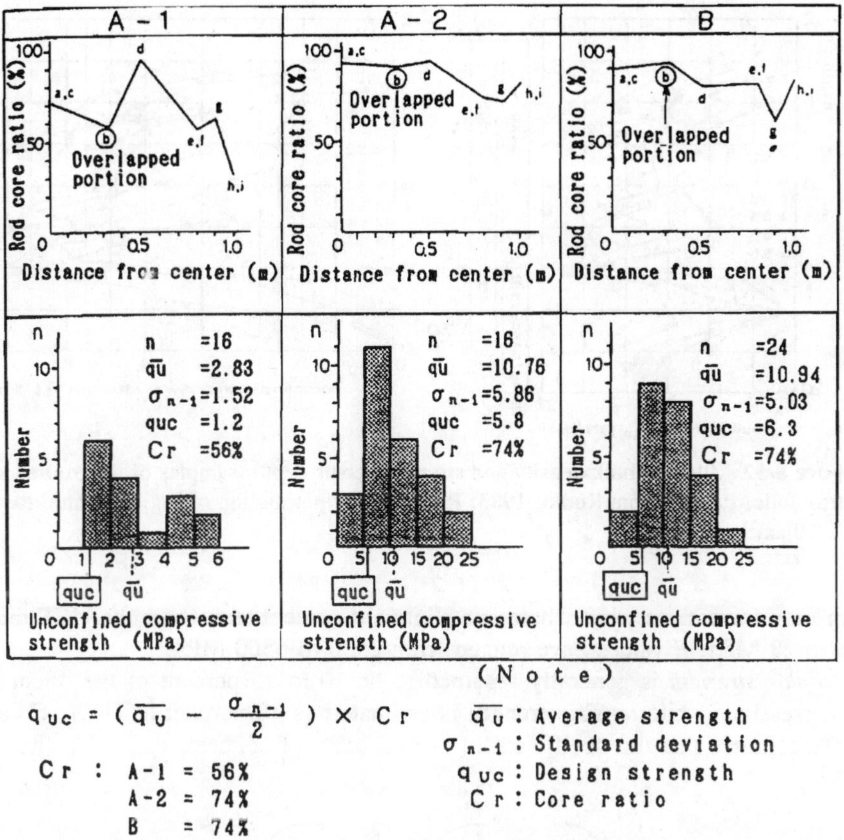

Figure 8-51 Core sampling rate and distribution of unconfined compressive strength, Japan. (From Ichihashi et al., 1992. Reproduced by permission of ASCE.)

(1992) described the data obtained from F1 soilcrete formed in silty sandy soils at Porto Tolle, Italy. Over 100 cored samples were examined and strength was related to bulk density, as shown in Figure 8-52.

Miyasaka et al. (1992) conducted tests on F3 soilcrete formed in loose alluvial silty clayey sand in Japan and recorded the data of Figure 8-53. The specified values for UCS (1 MPa), *cohesion* (0.3 MPa), and the *coefficient of subgrade reaction* (100 MN/m^2) were also surpassed.

The *deformability* of soilcrete is even more variable than strength, but in general the modulus/strength (*E*/*R*) ratio increases with strength (*R*) from about 300 to over 1000. Figure 8-54 shows secant modulus *E* at 40 percent peak strength, versus compressive strength from careful tests in various soilcretes from the Milan alluvials. Data from Trevi (1990) also support this high degree of variability. Figure 8-55 shows the results obtained on various sites in silty or sandy silty formation. For the same value of UCS, the *E* value may vary by a factor of 2 to 3. This also occurs

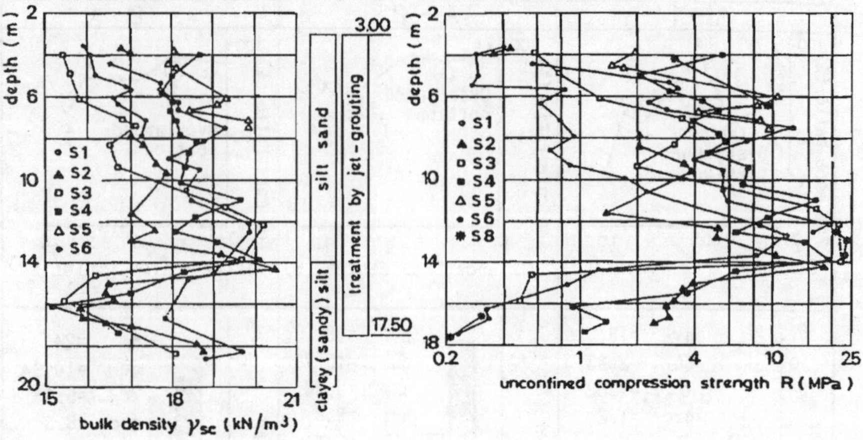

Figure 8-52 Plots of bulk density and strength versus depth (samples of jet grouted soil at Porto Tolle, Italy). (From Rodio, 1983. Reprinted by permission of Ing. Giovanni Rodio & C., Milan, Italy.)

in sandy formations where there is a slight silt content and where the UCS may be up to 28 MPa. *E* values have ranged from 2500 to 6500 MPa.

Shear strength is generally assumed to be 10 to 15 percent of the unconfined compressive, while *tensile* strength is estimated as 10 percent of UCS. Coomber (1985a) notes that the coefficient of *permeability* of soilcrete can be in the range

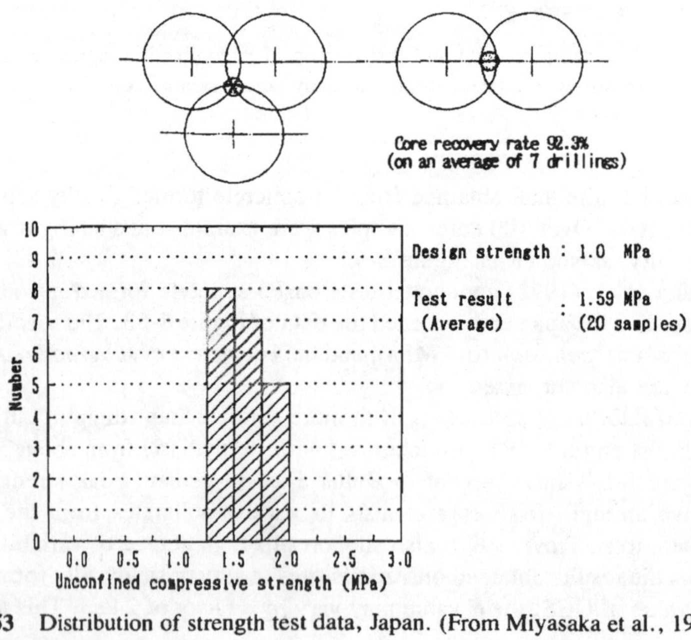

Figure 8-53 Distribution of strength test data, Japan. (From Miyasaka et al., 1992. Reproduced by permission of ASCE.)

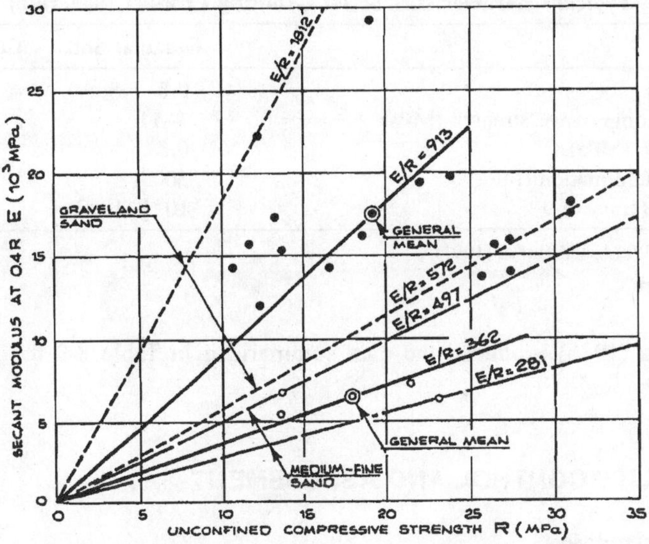

Figure 8-54 Milan Metro, Line 3: Italy. Plots of secant modulus versus strength (specimens of jet grouted alluvial soils). (From Gallavresi, 1992. Reproduced by permission of ASCE.)

10^{-6} to 10^{-9} cm/sec if bentonite is used, while Langbehn (1986) reminds that the overall permeability of a jet grout wall is "entirely contingent" on the continuity between adjacent columns or panels.

A synthesis of F3 work by Welsh and Burke (1991) provided the data of Table 8-8.

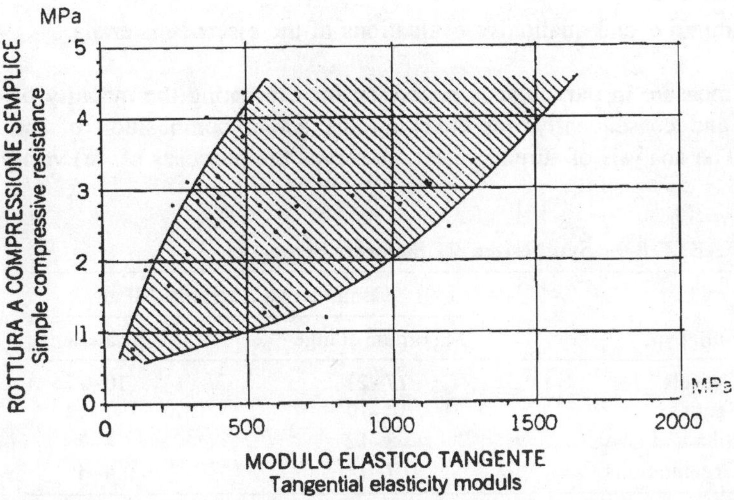

Figure 8-55 Values of the tangential elasticity modulus in relation to the values of unconfined compressive strength. Treatments in silty or silty-sandy soils (From Trevi, 1990.)

TABLE 8-8 Typical Characteristics of Jet Grouting Product (Soilcrete)

Parameter	Granular Soils	Cohesive Soils
Diameter (m)	0.8–1.8	0.5–1.5
Unconfined compressive strength (MPa)	5–11	1–5
Shear strength (MPa)	0.5	0.3
Modulus of deformation (MPa)	500	100
Permeability (cm/sec)	10^{-6}–10^{-7}	10^{-6}–10^{-7}

Source: From Welsh and Burke (1991).

Langbehn (1986) produced the data summarized in Table 8-9 for F1 grouting strengths.

8-5 QUALITY CONTROL AND ASSESSMENT

General Principles

In general the volume of grout injected is greater than the volume of voids in the natural soil. This reflects a primary role for the grout, in F1 and F2 applications, to act as a fracturing medium. It follows that during jet grouting, the excess fluid gives rise to soil displacements, and outflows of both grout and soil. Depending on the scale of the project, the following controls can be exercised:

- Precise leveling to check vertical movements on the surface or existing structures.
- Inclinometers to assess horizontal displacement at various distances from the treated area.
- Quantitative and qualitative evaluations of the ejected materials.

The last measure in particular is necessary for estimating the quantity of cement refused, and consequently, for checking the actual composition of the treated ground. The analysis of surplus materials serves the purposes of: (a) verifying the

TABLE 8-9 Synthesized F1 Soilcrete Strengths

	Compressive Strength (MPa)	
Soil Type	Maximum Range	Typical Range
Gravels	17–21	10–17
Sands	14–17	7–14
Silts and Clays	8–12	4–7
Organic Soils	0.7–7	0.4–4

Source: From Langbehn (1986).

design criteria; and (b) supplying data required to modify, if necessary, the jet grouting parameters.

Gallavresi (1992) lists the following options after treatment:

- Coring to provide laboratory specimens.
- Direct examination by special excavation.
- Static or dynamic penetration tests and pressuremeter tests to evaluate the increase in bearing capacities of intercolumn soils.
- Controls by sonic testing both to evaluate the improvement of the mechanical properties and to check the treatment uniformity and the possible overlapping of adjacent columns. In practice, vertical pipes are placed at preestablished spacings through which the transmitter and receiver are lowered.

Estimating the Composition of Cuttings and Soilcrete

Kauschinger et al. (1992b) described methods to estimate the composition of jet grouted bodies. The paper presented a detailed theoretical framework and laboratory experiments of potential use as guidelines when developing a qa/qc program. They discussed (a) development of mass balance equations that can be used to estimate the size and composition (soil, water, cement) of cuttings and soilcrete; (b) a test (heat of neutralization) that can be used to measure the concentration of cement in the cuttings and soilcrete; and (c) design charts (UCS, *E* with respect to w/c ratio, and cement factor) for samples of cemented Boston Blue Clay. A briefer guide to these researches was presented in an earlier work by Kauschinger and Welsh (1989), which concentrated on the heat of neutralization.

The major component of the cuttings that has to be determined is the amount of cement. Various techniques are available for estimating the amount of cement, but the heat of neutralization test (Queensland, 1978; Kauschinger and Hankour, 1989; U.S. Bureau of Reclamation, 1989) was determined to be simple, quick, and accurate. This test requires reagents (sodium acetate, glacial acetic acid) in certain proportions to be added to the cuttings. A digital thermometer is used to measure the temperature rise as the cement becomes neutralized. This rise in temperature is related to the cement content in the cuttings through the use of a calibration curve. Heat of neutralization experiments (Kauschinger and Hankour, 1989) were conducted on a mix with Boston Blue Clay. The calibration curves for estimating the cement content versus heat of neutralization for varying slurry densities (s/w = 1 : 1, 1 : 2, 1 : 5, 1 : 10) are plotted in Figure 8-56. The heat generated by the cement obviously increases in relation to the cement in the mixture, and is also a function of the density of the cuttings. Typically, jet grout cuttings contain about 40 percent cement, and therefore, the temperature rise above ambient room temperature would be about 15 to 20°C.

The value of the soil to water ratio (s/w) needed for input into Figure 8-56 can be calculated as follows:

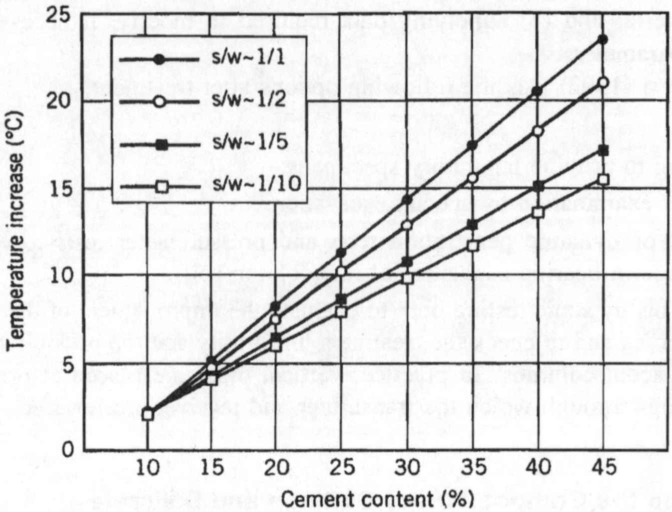

Figure 8-56 Heat of neutralization calibration curves for Boston Blue Clay. (From Kauschinger and Hankour, 1989.)

1. Obtain sample of cuttings, mix homogeneously, and determine unit weight of cuttings.
2. Use centrifuge to separate water, soil, cement. Measure amount of water, and determine water content of mixture at bottom of centrifuge tube. Possibly use microwave oven for water content determination for speed.
3. Calculate amount of water in cuttings from step 2.
4. Mix second sample homogeneously and perform heat of neutralization test.
5. From step 2 calculate amount of water in sample, from step 4 calculate amount of cement, and then the amount of soil. (Assumes volume of air in cuttings is negligible.)

The constituents of soil, water, and cement in the mixture calculated using the above procedure can be checked using the following equations:

Soil : Water (s : w) Ratio (Dimensionless):

$$(\text{sw}) = \frac{\gamma_{+}^{c}\ [1/\gamma_w + 1/(\gamma_w G_c)\ [\text{c/w}]] - [1 + \text{c/w}]}{[1 - \gamma_{+}^{c}/(\gamma_w G_s)]} \tag{8-1}$$

Weight of Water per Unit Volume:

$$W_w = \frac{\gamma_{+}^{c}}{\{\ 1 + \text{c/w} + \text{s/w}\}} \tag{8-2}$$

Weight of Cement per Unit Volume:

$$W_c = \frac{\gamma_+^c}{\left\{1 + \frac{s/w}{c/w} + \frac{1}{c/w}\right\}} \tag{8-3}$$

Weight of Soil per Unit Volume:

$$W_s = \frac{\gamma_+^c}{\left\{1 + \frac{c/w}{s/w} + 1/(s/w)\right\}} \tag{8-4}$$

where the units in Eqs. (8-1) to (8-4) are the same as the unit weight of the cuttings, γ_+^c, and the other variables are defined as:

γ_w = unit weight of water
G_c = specific gravity of cement (cement = 3 to 3.15)
G_s = specific gravity of soil (clay = 2.75)
c/w = cement to water weight ratio of the cuttings
s/w = soil to water weight ratio of cuttings

The sum of $W_w + W_c + W_s$ [Eqs. (8-1) to (8-4)] is equal to the density of the cuttings.

Quality Assessment by the Energy Approach

The aim of grouters has been to develop an effective monitoring procedure to permit a real-time quality control and provide a valuable basis for timely operational decisions. DePaoli et al. (1991) have recently reported on such an electronic monitoring system called PAPERJET.

A major feature of the system is the "energetic approach" for interpreting the parameters measured during drilling and grouting. This enables the determination of the in situ soil during drilling and the subsequent energy expended during the grouting. As a consequence, by means of the electronic measurement of pressures, flow rates, and volumes, the system makes it possible to energetically quantify the whole process.

In detail, the energy expended during the drilling and grouting phases is evaluated in the following.

The Specific Drilling Energy This energy, defined by Teale (1965), is a useful index for soil characterization, and is based on physical parameters playing a specific role during drilling. The excavation of a unit volume of soil or rock requires a certain energy, which has two components.

The first is the thrust F (kN) on the drilling tool, having the same section A (m^2) as the hole, for a unit downward displacement. This contribution is dimensionally a pressure, since the thrust for a unit displacement, divided by the excavated soil volume, corresponds to the pressure on the bottom of the hole.

The second is the torque T (kNm) of the power swivel, which is a function of the rotational speed S (rev/sec) and of the rate of penetration V (m/sec). Consequently, the specific drilling energy is expressed by:

$$E = \frac{F}{A} + \frac{2\pi ST}{AV} \quad (\text{kJ/m}^3) \tag{8-5}$$

This parameter characterizes a soil type from the energy viewpoint. The validity of Eq. (8-5) has been the subject of an experimental investigation by Rowlands (1971), who verified it under ideal drilling conditions, involving the following assumptions:

- The soil particles removed by the drilling bit are immediately removed by the drilling fluid, that is, without additional grinding at the bottom of the hole.
- The flushing action of the drilling fluid occurs without clogging or collar formation along the rods.
- There is no loss of energy along the rods due to friction against the sides of the hole, or vibrations.
- The wear conditions of the drilling bit are constant.

As a consequence, the use of the above equation requires that the drilling phase be carried out with some care so that "ideal" conditions may be verified: only if the drilling operations are conducted with the same procedure, is it possible to compare properly two different specific drilling energies. This condition can be verified only by monitoring the drilling.

The Specific Jet Grouting Energy For a unit length of column this energy depends mainly on:

Grout pressure	P (MPa)
Grout flow rate	Q (m^3/hr)
Withdrawal speed	V_t (m/hr)

and is expressed as:

$$E_s = \frac{P \cdot Q}{V_t} \quad (\text{MJ/m}) \tag{8-6}$$

When the F3 system is used, the jet grouting energy is calculated by adding together the energies of water and grout jetting, while the air contribution is neglected for simplicity.

The range of variability of the main parameters is summarized in Table 8-10. In

TABLE 8-10 Range of Jet Grouting Parameters[a]

		V_t (m/hr)		V_t (m³/hr)		Q (m³/hr)		P (MPa)		E (MJ/m)	
System	Fluid	Minimum	Maximum	Minimum	Maximum	Minimum	Maximum	Minimum	Maximum	Minimum	Maximum
F1	Grout	15	35	0.2	0.3	3.6	7.2	30	50	3	24
F2	Grout			0.5	1.5	5	9	30	50	8	110
	Air	4	18	10	90	150	360	0.7	1.2	—	—
F3	Grout			0.5	2	5	9	1	10	1	20
	Water	2.5	18	0.5	2	5	9	30	50	8	180
	Air			10	150	150	360	0.7	1.2	—	—

Source: From DePaoli et al. (1991).

[a] V_t = withdrawal speed; Q = grout flow rate; V = grout volume; P = grout pressure; E = grouting energy

the same table the energy values are presented for the pressures and volumes usually employed.

It is important to note that the PAPERJET acquisition system does not cause delays during construction. It involves:

- Transducers
- An acquisition unit
- A portable computer (p.c.) linked to a printer
- The various connections between the components

The transducers are placed on the drill rig, while the flowmeters are integrated into the injection equipment. The analog signals of these instruments are sent to the acquisition unit through the connection cables. This unit converts the analog signals into digital signals that are sent to the p.c. Both acquisition unit and p.c. support a dedicated software that performs all the required calculations and operations. The system provides:

- Real-time monitoring of drilling and jet grouting parameters and their visual display in graphical form.
- Printing in real time of all acquired numerical values.
- Real-time storage of the same data in magnetic support.
- Possibility of later plotting (and printing) the variation of the parameters with depth.

The parameters relevant to the drilling operations are: depth, withdrawal speed, withdrawal force, rotational speed, torque, grout, air and water pressures, grout and water flow rates, cumulated grout and water volumes, and grout and water specific volumes (per meter of column).

The jet grouting energy is calculated on the basis of the withdrawal speed, the grout and water pressures, and their flow rates.

The PAPERJET system enables the rig operator and the site engineers to observe, during the work, the main parameters both in graphical and numerical forms. The real-time monitoring of these quantities is useful since, if some mechanical malfunction occurs (such as a drop in pressure or flow rate, or clogging of the nozzles), the rig operator can immediately detect it on the display, and take the necessary actions. Equally, the site engineer is informed of any change in the soil condition, and can accordingly modify the jet grouting process. For example, if the system detects during drilling the presence of a cavity at a certain depth, the site engineer could decide to increase the treatment in that zone, for instance, by reducing the withdrawal speed. Also, the owner can monitor the execution of the work, confirming that it has been carried out correctly and within the design limits. In addition, the PAPERJET system permits the use of the data stored magnetically for the preparation of graphical outputs showing the variation of the measured quantities with depth.

The following case history provides an excellent illustration of the applicability of the technology.

The project for an overpass crossing the double track railway line at San Benedetto (Ancona, Italy) required consolidation by jet grouting of the bridge foundation materials. Owing to the proximity of the railway line, the design was very critical and required a thorough quality control to avoid damage during treatment. A comprehensive study was also carried out in order to define the most adequate geometry and treatment parameters, so as to obtain the best results from the consolidation viewpoint. The soil consisted of medium fine sands with silty interbeds of about 20-cm thickness, and the SPT tests classified the soil in a range varying from loose to medium sands. A trial was planned to acquire all necessary data for the optimization of the F2 and F3 jet grouting methods. The PAPERJET system, owing to its versatility, proved to be valuable in the following phases:

- During the instrumented drilling phase, it was possible to define the drilling specific energy of the natural soil for each column. In addition, it was possible

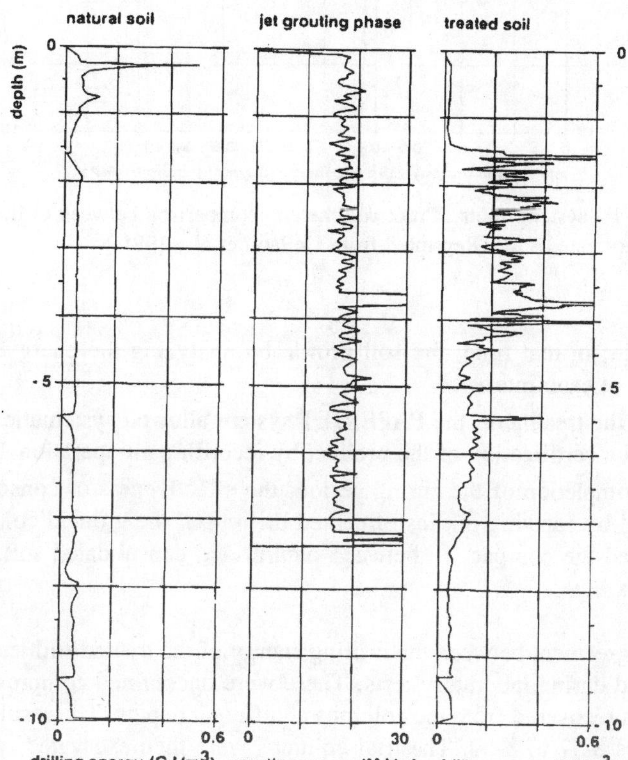

Figure 8-57 F2 system: plots of the drilling energy in natural and treated soil, and grouting energy. (Reprinted from DePaoli et al., 1991.)

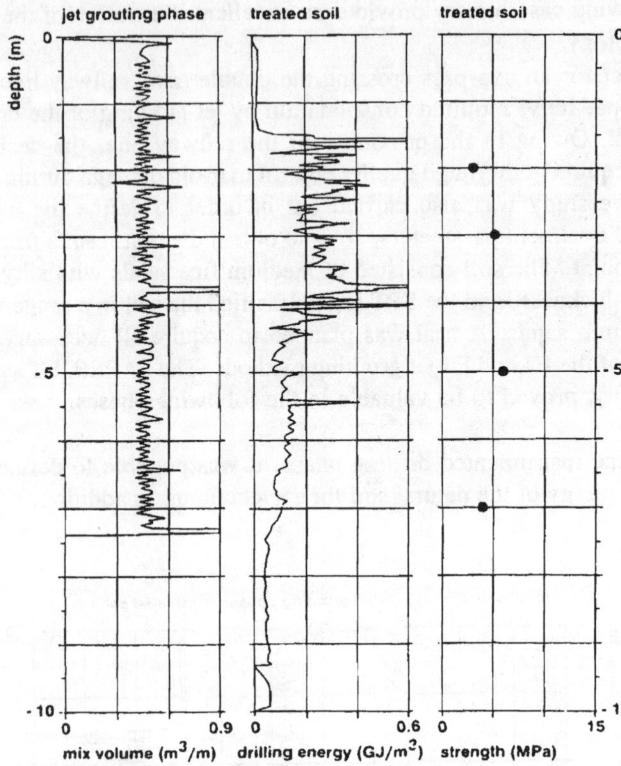

Figure 8-58 F2 system: plots of mix volume and comparison between drilling energy and strength of the treated soil. (Reprinted from DePaoli et al., 1991.)

to obtain, in real time, the soil profile by analyzing the energy required for drilling at each interval.

- During the treatment, the PAPERJET system allowed systematic quality control and a certification of the product by recording all operational parameters.
- After completion of the curing period, the effectiveness of consolidation was checked by carrying out instrumented drilling of the grouted columns, which permitted the compar... between natural and consolidated soil, in terms of energies.

A good agreement between the drilling energy of the treated soil and its strength was observed during laboratory tests. These were unconfined compressive tests on core samples recovered from the columns at different depths. The results are reported in Figures 8-57 to 8-60. The trial columns were then excavated to verify their geometry and homogeneity.

The columns executed by the F2 system appeared quite regular except in the peripheral area. In Figure 8-57, the graphs show the drilling specific energy in

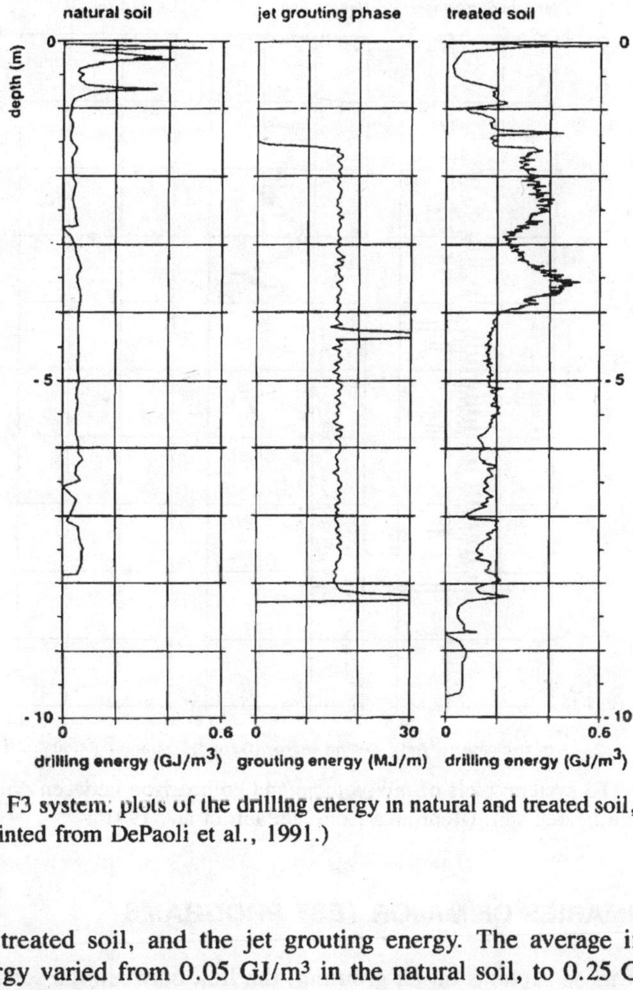

Figure 8-59 F3 system: plots of the drilling energy in natural and treated soil, and grouting energy. (Reprinted from DePaoli et al., 1991.)

natural and treated soil, and the jet grouting energy. The average increment of specific energy varied from 0.05 GJ/m^3 in the natural soil, to 0.25 GJ/m^3 in the column. The regularity of jet grouting energy confirmed that all the jet grouting operations had been correctly performed in accordance with the foreseen construction parameters.

Figure 8-58 displays the amount of grout employed for the column formation, the drilling energy of the treated soil, and consolidated soil strengths.

At this stage it seems too early to quantify the comparison between drilling energy and treated soil strength, but it appears that a good correlation exists between these two parameters. A further step would be the recording of sufficient data to create a family of significant statistical samples for different types of soil.

For the F3 work, an increase in drilling energy from 0.05 to 0.28 GJ/m^3 occurred and the agreement between drilling energy and treated soil strength was fairly good.

The large mass of data acquired with the PAPERJET system allowed the selection of the most suitable construction method, and defined the jet grouting operational parameters.

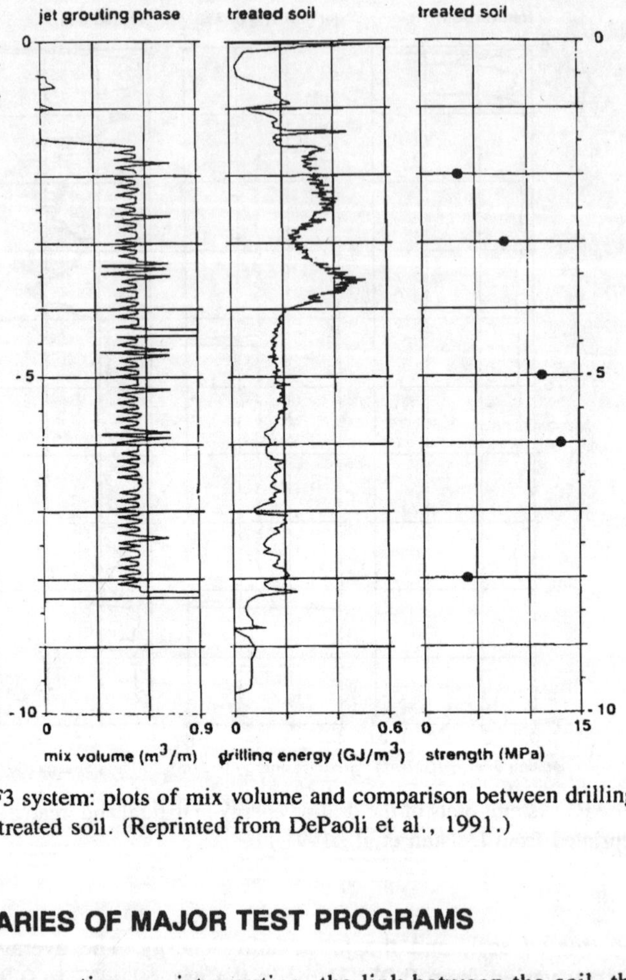

Figure 8-60 F3 system: plots of mix volume and comparison between drilling energy and strength of the treated soil. (Reprinted from DePaoli et al., 1991.)

8-6 SUMMARIES OF MAJOR TEST PROGRAMS

Throughout these sections on jet grouting, the link between the soil, the operating parameters, and the soilcrete volume and composition has been evident. Given the difficulties inherent in developing precise design methodologies, the value of the thoughtfully executed field test program has become paramount. The remainder of this section summarizes details from outstanding test programs conducted around the world. Some were executed as a necessary feature of a related major project, while others were conducted solely for research. They are described in the following sequence:

Location	Soils	Jet Grouting Method
Varallo Pombia, Italy	Sands	F1
Sao Paolo, Brazil	Sands	F1
Norfolk, Va.	Silty sand	F1
Volgodansk, C.I.S.	Silt	F1

Location	Soils	Jet Grouting Method
Singapore	Clay	F1
Sao Paulo, Brazil	Sand	F2
Osaka, Japan	Sand, clay	F3
Casalmaiocco, Italy	Sands and gravels	F1, F2, F3
Monte Olimpino, Italy	Peaty soils	F1, F2, F3

The list is not comprehensive, but provides an excellent cross section of methods, materials, and results.

Varallo Pombia, Italy

The F1 method was used (Rodio, 1983) at the Varallo Pombia test site to form 19 individual jet grout columns in moist alluvium (6 percent water content). The size and strength of the columns were determined for the soilcrete formed in the top 4 to 5 m, which corresponds to soil types ST1, ST2, and ST3 in Figure 8-61. Six of the jet grout columns were cored and samples of soilcrete were obtained from along the columns' central axis and near the outer perimeter of the column.

The unconfined compressive strength data for columns A1 and A14, listed in Table 8-11, were obtained from cores at the same elevation. The central axis strength is about 60 to 70 percent higher than the strength measured on specimens from the perimeter. The central axis and perimeter strength data for the other four columns (A7, A9, A13, A17) listed on this table were obtained within 0.1 m in elevation for samples taken from a particular column. Although the sampling locations were in close proximity, the ratio between central axis and perimeter strength varied widely between 0.4 and 1.9. Four of the six measurements had perimeter strengths higher than the central axis.

Six of the eight sets of strength measurements recorded in Table 8-11 were derived from samples formed using cement injections of about 140 to 145 kg/m of column. There appears to be a very large variability in the soilcrete strengths, between 3.6 to 21.1 MPa. However, when the size of the column is considered and the amount of cement injected per volume (kg/m^3) is calculated, then there is a clear linear relationship (Figure 8-62). The best linear relationship between unconfined strength (UCS) and injected cement factor (ICF) is: UCS = 0.6 (ICF) − 88, where UCS is measured in kg/cm^2 and ICF as kg/m^3.

Sao Paulo, Brazil (F1)

In 1982 Novatecna undertook extensive F1 field trials to examine how various drilling parameters interact with different soil types to produce a certain sized column (Guatteri et al., 1988). It was not the intent of the field study to obtain the optimum combination of drilling parameters, but rather, to understand how a single drilling parameter could be used to control the size and economy of F1 columns. Therefore, it was decided to examine what influence the lifting speed had upon the

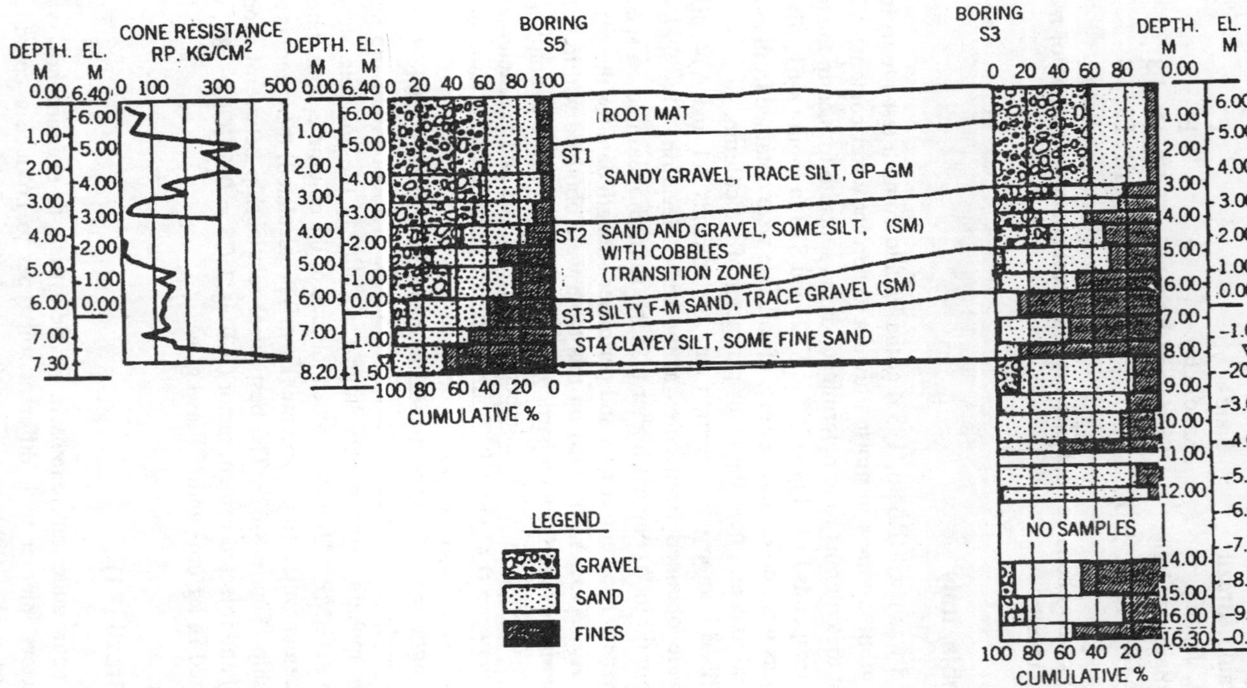

Figure 8-61 Generalized soil profile, grain size distribution, and cone resistance of alluvium at Varallo Pombia test site. (From Rodio, 1983. Reprinted by permission of Ing. Giovanni Rodio & C., Milan, Italy.)

TABLE 8-11 Variation of Unconfined Compressive Strength Along Central Axis and Outer Perimeter of Columns Formed at Varallo Pombia Trial Field

	Unconfined Compressive Strength		Cement Injected			
Column[a,b]	Central Axis (MPa)	Outer Perimeter (MPa)	kg	kg/m³	Ratio of Central: Perimeter	Average Column Diameter (m)
A1	9.7	5.7	145	288	0.6	0.80
A14[c]	4.8	3.2	140	178	0.7	1.00
A7	7.6	9.5	141	281	1.2	0.80
A7	7.6	3.1	141	281	0.4	0.80
A9	4.4	7.0	82	213	1.6	0.70
A13	3.6	6.9	143	324	1.9	0.75
A13[d]	21.1	7.7	143	431	0.4	0.65
A17	6.5	7.3	88	265	1.1	0.65

Source: From Rodio (1983).

[a]Samples for columns A1, A14 cored from same elevation.

[b]Samples for columns A7, A9, A13, A17 cored within 0.1 of each other in respective column.

[c]Prewash with water before jet grouting.

[d]Large boulders in column.

size of F1 columns formed in three soil types. However, since the lifting speed is only an indicator of the amount of time which the jet stream cuts the soil, it was decided for discussion purposes to also use the terms jet impact time and rod revolutions per step. The jet impact time is equal to the product of the lifting speed, lift increment per step, and number of nozzles. The jet impact time has units of

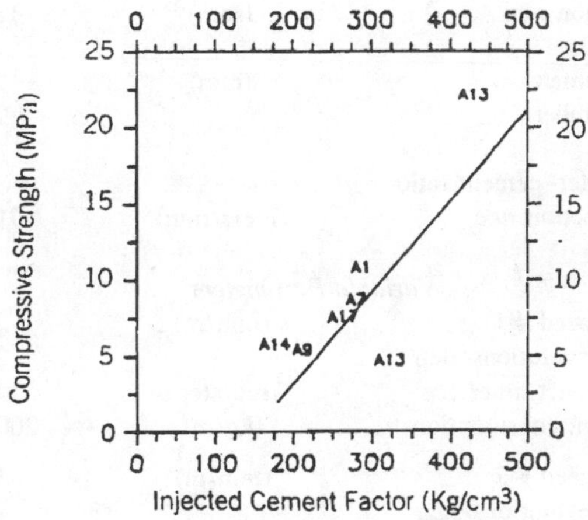

Figure 8-62 Soilcrete strength versus injected cement factor, F1 columns formed in moist sandy gravel, Varallo Pombia test site. (From Rodio, 1983. Reprinted by permission of Ing. Giovanni Rodio & C., Milan, Italy.)

time/step. The rod revolutions per step is the product of lifting speed, lift increment per step, and rod revolution rate.

The three soils selected for the field tests were sedimentary soils commonly found in the Sao Paulo area, and included sands, silty clay, and an organic clay obtained from the Sanegran Tunnel excavation. Three pits 10 × 10 × 5 m deep were excavated into the bank of the Tiete River above the tide line. All materials were end dumped from a truck into their respective pit and compacted. The highly plastic (LL = 60, PI = 40), soft organic clay was placed at its natural water content in 30-cm lifts and compacted. The silty clay and sand were compacted in a wet state. The in-place unit weights of the silty clay and sand were 102 and 113 pcf, respectively.

Twenty columns were formed in each pit. All 60 columns were made using the same nozzle pressure, rotation rate, size and number of nozzles, water-cement ratio, and grout injection rate as specified in Table 8-12. Within each pit 10 columns were drilled using a lifting speed of 2.9 min/m and then another 10 were formed using twice that lifting speed, and twice the amount of cement.

The measured size of the columns and average compressive strengths of the soilcrete are summarized in Table 8-13. The sand compressive strengths reported in Table 8-13 are based upon past experience with this material, and were not measured from columns formed during these trials. However, all other strengths recorded were obtained from soilcrete specimens formed during the field trials. These data support the following observations:

TABLE 8-12 Jet Grouting Parameters, F1 Test, Sao Paolo

Constant Parameters		
Pressure	(MPa)	30
Rotation rate	(rpm)	18
Nozzles		
Diameter	(mm)	2.2
Number		2
Grout		
Water–cement ratio		1–1
Injection rate	(liters/min)	91
Variable Parameters		
Lift speed #1	(min/m)	2.9
Rod revolutions/step		2
Jet impact time/step	(sec/step)	14
Cement consumption	(kg/m)	200
Lift speed #2	(min/m)	5.8
Rod revolutions/step		4
Jet impact time/step	(sec/step)	26
Cement consumption	(kg/m)	400

Source: From Guatteri et al. (1988).

TABLE 8-13 Results from F1 Test, Sao Paolo

Soil Type	Lifting Speed (min/m)	Column Diameter (cm)	Average Compressive Strength (MPa)
Clay	2.9	55–60	2
Clay	5.8	65–70	4
Silty clay	2.9	65–70	5
Silty sand	5.8	75–80	7
Sand	2.9	75–80	8
Sand	5.8	85–90	12

Source: From Guatteri et al. (1988)

1. Doubling the jet impact time from 14 to 26 sec did not have a significant influence upon the size of any F1 column formed. Doubling the jetting time resulted in a 15 to 25 percent increase in the diameter of the columns. It appeared that most of the cutting action of the jet was expended during the first two revolutions of the drill pipe.
2. The compressive strength for all soils was significantly increased by doubling the jet impact time, allowing double the weight of cement to be injected per meter of column. The increase in strength was double for the organic clay and about 50 percent more for the silty clay and sand.
3. The soil type played a significant role in determining the size of the column formed, and could account for up to 50 percent of the difference in the sizes of columns. As the amount of clay in the soil increased, the columns decreased in size. As shown in Table 8-13, the smallest columns were formed in the organic clay (0.55 to 0.60 m), intermediate-sized in silty clay (0.65 to 70 m), and largest in sand (0.85 to 0.90 m).
4. Finally, soilcrete made by mixing 1 to 1 water: cement grout with sand resulted in compressive strengths (8 MPa) about four times greater than comparable grout mixes in clay.

Norfolk, Virginia

The background to this project was reviewed under Vertical/Subvertical Applications in Section 8.3, while this account of the preliminary test program is also drawn from Andromalos and Gazaway (1989).

The F1 jet grout columns were installed individually in loose to dense silty fine to medium sands with a little gravel. SPT values were generally in the 40 blows/ft range. Thirty test columns were installed on shore near the subsequent work site. The operational variables included the number of nozzles, nozzle diameter lifting rate, cement injection, and grout flow rates. Pressures and grout mixes were varied to determine correlations between these factors and the quality and volume of the soilcrete.

One of the primary purposes of the initial test program was to optimize the soilcrete column diameter at the minimum cement injection rate. After each set of test columns was installed and allowed to set, the soil around each column was excavated and the minimum column diameter was determined.

Since the test program was conducted above the water table, several columns were installed in a compacted soil inside buried steel cylinders with a simulated head of water. These tests were performed in an effort to better correlate the test program data with anticipated actual soil conditions below the water table. A summary of the results of this test program are shown in Table 8-14.

In addition to the installation of the test columns, the test program included laboratory testing of various grout mixes, and the installation of supplementary test borings at the test program site and beneath the pier at several locations, to establish the similarity of soil types.

Analysis of data obtained in the test program indicated that the optimum combination of jet grouting construction parameters for this project was:

Pressure	40 MPa
Lift rate	0.18 to 0.30 m/min
Number of nozzles	2
Nozzle diameter	1.0 to 2.2 mm
Grout mixture	w/c = 1.4 (Type I Portland cement)
Injection rate	280 kg/m

Initial laboratory testing of the soilcrete samples obtained in the test section and early phases of production work consisted primarily of unconfined compressive tests. Figure 8-63 shows a plot of unconfined compressive strength with time for some of the initial testing performed on verification section samples with a 1.4 : 1 grout as well as that from the limited injection of a 1 : 1 grout. Apparently, the 1.4 : 1 grout at a lift rate of 0.18 m/min indicated a continuing upward trend, but project timing did not accommodate its verification. The 1 : 1 grout showed a much steeper curve and was much more readily acceptable from a technical standpoint, but the increased cement usage on a project of this size made the 1 : 1 grout unattractive financially. As a result, alternative verification testing methods were investigated.

Since the primary design consideration in this application was the shear strength of the soilcrete material, direct shear tests were performed to augment the previously performed unconfined compressive strength tests. While specific design details and results of long-term testing were not available for publication, it is known that direct shear tests verified that the soilcrete material installed with a 1.4 : 1 grout and a 0.18-m/min lift rate consistently satisfied minimum design criteria. Analysis of the above data also indicated that reliable test results could only consistently be obtained from samples that had attained an in situ cure time of 30 days.

As a result of the above observations and indications, construction verification procedures for the work were modified to allow this time period prior to sample

TABLE 8-14 Summary of Test Results, Norfolk, Virginia

Test Number[a]	Number of Jets	Jet Diameter (mm)	Injection Pressure		Grout Ratio	Average Flow (per min)		Lift Rate (per min)		Grout Quantity		Minimum Diameter	
			ksi	MPa	w–c[b]	gal	liter	ft	m	bags/ft	bags/m	in.	m
1	4	1.8	6.0	41	2–1	57	216	0.6	0.2	5.1	16.6	46	1.2
2	4	1.8	6.0	41	2–1	55	208	1.0	0.3	2.9	9.6	41	1.0
3	4	1.8	6.0	41	2–1	57	216	3.3	1.0	0.9	3.0	28	0.7
5	2	1.8	6.0	41	2–1	27	102	1.0	0.3	1.4	4.7	37	0.9
6	2	1.8	6.0	41	2–1	27	102	3.3	1.0	0.4	1.4	23	0.6
7	2	1.8	6.0	41	2–1	27	102	0.6	0.2	2.4	7.9	37	0.9
8	2	1.8	6.0	41	3–1	28	106	0.6	0.2	1.8	5.9	41	1.0
9	2	1.8	6.0	41	3–1	28	106	1.0	0.3	1.1	3.5	32	0.8
10	2	1.8	6.0	41	3–1	28	106	3.3	1.0	0.3	1.1	25	0.6
11	2	1.8	4.0	28	3–1	23	85	1.0	0.3	0.9	2.8	22	0.6
12	4	1.8	4.0	28	3–1	45	170	0.6	0.3	1.7	5.6	29	0.7
13[d]	4	1.8	4.0	28	2–1	50	189	0.6	0.1	3.2	10.4	40	1.0
14[d]	2	2.0	6.0	41	3–1	26	97	0.6	0.1	1.6	5.4	42	1.1
15[d]	4	1.8	6.0	41	3–1	51	193	0.6	0.1	3.2	10.6	60	1.5
16[d]	2	2.0	4.0	28	2–1	20	77	0.6	0.1	1.3	4.3	39	1.0
17[d]	2	2.0	4.0	28	3–1	21	80	0.6	0.1	1.3	4.4	39	1.0

(continued)

TABLE 8-14 *(Continued)*

Test Number[a]	Number of Jets	Jet Diameter (mm)	Injection Pressure		Grout Ratio	Average Flow (per min)		Lift Rate (per min)		Grout Quantity		Minimum Diameter	
			ksi	MPa	w–c[b]	gal	liter	ft	m	bags/ft	bags/m	in.	m
18[d]	4	1.8	4.0	28	3–1	51	193	0.6	0.1	3.2	10.6	39	1.0
19[d]	2	1.8	6.0	41	3–1	29	108	1.0	0.3	1.1	3.6	38	1.0
20[d]	2	1.8	6.0	41	3–1	29	108	0.6	0.1	1.8	5.9	39	1.0
21[d]	2	2.0	6.0	41	3–1	26	97	1.0	0.3	1.0	3.2	42	1.1
22[d]	2	2.2	6.0	41	3–1	31	117	1.0	0.3	1.2	3.9	45	1.1
23[d]	2	2.2	6.0	41	3–1	31	117	0.6	0.1	2.0	6.5	44	1.1
24	4	1.8	6.0	41	3–1	52	197	1.0	0.3	2.0	6.5	51	1.3
25	2	2.0	6.0	41	3–1	25	95	1.0	0.3	1.0	3.1	41	1.0
26	2	2.2	6.0	41	3–1	31	117	1.0	0.3	1.2	3.9	49	1.3
27	2	2.2	6.0	41	3–1	32	121	0.6	0.1	2.0	6.7	35	0.9
28	2	2.0	6.0	41	3–1	25	95	0.6	0.1	1.6	5.2	36	0.9
29[e]	2	1.8	6.0	41	3–1	28	106	0.6	0.1	1.8	5.8	30	0.8
30[e]	2	2.2	6.0	41	3–1	31	117	1.0	0.3	1.2	3.9	49	1.3

Source: From Andromalos and Gazaway (1989).

[a]Test numbers 1 through 12 performed in moist, fine sand, above water table. Test number 4 invalid—not reported.

[b]By volume.

[c]× 1,000.

[d]Test performed in 60-in.-diameter cylinders with simulated water head.

[e]Set of four connecting columns in alternating sequence 2-ft on centers.

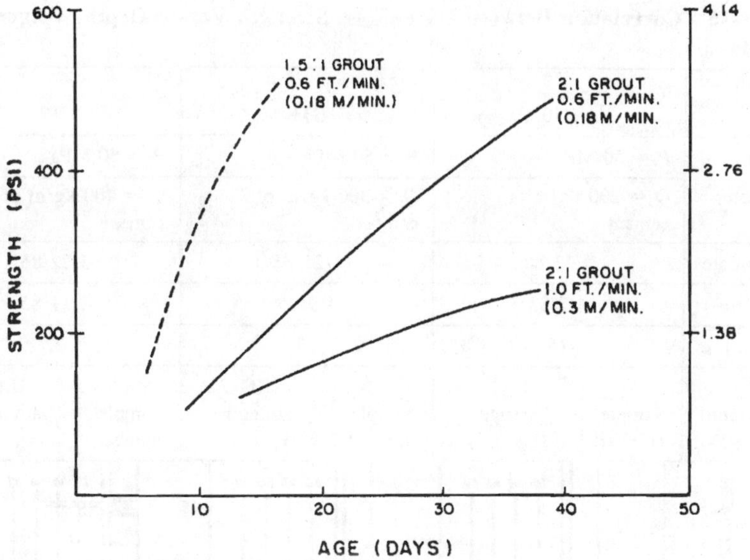

Figure 8-63 Plot of initial soilcrete strength gain trends, Norfolk, Virginia. (From Andromalos and Gazaway, 1989. Reproduced by permission of ASCE.)

retrieval, and to allow the use of direct shear tests to determine the adequacy of the installed soilcrete panel wall.

Volgodansk, C.I.S.

Twelve-story residential buildings were built on shallow mats founded over collapsible clayey silts. When a new reservoir system was placed into operation to service the area, the groundwater table rose and caused the buildings to experience about 50 cm of settlement (Rodio, 1985). The F1 system was used to underpin the buildings to a depth of about 18 m, with 70 percent of the 460 columns being formed under headroom of about 2.4 m. A trial was conducted prior to production work and eight of the test columns were exposed and their diameters measured. In general, the column diameters varied between 35 and 60 cm. The larger columns were produced using high pressures (50 MPa) and grout injection volumes of about 400 liters/m. The smaller diameter columns were desirable for underpinning the sensitive structure because only a small volume of soil was fluidized while jet grouting, and the hardened columns were the strongest possible for any of the three jet grout systems tested at the site. The F1 test columns were allowed to consolidate under their self-weight and age from 15 to 62 days. Results from laboratory tests on cored samples from columns C10, C11, and C12, along with the injection parameters, are listed in Table 8-15. For these three columns a pump pressure of 50 MPa was used and the lift speed varied so that 200, 300, and 400 kg of cement per meter of column could be injected. For column C10 (200 kg/m), 5 of the 8 samples had unconfined

TABLE 8-15 Correlation Between Unconfined Strength Versus Depth, Volgondansk Trial Field

Column Number	C10		C11		C12	
Injection parameter	P = 50 MPa		P = 50 MPa		P = 50 MPa	
	Q = 200 kg/m of cement		Q = 300 kg/m of cement		Q = 400 kg/m of cement	
Injection date	3/13/85		3/21/85		3/22/85	
Boring date	4/2/85		4/9/85		4/11/85	
Age (days)	15		19		22	
Depth (m)	Sample number	U.C. strength (kg/cm²)	Sample number	U.C. strength (kg/cm²)	Sample number	U.C. strength (kg/cm²)

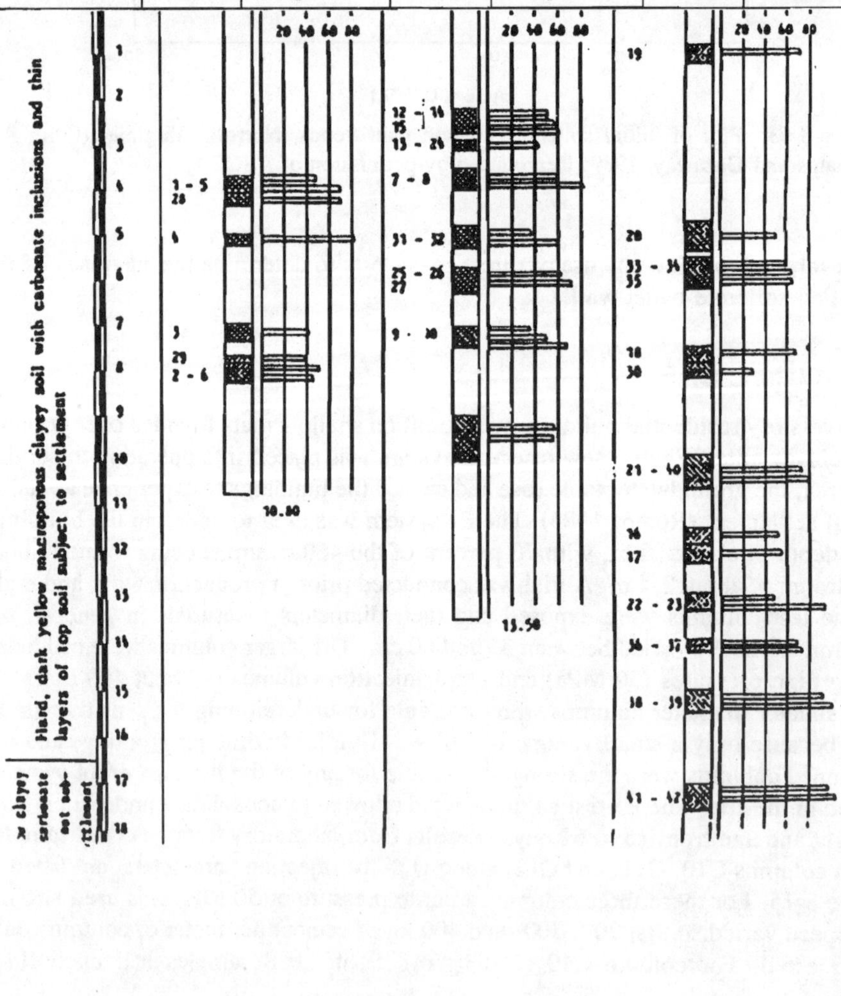

strengths between 4 and 5 MPa, whereas for column C12, the heaviest cement injection (400 kg/m) produced soilcrete generally above 6 MPa (5 of 7 samples). This represents a 15 to 30 percent increase in strength for a doubling of the cement content.

A total of 41 cores were taken from the test columns, and the values of unconfined compressive strength (UCS) versus depth are summarized in Figure 8-64. The average strength of the soilcrete in the collapsible silt (loess) was about 6 MPa, increasing to about 10 MPa at 16-m depth.

The correlation between the unconfined strength (R) and E/R ratio is shown in Figure 8-65. Typically, the values of E/R vary between 200 to 500 using total deformation data, and are about twice as large when E and R are obtained from the central portion of the sample. The values of E/R measured over the central portion of the sample were considered more accurate because the measurements were not as greatly influenced by the end restraint of the testing machine platens. Therefore, for an average unconfined strength of 6 MPa, the average Young's modulus should vary between 3000 and 6000 MPa.

Singapore

One of the first major applications of jet grouting for "block" consolidation from the surface was the project executed on Contract 106 of the Singapore Mass Transit Railway in 1983–1984. The job involved four tunnel sections with a total length of about 400 m, between Dhoby Ghaut and City Hall Stations, and was first described by Tornaghi and Perelli Cippo (1985) and Mongilardi and Tornaghi (1986).

The geology of Singapore Island is complex, but typically consists of beach sand

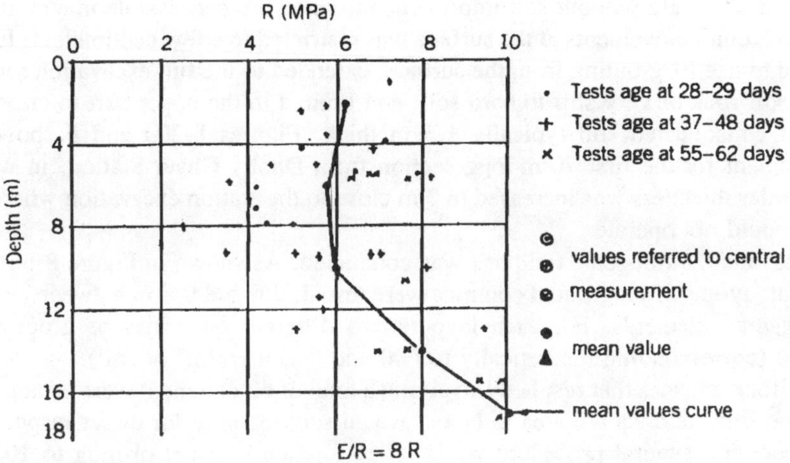

Figure 8-64 Variation of unconfined compressive strength with depth, Volgodansk, C.I.S. (From Rodio, 1985. Reprinted by permission of Ing. Giovanni Rodio & C., Milan, Italy.)

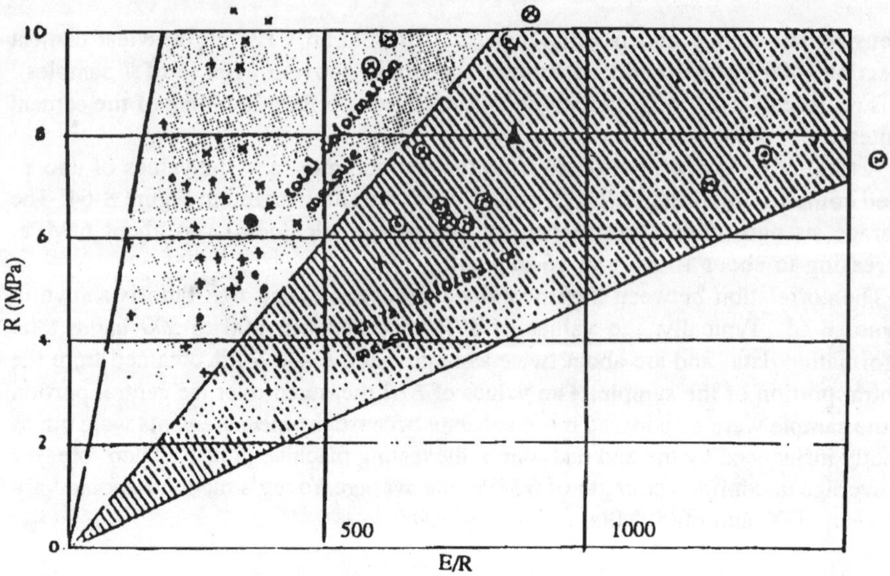

Figure 8-65 Correlation between unconfined compressive strength and E/R ratio, Volgodansk, C.I.S. (From Rodio, 1985. Reprinted by permission of Ing. Giovanni Rodio & C., Milan, Italy.)

and fill (3 to 5 m deep) that overlies very soft peaty clay, marine clay, and fluvial soils to combined depths of 15 m or more. The base of this sequence consists of stiff to hard cohesive soils or soft rock. Groundwater levels are only 1 to 2 m below surface.

In these hydrogeological conditions, even shield excavation could have been difficult and unsafe without soil improvement, since the permissible magnitude of vertical ground movements at the surface was restricted to a few centimeters. It was decided to use F1 grouting from the surface, extended to the full excavation section above soft rock or very stiff to hard soil, and limited in the upper part to create an arch of consolidated soil typically 1.5 m thick. Figures 8-20*a* and *b* show the arrangement for the first 70-m-long section from Dhoby Ghaut Station, in which the annular thickness was increased to 3 m close to the station excavation where the shield could not operate.

In advance, a full-scale field test was conducted. As shown in Figure 8-66, two different layouts of jet grouted columns were tested: 0.6 and 0.8 m between centers of staggered elements. For each layout, two different quantities of grout were injected (corresponding theoretically to 600 and 800 liters/m^3 of soil).

The four schemes that resulted (involving a total of 62 columns) were arranged to form the sides of a square area to be excavated subsequently for direct inspection. The following general procedure was applied to each scheme: drilling to 10.5-m depth, treatment from the bottom to 0.5-m depth, injecting a grout with a water/cement ratio of 1.6 at a pressure of 40 MPa.

The instrumentation previously installed consisted of two inclinometers to check

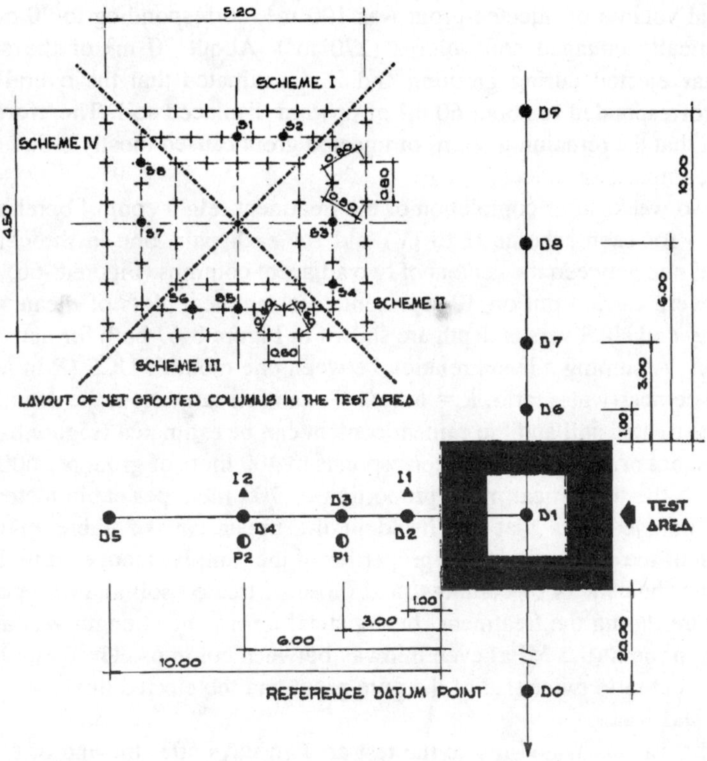

LEGEND

- ▲ CLINOMETER
- ◐ PIEZOMETER
- ● BOREHOLE
- ● DATUM POINT

SCHEME	SPAC. (m)	N. OF COLUMNS PER m² OF SOIL (n.)	VOLUME OF MIX PER L.m. OF COLUMN (l/m)	VOLUME OF MIX PER cu.m. OF SOIL (l/m³)
I	0.6	3.2	187	600
II	0.8	1.8	333	600
III	0.6	3.2	250	800
IV	0.8	1.8	444	800

Figure 8-66 Singapore Mass Transit System: general layout of Rodinjet test area. (From Tornaghi and Perelli Cippo, 1985.)

horizontal soil displacements, two piezometers to record pore pressure buildup and dissipation, and nine datum points to check vertical soil displacements with reference to a fixed point 20 m distant from the perimeter of the test area.

Inclinometer 1-1 (1 m from the axis of the external row of Scheme IV) recorded the maximum horizontal displacement of 23 cm at a depth of about 6 m; at a distance of 6 m, the displacements were less than 5 cm. The excess pore pressures recorded by the cells 6.5 m deep, and 3 m and 6 m from the perimeter of the treated area, were low (0.02 to 0.04 MPa) throughout. After completion of the treatment, the mean heave values were approximately 30 cm at 1 m distance, 17 cm at 3 m, 5 cm at 6 m, and 1 cm at 10 m.

The total volume of injected grout was 190 m³, corresponding to 70 percent of the theoretically engaged soil volume (270 m³). About 70 m³ of the soil-grout mixture was ejected during grouting and it is estimated that the overall surface upheaval corresponded to about 60 m³ of upward displaced soil. Therefore, it may be inferred that the remaining 60 m³ of injected grout caused mostly radial displacement and compression effects.

Some two weeks after completion of the treatment, eight control boreholes were drilled: two for each scheme (I to IV) and for each pair, one in the center of a column and one between the centers of two adjacent columns (Figure 8-66). Laboratory tests were carried out on 40 representative samples; plots of mean values of bulk density and UCS versus depth are shown in Figure 8-67 both for native and jet grouted soil. Assuming a mean relation between one month's UCS (R in MPa) and the overall cement/water ratio, $R = 6d(c/w)^2$, and full saturation, the actual composition of the treated soil and the cement content can be estimated (Figure 8-68). The average content of about 2 kN/m³ corresponds to 400 liters of grout per 600 liters of soil, close to the theoretical mean proportion of 700 liters per cubic meter of soil. The more detailed data that are listed in the upper part of Table 8-16 permit comparison of the composition and properties of the samples recovered in the center and between the centers of columns, and those of treated soil and of ejected soil-grout mixture during the treatment. In statistical terms, the strength was above the specified minimum (0.3 MPa) even midway between columns. The treated soil had a water/cement ratio twice that of the pure grout and the ejected mixture was richer in cement and water.

A test pit was excavated inside the test area 15 days after the end of treatment. There was overlapping of columns with significant discontinuities for Scheme II only. A spacing of 0.7 m was selected for the final design.

The field trial results were satisfactory in terms of the quality of soil improve-

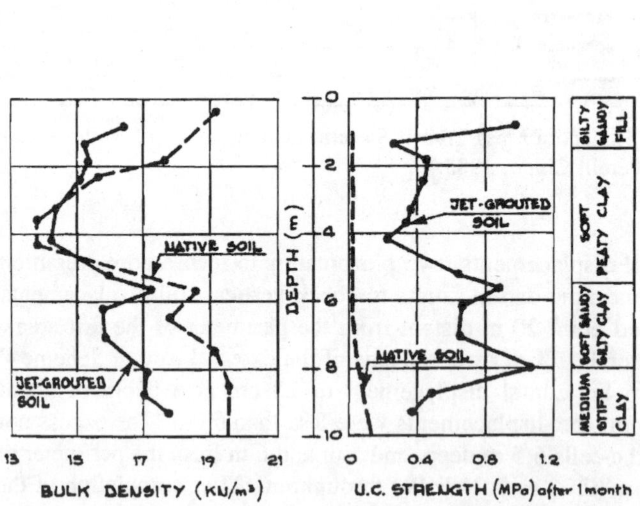

Figure 8-67 Plots of bulk density and strength versus depth (mean values recorded in Singapore test area on samples of jet grouted soil. (From Tornaghi and Perelli Cippo, 1985.)

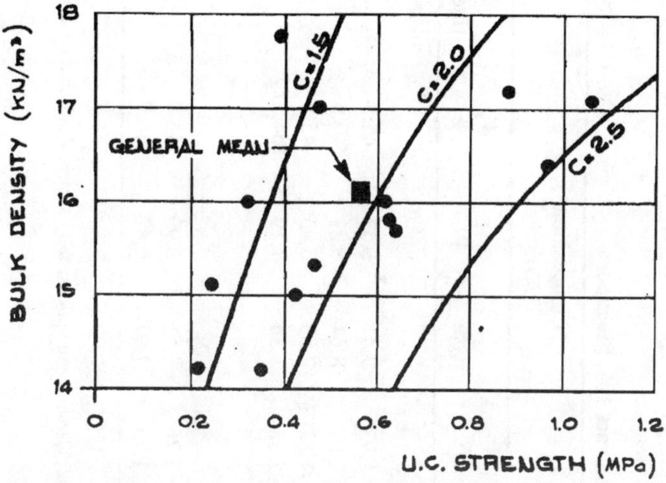

Figure 8-68 Estimate of cement content, kN/m^3, according to laboratory data recorded in Singapore Rodinjet test area. (From Tornaghi and Perelli Cippo, 1985.)

ment, but the magnitude of surface upheaval necessitated modification of the operating procedure, mainly to increase the volume of ejected material. Further testing led to a satisfactory solution of the problem by various adjustments of the drilling and jet grouting parameters. About 9400 m^3 of soil was subsequently treated by more than 10,000 columns. Ground movements, recorded daily by a close network of datum points, were kept within the safe limits (average 2 cm) by adjusting the sequence of grouting. Systematic controls by sampling, laboratory testing, and static cone penetration tests were used in monitoring the quality of the work during execution. An excellent performance during the subsequent shield tunneling was recorded.

Sao Paulo, Brazil (F2)

In order to quantify the effect of the additional air variables in the F2 system, Guatteri et al. (1988) conducted a trial in four phases from April to December, 1984.

The first two phases were performed to understand the role lifting speed played in the formation of the columns in cohesive soils. Phase three consisted of forming columns in sands, and in the last phase grout panels were experimented with, although not subsequently described.

The field trials were conducted in large pits (30 × 6 × 5 m deep), excavated in the banks of the Tiete River in the vicinity of the original F1 field trials [see the discussion of Sao Paulo (F1) earlier in Section ` 6]. During Phases I and II, the same soil type was used, and was typically silty (ay, with about 20 to 30 percent fine sand. The top 1.5 m of the fill was medium t , stiff in consistency; it was soft

TABLE 8-16 Data Recorded During Field Trial and Production Grouting, MRT Contract 106, Singapore

		Unit Weight (kN/m³)	UCS MPa	Composition[a] (kN/m⁻³)					
				C/W	*C*	*dS*	*W*	W_s (%)	V_g (liters/m³)
			Field Trial						
Jet grouted soil	Center columns	15.91	0.604	0.294	1.93	7.41	6.57	45.3	386
	Between columns	16.47	0.477	0.261	1.62	8.64	6.21	40.6	324
	Average	16.10	0.559	0.285	1.82	7.83	6.45	43.6	365
Ejected mixture		14.91	0.641	0.303	2.18	5.54	7.19	(64.3)	435
			Actual Treatment						
Jet grouted soil		15.31	1.090	0.395	2.76	5.56	6.99	43.0	550
Ejected mixture		16.14	0.519	0.272	1.75	7.97	6.42	44.0	350

Source: From Mongilardi and Tornaghi (1986).

[a]C = cement; W = water content of native soil; V_g = volume of grout per unit volume of treated soil; General assumptions: fully saturated soil and no drainage.

below 1.5 m to the bottom of the pit. The sand used during Phase III tests was a fine uniform sand, which was placed moist and in a loose state.

All F2 columns were formed using a nozzle pressure of 20 MPa and a single 3-mm nozzle, as listed in Table 8-17. The drill rods were rotated at about 18 rpm, and the air cone was delivered at 4 m^3/min at 0.7 MPa. The 1 : 1 water/cement grout was injected at about 70 liters/min.

Twenty-two columns were formed during Phase I. The only variable changed was the lifting speed, between 1.6 and 3 min/m. The faster lifting speed was selected to ensure that at least one full revolution of the drill pipe occurred before moving the jet to a higher elevation. Excavation of all 22 columns revealed nonhomogeneous, poorly cemented cylindrical bodies, with measured diameters between 1.8 and 2 m.

The solution to the nonhomogeneity of the column was to either use more than one jet, or jet longer. It was decided to investigate how extremely long jetting times, using one nozzle, would influence the size and strength of the columns. Therefore, during Phase II, six columns were formed using two longer lifting times, 12 and 18 min/m.

The results from the Phase II studies showed that no significant size increase resulted from jetting 4 to 6 times longer at a particular elevation; Phase II columns were still about the same size as the Phase I columns. However, the Phase II soilcrete bodies were much stronger and more homogeneous than in Phase I.

A total of 122 ea, 15-cm-high specimens were cut from the Phase II columns formed using the 12 min/m lifting speed. Samples were cut from the upper, middle, and lower third of the soilcrete body and were allowed to cure for 30 and 60 days. Both unconfined compressive and Brazilian tensile tests were performed. The results are summarized in Table 8-18, where it can be observed that the 30-day compressive strengths varied between 2.5 MPa at the top of the column to 1.4 MPa near the bottom. After 60 days of curing, the column tended to have a more uniform strength over its length, whereby the lower portion of the column experienced a 40 percent increase in compressive strength between 30 and 60 days. The results from the tensile strength tests were inconsistent: the strength in the upper and middle portion of the columns tended to decrease with age, while the strength in the lower third of the columns increased with age. This inconsistency is believed to be due to testing a small number (six samples) of specimens from each location at 30 days.

A more comprehensive set of strength data was reported by Guatteri and Teixeira (1987). Their study indicated that the average tensile strength of an F2 column, formed in a wide variety of soils, increased with time to about 10 to 15 percent of the average compressive strength.

Two columns were made in a loose sand during Phase III, as a reference to compare to those made in cohesive soils. Both columns were drilled using lifting times of 12 min/m and resulted in extremely uniform (1.9 to 2.1 m), and slightly larger columns than formed in cohesive soil. However, the sand soilcrete strength was significantly higher than in clay, which was consistent with the results from the F1 trials.

The results from Phase I and II tests indicated that the high-speed jet quickly cut

TABLE 8-17 Drilling Parameters Used During F2 Field Trials in Sao Paulo

	Constant Parameters	
Pressure	(MPa)	20
Rotation rate	(rpm)	18
Injectors		
Diameter	(mm)	3.0
Number		1
Grout		
Water-cement ratio		1–1
Injection rate	(liters/min)	70
Compressed air		
Pressure	(MPa)	0.7
Flow rate	(m^3/min)	4
	Phase I—Clayey Silt	
Lift speed #1	(min/m)	1.6
Rod revolutions/step		1.1
Jet impact time/step	(sec/step)	3.8
Cement consumption	(kg/m)	90
Lift speed #2	(min/m)	1.8
Rod revolutions/step		1.3
Jet impact time/step	(sec/step)	4.2
Cement construction	(kg/m)	100
Lift speed #3	(min/m)	2.0
Rod revolutions/step		1.3
Jet impact time/step	(sec/step)	4.2
Cement consumption	(kg/m)	100
Lift speed #4	(min/m)	3.0
Rod revolutions/step		2.2
Jet impact time/step	(sec/step)	7.2
Cement consumption	(kg/m)	175
	Phase II—Clayey Silt	
Lift speed #1	(min/m)	12
Rod revolutions/step		8.6
Jet impact time/step	(sec/step)	28.8
Cement consumption	(kg/m)	600
Lift speed #2	(min/m)	18.0
Rod revolutions/step		13.0
Jet impact time/step	(sec/step)	43.2
Cement consumption	(kg/m)	900

(continued)

TABLE 8-17 *(Continued)*

	Phase III—Sand	
Lift speed #1	(min/m)	12
Rod revolutions/step		8.6
Jet impact time/step	(sec/step)	28.8
Cement consumption	(kg/m)	600

Source: From Guatteri and Teixeira (1987).

the soil at a particular elevation, and only took about two revolutions (5 to 7 sec jet impact time) of the drill to form the column. However, this amount of time did not allow enough cement to be injected, nor was it sufficient time for the jet stream to properly mix the cut soil with the injected cement.

Osaka, Japan

Early F3 experiments were conducted in Japan, with the most comprehensive being reported by Yahiro et al. (1982) and Shibazaki and Ohta (1982).

The Yahiro in situ experiments were conducted at two different sites. The one at Koto-ku, Tokyo was essentially a loose silty fine sand site ($N = 10$). The log, shown in Figure 8-69, did not indicate if there was any groundwater present. However, it does show that the jet grout columns were constructed below grade, between Elevations -2 and -5 m. The second site, at Shiki-city in Saitama Prefecture, was a very soft clay site. Again, no water table location was reported in the drill log but the location of the jetted column is reported to be between Elevation -1 and -4 m. There were 9 jet columns constructed in the sandy deposit, while 13 columns were built in clay.

The Shibazaki tests were also conducted at two different sites: one sandy, the

TABLE 8-18 Average Compressive and Brazilian Strengths Obtained From Phase II F2 Columns

Column Portion Tested	Average Compressive Strength (MPa)				Average Brazilian Strength (MPa)			
	N^a	30 Day	N^a	60 Day	N^a	30 Day	N^a	60 Day
Upper	7	2.5	12	3.1	6	0.6	9	0.5
Middle	12	1.9	8	2.3	6	0.7	20	0.4
Lower	12	1.4	9	2.0	6	0.3	15	0.4

Source: From Guatteri and Teixeira (1987).

[a] *N* represents number of samples tested.

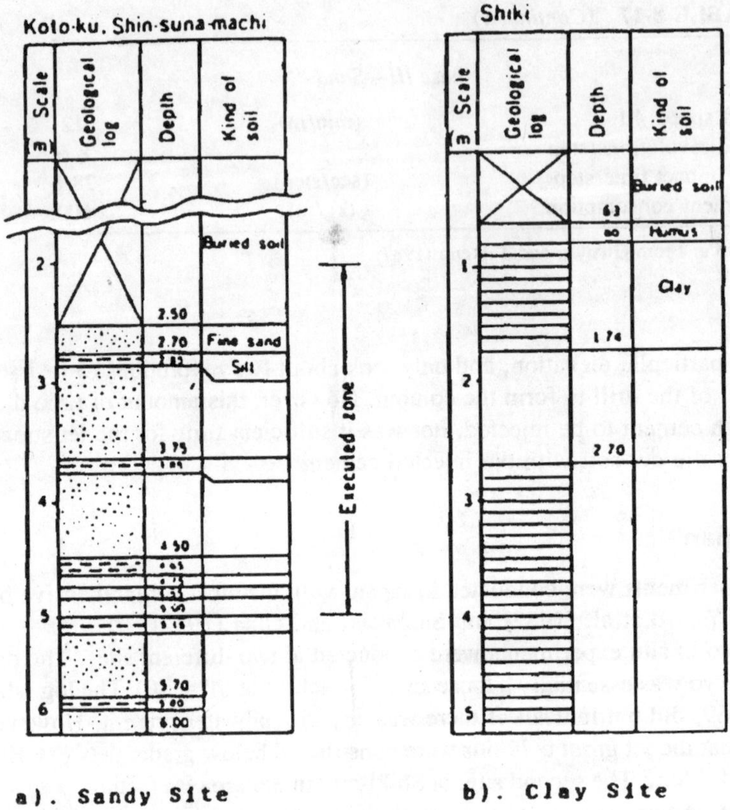

Figure 8-69 Geologic log for test sites in Japan used during research for column jet grouting. (*a*) Sandy site. (*b*) Clay site. (After Yahiro et al. 1982.)

other in clay. The sandy site consisted of 2.7 m of clayey soil, which overlay a layered deposit of fine sand and silt (Figure 8-70). Although the variation in the *N* values was reported for each soil, the scale was inadvertently omitted from Figure 8-70. In addition, the boring log does not indicate if groundwater was encountered. There were four columns constructed at the sandy site between Elevation −2 and −4 m. Figure 8-70 shows the geometries obtained for columns number 1 and 2. Shibazaki and Ohta (1982) indicated that five jet grouted columns were constructed at the clay site, but no boring logs were presented.

During the Shibazaki tests both the jet water pressure (40 MPa) and delivery rate (70 liters/min) were constant. In addition, the compressed air pressure was always 0.7 MPa and delivered at 1 m³/min. Shibazaki and Ohta examined what influence grout pressure and flow rate, along with lifting speed and rotation of the hydraulic monitor, had upon the resulting jet column. Yahiro et al. (1982) examined the same factors as Shibazaki, but in addition studied what influence changing the nozzle pressure from 20 to 40 MPa and flow rates (35 to 70 liters/min) had upon column

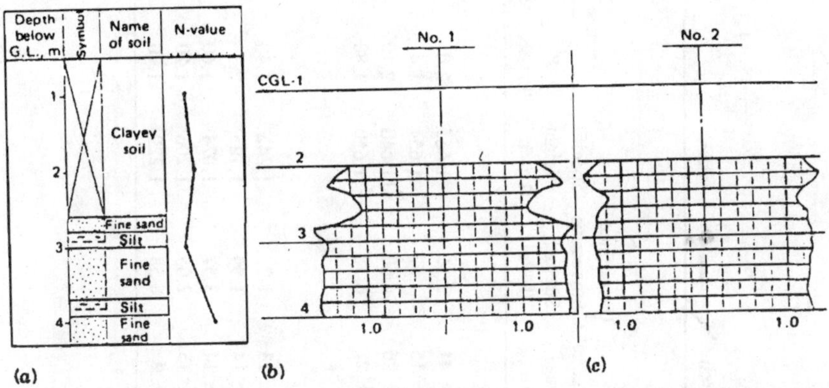

Figure 8-70 Boring log of test site and cross section of two jet grouted columns in sandy soil, Japan. (*a*) Boring log. (*b*) Column number 1. (*c*) Column number 2. (After Shibazaki and Ohta, 1982.)

formation. Yahiro et al. (1982) also employed compressed air at 0.5 MPa and 1 m^3/min.

Results from the Shibazaki experiments for both the sandy and clay site are summarized in Table 8-19. The following observations can be extracted from these data and the results reported by Yahiro et al. (1982).

The maximum column diameter for either the sand (3.40 m) or clay (3.24 m) occurred when both the lifting and rotational speed of the monitor was slowest (respectively, 5 cm/min and 5 rpm). Yahiro indicated that increasing the nozzle pressure from 20 to 40 MPa caused a 160 percent increase in the diameter of columns in sand, while for clays there was only a 50 percent increase.

Doubling the lifting speed from 5 to 10 cm/sec (columns number 7 and 8 on Table 8-19) resulted in nearly a 50 percent reduction in both the maximum and average size of a column jetted in clay. Yahiro reported that for sands the decrease was only 10 to 30 percent. When the rotational speed of the monitor was increased from 5 to 10 rpm, there was only a 15 percent decrease in the column diameter built in sand. No data were presented for clays.

Increasing the grout pressure to 30 MPa tended to cause a 30 percent increase in the size of the column constructed in clay (columns number 5 and 8). A similar observation was reported by Yahiro, but he indicated that the grout pressure was increased from 0 to 3 MPa, which is one-tenth the value reported by Shibazaki. Changing the grout pressure from 0 to 3 MPa had no significant effect on the diameter of columns jetted in sands.

Doubling the flow of grout equally did not have a significant effect upon the size of jet columns in sand (columns number 1 and 3), but Yahiro indicated that increasing the grout flow from 200 to 300 liters/min caused a 20 percent increase in the size of columns in clay.

Based on these results, Kauschinger and Perry (1986) concluded that the parameters listed in Table 8-20 should be used for F3 work in sands or clays.

TABLE 8-19 Parameters Used and Measurements Obtained During F3 Jet Grouting Research

	Process Specifications												
	Superhigh-Pressure Water		Hardening Agent		Compressed Air	Lifting	Revolving	Diameter		Unconfined Compressive	Specific	Seismic Wave	
Column Number	Pressure (MPa)	Flow Rate (liter/sec)	Pressure (MPa)	Flow Rate (liter/sec)	Pressure (kPa)	Speed (mm/sec)	Speed (rpm)	Maximum (cm)	Average (cm)	Strength (MPa)	Weight (kg/cm¹)	*P* (km/sec)	*S* (km/sec)
						Test 1—Sand							
1	40	1.2	0	1.2	700	0.8	5	340	280 ± 41	3.55	1.665	2.0	0.88
2			0	1.2	700	0.8	5	293	26[illegible] ± 15	2.52	1.654	1.4	0.64
3			0	2.5	700	0.8	5	293	256 ± 26	4.60	11.640	2.10	1.10
4			0	2.5	700	0.8	10	255	223 ± 21	6.78	1.650	1.80	0.85
						Test 2—Clay							
5			0	3.3	700	0.8	5	234	210 ± 13	1.73	1.242	1.40	0.50
6	40	1.2	30	3.3	700	1.25	5	214	179 ± 14	1.99	1.125	2.3	1.0
7			30	3.3	700	1.7	5	178	161 ± 14	3.51	1.074	1.60	0.69
8			30	3.3	700	0.8	5	324	302 ± 15	2.05	1.206	1.20	0.73
9			30	3.3	0	0.8	5	134	116 ± 9	6.58	1.229	1.70	1.0

Source: After Shibazaki and Ohta (1982).

TABLE 8-20 Recommended Parameters for Construction of F3 Columns

Water Jet	
Delivery pressure	40 MPa
Flow rate	50–70 liters/min
Nozzle diameter	2mm
Compressed Air Envelope	
Delivery pressure	0.7 MPa
Flow rate	1 m^3/min
Movement of Monitor	
Withdrawal rate	5 cm/min
Rotational speed	5 rpm
Grouting Parameters	
Delivery pressure	3.0 MPa
Rate of flow	200 liters/min
Estimated Column Diameter	
Sands	0.5–3.0 m
Clays	0.5–2.5 m

Source: From Kauschinger and Perry (1986).

Casalmaiocco, Italy

An extremely intensive field research program was conducted in sandy soils, of significantly higher water content than those at Varallo Pombia (Rodio, 1987). There were three distinct sandy soil types at the Casalmaiocco site (Figure 8-71). The upper 4 m of soil consisted of silt (ML) near the surface and silty fine sand beneath (SM). Between 4 and 6 m there were significant silt layers, with up to 70 percent silt in the soil horizon, 5 percent gravel, and 25 percent sand. Below 6 m the sandy soils became less silty (20 percent or less) and contained from 5 to 50 percent gravel. Probably the most important aspect of the Casalmaiocco trial was that all three jet grouting systems were used in the same soil conditions to investigate the variations in soilcrete strength. The soil water content with depth is shown in Figure 8-72, along with the measured strengths for a column formed using the F1 process. This column was formed using a constant cement injection of 200 kg/m. In general, the soilcrete strength varied inversely with the soil water content. In the top 4 m the soil water content was about 25 percent and the average soilcrete strength 10 MPa. Below 6 m the water content was typically 10 percent or less, but soilcrete strength usually over 30 MPa.

Obviously, the soil water content and soil type were strongly interrelated when

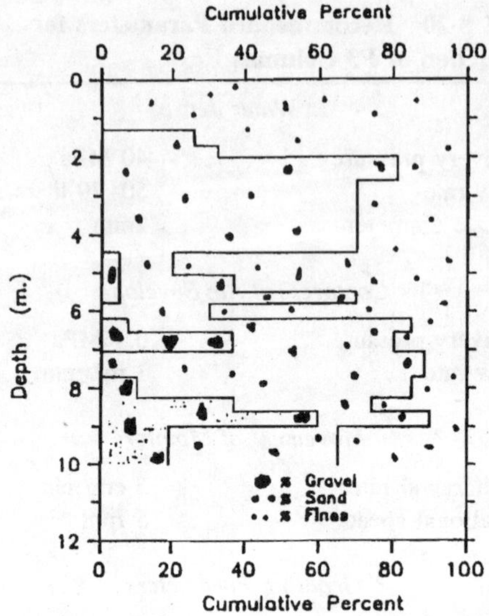

Figure 8-71 Soil stratigraphy: Boring S2, Casalmaiocco test site, Italy. (From Rodio, 1987. Reprinted by permission of Ing. Giovanni Rodio & C., Milan, Italy.)

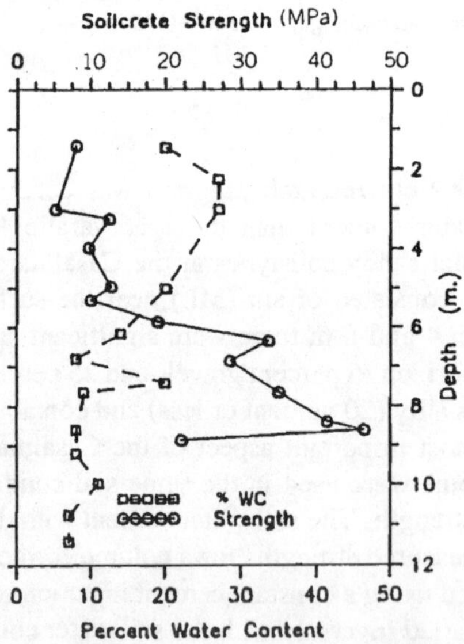

Figure 8-72 In situ water content, soilcrete strength variation versus depth; F1 jet grouting, Casalmaiocco test site, Italy. (From Rodio, 1987. Reprinted by permission of Ing. Giovanni Rodio & C., Milan, Italy.)

comparing the soil strata plotted in Figure 8-71 with the water content superimposed on Figure 8-72. The water content tended to be higher when the fines contents were higher. In the top 4 m, where the water content was about 20 to 30 percent, the fines varied between 20 to 75 percent. Below 6 m, there were only about 25 percent fines, some gravel, and therefore lower water content (usually less than 10 percent). The combined effect of lower water content and lower fines produced the concrete-like soilcrete below 6 m.

Attempts were made to compare soilcrete strength data, which were generated using similar ratios between injected cement (kg/m) and column cross-sectional area. The average column diameters measured for the single, double, and triple fluid systems, were respectively: 0.80, 2, and 2 m with corresponding cement injections of 200, 1000, and 1000 kg/m. Therefore, the amount of cement injected per volume (ICF) of soilcrete formed for each of the three systems was very similar, and varied between 320 to 395 kg/m^3. Lastly, the soilcrete strengths were all measured at about the same age, 220 to 240 days.

The unconfined strength data versus depth previously plotted in Figure 8-72 for F1 are presented in Figure 8-50 for ease in comparing soil type and water content to the soilcrete strengths produced by the three jet grouting systems. At almost every depth, F1 resulted in higher strengths than the other two jet grout methods. In the top 4 m, very weak soilcrete (2.5 MPa or less) was produced by the F2 and F3 systems, while F1 resulted in moderately strong soilcrete (5 to 12.5 MPa). Between 4 and 6 m there was a transition in the soilcrete strengths as the soil profile became more gravelly and the soil water content decreased. The highest soilcrete strength of 22.5 MPa was produced using F3. Below 6 m, F1 soilcrete strengths (30 to 40 MPa) were the second strongest soilcrete with central axis strengths varying between 15 and 30 MPa. F2 gave the weakest soilcrete with most strengths being between 7.5 and 15.0 MPa.

The average soilcrete strength (q_u) and standard deviations (Std Dev) calculated over the interval of 0 to 4 m, and 4 to 10 m are summarized in Table 8-21 for each of the three jet grout systems. The average column diameter measured over the exposed column length of about 4 m is also listed in Table 8-21. F1 again produced the strongest soilcrete for the two intervals examined, the average strengths being about five times and 50 percent greater than the other two systems at the shallower and deeper depths, respectively. In general, the variability of the soilcrete was less when the strengths were low, with the standard deviation being about 1 MPa. When the soilcrete had average strengths around 20 MPa, the standard deviation was about 10 MPa, that is, 63 percent of the strength values ranged between 10 to 30 MPa.

Also evident from Table 8-21 is the observation that, although the soilcrete produced using F1 was the strongest, it was at the expense of having a smaller diameter column. The column sizes measured for F1 were about half the size produced using the other two jet grout systems. However, the amounts of time spent jetting the F2 and F3 columns were about 3.2 and 4.6 times longer than the time spent forming the F1 column.

TABLE 8-21 Average Soilcrete Strengths, Column Sizes for Three Jet Grout Systems, Casalmaiocco Trial Field

Jet Grout System	Average Soilcrete Strength (MPa)				Average Column Diameter (m)
	0–4 m		4–10 m		
	q_u	Std. Dev.	q_u	Std. Dev.	
F1	8.6	2.7	27.6	11.9	0.80
F2	0.6	0.3	13.4	13.4	2.0
F3	1.6	0.6	19.7	9.3	2.0

Source: From Rodio (1987).

Monte Olimpino, Italy

A new railway route (Figure 8-73) flanks the existing line that was on highly compressible peaty soil mostly consisting of fibrous peat with a silty-clayey matrix (DePaoli et al., 1989; Gallavresi, 1992).

The thickness of this formation averaged 5 m. The old track, between Stations 40 + 495 and 40 + 925, built more than one century ago on similar soil, required frequent maintenance works due to ongoing settlements of the embankment: leveling surveys and remedial works were necessitated twice a year. For the new embankment foundation, the owner requested that the problem be resolved to ensure maintenance-free operation without even temporary interference to the adjacent operating line.

The peaty soil deposit had a thickness generally ranging between 4 and 6 m, and overlaid a sandy gravelly formation. Figure 8-74 shows the results of soil classification tests with depth (exploratory borehole S1). From data on absolute specific

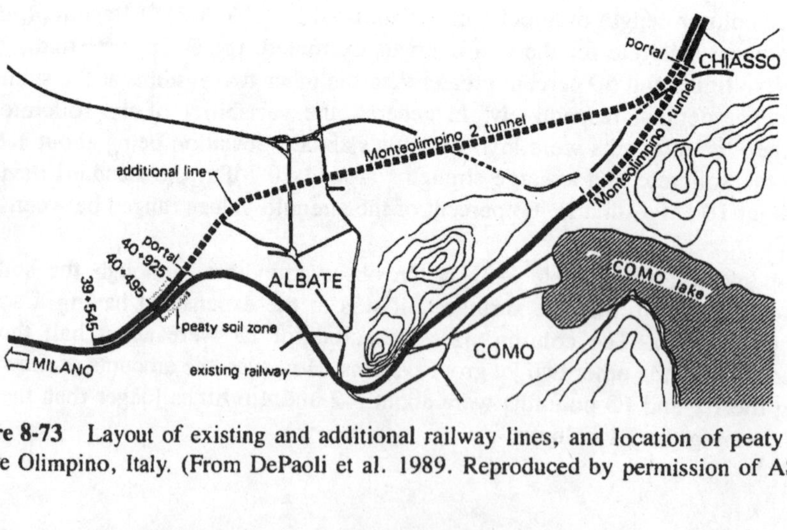

Figure 8-73 Layout of existing and additional railway lines, and location of peaty zone, Monte Olimpino, Italy. (From DePaoli et al. 1989. Reproduced by permission of ASCE.)

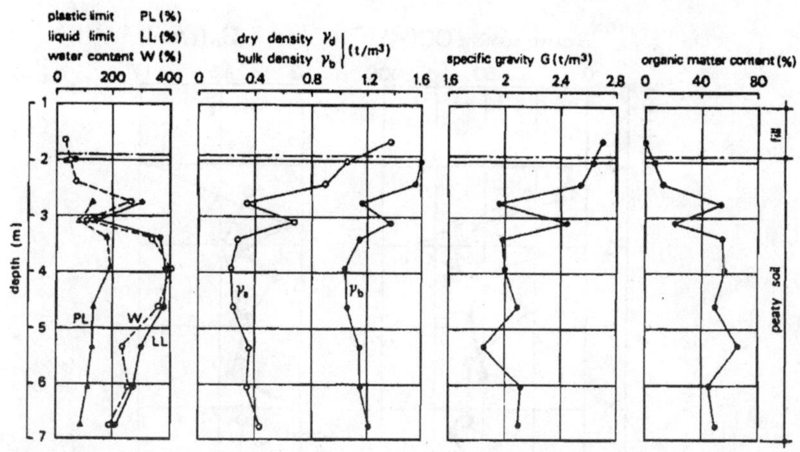

Figure 8-74 Characteristics of peaty soil with depth (borehole S1), Monte Olimpino, Italy. (From DePaoli et al., 1989. Reproduced by permission of ASCE.)

gravity, and assuming a specific gravity $G_m = 2.7$ for the mineral fraction, it was possible to calculate for the organic matter a specific gravity G_o between 1.3 and 1.5, as shown in Figure 8-75. The high compressibility was displayed by the range of odometer modulus when plotted against effective pressures: 0.15 to 0.30 MPa at overburden pressure. The rate of secondary compression was between 0.02 and 0.03.

Jet grouting was selected in preference to lime columns due to its potential for forming columns of larger diameter (0.6 to over 2 m), reducing therefore the working time per unit volume of treated soil, and providing a more homogeneous and effective improvement of mechanical properties. To explore the effectiveness of the jet grouting techniques, an extensive field trial was carried out.

A total of nine trial columns were formed: two F1, two F2, and five F3. During drilling, a "prewashing" treatment by the coaxial air-water jet was carried out in five columns in order to improve both the fracturing effect and partial replacement. In addition, two grout mixes were used (w/c = 0.7 and 0.85 by weight).

Approximately 15 days after grouting, the columns were exposed to a depth of 3.5 m to observe their dimensions. The measured diameters of columns within two depth ranges are shown in Table 8-22.

Unconfined compressive strengths were evaluated on cored specimens, and on samples trimmed from blocks recovered during excavation. Figure 8-76 shows the average strength values obtained as a function of the amount of cement injected per unit volume of treated soil.

The main conclusions drawn from the trial were:

- A design diameter of 2.1 m could be reliably assumed for columns formed by the F3 system with prewashing.

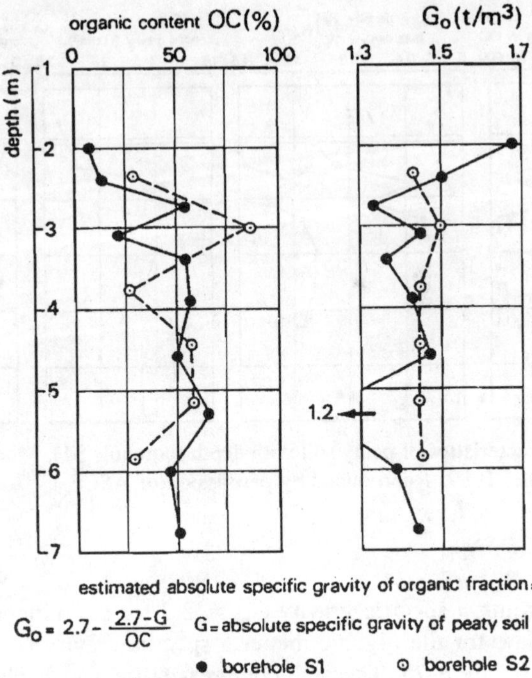

Figure 8-75 Content and estimated specific gravity of organic matter, Monte Olimpino, Italy. (From DePaoli et al., 1989. Reproduced by permission of ASCE.)

TABLE 8-22 Diameters and Depths of Columns, Monte Olimpino, Italy

Procedure	Column Number	Column Diameter (m)	
		From 0 to 1.50	From 1.50 to 3.50
F3	C_0	2.00	1.50
	C_1	2.20	1.50
	C_2	2.20	2.10
	C_3	2.40	2.10
	C_4	2.50	2.20
F2	C_5	1.80	1.60
	C_6	1.80	1.50
F1	C_7	0.80	0.70
	C_8	Not recorded	Not recorded

Source: From DePaoli et al. (1989).

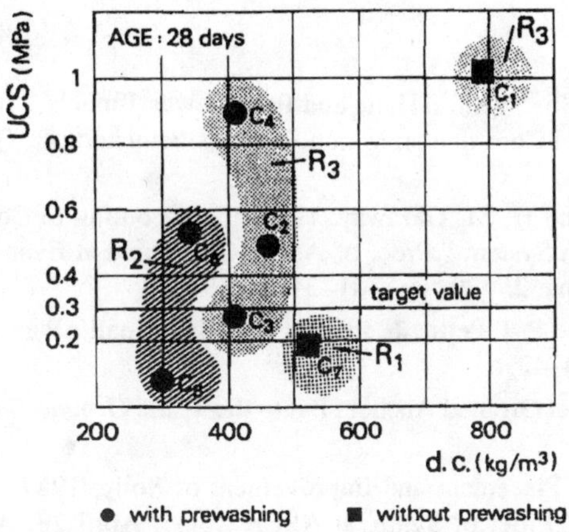

Figure 8-76 Relationship of column unconfined compressive strength (UCS), and dry cement injected per cubic meter of jet grouted soil (d.c.), Monte Olimpino, Italy. (From DePaoli et al., 1989. Reproduced by permission of ASCE.)

- A water-cement ratio of 0.7 and a cement content between 1.4 and 1.6 tonnes/m^3 were suggested.
- Since a fairly wide range of strengths might be possible, control tests on cored samples were required during the course of the work.
- A long-term strength target of 0.3 MPa was realistic.

The design was consequently based on columns 2.1 m in diameter, formed by F3 with prewashing, and arranged as shown in Figure 8-24. The distance between centers was 2.75 m along five longitudinal rows, 2.15 to 2.80 m apart. The columns, embedded 0.5 m into the underlying sandy gravelly formation, involved the treatment of about 60 percent of the peaty soil foundation.

8-7 OVERVIEW

Jet grouting is a relatively new technique that is enjoying tremendous growth around the world, even if the activity in the United States is currently and relatively less vibrant. It has reached the stage where sufficient fundamental information has been published on excellent field trials to give engineers a good starting point for preliminary designs and predictions of subsequent performance.

Increasingly the emphasis is changing to quality control and quality assurance on an industrial scale, and this will feature, as far as is practical, the use of automated drilling, grouting, and recording systems.

All these positive developments, supported by the increasingly impressive list of successful case histories, will continue to foster the popularity of this method for decades to come.

REFERENCES

Abramson, L. W., 1986. "Visit to Halic and Faith Sewer Tunnels," Internal Memorandum to Istanbul Rail Tunnel Consultants, March, Parsons Brinckerhoff, Quade and Douglas Inc., San Francisco, CA.

Andromalos, K. B. and H. M. Gazaway, 1989. "Jet Grouting to Construct a Soilcrete Wall Using a Twin Stem System," Proc. of ASCE Conf. held at Evanston, Ill., 2 vols., F. H. Kulhawy (Ed.), June 25–29, pp. 301–312.

Andromalos, K. B. and P. J. Pettit, 1986. "Jet Grouting: Snail's Pace of Adoption," *Civ. Eng. ASCE* Dec. pp. 40–43.

Anonymous, 1988. "Jet Grouted Anchors Put to the Test," *Ground Engineering,* Vol. 21, No. 6, pp. 16–17.

ASCE Committee on Placement and Improvement of Soils, 1987. "Soil Improvement—A Ten Year Update," Proc. of Symp. at ASCE Conv., April 28, Atlantic City, N.J., J. P. Welsh (Ed.), Special Pub. 12.

Aschieri, F., M. Jamiolkowski, and R. Tornaghi, 1983. "Case History of a Cut-Off Wall Executed by Jet Grouting," Proc. 8th European Conf. Soil Mechanics and Foundation Engineering, May 23–26, Helsinki, Vol. 1, pp. 121–126.

Broid, I. Z., M. F. Khasin, and V. S. Istomina, 1981. "Jet Grout Method and Cut-Off Wall Stability," Proc. 10th Int. Conf. on Soil Mechanics and Foundation Engineering, Stockholm, Sweden, pp. 397–399.

Bruce, D. A., 1988. "Developments in Geotechnical Construction Processes for Urban Engineering," *Civ. Eng. Practice,* Vol. 3, No. 1, Spring, pp. 49–97.

Bruce, D. A., C. H. Hall, and R. E. Triplett, 1992. "Structural Underpinning by Pinpiles," Proc. DFI Annual Meeting, October 21–23, New Orleans, 30 pp.

Burke, G. K., L. F. Johnsen, and R. A. Heller, 1989. "Jet Grouting for Underpinning and Excavation Support," Proc. ASCE Congr., Foundation Engineering: Current Principles and Practices, June 25–29, Evanston, Ill., pp. 291–300.

Coomber, D. B., 1985a. "Groundwater Control by Jet Grouting," Proc. 21st Reg. Conf. Engineering Group of Geological Society, Sept. 15–19, Sheffield, England, pp. 485–498.

Coomber, D. B., 1985b. "Tunnelling and Soil Stabilization by Jet Grouting." Proc., Tunnelling '85, March 10–15, Brighton, England, 7 pp.

DePaoli, B., R. Tornaghi, and D. A. Bruce, 1989. "Jet Grout Stabilization of a Peaty Layer to Permit Construction of a Railway Embankment in Italy," Proc. of ASCE Conf. held at Evanston, Ill., 2 vols., F. H. Kulhawy, Ed., June 25–29, pp. 272–290.

DePaoli, B., C. Stella, and A. Perelli Cippo, 1991. "A Monitoring System for the Quality Assessment of the Jet Grouting Process Through an Energy Approach," Proc. 4th Int. Conf. on Piling and Deep Foundations, April 7–12, Stresa, Italy.

Engineering News Record, 1974. "Grouting System Places 100-ft-deep Cutoff Wall," May 2, p. 16.

Engineering News Record, 1986. "Jet Grouting Doesn't Cut It," Feb. 27, p. 15.

Gallavresi, F., 1992. "Grouting Improvement of Foundation Soils," Proc. ASCE Conf. Grouting, Soil Improvement and Geosynthetics, Feb. 25–28, New Orleans, 2 vols., pp. 1–38.

Gazaway, H. N. and B. H. Jasperse, 1992. "Jet Grouting in Contaminated Soils," Proc.

ASCE Conf. Grouting, Soil Improvement and Geosynthetics, 2 vols., Feb. 25–28, New Orleans, pp. 206–214.

GKN Keller, 1985, London, England. Trade literature.

Guatteri, G. and A. Teixeira, 1987. "Improvement of Soft Clay by Jet Grouting Technology," Proc. Columbia Soil Mechanics and Foundation Engineering, CPMSIF, Carthagena, Columbia.

Guatteri, G., J. L. Kauschinger, A. C. Doria, and E. C. Perry, 1988. "Advances in the Construction and Design of Jet Grouting Methods in South America," Proc. 2nd Int. Conf. on Case Histories in Geotechnical Engineering, June 1–5, St. Louis, Mo., Paper No. 5.32.

Halliburton Services, 1987. Industrial Grouting Technical Data Sheets. Duncan, Okla.

Ichihashi, Y., M. Shibazaki, H. Kubo, M. Iji, and A. Mori, 1992. "Jet Grouting in Airport Construction," Proc. ASCE Conf., Grouting, Soil Improvement and Geosynthetics, New Orleans, Vol. 1, Feb. 25–28, pp. 182–193.

Imrie, A. S., W. F. Marcusson, and P. M. Byrne, 1988. "Seismic Cutoff," *Civ. Eng.*, Vol. 58, No. 12, pp. 50–53.

Kauschinger, J. L. and R. Hankour, 1989. "Heat and Neutralization of Boston Blue Clay," Geotechnical Engineering Report #189, Masters Thesis, Tufts University, Medford, Mass.

Kauschinger, J. L. and E. B. Perry, 1986. "Jet Grouting: State of the Art for Seepage Control," Interim Report to U.S. Army Crops of Engineers, Waterway Experiment Station.

Kauschinger, J. L. and J. P. Welsh, 1989. "Jet Grouting for Urban Construction," Proc. 1989 Geotechnical Lecture Series, Boston Society of Civil Engineering: Design Construction and Performance of Earth Support Systems, MIT, Cambridge, Mass., Nov. 18, 60 pp.

Kauschinger, J. L., E. B. Perry, and R. Hankour, 1992a. "Jet Grouting: State-of-the-Practice," Proc. ASCE Conf. Grouting, Soil Improvement and Geosynthetics, Feb. 25–28, New Orleans, 2 vols., pp. 169–181.

Kauschinger, J. L., R. Hankour, and E. B. Perry, 1992b. "Methods to Estimate Composition of Jet Grout Bodies," Proc. ASCE Conf. Grouting, Soil Improvement and Geosynthetics, Feb. 25–28, New Orleans, 2 vols., pp. 194–205.

Langbehn, W. K., 1986. "The Jet Grouting Method: Applications in Slope Stabilization and Landslide Repair," Master of Engineering Report, Dept. of Civil Engineering, Univ. of California, Berkeley, May 12, 72 pp.

Martin, D., 1986. "Lion City Leaps Ahead with MRT—Singapore Mass Rapid Transit," Tunnels and Tunnelling, July, pp. 14–24.

Miki, G., 1973. "Chemical Stabilization of Sandy Soils by Grouting in Japan," Proc. 8th Int. Conf. on Soil Mechanics and Foundation Engineering, p. 395.

Miki, G. and W. Nakanishi, 1984. "Technical Progress of the Jet Grouting Method and Its Newest Type," Proc. Int. Conf. In Situ Soil and Rock Reinforcement, Paris, Oct. 9–11, pp. 195–200.

Miyasaka, G., Y. Sasaki, T. Nagata, M. Shibazaki, M. Iji, and M. Yoda, 1992. "Jet Grouting for a Self-Standing Wall," Proc. ASCE Conf., Grouting, Soil Improvement and Geosynthetics, Feb. 25–28, New Orleans, Vol. 1, pp. 144–155.

Mongilardi, E. and R. Tornaghi, 1986. "Construction of Large Underground Openings and Use of Grouts," Proc. Int. Conf. on Deep Foundations, Beijing, Sept., 19 pp.

Nakanishi, W., 1974. "Method for Forming an Underground Wall Comprising a Plurality of Columns in the Earth and Soil Formation," U.S. Patent 3,800,544, Apr.

Nicholson, A. J., 1963. Discussion, ICE Conf. on Grouting and Drilling Muds in Engineering Practice. Butterworths, London, pp. 108–109.

Novatecna Consolidacoes E Construcoes S/A, 1988. "High Technology in Jet Grouting," Sao Paolo, Brazil, company brochure.

Otto, B. and A. Thut, 1991. "Means of In-Situ Measurements and Back Analysis," Proc. ASCE Geotechnical Engineering Congr., June 10–12, Boulder, Colo., Vol. I, pp. 160–172.

Pacchiosi Drill, 1978, Parma, Italy. Trade literature.

Parry-Davies, R., R. M. H. Bruin, G. Wardle, and M. G. Nixon, 1992. "Stabilization of Pier Foundation Using Jet Grouting Techniques," Proc. ASCE Conf., Grouting, Soil Improvement and Geosynthetics, Feb. 25–28, New Orleans, Vol. 1, pp. 156–168.

Queensland Main Roads Department, 1978. "Cement Content of Cement Treated Materials, Heat of Neutralization Test Method No. Q116B—1978," Brisbane, Australia.

Rodio & C., S.p.A., 1983. "Jet Grouting Test Results, Varallo Pombia RodinJet Trial Field," Rodio Internal Report No. L3052 and No. 19852, Nov.

Rodio & C., S.p.A., 1985. Volgodansk USSR LOESS Treatment by Rodinjet—Demonstrative Tests with Rodinjet 2 and 3 Techniques, Photographic Documentation, Report #20′311-1, Casalmaiocco, Italy.

Rodio & C., S.p.A., 1987. "Casalmaiocco Trial Field," Rodio Internal Report No. 20455/1, 20455/4, Apr.

Rowlands, D., 1971. "Some Basic Aspects of Diamond Drilling," Proc. 1st Australia New Zealand Conf. on Geomechanics, Melbourne.

Shibazaki, M. and S. Ohta, 1982. "A Unique Underpinning of Soil Solidification Utilizing Super-High Pressure Liquid Jet," Proc. ASCE Conf., Grouting in Geotechnical Engineering, Feb. 10–12, New Orleans, pp. 680–693.

Stella, C., G. Ceppi, and E. D'Appolonia, 1990. "Temporary Tunnel Support Using Jet-Grouting Cylinders," *J. Construction Engineering and Management*, Vol. 116, No. 1, Mar.

Teale, R., 1965. "The Concept of Specific Energy in Rock Drilling," *Int. J. Rock Mechanics Mining Science*, Pergamon Press.

Tornaghi, R. and A. Perelli Cippo, 1985. "Soil Improvement by Jet Grouting for the Solution of Tunnelling Problems," Proc. 4th Int. Symp. Min. and Met. Tunnelling, Mar. 10–15, Brighton, pp. 265–276.

Trevi, 1990, Cesena, Italy. Trade literature.

U.S. Bureau of Reclamation, 1989. "Procedure for Determining Cement Content of Soil-Cement (Heat of Neutralization Method)," Preliminary test procedure.

Welsh, J. P. and G. K. Burke, 1991. "Jet Grouting—Uses for Soil Improvement," Proc. ASCE Conf., Evanston, Ill., 2 Vols., June 25–29, pp. 334–345.

Welsh, J. P., R. M. Rubright, and D. B. Coomber, 1986. "Jet Grouting for Support of Structures," Proc. ASCE Conf., Grouting for the Support of Structures, Seattle, Wash., Apr. 16 pages.

Yahiro, T. and H. Yoshida, 1973. "Induction Grouting Method Utilizing High Speed Water Jet," Proc. 8th Int. Conf. on Soil Mechanics and Foundation Engineering.

Yahiro, T. and H. Yoshida, 1974. "On the Characteristics of High Speed Water Jet in the

Liquid and Its Utilization on Induction Grouting Methods," 2nd Int. Symp. on Jet Cutting Technology, Cambridge, England.

Yahiro, T., H. Yoshida, and K. Nishi, 1975. "The Development and Application of a Japanese Grouting System," *Water Power and Dam Construction*, Feb., p. 66.

Yahiro, T., H. Yoshida, and K. Nishi, 1982. "Soil Improvement Method Utilizing a High Speed and Air Jet (Column Jet Grout Method)," 6th Int. Symp. on Jet Cutting Technology, Surrey, England.

CHAPTER 9

REHABILITATION OF AGING ROCK SLOPES

9-1 INTRODUCTION

For the most part, knowledge of rock slope stability problems have been developed in connection with mine quarries, water resource projects, and transportation networks. However, rock slopes are frequently designed for any type of project situated in rocky terrain, including private residential and commercial developments and power facilities. The behavior mechanisms and design techniques for rock slopes are the same irrespective of facility type. However, the factors of safety used in design and analyses are different. For example, mining facility slopes are designed and maintained at the least cost for economic mining of the ore, in other words, a factor of safety near one. For transportation facilities, public safety and preservation of revenue producing arterials are an overriding consideration; the slopes are designed for a greater margin of safety in spite of higher costs, a factor of safety of two or three.

The rehabilitation of rock slopes requires detailed understanding of the site conditions, potential modes of failure, failure causes, slope condition rating methods, rehabilitation methods, and monitoring and maintenance methods. This information is given below with practical example problems when appropriate.

9-2 EXPLORATION METHODS

Geologic site exploration has three major components: review of existing data; site reconnaissance; and geotechnical investigation. Rarely are these steps avoided.

Work performed previously usually provides a very sound base from which new work can begin. A site visit is essential for getting a feel for the site and understanding the setting and environment that the rock slope and any investigation work will be within. Finally, additional geotechnical investigation work is usually required to provide more detail on the orientation and character of the intact rock and rock discontinuities (joints). Conclusions from the data gathered will be an essential part of the failure analyses and rehabilitation designs. The three major components of site exploration are described in more detail below.

Review of Existing Data

Data review should include all accessible documents generated for the facility including:

- Feasibility studies
- Previous geologic explorations
- Design drawings
- Specifications
- Supporting calculations
- Construction records
- Maintenance records
- Relevant published literature

Among other things, feasibility records will show the envisioned purpose of the facility, reasons for site selection, projected demands on the facility, and economic considerations such as dependence of neighboring facilities and cities. Previous geologic explorations are often quite useful for assessing general geology, nature of overburden materials, rock type(s), rock quality, and groundwater levels. Design documents will reflect how the original designer of the facility planned the geometrical layout, as well as excavations and slope support based on geological considerations. Construction records will show how easily these design concepts were implemented, how geologic conditions varied from the original assessments, and any unusual occurrences during construction. Most facilities have some form of maintenance program. Maintenance records should be analyzed to determine geologic similarities between areas of frequent instability and the project site vicinity. Finally, relevant published literature should be researched including geologic papers, geotechnical papers, and local newspaper accounts of events pertinent to construction, maintenance, or facility usage.

Site Reconnaissance

After or while background information is being collected and digested, members of the project team should go into the field to perform a detailed site reconnaissance. Spatial relationships of nearby cities and connecting transportation networks should be determined. Current facility usage patterns should be observed. The accuracy of maps and as-built drawings should be verified. The topography, geology, climatology, and general site conditions should be noted. If reliable topography maps do not exist, photogrammetric mapping may be required. Signs of imminent or previous rock slope instabilities should be investigated. Future geologic exploration methods and potential sites should be identified. Finally, preliminary rehabilitation concepts should be generated and checked to see whether site conditions are conducive to potential measures considered both from a constructibility standpoint as well as a user standpoint.

Geotechnical Investigation

After existing data have been reviewed, summarized, and then corroborated with field observations, the need to gather additional geologic information should be assessed. Usually, further exploration is required. The methods chosen will vary depending on the size of the site, accessibility, type of geologic formations, intended use of data, time constraints, and economic considerations. The geotechnical investigation may include:

- Geophysical surveys
- Borings
- Geologic mapping
- Stereographic data plots
- Geotechnical reports

Large-scale field tests are seldom conducted for slope stability problems but can include direct shear tests, dye tests, overcoring, packer permeability tests, and rockfall simulations (Wu, 1985).

Geophysical Surveys Geophysical surveys are sometimes conducted to establish depth of overburden soils, seismic velocities of the rock, rippability, depth of disturbance due to previous blasting, depth of weathering, and correlations with other strength parameters. A typical seismic velocity profile is shown in Figure 9-1.

The two most common geophysical methods used on slopes are seismic refraction methods and resistivity surveys. Seismic refraction consists of inducing a vibration in the ground with a hammer or explosives at the ground surface or in a borehole and then measuring, with transducers, the arrival times at various locations of elastic waves traveling through the ground. By plotting arrival times versus distances from the source of vibration, seismic velocities for differing zones and layers can be determined. Information about the thickness and character of rock, the rock discontinuities, fault zones, sinkholes, and the like can be inferred from these measurements.

Resistivity surveys are conducted usually with a line of four electrodes. The two outside electrodes induce a current in the ground, while the two inner electrodes are used to measure the voltage drop across a certain distance. Rock itself is for the most part nonconductive. Electricity is conducted by the water in the rock mass

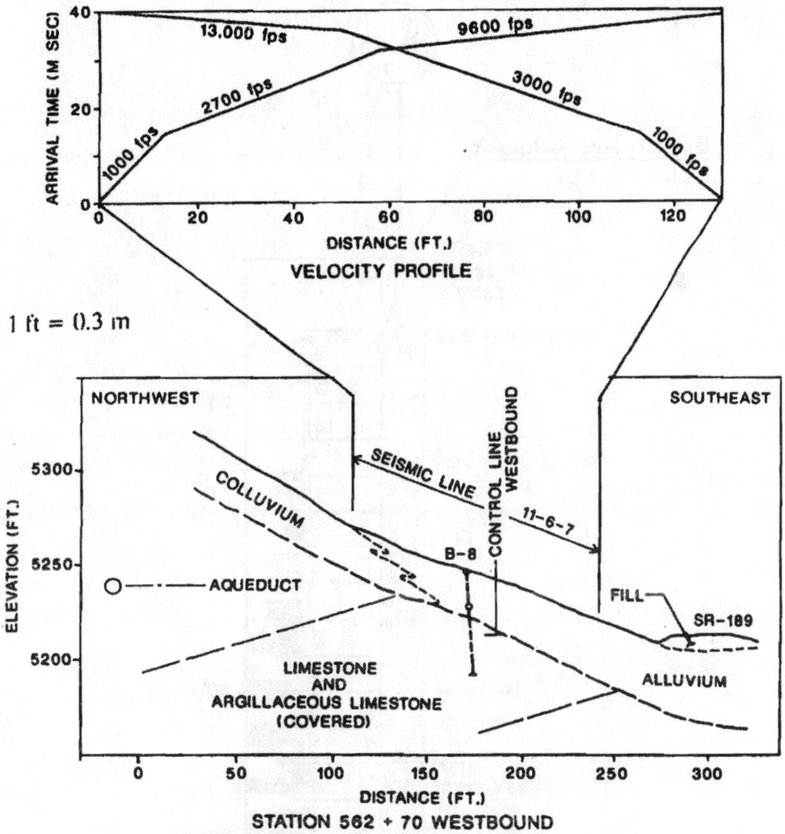

Figure 9-1 Typical seismic refraction results.

pores and joints. This is useful in locating highly fractured zones, changes in rock mass qualities, fault zones, and sinkholes.

Borings Depending on accessibility, borings can be drilled to obtain:

- Groundwater levels
- Rock quality designation (RQD) values
- Oriented core data
- Thickness/character of joints, shears, and faults
- Samples for laboratory testing
- Depth of weathering
- Depth of blast disturbance

Borings are usually drilled with a rotary wash drill rig, double- or triple-tube core barrels, and diamond studded core bits. Drilling fluid usually consists of water but

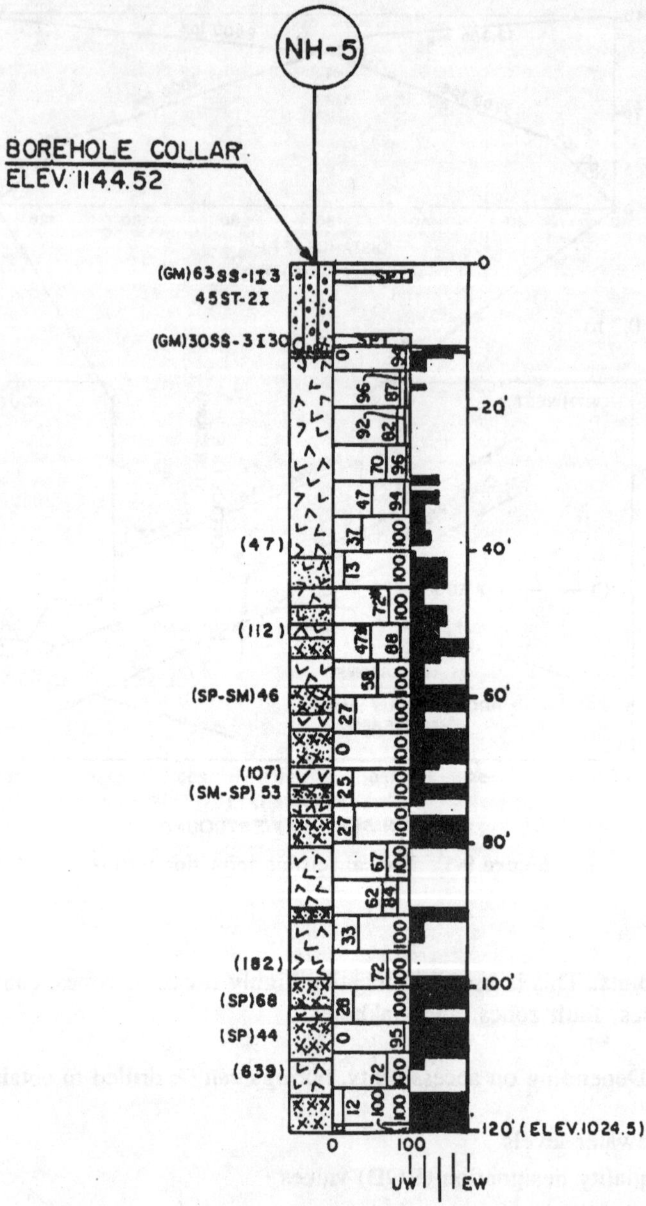

Figure 9-2 Typical rock core log.

air and foam can also be used. The size of the drill hole is usually NX (3 in. o.d., $2\frac{1}{8}$ in. i.d.), although hole sizes between $1\frac{7}{8}$ and $7\frac{3}{4}$ inches o.d. are possible. The core samples obtained are normally stored in boxes, photographed, logged, and tested. Data obtained from these operations are recorded on a core log. An example of what should be shown in the core log can be found in Figure 9-2.

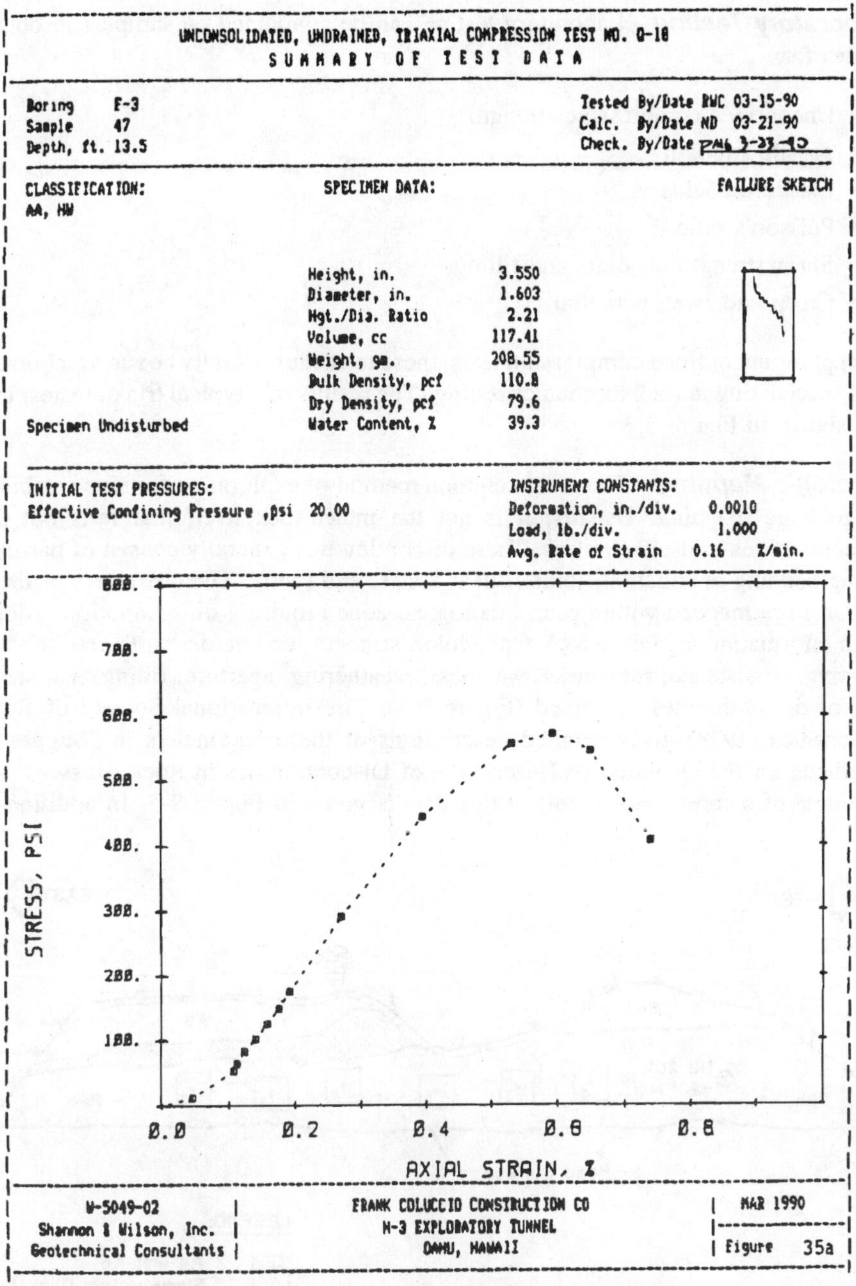

Figure 9-3 Typical triaxial shear test results.

Laboratory Testing Laboratory testing can be conducted on samples to obtain values for:

- Unconfined compressive strength
- Tensile strength
- Elastic modulus
- Poisson's ratio
- Shear strength of joints and fillings
- Creep and swell potential

Except for unconfined compression tests, these tests must usually be run at laboratories specializing in rock mechanics testing. The results of a typical triaxial shear test are shown in Figure 9-3.

Geologic Mapping The most common method of exploration for slope stability is geologic mapping. Usually, it is not the intact rock itself that fails but the discontinuities in the rock mass. These discontinuities generally consist of partings along bedding or foliation, joints, shear zones, and faults. The characters of these features are mapped within each lithological zone requiring differentiation. Along with information regarding rock type, color, strength, and hardness, the strike, dip, spacing, persistence, roughness, waviness, weathering, aperture, filling, and staining of discontinuities are noted (Figure 9-4). The International Society of Rock Mechanics (1978) gives detailed descriptions of these parameters in "Suggested Methods for the Quantitative Description of Discontinuities in Rock Masses." An example of a form used to collect this data is given in Figure 9-5. In addition to

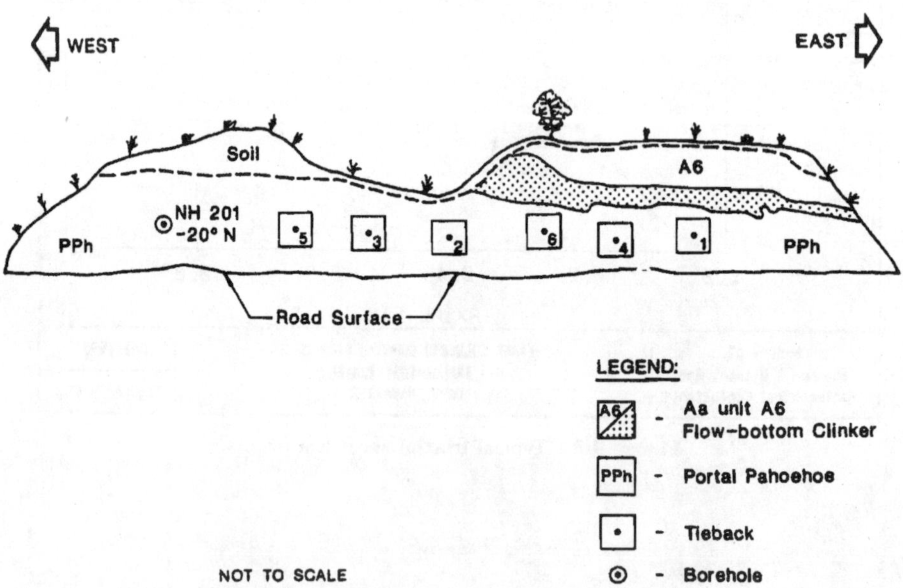

Figure 9-4 Typical geologic mapping results.

Name of project:
Site of survey:
Conducted by:
Date:

STRUCTURAL REGION

ROCK TYPE AND ORIGIN

DRILL CORE QUALITY R.Q.D.*

Excellent quality:	90 - 100%
Good quality:	75 - 90%
Fair quality:	50 - 75%
Poor quality:	25 - 50%
Very poor quality:	<25%

*R.Q.D. = Rock Quality Designation

WALL ROCK OF DISCONTINUITIES

Unweathered
Slightly weathered
Moderately weathered
Highly weathered
Completely weathered
Residual soil

GROUND WATER

INFLOW per 10 m of tunnel length litres/minute
or
WATER PRESSURE kPa
or
GENERAL CONDITIONS (completely dry, damp, wet, dripping or flowing under low/medium or high pressure:

STRENGTH OF INTACT ROCK MATERIAL

Designation	Uniaxial compressive strength, MPa	OR	Point-load strength index, MPa
Very high:	Over 250		>10
High:	100 - 250		4-10
Medium high:	50 - 100		2-4
Moderate:	25 - 50		1-2
Low:	5 - 25		< 1
Very low:	1 - 5		

SPACING OF DISCONTINUITIES

		Set 1	Set 2	Set 3	Set 4
Very wide:	Over 2 m				
Wide:	0,6 - 2 m				
Moderate:	200 - 600 mm				
Close:	60 - 200 mm				
Very close:	<60 mm				

NOTE: These values are obtained from a joint survey and not from borehole logs

STRIKE AND DIP ORIENTATIONS

Set 1 Strike: .... (average) (from .... to ....) Dip: .... (angle) .... (direction)
Set 2 Strike: .... (from .... to ....) Dip: ....
Set 3 Strike: .... (from .... to ....) Dip: ....
Set 4 Strike: .... (from .... to ....) Dip: ....

NOTE: Refer all directions to magnetic north.

(*a*)

CONDITION OF DISCONTINUITIES

PERSISTENCE (CONTINUITY)		Set 1	Set 2	Set 3	Set 4
Very low:	<1 m				
Low:	1 - 3 m				
Medium:	3 - 10 m				
High:	10 - 20 m				
Very high:	> 20 m				
SEPARATION (APERTURE)					
Very tight joints:	<0,1 mm				
Tight joints:	0,1 - 0,5 mm				
Moderately open joints:	0,5 - 2,5 mm				
Open joints:	2,5 - 10 mm				
Very wide aperture	> 10 mm				
ROUGHNESS (state also if surfaces are stepped, undulating or planar)					
Very rough surfaces:					
Rough surfaces:					
Slightly rough surfaces:					
Smooth surfaces:					
Slickensided surfaces:					
FILLING (GOUGE)					
Type:					
Thickness:					
Uniaxial compressive strength, MPa					
Seepage:					

MAJOR FAULTS OR FOLDS

Describe major faults and folds specifying their locality, nature and orientations.

GENERAL REMARKS AND ADDITIONAL DATA

NOTE:
(1) For definitions and methods consult ISRM document: '*Quantitative description of discontinuities in rock masses.*'
(2) The data on this form constitute the minimum required for engineering design. The geologist should, however, supply any further information which he considers relevant.

(*b*)

Figure 9-5 (a) and (b) Rock mapping data form. (Reprinted from: Bieniawski, Z. T., *Rock Mechanics Design in Mining and Tunneling*. 1984. A. A. Balkema, P. O. Box 1675. Rotterdam, Netherlands.)

geologic mapping, aerial photography can be studied for past landslides, fault traces, drainage features, and rock outcrops in the vicinity to be mapped.

9-3 STEREOGRAPHIC PROJECTION PLOTS

Rock discontinuity data from mapping and borings should be summarized and plotted on a polar stereographic projection. This graphic technique is used to find trends in field measurements. The technique is briefly discussed below. More detailed discussions on the methods and basis of stereographic projection are available in *Methods of Geological Engineering* by Goodman (1976), Hoek and Bray's (1981) *Rock Slope Engineering*, and *Hemispherical Projection Methods in Rock Mechanics* by Priest (1985).

In adapting stereographic projection to structural geology, the traces of planes on the surface of a reference sphere are used to define the dips and dip directions of the planes. The great circle that is traced out by the intersection of the plane and the reference sphere will define uniquely the inclination and orientation of the plane in space (Figure 9-6). Since the same information is given on both upper and lower parts of the sphere, only one of these need be used and, in engineering applications, the lower reference hemisphere is used for the presentation of data. In addition to the great circle, the inclination and orientation of the plane can also be defined by the pole of the plane. The pole is the point at which the surface of the sphere is pierced by the radial line which is normal to the plane. In order to communicate the information given by the great circle and the position of the pole on the surface of the lower hemisphere, a two-dimensional representation is obtained by projecting this information onto the horizontal or equatorial reference plane (Figure 9-7).

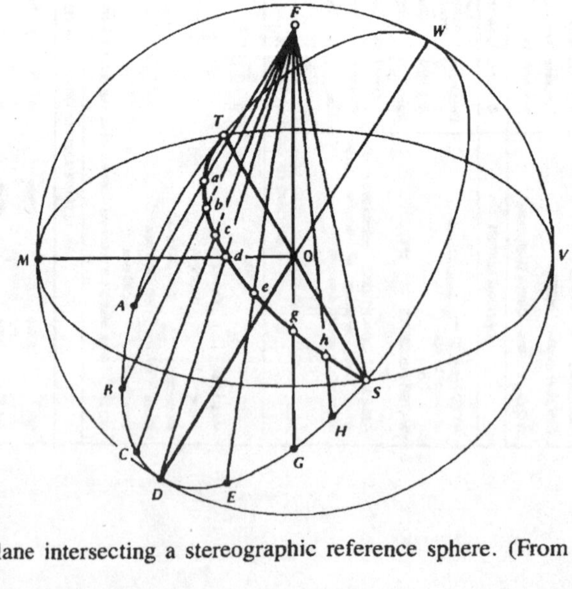

Figure 9-6 Plane intersecting a stereographic reference sphere. (From Goodman, 1980.)

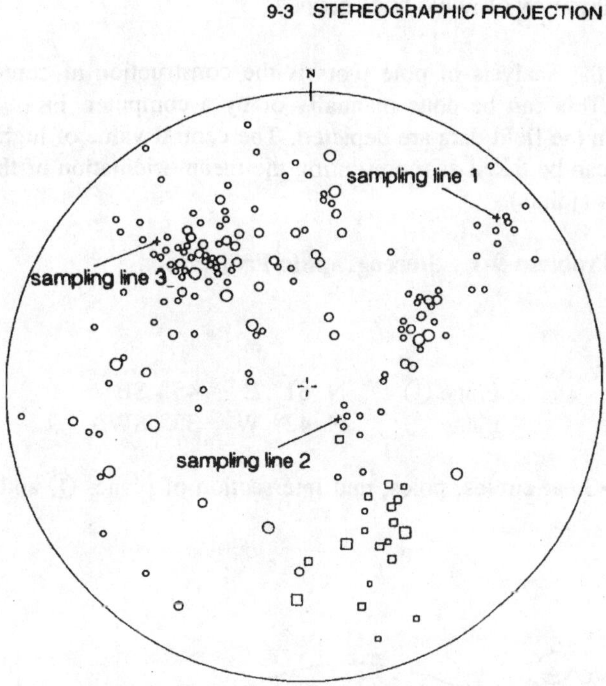

Figure 9-7 Lower hemisphere pole plot.

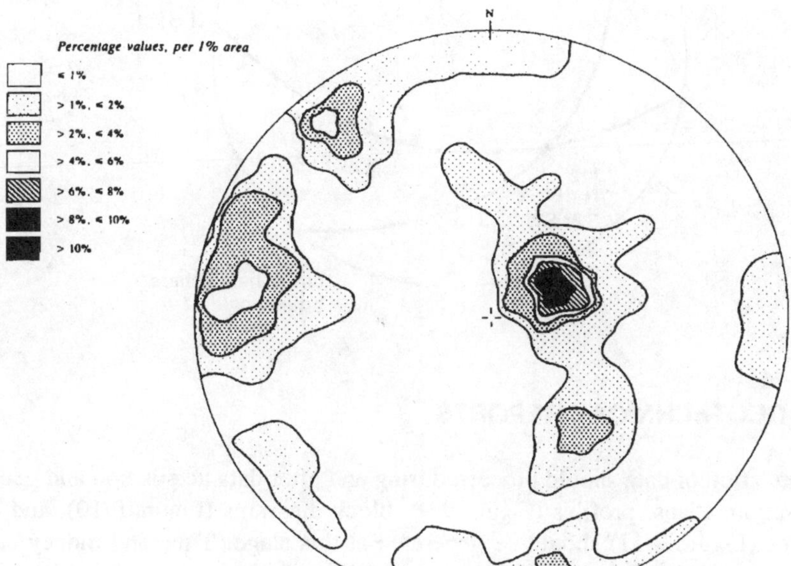

Figure 9-8 Lower hemisphere contour plot.

An aid in the analysis of pole plots is the construction of contour pole piots (Figure 9-8). This can be done manually or by a computer. From the contoured plots, trends in the field data are depicted. The central value of highest concentration of poles can be taken as representing the mean orientation of the given set of discontinuities (joints).

Example Problem 9-1 Stereographic Projections

Given

Plane ①	N 81° E	45° SE
Plane ②	N 42° W	33° SW

Required: Great circles, poles, and intersection of planes ① and ②.

Solution

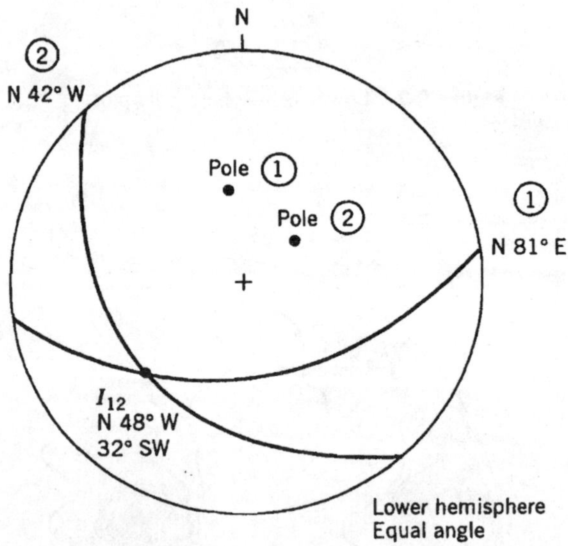

9-4 GEOTECHNICAL REPORTS

Interpretation of data should proceed during and after data acquisition and geologic exploration. Plans, profiles (Figure 9-9), block diagrams (Figure 9-10), and cross sections (Figure 9-11) should be generated at this stage. Time and money can be saved by considering potential future use of these documents and formatting them appropriately. Will they become part of construction contract documents, will they be used in a geotechnical report, or both?

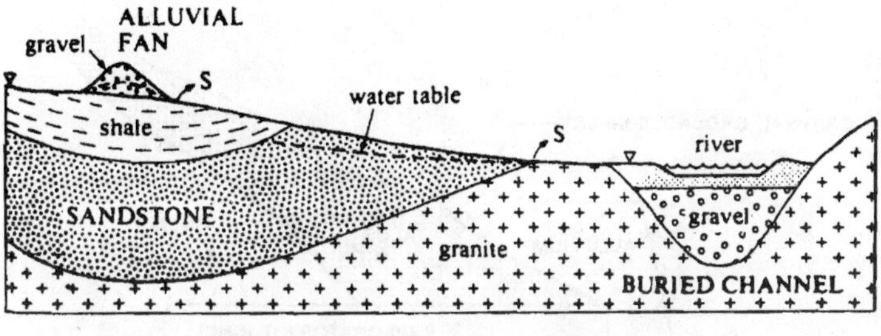

Figure 9-9 Typical geologic profile.

The types of geotechnical information that should be generated for a rock slope project fall into two categories. The first type is factual data; the second is interpretive data. These data should be prepared in two separate and distinct volumes that are clear and concise. The second volume can be made part of the contract document package.

Data Reports

The first volume or report should contain factual information:

- *Reconnaissance and Geological Background:* These data include regional geology, lithological and structural units, groundwater regime, and available references.
- *Boring Logs, Exploration, and Field Test Information:* This information includes quantitative data on depths, sampling, and recovery, drilling and sampling methods and equipment, sample and strata descriptions, drilling rates, drilling fluids, groundwater levels, in situ testing data, and presence of boulders and other obstructions.

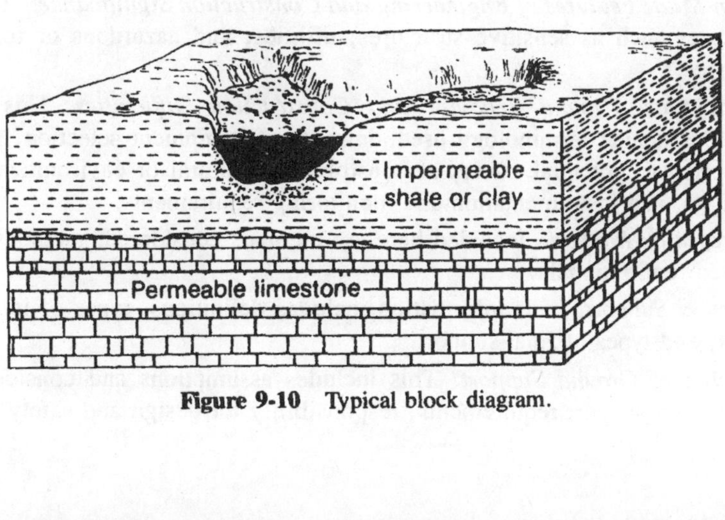

Figure 9-10 Typical block diagram.

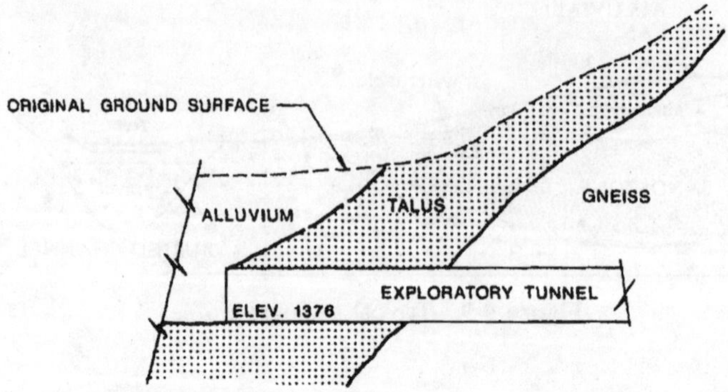

Figure 9-11 Typical geologic cross section.

- *Laboratory Test Information:* This includes diagrams and data sheets, correlations between various properties, and properties grouped by strata or lithology.
- *Reference Information.*

Interpretive Reports

The interpretive volume or report is usually referred to as the Geotechnical Design Summary Report and should contain the following:

- *Summary of Project.*
- *Sources of Information.*
- *Geologic Setting:* This should include regional geology.
- *Geologic Features of Engineering and Construction Significance:* Examples are engineering properties, bedrock weathering, geologic hazards, groundwater, and surface hydrology.
- *Man-Made Features of Engineering and Construction Significance:* These are features such as sensitive structures, utilities, and hazardous or toxic substances.
- *Anticipated Ground Behavior and Construction Difficulties:* This section should include terminology used, impacts of equipment selection, previous experience, potential effects of construction, cultural or environmental constraints, and instrumentation and monitoring requirements.
- *Excavation Method:* Included here are methods considered, methods not allowed, and rationale.
- *Ground Support:* This information includes definitions, types of initial support, and types of final support.
- *Design of Ground Support:* This includes assumptions and considerations, minimum support requirements, responsibility for design and safety, excava-

tion and support sequence, loading conditions, operational requirements, basis of analysis, and groundwater control.

- *Construction Specifications:* These include reasons for important or unusual requirements, special conditions, and allocation of risks.
- *Anticipated Quantities:* This section should list expected quantities along the project and at specific project features.

9-5 THEORETICAL BACKGROUND

Most rock slope stability problems are concerned with the jointing patterns and strength along joints of rock and the adverse effect that water between these joints tends to have. But to study the behavior of the rock mass, the intact strength must be known as well. And then, as for any engineering problems, one must decide how to relate imposed loads and stresses with the strength of the rock mass to determine the level of stability that is proposed or required. This introduces the concept of safety factors, which is discussed in more detail below.

Intact Rock

Rock slope analysis begins with knowing the type of rock and amount of weathering that has taken place since formation. These factors will determine to a large extent the intact strength of the rock. The strength of intact rock is important in that a slope becomes like a full-scale unconfined compressive test in the field because an unreinforced slope face provides no confinement to the rock. The intact strength is important for determining whether imposed loads on the slope are in excess of the unconfined strength. For very strong rock, the slope must be very high for unconfined compressive failure to be a potential problem. But for weak rock or a weak layer within a strong rock formation, failure could result from overstressing.

Another aspect of intact strength relates to movement along joints. If the normal stress levels holding a rock joint together are high enough, failure can occur along that joint by shearing through the asperities along the joint rather than sliding up and over the joint asperities. At high normal stresses, therefore, the shear strength along the joint will be higher for higher intact rock strengths, assuming no joint filling or joint weathering.

Discontinuous Rock Masses

Most rock mechanics problems are related to the behavior along the joints in the rock mass. This behavior is a function of joint orientation, spacing, length, roughness, planarity, weathering, filling, and strength. There are usually a number of joint sets present in the rock oriented at different strikes and dips. These joint sets are usually related to bedding planes, foliation, cooling cracks, stress relief fractures, tectonic stress and movement cracks (e.g., shears and faults), and other formational characteristics depending on the type and conditions of rock formation. For instance, sedimentary rocks tend to have orthogonal jointing patterns that are related to the bedding planes. Lava flows tend to have random jointing dependent on the time, location, and distance of flow and cooling history. Folded rocks will have variable jointing patterns depending on the axes of the folding and conditions prior to folding.

Jointing will control the slope stability behavior of the rock mass depending on the relative orientation of jointing with the orientation of the slope face and the tendency to form unstable blocks, slabs, and wedges. Generally speaking, if joints are steep and oriented toward the slope, stability along that joint set will be a problem. Or, if two or more joints intersect to form an unstable body along the intersection, this is of concern as well. Sometimes, it will not be immediately apparent which condition will control and use of stereographic projections will be quite handy in determining unstable joints or joint intersections.

Shear Strength of Joints

But just because a joint set or intersection dips into the excavation or slope does not necessarily mean that it will fail. One must consider the stresses along the joints and the strengths. Failure along one single joint is analogous to a block sliding on an inclined surface (Figure 9-12). If the rock joint has no cohesion, the block will move only when the inclination of the sliding surface reaches a certain critical value equal to the angle of internal friction of the joint surface. This simple analogy is only true when the joint has no cohesion and no water pressures are present. It is interesting to note that the critical angle is independent of the weight of the block. However, for most cases, this is not true and the weight of the block multiplied by the sin of the angle of inclination is the primary driving force as well as those due to water pressure, earthquake loads, and so on. The resisting forces are derived from the frictional resistance of the joint, which is a function of the weight of the block (the normal stress) and the cohesion of the joint (Figure 9-13).

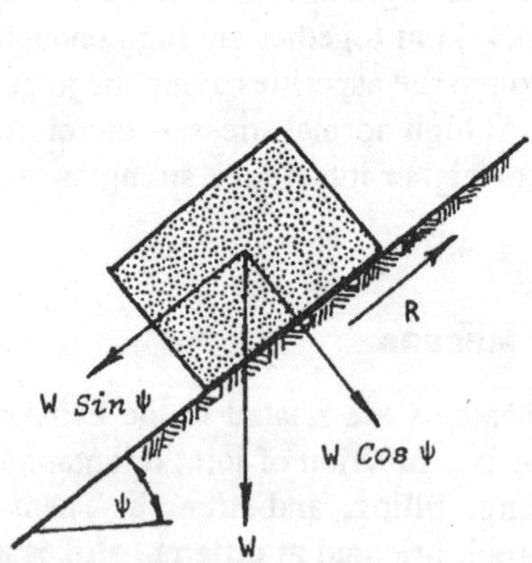

Figure 9-12 Block sliding on an inclined surface. (From Hoek and Bray, 1981.)

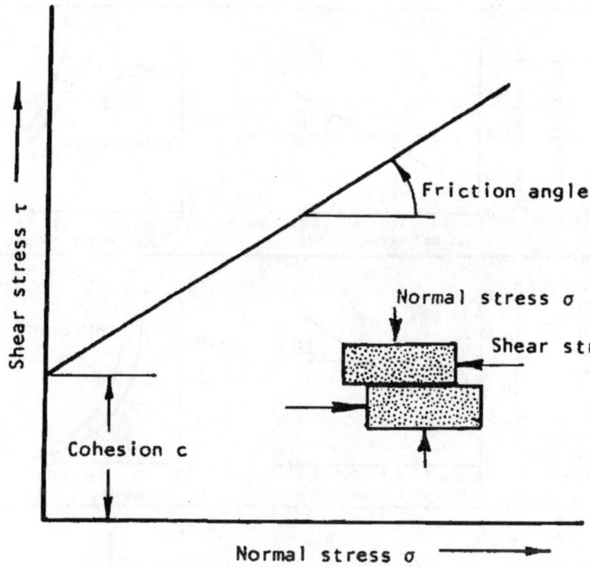

Figure 9-13 Shear strength of a rock joint. (From Hoek and Bray, 1981.)

Effects of Water

The effects of water on a rock block or wedge are not only confined to the water pressures imposed behind and around. The water also can impose uplift forces on the sliding surface itself, causing buoyancy and lubrication along the joint surface. The normal stress is reduced accordingly, which decreases the amount of frictional resistance that will be available to resist sliding. This is why the presence of water in a rock slope has such a dramatic effect and causes so many failures. It increases the driving forces and decreases the resisting forces at the same time.

Factors of Safety

The factor of safety of a slope is the ratio of resisting forces to driving forces. It is generally held by most practitioners that a factor of safety of 1.5 is adequate for most cases. Or in other words, the sum of resisting forces are 50 percent greater than the sum of driving forces. To back-calculate the strength of a filed slope, a factor of safety of 0.9 or 1.0 is usually input into the calculations. For some cases, temporary cuts or mine slopes for example, it may be adequate to use a factor of safety between 1.1 and 1.3. In other cases where the slopes are to be permanent and failure would have an extreme consequential cost or loss of life, it might be advisable to use 1.75 or 2.0. It must be pointed out that many slope stability calculations are based on uncertain characteristics and conditions. Parametric and probabilistic studies are usually necessary to obtain reliable results.

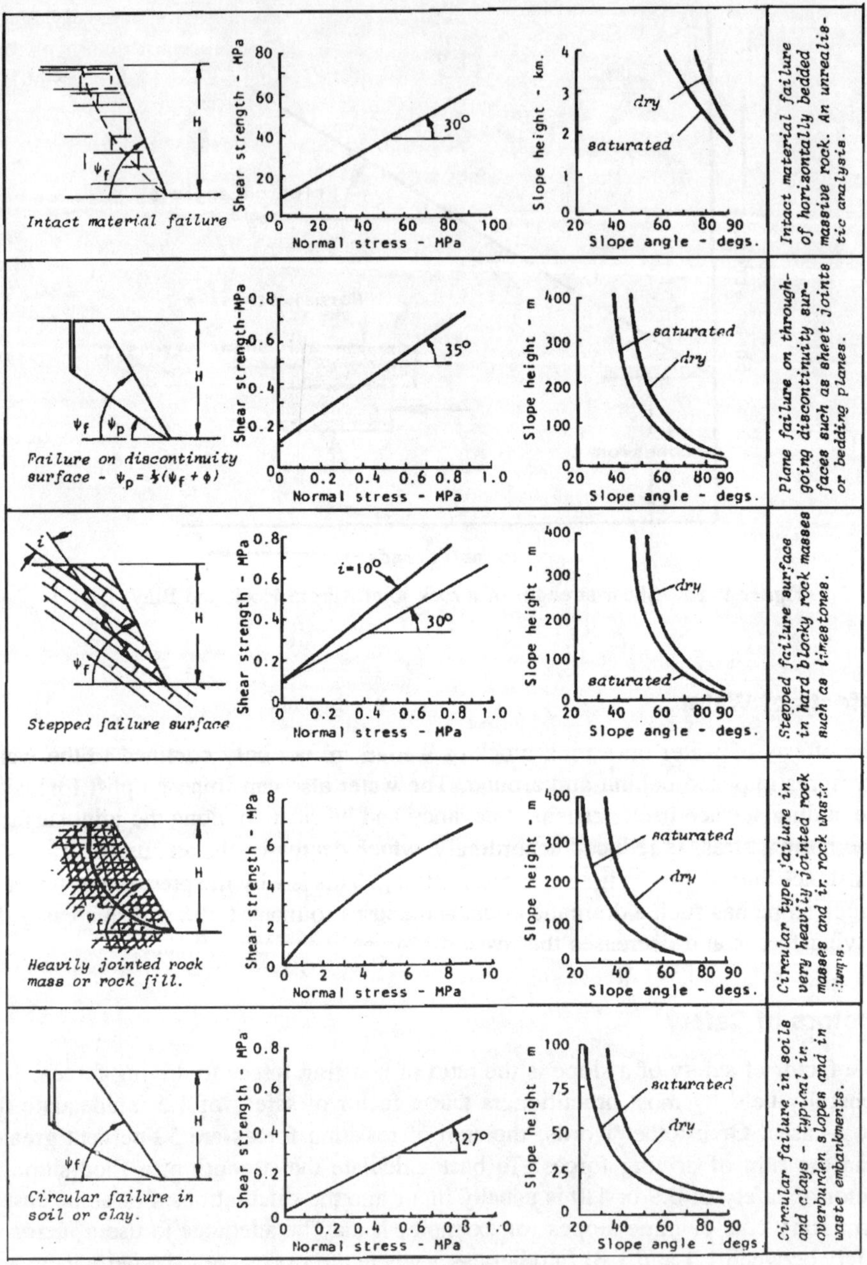

Figure 9-14 Geometrically stable rock slope conditions. (From Hock and Bray, 1981.)

If a slope is found to be unsafe according to the calculated factor of safety, steps can be taken to improve the stability of the slope. In addition to changing the geometrics of the slope (e.g., height, angle), redirecting water flow and rock bolting are two common methods for improving stability. These actions can be analyzed using the factor of safety concepts. Or similarly, factor of safety concepts can be used to determine the required amount of drainage or reinforcement.

9-6 STABILITY ANALYSES

From the above data acquisition and interpretation, potential failure modes can be ascertained using the orientation, height, and slope of an existing rock cut compared with the known geologic character of the rock mass (Figure 9-14). The predominant modes of rock slope failure are planar and wedge, although two-block, toppling, circular, and multiple plane failures occur frequently as well. Now with the advent of personal computers, computation of apparent safety factors for a variety of assumed conditions and probabilities has never been easier (Hendron et al., 1971; Kovari and Fritz, 1975; Abramson, 1985; Watts and West, 1985).

Unfortunately, our advancements made in computing factors of safety have not necessarily improved our ability to accurately model rock masses that are heterogeneous, inelastic, and discontinuous. The discontinuities in the rock mass are usually what control stability. The characteristics, strength, stress state, and stress history of discontinuities are still not clearly understood because of our inability to completely sample and test the rock mass. To further complicate the picture, the hydrologic characteristics of the rock mass, particularly the amount, pressure, and seepage of groundwater between discontinuities, is seldom clearly known and is often responsible for the onset of failure.

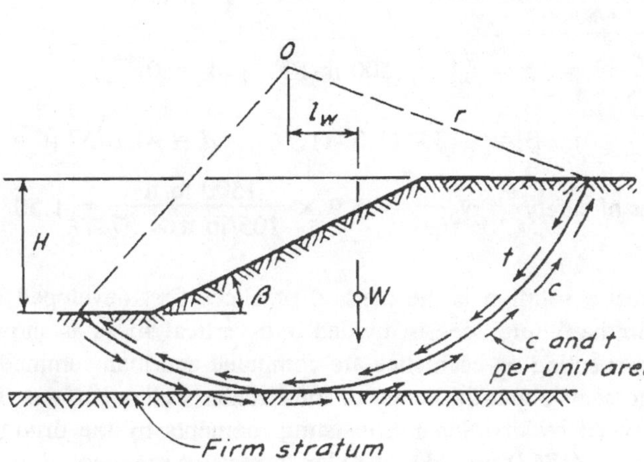

Figure 9-15 Circular failure surface.

Circular Analysis

If the rock is weathered or fractured to the point where it will behave like a soil, soil slope stability methods can be used. Many soil-like slopes will fail along a circular-shaped slip surface, as shown in Figure 9-15. If the soil is homogeneous and cohesive only, charts can be used to analyze the factor of safety based on the stability factor, which is a function of the height and angle of the slope and the cohesive strength of the soil (NAVFAC, 1982). If the soil possesses both frictional and cohesive components, Taylor (1948) provides charts for the solution of such problems.

Example Problem 9-2 Circular Failure Analysis

Given

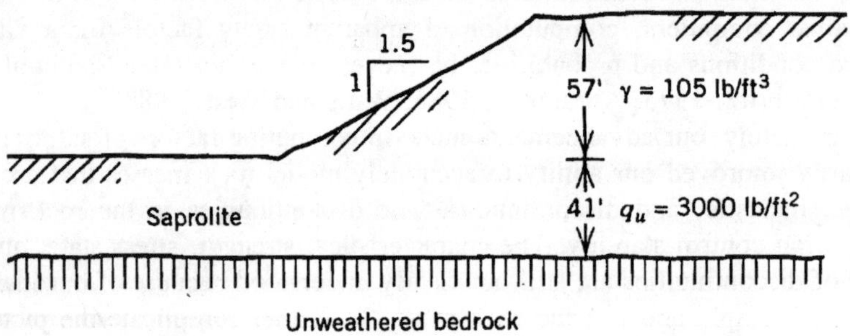

Required: Factor of safety against circular failure.

Solution: Use stability chart in NAVFAC (1982), page 7.1-319:

$$c = \frac{q_u}{2} = 1500 \text{ lb/ft}^2 \qquad \phi = 0$$

$$\beta = 33° \ (1.5\text{–}1) \qquad d = 41 \text{ ft}/57 \text{ ft} = 0.72$$

$$\text{Factor of safety} = N_0 \frac{c}{\gamma H} = 5.9 \times \frac{1500 \text{ lb/ft}^2}{105/\text{lb ft} \times 57} = 1.50$$

A more precise solution is the method of slices first developed by Fellenius (1939) wherein the failure zone is divided into vertical slices as shown in Figure 9-16. The forces acting on each slice are computed and then summed as resisting and driving moments about the center of the circular sliding surface. The factor of safety is computed by dividing the resisting moments by the driving moments. Terzaghi and Peck (1967) provide a tabular format for rapid solution of this problem. Janbu (1954), Bishop (1955), and Morgenstern and Price (1965) have proposed

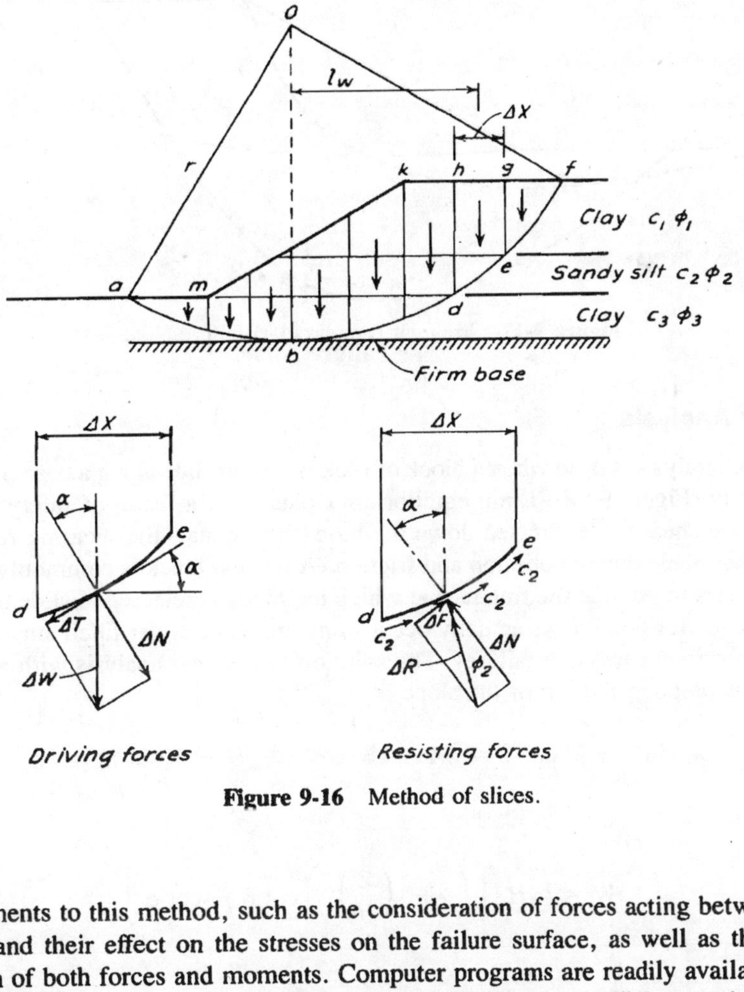

Figure 9-16 Method of slices.

refinements to this method, such as the consideration of forces acting between the slices and their effect on the stresses on the failure surface, as well as the equilibrium of both forces and moments. Computer programs are readily available that will solve soil slope stability problems using a number of different alternative methods.

Irregular Surface Analysis

Often slopes in soil-like materials do not fail along a circular-shaped surface. For a number of reasons including the presence of bedding, relic joints, sand or clay seams, vegetation, or previous slope movement, a noncircular or irregular failure surface can develop (Figure 9-17). Slope stability problems can still be solved using methods similar to those described above. Morgenstern and Price (1965) developed a method of analysis that treats a slip surface of arbitrary shape and satisfies all equilibrium requirements. It is easiest to solve this type of problem with the use of a computer.

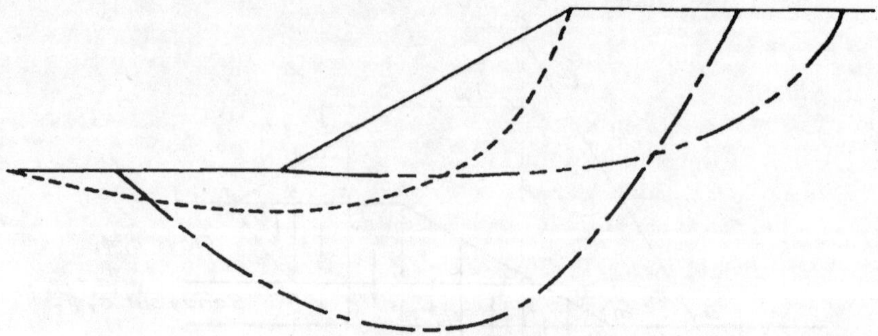

Figure 9-17 Irregular (noncircular) failure plane.

Planar Analysis

A planar analysis is used when a block of rock is free to slide along a base plane as a rigid body (Figure 9-18). Limit equilibrium exists (i.e., a factor of safety equal to one) if the shear force directed down the base plane equals the shearing resistance along that plane due to cohesion and friction. A tension crack is commonly used in the analysis to provide the free face at which the block is released to slide down the base plane. Tension cracks actually occur in nature and a water-filled tension crack often contributes to slope failures. The solution of a planar analysis with a tension crack intercepting the top of the slope is

$$W \sin \delta + V \cos \delta = SA + (W \cos \delta - U - V \sin \delta) \tan \phi \qquad (9\text{-}1)$$

where

$$W = \tfrac{1}{2}\gamma H^2 \left[\left(1 - \left(\frac{Z}{H} \right)^2 \right) \cot \delta - \cot \alpha \right] \qquad (9\text{-}2)$$

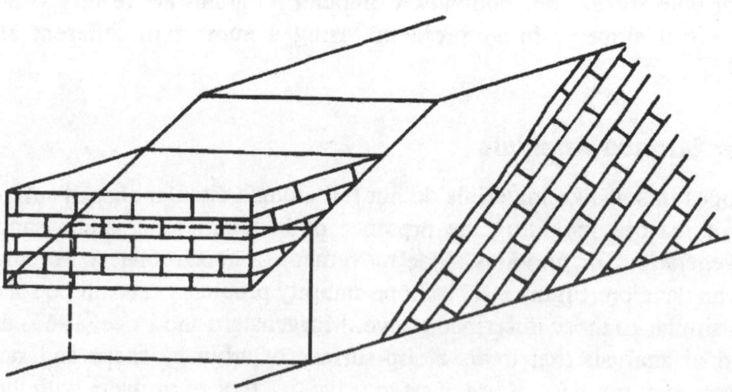

Figure 9-18 Planar analysis.

where W = weight of the sliding block
γ = unit weight of the rock
H = height of the slope
Z = depth of the tension crack
δ = dip of the sliding surface
α = dip of the slope
S = shear strength along sliding surface
A = area of sliding surface
ϕ = friction angle along sliding surface
U = water pressure along sliding surface
V = water pressure in tension crack

The solution of a planar analysis with a tension crack intercepting the slope is

$$W = \tfrac{1}{2}\gamma H^2\left[\left(1 - \frac{Z}{H}\right)^2 \cot\delta(\cot\delta\tan\alpha - 1)\right] \tag{9-3}$$

An example of a planar slide is shown in Figure 9-19.

Figure 9-19 Rock block sliding on planar surface.

Example Problem 9-3 Plane Failure Analysis

Given

Slope angle	$\alpha = 65°$
Sliding surface angle	$\delta = 50°$
Length of sliding surface	$A = 156.6$ ft
Height of slope	$H = 150$ ft
Depth of tension crack	$Z = 30$ ft
Angle of internal friction	$\phi = 10°$
Cohesion	$S = 1500$ lb/ft²
Rock density	$\gamma = 105$ lb/ft³

Required: Factor of safety when the tension crack is filled with water to a depth $Z_W = 10$ ft.

Solution

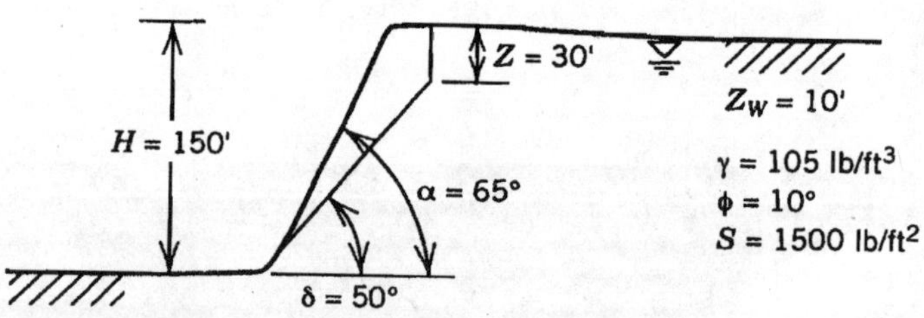

$$A = \frac{(H - Z)}{\sin \delta} = \frac{(150 \text{ ft} - 30 \text{ ft})}{\sin 50°} = 156.6 \text{ ft}$$

Crack intercepts the slope crest

$$W = \tfrac{1}{2}\gamma H^2 \left[\left(1 - \left(\frac{Z}{H}\right)^2\right) \cot \delta - \cot \alpha\right]$$

$$W = \tfrac{1}{2} \times 105 \times 150^2 \left[\left(1 - \left(\frac{30}{150}\right)^2\right) \cot 50° - \cot 65°\right] = 401^K$$

Factor of safety

$$\text{Factor of safety} = \frac{SA + (W \cos \delta - U - V \sin \delta)\tan \phi}{W \sin \delta + V \cos \delta}$$

$$U = \tfrac{1}{2}\gamma_W Z_W A = \tfrac{1}{2} \times 62.4 \times 10 \times 156.6 = 49^K$$

$$V = \tfrac{1}{2}\gamma_W Z_w^2 = \tfrac{1}{2} \times 62.4 \times 10^2 = 3^K$$

$$\text{Factor of safety} = \frac{1.5 \times 156.6 + (401 \cos 50° - 49 - 3 \sin 50°)\tan 10°}{401 \sin 50° + 3 \cos 50°}$$

$$= 0.88$$

Two-Block Analysis

A two-block analysis is similar to a planar slide analysis except two sliding blocks instead of one are involved. The toe of the slope is referred to as the "passive" block with a base plane angle flatter than the angle of internal friction (Figure 9-20). It would not slide on its own without the second or "active" block pushing on it with a base plane steeper than the angle of internal friction. By assuming that the two blocks are divided by a vertical plane, the problem of stability can be solved with the following equation:

$$F_b = \frac{W_1 \sin(\delta_1 - \phi_1)\cos(\delta_2 - \phi_2 - \phi_3) + W_2 \sin(\delta_2 - \phi_2)\cos(\delta_1 - \phi_1 - \phi_3)}{\cos(\delta_2 - \phi_2 - \Theta)\cos(\delta_1 - \phi_1 - \phi_3)} \quad (9\text{-}4)$$

where F_b = support force needed for equilibrium at angle Θ
W_1, W_2 = weight of the blocks
ϕ_1, ϕ_2, ϕ_3 = friction angles between blocks
δ_1, δ_2 = dip of the sliding surfaces

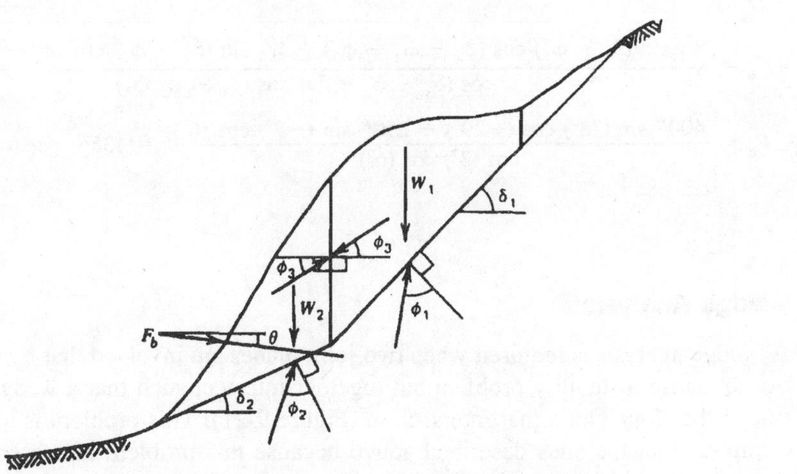

Figure 9-20 Two-block analysis. (From Goodman, 1980.)

Example Problem 9-4 Two-Block Failure Analysis

Given

$$W_1 = 400^K \qquad \delta_1 = 50^\circ$$
$$W_2 = 250^K \qquad \delta_2 = 15^\circ$$
$$\phi_1 = \phi_2 = \phi_3 = 22^\circ$$

Cohesion is neglected.

Required: Rock bolt force, F_b, for a factor of safety = 1.

Solution

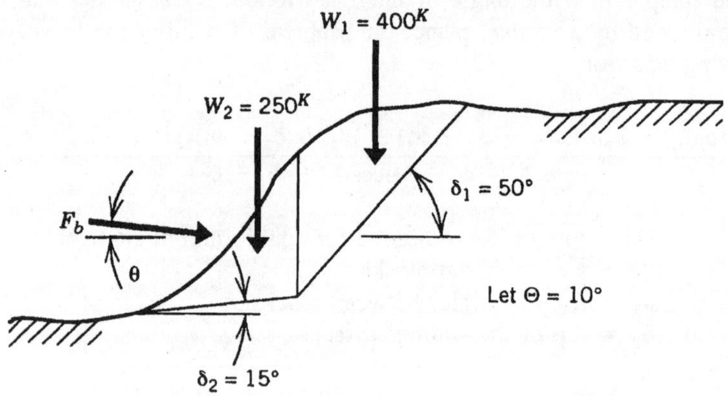

$$F_b = \frac{W_1 \sin(\delta_1 - \phi_1)\cos(\delta_2 - \phi_2 - \phi_3) + W_2 \sin(\delta_2 - \phi_2)\cos(\delta_1 - \phi_1 - \phi_3)}{\cos(\delta_2 - \phi_2 + \Theta)\cos(\delta_1 - \phi_1\ \phi_3)}$$

$$F_b = \frac{400^K \sin(28^\circ)\cos(-29^\circ) + 250^K \sin(-7^\circ)\cos(6^\circ)}{\cos(3^\circ)\cos(6^\circ)} = 135^K \text{ per foot of slope}$$

Wedge Analysis

A wedge analysis is required when two joint planes are involved that by themselves do not cause a stability problem but together intersect such that a wedge can slide out of the slope along that intersection (Figure 9-21). This problem is a little more complex than the ones described above because this problem has to be solved in three dimensions (Figure 9-22). Here is where a computer comes in handy and the listing of one such computer program is included in Abramson (1985). Rock bolt, psuedostatic seismic, and water forces are included in the solution as well as an

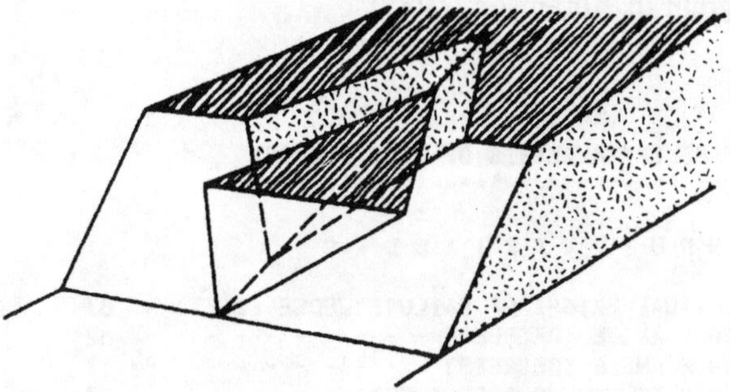

Figure 9-21 Wedge analysis.

interactive routine for changing the input parameters and performing sensitivity studies. Cohesion along the joint planes is ignored.

Example Problem 9-5 Wedge Failure Analysis

	Slope	Plane ①	Plane ②
Strike	N 63° W	N 81° E	N 42° W
Dip	N 82° SW	45° SE	33° SW
		$\phi^1 = 12°$	$\phi_2 = 27°$

Height of slope = 83 ft
Bench angle = 17°
Rock density = 135 lb/ft³
No groundwater or bolts

Required: Factor of safety against failure.

Solution

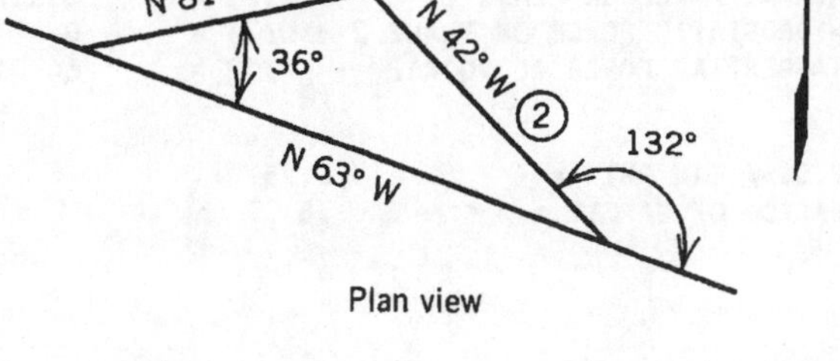

Plan view

$H = 83$ ft	$\beta_1 = 36°$	$\gamma_1 = 45°$	$\phi_1 = 12°$
$A = 82°$	$\beta_2 = 132°$	$\gamma_2 = 147°$	$\phi_2 = 27°$
$D = 17°$	$\gamma_{rock} = 135\ \text{lb/ft}^3$		

Use the program in Abramson (1985).

Results

```
STABILITY ANALYSIS OF ROCK SLOPES
*************************************

I N P U T   V A R I A B L E S

VERTICAL HEIGHT OF FAILURE WEDGE (FEET) -  83
SLOPE ANGLE (DEGREES) - - - - - - - - - -  82
BENCH ANGLE (DEGREES) - - - - - - - - - -  17
DEPTH TO GROUNDWATER (FEET) - - - - - - -  83

BETA  1 (DEGREES) -  36
LAMDA 1 (DEGREES) -  45
PHI   1 (DEGREES) -  12

BETA  2 (DEGREES) -  132
LAMDA 2 (DEGREES) -  147
PHI   2 (DEGREES) -  27

ANGLE OF INTERSECTION (DEGREES) -          31.10596

VOLUME OF WEDGE (CUBIC FEET) - -           968428.1

WEIGHT OF WEDGE (KIPS) - - - - - -        -130737.8

BOLT VECTOR -                  0           0             0

RESULTANT FORCE DUE TO WEDGE AND BOLTS - - -
 0           0            -130737.8

R DOT W1 =      92445.65
R DOT W2 =     -109645.8

WEDGE IS NOT LIFTED FROM EITHER PLANE

R DOT 1S12 =    51101.4
R DOT 2S12 =    23008.66

WEDGE SLIDES ALONG THE LINE OF INTERSECTION X12

NORMAL FORCE ON PLANE 1   - - - N(1) =     38414.93
HYDROSTATIC FORCE ON PLANE 1 - U(1) =      0
NORMAL FORCE ON PLANE 2   - - - N(2) =     85318.19
HYDROSTATIC FORCE ON PLANE 2 - U(2) =      0
TANGENTIAL FORCE ALONG X12   - - T12 =     64701.59


F.S. / TAN PHI   =             3.75
FACTOR OF SAFETY =             .8
```

Multiple Plane Analysis

Multiple plane problems become increasingly more complex to solve because of the multiple forces that must be treated on each plane and the increased number of

Figure 9-22 Rock wedge fallout.

failure modes. For instance, a block could rotate as well as slide in some cases (Figure 9-23). A method for solving such a problem may be found in Hendron et al. (1971).

Toppling

If the geometry of a block is such, it may tip over or "topple" before it begins to slide down an inclined plane (Figure 9-24). The variables considered are the shape

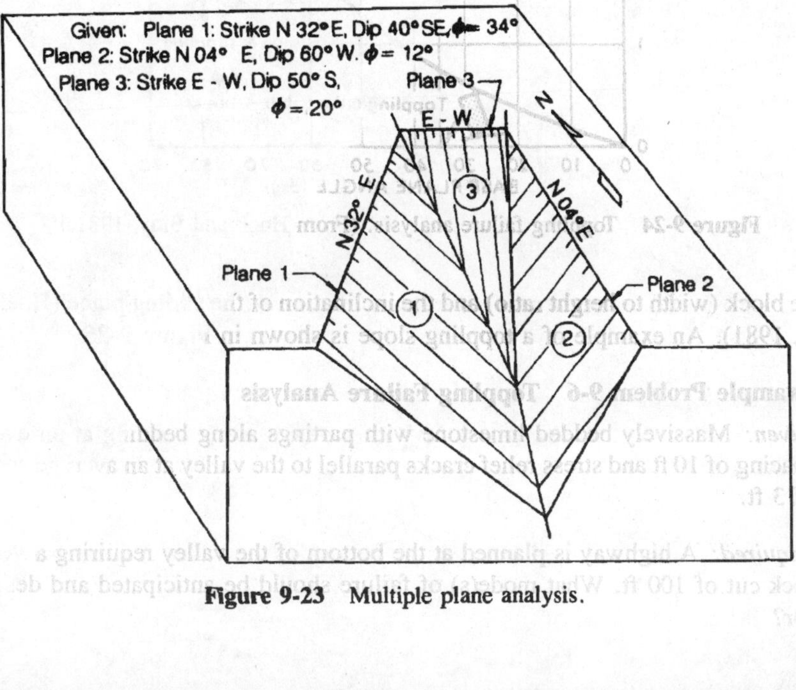

Figure 9-23 Multiple plane analysis.

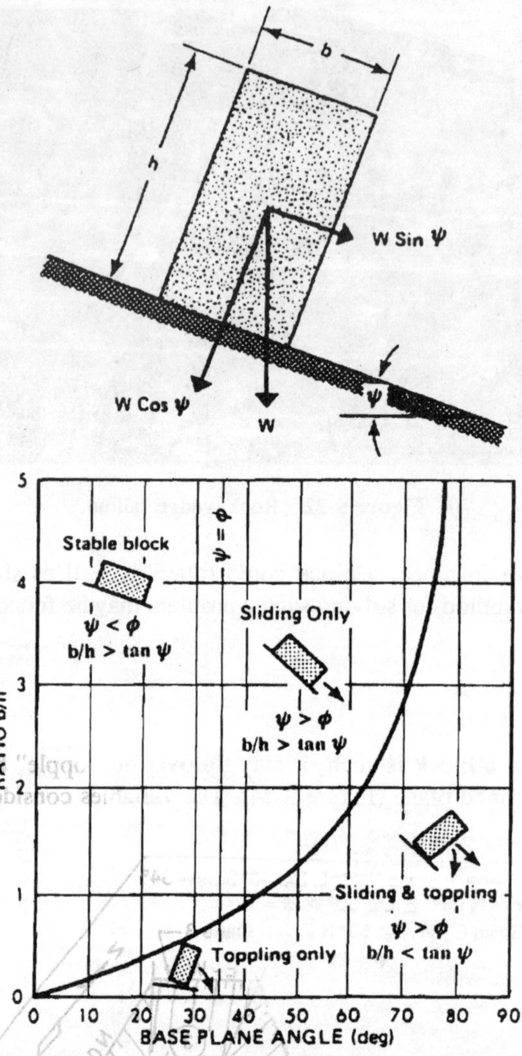

Figure 9-24 Toppling failure analysis. (From Hoek and Bray, 1981.)

of the block (width to height ratio) and the inclination of the sliding plane (Hoek and Bray, 1981). An example of a toppling slope is shown in Figure 9-25.

Example Problem 9-6 Toppling Failure Analysis

Given: Massively bedded limestone with partings along bedding at an average spacing of 10 ft and stress relief cracks parallel to the valley at an average spacing of 3 ft.

Required: A highway is planned at the bottom of the valley requiring a vertical rock cut of 100 ft. What mode(s) of failure should be anticipated and designed for?

Figure 9-25 Toppling failure conditions.

Solution: It depends on the bedding and friction angles. Consult Figure 9-24.

$$b/h = 3/10 = 0.33 = \tan\delta \qquad \delta = 18^\circ$$

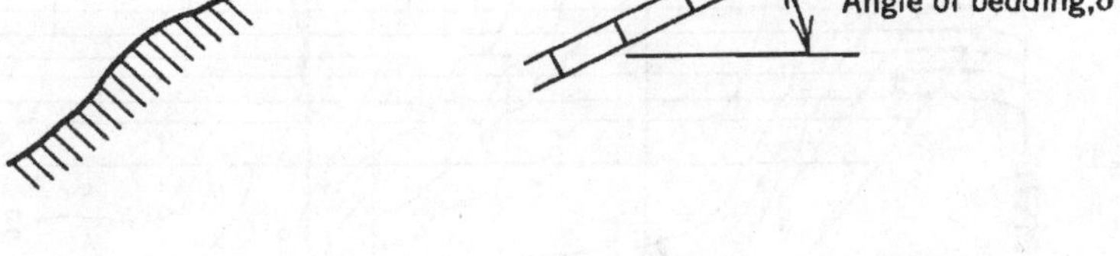

	$\delta > \phi$	$\delta < \phi$
$\delta > 18^\circ$	Sliding and toppling possible	Toppling possible
$\delta < 18^\circ$	Sliding possible	Stable condition

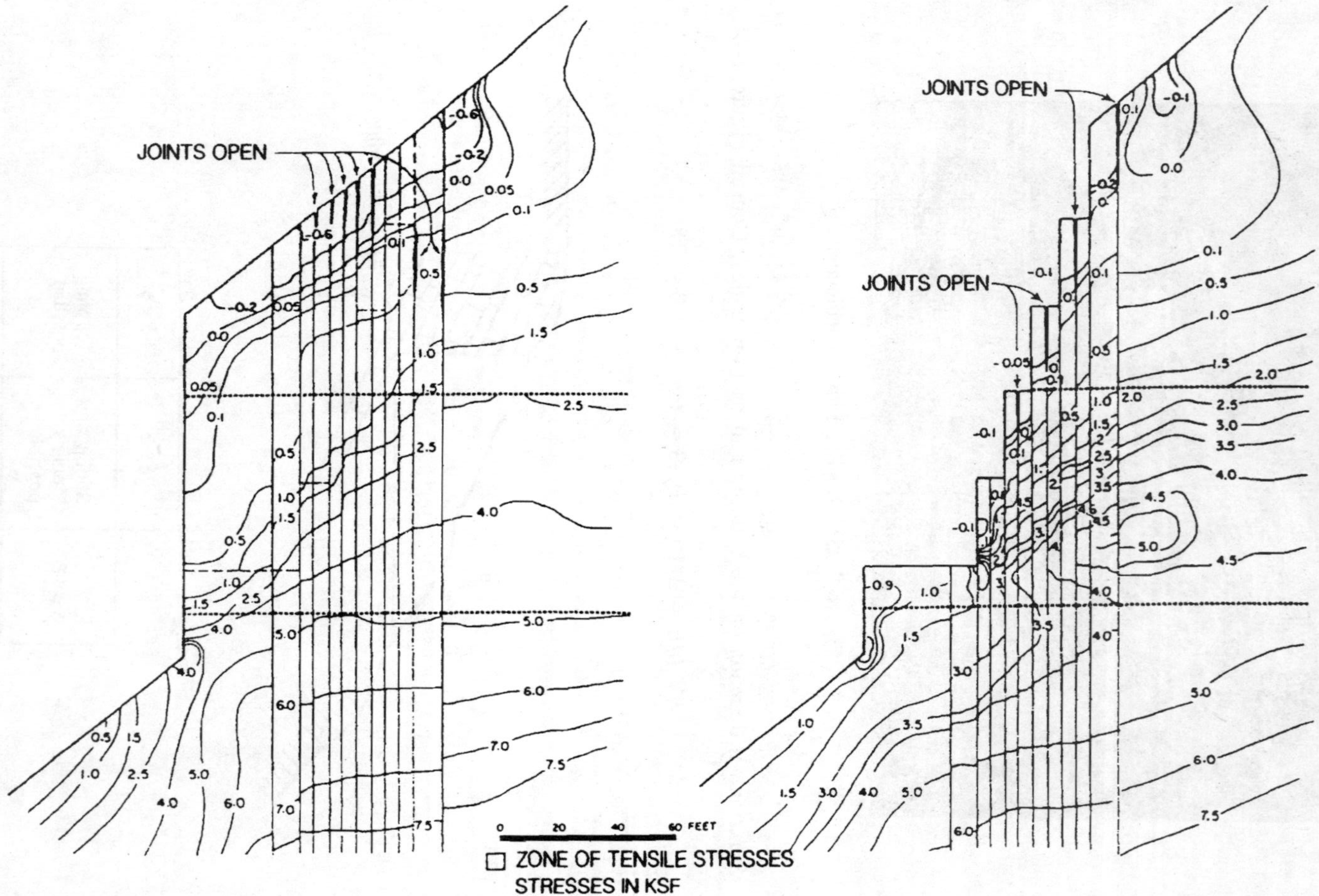

Figure 9-26 Finite element slope stability analysis.

Other Numerical Methods

Other numerical methods are sometimes used to solve rock slope stability problems. They primarily include key block analysis and finite element methods. Key block theory is used to analyze the rock blocks that form in a rock mass between discontinuities. Because of shape and orientation, some of these blocks are "key blocks," analogous to a key block in an archway. The other blocks are restrained from moving by this key block, and if the key block is held in place, so are the blocks it supports. More information on this approach can be found in Goodman and Shi (1985).

Another advanced method of analyzing rock slope stability problems is the finite element method. In this method, the rock mass is divided into discrete elements that have defined material properties, stiffnesses, and boundary conditions. A mesh is set up (Figure 9-26) with which the initial conditions can be set up as well as any changes in geometry (an excavation for instance) and material properties (due to weathering for instance). With this method, there are intrinsic difficulties in properly modeling the rock mass, especially the joints, and these difficulties usually outweigh the potential benefits of more sophisticated modeling. Excellent quality input data is required, and the results are highly sensitive to the modeling assumptions made.

9-7 POTENTIAL CAUSES OF FAILURE

Slopes do not fail because of the "modes of failure." Something causes this mode of failure to engage itself. Calculated factors of safety are essentially point measurements in time. Factors of safety change constantly during every season, every rainfall, every earthquake, and over the years as the rock slope ages due to exposure to the elements. The rate of change from stability to instability can be instantaneous, as in the case of an earthquake, or can take tens or hundreds of years, as in the case of slow degradation of a shale stratum undercutting an overlying massive sandstone bed (Figure 9-27). In most cases, there are a variety of weakening forces acting on the slope with different rates of degradation and relative importance. Sometimes, these phenomena are products of geologic transformation that began long before the facility was built. Stress relief jointing along river valleys is one common example.

Nevertheless, the principal causes of failure are often related to only a handful of natural processes:

- Weathering
- Hydrostatic pressure
- Freeze/thaw cycles
- Seismic events
- Creep
- Progressive failure

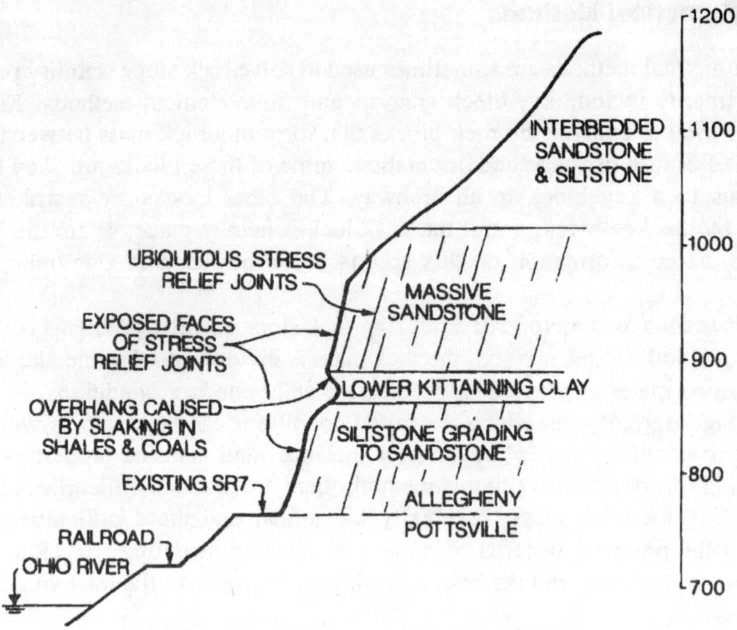

Figure 9-27 Undercutting of massive sandstone.

Progressive failure is a product of the other processes and their interrelationships. Additionally, human activities such as undercutting the toe of a slope, diverting surface runoff onto an unstable slope, or increasing the driving force on a slope by surcharge loading can also contribute to a failure. However, the theme of this chapter centers around aging slopes that are not being affected directly by human activities.

Generally, the rock mass starts off with one or more sets of throughgoing discontinuities. Construction of a facility frequently causes these discontinuities to open due to stress relief and/or blasting gas pressure and vibration. These effects are most pronounced at the slope face and become less pronounced within the rock mass. A rule-of-thumb is that the depth of loosening is equal to the height of cut or valley. Typically, instabilities occur within 50 ft of the slope face.

Weathering

Weathering is the processes by which intact rock is broken down into smaller pieces and weakened by disintegrative actions like grinding, shattering, or breaking (e.g., as a result of pore water freezing and expanding in the pores or joints); chemical alteration and decomposition; or dissolving of soluble minerals, called solutioning. All three processes (i.e., mechanical, chemical, and solution weathering) occur simultaneously but at different rates, depending on climate, topography, and composition of the host bedrock. Weathering can weaken intact rock as well as rock joints and joint fillings. Also, in interbedded formations, one type of rock interbed may be more prone to one type of weathering than another.

Hydrostatic Pressures and Freeze/Thaw Cycles

The water within the rock mass, in addition to facilitating weathering processes, tends to freeze during the winter and melt during the spring. As described before, freezing of the groundwater tends to push loosened rock blocks out, causing further loosening. Spring melt waters tend to impose hydrostatic forces on the bases and sides of discreet rock blocks, thereby increasing the magnitude of driving forces and decreasing the magnitude of resisting forces. This can occur within tension cracks and other open joints due to rainfall and snow melt whether a freeze cycle occurs or not.

Where there are strong seasonal climate variations, a high incidence of failures during the spring months usually occurs. This is due to joint water that has become frozen and expanded over the winter as it begins to melt. The water tends to jack loosened rock blocks and wedges during the winter, causing a transition from peak strength along discontinuities toward residual strength. During the spring, the ice melts, subjecting the rock to hydrostatic loading and increased lubrication along joints. As the strength along discontinuities decreases, the stress increases. Sooner or later the stress will exceed the strength and failure occurs (Figure 9-28). While this concept is easy to verbalize, our ability to analyze real field conditions and to predict failures is quite handicapped by the lack of accurate engineering parameters.

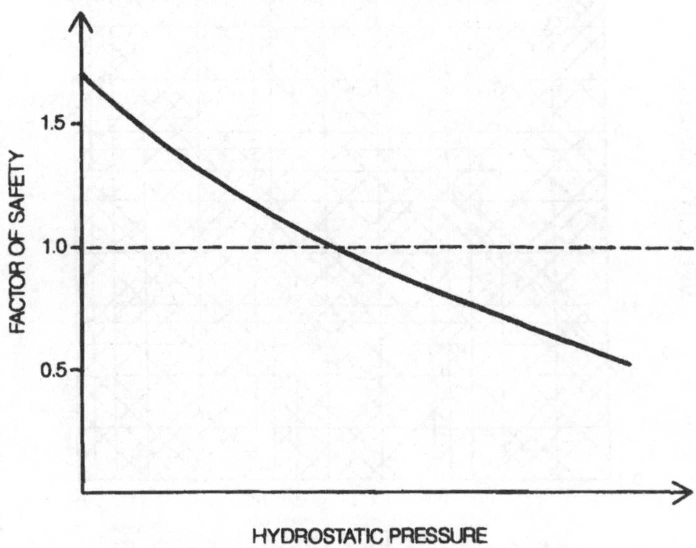

Figure 9-28 Failure due to hydrostatic pressure.

Seismic Events

Seismic events tend to cause dynamic displacements of a loosened rock mass. The collective effects of earthquakes on a slope can be progressive loosening, displacement, rotation of discrete blocks or wedges, or complete slope failure. Loosening could also possibly make the rock mass more stable by improving drainage and increasing interlocking. The possible effects of earthquakes on a rock slope must be evaluated the same way effects on other types of structures are evaluated. Namely, major faults in the vicinity must be identified and categorized by probable earthquake vibrations. Once the magnitude of probable ground vibrations is estimated, then the spectral response and effects on the slope can be estimated (Figure 9-29).

Creep

When clay minerals are prevalent in the rock mass, either as a constituent of intact rock or along discontinuities, then creep (displacement at constant load) can occur along open joints, particularly if the other processes described above are occurring also. It is known from direct shear tests that readjustment of stresses along joints

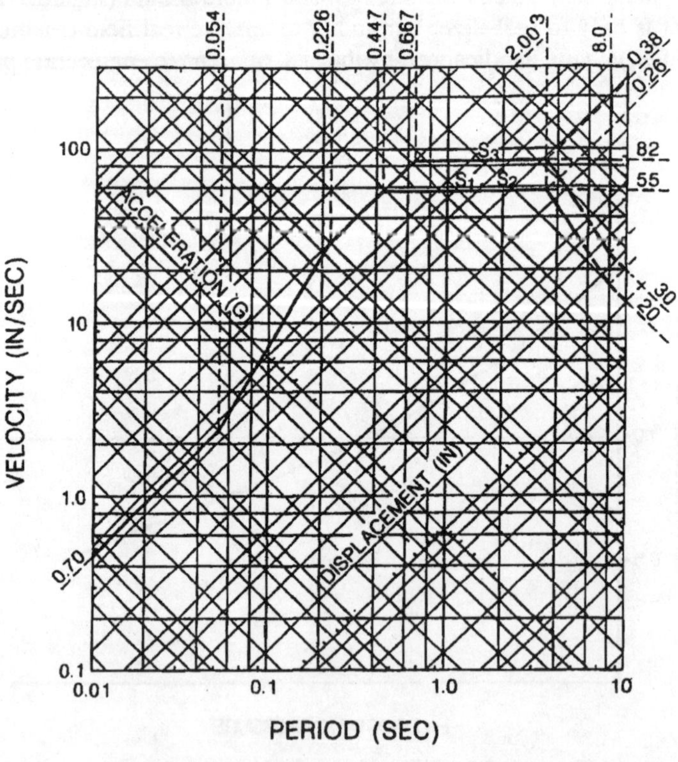

Figure 9-29 Earthquake spectral response.

occurs as irregularities are overridden, crushed, and/or sheared through. Therefore, it is easy to envision a block of rock creeping downslope, ever so slowly, redistributing its weight onto the most resistant surfaces until, finally, stress concentrations exceed the strength of potential support points and displacements accelerate until failure occurs (Figure 9-30).

Progressive Failure

The process by which most slope failures occur is called progressive failure, the compounding of events and processes that, over time, combine to cause ultimate instability of the slope. The other processes described above can be divided into short-term and long-term events. Single rainfall or runoff induced hydrostatic pressures, one freeze/thaw cycle, and earthquakes can be categorized as short-term events. Weathering, creep, and repetitive freeze/thaw cycles can be categorized as long-term events.

An aging rock slope is exposed to long-term events usually as soon as the rock cut is made. A myriad of factors will then determine which processes are most deleterious to the slope and establish the time rate of degradation. This is as true for new rock cuts as it is for ancient river valleys. The short-term events interplay with

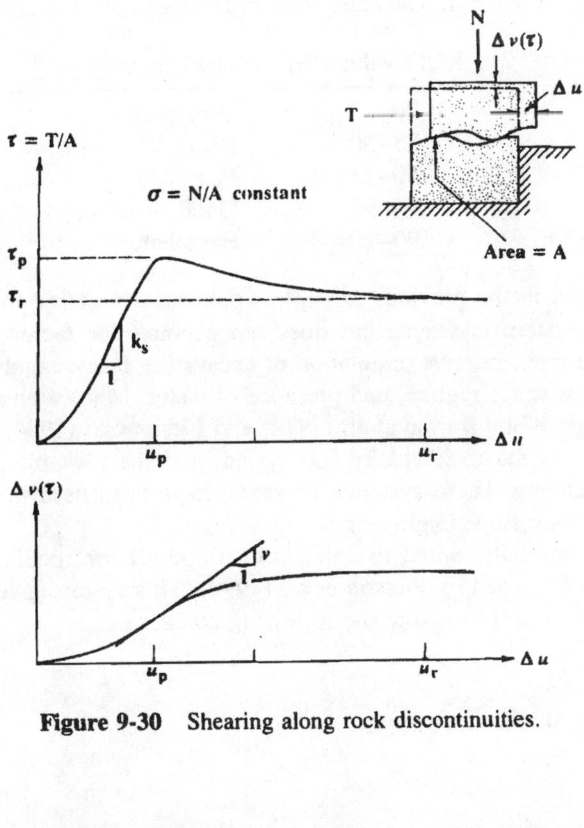

Figure 9-30 Shearing along rock discontinuities.

the long-term processes, usually resulting in localized accelerations or changes in the nature of ongoing degradation.

Tension cracks above the cut area are often a common sign of large-scale progressive failure. Small movements within the rock mass are usually magnified near the top of a slope. Scarps and separations at the top of the slope usually indicate that portions of the slope are moving. These openings can fill with water and increase progressive degradation. Often, tension cracks are not noted until after a failure due to hydrostatic loading in the crack. Sometimes, these cracks can be masked by overburden soils and vegetation. The presence of tension cracks should be a red flag in any slope condition assessment and the meaning and importance of such cracks cannot be overemphasized.

9-8 RATING SYSTEMS

Several rating systems are available in rock mechanics for characterizing the condition of a rock mass. The simplest and most common is the rock quality designation, or RQD (Deere and Deere, 1988). RQD is calculated by adding up the length of *NX*-size unweathered rock core pieces in a core run longer than 4 in. and dividing by the total length of the rock core run. So, for example, if there are 10 pieces of unweathered rock core longer than 4 in., say 4.5, 8, 7, 9, 13, 5.5, 7.6, 5.2, 4.3, and 6 in. long, of a 10-ft-long core run (120 in.), the RQD would equal 70.1 in. divided by 120 in., or 58 percent. The categories of the RQD system are as follows:

RQD Value (%)	RQD Category
0–25	Very poor
25–50	Poor
50–75	Fair
75–90	Good
90–100	Excellent

So, the core run in the previous example would be considered to be fair quality.

The RQD system is simple but does not account for factors like the intact strength of the rock, relative orientation of excavation faces, number of joint sets, joint conditions, stress regime, and presence of water. More sophisticated systems have been proposed by Barton et al. (1974) and Bieniawski (1984). These systems are referred to as the rock quality (Q) system and the rock mass rating (RMR) system, respectively. These systems, however, have been used in tunneling more often than for rock slope engineering.

A system especially suited to aging rock slopes is the rockfall hazard rating (RHR) system proposed by Pierson et al. (1990). This system gives relative point values ranging from 0 to 100 to the following categories:

- Slope height
- Catchment ditch effectiveness

TABLE 9-1 Rockfall Hazard Rating System

Catagory		Rating Criteria and Score			
		Points 3	Points 9	Points 27	Points 81
Slope Height		25 feet	50 feet	75 feet	100 feet
Ditch Effectiveness		Good catchment	Moderate catchment	Limited catchment	No catchment
Average, Vehicle Risk		25% of the time	50% of the time	75% of the time	100% of the time
Percent of Decision Sight Distance		Adequate sight distance, 100% of low design value	Moderate sight distance, 80% of low design value	Limited sight distance, 60% of low design value	Very limited sight distance, 40% of low design value
Roadway Width Including Paved Shoulders		44 feet	36 feet	28 feet	20 feet
Geologic Character	Case 1 Structural Condition	Discontinuous joints, favorable orientation	Discontinuous joints, random orientation	Discontinuous joints, adverse orientation	Continuous joints, adverse orientation
	Case 1 Rock Friction	Rough, irregular	Undulating	Planar	Clay infilling, or slicken sided
	Case 2 Structural Condition	Few differential erosion features	Occasional erosion features	Many erosion features	Major erosion features
	Case 2 Difference in Erosion Rates	Small difference	Moderate difference	Large difference	Extreme difference
Block Size		1 foot	2 feet	3 feet	4 feet
Quantity of Rockfall/Event		3 cubic yards	6 cubic yards	9 cubic yards	12 cubic yards
Climate and Presence of Water on Slope		Low to moderate precipitation; no freezing periods; no water on slope	Moderate precipitation or short freezing periods or intermittent water on slope	High precipitation or long freezing periods or continual water on slope	High precipitation and long freezing periods or continual water on slope and long freezing periods
Rockfall History		Few falls	Occasional falls	Many falls	Constant falls

Source: Pierson et al., 1990.

- Average vehicle risk (exposure in the fall zone)
- Percent of decision site distance
- Roadway width
- Geologic character
- Rockfall size
- Climate
- Presence of water
- Rockfall history

The criteria for assigning points to each of these categories are shown in Table 9-1. The total number of points for each slope is tallied. A total of 1000 is the worst score; low scores are thought to have the least rockfall hazard. From these totals, numerous slopes can be compared to determine if any need remediation and which have the most pressing urgency for remediation relative to the others.

9-9 REMEDIATION CRITERIA

By processing site conditions and geologic data, potential slope failure modes can be analyzed. With this information, potential causes of failure can be investigated to see which are present and their relative importance. Detailed surveys can be made with rating systems to ascertain which slopes are in the worst conditions. With this knowledge, a detailed slope condition assessment can be made. It is helpful to categorize the slope or slopes on a project according to some project specific criteria. The bases for these criteria could be:

- Schedule requirements
- Access requirements
- Type of repairs
- Type of equipment required
- Cost

Two bases can be combined in a matrix. For instance, one category might be inexpensive, easily accessed repairs. Another might be costly repairs requiring specialized equipment and shutdown of the facility.

Rock slope remediation must be planned based on certain criteria dealing with the costs of repairs, source of funding, schedule constraints, effects on the operation of the facility, design life of repairs, degree of safety desired, size of project, nature of construction personnel required, and type of construction contract. For instance, on a highway project, if the area requiring repair is localized, rehabilitation methods can be prescribed and implemented without undue commitment of funds and other resources. If, on the other hand, several locations or long stretches of roadway are being studied, the problem areas must be group classified and prioritized. Funds may be limited and effective utilization of these funds may be required.

The highest priorities should be given to choke points. Choke points are locations where rock slides could cause significant structural damage or economic dislocation due to long-term disruption of service. Examples would be:

- High rock slopes adjacent to high activity areas.
- Large areas where protracted postslide removal and repair could curtail use of the facility and have economically disruptive effects.

Lowest priorities should be given to slopes where minimal potential for structural damage, economic disruption, or personal injury exists. Within this category, slopes with significant annual maintenance costs should be considered for rehabilitation or reconstruction. Continued unrestrained deterioration may eventually lead to major slides involving large rock masses, which could cause substantial economic disruption.

9-10 REMEDIATION ALTERNATIVES

The methods available for slope rehabilitation are similar to those described by Piteau and Peckover (1978). There are typically seven repair methods used for rehabilitating a slope:

- Removal of unstable rock
- Catchment
- Flattening of the slope
- Buttresses
- Surface protection
- Reinforcement
- Drainage

These methods can be used singly or in combination.

Site specific conditions normally dictate whether to reinforce the rock or support it. Support methods are most commonly used to stabilize overhangs. Reinforcement is most commonly used to prevent ultimate sliding or rotational failure of potentially unstable rock masses along discontinuities. Also, surface protection using mesh or shotcrete can be used to prevent progressive raveling and attack by sunlight, air, and water.

Alternatively, relocation of the facility or surveillance may be preferable to repair for a variety of reasons. These alternatives, as well as reasons to defer repairs (do nothing), are discussed below.

Do Nothing

After a thorough analysis of the slope condition is made, it may be prudent to take no action at all. Potential reasons for this action could include:

- Insufficient funds
- Sufficient margin of safety
- Inaccessible site
- Intolerable disruption of service
- Warning/surveillance preferred to repair

However, it must be noted that none of these reasons would relieve the owner of liability for property damage and injury resulting from a rockfall.

While warning/surveillance is fairly simple, it does not mitigate the problem and can be very hazardous to workers. Rockfall warning signs are intended to be temporary until hazardous conditions are eliminated. Yet if rehabilitation is a low priority, the signs sometimes remain in place indefinitely, serving no realistically practical purpose. Electronic devices are sometimes, but not commonly, used as warning systems. These devices can include geotechnical instrumentation, electrical wire or fence, vibration monitors, and television cameras. The use of geotechnical instrumentation for rock slopes will be discussed in more detail later in this chapter. It is important to note, however, that eliminating potential rockfall hazards is preferable to monitoring and maintaining them for long periods of time.

Facility Relocation

An alternative to patrolling the facility is to relocate all or part of it away from the unstable area. Potential choices for relocation will depend on site conditions but may include moving the facility away from the rock slope onto an embankment, into an underground opening, or completely away from the general area (Figure 9-31). In most cases, it is more economical to rehabilitate the existing rock slope rather than to relocate the facility.

Removal of Unstable Rock

Removal of potentially unstable rock is typically necessary for slope rehabilitation whether it is to insure long-term performance or simply for worker safety. This may include removal of accumulated rock on benches, surface scaling by hand, and explosive removal of overhangs. Breakage and removal of the rock is normally done using conventional rock excavation equipment. Access to the slope is sometimes limited and may require hand-carried equipment and rappeling expertise. Alternatively, scaling can be done using a crane with specially designed "rake" or demolition ball.

Rock removal can be hazardous to the workers doing the work, as well as workers performing other tasks on-site. Additionally, vehicles and pedestrians pass-

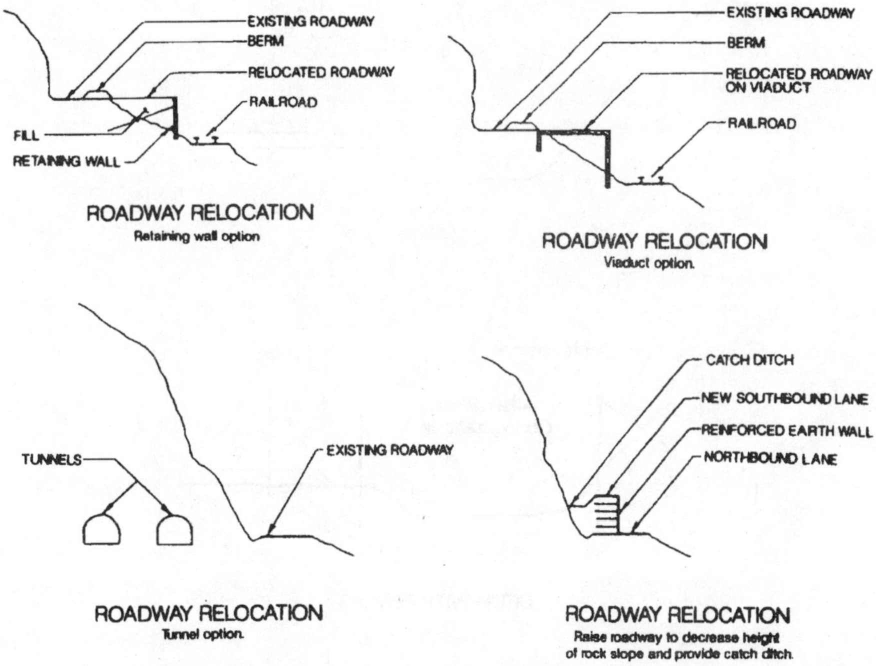

Figure 9-31 Relocation of a facility adjacent to a rock slope.

ing nearby may be endangered. Therefore, traffic must be stopped or diverted and on-site personnel must be informed about the nature and timing of overhead activities. Protection of existing structures adjacent to the work may also be necessary. All occupational safety regulations governing the facility should be adhered to.

Catchment

Regardless of which of the rehabilitation methods are chosen, there is usually the need for catchment of falling rock. Most slopes contain small pieces of rock that could loosen in the future but do not require extensive removal or reinforcement. Furthermore, just as the original design and construction does not eliminate future rockfall hazards completely, rehabilitation methods will not always be 100 percent effective due to the continual forces of nature. Catchment can consist of engineered benches, ditches, wide shoulders, berms, steel barriers, nets, fences, and concrete walls (Figure 9-32).

The type of catchment used depends largely on site conditions. Specifically, the height and angle of the slope and clearance between slope and the facility are important. This not only dictates how much space there is for catchment, but it also relates to anticipated paths of falling rocks. Obviously the catchment area must be somewhere along the paths of the rocks to be effective.

Generally, the flatter the slope, the more likely falling rocks will bounce and roll.

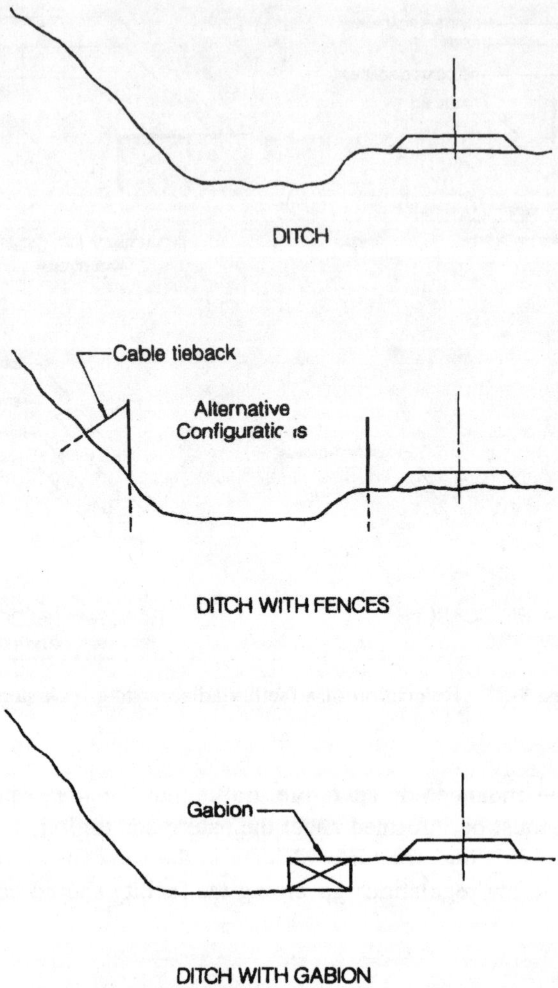

Figure 9-32 Catchment methods.

Simple rockfalls (i.e., with no bounce or roll) tend to occur on steep slopes and can be kept off of the facility with benches, ditches, and shoulders of appropriate widths and locations. Access should be provided to these areas because the catchment areas sometimes become filled with talus, forming a new slope that can eventually direct subsequent rockfalls onto the facility. The catchment area then must be accessed periodically to remove this debris, which can otherwise defeat the purpose of the catchment area.

For the flatter slopes where rockfalls tend to bounce and roll, a barrier is needed to deflect rocks away from the facility. Key considerations in the design of such barriers are height, location, and strength. Berms, steel barriers, fences, and concrete walls can be used for this purpose. These are normally used in conjunction with benches, ditches, and shoulders.

Example Problem 9-7 Catchment Area Design

Given: A highway is being designed that will require a 50-ft-high rock cut and a catchment area for rockfalls. There is enough right-of-way for a 12-ft-wide catchment area.

Required: How steep can the slope be with a 12-ft-wide catchment area?

Solution: Consult Schuster and Krizek (1978). The slope should be no steeper than 1 : 1 or 45°.

Flatten Slope

Generally speaking, the flatter a slope is, the more stable it is. The potential problems associated with flattening an existing slope are finding a practical place to start the excavation, acquiring additional property or right-of-way, and disposal of the spoil. The goals of flattening a slope are usually to increase the safety factor and eliminate the daylighting of unstable planes or wedges (Figure 9-33).

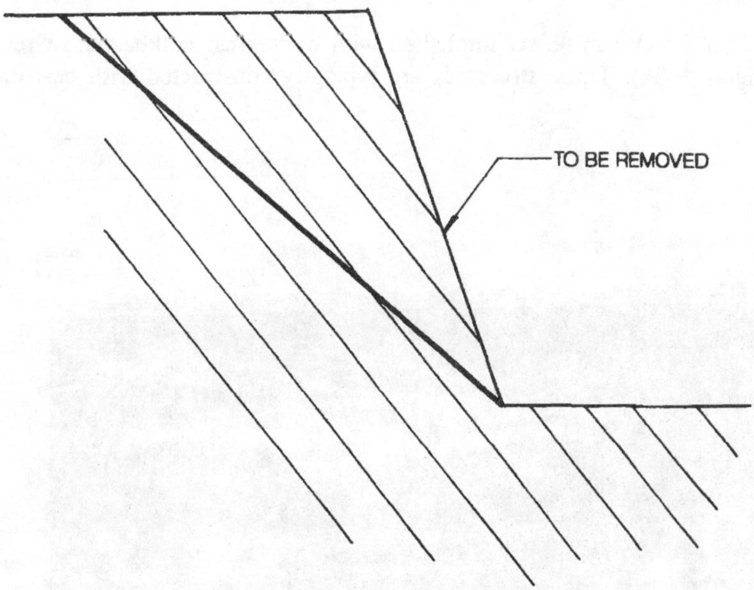

Figure 9-33 Flattening of a steep rock slope.

Example Problem 9-8 Flattening of a Rock Slope

Given: The slope given in Example Problem 9-5.

Required: How much must the slope be flattened to prevent a wedge failure?

Solution

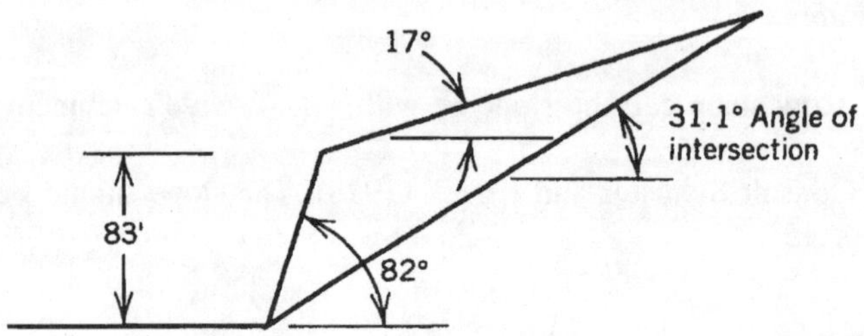

If the slope was made flatter than 31.1° (angle of intersection of planes ① and ②), a wedge failure would not be possible. The angle of intersection would not daylight in the slope.

In this example, it may not be economical to flatten the slope. Material removal would be about 450 yards³/ft of slope.

Buttresses

Support of the rock can be accomplished with buttresses, bulkheads, or retaining walls (Figure 9-34). These structures are typically constructed with cast-in-place

Figure 9-34 Rock slope buttress.

reinforced concrete, although stone or masonry can be used also. A new technique using the unstable rock as a "shot-in-place buttress" has been used in Tennessee by Moore (1986). Rock reinforcement is often used in combination with support methods.

Surface Protection

If space does not permit the construction of adequate catchment areas and barriers, potential rock debris must be held in place directly against the slope face. This can be accomplished using a net that is fastened to or over the slope. The netting commonly consists of chain link fence or gabion wire fabric held in place by rock bolts or cable tendons. Steel fabric, straps, or channels control rockfalls between reinforcement elements (dowels, bolts, or tiebacks).

Steel Mesh Steel mesh surface protection usually consists of chain link fence (Figure 9-35). Welded wire fabric can be used, especially if the possibility of using shotcrete also exists. Chain link fence interferes with the proper application of shotcrete. Another type of steel mesh is wire nets, which are discussed further below.

The main design components involved with specifying steel mesh are the wire size and the method and frequency of tie-downs. Standard sizes of chain link fence and welded wire fabric are available. Tie-downs usually consist of short rock dowels with the size, method of grouting, and number per square feet of slope specified. If rock dowels are being used to reinforce the slope in addition to the mesh, the mesh can be attached to the reinforcement dowels without additional tie-downs needed.

Wire Nets Wire nets have recently been introduced to the United States from Europe. Wire nets are similar to chain link fence except they have wider wire spacing and greater energy absorption characteristics.

Shotcrete Shotcrete is forced into open joints, fissures, seams, and irregularities in the rock surface and serves the same binding function as mortar in a stone wall (Figure 9-36). The adhesion of shotcrete to the rock surface, together with the shear strength of the shotcrete layer, provides resistance to the fallout of loose rock blocks as well as confinement of the rock mass. It will also seal rock that is prone to weathering due to exposure to the elements of nature.

Reinforcement

Rock reinforcement dowels, bolts, or tiebacks resist movement along joints and restrict block fallout and loosening. Tensioned reinforcement (bolts or tiebacks) will change the stress state around the slope face by inducing compressive stresses, which provide confinement, thereby improving the strength of the rock mass.

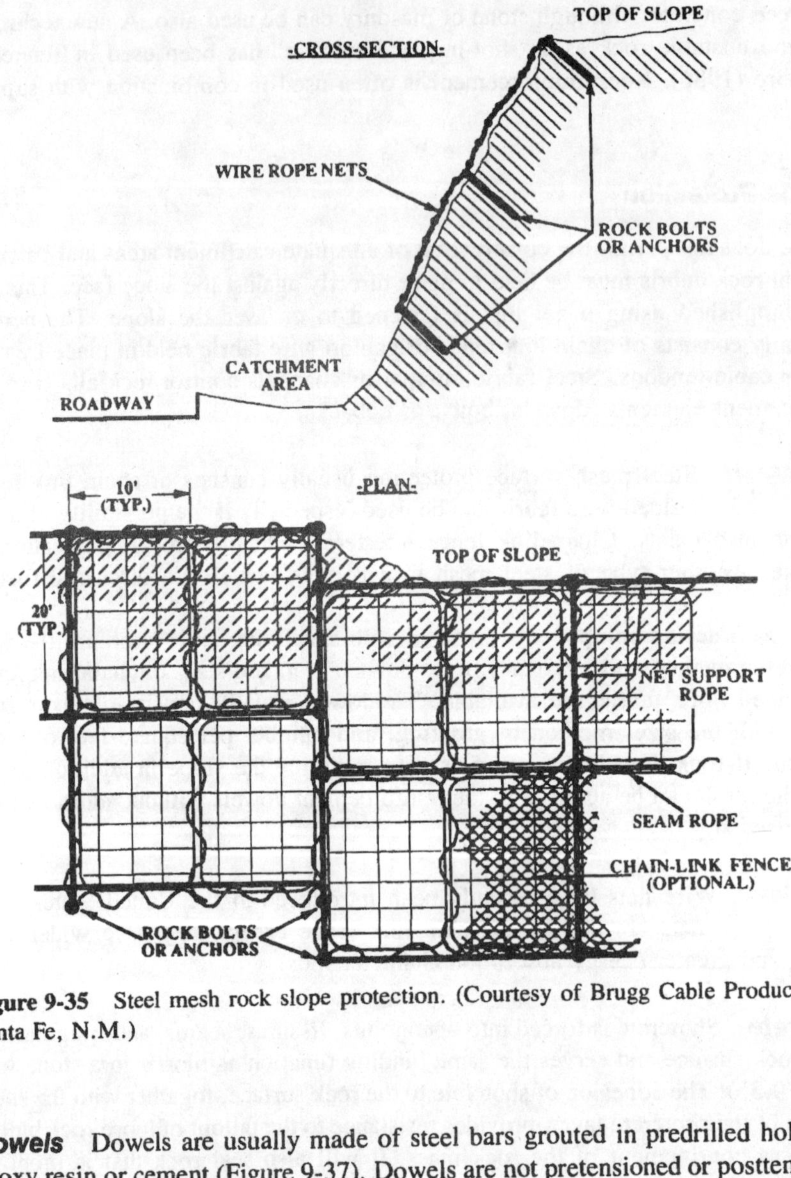

Figure 9-35 Steel mesh rock slope protection. (Courtesy of Brugg Cable Products, Inc., Santa Fe, N.M.)

Dowels Dowels are usually made of steel bars grouted in predrilled holes with epoxy resin or cement (Figure 9-37). Dowels are not pretensioned or posttensioned. The reason for this is that as the rock mass moves it will tend to tension the bar as it resists the movement. Reinforcing steel and high-strength bars are most commonly used. The bar diameter size can range between about 1 and $1\frac{1}{2}$ in. The hole size depends on the type of grout being used, the length of the bar (whether couplers are needed), and the type of rock.

Tiebacks Tiebacks are similar to dowels except that they are usually longer and are posttensioned. Steel strand is sometimes used instead of bars (Figure 9-38).

Figure 9-36 Shotcrete application on a slope.

Xanthakos (1991) provides a detailed discussion about the design, installation, and testing of tensioned ground anchors. The tensioning of the tieback provides additional confinement to the rock and increased shear strength across joints. Design forces of about 100 kips are not uncommon. The amount of confinement provided can be calculated by dividing the anchor force by the tieback spacing. For example, a 100-kip tieback on a 10 × 10 ft spacing would provide 1 ksf of confinement to the rock mass.

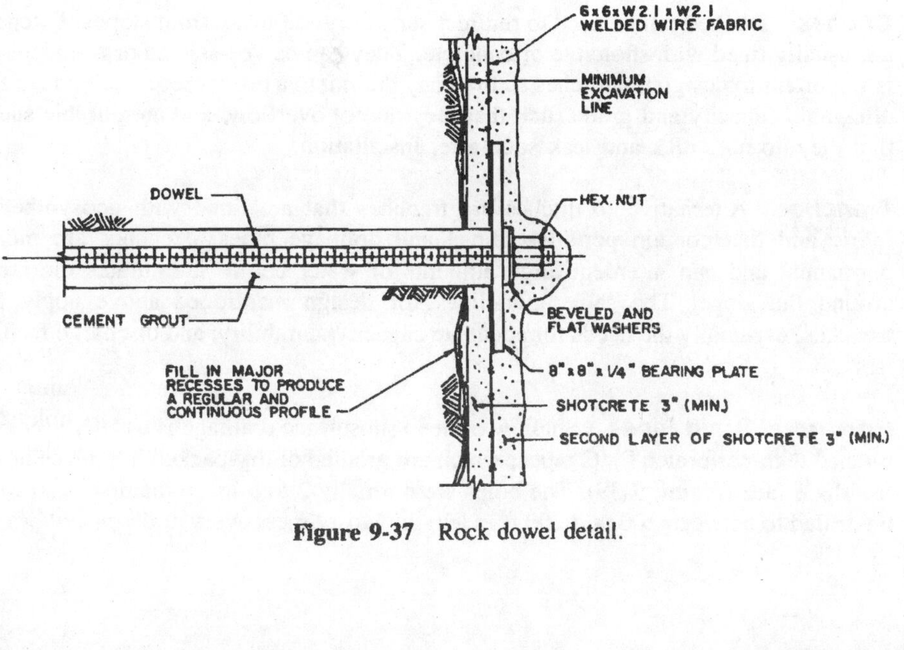

Figure 9-37 Rock dowel detail.

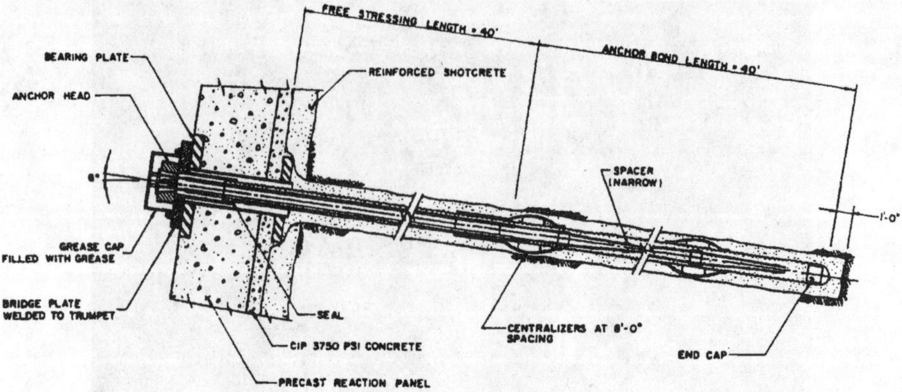

Figure 9-38 Tieback detail.

Drainage

To limit the amount of water that accumulates along the rock mass discontinuities, pressure relief drainage systems are used. These tend to prevent hydrostatic pressures from building up along discontinuities, and also limit the volume of water that can freeze in the winter causing rock block displacement due to expansion. The two general forms of drainage are surface and subsurface. Surface drainage refers to diversion of surface runoff away from tension cracks or open rock mass discontinuities near the slope face. This can be accomplished with dikes, ditches, or culverts. When discontinuity openings are particularly prone to water inflow, the gaps may be sealed and filled with grout until no significant void space remains.

Ditches Ditches can be used to redirect surface runoff away from slopes. Ditches are usually lined with shotcrete or concrete. They can be vee-shaped or rounded. It is important to design the ditches so that they drain into a proper receptacle, have an allowable capacity and grade such that they do not overflow, and are durable such that they do not crack and leak soon after installation.

Trenches Alternative to ditches are trenches that are lined with geosynthetic fabric and that contain pervious gravel and drainage pipes. Trenches are more permanent and can intercept large amounts of water before it saturates the area around the slope. The same guidelines for design mentioned above apply to trenches, especially the needs for adequate capacity, durability, and discharge facilities.

Horizontal Drain Pipes Shallow or deep subsurface drainage is usually accomplished with perforated PVC pipes, which are grouted or dry-packed into place near the slope face (Figure 9-39). The holes are normally 2 to 3 in. in diameter and can be drilled to between 100 and 200 ft in length. Most times, every drain pipe will not

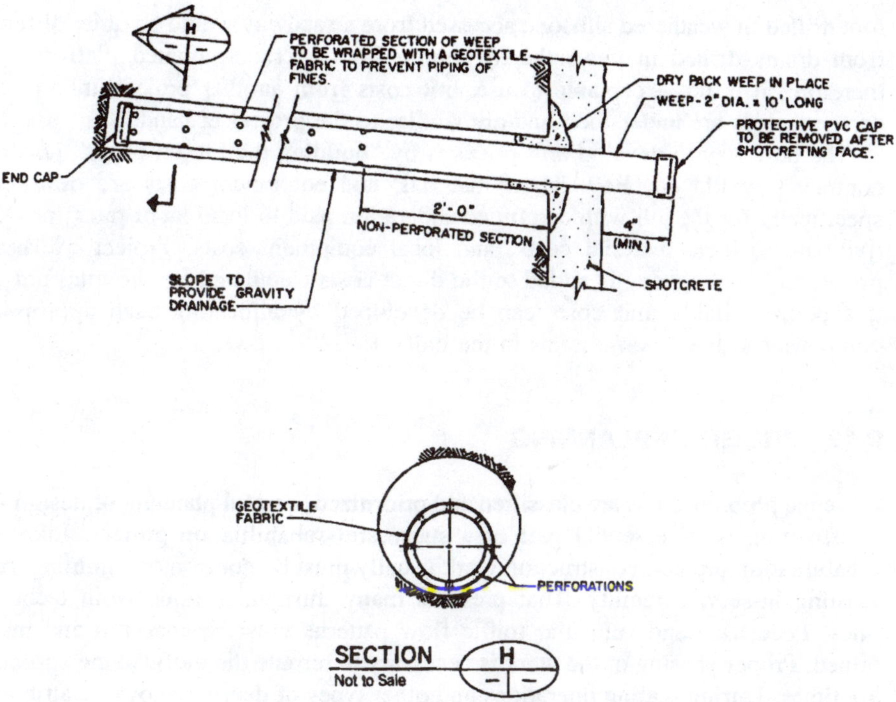

Figure 9-39 Horizontal drain pipe detail.

intercept water-bearing strata or discontinuities. A sufficient number must be installed to compensate for the "dry" ones. Also, they must be maintained in operating condition to be effective.

Tunnels and Other Methods Tunnels and other methods have been used to stabilize rock slopes. Although tunnels are seldom resorted to in stabilizing rock slopes because of the relatively high construction cost, they can be used as drainage galleries to reduce the quantity and rate of water flow at the face of the slope. Other exotic methods might include ground freezing, permanent dewatering systems, and grouted cutoff walls. The high cost of these methods are usually not warranted. Water flow is usually controlled by the methods described above at more reasonable prices.

9-11 ESTIMATION OF COSTS

The costs of rock slope repairs are often estimated on the basis of unit prices. This method is proper if the unit prices are generated on the basis of project-specific parameters. Unit prices taken from one project, however, may not necessarily represent the conditions of another. For instance, the price for horizontal drains per foot drilled in weathered siltstone accessed from a roadway would be quite different from drains drilled in unweathered granite from a crane-mounted platform. So therefore, it is not acceptable to use unit costs from another project unless cost-driven factors are understood and are similar to the project at hand.

The best way to develop unit prices is by "building the job on paper" just as a contractor would do. Raw labor, material, and equipment costs are developed specifically for the job with careful consideration paid to local labor rates, productivity rates, local material costs, and local equipment costs. Project overhead, profit, and contingency is added to the direct costs calculated for the total bid. At this point, reliable unit costs can be developed by allocating each appropriate component to the specific items in the bid list.

9-12 PROGRAM PLANNING

Once the problem areas are classified and prioritized, careful planning of design and construction is an essential part of a successful rehabilitation project. Like any rehabilitation project, construction work usually must be done on or adjoining to an existing in-service facility. That presents many difficulties aside from technical ones. Pedestrian and vehicular traffic flow patterns must be analyzed and maintained. Proper phasing of the work is necessary to reroute the traffic at the appropriate times. During scaling operations and other types of debris removal, traffic may have to be stopped temporarily or intermittently.

Classification of Problem Areas

For a small facility, the slope problems are probably confined to a limited area and remediation should be fairly straightforward. However, with a long linear facility such as a highway or railroad, it will probably be necessary to define major problem areas to focus remediation efforts on. Problem areas can be defined in different ways depending on what part of the remediation is a problem. For instance, the following might be problematic:

- Severe public danger
- High cost of remediation/limited funding
- Severe service disruption
- Weather constraints
- Accelerating deterioration
- Specialized equipment needed/long lead time

Problem areas are relative and must be assessed qualitatively for the most part, although cost is one quantitative criteria that might be used.

Prioritization of Remediation Program

Once the problem areas have been identified and all slopes have been categorized relative to the major problem areas, prioritization of the remediation program can be done. For large facilities, it is usually a program rather than a single repair that is being contemplated. The program might include a wide variety of activities that are carried out concurrently. One area might be selected for patrol and additional observation, one for major repair, and one for instrumentation, for instance. Work in other areas might be deferred until another time. Prioritization should have a sound basis and be flexible. Conditions could change over time, or slope remediation may turn out to be more or less serious than envisioned during study.

9-13 MONITORING AND MAINTENANCE

Geotechnical Instrumentation Monitoring

Long-term monitoring of slope performance can be accomplished using geotechnical instrumentation. In addition to performance monitoring, the instruments can be used to improve safety, cost economy, and design and construction adequacy. The two main parameters to be measured on slopes are deformation and water levels. Deformation can be measured with optical survey points, inclinometers, and extensometers. Water levels can be measured with observation wells and piezometers. All of these instruments can be read manually as well as by using automated, remote, monitoring systems.

The reliability of the geotechnical systems used is extremely important. Dunnicliff (1985) lists the basic ingredients of reliable instrumentation:

- Simplicity
- Self-verification
- Durability

In installing and monitoring the instruments, the program should provide for thorough planning, proper installation, regular care and maintenance, careful data collection, and timely data processing and interpretation. Making these ingredients part of a slope-monitoring program is essential for success.

When the instruments function properly and are read according to a suitable time schedule, a specific course of action in response to instrument readings is also necessary. The purpose of the instrumentation program is defeated if the data are collected but not evaluated on a timely basis. The personnel responsible for installing instruments and collecting data must understand what the readings mean, to whom they should be transmitted, and what magnitudes are cause for concern. Reading limits that should trigger quick response and the type of response required should be predetermined and understood by those who are charged with the responsibility for reaction. Some simple schedule for action, such as the following, should be in place:

Magnitude of Displacement	Required Action
Greater than 2 in.	Call chief engineer at once
1–2 in.	Read instruments weekly
Less than 1 in.	Read instruments monthly

Records of raw data, as well as processed results and plots, should be kept on a regular basis. Also at hand should be surveyed locations, installation dates, manufacturers' manuals, and project site information of events. This information should be easily accessible, organized, and up to date.

Slope Maintenance Programs

Few rock slopes are maintenance-free. Some type of observation and maintenance is usually required even after completion of a rehabilitation project. One reason for this is the fact that the slope will continue to be exposed to the forces of nature. Common types of slope maintenance include:

- Periodic patrolling
- Removal of accumulated debris on benches
- Removal of accumulated debris behind mesh
- Scaling
- Unclogging of drains
- Removal of ponding water

Maintenance can be done by owner forces or subcontractors. If it is done by the owner, the necessary equipment and personnel must be on hand. This also means that the owner takes on more of the responsibility for what happens in response to slope performance, including lost revenue and safety.

9-14 CASE HISTORIES

Oftentimes, rock slope failures occur due to the presence of water, weathering, progressive deterioration, or combinations of those three. Failure due to the presence of water can be caused by excess hydrostatic pressure on joint planes or ice jacking when the water between joint surfaces freezes and expands. Failure due to weathering generally occurs in fine-grained sedimentary rocks such as shale and joint fillings consisting of soil. Weathering causes a weakening of these materials such that over time the driving stresses overtake the resisting strengths. Progressive failure occurs over a prolonged period of time during which phenomena such as ice jacking and weathering are taking place, causing movement of rock blocks and weakening along joint planes a little at a time. The slope endures several cycles of these events before failure occurs. Some examples of these and other types of rock slope failures are given below.

Interstate 40 Sterling Mountain Portal Failure

On March 5, 1985, a massive rock slide involving roughly 14,000 yards^3 of rock debris occurred at the eastern portal area of the twin tunnels that carry Interstate 40 (I-40) through Sterling Mountain in North Carolina near the Tennessee state border (Abramson and Daly, 1986). The 150-ft canopy or tunnel extension, which had been built at the eastern end of the westbound tunnel to protect the interstate from falling rock, was destroyed and the eastbound lane was also completely blocked (Figure 9-40). A truck and a semitrailer just missed getting buried by the rock debris but, luckily, no deaths or private property damage occurred as a result of this failure.

The rock slope adjacent to the tunnel consists of good to excellent quality Longarm Quartzite underlain by poor to fair quality Longarm Quartzite interbedded with siltstone, slate, and phyllite. The Longarm Formation is one of four formations of the Snowbird Group, which is one of three regional lithologic units of the Ocoee Series. The Ocoee Series was formed from the metasediments that resulted from the erosion of the original Blue Ridge Mountains, which also formed most of the Great Smoky Mountains.

The knob of rock that failed was bounded by a fault trending nearly east-west, a major joint set trending northeast, and bedding that striking northwest. Because of the preexisting discontinuities in the rock mass, the abundant water flowing over

Figure 9-40 I-40 rock slope failure.

and through the slope (which accelerated weathering along the discontinuities), and the freeze/thaw pressures that developed every winter and spring season, the block of rock finally failed, destroying the westbound portal section of the tunnel.

Traffic was reinstated after the failure by diverting westbound traffic through the eastbound tunnel and eastbound traffic around the rock ridge. Existing data, geologic mapping, and core borings were used to characterize the rock mass for selection of remedial measures. These measures included demolition of badly damaged portions of the existing tunnel portal, construction of a new, cast-in-place concrete portal and retaining wall, the repair of damaged portions of the westbound tunnel, construction of a reinforced earth wall along the shoulder as a catchment area, and stabilization of the adjacent rock slopes with a combination of scaling, rock reinforcement, and drainage. Exploration of the site, design of remedial measures, and construction were completed within a total of eight months after the failure and prior to the approaching winter season, which begins in November.

The construction contract price was for $6 million and included:

- 17,000 yards^2 of slope scaling
- 93,000 linear ft of rock bolts
- 3400 yards^2 of steel mesh
- 3000 linear ft of horizontal drains
- 10,000 ft^2 of mechanically stabilized embankment (MSE)
- Portal reconstruction, tunnel repairs, and so on

Scaling and removal of rock overhangs was carried out first. The contractor piled hay around the existing structure for protection. Explosives were used to remove the rock overhangs. The blaster presplit a back line and then immediately fractured the freed blocks almost in midair. Scalers then rappelled down the side of the slope on ropes with pry bars, knocking loose any blocks or wedges that could conceivably be removed by hand. Traffic was stopped during scaling operations.

After scaling was complete, demolition of damaged structures and construction of new ones proceeded. Repair of the concrete tunnel arch began and excavation for new structures was completed. Shoring of the remaining tunnel was also required. The contractor erected reinforcing steel and poured concrete for new footings, retaining walls, and the portal structure while rock bolting was being carried out concurrently (Figure 9-41).

The rock slope was divided into areas according to the type of remediation required. In three areas, selective bolting was required using No. 9 reinforcing bars of varying lengths on an as-needed basis. Locations of these bolts were determined in the field. One area required pattern bolting and wire mesh. The most worrisome area was the one that included large rock blocks that could fail in the same manner as the original failure. Here, 40-ft-long, high-strength anchors were specified on a 10 × 15 ft pattern with 20-ft-long bolts to stabilize the surficial slabs in between. The resulting bolt pattern was 5 × 5 ft. All of this work was generally done off steel platforms hung from cranes.

Figure 9-41 I-40 remedial rock bolting.

Additional work included the MSE wall catchment area and rock fence above the new tunnel portal to collect any small-scale rockfalls that occur in the future. Also, horizontal drains were installed along the base of the slope to relieve future buildup of excess hydrostatic pressure within the rock mass.

Wire Netting Used in Montana and Nevada for Rockfall Protection

Three projects in Montana and Nevada utilized double-twisted, hexagonal, heavily galvanized mesh to mitigate rockfall problems along highways (Ciarla, 1986). Along Highway 15 in Montana, the highway was widened from two lanes to four lanes between Helena and Butte. Between Bernice and Basin, a detour road was required for local traffic during construction. Construction of this road required a 200-ft-high cut in a highly fractured rock formation. Ten thousand square yards of metallic mesh were installed with rock bolts and cables for $177,000 to protect traffic from rockfalls. The mesh was delivered in 13-ft-wide × 300-ft-long rolls. It was mounted using a crane by placing the roll at the base of the slope and unrolling it from the bottom to the top of the slope. One-inch-diameter rebar anchors and $\frac{3}{4}$-in.-diameter wire rope were used to fasten the mesh to the slope; workers utilized a crane-supported work cage. The anchors were installed on a 25-ft center-to-center pattern on the upper portion of the slope and a 50-ft pattern on the lower portion.

State Route 207 in Nevada was lowered, realigned, and widened to accommodate the heavy traffic between the area of Gardnerville and Minden and the Lake Tahoe area. Cuts in decomposed granite were required up to 125 ft in height. To protect the road bed, steel mesh tied to a 50 × 50 ft wire cable grid were used. Falling rocks thereby rolled or slid behind the mesh and were subsequently removed from the toe of the slope. Twenty thousand square yards of mesh was installed on the slope at a cost of $156,000.

The cut slopes along Highway 80 in Nevada near Sparks and Truckee require frequent maintenance and rock debris removal. To mitigate the threat of rockfalls onto the travel lanes, steel mesh has been installed. Near Sparks, 37,000 $yards^2$ were installed at a cost of $175,000. Near Truckee, 79,000 $yards^2$ of mesh were installed at a cost of $458,000.

Shot-in-Place Rock Buttress Used to Repair Landslide

A shot-in-place rock buttress was used to repair a block glide landslide in Tennessee along State Route 31 north of Mooresburg (Moore, 1986). The slide within an existing rock cut damaged the highway shoulder and deformed the roadway pavement. The rock mass consists of medium to thick beds of Clinch Formation quartzose sandstone with thin interbeds of shale and clay (Figure 9-42). The highway cut undercut the dipping beds of rock and exposed the bedding planes dipping into the highway.

As a result of weathering along joint surfaces and the infiltration of groundwater into the joints and bedding planes, movement was allowed to occur along the bedding planes toward the roadway. A 100-ft-long, 200-ft-wide, and 12-ft-thick

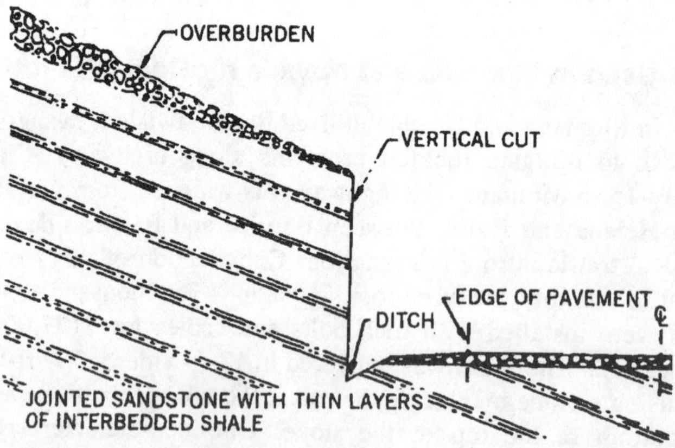

Figure 9-42 State Route 31 Highway rock slope failure subsurface conditions. (From Moore, 1986.)

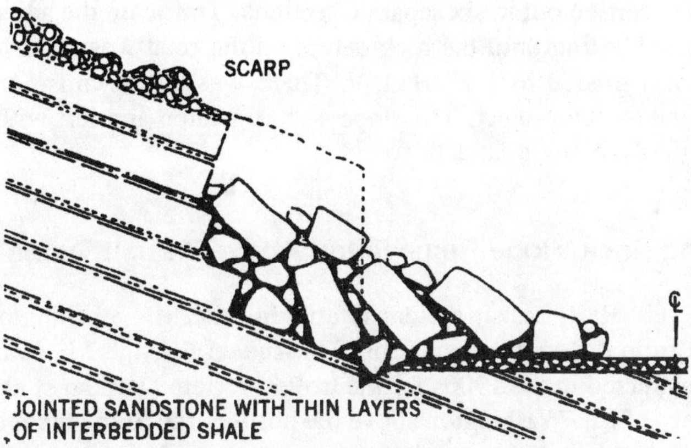

Figure 9-43 State Route 31 Block glide failure. (From Moore, 1986.)

rock block freed up and slid, forming a 5200 yard^3 "block glide" type landslide (Figure 9-43). A shot-in-place rock buttress was the chosen method of remediation.

Design parameters for the rock buttress included an angle of internal friction of 38° and a unit weight of 140 lb/ft^3. This was estimated to provide a factor of safety equal to 1.3 against future upslope movements. The base of the buttress was designed to be 25 ft wide with a slope of 23° up to the base of the scarp (Figure 9-44). A 14-ft-wide berm was left at the top of the buttress. The design blast pattern consisted of the following:

- Hole diameter = 2.5 in.
- Pattern = 5 × 7 ft
- Hole depth = 20 ft
- Explosive = 1.92 lb/linear ft of AnFo.

The construction cost for this work was $108,000.

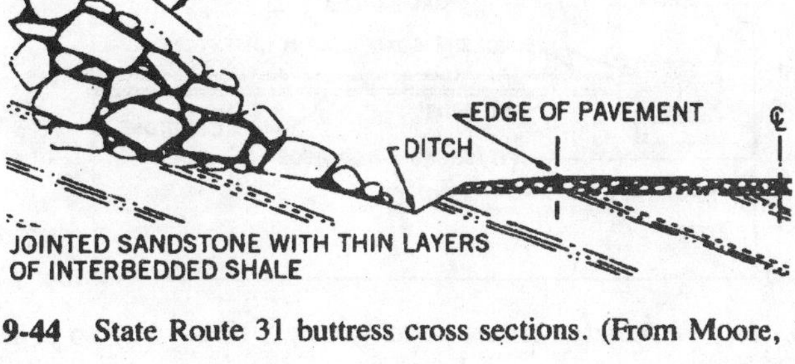

Figure 9-44 State Route 31 buttress cross sections. (From Moore, 1986.)

Blasting was carried out in six separate sections. Traffic on the adjacent highway was held during blasting until the rock debris on the road was removed. The final dressed face was graded to 1.25 H : 1 V. There was 2500 yards^3 of excess rock debris generated by the project. The slope was vegetated and left with no apparent signs of previously being a landslide.

Deteriorating Rock Slope Remediated Above Transit Tunnel Portal

Pittsburgh's Light Rail Transit system is utilizing the Mt. Washington Tunnel for access to and from the southern suburbs. This tunnel through Mt. Washington was originally constructed in the 1900s for the trolley system (Voytko et al., 1987). The 150-ft-high face of Mt. Washington above the north portal of the tunnel is composed of horizontally bedded shale, limestone, claystone, and sandstone (Figure 9-45). Differential weathering of the rock resulted in undercutting of the more resistant beds and overhangs. Jointing is near-vertical and orthogonal, forming columns of rock (Figure 9-46). Infiltration of groundwater between joint surfaces and subsequent freezing and thawing caused movement of these rock columns and concern about possible rockfalls that could damage the transit tunnel portal.

The sandstone overhang was repaired with rock bolts and a concrete buttress (Figure 9-47). Fifty 13-ft-long, $\frac{7}{8}$-in.-diameter, Grade 70, expansion anchor rock bolts were installed and postgrouted to stabilize the overhang. Rock bolts and formwork were then installed in preparation for pouring the concrete buttress. For

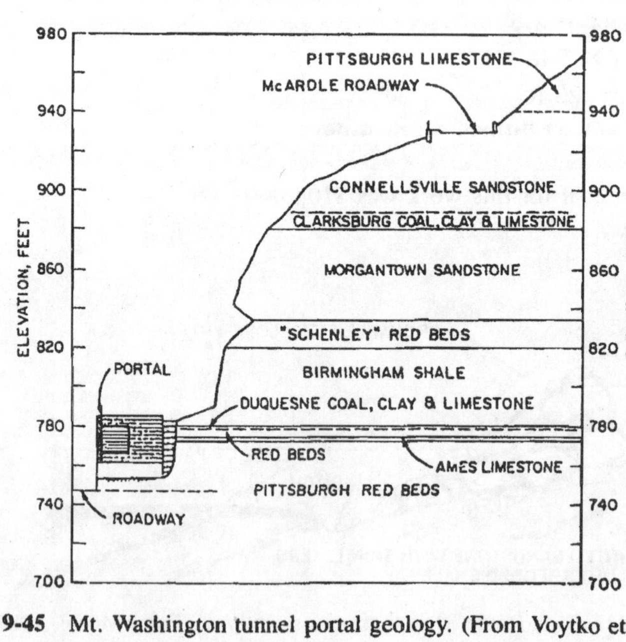

Figure 9-45 Mt. Washington tunnel portal geology. (From Voytko et al., 1987.)

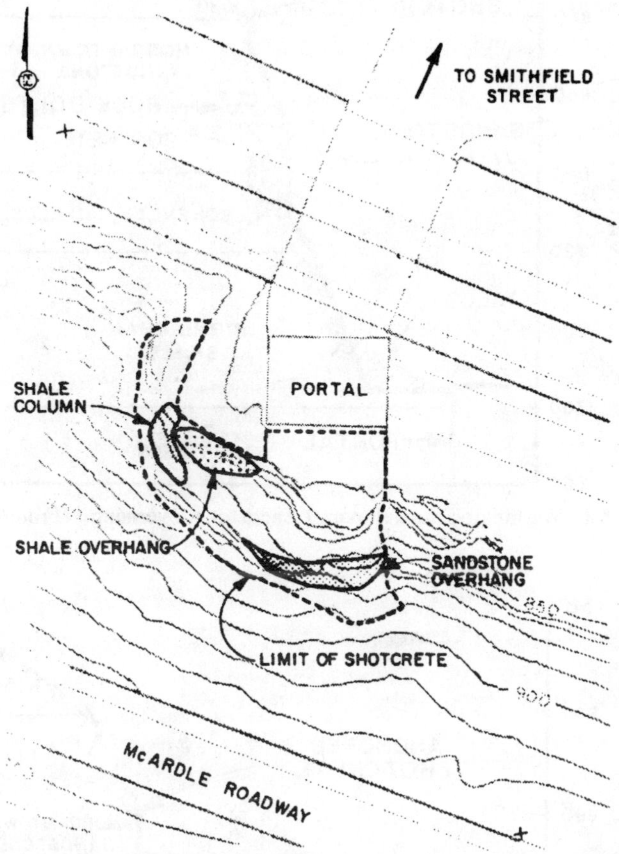

Figure 9-46 Mt. Washington tunnel portal jointing. (From Voytko et al., 1987.)

drainage, 4-in.-diameter pipes were installed to convey seepage from the rock to the face of the buttress.

The shale overhang was removed in sections using nonexplosive excavation methods. The overhang was removed to an existing predominant joint located behind the overhang to provide a relatively smooth final surface.

The shale column was intended to be removed in a controlled manner by initially being constrained by a wire mesh and cable restraining system (Figure 9-48) and subsequently being removed in small layers. However, the rock column failed before removal operations began and slid to the base of the cut without catastrophic consequences.

After the specialized work described above was completed, the area was scaled and reinforced with welded wire fabric (3 × 3–1.4 × 1.4) reinforced shotcrete with 2-in.-diameter weep holes spaced 15 ft apart. At the toe of the slope, grouted rip-rap was placed to protect the nonresistant red beds. To further protect the portal structure, a rock fence (Figure 9-49) and earth berm were constructed (Figure 9-50).

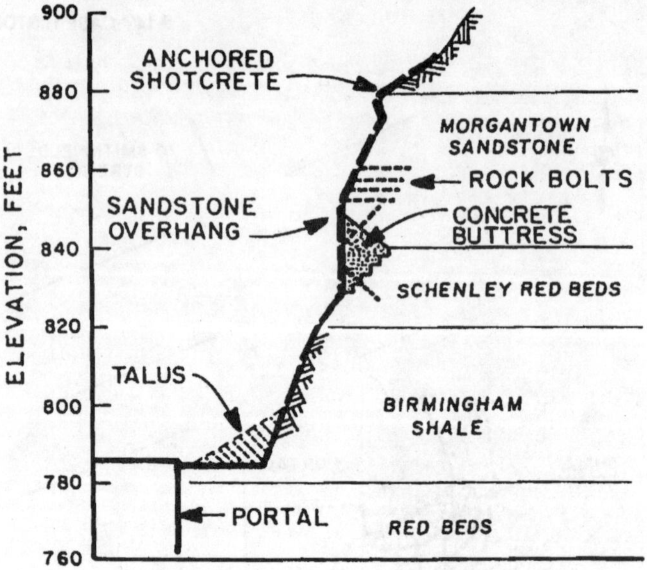

Figure 9-47 Mt. Washington tunnel portal sandstone overhang. (From Voytko et al., 1987.)

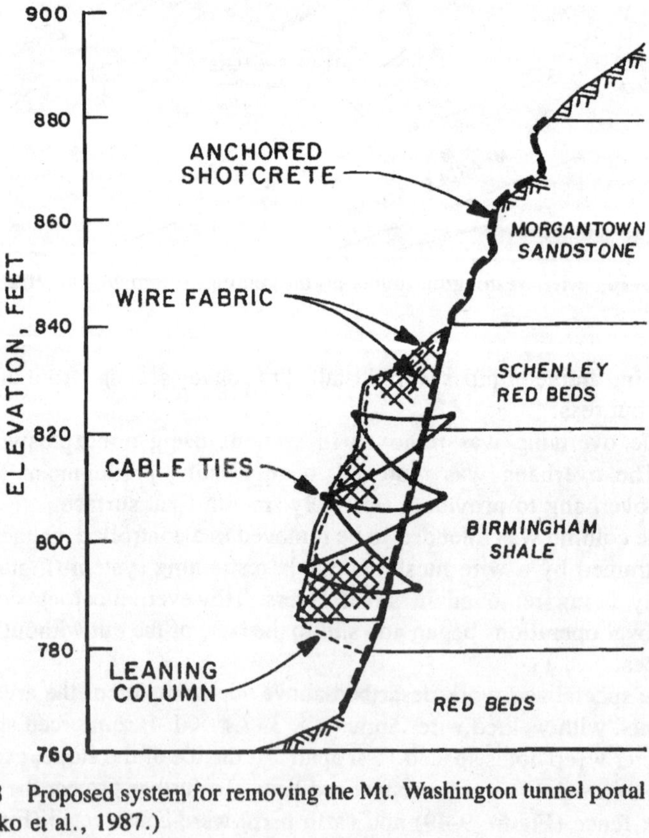

Figure 9-48 Proposed system for removing the Mt. Washington tunnel portal shale column. (From Voytko et al., 1987.)

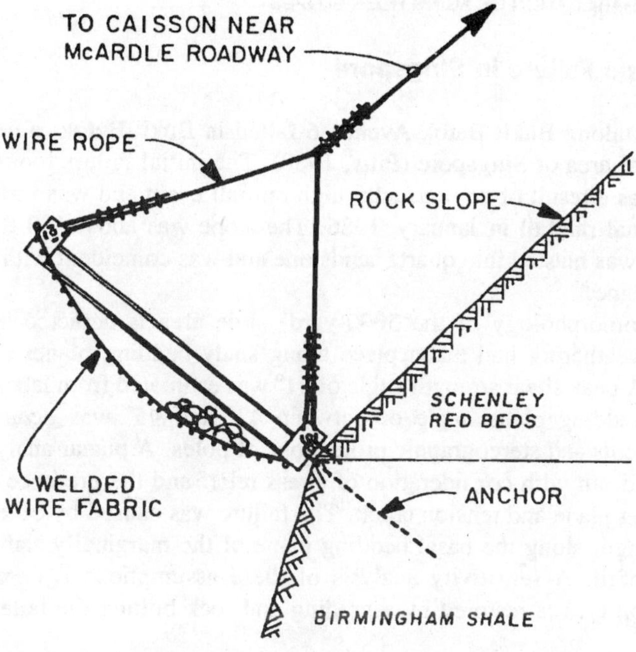

Figure 9-49 Mt. Washington tunnel portal rock fence. (From Voytko et al., 1987.)

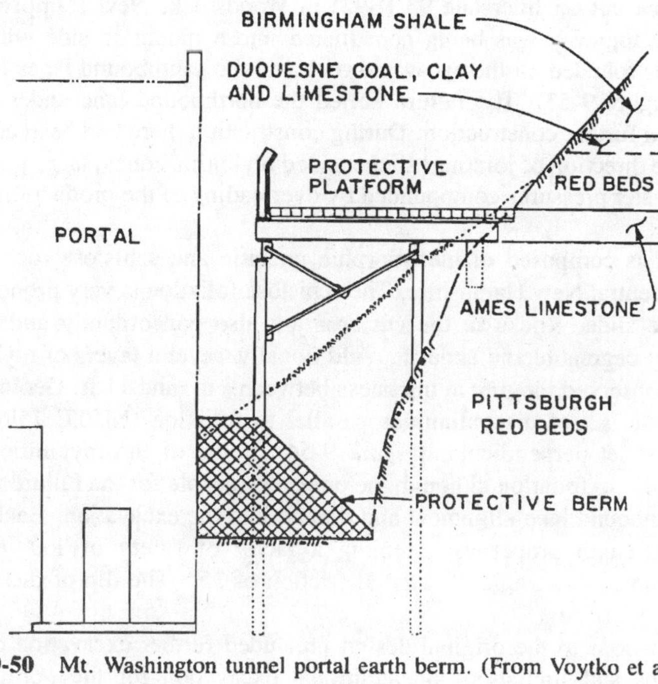

Figure 9-50 Mt. Washington tunnel portal earth berm. (From Voytko et al., 1987.)

Rock Slope Failure in Singapore

A rock cut along Bukit Batok Avenue 6 failed in Bukit Batok, a new town in the west-central area of Singapore (Pitts, 1988). The initial failure took place in October, 1985 as a result of an unusually high rainfall event and was further aggravated by additional rainfall in January, 1986. The slope was about 100 ft high and 45°. The slope was blasted into quartz sandstone and was coincident with a predominant bedding plane.

The geomorphology of the 5000-yard^3 slide area is depicted in Figure 9-51. Tropical weathering had taken place along shaly bedding planes and major joint surfaces. A peak shear strength angle of 41° was estimated from laboratory and field tests. A bedding plane angle of between 42° and 45° was deduced from joint measurements and stereographic projections of poles. A planar analysis of the slide was carried out with consideration of stress relief and the presence of water along the bedding plane and tension crack. The failure was caused by exceeding the peak shear strength along the basal bedding plane of the marginally stable slope due to heavy rainfall. A sensitivity analysis of these assumptions is depicted in Figure 9-52. Stability was restored by regrading and rock bolting the failed slope.

Woodstock Rock Slope Failure in New Hampshire

Approximately 17,000 yards^3 of rock slope failed along a recently excavated 1H : 8V rock cut on Interstate 93 (I-93) in Woodstock, New Hampshire (Fowler, 1976). The highway was being constructed into a mountain side with the northbound lanes founded on the excavated rock and the southbound lanes founded on a viaduct (Figure 9-53). The failure buried the northbound lane under construction and delayed further construction. During construction there had been concern about the adverse direction of jointing, slickensided mylonitic zones, large joint apertures, and high water pressures compounded by overloading of the production and presplit blast holes.

The cut is composed of metamorphic gneissic and schistose rock common to north and central New Hampshire. The schistose foliation is very pronounced in the area of the slide. Rocks in the cut area are also concordantly and discordantly intruded by pegmatite and andesite. Additionally, several layers of mylonitic materials were observed ranging in thickness between $\frac{1}{2}$ in. and 11 ft. Geologic mapping indicated one set of discontinuities parallel to foliation (N|50|E 75|NW) and the other major set perpendicular (Figure 9-54). Some of the mylonitic zones were perpendicular to foliation although the ones responsible for the failure were parallel to the northbound lane alignment and dipped into the excavation. Back-calculation of shear strength properties assuming a factor of safety of 1.0 indicated that the mylonite had an angle of internal friction of 35°. The dip of the failure plane was 38°.

Modifications to the original design precluded further excavation of the southbound lanes and limitations on additional excavation for the northbound lanes

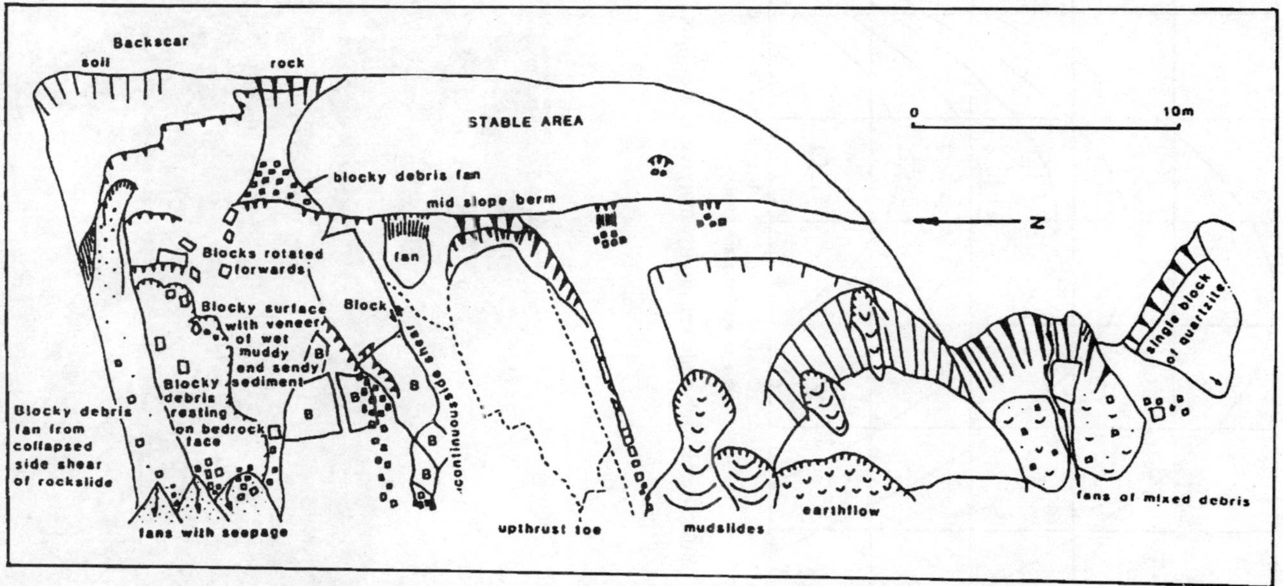

Figure 9-51 Geomorphology of the Bukit Batok rock slide. (From Pitts, 1988.)

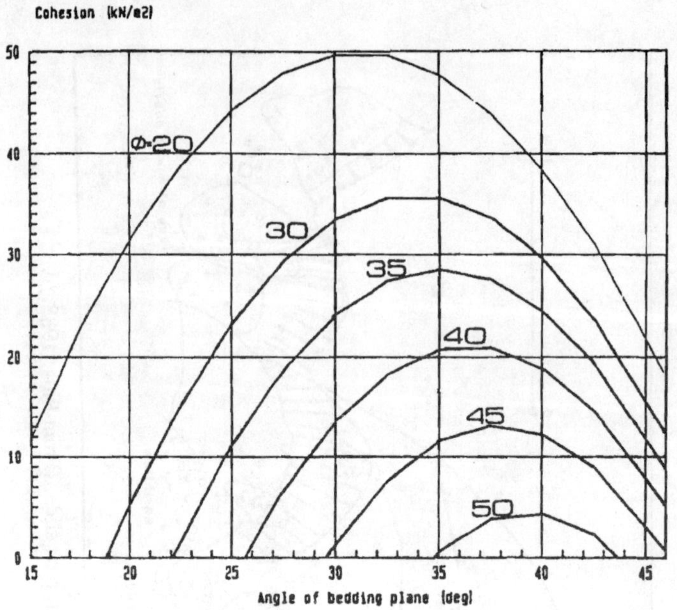

Figure 9-52 Bukit Batok rock slide sensitivity analysis. (From Pitts, 1988.)

Figure 9-53 Slope failure during construction of I-93 in Woodstock, New Hampshire. (From Fowler, 1976. Reproduced by permission of ASCE.)

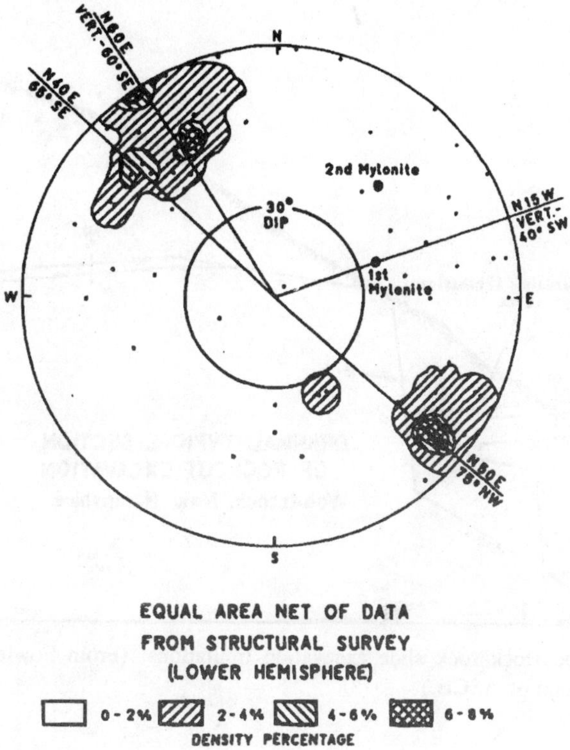

Figure 9-54 Woodstock rock slide jointing. (From Fowler, 1976. Reproduced by permission of ASCE.)

(Figure 9-55). The backslope was flattened to $\frac{1}{2}$H : 1V and the presplit line spacing was reduced from 36 in. to 18 in. with alternately loaded holes. During further excavation, another mylonitic zone was encountered and necessitated reinforcement of the slope with high-strength steel tendons, rock bolts, and horizontal drains (Figure 9-56).

Massive Rock Slide Derails Train in Pittsburgh

A historic case history involves the failure of 110,000 yards^3 of rock that slumped from the side of Brilliant Cut in Pittsburgh, Pennsylvania on March 20, 1941 (Hamel, 1971). This slide displaced three sets of railroad tracks and derailed a train. The cut is at the nose of a hill located at the junction of an abandoned river valley and the present valley of the Allegheny River (Figure 9-57). The rock mass consists of near-horizontal interbeds of sandstone, siltstone, claystone, limestone, shale, and coal (Figure 9-58).

The cut was originally made in the early 1900s for the railroad. The first major slide on record occurred in 1904. In 1930, the railroad was relocated further into the

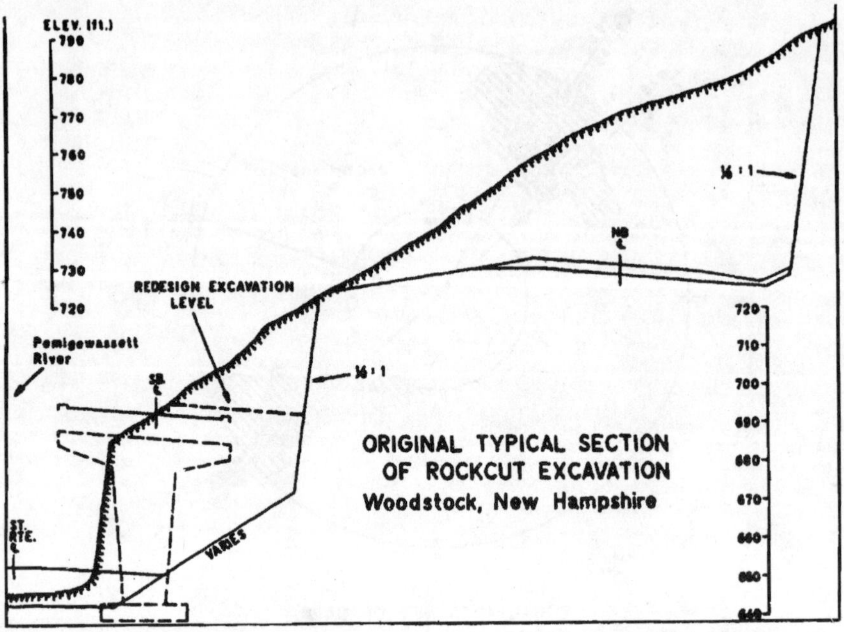

Figure 9-55 Woodstock rock slide excavation limitations. (From Fowler, 1976. Reproduced by permission of ASCE.)

slope, triggering two major rock siides in this area. During the 1930s, a 1-ft-wide joint opened up at the crest of the slope just above the Birmingham Shale. The joint was filled with concrete that, as time went on, had little effect on the continued movement. Finally, the slide in 1941 occurred and had a failure surface that generally followed the contact between the soft clay shale and indurated clay at about elevation 790. The upper part of the failure surface coincided with the crack in the Birmingham Shale.

Analysis showed that the rock slide was triggered by water pressure in the slope, which built up because natural drainage outlets in the slope were blocked by ice. The open vertical joint at the back of the slide occurred progressively over a period of several years and filled up with water, probably to the top of the slope. This joint was defined by stress relief jointing that had occurred as a result of downcutting by the rivers that formed the adjacent valleys. Back-calculated shear strength values after the slide resulted in estimates as follows:

Peak strengths	Cohesion = 0 to 2.3 kips/ft^2
	$\phi = 20°$ to $32°$
Residual strengths	Cohesion = 0
	$\phi = 11°$ to $16°$

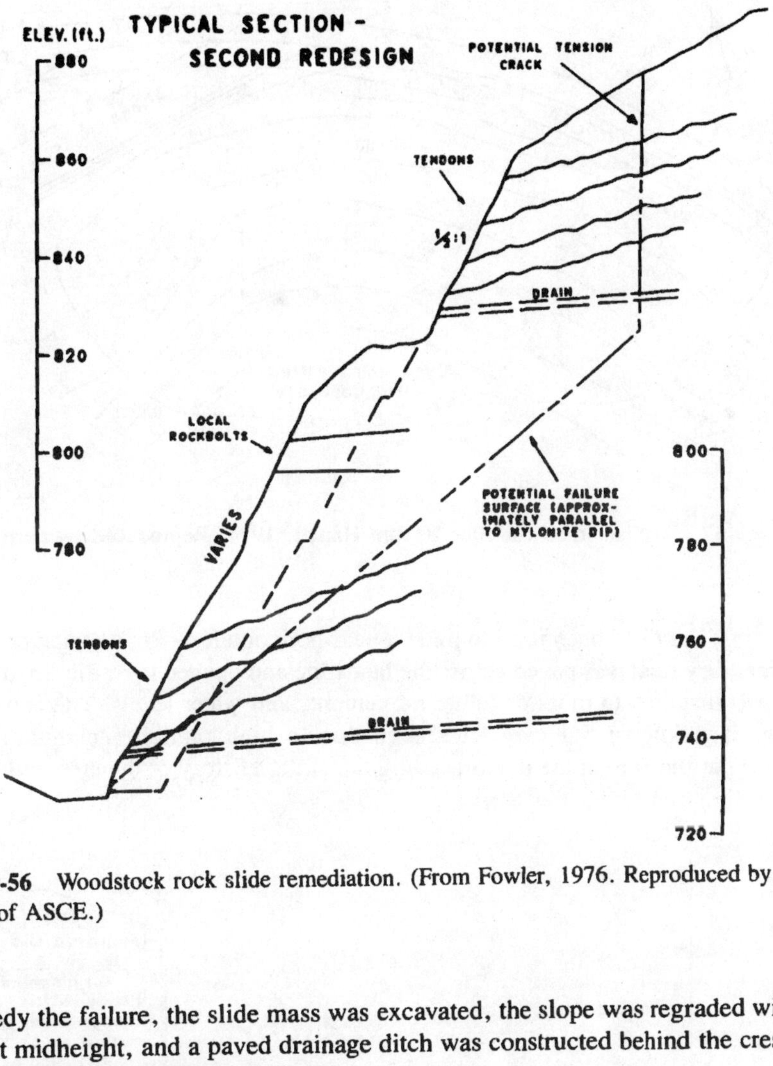

Figure 9-56 Woodstock rock slide remediation. (From Fowler, 1976. Reproduced by permission of ASCE.)

To remedy the failure, the slide mass was excavated, the slope was regraded with a bench at midheight, and a paved drainage ditch was constructed behind the crest of the slope to intercept surface drainage.

Cedar Canyon Landslide Destroys State Highway 14 in Utah

The Cedar Canyon landslide occurred about seven miles east of Cedar City, Utah, in March, 1989 (Harty, 1991). The 2-million-yard^3 landslide moved down the north-facing slope of Cedar Canyon, destroying about $\frac{1}{3}$ mile of Utah Highway 14 and threatening to block Coal Creek and cause flooding of the area. The slide surface was within the Cretaceous Tropic Formation shale, which is known to be susceptible to landsliding. The landslide moved downslope in a series of slumps and slides rather than as a single cohesive mass. Three abandoned coal mines underlay the

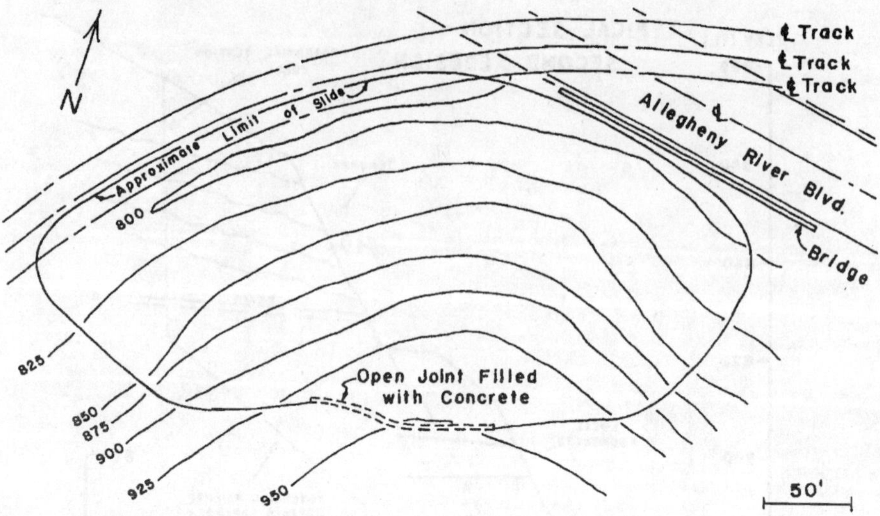

Figure 9-57 Brilliant rock cut slide. (From Hamel, 1971. Reproduced by permission of ASCE.)

slide area; they had been used to mine a near-horizontal, 4- to 5-ft-thick coal seam. A temporary road was paved across the landslide and opened to traffic. Instrumentation was installed to monitor future movements and water levels. The cause of the failure is unknown but may have been due to high rainfall accumulations that occurred at the site on the day preceding the slide. History of the area and the rock

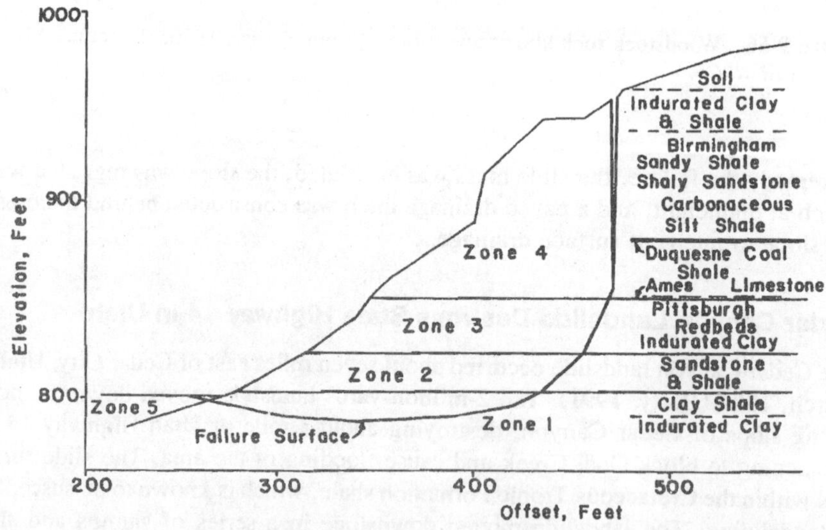

Figure 9-58 Brilliant rock cut geology. (From Hamel, 1771. Reproduced by permission of ASCE.)

formation suggests that the slope may have been in a weakened state prior to the rainfall due to highway and mine-related alterations to slope geometry and poor drainage.

REFERENCES

Abramson, L. W., 1985. "Rock Wedge Stability Analysis on a Personal Computer," Proceedings of the 26th U.S. Symp. on Rock Mechanics, Rapid City, S.Dak., Balkema, Rotterdam, pp. 675–682.

Abramson, L. W. and W. F. Daly, 1986. "Analysis and Rehabilitation of Aging Rock Slopes," 37th Annual Highway Geology Symp., Helena, Mont., pp. 56–86.

Barton, N., P. Lien, and J. Lunde, 1974. "Analysis of Rock Mass Quality and Support Practice in Tunneling, and a Guide for Estimating Support Requirements," Norwegian Geotechnical Inst. Internal Report No. 54206, June, 74 pp.

Bieniawski, Z. T., 1984. *Rock Mechanics Design in Mining and Tunneling*, A. A. Balkema, Rotterdam, Boston.

Bishop, A. W., 1955. "The Use of the Slip Circle in the Stability Analysis of Slopes," *Geotechnique*, Vol. 5, No. 1, pp. 7–17.

Ciarla, M., 1986. "Wire Netting for Rockfall Protection," 37th Annual Highway Geology Symp., Helena, Mont., 19 pp.

Deere, D. U. and D. W. Deere, 1988. "The Rock Quality Designation (RQD) Index in Practice," Rock Classification Systems for Engineering Purposes, ASTM Special Technical Publication No. 984, L. Kirkaldie, Ed., Philadelphia, pp. 91–101.

Dunnicliff, J., 1985. "Closing Remarks on Reliability of Geotechnical Instrumentation," Transportation Research Record No. 1004, Transportation Research Board, National Research Council, Washington, D.C., pp. 46–49.

Fellenius, W., 1939. *Erdstatische Berechnungen*, revised ed., W. Ernst u. Sohn, Berlin.

Fowler, B. K., 1976. "Construction Redesign—Woodstock Rockslide, N.H.," Specialty Conf. on Rock Engineering for Foundations and Slopes, Boulder, Colo., Vol. 1, ASCE, New York, pp. 386–403.

Goodman, R. E., 1976. *Methods of Geological Engineering in Discontinuous Rock*, West Publishing, St. Paul, Minn., Chapter 3, pp. 58–90.

Goodman, R. E., 1980. *Introduction to Rock Mechanics*, Wiley, New York, pp. 34 and 42.

Goodman, R. E. and Shi, G., 1985. *Block Theory and Its Application to Rock Engineering*, Prentice-Hall, Englewood Cliffs, N.J.

Hamel, J. V., 1971. "The Slide at Brilliant Cut," 13th Symp. on Rock Mechanics, Stability of Rock Slopes, E. J. Cording, Ed., ASCE, New York, pp. 487–510.

Harty, K. M., 1991. "Summary of Landslides in Utah, 1987–1990," 27th Symp. on Engineering Geology and Geotechnical Engineering, Utah State Univ., April, Logan, pp. 17-1–17-10.

Hendron, A. J., Jr., E. J. Cording, and A. K. Aiyer, 1971. "Analytical and Graphical Methods for the Analysis of Slopes in Rock Masses," U.S. Army Engineer Explosive Excavation Research Office, Livermore, Calif., Technical Report 36, NTIS No. AD-738-929, 148 pp.

Hoek, E. and J. W. Bray, 1981. *Rock Slope Engineering,* revised 3rd ed., Institution of Mining and Metallurgy, London, pp. 37–63 and 76.

International Society of Rock Mechanics, 1978. "Suggested Methods for the Quantitative Description of Discontinuities in Rock Masses," *Int. J. on Rock Mechanics, Mining Science and Geomechanics Abstract,* Vol. 14, pp. 319–368.

Janbu, N., 1954. "Application of Composite Slip Surfaces for Stability Analysis," Proc. Europ. Conf. on Stability Analysis of Earth Slopes, Stockholm, Sweden, Vol. 3, pp. 43–49.

Kovari, K. and P. Fritz, 1975. "Stability Analysis of Rock Slopes for Plane and Wedge Failure With the Aid of a Programmable Pocket Calculator," 16th U.S. Symp. on Rock Mechanics, ASCE, New York, pp. 25–34.

Moore, H. L., 1986. "The Construction of a Shot-in-Place Rock Buttress for Landslide Stabilization," 37th Annual Highway Geology Symp., Helena, Mont., pp. 137–157.

Morgenstern, N. R. and V. E. Price, 1965. "The Analysis of Stability of Slopes," *Geotechnique,* Vol. 15, No. 1, pp. 79–93.

Naval Facilities (NAVFAC) Engineering Command, 1982. *Soil Mechanics Design Manual 7.1,* NAVFAC DM-7.1, May, Department of the Navy, p. 7.1-319.

Pierson, L. A., S. A. Davis, and R. Van Vickle, 1990. "The Rockfall Hazard Rating System Implementation Manual," Oregon State Highway Division, Federal Highway Administration Report No. FHWA-OR-EG-90-01, Aug., 40 pp.

Piteau, D. R. and F. L. Peckover, 1978. "Engineering of Rock Slopes," *Landslides—Analysis and Control,* Transportation Research Board Special Report No. 176, National Academy of Sciences, Washington, D.C., Chapter 9, pp. 192–228.

Pitts, J., 1988. "Stability of a Rock Slope at Bukit Batok New Town, Singapore," 2nd Int. Conf. on Case Histories in Geotechnical Engineering, Univ. of Missouri—Rolla, S. Prakash, Ed., Vol. 1, pp. 115–121.

Priest, S. D., 1985. *Hemispherical Projection Methods in Rock Mechanics,* George Allen & Unwin Ltd., London.

Schuster, R. L., and R. J. Krizek, Eds. (1978). "Landslides—Analysis and Control," Transportation Research Board, Special Report No. 176, Washington, D.C., p. 220.

Taylor, D. W., 1948. *Fundamentals of Soil Mechanics,* Wiley, London.

Terzaghi, K. and R. B. Peck, 1967. *Soil Mechanics in Engineering Practice,* Wiley, New York.

Voytko, E. P., V. A. Scovazzo, and N. K. Cope, 1987. "Rock Slope Modifications Above the North Portal of the Mt. Washington Tunnel," 37th Annual Highway Geology Symp., Highway Construction in Unstable Topography, May, Pittsburgh, pp. 155–162.

Watts, C. F. and T. R. West, 1985. "Electronic Notebook Analysis of Rock Slope Stability at Cedar Bluff, Virginia," *Bull. of the Association of Engineering Geologists,* Vol. XXII, No. 1, pp. 67–85.

Wu, S., 1985. "Rockfall Evaluation by Computer Simulation," Transportation Research Rec. 1301, Transportation Research Board, National Research Council, Washington, D.C., pp. 1–5.

Xanthakos, P. P., 1991. *Ground Anchors and Anchored Structures,* Wiley, New York.

CHAPTER 10

VERTICAL SCREENS

10-1 INTRODUCTION

Vertical screens cover a broad area of protective or remedial systems, including continuous earth, semirigid, and rigid cutoff walls; plastic barriers and hot bituminous mastic inserted in narrow trenches; permeable treatment beds; synthetic membranes with overlapping or interlocking sheet-pile sections; and clay-cement grout injected under pressure into a preformed narrow slot. In general the intent is to provide essential control of groundwater movement where it is necessary to maintain the balance in the water supply, where the risk of pollution exists, and where deep excavations are contemplated. Examples are (a) cutoff walls built as seepage barriers beneath the main body of earth dams and canal embankments; (b) impervious curtains for pollution control in alluvial terrains, waste disposal sites, and industrial locations where groundwater flow must be diverted or contained; (c) landslide control by impeding percolation of groundwater; (d) the prevention of saline intrusion into the water supply; (e) groundwater and aquifer recharge schemes; and (f) special cutoffs built to control underground erosion. The presence of a vertical screen at a site means that impeded groundwater flow may cause an increase in upgradient hydraulic head with corresponding effects on the rate of vertical water movement. The associated hydrogeologic impact of a locally altered water table is thus a point of concern, and should be considered before applying this control.

The advantages of the method are fully explored if the vertical screen can satisfy the site requirements; namely: (a) it is not necessary to alter the groundwater flow or level during construction; (b) the insertion of the screen is not inhibited by site and ground conditions, even in very mobile formations; (c) the barrier can be made continuous, and in its final configuration it is flexible and can adjust to differential ground movement without cracking; and (d) the construction can be completed rapidly and at a low cost. In certain sites, vertical screens can be combined with or supplemented by other ground engineering techniques such as impervious blankets, sheet-pile walls, cofferdams, relief wells and surface drains, conventional dewatering and pumping, chemical grouting, and freezing, discussed in other chapters.

10-2 EARTH CUTOFFS

Cutoffs built with earth backfill are popular and relatively simple to construct, particularly if the materials are available at the site. A complete description of design and construction fundamentals is beyond the scope of this text, and the reader is referred to Xanthakos (1979). In the context of ground control and improvement, our concern is with the resistance of cutoffs to blowout failure, actual in situ permeability of earth backfills, and effect of pollutant infiltration on cutoff imperviousness and integrity.

Clay Mixes

Resistance of an earth backfill to water infiltration is markedly improved if the constituent materials are supplemented with fines, usually silt or clay, of a suitable type and source. Although the improved performance of a well-graded backfill is explicitly confirmed in field cases, sometimes designers tend to specify only the degree of watertightness, leaving backfill composition details to the specialist contractor.

The obvious function of a clay in an earth matrix justifies the analysis of its physical and flow properties in the design stage. The selected clay must form a successful blend with the bulk of the backfill, and also be comparatively inert and chemically inactive. Its physical and rheological characteristics will by and large determine the permeability and strength of the backfill in place, and its ability to resist displacement. The clay must also yield a blend that is compatible with the conditions of mixing, storage, placement, and self-compaction.

Initially the fine material must occupy and block the voids of the aggregate fraction without causing segregation or displacement of soil particles. It must also expand upon hydration and form a gel while at rest. This will reduce the bulk quantity while imparting to the backfill the ability to resist movement without the addition of rigidifying agents. Qualifying materials for this function usually are found among the naturally occurring clays, particularly those derived from sodium montomorillonite.

Sources of Clays Clays may be used as they occur naturally or may be treated and conditioned. Natural clays are preferred, especially if they are locally available. Economy is improved if the fine fraction is found in clayey and silty soil from the excavation properly blended with some sand and gravel, and then mixed with bentonite slurry. A mandatory treatment is to remove coarse particles and organic matter. Wet plastic clay is not considered suitable because it is difficult to mix with other grades. Where the source of natural clays is uncertain prepared clays should be selected, and among those sodium bentonite grades offer the best choice.

Marine or alluvial clays are preferred to glacial clays because the latter occasionally contain a high fraction of sand and silt that are erratic in distribution. Once acceptable grades are identified, the selection is based on the type of clay mineral expressed by the content of colloid-sized particles. These constitute the three basic groups of montmorillonite, illite, and kaolinite. The most active is the montmorillonite group because it exhibits a remarkable ability to swell by taking water molecules directly into the space lattice. Where natural clays are selected for either the backfill or the slurry, a standard laboratory analysis is mandatory to confirm the suitability of the material.

Physical Properties If the coarser fraction of in situ soil shows considerable variation in content, the particle size distribution of the clay is the most relevant physical property. A second index of suitability is the Atterberg liquid limit. If this is less than 60, the clay should not be considered unless it is practical and economical to remove the coarser fraction forming the skeleton of the soil to improve its fineness. If bentonite is the prime constituent of the clay, the liquid limit may range between 300 and 400, but this should not be the sole criterion for judging the flow properties.

Chemical Properties Relevant indexes are the surface activity, base-exchange capacity, and pH. In natural clays the base-exchange capacity is likely to vary considerably, prompting the same response in the backfill or the slurry. Since this variation is difficult to eliminate or control effectively, it implies acceptance of broader control limits for the backfill. This means that a relatively minor change in composition may result in a major change in the field properties of the backfill.

Clay suspensions usually tend to be slightly alkaline, and will flocculate with the addition of acidic agents. The pH should be not less than 7, which is the neutral point on the pH scale, but in most jobs the specified value of pH is 8 or greater.

Rheological Characteristics These include the yield stress, apparent and plastic viscosity, gel strength, filtration loss, and thixotropy (Xanthakos, 1979). The slurry must deposit a filter cake to keep the face stable and deter slurry loss to the ground, and then form a gel in the backfill to resist the applied gradient. Unless the clay exhibits thixotropy, it will not develop sufficient shear strength to resist blowout, and in this case rigidifying agents must be added.

Certain clays exhibit an unusual form of thixotropy whereby the gelling action is better promoted by agitation. In the field, this behavior will cause problems during the mixing and placement process. Efficient use of dispersants will remedy mild occurrence, but clays with strong tendency to gel upon agitation should not be considered.

Composition and Permeability of Backfill

The composition and permeability (in situ) of earth cutoffs vary widely with each project and application, but depend mainly on the gradation of available materials, and the mixing and placement procedures. Quoted in situ permeability is between 2 $\times 10^{-6}$ cm/sec (Katowicz, 1967) and 1×10^{-6} (Xanthakos and Bailey, 1975), but values as low as 5×10^{-8} cm/sec have been reported (La Russo, 1963). For a low permeability the backfill must contain a suitable clay fraction, sufficient to fill the pores created by the coarser particles. Good grading also imparts to the backfill low compressibility characteristics. Extra fines may be added to limit segregation of larger particles in the final position, but they may also enhance compressibility and subsequent consolidation.

Table 10-1 shows backfill gradation data from cutoffs in Europe and Australia. Relevant data from cutoffs built in the United States are shown in Table 10-2. Comparison of data from the two tables shows fairly good agreement. At least 10 percent of the backfill materials should pass the No. 200 sieve, and often the minimum percentage of fines is higher and close to 20 percent.

For the Calamus Dam the cutoff was constructed from suitable materials obtained from the slurry trench excavation and supplemented by a suitable borrow site. The clay was disked, pulverized, and processed so that 80 percent would pass the No. 4 sieve. The cutoff extends through deposits of dune sand, interbedded fine sand, coarse sand, and intermixing layers of peat, organic silts, soft silts and clays, and diatomaceous earth. These deposits are underlain by the Ogallala formation. Since the intent of the design was to obtain percolation reduction rather than total imperviousness, only a section of the cutoff penetrates the Ogallala formation.

Design Considerations

The design of a cutoff usually involves selection of a properly graded backfill to ensure the stipulated reduction in permeability. The cutoff must also resist displacement and blowout of the fines under the expected hydraulic head, and have low compressibility behavior. For long-term installations, durability and resistance to contamination have implicit importance and must be considered.

For preliminary purposes, a probable coefficient of permeability of the order of 2

TABLE 10-1 Typical Gradation Limits for Backfills in Europe and Australia

Screen Size, British Standard Sieve	Percentage Passing by Weight
3 in.	80–100
$\frac{3}{4}$ in.	40–100
$\frac{1}{4}$ in.	30–70
No. 25	20–50
No. 200	10–25

TABLE 10-2 Typical Gradation Limits for Backfills in the United States

U.S. Standard Sieve	Percentage Passing by Weight: Wanapum Development	Camanche Dike	Southport AWT Facilities, Indianapolis
3 in.	80–100	80–100	100
$\frac{3}{4}$ in.	40–100	60–100	55–100
No. 4	30–70	40–80	40–75
No. 30	20–50	20–60	25–50
No. 200	10–25	10–30	20–30

$\times 10^{-6}$ cm/sec (about 2 ft/yr) is reasonable. Blowout failure can initially be inferred by estimating the critical hydraulic gradient, expressed approximately by

$$i = \frac{6\tau_f}{\cos\theta} \frac{1}{D\gamma_f} \frac{1-n}{n} \tag{10-1}$$

where i = critical gradient
τ_f = shear strength of slurry in the soil pores
θ = tortuosity angle = 48° to 51° for viscous flow
D = mean size of upper and lower limits of gravel fraction
δ_f = weight of slurry
n = void ratio

Predictions made with the help of Eq. (10-1) are meaningful if: (a) the actual gel strength of the slurry can be accurately assumed (often taken as 2 or 3 times the 10-min gel strength); (b) the value of D as defined here is most relevant to blowout; and (c) the tortuosity angle θ, taken as 48° to 51°, is representative of in situ conditions for most earth backfills.

With the coefficient of permeability k and the critical gradient i tentatively estimated, we can check the total flow of water through the cutoff and compare it with the desired reduction in the percolation volume by reference to the relation

$$v = ki \tag{10-2}$$

where v is the discharge velocity, defined as the quantity of water that percolates in a unit time across a unit area. If the discharge velocity is too high, we may adjust the factors k and i by redesigning the mix or by increasing the cutoff thickness t_b. We should note that $i = \Delta h/t_b$, where Δh is the difference in the water level between upstream and downstream face of the cutoff. Equation (10-2) expresses what is known as Darcy's law.

The bentonite concentration in the slurry is based on appropriate control limits. The slurry must deposit a filter cake to isolate the trench from the ground environment, it must have sufficient shear strength to keep enough soil particles in suspension, and in the final position of the backfill its shear strength must resist blowout of fine particles.

The compressibility of earth materials in the cutoff must allow deformations to occur without cracking, yet it should not result in unwarranted movement. By and large, the percentage of coarse-grained particles has a decisive effect on both strength and compressibility. Strength increases whereas compressibility or plasticity decreases linearly with increasing percentage of the coarse-grained fraction. Compressibility limits therefore place corresponding limitations on the percentage of fines.

Blowout Tests

These are also called pressure tests, and are very useful if the procedures and the materials during the test represent the in situ conditions. A representative backfill is obtained by mixing bulk soil samples from the site with bentonite slurry. The tests have a two-fold function: they establish a probable hydraulic gradient at which failure by blowout will occur, and they allow estimation of the permeability of the backfill material.

Test Apparatus Figures 10-1 and 10-2 show the apparatus commonly used for a blowout test. A cylindrical tank, usually 28 to 30 in. high and 15 to 18 in. o.d., is provided with a drain, discharge pipe, and collector vessel and is connected to a water source to which air pressure can be applied.

A mix of clean saturated sand and gravel is placed in the tank to a height of 9 to 10 in. of the bottom. The tank is then filled with bentonite slurry that is allowed to

Figure 10-1 Steel tank and miscellaneous equipment for blowout test.

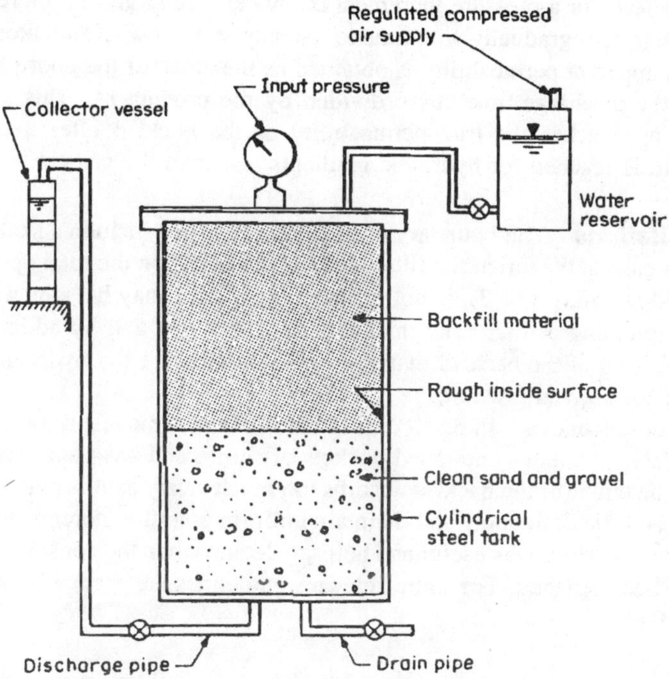

Figure 10-2 Schematic section of blowout apparatus.

stand until a seal is formed. The latter may be associated with surface filtration, deep filtration, or rheological blocking. Backfill is placed to fill the upper part of the tank, simultaneously displacing the slurry, and the lid is securely fastened and sealed to the top of the apparatus.

Pressure increments are applied to the system, and the resulting discharge of water is measured at suitable intervals. The critical hydraulic gradient is assumed to have been reached when a sudden or substantial increase in the flow rate occurs or when malfunctioning of the test becomes apparent. The presence and type of solids appearing in the collector vessel are also monitored and examined at the conclusion of the test.

The permeability of the backfill in the tank is estimated from the relation

$$k = \frac{Q}{tiA} \tag{10-3}$$

where k = coefficient of permeability, cm/sec
Q = discharge volume of water, cm^3
t = time over which the discharge is m asured, sec
i = hydraulic gradient
A = cross-sectional area of backfill normal to flow, cm^2

During the test, for a pressure increment the water discharge first increases rapidly and then decreases gradually with time to a steady-state flow (Xanthakos and Bailey, 1975). An apparent permeability is obtained as the slope of the chord between two points in the discharge-time curve divided by the product iA. This apparent permeability approaches the true permeability of the backfill after a consolidation equilibrium is reached for hydraulic gradients less than the critical.

Test Limitations The boundary conditions of the test include a seal, and probably a filter cake at the top of the filter layer whose effect on the total permeability, as determined from Eq. (10-3), is not isolated. This error may be somewhat compensated by assuming a filter cake thickness of 1 to 2 in., and by adding this to the actual thickness of the backfill material in the tank before the hydraulic gradient is computed for a given pressure.

Major discrepancies will result during the conduct of the test if: (a) the backfill is loosely placed, contains entrapped pockets of slurry, and has poor contact with the lining of the tank; (b) the backfill absorbs too much slurry as it is placed in the tank; and (c) it is difficult to apply and maintain uniform pressure increments. A previous experience with the test is useful and helps to decide when the consolidation equilibrium has been reached. For more relevant guidelines, reference is made to Xanthakos (1979).

Test Results Variations in the boundary and consolidation conditions between the testing apparatus and the field cutoff are incidental, yet relevant to the interpretation of results. However, the following conclusions are valid, and they are offered as general guidelines:

1. The presence of fines can reduce the permeability of an earth backfill to less than 1×10^{-6} cm/sec.
2. In the lower range of gradients the apparent permeability decreases with increased gradient, whereas in the upper range the permeability increases with increased gradient.
3. Immediate consolidation effects on the backfill in the tank have a corresponding influence on its true permeability for pressure increments of short duration. For long-duration pressure increments the measured permeability agrees fairly well with values obtained from standard permeability tests.
4. For cutoffs with gradation limits in the range of Tables 10-1 and 10-2, blowout failure appears to occur at gradients 30 to 40. For initial estimates a gradient 32 is admissible, and for a factor of safety 4 the approximate cutoff thickness is obtained by dividing the differential head by 8.

Examples of Blowout Tests Table 10-3 shows data and remarks from blowout tests carried out for the cutoff of Calamus Dam.

For test numbered 1 the base material was not wetted prior to adding slurry. After a seal was formed, the slurry was removed with scoops, and backfill was placed in

Test Number	Backfill Material and Composition	Base Density (lb/ft^3)	Estimated Blowout Gradient	Permeability Range (ft/yr)	Compressibility	Remarks
1	Poorly graded sand, 7% fines, 93% fine to medium sand; placement moisture = 7%; slurry added = 35%; slump = $4\frac{1}{2}$ in.	102 (dry)	≈ 32	4–10	≈ 12%	Satisfactory test; discharge flow-time curves consistent; no fines were observed in collector vessel; base gradation after test showed a 4% increase in fines
2	Silty sand—combination of fine and coarse sand; 20% fines, 71% coarse sand, 9% gravel; placement moisture = 11.3%; slurry added = 12%; slump = $3\frac{1}{2}$ in.	102 (wetted)	Unspecified but close to 8	5–12	≈ 8%	Test not satisfactory; failure related to base density; shear zone (cracks) developed due to differential settlement; piping developed along the cylinder walls; base showed a distinct zone containing visible bentonite fines
3	Backfill material same as in Test no. 2; placement moisture = 9.7%; slurry added = 15%; slump = $4\frac{1}{2}$ in.	108 (wetted)	Blowout not discernable	1.5–5.6	≈ 4%	Satisfactory test; discharge flow-time curve consistent, but erratic for i = 10–20; stabilized at $i > 20$, and became erratic again at i = 35–40; large crack in specimen propogated to base material small increase in base fines
4	Fine-coarse sand 40%; 40% gravel; 20% fines; placement moisture = 7.5%; slurry added = 10%; slump = 4 in.	108 (wetted)	> 55	1.9–3.6	≈ 3%	Satisfactory test; discharge flow-time curve consistent; only small cracks visible in backfill; full blowout not reached; coefficient of permeability initially erratic, then constant

3-in. lifts, compacted and rodded manually. From Figure 10-3, we note that the discharge versus time curve is consistent, and that the permeability pattern reveals a steady state. A permeability coefficient 4 to 10 ft/yr was estimated prior to blowout.

For test numbered 2 the base material was wetted before adding slurry to form a seal. The slurry was removed by pump, and the backfill was placed in 3-in. lifts and rodded by hand. At a gradient 8 to 10 malfunctioning of the test is apparent. At this stage differential settlement induced shear cracks extended through the backfill to the base. Piping occurred at the interface of the backfill and the cylinder wall. A small conical-shape zone of bentonite and silt fines was visible in the center of the base above the outlet orifice.

For test numbered 3 the base material was placed dry and then wetted and compacted by hand to increase its density to 108 pcf. The curve reversal observed at gradients 10 to 16 is an apparent anomaly that should not be related to blowout failure. The curve became erratic again at gradients exceeding 35, possibly indicating consolidation equilibrium by blowout failure.

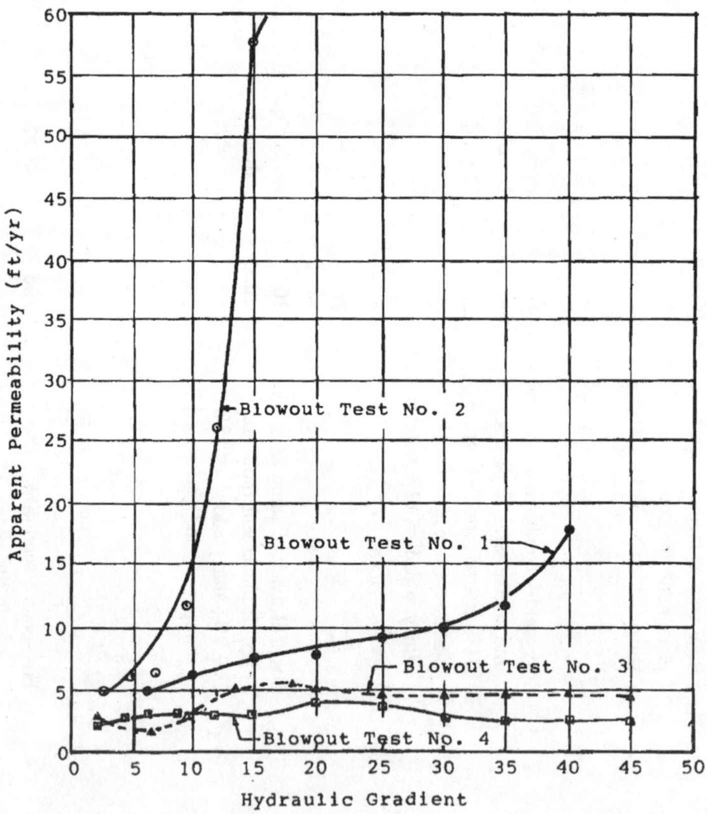

Figure 10-3 Apparent permeability versus hydraulic gradient, tests 1 to 4 of Table 10-3.

For test numbered 4 the material was placed as in test numbered 3. A slight anomaly observed for the gradient range 6 to 15 was reconciled, and produced a nearly steady permeability until blowout occurred at a gradient in excess of 55. Small cracks were observed on breakdown.

It appears from these tests that silty sand (20% fines, 70% fine to coarse sand, 10% fine gravel) would provide adequate reduction in permeability and reduce compressibility to 4 to 8 percent. A blowout gradient about 32 was considered reasonable for design purposes under a safety factor 4. Silty gravel with silty sand (20% fines, 40% fine to coarse sand, 40% fine to coarse gravel) would provide sufficient permeability reduction with least compressibility (about 3 percent) and a blowout gradient in excess of 50. Gravel, however, was not available at the site. Likewise, poorly graded sand (10% fines, 90% fine to medium sand) would provide some reduction in permeability, and increasing the fines to 20% would lower the permeability considerably. A blowout gradient close to 35 was reasonable, but a 12 percent compressibility was considered too high.

Certain problems experienced in blowout tests can be avoided by incorporating suitable details in the conduct of the test. For example localized piping often observed at the interface between the backfill material and the wall of the tank is prevented if suitable materials are used at this joint. A good connection is obtained if a mortar lining is applied to the wall and the perimeter is sealed at the top with a poured-in-place polyvinyl chloride rubber seal. Application of a silicone caulking material to roughen the wall is not guaranteed to be successful, and epoxy filler cement mixture has been found to be more effective.

Composite Permeability of Earth Cutoff

If we can support the assumption that the filter cake along the walls of the trench engages in the final cutoff function, the composite permeability of the backfill includes the contribution of all constituent materials. We can now write Darcy's law as a continuity equation (D'Appolonia, 1980):

$$v = ki = k\frac{\Delta h}{2t_c + t_b} = k_c\frac{\Delta h_c}{2t_c} = k_b\frac{\Delta h_b}{t_b} \tag{10-4}$$

where t = thickness

c = the filter cake

b = backfill

Δh = the head loss

Noting that t_b is much greater than t_c, we derive the following:

$$k = \frac{i_b}{\left(\frac{t_b}{k_b} = \frac{2t_c}{k_c}\right)} \tag{10-5}$$

The permeability of the backfill is now obtained from standard laboratory tests, whereas the ratio k_c/t_c may be determined approximately by simulating field conditions. D'Appolonia (1980) suggests a probable range of 5×10^{-9} to 25×10^{-9} cm/sec. The validity of this analysis remains in question, given the strong possibility that the filter cake will be damaged or scraped off the wall face during operations, and that the filter cake at the downstream face may be dehydrated and crack in the future.

Figure 10-4 shows plots for the relation between composite cutoff permeability and backfill and filter cake permeability expressed by Eq. (10-5), and assuming that only the upstream filter cake remains intact. We can see that the composite cutoff permeability is largely determined by the backfill alone when the backfill permeability is low, and the filter cake contribution becomes significant when the backfill permeability is relatively high. The overall cutoff permeability approaches an asymptotic limit of the order of 10^{-6} cm/sec imposed by the contribution of the filter cake. These results are valid for well-graded backfills enhanced by the presence of fines.

Effect of Water on Slump For a slump cone value of 2 to 5 in the water content in the backfill is between 25 and 30 percent. Figure 10-5 shows in graphical form the effect of water content on the slump of various backfill materials. Increasing the

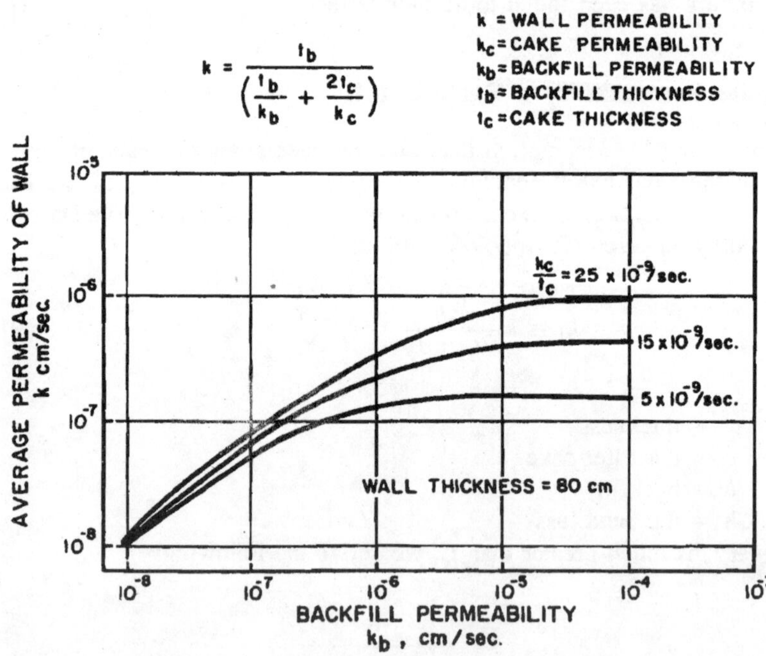

Figure 10-4 Theoretical relation between cutoff wall permeability and permeability of backfill and filter cake. (From D'Appolonia, 1980.)

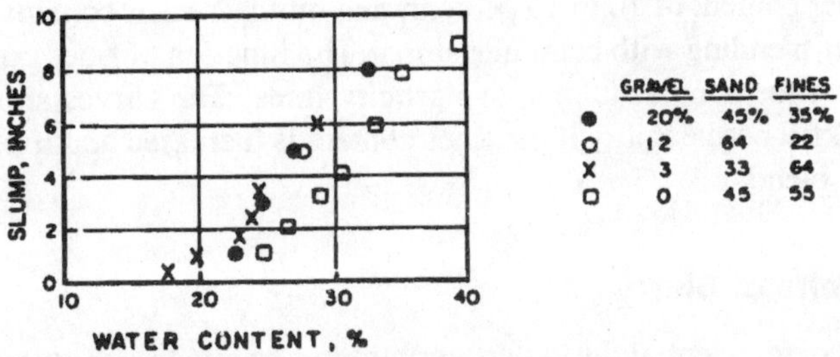

Figure 10-5 Effect of water content on the slump of backfill. (From D'Appolonia, 1980.)

water content increases the slump almost linearly. The backfill is usually mixed by blending the prepared materials with slurry containing a certain proportion of bentonite. The bentonite content in the mixed backfill is therefore controlled by the premix water content of the soil.

Effect of Gradation on Permeability Clearly the soil gradation has a marked effect on backfill permeability. D'Appolonia (1980) reports that gradations of coarse sand and gravel containing about 20 percent nonplastic fines have a permeability one or two orders of magnitude higher than gradations that contain more fines or plastic fines. The constituent determining backfill permeability is the small particle size that tends to block the voids in the backfill matrix. Other factors considered equal, natural soil with low permeability with produce backfill with lower permeability.

Figure 10-6 shows in graphical form the relations between backfill permeability and fines content (percentage passing a No. 200 sieve). The backfill materials have

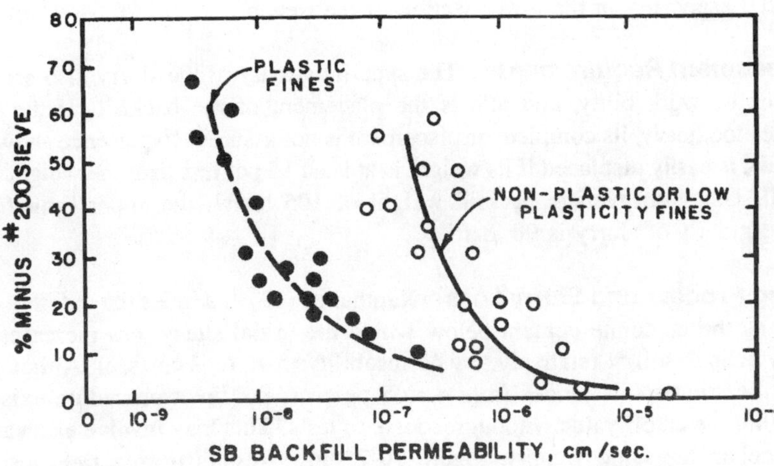

Figure 10-6 Effect of fine content on backfill permeability. (From D'Appolonia, 1980.)

a premix water content of 10 to 15 percent, and molded water content for a slump of 2 to 6 in. after blending with bentonite slurry. The function of fines content is shown for plastic and for nonplastic or low-plasticity fines. The curves show the distinct effect on backfill permeability if the fines content is increased and if plastic fines are added to the blend.

Control Limits of Slurry

Control limits are essential during construction to ensure trench stability, the continuity of the cutoff, and the composition and placement of the backfill. For a better understanding of these requirements, Xanthakos (1979) gives a summary of the physical and flow properties of slurries and also indicates the current test method. The relation between density and concentration depends on the specific gravity of bentonite. For a value of 2.2 and 6 percent concentration the slurry will have a specific gravity of 1.035, and a change of [illegible] percent in concentration will change the specific gravity of 0.005. Wyoming bentonite has a specific gravity of 2.5.

The colloid concentration C_c is usually expressed by weight. Thus

$$C_c = \frac{\text{lb bentonite}}{\text{100 lb of water}} \times 100 \qquad (10\text{-}6)$$

Other convenient expressions are kilograms per 100 kg of water, pounds per cubic foot of water, or bags per 600 liters of water. Expressing C_c as a weight ratio of bentonite and slurry (water plus weight of bentonite) is not advised.

Face Support Invariably, stability requirements in trench cutoffs indicate a slurry heavier than can normally be provided. The extra solids are taken from in situ soil added to the slurry to raise its initial density to 67 to 70 lb/ft³. As excavation continues soil from the trench is mixed with the slurry and raises its weight to 85 to 90 lb/ft³, especially in the lower section of the trench.

Displacement Requirements The specific gravity of the slurry also serves as an index of workability, and affects the placement of the backfill. If the slurry becomes too heavy, its complete displacement is not assured. Experience shows that the slurry is easily displaced if its weight is at least 15 pcf less than the weight of the backfill. Using an average backfill weight of 105 lb/ft³, the upper limit for the specific gravity of slurry is 90 lb/ft³.

Sealing Process and Slurry Loss Xanthakos (1979) defines the cutoff concentration as the bentonite content below which the initial slurry loss increases very sharply even in soil of relatively low permeability (5×10^{-3} cm/sec). Although for natural bentonites this ranges from 4 to 5 percent, no direct procedure exists for estimating the cutoff value without recourse to tests. This may involve excavating a test panel in the same vicinity. Figure 10-7 shows results from a field test, and evidently the fluid loss rises sharply if the bentonite concentration falls below 4.5 percent.

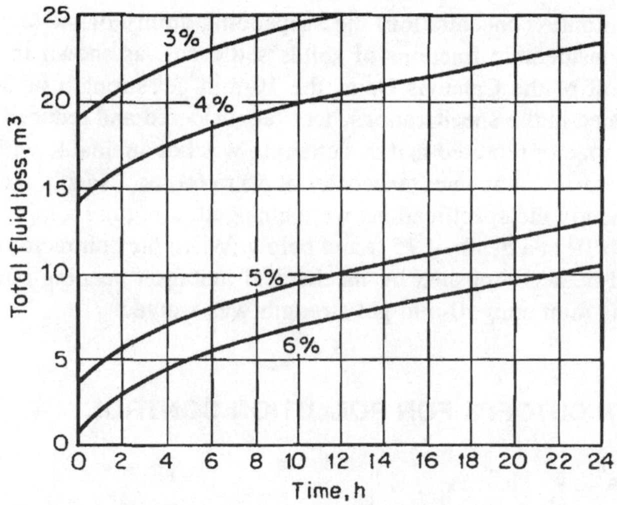

Figure 10-7 Variation in total fluid loss with time from a slurry-trench panel 15 × 100 × 2 ft. Ground permeability 5 × 10^{-3} cm/sec. (From Hutchinson et al., 1974.)

Suspension of Excavated Material Keeping enough soil particles in suspension ensures the specific gravity necessary for trench stability. A shear strength of 75 dyn/cm² or 15 lb/100 ft² will support sand particles 1 mm in size, which is the average particle size of coarse sand. According to experience, applying this limit to the 10-min gel strength gives good results. Below this limit (normally equivalent to

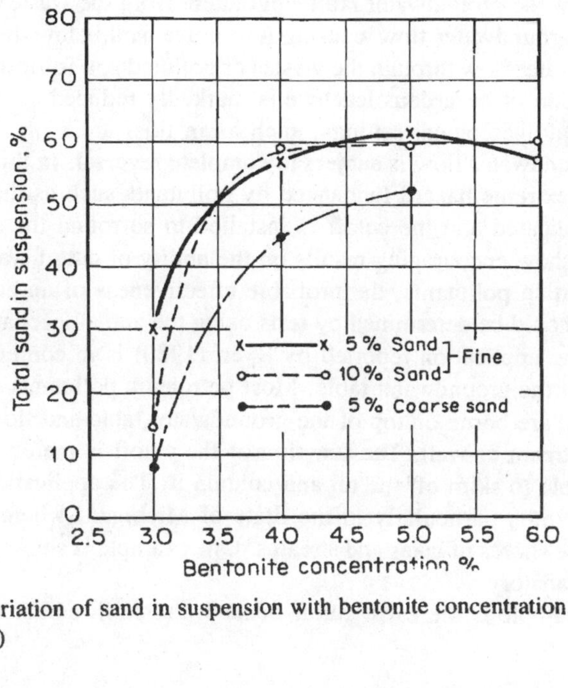

Figure 10-8 Variation of sand in suspension with bentonite concentration. (From Hutchinson et al., 1974.)

a 4 percent bentonite concentration) the suspending ability of the slurry is sharply reduced and considerable fractions of solids settle out, as shown in Figure 10-8.

For the cutoff of the Calamus Dam, the 10-min gel strength of 30 lb/100 ft^2, initially stipulated in the specifications, was reconsidered and reduced to 22 lb/100 ft^2 after the contractor reported that this strength was not attainable with the concentration that produced an apparent viscosity of 50 to 60 sec also stipulated as control limit. Subsequently the specifications were changed to require a minimum specific gravity of 80 lb/ft^3 at a depth of 15 ft and below. When the contractor demonstrated that this could be accomplished by mechanical agitation keeping soil particles in suspension, the minimum 10-min gel strength was waived.

10-3 EARTH CUTOFFS FOR POLLUTION CONTROL

Applications

Slurry earth cutoff walls have applications where it is necessary to provide pollution control. By enclosing or intercepting the site, municipal, industrial and chemical wastes can be contained. In recent years earth cutoffs have functioned as groundwater and leachate barriers around hazardous waste disposal sites. The alignment of the barrier depends on the direction and gradient of groundwater flow as well as on the location of the contaminated area.

When placed along the upgradient side of a contaminated site, the cutoff will probably cause the groundwater to flow around the waste area. The cutoff may be extended to penetrate an impervious layer for complete encapsulation. A barrier terminated below the groundwater table upgradient from the waste area can reduce the head of the groundwater flow, causing it to move well below the wastes. In this manner groundwater flow through the wastes or polluted soil is virtually eliminated, and the production of hazardous leachate is markedly reduced.

In certain hydrogeological settings, such as in tidal areas or along rivers, the direction of groundwater flow is subject to complete reversal. In this case, or in the presence of an extreme hazard (enhanced by pollutants such as dioxin), complete protection is indicated and the cutoff is installed to surround the site completely. Although tests show encouraging results on the ability of cutoff walls to withstand the effect of certain pollutants, the probable effectiveness of an earth barrier in a particular site should be determined by tests using the actual leachate from the site.

An interesting application reported by Ryan (1980) is in conjunction with oily wastes on top of the groundwater table. Most petroleum pollutants do not mix with groundwater but are borne on top of the groundwater table and flow laterally until they exit in a stream or wall. The function of the cutoff is merely to intersect the groundwater table to skim off the oil and contain it. The applications of this technique are numerous, particularly in the State of Michigan, where tank farms are located along the shores of lakes and streams. One example is shown in Figure 10-9, and is self-explanatory.

A second example is the earth cutoff beneath the earth embankment built as a

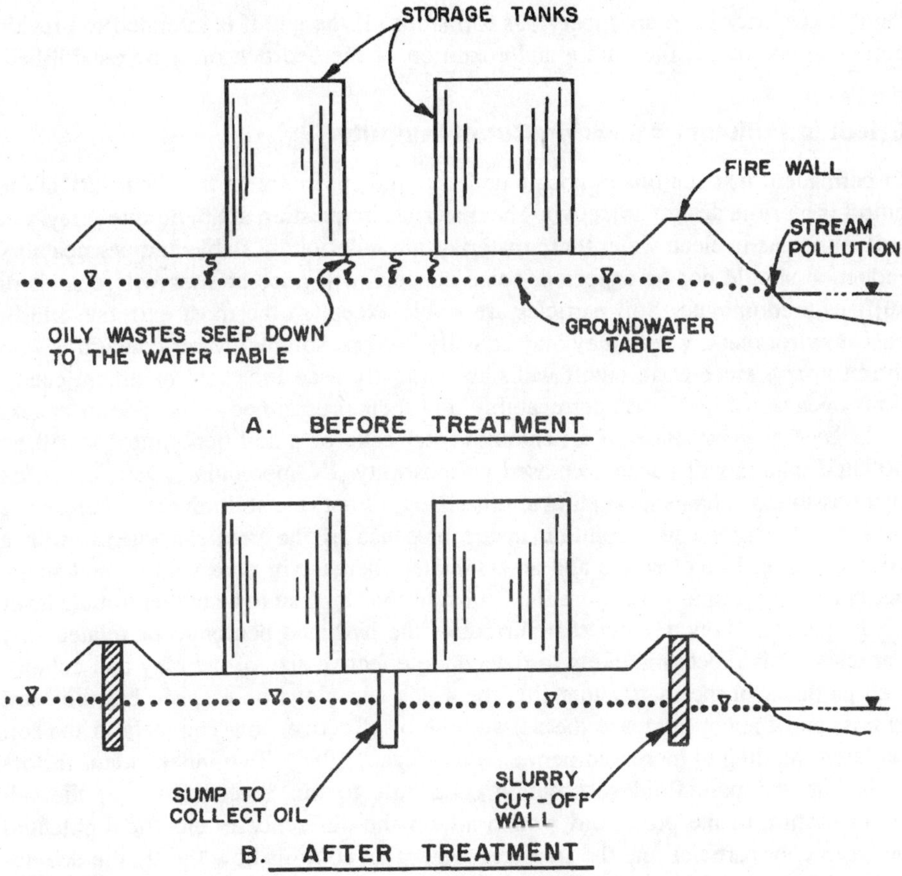

Figure 10-9 Typical application of cutoff wall to control groundwater pollution beneath oil storage tank.

containment dam for a waste management area in Elliot Lake, Ontario. This plan requires maintenance of an elevated groundwater table to minimize acid generation within the pyrite bearing tailings. This cutoff was designed to have a maximum permeability 1×10^{-6} cm/sec. The chemical compatibility of the soil-bentonite backfill with the acidic (pH = 3) drainage and its effect on permeability were also investigated. Results from tests indicated that the design criteria would be satisfied with locally available glacial tilt borrow (with a minimum of 30 percent passing the U.S. No. 200 sieve) supplemented by 2 percent (by weight) of sodium bentonite.

From the foregoing it follows that the use of an earth cutoff for pollution control must be preceded by a hydrogeologic and geotechnical investigation. The hydrogeologic study will determine the depth, rate, and direction of groundwater flow, and the chemical characteristics of the water. The geotechnical analysis will provide information on soil properties and data including permeability, stratification, and depth to bedrock or to an impervious formation. If the cutoff is intended to provide complete enclosure, the nature and condition of the bedrock must be established.

Effect of Pollutant Attack on Cutoff Integrity

In permanent installations in hostile underground environment the durability of the cutoff is a prime design objective. The materials in question are bentonite, clay, and soil. Whereas in clean water these materials are indefinitely stable and permeability reduction should not be expected, some concern will arise at sites where acids or sulfates predominate. Soil particles are stable except in the most extreme acid or basic environment, where they may actually become soluble. More concern is with the bentonite since earth cutoff walls are typically used in severe chemical conditions because of the lower permeability and their proclaimed resistance to attack.

In general permeation of a bentonite filter cake or a soil-bentonite backfill by polluted matter will cause increased permeability. D'Appolonia (1980) describes two possible mechanisms leading to this change. First, the soil minerals themselves may become prone to disolution in the presence of the permeant agent, with a corresponding loss of solids and an associated increase in pore volume and actual permeability. Secondly, the pore fluid substitution can lead to a smaller double layer of the partially bound water that surrounds the hydrated bentonite or related clay particles. This process is likely to reduce the effective size of the clay and colloid-size particles in the matrix that fills the voids and clogs pore space between soil grains, and thereby increase the net volume of effective flow channels in the soil skeleton, leading to increased permeability (Ryan, 1980). Two independent factors acting on the pore fluid substitution contribute to this mechanism: (a) the salt concentration of the pore fluid, which affects the difference in electrical potential between clay particles and the free pore water that controls how tightly the double-layer water is held; and (b) the sodium ions associated with sodium montmorillonite, which readily exchange with multivalent ions (e.g., calcium) carried in the polluting matter, a process also leading to a smaller double layer.

Reaction Time The time necessary to complete the changes associated with pore fluid substitution is relatively short. After a sample has been permeated by the advancing pollutant front of a volume of about twice the volume of the pore fluid in the sample, the initial pore fluid is largely leached out and the pollutant essentially comprises the new pore fluid. The time required to complete the cation exchange depends on the sodium cation exchange potential of the bentonite, as well as on the concentration of exchangeable cations carried by the invading pollutant. Sodium exhibits a tendency to readily exchange with multivalent cations in a calcium matrix, magnesium and heavy metals, prompting an exchange that is typically complete once an equivalent number of ions are supplied by the permeant to satisfy the total cation exchange capacity of bentonite.

Steady-State Conditions This stage is attained when the pore fluid is substituted and the cation exchange is completed. Steady-state conditions are characterized by constant permeability but at a higher value associated with the new pore fluid and the new cation montmorillonite. This behavior is shown in Figure 10-10, which expresses in graphical form permeability as a function of permeation time with calcium carbonate. In the plots the time scale is either the ratio of the leachate volume to the pore fluid volume, or the ratio of exchangeable cations released by the advancing permeant to the cations needed to complete the total cation exchange of bentonite.

Generalized Effects Pore fluid substitution is a condition that may lead to piping failure, especially if sufficient fines are not present in the blend. D'Appolonia (1980) reports laboratory tests on clean uniform sand mixed with 2 to 3 percent bentonite by weight. In the original state the mix had a permeability between 10^{-6} and 10^{-7} cm/sec. Prolonged permeation of the blend by calcium-rich water first led to a modest increase in permeability, but eventually a sudden piping failure occurred. If only a small amount of fines is present in the blend, the intergranular stresses are carried primarily by friction between sand grains, whereas bentonite expands to block the voids in the soil structure. The combination of cation exchange and higher salt concentration causes the breakdown between bound water and clay particles that can be forced out of the system in a process similar to blowout. A large number of long-term tests carried out to study the blowout effect

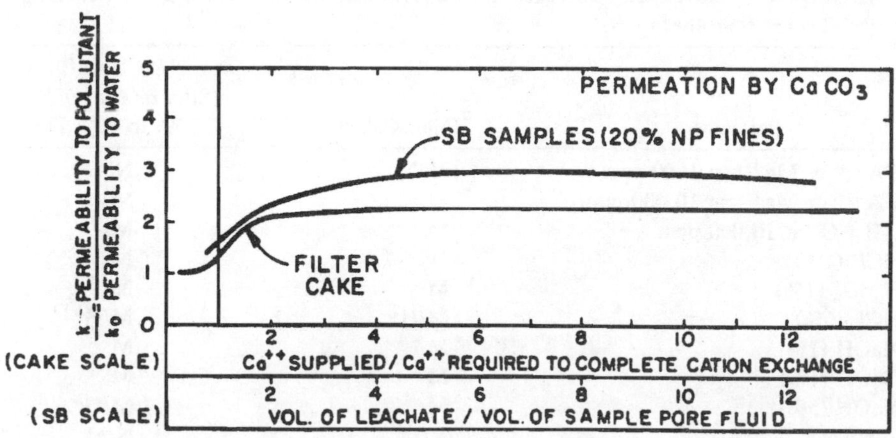

EXCHANGE CAPACITY OF CAKE (meq Ca^{++}/100g BENTONITE)	
BEFORE PERMEATION	AFTER PERMEATION
94	12

Figure 10-10 Effect of pollutant permeation on permeability of filter cake and soil-bentonite backfill.

manifested at high gradients (typically 100 to 200) confirm that piping by expulsion of the fines is prevented if a significant percentage of fines is present in the soil system. Sanning (1982) reports cases where the permeability of soil-bentonite mixes was found to increase in the presence of leachate but where the admixture of polymer compounds to the slurry reversed the process and prevented the breakdown of the supporting ability of the cutoff wall.

Summary of Effects Table 10-4 articulates the effects of advancing leachate front on the permeability of soil-bentonite mixes and filter cakes. In all cases listed, the blend contains significant plastic fines. The data give primarily a qualitative assessment of probable effects of pollutant permeation, and should not be used indiscriminately and without recourse to suitable tests representative of the in situ soil and the actual pollutant.

Based on a series of tests, D'Appolonia (1980) offers the following guidelines:

1. Well-graded soil-bentonite mixes with 30 percent or more plastic fines and about 1 percent bentonite are fairly resistant to contamination, and undergo only a modest increase in permeability under the effect of concentrated salt solutions at pH between 2 and 11. A similar effect has been observed with filter cakes formed with premium grades of natural bentonites. Granular soils

TABLE 10-4 Permeability Increase in Soil-Bentonite (SB) Mixes Due to Leaching with Various Pollutants

Pollutant	Filter Cake[a]	SB Backfill (Silty or Clayey Sand) 30% to 40% Fines
CA^{++} or Mg^{++} at 1000 ppm	N	N
CA^{++} or Mg^{++} at 10,000 ppm	M	M
NH_4NO_3 at 10,000 ppm	M	M
HCL (1%)	N	N
H_2SO_4 (1%)	M	N
HCL (5%)	M/H[b]	M/H[b]
NaOH (15)	M	M
CaOH (1%)	M	M
NaOH (5%)	M	M/H[b]
Sea water	N/M	N/M
Brine (SG = 1.2)	M	M
Acid mine drainage ($FeSO_4$; pH ~ 3)	N	N
Lignin (in CA^{++} solution)	N	N
Alcohol	H (failure)	M/H

[a]N = no significant effect, permeability increase by about a factor of 2 or less at steady state; M = moderate effect, permeability increase by factor of 2 to 5 at steady state; H = permeability increase by factor of 5 to 10.

[b]Significant dissolution likely.

without fines but with higher bentonite concentration undergo a marked permeability increase and even fail by piping.

2. A very low or a very high pH has considerable effect on permeability. Extremely basic solutions are likely to produce a greater permeability increase than strong acids. For example, amorphous silica becomes quite soluble in highly alkaline environments.
3. If a well-graded backfill contains the optimum content and type of fines, in situ stability can be ensured without the use of specially treated bentonite, since the associated advantages of the latter are uncertain. However, for backfill gradations without fines, the use of specially treated bentonites may offer some advantages.
4. Other relevant factors considered the same, backfills already contaminated by a pollutant are more stable than uncontaminated backfills in the sense that the relative change induced by subsequent permeation by a pollutant will be smaller.

By reference to Table 10-4 we notice that, among the chemicals tested, only alcohol was found to destroy the backfill completely. These results, however, are not general or conclusive, although they provide a reasonable index of cutoff behavior for a particular site and ground conditions.

Compressibility Characteristics

By reference to Table 10-3 we can note that the compressibility of most earth cutoffs is relatively high, and this allows the barrier to undergo deformations without cracking. In general a normal backfill will withstand compressive strains of several percent under in situ stress conditions without cracking (D'Appolonia and Ryan, 1979), and in the same context a cutoff can be designed for maximum flexibility under seismic effects.

Compressibility, and by analogy strength, is not a main concern for most seepage cutoffs. Exceptions are, however, noted when a cutoff is inserted underneath a dam, since in this case it must be compatible with the deformation characteristics of the surrounding ground to allow settlement of the dam above without shear effects. This implies that the cutoff must accommodate ground movement without cracking or changes that may affect permeability.

Consolidation of a backfill in situ under its own weight is restrained by the shear strength developed along the wall interface, and occurs for the most part as the backfill is placed. This imparts to the cutoff low compressibility characteristics in the final position. In terms of grading, compressibility depends mainly on the percentage of granular coarse particles, and low compressibility is obtained if these particles are contained in sufficient volume to result in direct interparticle contact that resists volume changes.

The relation between compressibility and fines content is shown in Figure 10-11 for soil-bentonite backfills. In this case compressibility is the compression ratio in a

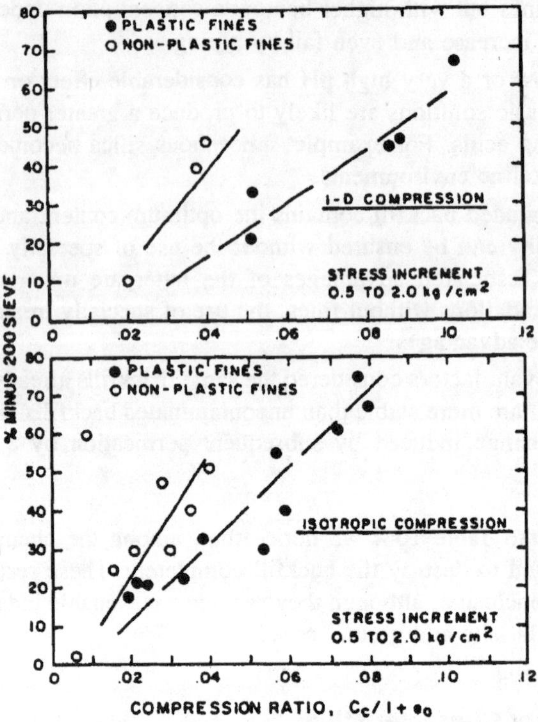

Figure 10-11 Compressibility versus fines content in soil-bentonite backfill. (From D'Appolonia, 1980.)

specific stress range, 0.5 to 2.0 kg/cm^2, which is typically the stress variation in most field applications. Data are given for one-dimensional and isotropic compression (D'Appolonia, 1980). As the fines content increases from 10 to 40 percent the compression ratio also increases almost linearly from 0.02 to 0.07, and this increase is accelerated with higher plasticity fines.

As we mentioned, low permeability and low compressibility have opposite requirements. A balanced design combines these two parameters in optimum terms if it can provide a granular structure where interstices are filled with fine-grained soil and bentonite, since the intent is to occupy void space without disrupting granular particle communication and contact. This establishes an upper limit in the fines content at about 30 percent. Where compressibility is not critical, a higher fines content is acceptable since the coarse fraction of the gradation curve becomes irrelevant.

The geomechanical behavior of soil-bentonite mixes indicates an almost perfectly plastic or strain-hardening material similar to a remolded soft sandy clay or clayey sand. Representative stress-strain curves for soil-bentonite mixes subjected to triaxial compression are shown in Figure 10-12 (D'Appolonia, 1980), together with typical effective stress paths in undrained shear. These samples fail by bulging,

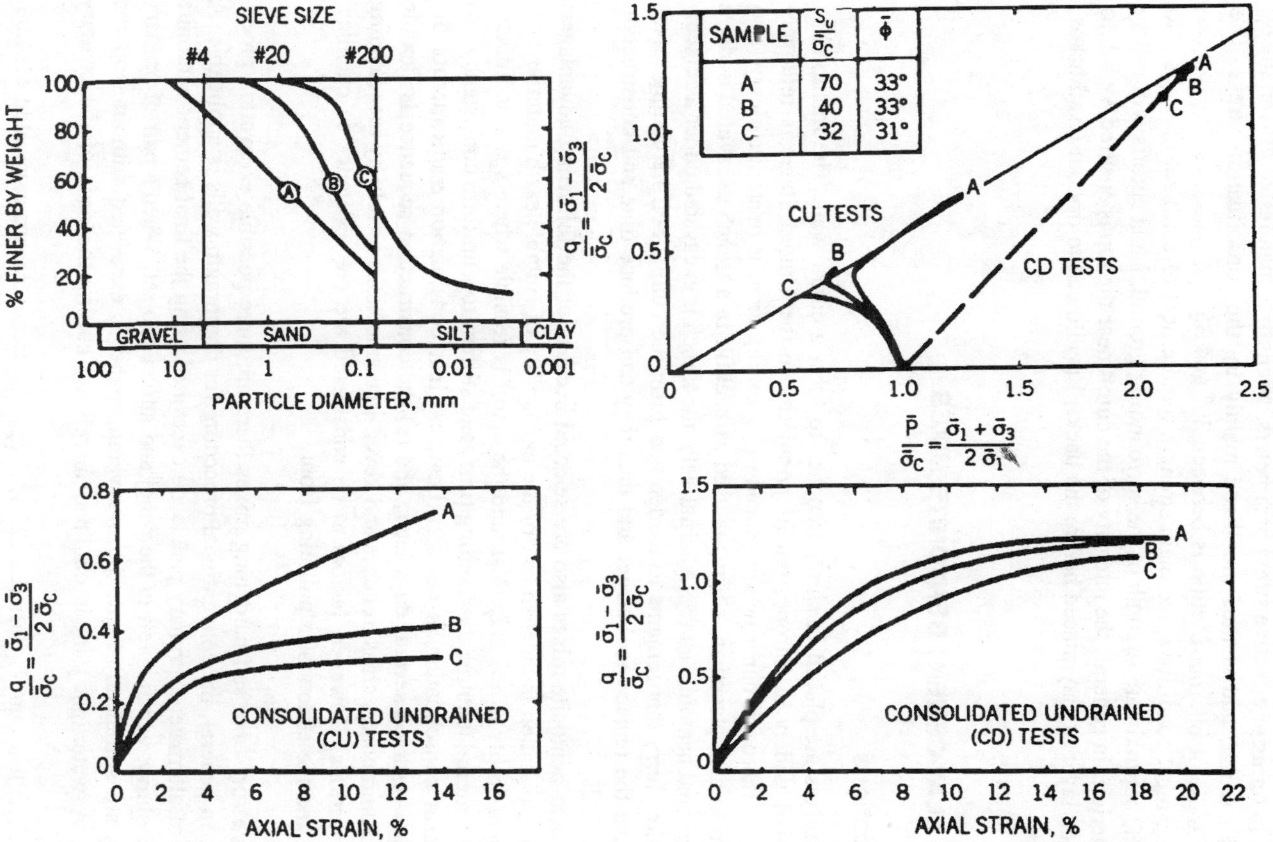

Figure 10-12 Stress-strain and strength behavior curves for soil-bentonite backfill mixes and gradient. (From D'Appolonia, 1980.)

and distinct shear planes appear only in heavily consolidated materials at large strains. In terms of effective stress parameters, strength is manifested essentially by internal friction, and is thus mobilized mainly by the consolidation stress. The quoted angle of internal friction is between 30° and 35°.

Under most conditions, the main strength requirement of earth cutoff walls is to attain the approximate strength of the surrounding ground. Most cutoffs will satisfy this criterion. In general, the integrity of the cutoff near the top is ensured by a 2- to 3-ft-thick layer of clay placed before the backfill is allowed to dry out (Xanthakos, 1979).

10-4 CLAY-CEMENT-BENTONITE MIXES

Applications

These mixes are placed in slurry trenches to form a cutoff wall. The method has been used widely in Europe, but its popularity in the United States is relatively recent. In the usual construction procedure, a setting cement-bentonite slurry (also referred to as replacement slurry) is used as backfill in a trench excavated under a conventional bentonite slurry. Alternatively, the trench is excavated under a cement-bentonite slurry that remains to harden and form the cutoff. Self-hardening mixes eliminate the replacement stage, and thus they can produce time and construction cost savings.

Cement-bentonite mixes also are selected because of the following advantages: (a) the application is independent of the availability of suitable earth materials; (b) site accessibility is not as critical, and the cutoff is feasible where space is limited; (c) the material requires short setting time, and stiffens to a uniform consistency; (d) long-term consolidation is not significant so that construction traffic across the trench is resumed several days later; and (e) the construction sequence is flexible and accommodates cross trenches and keyed-in connections merely by reexcavating short sections. However, factors to be considered are the resistance to chemical attack and the effects of a polluting front.

Excavation For self-hardening mixes there are three possible excavation procedures. In general, the conventional procedure for diaphragm walls is applicable. A series of alternate or primary panels are excavated using the final cement-bentonite mix as slurry, which is left in the panel and allowed to set. When a pair of primary panels are sufficiently hard, the intervening panel is excavated without stop-end tubes. A watertight joint is obtained merely by cutting the ends of the primary panel.

For shallow cutoffs (less than 25 ft deep), the trench can be excavated continuously. This will require a more rapid-setting slurry than for panel excavation. A mix that sets quickly inhibits self-circulation because of agitation due to excavation, and forms a uniform cutoff.

For very deep excavations ($>$ 100 ft or 30 m) or where chiseling is necessary to form a rock socket, it may not be possible to keep the mix fluid until the excavation is completed. In this case it is necessary to use a replacement slurry. This is normally injected at the base of the trench using a system of diffusers, and for complete displacement the new mix must be heavier than the excavation slurry. A common method of preparing this mix is by adding cement to slurry recovered from previous panels.

Fundamentals of Clay-Cement Mixes

In general cement is a material with adhesive and cohesive properties that make it capable of bonding mineral fragments into a compact whole. This characteristic embraces a wide spectrum of uses and applications. Clay and cement can be mixed in various proportions for use as base material in relatively flexible backfills or in special continuous curtains. The context in which we study these mixes here relates to the type of construction and does not apply to conventional grouting work.

The attainable strength, permeability, durability, and deformability are largely controlled by the relative proportions of the essential compounds. The rate of hardening and the associated development of strength are, however, slower, and the setting time is not well-defined. The thixotropic characteristics and gel strength of clay serve to keep the cement particles in suspension, and stop or limit the tendency for bleeding. During the hardening process the clay performs no other functions, and the final set is influenced largely by the ambient conditions.

The final strength of clay-cement mixes is much lower than in the all-cement structure, yet sufficient to satisfy typical field requirements. Field data show a wide strength range from 10 to 1000 lb/in^2 (70 kg/cm^2) based on the cement content (Greenwood and Raffle, 1963). In this strength range the cutoff can resist normal working pressure gradients and stresses caused by consolidation effects. The upper strength range is required in fissured rocks and open ground, where both hardening and deformability are equally important (Xanthakos, 1979).

Higher final strength implies lower clay content, usually a few percent of cement weight, indicating the use of active clay varieties such as bentonite. In these mixes the colloid is the agent giving rigidity to the base liquid of the cement matrix, and the cement still is the reaction product acting as a binder. This combination produces a relatively fluid cement suspension that can be freely placed and yet set to an adequate strength with minimum bleeding.

If the clay content is relatively high, the blend is referred to as clay-cement mix. In this configuration, the mixes are used for sealing coarse soils with permeability close to 0.1 cm/sec; injected grout screens are examples of this application. Strength is now less important, and the clay acts as a filler to increase the volume yield per unit cost of material. These mixes do not require the very active clays, and usually are prepared from locally available materials, modified when necessary with additives such as bentonite and sodium silicate for rapid setting.

Most clay-cement suspensions are anomalous systems. The gel that develops upon mixing resists displacement but also means a higher yield stress. If the mix must be injected, it will require higher injection pressure, particularly in fine soil. If the gel strength is too high and the necessary injection pressure is not available, the mix may not penetrate the pores to the extent desired. This behavior provides the basis for designing clay-cement mixes for injected screens. The flow properties deviate from the colloidal behavior of stable bentonite suspensions. Clay-cement systems are more viscous than clay-water or cement-water suspensions and have higher initial gel strength. If the pressure of clay increases the penetrability of the mix, this is because of the suspending action of the clay, rather than as a result of increase in fluidity, which inhibits the setting of cement particles and allows them to move faster and farther.

Typical characteristics of a clay-cement mix are shown in Table 10-5. This mix has a 7-day compressive strength of 5 lb/in^2. In strict terms the material is weak, yet its shear strength can sustain external gradients of 8 or higher in soil with permeability 1 cm/sec, or many times higher than gravitational gradients. A clay-cement mix of this class is suitable for coarse alluvial formations and sometimes rock fissures (Greenwood and Raffle, 1963). In this case fluidity is relatively unimportant since the mix is used only in soils sufficiently open to accept coarse cement particles without filtration. Where soil permeability is less than 0.1 cm/sec, clay-cement mixes are not effective in penetrating the soil pores because the cement particles are filtered out during injection. Conversely, the colloid size of fine clays stimulates penetration in soils with permeability 0.01 cm/sec and even lower.

Cement-Bentonite Mixes

The cement-bentonite mix of Table 10-5 has a reported seven-day compressive strength close to 650 lb/in^2 (about 46 kg/cm^2). This mix will fill cavities, rock fissures, and very open granular soil for permeability reduction and consolidation. Such a materials may be injected into soil voids with permeability close to 10 cm/sec under a pressure gradient 50. In similar ground the same mix will withstand a gradient close to 25 immediately after injection.

In conventional grout work the use of cement-bentonite blends is often disputed because of the practical difficulty of producing a thoroughly mixed material, the weakening and thickening effects, and the probable separation of the bentonite from

TABLE 10-5 Properties of a Clay-Cement Mix in Water Suspension

Type	Cement Content (% of Total Weight)	Clay Content (% of Weight of Cement)	Total Solid Content (% of Total Weight of Suspension)	Sp gr	Plastic Viscosity (cP)	Yield Strength (lb/100 ft^2)	Initial Shear (lb/100 ft^2)
Clay-cement	10	50	15	1.10	10	85	15
Cement-bentonite	75	5	80	1.50	90	270	125

the mix, causing cracks filled with bentonite only (Houlsby, 1990). Examples are reported in dam grouting (Deere, 1982). Although bentonite is used in many countries in grouting work for bleed control, it requires premixing with the use of high-speed, high-shear machines that are not always available at the site (see also Chapters 7 and 8).

Despite certain predispositions regarding the usefulness of the material, bentonite is added in small amounts to water-cement suspensions basically to extend the range of segregation-free and settlement-free performance. The dry cement and bentonite powder can be mixed together and then added to water, or a pregel of bentonite can be made and added to a cement slurry. Useful guidelines on mixing are given by Jefferis (1982). Conversion of bentonite into the calcium-exchanged form generally occurs through reaction with free lime released from cement, and the calcium bentonite is then flocculated by the excess of calcium cations. The flocs so formed lead to gelation that prevents sedimentation of the coarser cement particles. For fine cements, approximately 1.5 to 2 percent bentonite usually is sufficient to reverse the bleeding tendency of water-cement suspensions at water-cement ratios about 0.6.

Typically the presence of bentonite tends to lower the set strength since the cement particles are held apart in a matrix that inhibits consolidation. However, the 7- or 28-day strength of a water-cement mix containing a small fraction of bentonite (1 to 2 percent) is governed largely by the water-cement ratio as in ordinary water-cement mixes. This dependence is valid in very weak suspensions as those obtained with water-cement ratios as high as 20. The presence of bentonite enhances the formation of a cement gel which, although weak, is much stronger than any gel produced by bentonite alone.

In deciding on the quantity of bentonite, we should consider the following factors: (a) workability of mix (increasing the bentonite content results in increased stiffness, eventually making the mix unworkable); (b) set strength (overaddition of bentonite reduces the compressive strength of set cement markedly); (c) penetrability (particles with low specific gravity have restricted migrating tendencies, and a low gravity can result from increased bentonite concentration); and (d) stability toward sedimentation (at lower cement content more bentonite is needed to produce a system resistant to bleeding). For a preliminary evaluation of mix characteristics in cement-bentonite water blends, Figure 10-13 provides useful data.

Other Soil-Cement Mixes In certain areas the scarcity of natural impervious materials can make the cost of a barrier unreasonably high, higher than justified by the purpose of the project. In these cases granular soil-cement mixes are good substitutes, and successful applications are reported in earth embankments and dams (Holtz and Walker, 1962). The barrier is formed as a key-cutoff trench, slope paving, or core wall. Other uses reported by Zaffle (1970) are for impervious lining at the bottom of reservoirs.

Various types of soils can be blended with cement to produce a mix fairly watertight for most practical purposes. Shrinkage cracks in the final position increase seepage moderately from the uncracked condition. Results of laboratory

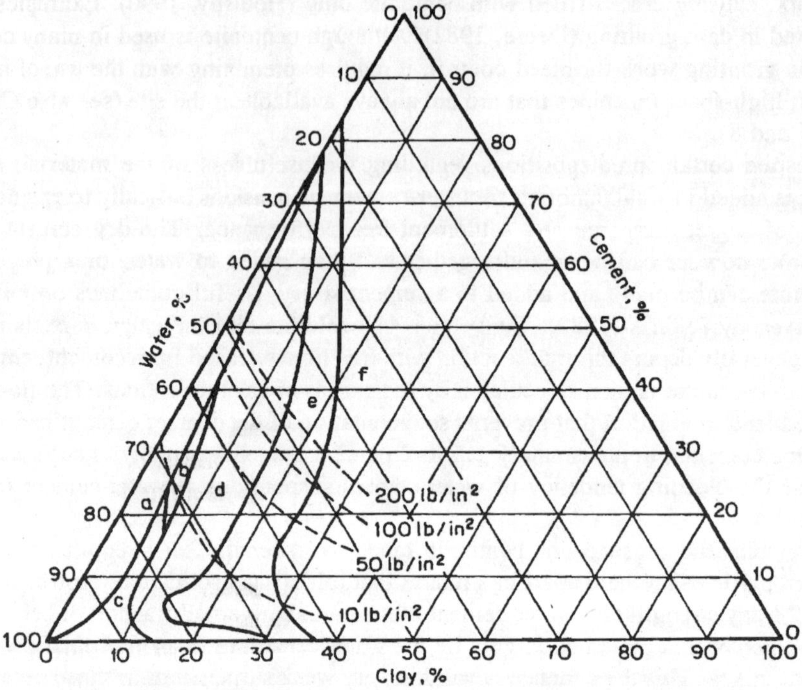

Figure 10-13 Bentonite (Fulbent 570)-cement compositions; *a* = unstable suspension: settles; *b* = temporarily stable suspension: settles before setting; *c* = clay-cement gels of low compressive strength; *d* = free-flowing, stable, and pumpable suspension; *e* = stable puttylike suspensions; *f* = solid unworkable mixes, normally powders; compressive strength on 2 × 2 in. cylinders. (From Jones, 1963.)

permeability tests on predominantly granular soil-cement materials show that substantial reduction in permeability is achieved if 7 to 10 percent of cement (by volume) is added to the blend (Zaffle, 1970). The conclusion is that the amount of cement normally required to stabilize the mix against freeze-thaw effects and wet-dry action will also reduce permeability to almost 1 ft/yr (1×10^{-6} cm/sec), which is the average permeability of earth cutoffs. The soils involved in these tests included fine and coarse sand, sandy loam, and loamy sand.

10-5 PLASTIC CONCRETE CUTOFFS

Applications

If a barrier is contemplated at a site that is not ideal in terms of soil engineering, the usual approach is to adapt the design to the actual in situ conditions. Whereas this decision is fairly simple, it implies a cutoff that must be flexible, deformable, durable, and of low compressibility but of adequate strength to withstand the worst credible conditions. Plastic concrete is a suitable material and also provides seepage control.

Plastic concrete attains an ultimate strength that does not exceed 1400 lb/in^2 (about 100 kg/cm^2) and often is much less. It also is defined according to the proportions of constituent materials, the specific results, or the plasticity of the set mix. In its final set, it is less stiff than conventional concrete but several times stiffer than earth backfill. This imparts to the cutoff an important behavior: high deformability that is not provided by rigid concrete, and strength that is not attained by earth cutoffs. This successful result is achieved without loss of watertightness.

Examples where a plastic concrete cutoff offers a good choice are sites where considerable additional overburden load is contemplated, such as the weight of a dam, or when extra hydraulic pressure is created, both placing extra strength requirements. Conversely, where relatively loose materials around a cutoff are expected to be squeezed and continue to deform until a new equilibrium is established, the design calls for a cutoff that can withstand the associated distortion. This geomechanical compatibility is satisfied if the barrier displays characteristics essentially similar to the behavior of the soil medium; hence the soil modulus is a relevant property and gives indication of the desired flexibility.

The foregoing criteria are not always met since it is not possible to establish a reliable measure of in situ conditions and changes that will occur with time. As an example, we may consider a relatively loose formation under an earth dam. Under increasing height the dam applies more load, the formation is compacted, and the stress on a soil element increases. After the dam is placed in service, saturation of the compacted layer is promoted by permeating water and higher hydraulic gradient. In these conditions the total deformation and settlement depend not only on the initial hydrogeologic conditions but also on subsequent changes. The composite system behavior becomes more complex if a cutoff is inserted beneath the dam causing changes in drainage and saturation from the upstream to the downstream side.

Selections of Modulus

For a plastic cutoff to be most functional the modulus of elasticity of the plastic concrete must approximate the soil modulus at the selected state of equilibrium, whereas the stress-strain curves for the two media must be substantially similar. The expectation of complete similarity is, however, erroneous and unwarranted, since in practice it is seldom if ever possible to duplicate the properties of a soil at any state or loading conditions. This effort is constrained by variations in soil elasticity, strength, and compressibility under loading, compaction, flow, saturation, and drainage.

The soil modulus, also quoted as Young's modulus, may be determined by plotting a stress-strain curve from an unconfined or, preferably, a confined (triaxial) compression test. It can also be extrapolated from plate-load tests. The value thus obtained is influenced by the confining stress, cycle of loading, void ratio, sample disturbance, stress history, and other factors. The soil modulus of cohesionless materials (pure sand) is affected very little by changes in moisture content and by particle size, but void ratio affects the E value to a greater extent, as does the deviator stress. Considerable evidence exists confirming that in soils with cohesion the elastic modulus determined in an unconfined compression test may include a substantial error, often by a factor of 4 to 5.

From the foregoing it follows that, even if a reliable measurement is available, choosing a soil modulus is of little practical value since the most relevant condition is the postconstruction period. For both granular and cohesive soils convenient assumptions are that the material is semiinfinite elastic, homogeneous, and isotropic. Apart from this compromise, it is impracticable (and often unnecessary) to make an exact estimation of soil modulus. It is more expedient to concentrate on the analysis of suitable plastic concrete mixes, focusing on economy and overall performance, and to match the elasticity of soil to the extent possible. For quick reference, the probable range of elasticity is shown in Tables 10-6 and 10-7 for various soil types, and from two different sources.

Elasticity and Strength of Plastic Concrete

A typical plastic concrete mix includes gravel, sand, clay, cement, and bentonite blended with water to produce a workable mass. The stress-strain diagram follows the curve shown in Figure 10-14. The curvature may cover the entire range as depicted in Figure 10-14*a* if the clay fraction predominates, or it may follow the more linear pattern of Figure 10-14*b* if the main constituents are gravel and sand. We see, therefore, that Young's modulus strictly applies to the straight portion of the curve or, if this portion is not present, to the tangent to the curve at the origin (also shown as initial tangent modulus). In addition, we distinguish the "secant modulus," represented by the slope of a line drawn from the origin to any point B on the curve; the "tangent modulus" referring to a line that is tangent to the curve at any point A; and the "chord modulus," represented by the slope of a line between any points B and C. The secant modulus is the most practical since it represents the actual deformation at the selected point.

TABLE 10-6 Young's Modulus for Repeated Loadings

Soil (1 atm Confining Pressure)	Young's Modulus, lb/in.^{2} Loose	Young's Modulus, lb/in.^{2} Dense
Screened crushed quartz, fine angular	17,000	30,000
Screened Ottawa sand, fine rounded	26,000	45,000
Ottawa standard sand, medium rounded	30,000	52,000
Screened sand, medium subangular	20,000	35,000
Screened crushed quartz, medium angular	18,000	27,000
Well-graded sand, coarse subangular	15,000	28,000

Source: From Chen (1948).

TABLE 10-7 Range of Values for Modulus of Elasticity *E*, for Selected Soils

Soil	E_s (lb/in.^{2})
Very soft clay	50–400
Soft clay	250–600
Medium clay	600–1,200
Hard clay	1,000–2,500
Sandy clay	4,000–6,000
Silty sand	1,000–3,000
Loose sand	1,500–3,500
Dense sand	7,000–12,000
Dense sand and gravel	14,000–28,000
Loess	14,000–18,000

Mixes that exhibit the behavior of Figure 10-14*b* are typical elastic-plastic materials. The mix is essentially elastic until the stress reaches the proportional limit, which defines the two regions of elastic and plastic behavior. The increase in strain while the load continues to act is caused by creep in the concrete, but the dependence of instantaneous strain on the speed of loading makes the demarcation between elastic and creep strains impractical. We thus choose an arbitrary distinction: we consider deformation occurring during loading elastic, and we regard subsequent increase in strain as creep. Unlike conventional concrete, plastic concrete mixes typically are allowed to reach the plastic stage in the ground.

Of practical interest is the flexural modulus determined from the deflection of a cantilever beam. This deformation represents the lateral translation and bending of a

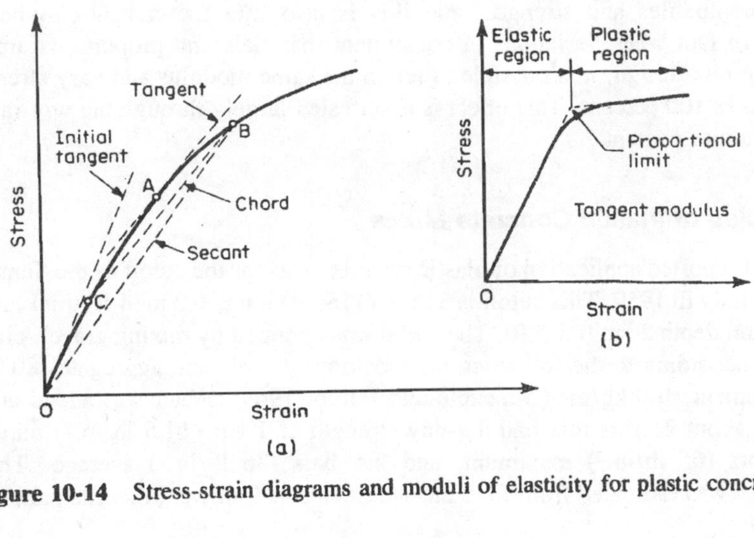

Figure 10-14 Stress-strain diagrams and moduli of elasticity for plastic concrete.

cutoff inserted in soil that becomes progressively stiffer with depth so that the deflection is zero at the bottom and maximum at the top. If this maximum deflection can be estimated or predicted, the elastic modulus can be calculated by flexural theory.

In the lower stress range plastic concrete behaves essentially elastically, but under constant load it shows a tendency to creep, although a clear distinction is not practical. While in the ground the mix is acted upon by forces gradually manifesting a steady loading state and change little thereafter.

Factors Affecting Strength and Modulus Elastic modulus is influenced by the same factors that affect strength, although to a lesser degree. These are the clay-bentonite content, the water-cement ratio, and the type and gradation of aggregate. Age increases modulus especially in richer mixes. Increasing the water-cement ratio causes a distinct reduction in the secant modulus for the usual range of plastic concrete. This range has a w/c ratio 1.5 to 3.0 or even higher.

The type of aggregate also has an effect on modulus, and since the deformability of concrete is partly the elastic deformation of the aggregate fraction, a higher modulus is obtained with stiffer aggregate. Limestone is more suitable than sandstone for decreased stiffness and increased plasticity. Aggregate grading influences modulus in the same way it affects strength: increased strength means increased modulus. Because of the high w/c ratio plastic concrete is not harsh or unworkable, and its elasticity responds to the elasticity of the fines.

The effect of free moisture on mixes of the same consistency and age is a higher modulus for wet than for dry concrete. Both modulus and strength are improved with longer curing periods, which typically are provided in slurry trench construction. The effect of bentonite on the set concrete is lower strength and higher plasticity.

Unlike the modulus of conventional concrete, which increases approximately with the square root of its strength, for plastic concrete no set relationship exists between modulus and strength, and this is also true for cement-clay-bentonite mixes. In fact when variations in constituent materials and proportions are introduced in mix design, it is possible to retain the same modulus and vary strength by as much as 100 percent. This effect is manifested largely through the w/c ratio and the bentonite content.

Examples of Plastic Concrete Mixes

The first reported application of plastic concrete was for the cutoff of the Santa Luce Dam in Italy in 1959. This cutoff is 354 m (1160 ft) long, 1.2 m (4 ft) thick, and has maximum depth 20 m (65.5 ft). The blend was obtained by mixing gravel, clay, and cement according to the following composition: gravel-sand aggregate, 80% clay, 20%; cement, 100 kg/m^3 (Dupeuple and Habib, 1969). Water was added at a w/c ratio of about 2. This mix had a 7-day strength of 1 bar (14.5 lb/in.^{2}) minimum, 4.45 bars (65 lb/in.^{2}) maximum, and 2.5 bars (36 lb/in.^{2}) average. The permeability was estimated from field measurements as follows: (a) at the panel center

$k_{min} = 1.4 \times 10^{-7}$ cm/sec, $k_{max} = 3 \times 10^{-6}$ cm/sec, $k_{ave} = 1 \times 10^{-6}$ cm/sec; (b) at the joints $k_{min} = 1.1 \times 10^{-6}$ cm/sec, $k_{max} = 1.9 \times 10^{-6}$ cm/sec, $k_{ave} = 1.6 \times 10^{-6}$ cm/sec.

A plastic concrete cutoff derives its watertightness largely from the presence of clay and colloid-size constituents. The same material disrupts the tendency of the coarse aggregate-cement structure to act as a brittle body, and imparts to the mix deformability without breaking.

In normally consolidated sedimentary soil the void ratio and water content usually decrease with depth; hence, the strength and soil modulus of most alluvial deposits also increase with depth. A plastic concrete cutoff is more compatible with these ground conditions if its modulus can exhibit the same behavior, and this is accomplished to a great extent if the mix contains a large proportion of sand.

Table 10-8 shows data from two plastic concrete cutoffs for dams (Little, 1974). The modulus of elasticity for this concrete is 1.5×10^4 kg/cm² (210,000 lb/in.²), or several times the modulus of any soil except rock. There is a marked difference in the composite permeability of the two cutoffs despite the similarities in material use and construction methods. The mix was placed in panels using tremie pipes. These panels have simple round tube joints enhancing leakage at these locations, and this may account for the variation in permeability.

Table 10-9 shows composition data from plastic mixes where sand is the only aggregate and represents the main bulk of the blend. The content of the fine fraction (clay and bentonite) is varied, but the cement ratio is the same for all mixes. The clay-bentonite ratio is 9, and the w/c ratio of about 2 is enough to produce a slump of about 15 cm (6 in.).

Figure 10-15 shows the 7-day stress strain curve for mix 1 of Table 10-9. The set concrete appears almost ideally elastic-plastic with a well-defined proportional limit obtained as the intersection of two tangents (point *R* on the graph). The material can withstand 6 to 7 percent strain without failing, although in this stress region a

TABLE 10-8 Materials and Properties of Plastic Concrete Cutoffs

	Balderhead Dam		Lluest Dam	
Material	kg/m³	%	kg/m³	%
Weight per unit volume	2039	100	1956	100
Bentonite	44	2.2	24	1.2
Water	400	19.6	405	20.7
Cement	195	9.6	227	11.6
Aggregate	1400	68.6	1300	66.5
Water-cement ratio	2.05		1.78	
Water-solids ratio	0.24		0.26	
Permeability *k* (cm/sec)	0.6×10^{-7}–2×10^{-7}		10^{-3}–10^{-4}	

Source: From Little (1974).

TABLE 10-9 Materials and Proportions of Plastic Mixes

Mix Number	Sand S (%)	Clay + Bentonite C + B (%)	Cement (% of S + C + B)	Water (% of S + C + B)
1	80	20	6	
2	83.3	16.7	6	10–12
3	85	15	6	
4	90	10	6	

Source: Dupeuple and Habib (1969).

permanent deformation is likely to occur. From the graph the 7-day initial tangent modulus is 250 bars (3600 lb/in.^{2}).

Figure 10-16 shows plots of the peak deviator stress at the rupture point for a confining stress zero or 5 bars (72.5 lb/in.^{2}) at 7 and 28 days as a function of the clay-bentonite content for the mixes of Table 10-9. The corresponding relation between modulus of elasticity and the clay-bentonite fraction for the same mixes and under the same conditions of confining stress and age is shown in Figure 10-17. The plots of Figures 10-16 and 10-17 distinctly show the effect of age, confining stress, and clay-bentonite content; all these factors appear to influence strength and modulus, but the effect on the latter occurs to a lesser extent. Quantitatively, the influence of clay-bentonite content approaches as asymptotic limit as this content reaches 20 percent.

The effect of bentonite on the development of strength also is shown in Table 10-10, summarizing data from three plastic concrete mixes. The water-cement ratio is constant and equal to 2, but the bentonite content is varied from 3.5 to 0. Both the 7- and 28-day strength are markedly reduced when bentonite is in the mix (Xanthakos, 1976b).

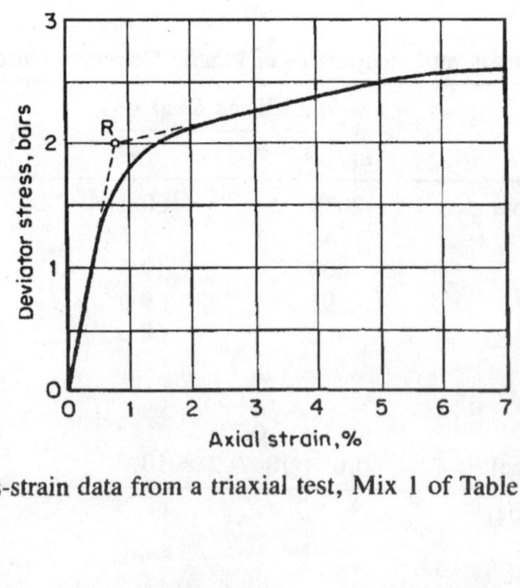

Figure 10-15 Stress-strain data from a triaxial test, Mix 1 of Table 10-9. (From Dupeuple and Habib, 1969.)

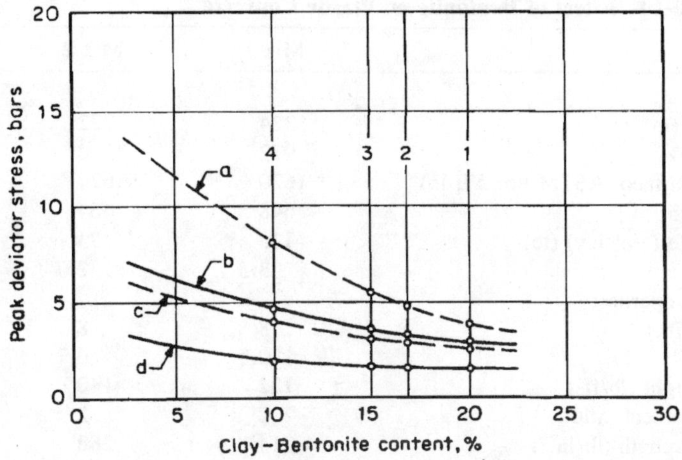

Figure 10-16 Relationship of strength to clay-bentonite content, mixes of Table 10-9; a = 28 days, confining stress of 5 bars (1 bar = 14.5 lb/in.^{2}); b = 7 days, confining stress of 5 bars; c = 28 days, 0 confining stress; d = 7 days, 0 confining stress. (From Dupeuple and Habib, 1969.)

Useful data have been provided by Kauschinger et al. (1991), following a long testing program performed to assess stress-strain characteristics and measure the permeability of plastic concrete. This work involved conventional shear strength tests for bentonite contents 0, 10, 20, and 40 percent, and the results were used to correlate the batch proportions with the unconfined compressive strength and the elastic modulus of the set mix. In addition, a series of CIUC and Q type triaxial tests were performed to study the effect of consolidation and confinement on the behavior of plastic concrete.

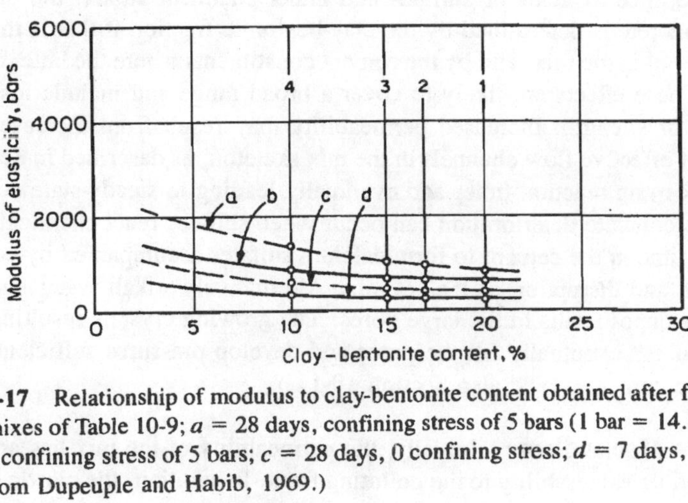

Figure 10-17 Relationship of modulus to clay-bentonite content obtained after first cycle of loading, mixes of Table 10-9; a = 28 days, confining stress of 5 bars (1 bar = 14.5 lb/in.^{2}); b = 7 days, confining stress of 5 bars; c = 28 days, 0 confining stress; d = 7 days, 0 confining stress. (From Dupeuple and Habib, 1969.)

TABLE 10-10 Effect of Bentonite on Plastic Concrete

	Mix 1	Mix 2	Mix 3
Materials:			
Cement (lb)	324	232	176
Sand (lb)	972	1390	1790
Stone (Romeo, ASTM no. 57, lb)	1620	1620	1600
Water (lb)	648	464	352
Bentonite (Volclay) (lb)	123	75	
%	3.5	2.0	0
Physical properties:			
Slump (in.)	8	8	8
Air (%)	0.5	0.7	1.6
Unit weight (lb/ft^3)	132	140	145.2
Water-cement ratio	2	2	2
7-day strength (lb/in.^{2})	214	280	313
28-day strength (lb/in.^{2})	375	407	539

Source: From Xanthakos (1976b).

Important conclusions from this program are that CIUC consolidation of plastic concrete can cause the undrained strength to increase by as much as tenfold over that measured in unconfined compression. The initial elastic modulus measured in a CIUC test is only 2 to 3 times greater than that measured in unconfined compression. Increased levels of bentonite (20 to 40 percent) cause the plastic concrete to behave as a ductile material. For the 40 percent bentonite mix, the strain at failure measured in a CIUC test was increased with setting time.

Durability of Plastic Concrete

In the presence of acids or sulfates and under chemical attack, the durability of plastic concrete is determined by the clay-bentonite fraction if this is the predominant material in the mix, and by the cement constituents where the latter control mix design. These effects are likely to cover a broad range and include loss of watertightness or strength. Increased permeability may result from increase in the net volume of effective flow channels in the mix skeleton, as described in Section 10-3, within a certain reaction time, and eventually leading to steady-state conditions.

Plastic concrete deterioration can occur when sulfates react chemically with the hydrated lime in the cement to form calcium sulfate, accompanied by considerable expansion and disruption of the set mix. Alternatively, alkali water entering concrete may deposit salts in the large pores. The growing crystals resulting from this deposition can eventually fill the pores and develop pressures sufficient to disrupt the plastic concrete. (See also Section 10-11.)

Available Mechanisms Initially, the permeability of the mix has an important bearing on its vulnerability to the polluting front. In addition, the plastic concrete as a whole contains voids since compaction is never complete. Pore space, as distinct from permeability, is measured by absorption; the two quantities are not necessarily related, and although absorption should not be used as a measure of quality of the plastic concrete, it gives an indication of mobility for an invading liquid front. Thus the flow of water through plastic concrete is fundamentally similar to flow through a porous body. Since, however, the mix is composed of particles connected over only a small fraction of their total surface, a part of the invading liquid is absorbed and a part continues to flow under an external gradient.

Polluted or acidic and chemical water can pass through large pores, construction joints, and cracks. This water can dissolve some of the readily soluble calcium hydroxide and other solids and cause appreciable errosion of the mix in the course of time. Damage can also be caused by surface errosion of the mix in direct exposure to organic acids, farm silage, polluting wastes, and other forms of chemical attack. Where these conditions are expected, tests should be carried out to determine their effect on the plastic concrete.

Resistance to sulfate attack is improved by the use of sulfate-resisting cement (ASTM type V). Resistance to disintegration caused by crystal growth requires a mix that is dense and impervious, and has a relatively low w/c ratio. This problem may be remedied in conventional concrete but is likely to persist in plastic concrete, particularly in the range of higher permeabilities and w/c ratios. (See also Section 10-11.).

10-6 PERMEABLE TREATMENT BEDS

Applications

In certain disposal sites pollution associated with emanating leachate may be intercepted or treated in place by the presence of a permeable treatment bed that physically and chemically can remove the contaminants. These beds may become saturated or plugged with time and thus need replacement. In this context they are considered temporary solutions. Several methods have been introduced for reducing the contaminant load present in groundwater, but are still considered to be state of the art and should be selected under competent technical advice.

Materials Relatively few materials are considered suitable for use in permeable beds. Among them are: (a) limestone or crushed shell; (b) activated carbon; (c) glauconitic greensands or zeolite; and (d) synthetic ion exchange resins.

A limestone bed is chemically appropriate where neutralization of acidic groundwater flow is needed, but it is also considered effective in removing certain metals such as cadmium, iron, and chromium (EPA, 1978a). Crushed shell has similar chemical characteristics, and in coastal regions where availability is good it can replace limestone effectively.

In sites contaminated with organic compounds, activated carbon is more applicable provided its cost is acceptable. Glauconitic greensand deposits reportedly exhibit good adsorption characteristics for certain heavy metals (Spoljaric and Crawford, 1979), and are accessible along the central Atlantic coastal region of New Jersey, Delaware, and Maryland.

Where natural materials are not available synthetic ion exchange resins are a good choice. These form effective beds that enhance the removal of heavy metal contaminants. However, they have high cost, and are often impractical because of short life and reactivation difficulties. For these reasons, they are recommended only where engineering and economic assessment prove their advantages for the intended use.

Design and Construction Considerations

Figure 10-18 shows a typical permeable treatment bed installation. The process involves trench excavation to intercept the flow of the contaminated groundwater, filling the trench with the selected materials, and capping the trench.

The trench must be long enough to contain the plume of the contaminated flow and deep enough to inhibit the groundwater from flowing underneath. The trench penetrates, therefore, an impermeable layer of bedrock. The width of the trench is

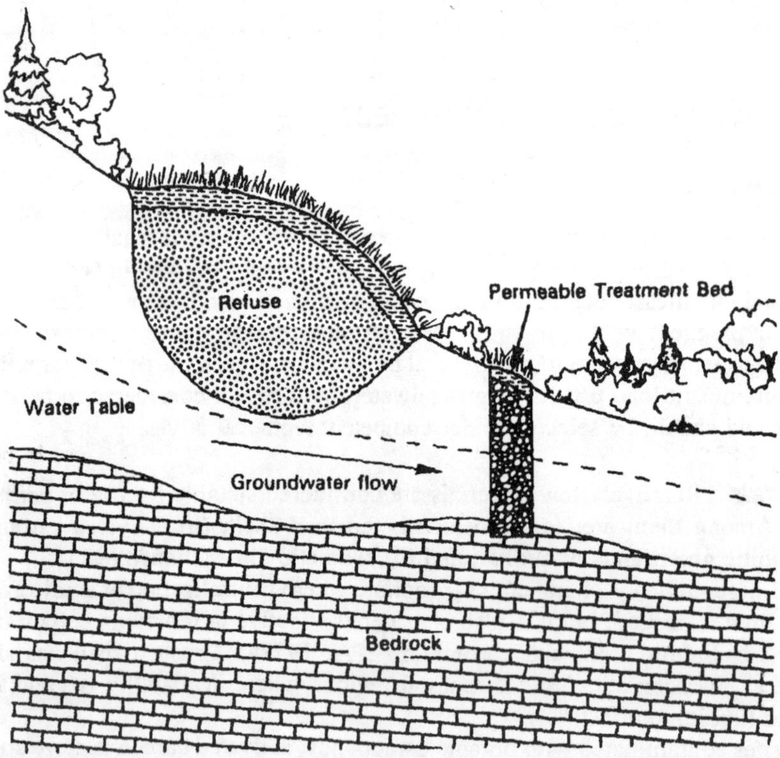

Figure 10-18 A permeable treatment bed installed to intercept *contaminated* groundwater.

decided from a consideration of the velocity of the flow, the permeability of the bed, and the contact time required for effective treatment.

Since the excavation is carried out in the dry, it requires temporary support. Invariably, the trench intercepts the groundwater before reaching an impermeable bed; hence dewatering is necessary during excavation. Groundwater pumped from the trench is likely to be contaminated and to require treatment.

For a properly designed bed, we begin by considering the mechanism of groundwater flow expressed by

$$v = \frac{ki}{n} \tag{10-7}$$

which is another form of Darcy's law,

where v = approach velocity or quantity of water through a unit area in a unit time
i = hydraulic gradient
k = permeability coefficient or proportionality constant of water at a given temperature flowing through a given material
n = effective porosity of the material.

The thickness of the bed is a function of required residence time and flow velocity of the groundwater through the bed, and is expressed as

$$w_b = (v_b)\,(t_c) \tag{10-8}$$

where w_b, v_b, and t_c are thickness, velocity, and time, respectively. The highest groundwater velocity measured in the immediate site should be used to give conservative results. Flow velocities within the bed and within the soil should be approximately equal to prevent disturbance in the natural groundwater flow. If the hydraulic gradients in the soil and in the bed are the same, this criterion is satisfied if the soil and the bed have the same permeability (Fair et al., 1966; Johnson Division, 1976).

Residence or contact time depends on the contamination level of groundwater and the treatment rendered by the material in the decontamination process. The residence time of contaminated groundwater flow through the bed must be long enough to satisfy optimum treatment conditions. This presupposes a knowledge of contaminated groundwater chemistry and a prediction of its response to treatment. In this context, an optimum contact time is determined on the basis of expected rates of interaction between groundwater and bed material.

Materials Analysis

Limestone Bed Crushed limestone is an effective low-cost landfill liner that enhances the attenuation of the migration of certain heavy metals from waste leachates. Pure limestone with high calcium content should be used in treatment beds. The particle size in the screen depends on the soil type containing the groundwater flow, and may vary from a typical gravel to a typical sand size. Besides treatment, optimum grading of limestone is necessary to minimize settling of the bed as the materials dissolve. In addition, smaller sand-sized particles prevent channeling through the bed and improve the contact between the contaminated flow and the bed.

The neutralizing action on acidic water flowing through a crushed limestone bed is reported by EPA (1978a). The results of tests show that the contact time necessary to change one pH unit (from 5 to 6, or from 6 to 7) varies from 8 to 15 days. However, little information is available on estimating the contact time needed for optimum removal of heavy metals. The contact time, together with the leachate concentration and pH, determines the efficiency of limestone in permeable treatment beds. The removal of metallic cations has been under study but conclusive results still are forthcoming. Thus, determining the optimum contact time for the removal process of heavy metals should be approached with caution.

Activated Carbon This material is indicated for the control of organic contaminants present in groundwater flow. Nonpolar organic compounds such as PCBs can be removed by activated carbon by adsorption resulting from Van de Waals and associated chemical attractive forces. Whereas the process enhances the removal of hydrocarbons, polar organic compounds such as alcohols and ketones may not be removed as effectively because of particular electrical changes. Activated carbon is also suggested for the removal of certain heavy metals, but in a metallic contaminants removal system the method is claimed to be impractical.

Glauconitic Greensands These are found along the Atlantic coastal plains, and have been widely used for the removal of several heavy metals from contaminated waters, although on a regional basis. Glauconite is a hydrous aluminosilicate clay mineral containing ferric iron and also rich in potassium. The material occurs as dark, light, or yellowish-green pellets 0.5 to 1 mm long, as casts of fossil shells, as coatings on other grains, and as a clayey matrix in coarser-grained sediments.

Spoljaric and Crawford (1979) report results of bench-scale studies showing that these materials had superior retention of heavy metal cations with leachate from the Pigeon Point landfill in Delaware. Highest removal efficiencies were reported for copper, mercury, nickel, arsenic, and cadmium, and these efficiencies improved further with increased contact time. In these tests contact time was estimated at 2 min, suggesting that with contact time of several days used in field assessments metal removal efficiency would be much higher. Since the method also was found to reduce odors, the conclusion was that adsorption of organics occurred as well.

These investigators suggest that the results of the tests indicated a high capacity of greensands for heavy metal cation retention, and this makes them a formidable material for permeable treatment beds. However, saturation points for heavy metal adsorption are yet to be determined, and the sorptive capacity of greensands has yet to be assessed through further experimentation. Since the method is an active technology, before it is considered for a given field program the designer should become familiar with current data and check the latest project status report.

Other Methods In the last decade, on-site treatment of contaminated soils has been enhanced by the development of separation and extraction techniques. These include, for example, treatment of refinery oil wastes, soil vapor extraction of organic chemicals, electro-reclamation processes, solvent extraction, and low energy extraction methods. These technologies emerged with the introduction of the SITE program, and several have been commercially demonstrated (Blank et al., 1990).

In situ stabilization/fixation approaches offer solutions without the need for excavation, but they have been associated with several problems. For example, a site undergoing a stabilization/fixation process cannot retain its flexibility in terms of future utilization. Furthermore, data on long-term aging effects of any stabilized/solidified matrix are scarce and at best incomplete, and this inhibits mitigation of risks. The low energy extraction process is a patented technology that can be used either on site or in a central location for discontaminating earth materials, sludges, and sediments containing organic pollutants. Laboratory-scale tests have been carried out with PCBs, and confirm the validity of the overall concept. Current plans include a pilot plant unit and a full size unit capable of processing 30 to 50 tons/hr. The process is based on solvent extraction using common hydrophilic and hydrophobic organic solvents to extract and further concentrate organic pollutants from any soil matrix.

Contingent Factors The final choice of a permeable treatment bed should be based on the analysis of factors that may determine effectiveness and efficiency in the long term. Among these are availability and cost of materials, life expectancy of the bed, potential plugging of the screen, solution-channeling through the bed, limitations in removing organic contaminants or polar organic compounds, and certain hazards associated with the operation.

Factors contributing to total cost are dewatering, and trench support and shoring. As a rule of thumb, for trenches up to 20 ft deep, permeable treatment beds compare favorably with the cost of cutoff walls and separation/extraction techniques. For deeper excavations the cost of temporary support escalates rapidly and deters an economical application in favor of other types of screens.

10-7 CEMENT-BENTONITE CUTOFFS (SOLIDIFIED WALLS)

Characteristics

These cutoffs, built according to procedures discussed in Section 10-4, are cement-bentonite mixes that contain no aggregate except a soil fraction mixed with the slurry in the process of excavation. The hardened wall exhibits flexibility that allows it to withstand differential horizontal and vertical movement without failing. In built-up areas the choice is attractive because it minimizes the need for materials, and the wall is completed with the shortest sequence of operations.

Once in the trench, the cement-bentonite mix becomes the final material and attains sufficient strength but without loss of elasticity. The quoted strength range is between 10 and 30 kg/cm^2, or 100 and 400 lb/in.^{2} (ICOS, 1973), and the modulus of elasticity between 200 and 500 kg/cm^2 (2800 to 7000 lb/in.^{2}). Apparently no set relation can be established between strength and modulus. If all relevant factors are introduced, such as bentonite content and incidental soil fraction, broad variations in elastic behavior should be expected. The material remains integral and free of cracks as long as it is buried in the ground and in a moist environment.

Installation and Panel Sequence The procedure described in Section 10-4 is illustrated in Figure 10-19. Element A is excavated in one pass and simultaneously filled with cement-bentonite slurry. The equipment is then moved to position B to repeat the same operation, and then returns to a position between A and B to complete the intervening tongue. As long as the mix in A is not excessively hard and the mix in B is stiff enough to be stable, the three elements blend together and form a jointless curtain. Likewise, the sequence is repeated with elements C, D, and so on.

Proportioning the Mix

Figure 10-20*a* shows a simple diagram that gives an indication of the principal regions in the bentonite-cement-water system. There is a limited range of compositions that will produce a satisfactory mix. If there is excess of bentonite or cement, the slurry will be too thick and unworkable. If there is insufficient cement, no set will occur, and if there is insufficient bentonite, the slurry will settle and release bleed water.

A typical mix contains: bentonite, 3 to 6 percent; cement, 15 to 20 percent; and aggregate mixed with the slurry, 5 to 10 percent. Jefferis (1983) gives the following material quantities required to make 1 m^3 of slurry: bentonite, 40 to 100 kg; cement, 80 to 350 kg; water, 850 to 950 kg. A retarder of the lignosulfite group is also added (usually 0.1 percent) to control the curing process and mainly to extend the curing time.

The limits on the quantity of bentonite are related to its quality, and define the practical range. If ordinary Portland cement is used, set will seldom occur at cement concentrations less than 100 kg/m^3, but this can be lower if sand and soil aggregate is added to strengthen the mix. At cement contents higher than 350 kg/m^3 the slurry normally will be too thick and set too rapidly to be of use. Also, in this range the mix will exhibit very high strength and brittle behavior, and thus be unable to accommodate ground movement. Interestingly, high cement ratio does not necessarily yield low permeability or improved durability.

Jefferis (1983) has introduced cutoff systems where part of the cement content is replaced by ground blast furnace slag. This replacement makes the slurry workable by extending the setting time but without adversely affecting the properties of the set mix. This addition produces higher strength and lower permeability, but it tends to increase brittleness.

The addition of fly ash to replace part of the cement will allow better control of

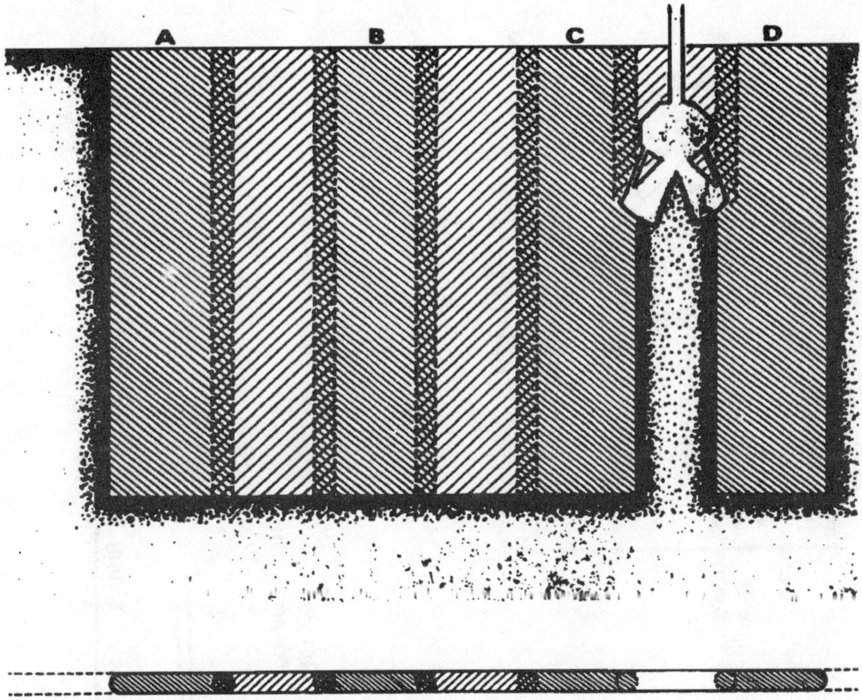

Figure 10-19 Construction sequence of a continuous flexible wall.

the setting time, but the resulting set may be too slow. Fly ash affects permeability to a lesser degree, but helps resist disintegration more effectively. It may therefore improve the resistance of the mix to aggressive chemicals.

Properties of Mix

From the foregoing composition it follows that the density (weight) of the set mix is considerably less than in the earth cutoff or for an all-concrete structure. For a water content 60 to 65 percent, the specific gravity of the mix is 1.25 to 1.30, corresponding to a density 80 lb/ft^3 or slightly higher.

Jefferis (1983) gives the properties of a typical cement-bentonite slurry as supplied to the trench, and on completion of the operation two days after excavation. These are as follows

	As supplied	After excavation
Density (kg/m^3)	1080	1250
Marsh viscosity (sec)	30–40	No flow
Apparent viscosity (cP)	15	> 130
Plastic viscosity (cP)	9	30–50
Gel strength 10 sec/10 min (Pa)	15/18	20/22
pH	12.5	12.5

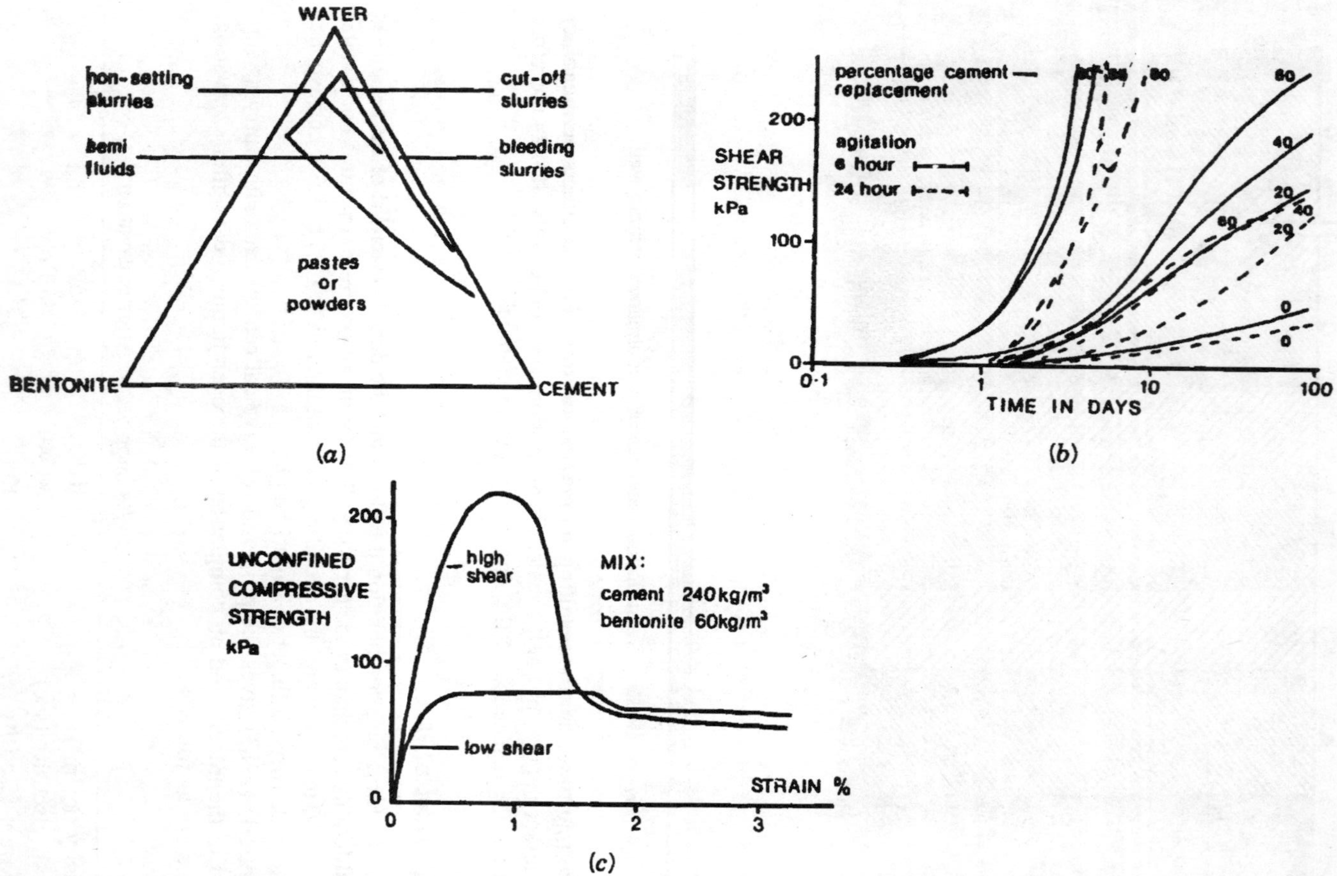

Figure 10-20 (*a*) Principal regions in the bentonite-cement-water system. (*b*) Shear strength development in slurries containing slag. (*c*) Effect of shear rate during mixing on behavior of set slurry.

Density Sufficient density ensures the hydrostatic force necessary to keep the trench stable (Xanthakos, 1979). Low density is unlikely in this case, since the addition of cement and soil from the excavation will raise it to adequate levels. Although the viscosity is raised for the mix in the trench, the gel strength remains relatively low, and this prevents high soil intake that would increase the density beyond the range 1300 kg/m^3 (80 lb/ft^3). For self-hardening slurry where no displacement is necessary, an upper limit on density does not need to be specified. For replacement slurries that must displace the excavation slurry, greater difference between the two densities means more complete displacement.

Rheological Characteristics The foregoing data show that toward the end of excavation the slurry becomes too thick for any flow in the Marsh funnel, and its apparent viscosity cannot even be measured. As long as the gel strength is low, the slurry remains workable and can be handled by the excavating equipment.

The 10-sec and 10-min gel strength show much less variation with time until the mix begins to set. Replacement of cement by fly ash or blast furnace slag has a minor effect on viscosity and gel strength, but the slurry tends to respond as pure bentonite slurry enhancing thixotropic behavior.

An interesting application is reported by Dutko et al. (1991) in conjunction with the use of fly ash in a cement-bentonite cutoff built to remediate seepage at an embankment dam in Ohio. Design and preconstruction evaluation indicated the potential of large slurry loss during trenching in the embankment (originally constructed of compacted boiler slag). This loss could result in trench instability with corresponding effects on the dam integrity. Laboratory tests were carried out using various cement-bentonite fly ash mixes, focusing on permeability and strength tests as well as on flow property tests. The results confirmed that the presence of fly ash in the mix would be beneficial in reducing slurry loss by improving the sealing process in the trench walls through an enhanced thixotropoic behavior.

pH For ordinary cement-bentonite slurries the pH is in the range 12 to 13, and this is the result of the cement presence in the mix. This is of no practical importance, except that for replacement slurries the excavation slurry should have a pH that does not exceed these values.

Filter Loss This is not a direct measure of the seal formation that normally is necessary to prevent excessive slurry loss toward the ground. Fluid losses in the trench for cement-bentonite slurries are very high in comparison with conventional slurries. Although this happens because the slurry enters the trench in a flocculated state, it still is capable of forming a filter cake and eventually of rheological blocking. Filter cake permeability calculated from filter loss tests indicates a probable range 1 to 4 $\times$ 10^{-6} cm/sec, which is sufficient for stability.

Bleeding Since the water-cement ratio is typically in the range 11:1 to 3:1, without bentonite the cement slurry will undergo rapid settlement of cement particles out of the suspension to produce free water. The presence of bentonite deters this tendency but some bleeding always occurs.

The bentonite concentration is important in controlling bleeding, as is the sequence and method of mixing. The bentonite should be prehydrated for at least 12 hr prior to mixing. Bleeding is also reduced by the use of high shear mixers to combine the cement and bentonite, and by longer mixing times. Continuous agitation reduces bleeding by keeping cement particles in suspension. A slurry properly designed should show a bleeding less than 1 percent.

Control of Setting Time

Figure 10-20*b* presents typical shear strength development curves for a mix containing 150 kg/m^3 cement and 50 kg/m^3 bentonite, with various percentages of cement replaced by slag. These data are for slurries setting after 6 and 24 hr of agitation. Continuous agitation reduces shear strength at all ages and for all stages of cement replacement. After 48 hr of agitation, slurries with small percentage of slag lose almost the entire setting ability. With high slag percentage, considerable setting potential (about 70 percent) still remains after seven days of agitation.

For unfinished excavations, keeping the slurry fluid and workable during operations is part of the problem. If the slurry is left undisturbed overnight, setting will begin, and as the second excavation shift starts the mix may be too thick and unworkable.

Agents added to the slurry to retard premature stiffening do not always produce the desired effect, particularly after the first 24-hr period. Conversely, certain retarders may even cause acceleration at this stage. Better results are obtained if the cement is prewet with part of the mixing water under a water-cement ratio 0.5, and then allowed to react before mixing it with bentonite slurry. In laboratory tests, sugar derivatives have been found to be the most effective retarders for this application, but many cause adverse effects on final strength.

Where suitable retarders are not available and long excavation periods are unavoidable, consideration should be given to some cement replacement by fly ash or slag. This should not inhibit shear strength development during the first 48 hr, but would help to ensure that if the slurry must be agitated for extended time, set will eventually occur.

Mixing Procedures

In general, a uniform slurry is obtained if the cement is added to the prehydrated bentonite while the latter is sheared in a mixer. Bentonite slurry should never be added to cement since this procedure will seldom produce a homogeneous mix. The level of shear attained in the mixer affects the properties of the mix, as shown in Figure 10-20*c*. The weakest mix is prepared in a low shear mixer working at 50 rpm. The stronger mix has the same proportions but is prepared in a high shear mixer running at 4000 rpm (Jefferis, 1983). The lower strength of the low shear mix is probably associated with the small modulus of undispersed cement, which acts as stress raiser and causes the mix to develop a lower average strength.

Mixes prepared in low shear conditions also exhibit higher bleeding, and in the set form higher permeability. However, high shear mixes are more sensitive to drying.

Strength and Permeability of Set Mix

The final properties of cement-bentonite mixes are largely controlled by the percentage of constituent materials. A broad range is obtained merely by varying the percentage of cement, bentonite, fly ash, and ground slag. If strength and elasticity parameters must be defined in a close range, the recommendation is to prepare trial batches using actual mix composition and carry out suitable tests to determine the properties of the set material.

Jefferis (1983) quotes typical 90-day shear strengths for cement-bentonite mixes containing 50 kg/m^3 of British bentonite and different levels of ordinary Portland cement. For cement content (kg/m^3) 100, 150, 200, and 230, the 90-day shear strength (kPa) is 6, 49, 80, and 180, respectively. Below 100 kg/m^3 cement content, set is unlikely to occur, and at higher cement content strength development increases and accelerates rapidly.

In practice, the goal is to produce a mix that has sufficiently low strength and elastic modulus to match the characteristics of the ground. In most cases, a typical requirement is that at 90 days the set mix should develop a strain at least 5 percent without cracking under a deviator stress 125 to 150 kPa. Field performance indicates that the mix can withstand large strains. Samples tested under confined drained conditions (usually most compatible with a cutoff) show that the mix may withstand strains up to 15 percent without cracking.

A main advantage of cement-bentonite mixes is their ability to attain low permeability. Quoted values typically are less than 10^{-6} cm/sec. Mixes made from ordinary Portland cement and fly ash generally have permeabilities 1 to 5 $\times$ 10^{-6} cm/sec at 7 days, dropping to 5 to 10 $\times$ 10^{-7} cm/sec at 90 days. If slag replacements are used ($>$ 30 percent), the permeability should be less than 10^{-7} cm/sec at 7 days and close to 10^{-9} cm/sec at 90 days, under a confining pressure 40 kPa.

Both confining pressure and age affect permeability, and this is amplified in weak mixes since these consolidate easily. Resistance to consolidation may account for observed increases (rather than decreases) in permeability with increasing cement content. Permeabilities should be quoted or specified in terms of a confining pressure.

Resistance to Deterioration

Permeability under High Gradients The functional ability of a cutoff often is assessed by monitoring its permeability under extended application of high gradients, and with actual or simulated groundwater. In these programs the gradients approach 500. Normally, there should be no increase in permeability with time unless the groundwater is chemically aggressive.

Drying In its final set, a cement-bentonite mix contains an average of 65 percent water. If dehydration is allowed to occur, substantial shrinkage will result. In the ground drying conditions are unlikely since water is held in the slurry by capillary forces unless the soil moisture is very low. Surface protection is, however, needed, and is provided in the form of a clay layer. The tendency to shrink is inhibited if graded fill is added to the mix to form a plastic concrete.

Chemical Attack Chemicals that have a damaging effect on cement are likely to limit the application of cement-bentonite mixes. Strong acids can cause rapid and complete disintegration, whereas alkalis and calcium or magnesium chlorides have only minor effects. In particular, calcium sulfate and magnesium sulfate cause rapid disintegration of unconfined materials. Slag replacement limits this attack slightly, but fly ash replacement reduces it significantly.

When confined samples are permeated by sulfate waters, no disruption occurs even at low confining pressures (20 kPa). The ability of the mix to withstand large strains without failure is characterized by its resistance to cracking under chemical attack causing local distortions, provided some confining pressure is available. During permeation with sulfates, a modest increase in permeability results and is associated with the effect on the microstructure of the cement.

Since different cements and bentonites have a different reaction to these effects, the exact chemistry of groundwater is important. If the risk of chemical attack exists, the bentonite concentration should be increased. Interestingly, a mix with permeability greater than 10^{-5} cm/sec will allow a continuous flow through the entire cutoff material, thus increasing the risk of erosion.

The Slurry Replacement Method

Because of the high viscosity of cement-bentonite mixes, a considerable quantity of slurry is lost either by escaping to the ground or by sticking to the cuttings. Slurry lost in this manner has been as high as 100 percent of the trench volume. If long interruptions are anticipated and the effect of retarders on delaying initial setting is uncertain, a replacement slurry should be considered. The two-slurry application offers the following advantages: (a) it eliminates strict adherence to schedule and thus allows better timing; (b) it is compatible with the development of material properties over a wider range; (c) soil intake from the excavation is better controlled; and (d) the initial excavation slurry can be prepared to ensure stable conditions during trenching.

The operation is facilitated by combining a suitable excavating equipment with an appropriate flow jet mixer. This arrangement allows the mixing stage of bentonite suspensions or cement-bentonite slurries to be completed simply by changing the switch on the same equipment, and also produces the desired uniform difference in density.

Design Principles

Stability of Filter Cake In addition to the hydrostatic thrust of the slurry, stability is helped by its shear strength. If the mix is placed in a single stage, the colloidal stability of the slurry in liquid form is controlled mainly by the effect of cement. This agent imparts to the slurry an uncertain action characterized by strong interaction with bentonite particles. It is likely that the slurry enters the trench in a flocculated state disrupted only by the agitation of the operation. Filtration of bentonite eventually is combined with deposited cement particles to form a cake, but these deposits are voluminous and porous, allowing the slurry to escape until rheological blocking stops the flow. The presence of any cake is thus inconsequential and irrelevant to the composite cutoff permeability.

Expected Performance When in place the material will exhibit expanding tendencies that, combined with dehydration, can lead to shrinkage cracks. As a constituent of a cement mix (dilute), the bentonite serves to produce a stable suspension, first by reducing settlement of cement particles during the hardening process, and then by increasing both the viscosity and the cohesion of the mix. For maximum effect, the bentonite should be fully hydrated.

In the original solidified state the mix has ample strength and is free of textural irregularities. As long as it is confined in the trench and in a continuous moist environment below ground level, the cutoff should be expected to perform satisfactorily. Field observations confirm that cutoffs of cement-bentonite blends respond elastically to excessive stress levels but without developing cracks typically observed in normal concrete. The mix requires, however, a high moisture content to retain its strength and elastic properties, but this behavior is yet to be articulated and better understood.

In situations that involve clean water the mix is indefinitely stable, and no reduction in permeability or overall performance should be expected. However, cement is known to perform as a poor and unstable agent in the presence of acids and sulfates.

Microstructure Analysis Figure 10-21 shows a photograph at microscopic scale of a cement-bentonite sample. The surface details of the vacuum-dry microstructure are presented at a magnification of 8000. The compact form should not be interpreted to mean material of low density; nonetheless this is true in this case.

Examination of three-dimensional assemblies of mineral particles within the 0.01- to 10-μm range in bentonite and cement-bentonite slurries has shown that particles of this size dominate the properties of the material to an extent of practical significance. The photograph of Figure 10-21 suggests that the cement-bentonite slurry provides an ideal environment for the growth of stable clusters and flocks of fibrous calcium silicate hydrate. These spiky crystals produce a card-house structure that apparently prevents the stacks of flat, platelike clay particles with wet surfaces from sliding apart to form a disconnected matrix.

The stereoscan picture also suggests that transformation of cement particles into

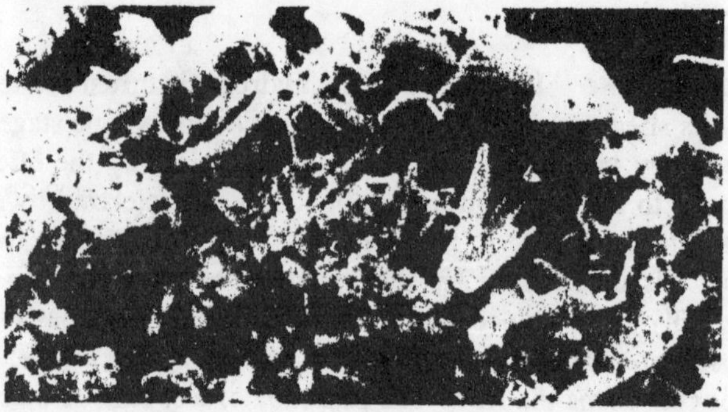

Figure 10-21 Photograph showing microstructure analysis of a cement-bentonite mix at a magnification of 8000 (ICOS).

clusters of fully developed spikes takes place under favorable conditions, and the structural skeleton of the thixotropic gel is arranged under minimum restraint. Once the skeleton of the products of hydration is in place, the gel can be assumed to be effectively trapped (ICOS, 1973). The strength of the composite system is thus much greater than the shear strength of bentonite gel and the strength of the spikes alone.

This interaction disappears, however, if the supporting action of bentonite is lost because the water was allowed to evaporate. The appearance of random large cracks on the surface indicates reversal from swelling to shrinking under the influence of progressive dehydration, and this process may lead to complete disintegration of the mix if dehydration is allowed to continue.

Experience with cement-bentonite mixes in cutoff walls thus far indicates that their properties display rather erratic variations, particularly under the combined effect of temperature and moisture. We can make, however, the following comments for this type of work: (a) higher water-cement ratios produce mixes that are weaker but of higher plasticity; (b) the presence of more soil and bentonite generally will reduce strength; (c) plasticity is decreased but strength is improved with time; and (d) neither strength nor plasticity show a definite correlation with fly ash content or bentonite concentration.

Examples

Solidified walls, also called grout walls, are widely used to protect urban excavations, without concern for the water table, and in hydraulic works. Applications are reported also for embankment dams.

An example is the Upper Peirce dam in Singapore, where a cutoff was installed directly beneath the main embankment (Little, 1974). When the fill in the rolled clay core reached a certain level, a 45-ft-deep slurry trench was excavated to the level of decomposed granite. The trench was filled with a cement-bentonite slurry that included a retarder to delay premature setting. The cutoff was processed in panels without stop-end tubes, but a key was formed by cutting into the preceding panel. Table 10-11 shows data for the mix. Evidently this is low-weight material, specific gravity 1.25 or unit weight 78 lb/ft^3, and has constituent proportions well within the range defined in the foregoing sections.

The use of fly ash in the mix (10.1 percent by weight) was necessary to improve its resistance to sulfate attack. By analogy, the same improvement is noted in mass or in plastic concrete, and covers a broad range of effects such as increased impermeability, reduced alkali-aggregate expansion, and better workability. Fly ash was used in the plastic concrete for the cutoff of Withens Clough dam, summarized in

TABLE 10-11 Mix Characteristics for the Cutoff of Upper Peirce Dam, Singapore

		Upper Peirce	
Weight per unit volume		1252.3/kg/m^3	100%
Lignosulfite		1.3	0.1
Bentonite	(B)	32	2.6
Water	(w)	841	67.1
Cement	(c)	252	20.1
Flyash	(Fa)	126	10.1
Aggregate	(A)	—	—
w/c		3.35	
w/solid		2.06	
w/(c + Fa)		2.23	
w/(c + Fa/2)		2.67	
B/(c + Fa)		0.09	
B/(c + Fa/2)		0.10	
Permeability k (cm/sec)		1.26×10^{-8}–2.07×10^{-8}	
Specimen diameter		76 mm	
Average axial strain at peak, ϵ_p		ϵ_p, %	σ_3, kN/m^2
14 days		—	—
14 days		—	—
28 days		—	—
28 days		—	—
90 days		3.38	197
90 days		3.83	296
Average peak deviator stress, q		q, kN/m^2	σ_3, kN/m^2
14 days		—	—
14 days		—	—
28 days		—	—
28 days		—	—
90 days		1808	197
90 days		2286	296

Source: From Little (1974).

TABLE 10-12 Fly Ash–Plastic Concrete Cutoff

Materials	Withens Clough Dam kg/m³	%
Weight per unit volume	1845	100
Bentonite (B)	25	1.3
Water (w)	409	22.3
Cement (c)	61	3.3
Fly ash (Fa)	300	16.3
Aggregate (A)	1050	56.9
w/c	6.7	
w/solids	0.28	
w/(c + Fa)	1.13	
B/(c + Fa)	0.07	

Source: From Little (1974).

Table 10-12. The water retained in this reservoir has a pH 3.8 (Little, 1974), suggesting aggressive conditions that could cause considerable damage to an unprotected cutoff.

A cement-bentonite wall was used for groundwater control at a construction site in Westmont, Illinois (Xanthakos, 1976a). The cutoff is 18 in. wide and extends into

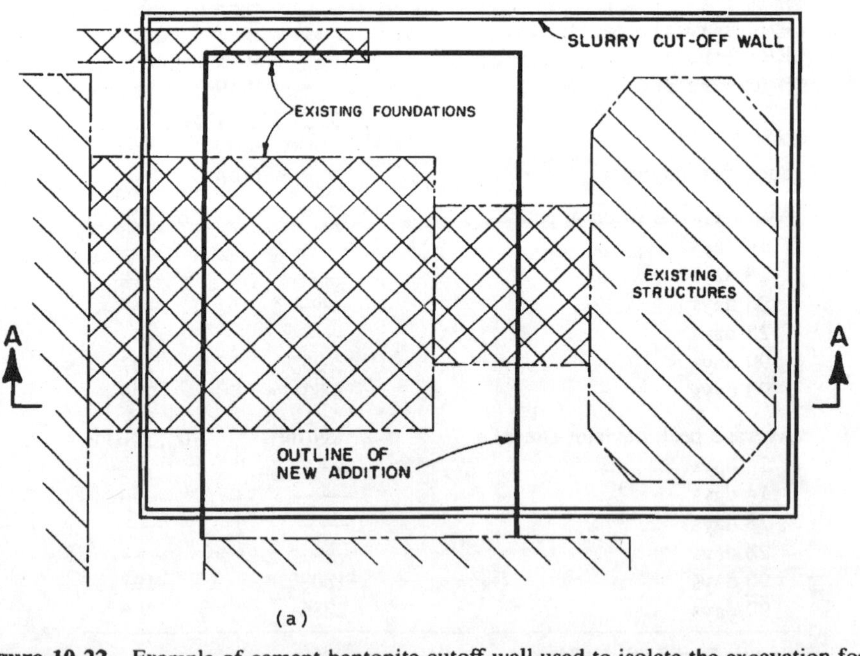

(a)

Figure 10-22 Example of cement-bentonite cutoff wall used to isolate the excavation for a new building. (*a*) Site plan. (*b*) Section of area to be protected. (*c*) Detail of cutoff wall.

a clay formation. Although 20-ft panels were initially specified, the contractor was allowed to construct the wall continuously after demonstrating that this would not affect trench stability.

The mix contained bentonite, 3 percent; cement, 17 percent; and water, 80 percent. Lignosulfite was also added, 0.1 percent by weight. Mixing with soil was allowed but not to exceed 10 percent. The cutoff was strong enough to withstand consolidation stresses caused by new construction, and since installation it has reduced water seepage into the basement sump by nearly 90 percent.

Urban Application Figures 10-22*a* and *b* show plan and section of a site with existing buildings and facilities. A new addition to the industrial plant requires temporary excavation. Among the associated problems is a soil profile of about 8 ft of rubble upper fill overlying 8 ft of sand over clay. Past construction has left massive concrete foundations that cover a major fraction of the total site, but the

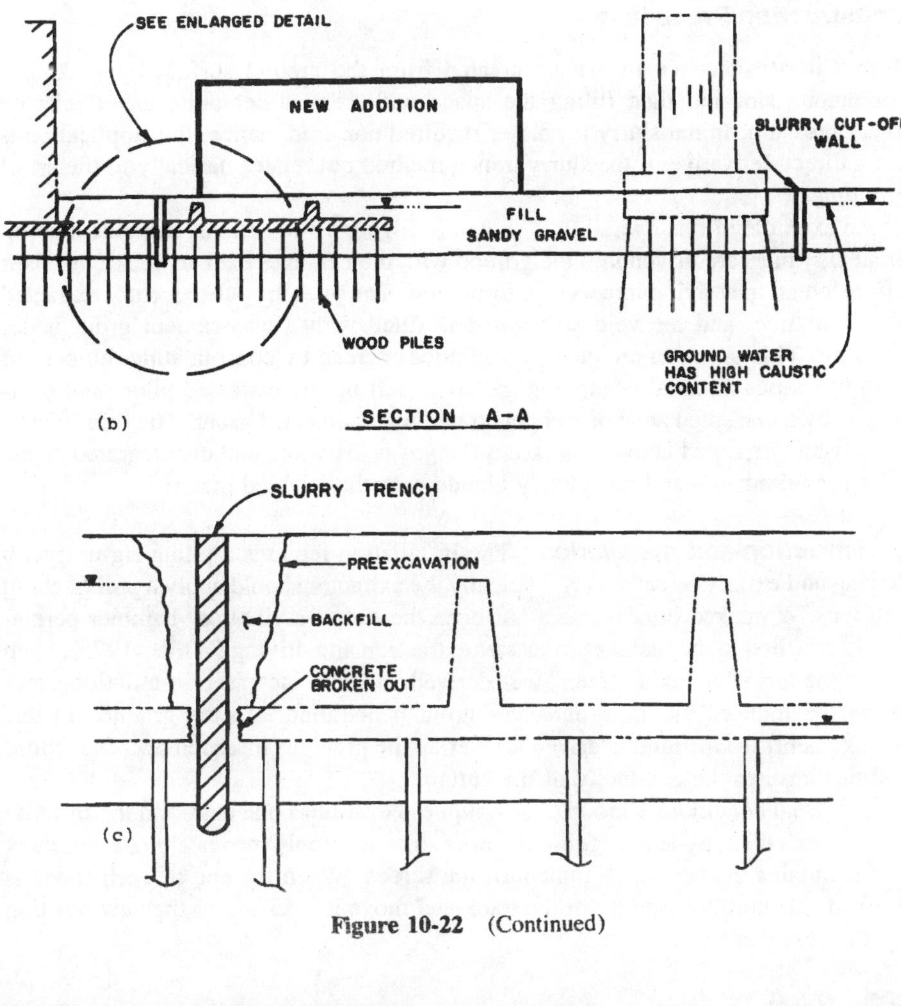

Figure 10-22 (Continued)

plan is to use the old foundations to the extent possible and without restricting the feasibility of new foundations. The groundwater is heavily polluted with caustic wastes, but lowering the water table would most likely affect the integrity of existing timber piles. Any water pumped from the ground would have to be treated before release. These difficulties are further compounded by restricted access and limited construction area.

In a practical scenario, the foregoing considerations would preclude an earth cutoff because of unavailability of suitable in situ backfill materials, stability problems, and constructability limitations. Conversely, a cement-bentonite cutoff was found chemically compatible with the high pH caustic groundwater, and was the obvious choice (Ryan, 1980a). As completed, the trench wall is 17 ft deep, and after the necessary preparation work it required 10 days to construct.

10-8 INJECTED SCREENS

Construction Procedure

A thin impervious screen can be created from the ground surface by making a continuous slot and then filling the space with cement-bentonite grout injected under pressure. Initial slurry is neither required nor used; hence, this application is not a direct derivative of the slurry-trench method but relates basically to the art of grouting.

An example of this operation is shown in Figure 10-23. In this instance a group of steel H piles are driven into the ground with their flanges back to back until their tips reach an underlying impervious formation. The piles are subsequently extracted one at a time, and the void so formed is filled with a clay-cement grout under pressure. This injection produces a continuous screen by consolidating the core of the impervious material occupying the space left by the extracted piles, and overlapped by a cemented zone of soil penetrated by the injected grout. The thin fillet of earth usually trapped between adjacent flanges is disrupted and disintegrated as the piles are withdrawn and completely blends with the injected material.

Construction and Installation The installation requires a piling rig unit with driving and extraction capability. Typically the extractor should apply a pull of about 150 tons. A marked improvement has been the use of a vibratory hammer permanently attached to the mandrel for easier extraction and driving (White, 1975), from which the term "vibrated screen" was derived. Besides increased installation capability, the induced vibration enhances grout penetration into the ground. In this arrangement, the mandrel is driven to overlap the previous insertion and extraction, and this ensures the continuity of the curtain.

In normal conditions a group of seven piles constitutes one unit, and the installation is carried out by leapfrogging the piles. The assembly moves along a rail track that maintains the correct alignment of the screen. When the end of each travel is reached, the unit is raised to lift the track and move it sideways to the new position for the next run.

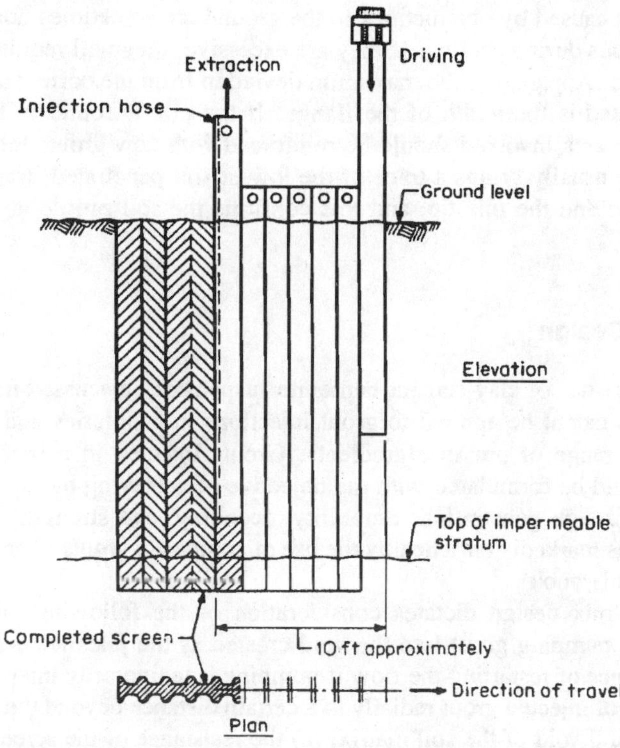

Figure 10-23 Method of installation of an injected-grout screen.

The piles should be heavy steel sections of sufficient flange and web thickness to resist distortion and ensure straight driving. A grout pipe is attached to the web. The selection of appropriate steel sections should take into account the depth and functional requirements of the screen as well as the driving conditions. In relatively soft and loose soils it is possible to drive 9-in.-deep sections without losing the overlapping. In dense sand and gravel the piles should be at least 15 in. deep, reinforced with cover plates for the entire length, and have the tops protected against repeated impact blows. The practical attainable depth for a continuous screen is 33 to 35 ft, but with special modifications of the equipment it can be increased to about 50 ft.

For one run all the piles in the group are installed by driving, and with the flanges back-to-back. The work then continues by simultaneously driving and pulling the piles so that for every pile extracted one pile is driven. The extraction begins by raising the pile 6 in. to create a void in the soil. With the grout hose connected, grout is injected under pressure until a surge of materials emerges at the top of the adjacent previously filled pile space. At this stage the lifting jacks and the injection pump are coordinated to work together, and by successive lifts the pile is withdrawn from the ground allowing the void to be filled with grout. Grout already placed should not be disturbed by vibration effects associated with pile driving at the far end, and a practical rule is to keep this distance to no less than 10 ft (3 m), which determines the number of pile units for the usual application.

Deviations caused by obstructions in the ground are sometimes unavoidable and become obvious during driving. If they are excessive, they will require supplementary injections. Apparently, the maximum deviation from the correct alignment that can be tolerated is the width of the flange. If the pile is found to be bent when extracted, the area involved should be reinforced with new grout. Interestingly, an extracted pile usually brings a trace of the lowest soil penetrated, trapped between the grout pipe and the pile tip, and this confirms the soil profile at the tip of the screen.

Grout Mix Design

The characteristics of clay-cement-bentonite suspensions discussed in Section 10-4 may to some extent be applied to grout injections in capillaries and previous soil over a wide range of pressure gradients. Grouts injected in a preformed screen likewise should be formulated with the objective of achieving the optimum combination of thorough penetration, durability, economy, and strength. The range of applications is markedly enriched by the use of composite grouts (discussed in other chapters of this book).

Optimum mix design dictates consideration of the following factors: (a) the feasibility of pumping grout into the void created as the piles are withdrawn, and the convenience of restarting the flow if pumping is temporarily interrupted; (b) the penetrability of injected grout radially to a certain distance beyond the artificial void and into natural void of the soil matrix; (c) the resistance of the screen to displacement under the service gradient; and (d) the reduction in permeability stipulated by the design.

These requirements are essentially satisfied if the clay-cement-water mix develops good penetrating characteristics, sets to a relatively plastic curtain that is resistant to softening, attains good set strength, and remains fairly watertight.

Proportioning of Materials Theoretically a cement-based grout is specified in terms of the relative proportions of water and cement. Quite frequently, the volumetric ratio is found convenient considering the fact that the dry volume of one bag of cement is about 1 ft^3, and the same bag occupies approximately $\frac{1}{2}$ ft^3 when added to water. For full hydration the w/c ratio is approximately 0.3:1 by weight, or 0.45:1 by volume, and water added in excess of this amount is intended to increase the flowability of the mix and improve the injection process. In the past, thin grouts were thought to improve penetration, but recent tests show that they also may exhibit numerous bleed paths when set. The trend is therefore toward the use of relatively thick grouts (Houlsby, 1982, 1985, 1990; Bozovic, 1985).

Among natural clays, marine or alluvial deposits are preferred since they are relatively free of coarse particles and have an Atterberg limit more than 60. The clay fraction of clay-cement grouts forms a gel that stabilizes the cement by reducing bleed but performs no significant chemical function in the final setting process. The clay should be credited for delaying setting and for a lower strength in the set grouts, although this strength clearly is far greater than is required to resist pressure gradients in normal water retention schemes.

In a typical grout mix, the blend contains 2 parts of clay to 1 part of cement, and the water is balanced to keep the mix flowable. The density of this material is slightly in excess of 100 lb/ft^3 (1.60 g/cm^3), and the 28-day crushing strength may range from 100 to 250 lb/in.^{2} (7 to 17.5 kg/cm^2).

The presence of bentonite in a cement grout is effective in reducing sedimentation of cement particles (see also Section 10-4), but careful proportioning of the cement and bentonite is essential to obtain the desired grout mixture. The data summarized in Figure 10-13 may be supplemented by Table 10-13, showing results obtained through the use of bentonite hydrated for two hours before mixing (Deere, 1982).

Flow Properties of Grout

For a relatively thin grout of clay and cement, the most relevant property to flow is the initial gel strength τ_s. Ignoring the curved portion of the flow diagram, the initial gel strength can be taken as the yield stress τ_0, and the flow is regarded as ideal plastic flow. In this case, the flow in a capillary of radius R is expressed in terms of the pressure gradient i_p as

$$\tau_s = \frac{R}{2} i_p \tag{10-9}$$

Actually, clay pastes and cement mixes used for this application are hardly expressed by Eq. (10-9), and their behavior is better described by Figure 10-24 (Marsland and Loudon, 1963). When the shear stress reaches the value τ_s, shear failure occurs near the wall of the capillary and the suspension moves forward as a plug (stage II). As the pressure gradient continues to increase, the diameter of the solid plug becomes progressively smaller (stage III) until the entire material in the capillary flows like a viscous fluid. The rate of flow increases thereafter linearly with the pressure gradient (stage IV).

Recent research by Lau and Crawford (1986) correlates the rheological properties with the cement type and the w/c ratio, and articulates several modes of

TABLE 10-13 Effect of Bentonite on Bleed

Grout Mix Water/Cement (weight)	Percentage of Sedimentation		
	0% Bentonite	2% Bentonite	4% Bentonite
3/1	62	40	22
2/1	52	20	6
1/1	16	3	1
0.66/1	5	0	0

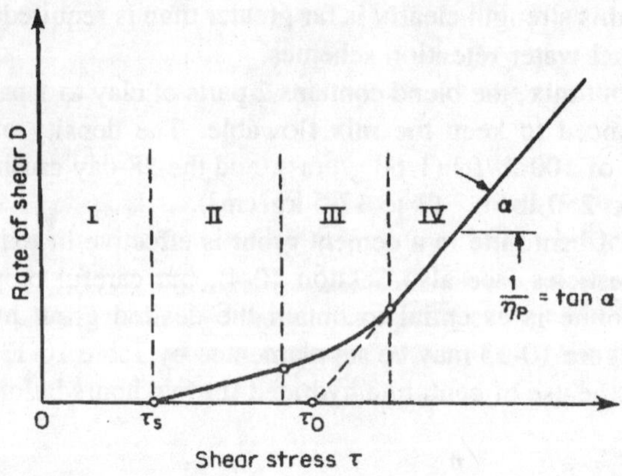

Figure 10-24 Flow curve for clay pastes and viscous grouts.

behavior. Thus for high early strength cement, a mix with w/c ≥ 2.5:1 (by weight) can be modeled as a Newtonian fluid; mixes with w/c between 2.5:1 and 0.75:1 behave essentially as Bingham fluids; and mixes thicker than 0.75:1 respond pseudoplastically. In any case the rheological properties can be markedly altered by the use of superplasticizers.

In the field the injection can be complicated by thixotropic thickening and changes in the gel structure occurring if the grout is allowed to rest. Tests of bentonite-cement suspensions reveal distinct anomalies in the pressure/rate of flow relationship during injection and with the pressure drop occurring at constant flow volume. Although there exists a minimum pressure that must be exceeded for flow to begin, there also exist distinct areas where an inverse relationship governs the pressure drop and volume flow rate. The same conclusion is also reached if the flow test is carried out at constant pressure and the volume flow rate is allowed to reach a steady state.

For impervious screens installed by this process the flow behavior is important in the low rates of shear. This is because the injection through the grout pipe must be correlated with the rate of pile withdrawal without damaging the slot, and usually this implies reduced shear rates. Clay-cement mixes behave as Bingham bodies only at low concentrations, and in this range the rate of shear can be related to apparent viscosity and yield stress. However, for mixes injected in preformed slots the clay-cement concentration is high enough to preclude Bingham flow, and also high enough to prompt rapid increases in apparent viscosity and yield stress. This complicates further the prediction of flow in the screen, and suggests the importance of capillary viscometer tests to obtain basic data before designing the mix.

Penetrability and Strength of Grout

Lateral penetration of the soil should always be expected, but the amount of fingering occurring in this fashion may be restricted by the high viscosity of the material.

Conversely, the available finite initial shear strength may lead to preferential filling of high-porosity areas in more erratic formations. Penetrability is limited if natural clays and other minerals, having larger particles in the final matrix, are used instead of purified bentonite.

If the grout contains even a small proportion of coarse particles, it may form a tight filter cake on the soil face near the injection source, and this will limit penetration. When the soil surrounding the slotted voids is steeply graded, particles in the grout larger than one-tenth of the soil particles tend to be trapped, thereby making penetration beyond the face impracticable. Since both permeability and critical filtering size depend on the average diameter of pore channels, the latter is correlated with the size of the largest particles in the grout. Clay-cement suspensions that contain many particles as large as 100 μm begin to form filter cakes in soils with permeability as high as 1/10 cm/sec (Scott, 1963). These can clog the pores and prevent further penetration.

Although the flow characteristics of unstable grout through a natural fracture system are basically dissimilar to the flow in a preformed artificial slot, essential physical similarities suggest that penetrability is affected by the viscosity, cohesion, specific gravity, tendency to settle, and tortuosity. The flow velocity as a function of the injection pressure is also relevant as it relates to the settling velocity of grout particles. Any tendency of the clay-cement suspension to form aggregated particles can be combined with the tendency to settle and block narrow paths and fractures.

Penetrability is thus complex. Studies of non-Newtonian fluids with fixed shear strength or for strength increasing with time have been made by Greenwood and Raffle (1961), Scott (1963), Marsland and Loudon (1963), and others (Houlsby, 1990). The results from these investigations should be used with caution since the clay-cement grouts typically used in these screens are structured to form a shear strength that is characteristic for a weak solid.

The strength and resistance to displacement developed by the set grout are favorable factors. Nonetheless, the screen can tolerate a relatively small differential ground movement. The tendency of the mix to extrude under a pressure head is resisted by the shear strength of the material over the internal surface of each void passage. A shear strength 1 to 2 $lb/in.^2$ (0.07 to 0.14 kg/cm^2) can resist hydraulic gradients close to 100 in soils with grain size 1 in. (0.54 cm). Resistance to displacement is further enhanced by the cohesion developed in granular soil penetrated by grout. The soil in this instance forms hard agglomerates that offer considerable resistance to deformation by shear.

Applications

Between 1956 and 1974 more than 5 million m^2 (50 million ft^2) of injected screens were built in Europe, by the so-called ETF process. The usual depth of these screens is from 16 to 30 ft, but machines have been developed that can operate at depths of 15 to 25 m (50 to 80 ft).

Based on field measurements and relevant data, about 1.3 ft^3 of void space produces 1 yd^2 of finished screen (9 ft^2) or 0.15 ft^3/ft^2 of screen. From actual grout take the grout volume is 3.8 $ft^3/yard^2$ of screen, or 0.42 ft^3/ft^2 of screen, including incidental penetration into the soil. About one- to two-tenths of that is used to fill the gap between flanges of adjacent piles. This consumption indicates that in average soil conditions the grout penetrates the ground by. several inches.

Figure 10-25 shows an exposed finished screen. Much of the soil had to be removed by pneumatic hammer in order to reveal the shape of the screen as formed. The loose soil fillet between steel sections is disintegrated and replaced by grout. The screen is 28 ft deep, and was produced by a series of seven H piles. The mix consists of clay, silt, and cement, and required an injection pressure of 20 lb/in.^{2} through a 1-in.-diameter pipe.

Applications are reported for sealing permeable layers to prevent inflow or underflow of groundwater that may cause ground instability, excessive pumping, or water loss. Natural candidate sites are therefore dikes, levees, dams, irrigation canals, cofferdams, and underground excavations. The screen may also be considered an underground barrier against flow of pollutants for major sources of groundwater pollution such as agricultural runoff, sewage, seawater intrusion, industrial wastes, petroleum, landfills and dumps, mining wastes, and certain chemical stockpiles.

Figure 10-25 Exposed impervious curtain built by injecting grout in preformed slots. Note the extent of penetration into the surrounding soil.

10-9 IMPERMEABLE MEMBRANES

The sensitivity of plastic concrete and cement-bentonite walls to certain pollutant attacks at contaminated sites can be remedied if the permanent encapsulation of the ground is ensured by a double-protection system. In addition to the basic slurry trench cutoff, the system may include a synthetic membrane inserted continuously along the basic center of the trench and before the mix is allowed to harden. The membrane will have to satisfy certain physical and mechanical requirements. Thus, it must (a) be completely watertight, including the base; (b) remain flexible to accommodate differential movement; (c) be suitable for installation in a variety of soil types; (d) have a simple installation procedure and ensure the absence of breaks in the screen; (e) offer resistance to chemical attack, vegetation growth, and rodent attack, and to decay caused by microorganisms; (f) be durable and have an extended effective life; and (g) be suitable for installation depths up to 30 to 40 m (100 to 130 ft).

Synthetic Membrane An example of a synthetic screen is shown in Figure 10-26. The Geolock system is a plastic screen consisting of high-density polyethylene (HDPE) extruded section in the form of a sheet pile wall member. Individual screen sections are locked together to ensure a fully watertight joint.

The cross section in Figure 10-26 shows the following distinct features:

1. A hammer-shaped bead that fits into the lock.
2. The main body, which consists of an HDPE liner, varying in width from 2 to 4 mm, welded to the end section of the screen.
3. The lock section, which slips onto the previous sheets to ensure continuity and complete seal.
4. A groove, which allows the insertion of a water stop that expands in place to seal the connection.

The sheet of the lock section usually is thicker than the main liner to provide a joint strength that can resist ground or wall deformation. Thus, if differential movement occurs, the lock section is not released but is stretched without affecting the watertightness of the joint.

Assembly and Handling The liner, labeled "2" in Figure 10-26, is welded onto the lock section by means of a hot air welding method. The sheet sections can be produced in any size or length, but storage and transport considerations dictate a

Figure 10-26 Geolock screen; typical cross section. (BACHY.)

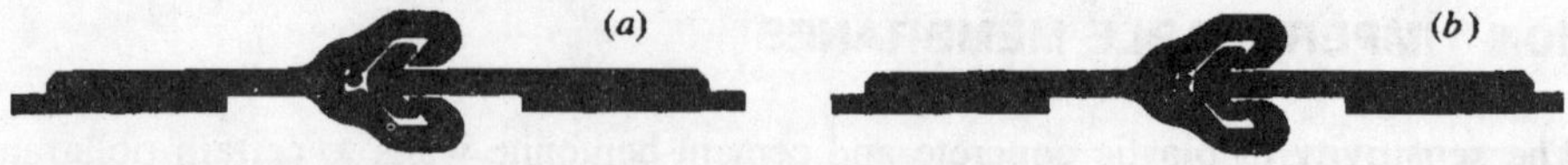

Figure 10-27 Water stop detail of Geolock screen. (*a*) Expansion profile before installation. (*b*) Final volume approximately after 40 hr. (BACHY.)

practical maximum length of 15 m (50 ft). Greater lengths can be provided at the site to suit the required depth by means of butt welds. However, inserting sections deeper than 15 m requires special procedures, and specially designed equipment and product modification. Friction against the lock or against the soil during installation must be considered in terms of temporary stresses induced in the board sections. Installation in a slurry trench will remedy potential stability problems and facillitate greater depths.

Water Stop Detail An important feature of the Geolock screen is the water stop joint that connects and seals adjoining sections. As shown in Figure 10-27, this consists of an expansion profile (water stop) that is fitted in the appropriate groove of the lock section. The expansion profile is made of a neoprene-based rubber with the following material properties: specific density, 1300 kg/m^3 (81 lb/ft^3); hardness shore $A = 52$; tensile strength, 2.9 N/mm^2 (420 lb/in.^{2}); elongation, 700 percent; and deformation, 32 percent.

The material is said to remain unaffected by chemical attack. Depending on the groundwater conditions, it can expand up to 10 times its original volume, and is accommodated with a built-in delay in operation to allow sufficient time for the installation of the screen before the swelling process beings. The time-expansion behavior is shown in Figure 10-28.

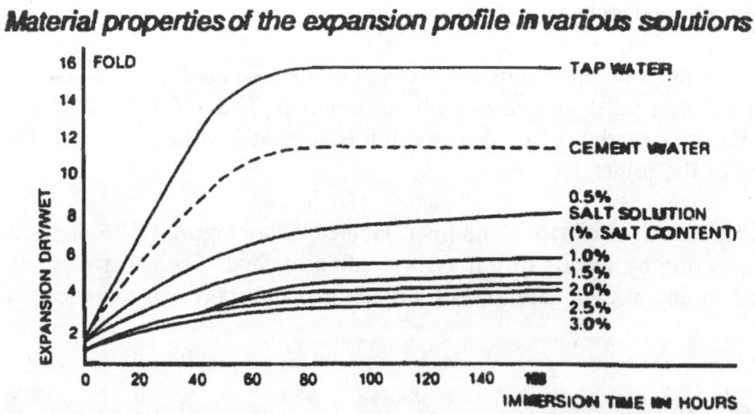

Figure 10-28 Characteristics of the expansion profile with time and in various solutions, Geolock screen. (BACHY.)

TABLE 10-14 Resistance of HDPE Screens to Chemical Attack

Attacking Chemical	Resistance[a]	Attacking Chemical	Resistance[a]
Aromatic compounds		Inorganic contamination	
Benzene	+	NH4	++
Ethylene benzene	++	Fluorine	++
Toulene	+	CN	++
Xylene	++	Sulfides	++
Phenol	++	Broom	0
		PO4	++
Polycyclic hydrocarbons			
Napthalene	++	Other sources of contamination	
Anthracene	++	Tetrahydroferane	+
Phenanthrene	++	Pyrides	++
Fluoranthene	++	Tetrahydrothiophene	++
Pyrene	++	Cyclohexamene	++
Benzopyrene	++	Styrene	++
		Petrol	++
Chlorinated hydrocarbons		Mineral oil	++
Aliphatic chlorinated hydrocarbons	++	Pesticides	
Chlorobenzenes	+	Organic chlorine compounds	++
Chlorophenol	++		
PCBs	++	Pesticides	++

Source: From BACHY (1989).

[a] ++ Good resistance/+ Average resistance/0 unsuitable.

TABLE 10-15 Material Properties, Geolock HDPE Screen

Sheet width	2 m
Thickness	2mm
Length	3–30 m
Tensile strength	34 kN/m
Tensile strength lock	50 kN/m
Specific density	960 kg/m^3
Melting temperature	122° C
E-modulus	800 N/mm^2
Elongation	600 %
Hardness shore *D*	59
Hydraulic permeability	10^{-13} m/sec

Source: From BACHY (1989).

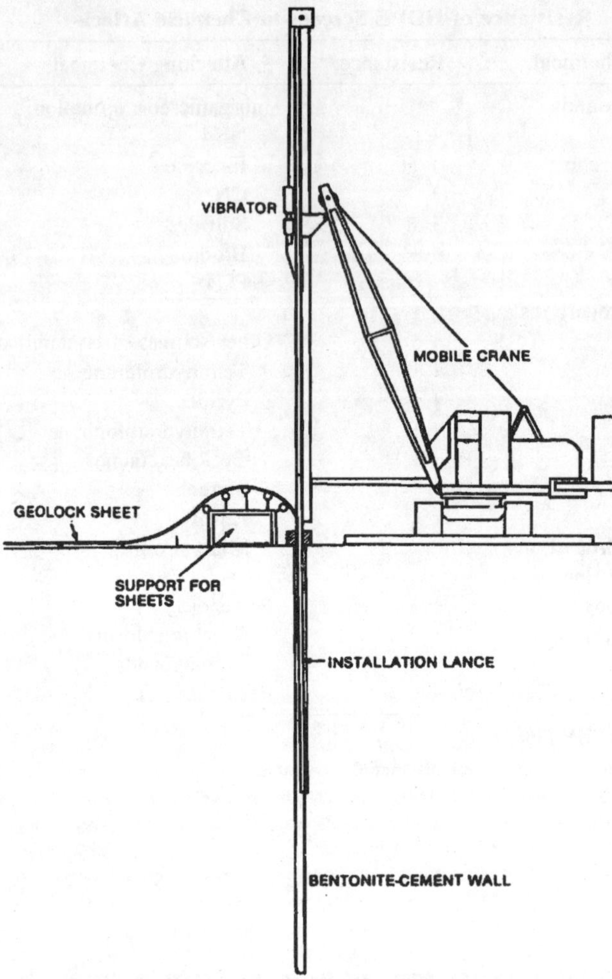

Figure 10-29 Typical installation procedure for HDPE screen in a cement-bentonite slurry wall. (BACHY.)

Chemical Resistance

High-density polyethelene is available in a wide variety of grades and characteristics with respect to chemical resistance. A general summary of average resistance to chemical attack is tabulated in Table 10-14. Requirements and assessment reflect criteria and standards used by most industrial codes, and apply to a 100 percent solution of the substance involved, although this high concentration normally would not be expected in actual polluted sites. Addition of carbon imparts to the material a greater resistance to the action of ultraviolent radiation.

The endurance of HDPE screens is determined by several factors, but symptoms of the aging process are manifested by: (a) temperature fluctuations; (b) tension fluctuations; (c) ultraviolet radiation; (d) physical attack; and (e) chemical attack. In situ conditions usually preclude the first four effects, leaving chemical attack as the prime factor that determines the durability and condition of the material. A suggested lifetime for the HDPE screen is close to 100 years in connection with the Dutch Delta Eastern Scheldt Works (BACHY, 1989). Material properties are summarized in Table 10-15.

Installation

The screen can be installed by a waterjet method. However, a main disadvantage in this case is the area of high permeability that occurs at the bottom of the sheet. This region must be injected with grout at a later stage to prevent excessive seepage.

A more effective installation is in conjunction with a cement-bentonite slurry wall where the sheets are installed along the center line. This construction is completely watertight, resistant to pollutant attack, and feasible in most soils and to greater depths. A typical installation procedure is shown in Figure 10-29.

Example of Environmental Application

Figure 10-30 shows a typical cross section and details of the dike at the Castle Peak (France) power station. The disposal areas for PFA hydraulically transported from the power station had to be completely isolated and protected to prevent environ-

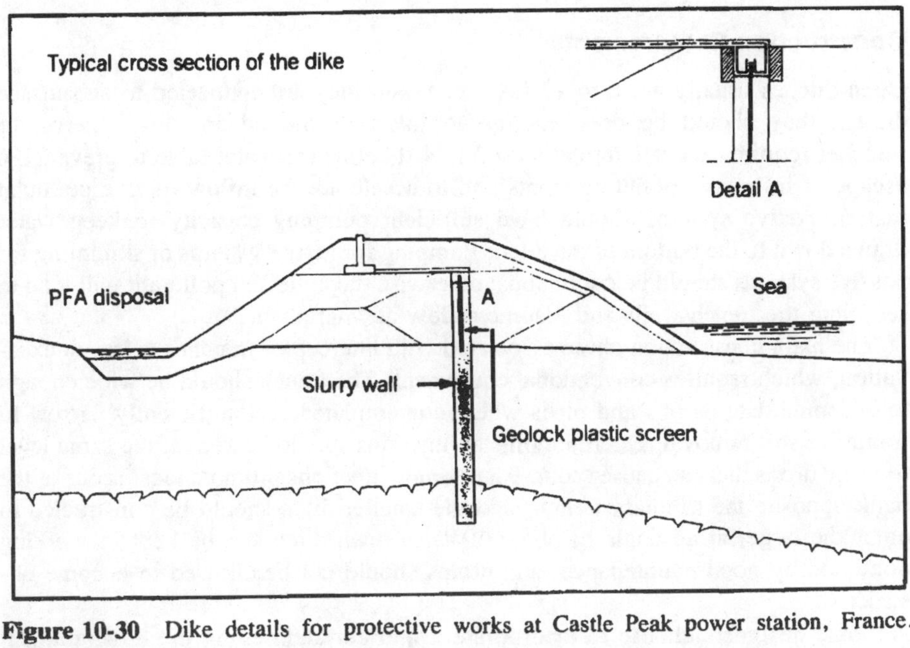

Figure 10-30 Dike details for protective works at Castle Peak power station, France. (BACHY.)

mental pollution of the seaside region, which would be particularly critical because of the proximity of oyster beds.

The area is protected by a continuous dike. Watertightness is ensured by a slurry wall penetrating into bedrock by a 20-cm socket. A plastic screen of HDPE materials, 4 m (13 ft) high, is inserted within the slurry wall and reduces permeability to inconsequential levels. The plastic screen is inserted 3.7 m in the slurry wall and extends 0.3 m into the concrete capping beam as shown in Detail A of Figure 10-30.

10-10 INTERCEPTOR TRENCHES

Interceptor trenches or ditches can be very effective in lowering the local water table and in controlling the direction of groundwater flow. They may be either active (pumped) or passive (gravity flow). Active systems have intermediately spaced vertical removal wells or a perforated, horizontal removal pipe (collector drain) located in the bottom of the trench. Active trenches usually are backfilled with a mix of coarse sand or gravel to ensure trench stability. Passive ditches normally are left open to accommodate the installation of a skimming pump for the collection and removal of the pollutant material. A typical application is the use of open ditches as interceptor drains to collect lateral surface seepage from a landfill, thus preventing it from percolating into the groundwater or flowing laterally to an area that should be protected. A second example is an open ditch used in certain cases to intercept subsurface collectors and carry the leachate to its ultimate disposal.

Construction Requirements

Open ditches usually are 6 to 12 ft deep. When they are connected to subsurface drains, they should be deep enough to intercept the underdrains. Interceptor trenches require excavation to at least 3 tp 4 ft below the water table to prevent the escape of inflowing polluting fronts and to accelerate the inflow of free pollutant matter. Active systems should have sufficient pumping capacity to keep water drawn down to the bottom of the ditch. Pumping for active systems or skimming for passive systems should be continuous; otherwise the collected pollutant will tend to seep into the trench walls and assume a flow downgradient.

The main construction phase associated with interceptor trenches is trench excavation, which requires conventional equipment. The trench should be wide enough to accommodate pumps and pipes where contemplated, yet sufficiently narrow to minimize soil removal. Ditch bottoms at junctions should be kept at the same level to avoid drops that can cause scour. Right-angle junctions promote local scour of the bank opposite the tributary trench, and the smaller ditch should be constructed to enter the larger at an angle of about 30°. An open ditch can be kept in working condition by good maintenance, and drains should not be allowed to become obstructed.

Some designers choose to incorporate a lining system along the downgradient face of the trench, usually an impermeable material such as polyethylene film, to stop floating pollutants from passing through. The main objection in this case is that the pollutant always tends to find its way around the ends of the barrier and penetrate the adjoining ground, while balancing the hydrostatic pressure on both sides of the film prompts the pollutant to float away.

Design Considerations

The purpose and operational characteristics of the ditch will determine the water level. Any flow in the trench should have a velocity compatible with the possibility of scouring of the bed and side slopes, and of sediment deposition. Factors affecting the flow velocity are the soil type, character of channel, well roughness, and anticipated sediment load.

The selection of side slopes is dictated by soil stability, taking into account possible groundwater pressures and vegetative cover. Side slope stability is improved by tamping or rolling. Trapezoidal cross sections are common since they are most efficient. In fine grained soils such as heavy clays, $\frac{1}{2}$ to 1 slopes are typical, but in coarser textured soils flatter slopes are necessary. Interceptor trenches are included in this review since in their final configuration they constitute physical control barriers.

In order to decide where to position the trench to intercept the expected flow, we must know the relationship between depth and flow together with the upgradient and downgradient influence of the trench. The upgradient influence may be determined from the following:

$$D_u = \frac{4}{3} H \tan \phi \tag{10-10}$$

where D_u = effective distance of drawdown upgradient, m
H = saturated thickness of the water-bearing strata not affected by drainage, m
ϕ = angle between the initial water table or ground surface and the horizontal plane

Likewise, a theoretical expression for the downgradient influence is as follows:

$$D_d = \frac{K \tan \phi}{q} (h_1 - h_2 + D_2) \tag{10-11}$$

where D_d = distance downgradient from the drain where the water table is lowered to the desired depth, m
K = hydraulic conductivity, m/day
q = drainage coefficient, m/day
h_1 = effective depth of drain, m
h_2 = desired depth to the water table after drainage, m
D_2 = distance from the ground surface to the water table before drainage at distance D_d downgradient, m

Advantages and Disadvantages

The most obvious advantage of interceptor trenches is the relative simplicity of the construction and the associated low cost. Other advantages are as follows:

1. Operating costs are relatively low since flow to underdrains is by gravity.
2. The trenches are useful in intercepting landfill side seepage and runoff without the use of impervious liners.
3. Higher rates of flow are attained by large wetted perimeter.
4. It is possible to monitor and recover pollutants, while the system produces much less residual fluid than well points.

The most obvious disadvantage is the requirement of continuous monitoring and maintenance. Other disadvantages are:

1. The system is not as efficient in poorly permeable soils.
2. The trench is not suited for deep disposal sites or impoundment, and may interfere with the use of land.
3. Very frequently, the need exists for additional safety/security measures.

10-11 HIGH-RESISTANCE NONCORROSIVE CUTOFFS

Pollution of alluvial terrains occurs in conjunction with a continuous groundwater surface but also with a series of local perched water tables developed within existing gravel lenses. The usual pollution scheme involves groundwater in direct connection with an aquifer or the application of considerable hydraulic heads, but movement of pollutants within an alluvial deposit may occur as a result of capillary and diffusive transport through what is known as a vadose zone. Polluting matter can be disseminated even in fine soil such as a silty-clay matrix.

It appears from the foregoing sections that the effects on the permeability and durability of plastic or rigid concrete cutoffs can vary widely but generally are caused by the disintegrating action traced to the factors discussed in the following sections.

Aggressive Water in Alkali Regions Sodium, potassium, and magnesium sulfates in alkali soil and water are usual causes of concrete deterioration (Xanthakos, 1979). The sulfates presumably react chemically with the hydrated lime in the cement paste to form calcium sulfate, a reaction followed by considerable expansion and disruption of the concrete. Alternatively, alkali water entering concrete may deposit salts in the larger pores, and the resulting growing crystals can fill the pores and eventually develop pressures sufficient to disrupt the concrete.

Disintegration of the first type is usually prevented by the use of sulfate-resisting cement (ASTM type V). Resistance to crystal growth effects is improved if the concrete mix is dense and impervious, has a relatively low water-cement ratio, and contains entrained air. Crystal growth is therefore more of a problem in plastic concrete than in normal concrete. The argument that these effects are moderated by the presence of filter cakes at the interface is not always valid since pollutants can also affect the colloidal stability and hence the permeability of a filter cake.

Leaching and Chemical Attack Water can pass through large pores, construction joints, and cracks in improperly constructed cutoffs, and can dissolve some of the readily soluble calcium hydroxide and other solids, thus causing errosion of concrete. Associated problems relate also to increases in permeability. These effects can be avoided if a wateright concrete is maintained.

Considerable damage can be caused by surface corrosion if concrete is directly exposed to organic acids, farm silage, polluting wastes, and other forms of chemical attack. Chemically active materials and substances can be harmful to unprotected concrete.

Fly Ash Mixes These have been prominently mentioned in the foregoing sections in connection with their remarkable ability to enhance resistance of concrete to sulfate attack. An example of fly ash mix is shown in Table 10-12. Tests on fly ash concrete confirm that this combination can provide reduction in permeability and simultaneously improve resistance to disintegration.

Fly ash is a pozzolan material consisting of fine solid particles of noncombustible ash carried out of a bed of solid fuel by the draft. Pozzolan is a silicious or aluminous substance that reacts chemically with slaked lime under moisture to form a cementlike material. Several commercial Portland-pozzolan cements show considerable resistance to sulfate attack. Justification for their use depends on the resulting long-term economy and improvement of the properties of mass concrete. This may include increased impermeability, reduced alkali-aggregate expansion, and improved workability. Inherent disadvantages in the use of pozzolan for structural concrete are slower strength development and lower resistance to deterioration caused by freezing and thawing, unless longer moist curing periods can be provided.

With mass concrete, that is, dams, sulfate attack is a minor problem, but it may be serious in concrete cutoffs of the usual thickness. On the other hand, the potentially detrimental effects of pozzolan on concrete are not necessarily adverse factors for cutoff walls since these are buried in the ground and are not subjected to weather changes.

Among the many test programs carried out to study the properties of fly ash concrete, mention is made of Dikeou (1970), and the National Ash Association (1971). Results from these tests show that fly ash consistently produces significant improvement in the sulfate resistance of the mix. These results are summarized by Xanthakos (1979).

Bituminous Mixes Among these, mastics are generally preferred because they melt more easily, have high plasticity and complete impermeability, and are highly resistant to sulfate attack. However, the many problems associated with slurry stability and control and also with the placement of bituminous mastic in continuous deep and narrow trenches under water or bentonite warrant ample assessment of the situation at hand. In spite of these difficulties, practice confirms the feasibility of inserting bituminous cutoff walls underground.

In many instances bituminous mastic backfills have been placed hot with the use of tremie pipes whose tip is lowered to within a few inches of the trench bottom. Whereas the material is sufficiently flowable and workable to fill the excavation by its own gravity, the main difficulty is associated with its flow in the tremie pipe.

Bituminous mixes are not affected by the corrosive action of underground water containing sodium chloride and magnesium sulfate. The mixes are generally prepared using 70 percent aggregate, 10 percent lime filler, and 20 percent bitumen. The binder should not necessitate excessively high temperature for placing, and once it is cooled, it should not be too fluid. A suitable temperature of the mix for placement is 160° to 180°C.

REFERENCES

Anton, W. F. and D. J. Dayton, 1972. "Camanche Dike 2 Slurry Trench Cutoff," Proc. Performance Earth Earth-Supported Struct. ASCE, Purdue Univ., Vol. 1, pp. 735–749.

BACHY, 1989. Geolock Screens, Brochure, Bachy Enterprise, Paris.

Bishop, A. W., 1963. "Discussion of Cutoff Efficiency," in *Grouts and Drilling Muds in Engineering Practice*, Butterworths, London.

Blank, Z., B. Rugg, and W. Steiner, 1990. "New Technology to Decontaminate PCB-Contaminated Sites, Status Report," Proc. EPA/AWMA Int. Symp., Cincinnati, pp. 325–341.

Bozovic, A., 1985. "Discussion of question 58; Foundation Treatment for Control of Seepage," 15th Congr. on Large Dams. Vol. 3, pp. 367–372.

D'Appolonia, D. J., 1980. "Soil-Bentonite Slurry Trench Cutoffs," *ASCE J. Geotech.*, Apr., pp. 399–417.

D'Appolonia, D. J. and C. R. Ryan, 1979. "Soil Bentonite Slurry Cut-off Walls," Geotechn. Exhibition and Techn. Conf., Mar., Chicago.

Deere, D. U., 1982. "Cement-bentonite grouting for dams," Proc. Conf. Grouting in Geotechnical Engineering, ASCE, New Orleans, pp. 279–300. Also vol. 2 (Discussion volume, only available to participants at Conf.), pp. 35, 36.

Deere, D. U. and G. Lombardi, 1985. "Grout slurries—Thick or thin?" Issues in Dam Grouting, ASCE, Denver, pp. 156–164.

Dikeou, J. T., 1970. "Fly Ash Increases Resistance of Concrete to Sulfate Attack," U.S. Bur. Recla. Res. Rep. 23.

Dupeuple, P., and P. Habib, 1969. A Plastic Concrete Cutoff, Proc. 7th Int. Conf. Soil Mech. Found. Eng., Spec. Sess. 14, 15, Mexico City, pp. 71–76.

Dutko, P., M. A. Khoury, and A. L. Harris, 1991. Tests and Field Measurements of Cement-Bentonite-Fly Ash Mixes, Slurry Walls, ASTM Symp., June 27–28, Atlantic City, N.J.

Fair, G. M., J. C. Geyer, and D. A. Okun, 1966. *Water and Wastewater Engineering*, Vol. 1, Wiley, New York.

FPS, 1973. "Specifications for Cast in Place Diaphragm Walling," Federations of Piling Specialists, London.

Fuchsberger, M., 1974. "Some Practical Aspects in Diaphragm Wall Construction," Proc. Diaphragm Walls Anchorages, Inst. Civ. Eng., London.

Greenwood, D. A., and J. F. Raffle, 1961. Non-newtonian Fluids, Proc. 5th Int. Conf. Soil Mech. Found. Eng., vol. 1, p. 789.

______ and ______, 1963. Formulation and Applications of Grouts Containing Clay, in "Grouts and Drilling Muds in Engineering Practice," Butterworths, London.

Holtz, W. G., and F. C. Walker, 1962. Soil-Cement as Slope Protection for Earth Dams, *J. Soil Mech. Found. Div.* ASCE, Dec., pp. 122–132.

Houlsby, A. C., 1982. "Optimum Water: Cement Ratios for Rock Grouting," Proc. Conf. on Grouting in Geotechnical Engineering, ASCE, pp. 317–331.

Houlsby, A. C., 1985. "Cement Grouting; Water Minimizing Practices," Issues in Dam Grouting, ASCE, pp. 34–75.

Houlsby, A. C., 1990. *Construction and Design of Cement Grouting*, Wiley, New York.

Hutchinson, M. T., et al., 1974. "The Properties of Bentonite Slurries Used in Diaphragm Walling and Their Control," Proc. Diaphragm Walls Anchorages, Inst. Civ. Eng., London.

ICOS, 1973. Economics of Cutoff Walls by New Slurry Method, International Construction, Milan.

Jefferis, S. A., 1972. "The Composition and Uses of Slurries in Civil Engineering Practice," Ph.D. Thesis, Univ. of London.

Jefferis, S. A., 1982. "Effects of Mixing on Bentonite Slurries and Grouts," Proc. Conf. on Grouting in Geotechnical Engineering, ASCE, New Orleans, pp. 62–76.

Jefferis, S. A., 1983. "Bentonite-Cement Slurries for Hydraulic Cutoffs," personal communications.

Johnson Division, UOP, Inc., 1976, "Groundwater and Wells," Edward F. Johnson, Inc., Saint Paul, Minn.

Katowicz, M. S., 1967. "The Design and Construction of the Bentonite Trench Cutoff in Khancoban Dam," Proc. 5th Aust.-N.Z. Conf. Soil Mech. Found. Eng., pp. 153–159.

Kauschinger, J. L., T. W. Kahl, and E. B. Kerry, 1991. "Stress-Strain-Strength Behavior and Permeability Measurements on Plastic Concrete," Slurry Walls, ASTM Symp., June 27–28, Atlantic City, N.J.

La Russo, R., 1963. "The Wanapum Development," Grouts and Drilling Muds in Engineering Practice, Butterworths, London.

Lau, D. and A. Crawford, 1986. "Grouting for the Underground Containment of Radioactive Waste," Univ. of Toronto, Dept. of Civ. Eng., Pub. 86-03, 126 pp.

Little, A. L., 1974. "In Situ Diaphragm Walls for Embankment Dams," Proc. Diaphragm Walls Anchorages, Inst. Civ. Eng., London.

Marsland, A., and A. G. Loudon, 1963. The Flow Properties and Yield Gradients of Bentonite Grouts in Sands and Capillaries, in "Grouts and Drilling Muds in Engineering Practice," Butterworths, London.

National Ash Association, 1971. "The VKR Lightweight Aggregate Plant and Quality Control Program for Fly Ash Utilization," NAA Rep. 3–71, Washington, D.C.

Nelson, F. and P. Sknonr'kov, 1949. "Seepage Through Saturated Media," Sovetska Nakne, Moskow.

Ryan, C. R., 1976. "Slurry Cut-Off Walls, Design and Construction," Proc. Slurry Wall Technical Course, Chicago, Apr.

Ryan, C. R., 1980a. "Slurry Cutoff Walls, Methods and Applications," Proc. Geotech. Conf., Mar., Chicago.

Ryan, C. R., 1980b. "Slurry Trench Cut-Offs to Halt Flow of Oil-Polluted Groundwater," Energy and Technology Conf. and Exhibition, ASME Petroleum Div., Feb., New Orleans.

Sanning, D. E., 1982. "Remedial Action of Disposal Sites," Office of Research and Development, U.S. Environmental Protection Agency, Cincinnati, Ohio, Report EPA-625/6-83-006.

Scott, R. A., 1963. Fundamental Considerations Governing the Penetrability of Grouts and Their Ultimate Resistance to Displacement, in "Grouts and Drilling Muds in Engineering Practice," Butterworths, London.

Spoljaric, N. and W. Crawford, 1979. "Removal of Contaminants from Landfill Leachates by Filtration Through Glauconitic Greensands," *Environmental Geology,* Vol. 2, No. 6, pp. 359–363.

Stanton, T. E., 1948. "Durability of Concrete Exposed to Sea Water and Alkali Soils," Calif. Experience, Proc. ACI, Vol. 44, pp. 821–847.

Tallard, G. R. and G. Caron, 1977. "Chemical Grouts for Soils, Vol. 1. Available Materials," Federal Highway Administration Report, FHWA-RD-77-50.

Terzaghi, K., 1955. "Evaluation of Coefficients of Subgrade Reaction," *Geotechnique*, Dec., pp. 295–326.

Tone Boring Co., 1980. "Plastic Cutoff Wall Systems," Intercompany Rep., Tokyo.

U.S. Environmental Protection Agency (US EPA), 1978a. "Guidance Manual for Minimizing Pollution from Waste Disposal Sites," Cincinnati, Ohio, EPA-600/2-78-142.

U.S. Environmental Protection Agency (US EPA), 1978b. Proc. of the 4th Annual Research Symp., March 6–8, San Antonio, Tex., EPA-600/9-78-016, pp. 282–298.

Xanthakos, P. P., 1976a. "Specifications for a Solidified Wall in Westmont, Illinois."

Xanthakos, P. P., 1976b. "Tests on the Effect of Bentonite on Plastic Concrete," unpublished report.

Xanthakos, P. P., 1979. *Slurry Walls,* McGraw-Hill, New York.

Xanthakos, P. P. and B. Bailey, 1975. "Report of Geotechnical Information, Slurry Trench Cutoff, Southport AWT Facilities," ATEC Associates, Indianapolis.

Zaffle, J. A., 1970. Soil-Cement for Water and Sewage Works, Soil-Cement Slope Protection. Portland Cement Assoc. Publ. C-24.

CHAPTER 11

ARTIFICIAL GROUND FREEZING

11-1 INTRODUCTION

Background

The first reported use of ground freezing as a method of stabilization was in conjunction with a mine shaft excavation in South Wales in 1862 (Maishman, 1975). Subsequently the process was patented in Germany by Poetsch in 1883. The basic method of circulating cooled brine through underground tubing is described in the patent as the "Poetsch Process," and remains the basic process in use today. The first use in the United States probably was in 1888 in conjunction with the construction of a mining shaft in Louisiana (Jumikis, 1966).

Primary use and development of the freezing method has been articulated in the mining industry, where excavations are selected based on ore location and related factors and have no reference to the economics and feasibility of engineered excavations. A similar siting problem has now become obvious for other forms of underground construction, attracting a variety of control techniques including freezing. Thus, in situ freezing for stabilization in both the mining and construction industries is applied in two basic modes: (a) as a supplementary or emergency technique for stabilizing ground excavations and installations using the more traditional support methods (underpinning, sheet piles, etc.); and (b) as a primary independent construction method for stabilizing underground openings.

Until recently most uses of the method have been as a supplement, exclusive of mining operations. However, in the last 20 years, in situ ground freezing as a primary method considered in the initial design has found increased uses related to the following factors: (a) increasing costs of conventional construction procedures compared to the cost of ground freezing; (b) expanded use of sites previously considered unsuitable; (c) development of design techniques with associated advances in the technology of ground freezing; (d) identification of optimum conditions for freezing applications with ample demonstration to potential users; and (e) improvement of a design methodology to eliminate much of the former overconservatism.

Basic Process

Essentially, the process of ground freezing involves removing the heat from the ground to cause a drop of subsurface temperature below the freezing point of moisture in the pore spaces. The frozen moisture acts as a cementing agent, binding the soil particles together and providing a structural support network in the soil mass. Heat is removed by circulating coolants through pipes installed from the surface into the zone to be frozen, and subsequently is transferred into the atmosphere.

In practice, a designed pattern of freezing pipes or "probes" is emplaced in the zone to be frozen. The probes are typically two pipes of different sizes, one within the other, so that the coolant can be pumped into one and extracted or allowed to escape from the other. In the soil, freezing progresses radially outward from the probe location, forming a frozen cylinder along the length of the probe. Adjoining cylinders eventually coalesce between probes to form a continuous wall or zone enclosing the area to be excavated with an impervious barrier sufficiently strong to be self-supported.

In closed systems the coolant is continuously circulated, cooled, and recirculated through the heat removal system, and this process is the most common technique used. Conversely, open systems are more direct, allowing the cooling to be accomplished by sublimating a solid or releasing pressurized liquefied gas to evaporate in the zone where cooling is desired. This permits the heat to be carried off directly to the atmosphere (Shuster, 1972). Intermediate systems are also possible, and allow repressurization and reuse of the gas.

11-2 SAND-ICE SYSTEMS

Mechanical Properties and Creep

The mechanical properties of saturated frozen sand depend mainly on the behavior of the ice matrix, volume concentration of sand particles, and temperature. Impurities such as air bubbles, salts, or organic matter will alter ice behavior, hence the behavior of the sand-ice system. At lower sand volume concentrations the behavior will be as in polycrystalline ice, whereas at higher sand volume concentrations interparticle friction and dilatancy of sand particles interact and become prominent. During deformation, adhesion between sand particles and the ice matrix produces high cohesion to the mass and may create an effect analogous to higher effective stresses when sand particles are in contact (Goughnour and Andersland, 1968).

Considerable data are yet to be derived about the interaction between sand grains and ice in a saturated frozen sand under stress. (See also subsequent sections.)

Gold (1963) has suggested that, for a given temperature, deformation rate is probably the major factor in determining what mechanisms are required for an ice grain to conform to the applied deformation. Glen (1963) states that creep rates for polycrystalline ice appear to be related to stress by a power law and that hydrostatic pressure does not affect flow. Butkovich and Landauer (1960) indicate that the rate process theory, associated with a hyperbolic sine stress dependence, probably best describes the experimental creep results on ice. Deformation rates of polycrystalline ice are summarized by Dillon and Andersland (1967), and appear to indicate that ice will creep under loads approaching zero stress.

Goughnour and Andersland (1968) have obtained useful data in describing or predicting the mechanical properties of sand-ice systems in a series of constant axial strain-rate and constant axial-stress creep tests on both polycrystalline ice and sand-ice samples.

Constant Axial Strain-Rate Tests Typical stress-strain curves for ice and sand-ice samples deformed at 0.0003 in./min and temperature equal to −12.03°C are shown in Figure 11-1 for various samples. The curves for samples 3, 13, and 18 are essentially similar, whereas samples 5 and 11 with high sand volume concentrations show considerable increase in strength at higher strains. These tests were carried out with no confining pressures.

Constant axial strain-rate tests on ice with 100-lb/in.^{2} confining pressures gave the results shown in Figure 11-2. For high strain rates the strength is increased, probably because volume change and accommodation cracking are restricted by the higher pressures. At lower strain rates ultimate strength remains about the same or

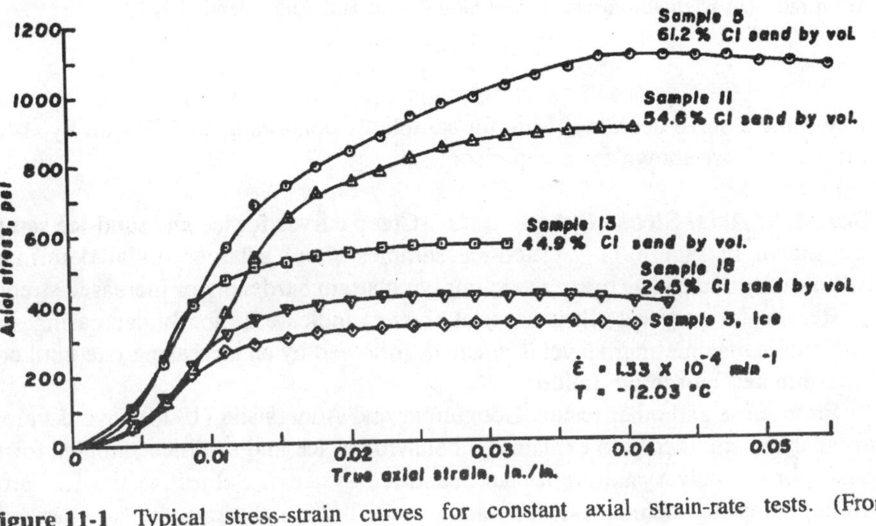

Figure 11-1 Typical stress-strain curves for constant axial strain-rate tests. (From Goughnour and Andersland, 1968.)

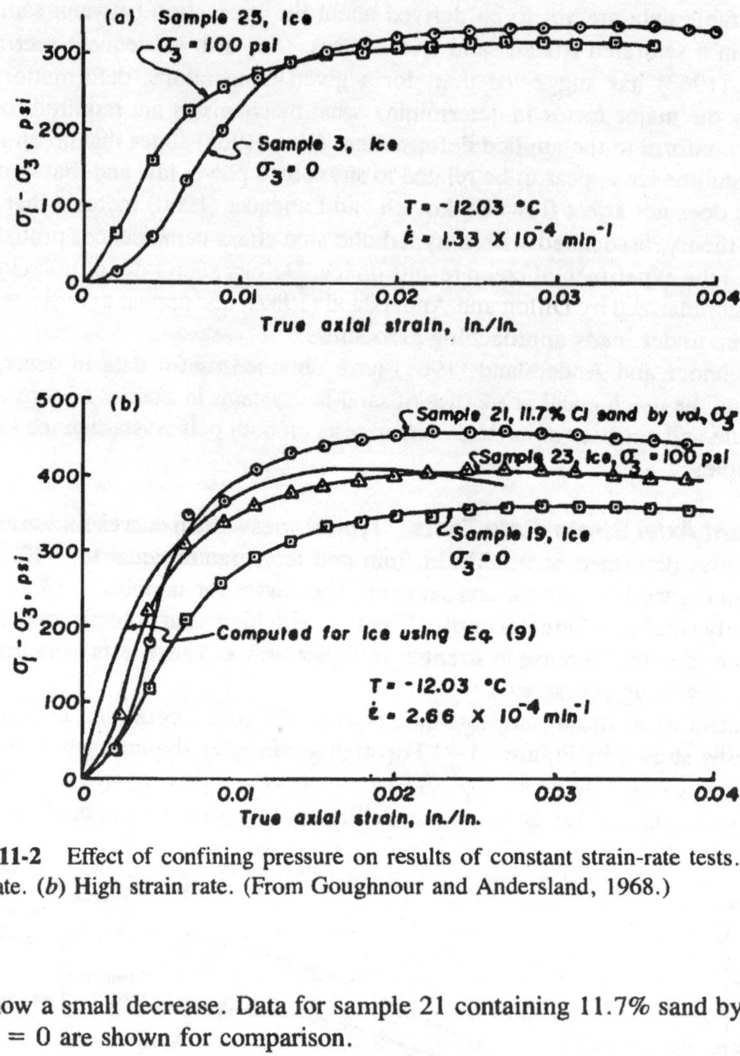

Figure 11-2 Effect of confining pressure on results of constant strain-rate tests. (*a*) Low strain rate. (*b*) High strain rate. (From Goughnour and Andersland, 1968.)

may show a small decrease. Data for sample 21 containing 11.7% sand by volume and $\sigma_3 = 0$ are shown for comparison.

Constant Axial Stress Creep Tests Creep curves for ice and sand-ice samples are shown in Figure 11-3. Sand-ice samples show a larger initial deformation followed by decreasing creep rates, implying strain hardening or increased strength.

Results for polycrystalline ice (not shown) indicate a rapidly decreasing strain rate until some minimum level is reached, followed by an increasing rate until creep is terminated by sample failure.

From these and other results Goughnour and Andersland (1968) have developed an analysis and theory to explain the behavior of ice and sand-ice samples for two basic forms: polycrystalline ice and sand-ice systems. Subject to the test procedures, sampling techniques, and sands used, the conclusions are summarized as follows.

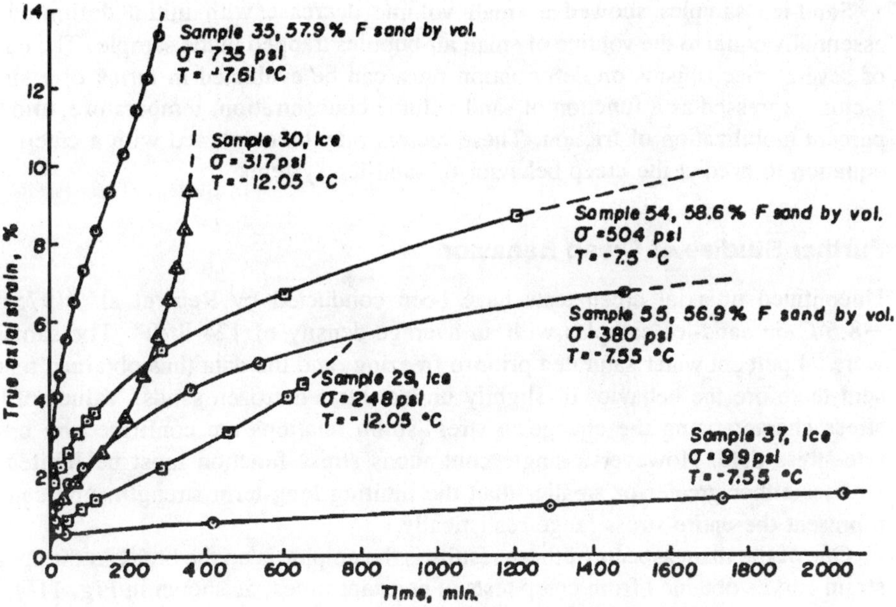

Figure 11-3 Creep curves for several ice and sand-ice samples. (From Goughnour and Andersland, 1968.)

Polycrystalline Ice The elastic modulus decreases with increasing strain. This is in agreement with the greater accommodation cracking and greater grain boundary distortion occurring on deformation (Gold, 1963). A small decrease in Young's modulus with colder temperatures is documented and may be the result of more extensive cracking.

The data suggest that the creep rate of polycrystalline ice depends on stresses, temperature, strain, and absorbed strain energy. Appropriate equations were developed for predicting creep rates, and include a strain-dependent hardening term and a strain-energy-dependent softening term. By extensions, these relations can be used to predict shear strength for constant axial strain-rate tests.

Ice samples in uniaxial compression show a small volume increase during initial deformation that may be associated with accommodation cracking, grain boundary sliding, and partial disruption of intergranular continuity. Application of confining pressure tends to hold the grains in more intimate contact, thus inhibiting grain boundary adjustments and increasing strength.

Sand-Ice System The presence of sand particles in the ice matrix alters behavior in several ways. At low sand volume concentrations, shear strength increase is a simple linear relation to the relative proportions of the sample. When a critical sand volume concentration is reached, about 40 percent, shear strength appears to increase rapidly whereas particle friction and dilatancy begin to contribute to the shear strength.

Sand-ice samples showed a small volume decrease with initial deformation, essentially equal to the volume of small air bubbles trapped in the sample. The effect of several mechanisms on deformation rates can be evaluated in terms of a stress factor, expressed as a function of sand volume concentration, temperature, and the percent mobilization of friction. These factors may be combined with a creep rate equation to predict the creep behavior of sand-ice systems.

Further Studies of Creep Behavior

Unconfined uniaxial creep tests have been conducted by Rein et al. (1975) at $-8.5°C$ on sand-ice samples with an average density of 124 lb/ft^3. The samples were 94 percent water saturated prior to freezing, and the data thus obtained represent therefore the behavior of slightly undersaturated frozen sands. Values of the stress characterizing the change in stress-strain relations are confirmed by creep rate-stress data. However a single continuous stress function must be limited to stresses either greater or smaller than the limiting long-term strength, and cannot represent the entire stress range realistically.

One of the main aspects of these results is the emphasis on the break in the stress-strain curves obtained from creep tests at constant times, as shown in Fig. 11-4. In these plots, stress is the independent variable but is plotted on the ordinate to provide consistency for data analysis.

The solid lines shown in Figure 11-4 have the smallest sum of the squares of deviations from the data for that time, consistent with restrictions requiring that the two sections of the curves intersect and that the curves do not intersect the stress axis at negative values. These relations are represented by

$$\epsilon - \epsilon_c = (\sigma - \sigma_c)\, h_1\,(t), \qquad \epsilon_f > \epsilon \geq \epsilon_c \tag{11-1}$$

$$\epsilon = \sigma^n\, h_2\,(t), \qquad \epsilon \leq \epsilon_c \tag{11-2}$$

where σ = stress

ϵ = strain

ϵ_f = strain at which tertiary creep begins

and the subscript c refers to the value of the variable at the point where there is a change in a stress-strain relation. The parameter n was found to be constant = 0.3 for these tests. The factors σ_c and ϵ_c may be functions of time, whereas the parameters $h_1\,(t)$ and $h_2\,(t)$ indicate functions of time.

These data suggest that the stress-strain behavior for long times of loading is articulated by two separate sections in the stress-strain curves. The change in the stress-strain relation becomes more apparent as the time of loading increases, or the average creep test rate decreases. Likewise, the stress σ_c, characterizing the change in the stress-strain relation, equals the stress σ_s, characterizing the transition from primary to secondary creep only when the characteristic stresses are near the limiting long-term strength. These two stresses are not equal at higher stress values.

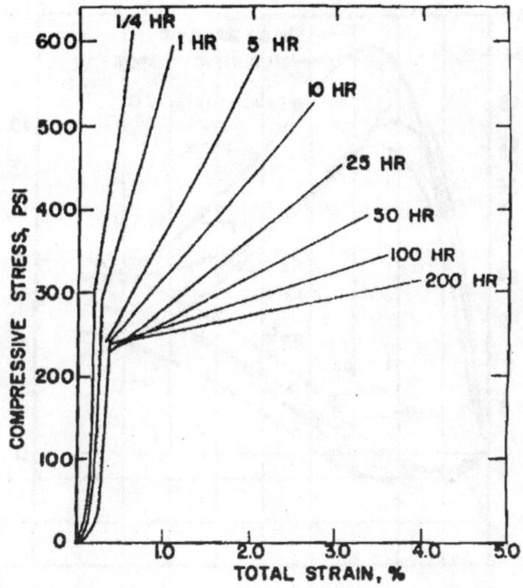

Figure 11-4 Summary of stress-strain data at different but constant times. (From Rein et al., 1975.)

Although the resulting conclusions are valid only for the test conditions, it appears that the stress characterizing the change in stress-strain relation can be obtained from stress-creep rate data in addition to stress-strain data. The limiting long-term strength can be obtained from the σ_c-time curve, in addition to the usual method of using σ_s and σ_{pr}-time curves, where σ_{pr} = tertiary creep stress, and σ_s = secondary creep stress.

Triaxial Tests of Frozen and Unfrozen Sands

Youssef and Hanna (1988) present results of an experimental program focusing on the behavior of frozen and unfrozen sands in triaxial testing, giving also emphasis to structures that are usually subjected to such changes in behavior because of seasonal temperature changes.

The high viscosity of intergranular ice imparts to the sand strength that combines its ice cohesion as well as its frictional components. This strength is time dependent. Unfrozen sand, on the other hand, is cohesionless material, and because of the low viscosity of the intergranular water its shear behavior is essentially time independent.

Triaxial testing of frozen sands is essentially in closed-system conditions since the intergranular ice is not free to move out of the sample during testing in shear, although the system exhibits volume changes. Triaxial testing of unfrozen sands may be of either type, drained or undrained.

Test results for frozen sands are shown in Figure 11-5. The short-term strength is influenced to a high degree by the applied strain rate ϵ_1 and the level of confining pressure σ_3. This strength is also a function of the physical properties, but mainly

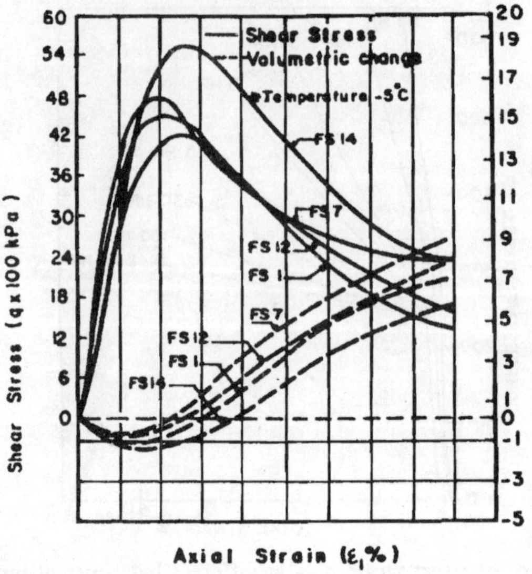

Figure 11-5 Test results for frozen sands; FS = test number. (From Youssef and Hanna, 1988.)

the initial void ratio and the degree of saturation. Increasing the confining pressure causes an increase in the shear strength, and this is in agreement with the foregoing results (Goughnour and Andersland, 1968) and with Chamberlain et al. (1972).

A higher degree of saturation results in a higher shear strength, attributed mainly to the increase in the contact area between sand particles and ice. This in turn causes intensification of the cementation bond. As can be seen from Figure 11-5, the volumetric change behavior is tested under frozen conditions. Initially the volume decreases with an increase in the axial strain, then it shows a rapid increase up to the failure strain, and continues to increase at a slower rate to the end of the test.

The dependence of uniaxial shear strength of frozen sand as a function of temperature and strain rate is presented by Parameswaran (1980), who documents that increasing the confining pressure or decreasing the temperature results in increasing the strength of frozen soil.

Test results of drained unfrozen sand are shown in Figure 11-6. Increasing the confining pressure increases the drained shear strength, and the failure strain varies from 2.95 to 8 percent, depending on the voids ratio after consolidation and the applied confining pressure. The volumetric strain at failure decreases with increasing confining pressure because of the decrease in the interlocking of the sand particles, and also with increasing porosity for the same confining pressure. The denser the sample, the higher the dilatancy observed.

Comparison Between Frozen and Unfrozen Sand A comparative study of frozen and unfrozen sand samples is illustrated in Figure 11-7, presenting curves of shear stress and strain for both frozen and water-saturated sands. Invariably the

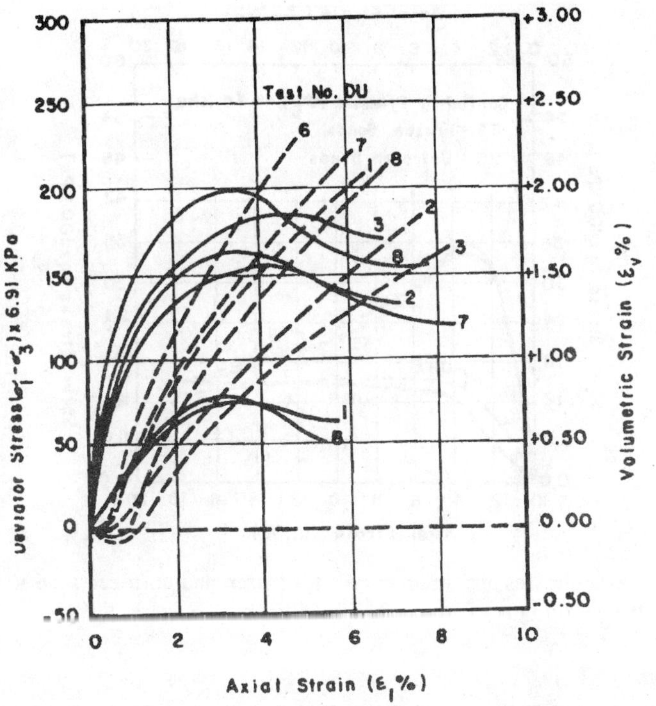

Figure 11-6 Test results for drained unfrozen sands. (From Youssef and Hanna, 1988.)

frozen samples have a much higher shear strength than unfrozen sand. Transformation of the water to solid state (ice) increases the brittleness of the samples.

The residual strength (at 20 percent strain or higher) of frozen sand approaches that of unfrozen soil. At higher strain levels (longer duration of loading) the contribution of the ice matrix to cohesion and friction will decrease to inconsequential values. Because of the high viscosity of the ice component, the strength of the frozen sample can be increased by increasing the strain rate during testing.

These tests also document the variation of the effective friction angle. The friction angle of frozen sand increased to about 48° and then decreased to 37° at a strain level 20 percent. At higher strain levels the friction angle approaches the ϕ value of unfrozen sand (35.4°) while cohesion approaches zero. This means that at higher strain levels the contribution of the ice matrix to the shear strength in terms of friction and cohesion appears to dissipate, and frozen sand tends to have the same shear as unfrozen undrained samples.

Volume Change Behavior From results of the same test program, Youssef and Hanna (1988) conclude that the apparent volume change behavior of frozen and unfrozen sands is similar. The volume first begins to decrease moderately and then increases progressively until the end of the test. However, the mechanisms of deformation are different for frozen and unfrozen samples. Data on volume change

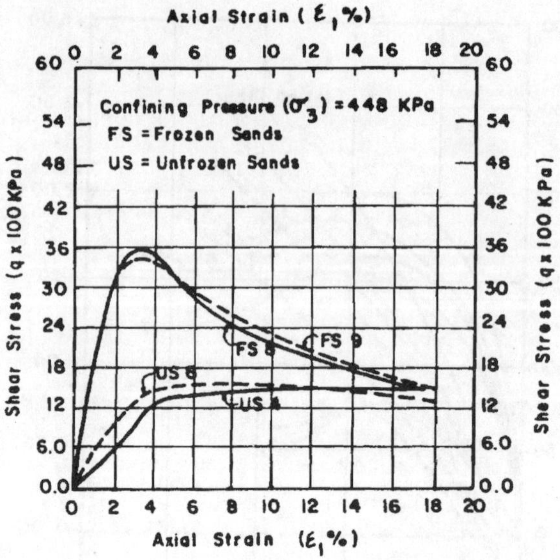

Figure 11-7 Shear stresses and strain curves for frozen and unfrozen sand. (From Youssef and Hanna, 1988.)

measurements have been presented by Goughnour and Andersland (1968), mentioned briefly in the foregoing sections, O'Connor (1975), and Lode et al. (1980). However, the mechanism controlling the behavior of the composite frozen material is not explained. Based on tests by Youssef (1984, 1985), it may be concluded that the initial volume decrease is due to the compressibility of both the frozen sample and the air bubbles entrapped in the system. Volume increase may be due to initiation and progress of cracks in the frozen soil.

Applications

The foregoing results demonstrate the advantages of freezing soils for construction purposes. For example, referring to Figure 11-7, it appears that at a temperature −5°C freezing the ground results in a shear strength increase by a factor of 2.5 or more. The strength of frozen sand increases because of the decrease in temperature. In practice, artificial ground freezing to low temperatures (−196°C can be achieved by the use of liquid nitrogen) can sharply increase the soil strength to a comparable rock strength level. Thus, a major advantage of ground freezing for construction purposes is the ability to make the ground self-supported.

Equally interesting is the volumetric strain behavior of thawed and frozen sands. Although this is apparently similar, the controlling mechanisms are different, and understanding this difference can have direct application in the control of foundation stability. In practice, in seasonal frozen areas above the frost line, building foundations will be subjected to thaw settlements resulting from changes in soil behavior from winter to summer, that is, from the frozen to unfrozen state. In these conditions foundations are constructed below the frost line. Likewise, in the construction of highways and roads, the material used in the subgrade should not be susceptible to frost in order to provide sufficient drainage. These applications are discussed in some detail in subsequent sections.

11-3 CLAY-ICE SYSTEMS

General Principles

The presence of a break in the stress-strain curves obtained from tests (shown in Figure 11-4) and the use of at least two stress functions to represent the entire stress range is also documented in tests of a frozen clay soil reported by Akili (1970). Typical stress-strain curves derived from constant stress-creep data are shown in Figure 11-8, and a definite break in the stress-strain curves is very pronounced. In these tests, however, the constant times are shorter than the ones of Figure 11-4.

Akili (1970) has investigated creep behavior of two frozen clay soils at a temperature range $-1°$ to $-22°C$ for the purpose of determining the limiting long-term strength of the frozen clay samples. The results demonstrate a limiting stress above which pseudoinstantaneous plastic strains are rather large with continuing deformation until failure. For applied stresses below this limiting stress, pseudoinstantaneous strains are small and creep rates decelerate until they become practically

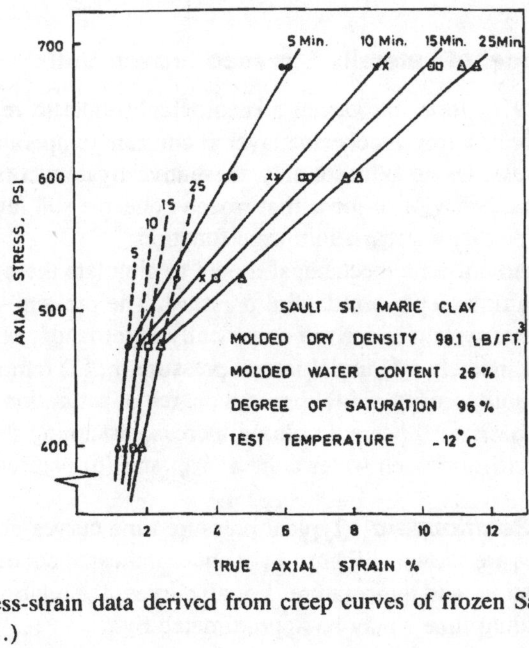

Figure 11-8 Stress-strain data derived from creep curves of frozen Sault St. Marie clay. (From Akili, 1971.)

zero. The magnitude of this limiting stress in these tests was nearly 70 percent of the magnitude of the ultimate strength of test specimens determined from constant axial-strain rate carried out at an average rate of about 5 percent/hr.

The significance of predicting the limiting long-term strength is in controlling creep instability by ensuring that applied stresses will not exceed the limiting long-term strength. However, as pointed out by Akili (1970), in several practical applications frozen soils can be subjected to different consecutive stress regimes rather than a constant stress level. In this case σ_∞ will not be constant but will tend to change since frozen soils are stress history dependent.

Another criterion of creep instability is the selection of a constant limiting strain for a given temperature and structure; if this is exceeded, failure results. The physical basis for this criterion is interpreted in terms of total damage occurred up to a certain level (Ladanyi, 1972). An acceptable limiting strain is a certain percentage of the total strain when tertiary creep is imminent. The amount of permanent strain at the onset of tertiary creep has been found to be approximately constant for a given temperature, structure, and test method (Ladanyi, 1972).

The foregoing brief review highlights certain basic problems when applying various methods to the analysis and design of ground freezing systems. The following are technical difficulties typically associated with stability analysis of frozen ground: (a) modeling the temperature-dependent rheological properties of frozen soils; (b) the initialization of stresses during the freezing process prior to excavation (a usual approach is to assume at rest earth pressure conditions during a gravity loading); and (c) the stress-strain characterization of the frozen soil in tension, a difficulty manifested mainly in cut-and-cover tunneling situations where high tensile stresses can occur.

Rheological Model of Laterally Stressed Frozen Soil

Aziz and Laba (1976) have introduced a rheological model to represent the time-dependent behavior of a frozen cohesive layer at constant temperature and under the effect of lateral stress. Using Windsor clay, these investigators correlated the stress-time and strain-time behavior of the actual frozen cohesive soil layer with a mathematical model to predict a stress-strain-time function.

These investigators used a mechanical model to simulate the behavior of frozen clay while under a time-dependent lateral pressure. The pressure-time relationship for the soil-ice-water system is derived empirically in terms of six obtainable influencing parameters, namely: (1) initial lateral pressure σ_i; (2) temperature of frozen soil T_F; (3) initial soil porosity n; (4) apparent degree of saturation S_i (ice saturation calculated on the basis of 9 percent volume increase assuming that all pore water converts into ice); (5) unfrozen water content W_u; and (6) elapsed time t.

Pressure-Time Relationship Typical pressure-time curves obtained for different initial pressures are shown in Figure 11-9 under indicated conditions of porosity, apparent ice saturation, and temperature. For all curves, the relationship between σ_t and the corresponding time t may be approximated by

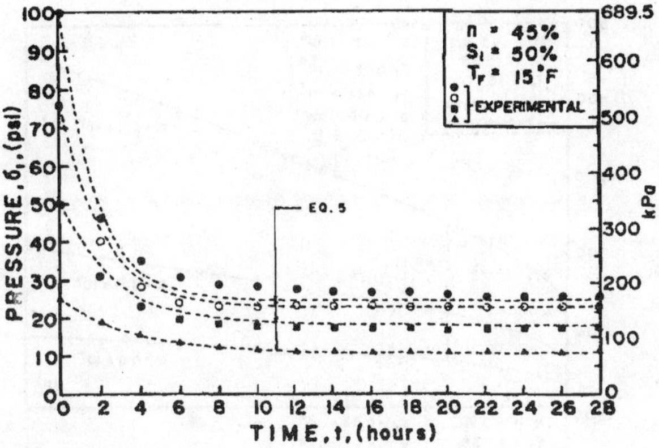

Figure 11-9 Pressure-time curves for four different initial pressures under indicated conditions. (From Aziz and Laba, 1976.)

$$\sigma_t = (\sigma_2 - \sigma_e) \exp(-ct) + \sigma_e \tag{11-3}$$

where σ_t = lateral pressure retained by the speciment at time t after the application of initial pressure

σ_i = initial lateral pressure

σ_e = long-term stable stress = $f(\sigma_i, T_F, S_i, n)$

c = time factor = $f(\sigma_i, T_F, S_i)$

t = time elapsed after the application of σ_i

The parameter σ_e can be expressed in terms of a long-term percentage reduction factor R_e as follows:

$$\sigma_e = \sigma_i\left(1 - \frac{R_e}{100}\right) \tag{11-4}$$

where R_e is derived empirically as suggested by Aziz and Laba (1976), who also give relations for the parameters W_u and S_i.

Stress-Strain-Time Relationship Using the pressure-time relationship for σ_t expressed by Eq. (11-3), the strain equations for model elements corresponding to the frozen cohesive system of this study (Windsor Clay; Ontario, Canada) are derived by Aziz and Laba as functions of the elastic constants E. With the initial application of pressure to the frozen soil system, high stress concentrations develop at the contact points between soil particles and ice crystals. Under excessive pressure, ice melts and flows from a higher pressure zone to a lower pressure zone, where it regelates. A flow of unfrozen water inside the frozen system occurs immediately after the initial pressure application. This flow decreases very rapidly with time due to the regelation of the melted ice.

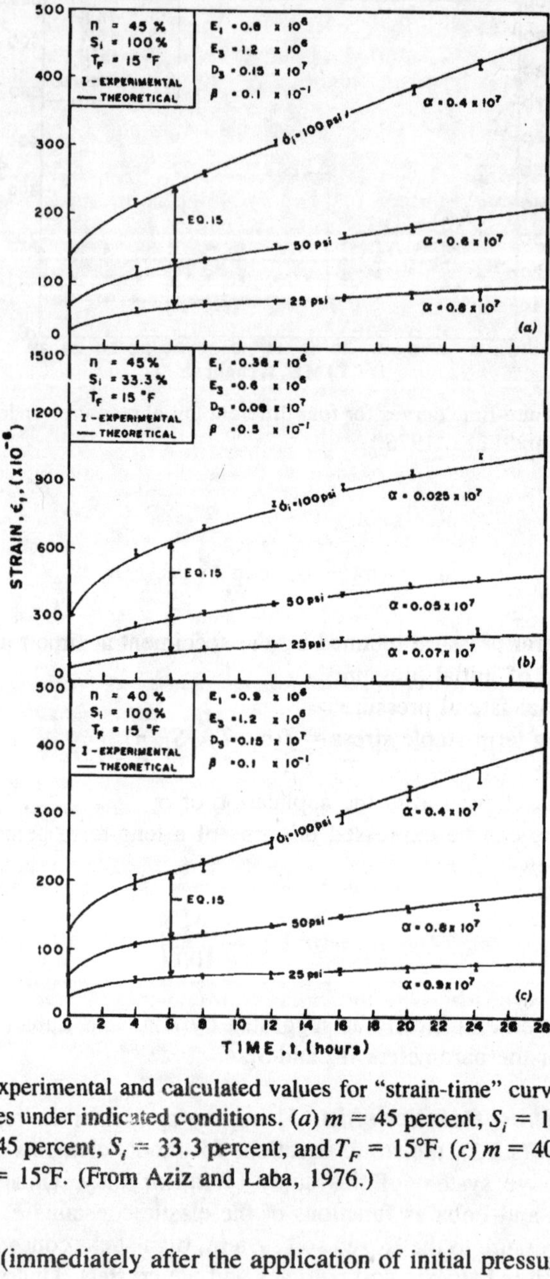

Figure 11-10 Experimental and calculated values for "strain-time" curves for three different initial pressures under indicated conditions. (*a*) $m = 45$ percent, $S_i = 100$ percent, and $T_F = 15°F$. (*b*) $m = 45$ percent, $S_i = 33.3$ percent, and $T_F = 15°F$. (*c*) $m = 40$ percent, $S_i = 100$ percent, and $T_F = 15°F$. (From Aziz and Laba, 1976.)

When $t = 0$ (immediately after the application of initial pressure σ_i), the initial strain is

$$\epsilon_0 = \sigma_i / E_1 \tag{11-5}$$

which is the instantaneous elastic deformation of the system under the applied initial pressure. When t is very large ($t = \infty$), the corresponding strain is

$$\epsilon_\infty = \frac{\sigma_e}{E_l} + \frac{\sigma_i - \sigma_e}{\alpha(c + \beta)} + \frac{\sigma_e}{\alpha\beta} + \frac{\sigma_e}{E_3} \tag{11-6}$$

where α and β are constant model coefficients (Aziz and Laba, 1976).

Strain-Time Curves Figures 11-10*a*, *b*, and *c* show three typical strain-time curves for frozen soil specimens that have the same values of initial soil porosity n, apparent degree of ice saturation, S_i, and temperature T_F, but that are subjected to three different initial pressures indicated on the curves. The two flanges of I-shaped symbols indicate two experimentally measured values of strain ϵ_t corresponding to a particular time t, attributed to the nonhomogeneous structure of the soil-ice-water system.

For the frozen cohesive soil layer under decreasing lateral pressure σ_t, the general tendency of strain increment with time is similar in all cases. All curves show an initial deformation ϵ_0 on the ordinate at $t = 0$ (immediately after the application of initial pressure), representing the elastic part of the total strain ϵ_t. Invariably, the strain rate is maximum at the beginning of the curve, but decreases with time and becomes nearly constant during the first 24 hr. From the curves, it appears that elastic strain occurs in the clay-ice-water system as soon as the initial pressure is applied. Plastic deformation starts thereafter and follows the general behavior of frozen soils (Tsytovich, 1960, 1966).

Aziz and Laba (1976) conclude that in a laterally stressed frozen cohesive soil layer of constant temperature, the induced stress decreases with time and is associated with simultaneous increase in lateral deformation. The rheological behavior of clay-ice-water systems is significantly influenced by the parameters σ_i, T_F, S_i, W_u, n, and t (time elapsed after the development of initial pressure).

11-4 SEISMIC EFFECTS AND DYNAMIC RESPONSE OF FROZEN GROUND

In general, fully frozen ground behaves well during earthquakes, and most engineering problems arise when saturated cohesionless soils are trapped beneath the frozen surface layer. However, results from investigations indicate the following: (a) dynamic stress-strain properties for coarse-grained soils can be nearly an order of magnitude greater than for fine-grained soils, and energy absorbing properties can vary significantly with soil type and composition; (b) in the range of void ratios 0.3 to infinity (ice) dynamic stress-strain properties for fully saturated soils can decrease by a factor of 5; (c) dynamic stress-strain properties of frozen soils increase with increasing degree of ice saturation; (d) the same properties decrease and damping properties increase with ascending temperatures; (e) the same properties decrease and damping properties increase with increasing axial strain amplitude from 10^{-30} to 10^{-10} percent; (f) the frequency of loading has only a minor effect on dynamic stress-strain properties, and the effect on damping properties may be important at low frequencies; and (g) confining pressure has an important effect on the dynamic properties of coarse-grained soils and a relatively unimportant effect on fine-grained soils.

Seismic Response and Dynamic Behavior

The behavior of frozen ground during an earthquake event becomes a prominent engineering feature in project development in the Northern Regions such as North Canada and Alaska, where the widespread presence of permafrost or perenially frozen ground is a typical geological feature. In some instances, moisture in the form of ice and unfrozen water may not be present. Unless ice is present, the dynamic engineering properties are not likely to be significantly different from those of the same soils in the unfrozen state. Since the response of frozen soils becomes important when ice is present, the terms permafrost, frozen ground, and frozen soil or rock all imply the presence of ice as a cementing agent.

Finn and Yong (1978) point out an important aspect of the dynamic response of frozen ground: what makes this prediction difficult, or is likely to lead to serious problems during the dynamic event, is not the frozen soil itself but the complex structural combinations of frozen and unfrozen soils in many regions of the North, that is, coastal areas, deltas, river valleys, and the margins of lakes.

Freezing Phenomena The structure of frozen soil has been studied by several investigators (Low et al., 1968a; 1968b; Hoekstra, 1969; Finn and Yong, 1978). Miller (1972) and Nakano and Froula (1973) have shown that the unfrozen water content of a soil is a hysteretic function of the negative temperature. The unfrozen water content is a more fundamental variable than temperature in controlling the behavior of frozen fine-grained soils, although all data on dynamic soil properties usually refer to negative temperatures without any correlation with temperature history or unfrozen water content.

The migration of pore water during freezing has an important influence on the dynamic response of frozen soil. When water freezes, its volume expands by about 9 percent. Fine-grained soils swell to accommodate the volume increase of frozen pore water, and additional water is attracted into the pores, leading to further volume changes and ice segregation. If temperature gradients are set up during freezing, differential thickness of absorbed films due to the differential rates of crystallization of ice will occur. The more mobile water in the thicker films migrates to the more frozen regions where the absorbed films are water deficient (Hoekstra, 1969). This migration pattern can lead to nonuniform water content distribution and ice segregation (Sheeran and Yong, 1975; Cary and Mayland, 1972).

In coarse-grained soils that have small specific surfaces and little attraction for water molecules, water is expelled from the freezing front (Mackay, 1975; McRoberts and Morgenstern, 1975). In closed systems this process can lead to high positive pore-water pressure (Miller, 1972; McRoberts and Morgenstern, 1975). Mackay (1977a,b) reports two field cases of very high pore-water pressures in sands caused by water expulsion from an advancing freezing front.

It appears that these phenomena are essential to the assessment of the dynamic response and liquefaction potential of saturated sands underlying thin permafrost during an earthquake. Even before the seismic event considerable positive pore-water pressures may already exist. In this case, a liquefaction analysis that did not consider the preexisting pore-water pressures could underestimate the liquefaction potential hazard.

Determination of Dynamic Properties Current procedures include ultrasonic methods and cyclic triaxial tests. Whereas existing data can be useful in providing good estimates of the dynamic properties of frozen soils, it is expedient to carry out an investigation of the effects of temperature, strain, and frequency on the dynamic behavior. These studies are commonly recommended not only on artificially frozen soils but also on frozen undisturbed field samples (Seed, 1976).

Frozen sands do not contain unfrozen water. Under a confining pressure the contact stresses between ice and sand increase, the ice melts, and the water migrates to areas of low stress and refreezes. In the latter case there are greater contact stresses between sand grains and hence a higher modulus. Silts and clays contain adsorbed water and the increased pressure does not lead to increased mineral grain-to-grain contact. Furthermore, the grains will not move much closer to each other in the absence of time for water expulsion to occur.

The dynamic properties of frozen soils and parameters affecting them are discussed in more detail in the following sections.

Seismic Exploration Methods for seismic exploration of frozen ground are reviewed by Finn and Yong (1978). They are used to determine the depth to frozen ground in continuous permafrost zones, to probe for bodies of frozen ground in unfrozen strata, to map structure in frozen overburden, to measure the seismic velocities of frozen soils in situ, and to measure the thickness of frozen ground. Reflection methods are confined to structures such as bedrock, or they are used for well logging. Refraction methods are used for exploring the shallower depths usually associated with engineering and construction activity.

The refraction method has its origin in the fact that seismic waves are refracted from a low velocity layer into a higher velocity layer and back again, so that a basic condition for successful application is that the velocity should increase with depth. This means that the thickness of a frozen layer cannot be determined directly by the refraction method when it is underlain by unfrozen soil with a lower seismic wave velocity. In this condition, only the velocity of the upper levels of the permafrost is obtained by the refraction method, whereas the average velocity required for computing the depth of deeper layers must be obtained by other procedures, such as well logging. More data on these techniques are provided by Barnes (1963), Roethlisberger (1972), and Hunter et al. (1976).

Field Data Existing data document the uncertainty regarding the characteristics of seismic motions in cold regions. This record covers spectra, attenuation of accelerations with distance, and the correlation of accelerations and velocities with earthquake magnitude, derived from instrumentation of unfrozen soils and rocks. In most sites, the ground stiffness increased with depth, whereas in permafrost below a certain depth the stiffness decreases because of increasing temperatures. There are limited quantitative data on the transmission characteristics of earthquake motions through permafrost. Since in North America permafrost often involves cohesionless material, its seismic wave velocities approach those of rock, suggesting that the seismic motions would have characteristics similar to rock.

An excellent report on seismic response of frozen ground is given by Ferrians (1966), and involves the ground cracking and landsliding in the Copper River Basin during the 1964 Alaska earthquake. When the earthquake occurred (March) the active layer was still frozen at about the maximum depth.

Most ground cracks in the basin occurred in coarse-grained deposits even though such deposits were thicker and stronger. The cracks were not localized in zones with linear patterns. On fairly level ground they were restricted to areas with the following prevailing conditions: (a) a frozen surficial layer existed; (b) perennially frozen ground was either absent or some distance below the surface with a substantial layer of unfrozen soil underlaying the frozen surface layer; and (c) the unfrozen soil was cohesionless and saturated. No cracks were noted in perennially frozen ground or when there was only a very thin layer of unfrozen coarse-grained soil between the active layer and the layer of perennially frozen ground.

The field data for the seismic response of ground in cold regions appear to indicate, therefore, that potentially dangerous conditions during earthquake are associated with layers of saturated unfrozen cohesionless soils sandwiched between a frozen active layer and the permafrost table. Solidly frozen grounds are likely to perform well during seismic events.

Seismic Response Analysis Analytical investigations are reported by Finn et al. (1977b) and by Singh and Donovan (1977a, 1977b). Mathematical models for frozen soils are presented by Schnabel et al. (1972); Lysmer et al. (1974; 1975); Streeter et al. (1976); Finn et al. (1977a); and Liou et al. (1977).

The dynamic response of saturated layers of cohesionless soils sealed by frozen surface layers has been studied by Finn et al. (1977b), using a nonlinear elastic mathematical model. The obvious effect of the frozen surface layer is to prevent drainage from the saturated cohesionless soils, and to explore this effect a dynamic analysis was carried out in terms of effective stress (Finn et al., 1977a). Two important features of the method are that: (a) the effect of increasing pore-water pressure during the earthquake on the dynamic properties of the unfrozen soil is continuously considered; and (b) drainage and redistribution of pore-water pressures under dynamically induced pore pressure gradients are incorporated in the analysis.

Effective stress dynamic analysis shows that the active layer plays a major role in the liquefaction potential of the frozen layer. If the permeability of the layer is low (k less than 0.0003 ft/sec), the effect of drainage or pore-water pressure redistribution is negligible. Drainage begins to have an appreciable effect on pore-water pressure and dynamic response when $k \geq 0.0003$ ft/sec. When drainage occurs at these higher permeability levels, the liquefaction potential is reduced in many cases. However, as the pore pressures are redistributed upwards to levels of lower effective stress, the level at which liquefaction first occurs may be closer to the surface than predicted with drainage ignored.

Parameter Effects on Dynamic Properties of Frozen Soils

Vinson (1978) discusses frozen soils considering the parameters that have been found to affect their dynamic properties. These parameters may be divided into two groups, as shown in Table 11-1, and they may be considered separately or be combined. There are a large number of combinations that would result in equivalent dynamic properties.

Examination of parameters that affect the dynamic response of frozen soils or frozen soil deposits is discussed by Vinson (1978) in considerable detail and will not be repeated here except for characteristic examples and a brief summary (see also the foregoing sections). Reference is also made to Bennett (1972); Bentley (1972, 1975); Roethlisberger (1972); and Mackay and Black (1973).

Coarse-grained soils (gravel and glacial till) generally have higher compression wave velocities than fine-grained soils (silts and clays). Furthermore, the compression wave velocities of material in the frozen state are much higher than in the unfrozen state.

The influence of density on the compression wave velocity of ice is shown in Figure 11-11. The compression wave velocity decreases markedly with decreasing density. The rate of decrease is somewhat greater for densities below 600 kg/m^3 than for densities above this value.

Results based on laboratory tests are reported by many investigators. Among these are Kaplar (1963; 1969); Nakano et al. (1972); Nakano and Arnold (1973); Stevens (1973; 1975); Vinson and Chaichanavong (1976); and Vinson et al. (1977). Useful references for laboratory tests using ultrasonic equipment are given by Vinson (1978). Interestingly, when examining the dynamic properties of frozen soil it should be emphasized that the structure of a sample reconstituted and frozen in the laboratory (artificial freezing) can differ markedly from the structure of a sample for a comparable soil type taken from a frozen ground deposit (naturally frozen soil). Accordingly, the properties of the two samples can differ significantly.

The dynamic stress-strain properties of frozen soil increase with descending temperature. The increase in longitudinal wave velocity is greatest for fine-grained

TABLE 11-1 Parameters Influencing Dynamic Properties of Frozen Soils

Field and/or Test Conditions Parameters	Material Parameters
Temperature	Material type and composition
Strain (or stress) amplitude of loading	Material density or void ratio
Frequency of loading	Ice content or degree of ice saturation
Confining pressure	Unfrozen water content
Duration of loading	Anisotropy

Source: From Vinson (1978).

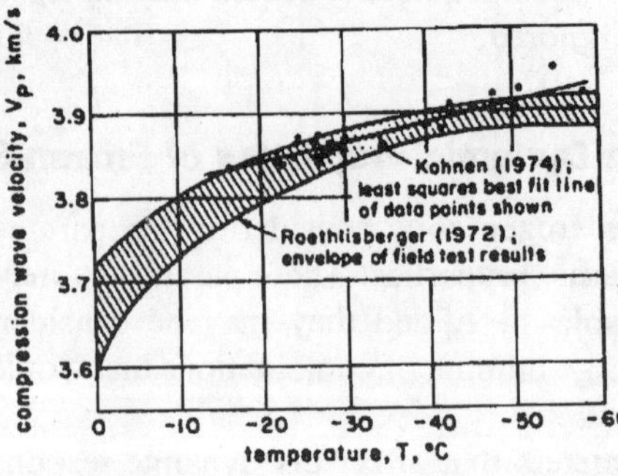

Figure 11-11 Compression wave velocity of ice versus temperature. (From Vinson, 1978.)

soils, particularly in the range 0° to −5°C, although there are some exceptions (Vinson, 1978). Stevens (1975) reports that dynamic moduli in the frozen state are generally more than two orders of magnitude greater than moduli in the unfrozen state. This corresponds to longitudinal wave velocities in the frozen state of more than one order of magnitude greater than velocities in the unfrozen state.

Figure 11-12 shows the influence of confining pressure on dynamic properties of five soil types and ice at two densities for various test and material conditions. For the four fine-grained soils the change in either the dynamic Young's modulus or the damping ratio over the range of confining pressure is very small. There is, however, an increase in dynamic modulus with increasing confining pressure for the coarse-grained sand and ice samples. The damping ratio for the coarse-grained sand and ice samples is not affected by confining pressure. Results of research by Vinson and Chaichanavong (1976) and Vinson et al. (1977) suggest that the relationships shown in Figure 11-12 are independent of the strain amplitude, frequency of loading, and the test temperature.

In general, all investigations confirm the influence of soil type on dynamic properties. For example, coarse-grained soils have higher dynamic stress-strain properties than fine-grained soils. Silts have been found to have higher dynamic stress-strain properties than clays. The wave velocities of gravelly sand are determined by Kaplar (1969) to be about three times the velocities of plastic clays. Likewise, energy-absorbing properties are higher for frozen clay than for frozen sand or silt, the latter two having similar energy-absorbing properties.

This brief review highlights the dynamic behavior of frozen soils, but it also demonstrates the significant range of soil types, materials, and field or test condition parameters that may affect the dynamic properties. The relative importance of these parameters is assessed by Vinson (1978).

Frozen Clay Under Cyclic Axial Loading

Cyclic triaxial tests on frozen Ontonagon clay have been carried out by Vinson et al. (1978) to determine values of dynamic Young's moduli and damping ratios. The

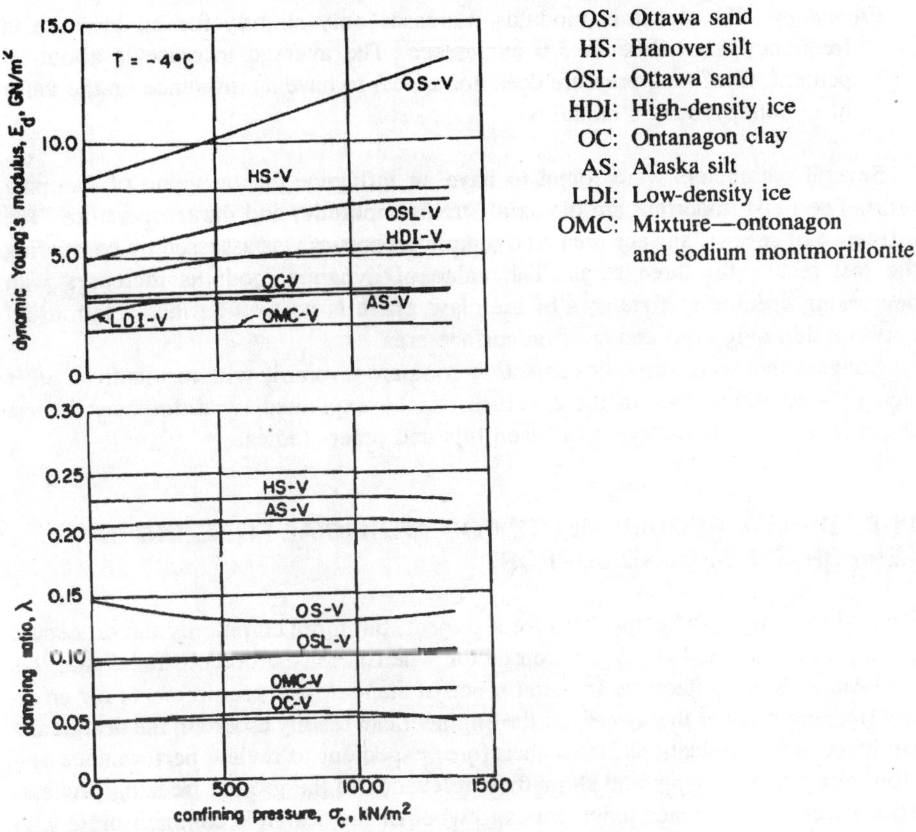

Figure 11-12 Dynamic Young's modulus and damping ratio of frozen soil versus confining pressure. (From Vinson, 1978.)

range of test conditions associated with this program was chosen to correspond to field conditions and loadings anticipated for frozen soil deposits subjected to strong motion earthquakes and low frequency dynamic loadings.

The value of dynamic Young's modulus of the frozen clay was determined to be in the range 50×10^3 to 870×10^3 lb/in.^{2} for the range of test conditions. Several parameters were found to have an influence on the dynamic modulus as follows:

Axial Strain Amplitude: The dynamic Young's modulus decreases with an increase in axial strain amplitude from 3.2×10^{-3} to 10^{-1} percent. In this case the average decrease is about 60 percent.

Temperature: The dynamic modulus decreases with ascending temperature from 14° to −30.2°F. The average decrease is about 60 percent.

Water Content: The dynamic modulus increases with increasing water content from 29 to 55 percent, giving an average increase of 40 percent.

Frequency: The dynamic modulus increases only slightly for an increase in frequency from 0.05 to 5.0 cycles/sec. The average increase is about 30 percent. Confining pressure does not appear to have an influence on the value of dynamic Young's modulus.

Several parameters were found to have an influence on the value of damping ratio. The most important are the axial strain amplitude, and the temperature. The influence of specific surface area on dynamic properties was assessed by comparing the test results for three clays. The value of dynamic modulus increases with decreasing specific surface area of the clay. There is no well-defined relationship between damping ratio and specific surface area.

Longitudinal wave velocities appear to compare favorably with results from other studies, and differences in these values can be explained by differences in test techniques and material types between this and other studies.

11-5 DESIGN REQUIREMENTS FOR ARTIFICIAL FREEZING IN TEMPORARY GROUND SUPPORT

Ground freezing may be specified for a project, but more commonly the procedure is proposed and detailed by the contractor when it can be demonstrated that this application is cost effective. In the absence of standardized requirements for artificial freezing, neither the owner nor the engineer can readily ascertain the design and the associated implications. It is, therefore, expedient to review performance and monitoring requirements and show their relevance to the ground freezing process. This review may enhance judgment as to whether the design is commensurate with project requirements.

Artificially frozen ground can be used where it is necessary to limit exterior groundwater drawdown; as temporary ground support before excavation; for various excavation configurations such as circular shafts, tunnels, deep basements, remedial work, and underpinning; and where safety and construction controls cannot be met by other cofferdam methods. Because of the considerable expense of energy and rental of the refrigeration plant, the economy of the method often depends on the duration of construction within the excavation.

Failure of an artificially frozen barrier because of marginal procedures or inadequate assessment of ground conditions can have serious implications, such as subsidence or structural damage within the influence zone of the excavation, along with associated legal implications and contractor claims. Catastrophic failures have been rare, but partial failures due to an unfrozen zone or unscheduled delays are common. Minimum performance requirements for a given project must therefore be established to meet the specific needs and as a function of the consequences of possible failure.

Relevant Data For proper design of a frozen ground wall, the soil exploration should be planned to meet the normal requirements of the project, including number and type of borings, undisturbed samples, groundwater data, and obstructions, since these may slow installation of the freezing system and thus affect construction cost and schedule.

The water content of the soils to be frozen is particularly relevant in cohesive soils. A large amount of energy must be removed to change pore water to ice, and freezing typically is achieved at a slower rate in high-water content cohesive soils. The density of soils to be frozen is also important. This parameter is readily determined by measuring and weighing undisturbed samples of cohesive soils. Densities of granular deposits are generally obtained with sufficient accuracy. The degree of saturation of granular soils above the groundwater table is also essential and should be determined.

The investigation program should include groundwater conditions to permit determination of the gradient across the site, as well as the grain size and permeability of each aquifer. These data are used in estimating seepage velocity through the soil pores. Temperature of the ground and groundwater should be determined. If ground freezing is considered in the design stage, the engineer may require undisturbed samples of critical strata for laboratory testing of both frozen and unfrozen strength and deformation. Frost heave and thaw consolidation tests will be useful if heave and settlement are predicted and may have adverse effects on existing or new structures (Lacy and Floess, 1988).

11-6 GENERAL DESCRIPTION OF FREEZING SYSTEMS

System Components

A ground freezing installation usually consists of a refrigeration plant and a distribution system for controlled circulation of coolant to the ground (Jumikis, 1966; Shuster, 1972; Powers and Maishman, 1981). In this section, only the most common freezing systems are briefly reviewed.

A typical refrigeration source is a conventional ammonia or freon plant, available in various capacities and usually trailer or skid mounted. The plant is powered by 100 to 300 HP motors providing freezing capacities between 40 and 120 tons of refrigeration (1 ton of refrigeration = 3.5 kW). Rated tonnage for ground freezing is dependent on brine temperature and is often based on cooling the circulating brine to −20°C. The evaporating temperature of the refrigerant in the chiller will be close to −25°C before this brine temperature is obtained. Lacy and Floess (1988) recommend that this relationship be established as a standard in artificial freezing construction. A refrigeration plant normally produces more than twice the rated tonnage during startup when the brine is warm and only 70 percent to less than 50 percent the rated value after the ground is frozen and brine temperatures approach practical lower limits. Rated tonnage also depends to a lesser degree on atmospheric conditions and refrigerant temperature.

Establishing the rated capacity of the refrigeration plant in the field is rather difficult. These plants are often modified and may have replacement components that are different from the initial assembly. Although the basic components are available from manufacturers, they are usually selected and assembled by a refrigeration specialist based on design and construction requirements.

Several plants can be combined if greater capacity is needed. A backup refrigeration unit should always be available during the entire excavation process to ensure the continuous stability of the frozen ground in the event of breakdown. A backup unit is also required during initial freeze if breakdown delays cannot be tolerated by project scheduling.

The refrigeration plant consists of a compressor, a condenser, and an evaporator, shown schematically in Figure 11-13. The compressor liquifies gaseous refrigerant as it is pressurized to several atmospheres. Pressurization raises the temperature of the refrigerant, which is cooled as it passes through water-cooled coils in the condenser. Next, the refrigerant passes through an expansion valve and is sprayed onto the coils of the evaporator. Coolant is chilled as it passes through the evaporator coils acting as a heat exchanger. The ammonia or freon gas then flows into the compressor where the cycle is repeated. The refrigeration plant is a closed system with the ammonia or freon refrigerant continuously circulated.

In a conventional ground freezing system the coolant is brine. This is a solution of calcium chloride and water that has a specific gravity 1.24 to 1.28. The brine is pumped into freeze pipes in the ground by means of a supply header. The chilled brine returns back through the annulus formed by the pipes extracting heat from the ground as it flows. Brine can be pumped directly from one freeze pipe into another if both ends of the freeze pipe are accessible, as in a tunnel. The normal brine is collected in a return header and rechilled at the refrigeration plant, and the cycle is repeated. The freeze pipes and headers form a closed system in which the brine is continuously circulated. Calcium chloride brine begins to gel at about −40°C. The system is simple and probably the most commonly used in ground freezing. Time required for freezing usually is weeks.

A system less frequently used entails the direct injection of a refrigerant, typically liquid nitrogen, into the freeze-pipe assembly where it evaporates. The result-

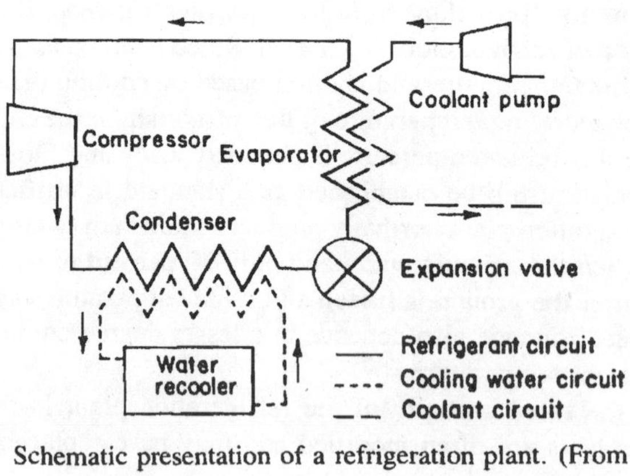

Figure 11-13 Schematic presentation of a refrigeration plant. (From Shuster, 1972.)

ing gas is released into the atmosphere, still at a very low temperature. The nitrogen system will freeze soil much faster than chilled brine, so that the freezing time is usually days rather than weeks.

Expandable refrigerants, such as nitrogen, are used mainly for small projects of short duration, or where emergency stabilization is needed. The main difficulty in this case is control of the open system. For example, unconfined venting of the refrigerant can result in a very irregular frozen ground zone. Liquid nitrogen has also been used as a refrigeration plant backup system to cool the brine. This application warrants ample control to avoid localized overcooling of the brine. Other freezing systems involve the recovery of evaporated refrigerants, and their subsequent reliquefication and recirculation. These are not in broad use today.

Review of Freezing Procedures

A wall of frozen ground is produced by positioning freeze pipes at a predetermined spacing along the line of the proposed excavation. Freeze pipes are normally made of metal in diameters of 80 to 100 mm. Larger pipes, up to 250 mm in diameter, are sometimes used, particularly when control of alignment is necessary.

Freeze pipes can be installed by either soil removal methods or soil displacement methods. Examples for horizontally installed pipe are: (a) rotary wet drilling with

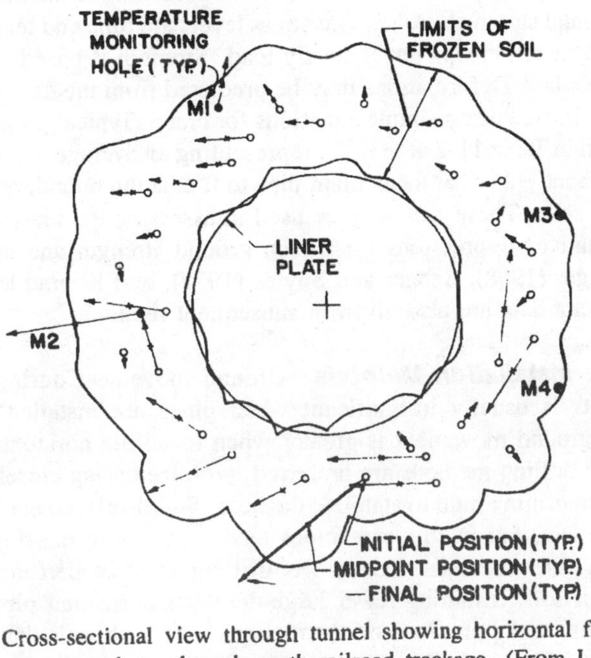

Figure 11-14 Cross-sectional view through tunnel showing horizontal freeze pipes and approximate frozen ground envelope beneath railroad trackage. (From Lacy and Floess, 1988.)

casing; (b) air track drilling with casing; (c) pipe jacking with interior soil removal; (d) the use of a pneumatic mole with following casing; (e) the use of a dry auger with casing; (f) jacking closed-end pipes; and (g) the use of a steerable larger diameter casing. Since it is easier to control and adjust the alignment of larger diameter pipe, it should be considered for horizontal freeze pipes because alignment control is more difficult in this case. Vertical pipes are normally installed in holes advanced with drilling mud or wet drilled with a casing.

After installation, the actual position of the pipes is determined using inclinometers for vertical holes and deflectometers for horizontal holes. Data thus obtained are used to check the spacing of adjacent freeze pipes with reference to the design values at any point along their length. If the spacing exceeds the design value, additional freeze pipes should be installed.

Figure 11-14 shows a cross section through a tunnel with the horizontal freeze pipes and the approximate frozen ground envelope at the critical location directly beneath overlying railroad tracks. Flow through individual pipes should be adjusted with valves to provide equal flow along the frozen ground structure. Bleed-off valves may be provided to remove air from the freeze-pipe system (see also subsequent sections).

11-7 DESIGN CONSIDERATIONS

Frozen Soil Walls Wall thickness is selected according to the limiting stresses in the frozen ground structure. Allowable stress levels are time and temperature dependent, and frozen soil creeps under steady load. Strength is based on plastic failure and Coulomb's law. Deformations may be predicted from the data of the foregoing sections or estimated using simple equations for creep. Typical properties for frozen soil are shown in Table 11-2 at $-10°C$, representing an average soil temperature that varies from about $-25°C$ at the coolant pipe to 0°C at the boundary between frozen and unfrozen soils. These data may be used in assessing the feasibility of a frozen ground alternative. Expressions for frozen ground strength and deformations are given by Sanger (1968), Sanger and Sayles (1978), and Konrad and Morgenstern (1980). Relevant data are also given in subsequent sections.

Freeze Pipe Installation Methods Ground movement during installation of these elements is usually insignificant when pipes are installed vertically. The potential for ground movement is greater when installing horizontal freeze pipes.

Rotary wet drilling methods are preferred, with the casing closely following the drill bit or with drilling mud to stabilize the hole. Soil displacement methods reduce ground settlement, but in this case pipes tend to be more misaligned than pipes installed with other methods. Rotary wet drilling must be performed carefully to prevent loss of soil. Installing fewer large-diameter horizontal pipes is preferable with long pipes, in the presence of obstructions, and with soils that are variable or dense. Smaller diameter pipes are suitable for lengths less than 100 ft in low-strength soils.

TABLE 11-2 Typical Frozen Ground Properties

	Sand[a]	Clay
Short-term strength		
Tsf	95–160	50–95
MPa	9.1–15.3	4.8–9.1
Stress causing failure at 60 days of load (%)	± 70	± 70
Allowable strength at 60 days (% of 1)	30–50	—
Elastic modulus		
Frozen soil		
Tsf	6,000	—
MPa	575	—
Unfrozen soil		
Tsf	500	—
MPa	48	—

Source: From Lacy and Floess (1988).

[a]Saturated soil (partially saturated soils have reduced strength).

Freeze Pipe Spacing Pipes for ground freezing are normally spaced 3 to 4 ft apart. A rule of thumb for smaller freeze pipes is to select a spacing-to-diameter ratio ≤ 13. This simple criterion is applicable to pipes that are 120 mm or less in diameter. Most specifications should include this requirement.

Brine Temperatures During freezing, these drop in the first several days of freezing and approach an equilibrium state between −20° and −30°C. A brine temperature −25°C or less is necessary to ensure that the soil is frozen rapidly, a condition minimizing frost heave and expediting construction. Temperature requirements will vary with strength requirements, and therefore with soil type and water content (see also subsequent sections).

Size of Refrigeration Plant In the United States, this component is normally measured in tons of refrigeration. Other options include horsepower, because the rated tonnage is dependent on several factors such as air temperature, relative humidity, and brine temperature.

Normally, 4 to 7 tons of refrigeration per 93m² of interior frozen ground surface is required to form the wall of shafts. This corresponds to about 0.013 to 0.025 tons of refrigeration per linear foot of freeze pipe. Tunnels sometimes have a double row of freeze pipes above the springline, and the refrigeration requirements for these cases are higher. Given the particular project, the refrigeration capacity depends on several factors, including desired speed of freezing, design temperature, and so on. Procedures are given by Sanger (1968), and are reviewed in subsequent sections. About 50 to 70 percent of the estimated tonnage requirement is normally needed to maintain the frozen ground after it is formed.

Special Considerations at Shallow Depth Sufficient moisture between the soil grains is necessary to form ice bonds. Saturated soil below groundwater usually attains high strength when frozen. Clay soils above the groundwater table are nearly saturated and normally develop high frozen strength. Silty soils near the water table usually have high moisture content as a result of capillarity, and will therefore develop high frozen strength.

At shallow depths, evaporation tends to dry the soil, and this may result in low frozen strength. Sand above the water table is normally too dry to form strong ice bonds when frozen, and typically may require additional moisture to develop adequate frozen strength. Wetting the soil surface from a ditch or using slotted PVC pipe can provide this extra moisture, but excessive application of water can delay formation of the frozen ground wall. Horizontal slotted pipes have been installed above tunnel alignment to provide extra moisture at sites with low groundwater table. In other instances, bentonite slurr, walls have been used to increase moisture in highly permeable cinder fill that hac low moisture content and where the water drained away too rapidly for the intended frozen strength.

11-8 BASIC DESIGN PARAMETERS

Basic design parameters necessary for a ground freezing program include the thermal, hydrological, and mechanical properties of the soil mass to be frozen. The effect of these parameters on the behavior of the ground must be analyzed in terms of performance criteria, cost factors, and time factors necessary to achieve the final design of the freezing plan.

Thermal Properties

Besides the initial subsurface temperature T_0, other thermal properties relevant to the design of a frozen structure and the freezing program are as follows.

Specific Heat The specific heat C of both the fluids and solids in the zone to be frozen, and conversely the ratio of the amount of heat required to change the temperature of a unit mass of material by one degree to the amount of heat required to raise the same mass of water by one degree, must be obtained. The usual approach is to use the term "heat capacity" for this quantity and to consider both a mass and a volumetric heat capacity term. Mass heat capacity C_m is taken as reference, and for water it is 1 cal/gm-°C or 1 BTU/lb-°F. Volumetric heat capacity C_v is sometimes more convenient, or

$$C_v = C_m \gamma_d \tag{11-7}$$

where γ_d is the dry unit weight of the material.

Frozen and unfrozen soils have different heat capacities. Moisture content w

(weight of water in percent of dry soil weight) is the major factor to be considered in calculating heat capacity.

The approximate volumetric capacity is as follows:

$$\text{Unfrozen soil} \quad C_u = \gamma_d C_{ms} + w\gamma_d \frac{C_{mw}}{100} \tag{11-8}$$

$$\text{Frozen soil} \quad C_f = \gamma_d C_{ms} + w\gamma_d \frac{C_{mi}}{100} \tag{11-9}$$

where γ_d = dry unit weight of soil, lb/ft^3
C_{ms} = mass heat capacity of dry soil
C_{mw} = mass heat capacity of pore water
C_{mi} = mass heat capacity of ice.

Typical values for dry unit weight and water content of soils are given in Table 11-3.

Latent Heat of Fusion The latent heat of fusion L of the pore water is the amount of heat that must be removed to convert the pore water to ice. Because latent heat is large compared to all other heat losses, it represents the most important factor in the freezing process. Removal of latent heat begins when the temperature of an element of soil is lowered to about 32°F. The temperature remains approximately at this level while water is converted to ice. In fine-grained, brine-saturated, or chemically contaminated soils this phase transition occurs over a temperature range rather than at a single point. Approximately 144 BTU are required to convert 1 lb of water into ice (or 80 cal/gm). For a body comprised of both solids and moisture, the latent heat of fusion is a function of the dry unit weight of the soil (γ_d) and the percent of water by dry weight (w), or

$$L = \gamma_d \, 1.44 \, w \ \text{BTU/ft}^3 \tag{11-10}$$

TABLE 11-3 Water Content and Dry Unit Weight of Typical Soils

Soil Type	Typical Values	
	w (% dry weight)	γ_d (pcf)
Silty or clayey well-graded sand and gravel	5	140
Clean well-graded sand and gravel	8	130
Well-graded sand	10	120
Poorly graded sand	15	110
Inorganic silt or fine sand and silt	15–25	110–85
Stiff to very stiff clay	20–30	95–80
Soft to medium clay	30–40	80–70

Since the variation in dry unit weight is small, the water content is of greater significance.

Thermal Conductivity This parameter, denoted by K, is defined as the quantity of heat transfer through a unit area in unit time under a unit thermal gradient. Thermal diffusivity (or temperature conductivity α) is the quotient of conductivity and volumetric heat capacity ($\alpha = K/C$). A summary of thermal conductivities for typical frozen and unfrozen soils is given by Kersten (1949).

Hydrologic Properties

These are interrelated with the bulk thermal properties and have a strong influence on the final design. The most important parameters are: (a) moisture content; (b) subsurface flow rates and direction of flow; (c) permeability of the soil; and (d) pore water chemistry.

Mechanical Properties

A summary of mechanical properties and creep for frozen sand and clay is presented in Sections 11-2 and 11-3, and includes results from tests of specific soil samples and test conditions.

A frozen soil mass is a viscoplastic material that will creep under applied stress. Normally, the creep rate, rather than ultimate strength, will control the design, but the latter is a useful index property in assessing creep.

Strength Tests for Design Tests are usually made using laboratory samples frozen to the temperature expected in the field. Problems ordinarily arise when heterogeneous deposits are encountered and the true in situ conditions are not represented in the laboratory investigation. In situ pressuremeter tests are used to assess deformation characteristics of soil after freezing (Shuster, 1975).

The strength of frozen ground may be more critical in certain cases, for example across the crown of a tunnel that is immediately beneath a heavy structure. For simple projects a general knowledge of the soils and their degree of saturation may be adequate, and design aides may be used to estimate frozen soil strength. If the design involves a critical frozen ground structure, it will be necessary to obtain undisturbed samples of in situ soils for laboratory testing.

From the foregoing it follows that the strength of frozen ground is a function of its temperature below freezing, considering also the fact that frozen soil creeps under load. Strength becomes greater at lower temperatures, but decreases with time of loading. As shown in Sections 11-2 and 11-3, the rate of creep deformation depends on the stress level.

A common method of obtaining frozen strength and modulus is by artificially freezing a small test section of the ground for in situ testing (Lacy et al., 1982). A soil auger is used to make a hole through the frozen ground, and a pressuremeter appropriate for rock testing is inserted into the hole and expanded against its sides.

Results of these tests correlate reasonably well with tests performed on soil samples frozen in the laboratory.

It is sometimes necessary to obtain the elastic modulus of both the frozen ground and the surrounding unfrozen zones to determine the distribution of unbalanced loads such as moving trains or vehicles above a frozen ground tunnel. If the frozen ground structure is close to the moving loads and the surface has heaved during freezing, the dynamic loads will be attracted to the stiff frozen soil section.

Analysis of stresses within a frozen ground structure can vary from relatively simple empirical methods to elaborate finite element techniques that give detailed contours of stress levels throughout the structure. Examples are presented in subsequent sections.

Creep The behavior of a viscoplastic frozen soil with time is dependent upon stress levels and temperature. Creep will increase with applied stresses and will decrease with temperatures below freezing (see also the foregoing sections and Figures 11-10, and 11-9). Typical behavior patterns are shown in Figures 11-15 and 11-16. Figure 11-15 shows the effect of increasing compressive stress on axial strain. Figure 11-16 shows strain increase with both higher stress and higher temperature. Stress is held constant for each of the three curves.

Point F in Figure 11-16 represents the line at which the rate of strain becomes increasingly greater with time. This phase is referred to as creep failure. Similar relationships are obtained for the effect of stress and temperature on time of creep failure. These show that as the temperature increases, the reciprocal of stress also

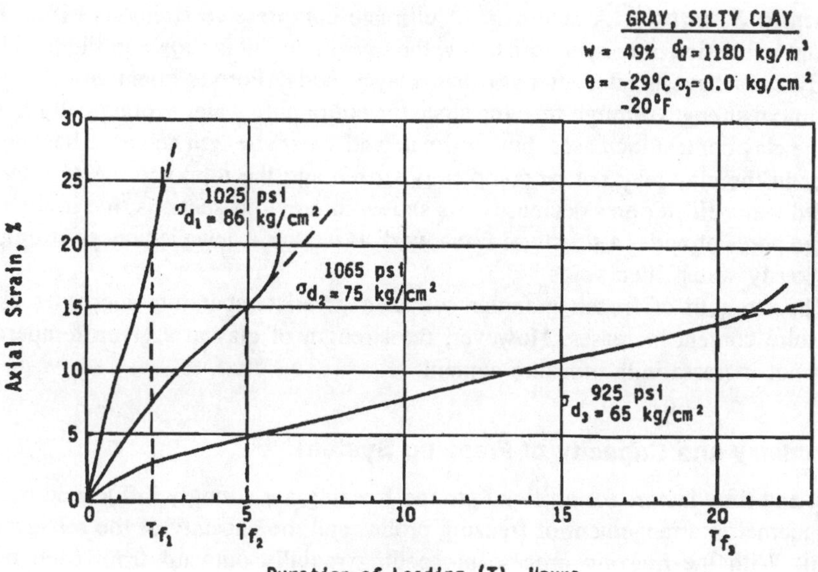

Figure 11-15 Strain versus time and loading for frozen soil. (From Shuster, 1972.)

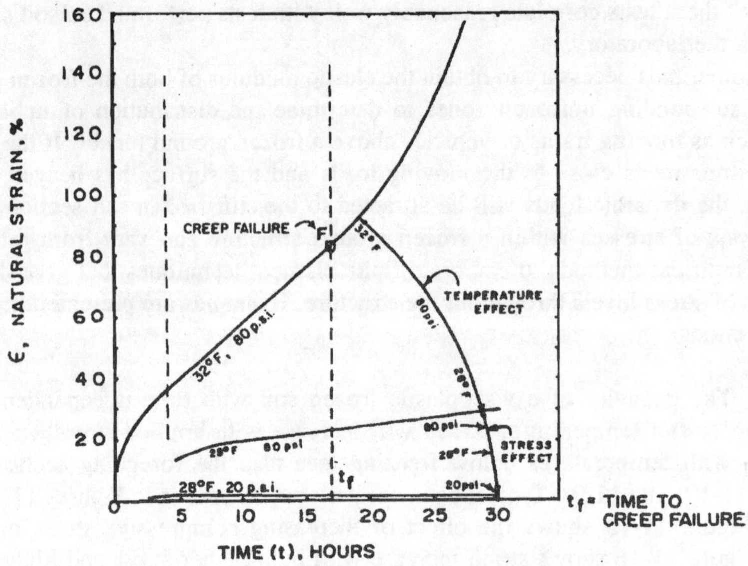

Figure 11-16 Creep curves for an organic silty clay with temperature influence. (From Sanger, 1968.)

increases, and also that the time to creep failure becomes exponentially longer as the reciprocal of stress increases.

Ultimate Strength A summary of ultimate compressive strengths of common soils as a function of temperature below the freezing point is shown in Figure 11-17. Sandy soils have greater strength than clayey soils. Porous granular soils attain maximum strength through freezing since the entire pore water is practically frozen. As the clay content increases, both ultimate and shear strength decrease because the water in the clay may not be completely frozen and the total volume of interconnected water-filled pores decreases. As shown in previous sections, ice in interconnected pores provides a structural framework as well as a new element of strength in previously water-filled voids.

The strength of frozen granular soil at a given temperature increases as the moisture content increases. However, the strength of clay at a given temperature does not increase with moisture content.

Geometry and Capacity of Freezing System

Cost and time factors for artificial ground freezing are strongly influenced by both the geometric arrangement of freezing probes and the capacity of the refrigeration plant. With the freezing process proceeding radially outward from each of the freezing probes, the rate of progress depends on: (a) the capacity of the equipment relative to the thermal load of all the combined probes and surface piping; (b) the

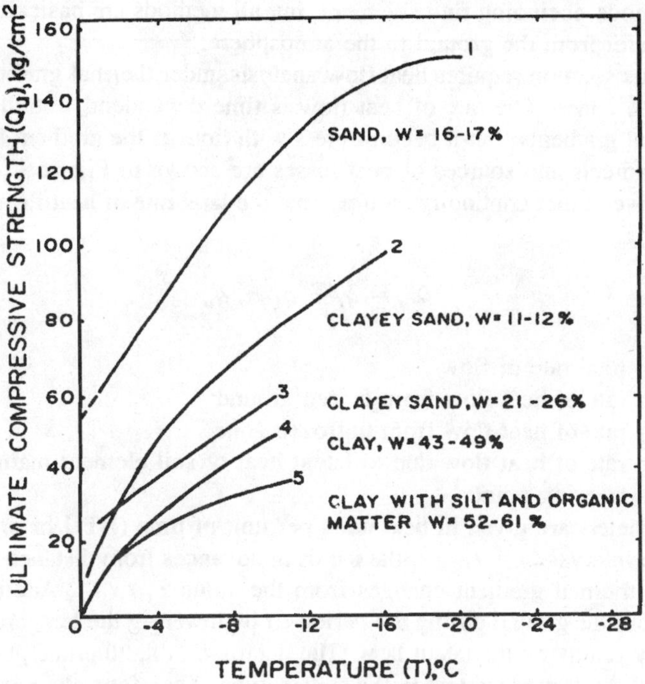

Figure 11-17 Ultimate short-term compressive strength as a function of temperature. (From Tsytovich, 1960.)

thermal gradient between the probe and surrounding materials; (c) the rate of heat transfer between the probe–frozen ground sector and the unfrozen soil mass; and (d) fringe losses at the freezing front due to lateral groundwater flow.

In designing the system, increased freezing rates can be obtained by decreasing freeze element spacing or by increasing the temperature differential through an increase of the capacity of the cooling equipment. Fringe losses are reduced as the radial freezing fronts converge between probes since the fronted areas between frozen and unfrozen masses are reduced and since thermal losses due to groundwater movement through the freezing mass are blocked.

11-9 GENERAL DESIGN APPROACH

Thermal Considerations

Background and Fundamentals The basic approach considers two main phases of operation as follows: (a) reducing the temperature of the soil mass to a level where the required frozen ground behavior is attained, and (b) maintaining all or some part of the frozen mass at a temperature where the mass will behave as predicted during construction activities. Theoretical analysis may be pursued by several methods, including finite element, but all methods are basically an exercise in heat transfer from the ground to the atmosphere.

A rigorous solution requires heat flow analysis under thermal gradients in frozen and unfrozen zones. The rate of heat flow is time dependent, initially high under steep thermal gradients, but it becomes less with time as the gradient becomes flat. Thermal gradients and sources of heat losses are shown in Figure 11-18.

At any given time, continuity requires that the total rate of heat flow must satisfy the following:

$$\Sigma_q = q_f + q_L + q_u \tag{11-11}$$

where Σ_q = total rate of flow
q_f = rate of heat flow from frozen ground
q_u = rate of heat flow from unfrozen zone
q_L = rate of heat flow due to latent heat of soil element maintained at the freezing point

These parameters are given in heat units per unit of time (BTU/hr or Cal/sec).

In a time interval $\Delta t = t_2 - t_1$ the ice front advances from distance z_i to distance z_2, and the thermal gradient changes from the value T_1 to T_2. Additional heat is removed from the ground during the period Δt by lowering the temperature from T_1 to T_2 and by removing the latent heat. The decrease in the thermal gradient during time interval Δt is most evident in the frozen zone. Therefore, the rate of total heat flow has also decreased.

The incremental heat loss during time Δt is

$$\Delta Q = \Delta Q_u + \Delta Q_L + \Delta Q_f \tag{11-12}$$

where ΔQ = incremental total heat loss
ΔQ_f and ΔQ_u = heat loss required to drop temperature from T_1 to T_2 in frozen and unfrozen zones, respectively
ΔQ_L = incremental latent heat loss

Rigorous Solutions These are rather complex because the thermal gradient changes with time. Examples are given by Sanger (1968). These procedures involve a closed form solution of the energy removal and time required to freeze a zone of a specified size. The mathematical requirements are complex, and the multiplicity of design variables makes this approach cumbersome except for very simple cases. Alternatively, computer models may be developed using finite element techniques if the project size warrants the cost. However, serious limitations may have to be accepted where it is necessary to design a freezing plan in heterogeneous deposits.

Simplified Solutions In this approach, the objective is to: (a) identify the zone to be frozen; (b) establish existing temperatures and temperatures after freezing; and (c) compute the amount of heat loss required to transfer the volume of soil in the

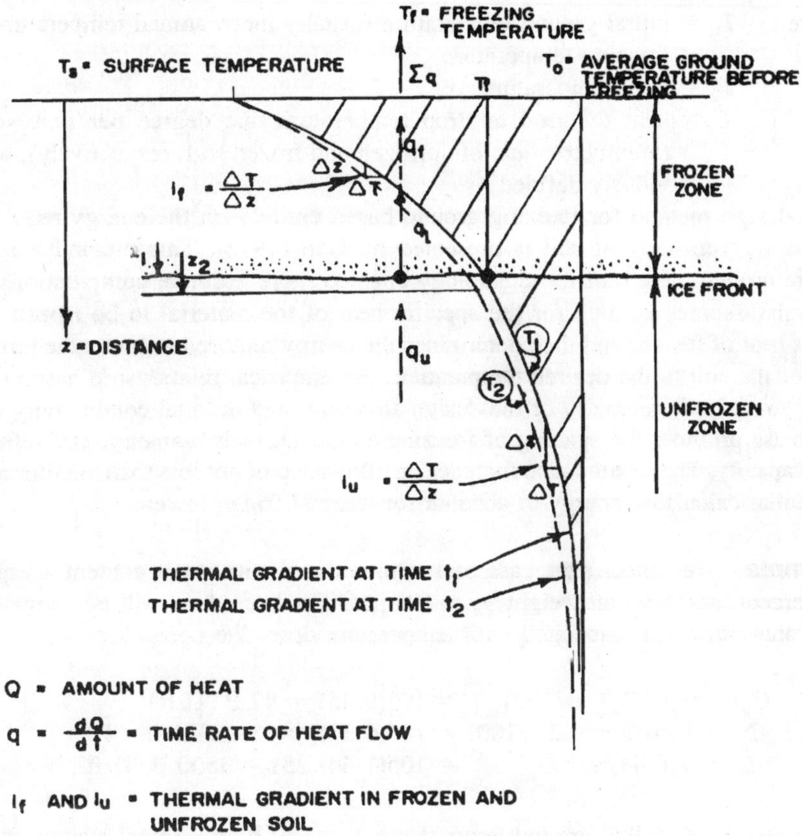

Figure 11-18 Heat flow under thermal gradient.

zone from the existing to the frozen condition. The temperature drop, hence heat loss at distances beyond the ice fronts, is neglected. However, for practical applications the heat loss within the frozen zone is large compared to the heat loss beyond the frozen zone.

We introduce the following notations:

Q_u = heat flow from soil, solids, and pore water required to drop temperature from initial soil temperature T_0 to the freezing T_f; Q_L = latent heat flow to transfer water to ice (occurs at constant temperature T_f); and Q_F = heat flow from soil, solids, and pore water required to drop temperature from freezing point T_f to the design subsurface temperature T_2. We can now write the total heat loss from a unit volume of soil as

$$
\begin{aligned}
Q_u &= C_u\,(T_0 - T_f) \\
Q_L &= \gamma_d\,(1.44)w \\
Q_f &= C_f\,(T_f - T_2)
\end{aligned}
\tag{11-13}
$$

where T_0 = initial ground temperature (usually mean annual temperature)
T_f = freezing temperature
T_2 = final temperature
C_u, C_f = heat required to drop temperature one degree per unit volume (volumetric heat of unfrozen and frozen soil, respectively), as previously defined

A design method for freezing ground based entirely on the energy required to freeze a given body of soil is presented by Gail (1972). This method was used before modern heat transfer technology enabled more accurate computations. The analysis assumes a value for the specific heat of the material to be frozen and a latent heat of fusion, and then determines the energy required to lower the temperature of the soil to the desired temperature. An empirical relationship based on the energy required, geometry of the design structure, and thermal conductivity of the soil mass provides the spacing of freezing elements, their diameter, and refrigeration capacity. The method also includes an allowance of not less than 100 percent of the initial calculated energy to account for thermal fringe losses.

Example We consider the case of a saturated soil with water content w equal to 25 percent, and dry unit weight γ_d = 105 pcf. The latent heat will be compared to the volumetric heat associated with temperature drop. We compute

$$\begin{aligned} C_u &= \gamma_d(0.2 + w/100) &&= 105(0.45) = 47 \text{ BTU/ft}^3/°\text{F} \\ C_f &= \gamma_d(0.2 + 0.5w/100) &&= 105(0.325) = 34 \text{ BTU/ft}^3/°\text{F} \\ L &= \gamma_d(144)w/100 &&= 105(1.44)(25) = 3800 \text{ BTU/ft}^3/°\text{F} \end{aligned}$$

If we assume an initial ground temperature T_0 = 50°F and a final average design temperature T_2 = 10°F, we can write

$$\begin{aligned} Q_u &= 47(50 - 32) = 840 \text{ BTU/ft}^3 \\ Q_L &= 3800 \text{ BTU/ft}^3 \\ Q_f &= 34(32 - 10) = 750 \text{ BTU/ft}^3 \end{aligned}$$

This simple example demonstrates the significance of latent heat relative to volumetric heat.

Groundwater Movement A rigorous approach to evaluate the influence of groundwater movement on the freezing process is presented by Hashemi and Sliepcevich (1973). The usual assumptions are: (a) the soil is homogeneous; (b) the latent heat of fusion is much greater than the specific heat (heat removal to further lower temperatures) of the frozen soil; and (c) the temperature varies only with time and radial position. Based on these assumptions, a closed solution is developed, but for a field application of multiple rows of closely spaced freeze pipes or temperatures below −40°C the solutions are not applicable.

Commentary The simplified version of the Gail approach allows practical but conservative estimates of energy requirements. Shuster (1972) gives an outline of the basic considerations and empirically supported theoretical correlations between various parameters. These correlations are based on a theory originally developed by Kamenskii (1971) for freezing with air convection.

Typical freezing times for various coolants are given in Figure 11-19 (laminar liquid coolant is about −15° to −40°C, and boiling liquid nitrogen about −175° to −190°C). The R factor suggests the important effect of freezing element spacing. Likewise, Figure 11-20 shows the effect of groundwater flow on freezing time. This time decreases with decreasing temperature of coolant, decreasing element spacing, and decreasing groundwater flow. A point to be emphasized is that the amount of energy and time required is governed mainly by the latent heat of fusion of the pore water.

Effect of Mechanical Parameters

The approach to structural design with frozen ground must consider the viscoplastic time and temperature-dependent behavior of the materials. Appropriate creep theories and rheological models are demonstrated in previous sections. These account

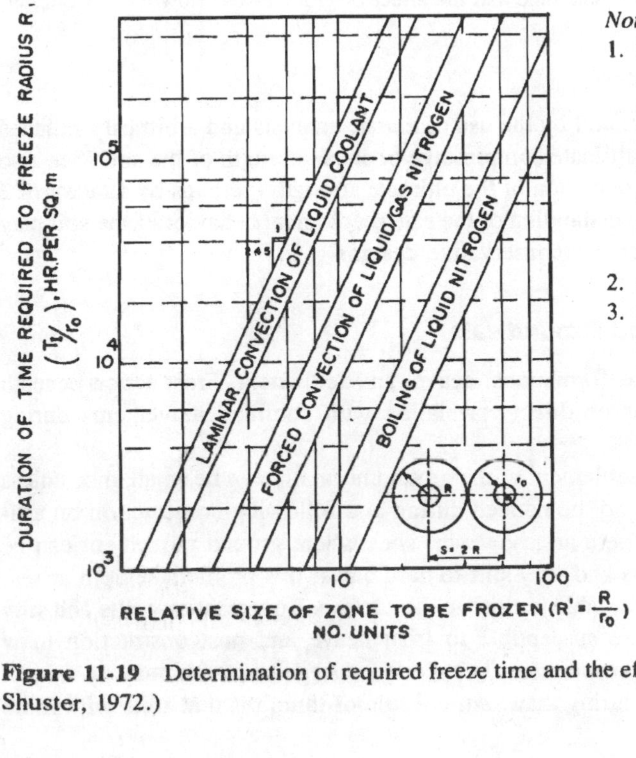

Notes:

1. Indicated bands represent normal range of observed field and laboratory results; however, results with forced convection of N_2 may vary more widely than indicated due to variables in control of the freezing process.
2. R, r_0 in meters.
3. T_I = time of active freezing (hr).

Figure 11-19 Determination of required freeze time and the effect of coolant types. (From Shuster, 1972.)

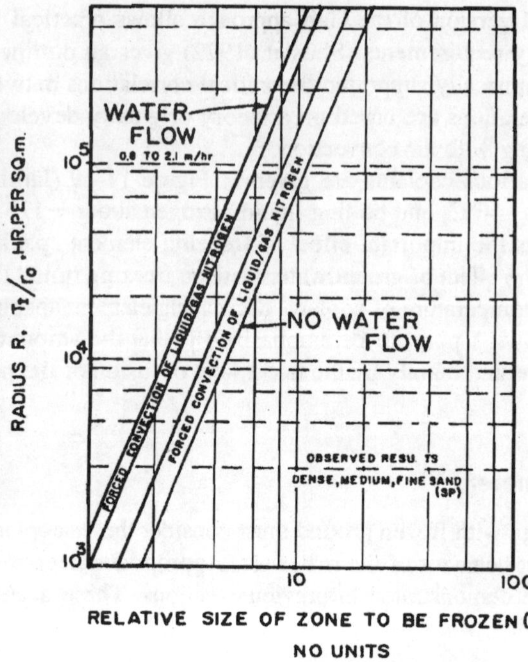

Notes:

1. Indicated bands represent normal range of observed field and laboratory results; however, results with forced convection of N_2 may vary more widely than indicated due to variables in control of the freezing process.
2. R, r_0 in meters.
3. T_t = time of active freezing (hr).

Figure 11-20 Duration of freeze time and the effect of groundwater flow. (From Shuster, 1972.)

for the creep of frozen ground by the use of elastic analysis and arbitrarily reduced values of the short-term ultimate compressive or shear strength of the soil (See also Figure 11-17). Arbitrary reduction of the ultimate strength (perhaps by a factor of 2 to 10) without explicit understanding of the true rheological behavior of the soil may equally produce unsafe or overconservative designs.

Ground Movement and Groundwater

This may involve lateral deformations, heave, and settlement. Frost action beneath unconfined pavement has no direct correlation with confined movements during ground freezing.

Frost heave and thaw settlement of the ground normally can be small and, unless carefully measured, may not be noticed during and following construction on artificially frozen ground. There are, however, sites where ground movement can be significant. In clean sands and very stiff to hard clays, this problem seldom arises, even when the groundwater table is close to the surface. Low plasticity silts and silty fine sands are much more susceptible to frost heave and postconstruction thaw settlement. Soft clays do not necessarily experience large heave but may be subject to significant settlement during thaw. An old rule of thumb is that soils with more than 3 percent by weight finer than 0.02-mm size are frost susceptible below pavement.

The Corps of Engineers criteria have been developed for nonsaturated soils in which the frost heave is associated with ice lens segregation. The primary mechanism of lens growth is by capillary migration of pore water to the ice lens. Clean, free-draining soils have insufficient fines to develop capillarity; hence they are not frost susceptible. The tendency of silty or clayey sands to freeze generally increases with the percentage passing 0.02 mm.

The direction of heat flow in a freezing system with vertical pipes is also perpendicular to the direction of the surface freezing. However, the geometry of groundwater, stratigraphic sequence, capillarity, and permeability relative to the freeze surface are different from the general frost heave model below pavement.

Soil movement from freezing is caused by two distinct but related phenomena. Frost heave, beyond the small volume increase resulting from the phase change of pore water to ice and the associated volume expansion, is caused by the formation of ice lenses along the freezing front that draw water from nearby, more permeable soils by negative water pressure (Konrad and Morgenstern, 1981). Thawing of incompressible soil after frost heave will result in settlement about equal to the frost heave caused by escaping thawed water from the ice lenses.

The magnitude of heave due to ice lensing can be estimated from laboratory tests. These are performed by subjecting undisturbed samples to a controlled negative heat gradient with vertical confining pressures that approximate in situ values (Nixon, 1982). The segregation potential method is used to estimate frost heave. The segregation potential of soil is determined by measuring the volume of water absorbed by the soil sample during freezing. The estimated magnitude of frost heave is function of the segregation potential and soil porosity.

The second phenomenon is manifested when compressible soils with natural moisture content significantly above the plastic limit are frozen (Chamberlain, 1980). Ice lenses form during freezing because of segregation of contained pore and film water within the clayey soils. If there is no external source of water, such ice lensing causes only small volume increase and heave. However, significant settlement will occur during thaw when water in the ice lenses escapes. The thawed soil finally reaches a lower water content and becomes denser than it was before freezing. Freezing, in effect, preconsolidates the soil between ice lenses. If the soil is also susceptible to frost heave from intake of external water, the two phenomena are additive. Thaw consolidation includes both settlement from frost heave and settlement from freeze-thaw preconsolidation.

Thaw consolidation implies that actual densification of soil occurs during freezing. The process becomes, however, evident only when the soil thaws. Then consolidation can be estimated by measuring thaw strain of samples frozen in a controlled negative heat gradient. In clay soil it can be estimated from the plastic limit and natural water content (Chamberlain, 1980).

Rapid freezing can be used to minimize, if not eliminate, ice segregation in soils. However, after a period of time when the ice front advance slows down or stagnates, the possibility for ice segregation and associated heave will exist. In these cases careful monitoring is essential, especially if structures are adjacent to the excavation.

Protection of Frozen Ground Structures Sunlight and high air temperatures can thaw frozen ground exposed during excavation. In such instances it is necessary to protect the frozen ground from deterioration. White plastic or canvas tarpaulins can be draped over exposed areas, or foam insulation sprayed on 50 to 75 mm thick and lightly reinforced with wire mesh anchored to the exposed frozen wall.

For a frozen shaft or tunnel the objective is to prevent surface water from contacting the frozen ground structure. Normal-temperature water flowing over the top of a shaft and into the excavation can rapidly cause the frozen ground to deteriorate. In this case it is necessary to divert drainage around the shaft and prevent inflow if there is a possibility of flooding during wet weather.

Groundwater Control Typically, artificial ground freezing is used where the groundwater table cannot be lowered, where existing structures are underlain by compressible deposits, or where watertight cofferdams cannot readily be installed. The design must therefore consider existing groundwater conditions. Frozen ground shafts normally penetrate an aquiclude below the subgrade, if practical. If the shaft must extend below the bedrock, freeze pipes must penetrate into the rock stratum to ensure a good seal. Bedrock within the shaft can be grouted to minimize water inflow according to rock characteristics (jointing and permeability).

The stability of soil subgrades in shafts is analyzed as in normal shaft construction. If necessary, underlying aquifers are depressurized by pumping before excavation. If no aquicludes are present, the shaft interior must be dewatered with wells. Groundwater inflow from below a frozen ground shaft resulting from dewatering system malfunction can cause rapid deterioration of frozen ground.

Flowing groundwater will impede formation of the ice wall. If the flow velocity through the soil pores exceeds about 1 to 2 m/day, formation of a continuous frozen ground wall may be inhibited (Takashi, 1969; Hashemi and Sliepcevich, 1973). This seepage velocity refers to the resultant actual seepage velocity in the soil pores. The groundwater gradient is also relevant to the freeze problems, since the seepage velocity in an aquifer with known gradient can be estimated if the permeability and porosity of the aquifer are also known.

Selection of Freezing System

Given the particular application, this phase must consider time, temperature, and cost. Generally, the lower the freezing temperature the higher the cost and the shorter the time. Basic elements of typical freezing systems are shown in Figure 11-21. Deviations and modifications from the typical refrigeration system are common and are usually considered proprietary designs among artificial ground freezing contractors.

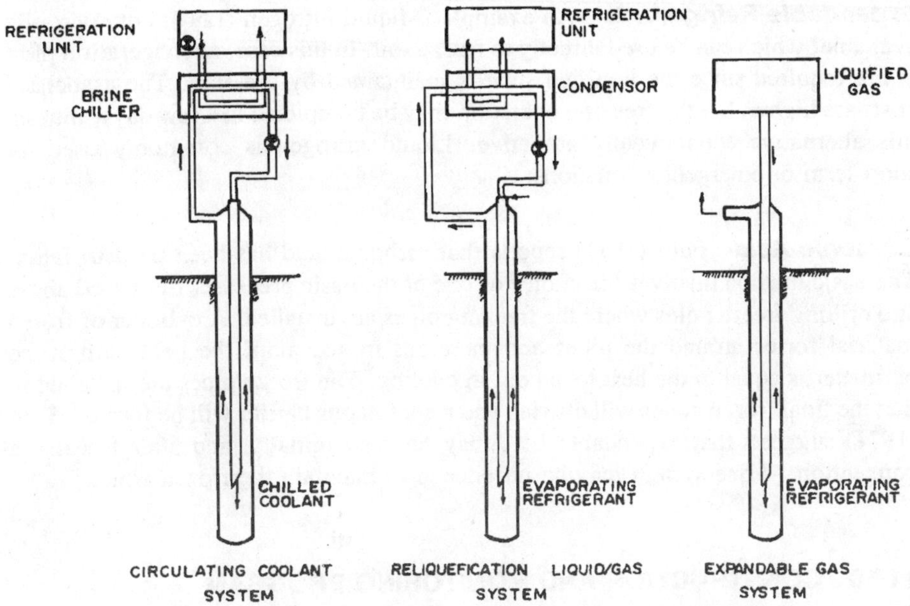

Figure 11-21 Basic refrigeration system components for artificial ground freezing.

A common method for soil freezing is the so-called Poetsch process. This system consists of an ammonia or freon primary refrigeration plant to chill a secondary coolant circulated into freeze pipes in the soil (see also Section 11-6 and Figure 11-13). Other freezing methods are available and may offer the advantage of a lower operating temperature at the soil interface and a quicker freezing time. These methods are normally more expensive but the associated time saving may justify the additional cost. More specifically, the alternatives to the Poetsch process involve the following.

On-Site Refrigeration Plant This system, together with the primary refrigerant pumped directly into the freezing pipes, has been tried using ammonia or freon. However, the system operates under a vacuum so that leaks are undetectable. With carbon dioxide the system operates under high pressure to keep the CO_2 liquid, hence expensive high-pressure plumbing is necessary.

Primary and Secondary Refrigerants This alternative uses a thermally cascaded system with a primary refrigerant that can produce low temperature, and a secondary refrigerant capable of transmitting this low temperature. A system using freon as the primary and CO_2 as the secondary coolant seems the most feasible for temperatures $-20°$ to $-55°C$. A disadvantage is that field control of the secondary refrigerant is more expensive. Improved procedures with required simplified control units can make this approach practical.

Expendable Refrigerants An example is liquid nitrogen (LN_2), commercially available, which can be used directly to freeze soil. In this case, a refrigeration plant is not required since the liquified state is maintained by pressure. The associated costs are higher, but the freezing operation may be completed in a few days, making this alternative economically attractive. Liquid nitrogen is commonly used for short-term or emergency situations.

Carbonic Acid Fujii (1971) reports that carbonic acid has been used in Japan. The basic method involves the choice of one of the basic processes discussed above and drilling freeze holes where the freezing pipes are installed. A cylinder of frozen material forms around the pipes and increases in size until the heat gain at the perimeter is equal to the heat taken out in cooling. The freeze pipes are installed so that the final frozen zones will overlap and a continuous barrier will be formed. Fujii (1971) suggests that expendable LN_2 may be used initially, and after freezing a conventional brine system may be introduced to maintain the frozen zone.

11-10 CONSTRUCTION AND MONITORING PROGRAM

General Approach

From the discussion of Sections 11-5 and 11-6 it follows that construction procedures for a ground freezing operation are relatively straightforward after the design has been selected. Modifications may be necessary as the process is under way to account for variations in freezing rate caused by variations in stratigraphy, groundwater movement, unforseen conditions, and freezing system departures from ideal design. Normally, the design, installation, and operation of a ground freezing system is undertaken by specialist contractors.

Freeze pipes are placed with spacing s and probe size r_0 according to time requirements and freeze wall thickness (see also Section 11-6, and Figures 11-19 and 11-20). Obtaining the required ice wall thickness is not necessarily a critical problem unless groundwater flows in excess of 6 ft/day are encountered. Frequently, low-temperature freezing techniques are used to overcome heat losses to the moving water above this range. As an example, a wall thickness in the range 8 to 15 ft will accommodate up to 100-ft depths; wall thicknesses of 8 to 12 ft may be needed below 100 ft. A large 50-ft-deep, excavation in sands and gravels has been supported by 5-ft thick straight ice walls surrounded by an elliptical 6-ft thick outer wall. A 9-ft thick wall for a 30-ft deep excavation in fine sand was used as an alternative to other support techniques during a cut-and-cover tunneling program.

Extra care is necessary when drilling the holes and placing the freeze pipes to ensure proper alignment. If one freeze pipe is out of line, closure of the freeze wall may not occur, resulting in a leak or stress concentration locally. Closure of the wall is critical prior to construction since, when the excavation begins and is dewatered, significant pressure gradients will occur across the frozen zone. Breaches in the freeze wall can lead to failure from groundwater inflow. In this case the excavation must be stopped, the partially completed hole allowed to flood, and freezing continued until the leak is closed.

The usual practice is to design the frozen structure so that either it bottoms in impervious stratum or a frozen bottom is part of the design. With the former procedure the freezing probes are commonly inserted several feet into the impervious zone to assure that watertight closure is accomplished.

Special Construction Problems Special details must be designed in areas with existing utilities and especially steam, water, and sewage. If frozen, these conduits can interrupt flow, and if not frozen they constitute a heat source with a potential leak in the freeze wall. One solution is to temporarily relocate the utilities, but where this is not feasible and freezing must proceed through the utilities, they can be insulated so that the 32°F isotherm remains in the insulation.

Concrete members whose thickness exceeds 1 ft may be poured directly against frozen earth. The warm concrete (usually 55 to 60°F) will thaw the surface of the frozen ground as it is placed, and the developing hydration heat will cause more thawing. Continued refrigeration will refreeze the thawed zone, eventually reaching and freezing the concrete. However, this will not occur until the concrete has attained its initial set. Normally, no special additives are required, but a somewhat richer concrete mix may be specified. Concrete placed against a freeze wall may be considered normal practice, and alternatively it is possible to place insulation of the frozen earth before concreting.

Monitoring

Monitoring of subsurface temperature is mandatory during construction of a frozen ground structure. This is more critical while the freeze walls are built and is intended to establish the progress rate and to ensure that no breaches exist. After the structure is completed, monitoring is somewhat relaxed but continuous through the excavation stage so that temperature-dependent ground strengths are known as the excavation proceeds.

For a chilled-brine system, both the brine flow and the brine temperature are normally monitored, as well as ground temperatures within and adjacent to the frozen ground wall. For shafts, the groundwater level inside the frozen ground ring should be monitored for a characteristic rise in interior water level, which indicates closure of the frozen cylinder. Heave and settlement of adjacent and new structures should be monitored by establishing survey points of the structures.

Brine Flow Brine flow in and out of freeze pipes or groups of freeze pipes is measured to identify blockage or air pockets in a pipe. Brine flow rates should be monitored after system startup to check for blockage and to tune the system. Thereafter, the flow rates could be checked periodically. Brine flow must also be checked before excavation for blockage and to ensure that brine is evenly distributed to all parts of the frozen wall.

Brine Temperature Both delivery and return temperatures should be monitored daily. The difference between brine delivery and return temperatures indicates the amount of heat transfer and will narrow as the frozen ground wall is developed.

It is also essential to monitor delivery and return temperature at each freeze pipe in order to locate inefficient or overcooled pipes. Groundwater flow that prevents freezing can be identified by large temperature differentials in individual pipes indicating the presence of a heat source.

Soil Temperature Measurement by thermostats or thermocouples in probe holes is the primary control system and temperature sensor. The probe holes are not installed until all freeze pipes have been completed and inclinometer surveys have been made in each hole to measure deviation from the intended position.

Temperatures should be monitored in each probe hole at given intervals, usually 1.5 to 7 m, with at least one thermocouple located in each soil stratum. Extra thermocouples can be positioned at critical locations or where closure is expected to take the longest. Measured temperatures are plotted versus radial distance from the freeze pipe in Figure 11-22. Similar data can be used to estimate the approximate thickness of the freeze wall with reference to the excavation line. Progressive expansion of frozen ground around freeze pipes for a shaft excavation is shown in Figure 11-23. The wall thickness will be minimal midway between freeze pipes; hence this point is where the probe holes should be located.

The thickness of artificially frozen ground can also be determined by frost indicators and test pits, usually suitable for shallow frozen structures. Test pits can

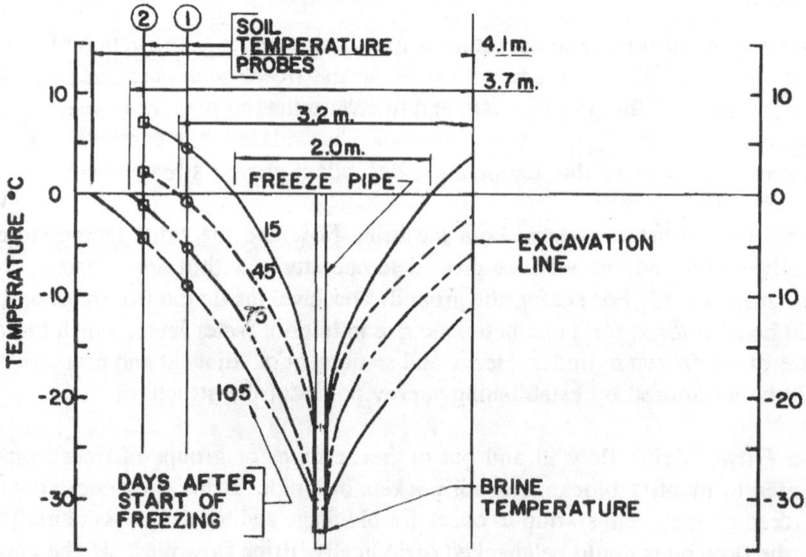

Figure 11-22 Temperature versus radial distance from freeze pipe. (From Lacy and Floess, 1988.)

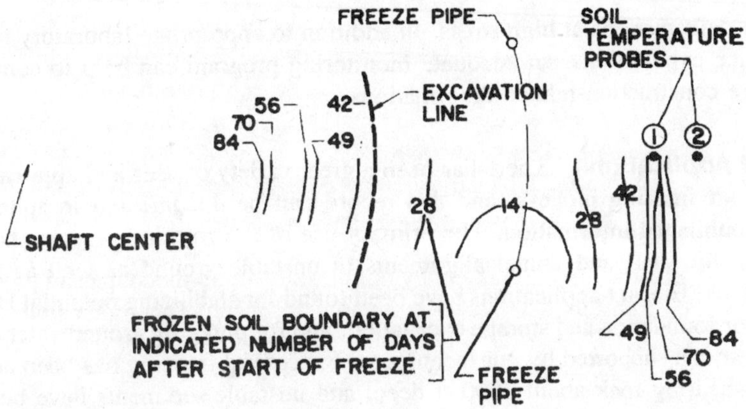

Figure 11-23 Progression of frozen ground around freeze pipe in shaft excavation. (From Lacy and Floess, 1988.)

be used to determine directly the limits of frozen ground in shallow excavations. The technique is particularly effective for examining the crown of shallow frozen ground tunnels, since test pits allow direct visual examination of the extent and condition of the frozen zone.

Probe holes may be located according to the special project conditions. For example, for a tunnel excavation the thickness and temperature along the crown would be of primary importance since this zone may be subjected to heavy surcharge loads. Likewise, where known groundwater flow conditions exist it is important to place probe holes on the upgradient side of the frozen barrier. Lacy and Floess (1988) give an example where temperature probes were located at the widest freeze-pipe spacing, which was on the downstream side of groundwater flow. Flowing groundwater caused a window in the ice wall, discovered only after beginning the excavation. This problem might have been identified earlier, preventing a blow-in condition, if temperature probes were placed on the upstream side of the shaft.

Settlement Ice lensing and the resulting heave and thaw consolidation and settlement are most severe in silty soils. These conditions warrant survey monitoring points that should be established before construction on all nearby structures and major utilities as well as new structures after their completion.

Surface heave may justify the importance of measurements during the freezing process. Although the largest deformations should be expected in fine-grained soils below the groundwater table, a heave monitoring program is indicated under most conditions and can be provided using conventional surface settlement platforms. If deformations related to heave are expected around sensitive structures, provisions must be included in the design to reduce their magnitude.

Lateral or settlement displacements associated with the construction process are more complex. As was shown in previous sections, a frozen zone of ground has much greater strength and stiffness than an unfrozen mass, but it may be subjected to creep deformations at high stress. In addition to appropriate laboratory tests and subsurface explorations, an adequate monitoring program can help to control and minimize construction-related deformations.

Typical Applications There has been a great variety of uses and applications of the ground freezing process, and this record can be documented in spite of the limited published information. The primary use of the freezing process, however, has been for shaft and tunnel alignments in unstable ground as well as in open excavations. Elegant applications have been found for stabilizing potential landslide zones, for foundation and storage excavation, and for plugging groundwater leakage in excavations supported by our techniques. Successful freezing has been achieved in water-bearing rock about 3000 ft deep, and unstable sediments have been successfully controlled at a depth approaching 2000 ft. General design approaches are presented in the following sections for shafts tunnels and open cut excavations.

11-11 SHAFT DESIGN AND CONSTRUCTION

Artificial freezing in shaft construction is intended to add structural strength, improve stability during excavation, and exclude groundwater. In this application, freezing provides the support mechanism for soils close to the water table for all types of shaft, and stabilizes water-saturated, low-strength rocks in deep mine shafts. Typical examples of the latter are the Blairmore formation in the Saskatchewan Potash Field and the based Permian sands of coal fields in Britain, which are partially cemented or uncemented sandstones. Freezing can also be beneficial in preventing water ingress from rocks of relatively high inherent strength as an alternate to grouting and pumping.

The foregoing discussion articulates the high degree of insensitivity of freezing to variable soil, water, and rock formations. A deep shaft can conceivably encounter unstable clay, silt, and sand overburden underlain by decomposed or fissured rocks and incompetent sandstones below the water table. These conditions may require a combination of the alternate ground control procedures discussed in other chapters, but may be dealt with a single freezing operation. A second advantage is that the water cutoff and structural ground support are in place before the shaft is excavated.

On the other hand, shaft freezing requires special technical considerations that may be actual constraints, summarized as follows:

1. The process is temporary, and unlike grouting no permanent improvement in strength and impermeability is introduced to the shaft zone. Where the shaft is used temporarily as access this constraint is inconsequential, but for permanent works shaft linings must be designed excluding freezing effects.
2. The degree of watertightness achieved by the permanent lining is not confirmed until after wall placement and dismantling of the freezing process.

Depending upon this performance, additional time and effort may be necessary to reduce leakage within the design requirements.

Design Principles

Initially, the design must address strength and deformation characteristics of frozen ground, and then the flow of heat through the ground before, during, and after freezing of its contained water. The "rigorous" treatment mentioned in Section 11-9 may not be possible for both phases (Sanger, 1968). Useful reference material that summarizes the theory and the design steps has been developed by Sanger and Sayles (1979).

Structural Aspects The frozen ground around an excavated shaft takes the shape of a hollow cylinder. Shafts deeper than 100 ft are normally designed as circular structures if freezing is required. Large, shallower shafts for heavy construction may have a rectangular cross section but the freeze wall installed will have an arched or circular configuration so that the stresses resisting collapse are compressive hoop stresses. The frozen soil will also rely on compressive properties when acting as load-bearing structure.

Two ratios are relevant to the thickness of the frozen cylinder. These are the expressions p/q_u and b/a, where p = external lateral ground pressure, q_u = uniaxial compressive strength of the frozen material, b = external radius of the frozen cylinder, and a = internal radius of the frozen cylinder. For soils and incompetent or weakly cemented granular rock, the relationship, given the thickness of the frozen cylinder, must also include the viscoelastic behavior of the material under load.

An expression still in use today was proposed by Domke (1915). It is derived from the concept of an inner plastic section of freeze wall surrounded by an outer section that behaves elastically, or

$$\frac{p}{q_u} = \tfrac{1}{2}\ln\left(\frac{b}{a}\right) + \tfrac{1}{2}\left(1 - \frac{a}{b}\right) \tag{11-14}$$

The required dimension $(b - a)$ can be found more readily from the following:

$$\frac{b - a}{a} = 0.29\left(\frac{p}{q_u}\right) + 2.3\left(\frac{p}{q_u}\right)^2 \tag{11-15}$$

Based on earlier work completed in Russia, Sanger (1968) has proposed a relationship that articulates the eventual plastic failure of a frozen soil with parameters obeying Coulomb's law, as follows:

$$\frac{b}{a} = \left[\left(\frac{p}{q_u}\{N - 1\}\right) + 1\right]^{\frac{1}{(N-1)}} \tag{11-16}$$

where N = flow factor = $\tan^2 (45° + \phi/2)$. For clays with $\phi = 0$, Eq. (11-16) simplifies to

$$\ln\left(\frac{b}{a}\right) = \frac{p}{q_u} \tag{11-17}$$

For cohesive materials with no internal friction, both Domke's and Sanger's formulas yield similar results for relatively thin freeze walls. For very deep shafts, other authors (Klein, 1980; Winter, 1980) consider Eqs. (11-14) and (11-17) too conservative and propose other calculation procedures.

Commentary Frozen soil-rock-water systems do not have a unique compressive strength. Failure may occur at various times after applying various levels of steady load, but typically it is preceded by deformation (see also Figure 11-15). For any specific stress level, the deformation will be less and the time to ultimate failure longer when the frozen material is colder (Akili, 1970). The creep effects discussed in the foregoing sections must be considered in a frozen clay cylinder under modest loads (construction shafts), and in more highly stressed sands and grossly fissured rocks (deep mine shafts). Since the short-term strength (hours after load application) may be several times greater than the long-term strength (after several months of loading), a relevant factor is the time lapse before the unsupported section of the freeze wall is permanently restrained by the structural lining.

A shallow shaft through frozen clay will thus require different design procedures, depending on the availability of rings of liner plate progressively installed to secure the shaft. For deeper shafts the freeze walls should be thicker and/or colder if shaft lining installation calls for long vertical cylinders of load-bearing frozen materials (Maishman, 1982). This has been the subject of extensive study here and in Europe. For example, because of the eventual extraction of coal adjacent to deep shafts frozen to 2000 ft in Germany, permanent shaft linings are designed as continuous flexible multilayer columns consisting of steel plate, asphalt, and concrete, and are built from the bottom up. Hegemann (1980) reports that a freeze wall 26 to 33 ft thick around a 30-ft-diameter excavation would be insufficient to prevent damaging creep effects if left unsupported during sinking (b/a = 3). This excavation was supplemented with concrete blocks and fiber chip board, giving a compound concrete/frozen support. By comparison, a shaft frozen to similar depth in Saskatchewan but permanently lined downwards, and with frozen surfaces exposed only for a few days, required a 15-ft-thick freeze wall around an excavation 24-ft in diameter (b/a = 2.25).

These examples demonstrate the physical dimensions of the frozen structure required for very deep shafts. By contrast, relatively shallow, above bedrock shafts up to 150 ft deep rarely require b/a ratios greater than 1.3.

Different design procedures might apply in a stratum consisting of a highly competent rock with water-filled pores or fissures. If rock stability does not pose a problem, freezing of the contained water must prevent its extrusion into the shaft opening by its own hydrostatic head. The associated support requirements are less than for mass stability, and they seldom need additional consideration in a freeze encompassing both conditions.

Thermal Aspects

The design strength of frozen soil must be provided at a chosen subfreezing temperature assumed to be an average temperature of the frozen cylinder. Actual temperatures vary, however, according to a pattern (see also Figures 11-22, 11-23) such that a minimum occurs at the freeze pipe (coolant temperature) and a maximum at the outside face of the frozen cylinder, as shown in Figure 11-24. Attaining the required thermal gradients within a specified period is the objective of thermal design. During the cool down heat is transmitted by conduction to the freeze pipes from three zones, each with different heat flow characteristics, as follows: (a) an inner frozen zone where soil and contained ice are cooled below freezing point; (b) a transitional zone where groundwater is converted to ice at its freezing temperature; and (c) an outer zone where unfrozen ground is cooled at temperatures above

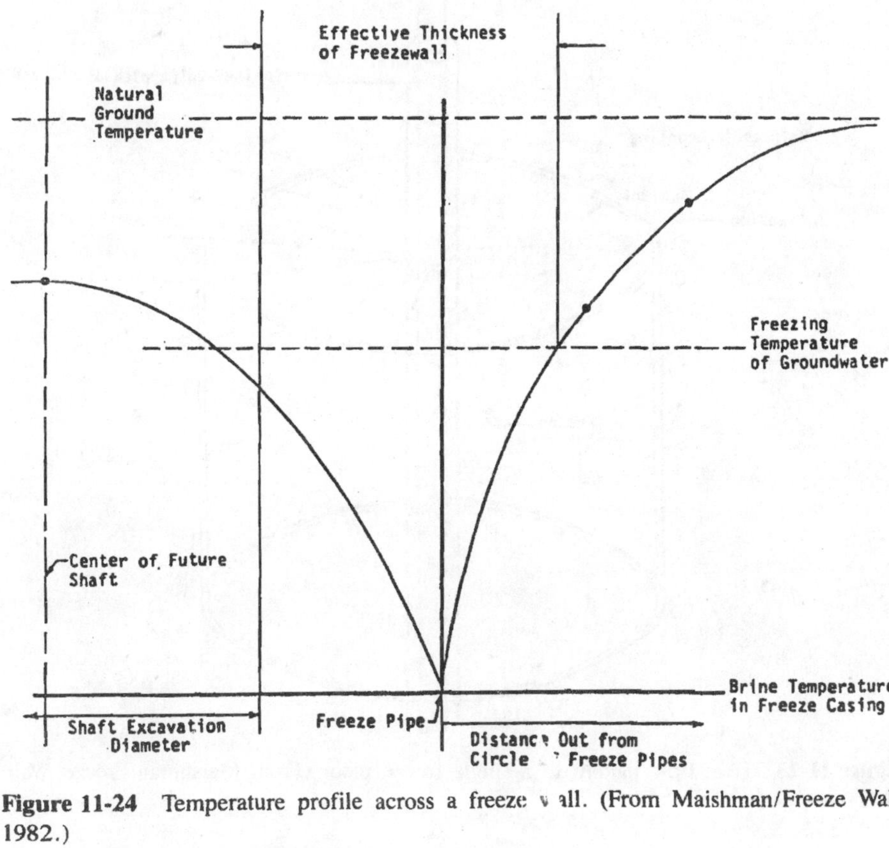

Figure 11-24 Temperature profile across a freeze wall. (From Maishman/Freeze Wall, 1982.)

freezing. The heat flow pattern around a single freeze pipe is shown in Figure 11-25.

The amount of heat to be extracted, rate of extraction, and two relevant time periods (time necessary to connect the freeze between adjacent pipes, and time necessary to extend the wall radially to its design thickness) have an obvious influence on freeze pipe location and refrigeration plant capacity (Sanger and Sayles, 1979).

The thermal design is usually a compromise where savings resulting from larger freeze hole spacing are offset by higher refrigeration capacity and longer freeze wall formation period. Optimum layout for shaft freezing operations usually has adjacent freeze pipes spaced 3 to 5 ft on centers, and allows a freezing time of 3 weeks to 3 months before excavation reaches the water table. Situations where freeze holes can diverge to create windows or weak spots in the wall should be avoided. For normal conditions, a minimum freeze wall thickness is about 4 ft for a small-diameter

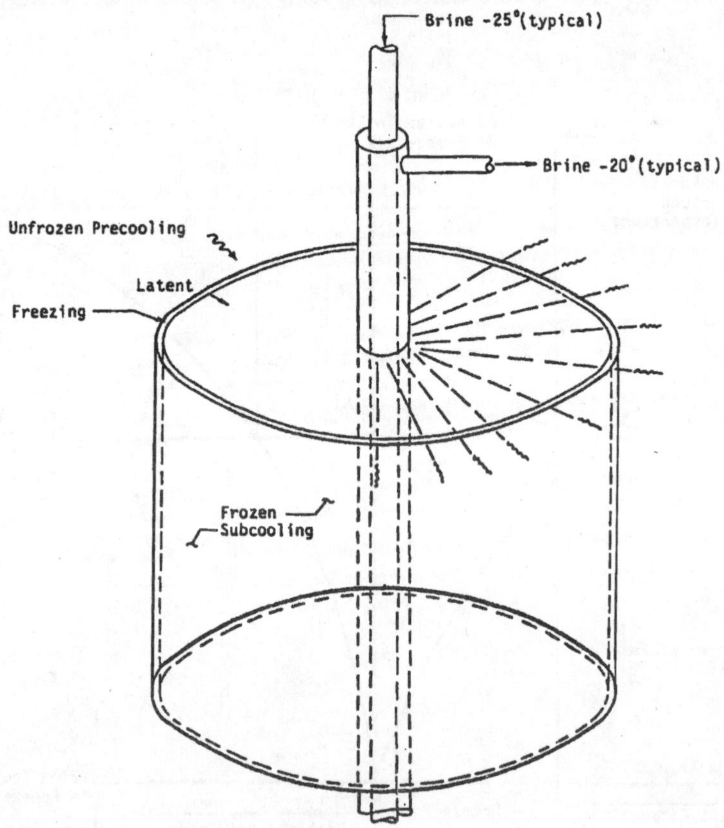

Figure 11-25 Heat flow pattern to a single freeze pipe. (From Maishman/Freeze Wall, 1982.)

shallow shaft. Likewise, the maximum thickness at the base of a single circular array around a deep shaft is about 16 ft, above which a double ring should be considered.

Relevant Design Data and System Components

Section 11-5 outlines the general design requirements of a freeze wall, and suggests a minimum exploratory program necessary to obtain and collect data relevant to this purpose. For a shaft excavation these data include: (a) the moisture content determined by a sufficient number of measurements; (b) geotechnical gradients and thermal characteristics such as unfrozen and frozen coefficients of heat conductivity and volumetric heat capacity; and (c) tests on frozen samples to determine their mechanical properties and long-term creep strength. The latter may be extrapolated from several tests to short-term failure considering the linear relationship between the log of time to failure and the reciprocal of applied stress (Shuster, 1972).

The first objective of shaft freeze design is to establish its depth. For this purpose, rock cores or undisturbed soil samples are useful. A low permeability bed below the problem areas is a suitable level to terminate the freeze pipes, and may occur above or below the base of the proposed shaft. With the cylindrical frozen structure formed, the groundwater contained within it is confined. This provides two advantages: (a) it facilitates monitoring of the closure by ensuring that there is no upward movement of groundwater due to hydrostatic imbalance when excavation begins; and (b) when the core is completely frozen (shafts terminating in a thick aquifer with no convenient lower cutoff stratum are often designed in this manner) having the freeze pipes below subgrade affords better protection. Alternatively, it may be possible to provide pressure relief by pumping the underlying aquifer, but this requires coordination between groundwater and freezing technologies.

Components of Freezing System Freezing systems and components are reviewed in Section 11-6. Hence, only details specific to shaft construction are discussed here.

The down-pipe method is routinely employed by freeze contractors for all depths of shaft construction. Freeze pipes consist of an outer steel shell (also referred to as casing) sealed at the lower end. Inside, there is a shorter open-ended tube permitting coolant from a cold delivery main to be circulated and returned to the surface. The casing is inserted and tested for leaks before the down-pipe and freeze head are installed.

The outer steel casing must withstand various stresses induced in the freezing process. Among these are tensile stresses that develop as the pipe is chilled because its coefficient of thermal contraction exceeds that of the frozen environment to which the casing adheres. Localized freeze pipe stresses can also be significant where several massive strata with different physical responses to freezing are penetrated. These conditions are more likely to arise in deep mine shafts. Oil wells cased into the arctic permafrost experience a similar physical environment (Goodman and

Wood, 1975; Ruedrich et al., 1976). Deep-shaft freezing through rock requires rotary drill rigs using appropriate oil field techniques.

For shallow shafts frozen in soils, special casing is not necessary. Standard mild steel pipe, 3 to 4 in. in nominal diameter, with lower seal plates is commonly used. It may be lowered in one piece by crane into freeze holes drilled, jetted, or augured. For depths exceeding about 50 ft it is advisable to check the vertical alignment of every freeze casing to ensure conformity with design.

For deep-shaft freezing, the casing is typically $4\frac{1}{2}$ to 6 in. in diameter, and is lowered in sections. Screwed connections are torqued up on the rig floor and each joint's pressure is tested using a bridge packer assembly.

Special considerations must address details such as ring mains and shaft top arrangement. Useful data and information are provided by specialist freeze contractors.

Freeze Plant The general schematic presentation of Figure 11-13 is supplemented by the details of Figure 11-26, showing the basic refrigeration cycle. In general, the mode of operation must conform with established practice. The system compresses a refrigerant in one or two stages by means of piston or rotary compressors, liquefies the high-pressure vapor at ambient temperature, and expands it in a heat exchanger where the cold energy released is transmitted to the circulating brine. The cycle is completed as shown.

Monitoring The general guidelines given in Section 11-10 are applicable in shaft construction. Thus temperatures are routinely taken using thermocouples inside a central observation wall. Alternatively, thermal monitor walls are placed externally and remain intact during freeze wall formation, maintenance, and thawing operations. Thermal monitors should be located so that temperature profiles may be obtained defining the thermal structure of the freeze wall at all stages (see also Figure 11-24).

Heave and settlement during the freeze/thaw cycle can vary for different jobs depending on the ground conditions (see also the foregoing sections). It may be insignificant or may approach several inches. In-shaft measurements will indicate whether shaft sinking and lining procedures are compatible with accepted rate and magnitude of movement. For large-diameter shafts, lateral accommodations to the changing stress system may be monitored in vertical holes inside or adjacent to the freeze wall. These holes are surveyed to a proper depth to obtain a static reference point. The associated technique is relatively new to ground freezing, but recent studies (Jones, 1980) suggest the importance of small movements occurring before excavation and their influence on design.

Other monitoring activities deal with plant performance. The useful output of the refrigeration plant is

$$TR = KQ\,\Delta t \tag{11-18}$$

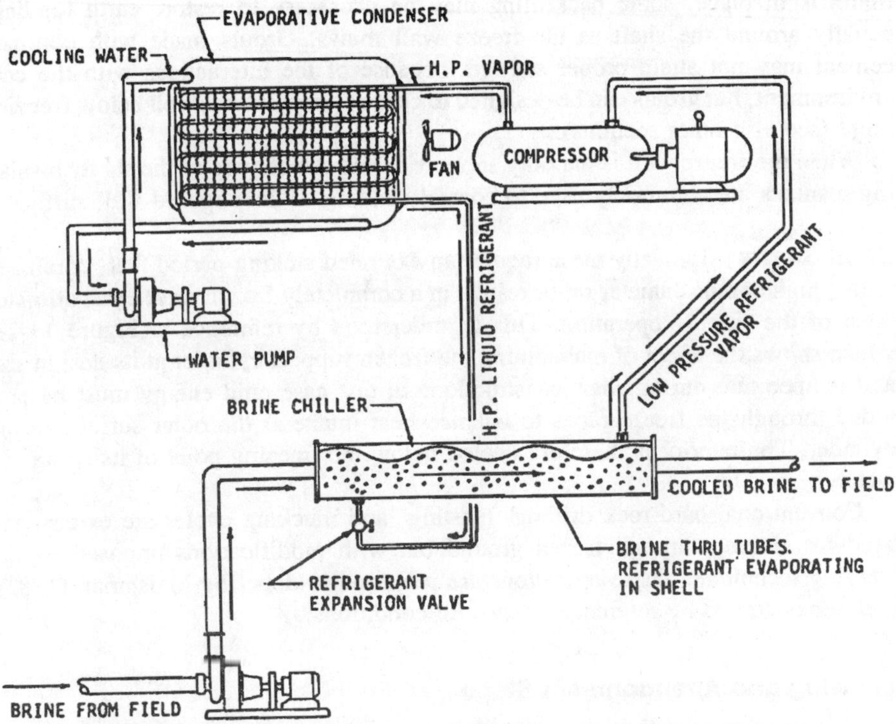

Figure 11-26 Basic refrigeration cycle. (From Maishman/Freeze Wall, 1982.)

where TR = tons refrigeration
Q = coolant flow in unit time
Δt = the difference in brine temperatures out/in measured at the heads of the freeze pipes
K = a constant determined by the physical characteristics of the brine

In addition to the basic brine flow, volume, and temperature measurements, it is useful to record refrigerant pressures, temperatures, motor amps, compression oil consumption, and so on, for effective mechanical control.

Shaft Construction Inside Frozen Earth

Shallow Shafts with Diameter > 20 ft Where the excavation can be completed within a few weeks, most of the earth removed can be carried out while the soil remains unfrozen. An internal observation wall can also be used to pump and drain most of the trapped water from granular formations, and just before the excavation beings. Earth removal may entail a combination of a backhoe or scraper and clamshell, as for a sheeted excavation. Some frozen ground may have to be trimmed from the sides for the correct installation of segmental linings. After the lining is in place, some backfilling may be necessary to restore earth loadings equally around the shaft as the freeze wall thaws. Grouts made with common cement may not attain proper strength because of the interactions with the cold environment, but grouts can be designed to cure at temperatures well below freezing point (see also other sections).

When the freeze wall is the only support during the service of the shaft, insulating blankets are commonly used to control spalling of the exposed wall surface.

Deep Shafts Typically these require an extended sinking period that, combined with a high depth/diameter ratio, results in a completely frozen shaft excavation for most of the sinking operation. This is understood by reference to Figure 11-27, which shows the effect of maintaining the frozen support cylinder at its design size and temperature during shaft construction. In this case cold energy must be provided through the freeze pipes to balance heat intake at the outer surface of the cylinder. The interior is therefore cooled below the freezing point of its contained water.

Conventional hard-rock drilling, blasting, and mucking cycles are extensively used for sinking through frozen ground but with modifications imposed by the freezing technique. These operations are reviewed in detail by Maishman (1982), and demonstrated by reference to specific conditions.

Thawing and Abandonment Stage

Three phases are distinguished: (a) degradation of the ice wall; (b) abandonment of shallow-shaft freezing through single-aquifer soils; and (c) abandonment of freeze pipes traversing several aquifers.

Degradation of the Ice Wall Natural thawing is a long process, usually taking months. Abandonment procedures can be planned according to technical requirements, geological conditions, and project scheduling. If the thawing phase is to be artificially supported, the brine is gradually heated after a period of neutral circulation, and the heating may also include the shaft ventilating system. During this phase, the temperature monitoring system remains in use.

Abandonment Through Single-Aquifer Soils In this case, the main objective is to fill voids created in the freeze pipe installation. Thawed soils eventually close minor annular space left between casing and hole through natural expansion. Hence, the freeze pipes are usually left in place after flushing out the brine, removing the down pipe, and filling with sand or tremied grout. This procedure does not require a complete thaw.

Abandonment of Freeze Pipes Traversing Several Aquifers In this case, aquifers must be separately sealed, and for deep freeze pipes surrounding a mine shaft in an advanced construction stage, the working environment is often unfavorable.

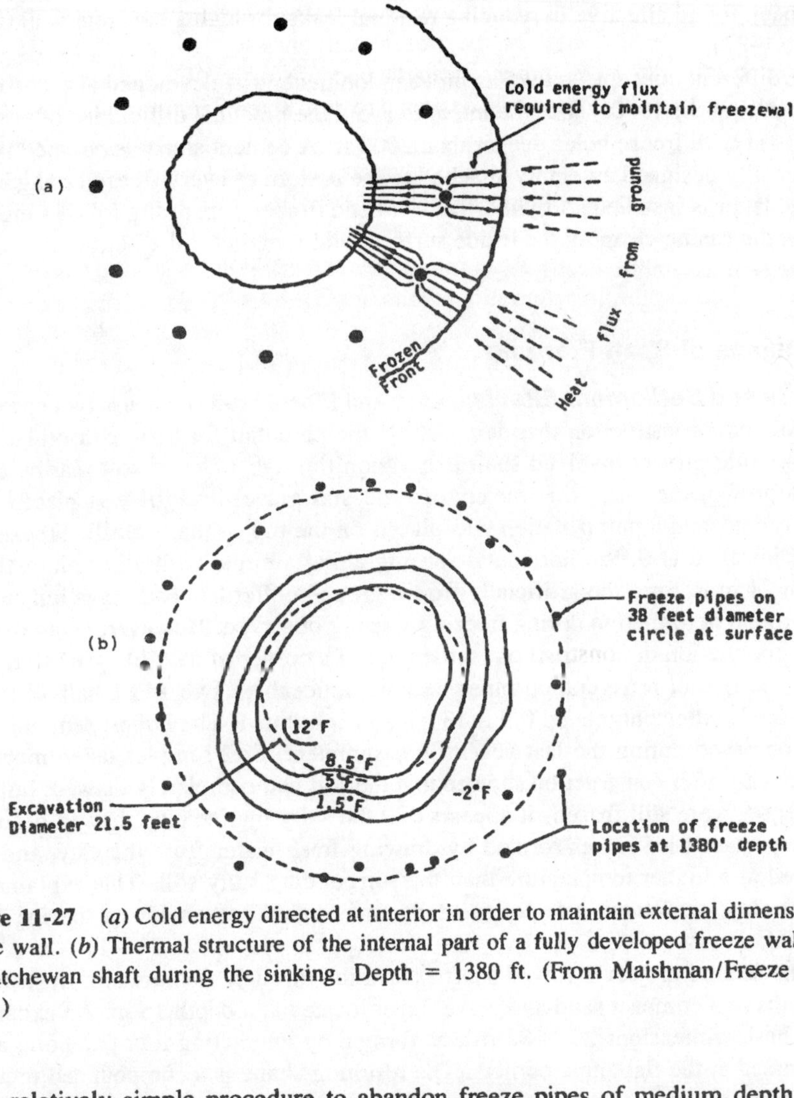

Figure 11-27 (*a*) Cold energy directed at interior in order to maintain external dimension of freeze wall. (*b*) Thermal structure of the internal part of a fully developed freeze wall in a Saskatchewan shaft during the sinking. Depth = 1380 ft. (From Maishman/Freeze Wall, 1982.)

A relatively simple procedure to abandon freeze pipes of medium depth (less than 1000 ft) is as follows. Following a period of slow warming through all pipes, a small group selected for immediate treatment has additional heat supplied so that the casing temperature will be elevated above freezing point for some time. The casing is pulled, steel drill pipe is lowered to the bottom of a hole where frozen walls are stable, and cement grout is placed until it fills the hole.

This procedure must be accomplished within a narrow range of temperature change, hence it is not always feasible for deep freeze pipes. After preliminary thawing, the deeper aquifers and adjacent confining beds are grouted through perforations or cuts made in the casings. Grouting takes place inside a decaying freeze wall and may reach the permanent shaft lining. In some cases, freeze hole grouting has been found effective in reducing residual leaks through a concrete wall (Maishman, 1982).

A different concept, known as preabandonment, was developed at potash shafts sunk during the 1960s, with the intent to avoid the practical difficulties of postabandonment with freeze holes deeper than 2000 ft. A cement slurry is pumped through a specially designed assembly attached to the bottom of every freeze casing as soon as the latter is installed. A cylindrical synthetic rubber wiper plug follows the slurry down the casing cleaning the inside surface, and remains at the bottom as a backup to the seal assembly.

Examples of Shaft Freezing

Heave and Settlement Effects Lacy and Floess (1988) discuss two case histories of shaft construction that demonstrate the potential for thaw consolidation.

The first project involved shaft excavation through 17 m of soft marine clay to the top of glacial till. Nine meters of sand and gravel backfill was placed in the excavation, and a pump station was placed on the top of the backfill. Freeze pipes were installed at 0.9-m horizontal spacing along a circle with 16-m diameter to a depth 21 m to form the artificially frozen ground cofferdam as excavation support.

Ground deformation during freezing was not observed. However, about one year after completion of construction, settlement of a corner of a building located 3.7 m from the line of refrigeration pipes became noticeable. Two and a half years later measured settlement totaled 0.9 m (3 ft), and continued. The largest settlement rate was observed during the last year of measurement. Soil samples taken more than three years after construction showed that the soil had completely thawed, but some ice lenses were still frozen. It appears that the salty marine clay thawed before the closely spaced ice lenses formed by drawing fresh water from the clay, and these thawed at a higher temperature than the surrounding salty soil. This explained the initial slow settlement rate followed by more rapid settlement as the ice lenses thawed.

The second project involved shaft excavation through medium stiff to stiff clays and silts to a compact sand and gravel layer located at a depth 15 m. An excavation with basic dimensions 29 × 33 m was formed by interacting four parabolic arches buttressed at the flat angle corners. The resulting shape is a compounded rectangle with curving walls between corners. During freezing, ground heaving was not observed. However, movement of groundwater from adjacent pumping caused partial wall failure, exposure of freeze pipes, and rupture of brine piping. Backfilling the hole with loose sand and refreezing the wall to a greater thickness extended the freezing period considerably. Since completion, a settlement of almost 80 mm has occurred, involving a shallow-supported retaining wall from a new structure across the line of the previously frozen ground. There is also evidence of pavement cracking and 230 mm of pavement settlement concentrated over the line of the frozen ground.

Innovative Applications An attractive solution in deep-shaft construction is to avoid freezing (often unnecessary) of overlying rocks. It may be feasible first to sink conventionally and then install freeze pipes from an underground chamber, although the method in this case may introduce counterproductive elements (Maishman, 1982).

An innovative solution using oil well directional drilling techniques was developed for the Boulby Mine in Yorkshire. The distribution mains were located in a shaft chamber 1800 ft below ground, but the freeze casings were previously set through holes drilled from the surface and uncovered during the chamber excavation. Because of the preabandonment procedure, artesian flows from the freeze holes did not occur. Surface locations of the drill rigs were established away from the permanent shaft top structures, and this allowed the freeze installation and initial shaft sinking to 1800 ft to proceed concurrently. This freezing operation was extended to 3200 ft, probably the deepest on record, but started almost at the lower half.

Freezing was used at a site in Wales to restore a long abandoned coal mine shaft to a safe condition. The shaft was flooded almost to the surface and contained a brick lining. Backfilling was not considered safe (similar backfilled shafts had collapsed suddenly when debris at the base of the shaft flowed away to adjacent mine openings), and dewatering might increase the load on the brick lining, induce settlement, and precipitate failure. Ground freezing by pipes inside the old shaft enabled access of 160 ft inside a 42-in diameter steel tube to install a reinforced concrete plug into the rock.

11-12 TEMPORARY TUNNEL SUPPORT BY ARTIFICIAL GROUND FREEZING

Temporary ground freezing for soft ground tunneling is an effective technique, particularly where more conventional systems, such as grouting or compressed air, are not feasible.

Whereas stratification and variation in permeability have little effect on freezing, these factors can affect a successful grouting operation. Fine sands and silts can be frozen, but are not susceptible to grouting because of the low permeability. Freezing also eliminates the hazards associated with compressed air such as blowouts and the effects of high pressures. Among the factors to be considered for artificial ground freezing in tunnel construction are the following.

Ground Movement As mentioned in the previous sections, this is associated with (a) frost expansion during the freezing period (see also Williams, 1968; Shuster, 1972); (b) stress relief during excavation (attributable to stress relaxation during excavation common to underground construction of this type); and (c) consolidation during the thawing period (see also Endo, 1969; Morgenstern and Smith, 1973; Poulos and Davis, 1974; Tsytovich, 1975).

Strength Factors The viscoplastic behavior of frozen soil has been discussed in the foregoing sections. Since frozen soil creeps under stress, its strength and deformation depend on both the internal friction and the cohesion between particles. This behavior is also strongly dependent on time, temperature, and stress level (see also Sayles, 1968, 1974; Vialov, 1965).

Cost Considerations The cost of artificial freezing is influenced by many variables; hence, it is not practical to assign a cost per cubic meter of material to be frozen. Major items to be mentioned are the ground conditions, freeze pipe spacing, available time for freezing, and time period over which the system must be in place. The cost of artificial freezing must also reflect potential savings to be realized by eliminating dewatering, compressed air, temporary support, and so on. When all these factors are combined with time parameters, freezing may be a competitive and economically attractive option. This is particularly true when the contractor can complete the freezing-related portion of a project within a short time period and thus minimize the associated energy costs.

Tunnel Freezing Studies

The applicability of the freezing method can be demonstrated by case studies. Jones and Brown (1978) present three case studies involving three tunnels constructed beneath multisets of railroad tracks. These tunnels are designed to support a Cooper E-80 engine loading (equivalent static loading = 5 kips/linear ft of track), applied on all tracks.

Georgia Tunnel This project is an arch-shaped tunnel 25 ft wide, in Atlanta. The tunnel is 98 ft long, and its crown is located approximately 12 ft beneath five sets of railroad tracks. The subsurface conditions consist of fill (50 ft thick) composed of loose silty fine sand, and with blow count 5 blows/ft. The unfrozen fill has a drained angle of friction 25° with no cohesion. Moisture content varies in the range 18 to 36 percent with an average of 23 percent. The average void ratio is 0.9, and the material has an average unit weight 90 lb/ft^3. The water table is located below the fill.

New York Tunnel This project is in upstate New York and involves a circular sewage outfall tunnel 11.5 ft in diameter. The tunnel is approximately 110 ft long and passes beneath three sets of mainline railroad tracks. The crown of the tunnel is 6 ft below the track level. The subsurface conditions consist of cinder and sand fill to the springline of the tunnel, with average moisture content 29 percent. The fill has a void ratio 1.2, and a dry unit weight 60 lb/ft^3. Typical penetration resistance is 2 blows/ft. Beneath the fill is a stratum of loose sand with some silt (average void ratio 1.5, moisture content 45 percent, and dry unit weight 64 pcf). Typical penetration resistance in the lower stratum is 2 to 3 blows/ft.

Washington Tunnel This project is a circular tunnel, 12.5 ft in diameter, 110 ft long, passing 9 ft beneath four sets of railroad tracks. The subsurface consists of sand and gravel with amounts of clay and silt to a depth 25 ft. The blow count in this layer varies from 2 to 50 blows/ft, with an average of 20. The average moisture content is 34 percent. Beneath this material is a thick silt layer with average penetration resistance 4 blows/ft and typical moisture content 60 to 80 percent.

Gradation characteristics for the three tunnel sites are shown in Table 11-4, based on representative samples.

Design Aspects of Case Studies

Two common freeze pipe configurations are a circular or elliptical frozen zone with the freeze pipes placed horizontally around the tunnel perimeter, and an arch-shape configuration with the freeze pipe placed vertically or inclined from the surface. The former type is shown in Figure 11-28. Because of the relatively small diameter and length of the New York and Washington tunnels and the accessibility from the tunnel end, a horizontal freeze pipe configuration was selected. The large size of the Georgia tunnel prohibited freezing around the entire periphery, and the system consisted of a combination of horizontal and vertical drilling with the tunnel floor left unfrozen.

The frozen soil thickness is basically dependent on loading conditions and freezing temperature. A 3-ft-thick zone of frozen soil was selected for the New York and Washington tunnels, and a 12-ft thickness was selected for the Georgia tunnel, to be verified by the analysis.

Laboratory Tests The design of tunnels typically requires evaluation of the effect of repetitive loading on the frozen zone. Since the actual train frequency is erratic, the usual approach is to obtain solutions for upper and lower bound frequencies. The lower bound is obtained by static creep tests simulating a train stopped above the tunnel for an extended time. Upper bound solutions involve repeated load triaxial tests at higher load frequencies than normally expected in the field.

Part of the testing program is intended to provide an indication of field behavior of the frozen soil, and more particularly the possibility of tensile stresses. In these studies, only nominal tensile stresses would be expected considering the small tunnels, and these would be below the tensile strength of the frozen soils.

Sample preparation, test equipment, procedure, and test results are summarized in detail by Jones and Brown (1978). Static creep tests on Georgia silty sand and

TABLE 11-4 Gradational Characteristics of Soils at Each Site

Tunnel	Material Type	D_{90} (mm)	D_{60} (mm)	D_{30} (mm)	D_{10} (mm)
New York	Cinder and sand fill	8.0	1.8	0.23	0.015
	Sand	0.3	0.16	0.047	0.0014
Georgia	Silty sand	1.5	0.3	0.09	0.009
Washington	Clayey sand	0.7	0.2	0.008	—
	Silt	0.03	0.003	—	—

Source: From Jones and Brown (1978).

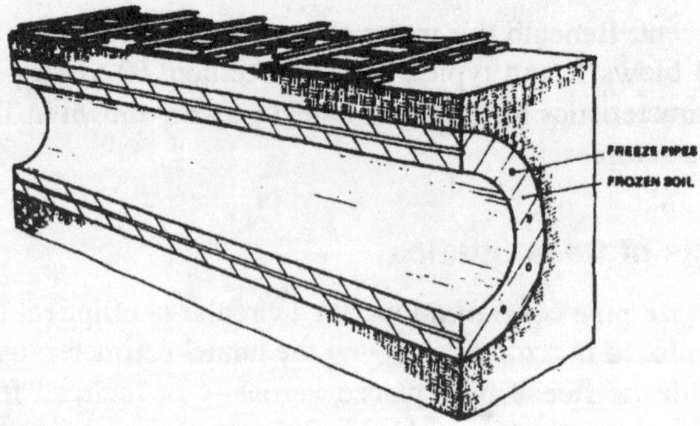

Figure 11-28 Schematic presentation of freeze pipe configuration for frozen soil tunnel. (From Jones and Brown, 1978.)

repeated load tests on remolded samples reveal an important point. There is little difference between the strain response from static tests and repeated-load tests for the low loading rate investigated, that is, creep effects are independent of cyclic loading.

A plot of the stress-strain curves obtained from the static creep and repeated load tests is shown in Figure 11-29 for the Georgia tunnel. Points on the curves are obtained by taking the ultimate strain developed with time under each given stress level. The hyperbolic representation of this curve is used in the stress analysis.

A sample of test results for the Washington tunnel is shown in Figure 11-30, presenting a plot of stress-strain curves obtained for various loading times. Repeated load tests were not considered necessary in this case. Static creep tests were

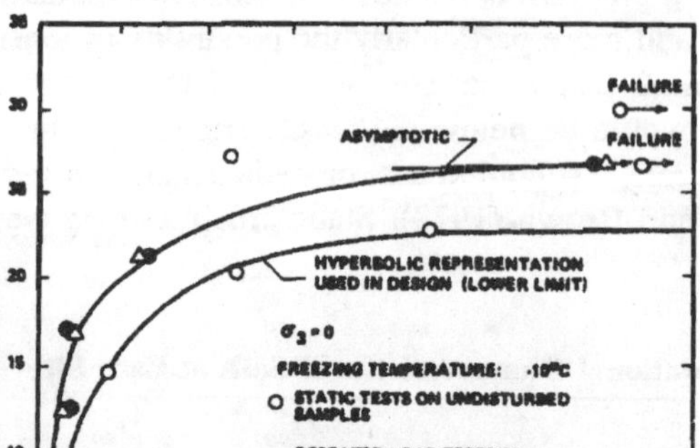

Figure 11-29 Stress-strain relationship for micaceous silty sand—Georgia Tunnel. (From Jones and Brown, 1978.)

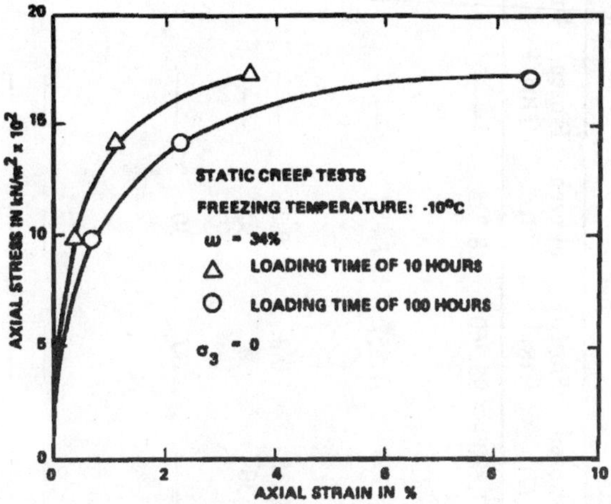

Figure 11-30 Stress-strain relationship for clayey sand—Washington Tunnel. (From Jones and Brown, 1978.)

performed on remolded samples of the clayey sand expected to have the lowest strength and highest deformability.

Summary of Frozen Soil Properties Frozen soil parameters and unfrozen soil properties used in the analysis are summarized in Table 11-5 for the three tunnels. Linear elastic stress-strain conditions are assumed for the Washington and New York tunnels. This is based on previous experience indicating that stresses in a 1-m-thick zone of frozen soil should be linearly proportional to strain and nominal in the tensile range. The elastic modulus is selected for stress levels 50 percent the ultimate stress, which is essentially in the linear range of triaxial test results shown in Figure 11-30.

Values of Poisson's ratio are extracted from available literature. Values of K_0 are obtained from the relationship $\mu/(1 - \mu)$, considered sufficient for the purpose of analysis.

The Georgia tunnel is much larger than most tunnels reported in the literature. Without a database for tunnel performance, Jones and Brown (1978) model the stress-strain characteristics of the frozen soil using a hyperbolic formulation (Kondner and Zelasko, 1963). The stress-strain relation is expressed for a loading time exceeding 100 hr, and obeys the hyperbolic formulation

$$E_t = \left[1 - \frac{R_f(1 - \sin\phi)(\sigma_1 - \sigma_3)}{2\sigma_3 \sin\phi + 2c\cos\phi}\right]^2 Kp_a\left(\frac{\sigma_3}{p_a}\right)^n \tag{11-19}$$

where E_t = tangent modulus
$R_f = (\sigma_1 - \sigma_3)_f/(\sigma_1 - \sigma_3)_{\text{ult}}$
ϕ = friction angle

TABLE 11-5 Summary of Soil Parameters Selected for Design

	Georgia Tunnel		New York Tunnel				Washington Tunnel			
	Micaceous Silty Sand		Cinder and Sand Fill		Fine Sand, Some Silt		Clayey Sand		Silt	
Data (1)	Frozen (2)	Unfrozen (3)	Frozen (4)	Unfrozen (5)	Frozen (6)	Unfrozen (7)	Frozen (8)	Unfrozen (9)	Frozen (10)	Unfrozen (11)
Modulus of elasticity, in kilonewtons per square meter	—	—	95,760—287,280	9,576—23,940	287,280	47,880	95,760	28,728	—	4,788
Unit weight, γ, in kilograms per cubic meter	1,842	1,842	1,442	1,442	1,506	1,506	1,857	1,857	—	1,440
Poissons ratio, μ	0.2	0.3	0.35	0.35	0.35	0.35	0.3	0.3	—	0.45
Angle of internal friction, φ, in degrees	0	25	—	30	0	30	0	30	—	0
Cohesion, C, in kilonewtons per square meter	1,101	0	622	—	622	—	30	30	—	0
Hyperbolic parameters										
K	3,000	100	—	—	—	—	—	—	—	—
n	0	0.7	—	—	—	—	—	—	—	—
R_f	0.94	0.71	—	—	—	—	—	—	—	—

Source: From Jones and Brown (1978).

Note: 1 ksf = 47.88 kN/m^2; 1 pcf = 16.02 kg/m^3.

c = cohesion
σ_3 = minor principal stress
σ_1 = major principal stress
K = modulus number
n = exponent determining the rate of variation of the initial modulus with σ_3
p_a = atmospheric pressure

For feasibility analysis, the frozen silty sand is assumed to have a cohesion equal to one-half the unconfined compressive strength and a friction angle 0°.

Structural Investigation

The tunnels studied in this review are transverse to the direction of train movement. Hence, a three-dimensional analysis of the frozen soil tunnels provides a more accurate presentation of stress and deformation behavior. Because of economic limitations, however, the investigations pursued a conservative two-dimensional model.

New York Tunnel The maximum available loading for this tunnel occurs with all tracks loaded with engines. Although in terms of statistical reliability these conditions should not be expected to occur, particularly for an extended time period, this loading is the design parameter. Given the small ratio of the tunnel diameter to the available loading distance along the tunnel axis, the train load is approximated by a large area load applied at the ground surface.

Frozen and unfrozen soil are modeled using linear elastic stress-strain relationships. The plane strain finite element model shown in Figure 11-31 is used to calculate tunnel stresses and deflections using varying modulus in the frozen cinder and sand fill. This analysis shows that varying the modulus of the frozen soil by more than 300% results in a change in the calculated maximum shear stress less than 10%.

The calculated maximum shear stress in the frozen soil is 7 kips/ft^2 and the corresponding major and minor principal stresses are 19.5 and 5.5 kips/ft^3, respectively, both in compression. The maximum shear stress occurs at the springline. The maximum tensile principal stress occurs at the crown and is 3.1 kips/ft^2 with a corresponding shear stress 1.1 kips/ft^2. The calculated differential deflection between top and bottom inverts of the tunnel varies from 0.3 to 0.8 in., depending on the particular combination of elastic moduli.

Washington Tunnel Plane strain analysis using linear elastic soil properties is also applicable to this tunnel, with a finite element mesh essentially similar to the New York tunnel model. The calculated maximum shear stress in the frozen zone is 7 kips/ft^2 with principal compressive stresses 14 8 and 0.7 kips/ft^2. Likewise, the maximum shear stress occurs at the springline. The maximum tensile principal stress is 5.5 kips/in.^{2} and occurs at the crown. The maximum calculated deflection is 0.7 in.

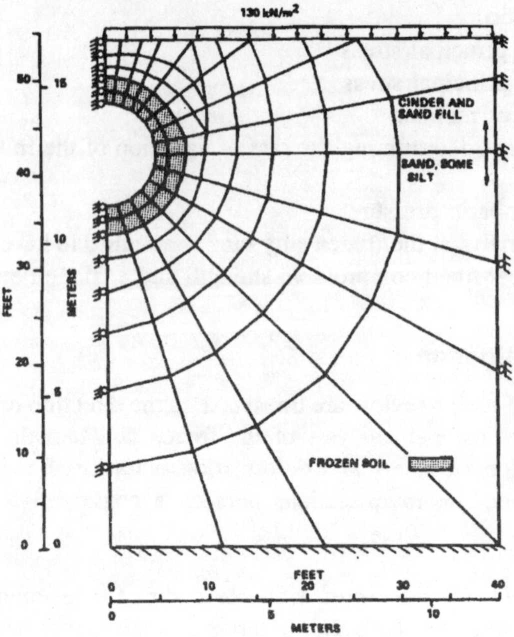

Figure 11-31 Finite element model for New York tunnel. (From Jones and Brown, 1978.)

Georgia Tunnel The finite element model for this tunnel, including structure geometry, is shown in Figure 11-32. The first step in the analysis is to determine the initial state of stress in the soil mats prior to stress changes caused by tunnel excavation. The initial state of stress includes the geostatic stresses and loading from trains on all five tracks. The three-dimensional stress distribution from the train loads is determined from the Boussinesq procedure for the case of line load. Superposition is then introduced to obtain the initial stress conditions due to geostatic stress and each of the five tracks being loaded simultaneously. The maximum stresses occur beneath the center track on a plane perpendicular to the tunnel, and are used as initial stresses in a plane strain finite element analysis. The procedure is used to calculate stresses induced by tunnel excavation by forcing the final shear and normal forces around the inside tunnel perimeter to be equal to zero for equilibrium.

Hyperbolic stress-strain modulus is used to represent both frozen and unfrozen soil. Since the analysis is based on long-term track loading, the design stress-strain relationship is taken to express ultimate conditions. The maximum shear stress calculated from this model is less than 9.4 kips/ft^2. However, fairly high tensile principal stresses are found to exist. For example, at point A in Figure 11-32, the principal stresses are calculated at 21.4 and 17.0 kips/ft^2, both in tension.

Commentary Reversal of movement resulting from ground heave due to freezing followed by noticeable settlement of supersaturated materials upon thawing can

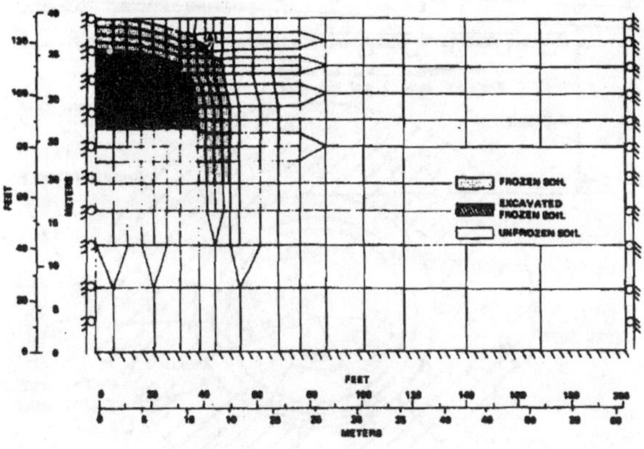

Figure 11-32 Finite element model and geometry of Georgia tunnel. (From Jones and Brown, 1978.)

produce more damage than movement due to heave or settlement alone. Displacement reversal can cause cosmetic and even structural damage in masonry structures.

Interestingly, the loading on underground structures is not a fixed concept, but results from the soil-structure interaction. Thus, the sequence of excavation and construction can influence the final medium-structure interaction and should be reflected in the analysis. The assumption of linear-elastic material behavior for the three tunnels subjected to low stress levels may be fully justified if the liner is placed simultaneously with the excavation, thus excluding time-dependent effects. Where the stresses in the soil medium are high and the permanent support is delayed, the time-dependent analysis may result in some stress relaxation and reduced final liner forces (see also Chapter 3). In addition, the two-dimensional plane strain finite element models may not be applicable to cross sections near the tunnel face, and may not account for deformations and stress changes experienced by the surrounding soil system before the liner is installed.

Observed Performance

New York Tunnel Additional studies on the tunnel were performed to supplement the stress analysis briefly described in the foregoing sections and the finite model shown in Figure 11-31. These studies incorporate the effects of freezing on the analysis process. Finite element modeling was used to study the volumetric expansion of the soil during freezing as cause of movement and stress changes in the unfrozen medium. The results indicated ground surface heaving up to 4 cm, expected during the freezing period. No movement was expected during the excavation period due to the very shallow overburden. The analysis indicated, however, approximately 3 cm of thaw consolidation (Chamberlain, 1980).

A minimum of 26 freeze pipes were used, as shown in Figure 11-33. Thermal considerations indicated a freeze pipe spacing/diameter ratio less than 13, and the selected frozen ground temperature between freeze pipes was $-10°C$ or below. A

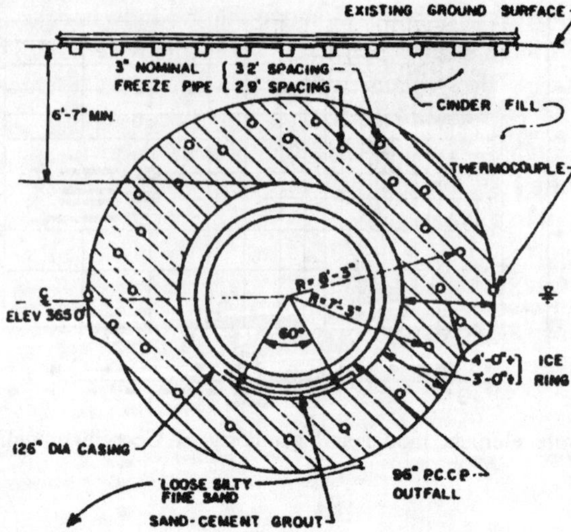

Figure 11-33 Cross section of New York tunnel showing freeze pipe arrangement and details. (From Lacy et al., 1982.)

design requirement was raising the moisture content of the cinder fill above the groundwater table (Lacy et al., 1982).

Construction Process The mining of the tunnel was followed by the installation of the liner plate. The annulus space between liner and excavation face was grouted with center mix. Two access shafts were excavated at each end of the tunnel, supported continuously with steel sheet piling. This limited groundwater lowering adjacent to the tracks, and pumping was confined to the cofferdam areas.

Monitoring of ground freezing was provided by thermocouples placed at four locations around the horizontal tunnel cylinder. Two test pits were required across the tunnel alignment adjacent to the tracks after the ground was frozen to examine visually the extent and quality of the frozen soil.

Construction Performance A practical problem was jacking horizontal freeze pipes about 125 ft through variable soil, and maintaining the accuracy of freeze pipe spacing. A second problem was associated with the upper part of the frozen cylinder being above the normal groundwater table. A third problem involved the actual mining, and resulted from the inherent geometry of a horizontal cylinder of frozen earth. Since it is not practical to prevent the tunnel face from freezing, mining may have to be carried out in frozen earth.

Maishman and Powers (1982) describe the solutions to these problems. The freeze pipes were installed within special heavy-wall steel sleeves, 8-in. in diameter, sufficiently stiff to resist deflection and misalignment. The sleeves were advanced with hydraulic pressure while an auger close to the cutting edge moved material back through the sleeve to the jacking pit.

The problem of unsaturated soil above the groundwater table was remedied by injecting fluid through sleeves perforated above the water table, also making provisions for inundating the ground surface between tracks. Bentonite slurry was injected through the perforated sleeves, creating (because of its viscosity) a mound in the groundwater table. The injected slurry was chilled with the refrigeration plant, and the lower temperature increased its viscosity. In this manner, the transition from saturated soil within the groundwater mound to frozen soil was continuous and with an accelerated formation.

Tunnel excavation in frozen ground required a special roadheader equipped with a cutter head capable of ripping the frozen earth efficiently.

Monitoring of Freezing The variation in temperature versus the distance from the average centerline of the frozen ring is shown in Figure 11-34. This pattern can be partly explained by variations in distance of the monitor points from the two freeze pipes. Figure 11-35 shows the variation in soil temperature with time for two typical monitor points, and the variation in brine temperature. The point closest to the freeze pipes responds rapidly as the second freeze unit is added and the brine temperature drops. Further from the freeze pipe the temperature drops more slowly. As the second unit is discontinued the soil temperature remains relatively constant.

Tunnel Excavation and Support Most of the tunnel excavation was carried out with a continuous rock mining machine giving an excavation progress more than 2 m per day. This rate was controlled by the placement of the liner plate rather than by the capability of equipment. No more than 1.6 m of exposed tunnel was allowed at any point. Grouting voids were properly found and hollow spots were filled with cement mix.

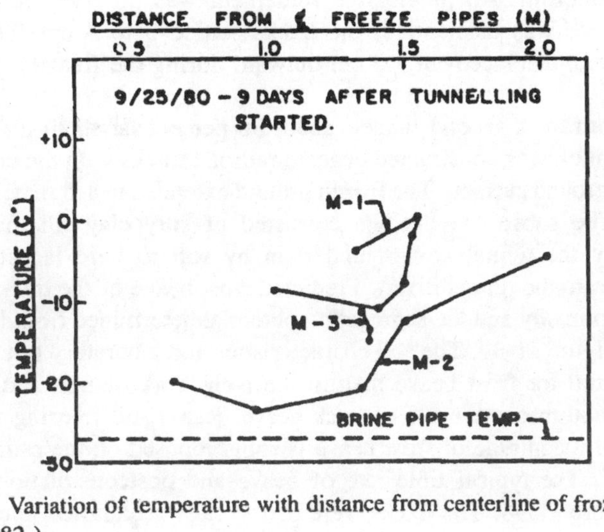

Figure 11-34 Variation of temperature with distance from centerline of frozen ring. (From Lacy et al., 1982.)

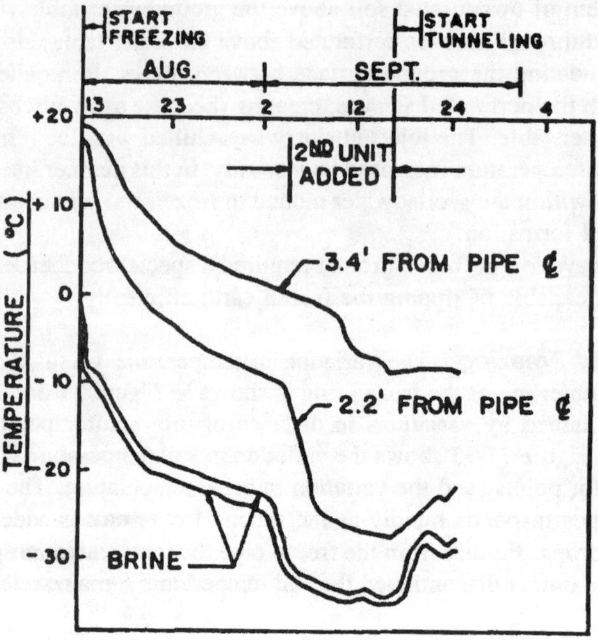

Figure 11-35 Soil temperature versus time. (From Lacy et al., 1982.)

Measured Displacement Figure 11-36 shows the variation in track level after making adjustments for reballasting the track. Remote readings show that the general area is settling at the rate of 2.3 cm/yr. During freeze pipe installation a settlement of 1.5 cm occurred, while the tracks experienced 6-cm heave during the period of artificial freezing. No measurable settlement was observed during tunneling. After the freeze was dismantled, the tracks settled 3 to 4 cm. Slope indicators showed a lateral displacement 1.3 cm outward during the freezing period.

Other Projects A second tunnel, about 30 percent larger in diameter than the New York tunnel, was constructed beneath railroad tracks with the crown about 3.3 m below the ground surface. The frozen ground extended to 1.5 m below the base of the tracks. The subsoil at the site consisted of silty clay fill that extended to approximately the tunnel crown, underlain by soft to hard lacustrine silty clay deposits down to the tunnel invert. Predicted frost heave of the tracks based on the range of soil porosity and the segregation potential determined from laboratory tests is shown in Figure 11-37. The soil characteristics and laboratory test data indicate a greater potential for frost heave than in the New York tunnel, and this was confirmed by actual measurements of track heave. Rapid soil freezing was necessary, and the actual magnitude of frost heave is superimposed on the estimated value in Figure 11-37. The typical time rate of heave and postconstruction settlement is shown in Figure 11-38. The tracks were periodically reballasted to compensate for these movements.

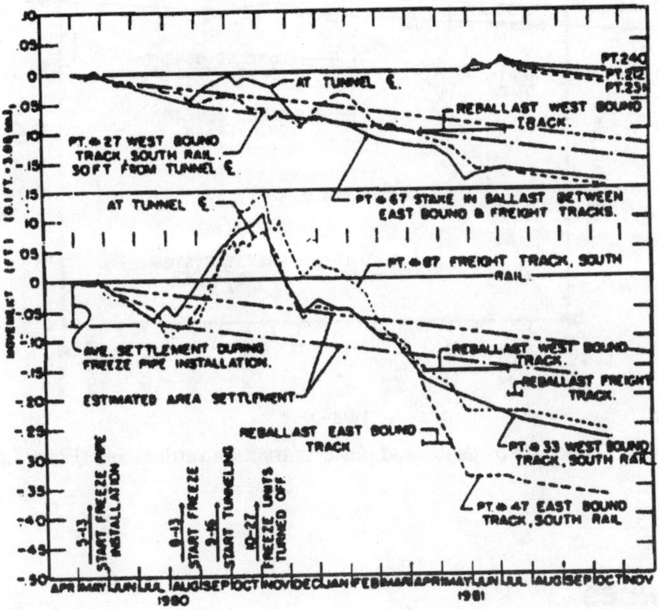

Figure 11-36 Variation of track level. (From Lacy et al., 1982.)

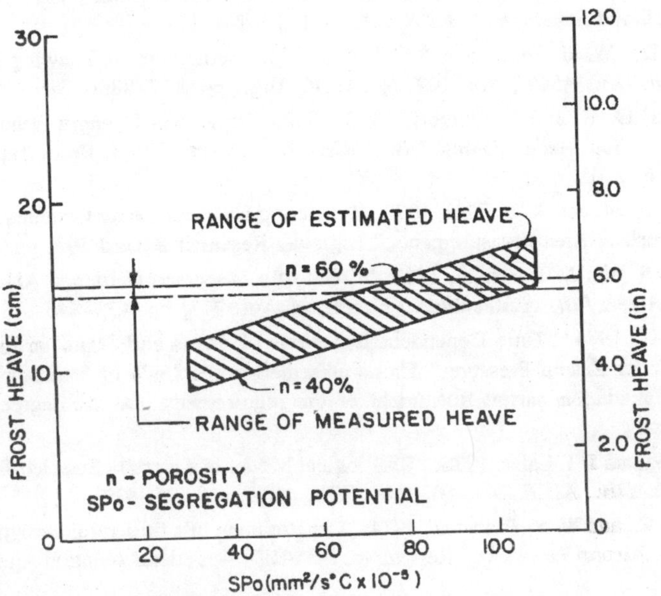

Figure 11-37 Estimated frost heave of tracks based on soil porosities and segregation potential. (From Lacy and Floess, 1988.)

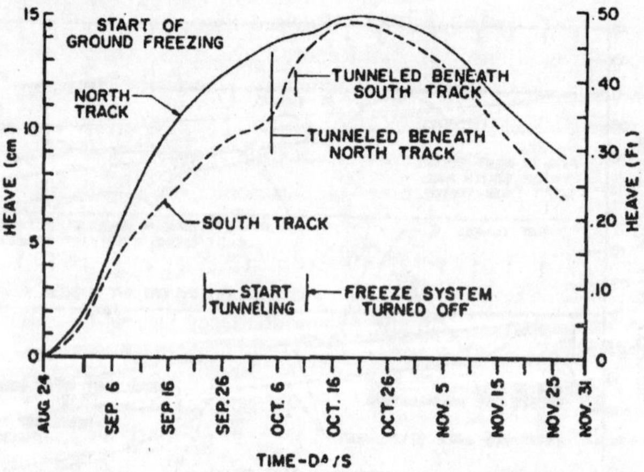

Figure 11-38 Time rate of heave and postconstruction settlement. (From Lacy and Floess, 1988.)

REFERENCES

Akili, W., 1966. "Stress Effect on Creep Rates of a Frozen Clay Soil from the Standpoint of Rate Process Theory," Thesis presented to Michigan State Univ., East Lansing, in partial fulfillment of the requirement for the degree of Doctor of Philosophy.

Akili, W., 1970. "On the Stress-Creep Relationship for a Frozen Clay Soil—Materials Research and Standards," *M.T.R.S.A.*, Vol. 10, No. 1.

Alkire, B.D., W. H. Haas, and J. J. Botz, 1976. "Settlement of Thawing Embankment," *Geotech. Div., ASCE,* Vol. 102, No. GT8, Aug., pp. 877–880.

Andersland, O. B. and I. AlNouri, 1970. "Time-Dependent Strength Behavior of Frozen Soils," *J. Soil Mech. Found. Div. ASCE,* Vol. 96, No. SM4, Proc. Paper 7406, July, pp. 1246–1265.

Anderson, D. M. and A. R. Tice, 1972. "Predicting Unfrozen Water Contents in Frozen Soils From Surface Area Measurements," Highway Research Record 393, pp. 12–18.

Ansell, G. S. and J. Weertman, 1959. "Creep of a Dispersed Hardened Aluminum Alloy," *Trans. Amer. Inst. of Mechanical Engineers,* Vol. 215, pp. 438–843.

Aziz, K. A., 1974. "Time Dependent Behaviour of Stress and Strain in Frozen Cohesive Soils Under Lateral Pressure," Thesis presented to the Univ. of Windsor, Windsor, Ontario, Canada, in partial fulfillment of the requirements for the degree of Doctor of Philosophy.

Aziz, K. A. and J. T. Laba, 1976. "Rheological Model of Laterally Stressed Frozen Soil," *J. Geotech. Div., ASCE,* Vol. 102, No. GT8, Aug., pp. 825–839.

Barker, W. R. and W. N. Brabston, 1974. "Development of a Structural Design Procedure for Flexible Airport Pavement," Report No. FFA-RD-74, Federal Aviation Agency, Washington, D.C.

Barnes, D. F., 1963. "Geophysical Methods for Delineating Permafrost," Proc., 1st Int. Conf. on Permafrost, National Academy of Sciences, National Research Council, Publ. No. 1287, pp. 349–355.

Berg, R. L. and T. C. Johnson, 1983. "Revised Procedure, for Pavement Design Under Seasonal Frost Conditions," Special Report 83-27, U.S. Army Corps of Engineers Cold Regions Research and Engineering Laboratory, Hanover, N.H.

Bergan, A. T. and C. L. Monismith, 1972. "Some Fatigue Considerations in the Design of Asphalt Concrete Pavements," *Proc., Canadian Tech. Asphalt Assoc.*, Vol. XVII.

Bennett, H. F., 1972. "Measurements of Ultrasonic Wave Velocities in Ice Cores from Greenland and Antarctica," Research Report 237, U.S. Army Cold Regions Research and Engineering Laboratory, Hanover, N.H., June.

Bentley, C. R., 1972. "Seismic Wave Velocities in Anisotropic Ice: A Comparison of Measured and Calculated Values in and Around the Deep Drill Hole at Bryd Station, Antarctica," *J. Geophysical Research,* Vol. 77, No. 23, pp. 4406–4420.

Bentley, C. R., 1975. "Advances in Geophysical Exploration of Ice Sheets and Glaciers," *J. Glaciology,* Vol. 15, No. 73, pp. 113–135.

Bono, N., 1986. "Highway Design for Frost Susceptible Soils," M.S. thesis, Tufts Univ., Medford, Mass.

Butkovich, T. R. and J. K. Landauer, 1960. "Creek of Ice at Low Stresses," Research Report 72, U.S. Army Snow Ice and Permafrost Research Establishment, Wilmette, Ill.

Cary, J. W. and H. F. Mayland, 1972. "Salt and Water Movement in Unsaturated Frozen Soil," *Soil Science Soc. Amer. Proc.*, Vol. 36, pp. 549–555.

Chamberlain, E. J., 1980. "Overconsolidation Effects of Ground Freezing," 2nd Int. Symp. on Ground Freezing, Trondheim, Norway.

Chamberlain, E. J. 1986. "Evaluation of Selected Frost Susceptibility Test Methods," CREL Report 86-14, U.S. Army Corps of Engineers Cold Regions Research and Engineering Laboratory, Hanover, N.H.

Chamberlain et al., 1972. "The Mechanical Behavior of Frozen Earth Materials Under High Pressure Triaxial Test Conditions," *Geotechnique,* Vol. 22.

Chamberlain, E. J., D. M. Cole, and T. C. Johnson, 1979. "Resilient Response of Two Frozen and Thawed Soils," *J. Geotech. Div., ASCE,* Vol. 105, No. GT2, Feb., pp. 257–271.

Chamberlain, E. J., P. N. Gaskin, D. Esch, and R. L. Berg, 1984. "Survey of Methods for Classifying Frost Susceptibility, In Frost Action and Its Control," Tech. Council on Cold Regions Engineering Monograph, ASCE, New York.

Cory, F. E., 1973. "Settlement Associated with the Thawing of Permafrost," Proc., 2nd Int. Conf. on Permafrost, North American Contribution, July, pp. 599–607.

Dillon, H. B. and O. B. Andersland, 1966. "Predicting Unfrozen Water Contents in Frozen Soils," *Canadian Geotech. J.*, Vol. III, No. 2, May, pp. 53–60.

Dillon, H. B. and O. B. Andersland, 1967. "Deformation Rates of Polycrystalline Ice," Proc. Int. Conf. on Physics of Snow and Ice, The Inst. of Low Temperature Science, Hokkaido Univ., Sapporo, Japan, Vol. 1, Part 1, pp. 313–327.

Domke, O., 1915. "Uber die Beanspruchung der Frostmauet beim Schachtabtenfen nach dem Gefrierverfahren," *Gluckauf,* Vol. 51.

Edgers, L. and N. Bono, 1985. "Highway Design for Frost Susceptible Soils: Laboratory and Field Data—Winchendon Soils," Department of Civil Engineering, Tufts Univ., Medford, Mass.

Edgers, L., L. Bedingfield, and N. Bono, 1987. "Field Evaluation of Criteria for Frost Susceptibility of Soils," Transportation Research Record 1190, TRB, National Research Council, Washington, D.C., pp. 73–85.

Endo, K., 1969. "Artificial Soil Freezing Method for Subway Construction," Japan Society of Civ. Engineers.

Esch, D. C., R. L. McHattie, and B. Connor, 1981. "Frost-Susceptibility Ratings and Pavement Structure Performance," in Transportation Research Record 809, TRB, National Research Council, Washington, D.C., pp. 27–34.

Evaluation of the Heave Stress System in Predicting Frost Susceptibility, 1982. Final Report, HPR Research Study R-122-7, Massachusetts Dept. of Public Works, Wellesley, Mass.

Ferrians, O. J., Jr., 1966. "Effects in the Copper River Basin Area," U.S. Geological Survey Prof. Paper 543-E, Supt. of Documents, U.S. Government Printing Office, Washington, D.C.

Finn, W. D. L. and R. N. Young, 1978. "Seismic Response of Frozen Ground," *J. Geotech. Div., ASCE,* Vol. 104, No. GT10, Oct., pp. 1225–1241.

Finn, W. D. L., K. W. Lee, and G. R. Martin, 1977a. "An Effective Stress Model for Liquefaction," *J. Geotech. Eng. Div., ASCE,* Vol. 103, No. GT6, Proc. Paper 13008, June, pp. 517–533.

Finn, W. D. L., R. N. Yong, and K. W. Lee. 1977b. "Liquefaction of Thawed Sandwich Layers," Annual Conv., Exposition, and Continuing Education Program, Oct. 17–21, San Francisco.

Fujii, T., 1971. "The Practical Application of Thermal and Freezing Methods to Soils Stabilization," Proc. 1st Australia-New Zealand Conf. on Geomechanics, Melbourne, pp. 337–343.

Gail, C. P., 1972. "Tunnel Driven Using Subsurface Freezing," *Civ. Eng., ASCE,* Vol. 42, No. 5, May, pp. 37–40.

Gail, C. P., 1973. "Subsurface Freezing as an Aid to Soft Ground Tunnel Construction," Paper presented at Univ. of Wisconsin—Extension Prof. Development Seminar on "Soft Ground Tunneling," April 24–25.

Glen, J. W., 1963. "The Rheology of Ice," *Ice and Snow,* W. D. Kingery, Ed., M.I.T. Press, Cambridge, Mass., pp. 3–7.

Gold, L. W., 1958. "Some Observations on the Dependence of Strain on Stress for Ice," *Canadian J. of Physics,* Vol. 36, No. 10, pp. 1265–1275.

Gold, L. W., 1963. "Deformation Mechanisms in Ice," *Ice and Snow,* W. D. Kingery, Ed., M.I.T. Press, Cambridge, Mass., pp. 8–27.

Goodman, M. A. and D. B. Wood, 1975. "A Mechanical Model for Permafrost Freeze—Back Pressure Behavior," *Society of Petroleum Engineers J.,* A.I.M.E., Aug., Paper S.P.E. 4589.

Goughnour, R. R., 1967. "The Soil-Ice System and the Shear Strength of Frozen Soils," thesis presented to Michigan State Univ. at East Lansing, in partial fulfillment of the requirements for the degree of Doctor of Philosophy.

Goughnour, R. R. and O. B. Andersland, 1968. "Mechanical Properties of a Sand-Ice System," Placement and Improvement of Soil to Support Structures, Proc., ASCE, Soil Mech. Found. Div., Aug. 26–28, Cambridge, Mass., pp. 287–320. Also in *J. Soil Mech. Found. Div., ASCE,* Vol. 94, No. SM4, Proc. Paper 6030, July, pp. 923–950.

Hashemi, H. T. and C. M. Sliepcevich, 1973. "Effect of Seepage Stream on Artificial Soil Freezing," *J. Soil Mech. and Found. Div., ASCE,* Vol. 99, No. SM3, Mar., pp. 267–287.

Haynes, F. D. et al., 1975. "Strain Rate Effect on the Strength of Frozen Silt," Research Report 350, U.S. Army Cold Regions Research and Engineering Laboratory, Hanover, N.H.

Hegemann, J., 1980. "A New Concept for Sinking Freeze Shafts to Great Depths," Int. Symp. on Ground Freezing, June, Trondheim, Norway.

Hoekstra, P., 1969. "The Physics and Chemistry of Frozen Soils," Special Report 103, Highway Research Board, pp. 78–90.

Hunter, J. A. M., et al., 1976. "The Occurrence of Permafrost and Frozen Sub-Bottom Materials in the Southern Beaufort Sea," Beaufort Sea Technical Report No. 22, Department of the Environment, Government of Canada, Apr., pp. 1–174.

Johnson, E. G., A. Phukan, and W. H. Haas, 1988. *Embankment Design and Construction in Cold Regions*, ASCE, New York, 180 pp.

Jones, J. S., 1980. "State-of-the Art Report: Engineering Practice in Artificial Ground Freezing," Int. Symp. on Ground Freezing, June, Trondheim, Norway.

Jones, J. S. and R. E. Brown, 1978. "Temporary Tunnel Support by Artificial Freezing," *J. Geotech. Div., ASCE*, Vol. 104, No. GT10, Oct., pp. 1257–1276.

Jumikis, A. R., 1966. *Thermal Soil Mechanics*, Rutgers Univ. Press, New Jersey, 267 pp.

Kamenskii, R. M., 1971. "Thermal Engineering Calculations of the Frozen Soil Watertight Cutoff Dams, Taking into Account the Mutual Influence of the Freezing Columns," published in Russian in *Gidrotechnicheskoi Stroitel'stvo*, No. 4, Apr., pp. 38–42.

Kaplar, C. W., 1963. "Laboratory Determination of the Dynamic Moduli of Frozen Soils and of Ice," Proc. 1st Int. Conf. on Permafrost, National Academy of Sciences, National Research Council, Publ. 1287, pp. 293–301.

Kaplar, C. W., 1969. "Laboratory Determination of Dynamic Moduli of Frozen Soils and of Ice," Research Report 163, U.S. Army Cold Regions Research and Engineering Laboratory, Hanover, N.H., Jan.

Kersten, M. S., 1949: "Final Report, Laboratory Research for the Determination of the Thermal Properties of Soils, Eng. Exp. Station, Univ. of Minnesota.

Khakimov, K. R., 1957. *Problems in the Theory and Practice of Artificial Freezing of Soil*, Academy of Sciences, Moscow, U.S.S.R.

Klein, J., 1980. "Compromise Cone—A Useful Form of Isotropic Yield Surface for Freeze Shaft Design," Int. Symp. on Ground Freezing, June, Trondheim, Norway.

Kohnen, H., 1974. "The Temperature Dependence of Seismic Waves in Ice," *J. Glaciology*, Vol. 13, No. 67, pp. 144–147.

Kondner, R. L. and J. S. Zelasko, 1963. "A Hyperbolic Stress-Strain Formulation for Sands," Proc. 2nd Pan-Am Conf., Soil Mech. Found. Eng., Vol. 1, Brazil.

Konrad, J. M. and N. R. Morgenstern, 1980. "A Mechanistic Theory of Ice Lens Formation in Fine Grained Soils," *Canadian Geotech. J.*, Vol. 17, pp. 473–486.

Konrad, J. M. and N. R. Morganstern, 1981. "The Segregation Potential of a Freezing Soil," *Canadian Geotech. J.*, Vol. 18, pp. 482–491.

Laba, J. T. and K. A. Aziz, 1972. "Pressure-Time Relationship in Laterally Stressed Frozen Granular Soils," Highway Research Record 393, pp. 79–87.

Lacy, H. S. and C. H. Floess, 1988. "Minimum Requirements for Temporary Support with Artificially Frozen Ground," Transportation Research Record 1190, TRB, National Research Council, Washington, D.C., pp. 46–56.

Lacy, H. S., J. S. Jones, and B. Gidlow, 1982. "A Case History of Tunnel Constructed by Ground Freezing," 3rd Int. Symp. on Ground Freezing, June 22–24, pp. 389–396.

Ladanyi, B., 1972. "An Engineering Theory of Creep of Frozen Soil," *Canadian Geotechnical J.*, Vol. 9, No. 1, Feb.

Lade, P. V., H. L. Jessberger, and N. Dickman, 1980. "Stress-Strain and Volumetric Behavior of Frozen Soils," presented at 2nd Int. Symp. on Ground Freezing, Trondheim, Norway.

Lenzin, P. A. and B. Briss, 1975. "Ground Stabilization: Review of Grouting and Freezing for Underground Openings," UILU-ENG-75-2017, FRAORD/D-75-95, Department of Civil Engineering, Univ. of Illinois at Urbana-Champaign, Urbana, Aug.

Liou, C. P., V. L. Streeter, and F. E. Richart, Jr., 1977. "Numerical Model for Liquefaction," *J. Geotech. Eng. Div. ASCE,* Vol. 103, No. GT6, Proc. Paper 12998, June, pp. 589–606.

Low, P. F., D. M. Anderson, and P. Hoekstra, 1968a. "Some Thermodynamic Relationships for Soils at or Below the Freezing Point. 1. Freezing Point Depression and Heat Capacity," *Water Resources Research,* Vol. 4, pp. 379–394.

Low, P. F., D. M. Anderson, and P. Hoekstra, 1968b. "Some Thermodynamic Relationships for Soils at or Below the Freezing Point. 2. Effects of Temperature and Pressure on Unfrozen Soil Water," *Water Resources Report,* Vol. 4, No. 3, pp. 541–544.

Lysmer, J. et al., 1974. "LUSH—A Computer Program for Complex Response Analysis of Soil-Structure Systems," Report No. EERC 74-4, Earthquake Engineering Research Center, Univ. of Calif., Berkeley, Calif., Apr.

Lysmer, J. et al., 1975. "FLUSH-A Computer Program for Approximate 3-D Analysis of Soil-Structure Interaction Problems," Report No. EERC 75-30, Earthquake Engineering Research Center, Univ. of Calif., Berkeley, Calif., Nov.

MacKay, J. R., 1975. "Freezing Processes at the Bottom of Permafrost, Tuktoyaktuk Peninsula Area, District of Mackenzie (107c)," Paper 75-1A, Report of Activities, Part A, Geological Survey of Canada, pp. 471–474.

MacKay, J. R., 1977a. "Permafrost Growth and Sub-permafrost Pore-water Expulsion, Tuktoyaktuk Peninsula, District of Mackenzie," Paper 77-1A, Report of Activities, Geological Survey of Canada.

MacKay, J. R., 1977b. "Pulsating Pingos, Tuktoyaktuk Peninsula," *Canadian J. Earth Sciences,* Vol. 14, No. 2, pp. 209–222.

Mackay, J. R. and R. F. Black, 1973. "Origin, Composition, and Structure of Perenially Frozen Ground and Ground Ice: A Review," North American Contribution to 2nd Int. Conf. on Permafrost, National Academy of Sciences, Washington, D.C., pp. 185–192.

Maishman, D., 1975. *Ground Freezing, Methods of Treatment of Unstable Ground* (in press), Butterworth.

Maishman, D., 1982. "Shaft Design and Construction for Underground Excavations," Univ. of Wisconsin, Ext. Div., Jan. 14–15.

Maishman, D. and J. P. Powers, 1982. "Ground Freezing for Tunnels—Three Case Histories," 3rd Int. Symp. on Ground Freezing, June 22–24, pp. 397–409.

McRoberts, E. C. and N. R. Morgenstern, 1975. "Pore-Water Expulsion during Freezing," *Canadian Geotech. J.,* Vol. 12, No. 1, pp. 131–141.

Miller, R. D., 1972. "Freezing and Heaving of Saturated and Unsaturated Soils," *Highway Research Record,* No. 393, pp. 1–11.

Morgenstern, N. R. and L. B. Smith, 1973. "Thaw Consolidation Tests on Remoulded Clays," *Canadian Geotech. J.,* Vol. 10, No. 1, pp. 25–40.

Nakano, Y. and R. Arnold, 1973. "Acoustic Properties of Frozen Ottawa Sand," *J. Water Resources Research,* Vol. 9, No. 1, Feb., pp. 178–184.

Nakano, Y. and N. H. Froula, 1973. "Sound and Shock Transmission in Frozen Soils," Permafrost: North American Contribution to 2nd Int. Conf. on Permafrost, Yakutsk, Siberia, U.S. National Academy of Sciences, pp. 359–369.

Nakano, Y., R. J. Martin, and M. Smith, 1972. "Ultrasonic Velocities of the Dilational and Shear Waves in Frozen Soils," *J. Water Resources Research,* Vol. 8, No. 4, Aug., pp. 1024–1030.

Nixon, J. F., 1982. "Field Frost Heave Predictions Using the Segregation Potential Concept," *Canadian Geotech. J.,* Vol. 19, pp. 526–529.

O'Connor, M. J., 1975. "Triaxial and Plane Strain Experiments on a Frozen Silt," Ph.D. thesis, Queen's Univ., Kingston, Ont., Canada.

Parameswaran, V., 1980. "Deformation Behavior and Strength of Frozen Sands," *Canadian Geotech. J.,* Vol. 17.

Poulos, H. G. and E. H. Davis, 1974. "Stresses and Displacement in Embankments and Slopes," *Elastic Solutions for Soil and Rock Mechanics,* Wiley, New York, pp. 199–228.

Powers, J. P. and D. Maishman, 1981. *Ground Freezing in Construction Dewatering: A Guide to Theory and Practice,* Wiley, New York, Chapter 20, pp. 349–359.

Rein, B. G., V. V. Hathi, and C. M. Sliepcevich, 1975. "Creep of Sand-Ice System," *J. Geotech. Div., ASCE,* Vol. 101, No. GT2, Feb., pp. 115–128.

Robinsky, E. I. and K. E. Bespflug, 1973. "Design of Insulated Foundations," *J. Geotech. Div. ASCE,* Vol. 99, No. SM9, Sept., pp. 649–667.

Roethlisberger, H., 1972. "Seismic Exploration in Cold Regions," Tech. Monograph II-A2a, Cold Regions Research and Engineering Laboratory, U.S. Army Corps of Engineers, Hanover, N.H., Oct.

Ruedrich, R. A., T. K. Perkins, J. A. Rochon, and S. A. Christman, 1976. "Casing Strain Resulting From Thawing of Prudhoe Bay Permafrost," Society of Petroleum Eng. Tech. Conf., New Orleans, Paper S.P.E. 6062.

Sanger, F. J., 1968. "Ground Freezing in Construction," *J. Soil Mech. Found Div. ASCE,* Jan.

Sanger, F. J. and F. H. Sayles, 1978. "Thermal and Rheological Computations for Artificially Frozen Ground Construction," Int. Symp. on Ground Freezing, March, Bochum, Germany.

Sayles, F. H., 1968. "Creep of Frozen Sands," U.S. Army Cold Regions Research and Engineering Laboratory Tech. Report 190, U.S. Army Cold Regions Research and Engineering Laboratory, Hanover, N.H., p. 1.

Sayles, F. H., 1974. "Triaxial Constant Strain Rate Tests and Triaxial Creep Tests on Frozen Ottawa Sand," Permafrost 2nd Int. Conf., North American Contribution, National Academy of Sciences, Washington, D.C., p. 382.

Schnabel, P. B., J. Lysmer, and H. B. Seed, 1972. "SHAKE—A Computer Program for Earthquake Response Analysis of Horizontally Layered Sites," Report No. EERC-72-12, Earthquake Engineering Research Center, Univ. of Calif., Berkeley, Calif., Dec.

Scott, R. F., 1969. "Freezing Process and Mechanics of Frozen Ground," Cold Region Science and Engineering Monography, Part 2, Cold Regions Research and Engineering Laboratory Mono II-D1, Hanover, N.H., Oct.

Seed, H. B., 1976. "Evaluation of Liquefaction Effects on Level Ground During Earthquakes," presented at the Sept. 27–Oct. 1 ASCE Annual Conv. and Exposition, Philadelphia, Pa. (Preprint 2752).

Sheeran, D. E. and R. N. Yong, 1975. "Salt and Water Re-distribution in Freezing Soils," Proc., Conf. on Soil-Water Problems in Cold Regions, American Geophysical Union, Calgary, Canada.

Shuster, J. A., 1975. "Controlled Freezing for Temporary Ground Support," Proc., North Amer. Rapid Excavation and Tunneling Conf., Vol. 2, Chicago, June 5–7.

Singh, S. and N. C. Donovan, 1977a. "Seismic Report of Frozen-Thawed Soil Systems," Proc., 6th Int. Conf. on Earthquake Engineering, New Delhi, India, pp. 611–616.

Singh, S. and N. C. Donovan, 1977b. "Seismic Behavior of Frozen-Thawed Profiles," ASCE Annual Conv., Exposition, and Continuing Education Program, Oct. 17–21, San Francisco.

Stevens, H. W., 1973. "Viscoelastic Properties of Frozen Soil Under Vibratory Loads," North American Contribution to 2nd Int. Conf. on Permafrost, National Academy of Sciences, Washington, D.C., pp. 400–409.

Stevens, H. W., 1975. "The Response of Frozen Soils to Vibratory Loads," Technical Report 265, U.S. Army Cold Regions Research and Engineering Laboratory, Hanover, N.H., June.

Streeter, V. L., E. B. Wylie, and F. E. Richart, 1974. "Soil Motion Computations by Characteristics Method," *J. Geotech. Eng. Div., ASCE*, Vol. 100, No. GT3, Proc. Paper 10410, Mar., pp. 247–263.

Sverdrup and Parcel and Associates, 1973. "Cut-and-Cover Tunneling Techniques," National Tech. Information Service, Report No. FHWA-RD-73-40, Springfield, Va.

Takashi, T., 1969. "Influence of Seepage Stream on the Joining of Frozen Soil Zones in Artificial Soil Freezing," in Special Report 103, HRB, National Research Council, Washington, D.C., pp. 273–286.

Thompson, E. G., and F. H. Sayles, 1972. "In Situ Creep Analysis of Room in Frozen Soil," *J. Soil Mech. Found. Div. ASCE*, Vol. 98, No. SM9, Proc. Paper 9209, Sept., pp. 899–915.

Tsytovich, N. A., 1960. "Bases and Foundations on Frozen Soil," Special Report 58, Highway Research Board.

Tsytovich, N. A., 1966. "Basic Mechanics of Freezing, Frozen and Thawing Soils," Technical Trans. 1239, National Research Council of Canada, Div. of Bldg. Research.

Tsytovich, N. A., 1975. *The Mechanics of Frozen Ground*, McGraw-Hill, New York.

Tsytovich, N. A., et al., 1965. "Consolidation of Thawing Soils," Proc. 6th Int. Conf. Soil Mech. Found. Eng., Vol. 1, Montreal, Canada, pp. 390–394.

Vialov, S. S., 1963. "Rheology of Frozen Soils," Proc., Permafrost Int. Conf., National Academy of Science–National Research Council, Publication 1287, Washington, D.C., p. 382.

Vialov, S. S., 1965. "The Strength and Creep of Frozen Soils and Calculations for Ice-Soil Retaining Structures," Translation 76, Cold Regions Research and Engineering Laboratory, Hanover, N.H.

Vinson, T. S., 1978. "Parameter Effects on Dynamic Properties of Frozen Soils," *J. Geotech. Div. ASCE*, Vol. 104, No. GT10, Oct., pp. 1289–1306.

Vinson, T. S. and T. Chaichanavong, 1976. "Dynamic Properties of Ice and Frozen Clay Under Cyclic Triaxial Loading Conditions," Report No. MSU-CE-76-4, Div. of Eng. Research, Michigan State Univ., East Lansing, Oct.

Vinson, T. S., R. Czajkowski, and J. Li, 1977. "Dynamic Properties of Frozen Cohesionless Soils Under Cyclic Triaxial Loading Conditions," Report No. MSU-CE-77-1, Div. of Eng. Research, Michigan State Univ., East Lansing, Jan.

Vinson, T. S., T. Chaichanavong, and R. Czajkowski, 1978. "Behavior of Frozen Clay under Cyclic Axial Loading," *J. Geotech. Div. ASCE*, Vol. 104, No. GT7, July, pp. 779–800.

Vyalov, et al., 1965: "Strength and Creep of Frozen Soils and Calculations for Ice-Soil Retaining Structures," Cold Regions Research and Engineering Laboratory, Transaction 76, Hanover, N.H., Sept.

Williams, P. J., 1964. "Unfrozen Water Content of Frozen Soils and Soil Moisture Suction," *Geotechnique*, London, Vol. XIV, No. 3, Sept., pp. 231–246.

Williams, P. J., 1968. "Ice Distribution in Permafrost Profiles," *Canadian J. Earth Sciences*, Vol. 5, No. 12, Dec.

Winter, H., 1980. "Creep of Frozen Shafts: A Semi-analytical Model," Int. Symp. on Ground Freezing, June, Trondheim, Norway.

Youssef, H., 1984. "Indirect Determination of Intergranular Stresses in Frozen Soils," Ph.D. thesis, Department of Civil Engineering, Ecole Polytechnique, Univ. of Montreal, Quebec, Canada.

Youssef, H., 1985. "Development of a New Triaxial Cell with Self-Cooling System (TC-WSCS) for Testing Ice and Frozen Soils," Presented at 4th Int. Symp. on Ground Freezing, Sapporo, Japan.

Youssef, H. and A. Hanna, 1988. "Behavior of Frozen and Unfrozen Sands in Triaxial Testing," Transportation Research Record 1190, TRB, National Research Council, Washington, D.C. pp. 57–64.

INDEX

Entries having an asterisk (*) have subentries not arranged alphabetically but in the order they appear in the pages.

Anchored walls, 127, 129, 140, 147, 184
Angle of internal friction, 246, 247, 254, 362, 363
Anisotropy, 173–175
Artesian water pressure, 7, 10, 20, 21, 42
Augered caissons, 87–88. *See also* Underpinning

Base heave, 170, 173, 175
Battered piles, 88–89. *See also* Underpinning
Bearing capacity, 244, 246, 247, 249, 255, 266–267, 273
Beaubourg Center, 181
Berms, 167–169. *See also* Bracing
Blasting:
- precision, 221
- restrictions, 219, 220
- safety, 219, 220

Bored pile walls, 218
Boylston Street, Boston, 431, 440
Brace stiffness, effect of, 163–165
Bracing, 162–170
Buildings, underpinning for, 99–102
Buoyancy forces, 5
Buttresses, 723, 728–729

*Cement-bentonite cutoffs:
- characteristics, 795
- proportioning, 796
- properties, 797–800
- setting time, 800
- mixing, 800
- strength, 801
- permeability, 801
- resistance, 801, 808
- design, 803
- microstructure analysis, 803, 804
- examples, 804–808

Circular slope failure, 702–704
Clay-cement-bentonite mixes:
- applications, 778
- background, 778
- characteristics, 779
- compressive strength, 780

Cohesion, 254, 363
Collapsible soils, 179–181, 240
Column jacking, 103, 104
Compaction, conventional, 234, 238, 245, 247, 255, 282
Compaction grouting, 234–237, 241, 242, 244, 245, 247, 250, 251, 283–284, 494
Compressed air tunneling, 16, 56–57
Cone of depression due to well pumping, 5, 19, 20
Cone penetrometer test, 238, 257, 258, 288–290, 389

Consolidation of clays:
  consolidation tests, 238, 257, 258–263
  drainage, 239, 246, 248, 254
  percent consolidation, 240, 262, 263
  rate, 248, 251, 261–264, 271–272, 277–280
  time factor, 240, 261–264
Coney Island, N.Y., 431, 444
Core drilling, 688
Creep of rock slopes, 715, 718
Cyclic stress ratio, 248, 268–270

Dams, 148–149, 216, 219
Darcy's law, 23, 34–37
Deep wells, 3, 16, 42, 44
Dewatering:
  case histories, 58–73
  discharge, 50–52
  effects, 52–54
  sumps, 3, 38, 39, 41
  trenches, 3, 38, 39
Degree of compaction, 238, 256
Densification of sands, 245, 246, 247, 253, 275
Diaphragm walls, *see* Slurry walls
Dilatometer testing, 290, 292
Direct sheer tests on rock, 718
Downdrag, 108
Drainage of soils, 1–74
Drawdown due to water pumping, 9, 20, 21, 25, 26, 28, 30
Driven piles, 84–87. *See also* Underpinning
Dynamic compaction, 234–237, 240, 241, 244, 245, 247, 250, 275, 282–283

*Earth cutoffs:
  clay mixes, 756
  backfill characteristics, 758
  design, 758
  blowout tests, 760
  permeability, 758, 765–768
  control limits, 768
  pollution control, 770
  compressibility, 775
Earth pressures, 345
Earth reinforcing, 471
Element walls, 125, 128–131
Embankments, 234, 238, 252
Equilibrium wall formula, 19, 20
ETF process, 813. *See also* Injected screens
Excavation support, 146–233, 336, 337, 338, 344

Fabric walls, 348
Field permeability tests, 12, 13
Fills, *see* Embankments
Flow nets, 23, 24, 35, 37
Foam drilling, 688
Freeze-thaw cycles of rock slopes, 715, 717

Geologic mapping, 686, 690–692
Geophysical testing, 290–291, 386–387
Geotechnical exploration, 10, 238, 254, 271, 276, 288–292, 389, 391, 684–692
Geotechnical instrumentation, 292–294, 346, 381–382, 735–736
Geotechnical reports, 694–697
Grain size distribution, 10, 12, 13, 14, 15, 16, 244, 254
Grillages, 94–95. *See also* Shoring
*Ground freezing, artificial:
  applications, 150, 827, 836, 872
  background, 57, 827, 828
  sand-ice systems, 828
  mechanical properties, 828, 856, 863
  creep, 828–833
  strain tests, 829
  tests on frozen and unfrozen sands, 833–836
  clay-ice systems, 837
  creep tests, 837
  strain-time curves, 841
  dynamic response of frozen ground, 841
  dynamic properties of frozen ground, 843–846
  cyclic axial loading, 846
  design for temporary support, 848
  freezing systems, 849, 866, 867
  system components, 849
  freezing procedures, 851
  design, 852, 859
  thermal properties, 854, 859
  hydrologic properties, 856
  system capacity and geometry, 858
  groundwater movement, 862
  monitoring, 869
Ground movement, 154, 156, 157, 160, 170, 175, 207–209, 864
Groundwater, effect of, 152–153, 156, 161, 414, 862, 864
Groundwater quality, 9
Grout curtains, 54, 148, 149, 217, 808
Grouted piles, 89–90. *See also* Underpinning

Header pipes, 41, 49
Horizontal drains, 38, 40, 41
Hydraulic conductivity, 7
Hydraulic gradient, 7, 9, 22, 44
Hydrofracture grouting, 493

*Impermeable membranes:
synthetic screens, 815
assembly, 815, 816
water stop, 816
chemical resistance, 818
installation, 819
examples and applications, 819
Inclined walls, 118, 140
*Injected screens:
construction, 808–810
mix design, 810
proportioning, 810–811
flow properties, 811–812
penetrability and strength, 812
applications, 813, 814
Interceptor trenches:
advantages, 822
construction, 820
design, 821

Jacked piles, 82–83. *See also* Underpinning
*Jet grouting:
background, 495, 580
one-fluid systems, 584
two-fluid systems, 585
three-fluid systems, 585
applications, 592
seepage barriers, 593
lateral retention, 593
bottom seals, 593
tunnel construction, 598, 620
structural underpinning, 605. *See also* Underpinning
slope stability, 612
design, 625
grout selection, 626
jet grouted sols, 632
control and assessment, 640
test programs, 650

Key block theory, 715

Lagging, 159
Landfills, 238, 240
Landslide stabilization, 331, 335, 340
Limit equilibriuim slope failure, 365
Liquefaction, 234, 238, 241, 244, 245, 248, 255, 268–270, 278

Mechanically stabilized embankment (MSE), 331, 333, 336, 341, 345, 348, 365
Micropiles, 340. *See also* Small-diameter elements
Minipiles, *see* Small-diameter elements
Modulus of soils, 252, 263, 264
Moisture content of soils, 238, 253, 256
Moisture-density curves, 239, 256
Mohr-Coulomb failure criteria, 363, 365
Monadnock Building, underpinning of, 100
Multiple rock joint plane failure, 710–711

Needle beams, 92. *See also* Shoring
Noncorrosive cutoffs:
aggressive water, 822
bituminous mixes, 823–824
fly ash mixes, 823
high resistance, 822
leaching and chemical attack, 823

Observation wells, 7, 11, 12, 19

Packer tests, 9, 12, 14
Perched groundwater, 7, 10, 28
Permanent dewatering, 4
Permeability of soils, 5, 7, 8, 9, 16, 17, 19, 20, 21, 26, 34, 36, 40, 246–248, 262, 272
Permeability tests, 9, 17
Permeable treatment beds:
applications, 791
design, construction, 792
materials, 793, 794
Permeameter, 17, 18
*Permeation grouting:
background, 483, 498
applications, 499
grouts, 501
design, 528
geotechnical parameters, 528
grout parameters, 531
grout selection, 540
drilling, 540
grouting, 557
performance, 566
geophysical methods, 566
core sampling, 568
laboratory tests, 569
hydraulic tests, 570
costs, 570
Piezometers, 7, 11, 238, 294
*Pin piles:
characteristics, 407
construction, 408, 473
applications, 408, 473. *See also* Underpinning
structural connections, 410–411
postgrouting, 412
design, 413, 473
testing, 414

*Pin piles (*Continued*)
  grout, 417
  bond, 421
  composite action, 421
  stability, 421, 473
  grout-ground bond, 422
  group effect, 426
  case histories, 431, 481
Plastic concrete cutoffs:
  applications, 782
  characteristics, 783
  durability, 790
  examples, 786–796
  modulus of elasticity, 784–786
  strength, 784, 786
Plasticity of soils, 254
Pokomake River Bridge, Md, 431, 461
Pore pressure, 233
Porosity of soils, 253–254
Postal Square, Washington, D.C., 431, 455
Prefounded columns, 120–124
Preloading, 238, 250, 277, 279, 280
Pressuremeter tests, 263, 264–266, 389
Pretest piles, 83–84. *See also* Underpinning
Planar rock joint failure, 704–707
Primary consolidation, 239
Proctor tests, 239, 256
Progressive failure of rock slopes, 715–716, 719–720
Pump hydraulics, 22, 29–33
Pump tests, 9, 10, 19, 21, 22, 26

Radius of influence, well pumping, 9, 20, 24, 34, 36
Rakers, 165, 169. *See also* Bracing
Recharge, water, 25, 50, 52
Rehabilitation of rock slopes, *see* Rock slope engineering
Relative density, 238, 250, 253, 255, 257
Retaining wall, repair, 341, 343
Reticulated walls, 102, 103, 114, 471. *See also* Small-diameter elements
Rib walls, 125–127
Rock bolts and dowels, 730–732
Rockfall catchment ditches, 725–727
Rockfall hazard rating system, 720–722
Rock quality designation (RQP), 687, 720
Rock slope engineering:
  case histories, 736–753
  drainage, 732–733
  failure modes, 715–720
  geometrically stable, 720
  hydrostatic pressure, 699, 715, 717
  joints and discontinuities, 697–698
  reinforcement, 729–732
  remediation, 723–733
  repair costs, 733–734
  safety factors, 699
  shear strength, 698
  stability analysis, 701–715
  theoretical background, 697–704
Root piles, 90–91, 102, 110. *See also* Reticulated walls

Sand cone method, 256
Sand drains, 43, 239, 248, 250, 251, 270, 277, 279–281, 284
Secant pile walls, 117
Seismic failiure of rock slopes, 715, 718, 719
Settlement (subsidence), 234, 238, 239, 241, 248, 250–252, 255, 261–266, 273
Settlement markers, 238, 292, 284
*Shafts (in frozen ground), *see also* Ground freezing
  temporary support, 872
  structural aspects, 873
  thermal aspects, 875
  system components, 877, 878
  monitoring, 878
  construction, 879, 880
  deep shafts, 880
  abandonment, 880–882
  examples, 882
Shear strength of clays, 249, 253, 254, 271
Shear strength tests, 238
Sheep's foot roller, 234
Sheet pile walls, steel, 54–56, 161–162
Shoring, 91–95
*Shotcrete:
  structural characteristics, 186
  steel-fiber reinforced, 188
  compressive strength, 189
  flexural strength, 190–193
  plate tests, 193
  rebound tests, 194
  pullout tests, 197
  tunnel support, 198, 209
  model tests, 200
  field observations, 200
  with rock bolts, 203
  bond, 206
  uses, 347, 381–382, 388, 729
Sinkholes, 234, 240, 241, 686
Slope stability, 246, 250, 253, 274, 332, 336, 338, 339, 356, 359, 362, 366
*Slurry walls:
  lateral protection, 3, 56, 115, 116, 119, 147
  permanent, 117

instrumentation, 119–120
Small-diameter elements:
pin piles, 407
type "A" walls, 471
Soil-cement walls, 224
Soil compaction and consolidation:
case histories, 294–330
combined methods, 244
costs, 244
Soil nailing:
active/resistant zones, 348
advantages/disadvantages, 332–335
bar size, 368, 370, 374, 385
bending stresses, 348, 363
bond stresses, 349, 369, 370
case histories, 392–402
computer programs, 366
configuration, 382, 471
construction, 336, 351, 384–389
corrosion protection, 350, 367, 378–381, 387–388
deflections, 350, 352–355
design, 356–384
design life, 350, 380
drilling, 385
drainage, 350, 355, 382–384, 388–389
driving, 384
facing, 331, 366, 367, 381, 388
failure modes, 349, 359, 361, 363, 365
grouting, 381, 386, 387–388
history, development, 342–343
internal/external stability, 348, 366, 371
length, 366–367, 370, 374
pull-out tests, 346, 361, 362, 364, 370, 389, 390
sheer stresses, 348, 369, 374
spacing, 367, 368, 370, 372
tensile stresses, 348, 349, 351, 364, 375, 378
testing, full scale, 346
theoretical background, 348–350
Soil doweling, 471
Soldier pile walls, 159, 164, 165, 184
Specific well capacity, 22, 24, 27
Standard penetration test, 11, 246–250, 252, 253, 268–269, 278, 288, 389
Stone columns, 235–237, 243–246, 248, 250, 252–253, 275, 290, 293
Storage capacity, 9, 22, 24
Storage coefficient, 9, 24, 26
Stereographic projections, 692–694
Stress paths, 153, 154, 155
Submersible pumps, 3, 4, 38, 43
Sump pumps, 4, 30
Surcharge loads, 131–140

Tiebacks, 332, 336, 339, 342, 380, 730–731
Time-drawdown, in water well pumping, 9
Time effects in excavations, 175
Temporary walls, 181
Toppling failure, 711–713
Transmissibility of water in soil, 9, 22, 24, 25, 26
Tresca's failure criteria, 364, 365
*Tunnel support, by ground freezing:
strength factors, 853
ground movement, 883
costs, 884
case studies, 884–885, 889
design aspects, 885
tests, 885–887
investigation, 889
performance, 891
Tunneling, underpinning, 102–104
Tunnel reinforcement, 331, 334, 339
Type "A" walls, 471
Two-block rock slope failure, 707–708

Unconfined compressive strength, 683, 697
*Underpinning:
remedial, 76, 95–99
precautionary, 76
grouping, 76, 77, 107–108
pit or pier, 78, 79–82, 99, 108, 119
maintenance, 81
pile, 82, 108, 119
jacked piles, 82–83
pretest piles, 83–89, 99
driven piles, 84–87
augered caissons, 87–88
battered piles, 88–89
grouted piles, 89–90
root piles, 90–91, 110
examples, 95–104
lateral, 115, 116, 124
by grouting, 150, 151, 605
Unit weight of soil, 256, 257
Uplift, 147, 149

Vibrocompaction, 234–237, 242–245, 247, 250, 273, 275, 277, 285–287
Void ratio, 247, 253, 255, 261

Weathering of rock slopes, 715–716
Wedge rock slope failure, 708–710
Well formulas, 35, 37
Well points:
eductors, 4, 16, 42
effective radius, 45
ejectors, 4, 16

Well points (*Continued*)
multiple stages, 4
plumbing, 48–50
pumps, 30, 41, 43, 45–48
spacing, 45–46
Wick drains, 43, 234–237, 239, 244, 250–252, 272, 277, 280–281, 284–285
Wire nets, 729